GEOREF
Thesaurus and Guide to Indexing
Second Edition

GEOREF
Thesaurus and Guide to Indexing
Second Edition

Edited by
Carol Heckman and others

Published by the
AMERICAN GEOLOGICAL INSTITUTE

Earlier Edition:
GeoRef Thesaurus and Guide to Indexing, First Edition
Copyright © 1977 American Geological Institute

GeoRef Thesaurus and Guide to Indexing, Second Edition
Copyright © 1978 American Geological Institute
All rights reserved

Library of Congress Catalog Card Number 78-65083
International Standard Book Number 0-913312-07-X
Printed in the United States of America

American Geological Institute
5205 Leesburg Pike
Falls Church, Virginia 22041

The Thesaurus was photocomposed on a Videocomp from a tape generated by SAMANTHA programs. Eterna bold and light fonts were used for the text.

Cover photograph: San Andreas Fault
This is on the Carrizo Plain, about half way between Los Angeles and San Francisco. (Photo by Robert E. Wallace, U.S. Geological Survey) Reprinted from *Geotimes*, February 1975

ERRATA: Please Note

The broader term of <u>radar methods</u> (P. 249) is <u>geophysical methods</u> and not <u>geophysical surveys</u>. Term entries on the following pages should read:

P. 112   <u>geophysical methods</u>
           NA radar methods

P. 113   <u>geophysical surveys</u>
           SA radar methods (delete <u>radar methods</u> as NA--
                              autoposting narrower term)

P. 249   <u>radar methods</u>
           SA geophysical surveys (delete <u>geophysical surveys</u>
                                    as BA--autoposted broader term)

CONTENTS

Introduction		v
Thesaurus, A-Z		1
Guide to Indexing, Introduction		351
List A	First order terms	353
List B	Second and third order terms for areas	395
List C	Commodities	401
List D	Element sets	405
List E	Geologic ages	408
List F	Fossils	409
Lise G	Terms appropriate to a large number of topics	427
List H	Igneous rocks	428
List I	Sedimentary rocks	431
List J	Metamorphic rocks	433
List K	Sedimentary structures	436
List L	Mineral groups	437
List M	Soils	440
List N	Sediments	447
List O	Areas (including main area terms)	448
Authorities		456
Abbreviations		456

INTRODUCTION

The second edition of the GeoRef Thesaurus and Guide to Indexing contains over 12,500 terms, of which approximately 3000 are new. There are geographic place names, systematic terms (rocks, fossils, minerals, etc.) and non-systematic terms (geologic features, processes, properties, materials, etc.).

The ANSI Z39 Standard, Guidelines for Thesaurus Structure, Construction and Use (1974), has been followed herein.[1]

SOURCE OF THE VOCABULARY

Since May 1978, GeoRef has included five segments, the first three of which are new additions:

(1) Bibliography and Index of North American Geology (1961-1970)
(2) Bibliography of Theses in Geology (1965-1966)
(3) Geophysical Abstracts (1966-1971)
(4) Bibliography and Index of Geology Exclusive of North America (1967-1968)
(5) Bibliography and Index of Geology (1969 to present)

The index terms and sets used in all five segments have much in common. They derive from a common source, the early volumes of the Bibliography and Index of North American Geology. This Thesaurus is based on that indexing vocabulary as it has evolved down the years. It displays, interrelates, and regularizes the vocabulary.

This edition has been derived from terms in GeoRef file Segments 4 and 5. Most of the terms in Segments 1-3 are also included, but we have been unable as yet to study the indexing in those Segments in detail in order to incorporate notes and cross-references for them. We intend to do so. Meanwhile, help in searching Segments 1-3 can be found in List A of the Guide.

INDEXING

Prior to 1977 when the first edition of the Thesaurus appeared, the Guide to Indexing was the only tool available for GeoRef indexers and searchers. The Guide, which consists of structured lists of terms, has been revised and is included herein, with its own Introduction.

GeoRef indexing is in three level sets (see the Introduction to the Guide), which follow definite rules which can be found under each first level term in the Thesaurus.

First and second level terms have been tightly controlled in GeoRef, and few changes have been made in terms on these levels. Level three has been

---

1. Available from the American National Standards Institute, 1430 Broadway, New York, N. Y., 10018. Price $4.50.

v

more open-ended to accommodate specific locations, minerals, etc. Beginning in 1978, level three has been shortened to a single term. The other terms which had been included in level three became supplemental index terms. This change does not affect searching. Its effect on the printed indexes is evident by a comparison between 1978 and earlier publications. Briefly, the 1978 indexes have a single term plus title on level three instead of a term string. Also in 1978 cross-references were added to the index and selected supplemental terms to the citations.

Over 1500 terms are valid on levels one and two. All are included herein. Also included are all systematic and geographic level three terms used over four times and all non-systematic terms used over 24 times.

AUTOPOSTING

If a term appears in the indexing of a document, it does not necessarily follow that its broader terms will appear. Certain broader terms *will* be added because they are called for by the set structures. For example, for most geographic, stratigraphic, and fossil narrower terms, one, but not every broader term is required by the set structures. Other broader terms are added to facilitate computer searching.

As of January 1978, selected broader terms are autoposted (automatically added) by the computer. For example, each time Kansas is entered by an editor/indexer, the computer adds United States.

Those broader terms which are autoposted are designated as BA or BZ terms. For example:

        apatite
          BA   phosphates
          BT   minerals

(Each time apatite is indexed, phosphates is autoposted, but minerals is not autoposted.)

Autoposting is used selectively for:

   a.   Areas (see List O in the Guide)
   b.   Age and stratigraphic terms, excluding formations (see List E)
   c.   Fossils (see List F)
   d.   Igneous, sedimentary, and metamorphic rocks (see Lists H, I, J)
   f.   Sediments (see List N)
   g.   Minerals (see List L)
   h.   Soils (see List M)
   i.   Non-systematic terms (in only a few cases, such as faults)

TERM RELATIONSHIPS

In term entries, relationships are indicated by the tags CO, UF, BT, BA, BX, BZ, NT, NA, NX, NZ, and SA.

a. Geographic Coordinates--

In use since September 1977, coordinates are assigned to the principal area(s) studied in a document.

A fix fielded format of thirty characters is used to represent coordinates. For example, the coordinates for Alabama, lat. N30°-N35°, long. W85°-W88°30' are given as:

> N300000N350000
> W0850000W0883000

Principal geographic areas and a few others have coordinates herein. Any area is eligible and coordinates will be added for more areas. Coordinates for areas are used as needed, whether or not they appear herein.

Coordinates are used <u>in addition to</u> applicable area terms. For more information see inside front cover.

b. Use/Used For--

The use/used for (UF) relationship indicates synonyms or alternate spellings of a term. For example:

> columbium
>   use niobium
>
> Ezan Cape
>   use Esan Cape

Also, inverted forms of multiword terms have been cross-referenced when appropriate. For example:

> control, erosian
>   use erosion control

c. Broader Term/Narrower Term--

The broader term/narrower term relationship is that of genus to species. The broader term represents a class of which the narrower term is a member. For example:

> olivine group
>   BA orthosilicates
>   silicates
>   BT minerals

The narrower term cannot be a part of the broader term, in a part-whole relationship, except in the case of geographic place names and geologic age terms.

Variants of BT-NT introduced in this edition are BA-NA, BX-NX, and BZ-NZ.

BA-NA indicates autoposting (see Autoposting above).

BX-NX shows logical broader terms on the same hierarchical level or from different hierarchies. For example:

>   Chautauqua County
>   BX   Kansas
>   New York
>   BT   United States

(Both Kansas and New York have a Chatauqua County.)

BZ-NZ is for a combination of BA-NA and BX-NX. For example:

>   Hawaii
>   BZ   Pacific Ocean
>   United States

(The broader terms are from different hierarchies and both are autoposted by Hawaii.)

d. <u>See Also</u>--

The see also (SA) indicates related terms or terms often appearing together in GeoRef indexing other than broader terms and narrower terms. Examples are:

>   Gallup Sandstone
>   BT   Upper Cretaceous
>   BT   Cretaceous
>   SA   Mesaverde Group
>   SA   New Mexico

>   geochronology
>   SA   absolute age

NOTES

A note may occur immediately after a term to indicate the following:

a. Date Term Became Valid--If no date is given, the term is valid for GeoRef Segments 4 and 5. If a date <u>is</u> given, this is the year in which the term began to be used in GeoRef. (It is not the year in which the term was added to the Thesaurus.)

b. Set Level--The set level on which the term is currently used may be given. For example:

>   boudinage
>   includes use on level 3 under sedimentary
>   structures (1) soft sediment deformation (2)...

(Boudinage can be found under soft sediment deformation, which is a second level term under sedimentary structures.)

All valid index terms can be used on level three. Unless otherwise specified in a note, a term cannot be used on level one or two. List A in the Guide shows the valid first level terms for all segments of GeoRef.

c. Set Structure--For each valid level one term, the Thesaurus shows how the sets are to be structured.

d. Usage--The phrase "includes use", which occurs frequently in the notes, suggests an example of current and significant GeoRef usage, but is not meant as a listing of all possibilities. It means "this is an important use", but not "this is the only use".

e. Reference to the Guide--Notes may refer to the lists in the Guide.

f. Previously Used Term--If a term has had 100 or more postings, but is no longer valid for indexing, a note appears under that term in the Thesaurus and a note to "also search" the term occurs under the preferred term entry.

   For invalid terms with a frequency of less than 100, there are notes to "also search" the term in the entry of the preferred Thesaurus term. There may be a UF also if it is appropriate.

g. Geography and Stratigraphy--Brief notes are given for most terms to assist the user in locating an area. Directions are abbreviated in these notes as N, E, S, and W.

h. Combined Terms--Usually only single terms occur on any level of a set. However, a few terms can occur together, separated by commas, such as mineral deposits, genesis. Explanatory notes are provided under each of these terms.

MULTILINGUAL THESAURUS

A multilingual thesaurus in geology is being developed by an IUGS/ICSU AB Working Group, of which G. N. Rassam, Chief Editor of GeoRef, is a member. One decision of this Group was not to use adjectives as terms. We have moved to conform to this decision (see Adjectives, below).

ADJECTIVES

Due to the Multilingual Thesaurus and to the dropping of term strings on level three (see under Indexing), several hundred adjectives have been replaced by nouns. For example:

andesitic	became	andesitic composition
disseminated	became	disseminated deposits
sublittoral	became	sublittoral environment

The only remaining group of adjectives are age terms (see List E).

TERM VALIDATION

The Thesaurus on computer tape is the source of direct access validation files currently in use to check the spelling, capitalization, and level of

terms. Index terms which don't validate can be accepted by overriding the validation. These become candidate Thesaurus terms when they have occurred five or more times.

ALPHABETIZATION

Terms are sorted word-by-word rather than letter-by-letter. Specifically:

- The sort is on letters, numbers, spaces and hyphens.
- Spaces sort before hyphens; hyphens sort before letters; letters sort before numbers.
- An open parenthesis is sorted as if it were a space.
- All characters, including punctuation marks, other than letters, numbers, hyphens and the open parenthesis, are squeezed out in the sort, i.e., are treated as if they did not exist.
- A few terms are identical except that one begins with a lower case letter, one with an upper case letter, e.g., alpine and Alpine. When this happens, the term with the lower case letter precedes the other term.

FEEDBACK

The computer file from which the Thesaurus was produced will be regularly updated for future editions. Users are encouraged to send in corrections and comments. Let us hear from you!

ACKNOWLEDGMENTS

The following petroleum companies provided a large portion of the support for the first edition of the Thesaurus:

        Amoco Production Company
        Atlantic Richfield Company
        Chevron Oil Field Research Company
        Exxon Production Research Company
        Getty Oil Company
        Phillips Petroleum Company
        Shell Development Company

In particular, we are grateful to William Baum (deceased) and Margaret McLean of ARCO.

Editorial work on this edition was done by Carol Heckman with assistance from G. N. Rassam, Chief Editor, and the editors/indexers. Work on the geographical terms for the first edition was done by Ed Moon. Jack Wolfire handled the programming and photocomposition. Lesa Warren and Hazel S. Kirby typed and formatted the Guide, Introduction and front matter.

                                              John Mulvihill
                                              Manager, GeoRef

A

aa
use aa lava

aa lava
Term introduced in 1978. Before 1978, also search aa AND lava.
UF aa
BT lava
SA pahoehoe
  volcanism

Aachen
City near the intersection of the Belgium and Netherland borders.
BT North Rhine-Westphalia
  Germany

Aaland
use Aland

Aalenian
Europe. Lowermost Middle or uppermost Lower Jurassic. Includes use on level 3 under age terms(1). See list E.
BA Jurassic
BT Mesozoic
SA Dogger
  Lower Jurassic
  Middle Jurassic

Aar Massif
BT Switzerland

Aar Valley
River valley in central and N Switzerland. Also search Aare Valley.
UF Aare Valley
BT Switzerland

Aare Valley
use Aar Valley

Aargau
Canton in N.
BT Switzerland

Abakan
Town on Yenisei River SW of Krasnoyarsk in Khakass Autonomous Oblast.
BT Russian Republic
  USSR

Abashiri
City in NE Hokkaido.
UF Abasiri
BT Hokkaido
  Japan

Abasiri
use Abashiri

Abee
Village in central.
BT Alberta
  Canada

Aberdeen
City on North Sea.
CO N565000N574500
  W0014500W0041000
BT Scotland
  Great Britain
  United Kingdom

Aberdeenshire
County in NE.
BT Scotland
  Great Britain
  United Kingdom

Aberystwyth
Town on Saint Georges Channel.
BT Wales
  Great Britain
  United Kingdom

Aberystwyth Grits
Upper Llandoverian. Form part of the Ystwyth Stage. N and central Wales and NW England.
BT Silurian
SA England
  United Kingdom
  Wales

Abilene
Arch in central Kansas.
UF Abilene Arch
BT Kansas
  United States

Abilene Arch
use Abilene

Abitibi
County on James Bay.
BT Quebec
  Canada

Abkhasia
use Abkhazia

Abkhazia
Abkhaz Autonomous Soviet Socialist Republic. Also search Abkhasia.
CO N422000N432000
  E0421500E0400000
UF Abkhasia
BT Georgian Republic
  USSR

ablation
Includes use on level 3 under sedimentation(1).
SA glaciers
  mass balance
  sedimentation
  wind erosion
  wind transport

Abo
use Turku

abrasion
UF mechanical erosion
SA detritus
  erosion
  glaciation
  grinding
  planation

abrasives
General. Includes use as level 3 commodity term under industrial minerals(1). See list C.
SA corundum
  diamonds
  diatomite
  garnet
  industrial minerals
  pumice
  silica

Abruzzi
Autonomous region on the Adriatic Sea.
BT Italy
NT Chieti
  Marsica

Absaroka Mountains
use Absaroka Range

Absaroka Range
Range of the Rocky Mountains. Index states as applicable. Also search Absaroka Range.
CO N433000N450000
  W1090000W1101500
UF Absaroka Mountains
BT United States
SA Beartooth Mountains
  Montana
  Rocky Mountains
  Wyoming

absarokite
Includes use on level 3 under igneous rocks(1) basalt family(2). See list H.
BA basalt family
BT igneous rocks

absolute age
For radiometric or radiogenic (isotopic) dating methods; for other methods (non-isotopic) see geochronology. Includes use on level 1; on level 2 under orogeny(1). If 1, term set options are:
dates
  material [rock group, rock type, mineral name (e.g. charcoal, granite, metamorphic rocks, shells, sediments)]
methods
  name of method [Ar/Ar, C-14, H-3, He-4/He-3, Io/Th, K/Ar, Pb-alpha, Pb/Pb, Pb/Th, Pb-210, Re/Os, Sr/Rb, Th/Th, Th/U, U/He, U/Pb, U-238/Pb-206, U-235/Pb-207, U/Pd, U/Th/Pb, uranium disequilibrium]
techniques
  subtopic [e.g. instruments, models, sample preparation, sampling]
topic [applications, bibliography, catalogs, interpretation, philosophy]
  subtopic
UF actual age (absolute age)
SA age
  Ar/Ar
  C-14
  changes of level
  charcoal
  dates
  diffusion
  geochemistry
  geochronology
  He-4/He-3
  Io/Th
  Io/U
  isochrons
  isotopes
  K/Ar
  new methods
  orogeny
  Pb-210
  Pb/Pb
  Pb/Th
  radioactive decay
  radiometric properties
  relative age
  Sr/Rb
  Th/Th
  Th/U
  time
  U-238/Pb-206
  U/He
  U/Pb
  U/Th/Pb
  uranium disequilibrium
  whole rock

absorption
Includes use on level 2 under aurora(1); on level 3 under spectroscopy(1) methods(2); on level 3 under geochemistry(1) processes(2).
SA absorption and scattering
  adsorption
  atomic absorption
  aurora
  emission spectroscopy
  geochemistry
  sorption
  spectroscopy

absorption and scattering
Includes use on level 2 under aeronomy(1).
SA absorption
  aeronomy
  aurora
  scattering

absorption spectroscopy
Not a valid index term. Use absorption under spectroscopy(1).

Abu Dhabi
Sheikdom. One of federation of 7 states at S end of Persian Gulf. Includes use on level 3 as an area term (list O).
BT United Arab Emirates
  Arabian Peninsula

Abukuma Mountains
NE of Utsunomiya in E central Honshu. Also search Abukuma Plateau.
UF Abukuma Plateau
BT Honshu
  Japan

Abukuma Plateau
use Abukuma Mountains

abundance
Includes use on level 2 under commodity terms (list C) and under chemical elements (list D); on level 2 under isotopes(1).
SA organic materials

abyssal cones
use submarine fans

abyssal environment
Term introduced in 1978. Before 1978, search environment.
SA deep-sea environment
  environment
  marine environment
  oceanography

abyssal fans
use submarine fans

abyssal plains
Includes use on level 3 under oceanography(1).
SA continental rise
  ocean basins
  ocean floors
  oceanography
  plains

Abyssinian Rift valley
use Ethiopian Rift

abyssolith
use batholiths

Ac
use actinium

Acadian
Provincial series, Canada. Includes use on level 3 under age terms(1). See list E.
BA Middle Cambrian
  Cambrian
BT Paleozoic
NT Lancara Formation
SA Acadian Phase
  Canada

Acadian Orogeny
use Acadian Phase

Acadian Phase
Use Acadian for the age. Before 1978, also search Acadian Orogeny. Also search orogeny AND Acadian.
UF Acadian Orogeny
BT Devonian
SA Acadian
  Antler Orogeny
  orogeny

Acantharina
Suborder. Includes use on level 2 under Radiolaria(1). See list F.
BA Porulosida
  Radiolaria
BT Invertebrata

acanthite
BA sulfides
BT minerals
SA argentite

Acanthodes
Genus. Includes use on level 3 under Pisces(1) Osteichthyes(2).

BA Osteichthyes
    Pisces
BT Vertebrata

acanthodians
use Acanthodii

Acanthodii
Subclass. Includes use on level 3 under Pisces(1) Placodermi(2). Also search acanthodians.
UF acanthodians
BA Placodermi
    Pisces
BT Vertebrata

acanthopores
SA Bryozoa

Acceglio
Village in S.
BT Italy
SA Piedmont

accelerograms
Includes use on level 3 under seismology(1). Also search acceleration; accelerographs; accelerometers.
SA engineering geology
    seismology
    seismometers

accessory minerals
Includes use on level 3 under petrology(1). Before 1978, search minerals AND accessory.
BT minerals
SA heavy minerals

Accomac County
use Accomack County

Accomack County
On the Delmarva Peninsula. Also search Accomac County.
UF Accomac County
BT Virginia
    United States

Accra
City on the Gulf of Guinea.
BT Ghana

accretion
Term used for sedimentation through 1977. After 1977, term includes use under planetology(1) as genetic concept for the Moon and planets.
SA deposition
    Moon
    planetology
    sedimentation

accumulation
Used as a general term.
SA deposition
    glaciers
    ice
    precipitation
    snow

accuracy
Used as a general term.
SA calibration
    corrections
    efficiency
    errors
    reliability

Acer
Genus. Includes use on level 3 under angiosperms(1) Dicotyledoneae(2).
BA Dicotyledoneae
    angiosperms
BT Plantae

Acheulean
use Acheulian

Acheulian
Archaeologic classification.
UF Acheulean
    Acheullian
BA Paleolithic

BT Cenozoic

Acheullian
use Acheulian

achondrites
Includes use on level 3 under meteorites(1).
BA meteorites
SA chondrites
    howardites
    stony irons

acid mine drainage
Before 1978, also search drainage AND mines.
SA drainage
    environmental geology
    mines
    pollution
    sulfuric acid

acidic
A valid level 2 index term through 1977. After 1977, use acidic composition on level 2 under igneous rocks(1).

acidic composition
Term introduced in 1978. Includes use on level 2 under igneous rocks(1). Before 1978, also search acidic.
SA acids
    composition
    igneous rocks
    pH

acidity
use pH

acids
SA acidic composition
    amino acids
    compounds
    fatty acids
    fulvic acids
    humic acids
    pH
    sulfuric acid

acmite
BA pyroxene group
    chain silicates
    silicates
BT minerals
SA aegirine

Aconcagua Province
Central Chile. Also search Aconcagua.
BT Chile
NT La Ligua

acoustic logging
use acoustical logging

acoustic methods
use acoustical methods

acoustic surveys
use acoustical surveys

acoustic waves
use acoustical waves

acoustical
A valid level 2 index term through 1977. After 1977, use acoustical logging on level 2 under well-logging(1).

acoustical logging
Not a valid index term from 1975 to 1977. After 1977, includes use on level 2 under well-logging(1). Before 1978, also search well-logging AND acoustic; acoustic logging.
UF acoustic logging
    logging, acoustical
BA well-logging
SA acoustical methods
    acoustical surveys

acoustical methods
Includes use on level 2 under geophysical methods(1). Also search

acoustic methods; acoustic.
UF acoustic methods
BA geophysical methods
SA acoustical logging
    acoustical surveys
    deep-tow methods
    echo sounding
    methods
    sonar methods

acoustical properties
Includes use on level 3 under engineering geology(1) or geophysical surveys(1).
SA engineering geology
    geophysical surveys
    physical properties
    properties

acoustical surveys
Includes use on level 2 under geophysical surveys(1). Also search acoustic surveys; acoustic.
UF acoustic surveys
BA geophysical surveys
BT surveys
SA acoustical logging
    acoustical methods
    sonar methods

acoustical waves
Also search acoustic AND waves; acoustical AND waves; sonic waves; sound waves.
UF acoustic waves
    sonic waves
    sound waves
SA waves

Acqui
Town in SE.
BT Italy
SA Piedmont

acquisition, data
use data acquisition

acritarchs
Hystrichosphaerids are included here and under Dinoflagellata. Includes use on level 2 under palynomorphs(1). See list F.
BA palynomorphs
NA Baltisphaeridium
NZ Hystrichosphaeridae
SA Dinoflagellata

actinium
Includes use on level 1 and 2 as a chemical element (list D).
UF Ac
SA elements

actinolite
BA amphibole group
    chain silicates
    silicates
BT minerals
SA actinolite facies
    asbestos

actinolite facies
Term introduced in 1978.
BT facies
SA actinolite
    metamorphic rocks

Actinopterygii
Includes use on level 3 under Pisces(1) Osteichthyes(2). See list F.
BA Osteichthyes
    Pisces
BT Vertebrata
NA Amiidae
SA Cyprinidae
    fish

action, frost
use frost action

activation analysis
Including field applications and instruments. Includes use on level 3 under chemical analysis(1) methods(2). Also search activation.

UF radioactivation analysis
SA analysis
    chemical analysis
    isotopes
    neutron activation analysis

activation energy
SA energy
    particles

active faults
BA faults

active layer
As of 1978, term is used on level 2 under permafrost(1).
UF annually thawed layer
    layer, active
    mollisol
SA permafrost
    soils

activity
Includes use on level 3 under geochemistry(1).
SA geochemistry

activity, igneous
use igneous activity

actual age (absolute age)
use absolute age

Acungui Group
BT Precambrian
SA Brazil
    Parana

Ada County
SW Idaho.
BT Idaho
    United States

Adak Island
In central part of Aleutian Islands SW of Alaska Peninsula.
BT Alaska
    United States

Adalia
use Antalya

Adamawa
Administrative region in N.
BT Cameroon

adamellite
Includes use on level 3 under igneous rocks(1) granite-granodiorite family(2). See list H.
BA granite-granodiorite family
BT igneous rocks

Adamello Massif
In Rhaetian Alps in N.
BT Italy

adamite
BA arsenates
BT minerals

Adamow Mine
In W central part of country.
BT Poznan
    Poland

Adams County
Index states as applicable.
BX Colorado
    Idaho
    Illinois
    Indiana
    Iowa
    Mississippi
    Nebraska
    North Dakota
    Ohio
    Pennsylvania
    Washington
    Wisconsin
BT United States

Adana
Province in S Anatolia. Also a city.
BT Turkey
    Middle East

adaptation
Includes use on level 3 under ecology(1) or paleoecology(1).
SA ecology
paleoecology
paleontology

Adavale Basin
S central.
BT Queensland
Australia

Addis Ababa
City in central part of country.
BT Ethiopia

Adelaide
City on Gulf Saint Vincent in SE.
BT South Australia
Australia

Adelaide Geosyncline
Also search Adelaide.
BT South Australia
Australia
SA geosynclines

Adelaide Island
Off W coast of Antarctic Peninsula S of Cape Horn.
BT Antarctica

Adelaidean
SE Australia.
UF Adelaidean system
BA Precambrian
SA upper Precambrian
upper Proterozoic

Adelaidean system
use Adelaidean

Adelie Coast
In the French Sector on the Indian Ocean side S of Australia. Also search Adelie Land.
UF Adelie Land
BT Antarctica

Adelie Land
use Adelie Coast

Aden
City on Gulf of Aden. Former British Colony.
BT Southern Yemen
Arabian Peninsula

Adirondack Anorthosite
NE New York.
BT Precambrian
SA New York

Adirondack Mountains
NE New York. Also search Adirondacks.
CO N430000N443000
W0733000W0750000
UF Adirondacks
BT New York
United States

Adirondacks
use Adirondack Mountains

Admire
Village in Lyon County in E central.
BT Kansas
United States

Adour Basin
River basin in extreme SW France. Also search Adour.
BT France

Adrar
Interior region in W central near Western Sahara border.
BT Mauritania

Adriatic Coast
use Adriatic region

Adriatic region
Index countries as applicable. Also search Adriatic Coast.
UF Adriatic Coast
BT Europe
SA Albania
Italy
Yugoslavia

Adriatic Sea
Between Italy on the W, and Albania and Yugoslavia on the E. Includes use on level 1 as an area term (list O). For term set options see list B.
BA Mediterranean Sea
SA Mediterranean region

adsorption
Includes use on level 3 under geochemistry(1) processes(2).
SA absorption
chromatography
clay mineralogy
desorption
geochemistry
processes
solution
sorption

adularia
BA feldspar group
framework silicates
silicates
BT minerals
SA aluminosilicates
K-feldspar
orthoclase

Adzhar
use Adzharistan

Adzharia
use Adzharistan

Adzharistan
Adzhar Autonomous Soviet Socialist Republic in SW Georgian Republic. Also search Adzhar; Adzharia.
CO N410000N411500
E0423000E0411500
UF Adzhar
Adzharia
BT Georgian Republic
USSR

Aegean Islands
Administrative region comprising islands in the Aegean Sea.
BT Greece
NT Chios
Cyclades
Dodecanese
Karpathos Island
Kos
Lesbos
Naxos
Rhodes
Samos
Santorin
Thera

Aegean Sea
Between Greece on the W and Turkey on the E. As of 1977 includes use on level 1 as an area term (list O). See list B for term set options.
CO N360000N410000
E0282000E0230000
BA Mediterranean Sea

aegerine
use aegirine

Aegina
Town and island in Saronic Gulf. Central Greece and Euboea.
BT Greece

aegirine
UF aegerine
BA pyroxene group
chain silicates
silicates
BT minerals
SA acmite

aenigmatite
UF enigmatite
BA chain silicates
silicates
BT minerals
SA rhonite

Aeolian Islands
use Lipari Islands

aeolotropy
use anisotropy

aerial photography
As of 1978, term is used on level 2 under remote sensing(1). Also search aerial; aerial photographs.
SA cartography
geomorphology
landform description
maps
photogeology
photogrammetry
photography
remote sensing

aerobic environment
Term introduced in 1978. Before 1978, search aerobic.
SA anaerobic environment
environment

aeromagnetic
Not a valid term for GeoRef. Use magnetic surveys and airborne methods.

aeromagnetic surveys
A valid term through 1973. After 1973, use airborne methods under geophysical surveys(1) magnetic surveys(2).

aeronomy
Usually out-of-scope for GeoRef. Includes use on level 1. Term set options are:
absorption and scattering
subtopic
composition
element or compound
densities and temperatures
element or compound
diffusion
mechanism
instruments
name of instrument
ionization
latitude
source of energy [e.g. cosmic rays, X-rays]
techniques
name of technique
tides
subtopic
turbulence
type
waves
type
winds
altitude by region [e.g. stratosphere, troposphere]
SA absorption and scattering
atmosphere
aurora
cosmic rays
densities and temperatures
diffusion
interplanetary space
ionization
ionosphere
magnetosphere
meteorology
stratosphere
tides
troposphere
turbulence
waves
winds

aerosols
Includes use on level 2 under meteorology(1).
SA air-sea interface
atmosphere
convection
meteorology
particles
radioactive tracers
turbidity

aeschynite
UF aschynite
eschinite
eschynite
BA oxides
BT minerals
SA niobates

Afar
Large desert region mostly in Ethiopia. Also search Danakil. Index Djibouti and/or Ethiopia as applicable.
UF Danakil
Dankalia
BT Africa
SA Djibouti
Ethiopia

Afar Depression
Part of Afar lying between the "Danakil Alps" and the Ethiopian highlands in NE Ethiopia. Section of the Great Rift Valley. Also search Afar; Afar Rift; Danakil; Danakil Depression.
UF Afar Rift
Danakil Depression
BT Ethiopia
SA Great Rift Valley

Afar Rift
use Afar Depression

Afars and Issas Territory
A valid level 1 term through 1977. After 1977, use Djibouti.

affinities
Includes use on level 2 or 3 under commodity terms (list C); used mostly on level 3 of fossil groups(1). See list F. Also search geochemical affinities.

Afghan-Tadzhik Basin
use Afghan-Tadzhik Depression

Afghan-Tadzhik Depression
Index Afghanistan and/or Tadzhikistan as applicable. Also search Afghan-Tadzhik; Afghan-Tadzhik Basin.
UF Afghan-Tadzhik Basin
BT Asia
SA Afghanistan
Tadzhik Depression
Tadzhikistan

Afghanistan
Includes use on level 1 as an area term (list O). For term set options see list B.
CO N290000N381500
E0750000E0600000
BA Asia
NT Altimur Mountains
Bamian
Dasht-i-Nawar
Ghazni
Kabul
Logar
Paktia
Wardak
SA Afghan-Tadzhik Depression
Amu Darya
Badakhshan
Hindu Kush
Murgab Basin
Pamirs
Turkestan

Africa
Includes use on level 1 or 2 as an area term (list O). For term set options see list B. To retrieve all documents, individual countries and physiographic regions should also be searched (see list O).

Africa ● airborne

CO S350000N370000
  E0510000E0180000
NT Afar
   African Platform
NA Algeria
   Angola
NT Atlas Mountains
NA Benin
NT Benue Valley
   Blue Nile
NA Botswana
   Burundi
   Cameroon
NT Cap Blanc
NA Cape Verde Islands
   Central Africa
   Central African Republic
   Chad
NT Chad Basin
NA Congo
NT Congo Basin
   Congo River
NA Djibouti
   East Africa
NT East African Rift
NA Egypt
   Equatorial Guinea
   Ethiopia
   Gabon
   Gambia
   Ghana
NT Gregory Rift
NA Guinea
   Guinea-Bissau
   Ivory Coast
NT Kalahari Desert
   Kasai River
NA Kenya
NT Lake Albert
   Lake Chad
   Lake Edward
   Lake Kariba
   Lake Kivu
   Lake Malawi
   Lake Natron
   Lake Tanganyika
   Lake Turkana
   Lake Victoria
   Lebombo Mountains
NA Lesotho
   Liberia
   Libya
NT Libyan Desert
   Limpopo Basin
   Logone River
NA Malagasy Republic
   Malawi
   Mali
NT Mali-Niger Syneclise
NA Mauritania
   Morocco
   Mozambique
NT Namaqualand
NA Niger
   Niger River
   Niger Valley
NA Nigeria
NT Nile River
   Nile Valley
   Nimba Mountains
NA North Africa
NT Nubia
   Orange River
   Reguibat Ridge
NA Rhodesia
   Rwanda
   Sahara
NT Sahel
NA Senegal
NT Senegal Basin
   Senegal River
NA Sierra Leone
   Somali Republic
   South Africa
   South-West Africa
   Southern Africa
   Sudan
   Swaziland
NT Tanezrouft
NA Tanzania
NT Taourirt
NA Togo
   Tunisia
   Uganda
NT Umba
NA Upper Volta
NT Volta Basin
NA West Africa
NT West African Shield
NA Zaire
NT Zambezi Valley
NA Zambia
SA African Plate
   Dwyka Series
   Gondwana
   Karroo System
   Katangan Orogeny
   Nubian Sandstone
   Tethys

African Plate
Includes the continent of Africa, adjoining areas of the Atlantic and Indian oceans, the Malagasy Republic, and much of E Mediterranean Sea.
SA Africa
   plate tectonics

African Platform
A vast platform of complex Precambrian and Paleozoic rocks, partly covered by Mesozoic and Tertiary sedimentary rocks, which underlies the entire continent.
BT Africa

aftershocks
Includes use on level 2 under earthquakes(1).
SA earthquakes
   focus
   foreshocks
   seismology

Aftonian
Interglacial interval. North America.
BA lower Pleistocene
   Pleistocene
   Quaternary
BT Cenozoic

Ag
use silver

Agades
Town in W central Niger. Also search Agadez.
UF Agadez
BT Niger

Agadez
use Agades

Agarak
Mining region in central.
BT Armenia
   USSR

agate
BA silica minerals
   framework silicates
   silicates
BT minerals
SA chalcedony
   gems
   quartz

age
Includes use as level 2 or 3 term appropriate to a large number of topics, e.g. on level 2 under ocean basins(1) and under ground water(1). See list G.
SA absolute age
   exposure age
   relative age

Ager Formation
Catalonia and the Pyrenees.
BT lower Eocene
   Eocene
   SA Spain

agglomerate
Includes use on level 3 under igneous rocks(1) or sedimentary petrology(1).
SA breccia
   conglomerate
   igneous rocks
   pyroclastics
   sedimentary petrology

agglutinates
SA fines
   Moon
   particles

aggradation
SA geomorphology
   processes

aggregate
Includes use as level 3 commodity term under construction materials(1). See list C.
BA construction materials
SA gravel
   pumice
   rocks
   sand

aggregation
A valid term through 1975. After 1975, use sedimentation.

Agly Massif
S part of country.
BT Pyrenees-Orientales
   France

agmatite
Includes use on level 3 under metamorphic rocks(1) migmatites(2). See list J.
BA migmatites
BT metamorphic rocks

Agnatha
Including ostracoderms. Includes use on level 2 under Pisces(1). See list F.
BA Pisces
BT Vertebrata
NA Heterostraci
   ostracoderms
SA fish

Agnes
A storm in 1972 on the East Coast of United States.
SA hurricanes

Agnostida
Includes use on level 2 under Trilobita(1). See list F.
BA Trilobita
BT Trolobitomorpha
   Arthropoda
   Invertebrata

Agnotozoic
use Proterozoic

agpaite
Includes use on level 3 under igneous rocks(1) syenite family(2). See list H.
BA syenite family
BT igneous rocks

Agricola Lake
BT Northwest Territories
   Canada

agricultural waste
Term introduced in 1978. Before 1978, search agricultural.
UF waste, agricultural
SA agriculture
   waste disposal

agriculture
As of 1978, term is used on level 2 under land use(1).
SA agricultural waste
   fertilization
   fertilizers
   land use
   soils
   utilization

Agto
Fishing settlement on small island in Davis Strait.
BT Greenland
   Arctic region

Agua Blanca Fault
Index Baja California and/or California.
SA Baja California
   California

Agulhas Bank
South of Cape of Good Hope.
BT Indian Ocean

Ahaggar
Volcanic upland in SE Algeria. Also search Ahaggar Mountains.
UF Ahaggar Mountains
   Hoggar Mountains
BT Algeria
SA Sahara

Ahaggar Mountains
use Ahaggar

ahermatypic taxa
Term introduced in 1978 on level 3 under fossil groups (list F). Before 1978, search ahermatypic.
UF taxa, ahermatypic
SA corals
   hermatypic taxa

Ahnet
Region in S central.
BT Algeria

Aichi
Prefecture in central Honshu. Capital is Nagoya.
BT Honshu
   Japan

Aigoual Massif
In Cevennes Mountains in S.
BT France

Aiguille Rouges
Alpine range in S.
BT France

Aiken
City in Aiken County in W.
BT South Carolina
   United States

Aiken County
W South Carolina.
BT South Carolina
   United States

aikinite
BA sulfides
BT minerals

Ain
Department in E France. Also a river.
BT France

air
As of 1978, term is used on level 2 under pollution(1).
SA atmosphere
   pollution

air-sea interface
Includes use on level 3 under oceanography(1) or meteorology(1) aerosols(2).
UF interface, air-sea
SA aerosols
   atmosphere
   oceanography

Aira Caldera
BT Kyushu
   Japan

airborne
A valid index term through 1977. After 1977, use airborne methods.

airborne methods
  Term introduced in 1978. Includes use on level 3 under geophysical surveys(1) magnetic methods(2). Before 1978, also search airborne.
  BA geophysical methods
  SA geophysical surveys
      magnetic methods
      magnetic surveys
      methods
      remote sensing

airglow
  Includes use on level 2 under ionosphere(1).
  SA atmosphere
      aurora
      emissions
      ionosphere
      luminescence

airy waves
  Term introduced in 1978. Includes use on level 2 under ocean waves(1).
  SA ocean waves
      waves

Aisne
  Department in N.
  CO N484500N501000
      E0041000E0010000
  BT France
  SA Oise River valley
      Somme River valley

Aix-en-Provence
  City in S part of country.
  BT Bouches-du-Rhone
      France

Aizu Basin
  Fukushima Prefecture in N central Honshu. Also search Aizu.
  BT Honshu
      Japan

Ajay River
  Index Indian states as applicable.
  BT India
  SA Bihar
      West Bengal

Ajman
  Sheikdom. One of federation of 7 states at S end of Persian Gulf.
  BT United Arab Emirates
      Arabian Peninsula

Ajmer
  City and district in central.
  BT Rajasthan
      India

akaganeite
  BA oxides
  BT minerals

Akbastau
  Village NE of Chimkent in S central.
  BT Kazakhstan
      USSR

Akchagylian
  Europe.
  BA upper Pliocene
      Pliocene
      Neogene
      Tertiary
  BT Cenozoic
  SA upper Tertiary

Akchatau
  Town in SE Karaganda Oblast in E central.
  BT Kazakhstan
      USSR

Akenobe Mine
  Hyogo Prefecture of which Kobe is the capital.
  BT Honshu
      Japan

Akera
  River in SW.
  BT Azerbaidzhan
      USSR

akermanite
  BA melilite group
      orthosilicates
      silicates
  BT minerals

Akhaltsikh
  use Akhaltsikhe

Akhaltsikhe
  City.
  UF Akhaltsikh
  BT Georgian Republic
      USSR

Akhtala
  Town in N.
  BT Armenia
      USSR

Akita
  City on Japan Sea in N.
  BT Honshu
      Japan

Akiyoshi
  Village in extreme SW.
  BT Honshu
      Japan

Akiyoshi Limestone
  SW Honshu.
  BT Permian
  SA Honshu
      Japan

Akjoujt
  Village in W central.
  BT Mauritania

Akron
  City in Summit County in NE.
  BT Ohio
      United States

Aksu
  Village in Taldy Kurgan Oblast in SE.
  BT Kazakhstan
      USSR

Aktyubinsk
  City and oblast in NW.
  BT Kazakhstan
      USSR

Akzhal
  Town in E Semipalatinsk Oblast in E.
  BT Kazakhstan
      USSR

Al
  use aluminum

Al-Kufrah
  use Kufra Basin

Al-Qusayr
  use Kosseir

Al-Quseir
  use Kosseir

Al-26
  Includes use on level 3 under isotopes(1).
  SA aluminum
      isotopes

Ala-Kul Lake
  use Alakol

Alabama
  Includes use on level 1 as an area term (list O). For term set options see list B.
  CO N300000N350000
      W0850000W0883000
  BA United States
  NT Alabama River
  NX Blount County
      Butler County
      Calhoun County
      Cherokee County
  NT Chilton County
  NX Clarke County
      Clay County
      Coffee County
  NT Coosa County
  NX Dallas County
  NT Escambia County
      Etowah County
  NX Fayette County
      Franklin County
      Greene County
  NT Highland Rim
  NX Jackson County
      Jefferson County
      Lawrence County
      Lee County
      Limestone County
      Macon County
      Madison County
      Marion County
      Marshall County
  NT Martin Lake
      Mobile
      Mobile County
  NX Monroe County
      Montgomery County
      Morgan County
      Perry County
      Pike County
      Randolph County
  NT Red Mountain
  NX Russell County
      Saint Clair County
  NT Talladega Front
      Tallapoosa County
      Tuscaloosa County
  NX Washington County
  SA Appalachian Basin
      Appalachian Plateau
      Bangor Limestone
      Birmingham
      Black Warrior Basin
      Brevard Zone
      Chattanooga Shale
      Chesterian
      Chickamauga Group
      Chipola Formation
      Citronelle Formation
      Claiborne Group
      Clayton Formation
      Cumberland Plateau
      Fernvale Formation
      Fort Payne Formation
      Gulf Coastal Plain
      Jackson Group
      Marianna Limestone
      Mississippi Embayment
      Monteagle Limestone
      Ocala Group
      Pennington Formation
      Piedmont
      Pottsville Group
      Red Mountain Formation
      Ripley Formation
      Rome Formation
      Saint Louis Limestone
      Sainte Genevieve Limestone
      Shady Dolomite
      Smackover Formation
      Talladega Group
      Tennessee River
      Tennessee Valley
      Tuscaloosa Formation
      Valley and Ridge Province
      Vicksburg Group
      Warsaw Formation
      Yazoo Clay

Alabama River
  Central and SW.
  BT Alabama
      United States

alabandine
  use alabandite

alabandite
  Also search alabandine.
  UF alabandine
      manganblende
  BA sulfides
  BT minerals

alabaster
  UF onyx marble
  BA sulfates
  BT minerals
  SA aragonite
      calcite
      gypsum

Alacran Reef
  Off north Yucatan coast.
  BT Gulf of Mexico

Alae Crater
  Kilauea volcano on island of Hawaii.
  BT Hawaii
      United States
  SA Kilauea

Alagoas
  State in NE Brazil. Also a city.
  BT Brazil
  SA Sao Francisco Basin
      Sergipe-Alagoas Basin

Alai Range
  One of the W ranges of the Tien Shan in SW Kirghizia. Also search Alai.
  UF Alay Range
  BT Kirghizia
      USSR
  SA Darvaza Range
      Hissar Range

Alaia
  use Alanya

Alais
  use Ales

Alaiye
  use Alanya

Alakol
  Lake in Taldy Kurgan Oblast E of Lake Balkhash in E Kazakhstan. Also search Ala-Kul Lake.
  UF Ala-Kul Lake
  BT Kazakhstan
      USSR

Alameda County
  On San Francisco Bay.
  CO N373000N375000
      W1213000W1222500
  BT California
      United States

Alaminos Canyon
  BT Gulf of Mexico

Aland
  Islands belonging to Finland. Also search Aland Islands; Aland Archipelago.
  UF Aaland
      Aland Archipelago
      Aland Islands
  BT Baltic Sea
      Atlantic Ocean

Aland Archipelago
  use Aland

Aland Islands
  use Aland

Alandroal
  Town in S central part of country.
  BT Evora
      Portugal

Alanya
  Town on the Gulf of Adalia in S Anatolia.
  UF Alaia
      Alaiye
      Alaya
  BT Turkey
      Middle East

Alaska
  Includes use on level 1 as an area term (list O). For term set options see list B.
  CO N510000N720000
      W1300000E1730000

Alaska • Aleutian Arc

BA United States
NT Adak Island
Alaska Peninsula
Alaska Range
Aleutian Islands
Amchitka Island
Anchorage
Augustine
Barrow
Birch Creek
Brooks Range
Cannikin
Chugach Mountains
Colville River
Colville River delta
Cook Inlet
Delta River
Denali Fault
Fairbanks
Fairweather Fault
Glacier Bay
Juneau
Juneau ice field
Katmai National Monument
Kenai Peninsula
Knik Arm
Kodiak Island
Martin River Glacier
Matanuska Valley
Meade Basin
Montague Island
Nome
North Slope
Nunivak Island
Point Barrow
Prince William Sound
Prudhoe Bay
Rat Island
Rat Islands
Seward Peninsula
Shemya Island
Sitka Sound
Susitna River basin
Trans-Alaska Pipeline
Valdez
Valley of Ten Thousand Smokes
Wrangell Mountains
Yukon-Tanana Upland
SA Arctic region
Bowser Formation
Chitistone Pass
Coast Mountains
Cordillera
Kenai Group
Lisburne Group
Prince of Wales Island
Rampart Group
Rocky Mountains
Sadlerochit Formation
Saint Elias Mountains
Tintina fault zone
Western U.S.
White Mountain
White River
Yukon River

Alaska Peninsula
SW Alaska.
BT Alaska
United States

Alaska Range
S central.
BT Alaska
United States

alaskite
Includes use on level 3 under igneous rocks(1) granite-granodiorite family(2). See list H.
BA granite-granodiorite family
BT igneous rocks

Alava
One of the Basque Provinces in N.
BT Spain
SA Basque Provinces

Alaverdi
City in N Armenia.
UF Allaverdy
BT Armenia
USSR

Alay Range
use Alai Range

Alaya
use Alanya

Alazani
Mining and oil producing area in E.
BT Georgian Republic
USSR

Alba-Iulia
Town in Alba County in SW.
BT Transylvania
Romania

Albacete
Province in SE.
BT Spain

Alban Hills
Part of the Lower Apennines SE of Rome.
BT Latium
Italy
SA Apennines

Albania
Includes use on level 1 as an area term (list O). For term set options see list B.
BA Europe
NT Albanides
Ionian Zone
SA Adriatic region
Balkan Peninsula
Mediterranean region

Albanides
Mountain area in W and SW.
BT Albania
SA Alps

Albany
City in Albany County in E.
BT New York
United States

Albany County
Index states as applicable.
BX New York
Wyoming
BT United States

Albatross Cordillera
use East Pacific Rise

albedo
Includes use on level 2 under ionosphere(1) for neutron albedo, magnetic albedo, and albedo of electromagnetic waves; on level 3 under interplanetary space(1) cosmic rays(2) and under magnetosphere(1) cosmic rays(2).
UF Bond albedo
SA cosmic rays
interplanetary space
ionosphere
magnetosphere
Moon
reflectivity
remote sensing

Albert Canyon
Village between Mount Revelstone and Glacier National Park in SE.
BT British Columbia
Canada

Alberta
Includes use on level 1 as an area term (list O). For term set options see list B.
CO N490000N600000
W1100000W1200000
BA Canada
NT Abee
Athabasca
Athabasca Glacier
Athabasca River
Banff
Banff National Park
Bow River Valley
Bruderheim
Calgary
Drumheller
Edmonton
Fort McMurray
Jasper National Park
Lake Louise
Leduc
Medicine Hat
Red Deer River
Smoky River
Strathcona Mine
Swan Hills
SA Altyn Limestone
Bearpaw Formation
Belly River Formation
Blairmore Group
Canadian Cordillera
Canadian Shield
Edmonton Formation
Elk Point Basin
Elk Point Group
Great Plains
Hector Formation
Keg River Formation
Kicking Horse River valley
Mannville Formation
Miette Group
Milk River Formation
Missouri River basin
North Saskatchewan River
Oldman Formation
Palliser Formation
Paskapoo Formation
Peace River
Slave Province
Souris River basin
South Saskatchewan River
Sweetgrass Arch
Viking Formation
Waterways Formation

Albian
Europe. Uppermost Lower Cretaceous or Middle Cretaceous of some authors. Above Aptian, below Cenomanian. Includes use on level 3 under age terms(1). See list E.
BA Cretaceous
BT Mesozoic
SA Lower Cretaceous
Middle Cretaceous

Albion Range
Cassia County in S.
BT Idaho
United States

albite
BA feldspar group
framework silicates
silicates
BT minerals
SA alkali feldspar
andesine
granite
plagioclase

albitite
Includes use on level 3 under igneous rocks(1) trachyte-phonolite family(2). See list H.
BA trachyte-phonolite family
BT igneous rocks

albitization
Includes use on level 3 under metasomatism(1) processes(2).
BA metasomatism
SA autometamorphism
processes
spilitization

albitophyre
Includes use on level 3 under igneous rocks(1) syenite family(2). See list H.
BA syenite family
BT igneous rocks

Alboran Sea
Between S Spain and Morocco in W.
BT Mediterranean Sea

Alborz Mountains
use Elburz

Albuquerque
City in Benalillo County in central.
BT New Mexico
United States

Alcoy
City in SE part of country.
BT Alicante
Spain

Alcudia Valley
S central part of country.
BT Ciudad Real
Spain

Aldabra Island
Atoll 250 miles NW of Malagasy Republic.
CO S092400S092400
E0462000E0462000
BT Seychelles
Indian Ocean

Aldan
City in SE Yakutia.
BT Russian Republic
USSR

Aldan Plateau
SE Yakutia.
CO N562000N581000
E1303000E1230000
BT Russian Republic
USSR

Aldan River
SE Yakutia. Flows into the Lena River.
BT Russian Republic
USSR

Aldan Shield
The Aldan Plateau area of SE Yakutia. Part of greater Angara Shield.
CO N540000N610000
E1350000E1230000
BT Russian Republic
USSR

Aldanian
Europe.
BA Lower Cambrian
Cambrian
BT Paleozoic

Aleksinac
Town in E Serbia.
UF Alexinats
Alexinatz
BT Serbia
Yugoslavia

Aleksod
Region in Ahaggar in SE.
BT Algeria

Alentejo
Former province in central and S.
BT Portugal

Ales
City in S part of country. Also search Alais.
UF Alais
BT Gard
France

Alessandria
City and province in SE.
BT Italy
SA Piedmont

Aleutian Arc
use Aleutian Islands

Aleutian Islands
Chain of volcanic islands extending in great curve from Alaska Peninsula in SW. Also search Aleutian Arc and Aleutians.
CO N510000N553000
W1610000E1720000
UF Aleutian Arc
Aleutians
BT Alaska
United States
SA Aleutian Trench
Amchitka Island
Rat Island
Rat Islands

Aleutian Ridge
Along S side of Aleutian chain.
BT Pacific Ocean
SA Aleutian Trench

Aleutian Trench
S of Aleutian Ridge and extending along the entire Aleutian chain.
BT Pacific Ocean
SA Aleutian Islands
Aleutian Ridge

Aleutians
use Aleutian Islands

Alexander Island
Large island off W coast of Antarctic Peninsula S of Cape Horn.
BT Antarctica

Alexandria
Governate. Also a city on the Mediterranean Sea.
BT Egypt

Alexandrian
North American provincial series. Lower Silurian (above Cincinnatian of Ordovician, below Niagaran).
UF Alexandrian Series
BA Lower Silurian
Silurian
BT Paleozoic

Alexandrian Series
use Alexandrian

alexandrite
BA oxides
BT minerals
SA chrysoberyl

Alexinats
use Aleksinac

Alexinatz
use Aleksinac

Alfisols
Includes use on level 3 under soils(1). See list M.
BA soils

Alfold
Mostly in central and E Hungary. Small part in NE Yugoslavia and W Romania. Index countries as applicable. Also search Great Hungarian Plain; Hungarian Basin; Hungarian Great Plain; Hungarian Plain.
UF Great Alfold
Great Hungarian Plain
Hungarian Basin
Hungarian Great Plain
Hungarian Plain
BT Europe
SA Hungary
Romania
Yugoslavia

algae
Includes use on level 1 and 2 as a fossil term (list F).
BT Plantae
NA calcareous algae
Charophyta
Chlorophyta
Chrysophyta
Coccolithophoraceae
Cyanophyta
Dasycladaceae
diatoms
Microcodium
nannofossils
Phaeophyta
Rhodophyta
NZ stromatolites
NA Tasmanites
SA algal biscuits
algal mats
algal mounds
bioherms
fungi
lichens
living culture
microorganisms
phytoplankton
plankton
Protista
Receptaculitaceae
zooxanthellae

algal banks
See list K.
BA biogenic structures
sedimentary structures
SA banks

algal biscuits
Includes use on level 3 under sedimentary structures(1) biogenic structures(2). See list K.
UF lake biscuits
water biscuits
BA biogenic structures
sedimentary structures
SA algae
girvanella

algal limestone
Common in early Paleozoic, especially in the Cambro-Ordovician of the Appalachian region of the United States. Before 1978, also search algal AND limestone.
BA limestone
carbonate rocks
BT sedimentary rocks

algal mats
Includes use on level 3 under sedimentary structures(1) biogenic structures(2). See list K.
BA biogenic structures
sedimentary structures
SA algae
stromatolites

algal mounds
Includes use on level 3 under sedimentary structures(1) biogenic structures(2). See list K.
UF mounds, algal
BA biogenic structures
sedimentary structures
SA algae
calcilutite
limestone

Algarve
Region in extreme S Portugal. Former province.
BT Portugal

Algau Alps
use Allgau Alps

Algeria
Includes use on level 1 as an area term (list O). For term set options see list B.
CO N190000N370000
E0120000W0090000
BA Africa
NT Ahaggar
Ahnet
Aleksod
Algiers
Annaba
Aures
Babor Range
Constantine
Dellys
Gour Oumelalen
Great Kabylia
Hassi Messaoud Field
Kabylia
Laghouat
Mouydir
Oran
Ougarta
Saoura
Tassili n'Ajjer
Tell
Tindouf
Tindouf Basin
SA Atlas Mountains
Mediterranean region
North Africa
Sahara
Tanezrouft
Taourirt
West Africa

Algerian Plain
use Balearic Basin

Algiers
City on Mediterranean Sea.
BT Algeria

Algoma
District on W Lake Superior.
BT Ontario
Canada

Algonkian
Includes use on level 3 as an age term (list E).
BA Precambrian
NT Baraboo Quartzite
SA Proterozoic
upper Precambrian

algorithms
Includes use on level 3 under automatic data processing(1) or mathematical geology(1).
SA automatic data processing
mathematical geology

Alicante
Province in SE Spain. Also a city on the Mediterranean Sea.
BT Spain
NT Alcoy

Alice Arm
Village on Observatory Inlet near Alaska border NE of Prince Rupert.
BT British Columbia
Canada

Alice Springs
Town in S central.
BT Northern Territory
Australia

alkali basalt
Includes use on level 3 under igneous rocks(1) alkali basalt family(2). See list H. Also search alkalic basalt; alkaline basalt.
UF alkalic basalt
alkaline basalt
BA alkali basalt family
BT igneous rocks
SA basalt
tholeiite

alkali basalt family
Includes use on level 2 under igneous rocks(1). See list H.
BT igneous rocks
NA alkali basalt
ankaratrite
basanite
crinanite
hawaiite
leucitite
melilitite
nepheline basalt
nephelinite
olivine nephelinite
spilite
tephrite
trachybasalt
trachydolerite
SA basalt family

alkali feldspar
BA feldspar group
framework silicates
silicates
BT minerals
SA albite
feldspar
K-feldspar

alkali gabbro family
Includes use on level 2 under igneous rocks(1). See list H.
BT igneous rocks
NA essexite
ijolite
melteigite
teschenite
urtite
SA foyaite
gabbro family
shonkinite

alkali granite
use alkalic granite

alkali metals
UF alkaline metals
BT metals
SA cesium
lithium
potassium
rubidium
sodium

alkali olivine basalt
BA basalt family
BT igneous rocks
SA basalt
olivine basalt

alkali syenite
Also search alkalic syenite.
UF alkalic syenite
BA syenite family
BT igneous rocks

alkalic
Valid level 2 term through 1977. After 1977, use alkalic composition on level 2 under igneous rocks(1).

alkalic basalt
use alkali basalt

alkalic composition
Term introduced in 1978. Includes use on level 2 under igneous rocks(1). Before 1978, also search alkalic AND igneous rocks.
SA calc-alkalic composition
composition
igneous rocks

alkalic granite
Also search alkali granite.
UF alkali granite
BA granite-granodiorite family
BT igneous rocks
SA granite

alkalic syenite
use alkali syenite

alkaline basalt
use alkali basalt

alkaline metals
use alkali metals

alkalinity
Includes use on level 3 under geochemistry(1).
SA geochemistry
pH

allanite
Also search orthite.
UF orthite
BA epidote group
orthosilicates

allanite • alpha rays

silicates
BT minerals

Allarechenskiy
Ore field in the Kola Peninsula in extreme NW European USSR. Also search Allarechenskoye.
UF Allarechenskoye
BT Russian Republic
  USSR
SA Kola Peninsula

Allarechenskoye
use Allarechenskiy

Allaverdy
use Alaverdi

Allegany County
Index states as applicable.
BX Maryland
  New York
BT United States

Alleghany County
Index states as applicable.
BX North Carolina
  Virginia
BT United States

Alleghany Mountains
use Allegheny Mountains

Alleghenies
use Allegheny Mountains

Allegheny County
W Pennsylvania.
BT Pennsylvania
  United States

Allegheny Front
Eastern slope of the Allegheny Mountains. Index states as applicable.
BT United States
SA Allegheny Mountains
  Maryland
  Pennsylvania
  Virginia
  West Virginia

Allegheny Group
Subdivided into three formations: Clarion, Kittanning, and Freeport. E Kentucky, W Maryland, E Ohio, W Pennsylvania, W Virginia, and West Virginia.
BT Middle Pennsylvanian
  Pennsylvanian
SA Kentucky
  Maryland
  Ohio
  Pennsylvania
  Virginia
  West Virginia

Allegheny Mountains
Ranges of the Appalachians constituting a part of the Allegheny Plateau. Index States as applicable.
UF Alleghany Mountains
  Alleghenies
BT United States
SA Allegheny Front
  Allegheny Plateau
  Appalachians
  Maryland
  Pennsylvania
  Virginia
  West Virginia

Allegheny Orogeny
Term introduced in 1978. An event which deformed the rocks of the Valley and Ridge province, and those of the adjacent Allegheny Plateau in the central and S Appalachians. Before 1978, also search Allegheny AND orogeny.
SA orogeny
  Pennsylvanian
  Permian

Allegheny Plateau
W section of the Appalachians extending from Cumberland Plateau on S to Mohawk Valley in New York on N and including the Allegheny Mountains. Index states as applicable.
BT United States
SA Allegheny Mountains
  Appalachian Plateau
  Appalachians
  Kentucky
  New York
  Ohio
  Pennsylvania
  Virginia
  West Virginia

Allen County
Index states as applicable.
BX Kansas
  Kentucky
  Ohio
BT United States

Allende
Point of impact in SE Chihuahua. Also search Allende Meteorite.
BT Chihuahua
  Mexico
SA meteorites

Allerod
Europe. An interval of late-glacial time following the Older Dryas and preceding the Younger Dryas.
BA Weichselian
  Quaternary
BT Cenozoic

allevardite
use rectorite

Allgaeu Alps
use Allgau Alps

Allgau Alps
Index Bavaria and/or Tyrol as applicable.
CO N471000N473000
  E0105000E0100000
UF Algau Alps
  Allgaeu Alps
BT Alps
  Europe
SA Austria
  Bavaria
  Germany
  Tyrol

Allier
Department in central.
CO N455000N464500
  E0040000E0023000
BT France
SA Limagne

allochthons
Includes use on level 3 under tectonics(1).
SA autochthons
  faults
  structural geology
  tectonics

alloclasite
BA sulfides
BT minerals

Allogromiina
Includes use on level 2 under foraminifera(1). See list F.
BA foraminifera
BT Invertebrata

allometry
SA fossils

allophane
BA sheet silicates
  silicates
BT minerals
SA clay minerals

alloys
SA base metals
  construction materials
  metallurgy
  metals
  native elements and alloys

alluvial deposits
use alluvium

alluvial fans
Includes use on level 3 under geomorphology(1) fluvial features(2).
UF detrital fan
  dry delta
  talus fan
SA Alluvial soils
  alluvium
  deposition
  fans
  fluvial features

alluvial plains
Includes use on level 3 under geomorphology(1).
UF river plain
SA deposition
  geomorphology
  paleogeography
  plains

Alluvial soils
Includes use on level 3 under soils(1). See list M.
BA soils
SA alluvial fans
  alluvium

alluvion
use alluvium

alluvium
Includes use on level 3 under sediments(1) clastic sediments(2). See list N.
UF alluvial deposits
  alluvion
BA clastic sediments
BT sediments
SA alluvial fans
  Alluvial soils
  clay
  colluvium
  eluvium
  floodplains
  gravel
  sand
  silt
  soil mechanics
  soils

Alma Dag
use Amanos Mountains

Alma River
In Crimea. Also search Alma.
BT Ukraine
  USSR

Alma-Ata
City in SE Kazakhstan.
UF Vernyi
  Vyernyi
BT Kazakhstan
  USSR

Almaden
Village in S central part of country.
BT Ciudad Real
  Spain

Almalyk
Town SE of Tashkent.
BT Uzbekistan
  USSR

almandine
UF almandite
BA garnet group
  orthosilicates
  silicates
BT minerals

almandite
use almandine

Almas Valley basin
River basin in W central.
BT Transylvania
  Romania

Almeria
Province in SE Spain. Also a city on the Mediterranean Sea.
BT Spain
NT Cerro del Hoyazo
  Sierra de Gador
  Sierra de los Filabres
  Velez Rubio
SA Andalusia

Almond Formation
In Mesaverde Group. Overlies Ericson Sandstone and underlies Lewis Shale. SW Wyoming.
BT Upper Cretaceous
  Cretaceous
SA Mesaverde Group
  Wyoming

Almora
Town and district in N.
BT Uttar Pradesh
  India

Alno
Island in Gulf of Bothnia.
CO N623000N623000
  E0173000E0173000
BT Sweden
SA Gulf of Bothnia

Alnus
Genus.
BA Dicotyledoneae
  angiosperms
BT Plantae

Alpago Basin
Belluno Province in N.
BT Veneto
  Italy

Alpena County
On Lake Huron in Lower Peninsula.
BT Michigan
  United States

Alpes-de-Haute Provence
Department in SE France. Before 1977, also search Basses-Alpes.
CO N433000N450000
  E0070000E0053000
UF Basses-Alpes
BT France
NT Castellane
  Digne
  Verdon Valley
SA Eoulx Basin
  Provence
  Provence Alps

Alpes-Maritimes
Department in extreme SE.
CO N432000N443000
  E0073000E0063000
BT France
NT Antibes
  Nice
  Saint-Vallier
SA Esterel
  Provence
  Provence Alps

alpha chalcocite
use digenite

Alpha Cordillera
Undersea feature N and NW of Queen Elizabeth Islands of Northwest Territories, Canada.
BT Arctic Ocean

alpha rays
Use when referring to the rays themselves. For application or methodology, use alpha ray-spectroscopy. Includes use on level 3 under spectroscopy(1) and under chemical element(1).
UF rays, alpha

SA alpha-ray spectroscopy
     cosmic rays
     spectroscopy

alpha-ray spectroscopy
Term introduced in 1978. Includes use on level 3 under spectroscopy (1) and under chemical element(1) analysis(2). Before 1978, also search alpha ray AND spectroscopy.
BT spectroscopy
SA alpha rays
     analysis
     gamma-ray spectroscopy
     radioactivity

alpine
A valid index term through 1977. Now use alpine environment, and use Alps for alpine regions.

Alpine
A valid term used for Alpine orogeny through 1977. After 1977, use Alpine Orogeny.

alpine environment
Term introduced in 1978. Includes use on level 3 under ecology(1). Before 1978, also search alpine. For alpine regions, use Alps.
SA Alps
     ecology
     environment

Alpine Fault
BT New Zealand

Alpine Foreland
use Prealps

Alpine Geosyncline
Term introduced in 1978. Before 1978, search geosynclines AND Alpine.
SA Alpine Orogeny
     geosynclines

Alpine Orogeny
Term introduced in 1978. Includes use on level 3 under orogeny(1) and tectonics(1). Before 1978, also search orogeny AND Alpine; Alpine structure.
UF Alpine structure
SA Alpine Geosyncline
     alpine-type
     Cenozoic
     Mesozoic
     orogeny
     tectonics

Alpine structure
use Alpine Orogeny

alpine-type
Used mostly as a general term on level 3 under mineral deposits, genesis(1).
SA Alpine Orogeny
     Alps
     ecology
     mineral deposits
     tectonics

Alps
Great mountain system of S central Europe. Index countries as applicable. Includes use on level 1 as an area term (list O). For term set options see list B.
BA Europe
NT Allgau Alps
     Bavarian Alps
     Carnic Alps
     Central Alps
     Cottian Alps
     Eastern Alps
     Julian Alps
     Karawanken
     Lepontine Alps
     Maritime Alps
     Otztal Alps
     Pennine Alps
     Rhaetian Alps
     Western Alps
     Zillertal Alps
SA Albanides
     alpine environment
     alpine-type
     Austria
     Bernese Alps
     Brianconnais Zone
     Dauphine Alps
     Dinaric Alps
     Dolomites
     France
     Gailtal Alps
     Germany
     Glarus Alps
     Hellenides
     Hohe Tauern
     Italy
     Ligurian Alps
     Limestone Alps
     Northern Limestone Alps
     Piedmont Alps
     Prealps
     Provence Alps
     Sau Alps
     Savoy Alps
     Sesia-Lanzo Zone
     Southern Alps
     Stubai Alps
     Swiss Alps
     Switzerland
     Vanoise
     Yugoslavia

Alpujarras
Mountainous area in S.
BT Spain

Alsace
Region and former province in NE.
BT France

alstonite
UF bromlite
BA carbonates
BT minerals

Alta
Village in extreme N part of country.
BT Finnmark
     Norway

Altai
Administrative Territory or Kray on the Kazakhstan and Mongolian borders.
UF Altay
BT Russian Republic
     USSR

Altai Mountains
Mountain system at meeting of borders of Altai Kray, Kazakhstan, Mongolia, and Sinkiang Uighur. Index countries as applicable. Also search Altai Range.
CO N443000N500000
     E1030000E0870000
UF Altai Range
     Altay Mountains
BT Asia
SA China
     Chuya Alps
     Gorny Altai
     Kalba Range
     Kuznetsk Alatau
     Mongolia
     Mongolian Altai
     USSR

Altai Range
use Altai Mountains

Altai-Sayan region
Region including adjoining Altai and Sayan mountains. Index Mongolia and/or USSR as applicable. Also search Altai-Sayan.
CO N500000N550000
     E1000000E0800000
BT Asia
SA Mongolia
     USSR

altaite
BA tellurides
     sulfides
BT minerals

Altamaha River
SE Georgia.
BT Georgia
     United States

Altan Teeli
Village in W Mongolia. Also search Altan Teli.
UF Altan Teli
     Dzereg
BT Mongolia

Altan Teli
use Altan Teeli

Altay
use Altai

Altay Mountains
use Altai Mountains

Altenberg
Town near Czechoslovak Border in S part of country.
BT Karl-Marx-Stadt
     East Germany

alteration
Formerly a general term appropriate to a large number of topics. As of 1976, use was restricted to any process having to do with changes in temperature and pressure, i.e. diagenesis, hydrothermal alteration, metasomatism, weathering, etc. The term is used to denote the whole spectrum of chemical and physical change. Includes use on level 2 under igneous rocks(1).
SA autometamorphism
     changes
     diagenesis
     fenitization
     hydrothermal alteration
     hydrothermal processes
     igneous rocks
     kaolinization
     leaching
     metasomatic rocks
     metasomatism
     processes
     serpentinization
     thermal alteration
     wallrock alteration
     weathering

Altimur Mountains
E part of country.
BT Afghanistan

Altin Tepe
use Altin-Tepe

Altin-Tepe
Ore bearing region. Also search Altin Tepe.
UF Altin Tepe
BT Dobruja
     Romania

altitude
Used as a general term. Also search elevation.
UF elevation

Alto Adige
Along with Trento comprises Trentino-Alto Adige autonomous region in NE.
BT Italy

Alto Alentejo
Former province in central.
BT Portugal

Altonian
Provincial series, New Zealand.
BA lower Miocene
     Miocene
     Neogene
     Tertiary
BT Cenozoic

Altoona
City in Blair County in W central.
BT Pennsylvania
     United States

Altyn Formation
use Altyn Limestone

Altyn Limestone
In Ravalli Group. Precambrian (Belt series). SW Alberta, SE British Columbia and NW Montana. Also search Altyn Formation.
UF Altyn Formation
BT Precambrian
SA Alberta
     British Columbia
     Montana
     Ravalli Group

Altyn-Topkan
Village in N.
BT Tadzhikistan
     USSR

alum
Also search potassium alum.
UF potash alum
     potassium alum
BA sulfates
BT minerals

alum rock
use alunite

alumina
Includes use on level 3 under geochemistry(1) or under commodities (list C). Also search aluminum oxide.
UF aluminum oxide
SA aluminum
     corundum
     geochemistry
     oxides

aluminosilicates
BA silicates
BT minerals
SA adularia
     amazonite
     bavenite
     berthierine
     bytownite
     carpholite
     chloritoid
     clinozoisite
     cookeite
     cordierite
     corrensite
     cymrite
     epidote
     euclase
     eucryptite
     glaucophane
     hydrobiotite
     hydrosodalite
     kaliophilite
     kalsilite
     lawsonite
     leifite
     maskelynite
     milarite
     mordenite
     offretite
     osumilite
     petalite
     prehnite
     pumpellyite
     sapphirine
     spodumene
     stilpnomelane
     sudoite

aluminum
Includes use on level 1 and 2 as a commodity term (list C) and as a chemical element (list D). Also

aluminum ● ammonium

search Al.
UF Al
SA Al-26
    alumina
    bauxite
    elements

aluminum oxide
use alumina

alumite
use alunite

alumohydrocalcite
BA carbonates
BT minerals

alumstone
use alunite

alunite
UF alum rock
    alumite
    alumstone
BA sulfates
BT minerals
SA jarosite

Alveolina
Includes use on level 3 under foraminifera(1) Alveolinellidae(2).
BA Alveolinellidae
    Miliolina
    foraminifera
BT Invertebrata

Alveolinellidae
Includes use as level 2 term under foraminifera(1). See list F.
BA Miliolina
    foraminifera
BT Invertebrata
NA Alveolina

Alwar
City in NE Rajasthan. Formerly a princely state.
UF Alwur
BT Rajasthan
    India

Alwur
use Alwar

Am
use americium

Amadeus Basin
SW Northern Territory and W Western Australia. Index state and/or territory as applicable.
CO S260000S230000
    E1340000E1290000
BT Australia
SA Northern Territory
    Western Australia

Amador County
E California.
BT California
    United States

Amagase
Village S of Osaka.
BT Honshu
    Japan

Amami
use Oshima

Amami-O-shima
Largest island in Amami-gunto Island group.
BT Ryukyu Islands
    Japan
SA O-shima

Amanos Mountains
Along SE Gulf of Alexandretta in SE Anatolia.
UF Alma Dag
    Gavur Mountains
BT Turkey
    Middle East

Amapa
Federal Territory on French Guiana border.
BT Brazil
NT Serra do Navio

Amargosa Desert
Index California and/or Nevada as applicable.
BT United States
SA California
    Nevada

amargosite
use bentonite

Amarjola
Region in the Rajmahal Hills in E.
BT Bihar
    India

Amasra
Town on the Black Sea in NW Anatolia.
BT Zonguldak
    Turkey

Amasra Basin
Coal basin in NW Anatolia.
BT Turkey
    Middle East

Amazon Basin
River drainage basin. Index countries as applicable. Also search Amazon; Amazon region.
UF Amazon region
BT South America
SA Bolivia
    Brazil
    Colombia
    Ecuador
    Peru
    Venezuela

Amazon region
use Amazon Basin

Amazon River
Largest river in the world if measured by volume of water carried. Rises in 2 major headstreams in the Peruvian Andes flowing N and then E across N Brazil into the Atlantic Ocean. Index countries as applicable. Also search Amazon.
BT South America
SA Brazil
    Colombia
    Peru

Amazonas
State in NW Brazil; a commissary in SE Colombia; a department in N Peru; and a territory in S Venezuela. Index countries as applicable.
CO S090000N020000
    W0570000W0740000
BT South America
SA Brazil
    Colombia
    Peru
    Venezuela

Amazonian Shield
use Brazilian Shield

amazonite
UF amazonstone
BA feldspar group
    framework silicates
    silicates
BT minerals
SA aluminosilicates
    microcline

amazonstone
use amazonite

Amba Dongar
Region.
BT Gujarat
    India

amber
UF bernstein
BA organic compounds
BT minerals
SA resins

amber mica
use phlogopite

Ambin Massif
SE part of country.
BT Savoie
    France

amblygonite
UF hebronite
BA phosphates
BT minerals
SA lithium
    montebrasite

Amblypoda
Includes use on level 2 under Mammalia(1). See list F.
BA Mammalia
BT Tetrapoda
    Vertebrata

Ambre Mountain
N part of country.
BT Malagasy Republic

Ambrosia Lake
BT New Mexico
    United States

Amchitka Island
In center of Aleutian Islands SW of Alaska Peninsula. Also search Amchitka.
BT Alaska
    United States
SA Aleutian Islands

Amelia
Mineral bearing region in Amelia County in S central.
BT Virginia
    United States

americium
Includes use on level 1 and 2 as a chemical element (list D).
UF Am
SA elements

Americus Limestone
use Americus Limestone Member

Americus Limestone Member
Of Foraker Limestone. E Kansas, SE Nebraska and central N Oklahoma. Also search Americus Limestone.
UF Americus Limestone
BT Permian
SA Kansas
    Nebraska
    Oklahoma

Ames Limestone
Of Conemaugh Formation. E Kentucky, SW Maryland, E Ohio, SW Pennsylvania, and N West Virginia.
BT Upper Pennsylvanian
    Pennsylvanian
SA Kentucky
    Maryland
    Pennsylvania
    West Virginia

amethyst
BA silica minerals
    framework silicates
    silicates
BT minerals
SA corundum
    gems
    quartz

Amiidae
Includes use on level 3 under Pisces(1) Osteichthyes(2).
BA Actinopterygii
    Osteichthyes
    Pisces
BT Vertebrata

amino acids
Includes use on level 2 under organic materials(1).
BA organic materials
SA acids
    fatty acids
    proteins
    racemization

Amisk Group
Considered as oldest rocks in Flin Flon Belt. Overlain by Missi Group.
BT Archean
SA Manitoba

Amisk Lake
Near Manitoba border in E.
BT Saskatchewan
    Canada

Amite River
Rises in SW Mississippi and flows S and E into Lake Maurepas in SE Louisiana. Index states as applicable.
BT United States
SA Louisiana
    Mississippi

Ammobaculites
Genus. Includes use on level 3 under foraminifera(1) Lituolacea(2).
BA Lituolacea
    Textulariina
    foraminifera
BT Invertebrata

Ammodiscacea
Includes use on level 2 under foraminifera(1). See list F.
BA Textulariina
    foraminifera
BT Invertebrata
NA Ammodiscidae
    Astrorhizidae

Ammodiscidae
Family. Includes use on level 3 under foraminifera(1) Ammodiscacea(2).
BA Ammodiscacea
    Textulariina
    foraminifera
BT Invertebrata

Ammonia
Genus. Includes use on level 3 under foraminifera(1) Rotaliacea(2).
BA Rotaliacea
    Rotaliina
    foraminifera
BT Invertebrata
NA Ammonia beccarii
SA ammonia compound

Ammonia beccarii
Includes use on level 3 under foraminifera(1) Rotaliacea(2).
BA Ammonia
    Rotaliacea
    Rotaliina
    foraminifera
BT Invertebrata

ammonia compound
Term introduced in 1978 to distinguish from fossil genus, Ammonia. Includes use on level 3 under geochemistry(1). Before 1978, also search ammonia.
SA Ammonia
    ammonium
    geochemistry

Ammonites
Includes use on level 3 under Mollusca(1) Cephalopoda(2).
BA Ammonoidea
    Cephalopoda
    Mollusca
BT Invertebrata
SA ammonoids
    aptychi
    Ceratites

ammonium
Includes use on level 3 under geochemistry(1).
SA ammonia compound

geochemistry
Ammonoidea
Subclass. Includes use on level 3 under Mollusca(1) Cephalopoda(2).
BA Cephalopoda
  Mollusca
BT Invertebrata
NA Ammonites
  ammonoids
  Baculites
  Bouleiceras
  Dactylioceratidae
  Goniatitida
  Hildoceratacea
  Hildoceratidae
  Perisphinctidae
  Scaphites

ammonoids
Includes use on level 3 under Mollusca(1) Cephalopoda(2).
BA Ammonoidea
  Cephalopoda
  Mollusca
BT Invertebrata
SA Ammonites

amorphous materials
General term introduced in 1978 for all materials not crystalline. Descriptive, not geologic term. Before 1978, also search amorphous AND materials.
SA materials

amosite
Commerical term.
BA amphibole group
  chain silicates
  silicates
BT minerals
SA anthophyllite
  asbestos
  gedrite

Amparo
City in E.
BT Sao Paulo
  Brazil

Amphibia
Includes use on level 1 and 2 as a fossil term (list F).
BT Tetrapoda
  Vertebrata
NA Labyrinthodontia
  Lepospondyli
  Lissamphibia

amphibole
Use amphibole group for group name.
BA amphibole group
  chain silicates
  silicates
BT minerals
SA hornblende

amphibole group
Includes use in combination with chain silicates (i.e. chain silicates, amphibole group) on level 2 under minerals(1). See list L.
BA chain silicates
  silicates
BT minerals
NA actinolite
  amosite
  amphibole
  anthophyllite
  arfvedsonite
  clinoamphibole
  crocidolite
  crossite
  cummingtonite
  ferrohastingsite
  gedrite
  glaucophane
  grunerite
  hastingsite
  holmquistite
  hornblende
  kaersutite
  magnesioriebeckite
  nephrite
  pargasite
  richterite
  riebeckite
  tirodite
  tremolite
  tschermakite

amphibolite
Includes use on level 3 under metamorphic rocks(1) amphibolites(2). See list J.
UF diabase amphibolite
BA amphibolites
BT metamorphic rocks
SA para-amphibolite

amphibolite facies
Includes use on level 3 under metamorphic rocks(1).
BT facies
SA epidote-amphibolite facies
  metamorphic rocks

amphibolites
Includes use on level 2 under metamorphic rocks(1). See list J.
BT metamorphic rocks
NA amphibolite
  orthoamphibolite
  para-amphibolite

Amphineura
Not a valid term for GeoRef. See Aplacophora and Polyplacophora under Mollusca(1).

Amphistegina
Genus. Includes use on level 3 under foraminifera(1) Orbitoidacea(2).
BA Orbitoidacea
  Rotaliina
  foraminifera
BT Invertebrata

amplitude
Do not use to indicate abundance. Includes use on level 3 under seismology(1).
SA elastic waves
  intensity
  seismology

Ampurdan
Coastal plain in Catalonia.
BT Gerona
  Spain

Amsterdam
City on IJsselmeer (former Zuider Zee). North Holland.
BT Netherlands

Amsterdam Island
Dependency of Malagasy Republic 2800 miles off SE African coast.
BT Indian Ocean

Amu Darya
River which rises in Pamirs and flows into Aral Sea. Index Afghanistan and Soviet republics as applicable.
UF Oxus
BT Asia
SA Afghanistan
  Tadzhikistan
  Turkmenia
  Uzbekistan

Amund Ringnes Island
One of the Queen Elizabeth Islands in District of Franklin.
BT Northwest Territories
  Canada
SA Queen Elizabeth Islands

Amur Basin
River basin covering an area including parts of Amur and Chita oblasts, and Khabarovsk Kray in the Russian Republic; parts of Inner Mongolia, and Manchuria in China; and E Mongolia. Index Manchuria, Mongolia, or Russian Republic as applicable. Also search Amur; Amur region.
CO N470000N540000
  E1410000E1210000
UF Amur region
BT Asia
SA Manchuria
  Mongolia
  Russian Republic
  Soviet Far East

Amur region
use Amur Basin

Amur River
Formed at junction of the Shilka and Argun rivers. After serving as the boundary between Manchuria and 2 oblasts of the Russian Republic, it flows NE across Khabarovsk Kray into the N end of the Tatar Strait. Index Manchuria and/or Russian Republic as applicable. Also search Amur.
BT Asia
SA Argun River
  Manchuria
  Russian Republic

Anabar Anteclise
use Anabar Shield

Anabar Bay
On the Laptev Sea in NW Yakutia. Also search Anabar.
BT Russian Republic
  USSR

Anabar Massif
use Anabar Shield

Anabar River
Flows into Anabar Bay in NW Yakutia. Also search Anabar.
BT Russian Republic
  USSR

Anabar Shield
Part of the Central Siberian Plateau in NW central Yakutia. Also search Anabar; Anabar Anteclise; Anabar Massif.
UF Anabar Anteclise
  Anabar Massif
BT Russian Republic
  USSR
SA Siberian Platform

Anacostia River basin
Index Maryland and/or District of Columbia as applicable. Also search Anacostia River.
BT United States
SA District of Columbia
  Maryland

Anadarko Basin
Index states as applicable.
BT United States
SA Kansas
  Oklahoma
  Texas

Anadyr Basin
River basin in Chukchi National Okrug in NE Siberia. Also search Anadyr.
BT Russian Republic
  USSR

Anadyr Range
Extends SE from Chaun Bay into W Chukchi Peninsula in extreme NE Siberia. Also search Anadyr and Chukchi Range.
UF Chukchi Range
BT Russian Republic
  USSR

anaerobic environment
Term introduced in 1978. Before 1978, search anaerobic.
SA aerobic environment
  environment
  sapropel

anaerobic taxa
Term introduced in 1978. Before 1978, search anaerobic.
UF taxa, anaerobic

analbite
BA feldspar group
  framework silicates
  silicates
BT minerals

analcime
Also search analcite.
UF analcite
BA framework silicates
  silicates
BT minerals
SA zeolite group

analcite
use analcime

analog simulation
Term introduced in 1978. Includes use on level 3 under mathematical geology(1). Also search analog AND automatic data processing.
SA analog techniques
  automatic data processing
  mathematical geology
  simulation

analog techniques
Term introduced in 1978. Before 1978, search analog AND automatic data processing.
SA analog simulation
  automatic data processing
  digital techniques
  techniques

analyses
A valid term through 1974. After 1974, use analysis.

analysis
Includes use on level 2 under name of element(1), under isotopes(1) and paleoecology(1). See list D. For methods, see chemical analysis(1), spectroscopy(1), X-ray diffraction analysis(1), and differential thermal analysis(1). For data, see material name. Before 1975, also search analyses.
SA activation analysis
  alpha-ray spectroscopy
  analytical methods
  atomic absorption
  autoradiography
  canonical analysis
  chemical analysis
  chemical composition
  chromatography
  cluster analysis
  colorimetry
  correspondence analysis
  data analysis
  differential thermal analysis
  discriminant analysis
  electron diffraction analysis
  electron paramagnetic resonance
  electron probe
  emission spectroscopy
  environmental analysis
  factor analysis
  finite difference analysis
  finite element analysis
  flame photometry
  fluid inclusions
  Fourier analysis
  gamma-ray spectroscopy
  infrared spectroscopy
  ion probe
  isotopes
  least-squares analysis
  major-element analyses
  microwave methods
  microwave spectroscopy

analysis ● andradite

minor-element analyses
modal analysis
Mossbauer spectroscopy
multispectral analysis
multivariate analysis
neutron activation analysis
neutron diffraction analysis
nuclear magnetic resonance
numerical analysis
optical spectroscopy
polarography
pollen analysis
qualitative analysis
quantitative analysis
radio-frequency spectroscopy
Raman spectroscopy
regression analysis
sampling
shape analysis
spectral analysis
spectroscopy
spherical harmonic analysis
statistical analysis
statistical methods
structural analysis
thermal analysis
thermogravimetric analysis
thermomagnetic analysis
trace-element analyses
trend-surface analysis
ultraviolet spectroscopy
vacuum fusion analysis
variance analysis
wet methods
X-ray analysis
X-ray diffraction analysis
X-ray fluorescence
X-ray spectroscopy

analytical
A valid level 2 index term through 1977. After 1977, use analytical methods on level 3 under level 1 terms.

analytical data
A valid term through 1975. After 1975, use data.

analytical methods
Not a valid term from 1975 to 1977. After 1977, includes use on level 3. Before 1978, also search analytical AND methods.
  SA analysis
        methods

Anantapur
City and district in SW.
  BT Andhra Pradesh
        India

Anantnag
Town and district in S central.
  BT Jammu and Kashmir
        India

Anapa
Town in Krasnodar Kray in the Northern Caucasus.
  BT Russian Republic
        USSR

Anapsida
Includes use on level 2 under Reptilia(1). See list F.
  BA Reptilia
  BT Tetrapoda
        Vertebrata
  NA Chelonia
        Pelomedusidae

Anarak
Town in central.
  BT Iran

Anasco Bay
On Mona Passage on the W coast.
  BT Puerto Rico
        West Indies

anastomosing streams
use braided streams

anatase
  UF octahedrite (mineral)
  BA oxides
  BT minerals
  SA brookite
        rutile

anatectite
use anatexite

anatexis
Includes use on level 3 under metamorphism(1).
  SA magmas
        metamorphism
        palingenesis

anatexite
  UF anatectite
  BA metamorphic rocks

Anatolia
Westernmost part of Asia. That part of Turkey which lies in Asia. Also search Asia Minor.
  CO N360000N420000
        E0360000E0260000
  UF Asia Minor
  BT Turkey
        Middle East

Ancenis
Town on the Loire River in W part of country.
  BT Loire-Atlantique
        France

Anchorage
City on Cook Inlet in S.
  BT Alaska
        United States

anchors
  SA engineering geology

ancient
A valid term through 1977 used with ice ages (i.e. ice ages, ancient) on level 2 under glacial geology(1). After 1977, use ancient ice ages on level 2.

ancient ice ages
Term introduced in 1978 on level 2 under glacial geology (1). Before 1978, search ice ages AND ancient.
  SA glacial geology

Ancona
City and province on the Adriatic Sea.
  BT Marches
        Italy

Andalusia
Eight Provinces in S and SW Spain: Almeria, Granada, Jaen, Malaga, Cadiz, Cordoba, Huelva, and Seville.
  BT Spain
  SA Almeria
        Andalusian
        Cadiz
        Cordoba
        Granada
        Huelva
        Jaen
        Malaga
        Seville

Andalusian
  BA Triassic
  BT Mesozoic
  SA Andalusia
        Betic Cordillera
        Spain

andalusite
Includes use as level 3 commodity term under ceramic materials(1). See list C.
  BA orthosilicates
        silicates
  BT minerals
  SA ceramic materials
        kyanite
        sillimanite

Andaman Basin
Between the Andaman and the Nicobar islands.
  BT Bay of Bengal
        Indian Ocean

Andaman Islands
E Bay of Bengal, N of Nicobar Islands. Andaman and Nicobar Territory.
  CO N103000N130000
        E0970000E0973000
  BT India

Andaman Sea
W of Burmese and Thai section of Malay Peninsula.
  BT Bay of Bengal
        Indian Ocean

Andean Geosyncline
Term introduced in 1978. Includes use on level 3 under geosynclines(1) evolution(2). Before 1978, search geosynclines AND Andean.
  SA Andean Orogeny
        Andes
        geosynclines

Andean Orogeny
Term introduced in 1978. Before 1978, search orogeny AND Andean.
  SA Andean Geosyncline
        Andes
        orogeny
        tectonics

Anderson County
Index states as applicable.
  BX Kansas
        Kentucky
        Tennessee
        South Carolina
        Texas
  BT United States

Andes
Mountain system extending along W coast from Venezuela and Colombia to Tierra del Fuego. Index countries as applicable. Includes use on level 1 as an area term (list O). For term set options see list B.
  UF Andes Mountains
  BA South America
  NA Central Andes
        Northern Andes
        Southern Andes
  SA Andean Geosyncline
        Andean Orogeny
        Argentina
        Bolivia
        Chile
        Colombia
        Eastern Cordillera
        Ecuador
        Peru
        Venezuela

Andes Mountains
use Andes

andesine
  BA feldspar group
        framework silicates
        silicates
  BT minerals
  SA albite
        anorthite
        plagioclase

andesite
Includes use on level 3 under igneous rocks(1) andesite-rhyolite family(2). See list H.
  BA andesite-rhyolite family
  BT igneous rocks
  SA andesitic composition
        meta-andesite
        pyroxene andesite

andesite basalt
Not a valid term for GeoRef. See basalt.

andesite porphyry
See list H.
  BA andesite-rhyolite family
  BT igneous rocks
  SA porphyry

andesite tuff
Also search andesitic tuff. See list H.
  UF andesitic tuff
  BA pyroclastics and glasses
  BT igneous rocks
  SA tuff

andesite-rhyolite family
Includes use on level 2 under igneous rocks(1). See list H.
  BT igneous rocks
  NA andesite
        andesite porphyry
        comendite
        dacite
        dacite porphyry
        dellenite
        latite
        liparite
        liparite porphyry
        pantellerite
        pyroxene andesite
        quartz latite
        quartz porphyry
        rhyodacite
        rhyolite
        rhyolite porphyry
        trachyandesite
        vitrophyre
  SA propylite

andesitic composition
Term introduced in 1978. Before 1978, search andesitic AND composition.
  SA andesite
        composition

andesitic tuff
use andesite tuff

Andhra Pradesh
State in central.
  BT India
  NT Anantapur
        Cuddapah
        Cuddapah Basin
        Guntur
        Khammam
        Kurnool
        Nellore
        Nellore mica belt
        Rajahmundry
        Srikakulam
        Visakhapatnam
  SA Aravalli System
        Cuddapah System
        Deccan Plateau
        Godavari River
        Godavari Valley
        Krishna
        Kurnool System
        Pranhita-Godavari Valley
        Raghavapuram Shales

Ando soils
use Andosols

Andorra
A principality in E Pyrenees between France and Spain. Includes use on level 1 as an area term (list O). For term set options see list B.
  BA Europe
  SA Pyrenees

Andosols
  UF Ando soils
  BA soils

andradite
  BA garnet group
        orthosilicates

silicates
BT minerals

Andriamena
Village in N central.
BT Malagasy Republic

Andros Island
Large island in W.
CO N233000N250000
W0773000W0783000
BT Bahamas
West Indies

Andrychow
Town in S part of country.
BT Krakow
Poland

Anegada Island
Northernmost island. British Virgin Islands.
BT West Indies
SA British Virgin Islands

anelastic material
use anelastic media

anelastic media
Includes use on level 3 under seismology. Before 1978, search anelastic.
UF anelastic material
SA media
seismology

anelasticity
SA deformation
elasticity
seismology

Angara
use Angara River

Angara River
Flows from Lake Baikal into Yenisei River. Formerly called Upper Tunguska in its lower course. Also search Angara.
UF Angara
Upper Tunguska
BT Russian Republic
USSR
SA Tunguska
Tunguska River

Angara-Lena Basin
Adjoining basin of two rivers in Irkutsk Oblast W and NW of Lake Baikal.
CO N520000N583000
E1090000E1020000
BT Russian Republic
USSR
SA Lena Basin

angiosperms
Includes use on level 1 and 2 as a fossil term (list F).
BT Plantae
NA Dicotyledoneae
Monocotyledoneae
SA Dryas
Florinites
fossil wood
gymnosperms
Pteropsida
sporangia

angle of dip
use dip

Anglesea
use Anglesey

Anglesey
Island and County in N Wales. Also search Anglesea.
CO N531000N533000
W0040000W0044000
UF Anglesea
BT Wales
Great Britain
United Kingdom

anglesite
BA sulfates
BT minerals

Angola
Peoples Republic of Angola. Includes use on level 1 as an area term (list O). For term set options see list B.
CO S180000S060000
E0230000E0120000
UF Portuguese West Africa
BA Africa
NT Cabinda
Catanda
Cuanza Basin
Cuanza-Sul
Luanda
Mossamedes
SA Congo Basin
Congo River
Kasai River
Zambezi Valley

Angora
use Ankara

Angouleme
City in W part of country.
BT Charente
France

Angra dos Reis
City in SW.
BT Rio de Janeiro
Brazil

Angren
City in Tashkent Oblast.
BT Uzbekistan
USSR

Angus
County on the North Sea. Formerly Forfar or Forfarshire.
CO N563000N570000
W0023000W0032500
UF Forfar
Forfarshire
BT Scotland
Great Britain
United Kingdom

anhydrite
Includes use on level 3 as a commodity term (list C) under gypsum(1); on level 3 under sediments(1) chemically precipitated sediments(2) and under sedimentary rocks(1) chemically precipitated rocks(2). See list I (sed. rocks) and list N (sediments).
UF cube spar
BA sulfates
BT minerals
SA cap rocks
chemically precipitated rocks
chemically precipitated sediments
evaporites
gypsum
sedimentary rocks
sediments

anhysteretic remanent magnetization
BA remanent magnetization
SA paleomagnetism

anilite
BA sulfides
BT minerals

animals
Used as general term, to distinguish from vegetation.
SA vegetation

Animas Mountains
Hidalgo County in extreme SW.
BT New Mexico
United States

Animikie Group
Includes Pokegama Quartzite, Biwabik Iron Formation Series, and Virginia Slate. N Michigan, NE Minnesota, Ontario, and N Wisconsin.
BT Precambrian
SA Biwabik Iron Formation
Gunflint Iron Formation
Michigan
Minnesota
Ontario
Wisconsin

anions
SA electrolytes
ion exchange
ions

Anisian
Europe. Above Scythian, below Ladinian. Includes use on level 3 under age terms(1). See list E.
BA Middle Triassic
Triassic
BT Mesozoic

anisotropic materials
Term introduced in 1978. Before 1978, search anisotropic.
SA anisotropy
materials

anisotropy
Includes use on level 3 under seismology(1).
UF aeolotropy
SA anisotropic materials
magnetic susceptibility
mechanical properties
seismology

Anjou
Old province in W.
BT France

Anjouan Island
One of the French Comoro Islands in the Mozambique Channel.
UF Johanna Island
BT Indian Ocean
SA Comoro Islands

Ankara
Province in W central Anatolia. Also a city formerly called Angora.
UF Angora
BT Turkey
Middle East

ankaramite
Includes use on level 3 under igneous rocks(1) basalt family(2) and ultramafic family(2). See list H.
BA basalt family
BT igneous rocks
SA ultramafic family

ankaratrite
Includes use on level 3 under igneous rocks(1) alkali basalt family(2). See list H.
BA alkali basalt family
BT igneous rocks
SA nepheline basalt
olivine nephelinite

ankerite
UF cleat spar
ferroan dolomite
BA carbonates
BT minerals
SA dolomite

Ann Arbor
City in Washtenaw County in SE Lower Peninsula.
BT Michigan
United States

Annaba
Department on the Mediterranean Sea. Also a city. Formerly called Bone.
BT Algeria

Annapolis Valley
River valley in SW Nova Scotia. Also search Annapolis.
BT Nova Scotia
Canada

Annecy
Town S of Lake Geneva in W.
BT Haute-Savoie
France

Annelida
Includes use on level 1 and 2 as a fossil term (list F).
BT Invertebrata
NA Chaetopoda
Polychaetia
SA scolecodonts
worms

Annonaceae
Family.
BA Dicotyledoneae
angiosperms
BT Plantae

Annot Sandstone
Priabonian (Bartonian). In the Alps of SE France.
BT Eocene
SA France

annual growth rings
use tree rings

annual meeting
Includes use on level 3 under associations(1).
SA associations

annual report
As of 1978, term is used on level 2 under surveys(1).
SA associations
current research
industry
progress report
report
surveys
symposia

annually thawed layer
use active layer

anomalies
Includes use as a level 2 or 3 term appropriate to a large number of topics, e.g. on level 2 under atmosphere(1), ecology(1), heat flow(1), and isostasy(1). See list G.
SA Bouguer anomalies
free-air anomalies
gravity anomalies
magnetic anomalies

Anomalinidae
Family. Includes use on level 3 under foraminifera(1) Cassidulinacea(2).
BA Cassidulinacea
Rotaliina
foraminifera
BT Invertebrata

anorthite
BA feldspar group
framework silicates
silicates
BT minerals
SA andesine
plagioclase

anorthoclase
BA feldspar group
framework silicates
silicates
BT minerals
SA orthoclase

anorthosite
Includes use on level 3 under igneous rocks(1) gabbro family(2). See list H.
BA gabbro family
BT igneous rocks

SA gabbroic anorthosite
anorthositic gabbro
See list H.
BA gabbro family
BT igneous rocks
SA gabbro

Ansbach
City in W.
BT Bavaria
West Germany

Antalya
Province in S Anatolia on Gulf of Adalia. Also a city formerly called Adalia.
UF Adalia
BT Turkey
Middle East

Antarctic Continent
use Antarctica

Antarctic Ocean
An arbitrary definition, considered the equivalent of Southern Ocean. Waters about Antarctica are sometimes called the Antarctic Ocean, but actually are only the parts of the Atlantic, Pacific and Indian Oceans south of approximately 55° S. Also search Antarctic. Index oceans as applicable. Includes use on level 1 or 2 as an area term (list O). If 1, see term set options under list B.
NT Bellingshausen Sea
Drake Passage
McMurdo Sound
Ross Sea
Scotia Sea
South Sandwich Islands
Weddell Sea
SA Antarctica
Atlantic Ocean
Indian Ocean
Pacific Ocean
subantarctic regions

Antarctic Peninsula
Peninsula jutting northward toward Cape Horn. Also search Antarctic and Palmer Peninsula.
UF Palmer Peninsula
BT Antarctica

Antarctic Plate
SA Antarctica
plate tectonics

Antarctic Platform
Central area which constitutes a high plateau. Also search Antarctic.
BT Antarctica

Antarctica
Continent centering on the South Pole. Includes use on level 1 or 2 as an area term (list O). If 1, see term set options under list B. To retrieve all documents, individual countries and physiographic regions should also be searched (see list O). Also search Antarctic.
CO S900000S610000
W0400000W0395900
UF Antarctic Continent
NT Adelaide Island
Adelie Coast
Alexander Island
Antarctic Peninsula
Antarctic Platform
Anvers Island
Arthur Harbor
Beacon Valley
Beardmore Glacier
Byrd Station
Coalsack Bluff
Deception Island
Don Juan Pond
Dry Valley Drilling Project
Dufek Intrusion
East Ongul Island
Ellsworth Land
Ellsworth Mountains
Enderby Land
Fosdick Mountains
Graham Land
Horlick Mountains
Hut Point Peninsula
Jones Mountains
King George Island
Livingston Island
Lutzow-Holm Bay
Marguerite Bay
Marie Byrd Land
McMurdo Ice Shelf
Mirnyy
Molodezhnaya Station
Ohio Range
Pensacola Mountains
Polar Cap
Polar Continental Shelf
Prince Charles Mountains
Queen Alexandra Range
Queen Maude Land
Queen Maude Range
Ross Dependency
Ross Ice Shelf
Ross Island
Shackleton Glacier
Shackleton Range
Sor-Rondane Mountains
South Orkney Islands
South Pole
South Shetland Islands
Taylor Glacier
Taylor Valley
Transantarctic Mountains
Victoria Land
Vostok Station
Wilkes Land
Wisconsin Range
Wright Valley
SA Antarctic Ocean
Antarctic Plate
Beacon Supergroup
Circum-Antarctic region
Falkland Islands
Ferrar Group
Fremouw Formation
Gondwana
Pacific-Antarctic Ridge
subantarctic regions
Yamato Mountains

Antelope Valley Limestone
In Pogonip Group. Central Nevada.
BT Ordovician
SA Lower Ordovician
Middle Ordovician
Nevada
Pogonip Group

anthophyllite
BA amphibole group
chain silicates
silicates
BT minerals
SA amosite
asbestos
cummingtonite
gedrite

Anthozoa
Includes use on level 2 under Coelenterata(1). See list F.
BA Coelenterata
BT Invertebrata
NA corals
Rugosa
Scleractinia
Tabulata
Tetracorallia
Zoantharia

anthracite
Includes use on level 3 under sedimentary rocks(1) organic residues(2). See list I.
BA organic residues
BT sedimentary rocks
SA coal
lignite
organic materials

Anthraconaia
Genus. Includes use on level 3 under Mollusca(1) Bivalvia(2).
BA Bivalvia
Mollusca
BT Invertebrata
SA Anthraconauta

Anthraconauta
Genus. Includes use on level 3 under Mollusca(1) Bivalvia(2).
BA Bivalvia
Mollusca
BT Invertebrata
SA Anthraconaia

anthraxolite
SA bitumens
coal
oil shale
organic materials

anthropogenic (activity)
use human activity

anthropology
Includes use on level 3.
SA fossil man
man

Anti-Atlas
Range of the Atlas Mountains SW of High Atlas Mountains in SW Morocco. Also search Anti-Atlas Mountains.
CO N290000N310000
W0060000W0090000
UF Anti-Atlas Mountains
BT Morocco
SA Atlas Mountains

Anti-Atlas Mountains
use Anti-Atlas

Antibes
Town on the Mediterranean Sea in extreme SE part of country.
BT Alpes-Maritimes
France

anticlinal
A valid term through 1977. After 1977, use anticlines.

anticlines
Not a valid index term from 1971 to 1977. Now use on level 3 under folds(1) style(2). Before 1978, also search folds AND anticlinal.
BA folds
SA anticlinoria
antiform folds
diapirs
domes
geanticlines
synclines

anticlinoria
Includes use on level 3 under folds(1) systems(2).
BA folds
SA anticlines
antiform folds
geanticlines
geosynclines
synclinoria
systems

Anticosti Island
In Gulf of Saint Lawrence.
BT Quebec
Canada

antidunes
Includes use on level 3 under sedimentary structures(1) bedding plane irregularities(2). See list K.
BA bedding plane irregularities
sedimentary structures
SA dunes
flame structures
sand waves

Antietam Formation
In Chilhowee Group. Maryland, SE Pennsylvania, Virginia, West Virginia.
BT Lower Cambrian
Cambrian
SA Chilhowee Group
Maryland
Pennsylvania
Virginia
West Virginia

antiform folds
Term introduced in 1978. Includes use on level 3 under folds(1) style(2). Before 1978, also search folds AND antiform.
BA folds
SA anticlines
anticlinoria
synform folds

Antigonish
Town and county W of Cape Breton Island.
BT Nova Scotia
Canada

antigorite
BA sheet silicates
silicates
BT minerals
SA serpentine group

Antilles
All of the islands of the West Indies except the Bahamas. Index Greater Antilles and/or Lesser Antilles.
BT West Indies
SA Caribbean region
Greater Antilles
Lesser Antilles

antimonates and antimonites
Includes use on level 2 under minerals(1). See list L.
UF antimonites, antimonates and
BT minerals

antimonides
Includes use on level 3 under minerals(1) sulfides(2). See list L.
BA sulfides
BT minerals
NA dyscrasite

antimonite
use stibnite

antimonites, antimonates and
use antimonates and antimonites

antimony
Includes use on level 1 and 2 as a commodity term (list C) and as a chemical element (list D).
UF Sb
SA elements
native elements and alloys

Antioquia
Department in NW Colombia. Also a city.
BT Colombia

antiperthite
BA feldspar group
framework silicates
silicates
BT minerals
SA perthite

Antler orogenic belt
use Antler Orogeny

Antler Orogeny
An orogeny which extensively deformed Paleozoic rocks of the Great Basin in Nevada. Named for relations in the Antler Peak Quadrangle near Battle Mountain. Before 1978, also search Antler; Antler orogenic belt.
UF Antler orogenic belt
BT Paleozoic

SA Acadian Phase
   Devonian
   Mississippian
   Nevada
   orogeny

Antlers Sands
In Trinity Group. SW Oklahoma and NE Texas.
BT Lower Cretaceous
   Cretaceous
SA Comanchean
   Oklahoma
   Texas
   Trinity Group

Antofagasta
Province in N Chile. Also a seaport N of Santiago.
BT Chile
NT Chuquicamata
SA Atacama Desert

Antong Java Rise
use Ontong Java Plateau

Antrim
County. Also a town.
CO N543000N551500
   W0054500W0063000
BT Northern Ireland
   United Kingdom

Antwerp
Province in N Belgium. Also a city at the mouth of the Western Scheldt River.
BT Belgium
NT Malines

Anura
Order. Includes use on level 3 under Amphibia(1) Lissamphibia(2).
BA Lissamphibia
   Amphibia
BT Tetrapoda
   Vertebrata

Anvers Island
Just off NW coast of the Antarctic Peninsula.
BT Antarctica

Anza Desert State Park
Occupies most of E San Diego County. Also search Anza.
BT California
   United States

Anzoategui
State in N extending from the Caribbean Sea to the Orinoco River.
BT Venezuela
NT Oficina

Aomori
City and prefecture in N.
BT Honshu
   Japan

Aosta Valley
use Valle d'Aosta

Aouelloul
A meteor crater near Chinguetti in W central Mauritania. Also search Aouelloul Crater.
UF Aouelloul Crater
BT Mauritania

Aouelloul Crater
use Aouelloul

Apache County
NE Arizona.
BT Arizona
   United States

Apalachicola
City at mouth of Apalachicola River on Apalachicola Bay in Franklin County in panhandle.
BT Florida
   United States

apatite
BA phosphates
BT minerals
SA carbonate apatite
   chlorapatite
   fluorapatite
   francolite
   hydroxylapatite
   pyromorphite

apatite ores
use phosphate

Apennine Front
Linear outer slopes of the Apennines. Also search Apennine AND front.
BT Italy
SA Apennines

Apennine Mountains
use Apennines

Apennine Range
use Apennines

Apennines
Mountain range extending the full length of the Italian peninsula. Includes use on level 1 as an area term (list O). For term set options see list B.
UF Apennine Mountains
   Apennine Range
BA Europe
SA Alban Hills
   Apennine Front
   Apuane Alps
   Italy
   Ligurian Apennines

aphanitic texture
Term introduced in 1978. Before 1978, search aphanitic.
SA igneous rocks
   textures

Aphebian
North America, middle Precambrian.
BA lower Proterozoic
   Proterozoic
BT Precambrian
NT Hurwitz Group
SA middle Precambrian

aplite
Includes use on level 3 under igneous rocks(1) granite-granodiorite family(2). See list H.
BA granite-granodiorite family
BT igneous rocks

apogranite
BA granite-granodiorite family
BT igneous rocks
SA granite

Apollo
Includes use on level 3 under Moon(1).
SA landing sites
   Moon
   remote sensing
   satellite methods

Apollo 9
Includes use on level 3 under remote sensing(1).
SA Moon
   remote sensing
   satellite methods

Apollo 11
Includes use on level 3 under Moon(1).
SA Moon
   remote sensing
   satellite methods

Apollo 12
Includes use on level 3 under Moon(1).
SA Moon
   remote sensing
   satellite methods

Apollo 13
Includes use on level 3 under remote sensing(1).
SA remote sensing
   satellite methods

Apollo 14
Includes use on level 3 under Moon(1).
SA Moon
   remote sensing
   satellite methods

Apollo 15
Includes use on level 3 under Moon(1).
SA Moon
   remote sensing
   satellite methods

Apollo 16
Includes use on level 3 under Moon(1).
SA Moon
   remote sensing
   satellite methods

Apollo 17
Includes use on level 3 under Moon(1).
SA Moon
   remote sensing
   satellite methods

apophyllite
BA sheet silicates
   silicates
BT minerals
SA zeolite

Appalachian Basin
Also search Appalachian Plateau. Index states as applicable.
BT United States
SA Alabama
   Appalachian Plateau
   Kentucky
   Maryland
   New York
   North Carolina
   Ohio
   Pennsylvania
   Tennessee
   Virginia
   West Virginia

Appalachian Mountains
use Appalachians

Appalachian Orogeny
use Appalachian Phase

Appalachian Phase
Term introduced in 1978. Includes use on level 3 under tectonics(1) structure(2). Before 1978, also search Appalachian structure; orogeny AND Appalachian; Appalachian Orogeny.
UF Appalachian Orogeny
   Appalachian structure
BT Permian
SA orogeny
   tectonics

Appalachian Plateau
Westernmost part of the Appalachians including the Allegheny and Cumberland plateaus. Index States as applicable. Also search Appalachian AND plateau.
BT United States
SA Alabama
   Allegheny Plateau
   Appalachian Basin
   Appalachians
   Cumberland Plateau
   Kentucky
   Maryland
   New York
   North Carolina
   Ohio
   Pennsylvania
   Tennessee
   Virginia
   West Virginia

Appalachian structure
use Appalachian Phase

Appalachians
Mountain system of eastern North America extending from Canadian maritime provinces to Alabama. Index countries as applicable. Includes use on level 1 as an area term (list O). For term set options see list B.
CO N330000N473000
   W0670000W0870000
UF Appalachian Mountains
BA North America
NT Blue Ridge Province
   Brevard Zone
   Carolina slate belt
NA Central Appalachians
   Northern Appalachians
   Southern Appalachians
SA Allegheny Mountains
   Allegheny Plateau
   Appalachian Plateau
   Blue Ridge Mountains
   Canada
   Catskill Mountains
   Cumberland Plateau
   Great Appalachian Valley
   Great Smoky Mountains
   Green Mountains
   Hudson Highlands
   United States

apparatus
Do not use for instruments. Includes use on level 3 under conodonts(1).
SA conodonts

appinite
Includes use on level 3 under igneous rocks(1) syenite family(2) and diorite family(2). See list H.
BA plutonic rocks
BT igneous rocks
SA diorite family
   syenite family

applications
Includes use as level 2 or 3 term appropriate to a large number of topics, e.g. on level 2 under mineral exploration(1). See list G. Also search application.

Apsheron Peninsula
Peninsula jutting into the Caspian Sea on E. Also search Apsheron.
BT Azerbaidzhan
   USSR

Apt
Town in SE part of country.
BT Vaucluse
   France

Aptian
Europe: Lower Cretaceous, or Lower and Middle Cretaceous. Above Barremian, below Albian. Includes use on level 3 under age terms(1). See list E.
BA Cretaceous
BT Mesozoic
SA Areado Formation
   Gargasian
   Lower Cretaceous
   Middle Cretaceous
   Zubair Formation

aptychi
Also search aptychus.
UF aptychus
SA Ammonites

aptychus
use aptychi

Apuane Alps
Division of Etruscan Apennines in E.
BT Tuscany
   Italy

SA Apennines

Apulia
Autonomous region comprising approximately the southern third of east coast.
UF Puglia
BT Italy
NT Bari
 Brindisi
 Foggia
 Gargano
 Lecce
 Murge
 Otranto
 Salentina Peninsula
 Taranto

Apuseni Mountains
A large mountain massif in WSW Transylvania. Also search Apuseni.
BT Romania
SA Muntii Metalici

aqueous solutions
Includes use on level 3 under geochemistry(1).
SA geochemistry
 solution
 solutions

Aquia Formation
In Pamunkey Group. Delaware, Maryland and E Virginia.
BT lower Eocene
 Eocene
SA Delaware
 Maryland
 Virginia

aquifer properties
A valid term through 1978. After 1978, see aquifers or specific property, e.g. transmissivity.

aquifers
Includes use on level 2 under ground water(1); on level 3 under hydrogeology(1).
BT ground water
NT unconfined aquifers
SA Darcy's law
 hydrogeology
 recharge
 transmissibility coefficient
 transmissivity
 water balance
 water supply

Aquitaine
A region of SW.
BT France

Aquitaine Basin
Large lowland in SW bounded by Pyrenees on S, Central Massif on N and NE, and Bay of Biscay on W.
CO N431500N460000
 E0020000W0013000
BT France

Aquitanian
Europe. Above Chattian (Oligocene), below Burdigalian. Includes use on level 3 under age terms(1). See list E.
BA lower Miocene
 Miocene
 Neogene
 Tertiary
BT Cenozoic

Ar
use argon

Ar-36
Includes use on level 3 under isotopes(1).
SA Ar-40/Ar-36
 argon
 isotopes

Ar-36/Ar-40
use Ar-40/Ar-36

Ar-37
Artificial. Includes use on level 3 under isotopes(1).
SA argon
 isotopes

Ar-38
Includes use on level 3 under isotopes(1).
SA argon
 isotopes

Ar-39
Artificial. Includes use on level 3 under isotopes(1).
SA Ar-40/Ar-39
 argon
 isotopes

Ar-39/Ar-40
use Ar-40/Ar-39

Ar-40
Includes use on level 3 under isotopes(1).
SA Ar-40/Ar-36
 Ar-40/Ar-39
 argon
 isotopes

Ar-40/Ar-36
Includes use on level 3 under isotopes(1). Also search Ar-36/Ar-40; Ar-36 AND Ar-40.
UF Ar-36/Ar-40
SA Ar-36
 Ar-40
 argon

Ar-40/Ar-39
Includes use on level 3 under isotopes(1). Also search Ar-39/Ar-40; Ar-39 AND Ar-40.
UF Ar-39/Ar-40
SA Ar-39
 Ar-40
 argon

Ar/Ar
Includes use on level 3 under absolute age(1) methods(2). Also search argon-argon.
SA absolute age
 argon
 methods

Arabia
use Arabian Peninsula

Arabian Desert
use Eastern Desert

Arabian Gulf
use Persian Gulf

Arabian Peninsula
Between Red Sea on the W; and the Persian Gulf, Gulf of Oman, and Arabian Sea on the E. Also search Arabia. Index countries as applicable. Includes use on level 1 and 2 as an area term (list O). For term set options see list B.
UF Arabia
BA Asia
NT Arabian Shield
NA Bahrain
 Kuwait
 Oman
 Qatar
 Saudi Arabia
 Southern Yemen
NT Timna
NA United Arab Emirates
 Yemen
SA Middle East

Arabian Plate
SA plate tectonics
 Saudi Arabia

Arabian Ridge
use Carlsberg Ridge

Arabian Sea
Between the Arabian Peninsula on the W and Pakistan and India in the E. As of 1977, includes use as a level 1 area term (list O). For term set options see list B.
BA Indian Ocean
NA Gulf of Aden
NT Gulf of Cambay
 Gulf of Oman
 Gulf of Tadjoura

Arabian Shield
NE part of the Ethiopian Shield. Index countries as applicable.
BT Arabian Peninsula
SA Saudi Arabia
 Southern Yemen
 Yemen

Arabian-Indian Midoceanic Ridge
use Carlsberg Ridge

Arabian-Indian Ridge
use Carlsberg Ridge

Arabic
Includes use on level 3 to indicate language of a document.

Aracaju
City on the Atlantic Ocean in NE part of country.
BT Sergipe
 Brazil

Aracena
Town near the Gulf of Cadiz.
BT Huelva
 Spain

Arachnida
Class. Includes use on level 2 under Arthropoda(1). See list F.
BA Chelicerata
 Arthropoda
BT Invertebrata

Arad
City and county near the Hungarian border.
BT Banat
 Romania

Arafura Sea
Sea between N Australia and Indonesia; 800 miles long by about 350 miles wide.
BT Pacific Ocean
SA Australia
 Indonesia

Aragats
A mountain in NW.
BT Armenia
 USSR

Aragon
Region and ancient kingdom in NE.
BT Spain

aragonite
BA carbonates
BT minerals
SA alabaster
 calcite
 pearls
 witherite

Aragua
State in N.
BT Venezuela

Aragvi
River in N central.
BT Georgian Republic
 USSR

Aral region
Index Soviet republics as applicable. Also search Aral; Aral Sea region.
UF Aral Sea region
BT USSR
SA Kazakhstan
 Turkmenia
 Uzbekistan

Aral Sea
Between Kazakhstan and Uzbekistan. Also search Aral.
CO N440000N460000
 E0620000E0580000
BT USSR

Aral Sea region
use Aral region

Arapahoe County
S and SE of Denver.
BT Colorado
 United States

Ararat
Mountain near Iranian and USSR borders.
BT Turkey
 Middle East

Arauco
Province in N.
BT Chile

Aravalli Range
Central.
BT Rajasthan
 India

Aravalli System
Includes Dabari, Matoon, and Udaipur formations.
BT Archean
SA Andhra Pradesh
 Gujarat
 India
 Rajasthan

Araxa
Town in W.
BT Minas Gerais
 Brazil

Araya Peninsula
W Sucre.
BT Venezuela
SA Sucre

Arbarastakh
Carbonatite complex in Yakutia.
BT Russian Republic
 USSR

Arbuckle Group
Subdivided into Fort Sill Limestone, Royer Marble, Signal Mountain Limestone, Chapman Ranch Dolomite, Wolf Creek Dolomite, Cool Creek Limestone, Alden Limestone, and West Spring Creek Formation. S Oklahoma.
BT Paleozoic
SA Cambrian
 Lower Ordovician
 Oklahoma
 Ordovician
 Upper Cambrian

Arbuckle Mountains
S central.
CO N340000N343000
 W0964500W0973000
BT Oklahoma
 United States

Arcachon Basin
Inlet of Bay of Biscay in SW part of country.
BT Gironde
 France
SA Bay of Biscay

arch dams
UF archdams
BT dams

Archaean
use Archean

Archaediscidae
Family. Includes use on level 3 under foraminifera(1) Fusulinina(2).
BA Fusulinina
 foraminifera
BT Invertebrata

Archaeocyatha
Includes use on level 1 and 2 as a

fossil term (list F).
UF cyathosponge
pleosponge
BT Invertebrata

Archaeogastropoda
Order. Includes use on level 3 under Mollusca(1) Gastropoda(2).
BA Gastropoda
Mollusca
BT Invertebrata

Archaeolithothamnium
Includes use on level 3 under algae(1) Rhodophyta(2).
BA Corallinaceae
Rhodophyta
algae
BT Plantae

archaeological sites
UF sites, archaeological
SA archaeology
artifacts
fossil man

archaeology
Includes use on level 3.
UF archeology
SA archaeological sites
artifacts
fossil man
man
paleoecology

Archaeomonadaceae
BA ebridians
Protista

Archaeopteryx
Genus. Includes use on level 3 under Aves(1) Archaeornithes(2).
BA Archaeornithes
Aves
BT Tetrapoda
Vertebrata

Archaeornithes
Includes use on level 2 under Aves(1). See list F.
BA Aves
BT Tetrapoda
Vertebrata
NA Archaeopteryx
SA Neornithes

Archangel
use Arkhangelsk

archdams
use arch dams

Archean
Includes use on level 1 and 2 as an age term (list E) as of 1978.
UF Archaean
BA lower Precambrian
Precambrian
NT Amisk Group
Aravalli System
Arunta Complex
Gangpur Series
Kalgoorlie System
Krivoy Rog Series
Rice Lake Group
Sausar Series
Singhbhum Granite
Willyama Complex
Yellowknife Group
SA Moldanubian

Archeocopida
Order. Includes use on level 2 under Ostracoda(1).
BA Ostracoda
BT Crustacea
Arthropoda
Invertebrata

archeology
use archaeology

arches
SA engineering geology

Arches
National Park in SE.
BT Utah
United States

Archosauria
Includes use on level 2 under Reptilia(1). See list F.
BA Reptilia
BT Tetrapoda
Vertebrata
NA Crocodilia
Ornithischia
Pteranodon
Pterosauria
Saurischia
SA dinosaurs

Archostemata
Includes use on level 3 under Insecta(1).
BA Insecta
BT Arthropoda
Invertebrata

arcs
Includes use on level 3 under aurora(1) electrical field(2).
SA aurora
electrical field

arcs, island
use island arcs

Arctic Archipelago
Large group of islands in the Arctic Ocean nearly coextensive with District of Franklin, Northwest Territories, Canada. Used as level 1 area term from 1977 to 1978. Now used only on level 3. Also search Arctic Islands.
UF Arctic Islands
BT Northwest Territories
Canada
SA Arctic region

Arctic Coastal Plain
SA Arctic region

arctic environment
Term introduced in 1978. Includes use on level 3 under ecology(1). Before 1978, also search arctic.
SA climate
ecology
environment
paleoclimatology
paleoecology

Arctic Islands
use Arctic Archipelago

Arctic Ocean
Extends from North Pole southward to approximately 70° N and is bounded by Alaska, Canada, Greenland, Norway, and the USSR. Includes use on level 1 or 2 as an area term (list O). If 1, see term set options under list B.
CO N700000N900000
W0400000W0395900
UF Arctic Sea
North Polar Sea
NT Alpha Cordillera
Barents Sea
Beaufort Sea
Beerenberg
Canada Basin
Chukchi Sea
East Siberian Sea
Eurasia Basin
Greenland Sea
NZ Jan Mayen
NT Kara Sea
Laptev Sea
Lomonsov Ridge
Mendeleyev Ridge
Svalbard
SA Arctic region

Arctic region
The Arctic Ocean and islands in it and adjacent to it plus mainland surfaces N of the Arctic circle. Index Alaska, Arctic Archipelago, Arctic Ocean, Greenland, and countries as applicable. Includes use on level 1 or 2 as an area term (list O). If 1, see term set options under list B.
NA Greenland
Spitsbergen
SA Alaska
Arctic Archipelago
Arctic Coastal Plain
Arctic Ocean
Canada
Finland
Hecla Hoek Formation
Jan Mayen
Norway
Soviet Arctic
subarctic regions
Sweden
USSR

Arctic Sea
use Arctic Ocean

arcuate faults
Before 1978, also search arcuate AND faults.
BA faults

Ardara Pluton
An igneous intrusion in NW part of country.
BT Donegal
Ireland

Ardeche
Department in S.
CO N441500N453000
E0050000E0034500
BT France
NT Largentiere
Privas
Vivarais

Ardennes
Wooded plateau region. Index countries as applicable.
CO N493000N503000
E0053000E0041000
UF Forest of Ardennes
BT Europe
NT Givet
Rocroi
Sedan
SA Belgium
France
Luxembourg

Ardnamurchan
Peninsula and parish in Argyll County in W.
BT Scotland
Great Britain
United Kingdom

arduinite
use mordenite

Areado Formation
BT Lower Cretaceous
Cretaceous
SA Aptian
Brazil
Minas Gerais

areal geology
Used for entries that might properly be placed under three or more of the second-order headings such as geomorphology, stratigraphy, structural geology. Includes use on level 2 under area terms(1). See list B.
SA bibliography
explanatory text
geology
guidebook
maps

areal studies
Includes use on level 2 under clay mineralogy(1).
UF studies, areal
SA clay mineralogy
mineral data

Arecibo
Town on the Atlantic Ocean W of San Juan.
BT Puerto Rico
West Indies

arenaceous texture
Term introduced in 1978. Includes use on level 3 under sedimentary rocks(1) and sediments(1). Before 1978, also search arenaceous AND specific sediment type or sedimentary rock.
UF arenarious texture
psammitic texture
sabulous texture
sandy texture
SA sand
sandstone
sedimentary rocks
sediments
textures

Arenal
Extinct volcano in NW.
BT Costa Rica

arenarious texture
use arenaceous texture

Arendal
A city on the Skagerrak.
BT Norway

arendalite
use epidote

Arenig
Two mountains in N Merioneth in N Wales. One is called Arenig Fach and the other Arenig Fawr.
BT Wales
Great Britain
United Kingdom

Arenigian
Europe. Above Tremadocian, below Llanvirnian. Includes use on level 3 under age terms(1). See list E.
BA Lower Ordovician
Ordovician
BT Paleozoic

arenite
Includes use on level 3 under sedimentary rocks(1) clastic rocks(2). See list I.
BA clastic rocks
BT sedimentary rocks
SA terrigenous materials

Arequipa
Department in S Peru. Also a city.
BT Peru

arfvedsonite
BA amphibole group
chain silicates
silicates
BT minerals

Argentera
Village in S.
BT Italy
SA Piedmont

Argentera Massif
S Piedmont.
BT Italy
SA Piedmont

Argentiere Glacier
In the Mount Blanc Massif of the Pennine Alps.
BT Haute-Savoie
France

Argentina
Includes use on level 1 as an area

Argentina ● Arkansas

term (list O). For term set options see list B.
  CO S550000S220000
      W0533000W0730000
  BA South America
  NT Barreal
      Buenos Aires Province
      Campo del Cielo
      Catamarca
      Chubut
      Corrientes
      Jujuy
      La Pampa
      La Rioja
      Mendoza
      Neuquen
      Neuquen Basin
      Pampas
      Pampean Mountains
      Patagonia
      Precordillera
      Puerto Deseado
      Rio Negro
      Salado Basin
      Salta
      San Luis
      Santiago del Estero
      Talacasto
      Tucuman
  SA Andes
      Baquero Formation
      Chaco
      Chanares Formation
      Cordoba
      Magdalena
      Paganzo Group
      Parana Basin
      Parana River
      Patagonian Andes
      Rio de la Plata
      San Juan
      Santa Cruz
      Santa Fe
      Tierra del Fuego
      Yacoraite Formation

Argentine Basin
E of Argentina.
  BT Atlantic Ocean

argentite
  UF argyrite
  BA sulfides
  BT minerals
  SA acanthite
      silver

Arges River
In S part of country.
  BT Romania

argillaceous rocks
use argillaceous texture

argillaceous texture
Term introduced in 1978. Includes use on level 3 under sediments(1) and sedimentary rocks(1). Before 1978, also search argillaceous AND specific sediment type or sedimentary rock. Also search argillaceous rocks.
  UF argillaceous rocks
  SA argillite
      clay
      clay minerals
      marl
      sedimentary rocks
      sediments
      shale
      textures

argillite
Includes use on level 3 under sedimentary rocks(1) clastic rocks(2). See list I.
  UF dropstone
  BA clastic rocks
  BT sedimentary rocks
  SA argillaceous texture
      terrigenous materials

Argo abyssal plain
Eastern.
  BT Indian Ocean

Argolis
Department in E.
  BT Peloponnesus
      Greece

argon
Includes use on level 1 and 2 as a chemical element (list D). Also search Ar.
  UF Ar
  BT noble gases
  SA Ar-36
      Ar-37
      Ar-38
      Ar-39
      Ar-40
      Ar-40/Ar-36
      Ar-40/Ar-39
      Ar/Ar
      elements
      helium
      isotopes
      K/Ar
      krypton
      neon
      radon
      xenon

Argovian
Europe. Substage in Great Britain: upper Jurassic (lower Lusitanian; above Oxfordian Stage, below Rauracian Substage). Includes use on level 3 under age terms(1). See list E.
  BA Upper Jurassic
      Jurassic
  BT Mesozoic

Argun
Village in central Tomsk Oblast in W central Siberia.
  BT Russian Republic
      USSR

Argun River
Rises in NW Manchuria and serves as part of boundary between NE China and USSR. Joins with Shilka River to form the Amur River. Index Manchuria and/or Russian Republic as applicable. Also search Argun.
  BT Asia
  SA Amur River
      Manchuria
      Russian Republic
      Shilka Valley

Argyll (County)
use Argyllshire

Argyllshire
County in W Scotland. Also search Argyll.
  UF Argyll (County)
  BT Scotland
      Great Britain
      United Kingdom

argyrite
use argentite

Ariake Bay
On Pacific Ocean in SE.
  BT Kyushu
      Japan

arid environment
Term introduced in 1978. Includes use on level 2 under land use. Before 1978, also search arid; before 1975 search arid regions.
  SA climate
      deserts
      ecology
      environment
      geomorphology
      land use
      semi-arid environment

arid regions
A valid term through 1974. After 1974, arid was used as an index term. After 1977, use arid environment.

Aridisols
Includes use on level 3 under soils(1). See list M.
  BA soils

Ariege
Department in S.
  CO N423500N432500
      E0021000E0005000
  BT France
  NT Castillon Massif
      Querigut Massif
      Saint-Girons
  SA Salat Valley

ariegite
Includes use on level 3 under igneous rocks(1) ultramafic family(2). See list H.
  BA ultramafic family
  BT igneous rocks

Aries Valley
River valley in W central Transylvania. Also search Aries.
  BT Transylvania
      Romania

Arikaree Group
Comprises Sharps, Monroe Creek, and Harrison Formations. Colorado, S South Dakota, SE Montana, and SE Wyoming.
  BT Miocene
  SA Colorado
      Montana
      South Dakota
      Wyoming

Arikareean
Provincial series, North America.
  BA Tertiary
  BT Cenozoic
  SA lower Miocene
      Neogene
      Paleogene
      upper Oligocene

Arita River
W Kii Peninsula S of Osaka. Also search Aritagawa.
  UF Aritagawa
  BT Honshu
      Japan

Aritagawa
use Arita River

Ariyalur
Town in S.
  BT Tamil Nadu
      India

Ariyalur Stage
S India.
  BT Upper Cretaceous
      Cretaceous
  SA India

Arize Massif
N of the central Pyrenees.
  BT France

Arizona
Includes use on level 1 as an area term (list O). For term set options see list B.
  CO N311500N370000
      W1090000W1150000
  BA United States
  NT Apache County
      Bisbee
      Black Mesa
      Black Mesa Basin
      Canyon Diablo
      Cochise County
      Coconino County
      Flagstaff
      Gila County
  NX Graham County
  NT Grand Canyon
      Holbrook
      Hopi Buttes Field
  NX Jerome
  NT Marble Canyon
      Maricopa County
      Mogollon Plateau
      Mohave County
      Naha test well
      Navajo County
      O'Leary Peak
      Petrified Forest National Park
      Phoenix
      Pima County
      Pinal County
      Salt River
      San Carlos Indian Reservation
      San Francisco Mountain
      San Francisco Peaks
      San Manuel
      Santa Catalina Mountains
  NX Santa Cruz County
  NT Santa Rita Mountains
      Sierra Ancha
      Sierrita Mountains
      Superstition Mountains
      Tombstone
      Tonto Basin
      Tucson
      Tucson Basin
      Tucson Mountains
      University of Arizona
      Verde Valley
      Yavapai County
      Yuma
  NX Yuma County
  SA Basin and Range Province
      Bidahochi Formation
      Chinle Formation
      Chuar Group
      Coconino Sandstone
      Colorado Plateau
      Colorado River
      Cutler Formation
      Four Corners
      Gila River
      Kaibab Formation
      Kayenta Formation
      Lake Mead
      Lake Powell
      Mancos Shale
      Martin Formation
      Mesaverde Group
      Moenkopi Formation
      Morrison Formation
      Navajo Indian Reservation
      Navajo Sandstone
      Paradox Basin
      Pedregosa Basin
      Redwall Limestone
      San Francisco Mountains
      San Juan Basin
      Sonoran Desert
      Southwestern U.S.
      Supai Formation
      Todilto Formation
      Toroweap Formation
      Unkar Group
      Verde Formation
      Virgin River valley
      White Mountains

Arkansas
Includes use on level 1 as an area term (list O). For term set options see list B.
  CO N330000N363000
      W0894000W0944000
  BA United States
  NT Barber County
      Batesville
  NX Benton County
  NT Boone County
      Calhoun County
      Carroll County
      Clark County
      Clay County

Columbia County
Crittenden County
Dallas County
NT Fayetteville
NX Franklin County
Fulton County
Grant County
Greene County
Jackson County
Jefferson County
Johnson County
Lafayette County
Lawrence County
Lee County
Lincoln County
Madison County
Marion County
Monroe County
Montgomery County
Nevada County
Newton County
Perry County
Phillips County
Pike County
Polk County
Pope County
Pulaski County
Randolph County
Scott County
Sevier County
Union County
Washington County
White County
SA Arkansas River
Arkansas River valley
Arkoma Basin
Atoka Formation
Bloyd Formation
Bossier Formation
Brassfield Formation
Buckner Formation
Chattanooga Shale
Cotton Valley Group
Desmoinesian
Everton Formation
Fayetteville Formation
Fernvale Formation
Jackfork Group
Johns Valley Formation
Midcontinent
Mississippi Embayment
Mississippi River
Mississippi Valley
Morrow Formation
Norphlet Formation
Ouachita Mountains
Ozark Mountains
Paluxy Formation
Pitkin Limestone
Red River
Red River valley
Saint Peter Sandstone
Saratoga Chalk
Savanna Formation
Schuler Formation
Smackover Formation
Smithville Formation
Stanley Group
Stuttgart
Tennessee Sandstone
Trinity Group
Washita Group
White River
Woodbine Formation

Arkansas City
Cowley County in S.
BT Kansas
   United States

Arkansas River
S central U.S., flows 1,450 mi. generally ESE from the Rocky Mts. of central Colorado, through Kansas, Oklahoma, and Arkansas, to Mississippi River N of Greenville, Mississippi. Index states as applicable. Also search Arkansas River valley.
BT United States
SA Arkansas
   Colorado
   Kansas
   Mississippi
   Oklahoma

Arkansas River valley
Index states as applicable.
BT United States
SA Arkansas
   Colorado
   Kansas
   Oklahoma

Arkhangelsk
City and oblast on the White Sea in NW Russian Republic. Also search Archangel.
UF Archangel
BT Russian Republic
   USSR

Arkoma Basin
Index states as applicable.
BT United States
SA Arkansas
   Oklahoma

arkose
Includes use on level 3 under sedimentary rocks(1) clastic rocks(2). See list I.
BA clastic rocks
BT sedimentary rocks
SA arkosic composition
   meta-arkose
   sandstone
   subarkose
   terrigenous materials

arkosic composition
Term introduced in 1978. Includes use on level 3 under sediments(1) and sedimentary rocks(1). Before 1978, also search arkosic AND specific sediment type or sedimentary rock.
SA arkose
   composition
   sedimentary rocks
   sediments

Arlan
Oil bearing region in Bashkiria in S Urals.
BT Russian Republic
   USSR

armalcolite
BA oxides
BT minerals
SA pseudobrookite

Armenia
Armenian Soviet Socialist Republic.
CO N390000N410000
   E0460000E0430000
BT USSR
NT Agarak
   Akhtala
   Alaverdi
   Aragats
   Arteni
   Dastakert
   Kadzharan
   Kafan
   Lake Sevan
   Megri
   Megrinskiy Pluton
   Pambak
   Razdan
   Sevan
   Shakhdag Range
   Shamlug
   Sisian
   Yerevan
   Zod
SA Caucasus
   European USSR
   Lesser Caucasus
   Oktemberyan Series
   Transcaucasia
   Zangezur

Armidale
Town in NE.
BT New South Wales
   Australia

armored mud balls
Includes use on level 3 under sedimentary structures(1) secondary structures(2). See list K.
UF mud balls, armored
BA secondary structures
   sedimentary structures
SA clay

Armorica
use Armorican Massif

Armorican Massif
NW part of country.
UF Armorica
BT France

Arno River Basin
N central.
BT Tuscany
   Italy

aromatic hydrocarbons
Term introduced in 1978. Before 1978, search aromatic or aromatics AND hydrocarbons.
BA organic materials
SA hydrocarbons

Arosa Bay
On the Atlantic Ocean in W between provinces of La Coruna and Pontevedra.
BT Galicia
   Spain

Arran
Island in Firth of Clyde in SW.
CO N552500N554500
   W0050500W0053000
BT Scotland
   Great Britain
   United Kingdom

arrays
Includes use on level 3 under seismology(1) or geophysical methods(1).
SA geophysical methods
   LASA
   seismology

arrival time
SA earthquakes
   elastic waves
   seismology
   time

Arrow Canyon Range
BT Nevada
   United States

arroyos
SA geomorphology

arsenates
Includes use on level 2 under minerals(1). See list L.
BT minerals
NA adamite
   austinite
NZ beudantite
NA conichalcite
   haidingerite
   legrandite
NZ mimetite
NA pharmacolite
   pharmacosiderite
   scorodite

arsenic
Includes use on level 1 and 2 as a commodity term (list C) and as a chemical element (list D).
UF As
SA arsenopyrite
   elements
   native elements and alloys

arsenical pyrites
use arsenopyrite

arsenides
Includes use on level 3 under minerals(1) sulfides(2). See list L.
BA sulfides
BT minerals
NA arsenopyrite
   cobaltite
   lollingite
   maucherite
   niccolite
   pararammelsbergite
   rammelsbergite
   safflorite
   skutterudite
   sperrylite

arsenites
Includes use on level 2 under minerals(1). See list L.
BT minerals
NA mixite

arsenopyrite
UF arsenical pyrites
   white pyrites
BA arsenides
   sulfides
BT minerals
SA arsenic
   lollingite

arsenosulfides
BA sulfides
BT minerals

Artemisia
Genus. Includes use on level 3 under angiosperms(1) Dicotyledoneae(2) or under palynomorphs(1) miospores(2).
BA Dicotyledoneae
   angiosperms
BT Plantae
SA miospores
   palynomorphs

Artemovsk
City in the Donets Basin.
UF Bakhmut
BT Ukraine
   USSR

Arteni
Region.
BT Armenia
   USSR

artesian waters
Includes use on level 2 under ground water(1). Before 1976, also search artesian waters and wells.
BT ground water
SA hydrogeology
   water

artesian waters and wells
A valid term through 1971. After 1972, use artesian waters.

Arthrodira
Subclass. Includes use on level 3 under Pisces(1) Placodermi(2).
BA Placodermi
   Pisces
BT Vertebrata

Arthropoda
Includes use on level 1 and 2 as a fossil term (list F).
BT Invertebrata
NA Chelicerata
   Crustacea
NT Insecta
NA Myriapoda
   Trilobitomorpha

Arthur Harbor
On Anvers Island. Off NW Antarctic Peninsula.
BT Antarctica

Atchison ● atolls

BT Kansas
United States

Atchison County
NE Kansas.
BT Kansas
United States

Atera Fault
Gifu Prefecture in central.
BT Honshu
Japan

Athabasca
Town in central Alberta.
UF Athabaska
BT Alberta
Canada

Athabasca Glacier
In Rocky Mountains between Jasper and Banff national parks near the British Columbia border.
BT Alberta
Canada

Athabasca River
Flows into Lake Athabasca.
UF Athabaska River
BT Alberta
Canada

Athabaska
use Athabasca

Athabaska River
use Athabasca River

Athens
City in Ohio; city in Georgia; and city in Central Greece and Euboea. Index Greece/Georgia/or Ohio as applicable.
SA Georgia
Greece
Ohio

Athgarh Sandstone
Sandstones of this formation are also grouped as part of Cuttack Stage. Part of Upper Gondwana sequence of the E coast.
BT Jurassic
SA India

Atlanta
City in Fulton County in N central.
BT Georgia
United States

Atlantic
Used primarily in Europe for an interval of post/glacial time following the Boreal and preceding the Subboreal. It corresponds to most of the Altithermal and the middle part of the Hypsithermal.
BA Holocene
Quaternary
BT Cenozoic

Atlantic Coastal Plain
Extends from the Maritime Provinces of Canada to Florida. Index states and provinces as applicable. Includes use on level 1 as an area term (list O). For term set options see list B.
UF East Coast
BA North America
SA Chesapeake Bay
Connecticut
Delaware
Florida
Georgia
Maine
Maryland
Massachusetts
New Brunswick
New Hampshire
New Jersey
New York
New York Bight
North Carolina
Nova Scotia
Rhode Island
South Carolina
United States
Virginia

Atlantic Ocean
Includes use on level 1 or 2 as an area term (list O). If 1, see term set options under list B.
CO S550000N750000
E0200000W0800000
NT Argentine Basin
Ascension Island
Atlantis
Atlantis fracture zone
Atlantis II
NA Azores
NT Baffin Bay
NA Baltic Sea
NT Barbados Ridge
Barracuda Ridge
Bay of Biscay
Bay of Fundy
NA Bermuda
NT Bermuda Platform
Bermuda Rise
Blake Plateau
Blake-Bahama Basin
Blake-Bahama Outer Ridge
Block Island Sound
Bouvet Island
NA Canary Islands
NT Caribbean Sea
Celtic Sea
Davis Strait
NA East Atlantic
English Channel
Faeroe Islands
Falkland Islands
NT Flemish Cap
Georges Bank
Gibbs fracture zone
Grand Banks
Great Bahama Bank
Great Meteor Seamount
Gulf of Guinea
Gulf of Maine
Gulf of Saint Lawrence
Gulf Stream
Hare Bay
Hatteras abyssal plain
NZ Iceland
NT Iceland-Faeroe Ridge
NA Irish Sea
NZ Jan Mayen
NT Kuno
Labrador Basin
Labrador Sea
Long Island Sound
NA Madeira
NT Maury Channel
Mid-Atlantic Ridge
NA North Atlantic
North Sea
Northeast Atlantic
Northwest Atlantic
NT Norwegian Sea
Puerto Rico Trench
Reykjanes Ridge
Rockall Bank
Rockall Plateau
Rockall Trough
Romanche fracture zone
Romanche Trench
Sable Island Bank
Saint Helena
Saint Paul Rocks
Sargasso Sea
Scotia Ridge
Scotian Shelf
NA Shetland Islands
NT Signy Island
NA South Atlantic
NT South Georgia
NA Southeast Atlantic
Southwest Atlantic
NT Strait of Gibraltar
Straits of Florida
Tristan da Cunha
Trou Sans Fond
Vema fracture zone
Walvis Ridge
NA West Atlantic
NT Wilkinson Basin
Wilmington Canyon
Yucatan Channel
SA Antarctic Ocean
Atlantic region
FAMOUS
Iapetus
Scotia Sea
South Sandwich Islands

Atlantic region
Term introduced in 1976. An artificial term used to indicate the coastal region immediately adjacent to the Atlantic Ocean. Includes land and the immediate littoral zone. Includes use on level 1 and 2 as an area term (list O). For term set options see list B.
SA Atlantic Ocean

Atlantis
Seamount SW of the Azores.
BT Atlantic Ocean

Atlantis fracture zone
Across the Mid-Atlantic Ridge in N.
BT Atlantic Ocean
SA Mid-Atlantic Ridge

Atlantis II
Seamounts off U.S. E coast roughly midway between Nova Scotia and Bermuda.
BT Atlantic Ocean

Atlantis II Deep
BT Red Sea

atlas
Includes use on level 3 under glossaries(1).
SA glossaries
maps
report

Atlas
use Atlas Mountains

Atlas Mountains
Mountain system of NW and N Africa extending from SW Morocco to NE Tunisia. Index countries as applicable. Also search Atlas AND appropriate area.
UF Atlas
BT Africa
SA Algeria
Anti-Atlas
High Atlas
Middle Atlas
Morocco
Tunisia

Atlin
Village near Alaska border NE of Juneau in NW.
BT British Columbia
Canada

atmosphere
Includes use on level 1(only for Earth); on level 2 or 3 under Moon(1) or other planets; on level 3 under environmental geology(1) pollution(2). For studies dealing with geochemical changes (short-term or during geologic time), observations related to interactions with phenomena of the Earth's surface, and other geophysical aspects. For changes in climate, see paleoclimatology. If 1, term set options are:
topic [age, changes, circulation, composition, evolution, general, genesis, instruments, observations, processes, properties, temperature]
subtopic
NA upper atmosphere
SA aeronomy
aerosols
air
air-sea interface
airglow
atmospheric precipitation
atmospheric pressure
biosphere
boundary interactions
circulation
climate
convection
convection currents
cycles
degassing
Earth
Ekman spiral
environmental geology
fallout
geochemistry
geophysics
glacial geology
humidity
hydrology
hydrosphere
ionosphere
magnetosphere
meteorology
Moon
ozone
paleoclimatology
planetology
pollution
sea water
storms
stratosphere
sulfur dioxide
troposphere
volcanology

atmospheric precipitation
As of 1978, term is used on level 2 under hydrology(1). Before 1978, also search hydrology AND precipitation.
SA atmosphere
evapotranspiration
hydrologic cycle
hydrology
meteorology
moisture
precipitation
raindrops
rainfall
retention
snow
water

atmospheric pressure
BT pressure
SA atmosphere

Atoka Formation
Sandstone members: Coody (Coata), Pope Chapel, Georges Fork, Dirty Creek, Webbers Falls, and Blackjack School. Subdivided, in Arkansas, to include Greenland Sandstone Member at base.
BT Atokan
Middle Pennsylvanian
Pennsylvanian
SA Arkansas
Oklahoma

Atokan
Provincial series, North America: lower Middle Pennsylvanian. Above Morrowan, below Desmoinesian. Includes use on level 3 under age terms(1). See list E.
BA Middle Pennsylvanian
Pennsylvanian
BT Paleozoic
NT Atoka Formation

atolls
Includes use on level 3 under

reefs(1).
BT reefs
SA bioherms
corals
islands
lagoons

atomic absorption
Includes use on level 3 under spectroscopy(1) methods(2); in combination with spectroscopy (i.e. spectroscopy, atomic absorption) under chemical element(1) analysis(2). See list D.
SA absorption
analysis
spectroscopy

atomic packing
Term introduced in 1978 for use under crystal structure(1), crystallography(1), or mineralogy(1). Before 1978, also search packing.
SA crystal structure
crystallography
mineralogy
packing

Atrypa
Genus. Includes use on level 3 under Brachiopoda(1) Articulata(2).
UF atrypoids
BA Atrypidae
Articulata
Brachiopoda
BT Invertebrata

Atrypidae
Family. Includes use on level 3 under Brachiopoda(1) Articulata(2).
BA Articulata
Brachiopoda
BT Invertebrata
NA Atrypa

atrypoids
use Atrypa

Atsumi Peninsula
S of Nagoya.
BT Honshu
Japan

attapulgite
use palygorskite

attenuation
Includes use on level 3 under seismology(1).
SA elastic waves
seismology

Atterberg limits
UF consistency limits
limits, Atterberg
SA deformation
plasticity
rock mechanics
soil mechanics

Attica
Ancient division and state. Currently a department. Central Greece and Euboea.
BT Greece

Attock
Town in NW.
BT Punjab
Pakistan

Aturia
Genus. Includes use on level 3 under Mollusca(1) Cephalopoda(2).
BA Nautiloidea
Cephalopoda
Mollusca
BT Invertebrata

Au
use gold

Aube
Department in N central.
BT France

Aubrac
Mountain range in S central.
BT France

Auckland
City in NW.
BT North Island
New Zealand

Auckland Islands
200 miles south of New Zealand.
BT Pacific Ocean

Aude
Department in S.
CO N423000N434000
E0031500E0014500
BT France
NT Corbieres
Mouthoumet Massif
Narbonne
Portel
Quillan
Salsigne Mine
SA Montagne Noire

augen gneiss
Includes use on level 3 under metamorphic rocks(1) gneisses(2). See list J.
BA gneisses
BT metamorphic rocks
SA gneiss

augite
BA pyroxene group
chain silicates
silicates
BT minerals
SA basalt
fassaite
omphacite
titanaugite

augitite
Includes use on level 3 under igneous rocks(1) ultramafic family(2). See list H.
BA ultramafic family
BT igneous rocks

Augusta
City in Kennebec County in S.
BT Maine
United States

Augusta County
Western Virginia.
BT Virginia
United States

Augustine
Volcano on Augustine Island in Kamishak Bay at mouth of Cook Inlet in S Alaska. Also search Augustine Volcano.
UF Augustine Volcano
BT Alaska
United States

Augustine Volcano
use Augustine

Augustow
Town in NE part of country.
BT Bialystok
Poland

aulacogens
Includes use on level 3 under tectonics(1).
SA grabens
tectonics
troughs

aureoles
Includes use on level 3 under metamorphism(1) and intrusions(1). Also search contact aureole; contact zone; metamorphic aureoles; metamorphic zone; thermal aureoles.
UF contact aureole
contact zone
exomorphic zone
metamorphic aureoles
metamorphic zone
thermal aureole
SA haloes
intrusions
metamorphism

Aures
Mountain massif in NE.
BT Algeria

aurora
Usually out-of-scope for GeoRef. Includes use on level 1 for special bibliographies; on level 2 under ionosphere(1). If 1, term set options are:
absorption
subtopic
auroral zone (including auroral oval)
subtopic
electrical field
topic [arcs, currents, electrojet, emmissions, particle precipitation]
emissions
name of element, wave length or subtopic
type of aurora or subtopic
type of emission or subtopic
magnetic field
topic [currents, electrojet, magnetic storms, micropulsations, type of aurora]
morphology
motions or subtopic
orientation or subtopic
synoptic studies or subtopic
time variations or subtopic
observations
type of aurora
theoretical studies
topic
SA absorption
absorption and scattering
aeronomy
airglow
arcs
astrophysics and solar physics
auroral oval
auroral zone
currents
electrical field
electrojet
emissions
intensity
ionosphere
magnetic field
magnetic storms
magnetosphere
meteorology
micropulsations
morphology
motions
solar wind
time variations

auroral oval
Includes use on level 3 under aurora(1).
SA aurora
auroral zone

auroral zone
Including auroral oval. Includes use on level 2 under aurora(1).
UF zone, auroral
SA aurora
auroral oval

Austin
City in Travis County in central.
BT Texas
United States

Austin Group
Comprises Ector Tongue, Bonham Clay, Blossom Sand, Brownstone Marl, Gober Tongue, Austin Chalk, and Burditt Marl. E Texas. Also search Austin Chalk.
BT Gulfian
Upper Cretaceous
Cretaceous
SA Texas

austinite
BA arsenates
BT minerals

Austral Islands
Group in French Oceania 330 miles S of Society Islands. Also search Austral.
UF Tubuai Islands
BT Pacific Ocean

Australasia
Islands of the South Pacific including Australia, New Zealand, New Guinea and adjacent islands. Term sometimes covers all Oceania. Index New Guinea and countries as applicable. Includes use on level 1 or 2 as an area term (list O). If 1, see term set options under list B.
NA New Guinea
New Zealand
Papua New Guinea
SA Australia
Oceania
Solomon Islands

Australia
Includes use on level 1 or 2 as an area term (list O). If 1, see term set options under list B. To retrieve all documents, individual countries and physiographic regions should also be searched (see list O).
CO S400000S100000
E1550000E1143000
NT Amadeus Basin
Australian Shield
Bass Strait
Canberra
Carpentaria Basin
Cooper Basin
Eucla Basin
Gambier Embayment
Georgina Basin
Giles Complex
Great Artesian Basin
Joseph Bonaparte Gulf
Murray Basin
Murray River
Musgrave Ranges
New England Batholith
NA New South Wales
Northern Territory
NT Nullarbor Plain
Officer Basin
Otway Basin
NA Queensland
NT Simpson Desert
Snowy Mountains
NA South Australia
NT Tasman Geosyncline
NA Tasmania
Victoria
Western Australia
SA Arafura Sea
Arunta Complex
Australasia
Baralaba Coal Measures
Bitter Springs Formation
Brockman Iron Formation
Broken River Formation
Bulli Seam
Gondwana
Hamersley Group
Hawkesbury Sandstone
Heemskirk Granite
Illawarra Coal Measures
Ipswich Coal Measures
Lilydale Limestone
Mersey River
Mersey Valley
Narrabeen Group
Oceania
Singleton Coal Measures

Australian Capital Territory
SE Australia, within state of New South Wales. Formerly called Federal Capital Territory.
- UF Federal Capital Territory
- BT New South Wales
     Australia

Australian Shield
SW and central.
- BT Australia

australite
- BA tektites
- SA meteorites

Australopithecinae
Subfamily. Includes use on level 3 under Mammalia(1) Primates(2).
- BA Primates
     Mammalia
- BT Tetrapoda
     Vertebrata

Australopithecus
Genus. Includes use on level 3 under Mammalia(1) Primates(2).
- BA Hominidae
     Primates
     Mammalia
- BT Tetrapoda
     Vertebrata

Austria
Includes use on level 1 as an area term (list O). For term set options see list B.
- CO N461500N490000
     E0171500E0093000
- BA Europe
- NT Burgenland
     Carinthia
     Enns Valley
     Gailtal Alps
     Hohe Tauern
     Koralpe Range
     Limestone Alps
     Lower Austria
     Salzburg
     Salzkammergut
     Semmering
     Styria
     Tauern Tunnel
     Totes Gebirge
     Tyrol
     Upper Austria
     Vorarlberg
     Wechsel
- SA Allgau Alps
     Alps
     Carnic Alps
     Central Alps
     Danube River
     Danube Valley
     Eastern Alps
     Hallstatt Limestone
     Inn Valley
     Karawanken
     Lake Constance
     Molasse Basin
     Morava River valley
     Otztal Alps
     Pannonia
     Pannonian Basin
     Rhaetian Alps
     Rhine Basin
     Rhine River
     Vienna Basin
     Wetterstein Limestone
     Zillertal Alps

authigenesis
Used for mineral genesis under diagenesis(1).
- SA authigenic minerals
     diagenesis
     minerals
     new minerals
     recrystallization
     sedimentary rocks

authigenetic minerals
use authigenic minerals

authigenic minerals
Also search authigenic AND minerals.
- UF authigenetic minerals
- BT minerals
- SA authigenesis

autochthons
- SA allochthons
     tectonics

autocorrelation
- BA statistical methods
- SA mathematical methods

automatic data processing
Includes use on level 1 and 2 (list A) for computer applications and machine modeling of geological data. Before 1972, also search computer methods; computer techniques. Term set options are:
     methods
          name of method or application
     topic [List B (except areal geology), or extraterrestrial geology, general, geophysics]
          subtopic
- UF data processing, automatic
- SA algorithms
     analog simulation
     analog techniques
     chemical analysis
     computer programs
     computers
     data
     data acquisition
     data analysis
     data bases
     data handling
     data processing
     data retrieval
     data storage
     digital simulation
     digital techniques
     Fortran
     Fortran IV
     Fourier analysis
     geology
     geophysics
     graphic display
     ground water
     information systems
     mathematical geology
     mathematical models
     microcomputers
     micropaleontology
     microprocessors
     minicomputers
     models
     nomograms
     programs
     punch cards
     remote sensing
     simulation
     statistical analysis
     trend-surface analysis
     well-logging

autometamorphism
Includes use on level 2 under metamorphism(1).
- BA metamorphism
- SA albitization
     alteration
     hydrothermal alteration
     igneous rocks
     mineral assemblages
     processes

autometasomatism
- BA metasomatism
- SA processes

autoradiography
Includes use on level 3 under geophysical methods(1) methods(2). Also search radiography.
- UF radiography
- SA analysis
     geophysical methods
     methods
     radioactivity

Autun
Town in E central part of country.
- BT Saone-et-Loire
     France

Autunian
Europe. Above Stephanian, below Saxonian. Includes use on level 3 under age terms(1). See list E.
- BA Lower Permian
     Permian
- BT Paleozoic
- SA Rotliegendes

autunite
- BA phosphates
- BT minerals

Auvergne
Historical region in S central.
- CO N443000N461500
     E0044500E0020000
- BT France

Auversian
Europe. Above Lutetian, below Bartonian. Includes use on level 3 under age terms(1). See list E.
- BA upper Eocene
     Eocene
     Paleogene
     Tertiary
- BT Cenozoic
- SA Priabonian

Auxerre
City in N central part of country.
- BT Yonne
     France

Avacha
Active volcano on S Kamchatka Peninsula. Also search Avacha Volcano.
- UF Avacha Volcano
- BT Russian Republic
     USSR

Avacha Volcano
use Avacha

avalanches
As of 1978, term is used on level 2 under geologic hazards(1).
- SA geologic hazards
     landslides
     slope stability

Avallon
Town in N central part of country.
- BT Yonne
     France

Avalon Peninsula
SW of Saint John's.
- CO N463000N481000
     W0523000W0541500
- BT Newfoundland
     Canada

Avalonian Orogenic Phase
use Avalonian Orogeny

Avalonian Orogeny
Term introduced in 1978. Used in connection with orogenic event near the end of Precambrian time along E border of North America. Named for the Avalon Peninsula in SE Newfoundland. Before 1978, search Avalonian AND orogeny.
- UF Avalonian Orogenic Phase
- BT Precambrian
- SA orogeny

Aves
Includes use on level 1 and 2 as a fossil term (list F).
- UF birds
- BT Tetrapoda
     Vertebrata
- NA Archaeornithes
     Neornithes
- SA eggs

Aves Island
Uninhabited Venezuelan island just East of Bonaire, Netherlands Antilles. Also search Aves in combination with Caribbean Sea.
- UF Bird Island
- BT Caribbean Sea

Aves Ridge
S of Aves Island in W.
- BT Caribbean Sea

Aveyron
Department in S central.
- CO N434500N450000
     E0033000E0014500
- BT France
- NT Decazeville
- SA Central Massif
     Rodez Trough
     Rouergue

Avila
Province W of Madrid. Also a city.
- BT Spain

Aviles
City on the Bay of Biscay.
- BT Oviedo
     Spain

Avonian
Europe. Includes use on level 3 under age terms(1). See list E.
- BA Lower Carboniferous
     Carboniferous
- BT Paleozoic
- SA Dinantian

Awamoan
Stage. New Zealand.
- UF Awamoan Stage
- BA lower Miocene
     Miocene
     Neogene
     Tertiary
- BT Cenozoic

Awamoan Stage
use Awamoan

awaruite
- BA native elements and alloys
- BT minerals

Awash Valley
River valley in E.
- BT Ethiopia

Awashima
Island in Sea of Japan.
- BT Honshu
     Japan

Axel Heiberg Island
Largest of Sverdrup Islands of District of Franklin.
- CO N780000N820000
     W0850000W0970000
- BT Northwest Territories
     Canada

axes, fold
use fold axes

axial surface
use axial-plane structures

axial-plane structures
Term introduced in 1978. Used for foliation or cleavage. Includes use on level 3 under foliation(1) style(2) and under structural analysis(1) theoretical studies(2). Before 1978, also search axial-plane; axial plane; axial plane foliation; axial planes.
- UF axial surface
- SA cleavage
     deformation
     flow cleavage
     folds

foliation
schistosity
slaty cleavage
structural analysis
structures

axinite
BA ring silicates
silicates
BT minerals
SA borosilicates

Ayacucho
Department in SW.
BT Peru

Ayrshire
County in SW.
CO N550000N555000
W0033000W0050000
BT Scotland
Great Britain
United Kingdom

Azalea Field
Midland County in W.
BT Texas
United States

Azerbaidzhan
Azerbaidzhan Soviet Socialist Republic in Transcaucasia. When referring to Iran, use Azerbaijan.
CO N383000N420000
E0500000E0450000
BT USSR
NT Akera
Apsheron Peninsula
Baku
Baku Archipelago
Bank
Dashkesan
Filizchay
Karabakh
Kedabek
Kirovabad
Kobystan
Lenkoran
Mugan Steppe
Nakhichevan
Neftechala
Neftyanyye Kamni
Ordubad
Peschanyy
Shemakha
Surakhany
SA Azerbaijan
Caspian Basin
Caucasus
European USSR
Greater Caucasus
Kura Lowland
Kura River
Lesser Caucasus
Transcaucasia
Zangezur

Azerbaijan
Used when referring to Iran. For USSR, use Azerbaidzhan.
BT Iran
SA Azerbaidzhan

Azof
use Azov

Azolla
Genus.
BA Filicopsida
pteridophytes
BT Plantae

Azores
Island group belonging to Portugal in E central North Atlantic Ocean. Includes use on level 1 as an area term (list O). For term set options see list B.
CO N370000N400000
W0250000W0310000
BA Atlantic Ocean
NT Sao Miguel Island
Terceira Island

Azov
Town near NE arm of Azov Sea in SW Rostov Oblast. Azov has also been used to index Azov Sea.
UF Azof
BT Russian Republic
USSR
SA Azov Sea

Azov region
Includes Azov Sea, the Crimea, the southern Ukraine bordering Azov Sea, and western section of the Northern Caucasus. Index Azov Sea and Soviet republics as applicable. Also search Azov; Azov-Kuban Basin.
CO N443000N470000
E0390000E0350000
UF Azov-Kuban Basin
BT USSR
SA Azov Sea
Russian Republic
Ukraine

Azov Sea
Shallow north arm of the Black Sea bounded on the SW by the Crimea, N by the Ukraine, and E by the Northern Caucasus. Also search Sea of Azov, and Azov.
CO N450000N470000
E0390000E0340000
UF Sea of Azof
Sea of Azov
BT USSR
SA Azov
Azov region
Black Sea

Azov-Kuban Basin
use Azov region

azurite
BA carbonates
BT minerals
SA smithsonite

B
use boron

B layer
use low-velocity layer

Ba
use barium

Babadag Lake basin
N Dobruja.
BT Dobruja
Romania

babingtonite
BA chain silicates
silicates
BT minerals

Babor Range
Coastal range of the Little Atlas Mountains in N Algeria. Also search Babor; Babors.
UF Babors
BT Algeria

Babors
use Babor Range

Bacau
City and county in central.
BT Moldavia
Romania

bacteria
Schizomycetes included here. Includes use on level 1 and 2 as a fossil term (list F); on level 3 under soils(1) biota(2) and under geochemistry(1).

UF schizmycetes
SA biology
biota
fungi
living culture
microorganisms
paleobotany
parasites
Plantae
Protista
soils

Baculites
Genus. Includes use on level 3 under Mollusca(1) Cephalopoda(2).
BA Ammonoidea
Cephalopoda
Mollusca
BT Invertebrata

Bad Deutsch Altenburg
Village on the Danube in E.
BT Lower Austria
Austria

Bad Gastein
Town in W central part of country.
BT Salzburg
Austria

Bad Pyrmont
Town in S.
BT Lower Saxony
West Germany

Badajoz
Province in W on Portuguese border. Also a city.
BT Spain

Badakhshan
Frontier province in NE Afghanistan and part of Gorno-Badakshan Autonomous Oblast in Tadzhikistan. Index Afghanistan and/or Tadzhikistan as applicable.
BT Asia
SA Afghanistan
Tadzhikistan

Badami Series
The Badami Group lies over the Kaladgi Group and is overlain by thick cover of Deccan Traps. Belgaum District in SW part of India.
BT Proterozoic
Precambrian
SA India
Mysore

baddeleyite
BA oxides
BT minerals

Baden
Former German state.
BT Baden-Wurttemberg
West Germany

Baden-Baden
City in West Germany. Also search Baden.
BT Baden-Wurttemberg
West Germany

Baden-Wuerttemberg
use Baden-Wurttemberg

Baden-Wurttemberg
A state of West Germany in SW. Also search Baden-Wuerttemberg.
UF Baden-Wuerttemberg
BT West Germany
NT Baden
Baden-Baden
Black Forest
Freiburg
Hegau
Heidelberg
Kaiserstuhl
Karlsruhe
Katzenbuckel
Neckar River
Swabian Alb
Weinheim
Wurttemberg
Wutach Valley
SA Franconia
Main River
Odenwald
Palatinate
Swabia

Badenian
Provincial series, Europe.
BA Miocene
Neogene
Tertiary
BT Cenozoic
SA lower Miocene
middle Miocene

Badkhyz
Steppe region in extreme S near Afghanistan and Iranian borders.
BT Turkmenia
USSR

Badlands
E of Black Hills in South Dakota and in SW North Dakota. Index states as applicable. A generic term for similar areas in other western states, South America and Asia. Geomorphologic rather than geographic.
BT United States
SA North Dakota
South Dakota

Badlands National Monument
Badlands area in Theodore Roosevelt National Memorial Park in SW.
BT South Dakota
United States

Baffin Bay
Arm of North Atlantic Ocean between Greenland and Baffin Island, Northwest Territories, Canada.
BT Atlantic Ocean

Baffin Island
Largest and most easterly of the Canadian Arctic Archipelago in District of Franklin.
CO N620000N733000
W0613000W0900000
BT Northwest Territories
Canada

Bagalkot
Town in NW.
BT Mysore
India

Bagh Beds
Includes Nimar Sandstone, Nodular Limestone, Deola Marl, Coralline Limestone.
BT Upper Cretaceous
Cretaceous
SA Gujarat
India
Madhya Pradesh

Bahama Islands
use Bahamas

Bahama Platform
use Blake Plateau

Bahamas
A chain of islands SE of Florida and N of Cuba. Includes use on level 1 as an area term (list O). For term set options see list B.
CO N204500N280000
W0720000W0793000
UF Bahama Islands
BA West Indies
NT Andros Island
Bimini
Exuma Sound
Northeast Providence Channel
Tongue of the Ocean
SA San Salvador

Bahariya Oasis
In the Libyan Desert in W.
BT Egypt

Bahia
State. Also old name for city of Salvador.
- BT Brazil
- NT Boquira Mine
  Curaca River basin
  Reconcavo Basin
  Salvador
  Tucano Basin
  Uaua
- SA Barreiras Formation
  Brazilian Shield
  Sao Francisco Basin

Bahrain
An archipelago in the W Persian Gulf. Includes use on level 1 as an area term (list O). For term set options see list B.
- UF Bahrein
- BA Arabian Peninsula
- SA Asia

Bahrein
use Bahrain

Baia Mare
Town in Maramures County in N Transylvania. Also search Baia-Mare.
- UF Baia-Mare
- BT Transylvania
  Romania

Baia-Mare
use Baia Mare

Baikal (Lake)
use Lake Baikal

Baikal Mountains
West shore of Lake Baikal mostly in Irkutsk Oblast. Also search Baikal; Baikal Range.
- UF Baikal Range
- BT Russian Republic
  USSR

Baikal Range
use Baikal Mountains

Baikal region
Lake Baikal, Baikal Mountains, and the surrounding Irkutsk Oblast and Buryat A.S.S.R. areas. Also search Baikal.
- CO N500000N570000
  E1130000E1020000
- BT Russian Republic
  USSR
- SA Lake Baikal

Baikalian Orogenic Phase
use Baikalian Phase

Baikalian Orogeny
use Baikalian Phase

Baikalian Phase
Term introduced in 1978. Widely used throughout the USSR for orogenies late in Precambrian time. Named after Lake Baikal in Siberia. Before 1978, search Baikalian AND orogeny; Baikalian Orogeny.
- UF Baikalian Orogenic Phase
  Baikalian Orogeny
- BT Cambrian
- SA orogeny
  Precambrian
  Russian Republic
  USSR

Bainbridge Formation
Characterized by two lithologically well-differentiated formations, St. Clair below, and Moccasin Springs above. SW Illinois and E Missouri.
- BT Niagaran
  Middle Silurian
  Silurian
- SA Illinois
  Missouri

Bairdia
Genus. Includes use on level 3 under Ostracoda(1) Podocopida(1).
- BA Bairdiidae
  Bairdiacea
  Podocopida
  Ostracoda
- BT Crustacea
  Arthropoda
  Invertebrata

Bairdiacea
Superfamily. Includes use on level 3 under Ostracoda(1) Podocopida(2).
- BA Podocopida
  Ostracoda
- BT Crustacea
  Arthropoda
  Invertebrata
- NA Bairdiidae

Bairdiidae
Family. Includes use on level 3 under Ostracoda(1) Podocopida(2).
- BA Bairdiacea
  Podocopida
  Ostracoda
- BT Crustacea
  Arthropoda
  Invertebrata
- NA Bairdia

Baita
Village in Maramures County, in N.
- BT Transylvania
  Romania

Baixo Alentejo
A region in S.
- BT Portugal

Baja California
A peninsula. Northern half is state of Baja California. Southern half is territory of Baja California Sur.
- CO N230000N323000
  W1090000W1170000
- BT Mexico
- NT Cerro Prieto
  Mexicali
  Ojo de Liebre Lagoon
  Todos Santos Bay
- SA Agua Blanca Fault
  Colorado River
  Colorado River delta
  Rosario Formation

Bajocian
Europe. Above Toarcian, below Bathonian. Includes use on level 3 under age terms(1). See list E.
- BA Middle Jurassic
  Jurassic
- BT Mesozoic
- SA Dogger

Bakal
City in Chelyabinsk Oblast in Southern Urals.
- BT Russian Republic
  USSR

Bakersfield
City in Kern County in S.
- BT California
  United States

Bakhchisarai
Town in S Crimea. Also search Bakhchisaray.
- UF Bakchisaray
- BT Ukraine
  USSR

Bakhchisaray
use Bakhchisarai

Bakhmut
use Artemovsk

Bakhtiari Formation
Synonymous with Bakhtiari Group. Comprises Upper and Lower Bakhtiari Formations.
- BT Pliocene
- SA Iran
  Iraq

Bakony Mountains
In NW central Hungary. Also search Bakony, and Transdanubian Central Mountains.
- UF Transdanubian Central Mountains
- BT Hungary

Baku
City on Apsheron Peninsula on Caspian Sea.
- BT Azerbaidzhan
  USSR

Baku Archipelago
Also search Baku.
- BT Azerbaidzhan
  USSR

Bala
England: Middle and Upper Ordovician. Comprises the Caradocian and Ashgillian. Includes use on level 3 under age terms(1). See list E.
- BA Ordovician
- BT Paleozoic
- SA Ashgillian
  Caradocian
  Middle Ordovician
  Upper Ordovician

Bala District
Urban district in NE Marioneth in NW Wales.
- BT Wales
  Great Britain
  United Kingdom

Balaghat
Town and district in central.
- BT Madhya Pradesh
  India

balance, mass
use mass balance

balance, water
use water balance

Balasore
Town and district on Bay of Bengal in N.
- BT Orissa
  India

Balaton (Lake)
use Lake Balaton

Balaton region
Area in W around Lake Balaton. Also search Balaton.
- BT Hungary
- SA Lake Balaton

Balcanoona
Village in E central.
- BT South Australia
  Australia

Bald Mountain
Name of peaks which occur in several states: central Colorado; central and SW central Idaho; E central Oregon; W South Dakota; NE Utah; N Wyoming and W central Wyoming. Index states as applicable.
- BT United States
- SA Colorado
  Idaho
  Oregon
  South Dakota
  Utah
  Wyoming

Balearic abyssal plain
use Balearic Basin

Balearic Basin
Between Balearic Islands and Sardinia.
- UF Algerian Plain
  Balearic abyssal plain
- BT Mediterranean Sea

Balearic Islands
Off E coast of Spain. Forms Spanish province of Baleares. Includes use on level 1 as an area term (list O). For term set options see list B.
- BA Mediterranean Sea
- NT Ibiza
  Majorca
  Minorca
  Palma

Balei
City in S central Chita Oblast E of Lake Baikal. Also search Baley.
- UF Baley
- BT Russian Republic
  USSR

Baley
use Balei

Bali
Island off E end of Java.
- BT Indonesia

Balkan Foreland
A stable area marginal to the Balkan Mountains of central.
- BT Bulgaria

Balkan Mountains
Range extending E and W across central.
- CO N423000N433000
  E0270000E0223000
- BT Bulgaria

Balkan Peninsula
SE Europe between the Adriatic and Ionian Seas on the W; the Mediterranean Sea on the S; and the Aegean and Black seas on the E. Includes use on level 1 as an area term (list O). For term set options see list B. Before 1978, also search Balkans.
- UF Balkan States
  Balkans
- BA Europe
- SA Albania
  Bulgaria
  Greece
  Romania
  Turkey
  Yugoslavia

Balkan States
use Balkan Peninsula

Balkans
use Balkan Peninsula

Balkhan
Two mountain ranges in Krasnovodsk Oblast in W: the Greater Balkhan Range and the Lesser Balkhan Range.
- BT Turkmenia
  USSR

Balkhash
City on N shore of Lake Balkhash.
- BT Kazakhstan
  USSR

Balkhash (Lake)
use Lake Balkhash

Balkhash region
Area around Lake Balkhash. Also search Balkhash.
- BT Kazakhstan
  USSR
- SA Lake Balkhash

ball and pillow
use ball-and-pillow

ball-and-pillow
Includes use on level 3 under sedimentary structures(1) soft sediment deformation(2). See list K. Also search ball and pillow.
- UF ball and pillow
- BA soft sediment deformation
  sedimentary structures
- SA pillow structure
  primary structures

balls, coal
use coal balls

Balmat
Village in Saint Lawrence County in N.
BT New York
    United States

Baltic Basin
use Baltic region

Baltic Coast
use Baltic region

Baltic Glaciation
Comprises Baltic region plus Norway which was completely covered by Pleistocene glaciation. Also search Baltic AND glaciation.
BT Pleistocene
SA Europe

Baltic region
Index Baltic Sea, countries and Soviet republics as applicable. Also search Baltic; Baltic Basin; Baltic Coast. Includes use on level 1 or 2 as an area term (list O). If 1, see term set options under list B.
CO N520000N680000
    E0310000E0100000
UF Baltic Basin
    Baltic Coast
SA Baltic Sea
    Denmark
    Estonia
    Europe
    Finland
    Germany
    Latvia
    Lithuania
    Peribaltic Syneclise
    Russian Republic
    Sweden

Baltic Sea
Enclosed by Denmark, Sweden, Finland, Soviet Union, Poland, and Germany. Includes use on level 1 as an area term (list O). For term set options see list B.
CO N550000N600000
    E0220000E1010000
BA Atlantic Ocean
NT Aland
    Gulf of Bothnia
    Gulf of Finland
    Gulf of Riga
    Kiel Bay
SA Baltic region

Baltic Shield
Index countries and Soviet republics as applicable.
BT Europe
SA Estonia
    Finland
    Latvia
    Norway
    Russian Republic
    Sweden

Baltic Syneclise
use Peribaltic Syneclise

Baltimore
City on Chesapeake Bay in Baltimore County.
BT Maryland
    United States

Baltimore County
Nearly surrounds city of Baltimore in N.
BT Maryland
    United States

Baltimore Gneiss
Unconformable below Glenarm Series. W Maryland, SE Pennsylvania, and central N Virginia.
BT Precambrian
SA Maryland
    Pennsylvania
    Virginia

Baltisphaeridium
BA acritarchs
    palynomorphs

Baluchistan
Province in W.
BT Pakistan
SA Sulaiman Range

Bambak
use Pambak

Bamble
Village on the Skagerrak.
BT Telemark
    Norway

Bambu
Village in NE Zaire.
UF Kilo
    Kilo-Mines
BT Zaire

Bambui Group
BT Proterozoic
    Precambrian
SA Brazil
    Minas Gerais

Bamian
Province in N central Afghanistan. Also a town.
BT Afghanistan

Banat
Region. Part of Voyvodina in Serbia and the Banat region of Romania. Index countries as applicable.
UF Banat of Temesvar
BT Europe
NT Arad
    Delinesti
    Mehadia
    Moldova Noua
    Resita
    Rusca Montana
    Sasca-Montana
    Semenic Mountains
    Severin
    Svinita
SA Getic Nappe
    Mehedinti Plateau
    Romania
    Yugoslavia

Banat of Temesvar
use Banat

Bancroft
Village between Ottowa and Georgian Bay in SE.
BT Ontario
    Canada

Band-e-Amir
A settlement in S part of country.
BT Fars
    Iran

Bandama River
Rises in N and flows into Gulf of Guinea.
BT Ivory Coast

banded gneiss
See list J.
BA gneisses
BT metamorphic rocks
SA gneiss

banded materials
Term introduced in 1978. Before 1978, also search banded.
SA materials

banded structures
Includes use on level 3 under igneous rocks(1) or metamorphic rocks(1). Also search banding; banded.
UF banding
SA igneous rocks
    metamorphic rocks
    petrology
    structures

banding
use banded structures

Banff
Town in Banff National Park near the British Columbia border in SW.
CO N570500N574500
    W0021500W0033000
BT Alberta
    Canada

Banff National Park
On the British Columbia border in SW.
BT Alberta
    Canada

Banffshire
County on Moray Firth in NE.
BT Scotland
    Great Britain
    United Kingdom

Bangalore
City and district in SE.
BT Mysore
    India

Bangka
An island between SE Sumatra and SW Borneo.
BT Indonesia

Bangkok
City on the Gulf of Siam in S.
BT Thailand

Bangla Desh
use Bangladesh

Bangladesh
Formerly East Pakistan. Includes use on level 1 as an area term (list O). For term set options see list B. Before 1973, also search East Pakistan or Pakistan AND east.
UF Bangla Desh
BA Asia
NT Magura
SA Bengal
    Brahmaputra River
    Ganges River
    Indian Plate
    Vindhyan Basin

Bangor Limestone
Includes Burgess Oolite and Rockwood Oolite, and Spout Spring Oolite near base. Alabama, central and E Tennessee, and NW Georgia.
BT Upper Mississippian
    Mississippian
SA Alabama
    Georgia
    Tennessee

Banja Luka
Town in N Bosnia and Herzegovina.
BT Yugoslavia

Bank
Town in SE.
BT Azerbaidzhan
    USSR

banks
BA biogenic structures
    sedimentary structures
SA algal banks

Banks Island
W District of Franklin.
CO N720000N743000
    W1150000W1250000
BT Northwest Territories
    Canada

Bankura
Town and district in W.
BT West Bengal
    India

Bannock County
SE Idaho.
BT Idaho
    United States

Banska Stiavnica
Town in S central.
BT Slovakia
    Czechoslovakia

Baoha
A town in N North Vietnam.
BT Vietnam

Baquero Formation
BT Lower Cretaceous
    Cretaceous
SA Argentina
    Santa Cruz

Bar
Town SW of Kiev in W Vinnitsa Oblast.
BT Ukraine
    USSR

Barabash Suite
Primorye Kray in Soviet Far East.
BT Upper Permian
    Permian
SA Russian Republic
    USSR

Baraboo
City in Sauk County in S central. Also a river.
BT Wisconsin
    United States

Baraboo Quartzite
Underlies Seeley Slate. Sauk and Columbia counties in S central Wisconsin.
BT Algonkian
    Precambrian
SA Wisconsin

Barakar
Town in W.
BT West Bengal
    India

Barakar Formation
use Barakar Stage

Barakar Stage
Lowest stage of the Damuda Series. Named for Barakar River in Raniganj coalfields of West Bengal. In N, NE, and E India. Also search Barakar Formation.
UF Barakar Formation
BT Lower Permian
    Permian
SA Damuda Series
    India

Baralaba Coal Measures
Permo-Carboniferous.
BT Paleozoic
SA Australia
    Carboniferous
    Pennsylvanian
    Permian
    Queensland

Baranya Mountains
use Mecsek Mountains

Baraolt
Town in SE.
BT Transylvania
    Romania

Baraolt Basin
SE Transylvania. Also search Baraolt.
BT Transylvania
    Romania

Barataria Bay
Inlet of Gulf of Mexico between Lafourche and Plaquemines parishes.
BT Louisiana
    United States
SA Gulf Coastal Plain

Barbados
Easternmost island of the Lesser Antilles. Includes use on level 1 as an

area term (list O). For term set options see list B.
CO N130000N132000
W0592000W0594500
BA West Indies
SA Lesser Antilles

Barbados Ridge
Extends N and S of Barbados with the island in the center.
BT Atlantic Ocean

Barber County
S on the Oklahoma border.
BT Arkansas
United States

Barberton
Town in SE near Swaziland border.
BT Transvaal
South Africa

Barberton Mountain Land
SE Transvaal.
BT Transvaal
South Africa

barbosalite
BA phosphates
BT minerals

Barcelona
Province in Catalonia in NE Spain. Also a city on the Mediterranean Sea.
BT Spain
NT Llogrebat River basin
Vich
SA Catalonia

bardiglio
use marble

Bardo Mountains
In SW part of country.
BT Wroclaw
Poland

Barents Sea
Part of Arctic Ocean N of Norway and USSR, and between Spitsbergen and Novaya Zemlya.
CO N700000N800000
E0670000E0100000
BT Arctic Ocean
NT Bear Island

Bari
City and province on the Adriatic Sea.
BT Apulia
Italy

Barinas
State in SW Venezuela. Also a city.
BT Venezuela

barite
Includes use on level 1 and 2 as a commodity term (list C).
BA sulfates
BT minerals

barium
Includes use on level 1 and 2 as a chemical element (list D). Also search Ba.
UF Ba
SA barylite
elements

Barmer
Town in W.
BT Rajasthan
India

Barnes ice cap
Central Baffin Island in District of Franklin.
BT Northwest Territories
Canada

Barnstable County
Coextensive with Cape Cod.
BT Massachusetts
United States

Baroda
City and former state in E.
BT Gujarat
India

barometry, geologic
use geologic barometry

Baronnies
Hilly region in SE part of country.
BT Drome
France

Barracuda Ridge
In the Anegada Passage area between the Virgin Islands and the N Leeward Islands.
BT Atlantic Ocean

Barrandian
Stage.
BA Middle Cambrian
Cambrian
BT Paleozoic

Barrandian area
use Barrandian Basin

Barrandian Basin
UF Barrandian area
BT Czechoslovakia

Barre Granite
An elongated, oval-shaped body that crops out in northwestern and west-central parts of East Barre quadrangle. Washington County, Vermont.
BT Devonian
SA Vermont

Barreal
Town in SW.
BT Argentina
SA San Juan

Barreiras Formation
BT Tertiary
SA Bahia
Brazil
Espirito Santo
Para
Pernambuco

Barreirinhas Basin
NE Maranhao.
BT Maranhao
Brazil

Barremian
Europe. Above Hauterivian, below Aptian. Includes use on level 3 under age terms(1). See list E.
BA Lower Cretaceous
Cretaceous
BT Mesozoic
SA Speeton Clay

barrier bars
use longshore bars

barrier islands
As of 1978, term is used on level 2 under shorelines(1).
SA barriers
bars
islands
shorelines

barrier reefs
BT reefs
SA fringing reefs
lagoons

barriers
Includes use on level 3 under geomorphology(1) or sedimentation(1).
SA barrier islands
embankments
geomorphology
sedimentation

Barrow
City at Point Barrow on the Arctic Ocean.
BT Alaska
United States

Barrow Island
Off NW coast.
BT Western Australia
Australia

bars
Includes use on level 3 under geomorphology(1) fluvial features(2) and shore features(2); on level 3 under sedimentary structures(1) planar bedding structures(2). See list K.
NT longshore bars
SA barrier islands
bedding plane irregularities
fluvial features
geomorphology
planar bedding structures
sedimentary structures
shore features

Barsakelmes
Region in Ust Urt Plateau in SW.
BT Kazakhstan
USSR

Barstovian
North America.
BA Miocene
Neogene
Tertiary
BT Cenozoic

Barstow
City in San Bernardino County in S.
BT California
United States

Barth
Town on the Baltic Sea.
BT East Germany

Bartlesville Sand
Washington County in NE Oklahoma.
BT Pennsylvanian
SA Oklahoma

Barton Beds
Overlain by the lower Headon Beds and rest on the upper Bracklesham Beds. Divided into lower, middle, and upper Barton Beds. Hampshire County in S England. Also search Barton.
BT Bartonian
Eocene
SA England
United Kingdom

Barton County
Index states as applicable.
BX Kansas
Missouri
BT United States

Bartonian
Europe. Above Auversian, below Ludian. Includes use on level 3 under age terms(1). See list E.
BA Eocene
Paleogene
Tertiary
BT Cenozoic
NT Barton Beds
SA Priabonian

Barwell
Meteorite. Town and parish in Leicester County in central England. Also search Barwell Meteorite.
BT England
Great Britain
United Kingdom
SA meteorites

barylite
BA ring silicates
silicates
BT minerals
SA barium

Bas-Dauphine
Region. Part of the historic province of Dauphine in SE France. Associated with the Dauphine Alps.
BT France
SA Dauphine Alps

Bas-Rhin
Department in Alsace in extreme NE.
BT France
NT Bouxwiller
Strasbourg
SA Saar Basin
Saar-Nahe Basin

basalt
Includes use on level 3 under igneous rocks(1) basalt family(2). See list H. Also search basaltic rocks; andesite basalt; andesitic basalt.
BA basalt family
BT igneous rocks
SA alkali basalt
alkali olivine basalt
augite
basaltic composition
columnar basalt
lava
melaphyre
metabasalt
nepheline basalt
pillow lava
tholeiitic basalt
tholeiitic composition
trap rock
trap rocks
volcanic rocks

basalt family
Includes use on level 2 under igneous rocks(1). See list H. Also search basalts.
BT igneous rocks
NA absarokite
alkali olivine basalt
ankaramite
basalt
columnar basalt
diabase
diabase porphyry
dolerite
melaphyre
mugearite
oceanite
olivine basalt
olivine diabase
olivine dolerite
olivine tholeiite
picrite
picrite porphyry
quartz diabase
quartz dolerite
shoshonite
tholeiite
tholeiitic basalt
tholeiitic dolerite
trap rocks
SA alkali basalt family

basalt glass
Not a valid term for GeoRef. Use volcanic glass and basalt.

basaltic
Not a valid term for GeoRef. Use basaltic composition or basaltic layer.

basaltic composition
Term introduced in 1978. Before 1978, search basaltic.
SA basalt
composition
igneous rocks

basaltic domes
use shield volcanoes

basaltic layer
Also search sima.
UF layer, basaltic
sima
SA granitic layer
lower crust

basaltic rocks
Not a valid term for GeoRef. Use ba-

salt family or basaltic composition.

basanite
Includes use on level 3 under igneous rocks(1) alkali basalt family (2). See list H.
BA alkali basalt family
BT igneous rocks

basculating faults
use wrench faults

base exchange
use ion exchange

base metals
As of 1978, term includes use on level 1 and 2 as a commodity (list C).
SA alloys
metals

base surges
Used mostly in sedimentation.
SA sedimentation
surges

base-level peneplain
use peneplains

Basel
Canton composed of demicantons of Basel-Land and Basel Stadt. Also a city.
UF Basle
BT Switzerland

basement
Includes use on level 3 under tectonics(1) structure(2). Also search basement rocks; basement structure.
UF basement rocks
basement structure
SA crust
igneous rocks
metamorphic rocks
Mohorovicic discontinuity
sedimentary cover
shields
tectonics

basement rocks
use basement

basement structure
use basement

Bashkiria
Bashkir Autonomous Soviet Socialist Republic in SE European USSR.
CO N520000N570000
E0600000E0530000
BT Russian Republic
USSR

Bashkirian
Europe.
UF Bashkirian Stage
BA Upper Carboniferous
Carboniferous
BT Paleozoic

Bashkirian Stage
use Bashkirian

basic hornfels

basic rocks
Not a valid term for GeoRef. Use mafic composition.

Basilicata
Autonomous region in S.
BT Italy
NT Matera
Potenza
SA Lucania

Basin and Range Province
Level 1 area term as of 1978. Physiographic province in W and SW United States characterized by a series of tilted fault blocks forming longitudinal, asymmetric ridges of mountains and broad intervening basins. Also search Basin and Range. Index states as applicable.
BA United States

SA Arizona
California
Idaho
Nevada
New Mexico
Oregon
Texas
Utah

basin range
Includes use on level 3 under tectonics(1) structure(2).
UF range, basin
SA basins
faults
tectonics

basins
Includes use on level 3 under folds(1) style(2) and under tectonics(1); on level 3 under paleogeography(1).
UF structural basins
SA basin range
coal basins
depressions
drainage basins
folds
intermontane basins
marginal basins
ocean basins
paleogeography
playas
sedimentary basins
sedimentation
tectonics

Baskunchak Series
S Urals.
SA Permian
Russian Republic
Triassic
USSR

Basle
use Basel

Basque Provinces
On the Bay of Biscay near the French border. Also search Basque. Index provinces as applicable.
BT Spain
SA Alava
Cantabrian Basin
Guipuzcoa
Vizcaya

Bass Basin
use Bass Strait

Bass Strait
Large channel separating Victoria from Tasmania and the Indian Ocean from the Tasman Sea. Also search Bass Basin.
UF Bass Basin
BT Australia

bassanite
BA sulfates
BT minerals

Basses-Alpes
use Alpes-de-Haute Provence

Basses-Pyrenees
use Pyrenees-Atlantiques

Bastar
Former Indian State. Now a district in SE.
BT Madhya Pradesh
India

bastnaesite
UF bastnasite
BZ carbonates
halides
BT minerals
SA fluorides

bastnasite
use bastnaesite

Bastogne
Town, E Luxembourg province.

BT Belgium

Basutoland
use Lesotho

Batchawana Bay
On Lake Superior N of Sault Sainte Marie.
BT Ontario
Canada

Batenev Ridge
In the Kuznetsk Alatau of the Khakass Autonomous Oblast SW of Krasnoyarsk.
BT Russian Republic
USSR

Batesville
City in Independence County in NE central.
BT Arkansas
United States

Bath
City and county borough near Bristol in W.
BT England
Great Britain
United Kingdom

batholiths
Includes use on level 2 under intrusions(1).
UF abyssolith
central granite
intrusive mountain
BA intrusions
SA dikes
domes
igneous rocks
laccoliths
lopoliths
plutons
stocks

Bathonian
Europe. Above Bajocian, below Callovian. Includes use on level 3 under age terms(1). See list E.
BA Middle Jurassic
Jurassic
BT Mesozoic
NT Great Oolite Series
SA Dogger

Bathurst
Region in NE.
BT New Brunswick
Canada

Bathurst Island
NW District of Franklin.
CO N750000N763000
W0973000W1023000
BT Northwest Territories
Canada

bathymetric maps
Term introduced in 1978. Before 1978, search maps AND bathymetric.
BA maps
SA bathymetry
ocean floors

bathymetric surveys
use bathymetry

bathymetry
Includes use on level 3 under marine geology(1) or geophysical surveys(1). Also search bathymetric surveys.
UF bathymetric surveys
SA bathymetric maps
echo sounding
geophysical surveys
marine geology
ocean floors
oceanography
paleo-oceanography
paleobathymetry

Baton Rouge
City on the Mississippi River in East Baton Rouge Parish.
BT Louisiana
United States

Battambang
Town in W.
BT Cambodia

Batum
use Batumi

Batumi
City on the Black Sea near the Turkish border. Also search Batum.
UF Batum
BT Georgian Republic
USSR

Bauchi Plateau
use Jos Plateau

Bauer Deep
In Southeast Pacific Basin N of Antarctica.
BT Pacific Ocean

Baumkirchen
Region in Inn Valley.
BT Tyrol
Austria

Bauru Formation
SA Brazil
Cretaceous
lower Tertiary
Minas Gerais
Sao Paulo
Tertiary
Upper Cretaceous

bauxite
Includes use on level 1 and 2 as a commodity term (list C).
SA aluminum
bauxitization
clastic rocks
clay minerals
laterites
sedimentary rocks

bauxitization
SA bauxite
laterites
laterization
mineral deposits
processes

Bavaria
State in SE.
CO N474000N503500
E0134500E0093000
BT West Germany
NT Ansbach
Bavarian Forest
Bayreuth
Berchtesgaden
Coburg
Eichstatt
Erlangen
Fichtelgebirge
Franconian Jura
Hagendorf
Kelheim
Kreuth
Middle Franconia
Munchberg Gneiss Massif
Munich
Nabburg
Neuburg
Nuremberg
Passau
Regensburg
Ries Crater
Sandelzhausen
Solnhofen
Spessart
Upper Bavaria
Upper Franconia
Upper Palatinate
Weissenburg
SA Allgau Alps

Bavarian Alps
Franconia
Inn Valley
Isar Valley
Main River
Odenwald
Palatinate
Rhon Mountains
Salzach River
Solnhofen Limestone
Swabia

Bavarian Alps
Range of the Central Alps along the Austro-German border. Index Bavaria and/or Tyrol as applicable.
BT Alps
      Europe
SA Bavaria
      Central Alps
      Tyrol

Bavarian Forest
A subsidiary forested mountain range of the Bohemian Forest in W Bavaria.
BT Bavaria
      West Germany

bavenite
BA sheet silicates
      silicates
BT minerals
SA aluminosilicates

Bay of Bengal
Part of Indian Ocean between India and the W coast of Burma.
BT Indian Ocean
NT Andaman Basin
      Andaman Sea

Bay of Biscay
Large inlet of the Atlantic Ocean between the W tip of Brittany and NW Spain. Also search Biscay; Biscay Bay.
UF Biscay (Bay)
      Biscay Bay
BT Atlantic Ocean
NT Cantabria Seamount
      Gulf of Gascony
SA Arcachon Basin

Bay of Bourgneuf
On Bay of Biscay off Vendee and Loire-Atlantique departments.
BT France

Bay of Fundy
Inlet between Maine and New Brunswick on the W and Nova Scotia on the E.
BT Atlantic Ocean
SA Minas Basin

Bay of Islands
Inlet of Gulf of Saint Lawrence off W coast.
BT Newfoundland
      Canada

Bay of Mont-Saint-Michel
use Bay of Saint-Michel

Bay of Naples
Inlet of Tyrrhenian Sea off the city of Naples.
BT Campania
      Italy

Bay of Plenty
Large inlet off NE coast of North Island, New Zealand.
BT Pacific Ocean

Bay of Saint Michel
use Bay of Saint-Michel

Bay of Saint-Michel
On English channel off SW corner of Normandy. Location of Mont-Saint-Michel. Also search Bay of Saint-Michel.
UF Bay of Mont-Saint-Michel
      Bay of Saint Michel
      Mont-Saint-Michel Bay
BT France

Bay of the Seine
Inlet of the English Channel off Normandy between Cotentin Peninsula on W to mouth of Seine River on E.
BT France

Bay Park
Town in Nassau County on Long Island.
BT New York
      United States

bayerite
Artificial.
BA oxides
BT minerals
SA gibbsite

Baylor County
SW of Wichita Falls in N.
BT Texas
      United States

Bayn Dzak
Region in the Gobi Desert.
BT Mongolia

Bayreuth
City in N.
BT Bavaria
      West Germany

bays
Includes use on level 3 under shorelines(1) or geomorphology(1) shore features(2).
SA fjords
      geomorphology
      inlets
      shore features
      shorelines

Be
use beryllium

Be-7
Includes use on level 3 under isotopes(1).
SA beryllium
      isotopes

Be-10
Includes use on level 3 under isotopes(1).
SA beryllium
      isotopes

beach erosion
Not a valid index term for GeoRef. Use beaches and erosion.

beach ridges
Includes use on level 3 under shorelines(1) or geomorphology(1) shore features(2).
SA geomorphology
      ridges
      shore features
      shorelines

beach rock
use beachrock

beaches
As of 1978, term is used on level 2 under shorelines(1).
SA coastal environment
      erosion
      geomorphology
      lacustrine features
      lagoons
      landform evolution
      paleoecology
      sedimentation
      shore features
      shorelines
      spits

beachrock
Includes use on level 3 under sedimentary rocks(1) carbonate rocks(2). See list I.

UF beach rock
BA carbonate rocks
BT sedimentary rocks

Beacon Group
use Beacon Supergroup

Beacon Supergroup
In Queen Maud Range along SW Ross Ice Shelf and in Victoria Land W of Ross Sea, both in Ross Dependency S of New Zealand. Also search Beacon Group.
UF Beacon Group
SA Antarctica
      Carboniferous
      Devonian
      Jurassic
      Mesozoic
      Paleozoic
      Permian
      Triassic

Beacon Valley
In S Victoria Land near SE Ross Sea in Ross Dependency S of New Zealand.
BT Antarctica

Bear Gulch Limestone Member
In Tyler Formation. Central N Montana. Also search Bear Gulch Limestone.
BT Mississippian
SA Montana

Bear Island
240 miles N of Norway. Along with Spitsbergen forms Svalbard island group.
BT Barents Sea
      Arctic Ocean
SA Svalbard

Bear Lake
Index states as applicable.
BT United States
SA Idaho
      Utah

Bear Province
A geological province in District of Mackenzie.
CO N630000N690000
      W1080000W1210000
BT Northwest Territories
      Canada

Bear River basin
Index states as applicable.
BT United States
SA Idaho
      Utah

Bear River Range
Index states as applicable.
BT United States
SA Idaho
      Utah

Bear Valley
Central California.
BT California
      United States

Bear-Slave Operation
Geochemical operation in Great Bear and Great Slave lakes region of District of Mackenzie.
BT Northwest Territories
      Canada

Beardmore Glacier
At NW end of Queen Maude Range just off Ross Ice Shelf in Ross Dependency S of New Zealand.
BT Antarctica
SA Queen Maude Range
      Ross Ice Shelf

bearing capacity
Includes use on level 3 under foundations(1).
SA capacity
      compaction
      foundations
      physical properties
      soil mechanics
      stress

Bearn
An historical region of SW.
BT France

Bearpaw Formation
In Montana Group. S Alberta; N, E, and S Montana; and Elk Basin region of central N Wyoming. Also search Bearpaw Shale.
UF Bearpaw Shale
BT Upper Cretaceous
      Cretaceous
SA Alberta
      Edmonton Formation
      Montana
      Montana Group
      Wyoming

Bearpaw Shale
use Bearpaw Formation

Beartooth Mountains
Index states as applicable. NE spur of Absaroka Range.
CO N445000N450000
      W1091500W1100000
BT United States
SA Absaroka Range
      Montana
      Wyoming

Beas River
One of the "Five Rivers" of the Punjab. Rises in Himachal Pradesh and flows into the Sutlej River. Index states as applicable.
BT India
SA Himachal Pradesh
      Punjab

Beata Ridge
Off south central coast of Hispaniola.
BT Caribbean Sea

Beatty
Underground nuclear explosion site. NW of Las Vegas.
BT Nevada
      United States

Beauce
Ancient region in N central. Index departments as applicable.
BT France
SA Eure-et-Loir
      Loir-et-Cher

Beaufort
City in Carteret County in E.
BT North Carolina
      United States

Beaufort County
On both sides of the Pamlico River in E.
BT North Carolina
      United States

Beaufort Formation
Underlies Castle Hayne Limestone or Yorktown Formation, or locally, an unnamed middle(?) Miocene unit; overlies Upper Cretaceous Pendee Formation. In E North Carolina. Also search Beaufort Series.
BT Paleocene
SA North Carolina

Beaufort Sea
Between N Alaska and the Arctic Archipelago of Canada.
BT Arctic Ocean

Beaumont
City in Jefferson County in E.
BT Texas
      United States

Beaver County
Index states as applicable.
BX Oklahoma

Pennsylvania
Utah
BT United States
Beaver Creek
Two Beaver Creeks. One in NE and one in W central.
BT Wyoming
United States
Beaver Lake
Garden County in western panhandle.
BT Nebraska
United States
Beaver River
Extreme SE.
BT Yukon Territory
Canada
Beaverhead County
Extreme SW.
BT Montana
United States
Beaverhead Formation
Unconformably overlies Kootenai, Dinwoody, Thaynes, Phosphoria, Quadrant and Madison. E central Idaho and SW Montana.
SA Cretaceous
Eocene
Idaho
Montana
Paleocene
Upper Cretaceous
Beaverhead Mountains
S part of Bitterroot Range. Index states as applicable.
BT United States
SA Bitterroot Range
Idaho
Montana
Beaverlodge
Lake N of Lake Athabasca.
BT Saskatchewan
Canada
Bechar
Town about 440 miles SW of Algiers.
BT Saoura
Algeria
Becraft Mountain
Columbia County in E.
BT New York
United States
bed forms
use bedforms
bed load
use bedload
bed-load
use bedload
Bedarieux
Town in S part of country.
BT Herault
France
bedded volcano
use stratovolcanoes
bedding
Includes use on level 3 under sedimentary structures(1) planar bedding structures(2) and under fractures(1) style(2). See list K.
BA planar bedding structures
sedimentary structures
SA bedding plane irregularities
columnar joints
cross-bedding
flaser bedding
fractures
graded bedding
laminations
lineation
massive bedding
rhythmic bedding
stratification
strike
bedding faults
Term introduced in 1978. Before 1978, search faults AND bedding.
BA faults
bedding plane irregularities
Includes use on level 2 under sedimentary structures(1). See list K.
BA sedimentary structures
NA antidunes
current markings
flute casts
fossil ice wedges
frost features
groove casts
grooves
megaripples
mounds
mud lumps
mudcracks
parting lineation
ripple marks
sand waves
scour casts
scour marks
shrinkage cracks
striations
tool marks
SA bars
bedding
dunes
ice wedges
imbrication
load casts
sand bodies
sole marks
bedding structures, planar
use planar bedding structures
bedding-plane slip
use flexural-slip
Bedford County
Index states as applicable.
BX Pennsylvania
Tennessee
Virginia
BT United States
Bedford Shale
Devonian and Mississippian(?): NE Kentucky, E Ohio, and SW Pennsylvania.
BT Paleozoic
SA Devonian
Kentucky
Mississippian
Ohio
Pennsylvania
Bedfordshire
County in SE.
CO N514500N522000
W0001000W0004000
BT England
Great Britain
United Kingdom
bedforms
Also search bed forms.
UF bed forms
SA fluvial features
hydrology
bedload
Also search bottom load.
UF bed load
bed-load
bottom load
SA streams
bedrock
Used as a general term.
SA outcrops
rocks
soils
beds, convoluted
use convoluted beds

beds, key
use key beds
beds, marker
use marker beds
Beemerville
Village in Sussex County in extreme N.
BT New Jersey
United States
Beerenberg
Extinct volcano on Norway's Jan Mayen Island.
BT Arctic Ocean
SA Jan Mayen
Behar
use Bihar
behavior
Used as a general term.
beidellite
BA sheet silicates
silicates
BT minerals
SA clay minerals
montmorillonite
Beil Limestone
use Beil Limestone Member
Beil Limestone Member
Of Lecompton Limestone. SW Iowa, E Kansas, SE Nebraska. Also search Beil Limestone.
UF Beil Limestone
BT Virgilian
Pennsylvanian
SA Iowa
Kansas
Lecompton Limestone
Nebraska
Beira Baixa
Former province in N central.
BT Portugal
Beirut
City on the Mediterranean Sea.
BT Lebanon
Middle East
Beius Basin
Bihor County in NW.
BT Transylvania
Romania
Bek-Budi
use Karshi
Belaya Gora
Village S of Arkhangelsk in Arkhangelsk Oblast in NW European USSR.
BT Russian Republic
USSR
Belchatow
Town in central part of country.
BT Lodz
Poland
Belcher Islands
Group in SE Hudson Bay.
BT Northwest Territories
Canada
belemnites
Includes use on level 3 under Mollusca(1) Cephalopoda(2).
BA Cephalopoda
Mollusca
BT Invertebrata
SA Belemnitidae
Belemnoidea
Belemnopsis
Belemnitidae
Family. Includes use on level 3 under Mollusca(1) Cephalopoda(2).
BA Cephalopoda
Mollusca
BT Invertebrata
SA belemnites
Belemnopsis

Belemnoidea
Extinct order of subclass Dibranchiata. Includes use on level 3 under Mollusca(1) Cephalopoda(2).
BA Cephalopoda
Mollusca
BT Invertebrata
SA belemnites
Belemnopsis
Belemnopsis
Includes use on level 3 under Mollusca(1) Cephalopoda(2).
BA Cephalopoda
Mollusca
BT Invertebrata
SA belemnites
Belemnitidae
Belemnoidea
Belfast
City in Co. Antrim and Co. Down.
BT Northern Ireland
United Kingdom
Belgaum
Town and district in W.
BT Mysore
India
Belgian Congo
use Zaire
Belgium
Includes use on level 1 as an area term (list O). For term set options see list B.
CO N493000N513000
E0063000E0023000
BA Europe
NT Antwerp
Bastogne
Brabant
Brabant Massif
Hainaut
Liege
Namur
SA Ardennes
Campine
European Platform
Grimmertingen
Limburg
Luxembourg Sandstone
Meuse River
Meuse Valley
North Sea Coast
Scheldt River
Stavelot-Venn Massif
Waterloo
Belgorod
City N of Kharkov in SE Russian Republic. Also an oblast.
BT Russian Republic
USSR
Belgrade
City on Danube River in N central Serbia.
UF Beograd
BT Serbia
Yugoslavia
Beli Isvor
use Beli Izvor
Beli Izvor
Village in E Bulgaria. Also search Beli Isvor.
UF Beli Isvor
BT Bulgaria
Belitung
use Billiton
Belize
Formerly British Honduras. Includes use on level 1 as an area term (list O). For term set options see list B. Before 1975, also search British Honduras.
BA Central America
NT Maya Mountains

Beljak
use Villach

Bell Canyon Formation
In Delaware Mountain Group. Includes Hegler Limestone, Pinery Limestone, Radar Limestone, Lamar Limestone and McCombs Limestone members. S New Mexico, and W Texas.
BT Guadalupian
  Permian
SA New Mexico
  Texas

Bell Creek Field
Oil and gas field.
BT Montana
  United States

Bell Island
In Conception Bay W of Saint John's.
BT Newfoundland
  Canada

Bellary
Town in E.
BT Mysore
  India

Belle Isle
Island constituting a salt dome on N shore of Atchafalaya Bay in Saint Mary Parish in S.
BT Louisiana
  United States

Belledonne
use Belledonne Massif

Belledonne Massif
Part of mountain range in SE France. Also search Belledonne; Belledonne Range.
UF Belledonne
  Belledonne Mountain Range
  Belledonne Range
BT France

Belledonne Mountain Range
use Belledonne Massif

Belledonne Range
use Belledonne Massif

Bellingshausen Sea
Inlet of South Pacific Ocean extending westward from Alexander Island off the W coast of Antarctic Peninsula to Thurston Island.
BT Antarctic Ocean

Belluno
Town and province in N.
BT Veneto
  Italy

Belly River Formation
Has been considered equivalent to Two Medicine Formation and Virgelle Sandstone of Blackfoot Indian Reservation, N Montana, and Judith River Formation, Claggett Formation, and Eagle Sandstone of central Montana.
BT Upper Cretaceous
  Cretaceous
SA Alberta
  Saskatchewan

Belomorsk
Town in Karelian A.S.S.R. in NW European USSR.
UF Soroka
BT Russian Republic
  USSR

Belorussia
use Byelorussia

Belozerka
Town in Kherson Oblast N of the Crimea.
BT Ukraine
  USSR

Belozero
Village W of Arkhangelsk in Arkhangelsk Oblast.
BT Russian Republic
  USSR

Belsk
Town in in NE part of country. Also search Bielsk Podlaski.
UF Bielsk
  Bielsk Podlaski
BT Bialystok
  Poland

Belt Basin
Cascade County in W central.
BT Montana
  United States

Belt Supergroup
Comprises Piegan Group, Missoula Group, and Ravalli Group. Also search Belt Series.
BT Precambrian
SA British Columbia
  Idaho
  Missoula Group
  Montana
  Ravalli Group
  Washington

Beltana
Village in E central.
BT South Australia
  Australia

belts
A valid term through 1978. After 1978, see specific terms, e.g. fold belts, metamorphic belts, etc.

belts, fold
use fold belts

belts, greenstone
use greenstone belts

belts, metamorphic
use metamorphic belts

belts, mobile
use mobile belts

belts, volcanic
use volcanic belts

Bembridge Marls
Defined to include Bembridge Oyster Beds, Lower Bembridge Marls, and Upper Bembridge Marls. Isle of Wight in English Channel S of Southampton.
BT Oligocene
SA England
  United Kingdom

benches
Includes use on level 3 under geomorphology(1) shore features(2).
SA erosion
  geomorphology
  shore features
  terraces

beneficiation
Used for commodity production technology.
SA dewatering
  flotation
  purification

Bengal
A region and former Indian province. Index Bangladesh and Indian states as applicable. Also search Bengal Basin.
UF Bengal Basin
BT Asia
SA Bangladesh
  Tripura
  West Bengal

Bengal Basin
use Bengal

Benham
Village in Harlan County in SE.
BT Kentucky
  United States

Beni Bouchera
Region in the Rif in NE.
BT Morocco

Benin
Was established in 1960 as Dahomey. Includes use on level 1 as an area term (List O). For term set options see list B. Before 1976, also search Dahomey.
BA Africa
SA Mali-Niger Syneclise
  Niger River
  Niger Valley
  West Africa

Benioff seismic zone
use Benioff zone

Benioff zone
Includes use on level 3 under plate tectonics(1) structure(2), and subduction(2); on level 3 under seismology(1).
UF Benioff seismic zone
  Benioff zones
  zone, Benioff
SA lithosphere
  plate tectonics
  plates
  sea-floor spreading
  seismology
  subduction
  subduction zones
  trenches

Benioff zones
use Benioff zone

Bennettitales
Includes use on level 2 under gymnosperms(1). See list F.
BA gymnosperms
BT Plantae
SA Ptilophyllum

Benoue Valley
use Benue Valley

Benson Mines
Also search Benson AND New York.
BT New York
  United States

benthic taxa
use benthonic taxa

benthonic
A valid index term through 1977. After 1977, use benthonic taxa.

benthonic taxa
Term introduced in 1977. Includes use on level 3 under fossil group(1). Before 1978, search benthonic; benthic.
UF benthic taxa
  taxa, benthonic
SA bottom features
  fossils
  marine geology
  paleoecology

Benton County
BX Arkansas
  Indiana
  Iowa
  Minnesota
  Mississippi
  Missouri
  Oregon
  Tennessee
  Washington
BT United States

Benton Formation
In Colorado Group. Comprising Graneros Shale, Greenhorn Limestone and Carlile Shale as members. Contains Codell Sandstone Member. E Colorado, South Dakota, Kansas, S Minnesota, SE Montana, Nebraska, NE New Mexico, E Wyoming.
BT Cretaceous
SA Carlile Shale
  Colorado
  Colorado Group
  Graneros Shale
  Greenhorn Limestone
  Kansas
  Lower Cretaceous
  Minnesota
  Montana
  Nebraska
  New Mexico
  South Dakota
  Upper Cretaceous
  Wyoming

bentonite
Includes use on level 1 as a commodity term (list C) only when of economic value; on level 3 under sediments(1) clastic sediments(2); on level 3 under sedimentary rocks(1) clastic rocks(2). See list I (sed. rocks) and list N (sediments).
UF amargosite
  mineral soap
  soap clay
  volcanic clay
BA clastic rocks
BT sedimentary rocks
SA clastic sediments
  clay mineralogy
  clay minerals
  metabentonite
  pyroclastics
  sheet silicates
  terrigenous materials
  tuff
  volcanic ash

Benue Trough
use Benue Valley

Benue Valley
River valley. Index countries as applicable. Also search Benue Trough.
UF Benoue Valley
  Benue Trough
  Binue Valley
BT Africa
SA Cameroon
  Nigeria

Beograd
use Belgrade

Beppu
City in NE.
BT Kyushu
  Japan

beraunite
BA phosphates
BT minerals

Berchtesgaden
Town in extreme SE.
BT Bavaria
  West Germany

Berea Sandstone
Devonian or Mississippian. NE Kentucky, S Michigan, Ohio, W Pennsylvania, and N West Virginia.
BT Paleozoic
SA Devonian
  Kentucky
  Michigan
  Mississippian
  Ohio
  Pennsylvania
  West Virginia

Beregovo
Town in Transcarpathian Oblast in extreme SE.
BT Ukraine
  USSR

Beresford Lake
In SW Manitoba.
BT Manitoba

Canada

beresite
Includes use on level 3 under igneous rocks(1). See list H.
BA granite-granodiorite family
BT igneous rocks
SA hypabyssal rocks
quartz porphyry

berg crystal
use quartz crystal

Bergamo
City in Bergamo Province in central.
BT Lombardy
Italy

Bergell Massif
In SE part of country.
BT Switzerland

Bergen
City on the North Sea in SW part of country.
BT Norway

Bergisch Gladbach
City in SW central.
BT North Rhine-Westphalia
West Germany

Bergisches Land
Region on right bank of the Rhine River.
BT West Germany

Bergstrasse
Name applied to the W slope of the Odenwald hills of SW.
BT West Germany

Berici Hills
Range of volcanic hills in Vicenza Province in W central.
BT Veneto
Italy

Bering land bridge
Massive Quaternary land link more than 1,000 miles wide which connected Asia and North America in the Bering Sea and Chukchi Sea area.
SA Bering Sea
Chukchi Sea

Bering Sea
Between NE Siberia and Alaska with the Aleutian Islands on the S and Bering Strait on the N. Includes use on level 1 as an area term (list O). For term set options see list B.
BA Pacific Ocean
NT Bowers Ridge
Bristol Bay
Saint Paul Island
SA Bering land bridge

Berkeley
City in Alameda County across San Francisco Bay from San Francisco.
BT California
United States

Berks County
SE central.
BT Pennsylvania
United States

Berkshire
County in S England.
CO N512000N514500
W0003000W0014000
BT England
Great Britain
United Kingdom

Berkshire Anticlinorium
use Berkshire Hills

Berkshire Highlands
use Berkshire Hills

Berkshire Hills
Hills in Berkshire County in W Massachusetts. Also search Berkshire Anticlinorium.
UF Berkshire Anticlinorium
Berkshire Highlands
Berkshire Massif
Berkshires
BT Massachusetts
United States

Berkshire Massif
use Berkshire Hills

Berkshires
use Berkshire Hills

Berlin
City partitioned into East Berlin and West Berlin since 1945. Index countries as applicable.
BT Germany
SA East Germany
West Germany

Bermuda
British colony comprising over 300 coral islands about 640 miles ESE of Cape Hatteras. Includes use on level 1 as an area term (list O). For term set options see list B. Also search Bermuda Platform; Bermuda Rise.
CO N321500N322500
W0644000W0645000
UF Bermuda Island
Bermudas
BA Atlantic Ocean
NT Harrington Sound

Bermuda Island
use Bermuda

Bermuda Platform
Part of Bermuda Rise.
BT Atlantic Ocean
SA Bermuda Rise

Bermuda Rise
Runs in SW-NE direction to the west of Bermuda.
BT Atlantic Ocean
SA Bermuda Platform

Bermudas
use Bermuda

Bern
Canton in W central Switzerland. Also a city. Also search Berne.
UF Berne
BT Switzerland
NT Biel Lake

Bernalillo County
Central.
BT New Mexico
United States

berndtite
BA sulfides
BT minerals

Berne
use Bern

Bernese Alps
N division of Central Alps in Bern and Valais cantons. Also search Bernese Oberland.
UF Bernese Oberland
BT Switzerland
SA Alps
Central Alps

Bernese Oberland
use Bernese Alps

Bernic Lake
BT Manitoba
Canada

bernstein
use amber

Berriasian
Europe. Lowermost Lower Cretaceous, above Portlandian (Jurassic), below Valanginian. Includes use on level 3 under age terms(1). See list E.
BA Lower Cretaceous
Cretaceous
BT Mesozoic
SA Neocomian

Berrien County
Index states as applicable.
BX Georgia
Michigan
BT United States

berryite
BA sulfosalts
BT minerals

berthierine
BA sheet silicates
silicates
BT minerals
SA aluminosilicates
chamosite

berthonite
use bournonite

bertrandite
BA ring silicates
silicates
BT minerals

Berwickshire
County in SE.
BT Scotland
Great Britain
United Kingdom

beryl
Includes use on level 3 under minerals(1); as level 3 commodity term under beryllium(1). See list C.
BA ring silicates
silicates
BT minerals
SA beryllium
emerald
gems

beryllium
Includes use on level 1 and 2 as chemical element (list D) and as a commodity term (list C).
UF Be
SA Be-10
Be-7
beryl
elements

Besancon
City near Swiss border.
BT Doubs
France

Beshchady Mountains
Section of the Carpathians. Index countries and Ukraine as applicable.
BT Europe
SA Carpathians
Czechoslovakia
Poland
Ukraine

Beskid Mountains
Both West and East Beskid mountains comprise mountain group of the Carpathians on NE border. Also search Beskids.
UF Beskids
BT Czechoslovakia
SA Carpathians

Beskids
use Beskid Mountains

Bessarabian
Pertains to Bessarabia, a former Romanian province which is now co-extensive with Moldavia, U.S.S.R.
BT Moldavia
USSR

Bet-Pak-Dala
Desert steppe W of Lake Balkhash. Also search Golodnaya Step.
CO N450000N470000
E0730000E0673000
UF Golodnaya Step
Hunger Steppe
BT Kazakhstan
USSR

betafite
Also search blomstrandite, ellsworthite, and hatchettolite.
UF blomstrandite
ellsworthite
hatchettolite
BA oxides
BT minerals
SA niobates
pyrochlore

betekhtinite
BA sulfides
BT minerals

Betic Cordillera
Extends along the N extremity of Andalusia between the Guadiana and Guadalaquivir rivers. It is a southern series of the Alpine System. Before 1978, also search Sierra Morena.
CO N380000N390000
W0030000W0060000
UF Cordillera Betica
Cordillera Marianica
BT Spain
SA Andalusian
Betic Zone
Prebetic Zone
Subbetic Zone

Betic Zone
Used in association with the Betic Cordillera in SW.
BT Spain
SA Betic Cordillera

Betsiboka Basin
Around town of Majunga on Bombetoka Bay. Before 1978, also search Majunga; Majunga Basin.
UF Majunga Basin
BT Malagasy Republic

Betts Cove
BT Newfoundland
Canada

Betula
Genus.
BA Dicotyledoneae
angiosperms
BT Plantae

betwixt mountains
use median masses

beudantite
BZ arsenates
sulfates
BT minerals

Bevier Coal
Bevier Formation comprises Bevier Coal with underlying gray clay and overlying black slate. SE Kansas.
BT Pennsylvanian
SA Desmoinesian
Kansas

Bewcastle
Village and parish in Cumberland County in NE.
BT England
Great Britain
United Kingdom

Bhandara
Town in extreme NE.
BT Maharashtra
India

Bhander Group
Bhander Series includes upper Bhander Sandstone, Sirbu shales, Middle Bhander Sandstone and Ganurgarh shales. Named after range of hills N of Narmada Valley. Also search Bhander Series; Bhander Limestone.
UF Bhander Limestone
Bhander Series
SA Cambrian

India
Madhya Pradesh
Precambrian

Bhander Limestone
use Bhander Group

Bhander Series
use Bhander Group

Bhilwara
Town in S.
BT Rajasthan
India

Bhuj
Town in Cutch in NW.
BT Gujarat
India

Bhuj Series
Consists of Zamia beds, Ptilophyllum beds, and Palmoxylon beds. Named after town of Bhuj in Cutch.
BT Middle Cretaceous
Cretaceous
SA Gujarat
India

Bhutan
Kingdom in E Himalayas between Tibet and NE India. Includes use on level 1 as an area term (list O). For term set options see list B.
BA Asia
SA Himalayas
Lesser Himalayas

Bi
use bismuth

Bialowieza
National park in E.
BT Poland

Bialystok
Province in extreme NE Poland. Also a city.
BT Poland
NT Augustow
Belsk
Suwalki

Biarritz
Town on the Bay of Biscay near Spanish border.
BT Pyrenees-Atlantiques
France

Biarritzian
Europe: middle to upper Eocene. Includes use on level 3 under age terms(1). See list E.
BA Eocene
Paleogene
Tertiary
BT Cenozoic
SA middle Eocene
upper Eocene

bibliography
Includes use on level 1 (list A); on level 2 under major disciplines; as a level 2 or 3 term appropriate to a large number of topics (list G). Covers major bibliographies on topics, areas or persons. If 1, term set options are:
topic [List B or extraterrestrial geology, general, geophysics]
subtopic
SA areal geology
associations
biography
catalogs
geology
glossaries
lexicons
mineralogy
oceanography
paleobotany
paleontology
petrology
research
review

sedimentary petrology
stratigraphy
structural geology
surveys
tectonophysics
volcanology

Bicaz Valley
River valley in Bacau County.
BT Moldavia
Romania

Bida
Region in W central.
BT Nigeria

Bidahochi Formation
Divided into six members for mapping purposes. Fifth member informally referred to as White Cone Member. In NE Arizona.
BT Pliocene
SA Arizona
New Mexico

bideauxite
BA halides
BT minerals

Biel Lake
In NW Bern.
UF Bienne Lake
Lake of Bienne
BT Bern
Switzerland

Biella
Town in Vercelli Province in N central.
BT Italy
SA Piedmont

Bielsk
use Belsk

Bielsk Podlaski
use Belsk

Bielsko
City in S part of country. Also search Bielsko-Biala.
UF Bielsko-Biala
BT Katowice
Poland

Bielsko-Biala
use Bielsko

Bienne Lake
use Biel Lake

Big Belt Mountains
Range of Rocky Mountains in W central.
BT Montana
United States
SA Rocky Mountains

Big Bend National Park
In Brewster County on the Mexican border in SW Texas. Also search Big Bend.
BT Texas
United States

Big Horn Basin
use Bighorn Basin

Big Horn County
Index states as applicable.
BX Montana
Wyoming
BT United States

Big Horn Mountains
use Bighorn Mountains

Big Pine Key
One of the principal islands in the Florida Keys.
BT Florida
United States
SA Florida Keys

Big Snowy Mountains
N central.
BT Montana
United States

Bigby-Cannon Limestone
Contemporary facies comprising this unit are Bigby Limestone and Cannon Limestone. Overlies Hermitage Formation and underlies Catheys Formation. Central Tennessee.
BT Middle Ordovician
Ordovician
SA Tennessee

Bighorn Basin
River basin. Index states as applicable. Also search Big Horn Basin.
UF Big Horn Basin
BT United States
SA Montana
Wyoming

Bighorn Mountains
Range of Rocky Mountains. Index states as applicable. Also search Big Horn Mountains.
CO N433000N450000
W1063000W1083000
UF Big Horn Mountains
BT United States
SA Montana
Rocky Mountains
Wyoming

Bihar
State in E India.
UF Behar
BT India
NT Amarjola
Bokaro
Chainpur
Dhanbad
Gaya
Hazaribagh
Jharia
Jharia coal field
Karanpura
Karanpura coal field
Karharbari
Monghyr
Mosaboni copper mines
Palamau
Rajgir
Rajmahal
Rajmahal Hills
Ramgarh coal field
Ranchi
Singhbhum
Sini
SA Ajay River
Damodar Valley
Damuda Series
Karharbari Stage
Kosi Basin
Panchet Series
Semri Series
Singhbhum Granite
Singhbhum shear zone
Son Valley

Bihar Mountains
use Bihor Mountains

Bihor
County in W.
BT Transylvania
Romania

Bihor Mountains
In W central Transylvania.
UF Bihar Mountains
BT Transylvania
Romania
SA Vladeasa Mountain

Biikzhal
Drill hole in Emba River region of Caspian Basin.
BT Kazakhstan
USSR

Bijapur
Town in NW.
BT Mysore
India

Bijawar System
Includes Majhauli Group, Bhitri Group, Lora Group, and Chanderdip Group.
BT Precambrian
SA India
Madhya Pradesh

Bikaner
A city and former state in NW.
BT Rajasthan
India

Bilaspur
Former Indian state. Now a district in N part of country.
BT Himachal Pradesh
India

Bilbao
City on the Bay of Biscay in the Basque country.
BT Vizcaya
Spain

Billings
City in Yellowstone county in S central.
BT Montana
United States

Billings County
In W North Dakota.
CO N464000N472000
W1030500W1034000
BT North Dakota
United States

Billiton
Island between Sumatra and SW Borneo.
UF Belitung
BT Indonesia

Bimini
Two small islands off S Florida coast.
UF Bimini Islands
BT Bahamas
West Indies

Bimini Islands
use Bimini

Bingham
Mining town in Salt Lake County in N central Utah.
BT Utah
United States

Bingham County
Before 1978, also search Bingham for the county.
BT Idaho
United States

Binghamton
City in Broome County on the Pennsylvania border.
BT New York
United States

Binn Valley
use Binnental

Binnatal
use Binnental

Binnen Tal
use Binnental

Binnental
Valley in E Valais. Also search Binn Valley, and Binnatal.
UF Binn Valley
Binnatal
Binnen Tal
BT Valais
Switzerland

Binue Valley
use Benue Valley

biocalcarenite
BA carbonate rocks
BT sedimentary rocks
SA calcarenite

biocenoses
Includes use on level 2 and 3 under fossil group(1). See list F. Also search biocenosis.
UF biocenosis
    biocoenoses
    life assemblage
SA communities
    paleontology
    thanatocenoses

biocenosis
use biocenoses

biochemistry
Includes use on level 2 under fossil group(1). See list F. Also search biogeochemistry.
SA biology
    chemistry
    fossils
    geochemistry
    paleobiology

biocirculation
Includes use on level 2 under ocean circulation(1).
SA ocean circulation

biocoenoses
use biocenoses

biofacies
Includes use on level 3 under ecology(1), paleoecology(1), sedimentation(1), or sediments(1).
UF biologic facies
SA assemblages
    biostratigraphy
    biota
    ecology
    facies
    sedimentation
    sediments

biogenic
A valid term through 1977. After 1977, use biogenic materials or biogenic structures.

biogenic structures
Includes use on level 2 under sedimentary structures(1). See list K.
BA sedimentary structures
NA algal banks
    algal biscuits
    algal mats
    algal mounds
    banks
    bioturbation
    burrows
    coprolites
    girvanella
    ichnofossils
    lebensspuren
    oncolites
NZ stromatolites
NA tracks
    trails
SA coal
    peat
    reefs
    structures

biogeochemical methods
Includes use on level 2 under mineral exploration(1). Before 1972, also search biogeochemical prospecting; biogeochemistry.
BT geochemical methods
SA geobotanical methods
    geochemistry
    methods
    mineral exploration

biogeochemical prospecting
A valid term through 1971. After 1971, use biogeochemical methods.

biogeochemistry
A valid term through 1978. After 1978, see biochemistry for fossils (list F), or biogeochemical methods under mineral exploration(1).

biogeography
Includes use on level 1 (list A); on level 2 under fossil groups (list F). To be used for descriptions of the geographic distribution of both fossil and modern organisms. If 1, term set options are:
age (List E; only where one age is discussed)
    area (or global)
fossil group [List F]
    age
topic [catalogs, concepts, distribution, evolution, interpretation, nomenclature, observations, patterns]
    subtopic
UF chorology
    phytogeography
    zoogeography
SA Boreal
    boreal region
    continental drift
    ecology
    faunal provinces
    fossils
    geography
    organisms
    paleobotany
    paleoecology
    paleogeography
    paleontology
    palynology
    range
    stratigraphy

biography
Includes use on level 1 (list A). This set is used only where there is an extensive bibliography. Term set options are:
general
    name of individual (last, first),
SA bibliography

bioherms
Includes use on level 3 under reefs(1).
UF organic mound
SA algae
    atolls
    biostromes
    calcareous composition
    corals
    reefs

biologic facies
use biofacies

biological processes
SA biology
    processes

biology
SA bacteria
    biochemistry
    biological processes
    biosphere
    biotypes
    ecology
    exobiology
    organisms
    paleobiology

biomass
Includes use on level 3 under ecology(1).
SA ecology
    organisms

biometry
Includes use on level 3 under fossil group(1) and under paleontology(1). See list F.
SA fossils
    mathematical geology
    paleontology
    statistical methods

biomicrite
Includes use on level 3 under sedimentary rocks(1) carbonate rocks(2).
BA carbonate rocks
BT sedimentary rocks
SA micrite

biopelite
use black shale

biosparite
BA carbonate rocks
BT sedimentary rocks
SA limestone

biosphere
SA atmosphere
    biology
    Earth
    hydrosphere
    lithosphere
    tectonosphere

biostratigraphy
Includes use on level 2 under fossil group(1). See list F.
SA biofacies
    diachronism
    fossils
    historical geology
    index fossils
    paleoecology
    paleontology
    range
    stratigraphy

biostromes
Includes use on level 3 under reefs(1).
SA bioherms
    corals
    reefs

biota
Includes use on level 2 under soils(1).
SA bacteria
    biofacies
    faunal studies
    floral studies
    Invertebrata
    microorganisms
    organisms
    soils
    Vertebrata

biotite
BA sheet silicates
    silicates
BT minerals
SA mica group

biotite gneiss
Includes use on level 3 under metamorphic rocks(1) gneisses(2). See list J.
BA gneisses
BT metamorphic rocks
SA gneiss

biotite granite
See list H.
BA granite-granodiorite family
BT igneous rocks
SA granite

biotite schist
Includes use on level 3 under metamorphic rocks(1) schists(2). See list J.
BA schists
BT metamorphic rocks
SA schist

bioturbation
Includes use on level 3 under sedimentary structures(1) biogenic structures(2). See list K.
BA biogenic structures
    sedimentary structures
SA burrows
    organisms
    sedimentation

biotypes
Term introduced in 1978. Includes use on level 3 under ecology(1) or paleoecology(1).
SA biology
    ecology
    paleoecology

Birch Creek
Village in E Alaska. Also a creek.
BT Alaska
    United States

Bird Island
use Aves Island

Bird River
Village in SE.
BT Manitoba
    Canada

Bird Spring Formation
In Spring Mountain area, Nevada, underlies Spring Mountain Formation and overlies Illipah Formation. Also search Bird Spring Group.
BT Paleozoic
SA California
    Lower Permian
    Mississippian
    Nevada
    Pennsylvanian
    Permian
    Upper Mississippian
    Utah

birds
use Aves

birefraction
use birefringence

birefringence
UF birefraction
    double refraction
SA extinction
    minerals
    optical properties
    polarization
    refraction
    refractive index

Birmingham
City in Jefferson County in central Alabama, and city in W central England. Index Alabama and/or England as applicable.
SA Alabama
    England

Birmingham University
In Birmingham in W central.
BT England
    Great Britain
    United Kingdom

birnessite
BA oxides
BT minerals

Biryusa
River SW of Tayshet in W Irkutsk Oblast.
BT Russian Republic
    USSR

Bisbee
City in Cochise County in SE.
BT Arizona
    United States

Biscay (Bay)
use Bay of Biscay

Biscay Bay
use Bay of Biscay

Biscayne Bay
S of Miami on E coast of Dade County.
BT Florida
    United States

Bishop Tuff
E central California. Bishop Tuff was erupted a short time after Sherwin glacial stage.

BT Pleistocene
  Quaternary
SA California

Bismarck Archipelago
Volcanic group of islands N and NE of E New Guinea.
BT Papua New Guinea
  Australasia
NT Rabaul Caldera
SA Melanesia
  New Britain
  New Ireland
  Oceania

Bismarck Sea
Part of W Pacific Ocean NE of New Guinea and NW of New Britain.
BT Pacific Ocean

bismuth
Includes use on level 1 and 2 as a chemical element (list D) and as a commodity term (list C).
UF Bi
SA elements
  heavy metals
  native elements and alloys

bismuthides
BA sulfides
BT minerals
NA froodite

bismuthinite
BA sulfides
BT minerals

Bison
Genus. Includes use on level 3 under Mammalia(1) Artiodactyla(2).
BA Bovidae
  Artiodactyla
  Mammalia
BT Tetrapoda
  Vertebrata
NA Bison latifrons
  Bison occidentalis

Bison latifrons
Includes use on level 3 under Mammalia(1) Artiodactyla(2).
BA Bison
  Bovidae
  Artiodactyla
  Mammalia
BT Tetrapoda
  Vertebrata
SA Ungulata

Bison occidentalis
Includes use on level 3 under Mammalia(1) Artiodactyla(2).
BA Bison
  Bovidae
  Artiodactyla
  Mammalia
BT Tetrapoda
  Vertebrata
SA Ungulata

Bistrita
City in Bistrita-Nasaud County in central.
BT Transylvania
  Romania

Bistrita Mountains
Range of the Carpathians in NE Romania.
UF Bistritei Mountains
BT Romania
SA Carpathians

Bistrita Valley
River valley in W central.
BT Moldavia
  Romania

Bistritei Mountains
use Bistrita Mountains

Bitkov
In W Ukraine.
UF Bitkow
  Bytkov
BT Ukraine
  USSR

Bitkow
use Bitkov

bitter spar
use dolomite

Bitter Springs Formation
Underlies Middle Cambrian sediments and overlies Heavitree Quartzite. SW and S central Northern Territory.
BT Proterozoic
  Precambrian
SA Australia
  Northern Territory

Bitterfeld
City in W.
BT Halle
  East Germany

Bitterroot Range
Range of the Rocky Mountains. Index states as applicable.
BT United States
SA Beaverhead Mountains
  Idaho
  Montana
  Rocky Mountains

bitumenite
use torbanite

bitumens
Includes use on level 1 as a commodity term (list C); on level 2 under organic materials(1).
BA organic materials
NA asphalt
SA anthraxolite
  bituminous shale
  hydrocarbons
  kerogen
  oil seeps
  petroleum

bituminous coal
Before 1978, also search coal AND bituminous.
UF soft coal
BA organic residues
BT sedimentary rocks
SA coal

bituminous shale
See list I.
BA clastic rocks
BT sedimentary rocks
SA bitumens
  shale

Bivalvia
Includes use on level 2 under Mollusca(1). See list F. Before 1976, also search Pelecypoda.
BA Mollusca
BT Invertebrata
NA Anthraconaia
  Anthraconauta
  Astarte
  Carbonicola
  Cardiidae
  Crassostrea virginica
  Daonella
  Didacna
  Edmondia
  Glycymeris
  Gryphaea
  Halobia
  Hippuritacea
  Inoceramidae
  Inoceramus
  Limnocardium
  Lucinidae
  Mactra
  Mercenaria
  Monotis
  Myophoria
  Mytilus
  Nuculanidae
  Nuculidae
  Ostreidae
  Pectinacea
  Pectinidae
  Rudistae
  rudists
  Tridacna
  Trigoniidae
  Unionidae

Biwa Lake
use Lake Biwa

Biwabik Iron Formation
In Animikie Group. In St. Louis County, NE Minnesota. Part of Mesabi Iron Range. Also search Biwabik Iron-Formation.
UF Biwabik Iron-Formation
BT Precambrian
SA Animikie Group
  Minnesota

Biwabik Iron-Formation
use Biwabik Iron Formation

bixbyite
UF partridgeite
BA oxides
BT minerals

Bjerkrem-Sogndal Massif
S part of country.
BT Norway

Bjurbole
Point of impact ENE of Helsinki. Also search Bjurbole Meteorite.
BT Finland
SA meteorites

Black earth
use Chernozems

Black Forest
Mountainous region in W.
BT Baden-Wurttemberg
  West Germany

Black Hills
Group of mountains primarily in SW South Dakota. Index states as applicable.
CO N431500N443500
  W1023000W1040500
BT United States
SA South Dakota
  Wyoming

black lead
use graphite

Black Mesa
Tableland region in NE.
CO N360000N370000
  W1090000W1110000
BT Arizona
  United States

Black Mesa Basin
In NE Arizona.
BT Arizona
  United States

Black Range
Extends N-S through parts of Grant and Sierra counties in SW.
BT New Mexico
  United States

Black River Group
Comprises Pamelia, Lowville, and Chaumont formations.
BT Middle Ordovician
  Ordovician
SA Blackriverian
  New York
  Pennsylvania

Black Rock Desert
Pershing County in NW central.
BT Nevada
  United States

Black Sea
Large inland sea bounded on the N and NE by the USSR, S by Turkey, and W by Bulgaria and Romania. Includes use on level 1 as an area term (list O). For term set options see list B.
CO N410000N470000
  E0420000E0280000
BA Eurasia
SA Azov Sea

Black Sea Basin
use Black Sea region

Black Sea Coast
use Black Sea region

Black Sea Lowland
use Black Sea region

Black Sea region
Index countries as applicable. Also search Black Sea Coast, Black Sea Lowland and Black Sea Basin.
CO N423000N473000
  E0400000E0300000
UF Black Sea Basin
  Black Sea Coast
  Black Sea Lowland
BT Eurasia
SA Bulgaria
  Romania
  Turkey
  USSR

black shale
Includes use on level 3 under sedimentary rocks(1) clastic rocks(2). Also search carbonaceous shale.
UF biopelite
  carbonaceous shale
BA clastic rocks
BT sedimentary rocks
SA shale
  terrigenous materials

Black Warrior Basin
Primarily in Alabama. Index states as applicable.
BT United States
SA Alabama
  Mississippi

Blackford County
E central.
BT Indiana
  United States

Blackhawk Formation
In Mesaverde Group. Six principal members are Desert, Grassy, Sunnyside, Kenilworth, Aberdeen, and Spring Canyon. Central E Utah.
BT Upper Cretaceous
  Cretaceous
SA Mesaverde Group
  Utah

Blackriverian
North America. Lower Mohawkian, below Trentonian. Includes use on level 3 as an age term (list E).
BA Middle Ordovician
  Ordovician
BT Paleozoic
SA Black River Group

Blacksburg
Town in Montgomery County in SW.
BT Virginia
  United States

Blackwater
River and falls in Tucker County in E.
BT West Virginia
  United States

Blaine County
Index states as applicable.
BX Idaho
  Montana
  Nebraska
  Oklahoma
BT United States

Blaine Formation
In Nippewalla, El Reno, or Pease River Group. W Oklahoma, central North and panhandle of Texas.
BT Permian
SA Oklahoma
Texas

Blairmore Group
Includes the Gladstone Formation, Beaver Mines Formation, and Mill Creek Formation in S Alberta.
BT Cretaceous
SA Alberta
Lower Cretaceous
Upper Cretaceous

Blake Basin
use Blake-Bahama Basin

Blake Outer Ridge
use Blake-Bahama Outer Ridge

Blake Plateau
E of N Florida and N of Grand Bahama Island. Also search Bahama Platform.
UF Bahama Platform
BT Atlantic Ocean

Blake Ridge
use Blake-Bahama Outer Ridge

Blake-Bahama Basin
Just to the E of the Blake Plateau E of North Florida. Also search Blake Basin.
UF Blake Basin
BT Atlantic Ocean

Blake-Bahama Outer Ridge
E of the Blake-Bahama Basin which is E of the Blake Plateau off N Florida.
UF Blake Outer Ridge
Blake Ridge
BT Atlantic Ocean

Blancan
North America. Covers an interval in the upper Pliocene above the Hemphillian age.
BA Cenozoic
SA Pleistocene
Pliocene
Quaternary
Tertiary
Texas

Blanco fracture zone
Off coast of Oregon. NE of E end of Mendocino Escarpment.
BT Pacific Ocean

blasting
Includes use on level 3 under engineering geology(1) or foundations(1).
SA engineering geology
excavations
explosions
foundations

Blastoidea
Includes use on level 2 under Echinodermata(1). See list F.
BA Crinozoa
Echinodermata
BT Invertebrata

blastomylonite
Includes use on level 3 under metamorphic rocks(1) mylonites(2). See list J.
BA mylonites
BT metamorphic rocks
SA mylonite

Bleiberg
Town in S part of country.
BT Carinthia
Austria

Bleikvassli
Ore bearing region.
BT Nordland
Norway

Blekinge
County on the Baltic Sea in extreme SE.
BT Sweden

Blind River
Town on Lake Huron.
BT Ontario
Canada

bloating shale
use shale

block
A valid index term through 1977. Now use block structures.

block clay
use melange

block faults
use block structures

Block Island Sound
Between Washington County, Rhode Island and Block Island.
BT Atlantic Ocean

block mountains
use block structures

block structures
Term introduced in 1978. Includes use on level 3 under faults(1) systems(2). Before 1978, also search faults AND block; block faults; block structure; block mountains.
UF block faults
block mountains
SA faults
horsts
normal faults
systems

blomstrandite
use betafite

Bloomsburg Formation
In Cayuga Group. Central and S Pennsylvania, W Maryland, W Virginia, and N West Virginia.
BT Upper Silurian
Silurian
SA Maryland
Pennsylvania
Virginia
West Virginia

Blount County
Index states as applicable.
BX Alabama
Tennessee
BT United States

Blow River
Flows into Mackenzie Bay in the extreme N.
BT Yukon Territory
Canada

Bloyd Formation
Shale consists of (ascending) Brentwood Limestone Member, Woolsey Member and an unnamed shale division. NW Arkansas and E Oklahoma.
BT Morrowan
Lower Pennsylvanian
Pennsylvanian
SA Arkansas
Oklahoma

blue algae
use Cyanophyta

blue chalcocite
use digenite

Blue Glacier
Mount Olympus, Olympic National Park in NW.
BT Washington
United States

blue lead
use galena

Blue Mountain Lake
In the Adirondack Mountains.
BT New York
United States

Blue Mountains
In W Idaho, NE Oregon, and SE Washington, and also a range in E Jamaica. Index Jamaica and states as applicable.
SA Idaho
Jamaica
Oregon
Washington

Blue Nile
Right headstream of the Nile River. Index countries as applicable.
BT Africa
NT Gezira
SA Ethiopia
Nile River
Sudan

Blue Ridge Mountains
E and SE range of the Appalachians. Index states as applicable. Also search Blue Ridge.
BT United States
SA Appalachians
Georgia
Grandfather Mountain
North Carolina
Virginia
West Virginia

Blue Ridge Province
Embraces the Appalachians E of the Great Appalachian Valley from South Mountain, Pa., southward including the Piedmont. Composed of some of the most ancient rocks of the Appalachians. Index states as applicable. Also search Blue Ridge.
BT Appalachians
North America
SA Georgia
Maryland
North Carolina
Pennsylvania
South Carolina
Virginia
West Virginia

blue-green algae
use Cyanophyta

Bluebell Mine
E shore of Kootenay Lake in SE.
BT British Columbia
Canada

blueschist
Includes use on level 3 under metamorphic rocks(1) schists(2). See list J.
BA schists
BT metamorphic rocks
SA glaucophane schist

blueschist facies
BT facies
SA greenschist facies
metamorphic rocks

bluff formation
use loess

Blyava
A section of the City of Mednogorsk in the S Urals near the Kazakh border.
BT Russian Republic
USSR

Bochnia
A mining town in S part of country.
BT Krakow
Poland

Bochum
City in the Ruhr Valley.
BT North Rhine-Westphalia
West Germany

Bodaibo
Town in NE Irkutsk Oblast.
BT Russian Republic
USSR

Bodega Head
Tip of peninsula sheltering Bodega Bay from the Pacific Ocean in Sonoma County.
BT California
United States

Bodensee
use Lake Constance

bodies, ore
use ore bodies

bodily tide
use Earth tides

bodily waves
use body waves

body waves
Used for P-waves or S-waves under seismology(1) when type of wave has not been indicated, or for collective use. Before 1976, also search seismic waves.
UF bodily waves
BT elastic waves
NT P-waves
S-waves
SA seismology

boehmite
BA oxides
BT minerals

Boeotia
A district in Attica Department. Central Greece and Euboea.
BT Greece

Boerzsoeny Mountains
use Borzsony Mountains

Bog soils
Includes use on level 3 under soils(1). See list M.
BA soils
SA Histosols
Hydromorphic soils
Intrazonal soils
peat
soil group

Bogota
City in central part of country.
BT Colombia

bogs
Include use on level 3 under geomorphology(1).
NT peat bogs
SA geomorphology
marshes
swamps

Boguszow
City in SW part of country.
BT Wroclaw
Poland

bohdanowiczite
BA sulfides
BT minerals

Bohemia
Region in westernmost part of country.
BT Czechoslovakia
NT Bohemian Massif
Central Bohemian Pluton
Chvaletice
Jachymov
Karlovy Vary
Kladno
Kutna Hora
Pilsen Basin
Plana
Prague
Pribram
Slany
Sobotka

Bohemia • Boreal

  Sokolov
  Sokolov Basin
  Teplice
SA Karkonosze Mountains
  Sudeten
  Sudeten Mountains
  Weisse Elster Basin
  White Mountain

Bohemian Massif
A dissected quadrangular plateau bounded by the Bohemian Forest, Erzgebirge, the Sudeten Mountains and the Bohemian-Moravian Heights.
CO N483000N501000
  E0163000E0123000
BT Bohemia
  Czechoslovakia
SA Central Bohemian Pluton

Boise
City in Ada County in WSW.
BT Idaho
  United States

Bojnice
Village in W central.
BT Slovakia
  Czechoslovakia

Bokaro
Coal field in central Bihar. Also search Bokaro coal field, Bokaro Coalfield, and Bokaro Field.
UF Bokaro coal field
  Bokaro Coalfield
  Bokaro Field
BT Bihar
  India

Bokaro coal field
use Bokaro

Bokaro Coalfield
use Bokaro

Bokaro Field
use Bokaro

Bolangir
Town and district in W.
BT Orissa
  India

Boleslawiec
Town is SW part of country.
BT Wroclaw
  Poland

Boliden
Village in NE.
BT Vasterbotten
  Sweden

Bolinas Lagoon
Off Bolinas Bay in SW Marin County.
BT California
  United States

Bolivar
State in SE.
BT Venezuela
NT Cartagena
  Imataca Complex
  Magdalena Delta

Bolivia
Includes use on level 1 as an area term (list O). For term set options see list B.
CO S230000S093000
  W0583000W0693000
BA South America
NT Cochabamba
  La Paz
  Oruro
NX Sucre
SA Amazon Basin
  Andes
  Chaco
  Eastern Cordillera
  Lake Titicaca
  Magdalena
  Santa Cruz

Bolivina
Genus. Includes use on level 3 under foraminifera(1) Buliminacea(2).
BA Buliminacea
  Rotaliina
  foraminifera
BT Invertebrata

Bolivinitidae
Family. Includes use on level 3 under foraminifera(2) Buliminacea(2).
BA Buliminacea
  Rotaliina
  foraminifera
BT Invertebrata

Bolivinoides
Genus. Includes use on level 3 under foraminifera(1) Buliminacea(2).
BA Buliminacea
  Rotaliina
  foraminifera
BT Invertebrata

Bologna
City and province in E central.
BT Emilia-Romagna
  Italy

Bolshezemelskaya Tundra
In Nenets National Okrug of Arkhangelsk Oblast. Also search Bol'shezemel'skaya Tundra.
BT Russian Republic
  USSR

Boltyshka Depression
Central Ukraine. Also search Boltyshka.
BT Ukraine
  USSR

Bolzano
City and province in N central.
BT Trentino-Alto Adige
  Italy

Bombay
City on the Arabian Sea.
BT Maharashtra
  India

Bonaire
Island off coast of Venezuela, 30 miles E of Curacao.
BT Netherlands Antilles
  Caribbean region

Bonaparte Basin
use Bonaparte Gulf basin

Bonaparte Gulf basin
N of Joseph Bonaparte Gulf and W of Northern Territory. Also search Bonaparte Basin.
UF Bonaparte Basin
BT Timor Sea
  Indian Ocean
SA Joseph Bonaparte Gulf

Bond albedo
use albedo

bonding
As of 1978, term is used on level 2 under crystal chemistry(1) and crystal structure(1).
SA cell dimensions
  crystal chemistry
  crystal structure
  order-disorder

Bondoc Peninsula
Between Ragay Gulf and Mompog Pass of S Quezon Province in S.
BT Luzon
  Philippine Islands

Bone Valley Formation
Unconformably overlain by Pleistocene terrace deposits ranging in age from Sunderland to Pamlico. S central Florida.
BT middle Pliocene
  Pliocene
SA Florida

bones
Includes use on level 3 under fossil groups(1) for type of material. See list F.
SA fossils
  jaws
  osteology
  paleontology
  skeletons
  skulls
  teeth
  Vertebrata

Bong Range
In N part of country.
BT Liberia

Bonin Islands
Group of 27 volcanic islands 600 miles S of Tokyo. Part of Tokyo Prefecture. Also search Ogasawara Islands.
UF Ogasawara Islands
BT Pacific Ocean

Bonner Formation
In Missoula Group. Precambrian (Belt series): In W Montana.
UF Bonner Quartzite
BT Precambrian
SA Missoula Group
  Montana

Bonner Quartzite
use Bonner Formation

Bonneterre Dolomite
use Bonneterre Formation

Bonneterre Formation
Includes Tom Sank limestone member at base. Also search Bonneterre Dolomite.
UF Bonneterre Dolomite
BT Upper Cambrian
  Cambrian
SA Missouri

Bonneville Salt Flats
Tooele County in NW.
BT Utah
  United States

book reviews
Includes use on level 1 for special indexes. Term set options are: topic [List B, extraterrestrial geology, general, geophysics]
  title
SA catalogs
  review
  textbooks

bookstone
use schist

Boone County
Index states as applicable.
BX Arkansas
  Illinois
  Indiana
  Iowa
  Kentucky
  Missouri
  Nebraska
  West Virginia
BT United States

Boothia Felix Peninsula
use Boothia Peninsula

Boothia Peninsula
N extremity of mainland Canada W of Gulf of Boothia in District of Franklin.
CO N690000N720000
  W0920000W0963000
UF Boothia Felix Peninsula
BT Northwest Territories
  Canada

Boquira Mine
Also search Boquira.
BT Bahia
  Brazil

Bor
Town in E.
BT Serbia
  Yugoslavia

boracite
BZ borates
  halides
BT minerals
SA chlorides

borates
Includes use on level 2 under minerals(1); on level 3 as a commodity term (list C) under boron(1); on level 3 under sediments(1) chemically precipitated sediments(2). See list L (minerals), list N (sediments).
BT minerals
NZ boracite
NA borax
  colemanite
NZ fluoborite
NA hulsite
  hydroboracite
  kaliborite
  kernite
  kotoite
  kurchatovite
  kurnakovite
  ludwigite
  preobrazhenskite
NZ sakhaite
NA solongoite
  suanite
  szaibelyite
  ulexite
  vimsite
NZ voltaite
NA vonsenite
SA chemically precipitated sediments
  evaporites

borax
BA borates
BT minerals

Borax Lake
BT California
  United States

Borborema
A plateau in NE part of country. Index states as applicable.
BT Brazil
SA Paraiba
  Rio Grande do Norte

Bordeaux
City in SW part of country.
BT Gironde
  France

Bordelais
Region in SW part of country.
BT Gironde
  France

Borden Group
In Kentucky. Comprises New Providence Formation, Brodhead Formation, Floyds Knob Formation, and Muldvaugh Formation. In S Indiana, comprises New Providence Shale, Locust Point, Carwood, Floyds Knob, and Edwardsville formations. S Indiana and E central Kentucky.
BT Mississippian
SA Indiana
  Kentucky
  Lower Mississippian
  Upper Mississippian

borderland, continental
use continental borderland

bore
use boreholes

Boreal
Includes use on level 3 as an age unit. Used primarily in Europe for an interval of postglacial time following

the Preboreal and preceding the Atlantic, during which the inferred climate was warm and dry.
BA Holocene
Quaternary
BT Cenozoic
SA biogeography
boreal region
paleoclimatology
Preboreal

boreal environment
Term introduced in 1978. Includes use on level 3 under ecology(1). Before 1978, also search boreal. (Boreal is used as an age unit).
SA boreal region
ecology
paleoclimatology
paleoecology

boreal region
Includes use on level 3 as a biogeographic term.
SA biogeography
Boreal
boreal environment

boreholes
Includes use on level 3 under well-logging(1). Before 1976, also search drill holes.
UF bore
SA cores
cuttings
dipmeter logging
drilling
electrical logging
heat flow
well-logging
wells

borings
As of 1978, use is restricted to indicate worm borings under ichnofossils(1). For well-logging, use boreholes or cores.
UF worm borings
SA burrows
cores
ichnofossils

Borislav
City in Lvov Oblast in W.
BT Ukraine
USSR

Borkut Deposit
Region of mineral deposits in Transcarpathian Oblast in W Ukraine. Also search Borkut.
BT Ukraine
USSR

Borneo
An island, the S two-thirds of which is known as Kalimantan, Indonesia, with the remainder comprising the two states of East Malaysia and the British protected sultanate of Brunei. Index Brunei, Kalimantan, and the states of East Malaysia as applicable.
BT Malay Archipelago
NT Brunei
SA East Malaysia
Kalimantan
Sabah
Sarawak

Borneo-Java Shelf
use Sunda Shelf

Bornholm
Island in W Baltic Sea S of Sweden. Also search Bornholm Island.
UF Bornholm Island
BT Denmark

Bornholm Island
use Bornholm

bornite
UF erubescite
horseflesh ore
purple copper ore
BA sulfides
BT minerals
SA copper

boron
Includes use on level 1 and 2 as a commodity term (list C) and as a chemical element (list D).
UF B
SA elements

boronatrocalcite
use ulexite

borosilicates
BA silicates
BT minerals
SA axinite
dravite

Borrego Mountain
In S California.
BT California
United States

Borrowdale Volcanic Series
Llanvirn and ? Llandeilo Series. Underlain by Skiddaw Slates and overlain by the Coniston Limestone Group. Lake District in NW England.
BT Ordovician
SA England
Llandeilian
Llanvirnian

Borshchev
City in Ternopol Oblast in W Ukraine. Also search Borshchov.
UF Borshchov
Borszczow
BT Ukraine
USSR

Borshchov
use Borshchev

Borsod Basin
Coal basin in NW Hungary. Also search Borsod.
BT Hungary

Borszczow
use Borshchev

Borzeta
Boreholes in W Carpathians.
BT Poland

Borzsony Mountains
Near Slovakian border in NW Nograd, Hungary. Also search Boerzsoeny Mountains.
UF Boerzsoeny Mountains
BT Hungary

Bos
Genus. Includes use on level 3 under Mammalia(1) Artiodactyla(2).
BA Bovidae
Artiodactyla
Mammalia
BT Tetrapoda
Vertebrata
NA Bos primigenius

Bos primigenius
Includes use on level 3 under Mammalia(1) Artiodactyla(2).
BA Bos
Bovidae
Artiodactyla
Mammalia
BT Tetrapoda
Vertebrata

Bosano
Region in NW.
BT Sardinia
Italy

Boshchekul
Village in Pavlodar Oblast in NE.

BT Kazakhstan
USSR

Bosnia
Region. Northern part of Republic of Bosnia and Herzegovina.
BT Yugoslavia

Boso Peninsula
use Chiba Peninsula

Bosphorus
use Bosporus

Bosporus
Strait connecting Black Sea with Sea of Marmara. Also search Bosphorus.
UF Bosphorus
BT Turkey
Middle East

Bossier Formation
In Cotton Valley Group. S Arkansas, W Louisiana, and E Texas.
BT Upper Jurassic
Jurassic
SA Arkansas
Cotton Valley Group
Louisiana
Texas

Boston
City in Suffolk County on Massachusetts Bay.
BT Massachusetts
United States

bostonite
Includes use on level 3 under igneous rocks(1) syenite family(2). See list N.
BA syenite family
BT igneous rocks

Bosumtwi Crater
S central.
BT Ghana

botany, paleo-
use paleobotany

Bothnian Sea
use Gulf of Bothnia

Botryococcus
Includes use on level 3 under algae(1) Chlorophyta(2).
BA Chlorophyta
algae
BT Plantae

Botswana
Before 1966 was British protectorate of Bechuanaland. Includes use on level 1 as an area term (list O). For term set options see list B.
BA Africa
SA Dwyka Series
Kalahari Desert
Karroo System
Limpopo Basin
Zambezi Valley

bottom currents
Papers are usually out of scope for GeoRef. Includes use on level 3 under ocean circulation.
BT currents
SA ocean circulation

bottom features
Includes use on level 2 under ocean floors(1); on level 3 under marine geology(1) and oceanography(1). See list B.
UF features, bottom
SA benthonic taxa
marginal basins
marine geology
mid-ocean ridges
ocean floors
submarine canyons
trenches
troughs

bottom load
use bedload

bottom water
SA oceanography
water

Botucatu Formation
Also search Botucatu Sandstone.
UF Botucatu Sandstone
SA Brazil
Mato Grosso
Mesozoic
Paleozoic
Paraguay
Permian
Sao Paulo
South America
Triassic

Botucatu Sandstone
use Botucatu Formation

Bou Azzer
Locality in Marrakesh region in W central Morocco. Also search Bou-Azzer.
UF Bou-Azzer
BT Morocco

Bou-Azzer
use Bou Azzer

Bouches-du-Rhone
Department on the Mediterranean Sea.
CO N431500N435500
E0054500E0041000
BT France
NT Aix-en-Provence
Carry-le-Rouet
Marseilles
Rhone Delta
Sainte-Victoire Mountain
SA Provence

boudinage
Includes use on level 3 under sedimentary structures(1) soft sediment deformation(2); on level 3 under deformation(1), and lineation(1) style(2); on level 3 under structural analysis(1). See list K.
UF pull apart structures
sausage structure
BA soft sediment deformation
sedimentary structures
SA boudins
deformation
lineation
melange
structural analysis

boudins
Includes use on level 3 under deformation(1), under lineation(1) style(2), and under structural analysis(1).
SA boudinage
deformation
lineation
melange
structural analysis

Bougainville
Largest of the Solomon Islands. ESE of the Bismarck Archipelago in the SW Pacific.
BT Papua New Guinea
Australasia
SA Solomon Islands

Bouguer
A valid term through 1976. After 1976, use Bouguer anomalies.

Bouguer anomalies
Includes use on level 3 under geophysical methods(1) and geophysical surveys(1). Also search Bouguer.
BT gravity anomalies
SA anomalies
free-air anomalies
geophysical methods

geophysical surveys

boulangerite
BA sulfantimonites
sulfosalts
BT minerals

Boulder
City in Boulder County in N central.
BT Colorado
United States

Boulder Batholith
BT Montana
United States

Boulder County
N central.
CO N395000N402000
W1050500W1053500
BT Colorado
United States

boulders
Includes use on level 3 under sediments(1) clastic sediments(2); on level 3 under geomorphology(1), glacial geology(1). Included use as sedimentary rock until 1977.
UF bowlder
BA clastic sediments
BT sediments
SA cobbles
erratics
geomorphology
glacial geology
gravel
rock glaciers
sedimentary rocks
terrigenous materials

Bouleiceras
Genus. Includes use on level 3 under Mollusca(1) Cephalopoda(2).
BA Ammonoidea
Cephalopoda
Mollusca
BT Invertebrata

Boulogne
City on the English Channel. Also search Boulogne-sur-Mer.
UF Boulogne-sur-Mer
BT Pas-de-Calais
France

Boulogne-sur-Mer
use Boulogne

Boulonnais
Old district in extreme N part of country.
BT Pas-de-Calais
France

boundary
Used for stratigraphic boundaries only.
SA stratigraphy
stratotypes

boundary interactions
Includes use on level 2 under meteorology(1).
UF interactions, boundary
SA atmosphere
meteorology

boundary layer
Includes use on level 2 under ocean circulation(1). Before 1976, also search boundary layer processes.
UF layer, boundary
SA ocean circulation
oceanography
turbulence

boundary layer processes
A valid term through 1975. After 1975, use boundary layer.

boundstone
BA carbonate rocks
BT sedimentary rocks

Bourges
City in central part of country.
BT Cher
France

Bourgogne
use Burgundy

bournonite
UF berthonite
endellionite
wheel ore
BA sulfantimonites
sulfosalts
BT minerals

Bouvet Island
Norwegian island 1600 miles SSW of Cape of Good Hope.
BT Atlantic Ocean

Bouxwiller
Town in extreme NE part of country.
BT Bas-Rhin
France

Bovidae
Family. Includes use on level 3 under Mammalia(1) Artiodactyla(2).
BA Ruminantia
Artiodactyla
Mammalia
BT Tetrapoda
Vertebrata
NA Bison
Bos

Bow River Valley
In S Alberta.
BT Alberta
Canada

bow shock waves
Includes use on level 2 under magnetosphere(1). Before 1976, also search bow shock.
SA intensity
magnetopause
magnetosheath
magnetosphere
shock waves
waves

Bowen
Town on the Coral Sea in the S Great Barrier Reef area.
BT Queensland
Australia

Bowen Basin
In E Queensland. Also search Bowen.
CO S250000S210000
E1500000E1480000
BT Queensland
Australia

Bowers Ridge
N of the Aleutian Islands.
BT Bering Sea
Pacific Ocean

Bowie Seamount
Just W of the Queen Charlotte Islands off central British Columbia.
BT Pacific Ocean

bowlder
use boulders

Bowser Formation
In Tuxedni Group, Middle(?) and Upper Jurassic: Central S Alaska.
BT Jurassic
SA Alaska
Middle Jurassic
Upper Jurassic

Box Elder County
Extreme NW.
BT Utah
United States

Boza Wola
Region in S.
BT Poland

Br
use bromine

Braarudosphaeridae
BA nannofossils
algae
BT Plantae

Brabant
Province in central.
BT Belgium
NT Louvain
University of Louvain

Brabant Massif
Also search Brabant.
BT Belgium

Brachiopoda
Includes use on level 1 and 2 as a fossil term (list F).
BT Invertebrata
NA Articulata
Inarticulata

Brachyura
Suborder. Includes use on level 3 under Arthropoda(1) Crustacea(2).
BA Decapoda
Crustacea
Arthropoda
BT Invertebrata

brackish water
Includes use on level 3 under ecology(1). Before 1978, also search brackish.
SA brackish-water environment
ecology
paleosalinity
salinity
salt water
sea water
water

brackish-water environment
Term introduced in 1978. Includes use on level 3 under ecology(1) name of fossil group(2). Before 1978, also search brackish or brackish water.
SA brackish water
ecology
environment
salinity
salt marshes
swamps
tidal flats

Brad
Town in Hunedoara County in SW.
BT Transylvania
Romania

Bradyodonti
Includes use on level 3 under Pisces(1) Chondrichthyes(2).
BA Elasmobranchii
Chondrichthyes
Pisces
BT Vertebrata

Braganca
District in NE Portugal. Also a city.
BT Portugal
SA Tras-os-Montes

Brahmaputra River
Rises in the Kailas Range of SW Tibet; it flows E across S Tibet, then S, SW, and again S through Bangladesh where it becomes the Jamuna River which merges with the Ganges River in the Ganges-Brhamaputra Delta on the Bay of Bengal. Index Tibet and countries as applcable.
BT Asia
SA Bangladesh
China
India
Tibet

braided streams
Includes use on level 3 under geomorphology(1) fluvial features(2). Also search braided channels.
UF anastomosing streams
BA streams
SA channels
fluvial features
geomorphology
sedimentation

Braila
City and county in NE.
BT Walachia
Romania

Bramsche
Town in SW.
BT Lower Saxony
West Germany

Branchiopoda
Class. Includes use on level 2 under Arthropoda(1). See list F.
BA Crustacea
Arthropoda
BT Invertebrata
NA Conchostraca

Brandenburg
Region and former state in central East Germany. Also a town on the Havel River.
BT East Germany

Brandenburg Stade
Europe. Term introduced in 1978. Before 1978, also search Brandenburg.
BA Weichselian
upper Pleistocene
Pleistocene
Quaternary
BT Cenozoic

Brandon
Region in Warwick County in central.
BT England
Great Britain
United Kingdom

Brandon Stade
Europe. Term introduced in 1978. Before 1978, also search Brandon.
BA Weichselian
upper Pleistocene
Pleistocene
Quaternary
BT Cenozoic

brannerite
BA oxides
BT minerals

Brasov
City and county in SE Transylvania. Also search Stalin.
UF Stalin
BT Transylvania
Romania

Brassfield Formation
Considered to consist of lower dolomitic limestone, including Whitfieldella bed, and Plum Creek Clay Member above. In central N Arkansas, S Indiana, SW Illinois, central Kentucky, SW Ohio, and S Tennessee. Also search Brassfield Limestone.
BT Lower Silurian
Silurian
SA Arkansas
Illinois
Indiana
Kentucky
Ohio
Tennessee

Bratislava
City on the Danube River in SW.
BT Slovakia
Czechoslovakia

braunite
    BA   orthosilicates
           silicates
    BT   minerals

Bravais lattice
    use lattice

bravoite
    BA   sulfides
    BT   minerals
    SA   pyrite

Brazil
    Includes use on level 1 as an area term (list O). For term set options see list B.
    CO   S340000N051500 W0340000W0740000
    BA   South America
    NT   Alagoas
           Amapa
           Bahia
           Borborema
           Brazilian Shield
           Ceara
           Espirito Santo
           Goias
           Guanabara
           Maranhao
           Mato Grosso
           Minas Gerais
           Para
           Paraiba
           Parana
           Parnaiba Basin
           Pelotas Basin
           Pernambuco
           Piaui
           Recife-Joao Pessoa
           Rio de Janeiro
           Rio Grande do Norte
           Rio Grande do Sul
           Rondonia
           Santa Catarina
           Sao Francisco Basin
           Sao Paulo
           Sergipe
           Sergipe-Alagoas Basin
           Serra do Mar
           Tocantins River region
    SA   Acungui Group
           Amazon Basin
           Amazon River
           Amazonas
           Areado Formation
           Bambui Group
           Barreiras Formation
           Bauru Formation
           Botucatu Formation
           Estrada Nova Formation
           Guiana Basin
           Guyana Shield
           Irati Formation
           Minas Series
           Natal
           Parana Basin
           Parana River
           Passa Dois Group
           Pirabas Formation
           Rio Bonito Formation
           Roraima Formation
           Santana Formation
           Serra Gerral Formation
           Tubarao Group
           Vitoria

Brazilian Shield
    The Brazilian uplands of N central, E central, and S. Index states as applicable.
    UF   Amazonian Shield
    BT   Brazil
    SA   Bahia
           Goias
           Maranhao
           Mato Grosso
           Minas Gerais
           Para
           Parana
           Santa Catarina
           Sao Paulo

Brazos River
    Central Texas.
    BT   Texas
           United States

Brazzaville
    City on the Congo River across from Kinshasa.
    BT   Congo

breakers
    use breaking waves

breaking strength
    use fracture strength

breaking waves
    Term introduced in 1978. Includes use on level 2 under ocean waves(1). Before 1978, also search waves AND breaking; breakers.
    UF   breakers
    BT   ocean waves
    SA   waves

breakwaters
    UF   water-break
    BT   marine installations
    SA   engineering geology
           shorelines

Breathitt Formation
    In Pottsville Group. In SE Kentucky.
    BT   Middle Pennsylvanian
           Pennsylvanian
    SA   Kentucky
           Pottsville Group

breccia
    Includes use on level 3 under sedimentary rocks(1) clastic rocks(2); on level 3 under faults(1) effects(2), and under stratigraphy(1), structural analysis(1), and Moon(1). See list I.
    UF   rubblerock
    BA   clastic rocks
    BT   sedimentary rocks
    SA   agglomerate
           breccia pipes
           brecciation
           conglomerate
           faults
           gouge
           gravel
           meteor craters
           microbreccia
           Moon
           sediments
           shear zones
           slickensides
           structural analysis
           suevite
           tectonics
           terrigenous materials

breccia pipes
    Includes use on level 3.
    BT   pipes
    SA   breccia
           diatremes
           intrusions
           plugs
           volcanic rocks
           volcanism
           volcanoes

brecciation
    SA   breccia
           processes

Brecknock (shire)
    use Brecknockshire

Brecknockshire
    County in SE Wales. Also search Brecknock.
    UF   Brecknock (shire)
           Breconshire
    BT   Wales
           Great Britain
           United Kingdom

Breconshire
    use Brecknockshire

Breidamerkurjokull
    Glacier.
    BT   Iceland

Brenham
    City in Washington County in SE central.
    BT   Texas
           United States

Brent Crater
    Meteor crater at Brent in Algonquin Provincial Park E of Georgian Bay.
    BT   Ontario
           Canada

Brescia
    City and province in E.
    BT   Lombardy
           Italy

Breslau
    use Wroclaw

Brest
    City on the Atlantic Ocean at W end of Brittany in Finistere, and a city in SW Byelorussia. Index Byelorussia and/or France.
    BT   Europe
    SA   Byelorussia
           France

Brest Basin
    Named after city of Brest in SW Byelorussia. Index Poland and Soviet republics as applicable. Also search Brest.
    BT   Europe
    SA   Byelorussia
           Poland
           Ukraine

Brevard fault zone
    use Brevard Zone

Brevard Zone
    Index states as applicable. Also search Brevard fault zone.
    UF   Brevard fault zone
    BT   Appalachians
    SA   Alabama
           Georgia
           North Carolina

Brewster County
    In SW Texas.
    BT   Texas
           United States

Brezno
    Town in central.
    BT   Slovakia
           Czechoslovakia

Briancon
    Town in SE part of country.
    BT   Hautes-Alpes
           France

Briancon Zone
    use Brianconnais Zone

Brianconnais Zone
    SE France and NW Italy. Index countries as applicable. Also search Brianconnaise Zone.
    UF   Briancon Zone
           Brianconnaise Zone
    BT   Europe
    SA   Alps
           Cretaceous
           Eocene
           France
           Italy
           Jurassic
           Mesozoic
           Paleocene
           Permian
           Tertiary

Brianconnaise Zone
    use Brianconnais Zone

briartite
    BA   sulfides
    BT   minerals

brick clays
    Not a valid term for GeoRef. Use clays.

Bridgerian
    North America.
    BA   Eocene
           Paleogene
           Tertiary
    BT   Cenozoic

bridges
    As of 1978, term is used on level 2 under foundations(1).
    SA   engineering geology
           foundations
           geomorphology
           highways

brief report
    A valid term through 1978. After 1978, use report, e.g. under associations(1) or symposia(1).

Brindisi
    City and province on the S Adriatic Sea.
    BT   Apulia
           Italy

brines
    Includes use on level 1 and 2 as a commodity term (list C).
    SA   bromine
           desalinization
           iodine
           salt
           salt water
           salt-water intrusion
           sea water

Brioverian
    Europe.
    BA   upper Proterozoic
           Proterozoic
           Precambrian

Brisbane
    City in extreme SE.
    BT   Queensland
           Australia

Bristol
    City on the Bristol Channel.
    BT   England
           Great Britain
           United Kingdom

Bristol Bay
    Between SW Alaskan mainland and Alaska Peninsula.
    BT   Bering Sea
           Pacific Ocean

Bristol Channel
    Arm of Atlantic Ocean. Index England and/or Wales.
    BT   United Kingdom
    SA   England
           Wales

britholite
    BZ   phosphates
           silicates
    BT   minerals

British Columbia
    Includes use on level 1 as an area term (list O). For term set options see list B.
    CO   N490000N600000 W1140000W1390000
    BA   Canada
    NT   Albert Canyon
           Alice Arm
           Atlin
           Bluebell Mine
           Cariboo Mountains
           Copper Mountain
           Crowsnest Pass
           Dogtooth Mountains

Endako
Esplanade Range
Fraser River
Guichon Creek Batholith
Harrison Lake
Hope
Kamloops
Kootenay Lake
Manning Park
Omineca Mountains
Pinchi Lake
Queen Charlotte Islands
Quesnel Lake
Revelstoke
Rossland
Saanich Inlet
Salmo
Selkirk Mountains
Shuswap Complex
Sifton Basin
Skeena Mountains
Slocan mining camp
Smithers
Sustut Basin
Tofino Basin
Topley Intrusions
Tulameen coal area
Vancouver Island
SA Altyn Limestone
Belt Supergroup
Burgess Shale
Canadian Cordillera
Coast Mountains
Columbia River
Columbia River basin
Cordillera
Juan de Fuca Strait
Kicking Horse River valley
Kootenay Formation
Lardeau Group
Liard River
Miette Group
Nahanni Formation
Nicola Group
Okanagan Valley
Pasayten Group
Peace River
Purcell System
Rocky Mountain Trench
Skagit Valley
Strait of Georgia
Vancouver
Windermere System

British Guiana
use Guyana

British Honduras
Not a valid index term for GeoRef since 1974. Use Belize.

British North Borneo
use Sabah

British Virgin Islands
Term introduced in 1978 for Virgin Islands excluding those of the United States. Before 1978, also search Virgin Islands for British Virgin Islands.
BT West Indies
SA Anegada Island
Lesser Antilles
United Kingdom
Virgin Islands

Brittany
Region and former province occupying peninsula between English Channel and Bay of Biscay.
BT France

Brive
Basin in S central part of country. Also search Brive-la-Gaillarde.
UF Brive-la-Gaillarde
BT Correze
France

Brive-la-Gaillarde
use Brive

Brno
City in S central.
BT Moravia
Czechoslovakia

Broach
City in S.
BT Gujarat
India

Broadlands
Geothermal field.
BT New Zealand

Broadwater County
SW central.
BT Montana
United States

brochantite
BA sulfates
BT minerals

Brocken Massif
In Harz Mountains in SE.
BT East Germany

Brockman Iron Formation
Wittenoom Gorge area in NW Western Australia.
BT Precambrian
SA Australia
Western Australia

Broegger Peninsula
use Brogger Peninsula

Brogger Peninsula
On the Island of Spitsbergen. Also search Broggerhalvoya.
UF Broegger Peninsula
Broggerhalvoya
BT Spitsbergen
Arctic region

Broggerhalvoya
use Brogger Peninsula

Broken Bay
Inlet of Pacific Ocean.
BT New South Wales
Australia

Broken Hill
City in W.
BT New South Wales
Australia

Broken River Formation
NE Queensland. Also search Broken River Group.
BT Paleozoic
SA Australia
Devonian
Queensland
Silurian

bromellite
BA oxides
BT minerals

bromine
Includes use on level 1 and 2 as a commodity term (list C) and as a chemical element (list D).
UF Br
SA brines
elements
salt

bromlite
use alstonite

Bronson Hill Anticlinorium
BT New Hampshire
United States

bronzite
BA pyroxene group
chain silicates
silicates
BT minerals
SA enstatite

bronzitite
Includes use on level 3 under igneous rocks(1) ultramafic family(2). See list H.
BA ultramafic family
BT igneous rocks
SA pyroxenite

brookite
BA oxides
BT minerals
SA anatase
rutile

Brooks Range
Between Yukon River and Arctic Ocean in N central.
BT Alaska
United States
SA Rocky Mountains

brown algae
use Phaeophyta

brown coal
Includes use on level 3 as commodity term under lignite(1) and on level 3 under sedimentary rocks(1) organic residues(2). See list C and list I.
BA organic residues
BT sedimentary rocks
SA coal
lignite
peat

Brown County
Index states as applicable.
BX Indiana
Kansas
BT United States

Brown forest soils
Includes use on level 3 under soils(1). See list M.
BA soils
SA Intrazonal soils
soil group

brown mica
use phlogopite

Brown soils
Includes use on level 3 under soils(1). See list M.
BA soils
SA Zonal soils

brucite
BA oxides
BT minerals

brucite marble
Includes use on level 3 under metamorphic rocks(1) marbles(2). See list J.
BA marbles
BT metamorphic rocks

Bruderheim
Village NE of Edmonton in central Alberta.
UF Bruederheim
BT Alberta
Canada

Bruederheim
use Bruderheim

Brule Formation
In White River Group. In NE Colorado, W Nebraska, W South Dakota, and E Wyoming.
BT Oligocene
SA Colorado
middle Oligocene
Nebraska
South Dakota
upper Oligocene
White River Group
Wyoming

Brunei
A sultanate under British protection on the South China Sea in N.
BT Borneo
Malay Archipelago
SA East Malaysia

Brunhes Epoch
Geomagnetic epoch.
UF Brunhes Normal
BA upper Pleistocene
Pleistocene
Quaternary
BT Cenozoic

Brunhes Normal
use Brunhes Epoch

Brunswick
City in Lower Saxony, West Germany, and in Glynn County, Georgia. Index Georgia and/or West Germany as applicable.
SA Georgia
West Germany

brushite
BA phosphates
BT minerals

Bryansk
City and oblast SW of Moscow.
BT Russian Republic
USSR

bryophytes
Mosses. Includes use on level 1 and 2 as a fossil term (list F).
UF mosses
BT Plantae
NA Hepaticae
Musci
SA paleobotany
thallophytes

Bryozoa
Includes use on level 1 and 2 as a fossil term (list F).
UF Polyzoa
BT Invertebrata
NA Cheilostomata
Cryptostomata
Ctenostomata
Cyclostomata
Cystoporata
Ectoprocta
Trepostomata
SA acanthopores

Bucegi Mountains
Group in the E Transylvanian Alps in S central.
BT Romania
SA Transylvanian Alps

Buchanan County
Index states as applicable.
BX Iowa
Missouri
Virginia
BT United States

Bucharest
City and administrative district in S central.
BT Walachia
Romania

Buckingham County
S central.
BT Virginia
United States

Buckner Formation
Subsurface. Overlies Smackover Limestone; underlies Cotton Valley Formation.
BT Upper Jurassic
Jurassic
SA Arkansas
Louisiana
Mississippi
Texas

Budapest
City on both sides of the Danube River in N central part of country.
BT Hungary

Budennovsk
use Prikumsk

Buenos Aires
City on the Rio de la Plata.
CO S410000S330000

W0570000W0640000
BT Federal District
    Argentina

Buenos Aires Province
In E part of country.
BT Argentina
NT Sauce Grande River valley

buergerite
BA ring silicates
BT minerals

Buffalo River
E of the Tennessee River in W central.
BT Tennessee
    United States

Bug region
Area including both the Southern Bug and the Western Bug rivers in SW European USSR and E Poland. Index Poland and Soviet republics as applicable.
BT Europe
SA Byelorussia
    Poland
    Ukraine

Bug River
The Southern Bug River in S Ukraine and the Western Bug River of the Ukraine, Byelorussia and E Poland. Index Poland and Soviet republics as applicable.
BT Europe
SA Byelorussia
    Poland
    Ukraine

building stone
Includes use as level 3 commodity term under construction materials(1), granite(1), sandstone(1), limestone(1) and marble(1). See list C.
UF stone, building
BA construction materials
SA dimension stone
    granite
    limestone
    marble
    sandstone

buildings
As of 1978, term is used on level 2 under foundations(1).
SA foundations
    structures

Bukantau
Mountains SE of Aral Sea in N.
BT Uzbekistan
    USSR

Bukhara
City and oblast in S near border of Turkmenia.
BT Uzbekistan
    USSR

Bukhara-Khiva
A region comprising neighboring former Khanates SE of the Aral Sea.
BT Uzbekistan
    USSR

Bukk Mountains
Southern spur of the Carpathians in NE.
BT Hungary
SA Carpathians

Bukovina
Region in E central Europe. Index Ukraine and/or Romania as applicable.
BT Europe
SA Romania
    Ukraine

Bukusu
Carbonatite complex.
BT Uganda

Bulawayan Group
Strong unconformities separate it from the overlying and underlying systems (Shamvaian and Sebakwian).
BT lower Precambrian
    Precambrian
SA Rhodesia

Bulgaria
Includes use on level 1 as an area term (list O). For term set options see list B.
CO N413000N441500
   E0284500E0221500
BA Europe
NT Balkan Foreland
   Balkan Mountains
   Beli Izvor
   Burgas
   Glavatsi
   Isker River
   Khaskovo
   Panagyurishte
   Pernik coal basin
   Pleven
   Plovdiv
   Rila Mountains
   Ruse
   Sakar Mountains
   Sofia
   Sredna Gora
   Varna
   Vidin
   Vratsa
SA Balkan Peninsula
   Black Sea region
   Danube Plain
   Danube River
   Danube Valley
   Dobruja
   Dobruja Basin
   Istranca Mountains
   Krajiste
   Macedonia
   Maritsa River
   Moesia
   Moesian Platform
   Osogovo Mountains
   Rhodope Mountains
   Struma River valley

Bulimina
Genus. Includes use on level 3 under foraminifera(1) Buliminacea(2).
BA Buliminacea
   Rotaliina
   foraminifera
BT Invertebrata

Buliminacea
Includes use on level 2 under foraminifera(1). See list F.
BA Rotaliina
   foraminifera
BT Invertebrata
NA Bolivina
   Bolivinitidae
   Bolivinoides
   Bulimina
   Gabonella
   Uvigerina
   Uvigerinidae
SA Protista

bulk modulus
UF incompressibility modulus
   modulus, bulk
   modulus of incompressibility
   volume elasticity
BA elastic constants
SA deformation
   elasticity
   shear modulus

Bulli Seam
Top of the Illawarra Coal Measures is marked by the Bulli Coal Seam. Coal seam in Sydney Basin in E New South Wales.
BT Permian
SA Australia
   Illawarra Coal Measures
   New South Wales

Buncombe County
In W North Carolina.
BT North Carolina
   United States

Bundelkhand
Region S of the Jumna River. Index states as applicable.
BT India
SA Madhya Pradesh
   Uttar Pradesh

Bundenbach
Area in Westerwald region in W.
BT Hesse
   West Germany

Bungonia Caves
In SE New South Wales.
BT New South Wales
   Australia

Bunol
Town in E part of country.
BT Valencia
   Spain

Bunter
Europe. Above Permian, below Muschelkalk. Includes use on level 3 under age terms(1). See list E. Also search Buntsandstein; Bunter Sandstone.
UF Bunter Sandstone
   Buntsandstein
BA Lower Triassic
   Triassic
BT Mesozoic
NT Voltzia Sandstone

Bunter Sandstone
use Bunter

Buntsandstein
use Bunter

Burbank Sand
Subsurface. Osage and Kay counties in N Oklahoma. Also search Burbank Sands.
UF Burbank Sands
BT Carboniferous
SA Kansas
   Oklahoma

Burbank Sands
use Burbank Sand

Burdekin Delta
On Upstart Bay on Pacific Ocean in Great Barrier Reef area in E.
BT Queensland
   Australia

Burdekin River
Flows into Upstart Bay on Pacific Ocean in Great Barrier Reef area in E.
BT Queensland
   Australia

Burdigalian
Europe. Above Aquitanian, below Helvetian. Includes use on level 3 under age terms(1). See list E.
BA lower Miocene
   Miocene
   Neogene
   Tertiary
BT Cenozoic

Burdwan
A city, district and division in central.
BT West Bengal
   India

Bureya
Town in SE Amur Oblast near the Manchurian border.
CO N480000N520000
   E1333000E1260000
BT Russian Republic
   USSR

Burgas
Province on the Black Sea in SE Bulgaria. Also a city.
BT Bulgaria

Burgenland
State on the Hungarian border in E.
BT Austria

Burgess Shale
BT Middle Cambrian
   Cambrian
SA British Columbia
   Canada

Burgos
Province in N Spain. Also a city.
BT Spain

Burgundy
Region and former province of E central and E France. Also search Bourgogne.
UF Bourgogne
BT France

Buriadia
Includes use on level 3 under gymnosperms(1) Coniferales(2).
BA Coniferales
   gymnosperms
BT Plantae

burial metamorphism
Term introduced in 1978. Includes use on level 2 under metamorphism(1). Before 1978, also search metamorphism AND burial.
BA metamorphism
SA geosynclines
   regional metamorphism

buried channels
Includes use on level 3 under geomorphology(1) fluvial features(2).
SA buried features
   buried valleys
   channels
   drift
   fluvial features
   geomorphology
   glacial geology
   paleogeography
   planar bedding structures

buried features
Term introduced in 1978. Before 1978, search buried.
UF features, buried
SA buried channels
   buried valleys

buried valleys
Includes use on level 3 under geomorphology(1), paleogeography(1), and under glacial geology(1).
SA buried channels
   buried features
   drift
   geomorphology
   glacial geology
   paleogeography
   valleys

Burke County
Index states as applicable.
BX Georgia
   North Carolina
   North Dakota
BT United States

Burleson County
SE central.
BT Texas
   United States

Burma
Includes use on level 1 as an area term (list O). For term set options see list B.
CO N090000N284500

E1014500E0920000
BA Asia
NT Shan State
SA Indochina
  Kama
  Phuket Group
  Siwalik System

Burnet County
Central.
BT Texas
  United States

Burnie
Town on Bass Strait in N.
BT Tasmania
  Australia

Burono
use Vourinos

Burro Mountain
BT California
  United States

burrows
Includes use on level 3 under sedimentary structures(1) biogenic structures(2). See list K.
BA biogenic structures
  sedimentary structures
SA bioturbation
  borings
  ichnofossils
  lebensspuren

bursts, rock
use rock bursts

Burundi
Formerly Urundi which was part of Belgian trust territory of Uranda-Urundi. Includes use on level 1 as an area term (list O). For term set options see list B.
BA Africa
SA East African Rift
  Lake Tanganyika
  Nile River

Buryat
Buryat Autonomous Soviet Socialist Republic. S and SE of Lake Baikal. Also search Buryatia, Buryat-Mongol, and Buryat-Mongolia.
UF Buryat-Mongol
  Buryat-Mongolia
  Buryatia
BT Russian Republic
  USSR

Buryat-Mongol
use Buryat

Buryat-Mongolia
use Buryat

Buryatia
use Buryat

Bushveld Complex
Igneous complex in E Transvaal. Also search Bushveld.
BT Transvaal
  South Africa

bustamite
BA chain silicates
  silicates
BT minerals

Butler County
Index states applicable.
BX Alabama
  Iowa
  Kansas
  Kentucky
  Missouri
  Nebraska
  Ohio
  Pennsylvania
BT United States

Butte
City in Silver Bow County in SW.
BT Montana

United States

Butte County
Index states as applicable.
BX California
  Idaho
  South Dakota
BT United States

Butte District
Mining district in the Butte area in SW Montana. Also search Butte.
BT Montana
  United States

Buzau
City and county in NE.
BT Walachia
  Romania

Buzau River
In NE Walachia.
BT Walachia
  Romania

Buzzards Bay
Inlet of Atlantic Ocean W of Cape Cod.
BT Massachusetts
  United States

Bydgoszcz
Province in N central Poland. Also a city.
BT Poland
NT Dobrzyn
  Grudziadz
  Inowroclaw
  Mogilno
  Torun

Byelorussia
Byelorussian Soviet Socialist Republic. Also search Belorussia.
CO N510000N560000
  E0330000E0240000
UF Belorussia
  White Russia
BT USSR
NT Byelorussian Massif
  Kletno
  Mezhrechye
  Minsk
  Mogilev
  Orsha
  Ostashkovichi
  Rechitsa
  Starobin
  Ustron
  Zhitkovichi
SA Brest
  Brest Basin
  Bug region
  Bug River
  Dnieper Basin
  Dnieper River
  Dvina River
  European USSR
  Neman River basin
  Polesye
  Pripet Basin
  Russian Plain
  Russian Platform
  Vyshkovo

Byelorussian Massif
S part of Lithuanian-Byelorussian Upland.
BT Byelorussia
  USSR

Byrd Station
In Marie Byrd Land SW of the Amundsen Sea on the Pacific Ocean side.
CO S795900S795900
  W1200100W1200100
BT Antarctica

Bystrzyca
River in SW part of country.
BT Wroclaw
  Poland

Bytkov
use Bitkov

Bytom
City in central.
BT Katowice
  Poland

bytownite
BA feldspar group
  framework silicates
  silicates
BT minerals
SA aluminosilicates
  plagioclase

C

C
use carbon

C-12
Includes use on level 3 under isotopes(1).
UF carbon-12
SA C-13/C-12
  carbon
  isotopes

C-12/C-13
use C-13/C-12

C-13
Includes use on level 3 under isotopes(1).
UF carbon-13
SA C-13/C-12
  carbon
  isotopes

C-13/C-12
Includes use on level 3 under isotopes(1). Also search C-12/C-13; C-12 AND C-13.
UF C-12/C-13
SA C-12
  C-13
  carbon
  geologic thermometry
  isotopes
  paleoclimatology
  stable isotopes

C-14
Includes use on level 3 under absolute age(1) methods(2). Before 1976, also search carbon-14.
SA absolute age
  carbon
  isotopes

Ca
use calcium

Cabaniss Formation
In Cherokee Group, SE Kansas, W Missouri, central and NE Oklahoma.
BT Desmoinesian
  Middle Pennsylvanian
  Pennsylvanian
SA Cherokee Group
  Kansas
  Missouri
  Oklahoma

Cabinda
A district and an enclave N of the mouth of the Congo River.
UF Kabinda
BT Angola

Cabo Ortegal
Cape in NW part of country in N.
BT La Coruna
  Spain

Cabo Rojo
Town in SW.
BT Puerto Rico
  West Indies

Cabrieres
Location in the Montagne Noire in S part of country.
BT Herault
  France

Caceres
Province in W on Portuguese border. Also a city.
BT Spain

Cache Creek
In NW California.
BT California
  United States

Cache La Poudre River
N central.
BT Colorado
  United States

Cache Valley
Index Idaho and/or Utah as applicable.
BT United States
SA Idaho
  Utah

cacoxenite
BA phosphates
BT minerals

Caddo County
SW of Oklahoma City.
BT Oklahoma
  United States

Cadiz
Province in SW on the Atlantic Ocean. Also a city.
BT Spain
NT Vejer de la Frontera
SA Andalusia
  Serrania de Ronda

cadmium
Includes use on level 1 and 2 as a chemical element (list D).
UF Cd
SA elements
  geochemistry
  heavy metals

cadmium blende
use greenockite

cadmium ocher
use greenockite

Cadomian Orogeny
Term introduced in 1978. Before 1978, search Cadomian AND orogeny.
SA Cambrian
  orogeny
  Precambrian

Caen
City in Normandy.
BT Calvados
  France

Caernarvon County
use Caernarvonshire

Caernarvonshire
County in NW Wales.
CO N524500N532000
  W0034500W0044500
UF Caernarvon County
BT Wales
  Great Britain
  United Kingdom

Cairngorm Mountains
Range of the Grampian Hills in highlands of central.
BT Scotland
  Great Britain
  United Kingdom

Cairo
Governorate in NE Egypt. Also a city on the Nile River.
BT Egypt

Caithness
County in extreme NE.
CO N581000N584000
W0030000W0035000
BT Scotland
Great Britain
United Kingdom

Calabria
An autonomous region in S Italy. The toe of the Italian boot.
CO N380000N401000
E0170000E0154500
BT Italy
NT Catanzaro
Cosenza
Reggio di Calabria
Sila Massif

Calabrian
Europe. Above Astian, below middle Pleistocene. Includes use on level 3 under age terms(1). See list E.
BA lower Pleistocene
Pleistocene
Quaternary
BT Cenozoic
SA Villafranchian

Calamites
Genus. Includes use on level 3 under pteridophytes(1) Sphenopsida(2).
BA Equisetales
Sphenopsida
pteridophytes
BT Plantae

Calatayud-Teruel Basin
NE Spain. Index provinces as applicable.
BT Spain
SA Saragossa
Teruel

Calaveras County
Central.
BT California
United States

Calaveras Fault
Central.
BT California
United States

calaverite
BA tellurides
sulfides
BT minerals
SA gold

calc-alkalic
A valid term through 1977. After 1977, use calc-alkalic composition(2) under igneous rocks(1).

calc-alkalic composition
Term introduced in 1978. Includes use on level 2 under igneous rocks(1).
SA alkalic composition
composition
igneous rocks

calc-schist
Includes use on level 3 under metamorphic rocks(1) schists(2). See list J.
BA schists
BT metamorphic rocks
SA schist

calc-sinter
use travertine

calc-tufa
use tufa

calcarenite
Includes use on level 3 under sedimentary rocks(1) carbonate rocks(2). See list I.
BA carbonate rocks
BT sedimentary rocks
SA biocalcarenite

limestone

calcareous
Valid index term through 1977. After 1977, use calcareous composition.

calcareous algae
Includes use on level 3 under algae(1). See list F.
BA algae
BT Plantae
SA stromatolites

calcareous clay
use marl

calcareous composition
Term introduced in 1978. Includes use on level 3 under sediments(1) and sedimentary rocks(1). Before 1978, also search calcareous AND specific sediment type or sedimentary rock.
SA bioherms
composition
corals
ooze
sedimentary rocks
sediments

calcareous nannofossils
use nannofossils

calcareous ooze
Not a valid term for GeoRef. Use calcareous composition and ooze.

calcareous sinter
use travertine

Calcareous soils
Before 1978, also search calcareous.
BA soils

calcareous tufa
use tufa

Calcasieu Parish
On the Texas border in SW.
BT Louisiana
United States

calcic composition
Term introduced in 1978. Includes use on level 2 under igneous rocks(1). Before 1978, also search igneous rocks AND calcic.
SA composition
igneous rocks

calcicrete
use calcrete

calcification
Term refers to fossils or soils.
SA diagenesis
fossils
processes
soils

calcilutite
Includes use on level 3 under sedimentary rocks(1) carbonate rocks(2). See list I.
BA carbonate rocks
BT sedimentary rocks
SA algal mounds
limestone

calciphyre
Includes use on level 3 under metamorphic rocks(1) marbles(2). See list J.
BA marbles
BT metamorphic rocks

Calcispherulidae
Family.
BA problematic fossils

Calcispongea
Includes use on level 2 under Porifera(1). See list F.
BA Porifera
BT Invertebrata
NA Sphinctozoa

calcite
Optical. Includes use on level 1 as a commodity term (list C).
BA carbonates
BT minerals
SA alabaster
aragonite
calcitization
chalk
dolomite
Iceland spar
light minerals
limestone
magnesian calcite
marble
tufa

calcitization
SA calcite
carbonatization
dedolomitization
diagenesis
dolomitization
processes

calcium
Includes use on level 1 and 2 as a chemical element (list D). Also search Ca.
UF Ca
SA elements

calcium carbonate
SA carbonates
geochemistry
lime

calcrete
Includes use on level 3 under sediments(1) chemically precipitated sediments(2); on level 3 under soils(1). See list N.
UF calcicrete
BA chemically precipitated sediments
BT sediments
SA caliche
conglomerate
duricrust
soils
weathering crust

calculation
A valid term through 1978. After 1978, see mathematical methods or mathematical models.

Calcutta
City on the Hooghly River near Bay of Bengal in S.
BT West Bengal
India

calderas
Includes use on level 3 under volcanology(1), and under geomorphology(1) volcanic features(2).
BT volcanic features
SA cauldrons
craters
geomorphology
volcanism
volcanoes
volcanology

Caledonia County
NE Vermont. Also search Caledonia.
BT Vermont
United States

Caledonian
A valid term through 1977. After 1977, use Caledonian Orogeny.

Caledonian Geosyncline
Term introduced in 1978. Before 1978, search geosynclines AND Caledonian.
SA Caledonian Orogeny
geosynclines

Caledonian Orogeny
Name commonly used for the early Paleozoic deformation in Europe which created an orogenic belt extending from Ireland and Scotland northwestward through Scandinavia. Before 1978, also search Caledonian.
BT Paleozoic
SA Caledonian Geosyncline
Caledonides
orogeny

Caledonides
Orogenic belt extending from Ireland and Scotland northwestward through Scandinavia, formed by the Caledonian Orogeny.
SA Caledonian Orogeny
Europe
Great Britain
Scandinavia

Calgary
City in SW.
BT Alberta
Canada

Calhoun County
Index states as applicable.
BX Alabama
Arkansas
Florida
Georgia
Illinois
Iowa
Michigan
Mississippi
South Carolina
Texas
West Virginia
BT United States

calibration
Used as a general term.
SA accuracy

caliche
Includes use on level 3 under sediments(1) chemically precipitated sediments(2); on level 3 under soils(1). See list N.
BA chemically precipitated sediments
BT sediments
SA calcrete
duricrust
soils
weathering crust

Calico Mountains
Small range in Mojave Desert.
BT California
United States

Caliente Range
BT California
United States

California
Includes use on level 1 as an area term (list O). For term set options see list B.
CO N323000N420000
W1141500W1243000
BA United States
NT Alameda County
Amador County
Anza Desert State Park
Bakersfield
Barstow
Bear Valley
Berkeley
Bodega Head
Bolinas Lagoon
Borax Lake
Borrego Mountain
Burro Mountain
NX Butte County
NT Cache Creek
Calaveras County
Calaveras Fault
Calico Mountains
Caliente Range
Calistoga

Cape Mendocino
Carmel
Carmel Bay
Cholame
Clear Lake
Coalinga
Colusa County
Contra Costa County
Coyote Creek Fault
Crestmore
Death Valley
Del Monte Beach
Del Norte County
Diablo Range
Eel River
El Centro
El Dorado County
Elk Hills Field
Elsinore Fault
NX Eureka
NT Fortuna
Fresno
Fresno County
Furnace Creek
Gabilan Range
Garlock Fault
Hayward
Hayward Fault
Hollister
NX Humboldt County
NT Imperial County
Imperial Valley
Inglewood
Inyo County
Inyo Mountains
Kern County
Kettleman Hills
La Jolla
NX Lake County
NT Lassen Peak
Lassen Volcanic National Park
Lompoc
Long Beach
Los Angeles
Los Angeles Basin
Los Angeles County
Madera County
Marin County
Mariposa County
Mendocino County
Merced County
Mill Creek
Modoc County
Modoc Plateau
Mojave Desert
Mono Basin
Mono County
Mono Craters
Mono Lake
Monterey
Monterey Bay
Monterey County
Mount Lyell
Napa County
Needles
NX Nevada County
NT Newport Bay
Newport-Inglewood Fault
Nicolaus
NX Orange County
NT Oroville Dam
Owens Valley
Oxnard
Pacoima Dam
Pajaro Valley
Palm Springs
Palo Alto
Palos Verdes Hills
Panamint Range
Parkfield
Pasadena
Peninsular Ranges
Perris
Pinnacles National Monument
Pisgah Crater
Placer County
Plumas County
Point Arena
Point Loma
Point Mugu
Point Reyes
Portola Valley
Preston Peak
Puente Hills
Rand Mountains
Riverside
Riverside County
Sacramento
Sacramento County
Sacramento Valley
Salinas
Salinas Valley
Salton Sea
Salton Trough
San Andreas Fault
San Benito County
San Bernardino
San Bernardino County
San Bernardino Mountains
San Clemente Island
San Diego
San Diego County
San Fernando
San Fernando Valley
San Francisco
San Francisco Bay
San Francisco County
San Francisco Peninsula
San Gabriel Mountains
San Giorgio Mountain
San Jacinto Fault
San Joaquin County
San Joaquin Valley
San Jose
San Juan Bautista
San Luis Obispo
San Luis Obispo County
San Mateo County
San Miguel Island
San Nicolas Island
San Pedro
San Pedro Valley
Santa Ana
Santa Ana Mountains
Santa Barbara
Santa Barbara Channel
Santa Barbara County
Santa Catalina Island
Santa Clara
Santa Clara County
Santa Clara Valley
NX Santa Cruz County
NT Santa Cruz Island
Santa Cruz Mountains
Santa Lucia Range
Santa Maria
Santa Monica Mountains
Santa Rosa
Santa Rosa Mountains
Santa Ynez Mountains
Sargent
Scripps Institution of Oceanography
Searles Lake
Shasta County
NX Sierra County
NT Simi Hills
Siskiyou County
Sonoma County
Southern California
Southern California Batholith
Stanislaus County
Sur fault zone
Sylmar Fault
Tehachapi Mountains
Temblor Range
The Geysers
Tomales Bay
Transverse Ranges
NX Trinity County
NT Trona Village
Tulare County
Tuolumne County
University of California
Ventura
Ventura Basin
Ventura County
Whittier Fault
Wilmington oil field
Yosemite National Park
Yuba County
SA Agua Blanca Fault
Amargosa Desert
Basin and Range Province
Bird Spring Formation
Bishop Tuff
Capistrano Formation
Cascade Range
Central Valley
Channel Islands
Coast Ranges
Colorado River
Cordillera
Franciscan Formation
Great Basin
Great Valley Sequence
Hat Creek Basalt
Hidden Valley Dolomite
Johnnie Formation
Kaibab Formation
Klamath Mountains
Lake Tahoe
Lost Burro Formation
Marca Shale Member
Moenkopi Formation
Monte Cristo Limestone
Monterey Formation
Moreno Formation
Noonday Dolomite
Pacific Coast
Pelona Schist
Pico Formation
Pogonip Group
Poway Conglomerate
Puente Formation
Punchbowl Formation
Rose Canyon Formation
Saint George Formation
San Diego Formation
San Lorenzo Formation
San Onofre Breccia
Santa Maria Formation
Sespe Formation
Shadow Mountains
Sierra Nevada
Sonoran Desert
Stirling Quartzite
Supai Formation
Tejon Formation
Truckee River
Tulare Formation
Twin Lakes
Umpqua Formation
Western U.S.
White Mountain
White Mountains
Wilmington
Wood Canyon Formation
Wyman Formation

californium
includes use on level 1 and 2 as chemical element (list D).
UF Cf
SA elements

Caliman (Mountains)
use Calimani Mountains

Calimani Mountains
Range in the Eastern Carpathians in E Transylvania. Also search Caliman, Calimani, and Calimanului Mountains.
UF Caliman (Mountains)
Calimanului Mountains
BT Transylvania
Romania
SA Carpathians
Eastern Carpathians

Calimanului Mountains
use Calimani Mountains

caliper logging
As of 1978, term is used on level 2 under well-logging(1). Before 1978, also search well-logging AND caliper.
UF logging, caliper
BA well-logging

Calistoga
City in Napa County NE of San Francisco Bay.
BT California
United States

Callander Bay
W edge of Lake Nipissing near city of North Bay E of Georgian Bay.
BT Ontario
Canada

Callianassa
Genus. Includes use on level 3 under Arthropoda(1) Crustacea(2).
BA Decapoda
Crustacea
Arthropoda
BT Invertebrata

Callipteris
Genus. Includes use on level 3 under pteridophytes(1) Filicopsida(2).
BA Filicopsida
pteridophytes
BT Plantae
SA gymnosperms
Pteridospermae

Callovian
Europe. Lowermost Upper Jurassic. Above Bathonian, below Oxfordian. Includes use on level 3 under age terms(1). See list E.
BA Upper Jurassic
Jurassic
BT Mesozoic
SA Dogger

Calpionella
Genus.
BA Tintinnidae
Protista

Calpionellidae
Includes use on level 3 under Protista(1) Tintinnidae(2). See list F.
BA Tintinnidae
Protista

Calpionellites
Genus.
BA Tintinnidae
Protista

Caltanissetta
City in central.
BT Sicily
Italy

Calvados
Department in NW.
CO N485000N493500
E0004000W0011500
BT France
NT Caen
May-sur-Orne
SA Normandy
Seine Estuary

Calvert Formation
In Chesapeake Group. Delaware, E Maryland, and Virginia.
BT middle Miocene
Miocene
SA Delaware
Maryland
Virginia

Camaguey
Province in W central Cuba. Also a city.
BT Cuba
West Indies

Camarines Norte
Province in SE.

BT Luzon
    Philippines

Cambay
City and former Indian state at N end of Gulf of Cambay.
BT Gujarat
    India

Cambay Basin
Gulf of Cambay area.
BT Gujarat
    India

Cambodia
Khmer Republic or Kampuchea. Includes use on level 1 as an area term (list O). For term set options see list B.
UF Kampuchea
    Khmer Republic
BA Asia
NT Battambang
SA Indochina

Cambrian
Includes use on level 1 as an age term (list E); on level 2 under paleoterms, e.g. paleoecology, paleomagnetism, paleogeography. Above Precambrian, below Ordovician.
BA Paleozoic
NT Baikalian Phase
    Lardeau Group
NA Lower Cambrian
    Middle Cambrian
NT Nama System
    Phuket Group
    Semri Series
    Tintic Quartzite
    Tons Member
NA Upper Cambrian
SA Arbuckle Group
    Assyntic Orogeny
    Bhander Group
    Cadomian Orogeny
    Dalradian
    Eocambrian
    Jacobsville Sandstone
    Kaimur Sandstone
    Knox Group
    Kurnool System
    Porsanger Dolomite Formation
    Talladega Group
    Vindhyan
    Wood Canyon Formation

Cambridge
City in Middlesex County in NE Massachusetts, and a city NNE of London. Also a former county in E England now a part of Cambridgeshire and Isle of Ely. Index England and/or Massachusetts as applicable.
CO N520000N524500
    E0003000W0001500
BX England
    Massachusetts

Cambridge Arch
Index states as applicable.
BT United States
SA Kansas
    Nebraska

Cambridge University
City of Cambridge NNE of London.
UF University of Cambridge
BT England
    Great Britain
    United Kingdom

Camelidae
Family. Includes use on level 3 under Mammalia(1) Artiodactyla(2).
BA Ruminantia
    Artiodactyla
    Mammalia
BT Tetrapoda
    Vertebrata

Cameron Parish
Extreme SW.
BT Louisiana
    United States

Cameroon
Includes former French trust territory and British trust territory of Southern Cameroons. Includes use on level 1 as an area term (list O). For term set options see list B.
CO N020000N130000
    E0170000E0080000
BA Africa
NT Adamawa
SA Benue Valley
    Chad Basin
    Lake Chad
    Logone River

Camp Century
On the Greenland Ice Sheet NE of Thule Air Base in NW.
BT Greenland
    Arctic region

Campania
An autonomous region in S.
CO N400000N413000
    E0154500E0134500
BT Italy
NT Bay of Naples
    Caserta
    Cilento
    Ischia
    Monte Somma
    Naples
    Phlegraean Fields
    Pozzuoli
    Roccamonfina
    Salerno
    Sorrento Peninsula
    Vesuvius
SA Lucania

Campanian
Europe. Above Santonian, below Maestrichtian. Includes use on level 3 under age terms(1). See list E.
BA Upper Cretaceous
    Cretaceous
BT Mesozoic
SA Rosario Formation
    Senonian

Campbell County
Index states as applicable.
BX Kentucky
    South Dakota
    Tennessee
    Virginia
    Wyoming
BT United States

Campeche
State on W part of Yucatan Peninsula.
BT Mexico
SA Gulf Coastal Plain
    Yucatan Peninsula

Campeche Bank
N of the Yucatan Peninsula.
BT Gulf of Mexico
SA Yucatan Shelf

Campine
Heathland area. Index countries as applicable.
BT Europe
SA Belgium
    Netherlands

Campo del Cielo
Point of impact N of village of Campo del Cielo in S Argentina. Also search Campo del Cielo Meteorite.
BT Argentina
SA Chaco
    meteorites

Campobasso
Town in S central part of country.
BT Italy

camptonite
Includes use on level 3 under igneous rocks(1) lamprophyre and carbonatite family(2). See list H.
BA lamprophyre and carbonatite family
BT igneous rocks

Canada
Includes use on level 1 or 2 as an area term (list O). If 1, see list B for term set options.
CO N420000N840000
    W0520000W1410000
BT North America
NA Alberta
NT Assiniboine River
    Assiniboine River valley
NA British Columbia
NT Canadian Cordillera
NA Canadian Shield
NT Carswell Structure
    Chaleur Bay
    Churchill Province
    Dunnage Melange
    Elk Point Basin
    Frontenac County
    Grenville Front
    Hudson Bay
    Hudson Bay Lowlands
    James Bay
    Kicking Horse River valley
NA Labrador
NT Lake Timiskaming
    Liard River
    Mackenzie Mountains
NA Manitoba
    Maritime Provinces
    New Brunswick
    Newfoundland
NT North Saskatchewan River
    Northumberland Strait
NA Northwest Territories
    Nova Scotia
    Ontario
NT Ottawa Valley
    Peace River
    Peel River
NA Prince Edward Island
    Quebec
NT Restigouche Estuary
    Rice Lake
    Richardson Mountains
NA Saskatchewan
NT Saskatchewan River
    Slave Province
    South Saskatchewan River
    Timiskaming
    Ungava
NA Yukon Territory
SA Acadian
    Appalachians
    Arctic region
    Burgess Shale
    Christopher Formation
    Cloridorme Formation
    Cypress Hills Formation
    Davidsville Group
    Espanola Formation
    Flinton Group
    Gander Lake Group
    Geological Survey of Canada
    Gowganda Formation
    Great Lakes
    Great Lakes region
    Great Plains
    Harbour Main Group
    Hudsonian Orogeny
    Huronian
    Keg River Formation
    Kenoran Orogeny
    Kootenay Formation
    Lardeau Group
    Leda Clay
    Mannville Formation
    Meguma Group
    Michelle Formation
    Miette Complex
    Miette Group
    Milk River Formation
    Missi Group
    Mississippi River basin
    Nahanni Formation
    Nicola Group
    Nipissing Diabase
    Onaping Formation
    Osler Series
    Palliser Formation
    Paskapoo Formation
    Peel Sound Formation
    Prince Albert Group
    Ramparts Formation
    Ravenscrag Formation
    Read Bay Formation
    Rice Lake Group
    Rocky Mountains
    Saint Lawrence River
    Shaunavon Formation
    Signal Hill Formation
    Sokoman Formation
    Taconic Allochthon
    Taconic Orogeny
    Western Interior
    Windsor Group
    Yellowknife Group

Canada Basin
N of Alaska and WNW of Canada's Queen Elizabeth Islands. Also search Canadian Basin.
UF Canadian Basin
BT Arctic Ocean

Canadian
Provincial series, North America. Above Croixian (Cambrian), below Champlainian. Includes use on level 3 under age terms(1). See list E.
BA Lower Ordovician
    Ordovician
BT Paleozoic

Canadian Basin
use Canada Basin

Canadian Cordillera
The Canadian Rocky Mountains. Index provinces as applicable.
BT Canada
SA Alberta
    British Columbia
    Cordillera
    Rocky Mountains

Canadian River
Index states as applicable.
UF South Canadian River
BT United States
SA Colorado
    New Mexico
    Oklahoma
    Texas

Canadian Shield
Index states and provinces as applicable. Introduced as a level 1 area term in 1978. Also search Canadian AND Shield.
CO N400000N720000
    W0565000W1210000
UF Laurentian Highlands
    Laurentian Plateau
    Precambrian Shield
BA Canada
    North America
SA Alberta
    Churchill Province
    Manitoba
    Michigan
    New York
    Northwest Territories
    Ontario
    Quebec
    Saskatchewan
    Wisconsin

canals
As of 1978, term is used on level 2 under waterways(1).

SA channels
streams
waterways

Cananea
City in N.
BT Sonora
Mexico

Canaries
use Canary Islands

Canary Islands
Spanish controlled island group off NW Africa. Includes use on level 1 as an area term (list O). For term set options see list B.
CO N274500N291500
W0130000W0180000
UF Canaries
BA Atlantic Ocean
NT Fuerteventura
Gomera
Grand Canary
Hierro
Lanzarote
Tenerife
SA Spain

Canavese Zone
Mountain region N of Turin.
BT Italy
SA Piedmont

Canberra
City between Sydney and Melbourne. Australian Capital Territory.
BT Australia

cancrinite
BZ carbonates
framework silicates
BT minerals

Candona
Genus. Includes use on level 3 under Ostracoda(1) Podocopida(2).
BA Cyprididae
Podocopida
Ostracoda
BT Crustacea
Arthropoda
Invertebrata

canfieldite
BZ sulfogermanates
sulfostannates
BA sulfosalts
BT minerals

Canidae
Family. Includes use on level 3 under Mammalia(1) Carnivora(2).
BA Carnivora
Mammalia
BT Tetrapoda
Vertebrata
NA Canis

Canis
Genus. Includes use on level 3 under Mammalia(1) Carnivora(2).
BA Canidae
Carnivora
Mammalia
BT Tetrapoda
Vertebrata

Cannikin
Name of site of nuclear explosion on Amchitka Island in the Aleutian Islands.
BT Alaska
United States

Canning Basin
Arid region in northern Western Australia, about 790 miles NNE of Perth.
CO S240000S180000
E1280000E1200000
UF Desert Basin
BT Western Australia
Australia

Cannon County
Central.
BT Tennessee
United States

Canon City
Fremont County in central.
BT Colorado
United States

Canon Diablo
use Canyon Diablo

canonical analysis
BA statistical methods
SA analysis
mathematical geology

Cantabria Knoll
use Cantabria Seamount

Cantabria Seamount
W Bay of Biscay. Also search Cantabria.
UF Cantabria Knoll
BT Bay of Biscay
Atlantic Ocean

Cantabrian Basin
Region in the Cantabrian Mountain area in N and NW Spain. Index Asturias, Basque Provinces, and Santander as applicable.
BT Spain
SA Asturias
Basque Provinces
Santander

Cantabrian Mountains
Range extending westward from the Pyrenees along the Bay of Biscay to the Galician Mountains.
CO N423000N433000
W0024000W0072000
BT Spain

Cantal
Department in S central.
CO N443000N453000
E0031500E0020500
BT France
NT Cantal Massif
Massiac
SA Central Massif

Cantal Massif
Part of Central Massif covering central.
BT Cantal
France
SA Central Massif

Canterbury
District in E central.
BT South Island
New Zealand

Canyon Diablo
Canyon in SE Coconino County in central Arizona. Also search Canon Diablo and Canyon Diablo Meteorite.
UF Canon Diablo
BT Arizona
United States
SA meteorites

Canyon Group
Includes Graford Formation, Winchell Limestone, Brad Formation, and Caddo Creek Formation. Central and N central Texas.
BT Upper Pennsylvanian
Pennsylvanian
SA Texas
United States

canyons
Includes use on level 3 under geomorphology(1). As of 1976, usage restricted to land canyons. Before 1976, also refers to submarine canyons.
UF land canyons
SA geomorphology
submarine canyons

Cap Blanc
Index countries as applicable.
BT Africa
SA Mauritania

cap rocks
Restricted to use for salt tectonics: an impervious layer or body of rock over a salt dome.
SA anhydrite
diapirs
gypsum
rocks
salt domes
salt tectonics

capacity
Used as a general term.
SA bearing capacity
dimensions
efficiency
exchange capacity
heat capacity
physical properties
volume

Capbreton
Town in SW.
BT Landes
France

Cape Breton Island
N part of province.
CO N453000N470000
W0593000W0613000
BT Nova Scotia
Canada

Cape Cod
SE Massachusetts. Coextensive with Barnstable County.
BT Massachusetts
United States

Cape Dyer
Village and cape on Davis Strait on E Baffin Island in District of Keewatin.
BT Northwest Territories
Canada

Cape Fear Arch
Extends NW from Cape Fear.
BT North Carolina
United States

Cape Hatteras
Promontory on Hatteras Island between Pamlico Sound and the Atlantic Ocean.
BT North Carolina
United States

Cape Kennedy
Formerly Cape Canaveral in Brevard County on the Atlantic Ocean. Site of John F. Kennedy Space Center.
BT Florida
United States

Cape Lookout
Headland at S end off Carteret County.
BT North Carolina
United States

Cape Mendocino
On W coast of Humboldt County. Extreme W part of state. Also search Mendocino.
UF Mendocino (Cape)
BT California
United States

Cape Province
Province in S.
BT South Africa
NT Cape Town
False Bay
Karroo Basin
Kimberley
Knysna
Langebaanweg
Molteno
Port Elizabeth
Postmasburg
SA Ecca Series
Namaqualand
Orange River
Vaal River

Cape Sable
S point of Cape Sable Island in extreme S.
BT Nova Scotia
Canada

Cape Town
City in extreme SW Cape Province.
UF Capetown
BT Cape Province
South Africa

Cape Valse
use False Cape

Cape Verde Islands
Former Portuguese overseas province W of Dakar, Senegal. Includes use on level 1 as an area term (list O). For term set options see list B.
BA Africa

Cape York Peninsula
Between Gulf of Carpentaria and Coral Sea in N.
BT Queensland
Australia

Capetown
use Cape Town

capillarity
UF capillary water
SA capillary pressure
soils

capillary pressure
BT pressure
SA capillarity

capillary water
use capillarity

Capistrano Formation
Subdivided to include Osa Member. Unconformably underlies Niguel Formation. Southern California between Santa Ana and Oceanside.
BT Neogene
Tertiary
SA California
lower Pliocene
upper Miocene

Capitan Formation
Massive reef limestone which grades basinward into upper Delaware Mountain Sandstone and lagoonward into Carlsbad Limestone. SE New Mexico and W Texas.
BT Guadalupian
Permian
SA New Mexico
Texas

Capnic
use Cavnic

capture
use stream capture

Carabobo
State on Caribbean Sea. Also a village.
BT Venezuela
NT Puerto Cabello

Caracas
City in N.
BT Federal District
Venezuela

Caracas Group
Middle and upper Mesozoic. Includes Las Brisas Formation, Antimano Formation, Tacagua Formation. In Cordillera de la Costa of Venezuela.
UF Caracas Series

BT Mesozoic
SA Federal District
middle Mesozoic
upper Mesozoic
Venezuela

Caracas Series
use Caracas Group

Caradocian
Europe. Below Ashgillian. Includes use on level 3 under age terms(1). See list E.
BA Upper Ordovician
Ordovician
BT Paleozoic
SA Bala

Caravaca
City in SE part of country.
BT Murcia
Spain

carbargilite
use coal

carbides
Includes use on level 3 under minerals(1). See list L.
BA native elements and alloys
BT minerals
SA cohenite
moissanite

carbohydrates
Includes use on level 2 under organic materials(1).
BA organic materials
NA cellulose

carbon
Includes use on level 1 and 2 as a chemical element (list D). Also search C.
UF C
SA C-12
C-13
C-13/C-12
C-14
carbonaceous composition
charcoal
diamond
diamonds
elements
hydrocarbons
isotopes
organic carbon
organic materials
stable isotopes

Carbon County
Index states as applicable.
BX Montana
Pennsylvania
Utah
Wyoming
BT United States

carbon dioxide
Includes use on level 3 under geochemistry(1), fluid inclusions(1), or phase equilibria(1). Also search CO2.
UF CO2
SA carbon monoxide
fluid inclusions
geochemistry
phase equilibria

carbon monoxide
UF CO
SA carbon dioxide

carbon-12
use C-12

carbon-13
use C-13

carbon-14
A valid term through 1976. After 1976, use C-14.

carbonaceous
A valid index term through 1977. After 1977, use carbonaceous chondrites under meteorites(1), and use carbonaceous composition as a general term referring to carbon and coal.

carbonaceous chondrites
Includes use on level 3 under meteorites(1). Before 1978, also search chondrites AND carbonaceous.
BA chondrites
BT meteorites

carbonaceous composition
A general term used with carbon and coal. Before 1978, also search carbonaceous.
SA carbon
coal
composition
organic materials

carbonaceous shale
use black shale

carbonado
BA native elements and alloys
BT minerals
SA diamonds
industrial minerals

carbonate
Restricted to use as a noun. Compound characterized by a fundamental anionic structure of CO3.
SA carbonate rocks
carbonate sediments
carbonates

carbonate apatite
Also search carbonate-apatites; tavistockite.
UF carbonate-apatites
tavistockite
BA phosphates
BT minerals
SA apatite
dahllite

carbonate compensation
use carbonate compensation depth

carbonate compensation depth
UF carbonate compensation depth of compensation
SA lysoclines

carbonate composition
As of 1978, term is used on level 2 under nodules(1).
SA composition
nodules

carbonate rocks
Includes use on level 2 under sedimentary rocks(1). See list I.
BT sedimentary rocks
NA beachrock
biocalcarenite
biomicrite
biosparite
boundstone
calcarenite
calcilutite
chalk
dolostone
grainstone
limestone
NZ marl
NA micrite
packstone
travertine
wackestone
SA carbonate
carbonate sediments
carbonates
carbonatite
dolomite
intraclasts
oolite
rocks
tufa

carbonate sediments
Includes use on level 2 under sediments(1). See list N.
BA sediments
SA carbonate
carbonate rocks
carbonates
coquina
dolomite
dolostone
gravel
limestone
oolite
sand
travertine
tufa

carbonate-apatites
use carbonate apatite

carbonates
Includes use on level 2 under minerals(1). See list L.
BT minerals
NA alstonite
alumohydrocalcite
ankerite
aragonite
artinite
azurite
NZ bastnaesite
NA calcite
NZ cancrinite
NA cerussite
chalybite
NZ dahllite
NA dawsonite
dolomite
gaylussite
huntite
hydromagnesite
Iceland spar
magnesian calcite
magnesite
malachite
NZ meionite
NA monohydrocalcite
nahcolite
nesquehonite
norsethite
NZ phosgenite
NA protodolomite
pyroaurite
rhodochrosite
NZ sakhaite
NA shortite
siderite
sjogrenite
smithsonite
NZ spurrite
NA strontianite
NZ synchisite
thaumasite
NA thermonatrite
NZ thomsonite
NA trona
vaterite
weddellite
weloganite
whewellite
witherite
SA calcium carbonate
carbonate
carbonate rocks
carbonate sediments
sodium carbonate

carbonation
use carbonatization

carbonatite
Includes use on level 3 under igneous rocks(1) lamprophyre and carbonatite family(2). See list H.
BA lamprophyre and carbonatite family
BT igneous rocks
SA carbonate rocks

carbonatite family, lamprophyre and
use lamprophyre and carbonatite family

carbonatization
Introduction of, or replacement by, carbonates. Also search carbonation.
UF carbonation
BA diagenesis
SA calcitization
dolomitization

Carbondale
City in Jackson County in S.
BT Illinois
United States

Carbondale Formation
In Kewanee Group. Illinois and W Kentucky.
BT Middle Pennsylvanian
Pennsylvanian
SA Illinois
Kentucky
Springfield Coal Member

Carbonicola
Genus. Includes use on level 3 under Mollusca(1) Bivalvia(2).
BA Bivalvia
Mollusca
Invertebrata

Carboniferous
Includes use on level 1 as an age term (list E); on level 2 under paleo- terms, e.g. paleoecology, paleogeography, paleomagnetism. Above Devonian, below Permian.
BA Paleozoic
NT Asturian Orogeny
Burbank Sand
Essen Beds
Johns Valley Formation
NA Lower Carboniferous
Middle Carboniferous
Namurian
Silesian
NT Springer Formation
Tubarao Group
NA Upper Carboniferous
SA Baralaba Coal Measures
Beacon Supergroup
Dwyka Series
Geirud Formation
Gondwana System
Karroo System
Kinderhookian
Mississippian
Paganzo Group
Pennsylvanian
Singleton Coal Measures
Talchir Formation

carbonification
use coalification

Cardiff
City in Glamorgan County on the Bristol Channel.
BT Wales
Great Britain
United Kingdom

Cardigan Bay
Large inlet of Saint George's Channel.
BT Wales
Great Britain
United Kingdom
SA Irish Sea

Cardigan County
use Cardiganshire

Cardiganshire
County on Cardigan Bay.
CO N520000N524000
W0034000W0044000
UF Cardigan County
BT Wales
Great Britain
United Kingdom

Cardiidae
Family. Includes use on level 3 under Mollusca(1) Bivalvia(2).
BA Bivalvia
 Mollusca
BT Invertebrata
NA Cardium

Cardium
Genus. Includes use on level 3 under Mollusca(1) Bivalvia(2).
BA Cardiidae
 Bivalvia
 Mollusca
BT Invertebrata
NA Cardium edule

Cardium edule
Includes use on level 3 under Mollusca(1) Bivalvia(2).
BA Cardium
 Cardiidae
 Bivalvia
 Mollusca
BT Invertebrata

Cariaco Basin
W of Cariaco Bay and between Isla la Tortuga and Venezuelan coast. Also search Cariaco Trench.
UF Cariaco Trench
BT Caribbean Sea

Cariaco Trench
use Cariaco Basin

Caribbean Mountain Range
Islands comprising the Antilles stretching from Florida to the coast of Venezuela constituting the remainder of several submerged Andean spurs meeting in Puerto Rico.
BT West Indies

Caribbean Plate
The Caribbean Sea including the Caribbean islands and across Central America to the Cocos plate just to the W.
SA Caribbean region
 plate tectonics

Caribbean region
Index Antilles, Caribbean Sea, Central America, Colombia, and Venezuela as applicable. Also search Caribbean. Includes use on level 1 or 2 as an area term (list O). If 1, see list B for term set options.
NA Netherlands Antilles
SA Antilles
 Caribbean Plate
 Caribbean Sea
 Central America
 Colombia
 Greater Antilles
 Hispaniola
 Lesser Antilles
 Venezuela
 Yucatan Peninsula

Caribbean Sea
Bounded by the West Indies on the N and E, South America on the S, and Central America on the W. Also search Caribbean. Includes use on level 1 or 2 as an area term (list O). If 1, see list B for term set options.
CO N090000N220000
 W0600000W0780000
BT Atlantic Ocean
NT Aves Island
 Aves Ridge
 Beata Ridge
 Cariaco Basin
 Cayman Trough
 Colombian Basin
 Grand Cayman Island
 Gulf of Cariaco
 Los Testigos
 Nicaragua Rise
 Venezuelan Basin
 Yucatan Basin
SA Caribbean region

Cariboo Mountains
Range of the Rocky Mountains in E central.
BT British Columbia
 Canada
SA Rocky Mountains

Carinthia
State in S.
CO N463000N470000
 E0150000E0123000
BT Austria
NT Bleiberg
 Krappfeld
 Sau Alps
 Villach
 Wolfsberg

Carixian
Europe.
BA Lower Jurassic
 Jurassic
BT Mesozoic

Carlile Shale
In Colorado Group. E Colorado, Kansas, SE Montana, Nebraska, NE New Mexico, South Dakota, E Wyoming.
BT Upper Cretaceous
 Cretaceous
SA Benton Formation
 Colorado
 Colorado Group
 Kansas
 Montana
 Nebraska
 New Mexico
 South Dakota
 Wyoming

Carlin
Town in Elko County in NE.
BT Nevada
 United States

Carlin Mine
NE Nevada. Also search Carlin.
BT Nevada
 United States

Carlsbad
City in Eddy County in SE. Use Karlovy Vary for town in Czechoslovakia.
BT New Mexico
 United States

Carlsbad Caverns
In SW Eddy County just N of Texas border.
BT New Mexico
 United States

Carlsberg Ridge
Between Somalia and the Seychelles in the W and the Laccadive and Maldive islands in the E. Also search Arabian-Indian Midoceanic Ridge and Arabian-Indian Ridge.
UF Arabian Ridge
 Arabian-Indian Midoceanic Ridge
 Arabian-Indian Ridge
BT Indian Ocean

Carmarthenshire
County in W on Bristol Channel.
UF Carmethen County
BT Wales
 Great Britain
 United Kingdom

Carmel
City in Monterey County.
UF Carmel-by-the-Sea
BT California
 United States

Carmel Bay
NW Monterey County.
BT California
 United States

Carmel-by-the-Sea
use Carmel

Carmethen County
use Carmarthenshire

Carmona
City in SW part of country.
BT Seville
 Spain

carnallite
BA halides
BT minerals
SA chlorides

Carnarvon Basin
S of Lake McCleod in Carnarvon region in W.
CO S270000S210000
 E1160000E1143000
BT Western Australia
 Australia

Carnegie Ridge
Just S of the equator between Ecuador and Galapagos Islands.
BT Pacific Ocean

Carnian
Europe. Above Ladinian, below Norian. Includes use on level 3 under age terms(1). See list E.
BA Upper Triassic
 Triassic
BT Mesozoic
SA Hallstatt Limestone

Carnic Alps
Range of Eastern Alps. Index countries as applicable.
BT Alps
 Europe
SA Austria
 Eastern Alps
 Italy
 Karawanken

Carnivora
Includes use on level 2 under Mammalia(1). See list F.
BA Mammalia
BT Tetrapoda
 Vertebrata
NA Canidae
 Felidae
 Hyaenidae
 Mustelidae
 Ursidae

carnotite
BA vanadates
BT minerals

Carolina slate belt
W North Carolina and NW South Carolina. Index states as applicable.
BT Appalachians
SA North Carolina
 South Carolina

Caroline Islands
Central and W island group of U.S. Trust Territory on the Pacific Islands.
UF Carolines
BT Pacific Ocean
SA Micronesia

Carolines
use Caroline Islands

carotenoids
BA organic materials
SA pigments

Carpathian Foredeep
Elongated depressions bordering Carpathians. Index countries and Ukraine as applicable. Also search Carpathian AND foredeep.
BT Europe
SA Carpathians
 Czechoslovakia
 Poland
 Romania
 Ukraine

Carpathian Foreland
Stable area marginal to Carpathians. Index countries and Ukraine as applicable. Also search Carpathian AND foreland.
BT Europe
SA Carpathians
 Czechoslovakia
 Poland
 Romania
 Subcarpathians
 Ukraine

Carpathian Mountains
use Carpathians

Carpathians
Mountain system of E and central Europe enclosing the Alfold. Index Ukraine and countries as applicable. Includes use on level 1 as an area term (list O). For term set options see list B. Before 1973, also search Carpathian Mountains.
CO N460000N495000
 E0280000E0180000
UF Carpathian Mountains
BA Europe
NT Eastern Carpathians
 Tatra Mountains
 Western Carpathians
SA Beshchady Mountains
 Beskid Mountains
 Bistrita Mountains
 Bukk Mountains
 Calimani Mountains
 Carpathian Foredeep
 Carpathian Foreland
 Czechoslovakia
 Hungary
 Low Tatra Mountains
 Matra Mountains
 Poland
 Rodna Mountains
 Romania
 Spis-Gemer
 Subcarpathians
 Tokaj-Eperjes Mountains
 Transylvanian Alps
 Ukraine
 Ukrainian Carpathians
 Vrancea

Carpentaria Basin
Region around Gulf of Carpentaria. Index Northern Territory and/or Queensland as applicable.
CO S190000S150000
 E1430000E1400000
BT Australia
SA Northern Territory
 Queensland

Carpentarian
SE Australia.
BA middle Precambrian
 Precambrian

carpholite
BA chain silicates
 silicates
BT minerals
SA aluminosilicates
 ferrocarpholite

Carriacou
Largest island of the Grenadines in the Windward Islands.
BT West Indies
SA Grenada
 Windward Islands

Carroll County
Index states as applicable.
BX Arkansas
 Georgia
 Illinois
 Indiana
 Iowa
 Kentucky
 Maryland

    Mississippi
    Missouri
    New Hampshire
    Ohio
    Tennessee
    Virginia
  BT United States

carrollite
  UF sychnodymite
  BA sulfides
  BT minerals
  SA linnaeite

Carry-le-Rouet
  A section in S.
  BT Bouches-du-Rhone
      France

carst
  use karst

Carswell Structure
  Lake Athabaska area of NW Saskatchewan, and the Baker Lake area in District of Keewatin of the Northwest Territories. Index province and territory as applicable.
  BT Canada
  SA Northwest Territories
      Saskatchewan

Cartagena
  City on the Caribbean Sea.
  BT Bolivar
      Colombia

Carter County
  Index states as applicable.
  BX Kentucky
      Missouri
      Montana
      Oklahoma
      Tennessee
  BT United States

Cartersville
  City in Bartow County in NW.
  BT Georgia
      United States

cartography
  For methods, instruments, programs. Includes use on level 2 under maps(1). Before 1972, also search geological exploration; before 1976 also search geologic mapping; before 1978 also search mapping.
  UF chartology
  SA aerial photography
      geodesy
      geomorphology
      legend
      leveling
      maps
      photogeology
      photogrammetry
      topography
      triangulation

Casablanca
  City on the Atlantic Ocean.
  BT Morocco

Casapalca
  Town NE of city of Lima.
  BT Lima
      Peru

Cascade Mountains
  use Cascade Range

Cascade Range
  Northern continuation of the Sierra Nevada Mountains. Index states as applicable. Also search Cascade Mountains and Cascades.
  CO N400000N510000
      W1203000W1230000
  UF Cascade Mountains
      Cascades
  BT United States
  SA California
      Mount Rainier
      Oregon
      Sierra Nevada
      Washington

cascades
  Includes use on level 3 under hydrology(1) or geomorphology(1) fluvial features(2).
  SA fluvial features
      geomorphology
      hydrology
      streams

Cascades
  use Cascade Range

Cascadia Basin
  Off Washington coast.
  BT Pacific Ocean

Cascadia Channel
  Off coast of Oregon SW of Cascadia Basin.
  BT Pacific Ocean

case histories
  Includes use as a general term.
  UF histories, case

case studies
  As of 1978, term is used on level 2 under pollution(1), rock mechanics(1), and soil mechanics(1).
  UF studies, case
  SA pollution
      rock mechanics
      soil mechanics

Caserta
  City and province NE of Naples.
  BT Campania
      Italy

Caseyville Formation
  In McCormick Group. SE Illinois and W Kentucky.
  BT Lower Pennsylvanian
      Pennsylvanian
  SA Illinois
      Kentucky

Cashel
  Town at the base of the Rock of Cashel in S central part of country.
  BT Tipperary
      Ireland

Casper
  City in Natrona County in central.
  BT Wyoming
      United States

Casper Formation
  Includes all beds from top of Madison Limestone to base of Permian (Opeche) red shales. In SE Wyoming.
  BT Paleozoic
  SA Pennsylvanian
      Permian
      Wyoming

Caspian Basin
  Region including the Caspian Sea and surrounding areas. Index Caspian Sea, Soviet republics and Iran as applicable. Also search Caspian; Caspian region.
  UF Caspian region
  BT Eurasia
  SA Azerbaidzhan
      Caspian Depression
      Caspian Sea
      Iran
      Kazakhstan
      Russian Republic
      Turkmenia

Caspian Depression
  That part of the Caspian Basin lying N of the Caspian Sea; it is a large lowland area which was submerged by the Caspian Sea at one time. Index Soviet republics as applicable. Also search Caspian; Caspian Lowland.
  UF Caspian Lowland
  BT USSR
  SA Caspian Basin
      Kazakhstan
      Russian Republic

Caspian Lowland
  use Caspian Depression

Caspian region
  use Caspian Basin

Caspian Sea
  Largest inland sea in the world. Index Iran and/or the USSR as applicable. Also search Caspian. Includes use on level 1 as an area term (list O). For term set options see list B.
  CO N360000N480000
      E0530000E0470000
  BA Eurasia
  SA Caspian Basin
      Iran
      Kara-Bogaz-Gol
      Neftyanyye Kamni
      USSR

Cass County
  Index states as applicable.
  BX Illinois
      Indiana
      Iowa
      Michigan
      Minnesota
      Missouri
      Nebraska
      North Dakota
      Texas
  BT United States

Cassidulinacea
  Includes use on level 2 under foraminifera(1). See list F.
  BA Rotaliina
      foraminifera
  BT Invertebrata
  NA Anomalinidae

cassiterite
  BA oxides
  BT minerals

Castellane
  Village in SE.
  BT Alpes-de-Haute Provence
      France

Castellon
  use Castellon de la Plana

Castellon de la Plana
  Province on Gulf of Valencia in E Spain. Also a city. Search Castellon.
  UF Castellon
  BT Spain
  SA Maestrazgo

Castile
  Region and ancient kingdom of central and N central part of country consisting of 13 modern provinces.
  BT Spain

Castile Formation
  Formation, as now defined, includes beds in lower part of Ochoa Series that are confined in extent to Delaware basin and overlap sloping surface of Capitan Limestone along its margins. SE New Mexico, and W Texas.
  BT Permian
  SA New Mexico
      Texas

Castillon Massif
  In Ariege Pyrenees in S.
  BT Ariege
      France

Castle Hayne
  Village in New Hanover County in SE.
  BT North Carolina
      United States

Castle Hayne Limestone
  Underlies Yorktown Formation near Pollicksville, North Carolina. Coastal Plain of North Carolina and South Carolina.
  BT Eocene
  SA middle Eocene
      North Carolina
      South Carolina
      upper Eocene

Castleton
  Town in Rutland County in S central.
  BT Vermont
      United States

Castoridae
  Family. Includes use on level 3 under Mammalia(1) Rodentia(2).
  BA Rodentia
      Mammalia
  BT Tetrapoda
      Vertebrata

casts, flute
  use flute casts

casts, groove
  use groove casts

casts, load
  use load casts

cataclasites
  Includes use on level 2 under metamorphic rocks(1). See list J.
  BT metamorphic rocks

catagenesis
  Also search katagenesis.
  UF katagenesis
  SA sedimentary rocks

catalogs
  Includes use on level 1 (list A); as a level 2 or 3 term appropriate to a large number of topics, e.g. on level 2 under paleontology(1). See list G. Also search catalog. If 1, term set options are:
  topic [list B, extraterrestrial geology, general, geophysics]
      subtopic (e.g. language if not English)
  SA associations
      bibliography
      book reviews
      collections
      education
      geology
      glossaries
      lexicons
      maps
      mineralogy
      museums
      oceanography
      paleobotany
      paleontology
      petrology
      sedimentary petrology
      stratigraphy
      structural geology
      surveys
      volcanology

Catalonia
  Historical region in NE Spain. Index provinces as applicable.
  CO N403000N423000
      E0033000E0000000
  BT Spain
  SA Barcelona
      Gerona
      Lerida
      Tarragona

catalysis
  SA chemical analysis
      geochemistry

Catamarca
  Province in NW Argentina. Also a city.

Catamarca ● cement

CO S300000S250000
W0650000W0690000
BT Argentina

Catanda
Village in E.
BT Angola

Catanzaro
City in central.
BT Calabria
Italy

catastrophic
A valid term through 1977. After 1977, use catastrophic waves on level 2 under ocean waves(1).

catastrophic advance
use glacier surges

catastrophic waves
Term introduced in 1978. Includes use on level 2 under ocean waves(1). Before 1978, also search catastrophic AND ocean waves.
BT ocean waves
SA tsunamis
waves

catastrophies
Term in 1978. Includes use on level 2 under geologic hazards(1).
SA destruction
geologic hazards

catastrophism
SA uniformitarianism

catchments
use drainage basins

catenas
Soil association of a given area developed from common parent material.
SA soils

cathodoluminescence
SA luminescence

cation exchange
A valid term through 1978. After 1978, use ion exchange or cation exchange capacity, e.g. under geochemistry(1) or soils(1).

cation exchange capacity
Includes use on level 3 under geochemistry(1) processes(2). Also search cation exchange.
SA clay mineralogy
exchange capacity
geochemistry
ion exchange
ions

cations
A valid term through May 1978. After May 1978, use ions.

Catoctin Formation
Divided into Warrenton Agglomerate Member at base and basalt flows now altered to metabasalt or "greenstone" schist. W Maryland, N Virginia, and NE West Virginia.
BT Precambrian
SA Maryland
Virginia
West Virginia

Catron County
In W on Arizona border.
BT New Mexico
United States

Catskill Delta
Devonian age complex. Index states as applicable.
BT United States
SA Maryland
New York
Pennsylvania
West Virginia

Catskill Formation
In Susquehanna Group.
BT Paleozoic
SA Devonian
Lower Mississippian
Maryland
Middle Devonian
Mississippian
New York
Pennsylvania
Upper Devonian
Virginia

Catskill Mountains
Group of the Appalachians in SE.
BT New York
United States
SA Appalachians

cattierite
BA sulfides
BT minerals

Caucasian Foreland
use Caucasus Foreland

Caucasus
Mountain system between the Black and Caspian Seas which separates the Northern Caucasus from Transcaucasia. Also search Caucasus Mountains. Index Greater Caucasus, Lesser Caucasus, and Soviet republics as applicable.
CO N410000N440000
E0480000E0400000
UF Caucasus Mountains
BT USSR
SA Armenia
Azerbaidzhan
Georgian Republic
Greater Caucasus
Lesser Caucasus
Russian Republic

Caucasus Foreland
The Northern Caucasus or Ciscaucasia area which includes the northern slopes and foothills of the Caucasus plus the Kuban Steppes to the north. Also search Caucasian Foreland.
UF Caucasian Foreland
BT Russian Republic
USSR
SA Northern Caucasus

Caucasus Mountains
use Caucasus

cauldrons
SA calderas
volcanic features
volcanism
volcanology

causes
Includes use as level 2 or 3 term appropriate to a large number of topics, e.g. on level 2 under orogeny(1). See list G.

Causses
An arid Jurassic limestone plateau of S Central Massif in S and SW France. Also search South Massif.
UF South Massif
BT France
SA Central Massif

caustobiolith
Includes use on level 3 under sedimentary rocks(1) organic residues(2). See list I. Also search caustobioliths.
BA organic residues
BT sedimentary rocks
SA organic materials

Cauvery Basin
River basin in S part of country. Index states as applicable.
UF Kaveri Basin
BT India
SA Mysore
Tamil Nadu

Caux
Chalky tableland in N.
BT Seine-Maritime
France

Cavellinidae
Family. Includes use on level 3 under Ostracoda(1) Podocopida(2).
BA Podocopida
Ostracoda
BT Crustacea
Arthropoda
Invertebrata
SA Healdiidae

caverns
Icludes use on level 3 under geomorphology(1) solution features(2).
SA caves
geomorphology
solution features

caves
Includes use on level 3 under geomorphology(1) solution features(2) and volcanic features(2).
SA caverns
geomorphology
karst
solution
solution cavities
solution features
speleology
speleothems
volcanic features

cavities
Used as a general term through 1978. After 1978, see specific terms, e.g. underground space or solution cavities.

cavities, solution
use solution cavities

Cavnic
Village in Maramures County in N Transylvania.
UF Capnic
BT Transylvania
Romania

Cayman Trench
use Cayman Trough

Cayman Trough
South of the Cayman Islands and Cuba. Also search Cayman Trench.
UF Cayman Trench
BT Caribbean Sea

Caytoniales
Includes use on level 2 under gymnosperms(1). See list F.
BA gymnosperms
BT Plantae
NA Pachypteris

Cayugan
Provincial series, North America. Above Niagaran, below Ulsterian (Devonian). Includes use on level 3 under age terms(1). See list E.
BA Upper Silurian
Silurian
BT Paleozoic
NT Williamsport Sandstone

Cd
use cadmium

Ce
use cerium

Ceara
State in NE.
BT Brazil
NT Jaguaribe

Cebu
Island in Visayan Islands in central.
BT Philippine Islands

Cedar City Formation
BT Devonian
SA Missouri

Cedar Mountain Formation
Together with Buckhorn Conglomerate comprises Cedar Mountain Group. NW Colorado and central Utah.
BT Lower Cretaceous
Cretaceous
SA Colorado
Utah

Cedar Valley Formation
Includes Linwood, Littleton, Coralville, Solon and Rapid members. E Iowa, SW Illinois, and SE Minnesota.
BT Middle Devonian
Devonian
SA Illinois
Iowa
Minnesota

celadonite
BA sheet silicates
silicates
BT minerals
SA glauconite
mica group

Celebes
Island E of Borneo. Also search Sulawesi, the official Indonesian name.
UF Sulawesi
BT Indonesia

Celebes Sea
Bounded on N by Philippines, W by Borneo, and S by island of Celebes. Also search Celebes. Includes use on level 1 as an area term (list O). For term set options see list B.
BA Pacific Ocean
SA Darvel Bay
Malay Archipelago

celestite
BA sulfates
BT minerals

Celje
Town in NW part of country.
BT Slovenia
Yugoslavia

cell dimensions
As of 1978, term is used on level 2 under crystal structure(1).
SA bonding
crystal structure
dimensions
lattice
minerals

cell, unit
use unit cell

cells, convection
use convection cells

cellulose
BA carbohydrates
organic materials

celsian
BA feldspar group
framework silicates
silicates
BT minerals
SA paracelsian

Celtic Sea
Body of water over the continental shelf between the S coast of Ireland and Brittany. It is arbitralily divided from the English Channel by a line from Land's End to Ushant Island off Brittany.
BT Atlantic Ocean

cement
Term restricted to use as the material. Use cementation for the process under sedimentary rocks(1).
SA cement materials
cementation
concrete

construction materials
sedimentary rocks
sediments

cement materials
Includes use as level 3 commodity term under limestone(1). See list C.
BA construction materials
SA cement
    limestone
    materials

cementation
Includes use on level 3 under sedimentation(1) diagenesis(2); on level 3 under sedimentary rocks(1) diagenesis(2). See list I.
BA diagenesis
SA cement
   clastic sediments
   consolidation
   deposition
   lithification
   sedimentation

Ceneri Zone
In Southern Alps. Index countries as applicable.
BT Europe
SA Italy
   Paleozoic
   Switzerland

Cenomanian
Europe. Lowermost Upper Cretaceous, or Middle Cretaceous. Above Albian, below Turonian. Includes use on level 3 under age terms(1). See list E.
BA Cretaceous
BT Mesozoic
SA Chalk
   Middle Cretaceous
   Upper Cretaceous

Cenozoic
Includes use on level 1 as an age term (list E); on level 2 under paleo-terms, e.g. paleoecology, paleogeography, paleomagnetism.
BA Phanerozoic
NA Blancan
NT Leda Clay
   Loup Fork Group
NA lower Cenozoic
   Matuyama Epoch
NT Osaka Group
NA Paleolithic
   Quaternary
NT San Juan Formation
   Santa Maria Formation
   Siwalik System
NA Tertiary
NT Tulare Formation
   Uonuma Group
NA upper Cenozoic
NT Verde Formation
NA Villafranchian
SA Alpine Orogeny
   Ferrar Group

centers, color
use color centers

centers, spreading
use spreading centers

central
General term used after any of the main geographic terms (list O).
SA east-central
   north-central
   south-central
   west-central

Central Africa
Term introduced in 1978. Before 1978, search Africa AND central.
BA Africa
SA North Africa
   Southern Africa

Central African Republic
Formerly French territory of Ubangi-Shari. Became independent in 1960. Includes use on level 1 as an area term (list O). For term set options see list B.
CO N030000N120000
   E0273000E0140000
BA Africa
NT Koto
SA Congo Basin

Central Alps
Ranges of the Alps extending from the Great Saint Bernard Pass in the W to Innsbruck and the Brenner Pass in the E. Index countries as applicable.
CO N460000N470000
   E0113000E0090000
BT Alps
   Europe
SA Austria
   Bavarian Alps
   Bernese Alps
   Glarus Alps
   Italy
   Lepontine Alps
   Pennine Alps
   Rhaetian Alps
   Switzerland
   West Germany

Central America
Index countries as applicable. Includes use on level 1 or 2 as an area term (list O). If 1, see list B for term set options.
NA Belize
   Costa Rica
   El Salvador
   Guatemala
   Honduras
   Nicaragua
   Panama
NT Panama Canal Zone
SA Caribbean region

Central Andes
Term introduced in 1978. Before 1978, also search Andes AND central.
BA Andes
BT South America
SA Northern Andes
   Southern Andes

Central Appalachians
Term introduced in 1978. Before 1978, also search Appalachians AND central.
BA Appalachians
BT North America
SA Northern Appalachians
   Southern Appalachians

Central Asia
Region known as Soviet Central Asia. Index Soviet republics as applicable.
CO N350000N440000
   E0900000E0540000
BT USSR
SA Kazakhstan
   Kirghizia
   Tadzhikistan
   Turkmenia
   Uzbekistan

Central Basin
Central.
BT Tennessee
   United States

Central Bohemian Pluton
Part of the Bohemian Massif in central.
BT Bohemia
   Czechoslovakia
SA Bohemian Massif

Central Europe
CO N450000N550000
   E0200000E0070000
BA Europe

central granite
use batholiths

Central Massif
An old, eroded plateau region in S central part of country. Index departments as applicable. Also search Massif Central.
CO N440000N470000
   E0050000E0004500
UF Massif Central
BT France
SA Aveyron
   Cantal
   Cantal Massif
   Causses
   Haute-Loire
   Montagne Noire
   Morvan
   Sidobre Massif

Central Polish Glaciation
The Elster and Saale stages of glaciation both of which covered central Poland. They were respectively the oldest and the middle stage of 3 stages which covered northern Europe.
BT Poland
SA Saalian

Central Rocky Mountains
Term introduced in 1978. Before 1978, also search Rocky Mountains AND central.
BA Rocky Mountains
BT North America
SA Northern Rocky Mountains
   Southern Rocky Mountains

Central Valley
Valley of Sacramento and San Joaquin rivers in California, and a 600 mile long valley between the Andes and the Coastal Range in central Chile. Index California and/or Chile as applicable. Before 1978, also search Great Valley for the valley in California.
SA California
   Chile
   Sacramento Valley
   San Joaquin Valley

centrallasite
use gyrolite

Centre County
Central.
BT Pennsylvania
   United States

centrum (seismology)
use focus

Cephalonia
Island in center of Ionian Islands in Ionian Sea W of Greek mainland. Also search Kefallinia.
UF Kefallinia
BT Ionian Islands
   Greece

Cephalopoda
Includes use on level 2 under Mollusca(1). See list F.
BA Mollusca
BT Invertebrata
NA Ammonoidea
   belemnites
   Belemnitidae
   Belemnoidea
   Belemnopsis
   Ceratites
   Coleoidea
   Nautiloidea

ceramic materials
Includes use on level 1 as a commodity term (list C).
SA andalusite
   clay mineralogy
   kyanite
   materials
   refractory materials
   sillimanite

Ceratites
Genus. Includes use on level 3 under Mollusca(1) Cephalopoda(2).
BA Cephalopoda
   Mollusca
BT Invertebrata
SA Ammonites

Ceratolithus
BA Coccolithophoraceae
   algae
BT Plantae

Cerithiidae
Family. Includes use on level 3 under Mollusca(1) Gastropoda(2).
BA Gastropoda
   Mollusca
BT Invertebrata
NA Cerithium

Cerithium
Genus. Includes use on level 3 under Mollusca(1) Gastropoda(2).
BA Cerithiidae
   Gastropoda
   Mollusca
BT Invertebrata

cerium
Includes use on level 1 and 2 as a chemical element (list D).
UF Ce
SA elements
   lanthanum
   rare earths

cerolite
BA sheet silicates
   silicates
BT minerals
SA serpentine group

Cerozem
use Sierozems

Cerro de Pasco
Town in W central part of country.
BT Peru

Cerro del Hoyazo
In the Sierra Alhamilla area in SE part of country.
BT Almeria
   Spain

Cerro Prieto
Geothermal area near Mexicali in N.
BT Baja California
   Mexico

cerussite
BA carbonates
BT minerals

Cervidae
Family. Includes use on level 3 under Mammalia(1) Artiodactyla(2).
BA Ruminantia
   Artiodactyla
   Mammalia
BT Tetrapoda
   Vertebrata
NA Cervus

Cervus
Genus. Includes use on level 3 under Mammalia(1) Artiodactyla(2).
BA Cervidae
   Ruminantia
   Artiodactyla
   Mammalia
BT Tetrapoda
   Vertebrata

cesium
Includes use on level 1 and 2 as a chemical element (list D). Also

search Cs.
UF Cs
SA alkali metals
 Cs-137
 elements

Cetacea
Includes use on level 2 under Mammalia(1). See list F.
BA Mammalia
BT Tetrapoda
 Vertebrata

Cette
use Sete

Cevennes
Mountain Range in S.
BT France

Ceylon
A valid term through 1973. After 1973, use Sri Lanka.

Cf
use californium

chabazite
BA zeolite group
 framework silicates
 silicates
BT minerals

Chablais
Limestone massif of Savoy Alps and region in SE part of country.
BT Haute-Savoie
 France

Chaco
Region. Index countries as applicable.
UF Gran Chaco
BT South America
SA Argentina
 Bolivia
 Campo del Cielo
 Paraguay

Chad
Includes use on level 1 as an area term (list O). For term set options see list B.
BA Africa
NT Kanem
 Tibesti Massif
SA Chad Basin
 Lake Chad
 Logone River
 Sahara
 Sahel

Chad Basin
Index countries as applicable.
BT Africa
SA Cameroon
 Chad
 Lake Chad
 Niger
 Nigeria

Chadak
Village in E.
BT Uzbekistan
 USSR

Chadobets Uplift
SE Krasnoyarsk Kray in S Siberia.
BT Russian Republic
 USSR

Chadron Arch
NW Nebraska and SW South Dakota. Index states as applicable.
BT United States
SA Nebraska
 South Dakota

Chadron Formation
In White River Group. NE Colorado, W Nebraska, W South Dakota, and E Wyoming.
BT lower Oligocene
 Oligocene
SA Chadronian
 Colorado
 Nebraska
 South Dakota
 White River Group
 Wyoming

Chadronian
North America.
BA lower Oligocene
 Oligocene
 Paleogene
 Tertiary
BT Cenozoic
SA Chadron Formation

Chaetetidae
Family.
BA Tabulata
 Anthozoa
 Coelenterata
BT Invertebrata

Chaetopoda
Includes use on level 3. Contains Oligochaetia and Polychaetia.
BA Annelida
BT Invertebrata
SA Polychaetia
 worms

Chaffee County
Central.
BT Colorado
 United States

Chain Deep
BT Red Sea

chain silicates
Includes use on level 2 under minerals(1); in combination with amphibole group and pyroxene group (i.e. chain silicates, amphibole group) to form terms on level 2 under minerals(1). See list L.
UF inosilicates
BA silicates
BT minerals
NA aenigmatite
 amphibole group
 babingtonite
 bustamite
 carpholite
 deerite
 ferrocarpholite
 howieite
 neptunite
 pectolite
 plancheite
 pyroxene group
 pyroxmangite
 rhodonite
 rhonite
 scawtite
 shattuckite
 tobermorite
 wollastonite
 xonotlite
SA asbestos

Chaine des Puys
use Monts Dome

Chainpur
Point of impact in W Bihar. Also search Chainpur Meteorite.
BT Bihar
 India
SA meteorites

chalcanthite
BA sulfates
BT minerals

chalcedony
BA silica minerals
 framework silicates
 silicates
BT minerals
SA agate
 gems
 quartz

chalcocite
BA sulfides
BT minerals
SA copper
 copper sulfides
 digenite

chalcopyrite
UF copper pyrites
BA sulfides
BT minerals
SA copper

chalcostibite
BA sulfantimonites
 sulfosalts
BT minerals

Chaleur Bay
Inlet of Gulf of Saint Lawrence between New Brunswick and the Gaspe Peninsula. Index provinces as applicable.
BT Canada
SA New Brunswick
 Quebec

chalk
Includes use as level 3 commodity term under limestone(1); on level 3 under sedimentary rocks(1) carbonate rocks(1). See list C (commodities) and list I (sed. rocks). For stratigraphic term, use Chalk.
BA carbonate rocks
BT sedimentary rocks
SA calcite
 Chalk
 limestone

Chalk
Europe. Includes use on level 3 under age terms(1). For commodity or sedimentary rock, see chalk.
BT Upper Cretaceous
 Cretaceous
 Mesozoic
SA Cenomanian
 chalk
 Senonian
 Turonian

Challenger Knoll
BT Gulf of Mexico

chalmersite
use cubanite

chalybite
BA carbonates
BT minerals
SA siderite

Chama Basin
River basin in N.
BT New Mexico
 United States

Chamba
Town and district in N part of country.
BT Himachal Pradesh
 India

chambers, magma
use magma chambers

chamosite
BA sheet silicates
 silicates
BT minerals
SA berthierine
 chlorite group

Champlain Valley
Lake Champlain area between Adirondack Mountains and Green Mountains and extending into S Quebec. Index Quebec and states as applicable.
SA Lake Champlain
 New York
 Quebec
 Vermont

Champlainian
Provincial series, North America. Above Canadian. Includes use on level 3 under age terms(1). See list E.
BA Middle Ordovician
 Ordovician
BT Paleozoic
SA Trentonian

Chanares
Village in central.
BT Cordoba
 Argentina

Chanares Formation
BT Triassic
SA Argentina
 La Rioja

Chandigarh
City in E.
BT Punjab
 India

Chandler wobble
Includes use on level 3 under Earth(1) motions(2).
SA Earth
 motions

Changai Mountains
use Hangay Mountains

changes
Includes use as level 2 or 3 term appropriate to a large number of topics, e.g. on level 2 under paleoecology(1). See list G.
SA alteration
 conversion
 transformations

changes of level
Used mostly under stratigraphy for sea-level (sometimes lake-level) changes of the upper Cenozoic. Includes use on level 1 (list A). Term set options are:
topic [age, causes, concepts, correlation, detection, evolution, extent, genesis, interpretation, mechanism, observations, patterns, processes, rates]
 subtopic (no area term)
UF level, changes of
SA absolute age
 continental shelf
 eustacy
 geodesy
 geomorphology
 glacial geology
 glaciation
 isostasy
 neotectonics
 paleoecology
 paleogeography
 reefs
 regression
 sedimentation
 shorelines
 stratigraphy
 submergence
 tectonics
 tectonophysics
 terraces
 transgression

Changwat Nakhon Phanom
Province in NE Thailand. Nakhon Phanom is also a city.
BT Thailand

channel geometry
For channel cross-section only. Includes use on level 3 under geomorphology(1) and hydrology(1).
SA channels
 fluvial features
 geometry
 geomorphology
 hydrology
 rivers
 rivers and streams

Channel Islands
The Santa Barbara Islands off SW California coast between Santa Barbara and San Diego, and three British islands in the English Channel off the W coast of Normandy. Index California and/or England as applicable.
CO N491000N494500
W0020000W0024500
SA California
England

channel order
use stream order

channels
As of 1978, term is used on level 2 under waterways(1).
SA braided streams
buried channels
canals
channel geometry
continental slope
fluvial features
geomorphology
grooves
gullies
lava channels
levees
planar bedding structures
rivers
sedimentary structures
sedimentation
sinuosity
streams
tidal channels
troughs
waterways

chaoite
BA native elements and alloys
BT minerals
SA meteorites

Char
Rock belt in the Lake Zaysan and Kalba Mountain areas of NE.
BT Kazakhstan
USSR

Characeae
Family.
BA Chlorophyta
algae
BT Plantae

characterization
Includes use on level 3 under soils(1) water regimes(2).
SA soils
water regimes

charcoal
Includes une on level 3 under coal(1) as a commodity.
SA absolute age
carbon
coal

Chardzhov
City and oblast in E.
BT Turkmenia
USSR

Charente
Department in W France. Also a river.
BT France
NT Angouleme
Saint-Severin

Charente-Inferieure
use Charente-Maritime

Charente-Maritime
Department in W France.
UF Charente-Inferieure
BT France

Charentes
Region in W and W central.
BT France

Charleston
City in Charleston County in SE South Carolina, and a city in Kanawha County, West Virginia. Index states as applicable.
BX South Carolina
West Virginia
BT United States

Charleston County
Along the Atlantic Ocean in SE.
BT South Carolina
United States

Charlevoix
Cryptoexplosion or impact feature in Quebec. Also a city on Lake Michigan. Index Quebec or Michigan as applicable. Aslo search Charlevoix structure.
SA Michigan
Quebec

Charlevoix County
N Lower Peninsula.
BT Michigan
United States

Charlotte County
In extreme S New Brunswick. Also counties in Florida and Virginia, U.S. Index New Brunswick and states as applicable.
SA Florida
New Brunswick
Virginia

charnockite
Includes use on level 3 under igneous rocks(1) granite-granodiorite family(2). See list H.
BA granite-granodiorite family
BT igneous rocks

Charophyta
Including chara. Includes use on level 2 under algae(1). See list F. Also search charophytes.
UF charophytes
BA algae
BT Plantae

charophytes
use Charophyta

Charters Towers
City just E of the Great Dividing Range in NE central.
BT Queensland
Australia

chartology
use cartography

Chase County
Index states as applicable.
BX Kansas
Nebraska
BT United States

Chase Group
Includes Odell Formation, Nolans Formation, Wreford Limestone, Matfield Shale, Doyle Shale, Winfield Limestone, Herrington Limestone, Fort Riley Limestone and Barneston Limestone. E Kansas, N central Oklahoma, and SE Nebraska.
BT Permian
SA Fort Riley Limestone
Kansas
Nebraska
Oklahoma
Wreford Limestone

Chateaulin Basin
In extreme W Brittany.
BT Finisterre
France

Chatham County
Index states as applicable.
BX Georgia
North Carolina
BT United States

Chatham Rise
500 miles W of New Zealand.
BT Pacific Ocean

Chatkal Range
Branch of NW Tien Shan. Also search Chatkal.
BT Kirghizia
USSR
SA Tien Shan

Chattanooga Shale
Consists of lower Dowelltown Member and upper Gassaway Member. In some areas, includes Hardin Sandstone Member below the Dowelltown. N Alabama, Arkansas, NW Georgia, Illinois, E and W Kentucky, NE Mississippi, Missouri, Oklahoma, and Tennessee.
BT Paleozoic
SA Alabama
Arkansas
Devonian
Georgia
Illinois
Kentucky
Mississippi
Mississippian
Missouri
Oklahoma
Tennessee
Upper Devonian

Chattian
Europe. Above Rupelian, below Aquitanian (Miocene). Includes use on level 3 under age terms(1). See list E.
BA upper Oligocene
Oligocene
Paleogene
Tertiary
BT Cenozoic

Chaturi
use Chiatura

Chautauqua County
Index states as applicable.
BX Kansas
New York
BT United States

Chaves County
In SE New Mexico.
BT New Mexico
United States

Chaya Massif
A gabbro-peridotite massif in northern Baikal region.
BT Russian Republic
USSR

Chazyan
North America. Below Mohawkian, above lower Ordovician. Includes use on level 3 under age terms(1). See list E.
BA Middle Ordovician
Ordovician
BT Paleozoic

Chechen-Ingush
Autonomous Soviet Socialist Republic on N slopes of Caucasus.
CO N423000N440000
E0463000E0443000
UF Checheno-Ingush
BT Russian Republic
USSR

Checheno-Ingush
use Chechen-Ingush

Chedabucto Bay
Inlet of Atlantic Ocean in NE just S of Cape Breton Island.
BT Nova Scotia
Canada

Cheilostomata
Includes use on level 2 under Bryozoa(1). See list F.
BA Bryozoa
BT Invertebrata

Cheju Island
In East China Sea S of mainland.
BT Korea

Chekalin
Town in W Tula Oblast in SW European USSR. Also search Likhvin.
UF Likhvin
BT Russian Republic
USSR

Cheleken
Town on the Caspian Sea in Krasnovodsk Oblast.
BT Turkmenia
USSR

Cheleken Peninsula
In Krasnovodsk Oblast on Caspian Sea.
BT Turkmenia
USSR

Chelicerata
Subphylum. Use only for taxa not included in Merostomata and Arachnida. Includes use on level 2 under Arthropoda(1). See list F.
BA Arthropoda
BT Invertebrata
NA Arachnida
Merostomata

Chelonia
Order and genus. Includes use on level 3 under Reptilia(1) Anapsida(2).
BA Anapsida
Reptilia
BT Tetrapoda
Vertebrata

Chelyabinsk
City and oblast in Southern Urals.
BT Russian Republic
USSR

chemical
A valid term through 1977. After 1977, use more specific term, e.g. chemical composition, or chemical properties, or chemical analysis, or chemical weathering.

chemical analysis
Use only for methodology. Includes use on level 2 under minerals(1). For data, see appropriate material. Includes use on level 1 (list A). Also search chemical analyses; analysis AND chemical. Term set options are:
methods
name of method [activation analysis, chromatography, colorimetry, electrolytic analysis, infrared spectroscopy, major-element analyses, minor-element analyses, polarography, spectroscopy, trace-element analyses, vacuum fusion analysis, volumetric analysis, wet methods, X-ray fluorescence]
techniques
subtopic [e.g. sample preparation, titration, material, reagent, etc.]
SA activation analysis
analysis
automatic data processing
catalysis
chemical composition
chromatography
clay mineralogy
color
colorimetry
crystal chemistry
crystallography
decrepitation
differential thermal analysis

diffusion
electron microscopy
electron probe
emission spectroscopy
flame photometry
geochemical methods
geochemistry
geochronology
igneous rocks
inclusions
infrared spectra
ion probe
ion probe data
isotopes
major-element analyses
minerals
minor-element analyses
neutron activation analysis
organic materials
photometry
polarography
qualitative analysis
quantitative analysis
sample preparation
solubility
spectroscopy
standard materials
thermal analysis
thermogravimetric analysis
trace-element analyses
vacuum fusion analysis
wet methods
X-ray analysis
X-ray diffraction analysis
X-ray fluorescence

chemical composition
To discuss data and results of methods. Includes use on level 3 under chemical analysis(1) methods(2) and under soils(1) composition(2). Also search chemical AND composition.
  SA analysis
      chemical analysis
      chemical methods
      chemistry
      composition
      gas chromatography
      soils

chemical elements
  use elements

chemical explosions
As of 1978, term is used on level 2 under explosions(1)
  BT explosions

chemical fossils
  SA fossils
      ichnofossils
      problematic fossils

chemical magnetization
  use chemical remanent magnetization

chemical methods
Includes use on level 3 under soils(1) analysis(2). Also search chemical.
  SA chemical composition
      geochemical methods
      methods
      sample preparation
      soils

chemical properties
Includes use under commodity terms (list C).
  SA coal
      geochemistry
      properties
      thermochemical properties
      valency

chemical reactions
  use reactions

chemical remanence
  use chemical remanent magnetization

chemical remanent magnetization
Includes use on level 3 under paleomagnetism(1). Also search chemical magnetization.
  UF chemical magnetization
      chemical remanence
      CRM
      crystallization magnetization
      crystallization remanent magnetization
  BA remanent magnetization
  SA magnetization
      paleomagnetism

chemical weathering
Includes use on level 3 under weathering(1). Also search decomposition.
  UF decomposition
  BT weathering

chemically precipitated rocks
Includes use on level 2 under sedimentary rocks(1). See list I.
  BT sedimentary rocks
  NA chert
      evaporites
      flint
      iron formations
      ironstone
      itabirite
      jasperoid
      jaspilite
      phosphate rocks
      phosphorite
      taconite
      tufa
  SA anhydrite
      chemically precipitated sediments
      dolomite
      gypsum
      halite
      iron-rich composition
      novaculite
      rocks
      siliceous sinter
      travertine

chemically precipitated sediments
Includes use on level 2 under sediments(1). See list N.
  BA sediments
  NA calcrete
      caliche
      duricrust
      ferricrete
      silcrete
      siliceous sinter
      weathering crust
  SA anhydrite
      borates
      chemically precipitated rocks
      dolostone
      evaporites
      gypsum
      halite
      iron-rich composition
      siliceous composition
      travertine
      tufa

chemistry
Includes use on level 2 under soils(1).
  SA biochemistry
      chemical composition
      crystal chemistry
      geochemistry
      hydrochemistry
      soils

Chenango River
In S central.
  BT New York
      United States

Cheniers
Village in central part of country.
  BT Creuse •
      France

Cher
Department in central France. Also a river.
  BT France
  NT Bourges

Cheremkhovo Basin
  use Irkutsk Basin

Cheremkovo coal basin
  use Irkutsk Basin

Cherkasi
  use Cherkassy

Cherkassy
City and oblast in central Ukraine.
  UF Cherkasi
  BT Ukraine
      USSR

Chernigov
City in Chernigov Oblast in N.
  BT Ukraine
      USSR

Chernozems
Includes use on level 3 under soils(1). See list M. Also search Black earth.
  UF Black earth
      Chernozyom
      Tchornozem
      Tschernosem
      Tschernosiom
  BA soils
  SA Chestnut soils
      soil group
      Zonal soils

Chernozyom
  use Chernozems

Cherokee County
Index states as applicable.
  BX Alabama
      Georgia
      Iowa
      Kansas
      North Carolina
      Oklahoma
      South Carolina
      Texas
  BT United States

Cherokee Group
Divided into Riverton, Warner, Rowe, Dry Wood, Bluejacket, Seville, Weir, Tebo, Scammon, Mineral, Robinson, Branch, Fleming, Croweburg, Verdigris, Bevier, Lagonda, Mulky, and Excello formations. SW Iowa, E Kansas, W Missouri, and SE Nebraska.
  BT Desmoinesian
      Middle Pennsylvanian
      Pennsylvanian
  SA Cabaniss Formation
      Iowa
      Kansas
      Missouri
      Nebraska

chert
Includes use on level 3 under sedimentary rocks(1) chemically precipitated rocks(2). See list I. Also search hornstone.
  UF hornstein
  BA chemically precipitated rocks
  BT sedimentary rocks
  SA flint
      jasper

chertification
  BA diagenesis
  SA silicification

Chesapeake Bay
Inlet of Atlantic Ocean. Index states as applicable.
  BT United States
  SA Atlantic Coastal Plain
      Maryland
      Virginia

Cheshire
County in NW.
  CO N530000N533000
      W0020000W0032000
  BT England
      Great Britain
      United Kingdom

Chesil Beach
On the English Channel in Dorset.
  BT England
      Great Britain
      United Kingdom

Chester County
Index states as applicable.
  BX Pennsylvania
      South Carolina
      Tennessee
  BT United States

Chester Series
  use Chesterian

Chesterian
Provincial series, North America. Formations included in standard section: Aux Vases Sandstone, Renault Limestone, Bethel Sandstone, Point Creek Formation. N Alabama, Illinois, S Indiana, Kentucky, E Missouri, and Tennessee. Includes use on level 3 under age terms(1). See list E.
  UF Chester Series
  BA Upper Mississippian
      Mississippian
  BT Paleozoic
  SA Alabama
      Illinois
      Indiana
      Kentucky
      Meramecian
      Missouri
      Morrowan
      Springer Formation
      Tennessee

Chestnut soils
Includes use on level 3 under soils(1). See list M.
  BA soils
  SA Chernozems
      soil group
      Zonal soils

Cheviot Hills
Range of hills along the English-Scottish border. Index England and/or Scotland as applicable.
  BT United Kingdom
  SA England
      Scotland

chevkinite
  BA orthosilicates
      silicates
  BT minerals

chevron folds
Term introduced in 1978. Includes use on level 3 under folds(1). Before 1978, also search folds AND chevron.
  BA folds
  SA kink folds

Cheyenne County
Index states as applicable.
  BX Colorado
      Kansas
      Nebraska
  BT United States

Cheyenne Sandstone
Underlies Kiowa Shale Member; overlies Morrison Formation. SW Kansas, SE Colorado, and W Ok-

lahoma.
BT Lower Cretaceous
    Cretaceous
SA Colorado
    Kansas
    Oklahoma

Chhindwara
Town in central.
BT Madhya Pradesh
    India

Chhindwara District
Central.
BT Madhya Pradesh
    India

Chiapas
State in SE.
BT Mexico

Chiatura
City in central Georgian Republic, USSR. Also search Chiaturi.
UF Chaturi
BT Georgian Republic
    USSR

Chiayi-Hsinying area
BT Taiwan

Chiba
City and prefecture SE of Tokyo.
BT Honshu
    Japan

Chiba Peninsula
Between Tokyo Bay and Sagami Sea. Also search Boso Peninsula, and Chiba.
UF Boso Peninsula
BT Honshu
    Japan

Chibougamau
Village SE of Lake Mistassini in central.
BT Quebec
    Canada

Chicago
City on Lake Michigan in Cook County.
BT Illinois
    United States

Chichibu
City NW of Tokyo in central.
BT Honshu
    Japan

Chichibu Belt
BT Honshu
    Japan

Chichibu Mine
Central.
BT Honshu
    Japan

Chickamauga Group
Includes limestone strata of Middle and Upper Ordovician age. N Alabama, NW Georgia, E Tennessee, and SW Virginia. Also search Chickamauga Limestone.
BT Ordovician
SA Alabama
    Georgia
    Middle Ordovician
    Tennessee
    Upper Ordovician
    Virginia

Chickasawhay Formation
In Limestone Creek Group. In SE Mississippi.
BT upper Oligocene
    Oligocene
SA Mississippi

Chickmagalur
Village in SW.
BT Mysore
    India

Chieti
Town in E.
BT Abruzzi
    Italy

Chigoku Sammayuku Mountains
use Chugoku

Chihuahua
State in northern Mexico. Also a city.
CO N254500N320000
    W1030000W1080000
BT Mexico
NT Allende
SA Chihuahua tectonic belt
    Pedregosa Basin
    Rio Grande
    Rio Grande Rift
    Rio Grande Valley
    Sierra Madre Occidental

Chihuahua tectonic belt
Index Mexican and U. S. states as applicable.
SA Chihuahua
    Coahuila
    New Mexico
    Rio Grande Rift
    Texas

Chile
Includes use on level 1 as an area term (list O). For term set options see list B.
CO S560000S174500
    W0670000W0760000
BA South America
NT Aconcagua Province
    Antofagasta
    Arauco
    Atacama
    Atacama Desert
    Coquimbo
    Magallanes
    Santiago
    Tarapaca
    Valparaiso
SA Andes
    Central Valley
    Patagonia
    Patagonian Andes
    Tierra del Fuego

Chile Ridge
Runs SE-NW off Chilean coast between 80° W-105° W. Also search Chile Rise.
UF Chile Rise
BT Pacific Ocean

Chile Rise
use Chile Ridge

Chilhowee Group
Includes Loudoun Formation, Weverton Sandstone, Harpers Shale, Antietam Sandstone, Unicoi Formation, Hampton Shale, Erwin Quartzite, Vann Quartzite, Sandsuck Shale, Cochran Quartzite, Nebo Quartzite, Murry Shale, Hesse Quartzite, Shady Dolomite, Rome Formation, Nichols Shale, Helenmode Formation.
BT Lower Cambrian
    Cambrian
SA Antietam Formation
    Maryland
    North Carolina
    Shady Dolomite
    Tennessee
    Virginia

Chilton County
Central.
BT Alabama
    United States

China
Peoples Republic of China. Includes use on level 1 as an area term (list O). For term set options see list B.
CO N200000N530000
    E1350000E0740000
BA Asia
NT Han River basin
    Heilungkiang
    Inner Mongolia
    Manchuria
    Shansi
    Shensi
    Sinkiang Uighur
    Szechwan
    Tibet
    Yunnan
SA Altai Mountains
    Brahmaputra River
    Far East
    Khanka Lake

China Sea
Bordering on China and divided by Taiwan into East China Sea to N and South China Sea extending to Malaysia in the S. Index seas as applicable.
BT Pacific Ocean
SA East China Sea
    South China Sea

Chingis-Tau
Mountain range in Semipalatinsk Oblast in E Kazakhstan. Also search Chingiz Range.
UF Chingiz Range
BT Kazakhstan
    USSR

Chingiz Range
use Chingis-Tau

Chinkuashih
Village in N Taiwan.
UF Kinnashih
BT Taiwan

Chinkuashih Mine
N Taiwan. Also search Chinkuashih.
UF Kinwashih Mine
BT Taiwan

Chinle Formation
In Dockum Group. N Arizona, SW Colorado, SE Nevada, N New Mexico, S Utah.
BT Upper Triassic
    Triassic
SA Arizona
    Colorado
    Dockum Group
    Nevada
    New Mexico
    Utah

Chios
City and island in Aegean Sea off W coast of Turkey.
UF Khios
BT Aegean Islands
    Greece

Chipola Formation
In Alum Bluff Group. SE Alabama and NW Florida.
BT lower Miocene
    Miocene
SA Alabama
    Florida

Chirchik
City in Tashkent Oblast in E.
BT Uzbekistan
    USSR

Chirchik River
Flows into the Syr Darya at Chinaz. Index Soviet republics as applicable. Also search Chirchik.
BT USSR
SA Kazakhstan
    Uzbekistan

Chiroptera
Includes use on level 2 under Mammalia(1). See list F.
BA Mammalia
BT Tetrapoda

Vertebrata

Chita
City and oblast E of Lake Baikal on the Mongolian and Manchurian borders.
BT Russian Republic
    USSR

Chitaldroog
use Chitradurga

Chitaldrug
use Chitradurga

Chitaldrug schist belt
Crystalline rocks in this area are considered to be Dharwars. Also search Chitaldrug.
BT Mysore
    India
SA Dharwars

Chitinozoa
Includes use on level 2 under palynomorphs(1). See list F. Also search chitinozoans.
UF chitinozoans
BA palynomorphs

chitinozoans
use Chitinozoa

Chitistone Pass
In the St. Elias Mountains. Index Alaska and/or Yukon Territory as applicable.
SA Alaska
    Yukon Territory

Chitradurga
Town in E central Mysore. Also search Chitaldrug.
UF Chitaldroog
    Chitaldrug
BT Mysore
    India

Chitral
Town and river.
BT North-West Frontier Province
    Pakistan

Chittenden County
On Lake Champlain in NW.
BT Vermont
    United States

Chiuzbaia
Region in Maramures County in N.
BT Transylvania
    Romania

Chkalov
use Orenburg

chkalovite
BA framework silicates
    silicates
BT minerals

chlorapatite
BA phosphates
BT minerals
SA apatite

chloride ion
Term introduced in 1978. Before 1978, search chloride.
SA chlorides
    ions

chlorides
Includes use on level 3 under minerals(1) halides(2). See list L.
BA halides
BT minerals
SA atacamite
    boracite
    carnallite
    chloride ion
    eudialyte
    halite
    kainite
    meionite
    mimetite
    parafacamite

phosgenite
pyromorphite
rinneite
sodium chloride
sylvite
vanadinite

chlorine
Includes use on level 1 and 2 as a chemical element (list D). Also search Cl.
UF   Cl
SA   elements

chlorite
BA   sheet silicates
     silicates
BT   minerals
SA   chlorite group
     chloritization

chlorite group
Includes use in combination with sheet silicates (i.e. sheet silicates, chlorite group) on level 2 under minerals(1). See list L.
BA   sheet silicates
     silicates
BT   minerals
SA   chamosite
     chlorite
     clinochlore
     cookeite
     kammererite
     sudoite

chlorite schist
Includes use on level 3 under metamorphic rocks(1) schists(2). See list J.
BA   schists
BT   metamorphic rocks
SA   schist

chloritization
Includes use on level 3 under metasomatism(1).
BA   metasomatism
SA   chlorite
     processes

chloritoid
BA   orthosilicates
     silicates
BT   minerals
SA   aluminosilicates

chloropal
use   nontronite

chlorophaeite
BA   silicates
BT   minerals

chlorophyll
BA   organic materials

Chlorophyta
Green algae and desmids are included here. Includes use on level 2 under algae(1). See list F.
UF   green algae
BA   algae
BT   Plantae
NA   Botryococcus
     Characeae
     Codiaceae
     desmids

Choctawhatchee Bay
Inlet of Gulf of Mexico in the Florida panhandle in NW.
BT   Florida
     United States

Cholame
Town in San Luis Obispo County on the Pacific in S.
BT   California
     United States

Chondrichthyes
Selachians are included here. Includes use on level 2 under Pisces(1). See list F.
BA   Pisces
BT   Vertebrata
NA   Elasmobranchii
SA   fish

chondrites
Includes use on level 3 under meteorites(1).
BA   meteorites
NA   carbonaceous chondrites
     enstatite chondrites
SA   achondrites
     chondrules
     stony irons

chondrules
Includes use on level 3 under meteorites(1).
SA   chondrites
     meteorites

Choptank Formation
In Chesapeake Group. E Maryland and Virginia.
BT   middle Miocene
     Miocene
SA   Maryland
     Virginia

Choptank River
Index states as applicable.
BT   United States
SA   Delaware
     Maryland

Chor
Village in SE part of country.
BT   Sind
     Pakistan

Chordata
Includes use on level 1 and 2 as a fossil term (list F). Term is to be used only when more specific terms do not apply.
SA   Hemichordata
     Vertebrata

chorology
use   biogeography

Chorzow
City in S part of country.
BT   Katowice
     Poland

Choybalsan
Town on Kerulen River in E Mongolia. Also search Kerulen.
UF   Kerulen
BT   Mongolia
     Asia

Christchurch
City on Pacific Ocean in E.
BT   South Island
     New Zealand

Christiansund
use   Kristiansund

Christmas Island
An external territory of Australia, about 225 miles S of W end of Java.
BT   Indian Ocean

Christmas Mountains
In Big Bend region in SW.
BT   Texas
     United States

Christopher Formation
Ellef Ringnes Island in Franklin District.
BT   Lower Cretaceous
     Cretaceous
SA   Canada
     Northwest Territories

chromates
Includes use on level 2 under minerals(1). See list L.
BT   minerals
NA   crocoite
     hemihedrite

chromatography
Includes use on level 3 under chemical analysis(1) methods(2).
SA   adsorption
     analysis
     chemical analysis
     gas chromatography
     spectroscopy

chrome diopside
UF   chrome-diopside
BA   pyroxene group
     chain silicates
     silicates
BT   minerals
SA   diopside

chrome spinel
Also search picotite.
UF   picotite
BA   oxides
BT   minerals
SA   spinel

chrome-diopside
use   chrome diopside

chromite
Includes use on level 1 and 2 as a commodity term (list C).
BA   oxides
BT   minerals
SA   chromitite

chromitite
Includes use on level 3 under igneous rocks(1) ultramafic family(2). See list H.
BA   ultramafic family
BT   igneous rocks
SA   chromite

chromium
Includes use on level 1 and 2 as a chemical element (list D). Also search Cr.
UF   Cr
SA   elements

chronology
Used as a general term.
SA   geochronology
     time

chronostratigraphy
Includes use under stratigraphy(1) or individual ages (list E).
SA   stratigraphy

chrysoberyl
BA   oxides
BT   minerals
SA   alexandrite

chrysocolla
BA   sheet silicates
     silicates
BT   minerals

Chrysophyta
Including golden-brown algae. Includes use on level 2 under algae(1). See list F.
UF   golden-brown algae
BA   algae
BT   Plantae

chrysotile
Also search chrysotile asbestos.
BA   sheet silicates
     silicates
BT   minerals
SA   serpentine group

Chu
Town and river in Dzhambul Oblast in S.
BT   Kazakhstan
     USSR

Chu-Ili Mountains
Branch of the Tien Shan in Dzhambul Oblast in S.
BT   Kazakhstan
     USSR
SA   Tien Shan

Chuar Group
Grand Canyon Series. Grand Canyon region in N Arizona.
BT   Precambrian
SA   Arizona

Chubut
Province in S.
CO   S460000S420000
     W0640000W0730000
BT   Argentina
SA   Patagonia

Chuckanut Formation
Overlies schists and igneous rocks. San Juan Islands in NW Washington.
BT   Eocene
SA   Washington

Chudleigh
Town in Devon in SW.
BT   England
     Great Britain
     United Kingdom

Chugach Mountains
Along S coast.
BT   Alaska
     United States

Chugoku
Mountains in SW Honshu.
UF   Chigoku Sammayuku Mountains
BT   Honshu
     Japan

Chukchi Peninsula
NE extremity of Siberia in Chukchi National Okrug of Magadan Oblast. Also search Chukchi, and Chukotka.
CO   N630000N700000
     W1700000E1590000
UF   Chukot Peninsula
     Chukotka (Peninsula)
     Chukotsk Peninsula
     Chukotski Peninsula
     Chukotskiy Peninsula
BT   Russian Republic
     USSR
SA   Okhotsk-Chukchi

Chukchi Range
use   Anadyr Range

Chukchi Sea
N of Bering strait. Between Alaska and Siberia. Also search Chukchi.
BT   Arctic Ocean
SA   Bering land bridge

Chukot Peninsula
use   Chukchi Peninsula

Chukotka (Peninsula)
use   Chukchi Peninsula

Chukotsk Peninsula
use   Chukchi Peninsula

Chukotski Peninsula
use   Chukchi Peninsula

Chukotskiy Peninsula
use   Chukchi Peninsula

Chuma River
use   Uda River

Chuquicamata
Mining settlement in N part of country.
BT   Antofagasta
     Chile

Church Stretton
Urban district in S central Shropshire in W.
BT   England
     Great Britain
     United Kingdom

Churchill
Town on Hudson Bay.
BT   Manitoba
     Canada

Churchill County
In W Nevada.
BT Nevada
United States

Churchill Falls
On Churchill River in SW and S Labrador. Also search Grand Falls.
UF Grand Falls
BT Labrador
Canada

Churchill Province
That part of the Canadian Shield lying in Canada. Index Labrador and provinces as applicable.
CO N540000N720000 W0600000W1120000
BT Canada
SA Canadian Shield
Labrador
Manitoba
Northwest Territories
Ontario
Quebec
Saskatchewan

Chuya
Village in NE Irkutsk Oblast near the SW Yakut A.S.S.R. border.
BT Russian Republic
USSR

Chuya Alps
Range in Altai Mountains near the Mongolian and Sinkiang-Uighur borders.
BT Russian Republic
USSR
SA Altai Mountains

Chuya Basin
River basin in Altai Kray near the Mongolian and Sinkiang-Uighur borders.
BT Russian Republic
USSR

Chvaletice
Mine site in E.
BT Bohemia
Czechoslovakia

Cibicides
Includes use on level 3 under foraminifera(1) Orbitoidacea(2).
BA Orbitoidacea
foraminifera
BT Invertebrata

Cicatricosisporites
BA miospores
palynomorphs

Cieszyn
City in S part of country.
BT Katowice
Poland

Cilento
Area in SW part of country.
BT Campania
Italy

Cima d'Asta
Peak in the Dolomites in E.
BT Trentino-Alto Adige
Italy
SA Dolomites

Cimmaron County
In extreme W of panhandle.
BT Oklahoma
United States

Cimmerian
Black Sea area. Above Meotian, below Kuyalnikian. Includes use on level 3 under age terms(1). See list E.
UF Kimmerian
BA Pliocene
Neogene
Tertiary
BT Cenozoic

Cincinnati
City on the Ohio River in Hamilton County in SW.
BT Ohio
United States

Cincinnati Arch
Index states as applicable.
BT United States
SA Indiana
Kentucky
Ohio

Cincinnatian
Provincial series, North America. Above Champlainian, below Alexandrian (Silurian). Includes use on level 3 under age terms(1). See list E.
BA Upper Ordovician
Ordovician
BT Paleozoic

cinder cones
UF cones, cinder
BT volcanic features
SA volcanism

cinerite
Includes use on level 3 under sedimentary rocks(1) clastic rocks(2). See list I.
BA clastic rocks
BT sedimentary rocks
SA terrigenous materials

cinnabar
BA sulfides
BT minerals
SA mercury

circulation
A valid level 1 term through 1977 used in combination with oceans (i.e. oceans, circulation). After 1977, includes use under meteorology(1).
SA atmosphere
climate-induced circulation
meteorology
ocean circulation
ocean waves
paleocirculation
permeability
springs
thermal circulation
thermohaline circulation
winds

Circum-Antarctic region
SA Antarctica

Circum-Pacific Belt
use Circum-Pacific region

Circum-Pacific region
Term introduced in 1978. The great-circle belt that borders the Pacific Ocean along the continental margin of Asia and the Americas. Before 1978, also search Circum-Pacific Belt; Circum-Pacific.
UF Circum-Pacific Belt
BT Pacific Ocean
SA Pacific region

cirques
Includes use on level 3 under glacial geology(1) glacial features(2).
BT glacial features
SA erosion
eskers
glacial geology
glacial lakes
kames

Cirripedia
Includes use on level 2 under Arthropoda(1). See list F.
BA Crustacea
Arthropoda
BT Invertebrata

Ciscaucasia
use Northern Caucasus

cistern rock
use laccoliths

Citronelle Formation
Recognized to be equivalent of Willis Formation of Texas and Louisiana. Graham Ferry Formation of Pliocene and Pleistocene age disconformably underlies Citronelle Formation. Gulf Coastal Plain from W Florida and S Georgia to E Texas inclusive.
BT Pliocene
SA Alabama
Florida
Georgia
Louisiana
Mississippi
Texas

Ciudad de Valles
use Valles

Ciudad Real
Province in S central.
BT Spain
NT Alcudia Valley
Almaden
SA Montes de Toledo

Ciudad Trujillo
use Santo Domingo

civil engineering
UF engineering, civil
SA engineering geology

Cl
use chlorine

Cladophlebis
BA miospores
palynomorphs

Claiborne Group
Includes Carrizo Sand, Reklaw Formation, Queen City Sand, Weches Glauconitic Marl, Viesca Glauconitic Marl, Sparta Sand, Stone City Beds, Crockett Formation, Landrum Shale, Spiller Sand, Yegua Formation, Gosport Sand, Tallahatta Formation, Zilpha Clay, Winona Formation, Wautubbee Formation, Cockfield Formation, Avon Park Limestone, Tallahassee Limestone, Lake City Limestone, Lisbon Formation, Neshoba Sand, Cook Mountain Formation, Meridian Formation, Kosciusko Formation, Cane River Formation. Gulf Coastal Plain from Georgia to S Texas.
BT middle Eocene
Eocene
SA Alabama
Cook Mountain Formation
Florida
Georgia
Louisiana
Mississippi
Queen City Formation
Texas

Claiborne Parish
In NW Louisiana.
BT Louisiana
United States

Clare
County on the Atlantic in W.
BT Ireland

Clark County
Index states as applicable.
BX Arkansas
Idaho
Illinois
Indiana
Kansas
Kentucky
Missouri
Nevada
Ohio
South Dakota
Washington
Wisconsin
BT United States

Clarke County
Index states as applicable.
BX Alabama
Georgia
Iowa
Mississippi
Virginia
BT United States

Clarno
Village in Wheeler County in N central.
BT Oregon
United States
SA Clarno Formation

Clarno Formation
Subdivided into four units. N central Oregon.
BT Eocene
SA Clarno
Oregon

classification
Includes use as level 2 or 3 term appropriate to a large number of topics, e.g. on level 2 under geochemistry(1) and under lava(1). See list G.
SA definition
identification
nomenclature
terrain classification

Classopollis
BA miospores
palynomorphs

clastic dikes
Includes use on level 3 under sedimentary structures(1) soft sediment deformation(2). See list K.
BA soft sediment deformation
sedimentary structures
SA dikes
sandstone dikes

clastic rocks
Valid index term for GeoRef since 1976. Before 1976, search clastics AND terrigenous or clastics AND nonterrigenous in combination with sedimentary rocks. From 1976 through 1977, search clastic rocks AND terrigenous or nonterrigenous. After 1977, use clastic rocks on level 2 under sedimentary rocks(1). Also search clastic.
BT sedimentary rocks
NA arenite
argillite
arkose
bentonite
bituminous shale
black shale
breccia
cinerite
claystone
conglomerate
contourite
diamictite
diatomaceous earth
diatomite
eolianite
fanglomerate
flysch
gaize
graywacke
NZ marl
NA microbreccia
mixtite
molasse
mudstone
novaculite
orthoquartzite
porcellanite
pyroclastics
radiolarite

       red beds
       sandstone
       saprolite
       shale
       siltstone
       sparagmite
       spongolite
       subarkose
       subgraywacke
       tillite
       tilloid
       tonstein
       turbidite
  SA bauxite
       clastic sediments
       fragments
       greensand
       nonterrigenous materials
       siliceous composition
       terrigenous materials
       wildflysch

clastic sediments
Valid index term for GeoRef since 1976. Before 1976 search clastics AND terrigenous or clastics AND nonterrigenous. Before 1978 included use in combination with terrigenous or nonterrigenous (i.e. clastic sediments, terrigenous) on level 2 under sediments(1). After 1978, use clastic sediments on level 2. Also search clastic.
  BA sediments
  NA alluvium
       boulders
       clay
       cobbles
       colluvium
       coquina
       diamicton
       drift
       dust
       eluvium
       flint clay
       gravel
       loess
       mud
       ooze
       outwash
       pebbles
       proluvium
       quartz sand
       residuum
       sand
       shingle
       silt
       till
  SA bentonite
       cementation
       clastic rocks
       flysch
       gaize
       greensand
       nonterrigenous materials
       silcrete
       terrigenous materials
       volcanic ash
       wildflysch

clastics
A level 2 term through 1975 used in combination with terrigenous or nonterrigenous to indicate sedimentary rocks or sediments (i.e. clastics, terrigenous or clastics, nonterrigenous). After 1975 through 1977, clastic rocks and clastic sediments were used on level 2 in combination with terrigenous or nonterrigenous (i.e. clastic rocks, terrigenous). After 1977, use clastic rocks or clastic sediments on level 2 under respective level 1 term.

clasts
Includes use on level 3 under sedimentary rocks(1) or sediments(1).
  SA fragments
       sedimentary rocks
       sediments

Clatsop County
Extreme NW.
  CO N454500N461500
       W1232000W1240200
  BT Oregon
       United States

clay
When of economic value, use clays. Includes use on level 3 under sediments(1) clastic sediments(2). See list N.
  BA clastic sediments
  BT sediments
  SA alluvium
       argillaceous texture
       armored mud balls
       clay mineralogy
       clays
       claystone
       fireclay
       flint clay
       fuller's earth
       kaolin
       loam
       marl
       mud
       mud lumps
       silt
       terrigenous materials
       underclay

clay (soil)
use Clay soils

Clay County
Index states as applicable.
  BX Alabama
       Arkansas
       Florida
       Georgia
       Illinois
       Iowa
       Kansas
       Kentucky
       Minnesota
       Mississippi
       Missouri
       Nebraska
       North Carolina
       South Dakota
       Tennessee
       Texas
       West Virginia
  BT United States

clay mineralogy
Includes use on level 1 (list A). Term set options are:
    areal studies
        area
    experimental studies
        chemical property
        material
        mineral
        physical property
        subtopic
    mineral data
        mineral name
  SA adsorption
       areal studies
       bentonite
       cation exchange capacity
       ceramic materials
       chemical analysis
       clay
       clay minerals
       clays
       crystal chemistry
       crystal growth
       crystal structure
       crystallinity
       crystallography
       differential thermal analysis
       electron microscopy
       flocculation
       fuller's earth
       glauconite
       ion exchange
       kaolin
       mineral data
       mineralogy
       minerals
       paragenesis
       sedimentary petrology
       sedimentary rocks
       sediments
       sheet silicates
       silicates
       soils
       spectroscopy
       thermal analysis
       transformations
       vermiculite
       weathering
       X-ray analysis

clay minerals
Includes use in combination with sheet silicates (i.e. sheet silicates, clay minerals) on level 2 under minerals(1). See list L.
  BA sheet silicates
       silicates
  BT minerals
  SA allophane
       argillaceous texture
       bauxite
       beidellite
       bentonite
       clay mineralogy
       corrensite
       dickite
       differential thermal analysis
       halloysite
       hectorite
       hisingerite
       hydromica
       illite
       imogolite
       kaolin
       kaolinite
       kaolinization
       metabentonite
       metakaolin
       mixed-layer minerals
       montmorillonite
       nacrite
       nontronite
       palygorskite
       rectorite
       saponite
       sepiolite
       smectite
       stevensite
       tosudite
       vermiculite

Clay soils
  UF clay (soil)
  BA soils

clay stone
use claystone

clays
Includes use as a commodity term on level 1 (list C). Use only when of economic value, otherwise use clay. Also search brick clays.
  SA clay
       clay mineralogy
       engineering geology
       flint clay
       fuller's earth
       kaolin
       quick clay
       refractory materials
       shale
       soil mechanics

claystone
Includes use on level 3 under sedimentary rocks(1) clastic rocks(2). See list I.
  UF clay stone
  BA clastic rocks
  BT sedimentary rocks
  SA clay
       mudstone
       terrigenous materials

Clayton Formation
In Midway Group. S Alabama, SW Georgia, NE Mississippi, SE Missouri, and S Tennessee.
  BT Paleocene
  SA Alabama
       Georgia
       Mississippi
       Missouri
       Tennessee

Clear Creek
River in N central.
  BT Colorado
       United States

Clear Creek County
N central.
  BT Colorado
       United States

Clear Lake
Lake County in NW central.
  BT California
       United States

Clearwater County
Index states as applicable.
  BX Idaho
       Minnesota
  BT United States

cleat spar
use ankerite

cleavage
Includes use on level 3 under foliation(1) and under analysis(1). Before 1978, term also refers to cleavage folds. After 1978, when discussing folds, use cleavage folds.
  SA axial-plane structures
       cleavage folds
       crystal growth
       crystallography
       crystals
       flow cleavage
       folds
       foliation
       fracture cleavage
       lineation
       minerals
       schistosity
       slaty cleavage
       slip cleavage
       structural analysis

cleavage folds
Term introduced in 1978. Includes use on level 3 under folds(1). Before 1978, also search cleavage AND folds.
  UF shear-cleavage folds
  BA folds
  SA cleavage

Clermont
City in N part of country.
  BT Oise
       France

Clermont-Ferrand
City in S central part of country.
  BT Puy-de-Dome
       France

Cleveland
City on Lake Erie in Cuyahoga County in N.
  BT Ohio
       United States

cliff of displacement
use fault scarps

cliffs
Includes use on level 3 under geomorphology(1) erosion features(2), fluvial features(2), and shore features(2).

SA erosion features
    fault scarps
    fluvial features
    geomorphology
    gullies
    scarps
    shore features
    slopes
    talus slopes
climate
Includes use on level 3 under soils(1) genesis(2) as type of factor.
SA arctic environment
    arid environment
    atmosphere
    factors
    humidity
    meteorology
    paleoclimatology
    precipitation
    soils
    storms
climate-induced circulation
Term introduced in 1978. Includes use on level 2 under ocean circulation(1). Before 1978, also search oceans AND circulation AND climate induced or climate-induced.
SA circulation
    ocean circulation
climatology, paleo-
use paleoclimatology
Climax
Village in Lake County in central.
BT Colorado
    United States
clinoamphibole
BA amphibole group
    chain silicates
    silicates
BT minerals
clinochlore
BA sheet silicates
    silicates
BT minerals
SA chlorite group
clinoenstatite
BA pyroxene group
    chain silicates
    silicates
BT minerals
clinohumite
BZ halides
    orthosilicates
BT minerals
SA fluorides
    humite group
    titanoclinohumite
clinohypersthene
BA pyroxene group
    chain silicates
    silicates
BT minerals
clinoptilolite
BA zeolite group
    framework silicates
    silicates
BT minerals
SA heulandite
clinopyroxene
BA pyroxene group
    chain silicates
    silicates
BT minerals
SA pigeonite
clinopyroxenite
BA ultramafic family
BT igneous rocks
SA pyroxenite
clinozoisite
BA epidote group
    orthosilicates

    silicates
BT minerals
SA aluminosilicates
    zoisite
Clinton Group
Includes Willowvale Shale, Rose Hill Formation, Keefer Sandstone, Rochester Shale, Osgood Formation, Laurel Formation, Waldron Formation, Reynales Formation, Irondequoit Formation, Decew Formation, Neahga Formation, Thorold Formation, Maplewood Formation, Furnaceville Formation, Sodus Formation, Williamson Formation. New York to NE Tennessee; also Michigan.
BT Middle Silurian
    Silurian
SA Maryland
    Michigan
    New York
    Pennsylvania
    Virginia
    West Virginia
clintonite
Also search xanthophyllite.
UF xanthophyllite
BA sheet silicates
    silicates
BT minerals
SA mica group
Clitheroe
City E of Liverpool in E Lancashire.
BT England
    Great Britain
    United Kingdom
Cloridorme Formation
Gaspe Peninsula.
BT Middle Ordovician
    Ordovician
SA Canada
    Quebec
closed fractures
Term introduced in 1978. Includes use on level 3 under fractures(1). Before 1978, also search fractures AND closed.
BA fractures
closed systems
Used as a general term.
SA open systems
    systems
Cloud County
N central.
BT Kansas
    United States
clouds
Includes use on level 3 under meteorology(1) water(2).
SA droplets
    meteorology
    raindrops
    water
Clovis
City in Curry County on the Texas border in E.
BT New Mexico
    United States
Cluj
City in Cluj County in central.
BT Transylvania
    Romania
cluster analysis
Includes use on level 3 under mathematical geology(1).
BA statistical methods
SA analysis
    mathematical geology
    statistical analysis
Cm
use curium

Co
use cobalt
CO
use carbon monoxide
Coahuila
State in NE.
CO N243000N295000 W1000000W1040000
BT Mexico
NT Parras Basin
SA Chihuahua tectonic belt
    Difunta Group
    Rio Grande
    Rio Grande Rift
    Rio Grande Valley
    Sierra Madre Oriental
    Zuloaga Limestone
coal
Includes use on level 1 as a commodity term (list C); on level 3 under sedimentary rocks(1) organic residues(2). Also search carbargilite.
UF carbargilite
BA organic residues
BT sedimentary rocks
NT humic coal
SA anthracite
    anthraxolite
    biogenic structures
    bituminous coal
    brown coal
    carbonaceous composition
    charcoal
    chemical properties
    coal balls
    coal basins
    coal fields
    coal measures
    coal seams
    coalification
    energy sources
    exinite
    fusinite
    gasification
    inertinite
    lignite
    lithification
    macerals
    micrinite
    open-pit mining
    organic materials
    peat
    rank
    resinite
    sapropelite
    sporinite
    vitrain
    vitrinite
    volatiles
coal balls
Includes use on level 3 under coal(1) as a commodity.
UF balls, coal
SA coal
    coal seams
    concretions
coal basins
Includes use on level 3 under coal(1) as a commodity.
SA basins
    coal
    coal fields
    coal measures
coal beds
use coal seams
coal clay
use underclay
Coal County
SE central.
BT Oklahoma
    United States
coal fields
Includes use on level 3 under coal(1) as a commodity. Also search

coalfields.
UF coalfields
    fields, coal
SA coal
    coal basins
coal measures
Includes use on level 3 under coal(1) as a commodity.
SA coal
    coal basins
    Coal Measures
    coal seams
    sedimentary rocks
Coal Measures
Europe. Includes use on level 3 under age terms(1). See list E.
BA Upper Carboniferous
    Carboniferous
BT Paleozoic
SA coal measures
coal seams
Includes use on level 3 under coal(1) as a commodity. Also search coal beds; coal deposits; coal strata.
UF coal beds
    seams, coal
SA coal
    coal balls
    coal measures
    underclay
coalfields
use coal fields
coalification
Includes use on level 3 under coal(1) as a commodity.
UF carbonification
    incarbonization
    incoalation
SA coal
    diagenesis
    processes
Coalinga
City in Fresno County in S central.
BT California
    United States
Coalsack Bluff
Site of paleontological investigations in Beardmore Glacier area around NW end of Queen Maud Range off the Ross Ice Shelf in Ross Dependency.
BT Antarctica
coast
A valid term through 1978. After 1978, use coastal environment.
Coast Mountains
Extend from Yukon Territory to the Fraser River serving as border between NW British Columbia and the Alaskan panhandle. Index Alaska, British Columbia, and Yukon Territory as applicable.
SA Alaska
    British Columbia
    Yukon Territory
Coast Ranges
Mountains extending along the Pacific Ocean from Washington to the Mexican border. Index states as applicable.
BT United States
SA California
    Diablo Range
    Gabilan Range
    Klamath Mountains
    Oregon
    Peninsular Ranges
    Santa Cruz Mountains
    Santa Lucia Range
    Washington
coastal
A valid term through 1977. After 1977, use coastal environment.

coastal environment
Term introduced in 1978. Includes use on level 3 under sedimentation(1). Before 1978, also search coastal; search coast.
SA beaches
    coastal plains
    environment
    inlets
    lagoons
    playas
    sedimentation
    shorelines
    spits
    tidal flats

coastal plains
Includes use on level 3 under geomorphology(1).
SA coastal environment
    geomorphology
    plains

cobalt
Includes use on level 1 and 2 as a commodity term (list C) and as a chemical element (list D). Also search Co.
UF Co
SA cobaltite
    elements
    heavy metals

Cobalt
Mining town in SE Ontario.
BT Ontario
    Canada

cobalt pyrites
use linnaeite

cobaltite
BA arsenides
    sulfides
BT minerals
SA cobalt

Cobb Seamount
Aproximately 550 miles W of Washington.
BT Pacific Ocean

cobbles
Includes use on level 3 under sediments(1) clastic sediments(2). See list N.
BA clastic sediments
BT sediments
SA boulders
    gravel
    pebbles
    shingle
    terrigenous materials

Cobequid Bay
Eastern arm of Minas Basin off Bay of Fundy.
BT Nova Scotia
    Canada
SA Minas Basin

Coburg
City in N.
BT Bavaria
    West Germany

Coccolithophoraceae
Includes use on level 2 under algae(1). See list F. Also search coccolithophores; Coccolithophoridae.
UF coccolithophores
    Coccolithophoridae
BA algae
BT Plantae
NA Ceratolithus
    Coccolithus
SA coccoliths
    discoasters
    nannofossils

coccolithophores
use Coccolithophoraceae

Coccolithophoridae
use Coccolithophoraceae

coccoliths
SA Coccolithophoraceae
    Coccolithus
    ooze
    sediments

Coccolithus
BA Coccolithophoraceae
    algae
BT Plantae
SA coccoliths

Cochabamba
Department in central Bolivia. Also a city.
BT Bolivia

Cochise County
Extreme SE.
BT Arizona
    United States

Cochrane
Town and district S of James Bay.
BT Ontario
    Canada

Coconino County
N central.
CO N341500N370000
    W1104500W1132000
BT Arizona
    United States

Coconino Sandstone
In Aubrey Group. N Arizona, S Utah, and SE Nevada.
BT Permian
SA Arizona
    Nevada
    Utah

Cocos Plate
In the Pacific Ocean W of the Caribbean Plate, E and S of the Pacific Plate, and N of the Nazca Plate which is S of the Galapagos Islands.
BT Pacific Ocean
SA plate tectonics

Cocos Ridge
Between Costa Rica and the Galapagos Islands.
BT Pacific Ocean

coda waves
Includes use on level 3 under seismology(1).
SA seismology
    waves

Codiaceae
Family.
BA Chlorophyta
    algae
BT Plantae
NA Halimeda

Cody Shale
In Colorado Group or Montana Group. Bighorn Basin in N Wyoming.
BT Upper Cretaceous
    Cretaceous
SA Colorado Group
    Montana Group
    Wyoming

coefficient, correlation
use correlation coefficient

coefficient of permeability
use hydraulic conductivity

coefficient of transmissibility
use transmissibility coefficient

coefficients, partition
use partition coefficients

Coelenterata
Includes use on level 1 and 2 as a fossil term (list F).
BT Invertebrata
NA Anthozoa
    Hydrozoa
    Octocorallia
    Scyphozoa
    Stromatoporoidea
SA corals
    hermatypic taxa
    Receptaculitaceae

Coelodonta antiquitatis
Includes use on level 3 under Mammalia(1) Perissodachyla(2).
BA Perissodachyla
    Mammalia
BT Tetropoda
    Vertebrata

coercive force
use coercivity

coercivity
Before 1978, also search coercive force.
UF coercive force
SA magnetic field
    paleomagnetism
    remanent magnetization

coesite
BA silica minerals
    framework silicates
    silicates
BT minerals

Coeur d'Alene
City in Kootenai County in N.
BT Idaho
    United States

Coeur d'Alene River
Flows into Coeur d'Alene Lake in Kootenai County in N.
BT Idaho
    United States

coexisting
A valid term through 1977. After 1977, use coexisting minerals or coexisting materials.

coexisting materials
Term introduced in 1978. Before 1978, search coexisting. If the material is mineral, use coexisting minerals.
SA coexisting minerals
    materials

coexisting minerals
Term introduced in 1978. Before 1978, also search coexisting AND minerals.
BT minerals
SA coexisting materials

Coeymans Formation
Includes Stormville Sandstone Member, Elbow Ridge Sandstone Member. W Maryland, E New York, E Pennsylvania, W Virginia, N West Virginia, and New Jersey.
BT Lower Devonian
    Devonian
SA Maryland
    New York
    Pennsylvania
    Virginia
    West Virginia

Coffee County
Index states as applicable.
BX Alabama
    Georgia
    Tennessee
BT United States

coffinite
BA orthosilicates
    silicates
BT minerals

cohenite
Meteorite mineral.
BA native elements and alloys
BT minerals
SA carbides
    meteorites

cohesion
use shear strength

cohesive materials
Introduced in 1978 as general term mostly used in engineering geology. Before 1978, also search cohesive.
SA engineering geology
    materials
    rock mechanics
    soil mechanics

coiling
SA fossils
    morphology

Coimbatore
City in E.
BT Tamil Nadu
    India

Coimbra
District in W Portugal. Also a city.
BT Portugal

Cojedes
State in NW central.
BT Venezuela
NT Tinaquillo

Colchis
Ancient country on the Black Sea S of the Caucasus.
BT Georgian Republic
    USSR

Coldwater
Point of impact in Comanche County in S Kansas.
BT Kansas
    United States
SA meteorites

colemanite
BA borates
BT minerals

Coleoidea
Subclass. Includes use on level 3 under Mollusca(1) Cephalopoda(2).
BA Cephalopoda
    Mollusca
BT Invertebrata
SA Decapoda

Coleoptera
Includes use on level 2 under Insecta(1). See list F.
BA Insecta
BT Arthropoda
    Invertebrata

Coles Bay
BT Tasmania
    Australia

Colfax County
Index states as applicable.
BX Nebraska
    New Mexico
BT United States

collagen
BA organic materials
SA proteins

collapse structures
SA land subsidence
    slump structures
    structures

collecting
Includes use on level 3 under minerals(1).
SA collections
    mineral collecting
    mineralogy

collections
Usage restricted to collections in museums, etc. Includes use on level 2 under meteorites(1); on level 3 under fossil names (list F), mineral names (list L), and paleontology(1). Also use collecting on level 3 under minerals(1).

SA associations
    catalogs
    collecting
    geology
    mineral collecting
    museums
    paleontology

college level
  A valid term through 1977. After 1977, use college-level education.

college-level education
  Term introduced in 1978 on level 3 under education(1). Before 1978, search college level.
  SA education
      elementary school
      high school
      junior high school

Collenia
  Genus.
  BA stromatolites
      algae
  BT Plantae
  SA Cyanophyta
      ichnofossils

collision
  Includes use on level 3 under plate tectonics(1) structure(2), concepts(2) and area terms (list O).
  SA plate tectonics

colloidal materials
  Term introduced in 1978. Before 1978, search colloidal; colloids.
  UF colloids
  NT gels
  SA flocculation
      materials

colloids
  use colloidal materials

colloquia
  use symposia

colluvium
  Includes use on level 3 under sediments(1) clastic sediments(2). See list N.
  BA clastic sediments
  BT sediments
  SA alluvium
      erosion features
      geomorphology
      soils
      talus slopes
      terrigenous materials

Cologne
  City on the left bank of Rhine River in W.
  BT North Rhine-Westphalia
      West Germany

Colombia
  Includes use on level 1 as an area term (list O). For term set options see list B.
  CO S040000N121500
      W0670000W0790000
  BA South America
  NT Antioquia
      Bogota
      Guajira Peninsula
      Magdalena River
      Magdalena Valley
      Sabana de Bogota
      Santa Marta
      Sierra Nevada de Santa Marta
  NX Sucre
  SA Amazon Basin
      Amazon River
      Amazonas
      Andes
      Caribbean region
      Eastern Cordillera
      Llanos
      Magdalena
      Orinoco River
      Santander

Colombia Basin
  use Colombian Basin

Colombian Basin
  N of Colombia in S Caribbean. Also search Colombia Basin.
  UF Colombia Basin
      Colombian Plain
  BT Caribbean Sea

Colombian Plain
  use Colombian Basin

Colon Archipelago
  use Galapagos Islands

colonial taxa
  Term introduced in 1978. Before 1978, also search colonial; colonies.
  UF colonies
      taxa, colonial

colonies
  use colonial taxa

color
  Includes use on level 3 under soils(1) morphology(2).
  SA chemical analysis
      colorimetry
      soils

color centers
  UF centers, color
  SA crystal structure
      crystallography
      mineralogy
      optical properties

Colorado
  Includes use on level 1 as an area term (list O). For term set options see list B.
  CO N370000N410000
      W1020000W1090000
  BA United States
  NX Adams County
  NT Arapahoe County
      Boulder
      Boulder County
      Cache La Poudre River
      Canon City
      Chaffee County
  NX Cheyenne County
  NT Clear Creek
      Clear Creek County
      Climax
      Cortez
      Crested Butte
      Cripple Creek
  NX Custer County
  NT Denver
      Dillon
  NX Douglas County
  NT Eagle County
      Eaton
  NX El Paso County
  NT Elk Mountains
  NX Fremont County
  NT Front Range
      Front Range urban corridor
  NX Garfield County
  NT Gilpin County
      Golden
      Golden Horn Batholith
      Gore Range
  NX Grand County
  NT Grand Junction
      Greeley
      Gunnison County
      Henderson
  NX Jackson County
      Jefferson County
      Kiowa County
      Lake County
  NT Larimer County
      Las Animas County
      Leadville
  NX Lincoln County
  NT Loveland
      Mesa County
  NX Mineral County
      Morgan County
  NT Mosquito Range
  NX Otero County
  NT Ouray County
  NX Park County
      Phillips County
  NT Piceance Creek basin
      Pikes Peak
      Pikes Peak Batholith
      Pitkin County
      Pueblo County
      Rangely
      Rangely Anticline
      Rangely Field
      Rifle
      Rio Blanco County
      Saguache County
      Salida
  NX San Juan County
  NT San Juan Mountains
      San Juan volcanic field
      Sawatch Range
  NX Sedgwick County
  NT Silverton Caldera
      Spanish Peaks
  NX Summit County
  NT Vail Pass
  NX Washington County
  NT Weld County
      Wet Mountains
      White River Plateau
  NX Yuma County
  SA Arikaree Group
      Arkansas River
      Arkansas River valley
      Bald Mountain
      Benton Formation
      Brule Formation
      Canadian River
      Carlile Shale
      Cedar Mountain Formation
      Chadron Formation
      Cheyenne Sandstone
      Chinle Formation
      Colorado Group
      Colorado Plateau
      Colorado River
      Cutler Formation
      Dakota Formation
      Denver Basin
      Dockum Group
      Duchesne River Formation
      Fort Hays Limestone Member
      Fort Union Formation
      Four Corners
      Fox Hills Formation
      Frontier Formation
      Graneros Shale
      Great Plains
      Green River
      Green River basin
      Green River Formation
      Greenhorn Limestone
      Kayenta Formation
      Kiowa Formation
      Lance Formation
      Laney Shale Member
      Laramie Formation
      Laramie Mountains
      Leadville Formation
      Lewis Shale
      Loup Fork Group
      Madison Group
      Mancos Shale
      Manitou Formation
      Medicine Bow Mountains
      Menefee Formation
      Mesaverde Group
      Midcontinent
      Minturn Formation
      Missouri River basin
      Moenkopi Formation
      Montana Group
      Morrison Formation
      Navajo Sandstone
      Niobrara Formation
      Ogallala Formation
      Paradox Basin
      Paradox Member
      Pierre Shale
      Raton Basin
      Rio Grande
      Rio Grande Valley
      San Jose Formation
      San Juan Basin
      San Juan Formation
      San Juan River
      San Luis Valley
      Sangre de Cristo Mountains
      Smoky Hill Chalk Member
      South Platte River valley
      Sundance Formation
      Sunlight
      Todilto Formation
      Twilight Gneiss
      Twin Lakes
      Uinta Basin
      Wasatch Formation
      Washakie Basin
      Weber Sandstone
      White Limestone
      White River Group
      Yule Marble

Colorado Group
  Includes Thermopolis Shale, Mowry Shale, Warm Creek Shale, Graneros Shale, Greenhorn Limestone, Carlile Shale, Niobrara Formation, Belle Fourche Shale, Marias River Formation, Skull Creek Shale, Fall River Sandstone, Newcastle Sandstone, and Telegraph Shale.
  BT Cretaceous
  SA Benton Formation
      Carlile Shale
      Cody Shale
      Colorado
      Frontier Formation
      Graneros Shale
      Greenhorn Limestone
      Idaho
      Iowa
      Kansas
      Lower Cretaceous
      Montana
      Mowry Shale
      Nebraska
      New Mexico
      Niobrara Formation
      North Dakota
      South Dakota
      Upper Cretaceous
      Wyoming

Colorado Plateau
  Index states as applicable. Introduced as level 1 area term in 1978.
  BA United States
  SA Arizona
      Colorado
      New Mexico
      Utah

Colorado River
  Rises in N Colorado and flows SW, then W through the Grand Canyon, and finally S into the Gulf of California. Index Mexican and U. S. states as applicable.
  SA Arizona
      Baja California
      California
      Colorado
      Nevada
      Sonora
      Utah

Colorado River delta
  At N tip of Gulf of California. Index states as applicable.
  BT Mexico
  SA Baja California
      Sonora

coloradoite
- BA tellurides
  - sulfides
- BT minerals

colorimetry
Includes use on level 3 under chemical element(1) analysis(2) and under chemical analysis(1) methods(2). See list D.
- SA analysis
  - chemical analysis
  - color
  - spectroscopy

Columbia
Name of cities in Missouri and South Carolina. Also cities in Illinois, Kentucky, Louisiana, and Tennessee. Town in Connecticut. Index states as applicable.
- BX Connecticut
  - Illinois
  - Kentucky
  - Louisiana
  - South Carolina
  - Tennessee
- BT United States

Columbia Basin
use Columbia River basin

Columbia County
Index states as applicable.
- BX Arkansas
  - Florida
  - Georgia
  - New York
  - Oregon
  - Pennsylvania
  - Washington
  - Wisconsin
- BT United States

Columbia Plateau
A level 1 area term as of 1978. Lava basin. Index states as applicable. Also search Columbia River plateau.
- CO N440000N483000
  - W1150000W1200000
- UF Columbia River plateau
- BA United States
- SA Idaho
  - Oregon
  - Washington

Columbia River
Rises in SE British Columbia and flows S and then W into the Pacific Ocean on the Oregon-Washington border. Also search Columbia AND appropriate area. Index British Columbia and states as applicable.
- CO N453000N481500
  - W1185400W1241000
- SA British Columbia
  - Oregon
  - Washington

Columbia River Basalt
Miocene and Pliocene (?). N Idaho, Oregon, and Washington. Also search Columbia River Group.
- BT Tertiary
- SA Idaho
  - Miocene
  - Oregon
  - Pliocene
  - Washington

Columbia River basin
Drainage basin. Index British Columbia, and states as applicable. Also search Columbia Basin.
- UF Columbia Basin
- SA British Columbia
  - Idaho
  - Montana
  - Nevada
  - Oregon
  - Utah
  - Washington
  - Wyoming

Columbia River estuary
Tidal mouth of the Columbia River. Index states as applicable.
- BT United States
- SA Oregon
  - Washington

Columbia River plateau
use Columbia Plateau

columbite
- BA oxides
- BT minerals
- SA tantalates

columbium
use niobium

Columbus
City in Franklin County in central Ohio; a city in Muscogee County in W Georgia. Index states as applicable.
- BX Georgia
  - Ohio
- BT United States

Columbus Limestone
Comprises Bellepoint, Eversole (?), Delhi, Marblehead, and Venice members. Named for exposure at Columbus in Franklin County, Ohio.
- BT Middle Devonian
  - Devonian
- SA Ohio

columnar basalt
Before 1978, also search columnar AND basalt.
- BA basalt family
- BT igneous rocks
- SA basalt

columnar jointing
use columnar joints

columnar joints
Term introduced in 1978. Includes use on level 3 under fractures(1) style(2). Before 1978, also search fractures AND columnar.
- UF columnar jointing
- BA fractures
- SA bedding
  - joints

Colusa County
N central.
- BT California
  - United States

Colville River
Flows E along N slope of Brooks Range and then N into Arctic Ocean.
- BT Alaska
  - United States

Colville River delta
On Arctic Ocean in N.
- BT Alaska
  - United States

Comanche County
Index states as applicable.
- BX Kansas
  - Oklahoma
  - Texas
- BT United States

Comanchean
Provincial series, North America. Lower and Upper Cretaceous: above Coahuilan, below Gulfian. Includes use on level 3 under age terms(1). See list E.
- UF Comanchian
- BA Cretaceous
- BT Mesozoic
- NT Edwards Formation
  - Fredericksburg Group
  - Mentor Beds
  - Paluxy Formation
  - Trinity Group
- SA Antlers Sands
  - Lower Cretaceous
  - Upper Cretaceous

Comanchian
use Comanchean

combustion
- SA geochemistry
  - processes

comendite
Includes use on level 3 under igneous rocks(1) andesite-rhyolite family(2). See list H.
- BA andesite-rhyolite family
- BT igneous rocks
- SA pantellerite

comets
Includes use on level 2 under interplanetary space(1).
- SA interplanetary space
  - meteorites
  - meteors
  - solar system
  - solar wind
  - tektites

commencements, sudden
use sudden commencements

common mica
use muscovite

common salt
use halite

common-depth-point method
- SA methods
  - seismic methods

communities
Includes use on level 3 under ecology(1) and paleoecology(1).
- SA assemblages
  - biocenoses
  - ecology
  - paleoecology

Como
City, lake and province in NW.
- BT Lombardy
  - Italy

Comoro Islands
Group of volcanic islands in N Mozambique Channel between Mozambique and Malagasy Republic. An overseas territory of France. As of 1977 includes use as a level 1 area term (list O). See list B for term set options.
- BA Indian Ocean
- SA Anjouan Island

compaction
Includes use on level 3 diagenesis(1) and soil mechanics(1).
- SA bearing capacity
  - consolidation
  - diagenesis
  - lithification
  - porosity
  - processes
  - soil mechanics

comparison
Used as a general term through May 1978. No longer a valid term for Georef.

compensation
Includes use on level 2 under isostasy(1).
- SA crust
  - isostasy

competence
As of 1978, the term is restricted to hydraulics. The ability of a current of water or wind to transport detritus.
- SA competent materials
  - hydraulics
  - transport
  - water
  - winds

competent materials
Term introduced in 1978.
- SA competence
  - deformation
  - materials
  - rock mechanics
  - soil mechanics

complexes
Not to be used for structural complexes under tectonics(1). Includes use on level 3 under igneous rocks(1) and metamorphic rocks(1).
- SA igneous rocks
  - massifs
  - metamorphic rocks
  - metamorphism
  - ring complexes
  - structural complexes

complexing
- SA geochemistry
  - metals
  - organic materials
  - processes

components
Used as a general term.

composite cone
use stratovolcanoes

composite volcano
use stratovolcanoes

composition
Includes use as level 2 or 3 term appropriate to a large number of topics, e.g. on level 2 under meteorology(1). See list G. Also search content.
- UF content
- SA acidic composition
  - alkalic composition
  - andesitic composition
  - arkosic composition
  - basaltic composition
  - calc-alkalic composition
  - calcareous composition
  - calcic composition
  - carbonaceous composition
  - carbonate composition
  - chemical composition
  - dolomitic composition
  - ferromanganese composition
  - ferruginous composition
  - gabbroic composition
  - glauconitic composition
  - granitic composition
  - iron-rich composition
  - mafic composition
  - mineral composition
  - phosphate composition
  - potassic composition
  - rhyolitic composition
  - saline composition
  - siliceous composition
  - tholeiitic composition
  - ultrabasic composition
  - ultramafic composition

compounds
Used as a general term.
- SA acids
  - organic compounds

compressibility
Includes use on level 3 under soil mechanics(1).
- UF modulus of compression
- SA compression
  - elasticity
  - physical properties
  - plasticity
  - soil mechanics

compression
Includes use on level 3 under deformation(1) field studies(2).
- SA compressibility
  - deformation
  - flow cleavage
  - lithification

pressure
tension
yield strength

compressional wave
use P-waves

compressive strength
SA deformation
rock mechanics
soil mechanics
strength

computer methods
A valid term through 1971. After 1971, use automatic data processing.

computer programs
Includes use on level 3 under automatic data processing(1).
NT Fortran
Fortran IV
SA automatic data processing
programs

computers
For documents discussing the machine itself. Includes use on level 3 under automatic data processing(1).
SA automatic data processing
microcomputers
minicomputers
programs

concentration
Used as a a general term.
SA localization
saturation
solubility
solution

concentric folds
Term introduced in 1978. Includes use on level 3 under folds(1). Before 1978, also search concentric AND folds; search parallel AND folds.
UF parallel folds
BA folds
SA similar folds

concentric fractures
Term introduced in 1978. Includes use on level 3 under fractures(1) style(2). Before 1978, also search fractures AND concentric.
BA fractures

concepts
Includes use as level 2 or 3 term appropriate to a large number of topics, e.g. on level 2 under geochemistry(1) and mineral exploration(1). See list G.

conchiolin
BA organic materials
SA proteins

Conchostraca
Order. Includes use on level 3 under Arthropoda(1) Branchiopoda(2).
BA Branchiopoda
Crustacea
Arthropoda
BT Invertebrata

concrete
Includes use on level 3 under engineering geology(1).
BA construction materials
SA cement
engineering geology

concretions
Includes use on level 3 under sedimentary structures(1) secondary structures(2). See list K.
BA secondary structures
sedimentary structures
SA coal balls
cone-in-cone
geodes
nodules
septaria

condensates
Liquid hydrocarbon that emanates from a gas well or from the gas-cap of an oil well.
SA heavy oil
natural gas
petroleum

condensation
Includes use under Moon(1) and planetology(1) for a genetic concept.
SA geochemistry
Moon
planetology
planets

conditions, P-T
use P-T conditions

conductivity
Includes use on level 3 under heat flow(1). A level 2 term under heat flow(1) until 1976; now use thermal conductivity(2).
SA electrical conductivity
electrical logging
geothermal gradient
heat flow
hydraulic conductivity
measurement
resistivity
thermal conductivity

Condylarthra
Includes use on level 2 under Mammalia(1). See list F.
BA Mammalia
BT Tetrapoda
Vertebrata
SA Ungulata

cone-in-cone
Includes use on level 3 under sedimentary structures(1) secondary structures(2). See list K.
BA secondary structures
sedimentary structures
SA concretions
septaria

Conemaugh
River in SW.
BT Pennsylvania
United States

Conemaugh Group
Divided into nine parts: Mahoning, Buffalo, Saltsburg, Grafron, Barton, Morgantown, Lonaconing, Connellsville, and one not yet named. Also includes Nadine Limestone, Woods Run Limestone, Carnahan Run Shale, and Brush Creek Limestone. W Maryland, E Ohio, Pennsylvania, N Virginia, and W west Virginia. Also search Conemaugh Formation.
BT Pennsylvanian
SA Maryland
Ohio
Pennsylvania
Virginia
West Virginia

cones, cinder
use cinder cones

cones, shatter
use shatter cones

conferences
Not a valid term for GeoRef. See under associations and under symposia.

configuration
Includes use on level 2 under magnetosphere(1).
SA Earth
magnetosphere

congelifraction
Also search frost shattering; frost splitting; frost weathering; frost wedging; gelifraction.

UF frost bursting
frost shattering
frost splitting
frost weathering
frost wedging
gelifraction
gelivation
SA cryoturbation
frost action
frost heaving
periglacial features
rock mechanics
soil mechanics

congeliturbation
use cryoturbation

conglomerate
Includes use on level 3 under sedimentary rocks(1) clastic rocks(2). See list I.
UF conglomerite
BA clastic rocks
BT sedimentary rocks
SA agglomerate
breccia
calcrete
ferricrete
metaconglomerate
terrigenous materials

conglomerite
use conglomerate

Congo
This term is now limited to the present Congo (People's Republic of the Congo). Formerly it also included what is now Zaire. Includes use on level 1 as an area term (list O). For term set options see list B.
UF People's Republic of the Congo
BA Africa
NT Brazzaville
SA Congo Basin
Congo River
Zaire

Congo (River)
use Congo River

Congo Basin
River drainage basin. Index countries as applicable.
BT Africa
SA Angola
Central African Republic
Congo
Zaire
Zambia

Congo River
Rises as the Lualaba River in SE Zaire and flows N, then W, and finally SW into the Atlantic Ocean. Index countries as applicable.
UF Congo (River)
Kongo River
Zaire River
BT Africa
SA Angola
Congo

congresses
Not a valid term for GeoRef. See under associations and under symposia.

Coniacian
Europe. Above Turonian, below Santonian. Includes use on level 3 under age terms(1). See list E.
BA Upper Cretaceous
Cretaceous
BT Mesozoic
SA Senonian

conical fractures
Term introduced in 1978. Includes use on level 3 under fractures(1) style(2). Before 1978, also search fractures AND conical.

BA fractures

conichalcite
BA arsenates
BT minerals

Coniferae
Treated variously as a family, order or class coextensive with order Coniferales.
BA Coniferales
gymnosperms
BT Plantae

Coniferales
Includes use on level 2 under gymnosperms(1). See list F.
BA gymnosperms
BT Plantae
NA Buriadia
Coniferae
Cupressaceae
Pinaceae
Taxodiaceae
Taxodium
SA Dadoxylon

conjugate folds
Term introduced in 1978. Includes use on level 3 under folds(1) style(2). Before 1978, search folds AND conjugate.
BA folds
SA joints

connate waters
Includes use on level 2 under ground water(1). Also search formation water; formation waters.
UF formation waters
fossil water
fossilized brine
native water
SA ground water
pore water
water

Connecticut
Includes use on level 1 as an area term (list O). For term set options see list B.
CO N410000N420300
W0714800W0734400
BA United States
NX Columbia
NT Hartford County
Litchfield County
NX Middlesex County
NT Roaring Brook Valley
Suffield
NX Windham County
Woodbury
SA Atlantic Coastal Plain
Connecticut River
Connecticut Valley
Eastern U.S.
Manhattan Formation
New England
New York City Group
Newark Group
Westerly Granite

Connecticut River
Rises in N New Hampshire and flows S emptying into Long Island Sound. Index states as applicable.
BT United States
SA Connecticut
Massachusetts
New Hampshire
Vermont

Connecticut River valley
use Connecticut Valley

Connecticut Valley
River valley. Index states as applicable. Also search Connecticut, and Connecticut River valley.
CO N413000N452000
W0711500W0724500
UF Connecticut River valley
BT United States

SA Connecticut
   Massachusetts
   New Hampshire
   Vermont

Connemara
Barren, mountainous coastal region in County Galway on the Atlantic Ocean.
CO N531500N534000
   W0090000W0102000
BT Ireland

conodonts
Includes use on level 1 and 2 as a fossil term (list F).
NA Hindeodella
   Icriodus
   Ozarkodina
   Palmatolepis
   Panderodus
   Polygnathus
   Spathognathodus
SA apparatus

Conrad discontinuity
Includes use on level 3 under crust(1).
SA crust
   discontinuities
   Earth
   velocity

conservation
Level 1 term as of 1978. Includes use on level 2 under impact statements(1) and land use(1). Used for environmental conservation of natural resources. If 1, term set options are:
environment
   beaches
   bogs
   caves
   ecosystems
   marshes
   reefs
   shorelines
natural resources
   coal
   energy sources
   forests
   ground water
   metals
   mineral resources
   underground space
   water resources
   wetlands
topic [experimental studies, impact statements, methods, programs, surveys]
   subtopic
SA deforestation
   depletion
   environmental geology
   erosion control
   impact statements
   land leases
   land use
   natural resources
   pollution
   reclamation
   resources
   soils
   urbanization
   waste disposal

consistency
Includes use on level 3 under soils(1) morphology(2).
SA morphology
   particles
   soils

consistency limits
use Atterberg limits

consolidation
Includes use on level 3 under diagenesis(1). Before 1978, also search solidification.

UF solidification
SA cementation
   compaction
   diagenesis
   lithification
   overconsolidated materials
   soil mechanics

Constance Lake
use Lake Constance

constant, dielectric
use dielectric constant

Constanta
County. Also a city on the Black Sea.
UF Constantsa
BT Dobruja
NT Romania

Constantine
Department in NE Algeria. Also a city.
BT Algeria
NT Guelma
   Hodna Basin

constants, elastic
use elastic constants

Constantsa
use Constanta

construction
As of 1978, term is used on level 2.
SA construction materials
   dams
   dredging
   foundations
   highways
   marine installations
   reservoirs
   shorelines
   tunnels
   underground installations

construction materials
Includes use on level 1 as commodity term (list C).
NA aggregate
   building stone
   cement materials
   concrete
   dimension stone
   trap rock
SA alloys
   cement
   construction
   granite
   gravel
   limestone
   marble
   materials
   perlite
   refractory materials
   sand
   sandstone
   shale
   trap rocks

contact
Includes use on level 2 under intrusions(1). Before 1977, term used in combination with metamorphism. After 1977, use contact metamorphism.
SA contact metamorphism
   igneous rocks
   intrusions
   metamorphism

contact aureole
use aureoles

contact metamorphism
Not a valid term from 1973 through 1977. After 1977, includes use on level 2 under metamorphism. Also search metamorphism AND contact.
BA metamorphism
SA contact

contact zone
use aureoles

contamination
Includes use on level 2 under ground water(1); on level 3 under hydrogeology(1).
SA assimilation
   ground water
   hydrogeology
   impurities
   magmas
   pollution
   purification
   salt-water intrusion

contemporaneous faults
use growth faults

content
use composition

conterminous regions
Used to indicate regions with a common boundary. Used especially for the United States.
SA United States

continental
A valid term through 1977. After 1977, use continental borderland, continental type (for the crust), or continental rise.

continental borderland
The area of the continental margin between the shoreline and the continental slope which is topographically more complex than the continental shelf.
UF borderland, continental
SA continental margin
   continental shelf
   continental slope

continental crust
use continental type

continental displacement
use continental drift

continental drift
For general discussions of "classical" drift concepts as well as plate tectonic reconstructions. Includes use on level 1 (list A); on level 3 under plate tectonics(1). Also search continental AND drift. Term set options are:
area [use terms like Gondwana, Iapetus, Laurasia, Mesogaea, Paleoafrica, Pangaea, Paratethys, Tethys]
   locality
topic [causes, concepts, evolution, general, genesis, interpretation, mechanism, paleobotany, paleoclimatology, paleomagnetism, paleontology, patterns]
   subtopic (no area term)
UF continental displacement
   continental migration
   displacement theory
   epeirophoresis theory
   Wegener hypothesis
SA biogeography
   continents
   drift
   geosynclines
   Gondwana
   Laurasia
   mantle
   Mohorovicic discontinuity
   ocean basins
   paleogeography
   paleomagnetism
   paleontology
   Pangaea
   Paratethys
   plate tectonics
   polar wandering
   reconstruction
   sea-floor spreading
   tectonics
   tectonophysics
   Tethys
   volcanology

continental genesis
use continental type

continental margin
Use term to denote both shelf and slope. Includes use on level 3 under area sets (list B) and under plate tectonics(1) or tectonics(1). Also search continental margins; continental AND margin.
UF margin, continental
SA continental borderland
   continental rise
   continental shelf
   continental slope
   continents
   cratons
   island arcs
   ocean floors
   plate tectonics
   submarine canyons
   tectonics
   tectonophysics

continental migration
use continental drift

continental platform
use continental shelf

continental rise
Includes use on level 3 under continental slope(1) or ocean floors(1).
UF rise, continental
SA abyssal plains
   continental margin
   continental slope
   nepheloid layer
   ocean floors

continental shelf
For studies of geological processes taking place on the shelf. Restricted to the modern shelf area. Includes use on level 1 (list A). Also search continental AND shelf. Term set options are:
topic [List B except for areal geology]
   subtopic (no area term)
UF continental platform
   shelf, continental
SA changes of level
   continental borderland
   continental margin
   continental slope
   continents
   currents
   inner shelf
   marine geology
   ocean circulation
   ocean floors
   oceanography
   offshore
   outer shelf
   Polar Continental Shelf
   reefs
   sedimentation
   shelf environment
   slope environment
   slumping
   submarine canyons

continental slope
Includes use on level 1 (list A). For modern slope area. Also search continental AND slope. Term set options are:
topic [List B except for areal geology]
   subtopic (no area term)
UF slope, continental
SA channels
   continental borderland
   continental margin
   continental rise
   continental shelf
   continents

inner slope
marginal basins
marine geology
ocean circulation
ocean floors
oceanography
offshore
outer slope
sedimentation
slope environment
slopes
slumping
submarine canyons

continental type
Used in relation to crust. Before 1978, search continental; continental crust; genesis AND continental; search continental AND crust.
UF   continental crust
       continental genesis
SA   crust
       plate tectonics

continents
Includes use on level 3 when discussing continents as units (e.g. geometry of a continent). See continental drift(1) or topic under area terms, list B.
SA   continental drift
       continental margin
       continental shelf
       continental slope
       cratons
       Earth
       epeirogeny
       island arcs
       islands
       isostasy
       marginal seas

contour maps
BA   maps
SA   structure contour maps
       topographic maps

contourite
Includes use on level 3 under sedimentary rocks(1) clastic rocks(2). See list I.
BA   clastic rocks
BT   sedimentary rocks
SA   terrigenous materials

Contra Costa County
On San Francisco Bay.
BT   California
       United States

contraction
Used as a general term.
SA   processes

control, erosion
use erosion control

control, production
use production control

controls
Includes use on level 2 under mineral deposits, genesis(1) and sedimentation(1).
SA   geochemical controls
       hydrogeological controls
       lithologic controls
       mechanical controls
       mineral deposits
       paleogeographic controls
       sedimentation
       slope stability
       stratigraphic controls
       structural controls
       waste disposal
       waterways

convection
Includes use on level 2 and 3 under meteorology(1) and under tectonophysics(1); on level 3 under plate tectonics(1).
SA   aerosols
       atmosphere
       convection cells
       convection currents
       heat flow
       mantle
       meteorology
       oceanography
       plate tectonics
       tectonophysics

convection cells
Includes use on level 3 under plate tectonics(1).
UF   cells, convection
SA   convection
       convection currents
       mantle
       plate tectonics

convection currents
Includes use on level 3 under Earth(1), under tectonophysics(1), and under plate tectonics(1).
BT   currents
SA   atmosphere
       convection
       convection cells
       density
       mantle
       plate tectonics
       tectonophysics

convergence
Includes use on level 3 under plate tectonics(1).
SA   plate tectonics

Converse County
E central.
BT   Wyoming
       United States

conversion
Used as a general term.
SA   changes

convoluted beds
Includes use on level 3 under sedimentary structures(1) soft sediment deformation(2) and turbidity current structures(2). See list K.
UF   beds, convoluted
       convoluted deformation
BA   soft sediment deformation
       sedimentary structures
SA   slump structures
       turbidity current structures

convoluted deformation
use convoluted beds

Coo
use Kos

Cooch Behar
City just N of N Bangladesh. A former state.
BT   West Bengal
       India

Cook County
Index states as applicable.
BX   Georgia
       Illinois
       Minnesota
BT   United States

Cook Inlet
Arm of the Pacific Ocean W of Kenai Peninsula in S.
BT   Alaska
       United States

Cook Islands
Group of 15 islands W of French Polynesia and E of Samoa and Tonga islands.
UF   Southern Cook Islands
BT   Pacific Ocean
SA   Polynesia

Cook Mountain Formation
In Claiborne Group. NW Louisiana, S and E Texas.
BT   middle Eocene

Eocene
SA   Claiborne Group
       Louisiana
       Texas

cookeite
BA   sheet silicates
       silicates
BT   minerals
SA   aluminosilicates
       chlorite group
       lepidolite

cooling
Used as a general term.
SA   temperature

Cooma
Town in S.
BT   New South Wales
       Australia

Cooper Basin
Also search Cooper's Creek basin. Index states as applicable.
CO   S300000S220000
       E1430000E1370000
UF   Cooper's Creek basin
BT   Australia
SA   Queensland
       South Australia

cooperation, international
use international cooperation

Cooper's Creek basin
use Cooper Basin

coordinate systems
Includes use on level 2 under magnetosphere(1).
SA   magnetosphere
       systems

coordinates
Includes use on level 3 under Moon(1) or maps(1). Use geodetic coordinates under geodesy(1). For geographic coordinates, see specific area.
SA   geodesy
       geodetic coordinates
       maps
       Moon

coordination
As of 1978, term is used on level 2 under crystal chemistry(1).
SA   crystal chemistry

Coorong Lagoon
Off Lacepede Bay of Indian Ocean in SE South Australia. Also search Coorong.
UF   The Coorong
BT   South Australia
       Australia

Coos Bay
Inlet on coast of Coos County on the Pacific Ocean in SW.
CO   N431500N433500
       W1241000W1243000
BT   Oregon
       United States

Coos County
On the Pacific Ocean in SW.
CO   N424000N434000
       W1234000W1243500
BT   Oregon
       United States

Coosa County
Central.
BT   Alabama
       United States

Copenhagen
County on Zealand Island in E Denmark. Also a city.
UF   Kobenhavn
BT   Denmark

Copepoda
Class. Includes use on level 2 under

Arthropoda(1). See list F.
BA   Crustacea
       Arthropoda
BT   Invertebrata

copiapite
BA   sulfates
BT   minerals

Copiapo
Town in N part of country. Also a river.
UF   San Francisco de la Selva
       Selva
BT   Atacama
       Chile

copper
Includes use on level 1 and 2 as a commodity term (list C) and as a chemical element (list D). Also search Cu.
UF   Cu
SA   bornite
       chalcocite
       chalcopyrite
       elements
       enargite
       heavy metals
       native elements and alloys
       porphyry copper

Copper Canyon
Canyon in Lander County in N central.
BT   Nevada
       United States

Copper Harbor Conglomerate
Upper Keweenawan. Keweenaw County on the Upper Peninsula, NW Michigan.
BT   Precambrian
SA   Michigan

Copper Mountain
Village in S central.
BT   British Columbia
       Canada

copper pyrites
use chalcopyrite

copper sulfides
BA   sulfides
BT   minerals
SA   chalcocite
       covellite

Copperbelt
Area.
BT   South Africa

Coppermine River
Flows into Coronation Gulf in N District of Mackenzie.
BT   Northwest Territories
       Canada

coprolites
Includes use on level 1 and 2 as a fossil term(list F); on level 3 under sedimentary structures(1) biogenic structures(2). See list K.
BA   biogenic structures
       sedimentary structures
SA   Invertebrata

coquimbite
BA   sulfates
BT   minerals

Coquimbo
Province in central Chile. Also a city.
BT   Chile

coquina
BA   clastic sediments
BT   sediments
SA   carbonate sediments
       terrigenous materials

coral pinnacle
use pinnacle reefs

coral reefs
A valid term through 1974. After

1974, use reefs.

Coral Sea
Between Queensland, Australia, on the W and New Hebrides and New Caledonia on the E. As of 1977, includes use on level 1 as an area term (list O). For term set options see list B.
BA Pacific Ocean
NT Coral Sea Basin
    Great Barrier Reef

Coral Sea Basin
Between NE Queensland and Solomon Islands.
BT Coral Sea
    Pacific Ocean

Corallinaceae
Includes use on level 3 under algae(1) Rhodophyta(2). See list F.
BA Rhodophyta
    algae
BT Plantae
NA Archaeolithothamnium
    Lithophyllum
    Lithothamnium

corals
Includes use on level 3 under Coelenterata(1) Anthozoa(2). See list F.
BA Anthozoa
    Coelenterata
BT Invertebrata
SA ahermatypic taxa
    atolls
    bioherms
    biostromes
    calcareous composition
    Coelenterata
    hermatypic taxa
    reefs
    Rugosa
    Scleractinia
    Tabulata

Corbieres
Outliers of the E Pyrenees in S Aude. Also search Corbieres Mountains.
UF Corbieres Mountains
BT Aude
    France

Corbieres Mountains
use Corbieres

Cordaitales
Includes use on level 2 under gymnosperms(1). See list F.
BA gymnosperms
BT Plantae
NA Cordaites
SA Dadoxylon

Cordaites
Genus.
BA Cordaitales
    gymnosperms
BT Plantae

cordierite
BA ring silicates
    silicates
BT minerals
SA aluminosilicates

Cordillera
Area. Index states and provinces as applicable.
BT North America
SA Alaska
    British Columbia
    California
    Canadian Cordillera
    Cordilleran
    Idaho
    Montana
    Nevada
    Oregon
    Washington
    Yukon Territory

Cordillera Betica
use Betic Cordillera

Cordillera de la Costa
Along Caribbean Sea in central N.
BT Venezuela

Cordillera Marianica
use Betic Cordillera

Cordillera Oriental
use Eastern Cordillera

Cordilleran
Includes use on level 3 under geosynclines(1) evolution(2).
SA Cordillera
    geosynclines

Cordilleran Geosyncline
Term introduced in 1978. Before 1978, search geosynclines AND Cordilleran.
SA Cordilleran Orogeny
    geosynclines

Cordilleran Orogeny
Term introduced in 1978. Includes use on level 3 under orogeny(1). Before 1978, search Cordilleran AND orogeny.
SA Cordilleran Geosyncline
    Laramide Orogeny
    orogeny
    tectonics

Cordoba
Province. Also a city. Index countries as applicable.
CO S350000S290000
    W0630000W0660000
NT Chanares
    Los Pedroches
    Valsequillo
SA Andalusia
    Argentina
    Spain

core
Used for the core of the Earth. For that of other planets and Moon, see appropriate set. Includes use on level 1 (list A); on level 2 under Earth(1), and under seismology(1). Term set options are:
topic [composition, concepts, evolution, general, genesis, interpretation, processes, properties, shape, structure, temperature, theoretical studies]
    subtopic
NA inner core
    outer core
SA crust
    Earth
    geophysics
    heat flow
    heat sources
    interior
    lithosphere
    magnetic field
    magnetohydrodynamics
    mantle
    Mohorovicic discontinuity
    seismology
    tectonophysics
    transition zones

cores
Includes use on level 3 under well-logging(1). Before 1978, also search borings.
UF drill cores
SA boreholes
    borings
    cuttings
    well-logging
    wells

Corinth
City on Gulf of Corinth in Corinth Department in NE.
BT Peloponnesus
    Greece

Coriolis force
Includes use on level 2 under ocean circulation(1).
UF force, Coriolis
    geostrophic force
SA Earth
    ocean circulation

Cork
County on the Atlantic Ocean. Also a city. Search County Cork.
CO N513000N522000
    W0081500W0103000
UF County Cork
BT Ireland

Cormeilles-en-Parisis
Town just N of Paris.
BT Val-d'Oise
    France

Cornwall
County in extreme SW.
CO N500000N505000
    W0041500W0063000
BT England
    Great Britain
    United Kingdom

Cornwallis Island
One of the Parry Islands in central District of Franklin.
CO N743000N753000
    W0933000W0970000
BT Northwest Territories
    Canada

Coromandel Coast
SE coast from Point Calimere in the S to mouths of the Krishna River in Andhra Pradesh. Also search Coromandel.
BT India

Coromandel Peninsula
ENE of Auckland. Also search Coromandel.
BT North Island
    New Zealand

corona
Includes use on level 2 under astrophysics and solar physics(1).
SA astrophysics and solar physics
    plasma instabilities
    solar wind
    Sun

coronadite
BA oxides
BT minerals
SA hollandite

Coronation Mine
E central.
BT Saskatchewan
    Canada

Corophioides
BA ichnofossils

Corpus Christi Bay
Inlet of Gulf of Mexico between Neuces and Patricio counties in SE.
BT Texas
    United States

corrections
Used as a general term.
SA accuracy
    errors

correlation
Formerly a general term appropriate to a large number of topics (list G). As of 1976, the term was restricted to stratigraphical correlation. Includes use on level 2 under stratigraphy(1).
SA stratigraphy

correlation coefficient
UF coefficient, correlation
SA statistical analysis
    statistical methods

corrensite
BA sheet silicates
    silicates
BT minerals
SA aluminosilicates
    clay minerals

correspondence analysis
BA statistical methods
SA analysis
    statistical analysis

Correze
Department in S central.
CO N445000N454500
    E0023000E0011000
BT France
NT Brive

Corrientes
Province between the Parana and the Uruguay rivers in NE Argentina. Also a city.
CO S300000S270000
    W0560000W0600000
BT Argentina

corrosion
Includes use under geomorphology(1). Chemical meaning also used.
SA destruction
    erosion
    geomorphology

Corsica
Island in the Mediterranean Sea and a department of France. Includes use on level 1 as an area term (list O). For term set options see list B.
CO N411500N430000
    E0093000E0083000
BA France
    Europe
SA Mediterranean region

Cortez
City in Montezuma County in extreme SW.
BT Colorado
    United States

Cortez Mountains
Eureka County in central Nevada. Also search Cortez.
BT Nevada
    United States

Cortina
use Cortina D'Ampezzo

Cortina D'Ampezzo
Town in Dolomites in N Veneto.
UF Cortina
BT Veneto
    Italy

corundum
Includes use on level 1 as a commodity term (list C).
BA oxides
BT minerals
SA abrasives
    alumina
    amethyst
    sapphire

Corynexochida
Includes use on level 2 under Trilobita(1). See list F.
BA Trilobita
BT Trilobitomorpha
    Arthropoda
    Invertebrata

Cos
use Kos

cosalite
BA sulfobismuthites
    sulfosalts
BT minerals

Cosenza
Town and province in W central.
BT Calabria
Italy

cosmic dust
Includes use on level 2 under planetology(1).
UF zodiacal dust
SA asteroids
dust
interplanetary dust
interplanetary space
meteorites
Moon
particles
planetology

cosmic rays
Includes use on level 2 under interplanetary space(1) and magnetosphere(1); on level 3 under aeronomy(1) ionization(2).
UF rays, cosmic
SA aeronomy
albedo
alpha rays
asteroids
cutoff rigidities
electromagnetic radiation
exposure age
extraterrestrial geology
gamma rays
intensity
interplanetary space
ionization
magnetic field
magnetosphere
Moon
particle track
particles
protons
solar cycles
solar flares
spallation

cosmochemistry
Includes use on level 3 under asteroids(1).
SA asteroids
elements
geochemistry
planetology

cosmogenic elements
Term introduced in 1978. Before 1978, search cosmogenic; cosmogenic AND isotopes.
SA elements
isotopes

cosmolites
use meteorites

Costa Rica
Includes use on level 1 as an area term (list O). For term set options see list B.
BA Central America
NT Arenal

Cote d'Or
use Cote-d'Or

Cote-d'Or
Department in E central France. Also search Cote d'Or.
UF Cote d'Or
BT France
NT Dijon
Pouilly-en-Auxois
SA Morvan

Cotentin Peninsula
Peninsula in Normandy jutting into the English Channel in N.
BT Manche
France

Cotes-du-Nord
Department on the English Channel in N Brittany.
CO N480000N490000
W0020000W0034500
BT France

Cotswold Hills
Range of hills in Gloucestershire in W central England. Also search Cotswolds.
CO N514500N521500
W0011500W0014500
UF Cotswolds
BT England
Great Britain
United Kingdom

Cotswolds
use Cotswold Hills

Cottbus
District in SE. Also a city.
BT East Germany
NT Doberlug
SA Saxony
Upper Lusatia

Cottian Alps
Division of Western Alps. Index countries as applicable.
BT Alps
Europe
SA Dauphine Alps
France
Italy
Piedmont Alps
Western Alps

Cotton
County on the Texas border in SW.
BT Oklahoma
United States

Cotton Valley Group
Includes Schuler Formation, and Bossier Formation. S Arkansas, N Louisiana, W Mississippi, and E Texas.
BT Upper Jurassic
Jurassic
SA Arkansas
Bossier Formation
Louisiana
Mississippi
Schuler Formation
Texas

Cottonwood Limestone
In Council Grove Group. E Kansas, SE Nebraska, and central northern Oklahoma. Also search Cottonwood Limestone Member.
UF Cottonwood Limestone Member
BT Permian
SA Kansas
Nebraska
Oklahoma

Cottonwood Limestone Member
use Cottonwood Limestone

Coulombs' modulus
use shear modulus

country rocks
For commodities (list C), host rocks should be used. Includes use on level 3 under intrusions(1).
SA host rocks
igneous rocks
intrusions
rocks

County Cork
use Cork

Coupon Bight
Florida keys area.
BT Florida
United States

Courland Spit
Narrow sandspit on the Baltic Sea in both Kaliningrad Oblast and Lithuania. Index Soviet republics as applicable.
UF Kurland Spit

BT USSR
SA Lithuania
Russian Republic

Couvinian
Europe. Above Emsian, below Givetian. Includes use on level 3 under age terms(1). See list E.
BA Middle Devonian
Devonian
BT Paleozoic
SA Eifelian

Covasna
County in SE.
BT Transylvania
Romania

covelline
use covellite

covellite
UF covelline
BA sulfides
BT minerals
SA copper sulfides

cover, sedimentary
use sedimentary cover

Cowley County
On the Oklahoma border in S.
BT Kansas
United States

Cowlitz County
On the Columbia River in SW.
BT Washington
United States

Coyote Creek Fault
W California.
BT California
United States

CO2
use carbon dioxide

Cr
use chromium

crabs
SA Crustacea

cracks
Includes use on level 3 under rock mechanics(1).
NT microcracks
SA fissures
fractures
open fractures
rock mechanics
shrinkage cracks

Cracow
use Krakow

Crai
Forest in Apuseni Mountains in WSW.
BT Transylvania
Romania

crandallite
BA phosphates
BT minerals

Crassostrea virginica
Includes use on level 3 under Mollusca(1) Bivalvia(2).
BA Bivalvia
Mollusca
BT Invertebrata

Crater Lake
In Cascade Mountains in Klamath County in SW.
BT Oregon
United States

cratering
Includes use on level 3 under earthquakes(1), geomorphology(1), Moon(1), seismology(1), and volcanology(1).
SA craters
cryptoexplosion features
impact features

craters
Includes use on level 3 under geomorphology(1) impact features(2) and landform description(2); on level 3 under volcanology(1) and meteor craters(1).
SA calderas
cratering
impact features
impacts
maars
meteor craters
microcraters
Moon
volcanology

cratons
Includes use on level 3 under tectonics(1).
SA continental margin
continents
crust
shields
tectonics

creep
As of 1978, term is used on level 2 under permafrost(1) and slope stability(1).
UF creeping
SA deformation
elastic limit
elastic strain
etching
landslides
mass movements
mass wasting
mechanical properties
permafrost
slope stability
slumping
soils
solifluction
stress

creeping
use creep

Creil
Town N of Paris.
BT Oise
France

crenulation cleavage
use slip cleavage

Creodonta
Order. Includes on level 2 under Mammalia(1).
BA Mammalia
BT Tetrapoda
Vertebrata

Crested Butte
Town and Peak in Gunnison County in W central.
BT Colorado
United States

Crestmore
Area.
BT California
United States

Cretaceous
Includes use on level 1 as an age term (list E); on level 2 under paleoterms, e.g. paleoecology, paleogeography, paleomagnetism. Above Jurassic, below Tertiary.
BA Mesozoic
NA Albian
Aptian
NT Benton Formation
Blairmore Group
NA Cenomanian
NT Colorado Group
NA Comanchean
NT Dakota Formation
Eureka Sound Formation
Graneros Shale
NA Lower Cretaceous
NT Lower Greensand

          Mancos Shale
  NA Middle Cretaceous
  NT Mishash Formation
          Nubian Sandstone
          Potomac Group
          Robles Formation
          San Felipe Formation
          Santana Formation
          Shiranish Formation
  NA Upper Cretaceous
  NT Viking Formation
  NA Vraconian
  NT Washita Group
          Weald Clay
          Whitemud Formation
          Yacoraite Formation
  SA Bauru Formation
          Beaverhead Formation
          Brianconnais Zone
          Deccan Traps
          Difunta Group
          Ewekoro Formation
          Fort Union Formation
          Franciscan Formation
          Gondwana System
          Great Valley Sequence
          Intertrappean Beds
          Ionian Zone
          Kootenay Formation
          Laramide Orogeny
          Maiolica Limestone
          Mardin Formation
          Muro Group
          Niniyur Group
          Paskapoo Formation
          Scaglia Formation
          Serra Gerral Formation
          Shimanto Group
          Tal Formation
          Tertiary

Crete
Administrative region and island in E Mediterranean Sea.
  CO N345500N354500
          E0263000E0231500
  BT Greece
  SA Hellenides

Creuse
Department in central France. Also a river.
  BT France
  NT Cheniers

Cricetidae
Family. Includes use on level 3 under Mammalia(1) Rodentia(2).
  BA Rodentia
          Mammalia
  BT Tetrapoda
          Vertebrata
  NA Pliomys

Crimea
Peninsula and oblast jutting into Black Sea. Former A.S.S.R. which was downgraded to an oblast in 1945 because of collaboration with Nazis during World War II occupation.
  CO N440000N460000
          E0370000E0320000
  BT Ukraine
          USSR

Crimean Mountains
Range in S Crimea along Black Sea Coast.
  BT Ukraine
          USSR

Crimean Plain
Dry, level steppe covering the northern 80% of Crimean Peninsula.
  BT Ukraine
          USSR

crinanite
Includes use on level 3 under igneous rocks(1) alkali basalt family(2).
See list H.
  BA alkali basalt family
  BT igneous rocks

Crinoidea
Includes use on level 2 under Echinodermata(1). See list F.
  BA Crinozoa
          Echinodermata
  BT Invertebrata
  NA Delocrinus
          Inadunata

Crinozoa
Includes use on level 2 under Echinodermata(1). See list F.
  BA Echinodermata
  BT Invertebrata
  NA Blastoidea
          Crinoidea
          Cystoidea
          Eocrinoidea

Cripple Creek
City in Teller County in central.
  BT Colorado
          United States

Crisana-Maramures
Historical province in NW.
  BT Transylvania
          Romania

cristobalite
  BA silica minerals
          framework silicates
          silicates
  BT minerals
  SA quartz
          tridymite

Crittenden County
Index states as applicable.
  BX Arkansas
          Kentucky
  BT United States

CRM
use chemical remanent magnetization

Croatia
Constituent republic in N.
  CO N423000N464500
          E0191500E0153000
  BT Yugoslavia
  NT Limski Channel
          Northern Limestone Alps
          Split
          Velebit Mountains
  SA Istria

crocidolite
  BA amphibole group
          chain silicates
          silicates
  BT minerals
  SA riebeckite

Crocodile River basin
use Limpopo Basin

Crocodilia
Order. Includes use on level 3 under Reptilia(1) Archosauria(2).
  BA Archosauria
          Reptilia
  BT Tetrapoda
          Vertebrata
  NA Teleosauridae

crocoite
  BA chromates
  BT minerals

Cromerian
Glaciation. Europe. Includes use on level 3 as an age term (list E).
  BA upper Pleistocene
          Pleistocene
  BT Quaternary
  NT South Polish Glaciation

Crook County
Index states as applicable.
  BX Oregon
          Wyoming
  BT United States

Crooked Creek Formation
In Meade Group. Southwestern Kansas and northwestern Oklahoma.
  BT Pleistocene
  SA Kansas
          Oklahoma

crops, field
use field crops

cross folds
use superposed folds

cross fractures
Term introduced in 1978. Before 1978, search cross AND fractures.
  BA fractures
  SA cross joints
          joints

cross joints
Term introduced in 1978. Includes use on level 3 under fractures(1) style(2). Before 1978, also search joints AND cross.
  BT joints
  SA cross fractures
          fractures

cross-bedding
Includes use on level 3 under sedimentary structures(1) planar bedding structures(2). See list K.
  BA planar bedding structures
          sedimentary structures
  SA bedding
          cross-laminations
          cross-stratification
          laminations

cross-laminations
Includes use on level 3 under sedimentary structures(1) planar bedding structures(2). See list K.
  UF diagonal lamination
  BA planar bedding structures
          sedimentary structures
  SA cross-bedding
          cross-stratification
          laminations
          ripple drift-cross laminations

cross-stratification
Includes use on level 3 under sedimentary structures(1) planar bedding structures(2). See list K.
  BA planar bedding structures
          sedimentary structures
  SA cross-bedding
          cross-laminations
          stratification

crossite
  BA amphibole group
          chain silicates
          silicates
  BT minerals
  SA glaucophane

Crossopterygii
Order. Includes use on level 3 under Pisces(1) Osteichthyes(2).
  BA Osteichthyes
          Pisces
  BT Vertebrata
  NA Rhipidistia

Crowsnest Pass
In SE British Columbia.
  BT British Columbia
          Canada

Crozet Islands
Five small French islands SE of South Africa.
  BT Indian Ocean

Crozon Peninsula
S of Brest in Brittany in W.
  BT Finistere
          France

crust
Refers to crust of Earth. For crust of other planets and Moon(1), see appropriate set. Includes use on level 1 (list A); on level 2 under seismology(1); on level 3 under plate tectonics(1) structure(2). Also search crusts. If 1, term set options are: topic [age, anomalies, composition, concepts, evolution, genesis, interpretation, observations, processes, properties, structure, theoretical studies, thickness]
          subtopic (no area term)
  NA lower crust
          upper crust
  SA asthenosphere
          basement
          compensation
          Conrad discontinuity
          continental type
          core
          cratons
          diastrophism
          discontinuities
          Earth
          earthquakes
          epeirogeny
          geophysics
          geosynclines
          geothermal gradient
          heat flow
          heat sources
          isostasy
          lithosphere
          low-velocity zones
          magmas
          mantle
          mobile belts
          Mohorovicic discontinuity
          neotectonics
          ocean basins
          oceanic type
          paleomagnetism
          plate tectonics
          plates
          sea-floor spreading
          seismic surveys
          seismology
          shields
          tectonics
          tectonophysics
          thickness
          undation
          volcanology

crust, weathering
use weathering crust

Crustacea
Superclass. Exclude Ostracoda. Use for groups of taxa; also use for recent classes Cephalocarida(3), Mystacocarida(3), and Branchiuria(3). Includes use on level 2 under Arthropoda(1). See list F.
  BA Arthropoda
  BT Invertebrata
  NA Branchiopoda
          Cirripedia
          Copepoda
          Decapoda
          Malacostraca
  NT Ostracoda
  SA crabs
          Trilobita

crustal shortening
Before 1978, also search shortening AND crustal.
  UF shortening, crustal
  SA gravity sliding
          plate tectonics
          tectonics
          thrust faults

crustal structure
A valid term through 1978. After 1978, see structure(2) under crust(1).

crustal studies
A valid term through 1978. After 1978, see crust(1).

Cruziana
BA ichnofossils
SA Trilobita

cryokarst
use thermokarst

cryolite
Includes use as level 3 commodity term under fluorspar(1). See list C.
BA halides
BT minerals
SA fluorspar

cryopedology
Includes use on level 3 under glacial geology(1) periglacial features(2).
SA engineering geology
    frost action
    glacial geology
    periglacial features
    permafrost

cryoturbation
UF congeliturbation
   frost churning
   frost stirring
   geliturbation
SA congelifraction
   frost action
   frost heaving
   periglacial features
   rock mechanics
   soil mechanics

cryptoexplosion features
Includes use on level 2 under geomorphology(1). Before 1973, also search cryptoexplosion structures.
UF features, cryptoexplosion
SA astroblemes
   cratering
   explosions
   geomorphology
   impact features
   impactite
   impacts
   meteor craters
   shatter cones
   suevite

cryptoexplosion structures
A valid term through 1972. After 1972, use cryptoexplosion features.

Cryptolithus
Genus. Includes use on level 3 under Trilobita (1) Ptychopariida(2).
BA Ptychopariida
   Trilobita
BT Trilobitomorpha
   Arthropoda
   Invertebrata

cryptomelane
BA oxides
BT minerals

cryptoperthite
BA feldspar group
   framework silicates
   silicates
BT minerals
SA perthite

Cryptostomata
Includes use on level 2 under Bryozoa(1). See list F.
BA Bryozoa
BT Invertebrata
NA Fenestellidae
   Rhabdomesidae

crystal chemistry
Used for the relations among chemical composition, structure and properties of crystals. Includes use on level 1 (list A) and on level 2 under minerals(1). Term set options are:
mineral group (List L)
   mineral species
topic (if more than one single specific mineral; e.g. bonding, coordination, ion exchange, order-disorder, partitioning, phase equilibria)
   subtopic
UF stereochemistry
SA bonding
   chemical analysis
   chemistry
   clay mineralogy
   coordination
   crystal growth
   crystal structure
   crystallography
   electron microscopy
   geochemistry
   ion exchange
   isomorphism
   lattice parameters
   mineralogy
   minerals
   order-disorder
   partition coefficients
   partitioning
   phase equilibria
   substitution
   transformations

crystal field
Includes use on level 3 under minerals(1).
UF field, crystal
SA minerals

crystal form
UF form, crystal
SA crystal growth
   habit

crystal growth
Used for studies on natural or artificial growth of crystals. Includes use on level 1 (list A). Term set options are:
mineral group (list L)
   mineral species
topic (if more than one single specific mineral; e.g. crystal form, mechanism, phase equilibria, synthesis, twinning)
   subtopic
SA clay mineralogy
   cleavage
   crystal chemistry
   crystal form
   crystal structure
   crystallography
   electron microscopy
   epitaxy
   fluid inclusions
   geochemistry
   growth
   growth spirals
   habit
   inclusions
   intergrowths
   lattice parameters
   mineralogy
   minerals
   nucleation
   overgrowths
   phase equilibria
   single-crystal method
   synthesis
   twinning
   zoning

crystal habit
use habit

crystal lattice
use lattice

crystal, quartz
use quartz crystal

Crystal River Formation
In Ocala Group. N and W Florida.
BT Eocene
SA Florida
   Ocala Group

crystal structure
Used for the internal structure of the crystal. Includes use on level 1 (list A) and level 2 under minerals(1). Also search atomic structure. Term set options are:
mineral group (list L)
   mineral species
topic (if more than one single specific mineral; e.g. bonding, cell dimensions, defects, refinement)
   subtopic
UF crystalline structure
SA atomic packing
   bonding
   cell dimensions
   clay mineralogy
   color centers
   crystal chemistry
   crystal growth
   crystallography
   defects
   dimorphism
   etching
   exsolution
   growth spirals
   inclusions
   ion exchange
   lattice
   lattice parameters
   metamict
   metamictization
   mineralogy
   minerals
   molecular structure
   perovskite structure
   polyhedra
   polymorphism
   polytypism
   refinement
   single-crystal method
   space groups
   structure
   superstructure
   unit cell
   X-ray analysis

crystalline limestone
use marble

crystalline rocks
Used when rocks cannot be distinguished as igneous or metamorphic.
SA igneous rocks
   metamorphic rocks
   rocks

crystalline schist
use schist

crystalline structure
use crystal structure

crystallinity
Degree to which a clay mineral, such as illite, is crystalline.
SA clay mineralogy
   diagenesis
   illite

crystallites
SA magmas
   volcanic glass

crystallization
Includes use on level 3 under intrusions(1) or magmas(1).
SA crystals
   differentiation
   fractional crystallization
   genesis
   granitization
   intergrowths
   intrusions
   lithification
   magmas
   precipitation
   recrystallization

crystallization magnetization
use chemical remanent magnetization

crystallization remanent magnetization
use chemical remanent magnetization

crystallography
Treated as a whole. Used for general studies on the discipline of crystallography. Includes use on level 1 (list A). Term set options are:
topic [automatic data processing, bibliography, catalogs, classification, concepts, education, experimental studies, general, history, instruments, methods, nomenclature, objectives, observations, philosophy, practice, principles, symposia, textbooks, theoretical studies]
   subtopic
SA atomic packing
   chemical analysis
   clay mineralogy
   cleavage
   color centers
   crystal chemistry
   crystal growth
   crystal structure
   crystals
   electron microscopy
   geochemistry
   growth spirals
   holography
   intergrowths
   lattice parameters
   mineralogy
   minerals
   monoclinic system
   paragenesis
   petrology
   phase equilibria
   pleochroism
   reflectivity
   spectroscopy
   standard materials
   X-ray analysis

crystals
Used as a general term. Includes use on level 3 under minerals(1).
SA cleavage
   crystallization
   crystallography
   intergrowths
   minerals
   phenocrysts

Cs
use cesium

Cs-137
Includes use on level 3 under isotopes(1).
SA cesium
   isotopes

Ctenostomata
Includes use on level 2 under Bryozoa(1). See list F.
BA Bryozoa
BT Invertebrata

Cu
use copper

Cuanza Basin
River basin in W central Angola.
UF Kwanza Basin
BT Angola

Cuanza-Sul
District on the South Atlantic in W.
BT Angola

Cuba
Includes use on level 1 as an area term (list O). For term set options see list B.
CO N195000N231500
   W0740000W0850000
BA West Indies
NT Camaguey

Havana
Isle of Pines
Las Villas
Oriente
Pinar del Rio
SA Greater Antilles

cubanite
UF chalmersite
BA sulfides
BT minerals

cube spar
use anhydrite

Cuddalore Sandstone
use Cuddalore Series

Cuddalore Sandstones
use Cuddalore Series

Cuddalore Series
Late Miocene-Pliocene. Also search Cuddalore Sandstone, and Cuddalore Sandstones.
UF Cuddalore Sandstone
    Cuddalore Sandstones
BT Tertiary
SA India
    Miocene
    Pliocene
    Tamil Nadu
    upper Miocene

Cuddapah
Town in S central.
BT Andhra Pradesh
    India

Cuddapah Basin
S central Andhra Pradesh. Also search Cuddapah.
BT Andhra Pradesh
    India

Cuddapah System
Now included in Purana Group. Subdivided into Kistna Series, Nallamalai Series, Cheyair Series and Papaghni Series. Also search Cuddapah.
BT Precambrian
SA Andhra Pradesh
    India

Cue
Town in W central.
BT Western Australia
    Australia

Cuenca
Province in E central Spain. Also a city.
BT Spain
SA Serrania de Cuenca

Cuisian
Europe. Above Ypresian, below Lutetian. Includes use on level 3 under age terms(1). See list E.
BA lower Eocene
    Eocene
    Paleogene
    Tertiary
BT Cenozoic

Culberson County
In W Texas.
BT Texas
    United States

Culm
Provincial Series, Europe.
BA Lower Carboniferous
    Carboniferous
BT Paleozoic

culture, living
use living culture

Cumana
City on the Caribbean Sea.
BT Venezuela
SA Sucre

Cumberland
County in NW England. Cities in Kentucky and Maryland. Town in Rhode Island. Village in central Virginia. Index countries and states as applicable.
CO N541500N551500
    W0021500W0033000
SA England
    Kentucky
    Maryland
    Rhode Island
    Virginia

Cumberland County
Index states as applicable.
BX Illinois
    Kentucky
    Maine
    New Jersey
    North Carolina
    Pennsylvania
    Tennessee
    Virginia
BT United States

Cumberland Mountains
use Cumberland Plateau

Cumberland Peninsula
On easternmost Baffin Island on Davis Strait in District of Franklin. Also search Cumberland AND peninsula.
BT Northwest Territories
    Canada

Cumberland Plateau
Southwesternmost division of the Appalachians. Also search Cumberland AND plateau. Index states as applicable.
UF Cumberland Mountains
BT United States
SA Alabama
    Appalachian Plateau
    Appalachians
    Kentucky
    Tennessee
    Virginia
    West Virginia

cummingtonite
BA amphibole group
    chain silicates
    silicates
BT minerals
SA anthophyllite

cumulates
SA igneous rocks
    magmas

Cupressaceae
Family.
BA Coniferales
    gymnosperms
BT Plantae

cuprite
BA oxides
BT minerals

Curaca River basin
Just S of the Sao Francisco River in NE.
BT Bahia
    Brazil

Curacao
Island in the Netherlands Antilles just N of W Venezuela.
BT Netherlands Antilles
SA West Indies

Curie point
Includes use on level 3 under heat flow(1) or paleomagnetism(1). Also search Curie temperature.
UF Curie temperature
SA heat flow
    magnetic properties
    magnetic susceptibility
    paleomagnetism
    thermoremanent magnetization

Curie temperature
use Curie point

curium
Includes use on level 1 and 2 as chemical element (list D).
UF Cm
SA elements

current directions
Includes use on level 3 under ocean circulation(1) or sedimentation(1).
UF directions, current
SA currents
    ocean circulation
    sedimentation

current lineations
use parting lineation

current markings
Includes use on level 3 under sedimentary structures(1) bedding plane irregularities(2). See list K.
UF markings, current
BA bedding plane irregularities
    sedimentary structures
SA flute casts
    scour marks
    tool marks

current partings
use parting lineation

current research
As of 1978, term is used on level 2 under surveys(1).
SA annual report
    associations
    geology
    progress report
    report
    research
    surveys

currents
Includes use on level 2 under ionosphere(1) and ocean circulation(1); on level 3 under aurora(1) electrical field(2) and magnetic field(2); on level 3 under meteorology(1) electrical phenomena(2).
NT bottom currents
    convection currents
    density currents
    longshore currents
    turbidity currents
SA aurora
    continental shelf
    current directions
    electrical field
    electrical phenomena
    electrojet
    ionosphere
    lightning
    magnetic field
    meteorology
    ocean circulation
    paleocurrents
    upwelling

curricula
Includes use on level 3 under education(1).
SA education

Curry County
Index states as applicable.
BX New Mexico
    Oregon
BT United States

curves, traveltime
use traveltime curves

Custer County
Index states as applicable.
BX Colorado
    Idaho
    Montana
    Nebraska
    Oklahoma
    South Dakota
BT United States

cut and fill
Includes use on level 3 under sedimentary structures(1) planar bedding structures(2). See list K.
BA planar bedding structures
    sedimentary structures

Cutch
District in E Gujarat. Also search Kutch. Rann of Cutch is to the N and NE.
UF Kutch
BT Gujarat
    India

Cutler Formation
Includes Halgaito Tongue, Cedar Mesa Sandstone Tongue, Organ Rock Tongue, White Rim Sandstone Member, DeChelly Sandstone Member, Hoskinnini Tongue, White River Sandstone, Rico transition facies. NE Arizona, SW Colorado, NW New Mexico, and SE Utah.
BT Permian
SA Arizona
    Colorado
    New Mexico
    Utah

cutoff rigidities
Includes use on level 3 under interplanetary space(1) cosmic rays(2) and under magnetosphere(1) cosmic rays(2).
UF rigidities, cutoff
SA cosmic rays
    interplanetary space
    magnetosphere

Cuttack
City in E.
BT Orissa
    India

cuttings
Includes use on level 3 under well-logging(1).
UF drill cuttings
SA boreholes
    cores
    well-logging
    wells

Cuvier abyssal plain
Off Western Australia. Also search Cuvier.
BT Indian Ocean

Cyanophyta
Blue algae, blue-green algae and Schizophyta are included here. Includes use on level 2 under algae(1). See list F.
UF blue algae
    blue-green algae
    Schizophyta
BA algae
BT Plantae
SA Collenia

cyathosponge
use Archaeocyatha

Cycadales
Includes use on level 2 under gymnosperms(1). See list F.
BA gymnosperms
BT Plantae
NA Ptilophyllum
SA Pachypteris
    Taeniopteris

Cyclades
Group of Islands in S Aegean Sea.
BT Aegean Islands
    Greece
SA Thera

cycles
Includes use on level 2 under geochemistry(1) and paleoclimatology(1).
SA atmosphere

cyclic processes
cyclothems
geochemistry
hydrologic cycle
hydrology
paleoclimatology
solar cycles

cyclic
A valid level 2 term through 1977. After 1977, use cyclic processes on level 2 under sedimentation(1).

cyclic loading
Includes use on level 3 under engineering geology(1).
SA engineering geology
  loading

cyclic processes
Term introduced in 1978. Includes use on level 2 under sedimentation(1). Before 1978, also search sedimentation AND cyclic.
SA cycles
  cyclothems
  processes
  sedimentation

cyclosilicates
use ring silicates

Cyclostomata
Includes use on level 2 under Bryozoa(1). See list F.
BA Bryozoa
BT Invertebrata
SA Cystoporata

cyclothems
Includes use on level 3 under sedimentary structures(1) planar bedding structures(2). See list K.
BA planar bedding structures
  sedimentary structures
NA megacyclothems
SA cycles
  cyclic processes
  rhythmite
  sedimentation

cylindrical folds
Term introduced in 1978. Before 1978, search folds AND cylindrical.
UF cylindroidal fold
BA folds

cylindrical structures
Includes use on level 2 under sedimentary structures(1). See list K.
BA sedimentary structures
SA structures

cylindrite
BA sulfantimonates
  sulfosalts
BT minerals

cylindroidal fold
use cylindrical folds

cymrite
BA sheet silicates
  silicates
BT minerals
SA aluminosilicates

Cynodontia
Infraorder. Includes use on level 3 under Reptilia(1) Synapsida(2).
BA Therapsida
  Synapsida
  Reptilia
BT Tetrapoda
  Vertebrata

Cypress Hills Formation
BT Oligocene
SA Canada
  Saskatchewan

Cyprididae
Family.
BA Podocopida
  Ostracoda
BT Crustacea
  Arthropoda
  Invertebrata
NA Candona

Cyprinidae
Family. Includes use on level 3 under Pisces(1) Osteichthyes(2).
BA Osteichthyes
  Pisces
BT Vertebrata
SA Actinopterygii
  Teleostei

Cyprus
Island nation in the Mediterranean Sea. Includes use on level 1 as an area term (list O). For term set options see list B.
CO N343000N354000
  E0343000E0323000
BA Middle East
NT Troodos Massif
SA Mediterranean region
  Near East

Cyrenaica
Easternmost part of country. Former province under the Italians.
BT Libya

cyrtolite
BA orthosilicates
  silicates
BT minerals
SA zircon

Cyrtospirifer
Includes use on level 3 under Brachiopoda(1) Articulata(2).
BA Articulata
  Brachiopoda
BT Invertebrata

Cystoidea
Includes use on level 2 under Echinodermata(1). See list F.
BA Crinozoa
  Echinodermata
BT Invertebrata

Cystoporata
BA Bryozoa
BT Invertebrata
SA Cyclostomata

Cytheracea
Superfamily. Includes use on level 3 under Ostracoda(1) Podocopida(2).
BA Podocopida
  Ostracoda
BT Crustacea
  Arthropoda
  Invertebrata
NA Cytheridae
  Leptocythere
  Trachyleberididae

Cytherella
Genus. Includes use on level 3 under Ostracoda(1) Podocopida(2).
BA Cytherellidae
  Podocopida
  Ostracoda
BT Arthropoda
  Crustacea
  Invertebrata

Cytherellidae
Family. Includes use on level 3 under Ostracoda(1) Podocopida(2).
BA Podocopida
  Ostracoda
BT Crustacea
  Arthropoda
  Invertebrata
NA Cytherella
  Cytherelloidea

Cytherelloidea
Genus. Includes use on level 3 under Ostracoda(1) Podocopida(2).
BA Cytherellidae
  Podocopida
  Ostracoda
BT Crustacea
  Arthropoda
  Invertebrata

Cytheridae
Family. Includes use on level 3 under Ostracoda(1) Podocopida(2).
BA Cytheracea
  Podocopida
  Ostracoda
BT Crustacea
  Arthropoda
  Invertebrata

Czechoslovakia
Includes use on level 1 as an area term (list O). For term set options see list B.
CO N473000N510000
  E0224500E0120000
BA Europe
NT Barrandian Basin
  Beskid Mountains
  Bohemia
  Moravia
  Slovakia
  Sudeten
SA Beshchady Mountains
  Carpathian Foredeep
  Carpathian Foreland
  Carpathians
  Danube River
  Danube Valley
  Elbe River
  Elbe Valley
  Erzgebirge
  Isergebirge
  Izera Mountains
  Moldanubian
  North Sudetic Basin
  Oder Valley
  Silesia
  Sniezník
  Subcarpathians
  Sudetic Basin
  Tatra Mountains
  Vienna Basin
  Western Carpathians

Czekanowskiales
BA Ginkgoales
  gymnosperms
BT Plantae

Czestochowa
City in N.
BT Katowice
  Poland

Czestochowa-Zawierce Basin
In NE Katowice.
BT Katowice
  Poland

D

D-region
Includes use on level 2 under ionosphere(1).
SA E-region
  F-region
  ionosphere

D/H
SA deuterium
  hydrogen
  isotopes
  stable isotopes
  tracer experiments
  tracers

Dacht-e-Nawar
use Dasht-i-Nawar

Dacian
Europe. Includes use on level 3 under age terms(1). See list E.
BA Pliocene

Anthropoda
Invertebrata

Cytheridae
Family. Includes use on level 3 under Ostracoda(1) Podocopida(2).
BA Cytheracea
  Podocopida
  Ostracoda
BT Crustacea
  Arthropoda
  Invertebrata

Neogene
Tertiary
BT Cenozoic

Dacian Basin
Region roughly covering modern Romania.
BT Romania

dacite
Includes use on level 3 under igneous rocks(1) andesite-rhyolite family(2). See list H.
BA andesite-rhyolite family
BT igneous rocks

dacite porphyry
BA andesite-rhyolite family
BT igneous rocks
SA porphyry

Dacryoconarida
Order.
BA problematic fossils
SA Mollusca

Dactylioceratidae
Family. Includes use on level 3 under Mollusca(1) Cephalopoda(2).
BA Ammonoidea
  Cephalopoda
  Mollusca
BT Invertebrata

Dade County
Index states as applicable.
BX Florida
  Georgia
  Missouri
BT United States

Dadoxylon
Genus. Includes use on level 3 under gymnosperms(1).
BA gymnosperms
BT Plantae
SA Coniferales
  Cordaitales
  Ginkgoales
  Glossopteris

Dagestan
Autonomous Soviet Socialist Republic on W shore of Caspian Sea. Also search Daghestan.
CO N410000N450000
  E0470000E0460000
UF Daghestan
BT Russian Republic
  USSR

Daghestan
use Dagestan

dahllite
BZ carbonates
  phosphates
BT minerals
SA carbonate apatite

Dahomey
A valid index term through 1976. After 1976, use Benin.

daily variation
use diurnal variations

Dakar
City on the Atlantic Ocean.
BT Senegal

Dakota Formation
E Colorado, NE New Mexico, North Dakota, Kansas, Minnesota, SE Montana, Nebraska, South Dakota, W Oklahoma, E Wyoming. Also search Dakota Sandstone; Dakota Group.
UF Dakota Group
  Dakota Sandstone
BT Cretaceous
SA Colorado
  Kansas
  Lower Cretaceous
  Minnesota
  Montana
  Nebraska

New Mexico
North Dakota
Oklahoma
South Dakota
Upper Cretaceous
Wyoming

Dakota Group
use Dakota Formation

Dakota Sandstone
use Dakota Formation

Dalarna
Region in W central Sweden. Also search Dalecarlia.
UF Dalarne
    Dalecarlia
BT Sweden

Dalarne
use Dalarna

Dalat
Town NE of Saigon.
BT Vietnam

Dalecarlia
use Dalarna

Dalgaranga
Region.
BT Western Australia
    Australia

Dallas
City in Dallas County in NE central.
BT Texas
    United States

Dallas County
Index states as applicable.
BX Alabama
    Arkansas
    Iowa
    Missouri
    Texas
BT United States

Dalmatia
Region, including many islands, which extends along the Adriatic Sea from Zadar on the N to near the Albanian border on the S. Sometimes the name is applied to most of the Yugoslav coast.
BT Yugoslavia

Dalradian
Great Britain, especially Scotland.
SA Cambrian
    Precambrian

damage, radiation
use radiation damage

Damara System
Includes Khomas Series, Marble Series, and Quartzite Series, Chuos Tillite, and Otavi Series. In W Damaraland region. NW of Walvis Bay.
BT Proterozoic
SA South-West Africa

Damascus
City in SW.
BT Syria

Damodar Valley
River valley. Index states as applicable.
BT India
SA Bihar
    West Bengal

Dampier Ridge
NE of Sidney in N.
BT Tasman Sea
    Pacific Ocean

damping
SA elastic waves
    seismology

dams
Level 1 term as of 1978 used for geologic studies on dams. Includes use on level 2 under foundations(1).
If level 1, term set options are:
construction
    subtopic
design
    subtopic
foundations
    subtopic
site exploration
    subtopic
NT arch dams
SA construction
    design
    embankments
    engineering geology
    foundations
    grouting
    land subsidence
    Oroville Dam
    reservoirs
    rock mechanics
    site exploration
    slope stability
    soil mechanics
    water storage

dams and damsites
Not a valid index term for GeoRef. Use dams.

Damuda Series
Forms middle part of lower Gondwanas. Subdivided into three stages: Raniganj (series), Ironstone shales, and lower Damudas, afterwards called Barakars.
BT Permian
SA Barakar Stage
    Bihar
    India
    Karharbari Stage
    Madhya Pradesh
    Raniganj Stage

Dan River basin
Index states as applicable.
BT United States
SA North Carolina
    Virginia

Danakil
use Afar

Danakil Depression
use Afar Depression

danburite
BA framework silicates
    silicates
BT minerals

Dangerous Islands
use Tuamotu Islands

Danian
Europe. Above Maestrichtian (Cretaceous), below Montian. Includes use on level 3 under age terms(1). See list E.
BA lower Paleocene
    Paleocene
    Paleogene
    Tertiary
BT Cenozoic
NT Niniyur Group

Daniel's Harbour
Village on upper Gulf of Saint Lawrence in NW.
BT Newfoundland
    Canada

Dankalia
use Afar

Danube Delta
On the Black Sea primarily in Romania. Index Romania and/or Ukraine as applicable. Also search Danube River delta.
UF Danube River delta
BT Europe
SA Romania
    Ukraine

Danube Plain
Covers the Walachian Plain and the N Bulgaria lowlands. Also search Danube. Index countries as applicable.
BT Europe
SA Bulgaria
    Romania

Danube River
Flows E from the Black Forest and turns S just N of Budapest. It then flows S, SE, N, and finally E into the Black Sea. Index Ukraine and countries as applicable. Also Search Danube.
UF Donau River
    Duna River
    Dunai River
    Dunarea River
    Dunau River
SA Austria
    Bulgaria
    Czechoslovakia
    Germany
    Hungary
    Romania
    Ukraine
    Yugoslavia

Danube River delta
use Danube Delta

Danube Stage
Term introduced in 1978. Europe.
BA lower Pleistocene
    Pleistocene
    Quaternary
BT Cenozoic

Danube Valley
River valley. Index Ukraine and countries as applicable. Also search Danube.
BT Europe
SA Austria
    Bulgaria
    Czechoslovakia
    Germany
    Hungary
    Romania
    Ukraine
    Yugoslavia

Danville
City in Pittsylvania County on the North Carolina border.
BT Virginia
    United States

Danzig
use Gdansk

Daonella
Includes use on level 3 under Mollusca(1) Bivalvia(2).
BA Bivalvia
    Mollusca
BT Invertebrata

Darasun
Town in S central Chita Oblast. N of Mongolia.
BT Russian Republic
    USSR

Darcy's law
SA aquifers
    fluid phase
    gases
    ground water
    hydrodynamics
    permeability
    petroleum
    porosity
    reservoir properties

Dare County
In E between Albermarle and Pamlico Sounds.
BT North Carolina
    United States

Darfur
Province in W.
BT Sudan

Darien
E part of isthmus between Gulf of Darien on E and Gulf of San Miguel on W.
BT Panama
    Central America

Darjeeling
Town, hill station, and district in extreme N near Sikkim border.
UF Darjiling
BT West Bengal
    India

Darjiling
use Darjeeling

Darling Downs
Tableland in SE.
BT Queensland
    Australia

Darling Range
In SW.
BT Western Australia
    Australia

Darmstadt
City in S.
BT Hesse
    West Germany

Dartmoor
Tableland in Devonshire in SW.
BT England
    Great Britain
    United Kingdom

Darvaz
Peak in the W Pamirs in central Tadzhikistan.
UF Kaganovich Peak
BT Tadzhikistan
    USSR

Darvaza
Town in Ashkhabad Oblast in S.
BT Turkmenia
    USSR

Darvaza Range
Branch of Alai Range N of Afghanistan in S central.
BT Tadzhikistan
    USSR
SA Alai Range

Darvel Bay
Inlet of Celebes Sea in Sabah in East Malaysia.
BT Malaysia
SA Celebes Sea

Darwin
City on the Timor Sea in N.
BT Northern Territory
    Australia

Darwin glass
Includes use on level 3 under tektites(1).
UF queenstownite
BA tektites
SA glasses
    impactite

Darwinula
Includes use on level 3 under Ostracoda(1) Podocopida(2).
BA Podocopida
    Ostracoda
BT Crustacea
    Arthropoda
    Invertebrata

Dashkesan
City in W.
BT Azerbaidzhan
    USSR

Dasht-e Bayaz
use Dasht-e-Bayaz

Dasht-e-Bayaz
Region in NE part of country. Locale of earthquake in 1968. Also search Dasht-e Bayaz.
UF Dasht-e Bayaz
BT Iran

Dasht-e-Lut Basin
use Lut Desert

Dasht-e-Nawar
use Dasht-i-Nawar

Dasht-i-Lut
use Lut Desert

Dasht-i-Nawar
Volcanic region in central Afghanistan. Also search Dacht-e-Nawar.
UF Dacht-e-Nawar
Dasht-e-Nawar
BT Afghanistan

Dastakert
Village in S.
BT Armenia
USSR

Dasycladaceae
Includes use on level 2 under algae(1). See list F.
BA algae
BT Plantae

data
Used as a general term, e.g. on level 3 under automatic data processing(1). Before 1975 also search analytical data.
SA automatic data processing
data acquisition
data analysis
data bases
data retrieval
electron microscopy data
electron probe data
ion probe data
mineral data
new data
SEM data
TEM data
X-ray data

data acquisition
Term introduced in 1978.
UF acquisition, data
SA automatic data processing
data
data storage

data analysis
Term introduced in 1978. Includes use on level 3 under automatic data processing(1).
SA analysis
automatic data processing
data

data bases
Also search data systems.
UF data systems
SA automatic data processing
data
data handling
data processing
data retrieval
data storage
information systems

data handling
Includes use on level 3 under automatic data processing(1). Term introduced in 1978.
UF handling, data
SA automatic data processing
data bases
data processing
data retrieval
information systems

data processing
UF processing, data
SA automatic data processing
data bases
data handling
information systems

data processing, automatic
use automatic data processing

data retrieval
Includes use on level 3 under automatic data processing(1).
UF retrieval, data
SA automatic data processing
data
data bases
data handling
data storage
information systems

data storage
Includes use on level 3 under automatic data processing(1).
SA automatic data processing
data acquisition
data bases
data retrieval
information systems
storage

data systems
use data bases

dates
Includes use on level 2 under absolute age(1).
SA absolute age
geochronology

dating, fission-track
use fission-track dating

datolite
BA orthosilicates
silicates
BT minerals

Dauphine
Historical region and former province in SE.
BT France

Dauphine Alps
W offshoot of Cottian Alps in SE.
BT France
SA Alps
Bas-Dauphine
Cottian Alps
Pelvoux Massif
Vercors

Davenport
City on the Mississippi River in Scott County.
BT Iowa
United States

Davidson
Town in Mecklenburg County on the South Carolina border in S.
BT North Carolina
United States

Davidson County
Index states as applicable.
BX North Carolina
Tennessee
BT United States

Davidsville Group
In NE Newfoundland.
BT Ordovician
SA Canada
Newfoundland

Davis Mountains
Jeff Davis County in W.
BT Texas
United States

Davis Strait
Between SW Greenland and Baffin Island.
BT Atlantic Ocean

Davos
Town consisting of Davos Platz and Davos Dorf in N central.
BT Graubunden
Switzerland

Dawson
City in W central.
BT Yukon Territory
Canada

dawsonite
BA carbonates
BT minerals

De Twente
use Twente

Dead Sea
A salt lake lying in part of the Great Rift Valley and constituting the lowest point on the Earth's surface. Also search Dead Sea Rift. Index countries as applicable.
BT Middle East
SA Dead Sea Rift
Great Rift Valley
Israel
Jordan

Dead Sea Rift
BT Middle East
SA Dead Sea
Israel
Jordan

Deadwood Formation
In N Black Hills, conformably underlies Aladdin Sandstone. SE Montana, W South Dakota (Black Hills), and E Wyoming.
BT Paleozoic
SA Lower Ordovician
Montana
South Dakota
Upper Cambrian
Wyoming

death assemblages
use thanatocenoses

Death Valley
Lowest point in United States. In Inyo County near the Nevada border.
CO N354500N370000
W1162000W1173000
BT California
United States

Debrecen
City in E part of country.
BT Hungary

debris
Used as a general term.
SA detritus
geomorphology
regolith

debris flows
As of 1978, term is used on level 2 under slope stability(1).
SA flows
slope stability

debris slopes
use talus slopes

Decapoda
Order. Includes use on level 3 under Arthropoda(1) Crustacea(2). Also search decapods.
UF decapods
BA Crustacea
Arthropoda
BT Invertebrata
NA Brachyura
Callianassa
SA Coleoidea
Ophiomorpha
Ophiomorpha nodosa

decapods
use Decapoda

Decatur County
In NW Kansas.
BT Kansas
United States

Decaturville
Village in Camden County in central.
BT Missouri
United States

decay, radioactive
use radioactive decay

Decazeville
Town in S central part of country.
BT Aveyron
France

Deccan Intertrappean
use Intertrappean Beds

Deccan Intertrappean Beds
use Intertrappean Beds

Deccan Intertrappean Series
use Intertrappean Beds

Deccan Plateau
Triangular tableland between the Eastern and Western Ghats and between the Narmada River on the N and the Krishna River on the S. Also search Deccan. Index states as applicable.
UF Dekkan Plateau
BT India
SA Andhra Pradesh
Indian Shield
Madhya Pradesh
Maharashtra
Mysore
Orissa

Deccan Trap
use Deccan Traps

Deccan Traps
Upper Cretaceous to lower Eocene. Named after the step-like aspect of the flat topped hills of The Deccan Plateau. Also search Deccan Trap.
UF Deccan Trap
SA Cretaceous
Eocene
India
lower Eocene
Upper Cretaceous

Deception Island
One of Britain's South Shetland Islands off N Antarctic Peninsula.
BT Antarctica
SA South Shetland Islands

Dechenellidae
Includes use on level 3 under Trilobita(1).
BA Trilobita
BT Trilobitomorpha
Arthropoda
Invertebrata

declination
Includes use on level 3 under Earth(1), geochronology(1), or paleomagnetism(1).
SA Earth
geochronology
magnetic field
paleomagnetism

decollement
Includes use on level 3 under folds(1) mechanics(2).
UF detachment
SA deformation
disharmonic folds
faults
flexural-slip
folds
mechanics
tectonics

decomposition
use chemical weathering

deconvolution
SA elastic waves
filters
filtration
seismology

decrepitation
Includes use on level 3 under chemical analysis(1) or mineralogy(1).
SA chemical analysis
    geologic thermometry
    mineralogy
    sample preparation

dedolomitization
SA calcitization
    diagenesis
    dolomitization
    processes

deep earthquakes
use deep-focus earthquakes

deep faults
use deep-seated structures

deep focus
A valid term through 1977. After 1977, use deep-focus earthquakes.

Deep Sea Drilling Project
Includes use on level 3 under individual oceans. Also search DSDP.
UF DSDP
SA Leg 18
    Leg 25
    Leg 27
    Leg 45
    Leg 46
    Leg 48
    ocean basins
    ocean floors
    oceanography

deep seismic sounding
Includes use on level 3 under seismology(1), geophysical surveys(1), and geophysics(1). Also search DSS; seismic sounding.
UF DSS
    seismic sounding
SA deep sounding
    explosions
    geophysical surveys
    geophysics
    seismology
    sounding

deep sounding
Includes use on level 3 under geophysical surveys(1) or geophysical methods(1).
SA deep seismic sounding
    echo sounding
    geophysical methods
    geophysical surveys
    sounding

deep structure
use deep-seated structures

deep-focus earthquakes
Term introduced in 1978. Includes use on level 3 under earthquakes(1). Before 1978, search earthquakes AND deep focus.
UF deep earthquakes
SA earthquakes
    focus
    intermediate-focus earthquakes
    shallow-focus earthquakes

deep-sea
A valid term through 1977. After 1977, use deep-sea environment.

deep-sea environment
Term introduced in 1978. Before 1978, search deep-sea or deep sea.
SA abyssal environment
    environment

deep-sea fans
use submarine fans

deep-seated structures
Term introduced in 1978. Includes use on level 3 under tectonics(1). Before 1978, search deep structure; deep-seated. Also search deep structure; deep faults.
UF deep faults
    deep structure
SA structures
    tectonics

deep-tow methods
Term introduced in 1978. Before 1978, search deep-tow. Includes use on level 3 under geophysical methods(1).
BA geophysical methods
SA acoustical methods
    methods

deepening
use dredging

deerite
BA chain silicates
    silicates
BT minerals

defects
As of 1978, term is used on level 2 under crystal structure(1).
SA crystal structure
    deformation
    destruction

definition
Used as a general term.
SA classification
    identification
    nomenclature
    photogrammetry

Deflandrea
BA Dinoflagellata
    palynomorphs

deforestation
SA conservation
    ecology
    forestry
    forests
    reclamation

deformation
Used for the process of folding, faulting, shearing, compression, or extension of the rocks as a result of various forces. Only small-scale deformations considered here. See tectonics(1) for large scale. Includes use on level 1 (list A) and level 2 under rock mechanics(1) and soil mechanics(1). Term set options are: experimental studies
    effects (e.g. creep, elastic strain, flow lines, fractures, kink-band structures, plastic flow, recrystallization, relaxation, twin-gliding) or name of modulus or limit (e.g. compression, elastic limit, fracture strength, shock, tension, torsion, yield strength) or theory (e.g. elasticity, plasticity, viscoelasticity, viscosity) or material
field studies
    effects or name of modulus or limit or material
theoretical studies
    effects or name of modulus or limit or material
SA anelasticity
    Atterberg limits
    axial-plane structures
    boudinage
    boudins
    bulk modulus
    competent materials
    compression
    compressive strength
    creep
    decollement
    defects
    diastrophism
    dilatancy
    dislocations
    distortion
    ductility
    elastic constants
    elastic limit
    elastic materials
    elastic properties
    elastic strain
    elastic waves
    elasticity
    engineering geology
    enterolithic folds
    extension
    extensometers
    fabric
    failure
    faults
    field studies
    finite strain
    flexure
    flow
    flow lines
    folds
    foliation
    fracture strength
    fractures
    geodesy
    geophysics
    Hooke's law
    isoclinal folds
    kink-band structures
    lamellae
    lineation
    mechanical properties
    melange
    metamorphic rocks
    metamorphism
    meteor craters
    microfractures
    mobilization
    mylonitization
    orogeny
    petrofabrics
    plastic flow
    plasticity
    plate tectonics
    pressure
    recrystallization
    rheology
    rock mechanics
    salt tectonics
    seismology
    shatter cones
    shear
    shear stress
    shock waves
    soil mechanics
    strain
    strainmeters
    strength
    stress
    structural analysis
    tectonics
    tectonite
    tectonophysics
    tensile strength
    tension
    torsion
    triaxial tests
    viscoelasticity
    viscosity
    yield strength

degasification
use degassing

degassing
Includes use on level 3 under atmosphere(1) or mantle(1). Also search degasification; outgassing.
UF degasification
SA atmosphere
    mantle

deglaciation
Includes use on level 3 under glacial geology(1) glaciation(2).
SA glaciation
    glaciers

degradation
Includes use on level 3 under geomorphology(1), soils(1), or weathering(1). Also search deterioration.
UF deterioration
SA denudation
    erosion
    geomorphology
    leaching
    soils
    transport
    weathering

Dehra Dun
City in N Uttar Pradesh. Also search Dehradun.
UF Dehradun
BT Uttar Pradesh
    India

Dehradun
use Dehra Dun

dehydration
Includes use on level 3 under geochemistry(1), minerals(1), or sediments(1).
UF drying
SA dewatering
    geochemistry
    hydration
    minerals
    sediments

Deimos
SA Mars
    Phobos
    satellites

Dekkan Plateau
use Deccan Plateau

Del Monte Beach
On Monterey Peninsula in Monterey County in W.
BT California
    United States

Del Norte County
Extreme NW.
BT California
    United States

Del Rio
City in Val Verde County on the Rio Grande in SW.
BT Texas
    United States

delafossite
BA oxides
BT minerals

Delaware
Includes use on level 1 as an area term (list O). For term set options see list B.
CO N382700N395000
    W0750500W0754600
BA United States
NX Sussex County
SA Aquia Formation
    Atlantic Coastal Plain
    Calvert Formation
    Choptank River
    Delaware Bay
    Delaware River
    Delaware River basin
    Delmarva Peninsula
    Eastern U.S.
    Magothy Formation
    Marshalltown Formation
    Monmouth Group
    Navesink Formation
    Newark Group
    Potomac Group
    Raritan Formation
    Saint Marys Formation
    Wicomico Formation
    Wilmington
    Wissahickon Formation

Delaware Basin
SE New Mexico and SW Texas. In-

dex states as applicable.
BT United States
SA New Mexico
   Texas

Delaware Bay
Arm of Atlantic Ocean. Index states as applicable.
BT United States
SA Delaware
   New Jersey

Delaware County
Index states as applicable.
BX Indiana
   Iowa
   New York
   Ohio
   Oklahoma
   Pennsylvania
BT United States

Delaware Limestone
Overlies Columbus Limestone; underlies Olentangy Shale. Central and N Ohio.
BT Middle Devonian
   Devonian
SA Ohio

Delaware River
Flows from S New York SE into Delaware Bay. Index states as applicable.
BT United States
SA Delaware
   New Jersey
   New York
   Pennsylvania

Delaware River basin
Index states as applicable.
BT United States
SA Delaware
   New Jersey
   New York
   Pennsylvania

Delhi
Territory in N central part of country. Also a city NNE of New Delhi.
BT India
NT New Delhi

Delhi System
Included Ajabgarh Series, Hornstone Breccia, Kushalgarh Limestone, Alwar Series, Raialo limestones and quartzes. Hornstone Breccia and Kushalgarh Limestone later confined to NE Rajputana, and Raialos excluded from Delhis to form separate unit. Named after city of Delhi.
BT Precambrian
SA Haryana
   India

Delinesti
Village in Caras-Severin County in W part of country.
BT Banat
   Romania

dellenite
Includes use on level 3 under igneous rocks(1) andesite-rhyolite family(2). See list H.
BA andesite-rhyolite family
BT igneous rocks
SA rhyodacite

Dellys
City on Mediterranean Sea E of Algiers.
BT Algeria

Delmarva Peninsula
Between Chesapeake and Delaware bays. Index states as applicable.
BT United States
SA Delaware
   Maryland
   Virginia

Delmas
Village in W central.
BT Saskatchewan
   Canada

Delocrinus
Genus. Includes use on level 3 under Echinodermata(1) Crinoidea(2).
BA Crinoidea
   Crinozoa
   Echinodermata
BT Invertebrata

Delta Area
BT Netherlands

Delta River
In E Alaska.
BT Alaska
   United States

deltaic
A valid term through 1977. After 1977, use deltas.

deltas
Includes use on level 3 under geomorphology(1) fluvial features(2) and shore features(2); on level 3 under ecology(1) and under paleoecology(1).
SA ecology
   fluvial features
   geomorphology
   paleoecology
   sedimentation
   shore features

demagnetization
Includes use on level 3 under geophysics(1) or paleomagnetism(1).
SA geophysics
   magnetization
   paleomagnetism
   remanent magnetization

Demospongea
Includes use on level 2 under Porifera(1). See list F.
BA Porifera
BT Invertebrata

Denali Fault
Mount McKinley area in S central.
BT Alaska
   United States

Denbighshire
County in N.
BT Wales
   Great Britain
   United Kingdom

dendrograms
SA diagrams
   statistical analysis

Dendroidea
Includes use on level 2 under Graptolithina(1). See list F.
BA Graptolithina
BT Invertebrata
SA Hemichordata

Dengizkul
Lake SW of Bukhara.
BT Uzbekistan
   USSR

denitrification
Includes use on level 3 under geochemistry(1).
SA geochemistry

Denmark
Includes use on level 1 as an area term (list O). For term set options see list B.
CO N541500N580000
   E0130000E0080000
BA Europe
NT Bornholm
   Copenhagen
   Helsingor
   Vendsyssel
   Zealand
SA Baltic region
   European Platform
   Fennoscandia
   Jutland
   North Sea Coast
   Oresund
   Scandinavia
   Zechstein

densities and temperatures
Includes use on level 2 under aeronomy(1).
SA aeronomy
   density
   ion densities and temperatures
   temperature

density
Includes use on level 3 under Earth(1). Also search densities.
SA convection currents
   densities and temperatures
   ion densities and temperatures
   isostasy
   sea water
   specific gravity

density currents
BT currents
SA turbidity currents

Dentalium
Genus. Includes use on level 3 under Mollusca(1) Scaphopoda(2).
BA Scaphopoda
   Mollusca
BT Invertebrata

denudation
Includes use on level 3 under geomorphology(1) processes(2).
SA degradation
   erosion
   erosion features
   geomorphology
   water erosion
   weathering

Denver
City in Denver County in NE central.
BT Colorado
   United States

Denver Basin
Primarily in central and NE Colorado. Index states as applicable.
UF Denver Julesberg Basin
BT United States
SA Colorado
   Nebraska
   Wyoming

Denver Julesberg Basin
use Denver Basin

depletion
Used as a general term.
SA conservation
   migration of elements
   natural resources
   petroleum
   reclamation

deposition
Includes use on level 2 under sedimentation(1); on level 3 under glacial geology(1) glaciation(2).
SA accretion
   accumulation
   alluvial fans
   alluvial plains
   cementation
   diagenesis
   glacial geology
   glaciation
   ice movement
   reservoirs
   reworking
   sedimentary processes
   sedimentary structures
   sedimentation
   sediments
   stratification
   uniformitarianism

depositional fault
use growth faults

depositional remanent magnetization
Also search detrital remanent magnetization.
UF detrital remanent magnetization
BA remanent magnetization
SA paleomagnetism

deposits
Includes use on level 2 (for nonmetals) under commodities (list C).
SA disseminated deposits
   lead-zinc deposits
   lenses
   massive deposits
   mineral deposits
   ore deposits
   roll-type deposits
   stockwork deposits
   stratabound deposits
   stratiform deposits

depressions
Includes use on level 3 under geomorphology(1).
SA basins
   folds
   geomorphology

depth
Includes use on level 2 under Mohorovicic discontinuity(1).
SA dimensions
   Mohorovicic discontinuity

depth of compensation
use carbonate compensation depth

Derby
City in Dereyshire in central.
BT England
   Great Britain
   United Kingdom

Derbyshire
County in central.
CO N524500N533000
   W0011500W0020000
BT England
   Great Britain
   United Kingdom

Des Moines Series
use Desmoinesian

desalination
use desalinization

desalinization
Includes use on level 3 under ground water(1) or water resources(1). Also search desalination.
UF desalination
SA brines
   ground water
   leaching
   salinity
   salt
   sea water
   soils
   treatment

Descartes
Village in NW part of country.
BT Oran
   Algeria

descloizite
BA vanadates
BT minerals

description
Used as a general term.
SA landform description
   morphology

Desert Basin
use Canning Basin

desert pediplains
use pediplains

desert plains • diagenesis 78

desert plains
  use pediplains
Desert soils
  Includes use on level 3 under soils(1). See list M.
  BA soils
  SA soil group
      Zonal soils
desertification
  SA ecology
deserts
  Includes use on level 3 under geomorphology(1) eolian features(2); on level 3 under ecology(1).
  SA arid environment
      ecology
      eolian features
      geomorphology
      playas
desiccation
  SA ecology
      moisture
      pore water
      soils
design
  As of 1978, term is used on level 2.
  SA dams
      development
      engineering geology
      foundations
      instruments
      marine installations
      nuclear facilities
      reservoirs
      shorelines
      tunnels
      underground installations
      waterways
desmids
  Includes use on level 3 under algae(1) Chlorophyta(2). See list F.
  BA Chlorophyta
      algae
  BT Plantae
desmine
  use stilbite
Desmoinesian
  Comprises Cherokee Group, Marmaton Group, Cabaniss Group, and Krebs Group. Subdivided into Venteran and Cygnian substages. Also search Des Moines Series; Desmoinesian Series.
  UF Des Moines Series
      Desmoinesian Series
  BA Middle Pennsylvanian
      Pennsylvanian
  BT Paleozoic
  NT Cabaniss Formation
      Cherokee Group
      Labette Shale
      Marmaton Group
      Savanna Formation
      Wewoka Formation
  SA Arkansas
      Bevier Coal
      Francis Creek Shale
      Iowa
      Kansas
      Missouri
      Nebraska
      Oklahoma
Desmoinesian Series
  use Desmoinesian
Desmostylia
  Includes use on level 2 under Mammalia(1). See list F.
  BA Mammalia
  BT Tetrapoda
      Vertebrata
Desolation Islands
  use Kerguelen Islands

desorption
  SA adsorption
      geochemistry
destruction
  Used as a general term.
  SA catastrophies
      corrosion
      defects
      geologic hazards
detachment
  use decollement
detection
  Includes use as a level 2 or 3 term appropriate to a large number of topics, e.g. on level 2 under earthquakes(1). See list G.
detergents
  Includes use on level 3 under environmental geology(1) waste disposal(2).
  SA environmental geology
      liquid waste
      waste disposal
      waste water
deterioration
  use degradation
detrital fan
  use alluvial fans
detrital remanent magnetization
  use depositional remanent magnetization
detritus
  Includes use on level 3 under sediments(1) or sedimentary rocks(1). Also search detrital; detrital minerals.
  SA abrasion
      debris
      erosion
      weathering
Detroit River Group
  Includes Sylvania Sandstone, Amherstburg Formation, Lucas Formation. SE Michigan, N Ohio, and W Ontario.
  BT Middle Devonian
      Devonian
  SA Michigan
      Ohio
      Ontario
deuterium
  Includes use on level 1 and 2. See list D (chemical elements). Also search H-2.
  UF H-2
  SA D/H
      elements
      hydrogen
      isotopes
      stable isotopes
Deux-Sevres
  Department in W.
  CO N460000N471000
      E0001500W0010000
  BT France
  SA Poitou
development
  Used as a general term.
  SA design
      economic geology
      exploration
      ground water
Devensian
  Europe.
  BA upper Pleistocene
      Pleistocene
      Quaternary
  BT Cenozoic
deviation, standard
  use standard deviation

Devils Icebox
  Region in Boone County in central.
  BT Missouri
      United States
devitrification
  SA geochemistry
      glasses
      volcanic glass
Devoluy
  Limestone range of the Dauphine Pre-Alps in SE.
  BT France
Devon Island
  N of Baffin Island off Baffin Bay in District of Franklin. Also search Devon.
  CO N743000N763000
      W0793000W0970000
  UF North Devon Island
  BT Northwest Territories
      Canada
Devonian
  Includes use on level 1 as an age term (list E); on level 2 under paleoterms, e.g. paleoecology, paleogeography, paleomagnetism. Above Silurian, below Carboniferous.
  BA Paleozoic
  NT Acadian Phase
      Barre Granite
      Cedar City Formation
      Genesee Group
      Heemskirk Granite
      Hunsruck Shale
      Keg River Formation
      Levis Shale
      Lilydale Limestone
      Lost Burro Formation
  NA Lower Devonian
  NT Martin Formation
  NA Middle Devonian
  NT Miette Complex
      Muth Quartzite
  NA Old Red Sandstone
  NT Ramparts Formation
      Tor Formation
      Traverse Group
  NA Upper Devonian
  NT Upper Old Red Sandstone
      Waterways Formation
  SA Antler Orogeny
      Beacon Supergroup
      Bedford Shale
      Berea Sandstone
      Broken River Formation
      Catskill Formation
      Chattanooga Shale
      Galway Granite
      Geirud Formation
      Hercynian Orogeny
      Hidden Valley Dolomite
      Hunton Group
      Peel Sound Formation
      Read Bay Formation
      Ringerike Sandstone
      Rondout Formation
      Talladega Group
      Woodford Shale
Devonshire
  County in SW England.
  CO N501500N511500
      W0030000W0043000
  BT England
      Great Britain
      United Kingdom
dewatering
  SA beneficiation
      dehydration
      separation
Dewey County
  Index states as applicable.
  BX Oklahoma
      South Dakota
  BT United States

dextral faults
  use right-lateral faults
Dhanbad
  Town in SE.
  BT Bihar
      India
Dharwar
  City in NW Mysore. Now part of the joint municipality of Hubli-Dharaar.
  BT Mysore
      India
Dharwars
  "Sub metamorphic" crystalline rocks of the Dharwar System, named after city of Dharwar in Mysore, but occurring throughout the country.
  SA Chitaldrug schist belt
      India
Dhenkanal
  Town and district in E central.
  BT Orissa
      India
diabase
  Includes use on level 3 under igneous rocks(1) basalt family(2). See list H.
  BA basalt family
  BT igneous rocks
  SA dolerite
      metadiabase
      olivine diabase
      quartz diabase
      trap rock
      trap rocks
diabase amphibolite
  use amphibolite
diabase porphyry
  BA basalt family
  BT igneous rocks
  SA porphyry
Diablo Bolson
  use Diablo Platform
Diablo Platform
  Extreme W Texas.
  UF Diablo Bolson
  BT Texas
      United States
Diablo Range
  One of the Coast Ranges in W central California.
  BT California
      United States
  SA Coast Ranges
diachronism
  SA biostratigraphy
      stratigraphy
diagenesis
  Level 1 term as of 1978, used for all the chemical, physical and biological changes, modifications, or transformations undergone by a sediment after its initial deposition, and during and after its lithification, exclusive of weathering and metamorphism. Includes use on level 2 under sedimentary rocks(1) and sediments(1). If level 1, term set options are:
  controls
      subtopic
  dolomitization
      subtopic
  effects
      subtopic
  geochemistry
      subtopic
  indicators
      subtopic
  materials
      subtopic
  mechanism
      subtopic
  processes
      subtopic

sedimentation
    subtopic
topic [e.g. environment, experimental studies, models, theoretical studies]
    subtopic
NA carbonatization
    cementation
    chertification
    early diagenesis
    late diagenesis
    syngenesis
SA alteration
    authigenesis
    calcification
    calcitization
    coalification
    compaction
    consolidation
    crystallinity
    dedolomitization
    deposition
    dolomitization
    halmyrolysis
    indicators
    lithification
    metamorphism
    metasomatism
    phosphatization
    pressure solution
    processes
    sedimentary petrology
    sedimentary rocks
    sedimentary structures
    sedimentation
    sediments
    silicification
    weathering

diagonal lamination
use cross-laminations

diagonal-slip faults
Term introduced in 1978. Before 1978, search faults AND diagonal-slip.
BA faults

diagrams
Used as a general term.
SA dendrograms
    legend
    maps
    phase diagrams
    pollen diagrams

diallagite
Includes use on level 3 under igneous rocks(1) ultramafic family(2). See list H.
BA ultramafic family
BT igneous rocks

dialogite
use rhodochrosite

diamictite
Includes use on level 3 under sedimentary rocks(1) clastic rocks(2). See list I. Also search mixtite.
BA clastic rocks
BT sedimentary rocks
SA mixtite
    terrigenous materials

diamicton
UF symmicton
BA clastic sediments
BT sediments
SA till

diamond
When of economic value, use diamonds. Includes use on level 3 under minerals(1) native elements and alloys(2). See list L.
BA native elements and alloys
BT minerals
SA carbon
    diamonds
    gems
    graphite

kimberlite

diamonds
Use only when of economic value. Includes use on level 1 as a commodity term (list C);on level 2 under placers(1).
SA abrasives
    carbon
    carbonado
    diamond
    gems
    graphite
    kimberlite
    minerals
    placers

diaphorite
use ultrabasite

diaphthoresis
use retrograde metamorphism

diapiric fold
use diapirs

diapirism
SA diapirs
    intrusions
    salt domes
    salt tectonics

diapirs
Includes use on level 3 under folds(1) style(2) and under salt tectonics(1).
UF diapiric fold
    piercement
    piercing fold
SA anticlines
    cap rocks
    diapirism
    domes
    folds
    salt
    salt domes
    salt tectonics

diaspore
BA oxides
BT minerals

diastrophism
General term used for large scale features. Includes use on level 3 under orogeny(1) and tectonics(1).
SA crust
    deformation
    epeirogeny
    orogeny
    tectonics

diatomaceous earth
Includes use on level 3 under sedimentary rocks(1) clastic rocks(2). See list I. Also search tripolite.
UF earth, diatomaceous
    tripolite
BA clastic rocks
BT sedimentary rocks
SA diatomite
    diatoms
    nonterrigenous materials

diatomite
Includes use on level 1 as a commodity term (list C); on level 3 under sedimentary rocks (1) clastic rocks(2). See list I.
BA clastic rocks
BT sedimentary rocks
SA abrasives
    diatomaceous earth
    nonterrigenous materials

diatoms
Includes use on level 2 under algae(1). See list F.
BA algae
BT Plantae
NA Melosira
    Navicula
    Nitzschia
SA diatomaceous earth

frustules
phytoplankton
plankton

diatremes
Includes use on level 2 under intrusions(1).
BA intrusions
SA breccia pipes
    pipes
    volcanic features

Dickenson County
In SW Virginia.
BT Virginia
    United States

Dickinson County
Index states as applicable.
BX Iowa
    Kansas
    Michigan
BT United States

dickite
BA sheet silicates
    silicates
BT minerals
SA clay minerals

Dicotyledoneae
Includes use on level 2 under angiosperms(1). See list F. Also search dicotyledons.
UF dicotyledons
BA angiosperms
BT Plantae
NA Acer
    Alnus
    Annonaceae
    Artemisia
    Betula
    Ericaceae
    Euphorbiaceae
    Fagus
    Juglandaceae
    Juglans
    Lauraceae
    Leguminosae
    Quercus
    Ulmaceae
SA Monocotyledoneae

dicotyledons
use Dicotyledoneae

Dicroidium
Includes use on level 3 under pteridophytes(1) Filicopsida(2).
BA Filicopsida
    pteridophytes
BT Plantae
SA gymnosperms
    Pteridospermae

dictionaries
Includes use on level 3 under lexicons(1).
SA glossaries
    lexicons

Dictyocha
BA Silicoflagellata
    Protista

Didacna
Genus. Includes use on level 3 under Mollusca(1) Bivalvia(2).
BA Bivalvia
    Mollusca
BT Invertebrata

Didymograptus
Genus.
BA Graptoloidea
    Graptolithina
BT Invertebrata

dielectric constant
UF constant, dielectric
SA dielectric properties
    electrical field
    electrical methods

dielectric properties
SA dielectric constant
    electrical field
    electrical methods
    properties

Dieppe
City on English Channel.
BT Seine-Maritime
    France

differential thermal analysis
Used on level 1 for methodology through 1977. After 1977, includes use on level 2 under thermal analysis(1).
UF thermography
SA analysis
    chemical analysis
    clay mineralogy
    clay minerals
    DTA data
    electron microscopy
    sample preparation
    spectroscopy
    thermal analysis
    thermogravimetric analysis
    X-ray diffraction analysis

differential thermal analysis data
use DTA data

differential weathering
Before 1978, also search weathering AND differential.
BT weathering

differentiation
Includes use on level 2 under magmas(1); on level 3 under intrusions(1) petrology(2) and under igneous rocks(1) petrology(2); on level 3 under deformation(1) and under Earth(1).
SA crystallization
    fractional crystallization
    horizon differentiation
    igneous rocks
    intrusions
    laccoliths
    magmas
    plugs
    sills
    stocks

diffraction
Includes use on level 2 under ocean waves(1); on level 3 under seismology(1).
SA ocean waves
    reflection
    refraction
    seismology

diffusion
Includes use on level 2 under aeronomy(1), meteorology(1), and ocean circulation(1); on level 3 under geochemistry(1) processes(2). See also under specific elements. Included use on level 2 under oceanography(1) until 1976.
SA absolute age
    aeronomy
    chemical analysis
    dissipation
    geochemistry
    ion exchange
    meteorology
    ocean circulation
    oceanography
    permeability
    polarography
    processes

diffusivity, thermal
use thermal diffusivity

Difunta Group
SA Coahuila
    Cretaceous
    lower Paleocene

Mexico
Paleocene
Upper Cretaceous

digenite
UF alpha chalcocite
blue chalcocite
BA sulfides
BT minerals
SA chalcocite

Digha
Coastal region along the Bay of Bengal.
BT West Bengal
India

digital simulation
Term introduced in 1978. Before 1978, search digital AND simulation.
SA automatic data processing
digital techniques
simulation

digital techniques
Term introduced in 1978. Before 1978, also search digital AND automatic data processing.
SA analog techniques
automatic data processing
digital simulation
techniques

Digne
City in SE part of country.
BT Alpes-de-Haute Provence
France

Dijon
City in E central part of country.
BT Cote-d'Or
France

dike swarms
Includes use on level 3 under intrusions(1).
SA dikes
igneous rocks
intrusions
ring dikes
swarms

dikes
Includes use on level 2 under intrusions(1).
BA intrusions
NT ring dikes
SA batholiths
clastic dikes
dike swarms
igneous rocks
laccoliths
ring complexes
ring structures
sills

dilatancy
Includes use on level 3 under deformation(1), seismology(1), earthquakes(1), and geophysics(1).
SA deformation
earthquakes
expansion
physical properties
seismology

dilatation
use dilation

dilatational wave
use P-waves

dilation
Used as a general term.
UF dilatation

Dillon
Village in Summit County in N central.
BT Colorado
United States

dimension stone
Includes use as level 3 commodity term under construction materials(1), granite(1), sandstone(1), limestone(1), and marble(1). See list C.
UF stone, dimension
BA construction materials
SA building stone
granite
limestone
marble
sandstone

dimensions
SA capacity
cell dimensions
depth
grain size
grains
shape
size

Dimitrovo coal basin
use Pernik coal basin

dimorphism
Includes use on level 3 under crystal structure(1), minerals(1), and fossils (list F). Also search polymorphism.
SA crystal structure
minerals
polymorphism
sexual dimorphism

Dinant
Town in S part of country.
BT Namur
Belgium

Dinant Basin
S part of country.
BT Namur
Belgium

Dinantian
Europe. Includes Tournaisian and Visean. Includes use on level 3 under age terms(1). See list E.
BA Lower Carboniferous
Carboniferous
BT Paleozoic
SA Avonian
Tournaisian
Visean

Dinaric Alps
SE division of Eastern Alps along Adriatic Sea. Also search Dinaric.
CO N430000N450000
E0180000E0150000
BT Yugoslavia
SA Alps
Eastern Alps
Northern Limestone Alps
Velebit Mountains
Zlatibor Mountains

Dingle Peninsula
Juts into Atlantic in W County Kerry.
CO N520500N521500
W0094500W0103000
BT Ireland

Dinoflagellata
Including peridineans. Hystrichosphaerids are included here and under acritarchs. Includes use on level 2 under palynomorphs(1). See list F. Also search dinoflagellates; Dinophyceae.
UF dinoflagellates
Dinophyceae
peridineans
BA palynomorphs
NA Deflandrea
NZ Hystrichosphaeridae
NA Peridinium
SA acritarchs
plankton
Pyrrhophyta

dinoflagellates
use Dinoflagellata

Dinophyceae
use Dinoflagellata

Dinosauria
use dinosaurs

dinosaurs
Includes use on level 3 under Reptilia(1). See list F. Also search Dinosauria.
UF Dinosauria
BA Reptilia
BT Tetrapoda
Vertebrata
SA Archosauria
Hadrosauridae

Diois
Massif in SE part of country.
BT Drome
France

diopside
BA pyroxene group
chain silicates
silicates
BT minerals
SA chrome diopside

diorite
Includes use on level 3 under igneous rocks(1) diorite family(2). See list H.
BA diorite family
BT igneous rocks
SA metadiorite
microdiorite
quartz diorite

diorite family
Includes use on level 2 under igneous rocks(1). See list H.
BT igneous rocks
NA diorite
ferrodiorite
mangerite
microdiorite
monzodiorite
quartz diorite
syenodiorite
tonalite
trondhjemite
SA appinite
orbicular texture

diorite porphyry
BA volcanic rocks
BT igneous rocks
SA porphyry

dip
As of 1978, term is used on level 3 as a noun. Before 1978, included use for dip faults and dip fractures.
UF angle of dip
SA dip-slip faults
dipmeter logging
orientation
strike
structural analysis

dip faults
Term introduced in 1978. Before 1978, search faults AND dip.
BA faults

dip fractures
Term introduced in 1978. Before 1978, search dip AND fractures.
BA fractures

dip-slip faults
Term introduced in 1978. Includes use on level 3 under faults(1) displacements(2). Before 1978, search faults AND dip-slip.
BA faults
SA dip
displacements

Diplograptina
Includes use on level 2 under Graptolithina(1). See list F.
BA Graptoloidea
Graptolithina
BT Invertebrata

dipmeter logging
Term introduced in 1978. Includes use on level 2 under well-logging(1). Before 1978, also search well-logging AND dipmeter or dip-meter.
UF logging, dipmeter
BA well-logging
SA boreholes
dip

dipole moment
SA magnetic field
paleomagnetism

Diptera
Includes use on level 2 under Insecta(1). See list F.
BA Insecta
BT Arthropoda
Invertebrata
SA Raphidiodea

dipyrite
use pyrrhotite

directions, current
use current directions

directory
Includes use on level 3 under commodity terms(1), e.g. under mineral resources(1). See list C.
SA guidebook

discharge
Includes use on level 3 under ground water(1), or hydrology(11). Also search effluents.
UF effluents
SA ground water
hydrology
recharge
runoff
water recovery

Discoasteridae
use discoasters

discoasters
Includes use on level 3 under algae(1) nannofossils(2). See list F. Also search discoaster; Discoasteridae.
UF Discoasteridae
BA nannofossils
algae
BT Plantae
NA Ethmodiscus
SA Coccolithophoraceae
Sphenolithus

Discocyclina
Genus. Includes use on level 3 under foraminifera(1) Orbitoidacea(2).
BA Orbitoidacea
Rotaliina
foraminifera
BT Invertebrata

discontinuities
Includes use on level 3 under crust(1), mantle(1), or seismology(1). Use for interface in seismology.
SA Conrad discontinuity
crust
mantle
Mohorovicic discontinuity
seismology

Discorbacea
Includes use on level 2 under foraminifera(1). See list F.
BA Rotaliina
foraminifera
BT Invertebrata

discordant folds
Term introduced in 1978. Includes use on level 3 under folds(1) orientation(2). Before 1978, also search folds AND discordant.
BA folds
SA longitudinal orientation
orientation

discoveries
Used as a general term.
SA mineral deposits
    placers

Discovery Bay
Village on central N coast.
BT Jamaica
    West Indies

Discovery Deep
Also search Discovery AND deeps.
BT Red Sea

discriminant analysis
Includes use on level 3 under mathematical geology(1).
UF discriminant function analysis
BA statistical methods
SA analysis
    mathematical geology

discriminant function analysis
use discriminant analysis

disharmonic folds
Term introduced in 1978. Includes use on level 3 under folds(1) style(2). Before 1978, also search folds AND disharmonic.
BA folds
SA decollement
    flexural-slip
    harmonic folds

Disko Island
Davis Strait just off SW Greenland. Also search Disko.
CO N683000N702000
    W0520000W0550000
BT Greenland

dislocations
Includes use on level 3 under deformation(1), faults(1), or seismology(1).
SA deformation
    displacements
    faults
    seismology
    structural geology

dispersal
use dispersion

dispersion
Used as general term. Also search dispersal.
UF dispersal
SA dissipation
    homogenization
    primary dispersion
    secondary dispersion

dispersion patterns
Includes use on level 3 under mineral exploration(1) geochemical methods(2).
SA geochemical methods
    mineral exploration
    patterns

displacement theory
use continental drift

displacements
Includes use on level 2 under faults(1).
SA dip-slip faults
    dislocations
    drag folds
    faults
    nappes
    overthrust faults
    reactivation
    reverse faults
    transcurrent faults
    transform faults
    wrench faults

display, graphic
use graphic display

disposal, waste
use waste disposal

disseminated deposits
Term introduced in 1978. Before 1978, search disseminated AND ore deposits; search deposits AND disseminated.
SA deposits
    mineral deposits
    ore deposits

dissipation
Used as a general term.
SA diffusion
    dispersion
    scattering

dissociation
Includes use on level 3 under geochemistry(1) processes(2).
SA electrolysis
    geochemistry
    processes

dissolution
A valid term through May 1978. After May 1978, use solution or solutions, e.g. under geochemistry(1).

dissolved materials
Term introduced in 1978. Before 1978, search dissolved.
SA materials
    suspended materials

Distephanus
BA Silicoflagellata
    Protista

distortion
Includes use on level 3 under deformation(1).
SA deformation

distribution
Includes use as a level 2 or 3 term appropriate to a large number of topics, e.g. on level 2 under faults(1), folds(1), fractures(1) and under ocean circulation(1). See list G.
SA size distribution

District of Columbia
Includes use on level 1 as an area term (list O). For term set options see list B.
BA United States
NT Rock Creek Park
    Smithsonian Institution
SA Anacostia River basin
    Potomac River
    Potomac River basin
    Wicomico Formation

District of Franklin
The northernmost district.
CO N630000N850000
    W0600000W1250000
UF Franklin District
BT Northwest Territories
    Canada
SA Eureka Sound Formation

District of Keewatin
In SE Northwest Territories. Before 1978, also search Keewatin.
CO N600000N690000
    W0830000W1020000
UF Keewatin District
BT Northwest Territories
    Canada
SA Keewatin

District of Kenora
Extreme N, NW, and W.
BT Ontario
    Canada

District of Mackenzie
Central and W Northwest Territories. Also search Mackenzie; Mackenzie District.
CO N600000N710000
    W1020000W1370000
UF Mackenzie District
BT Northwest Territories
    Canada

Distrito Federal
use Federal District

disturbances
Term is restricted to extraterrestrial phenomena. Includes use on level 2 under ionosphere(1).
SA intensity
    ionosphere
    magnetic storms
    magnetosphere
    solar activity
    solar cycles
    solar flares
    substorms
    sudden commencements
    variations

Ditrau
Town in Mures County in SW central.
BT Transylvania
    Romania

Dittonian
In Old Red Sandstone, England: upper Gedinnian; above Downtonian. Includes use on level 3 under age terms (1). See list E.
BA Lower Devonian
    Devonian
BT Paleozoic
SA Old Red Sandstone

diurnal variations
Also search diurnal.
UF daily variation
SA Earth
    magnetic field
    variations

diversity
Includes use on level 3 under fossil groups (list F).
SA species
    species diversity

Dixie Valley
Churchill County in W.
BT Nevada
    United States

Djadjerud Valley
BT Iran

Djadokhta Formation
In the Gobi Desert.
BT Upper Cretaceous
    Cretaceous
SA Mongolia

Djibouti
Became independent June 27, 1977. Formerly French Territory of the Afars and Issas. Capital city also named Djibouti. As of 1978, includes use on level 1 as an area term. Before 1978, also search Afars and Issas Territory.
UF French Somaliland
BA Africa
SA Afar
    Red Sea Basin

djurleite
BA sulfides
BT minerals

Djursland
Peninsula of Arhus County jutting into the Kattegat in E.
BT Jutland
    Denmark

Dnepropetrovsk
City and oblast in E central.
BT Ukraine
    USSR

Dnieper Basin
River drainage basin. Index Soviet republics as applicable. Also search Dnieper; Dnieper region.
UF Dnieper region

BT USSR
SA Byelorussia
    Dnieper-Donets Basin
    Russian Republic
    Ukraine

Dnieper region
use Dnieper Basin

Dnieper River
Rises in S Valdai Hills and flows S, then SE into a big bend at Dnepropetrovsk, and finally into the Black Sea near Kherson. Index Soviet republics as applicable. Also search Dnieper.
BT USSR
SA Byelorussia
    Russian Republic
    Ukraine

Dnieper-Donets Basin
Region comprising the Donets Basin and that part of the Dnieper Basin in the nearby Dnepropetrovsk area. Index Soviet republics as applicable. Also search Dnieper-Donets; Dnieper-Donets Depression; Dnieper-Donets region.
CO N460000N520000
    E0420000E0300000
UF Dnieper-Donets Depression
    Dnieper-Donets region
BT USSR
SA Dnieper Basin
    Donets Basin
    Russian Republic
    Ukraine

Dnieper-Donets Depression
use Dnieper-Donets Basin

Dnieper-Donets region
use Dnieper-Donets Basin

Dniester River
Flows into Black Sea SW of Odessa. Index Soviet republics as applicable. Also search Dniester, Dniester region, Dniester Valley.
BT USSR
SA Moldavia
    Ukraine

Dniester-Prut Interfluve
Area between the two rivers. Includes all of Moldavia which formerly was Bessarabia. Index Soviet republics as applicable.
BT USSR
SA Moldavia
    Ukraine

Doberlug
Town in SE part of country.
BT Cottbus
    East Germany

Doboy Sound
McIntosh County on the Atlantic Ocean.
BT Georgia
    United States

Dobrogea
use Dobruja

Dobrudja
use Dobruja

Dobrudzha
use Dobruja

Dobruja
Region on the Black Sea. Also search Dobrogea which is the Romanian name.
UF Dobrogea
    Dobrudja
    Dobrudzha
BT Romania
NT Altin-Tepe
    Babadag Lake basin
    Constanta
    Mangalia

Sulina
Tulcea
SA Bulgaria

Dobruja Basin
Black Sea coastal strip S of the Danube river including NE Bulgaria. Index countries as applicable. Also search Dobruja.
BT Europe
SA Bulgaria
Romania

Dobrzyn
Town in N central part of country.
BT Bydgoszcz
Poland

Dockum Group
Includes Sloan Canyon Formation, Sheep Pen Canyon Formation, Tecovas Shale, Santa Rosa Sandstone, Chinle Formation, Pierce Canyon Redbeds, Baldy Hill Formation, Travesser Formation. Colorado, Kansas, Oklahoma, New Mexico, and W Texas.
BT Upper Triassic
Triassic
SA Chinle Formation
Colorado
Kansas
New Mexico
Oklahoma
Texas

Docodonta
Includes use on level 2 under Mammalia(1). See list F.
BA Mammalia
BT Tetrapoda
Vertebrata

Doda
Village in SW.
BT Jammu and Kashmir
India

Dodecanese
Group of islands in SE Aegean Sea.
BT Aegean Islands
Greece
SA Rhodes

Dogger
Europe. Above Lias, below Malm. Includes use on level 3 under age terms(1). See list E.
BA Middle Jurassic
Jurassic
BT Mesozoic
SA Aalenian
Bajocian
Bathonian
Callovian

Dogger Bank
Submerged sand bank.
BT North Sea
Atlantic Ocean

Dogo
Largest of Oki Islands off W coast of Honshu in Japan Sea.
BT Honshu
Japan
SA Oki Islands

Dogtooth Mountains
In SE British Columbia.
BT British Columbia
Canada

dolerite
Includes use on level 3 under igneous rocks(1) basalt family(2). See list H.
BA basalt family
BT igneous rocks
SA diabase
metadolerite
olivine dolerite
quartz dolerite
tholeiitic dolerite

Dolina
City in Lvov Oblast in W.
BT Ukraine
USSR

dolinen
use dolines

dolines
UF dolinen
SA hydrogeology
karst
sinkholes

dolomite
Includes use on level 3 under minerals(1) carbonates(2). For sediments, use carbonate sediments, dolomitic composition. To indicate the rock made of dolomite, use dolostone. Before 1976, also search dolomite for the commodity, rock, and the sediment. See list I (sed. rocks), list L (minerals), and list N (sediments).
UF bitter spar
magnesian spar
BA carbonates
BT minerals
SA ankerite
calcite
carbonate rocks
carbonate sediments
chemically precipitated rocks
dolomitic composition
dolomitization
dolostone
evaporites
light minerals
marble
marbles

Dolomite Alps
use Dolomites

dolomite marble
A valid metamorphic rock term through 1977. After 1977, use marble AND dolomite.

Dolomites
Mountain range of the Eastern Alps between the Adige and Piave river valleys in NE part of country. Index autonomous regions as applicable.
CO N460000N464500
E0121500E0113000
UF Dolomite Alps
BT Italy
SA Alps
Cima d'Asta
Eastern Alps
Latemar Massif
Lessini Mountains
Trentino-Alto Adige
Veneto

dolomitic composition
Term introduced in 1978. Includes use on level 3 under sediments(1) and sedimentary rocks(1). Before 1978, also search composition AND dolomitic.
SA composition
dolomite
dolomitization
dolostone
sedimentary rocks
sediments

dolomitite
use dolostone

dolomitization
As of 1978, term is used on level 2 under diagenesis(1).
SA calcitization
carbonatization
dedolomitization
diagenesis
dolomite
dolomitic composition
dolostone
limestone

metasomatism
processes
sedimentary rocks
sedimentation
sediments

dolostone
Includes use on level 1 as a commodity (list C); on level 3 under sedimentary rocks(1) carbonate rocks(2). To be used for rocks made of dolomite (mineral). For sediments, use carbonate sediments, dolomitic composition. See list I (sed. rocks) and list N (sediments). Before 1976, also search dolomite for the rock, sediment, and commodity.
UF dolomitite
BA carbonate rocks
BT sedimentary rocks
SA carbonate sediments
chemically precipitated sediments
dolomite
dolomitic composition
dolomitization

domain structure
Includes use on level 3 under paleomagnetism(1).
SA magnetic domains
paleomagnetism
structure

domains
Used as a general term.
SA environment
facies
lithology
magnetic domains
sedimentary petrology

Domerian
Europe. Above Charmouthian, below Whitbian. Includes use on level 3 under age terms(1). See list E.
BA Lower Jurassic
Jurassic
BT Mesozoic

domes
Includes use on level 3 under folds(1) style(2).
BA folds
SA anticlines
batholiths
diapirs
geomorphology
intrusions
laccoliths
salt domes
uplifts

Dominican Republic
Occupies eastern two thirds of island of Hispaniola in the Greater Antilles. Includes use on level 1 as an area term (list O). For term set options see list B.
BA Hispaniola
West Indies
NT Santo Domingo
SA Hispaniola

Don Basin
River basin in central S and S Russian Republic. Also search Don.
BT Russian Republic
USSR
SA Don River

Don Juan Pond
In Wright Valley on W shore of Ross Sea near Ross Ice Sheet in N Victoria Land.
BT Antarctica

Don River
Rises SE of Tula and flows SE into a big bend near the Volga River then flows SW into the Azov Sea at Rostov. Also search Don.
BT Russian Republic

USSR
SA Don Basin

Dona Ana County
On the Mexican border.
BT New Mexico
United States

Donau River
use Danube River

Donbas
use Donets Basin

Donbass
use Donets Basin

Donegal
County. W of Northern Ireland.
BT Ireland
NT Ardara Pluton
Rosses Granite

Donets Basin
Industrial region in plain of Donets River primarily in E Ukraine. Also search Donbas, Donbass, and Donets.
CO N470000N483000
E0400000E0373000
UF Donbas
Donbass
BT USSR
SA Dnieper-Donets Basin
Novorayskoe Formation
Russian Republic
Ukraine

Doniphan County
Extreme NE.
BT Kansas
United States

Donjek Glacier
In SW Yukon Territory.
BT Yukon Territory
Canada

Donjek River
In SW Yukon Territory.
BT Yukon Territory
Canada

Doornik
use Tournai

Dordogne
Department in SW central.
CO N443000N454500
E0021500W0010000
BT France

Dore Lake Complex
BT Quebec
Canada
SA Precambrian

Dorog
Town NW of Budapest.
BT Hungary

Dorog Basin
Also search Dorog.
BT Hungary

Dorpat
use Tartu

Dorset
County on the English Channel. Also search Dorsetshire.
CO N503000N511000
W0014500W0030000
UF Dorsetshire
BT England
Great Britain
United Kingdom

Dorsetshire
use Dorset

double refraction
use birefringence

double-refracting spar
use Iceland spar

Doubs
Department in E France. Also a river in France and Switzerland.

BT France
NT Besancon

Douglas County
Index states as applicable.
BX Colorado
Georgia
Illinois
Kansas
Minnesota
Missouri
Nebraska
Nevada
Oregon
South Dakota
Washington
Wisconsin
BT United States

Douglas Group
Includes Stranger Formation, and Lawrence Formation. SW Iowa, E Kansas, NW Missouri, and SE Nebraska.
BT Virgilian
Upper Pennsylvanian
Pennsylvanian
SA Iowa
Kansas
Lawrence Formation
Missouri
Nebraska

Downtonian
In Old Red Sandstone, England. Uppermost Silurian or lowermost Devonian; Schmidt, 1960–lowermost Gedinnian (Lower Devonian). Includes use on level 3 under age terms(1). See list E.
BA Old Red Sandstone
Devonian
BT Paleozoic
SA Lower Devonian
Upper Silurian

Drac Valley
River valley in SE.
BT France

drag folds
Term introduced in 1978. Includes use on level 3 under folds(1) orientation(2). Before 1978, also search drag AND folds.
BA folds
SA displacements
faults
orientation

drainage
Includes use on level 3 under soils(1) water regimes(2).
SA acid mine drainage
drainage basins
drainage patterns
dredging
lakes
seepage
soils
stream order
streams
valleys
water
water regimes

drainage basins
Includes use on level 3 under geomorphology(1) fluvial features(2). Also search catchments.
UF catchments
feeding ground
SA basins
drainage
drainage patterns
fluvial features
geomorphology
precipitation
rivers
rivers and streams
storage

valleys
water balance
water yield
watersheds

drainage changes
A valid term through 1974. After 1974, see landform evolution and fluvial features under geomorphology(1).

drainage networks
use drainage patterns

drainage patterns
Includes use on level 3 under geomorphology(1) fluvial features(2).
UF drainage networks
SA drainage
drainage basins
fluvial features
geomorphology
patterns
rivers

Drake Passage
Strait between Cape Horn and South Shetland Islands N of Antarctica.
UF Drake Strait
BT Antarctic Ocean

Drake Strait
use Drake Passage

Drammen
City on a branch of Oslo Fjord.
BT Norway

drape folding
use drape folds

drape folds
Term introduced in 1978. Also search drape folding.
UF drape folding
BA folds

dravite
BA ring silicates
silicates
BT minerals
SA borosilicates
tourmaline

drawdown
Includes use on level 3 under ground water(1) levels(2). Also search pumping.
SA ground water
hydraulics
levels
reservoirs

dredging
Includes use on level 3 under engineering geology(1) or mining geology(1).
UF deepening
SA construction
drainage
engineering geology
excavations
mining geology
ocean floors

Dresden
District in East Germany. Also a city.
BT East Germany
SA Saxony
Upper Lusatia

drift
Includes use on level 3 under sediments(1) clastic sediments(2). See list N.
BA clastic sediments
BT sediments
SA buried channels
buried valleys
continental drift
glacial transport
littoral drift
loess
moraines
terrigenous materials

till

drill cores
use cores

drill cuttings
use cuttings

drill holes
A valid term through 1975. After 1975, use boreholes.

drilling
Used for the process. Includes use as level 3 term under commodities (list C), e.g. petroleum(1) or under well-logging(1) or engineering geology(1).
SA boreholes
engineering geology
petroleum
well-logging
wells

Drome
Department in SE France. Also a river.
CO N441500N452500
E0055000E0043500
BT France
NT Baronnies
Diois
Valence
SA Vercors

Dronning Maud Land
use Queen Maude Land

droplets
Includes use on level 3 under meteorology(1) water(2).
SA clouds
meteorology
raindrops
water

dropstone
use argillite

Drumheller
City in central.
BT Alberta
Canada

drumlins
Includes use on level 3 under glacial geology(1) glacial features(2).
BT glacial features
SA glacial geology
moraines
till

dry delta
use alluvial fans

Dry Valley Drilling Project
In different locations such as Lake Vanda, Lake Vida, New Harbor, Ross Island, and McMurdo Sound off SW Ross Ice Sheet in the Victoria Land area of Ross Dependency.
BT Antarctica

Dryas
Genus. Before 1978, term also included for the intervals of late-glacial time.
BA miospores
palynomorphs
SA angiosperms
Older Dryas
Oldest Dryas
Younger Dryas

drying
use dehydration

DSDP
use Deep Sea Drilling Project

DSS
use deep seismic sounding

DTA data
Term introduced on level 3 under thermal analysis(1).
UF differential thermal analysis data

SA differential thermal analysis
endothermic reactions
thermal analysis

Dubai
Emirate. One of federation of 7 states at S end of Persian Gulf. Includes use on level 3 as an area term (list O).
BT United Arab Emirates
Arabian Peninsula

Dublin
County on Irish Sea. Also a city.
BT Ireland

Duchesne County
In NE Utah.
BT Utah
United States

Duchesne River Formation
Unconformably underlies Bishop Conglomerate; unconformably overlies Uinta and Uinta (?) Formation. NW Colorado and NE Utah.
BT Paleogene
Tertiary
SA Colorado
Eocene
Oligocene
Utah

Ducktown
Town in Polk County in extreme SE.
BT Tennessee
United States

ductility
SA deformation
rock mechanics

Dudley
Town SW of Birmingham in central.
BT England
Great Britain
United Kingdom

Dufek Intrusion
In the Dufek Massif area of Forrestal Range in the Pensacola Mountains of Edith Ronne Land S of the Weddell Sea.
BT Antarctica

dufrenoysite
BA sulfarsenites
sulfosalts
BT minerals

Dukla
Town in SE part of country.
BT Rzeszow
Poland

Duluth
City at extreme W end of Lake Superior in Saint Louis County.
BT Minnesota
United States

Duluth Complex
In NE Minnesota.
BT Minnesota
United States
SA Precambrian

dumortierite
BA orthosilicates
silicates
BT minerals

Duna River
use Danube River

Dunai River
use Danube River

Dunantul
use Transdanubia

Dunarea River
use Danube River

Dunau River
use Danube River

Dundee
City on Firth of Tay on North Sea.

Dundee • Earth

BT Scotland
    Great Britain
    United Kingdom

dune rock
use eolianite

Dunedin
City on Pacific Ocean in SE.
BT South Island
    New Zealand

dunes
Includes use on level 3 under geomorphology(1) eolian features(2) and under sedimentary structures(1) bedding plane irregularities(2). See list K.
SA antidunes
    bedding plane irregularities
    eluvium
    eolian features
    geomorphology
    sand waves
    sedimentary structures

dunite
Includes use on level 3 under igneous rocks(1) ultramafic family(2). See list H.
BA ultramafic family
BT igneous rocks
SA peridotite

Dunkard Basin
Creek basin in SW.
BT Pennsylvania
    United States

Dunkard Group
Includes Washington and Greene formations. W Maryland, E Ohio, SW Pennsylvania, and N West Virginia. Also search Dunkard Series.
BT Paleozoic
SA Maryland
    Ohio
    Pennsylvania
    Pennsylvanian
    Permian
    West Virginia

Dunkerque
use Dunkirk

Dunkirk
City at S end of North Sea.
UF Dunkerque
BT Nord
    France

Dunnage Melange
BT Canada

Duplin Formation
Base of Duplin is an unconformity. Florida, E Georgia, E North Carolina, and E South Carolina.
BT upper Miocene
    Miocene
SA Florida
    Georgia
    North Carolina
    South Carolina

durain
Includes use on level 3 under sedimentary rocks(1) organic residues(2). See list I.
BA organic residues
BT sedimentary rocks

Durance
Village in SW part of country.
BT Lot-et-Garonne
    France

Durance Basin
River basin in SE France. Also search Durance.
BT France

Durango
State in W central Mexico. Also a city.
BT Mexico

NT Mapimi
SA Sierra Madre Occidental

Durban
City on the Indian Ocean.
BT Natal
    South Africa

Durham
City in Durham County in NE.
CO N543000N550000
    W0011000W0022000
BT England
    Great Britain
    United Kingdom

duricrust
Includes use on level 3 under sediments(1) chemically precipitated sediments(2); on level 3 under soils(1) and under weathering(1). See list N.
BA chemically precipitated sediments
BT sediments
SA calcrete
    caliche
    ferricrete
    soils
    weathering
    weathering crust

Dushambe
use Dushanbe

Dushanbe
City in W Tadzhikistan.
UF Dushambe
    Stalinabad
BT Tadzhikistan
    USSR

dust
Includes use on level 3 under sediments(1) clastic sediments(2). See list N.
BA clastic sediments
BT sediments
SA cosmic dust
    dust storms
    interplanetary dust
    particles
    terrigenous materials
    volcanic ash

dust storms
Also search dust clouds.
SA dust
    storms

Dutch Guiana
use Surinam

Dutch New Guinea
use Irian Jaya

Dutchess County
On Connecticut border in SE.
BT New York
    United States

Duval County
In S Texas.
BT Texas
    United States

Dvina River
The Western Dvina River which flows into the Gulf of Riga and the Northern Dvina in the White Sea Basin. Index Soviet republics as applicable.
BT USSR
SA Byelorussia
    Latvia
    Russian Republic

Dwyka Series
Of the Karroo System. Includes Upper Shales, Conglomerate (now Tillite), and Lower Shales. Also search Dwyka Tillite.
BT upper Paleozoic
    Paleozoic
SA Africa

    Botswana
    Carboniferous
    Karroo System
    Permian
    South Africa
    South-West Africa

Dy
use dysprosium

dynamic
A valid term through 1977. After 1977, use dynamic metamorphism on level 2 under metamorphism(1).

dynamic metamorphism
Term introduced in 1978. Includes use on level 2 under metamorphism(1). Before 1978, also search metamorphism AND dynamic.
BA metamorphism
SA regional metamorphism
    retrograde metamorphism

dynamics
As of 1978, term is used on level 2 under shorelines(1).
SA shorelines

dyscrasite
BA antimonides
    sulfides
BT minerals

dysprosium
Includes use on level 1 and 2 as chemical element (list D).
UF Dy
SA elements
    rare earths

Dzereg
use Altan Teeli

Dzhagdy Range
NW of Komsomolsk in Khabarovsk Kray in Soviet Far East.
BT Russian Republic
    USSR

Dzhezkazgan
Town in Karaganda Oblast in E central.
BT Kazakhstan
    USSR

Dzhida River
S of Lake Baikal in Buryat A.S.S.R.
BT Russian Republic
    USSR

Dzhungaria
Region in N Sinkiang Uighur. Also search Dzungaria.
UF Dzungaria
    Zungaria
BT Sinkiang Uighur
    China

Dzhungarian Alatau
Northernmost branch of the Tien Shan. Index Kazakhstan and/or Sinkiang Uighur as applicable. Also search Dzungar Alatau, Dzungarian Ala-Tau and Dzungarian Alatau.
CO N445000N453000
    E0823000E0784500
UF Dzungar Alatau
    Dzungarian Ala-Tau
    Dzungarian Alatau
BT Asia
SA Kazakhstan
    Sinkiang Uighur

Dzirula Massif
Central.
BT Georgian Republic
    USSR

Dzungar Alatau
use Dzhungarian Alatau

Dzungaria
use Dzhungaria

Dzungarian Ala-Tau
use Dzhungarian Alatau

Dzungarian Alatau
use Dzhungarian Alatau

E

E-region
Includes use on level 2 under ionosphere(1).
SA D-region
    F-region
    ionosphere

Eagle County
W central.
BT Colorado
    United States

early diagenesis
Term introduced in 1978. Includes use on level 3 under diagenesis(1).
BA diagenesis
SA late diagenesis

Earth
Treated as a whole. For general concepts or studies. Includes use on level 1(list A). Term set options are: topic [age, composition, concepts, evolution, general, genesis, gravity field, interior, magnetic field, motions, observations, processes, properties, shape, structure, temperature, theoretical studies] subtopic [e.g. Chandler wobble, convection currents, density, differentiation, Earth tides, expansion, free oscillations, melting, rotation]
SA asthenosphere
    atmosphere
    biosphere
    Chandler wobble
    configuration
    Conrad discontinuity
    continents
    core
    Coriolis force
    crust
    declination
    diurnal variations
    Earth tides
    expansion
    figure of Earth
    free oscillations
    geodesy
    geodynamics
    geography
    geoid
    geophysics
    geothermal gradient
    geothermal processes
    gravity field
    heat flow
    hydrosphere
    interior
    isostasy
    lithosphere
    low-velocity layer
    magnetic field
    magnetosphere
    mantle
    Mohorovicic discontinuity
    Moon
    motions
    oscillations
    paleomagnetism
    planetology
    planets
    plate tectonics
    pulsations
    secular variations
    seismology
    shape

solar system
spherical harmonic analysis
surface water
tectonophysics
terrestrial planets
tilt
variations

earth, diatomaceous
   use diatomaceous earth

earth flows
   use earthflows

earth pressure
As of 1978, term is used on level 2 under soil mechanics(1).
SA pressure
     soil mechanics

Earth tides
Includes use on level 3 under Earth(1) processes(2); under geophysics(1) and ocean circulation(1).
UF bodily tide
SA Earth
     geodesy
     geophysics
     ocean circulation
     tectonophysics
     tides
     underground installations

Earth-current methods
Includes use on level 2 under geophysical methods(1).
UF telluric methods
BA geophysical methods
SA Earth-current surveys
     electrical field
     magnetic field
     magnetotelluric methods
     methods

Earth-current surveys
Includes use on level 2 under geophysical surveys(1).
UF telluric surveys
BA geophysical surveys
BT surveys
SA Earth-current methods
     magnetotelluric surveys

earthflows
As of 1978, term is used on level 2 under slope stability(1). Also search earth flows.
UF earth flows
SA mudslides
     slope stability

earthquake magnitude
   use magnitude

earthquake record
   use seismograms

earthquake sea wave
   use tsunamis

earthquakes
Used on level 1 and 2 for studies emphasizing individual earthquakes. Includes use on level 2 under geologic hazards(1), mantle(1) and seismology(1). If level 1, term set options are:
topic [aftershocks, causes, classification, detection, effects, epicenters, focus, genesis, intensity, magnitude, mechanism, prediction]
    subtopic [e.g. deep-focus earthquakes, intermediate-focus earthquakes, shallow-focus earthquakes, or year, for specific earthquake] (no area term)
SA aftershocks
     arrival time
     crust
     deep-focus earthquakes
     dilatancy
     elastic waves
     engineering geology
     epicenters
     explosions
     faults
     fluid injection
     focal mechanism
     focus
     foundations
     geologic hazards
     ground motion
     intensity
     intermediate-focus earthquakes
     isoseismic maps
     magnitude
     mantle
     microearthquakes
     microseisms
     modified Mercalli scale
     Mohorovicic discontinuity
     moonquakes
     nuclear facilities
     periodicity
     plate tectonics
     prediction
     Q
     reservoirs
     rock mechanics
     seismic sources
     seismograms
     seismographs
     seismology
     seismometers
     shallow-focus earthquakes
     slope stability
     soil mechanics
     stick-slip
     surface waves
     swarms
     teleseismic signals
     tiltmeters
     tsunamis

earthworks
As of 1978, term is used on level 2 under foundations(1).
SA foundations

east
General term used after any of the main geographic terms (list O).
UF eastern
SA east-central
     northeast
     southeast

East Africa
BA Africa
SA North Africa

East African Rift
Part of Great Rift Valley. Index countries and lakes as applicable. Also search East African Rift system; East African Rift valley; East African Rift zone.
UF East African Rift system
     East African Rift valley
BT Africa
SA Burundi
     Ethiopia
     Great Rift Valley
     Kenya
     Lake Albert
     Lake Edward
     Lake Kivu
     Lake Malawi
     Lake Tanganyika
     Lake Turkana
     Malawi
     Mozambique
     Rwanda
     Tanzania
     Uganda

East African Rift system
   use East African Rift

East African Rift valley
   use East African Rift

East Anglia
Region including counties of Norfolk and Suffolk on the North Sea in extreme E.
BT England
     Great Britain
     United Kingdom

East Atlantic
Term introduced in 1978. Before 1978, also search Atlantic Ocean AND east or eastern.
BA Atlantic Ocean
SA Northeast Atlantic
     Northwest Atlantic
     South Atlantic
     Southeast Atlantic
     Southwest Atlantic
     West Atlantic

East Carpathians
   use Eastern Carpathians

east central
   use east-central

East China Sea
Enclosed by China, Korea, Japan, and Taiwan. As of 1977, includes use on level 1 as an area term (list O).
UF Eastern Sea
BA Pacific Ocean
SA China Sea

East Coast
   use Atlantic Coastal Plain

East European Plain
   use Russian Plain

East European Platform
Not a valid term for GeoRef. Use Russian Platform.

East Frisian Islands
In North Sea off NW.
BT Lower Saxony
     West Germany

East Germany
Officially German Democratic Republic or Deutsche Demokratische Republik. In N central Europe, bounded on N by the Baltic Sea, on E by Poland, on S by Czechoslovakia and West Germany, and on W by West Germany. Introduced as a level 1 area term in 1978.
CO N503000N543000
     E0150000E0100000
BA Germany
     Europe
NT Barth
     Brandenburg
     Brocken Massif
     Cottbus
     Dresden
     Erfurt
     Gera
     Halle
     Karl-Marx-Stadt
     Kyffhauser Range
     Leipzig
     Magdeburg
     Mecklenburg
     Potsdam
     Rugen Island
     Ruhla
     Saxony
     Saxony-Anhalt
     Saxony-Thuringia
     Schwerin
     Steinach
     Thuringia
     Thuringian Basin
     Thuringian Forest
     Unstrut River
     Upper Lusatia
     Weimar
     Weissenberg
SA Berlin
     Elbe River
     Elbe Valley
     Erzgebirge
     Frankfurt
     Harz Foreland
     Harz Mountains
     Harz region
     Lusatia
     North German Plain
     Oder Valley
     Pomerania
     Saale River
     Silesia
     Weisse Elster Basin
     Werra
     Zechstein

East Greenland
Term introduced in 1978. Before 1978, search Greenland AND east.
BA Greenland
SA South Greenland

East Indian Ocean
Term introduced in 1978. Before 1978, also search Indian Ocean AND east or eastern.
BA Indian Ocean

East Indian Ridge
Term introduced in 1978. Before 1978, search Mid-Indian Ridge AND east.
BT Mid-Indian Ridge
     Indian Ocean

East Lothian
County on S bank of Firth of Forth.
CO N555000N560500
     W0022000W0030000
UF Hardington County
     Hardingtonshire
BT Scotland
     Great Britain
     United Kingdom

East Malaysia
That part of the Federation of Malaysia which is comprised of the two states of Sabah and Sarawak on the island of Borneo.
BA Malaysia
SA Borneo
     Brunei
     Sabah
     Sarawak

East Mediterranean
Term introduced in 1978. Before 1978, search Mediterranean Sea AND east.
BA Mediterranean Sea
SA West Mediterranean

East Midlands
Eastern part of the midlands of central.
BT England
     Great Britain
     United Kingdom

East Ongul Island
In E Lutzow-Holm Bay off Prince Olav Coast in Norwegian Sector on Indian Ocean side.
BT Antarctica

East Pacific
Term introduced in 1978. Before 1978, search Pacific Ocean AND east or eastern.
BA Pacific Ocean
SA Equatorial Pacific
     North Pacific
     Northeast Pacific
     Northwest Pacific
     South Pacific
     Southeast Pacific
     Southwest Pacific
     West Pacific

East Pacific Rise
In S Pacific. Also search Albatross Cordillera.
UF Albatross Cordillera
     Easter Island Cordillera

BT Pacific Ocean
East Pakistan
A valid term through 1972. After 1972, use Bangladesh. Also search Pakistan AND east.
East Rudolf
Basin in N.
BT Kenya
East Sayan
use Eastern Sayan
East Siberian Sea
Between New Siberian Islands and Wrangel Island N of Magadan Oblast and NE Yakutia.
BT Arctic Ocean
East Texas Field
Oil field in E.
BT Texas
    United States
east-central
General term used after any of the main geographic terms (list O).
UF east central
SA central
    east
Easter Island
About 2,000 miles W of the Chilean coast.
BT Pacific Ocean
SA Polynesia
Easter Island Cordillera
use East Pacific Rise
eastern
use east
Eastern Alps
Ranges of the Alps extending from Innsbruck in the E to near Vienna and including the Dinaric Alps of Yugoslavia. Index countries as applicable.
BT Alps
    Europe
SA Austria
    Carnic Alps
    Dinaric Alps
    Dolomites
    Hohe Tauern
    Julian Alps
    Karawanken
    Otztal Alps
    Salzkammergut
    Wechsel
    Wiener Wald
    Yugoslavia
    Zillertal Alps
Eastern Carpathians
Ranges of the Carpathians in SW Ukraine and Moldavia. Index Ukraine and/or Romania as applicable. Also search East Carpathians.
UF East Carpathians
BT Carpathians
    Europe
SA Calimani Mountains
    Haghimas Syncline
    Romania
    Tulghes Series
    Ukraine
Eastern Cordillera
Eastern range of the Andes. Index countries as applicable. Also search Cordillera Oriental.
UF Cordillera Oriental
BT South America
SA Andes
    Bolivia
    Colombia
    Ecuador
    Peru
Eastern Desert
Between Nile on W and Gulf of Suez and Red Sea on E. Also search Arabian Desert.
UF Arabian Desert
BT Egypt
Eastern Ghats
Low mountain range extending about 500 miles along SE and E coast as far N as Mahanadi River.
BT India
SA Ghats
Eastern Goldfields
BT Western Australia
    Australia
Eastern Hemisphere
Used when discussing many large areas too numerous to mention. Includes use on level 1 and 2 as an area term (list O). If 1, see term set options under list B.
SA Northern Hemisphere
    Southern Hemisphere
    Western Hemisphere
Eastern Sayan
Term introduced in 1978. Before 1978, use Sayan AND east or eastern.
UF East Sayan
BA Sayan
BT USSR
Eastern Sea
use East China Sea
Eastern U.S.
Term introduced in 1978. Includes use as a level 1 area term. Before 1978, also search United States AND east or eastern.
BA United States
SA Connecticut
    Delaware
    Florida
    Georgia
    Maine
    Maryland
    Massachusetts
    New Hampshire
    New Jersey
    New York
    North Carolina
    Pennsylvania
    Rhode Island
    South Carolina
    Vermont
    Virginia
Eaton
Town in Weld County in NE.
BT Colorado
    United States
Ebino
Region.
BT Kyushu
    Japan
ebridians
Includes use on level 2 under Protista(1). See list F.
BA Protista
NA Archaeomonadaceae
Ebro Basin
River basin in NE extending from Santander Province to Catalonia. Also search Ebro River basin.
UF Ebro River basin
BT Spain
Ebro River
Rises in the Cantabrian Mountains in N and flows SE into the Mediterranean Sea SW of Barcelona.
BT Spain
Ebro River basin
use Ebro Basin
Ecca Series
Karroo Basin.
BT Permian
SA Cape Province
    South Africa
eccentricity
Used as a general term.
SA geometry
Echinodermata
Includes use on level 1 and 2 as a fossil term (list F).
BT Invertebrata
NA Asterozoa
    Crinozoa
    Echinozoa
    Homalozoa
    Pelmatozoa
SA Receptaculitaceae
    sclerites
Echinoidea
Includes use on level 2 under Echinodermata(1). See list F.
BA Echinozoa
    Echinodermata
BT Invertebrata
Echinozoa
Includes use on level 2 under Echinodermata(1). See list F.
BA Echinodermata
BT Invertebrata
NA Echinoidea
    Edrioasteroidea
    Holothuroidea
Echo Bay
Village just E of Sault Ste. Marie in S Ontario.
BT Ontario
    Canada
echo sounding
Includes use on level 3 under geophysical methods(1) acoustical methods(2), or geophysical surveys(1).
BA geophysical methods
SA acoustical methods
    bathymetry
    deep sounding
    geophysical surveys
    oceanography
    sonar methods
    sounding
eclogite
Includes use on level 3 under metamorphic rocks(1) granulites(2). See list J.
BA granulites
BT metamorphic rocks
ecology
Includes use on level 1 (list A); on level 2 under fossil groups (list F). Used for the study of relationships between organisms and their environments, including the study of communities, patterns of life, natural cycles, relationships of organisms to each other, biogeography, and population changes. If 1, term set options are:
name of fossil group (List F) (for Recent fauna and flora)
    type of environment [e.g. arctic environment, alpine environment, boreal environment, brackish-water environment, coastal environment, coastal swamps, deltas, eolian environment, estuaries, glacial environment, lagoons, lakes, marine environment, marshes, reefs, shorelines, terrestrial environment, tidal environment]
topic [analysis, changes, concepts, evolution, interpretation, observations, processes]
    subtopic (no area term)
SA adaptation
    alpine environment
    alpine-type
    arctic environment
    arid environment
    biofacies
    biogeography
    biology
    biomass
    biotypes
    boreal environment
    brackish water
    brackish-water environment
    communities
    deforestation
    deltas
    desertification
    deserts
    desiccation
    ecosystems
    environment
    environmental geology
    estuaries
    eutrophication
    fauna
    flora
    fresh water
    fresh-water environment
    geomorphology
    glacial environment
    grasslands
    habitat
    human ecology
    humid environment
    hydrology
    impact statements
    lagoons
    lakes
    littoral environment
    mangrove swamps
    marine environment
    microfauna
    paleoecology
    peat bogs
    ponds
    reefs
    semi-arid environment
    shallow-water environment
    soils
    swamps
    taiga environment
    terrestrial environment
    tidal environment
    tundra
    vegetation
    weathering
    wetlands
economic geology
For the discipline as a whole. See also specific commodities (list C). Includes use on level 1 (list A); on level 2 under area terms (list B). If 1, term set options are:
topic [applications, bibliography, catalogs, classification, concepts, education, experimental studies, general, history, instruments, interpretation, methods, nomenclature, objectives, philosophy, practice, principles, symposia, textbooks, theoretical studies]
    subtopic
SA development
    economics
    energy sources
    gems
    geology
    industrial minerals
    industry
    legislation
    metallogenic provinces
    metallogeny
    metals
    mineral deposits
    mineral economics
    mineral exploration
    mineral resources
    mining geology
    nonmetals
    oil and gas fields

ore minerals
paragenesis
water resources

economic geology maps
Term introduced in 1978. Before 1978, search maps AND economic geology.
BA maps

economics
Includes use on level 2 under commodity terms (list C) for general treatments not tied to specific areas.
SA economic geology
mineral economics

ecosystems
Includes use on level 3 under ecology(1).
SA ecology
environment

ecoulement
use gravity sliding

Ecrins-Pelvoux Massif
use Pelvoux Massif

ectinite
Includes use on level 3 under metamorphic rocks(1) migmatites(2). See list J.
BA migmatites
BT metamorphic rocks

Ectoprocta
BA Bryozoa
BT Invertebrata

Ector County
In W Texas.
BT Texas
United States

Ecuador
Includes use on level 1 as an area term (list O). For term set options see list B.
CO S050000N013000
W0750000W0810000
BA South America
SA Amazon Basin
Andes
Eastern Cordillera
Galapagos Islands

Eddy County
Index states as applicable.
BX New Mexico
North Dakota
BT United States

Eden Shale
Includes Economy Formation, Southgate Formation, McMicken Formation. S Indiana, central Kentucky, and SW Ohio.
BT Edenian
Upper Ordovician
Ordovician
SA Indiana
Kentucky
Ohio
Upper Ordovician

Edenian
North America. Above Mohawkian, below Maysvillian. Includes use on level 3 under age terms(1). See list E.
BA Upper Ordovician
Ordovician
BT Paleozoic
NT Eden Shale

Edentata
Includes use on level 2 under Mammalia(1). See list F.
BA Mammalia
BT Tetrapoda
Vertebrata

Edinburgh
City near the Firth of Forth in SE.
BT Scotland

Great Britain
United Kingdom

edingtonite
BA zeolite group
framework silicates
silicates
BT minerals

Edmondia
BA Bivalvia
Mollusca
BT Invertebrata

Edmonton
City in central.
BT Alberta
Canada

Edmonton Formation
Now considered equivalent of either Bearpaw Shale or Fox Hills Sandstone of Montana.
BT Upper Cretaceous
Cretaceous
SA Alberta
Bearpaw Formation
Fox Hills Formation
Northwest Territories
Saskatchewan

Edo
use Tokyo

Edrioasteroidea
Includes use on level 2 under Echinodermata(1). See list F.
BA Echinozoa
Echinodermata
BT Invertebrata

education
Includes use on level 1 (list A) for general discussions of methodology and curricula in education; on level 2 under major disciplines; as a level 2 or 3 term appropriate to a large number of topics (list G). If 1, term set options are:
topic [List B (except for areal geology), or extraterrestrial geology, general, geophysics]
subtopic (college-level education, curricula, elementary geology, elementary school, graduate-level education, high school, junior high school, materials, methods, objectives, popular and elementary geology, popular geology, vocational school)
SA associations
catalogs
college-level education
curricula
elementary geology
elementary school
general
geology
glossaries
graduate-level education
high school
historical geology
junior high school
materials
museums
oceanography
petrology
popular geology
programs
stratigraphy
textbooks

Edwards County
Index states as applicable.
BX Illinois
Kansas
Texas
BT United States

Edwards Formation
Overlies Walnut Clay and underlies Georgetown Limestone of Washita

Group. In S Texas. Also search Edwards Limestone.
BT Comanchean
Cretaceous
SA Lower Cretaceous
Texas

Edwards Plateau
Highland region of W central Texas.
CO N300000N313000
W0990000W1020000
BT Texas
United States

Eel River
Rises in N Mendocino County in NW and flows NW into the Pacific Ocean.
BT California
United States

Eemian
Europe.
BA upper Pleistocene
Pleistocene
Quaternary
BT Cenozoic

effects
Includes use as a level 2 or 3 term appropriate to a large number of topics, e.g. on level 2 under earthquakes(1) and under faults(1). See list G.
SA geomorphologic effects

efficiency
Used as a general term.
SA accuracy
capacity
streams

effluents
use discharge

Eger
City NE of Budapest.
BT Hungary

Egerian
Europe.
BA upper Oligocene
Oligocene
Paleogene
Tertiary
BT Cenozoic

Eggenburgian
Europe.
BA lower Miocene
Miocene
Neogene
Tertiary
BT Cenozoic

eggs
Includes use on level 3 under Aves(1) or Reptilia(1).
SA Aves
Reptilia

eggstone
use oolite

Egypt
Includes use on level 1 as an area term (list O). For term set options see list B.
CO N220000N320000
E0353000E0250000
BA Africa
NT Alexandria
Aswan
Bahariya Oasis
Cairo
Eastern Desert
Fayum
Fayum Depression
Kharga Oasis
Kosseir
Nile Delta
Safaga
Sinai
Suez Canal
Western Desert

SA Levant
Libyan Desert
Mediterranean region
Middle East
Near East
Nile River
Nile Valley
Nubia
Nubian Sandstone
Red Sea Basin
Sahara

Eh
Includes use on level 3 under geochemistry(1). Also search oxidation-reduction; oxidation-reduction potential; redox; redox potential.
UF oxidation-reduction potential
redox
redox potential
SA geochemistry
oxidation
oxidation zone
reduction

Ehime
Prefecture.
BT Shikoku
Japan

Eichkogel
Region in NE part of country.
BT Lower Austria
Austria

Eichstatt
Town in central.
BT Bavaria
West Germany

Eifel
Hilly region in W between Rhine and Moselle rivers. Also search Eifel Mountains.
UF Eifel Mountains
BT Rhineland-Palatinate
West Germany

Eifel Mountains
use Eifel

Eifelian
Europe. Above Emsian, below Givetian. Includes use on level 3 under age terms(1). See list E.
BA Middle Devonian
Devonian
BT Paleozoic
SA Couvinian

eigenvalues
SA mathematical methods

Eilat
use Elath

Eilenriede
Forest area E of Hanover.
BT Lower Saxony
West Germany

Eire
use Ireland

ejecta
Includes use on level 3 under volcanology(1) or mud volcanoes(1).
SA lapilli
mud volcanoes
pyroclastics
volcanic ash
volcanology

ekanite
BA ring silicates
silicates
BT minerals

Ekaterinburg
use Sverdlovsk

Ekibastuz
Town in Pavlodar Oblast in NE.
BT Kazakhstan
USSR

Ekman spiral
　Includes use on level 2 under ocean circulation(1).
　SA　atmosphere
　　　ocean circulation
　　　patterns
　　　winds

El Alboran Island
　Spanish islet 135 miles E of Gibraltar in SW.
　BT　Mediterranean Sea

El Centro
　City in Imperial County in Imperial Valley in S.
　BT　California
　　　United States

El Dorado County
　SW of Lake Tahoe in N.
　BT　California
　　　United States

El Kharga
　use Kharga Oasis

El Paso
　City in El Paso County in extreme W.
　BT　Texas
　　　United States

El Paso County
　Index states as applicable.
　BX　Colorado
　　　Texas
　BT　United States

El Salvador
　Includes use on level 1 as an area term (list O). For term set options see list B.
　BA　Central America
　SA　San Salvador

Elasmobranchii
　Subclass. Includes use on level 3 under Pisces(1) Chondrichthyes(2).
　BA　Chondrichthyes
　　　Pisces
　BT　Vertebrata
　NA　Bradyodonti
　　　Selachii

elastic constants
　Includes use on level 3 under deformation(1) or geophysics(1). Also search elastic moduli; modulus of elasticity; elastic modulus.
　UF　constants, elastic
　　　elastic moduli
　　　modulus of elasticity
　NA　bulk modulus
　　　Poisson's ratio
　　　shear modulus
　　　Young's modulus
　SA　deformation
　　　elastic properties
　　　elasticity
　　　geophysics

elastic limit
　Includes use on level 3 under deformation(1) field studies(2).
　UF　limit, elastic
　SA　creep
　　　deformation
　　　elastic properties
　　　elastic strain
　　　elasticity
　　　fracture strength
　　　Hooke's law
　　　mechanical properties
　　　strain
　　　viscoelasticity
　　　yield strength

elastic materials
　Introduced in 1978 as a general term, mostly under deformation(1). Before 1978, also search elastic AND materials.
　SA　deformation
　　　elasticity

　　　explosions
　　　geophysics
　　　materials
　　　plastic materials
　　　rock mechanics

elastic moduli
　use elastic constants

elastic properties
　Includes use on level 3 under deformation(1) and under material name; on level 3 under seismology(1).
　SA　deformation
　　　elastic constants
　　　elastic limit
　　　elastic strain
　　　elasticity
　　　geophysics
　　　Hooke's law
　　　mechanical properties
　　　properties
　　　seismology
　　　tensile strength

elastic strain
　Includes use on level 3 under deformation(1) and under engineering geology(1).
　SA　creep
　　　deformation
　　　elastic limit
　　　elastic properties
　　　elasticity
　　　engineering geology
　　　Hooke's law
　　　strain
　　　viscoelasticity

elastic waves
　Do not use seismic waves. Includes use on level 2 under seismology(1) and under mantle (1); on level 3 under earthquakes(1). Before 1976, also search seismic waves.
　NT　body waves
　　　surface waves
　SA　amplitude
　　　arrival time
　　　asthenosphere
　　　attenuation
　　　damping
　　　deconvolution
　　　deformation
　　　earthquakes
　　　explosions
　　　focus
　　　foreshocks
　　　long-period waves
　　　Love waves
　　　mantle
　　　Mohorovicic discontinuity
　　　propagation
　　　Q
　　　Rayleigh waves
　　　scattering
　　　seismic methods
　　　seismic sources
　　　seismology
　　　SH-waves
　　　traveltime
　　　traveltime curves
　　　Vibroseis
　　　waves

elasticity
　As of 1978, term is used on level 2 under rock mechanics(1) and soil mechanics(1).
　SA　anelasticity
　　　bulk modulus
　　　compressibility
　　　deformation
　　　elastic constants
　　　elastic limit
　　　elastic materials
　　　elastic properties
　　　elastic strain
　　　engineering geology
　　　mechanical properties

　　　plasticity
　　　Poisson's ratio
　　　rigidity
　　　rock mechanics
　　　shear modulus
　　　soil mechanics
　　　strain
　　　stress
　　　viscoelasticity
　　　yield strength
　　　Young's modulus

Elat
　use Elath

Elath
　City at head of Gulf of Aqaba. Also search Eilat and Elat.
　UF　Eilat
　　　Elat
　BT　Israel

Elba
　Island in Mediterranean Sea between Corsica and mainland of Italy.
　BT　Tuscany
　　　Italy

elbaite
　BA　ring silicates
　　　silicates
　BT　minerals
　SA　tourmaline

Elbe River
　Rises in N Bohemia and flows NW through Hamburg into the North Sea. Index countries as applicable. Also search Elbe.
　BT　Europe
　SA　Czechoslovakia
　　　East Germany
　　　West Germany

Elbe Valley
　Index countries as applicable. Also search Elbe.
　BT　Europe
　SA　Czechoslovakia
　　　East Germany
　　　West Germany

Elbingerode
　Town in W part of country.
　BT　Magdeburg
　　　East Germany

Elbistan
　Town in S central part of Anatolia.
　BT　Turkey

Elborus
　use Elbrus

Elbrus
　Peak in a N subsidiary spur of the main range of the Caucasus. Highest peak in Europe.
　UF　Elborus
　　　Elbrus Mountain
　　　Mount Elbrus
　　　Mt. Elbrus
　BT　Russian Republic
　　　USSR

Elbrus Mountain
　use Elbrus

Elburz
　Range in N parallel to S shore of Caspian Sea. Also search Elburz Mountains.
　UF　Alborz Mountains
　　　Elburz Mountains
　BT　Iran

Elburz Mountains
　use Elburz

Eldorado
　Village in NE.
　BT　Saskatchewan
　　　Canada

electric fields
　A valid term through 1974. After

1974, use electrical field.

electrical
　A valid term through 1977. After 1977, use electrical logging on level 2 under well-logging(1) or electrical conductivity under geophysical methods(1).

electrical conductivity
　Includes use on level 3 under geophysical methods(1).
　SA　conductivity
　　　electrical field
　　　geophysical methods
　　　resistivity
　　　thermal conductivity

electrical field
　Includes use on level 2 under aurora(1), ionosphere(1), and magnetosphere(1). Before 1974, also search electric fields.
　UF　field, electrical
　SA　arcs
　　　aurora
　　　currents
　　　dielectric constant
　　　dielectric properties
　　　Earth-current methods
　　　electrical conductivity
　　　electromagnetic field
　　　electromagnetic waves
　　　emissions
　　　induced polarization
　　　ionosphere
　　　magnetic field
　　　magnetosphere
　　　magnetotelluric methods

electrical logging
　Not a valid term from 1971 through 1977. Now includes use on level 2 under well-logging(1). Before 1978, also search well-logging AND electrical.
　UF　logging, electrical
　BA　well-logging
　SA　boreholes
　　　conductivity
　　　resistivity

electrical methods
　Includes use on level 2 under geophysical methods(1).
　BA　geophysical methods
　SA　dielectric constant
　　　dielectric properties
　　　electrical properties
　　　electrical surveys
　　　electromagnetic methods
　　　homogeneity
　　　induced polarization
　　　inhomogeneity
　　　methods
　　　resistivity

electrical phenomena
　Includes use on level 2 under meteorology(1).
　UF　phenomena, electrical
　SA　currents
　　　lightning
　　　meteorology
　　　raindrops
　　　thunderstorms

electrical properties
　Includes use on level 3 under material name.
　SA　electrical methods
　　　electrical surveys
　　　properties
　　　surface properties

electrical surveys
　Includes use on level 2 under geophysical surveys(1).
　BA　geophysical surveys
　BT　surveys
　SA　electrical methods
　　　electrical properties

electromagnetic logging
electromagnetic methods
induced polarization
resistivity
self-potential methods

electrochemical properties
Includes use on level 3 under geochemistry(1) properties(2).
SA geochemistry
properties

electrodes
Used as a general term.
SA electrolysis

electrojet
Includes use on level 2 under ionosphere(1); on level 3 under aurora(1) electrical field(2) and magnetic field(2).
SA aurora
currents
equatorial region
ionosphere
magnetic field
polar regions

electrolysis
Includes use on level 3 under geochemistry(1) processes(2).
SA dissociation
electrodes
geochemistry
polarography
processes
solution

electrolytes
SA anions
geochemistry
ion exchange
ions

electromagnetic
A valid term through 1977. After 1977, use electromagnetic logging on level 2 under well-logging(1).

electromagnetic field
Includes use on level 3 under geophysical methods(1) or geophysical surveys(1).
UF field, electromagnetic
SA electrical field
geophysical methods
geophysical surveys
magnetic field
magnetotelluric methods

electromagnetic logging
Term introduced in 1978. Includes use on level 2 under well-logging(1). Before 1978, also search well-logging AND electromagnetic.
UF logging, electromagnetic
BA well-logging
SA electrical surveys
electromagnetic methods
geophysical methods
geophysical surveys

electromagnetic methods
Includes use on level 2 under geophysical methods(1).
UF electromagnetic prospecting
BA geophysical methods
SA electrical methods
electrical surveys
electromagnetic logging
electromagnetic surveys
induction
magnetotelluric methods
magnetotelluric surveys
methods
mineral exploration

electromagnetic prospecting
use electromagnetic methods

electromagnetic radiation
Includes use on level 2 under astrophysics and solar physics(1) and under interplanetary space(1).
SA astrophysics and solar physics
cosmic rays
electromagnetic waves
gamma rays
interplanetary space
luminescence
magnetic field
particles
radiation
radio-wave methods
radioactivity
solar wind
whistlers
X-rays

electromagnetic surveys
Includes use on level 2 under geophysical surveys(1).
BA geophysical surveys
BT surveys
SA electromagnetic methods
magnetic field
magnetotelluric surveys

electromagnetic waves
Includes use on level 2 under meteorology(1).
SA electrical field
electromagnetic radiation
magnetic field
meteorology
radio-wave methods
waves

electron content
Includes use on level 2 under ionosphere(1).
SA ion densities and temperatures
ionosphere

electron diffraction analysis
Includes use on level 3 under appropriate material.
SA analysis
electrons
neutron diffraction analysis
X-ray diffraction analysis

electron microprobe
use electron probe

electron microscopy
Includes use on level 1 (list A) for methodology, not for data. If 1, term set options are:
applications
  subtopic
instruments
  subtopic
methods
  subtopic
techniques
  subtopic
UF microscopy, electron
SA chemical analysis
clay mineralogy
crystal chemistry
crystal growth
crystallography
differential thermal analysis
electron microscopy data
scanning method
scattering
SEM data
spectroscopy
structural analysis
TEM data
thermal analysis
thin sections
transmission method
X-ray analysis

electron microscopy data
Term introduced in 1978. Used to distinguished electron microscopy as a method from the data.
SA data
electron microscopy

electron paramagnetic resonance
As of 1978, term is used only for methodology. For data, see EPR spectra. Includes use on level 3 under spectroscopy(1) methods(2); in combination with spectroscopy (i.e. spectroscopy, electron paramagnetic resonance) on level 3 under chemical element(1) analysis(2). See list D. Also search electron spin resonance.
UF electron spin resonance
SA analysis
EPR spectra
methods
nuclear magnetic resonance
resonance
spectroscopy

electron paramagnetic resonance spectra
use EPR spectra

electron probe
As of 1978, term is restricted to methodology. For data, see electron probe data. Includes use on level 3 under spectroscopy(1) methods(2); chemical analysis(1) methods (2); in combination with spectroscopy (i.e. spectroscopy, electron probe) on level 3 under chemical element(1) analysis(2). See list D. Also search electron microprobe.
UF electron microprobe
electron probe analysis
probe, electron
SA analysis
chemical analysis
electron probe data
ion probe
methods
spectroscopy

electron probe analysis
use electron probe

electron probe data
Term introduced in 1978. Before 1978, search electron probe.
UF electron probe spectra
SA data
electron probe

electron probe spectra
use electron probe data

electron spin resonance
use electron paramagnetic resonance

electrons
Used as a general term.
SA electron diffraction analysis
neutrons
particles
protons

elementary geology
Term introduced in 1978. Includes use on level 3 under education(1). Before 1978, search popular and elementary geology.
SA education
geology
popular geology

elementary school
Includes use on level 3 under education(1).
SA college-level education
education
high school
junior high school

elements
Includes use on level 3 under geochemistry(1). See list D. Also search chemical elements.
UF chemical elements
SA actinium
aluminum
americium
antimony
argon
arsenic

barium
beryllium
bismuth
boron
bromine
cadmium
calcium
californium
carbon
cerium
cesium
chlorine
chromium
cobalt
copper
cosmochemistry
cosmogenic elements
curium
deuterium
dysprosium
erbium
europium
fluorine
gadolinium
gallium
geochemistry
germanium
gold
hafnium
helium
holmium
indium
iodine
iridium
iron
isotopes
krypton
lanthanum
lead
lithium
lithophile elements
lutetium
magnesium
major elements
major-element analyses
manganese
mercury
metals
migration of elements
minor elements
minor-element analyses
molybdenum
neodymium
neon
neptunium
nickel
niobium
nitrogen
nonmetals
osmium
oxygen
palladium
phosphorus
platinum
plutonium
polonium
potassium
praesodymium
promethium
protactinium
radium
radon
rare earths
rhenium
rhodium
rubidium
ruthenium
samarium
scandium
selenium
siderophile elements
silicon
silver
sodium
strontium
sulfur
tantalum

technetium
tellurium
terbium
thallium
thorium
thulium
tin
titanium
trace elements
trace-element analyses
tritium
uranium
vanadium
volatile elements
xenon
ytterbium
yttrium
zinc
zirconium

Elephantidae
Includes use on level 3 under Mammalia(1) Proboscidea(2).
BA Proboscidea
    Mammalia
BT Tetrapoda
    Vertebrata
NA Elephas
    Mammuthus
    Palaeoloxodon naumanni
    Stegodon

Elephas
Genus. Includes use on level 3 under Mammalia(1) Proboscidea(2).
BA Elephantidae
    Proboscidea
    Mammalia
BT Tetrapoda
    Vertebrata
NA Elephas antiquus

Elephas antiquus
Includes use on level 3 under Mammalia(1) Proboscidea(2).
BA Elephas
    Elephantidae
    Proboscidea
    Mammalia
BT Tetrapoda
    Vertebrata

elevation
use altitude

ELF
Includes use on level 3 under ionosphere(1) wave propagation(2).
SA HF
    ionosphere
    wave propagation
    waves

Elk Hills Field
U. S. naval oil reservation in Kern County in S California. Also search Elk Hills, and Elk Hills Oil Field.
UF Elk Hills oil field
BT California
    United States

Elk Hills oil field
use Elk Hills Field

Elk Mountains
Range of Rocky Mountains in W central.
BT Colorado
    United States
SA Rocky Mountains

Elk Point Basin
Index provinces as applicable.
BT Canada
SA Alberta
    Manitoba
    Saskatchewan

Elk Point Group
Includes Ashern Formation, Elm Point Formation, Winnipegosis Formation, and Prairie Formation.
BT Middle Devonian

Devonian
SA Alberta
    Manitoba
    Montana
    North Dakota
    Prairie Evaporite
    Saskatchewan
    South Dakota
    Winnipegosis Formation

Elko County
In NE Nevada.
BT Nevada
    United States

Ellef Ringnes Island
One of the Sverdrup Islands in District of Franklin.
CO N773000N790000
    W0980000W1060000
BT Northwest Territories
    Canada

Ellesmere Island
W of NW Greenland in District of Franklin.
CO N760000N840000
    W0600000W0920000
BT Northwest Territories
    Canada

Elliot Lake
Town in S central.
BT Ontario
    Canada

Elliott County
In E Kentucky.
BT Kentucky
    United States

ellipticity
General property.
SA geodesy
    geometry

Ellis County
Index states as applicable.
BX Kansas
    Oklahoma
    Texas
BT United States

Ellsworth County
Central.
BT Kansas
    United States

Ellsworth Highland
use Ellsworth Land

Ellsworth Land
High plateau just S and SW of the Antarctic Peninsula. Also search Ellsworth Highland.
UF Ellsworth Highland
    James W. Ellsworth Land
BT Antarctica

Ellsworth Mountains
Range S of Ellsworth Land and SW of the Antarctic Peninsula.
BT Antarctica

ellsworthite
use betafite

elongate minerals
Includes use on level 3 under lineation(1) style(2).
BT minerals
SA lineation
    structural analysis

Elphidium
Genus. Includes use on level 3 under foraminifera(1) Rotaliacea(2).
BA Rotaliacea
    Rotaliina
    foraminifera
BT Invertebrata

Elsinore Fault
Extending from a point E of Los Angeles to a point E of San Diego.
BT California

United States

eluvium
Includes use on level 3 under sediments(1) clastic sediments(2). See list N.
BA clastic sediments
BT sediments
SA alluvium
    dunes
    soils
    terrigenous materials

Elvas
City near the Spanish border in S part of country.
BT Portugal

Ely
Town in White Pine County, Nevada, and Saint Louis County, Minnesota. Index states as applicable.
BX Minnesota
    Nevada
BT United States

Emba
Town in Aktyubinsk Oblast in NW.
CO N470000N490000
    E0560000E0523000
BT Kazakhstan
    USSR

Emba River
Flows into NE Caspian Sea. There are oil fields in its lower course. Also search Emba.
BT Kazakhstan
    USSR

embankments
As of 1978, term is used on level 2 under highways(1) and slope stability(1).
SA barriers
    dams
    highways
    levees
    reservoirs
    slope stability
    slopes

emerald
Variety of beryl.
UF smaragd
BA ring silicates
    silicates
BT minerals
SA beryl
    gems

Emery
Town in Emery County in central.
BT Utah
    United States

Emery County
Central.
BT Utah
    United States

Emilia-Romagna
Autonomous region in N Italy. Also search Emilia.
BT Italy
NT Bologna
    Modena
    Parma
    Piacenza
    Romagna
    Taro Valley
    Vigarano
SA Po River
    Po Valley

emission
A valid term through 1977 used to denote the method. After 1977, use emission spectroscopy.

emission spectroscopy
Not a valid term from 1971 through 1977. After 1977, includes use on level 3 under spectroscopy(1) and under chemical element(1) analysis(2). Before 1978, also search spectroscopy AND emission.
BT spectroscopy
SA absorption
    analysis
    chemical analysis
    emissions

emission, thermal
use thermal emission

emissions
Includes use on level 2 and 3 under aurora(1).
SA airglow
    aurora
    electrical field
    emission spectroscopy
    solar flares
    VLF

Emperor Seamount Chain
use Emperor Seamounts

Emperor Seamounts
SE of Kamchatka Peninsula. Also search Emperor Seamount Chain.
UF Emperor Seamount Chain
BT Pacific Ocean

emplacement
Includes use on level 3 under intrusions(1).
SA igneous rocks
    intrusions
    magmas
    plutons
    salt tectonics

emplectite
BA sulfobismuthites
    sulfosalts
BT minerals

Emporia
City in Lyon County in E central.
BT Kansas
    United States

Ems River
Flows into the North Sea. Index West German states as applicable. Also search Ems.
BT West Germany
SA Lower Saxony
    North Rhine-Westphalia
    Weser-Ems

Emsian
Europe. Above Siegenian, below Eifelian. Includes use on level 3 under age terms(1). See list E.
BA Lower Devonian
    Devonian
BT Paleozoic

Emsland
Swampy region between Ems River and Netherlands border.
BT Lower Saxony
    West Germany

en echelon faults
Term introduced in 1978. Includes use on level 3 under faults(1) patterns(2) and folds(1). Before 1978, also search faults AND en echelon.
BA faults
SA systems

en echelon folds
Term introduced in 1978. Before 1978, search folds AND en echelon.
BA folds

enargite
BA sulfosalts
BT minerals
SA copper

Enari
use Inari

Encounter Bay
Inlet of Indian Ocean SE of Adelaide.

BT South Australia
Australia

encroachment (ground water)
use salt-water intrusion

Encruzilhada
use Encruzilhada do Sul

Encruzilhada do Sul
City in SE Rio Grande do Sul. Also search Encruzilhada.
UF Encruzilhada
BT Rio Grande do Sul
Brazil

Endako
Village in central.
BT British Columbia
Canada

endellionite
use bournonite

endellite
use halloysite

endemic taxa
Includes use on level 3 under fossil groups (list F) as applicable. Also search endemism.
UF endemism
taxa, endemic

endemism
use endemic taxa

enderbite
Includes use on level 3 under igneous rocks(1) granite-granodiorite family(2). See list H.
BA granite-granodiorite family
BT igneous rocks

Enderby Land
Semicircular projection of land on the Indian Ocean side claimed by Australia.
BT Antarctica

endogene processes
Includes use on level 3 under mineral deposits, genesis(1) processes(2).
SA mineral deposits
processes

endothermic reactions
Term introduced in 1978. Includes use on level 3 under thermal analysis(1). Before 1978, also search endothermic.
SA DTA data
reactions
thermal analysis

Endothyra
Genus. Includes use on level 3 under foraminifera(1) Fusulinina(2).
BA Fusulinina
foraminifera
BT Invertebrata

energy
Used as a general term. Does not include energy sources.
SA activation energy
energy sources
fission
free energy
geothermal energy
kinetics
nuclear energy
solar energy

energy sources
Includes use on level 1 as a commodity term (list C). See also names of appropriate commodities.
UF sources, energy
SA coal
economic geology
energy
geothermal energy
heat flow
natural gas
nuclear energy

petroleum
power plants
resources
solar energy
water resources

Enewetak
use Eniwetok Atoll

engineering, civil
use civil engineering

engineering geology
Used for geology as applied to engineering practice, especially mining and civil engineering. See appropriate features and processes under geomorphology(1). Includes use on level 1 (list A); on level 2 under area terms (list B). If 1, term set options are:
field studies
subtopic
materials, properties
(for material not covered under rock mechanics and soil mechanics)
material (type of rock, clays, rocks, sediments, soils) or property (e.g. elastic strain)
methods
type of method (cartography, photogeology, photogrammetry)
topic [bibliography, education, experimental studies, feasibility studies, frost action, maps, petroleum engineering, practice, site exploration, surveys, symposia, techniques]
subtopic or area
UF geologic engineering
SA accelerograms
acoustical properties
anchors
arches
blasting
breakwaters
bridges
civil engineering
clays
cohesive materials
concrete
cryopedology
cyclic loading
dams
deformation
design
dredging
drilling
earthquakes
elastic strain
elasticity
environmental geology
excavations
explosions
foundations
fractures
frost action
gas storage
geodesy
geologic hazards
geology
geomorphology
geophysical methods
geophysical surveys
geophysics
granular materials
ground water
grouting
highways
hydraulics
hydrology
impact statements
industry
injection
inventory
land subsidence
land use
landfills

landslides
layered media
legislation
liquid waste
load pressure
maps
marine installations
mass wasting
materials
mechanical properties
microfractures
mining geology
nuclear explosions
nuclear facilities
overconsolidated materials
penetration
permafrost
permeability
petroleum engineering
photogeology
photogrammetry
plastic materials
Poisson's ratio
pore pressure
quick clay
radioactive waste
radioactivity
remote sensing
reservoirs
rock mechanics
sedimentation
seepage
settlement
shear stress
shorelines
site exploration
slope stability
soil mechanics
soils
solid waste
solifluction
stability
strain
structural analysis
submarine installations
subsurface reservoirs
surface reservoirs
surface waves
surveys
tensile strength
thermal properties
tunnels
underground installations
uniaxial tests
waste disposal
water storage
waterways
weathering
wells

engineering, petroleum
use petroleum engineering

engineering properties
Includes use on level 2 under soils(1).
SA foundations
highways
permafrost
properties
sediments
soil mechanics
soils

England
Includes use on level 1 as an area term (list O). For term set options see list B.
CO N500000N554500
E0013000W0063000
BT Great Britain
United Kingdom
BA Europe
NT Barwell
Bath
Bedfordshire
Berkshire
Bewcastle
Birmingham University

Brandon
Bristol
NX Cambridge
NT Cambridge University
Cheshire
Chesil Beach
Chudleigh
Church Stretton
Clitheroe
Cornwall
Cotswold Hills
Dartmoor
Derby
Derbyshire
Devonshire
Dorset
Dudley
Durham
East Anglia
East Midlands
Essex
Exeter
Fawley
Folkestone
Forest of Dean
Freshwater
Gibraltar Point
Gloucestershire
Hampshire Basin
Hertfordshire
Holderness
Horsham
Howgill Fells
Humber Estuary
Ingleborough
Ingleton
Isle of Wight
Lake District
Lancashire
Land's End
Leicestershire
Lincolnshire
Liverpool Bay
London
London Basin
Malvern
Malvern Hills
Meldon Aplite
Mendip Hills
Midlands
Morecambe Bay
Newcastle
Northamptonshire
Northumberland
Norwich
Nottingham
Nottinghamshire
Okehampton
Oxford
Oxfordshire
Padstow
Peak District
Pennines
Piltdown
Sheffield
Shropshire
Staffordshire
Stonehenge
Suffolk
Surrey
Sussex
Tamar Valley
Teign Valley
Thames Estuary
Thames River
The Weald
Torquay
Tyne River
Warwickshire
Wenlock Edge
Westmorland
Wiltshire
Wolverhampton
Worcestershire
Yorkshire
SA Aberystwyth Grits
Barton Beds

Bembridge Marls
Birmingham
Borrowdale Volcanic Series
Bristol Channel
Channel Islands
Cheviot Hills
Cumberland
Esk Trough
Great Oolite Series
Great Scar Limestone
London Clay
Lower Greensand
Mersey River
Mersey Valley
North Sea Coast
Oxford Clay
Severn Valley
Shap Granite
Skiddaw Slates
Somerset
Speeton Clay
Trent Valley
Weald Clay
Welsh Borderland
Wenlock Limestone
Wye Valley
York

English
Includes use on level 3 to indicate language of document.

English Channel
Strait between England and France connecting Atlantic Ocean with the North Sea. Includes use on level 1 as an area term (list O). For term set options see list B.
CO N480000N510000
    E0020000W0060000
BA Atlantic Ocean

enigmatite
use aenigmatite

Enisei Ridge
use Yenisei Ridge

Enisei River
use Yenisei River

Eniwetok
use Eniwetok Atoll

Eniwetok Atoll
Part of the U. S. Trust Territory of the Pacific Islands. Also search Eniwetok.
UF Enewetak
    Eniwetok
BT Marshall Islands
    Micronesia

Enna
City and province in central.
BT Sicily
    Italy

Enns Valley
River Valley. Index Austrian states as applicable. Also search Enns.
BT Austria
SA Lower Austria
    Upper Austria

enrichment
Includes use on level 3 under geochemistry(1), metasomatism(1) or weathering(1).
SA geochemistry
    hypogene processes
    metasomatism
    migration of elements
    mineral deposits
    supergene processes
    weathering

enstatite
BA pyroxene group
    chain silicates
    silicates
BT minerals
SA bronzite

enstatite chondrites
BA chondrites
    meteorites

Enteletacea
Superfamily. Includes use on level 3 under Brachiopoda(1) Orthida(2).
BA Orthida
    Articulata
    Brachiopoda
BT Invertebrata
NA Schizophoria

enterolithic folding
use enterolithic folds

enterolithic folds
Term introduced in 1978. Also search enterolithic; enterolithic folding.
UF enterolithic folding
BA folds
SA deformation

enthalpy
Includes use on level 3 under geochemistry(1). Also search heat content.
UF heat content
SA entropy
    geochemistry
    thermodynamic properties
    thermodynamics

Entisols
Includes use on level 3 under soils(1). See list M.
BA soils

entropy
Includes use on level 3 under phase equilibria(1).
SA enthalpy
    phase equilibria
    sediments
    thermodynamic properties
    thermodynamics

environment
Includes use as a level 2 or 3 term appropriate to a large number of topics, e.g. on level 2 under sedimentation(1). See list G.
SA abyssal environment
    aerobic environment
    alpine environment
    anaerobic environment
    arctic environment
    arid environment
    brackish-water environment
    coastal environment
    deep-sea environment
    domains
    ecology
    ecosystems
    environmental analysis
    environmental geology
    fresh-water environment
    geomorphology
    glacial environment
    glaciofluvial environment
    habitat
    high-energy environment
    humid environment
    interglacial environment
    intertidal environment
    littoral environment
    low-energy environment
    marine environment
    nearshore environment
    paleoenvironment
    pelagic environment
    periglacial environment
    postglacial environment
    reclamation
    sebkha environment
    sedimentation
    semi-arid environment
    shallow-water environment
    shelf environment
    slope environment
    subaerial environment
    sublittoral environment
    submarine environment
    subtidal environment
    subtropical environment
    supratidal environment
    taiga environment
    temperate environment
    terrestrial environment
    tidal environment
    tropical environment

environmental analysis
For papers discussing the paleoenvironmental implications of several rock types and for papers whose main purpose is to determine environment. Includes use on level 2 under sedimentary rocks(1), sediments(1), and sedimentary structures(1). Also search environmental analyses.
SA analysis
    environment
    environmental geology
    paleoenvironment
    sedimentary rocks
    sedimentary structures
    sedimentation
    sediments

environmental geology
Includes use on level 1 (list A); on level 2 under area terms (list B). Used for studies involving the collection, analysis, and application of geologic data and principles to problems created by human occupancy and use of the physical environment. If 1, term set options are:
topic [concepts, education, maps, surveys, symposia]
    subtopic [controls, floods, human ecology, regional planning, urban planning]
SA acid mine drainage
    atmosphere
    conservation
    detergents
    ecology
    engineering geology
    environment
    environmental analysis
    floods
    geochemistry
    geologic hazards
    geology
    geomorphology
    ground water
    human ecology
    hydrology
    impact statements
    industrial waste
    industry
    injection
    inventory
    land use
    landfills
    legislation
    liquid waste
    maps
    marine geology
    medical geology
    mineral resources
    mining geology
    natural resources
    pesticides
    pollution
    radioactive waste
    reclamation
    regional planning
    remote sensing
    soils
    solid waste
    strip mining
    surface water
    surveys
    toxic materials
    toxicity
    trace metals
    urban planning
    urbanization
    waste disposal
    waste water
    water
    water resources
    waterways

enzymes
BA organic materials

Eocambrian
Uppermost Precambrian(?)/lowermost Cambrian(?).
BA Precambrian
SA Cambrian
    Infracambrian
    Riphean
    Vendian

Eocene
World. Above Paleocene, below Oligocene. Includes use on level 1 as an age term (list E); on level 2 under paleo- terms, e.g. paleoecology, paleogeography, paleomagnetism.
BA Paleogene
    Tertiary
BT Cenozoic
NT Annot Sandstone
NA Bartonian
    Biarritzian
    Bridgerian
NT Castle Hayne Limestone
    Chuckanut Formation
    Clarno Formation
    Crystal River Formation
    Golden Valley Formation
    Green River Formation
NA lower Eocene
    Lutetian
    middle Eocene
NT Mirador Formation
    Nanjemoy Formation
NA Narizian
NT Punta Mosquito Formation
    Pyrenean Orogeny
NA Refugian
NT Rose Canyon Formation
    Swauk Formation
    Umpqua Formation
NA upper Eocene
NT Werillup Formation
    Yamhill Formation
SA Beaverhead Formation
    Brianconnais Zone
    Deccan Traps
    Duchesne River Formation
    Floridan Aquifer
    Intertrappean Beds
    Paskapoo Formation
    San Lorenzo Formation
    Scaglia Formation
    Sespe Formation
    Sinjar Formation
    Twin River Formation
    Wasatch Formation

Eocrinoidea
Class. Includes use on level 2 under Echinodermata(1). See list F.
BA Crinozoa
    Echinodermata
BT Invertebrata

eolian
A valid term through 1977 for type of environment. After 1977, see wind transport.

eolian features
Includes use on level 2 under geomorphology(1).
UF features, eolian
SA deserts
    dunes
    geomorphology
    loess

wind erosion
wind transport
winds

eolianite
Includes use on level 3 under sedimentary rocks(1) clastic rocks(2). See list I. Also search dune rocks.
UF dune rock
BA clastic rocks
BT sedimentary rocks
SA terrigenous materials

eosphorite
BA phosphates
BT minerals

Eoulx Basin
Index departments as applicable.
BT France
SA Alpes-de-Haute Provence
Hautes-Alpes

epeirogenesis
A valid term through 1974. After 1974, use epeirogeny.

epeirogeny
Used for specific epeirogenic activity. For general treatment, see tectonics. Includes use on level 1 (list A). Before 1975, also search epeirogenesis. Term set options are: age [list E]
   area
SA continents
   crust
   diastrophism
   eustacy
   isostasy
   neotectonics
   orogeny
   tectonics
   uplifts

epeirophoresis theory
use continental drift

ephemeral streams
BA streams
SA fluvial features

epicenters
Includes use on level 2 under earthquakes(1).
SA earthquakes
   focus
   seismology

epidiorite
Includes use on level 3 under metamorphic rocks(1) schists(2). See list J.
BA schists
BT metamorphic rocks

epidote
UF arendalite
   pistacite
BA epidote group
   orthosilicates
   silicates
BT minerals
SA aluminosilicates

epidote group
Includes use in combination with orthosilicates (i.e. orthosilicates, epidote group) on level 2 under minerals(1). See list L.
BA orthosilicates
   silicates
BT minerals
NA allanite
   clinozoisite
   epidote
   piemontite
   tanzanite
   zoisite

epidote-amphibolite facies
BT facies
SA amphibolite facies
   metamorphic rocks

epigene processes
Includes use on level 3 under mineral deposits, genesis(1) processes(2). Also search epigenesis; epigenic processes.
UF epigenic processes
SA mineral deposits
   processes

epigenic processes
use epigene processes

Epirus
Administrative region in NW.
BT Greece
SA Pindus Mountains

epistilbite
BA zeolite group
   framework silicates
   silicates
BT minerals

epitaxy
Includes use on level 3 under crystal growth(1).
SA crystal growth
   twinning

epithermal processes
Also search epithermal, or epithermal AND processes.
SA hydrothermal processes
   mineral deposits
   processes

EPR spectra
Term introduced in 1978. Before 1978, also search electron paramagnetic resonance for data. After 1978, use electron paramagnetic resonance only for methodology.
UF electron paramagnetic resonance spectra
SA electron paramagnetic resonance spectra

epsomite
BA sulfates
BT minerals

equations
Used as a general term.
SA equations of state
   mathematical geology
   statistical methods

equations of state
Used as a general term, e.g. on level 3 under geochemistry(1) or phase equilibria(1).
SA equations
   geochemistry
   phase equilibria

equatorial
A valid term through 1977. After 1977, use equatorial regions.

Equatorial Guinea
Formerly Spanish Guinea. Comprises province of Rio Muni on the mainland and the province of Fernando Po consisting of the islands of Fernando Po and Annobon. Includes use on level 1 as an area term (list O). For term set options see list B. Before 1976, also search Spanish Guinea.
BA Africa
NT Fernando Po

Equatorial Pacific
Term introduced in 1978. Before 1978, search Pacific Ocean AND equatorial.
BA Pacific Ocean
SA East Pacific
   North Pacific
   Northeast Pacific
   Northwest Pacific
   South Pacific
   Southeast Pacific
   Southwest Pacific
   West Pacific

equatorial region
Term introduced in 1978. Includes use on level 3 under area terms. Before 1978, also search equatorial.
SA electrojet
   ionosphere

Equidae
Family. Includes use on level 3 under Mammalia(1) Perissodactyla(2).
BA Perissodactyla
   Mammalia
BT Tetrapoda
   Vertebrata
NA Equus
   Hipparion
SA Ungulata

equilibrium
Used as a general term.
SA phase equilibria

Equisetales
Order.
BA Sphenopsida
   pteridophytes
BT Plantae
NA Calamites

Equus
Genus. Includes use on level 3 under Mammalia(1) Perissodactyla(2).
BA Equidae
   Perissodactyla
   Mammalia
BT Tetrapoda
   Vertebrata

Er
use erbium

Er Rif
use Rif

Er Riff
use Rif

Eratosthenian
Term introduced in 1978. Lunar stratigraphy. Eratosthenian rocks are older than those of the Copernican system but younger than those of the Imbrian System.
SA Moon

erbium
Includes use on level 1 and 2 as a chemical element (list D).
UF Er
SA elements
   rare earths

Erevan
use Yerevan

Erfurt
District in SW East Germany. Also a city.
BT East Germany
SA Ruhla
   Thuringia
   Thuringian Forest
   Unstrut River
   Weimar

Erianfels
use granulite

Ericaceae
Family.
BA Dicotyledoneae
   angiosperms
BT Plantae

Erie County
Index states as applicable.
BX New York
   Ohio
   Pennsylvania
BT United States

Erimo Seamount
Off S coast of Hokkaido, Japan.
BT Pacific Ocean

erionite
BA zeolite group
   framework silicates
   silicates
BT minerals

Erivan
use Yerevan

Erlangen
City in N central.
BT Bavaria
   West Germany

erodability
use erodibility

erodibility
Rate of erosion.
UF erodability
   erosibility
SA erosion

Eromanga Basin
In SW Queensland.
BT Queensland
   Australia

erosibility
use erodibility

erosion
Includes use on level 2 under soils(1); on level 3 under geomorphology(1) processes(2) and under glacial geology(1) glaciation(2).
NT grinding
   nivation
   water erosion
   wind erosion
SA abrasion
   beaches
   benches
   cirques
   corrosion
   degradation
   denudation
   detritus
   erodibility
   erosion control
   erosion features
   erosion surfaces
   etching
   exfoliation
   geologic hazards
   geomorphology
   glacial geology
   glaciation
   ice movement
   landform evolution
   landslides
   mass wasting
   mobility
   mudflows
   planation
   potholes
   processes
   sediment yield
   shorelines
   slope stability
   slumping
   soils
   talus slopes
   valleys
   waterways
   weathering

erosion control
Includes use on level 3 under soils(1) conservation(2).
UF control, erosion
SA conservation
   erosion
   soils
   water erosion
   wind erosion

erosion features
Includes use on level 2 under geomorphology(1).
UF features, erosion
NT erosion surfaces
   potholes

SA cliffs
    colluvium
    denudation
    erosion
    geomorphology
    gullies
    highlands
    inselbergs
    klippen
    pediments
    peneplains
    talus slopes
    terraces
    valleys
    water erosion

erosion surfaces
Includes use on level 3 under geomorphology(1) erosion features(2). Also search planation surfaces.
UF planation surfaces
   surfaces, erosion
BT erosion features
SA erosion
   geomorphology
   pediments
   pediplains
   peneplains
   planation

erratic blocks
use erratics

erratic boulders
use erratics

erratics
Includes use on level 3 under glacial geology(1). Also search erratic blocks; erratic boulder(s).
UF erratic blocks
   erratic boulders
SA boulders
   glacial geology
   glacial transport
   glaciers

errors
Used as a general term.
SA accuracy
   corrections
   reliability

ERTS
A valid term through 1977. After 1977, use Landsat.

ERTS-1
A valid term through 1977. After 1977, use Landsat.

erubescite
use bornite

eruptions
Includes use on level 3 under volcanology(1) volcanism(2).
SA lava
   pyroclastics
   volcanism
   volcanoes
   volcanology

Ervine Creek Limestone
Of Deer Creek Limestone. SW Iowa, NE Kansas, NW Missouri and SE Nebraska.
UF Ervine Creek Limestone Member
BT Virgilian
   Upper Pennsylvanian
   Pennsylvanian
SA Iowa
   Kansas
   Missouri
   Nebraska

Ervine Creek Limestone Member
use Ervine Creek Limestone

Erzgebirge
Mountain range on the border of East Germany and Bohemia. Index countries as applicable. Also search Krusne Hory and Krusny Hory Mountains.
UF Krusne Hory
   Krusnehory Mountains
   Krusny Hory Mountains
BT Europe
SA Czechoslovakia
   East Germany

Esan Cape
At E entrance of Tsugaru Strait in SW Hokkaido. Also search Esan.
UF Ezan Cape
BT Hokkaido
   Japan

Escambia County
In S Alabama.
BT Alabama
   United States

escarpments
use scarps

eschinite
use aeschynite

eschynite
use aeschynite

Escondido Formation
In Navarro Group. In S Texas.
BT Gulfian
   Upper Cretaceous
   Cretaceous
SA Texas

Esk Trough
Index England and/or Scotland as applicable. Also search Esk.
BT United Kingdom
SA England
   Scotland

Eskdale
Village in Kanawha County in W.
BT West Virginia
   United States

eskebornite
BA selenides
   sulfides
BT minerals

eskers
Includes use on level 3 under glacial geology(1) glacial features(2).
UF os
BT glacial features
SA cirques
   glacial geology
   kames

Eskisehir
Province in NW central Anatolia. Also a city.
BT Turkey
   Middle East

Eskridge Shale
In Council Grove Group. E Kansas, SE Nebraska, and central N Oklahoma.
BT Permian
SA Kansas
   Nebraska
   Oklahoma

Esmeralda County
In SE Nevada.
BT Nevada
   United States

Espanola Formation
Of the Quirke Lake Group. Divisible into three members. S central Ontario.
BT Precambrian
SA Canada
   Ontario

Espirito Santo
State on the Atlantic Ocean NE of Rio de Janeiro.
BT Brazil
SA Barreiras Formation

Esplanade Range
Range of Selkirk Mountains in SE.
BT British Columbia
   Canada
SA Selkirk Mountains

Essei
use Yessey

Essen Beds
Ruhr District. Also search Essen.
BT Carboniferous
SA North Rhine-Westphalia
   West Germany
   Westphalian

Essex
County on the North Sea NE of London.
CO N513000N521000
   E0011500E0000000
BT England
   Great Britain
   United Kingdom

Essex County
Index states as applicable.
BX Massachusetts
   New Jersey
   New York
   Vermont
   Virginia
BT United States

essexite
Includes use on level 3 under igneous rocks(1) alkali gabbro family(2). See list H.
BA alkali gabbro family
BT igneous rocks

Esterel
Mountainous forested region. Index departments as applicable.
BT France
SA Alpes-Maritimes
   Var

Esterhazy
Town near the Manitoba border in SE.
BT Saskatchewan
   Canada

Estevan
Town near the North Dakota border in SE.
BT Saskatchewan
   Canada

estimation
Used as a general term through 1977. After 1977, see evaluation.

Estonia
Soviet Socialist Republic.
CO N580000N593000
   E0270000E0220000
BT USSR
NT Tallin
   Tartu
SA Baltic region
   Baltic Shield
   European USSR
   Russian Plain
   Russian Platform

Estrada Nova Formation
Formation in Brazil: Upper Permian, Permian, and Triassic.
SA Brazil
   Goias
   Parana
   Permian
   Santa Catarina
   Sao Paulo
   Triassic
   Upper Permian

Estremadura
Region in W central.
BT Spain

estuaries
Includes use on level 3 under geomorphology(1) fluvial features(2) or shore features(2); on level 3 under ecology(1) and paleoecology(1); on level 3 under sedimentation(1) for type of environment. Also search estuary.
SA ecology
   fjords
   fluvial features
   geomorphology
   paleoecology
   rivers
   rivers and streams
   sedimentation
   shore features
   waterways

estuarine
A valid term through 1977. After 1977, use estuaries.

etching
Includes use on level 3 under crystal structure(1) or geomorphology(1).
SA creep
   crystal structure
   erosion
   geomorphology
   landform evolution
   mass wasting
   weathering

ethane
BA organic materials
SA hydrocarbons
   methane

Ethiopia
Includes use on level 1 as an area term (list O). For term set options see list B.
CO N080000N180000
   E0480000E0330000
BA Africa
NT Addis Ababa
   Afar Depression
   Awash Valley
   Ethiopian Rift
   Fantale
   Harar
   Omo
   Omo River
   Omo Valley
   Tigre
SA Afar
   Blue Nile
   East African Rift
   Lake Turkana
   Nile River
   Red Sea Basin

Ethiopian Rift
That part of the East African Rift extending NE-SW from the Red Sea to Lake Turkana. Also search Ethiopian Rift system and Ethiopian Rift valley.
UF Abyssinian Rift valley
   Ethiopian Rift system
   Ethiopian Rift valley
BT Ethiopia

Ethiopian Rift system
use Ethiopian Rift

Ethiopian Rift valley
use Ethiopian Rift

Ethmodiscus
Includes use on level 3 under algae(1) nannofossils(2).
BA discoasters
   nannofossils
   algae
BT Plantae

Etna
use Mount Etna

Etowah County
In NE Alabama.
BT Alabama
   United States

ettringite
  BA sulfates
  BT minerals

Eu
  use europium

Eua Island
  One of the Tongapu group in S Tonga. Also search Eua.
  BT Tonga
      Pacific Ocean

Euboea
  Island in the Aegean Sea.
  UF Evvoia
  BT Greece

Eucla Basin
  Index states as applicable. Also search Eucla.
  CO S320000S300000
      E1300000E1250000
  BT Australia
  SA South Australia
      Western Australia

euclase
  BA orthosilicates
      silicates
  BT minerals
  SA aluminosilicates

eucrite
  Includes use on level 3 under igneous rocks(1) gabbro family(2). See list H.
  BA gabbro family
  BT igneous rocks
  SA howardites

eucryptite
  BA orthosilicates
      silicates
  BT minerals
  SA aluminosilicates

eudialite
  use eudialyte

eudialyte
  Also search eudialite.
  UF eudialite
  BZ halides
      ring silicates
  BT minerals
  SA chlorides

Euganean Hills
  Range of hills in Padova Province in S central.
  BT Veneto
      Italy

eugeosynclines
  Includes use on level 3 under geosynclines(1) or tectonics(1).
  UF pliomagmatic zone
  SA geosynclines
      miogeosynclines
      tectonics

Eulengebirge
  use Sowie Mountains

eulite
  BA pyroxene group
      chain silicates
      silicates
  BT minerals

eulysite
  Includes use on level 3 under igneous rocks(1) ultramafic family(2). See list H.
  BA ultramafic family
  BT igneous rocks

Euphorbiaceae
  Family.
  BA Dicotyledoneae
      angiosperms
  BT Plantae

Euphrates River
  Rises in E central Turkey and at its confluence with the Tigris River it forms the Shatt-al-Arab which flows into the Persian Gulf. Index countries as applicable.
  BT Asia
  SA Iraq
      Syria
      Turkey

Eurasia
  Land mass comprising the continents of Europe and Asia. Index continents as applicable. Includes use on level 1 or 2 as an area term (list O). If 1, see term set options under list B.
  NA Black Sea
  NT Black Sea region
      Caspian Basin
  NA Caspian Sea
  SA Asia
      Eurasian Plate
      Europe
      Laurasia
      USSR

Eurasia Basin
  Between the North Pole and the major islands lying N of the Eurasian land mass. Also search Eurasian Basin.
  UF Eurasian Basin
  BT Arctic Ocean

Eurasian Basin
  use Eurasia Basin

Eurasian Plate
  That part of the Earth's crust comprising Eurasia, including the British Isles, which "floats" independently.
  SA Eurasia
      plate tectonics

Eure
  Department in NW.
  BT France
  SA Eure Valley
      Normandy
      Seine Estuary

Eure Valley
  River valley in NW France. Index departments as applicable.
  BT France
  SA Eure
      Eure-et-Loir
      Orne

Eure-et-Loir
  Department in NW central.
  BT France
  SA Beauce
      Eure Valley

Eureka
  Village in Nevada. Cities in California, Kansas, and Utah. Index states as applicable.
  BX California
      Kansas
      Utah
      Nevada
  BT United States

Eureka County
  N central Nevada. Also search Eureka AND Nevada.
  CO N391000N410000
      W1154500W1163500
  BT Nevada
      United States

Eureka Sound Formation
  BT Cretaceous
  SA District of Franklin
      Northwest Territories

Europe
  Includes use on level 1 or 2 as an area term (list O). If 1, see term set options under list B. To retrieve all documents, individual countries and physiographic regions should also be searched (see list O).
  CO N350000N710000
      E0750000W0250000
  NT Adriatic region
  NA Albania
  NT Alfold
  NA Alps
      Andorra
      Apennines
  NT Ardennes
      Arve Valley
  NA Austria
      Balkan Peninsula
  NT Baltic Shield
      Banat
  NA Belgium
  NT Beshchady Mountains
      Brest
      Brest Basin
      Brianconnais Zone
      Bug region
      Bug River
      Bukovina
  NA Bulgaria
  NT Campine
      Carpathian Foredeep
      Carpathian Foreland
  NA Carpathians
  NT Ceneri Zone
  NA Central Europe
      Czechoslovakia
  NT Danube Delta
      Danube Plain
      Danube Valley
  NA Denmark
  NT Dobruja Basin
      Elbe River
      Elbe Valley
      Erzgebirge
      European Platform
      Fennoscandia
  NA Finland
      France
  NT Frankfurt
      Galicia
  NA Germany
  NT Gibraltar
  NA Greece
  NT Harz Foreland
      Harz Mountains
      Harz region
  NA Hungary
  NT Iberian Peninsula
  NZ Iceland
  NT Inn Valley
  NA Ireland
  NT Isar Valley
      Isergebirge
  NA Italy
  NT Izera Mountains
      Jura Mountains
      Jutland
      Karkonosze Mountains
      Karst region
      Krajiste
      Lago Maggiore
      Lake Constance
      Lake Geneva
      Lapland
      Liechtenstein
      Limburg
      Lower Rhine Basin
      Lusatia
  NA Luxembourg
  NT Macedonia
      Maritsa River
      Meuse River
      Meuse Valley
      Moesia
      Moesian Platform
      Molasse Basin
      Moldavia
  NA Monaco
  NT Monte Rosa
      Morava River valley
      Moselle River
      Moselle Valley
      Navarre
  NA Netherlands
  NT North German Plain
      North Sea Coast
      North Sudetic Basin
  NA Norway
  NT Oder Valley
      Oresund
      Osogovo Mountains
      Pannonia
      Pannonian Basin
      Peribaltic Syneclise
  NA Poland
  NT Pomerania
  NA Portugal
  NT Prut River
  NA Pyrenees
  NT Rhine Basin
      Rhine Graben
      Rhine River
      Rhine Valley
      Rhodope Mountains
      Rhone River
      Rhone Valley
  NA Romania
  NT Roztocze
      Saale River
      Saar Basin
      Saar-Nahe Basin
      Salzach River
      Savoy
  NA Scandinavia
  NT Scheldt River
      Silesia
      Simplon region
      Siret River
      Sneznik
      Somes Basin
  NA Spain
  NT Spis
      Stavelot-Venn Massif
      Struma River valley
      Subcarpathians
      Sudeten Mountains
      Sudetic Basin
      Sulitjelma
  NA Sweden
      Switzerland
  NT Tagus Basin
      Tagus River
      Thrace
      Tisza River
      Tokaj-Eperjes Mountains
  NA United Kingdom
  NT Vardar River
      Variscides
      Vienna Basin
      Weisse Elster Basin
      Werra
  NA Western Europe
      Yugoslavia
  SA Baltic Glaciation
      Baltic region
      Caledonides
      Eurasia
      Faeroe Islands
      Tethys

European
  A valid term through 1977 used in combination with USSR. After 1977, use European USSR.

European Platform
  Ancient platform of Precambrian crystalline rocks overlain by sedimentary deposits of the North European Plain extending from the Low Countries to the Urals. Index countries as applicable.
  BT Europe
  SA Belgium
      Denmark
      Germany
      Netherlands
      Poland
      USSR

European USSR
  Term introduced in 1978. Index

Soviet republics as applicable. Before 1978, also search USSR AND European.
  CO N400000N700000
      E0750000E0200000
  BT USSR
  SA Armenia
      Azerbaidzhan
      Byelorussia
      Estonia
      Georgian Republic
      Latvia
      Lithuania
      Moldavia
      Russian Republic
      Ukraine
europium
  Includes use on level 1 and 2 as a chemical element (list D).
  UF Eu
  SA elements
      rare earths
Euryapsida
  Includes use on level 2 under Reptilia(1). See list F.
  BA Reptilia
  BT Tetrapoda
      Vertebrata
  NA Plesiosauria
Eurypterida
  Includes use on level 3 under Arthropoda(1) Merostomata(2). See list F.
  BA Merostomata
      Chelicerata
      Arthropoda
  BT Invertebrata
eustacy
  Includes use on level 3 under isostasy(1). Before 1976, also search eustatism. Also search eustasy.
  UF eustasy
  SA changes of level
      epeirogeny
      isostasy
eustasy
  use eustacy
eutrophication
  Includes use on level 3 under ecology(1) or hydrology(1).
  SA ecology
      hydrogen sulfide
      hydrology
      lakes
euxenite
  BA oxides
  BT minerals
  SA niobates
      tantalates
evaluation
  Includes use on level 2 under mining geology(1).
  SA mining geology
Evangeline Parish
  S central.
  BT Louisiana
      United States
evaporation
  Includes use on level 3 under hydrology(1). Also search vaporization.
  UF vaporization
  SA evapotranspiration
      hydrology
      sublimation
      water vapor
evaporites
  Includes use on level I as a commodity term (list C); on level 3 under sedimentary rocks (I) chemically precipitated rocks (2). Includes use on level 3 under sediments(1) chemically precipitated sediments(2) for treatment of 2 or more of the following types of sediments: anhydrite; borates; dolomite; gypsum; halite. See list I (sed. rocks) and list N (sediments).
  BA chemically precipitated rocks
  BT sedimentary rocks
  SA anhydrite
      borates
      chemically precipitated sediments
      dolomite
      gypsum
      halite
      salt
      sediments
      sodium chloride
evapotranspiration
  Includes use on level 3 under hydrology(1).
  SA atmospheric precipitation
      evaporation
      hydrology
Everglades
  Vast swampy region in six counties lying S of Lake Okeechobee in S.
  BT Florida
      United States
Everton Formation
  In Buffalo River Group. Includes Calico Rock Sandstone Member, Kings River Member, Newton Sandstone Member. N Arkansas, and S Missouri.
  BT Middle Ordovician
      Ordovician
  SA Arkansas
      Missouri
evolution
  Includes use as a level 2 or 3 term appropriate to a large number of topics, e.g. on level 2 under paleontology(1), tectonics(1) and under fossil groups (list F). See list G (general terms).
  SA landform evolution
      natural selection
      species diversity
Evora
  District in SE central.
  BT Portugal
  NT Alandroal
Evros River
  use Maritsa River
Evvoia
  use Euboea
Ewekoro
  Region in SW.
  BT Nigeria
Ewekoro Formation
  NW, S, SW, and W Nigeria.
  SA Cretaceous
      lower Tertiary
      Nigeria
      Tertiary
      Upper Cretaceous
excavations
  As of 1978, term is used on level 2 under tunnels(1) and underground installations(1). Used under engineering geology(1) through 1977.
  SA blasting
      dredging
      engineering geology
      explosions
      rock mechanics
      slope stability
      tunnels
      underground installations
Excello Shale
  SE Kansas, W and N Missouri, and NE Oklahoma.
  BT Pennsylvanian
  SA Kansas
      Missouri
      Oklahoma
exchange capacity
  Includes use on level 3 under geochemistry(1) properties(2).
  SA capacity
      cation exchange capacity
      geochemistry
      ion exchange
      ions
Exeter
  City in Devonshire in SW.
  BT England
      Great Britain
      United Kingdom
exfoliation
  SA erosion
      geomorphology
      weathering
exhalative processes
  Includes use on level 3 under mineral deposits, genesis(1).
  SA mineral deposits
      processes
exine
  SA miospores
      palynomorphs
      pollen
      spores
      sporopollenin
exinite
  BA macerals
  SA coal
      organic residues
      sedimentary rocks
exobiology
  SA biology
      planetology
exogene processes
  Includes use on level 3 under mineral deposits, genesis(2) processes(2).
  SA mineral deposits
      processes
exogenous inclusions
  use xenoliths
exogeology
  use extraterrestrial geology
exomorphic zone
  use aureoles
expansion
  Includes use on level 3 under Earth(1).
  SA dilatancy
      Earth
expansive materials
  Term introduced in 1978. Before 1978, also search expansive soils.
  UF expansive soils
  SA materials
      soils
expansive soils
  use expansive materials
expeditions
  Used as a general term. Includes use on level 3 under mining geology(1) and associations(1).
  SA associations
      mining geology
experimental studies
  Includes use as a level 2 or 3 term appropriate to a large number of topics, e.g. on level 2 under meteorology(1), folds(1), fractures(1), heat flow(1), and under oceanography(1). See list G.
  UF studies, experimental
  SA laboratory studies
      theoretical studies
experiments, tracer
  use tracer experiments
explanatory text
  Includes use on level 3 under areal geology(1); on level 3 under maps(1). See list B.
  UF text, explanatory
  SA areal geology
      guidebook
      maps
exploitation
  A valid term through 1978. After 1978, see mining, mining geology, or production.
exploration
  Includes use on level 2 or 3 under commodity terms(1), e.g. on level 2 under energy sources(1) and placers(1). See list C. Before 1972, also search geological exploration.
  SA development
      geochemical methods
      geochemical prospecting
      geodesy
      geological methods
      geophysical methods
      mineral exploration
      mineral resources
      Moon
      ore deposits
      photogeology
      prospecting
      site exploration
      speleology
      surveys
Explorer 35
  SA Moon
      satellite methods
explosion phenomena
  A valid term through 1972. After 1972, see cryptoexplosion features(2) and impact features(2) under geomorphology(1). See explosions(1) and nuclear explosions(3) under engineering geology(1).
explosions
  A level 1 term as of 1978. Includes use on level 2 under geologic hazards(1). Used for geotechnical and seismological studies on the effects of explosions. Before 1973, also search explosion phenomena. If 1, term set options are:
      applications
          subtopic
      chemical explosions
          subtopic
      detection
          subtopic
      effects
          subtopic
      elastic waves
          subtopic
      excavations
          subtopic
      experimental studies
          subtopic
      materials, properties
          subtopic
      nuclear explosions
          subtopic
      site exploration
          subtopic
      theoretical studies
          subtopic
  NT chemical explosions
      nuclear explosions
  SA blasting
      cryptoexplosion features
      deep seismic sounding
      earthquakes
      elastic materials
      elastic waves
      engineering geology
      excavations
      foundations
      geologic hazards

land subsidence
marine installations
mining geology
rock mechanics
seismic sources
seismic surveys
seismology
shatter cones
site exploration
slope stability
soil mechanics

exposure age
Includes use on level 2 under geochronology(1). Before 1975, also search exposure ages.
SA age
cosmic rays
geochronology

exsolution
Includes use on level 3 under crystal structure(1). Also search unmixing.
SA crystal structure
solution

extension
Used to indicate the mechanics of deformation. Before 1978, also used to indicate style of structure. Now use extension faults or extension fractures for style of structure.
SA deformation
extension faults
extension fractures

extension faults
Before 1978, also search extension AND faults.
BA faults
SA extension

extension fractures
Term introduced in 1978. Before 1978, search extension AND fractures.
BA fractures
SA extension

extensometers
SA deformation
instruments
rock mechanics
strainmeters
stress

extent
Includes use on level 2 or 3 appropriate to a large number of topics, e.g. on level 2 under orogeny(1). See list G.

extinct
A valid term through 1977 used in combination with lakes (i.e. lakes, extinct). After 1977, use extinct lakes. When referring to taxa, use extinct taxa.

extinct lakes
Includes use on level 3 under glacial geology(1) glacial features(2) and periglacial features(2); on level 3 under geomorphology(1) lacustrine features(2). Before 1976, also search extinct AND lakes.
SA geomorphology
glacial features
glacial geology
glacial lakes
lacustrine features
lakes
periglacial features

extinct taxa
Term introduced in 1978. Includes use on level 3 under fossil groups(1). Before 1978, also search extinct.
UF taxa, extinct

extinction
Includes use on level 3 under fossil groups (list F) and for optical properties under minerals(1). Also search mass extinction.
SA birefringence
minerals
optical properties

extraction
Used for mining geology through May 1978. After May 1978, use production under mining geology(1).

extraterrestrial geology
Includes use on level 2 under automatic data processing(1), catalogs(1), and symposia(1). Term is mainly used to describe a catgory, and is not frequently used in indexing. See also names of planets.
UF exogeology
SA astrophysics and solar physics
cosmic rays
geology
landing sites
meteorites
planetology
planets
solar system

extrusive rocks
Not a valid index term through 1977. After 1977, includes use on level 3 under lava (1) and volcanology (1). Before 1978, also search extrusive.
SA igneous rocks
lava
rocks
volcanology

Exuma Sound
Body of water SE of Nassau.
BT Bahamas
West Indies

Eyre Peninsula
Between Great Australian Bight and Spencer Gulf.
BT South Australia
Australia

Ezan Cape
use Esan Cape

F
use fluorine

F-region
Includes use on level 2 under ionosphere(1).
SA D-region
E-region
ionosphere

fabric
Includes use on level 3 under igneous rocks(1), metamorphic rocks(1), sediments(1), and sedimentary rocks(1) textures(2).
SA deformation
igneous rocks
lineation
metamorphic rocks
orientation
petrofabrics
preferred orientation
sedimentary rocks
sediments
structural analysis
tectonite

fabric analysis
A valid term through 1972. After 1972, use structural analysis or fabric.

facies
Includes use on level 2 under metamorphic rocks(1).
NT actinolite facies
amphibolite facies
blueschist facies
epidote-amphibolite facies
granulite facies
greenschist facies
prehnite-pumpellyite facies
zeolite facies
SA biofacies
domains
grade
isograds
lithofacies
metamorphic rocks
microfacies

facilities, nuclear
use nuclear facilities

factor analysis
BA statistical methods
SA analysis
mathematical geology

factors
Includes use on level 3 under soils(1) genesis(2).
SA climate
organisms
parent materials
soils
time

Faeroe Islands
Self governing Danish island group between Iceland and the Shetland Islands. Includes use on level 1 as an area term (list O). For term set options see list B.
UF Faeroes
BA Atlantic Ocean
SA Europe
Kuno

Faeroe-Iceland Ridge
use Iceland-Faeroe Ridge

Faeroes
use Faeroe Islands

Fagaras Mountains
Highest range in Transylvanian Alps. Index Transylvania and/or Walachia as applicable.
BT Romania
SA Transylvania
Transylvanian Alps
Walachia

Fagus
Genus.
BA Dicotyledoneae
angiosperms
BT Plantae
SA gymnosperms

failure
As of 1978, term is used on level 2 under rock mechanics(1) and slope stability(1). Also search rock failure.
UF rock failure
SA deformation
faults
fractures
rock mechanics
slope stability
stress
tensile strength

Fair Isle
Most southerly of the Shetland Islands.
BT Scotland
Great Britain
United Kingdom
SA Shetland Islands

Fairbanks
City in E central.
BT Alaska
United States

Fairfield
Village in Utah County in N central.
BT Utah
United States

Fairview Formation
In Maysville Group. Includes Mount Hope Member, Fairmount Member. SE Indiana, N central Kentucky, and SW Ohio.
BT Upper Ordovician
Ordovician
SA Indiana
Kentucky
Ohio

Fairview Peak
Central.
BT Nevada
United States

Fairweather Fault
In the Fairweather Range area near the British Columbia border in SE.
BT Alaska
United States

Faiyum
use Fayum

Falcon
State on the Caribbean Sea in NW.
BT Venezuela
NT Paraguana Peninsula

Falkenau
use Sokolov

Falkland Islands
British colony 300 miles E of Straits of Magellan. Islands are claimed by Argentina. Includes use on level 1 as an area term (list O). For term set options see list B.
BA Atlantic Ocean
SA Antarctica
South Georgia
South Orkney Islands
South Sandwich Islands

Falknov
use Sokolov

Fall River County
Extreme SW.
BT South Dakota
United States

Fall River Formation
In Inyan Kara Group. Includes Keyhole Sandstone Member. W South Dakota and NE Wyoming.
BT Lower Cretaceous
Cretaceous
SA South Dakota
Wyoming

fallout
Restricted to use for radioactive fallout.
SA atmosphere
isotopes
pollution
radioactivity

False Bay
E of Cape Good Hope.
BT Cape Province
South Africa

False Cape
SW tip of Dolak Island off S coast.
UF Cape Valse
Kaap Valsch
BT Irian Jaya
Indonesia

famatinite
BA sulfantimonates
sulfosalts
BT minerals
SA luzonite

Famennian
Europe. Above Frasnian, below

Tournaisian (Carboniferous). Includes use on level 3 under age terms(1). See list E.
BA Upper Devonian
 Devonian
BT Paleozoic

FAMOUS
French-American expedition to the Mid-Atlantic Ridge, using submersibles.
SA Atlantic Ocean
 Mid-Atlantic Ridge
 submersibles

fanglomerate
Includes use on level 3 under sedimentary rocks(1) clastic rocks(2). See list I.
BA clastic rocks
BT sedimentary rocks
SA terrigenous materials

Fanning Island
One of the Line Islands S of Hawaii. Included in British colony of Gilbert and Ellice Islands.
BT Pacific Ocean
SA Line Islands

fans
Includes use on level 3 under glacial geology(1) periglacial features(2).
SA alluvial fans
 geomorphology
 glacial geology
 periglacial features
 submarine fans

Fantale
In Ethiopian Rift valley in W Ethiopia. Also search Fantale Volcano.
UF Fantale Volcano
BT Ethiopia

Fantale Volcano
use Fantale

Far East
Easternmost Asia along the Pacific Ocean. Index countries and regions as applicable. SE Asian countries are sometimes included. Includes use on level 1 as an area term (list O). For term set options see list B.
CO N320000N600000
 E1450000E1200000
BA Asia
SA China
 Indochina
 Indonesia
 Japan
 Korea
 Malay Archipelago
 Malaysia
 Mongolia
 Pacific region
 Soviet Far East
 Taiwan
 USSR

Faristan
use Fars

Fars
Province in SW Iran.
UF Faristan
BT Iran
NT Band-e-Amir

farside
SA Moon

farsundite
Includes use on level 3 under igneous rocks(1) granite-granodiorite family(2). See list H.
BA granite-granodiorite family
BT igneous rocks

Farther India
use Indochina

fassaite
BA pyroxene group
 chain silicates
 silicates
BT minerals
SA augite

fatty acids
Includes use on level 2 under organic materials(1).
BA organic materials
SA acids
 amino acids

faujasite
BA zeolite group
 framework silicates
 silicates
BT minerals

fault escarpment
use fault scarps

fault ledge
use fault scarps

fault planes
UF planes, fault
SA faults

fault scarps
Includes use on level 3 under faults(1).
UF cliff of displacement
 fault escarpment
 fault ledge
SA cliffs
 faults
 scarps
 slopes

fault zones
Includes use on level 3.
UF zones, fault
SA faults
 lineaments
 plate tectonics
 rift zones

faults
To be used for studies primarily stressing individual faults or systems of faults. For relationships with other structures, see tectonics. Includes use on level 1 (list A); on level 2 under structural analysis(1), geologic hazards(1), and nuclear facilities(1). Also search fault. If level 1, term set options are:
 displacements
  subtopic [diagonal-slip faults, dip-slip faults, gravity faults, normal faults, overthrust faults, reverse faults, strike-slip faults, thrust faults, transcurrent faults, transform faults, wrench faults]
 distribution
  subtopic (e.g. topic or area)
 effects
  subtopic [breccia, gouge, mullions, mylonite, shear zones, slickensides, ultramylonite]
 extent
  subtopic
 mechanics
  subtopic [e.g. compression, flexure, shear, stick-slip]
 orientation
  subtopic [arcuate faults, bedding faults, dip faults, longitudinal orientation, oblique orientation, strike faults, transverse faults]
 patterns
  subtopic [en echelon faults, parallel faults, peripheral faults, radial faults]
 systems
  subtopic [block structures, grabens, horsts, rift zones, step faults]
 theoretical studies
  subtopic
NA active faults
 arcuate faults
 bedding faults
 diagonal-slip faults
 dip faults
 dip-slip faults
 en echelon faults
 extension faults
 growth faults
 hinge faults
 low-angle faults
 normal faults
 overthrust faults
 parallel faults
 peripheral faults
 radial faults
 reverse faults
 right-lateral faults
 step faults
 strike faults
 strike-slip faults
 thrust faults
 transcurrent faults
 transform faults
 transverse faults
 underthrust faults
 wrench faults
SA allochthons
 basin range
 block structures
 breccia
 decollement
 deformation
 dislocations
 displacements
 drag folds
 earthquakes
 failure
 fault planes
 fault scarps
 fault zones
 folds
 foliation
 fracture zones
 fractures
 geologic hazards
 gouge
 grabens
 horsts
 klippen
 lineaments
 mechanics
 mylonite
 mylonitization
 nappes
 neotectonics
 nuclear facilities
 oblique orientation
 orientation
 orogeny
 plate tectonics
 reactivation
 rift valleys
 salt tectonics
 scarps
 seismology
 shear zones
 slickensides
 stick-slip
 strike
 structural analysis
 structural complexes
 structural geology
 systems
 tectonics
 tectonophysics

fauna
Includes use on level 3 under ecology(1). Use fossiliferous materials under sedimentary rocks(1).
SA ecology
 faunal list
 faunal provinces
 faunal studies
 flora
 fossiliferous materials
 fossils
 microfauna
 microfossils
 microorganisms
 paleontology

faunal list
UF list, faunal
SA fauna
 floral list
 fossils

faunal provinces
Includes use on level 3 under biogeography(1).
UF provinces, faunal
SA assemblages
 biogeography
 fauna
 paleontology

faunal studies
Includes use on level 2 under fossil group(1) for many classes or orders, or for genera. See list F.
UF studies, faunal
SA biota
 fauna
 floral studies
 fossils
 microorganisms

Favosites
Genus. Includes use on level 3 under Coelenterata(1) Tabulata(2) or Anthozoa(2).
BA Favositidae
 Tabulata
 Anthozoa
 Coelenterata
BT Invertebrata
SA Palaeofavosites

Favositidae
Family. Includes use on level 3 under Coelenterata(1) Tabulata(2) or Anthozoa(2).
BA Tabulata
 Anthozoa
 Coelenterata
BT Invertebrata
NA Favosites
 Palaeofavosites

Fawley
Town and parish SSE of Southampton in S.
BT England
 Great Britain
 United Kingdom

fayalite
BA olivine group
 orthosilicates
 silicates
BT minerals

Fayette County
Index states as applicable.
BX Alabama
 Georgia
 Illinois
 Indiana
 Iowa
 Kentucky
 Ohio
 Pennsylvania
 Tennessee
 Texas
 West Virginia
BT United States

Fayetteville
City in Washington County in NW.
BT Arkansas
 United States

Fayetteville Formation
Includes Mayes Limestone Member. N Arkansas, S Missouri; and NE, central, and E Oklahoma.
BT Upper Mississippian
 Mississippian
SA Arkansas
 Missouri

Oklahoma

Fayum
Province in Upper Egypt, SW of Cairo.
UF Faiyum
    Fayyum
BT Egypt

Fayum Depression
Bed of ancient Lake Moeris in N Upper Egypt SW of Cairo. Also search Fayum.
BT Egypt

Fayyum
use Fayum

Fe
use iron

Fe-55
SA iron
    isotopes

Fe-57
SA iron
    isotopes

feasibility studies
As of 1978, term is used on level 2. Also search feasibility.
UF studies, feasibility
SA highways
    marine installations
    nuclear facilities
    reservoirs
    tunnels
    underground installations

features, bottom
use bottom features

features, buried
use buried features

features, cryptoexplosion
use cryptoexplosion features

features, eolian
use eolian features

features, erosion
use erosion features

features, fluvial
use fluvial features

features, frost
use frost features

features, glacial
use glacial features

features, impact
use impact features

features, lacustrine
use lacustrine features

features, periglacial
use periglacial features

features, shore
use shore features

features, solution
use solution features

features, surface
use surface features

features, volcanic
use volcanic features

Federal Capital Territory
use Australian Capital Territory

Federal District
Area of which Mexico City is a part in S central Mexico. Also search Distrito Federal.
UF Distrito Federal
BT Mexico
NT Buenos Aires
    Caracas
    Mexico City
    Santiaguito
SA Caracas Group
    Valley of Mexico

Federated Shan States
use Shan State

Fedorov stage
use universal stage

feeding ground
use drainage basins

feldspar
Includes use on level 1 as a commodity term (list C); on level 3 under minerals(1) framework silicates, feldspar group(2). See list L. Through 1977, also search each individual feldspar mineral to retrieve all feldspars. As of 1978, individual feldspar minerals are autoposted to feldspar group.
BA feldspar group
    framework silicates
    silicates
BT minerals
SA alkali feldspar
    feldspathization
    light minerals

feldspar group
Includes use in combination with framework silicates (i.e. framework silicates, feldspar group) on level 2 under minerals(1). See list L.
BA framework silicates
    silicates
BT minerals
NA adularia
    albite
    alkali feldspar
    amazonite
    analbite
    andesine
    anorthite
    anorthoclase
    antiperthite
    bytownite
    celsian
    cryptoperthite
    feldspar
    hyalophane
    K-feldspar
    labradorite
    maskelynite
    microcline
    microperthite
    moonstone
    myrmekite
    oligoclase
    orthoclase
    paracelsian
    peristerite
    perthite
    plagioclase
    sanidine

feldspathization
Includes use on level 3 under metasomatism(1).
BA metasomatism
SA feldspar
    granitization
    metamorphism

Felidae
BA Carnivora
    Mammalia
BT Tetrapoda
    Vertebrata

felsite
Includes use on level 3 under igneous rocks(1) granite-granodiorite family. See list H.
BA granite-granodiorite family
BT igneous rocks

Fenestellidae
Family. Includes use on level 3 under Bryozoa(1) Cryptostomata(2).
BA Cryptostomata
    Bryozoa
BT Invertebrata

fenite
Includes use on level 3 under metasomatic rocks(1). Before 1978, included use on level 3 under metamorphic rocks(1).
BA metasomatic rocks
SA fenitization
    metasomatism

fenitization
Includes use on level 3 under metasomatism(1).
BA metasomatism
SA alteration
    fenite

Fennoscandia
Geological usage for that part of N Europe consisting of Denmark, Finland, Norway, and Sweden. Index countries as applicable.
BT Europe
SA Denmark
    Finland
    Norway
    Sweden

ferberite
BA tungstates
BT minerals

Fergana
Not a valid term for GeoRef. Use Fergana Basin.

Fergana Basin
Mountain enclosed steppe and desert region. Index Soviet republics as applicable. Also search Fergana and Fergana Valley.
CO N394500N413000
    E0730000E0693000
UF Fergana Valley
BT USSR
SA Kirghizia
    Tadzhikistan
    Uzbekistan

Fergana Valley
use Fergana Basin

Fergus County
Central.
BT Montana
    United States

fergusonite
BA oxides
BT minerals
SA niobates
    tantalates

Fermanagh
County in W.
BT Northern Ireland
    United Kingdom

Fernando Po
Province in Bight of Biafra. Also an island.
UF Fernando Poo
BT Equatorial Guinea

Fernando Poo
use Fernando Po

ferns
SA Filicopsida
    Gleicheniaceae
    pteridophytes

Fernvale Formation
In Richmond Group and Patterson Ranch Group. N Arkansas, SW Illinois, SE Missouri, and W Tennessee, NW Alabama, central E and NE Oklahoma.
BT Upper Ordovician
    Ordovician
SA Alabama
    Arkansas
    Illinois
    Missouri
    Oklahoma
    Richmond Group
    Tennessee

Ferrar Group
In Victoria Land W of Ross Sea on Pacific Ocean side.
BT Jurassic
SA Antarctica
    Cenozoic
    Mesozoic
    Paleozoic

ferric iron
Before 1978, search iron AND ferric.
SA iron

ferricrete
See list N.
BA chemically precipitated sediments
BT sediments
SA conglomerate
    duricrust

ferrierite
BA zeolite group
    framework silicates
    silicates
BT minerals

ferrimolybdite
BA molybdates
BT minerals

ferroan dolomite
use ankerite

ferrocarpholite
BA chain silicates
    silicates
BT minerals
SA carpholite

ferrodiorite
Also search ferrogabbro.
UF ferrogabbro
BA diorite family
BT igneous rocks

ferrogabbro
use ferrodiorite

ferrohastingsite
BA amphibole group
    chain silicates
    silicates
BT minerals
SA hastingsite

ferromanganese
A valid term through 1977. After 1977, use ferromanganese composition on level 2 under nodules(1).

ferromanganese composition
Term introduced in 1978. Includes use on level 2 under nodules(1). Before 1978, also search ferromanganese.
SA composition
    iron
    manganese
    nodules

Ferron Sandstone
use Ferron Sandstone Member

Ferron Sandstone Member
Of Mancos Shale. Central E Utah.
UF Ferron Sandstone
BT Upper Cretaceous
    Cretaceous
SA Mancos Shale
    Utah

ferropseudobrookite
BA oxides
BT minerals
SA pseudobrookite

ferrosilicon
BA native elements and alloys
BT minerals
SA silicides

ferrosilite
BA pyroxene group
    chain silicates
    silicates
BT minerals

ferrospinel
use hercynite

ferrous iron
Before 1978, search iron AND ferrous.
SA iron

ferruginous composition
Term introduced in 1978. Includes use on level 3 under sediments(1) and sedimentary rocks(1). Before 1978, also search ferruginous.
SA composition
iron
iron-rich sediments
red beds
sedimentary rocks
sediments

ferruginous quartzite
BA metamorphic rocks
SA quartzite

fertilization
Includes use on level 3 under soils(1) utilization(2).
SA agriculture
fertilizers
nutrients
soils
treatment
utilization

fertilizers
Includes use on level 2 under soils(1).
SA agriculture
fertilization
nitrogen
nutrients
phosphorus
soil management
soils
yields

Fezzan
Desert region in SW.
BT Libya

fibrolite
use sillimanite

Fichtelgebirge
Mountain range in NE.
BT Bavaria
West Germany

field crops
Includes use on level 3 under soils(1) utilization(2) and yields(2).
UF crops, field
SA soils
tillage
utilization
yields

field, crystal
use crystal field

field, electrical
use electrical field

field, electromagnetic
use electromagnetic field

field, gravity
use gravity field

field, magnetic
use magnetic field

field studies
Includes use on level 2 under deformation(1), and soils(1).
UF studies, field
SA deformation
sampling
soils

fields, coal
use coal fields

fields, geothermal
use geothermal fields

fields, giant
use giant fields

fields, lava
use lava fields

fields, oil and gas
use oil and gas fields

Fife
County N of Firth of Forth in E.
BT Scotland
Great Britain
United Kingdom

Fig Tree Group
use Fig Tree Series

Fig Tree Series
In Barberton Mountain Land in SE Transvaal. Also search Fig Tree Group.
UF Fig Tree Group
BT Precambrian
SA South Africa
Transvaal

figure of Earth
Includes use on level 2 under geodesy(1).
SA Earth
geodesy
geoid

Fiji
Independent state consisting of island group between New Caledonia and Samoa in S Pacific Ocean. Includes use on level 1 as an area term (list O). For term set options see list B.
UF Fiji Islands
BA Melanesia
NT Suva
Tavua
Vatukoula
Viti Levu

Fiji Islands
use Fiji

Fiji Plateau
Just E of Fiji in S.
BT Pacific Ocean

Filicales
use Filicopsida

Filicopsida
Including Filicales. Includes use on level 2 under pteridophytes(1). See list F. Also search Filicales.
UF Filicales
BA pteridophytes
BT Plantae
NA Azolla
Callipteris
Dicroidium
Gleicheniaceae
Pecopteris
NZ Sphenopteris
NA Taeniopteris
SA ferns
Juglans

Filizchay
Ore bearing region.
BT Azerbaidzhan
USSR

Fillmore Formation
Overlies House Limestone; underlies Wahwah Limestone. E Nevada and W central Utah. Also search Fillmore Limestone.
BT Lower Ordovician
Ordovician
SA Nevada
Utah

filters
Includes use on level 3 under seismology(1). Also search filtering.
SA deconvolution
filtration
seismology

filtration
As of 1978, term is restricted to use with sediments, i.e. for process of removing suspended material from a liquid.
SA deconvolution
filters
processes
sediments

fines
Includes use on level 3 under Moon(1).
SA agglutinates
grain size
mining
Moon
orange material
particles
sediments

Finger Lakes
Group of long narrow lakes in W.
BT New York
United States

Finistere
Department on W tip of Brittany.
CO N474000N484000
W0034000W0050000
BT France
NT Chateaulin Basin
Crozon Peninsula
Morlaix

finite difference analysis
Before 1978, also search finite difference method; finite difference.
UF finite difference method
SA analysis
mathematical geology

finite difference method
use finite difference analysis

finite element analysis
Includes use on level 3 under mathematical geology(1). Also search finite-element analysis.
UF finite-element analysis
BA statistical methods
SA analysis
mathematical geology

finite strain
SA deformation
rock mechanics
soil mechanics
strain
stress

finite-element analysis
use finite element analysis

Finland
Includes use on level 1 as an area term (list O). For term set options see list B.
CO N594500N700000
E0314500E0190000
BA Europe
NT Bjurbole
Haveroe
Inari
Lake Lappajarvi
Outokumpu
Tampere
Turku
Ylojarvi
SA Arctic region
Baltic region
Baltic Shield
Fennoscandia
Gulf of Bothnia
Gulf of Finland
Lapland

Finney County
W central.
BT Kansas
United States

Finnmark
County in N.
BT Norway
NT Alta
Porsang Fjord
Seiland
Tana Fjord
Varanger Fjord
Varanger Peninsula
SA Porsanger Dolomite Formation

fiord
use fjords

Fiordland National Park
SW South Island. Also search Fiordland.
UF Sounds National Park
BT South Island
New Zealand

fire clay
use fireclay

Fire Island
Long, narrow sand spit off S Long Island.
BT New York
United States

fire-clay
use fireclay

fireclay
Also search fire clay; fire-clay; refractory clay.
UF fire clay
fire-clay
refractory clay
SA clay
refractory materials
underclay

Firenze
use Florence

firn
Includes use on level 3 under glacial geology(1) glacial features(2), glaciers(2) and periglacial features(2).
SA glacial features
glacial geology
glaciers
periglacial features
snow

Firth of Clyde
An estuary of the Clyde River in SW Scotland. Also search Clyde River.
BT Scotland
Great Britain
United Kingdom

Firth of Tay
use Tay Estuary

fish
SA Actinopterygii
Agnatha
Chondrichthyes
Holostei
Osteichthyes
Pisces
Placodermi
Teleostei

Fiskenaesset
Settlement on SW coast.
BT Greenland
Arctic region

fission
UF nuclear fission
SA energy
fission tracks
fission-track dating
fusion
particles

fission tracks
Used for the phenomena. For method, use fission-track dating.
SA fission
fission-track dating
particle track
radiation damage
radioactivity
tracks

fission-track
A valid term through 1977. After 1977, use fission-track dating.

fission-track dating
Term introduced in 1978. Includes use on level 2 under geochronology(1). Before 1978, also search fission-track; fission tracks; fission-track method.
UF dating, fission-track
    fission-track method
    spontaneous fission-track dating
SA fission
    fission tracks
    geochronology
    methods
    radiation damage

fission-track method
use fission-track dating

fissures
Includes use on level 3 under fractures(1).
SA cracks
    fractures
    joints
    microfractures

fixation
Used in geochemistry, e.g. for fixation of ions.
SA geochemistry
    ions

fjords
Includes use on level 3 under geomorphology(1) shore features(2); on level 3 under glacial geology(1) glacial features(2).
UF fiord
    fyord
SA bays
    estuaries
    geomorphology
    glacial features
    glacial geology
    glaciation
    shore features

Flack Lake
BT Ontario
   Canada

Flagstaff
City in Coconino County in N central.
BT Arizona
   United States

flame emission spectrometry
use flame photometry

flame photometry
Includes use on level 3 under spectroscopy(1) methods(2); on level 3 under chemical element(1) analysis(2). See list D.
UF flame emission spectrometry
SA analysis
   chemical analysis
   photometry
   quantitative analysis
   spectroscopy

flame structures
Includes use on level 3 under sedimentary structures(1) soft sediment deformation(2). See list F.
BA soft sediment deformation
   sedimentary structures
SA antidunes
   load casts
   structures

Flandrian
Europe.
BA Holocene
   Quaternary
BT Cenozoic

flares
Term used on level 2 through 1978. Now use solar flares.

flaser bedding
Includes use on level 3 under sedimentary structures(1) planar bedding structures(2). See list K.
BA planar bedding structures
   sedimentary structures
SA bedding

Flathead Lake
NW Montana.
BT Montana
   United States

Flathead Sandstone
In Wyoming, unconformably overlies Precambrian granite and underlies Gallatin Limestone and Gros Ventre Formation. Montana and NW Wyoming.
BT Middle Cambrian
   Cambrian
SA Montana
   Wyoming

Fleming Formation
In Grand Gulf Group. Includes Lena, Carnahan Bayou, Dough Hills, Williamson Creek, Castor Creek, and Blounts Creek members. Also considered a group name including Oakville and Cuero formations. E Texas, and W Louisiana.
BT Miocene
SA Louisiana
   Texas

Flemish Cap
A marine bank NE of Grand Banks off Newfoundland.
BT Atlantic Ocean

Fleurieu Peninsula
S of Adelaide between Gulf Saint Vincent and Indian Ocean.
BT South Australia
   Australia

flexural-slip
Includes use on level 3 under folds(1) mechanics(2).
UF bedding-plane slip
SA decollement
   disharmonic folds
   folds
   mechanics

flexure
Used only to indicate the mechanics of deformation. Before 1978, also used to indicate style of structure. Now use flexure folds or growth faults for style of structure.
SA deformation
   flexure folds
   growth faults

flexure faults
use growth faults

flexure folds
Before 1978, also search flexure AND folds.
BA folds
SA flexure
   monoclines

Flin Flon
Town near Saskatchewan border in W.
BT Manitoba
   Canada

Flinders Island
Largest island of the Furneaux Islands off NE Tasmania.
BT Tasmania
   Australia

Flinders Range
use Flinders Ranges

Flinders Ranges
Between Lake Frome and Lake Torrens in E central South Australia. Also search Flinders Range.
CO S323000S293000
   E1393000E1380000
UF Flinders Range
BT South Australia
   Australia

flint
Includes use on level 3 under sedimentary rocks(1) chemically precipitated rocks(2). See list I.
BA chemically precipitated rocks
BT sedimentary rocks
SA chert

flint clay
Includes use on level 3 under sediments(1) clastic sediments(2). See list N.
BA clastic sediments
BT sediments
SA clay
   clays
   refractory materials
   terrigenous materials

Flint Creek Range
Range of Rocky Mountains in SW.
BT Montana
   United States
SA Rocky Mountains

Flint Hills
Index states as applicable.
CO N370000N383000
   W0963000W0965000
BT United States
SA Kansas
   Oklahoma

Flinton Group
E Ontario.
BT Precambrian
SA Canada
   Ontario

flocculation
SA clay mineralogy
   colloidal materials
   geochemistry
   suspended materials

flokite
use mordenite

flood plains
use floodplains

flood tuff
use ignimbrite

floodplains
Includes use on level 3 under geomorphology(1), fluvial features(2). Also search flood plains.
UF flood plains
SA alluvium
   floods
   fluvial features
   geomorphology
   meanders
   plains
   rivers
   terraces

floods
As of 1978, term is used on level 2 under geologic hazards(1) and waterways(1).
SA environmental geology
   floodplains
   geologic hazards
   reservoirs
   rivers and streams
   surface water
   watersheds
   waterways

flora
Includes use on level 3 under ecology(1). Use fossiliferous materials under sedimentary rocks(1).
SA ecology
   fauna
   floral list
   floral provinces
   floral studies
   fossiliferous materials
   fossils
   microfossils
   microorganisms
   paleobotany
   Plantae

floral list
SA faunal list
   flora

floral provinces
UF provinces, floral
SA flora

floral studies
Includes use on level 2 under fossil group(1) for many classes or orders, or for genera. See list F.
UF studies, floral
SA biota
   faunal studies
   flora
   fossils
   microorganisms
   paleobotany
   palynomorphs
   Plantae
   vegetation

Florence
City in E central Tuscany. Also search Firenze.
UF Firenze
BT Tuscany
   Italy

florencite
BA phosphates
BT minerals

Florida
Includes use on level 1 as an area term (list O). For term set options see list B.
CO N243000N310000
   W0800000W0873000
BA United States
NT Apalachicola
   Big Pine Key
   Biscayne Bay
NX Calhoun County
NT Cape Kennedy
   Choctawhatchee Bay
NX Clay County
   Columbia County
NT Coupon Bight
NX Dade County
NT Everglades
   Florida Keys
   Floridan Aquifer
NX Franklin County
   Hamilton County
   Jackson County
   Jefferson County
NT Key Largo
NX Lafayette County
   Lake County
NT Lake Okeechobee
NX Lee County
   Madison County
   Marion County
NT Miami
NX Monroe County
   Nassau County
   Orange County
NT Palm Beach County
   Pensacola
   Pinellas County
NX Polk County
   Putnam County
NT Saint Johns River basin
   Tampa Bay
NX Taylor County
   Union County
   Walton County

Florida ● fluvial features

Washington County
SA Atlantic Coastal Plain
    Bone Valley Formation
    Charlotte County
    Chipola Formation
    Citronelle Formation
    Claiborne Group
    Crystal River Formation
    Duplin Formation
    Eastern U.S.
    Gulf Coastal Plain
    Hawthorn Formation
    Jackson Group
    Key Largo Limestone
    Leon
    Marianna Limestone
    Miami Limestone
    Ocala Group
    Okefenokee Swamp
    Panhandle
    Suwannee Limestone
    Vicksburg Group
    Wicomico Formation
    Wilcox Group

Florida Bay
Body of water between S tip of Florida and Florida Keys.
BT Gulf of Mexico

Florida Keys
150 mile chain of coral limestone islands extending SW off S tip of Florida.
CO N243000N252000
    W0815000W0830000
BT Florida
    United States
SA Big Pine Key

Florida Mountains
Luna County in SW.
BT New Mexico
    United States

Florida Strait
use Straits of Florida

Florida Straits
use Straits of Florida

Floridan Aquifer
N and central Florida. Below the confining bed consists from top to bottom, of limestone in the bottom part of the Hawthorn Formation and limestone, dolomite, and dolomitic limestone in formations of Eocene age that include the Ocala Group, the Avon Park Limestone, the Lake City Limestone, and, in part, the Oldsmar Limestone.
BT Florida
SA Eocene
    Ocala Group

Florinites
BA miospores
    palynomorphs
SA angiosperms

flotation
Includes use on level 3 under commodities (list C).
SA beneficiation

flow
Used only to indicate the mechanics of deformation. Before 1978, also used to indicate style of structure. Now use flow folds for style of structure.
SA deformation
    flow folds
    flows
    heat flow
    plastic flow
    rheology
    viscosity

flow (volcanic)
use lava flows

flow cleavage
Includes use on level 3 under foliation(1) genesis(2).
BA foliation
SA axial-plane structures
    cleavage
    compression
    schistosity
    slaty cleavage
    structural analysis

flow folds
Before 1978, also search flow AND folds.
UF flowage fold
BA folds
SA flow

flow lines
Includes use on level 3 under deformation(1) field studies(2).
UF lines, flow
SA deformation
    lineation
    plastic flow
    rheology

flow mechanism
Includes use on level 2 under lava(1).
SA lava
    lava flows
    mechanism
    rheology
    viscosity

flow regime
Includes use on level 2 under sedimentation(1). Also search streamflow.
SA ground water
    hydraulic conductivity
    hydrographs
    hydrostatic pressure
    roughness
    sedimentation
    stream transport
    transport

flow structures
Includes use on level 3 under sedimentary structures(1) soft sediment deformation(2) and turbidity current structures(2). See list K.
BA soft sediment deformation
    sedimentary structures
SA sedimentation
    structures
    turbidity current structures

flowage fold
use flow folds

flows
When discussing lava, use lava flows.
SA ash flows
    debris flows
    flow
    hydraulics
    lava flows
    mass movements
    velocity

Floyd County
Index states as applicable.
BX Georgia
    Indiana
    Iowa
    Kentucky
    Texas
    Virginia
BT United States

fluctuations
Used as a general term. Includes use on level 3 under paleoclimatology(1).
SA paleoclimatology

fluid inclusions
A level 1 term as of 1978. Used for gaseous or liquid inclusions found in crystals. Before 1978, included use on level 2 under inclusions(1). If 1, term set options are:
topic [analysis, changes, composition, detection, experimental studies, genesis, geochemistry, geologic barometry, geologic thermometry, interpretation, paleosalinity, temperature, theoretical studies]
    type of inclusion [e.g. ammonia, carbon dioxide, gases, methane] or subtopic
SA analysis
    carbon dioxide
    crystal growth
    gases
    geochemistry
    geologic barometry
    geologic thermometry
    host materials
    host rocks
    hydrothermal solutions
    igneous rocks
    inclusions
    mineral deposits
    mineral inclusions
    minerals
    paleosalinity
    petrology
    rocks
    temperature
    xenoliths

fluid injection
Includes use on level 3 under earthquakes(1) or waste disposal(1). Also search injection; injection wells.
SA earthquakes
    injection
    waste disposal

fluid phase
Term introduced in 1978. Includes both gaseous phase and liquid phase. Includes use as a general term under geochemistry(1). Also search fluids; fluid; liquid.
UF fluids
SA Darcy's law
    gaseous phase
    geochemistry
    liquid phase
    solid phase

fluid pressure
SA pressure

fluids
use fluid phase

fluids, ore-forming
use ore-forming fluids

fluoborite
BZ borates
    halides
BT minerals
SA fluorides

fluor-phlogopite
BA sheet silicates
    silicates
BT minerals
SA mica group
    phlogopite

fluorapatite
BA phosphates
BT minerals
SA apatite

fluorescence
SA luminescence
    optical properties
    thermoluminescence
    X-ray fluorescence

fluoride ion
Term introduced in 1978. Before 1978, search fluoride.
SA fluorides
    fluorine
    ions

fluorides
See list L.
BA halides
BT minerals
SA bastnaesite
    clinohumite
    fluoborite
    fluoride ion
    humite
    leifite
    lepidolite
    minyulite
    neighborite
    norbergite
    prosopite
    sarcopside
    synchisite
    thomsenolite
    topaz
    triplite
    villiaumite
    weberite
    zinnwaldite

fluorine
Includes use on level 1 and 2 as a chemical element (list D).
UF F
SA elements
    fluoride ion

fluorite
Includes use on level 3 under minerals(1) halides(2). When of economic value, use fluorspar.
BA halides
BT minerals
SA fluorite structure
    fluorspar

fluorite structure
Term introduced in 1978.
SA fluorite

fluorspar
Use only when of economic value; otherwise use fluorite. Includes use on level 1 and 2 as a commodity term (list C).
SA cryolite
    fluorite

flute casts
Includes use on level 3 under sedimentary structures(1) bedding plane irregularities(2). See list K.
UF casts, flute
BA bedding plane irregularities
    sedimentary structures
SA current markings
    load casts
    scour casts
    scour marks
    soft sediment deformation
    sole marks
    turbidity current structures

fluvial
A valid term through 1977. After 1977, use streams and refer to fluvial features, or see rivers and streams(2) under geomorphology(1).

fluvial features
Includes use on level 2 under geomorphology(1).
UF features, fluvial
NT meanders
    waterfalls
SA alluvial fans
    bars
    bedforms
    braided streams
    buried channels
    cascades
    channel geometry
    channels
    cliffs
    deltas
    drainage basins

drainage patterns
ephemeral streams
estuaries
floodplains
geomorphology
mud banks
overwash
rivers
rivers and streams
sedimentary structures
shoals
sinuosity
stream order
streams
terraces
valleys

fluvial transport
use stream transport

fluvioglacial environment
use glaciofluvial environment

flux
General term. Includes use on level 2 under meteorites(1).
SA heat flux
    meteorites
    radioactivity

fly ash
use ash

flysch
Includes use on level 3 under sedimentary rocks(1) clastic rocks(2); on level 3 under sediments(1) clastic sediments(2). See list I (sed. rocks) and list N (sediments).
BA clastic rocks
BT sedimentary rocks
SA clastic sediments
    terrigenous materials
    wildflysch

Foaming Sea
UF Mare Spumans
BT Moon

focal mechanism
Includes use on level 3 under earthquakes(1). Also search focus AND mechanism.
SA earthquakes
    focus
    mechanism

foci
use focus

focus
Includes use on level 2 under earthquakes(1). Also search hypocenters; foci.
UF centrum (seismology)
    foci
    hypocenter
    seismic focus
SA aftershocks
    deep-focus earthquakes
    earthquakes
    elastic waves
    epicenters
    focal mechanism
    intermediate-focus earthquakes
    seismology
    shallow-focus earthquakes

Foggia
City and province in N.
BT Apulia
    Italy

fold axes
Includes use on level 3 under structural analysis(1) theoretical studies(2).
UF axes, fold
SA folds
    structural analysis

fold belts
Includes use on level 3 under tectonics(1). Also search belts; foldbelts.
UF belts, fold
    foldbelts
SA tectonics

foldbelts
use fold belts

folding
use folds

folds
Used for studies on folds and not those on several types of structure. Includes use on level 1 (list A); on level 2 and 3 under structural analysis(1). Also search folding. If 1, term set options are:
distribution
    subtopic
experimental studies
    subtopic
geometry
    subtopic [cylindrical folds, plane cylindrical folds, plane noncylindrical folds]
mechanics
    subtopic [compaction, decollement, flexural-slip, flexure, flow, kink-band structures, shear]
orientation (attitude of fold elements with respect to external coordinates)
  a. (folds defined on the basis of orientation in relation to the geographic horizontal plane): horizontal orientation, inclined folds, nappes, normal folds, overturned folds, plunging folds, recumbent folds, vertical orientation]
  b. (orientation of folds relative to spatially associated macroscopic structures such as large folds, fold systems, and orogenic zones): [discordant folds, drag folds, longitudinal orientation, oblique orientation, superposed folds]
style
    subtopic (anticlines, antiform folds, asymmetric folds, basins, chevron folds, cleavage folds, concentric folds, conjugate folds, convolute folds, cross folds, diapirs, disharmonic folds, domes, harmonic folds, intrafolial folds, isoclinal folds, kink folds, monoclines, polyclinal folds, ptygmatic folds, similar folds, symmetric folds, synclines, synform folds, troughs]
systems
    subtopic [anticlinoria, en echelon folds, synclinoria], (area)
theoretical studies
    subtopic
UF folding
NA anticlines
    anticlinoria
    antiform folds
    asymmetric folds
    chevron folds
    cleavage folds
    concentric folds
    conjugate folds
    cylindrical folds
    discordant folds
    disharmonic folds
    domes
    drag folds
    drape folds
    en echelon folds
    enterolithic folds
    flexure folds
    flow folds
    harmonic folds
    inclined folds
    intrafolial folds
    isoclinal folds
    kink folds
    monoclines
    normal folds
    overturned folds
    plunging folds
    ptygmatic folds
    recumbent folds
    similar folds
    superposed folds
    symmetric folds
    synclines
    synclinoria
    synform folds
SA axial-plane structures
    basins
    cleavage
    decollement
    deformation
    depressions
    diapirs
    faults
    flexural-slip
    fold axes
    foliation
    fractures
    geometry
    horizontal orientation
    kink-band structures
    lineation
    longitudinal orientation
    mechanics
    metamorphic rocks
    metamorphism
    nappes
    neotectonics
    oblique orientation
    orientation
    orogeny
    salt domes
    salt tectonics
    strike
    structural analysis
    structural complexes
    structural geology
    style
    systems
    tectonics
    tectonophysics
    vertical orientation

foliation
Includes use on level 1 (list A); on level 2 and 3 under structural analysis(1). Used for planar arrangements of textural or structural features in any type of rock. If 1, term set options are:
genesis
    subtopic [flow cleavage, shear cleavage], (area)
interpretation
    subtopic (correlation with movement, with strain, with stress)
style
    subtopic [axial-plane structures, cleavage, fracture cleavage, lamination, schistosity, slaty cleavage, slip cleavage], (area)
NA flow cleavage
    fracture cleavage
    schistosity
    slaty cleavage
    slip cleavage
SA axial-plane structures
    cleavage
    deformation
    faults
    folds
    fractures
    intrusions
    laminations
    lineation
    metamorphic rocks
    metamorphism
    neotectonics
    structural analysis
    structural geology
    style
    tectonics
    tectonophysics

Folkestone
City on Strait of Dover in SE.
BT England
    Great Britain
    United Kingdom

Folldal
Village in SE part of country.
BT Norway

Fontainebleau
Town in N central part of country.
BT Seine-et-Marne
    France

foraminifera
Includes use on level 1 and 2 as a fossil term (list F).
UF Foraminiferida
BT Invertebrata
NA Allogromiina
    Fusulinina
    Miliolina
    Rotaliina
    Textulariina
SA micropaleontology
    plankton
    Protista
    Receptaculitaceae

Foraminiferida
use foraminifera

force, Coriolis
use Coriolis force

foreshocks
SA aftershocks
    elastic waves
    shock waves

Forest City Basin
Index states as applicable.
BT United States
SA Iowa
    Kansas
    Missouri
    Nebraska

Forest of Ardennes
use Ardennes

Forest of Dean
Gloucestershire in W central.
BT England
    Great Britain
    United Kingdom

forestry
Includes use on level 3 under soils(1) utilization(2).
SA deforestation
    forests
    soils
    utilization

forests
Includes use on level 3 under soils(1) yields(2).
SA deforestation
    forestry
    soils
    yields

Forfar
use Angus

Forfarshire
use Angus

form, crystal
use crystal form

formation waters
use connate waters

formations, iron
use iron formations

Formosa
Not a valid index term for GeoRef since 1976. Use Taiwan.

Formosa Strait
Channel between Fukien Province of China and Taiwan. Also search Taiwan Strait.
UF Taiwan Strait
BT Pacific Ocean

formula
Term restricted to use for chemical formulas, e.g. for minerals.
SA minerals

forsterite
BA olivine group
   orthosilicates
   silicates
BT minerals

Fort Churchill
Destroyed wooden fort at town of Churchill on Hudson Bay in NE.
BT Manitoba
   Canada

Fort Good Hope
Trading station on Mackenzie River in NW District of Mackenzie.
UF Good Hope
BT Northwest Territories
   Canada

Fort Gouraud
Village near Spanish Sahara border in NW central Mauritania. Also search Idjil.
UF Idjil
BT Mauritania

Fort Hays Limestone
use Fort Hays Limestone Member

Fort Hays Limestone Member
Of Niobrara Formation. E Colorado, W Kansas, NE New Mexico, and SE South Dakota. Also search Fort Hayes Limestone.
UF Fort Hays Limestone
BT Upper Cretaceous
   Cretaceous
SA Colorado
   Kansas
   New Mexico
   Niobrara Formation
   South Dakota

Fort McMurray
Town on Athabaska River in NE central.
BT Alberta
   Canada

Fort Payne Chert
use Fort Payne Formation

Fort Payne Formation
Includes Greasy Creek facies, Short Mountain facies. N and E Alabama, NW Georgia, Kentucky, NE Mississippi, and Tennessee. Also search Fort Payne Chert.
UF Fort Payne Chert
BT Lower Mississippian
   Mississippian
SA Alabama
   Georgia
   Kentucky
   Mississippi
   Tennessee

Fort Riley Limestone
In Chase Group. E Kansas, SE Nebraska, and central N Oklahoma. Also search Fort Riley Limestone Member.
UF Fort Riley Limestone Member
BT Permian
SA Chase Group
   Kansas
   Oklahoma

Fort Riley Limestone Member
use Fort Riley Limestone

Fort Ross
Trading post on Somerset Island, W of Baffin Island, in District of Franklin.
BT Northwest Territories
   Canada

Fort Ternan
Village in SW part of country. Nyanza province.
BT Kenya

Fort Union Formation
Includes Lebo Shale Member, Tongue River Member, Tullock Member, Ludlow Member, Sentinel Butte Member. NW Colorado, Montana, North Dakota, NW South Dakota, and Wyoming.
SA Colorado
   Cretaceous
   Montana
   North Dakota
   Paleocene
   South Dakota
   Upper Cretaceous
   Wyoming

Fort Worth Basin
N central.
BT Texas
   United States

Forth Valley
River valley in S central.
BT Scotland
   Great Britain
   United Kingdom

Forties Field
SA North Sea
   oil and gas fields

Fortran
BT computer programs
SA automatic data processing

Fortran IV
Includes use on level 3 under automatic data processing(1).
BT computer programs
SA automatic data processing
   mathematical geology

Fortuna
Town in Humboldt County in NW.
BT California
   United States

Fosdick Mountains
N of Edsel Ford Range in W Marie Byrd Land near the Pacific Ocean.
BT Antarctica

Fossa Magna
Structural trench crossing the mountain ranges of Honshu.
BT Honshu
   Japan

fossil
A valid index term through 1977 used in combination with man (i.e. man, fossil) on level 1 and 2. After 1977, use fossil man on level 1 and 2.

fossil assemblages
use assemblages

fossil ice wedges
UF ice wedges, fossil
   ice-wedge cast
   ice-wedge fill
   ice-wedge pseudomorph
   wedges, fossil ice
BA bedding plane irregularities
   sedimentary structures
SA frost action
   frost features
   ice wedges
   periglacial features

fossil man
Term introduced in 1978 as valid on level 1 and 2. Used mainly for archaeological and anthropological studies. See Mammalia(1) Primates(2) for taxonomical studies. Before 1978, search man AND fossil.
SA anthropology
   archaeological sites
   archaeology
   artifacts
   Mammalia
   man
   Neanderthal
   Primates
   Vertebrata

fossil meteorite craters
use astroblemes

fossil soils
use Paleosols

fossil water
use connate waters

fossil wood
Not silicified wood. Includes use on level 3 under Plantae(1), gymnosperms(1), and under angiosperms(1). See list F.
SA angiosperms
   gymnosperms
   Plantae
   wood

fossiliferous materials
Term introduced in 1978. Includes use on level 3 under sediments(1) and sedimentary rocks(1). Before 1978, also search fossiliferous AND specific sediment type or sedimentary rocks.
SA fauna
   flora
   fossilization
   fossils
   living materials
   materials
   mollusks
   sedimentary rocks
   sediments

fossilization
Includes taphonomy. Includes use on level 2 under fossil group(1); on level 2 under paleontology(1). See list F.
SA fossiliferous materials
   fossils
   ichnofossils
   paleobotany
   paleontology
   problematic fossils
   silicification
   taphonomy

fossilized brine
use connate waters

fossils
A valid level 1 term through 1977 used in combination with problematic (i.e. fossils, problematic). After 1978, includes use on level 3. Use problematic fossils on level 1 and 2.
SA allometry
   assemblages
   benthonic taxa
   biochemistry
   biogeography
   biometry
   biostratigraphy
   bones
   calcification
   chemical fossils
   coiling
   fauna
   faunal list
   faunal studies
   flora
   floral studies
   fossiliferous materials
   fossilization
   functional morphology
   habitat
   histology
   ichnofossils
   index fossils
   jaws
   locomotion
   microfossils
   micropaleontology
   nannofossils
   neotypes
   ontogeny
   otoliths
   paleobiology
   paleobotany
   paleoclimatology
   paleontology
   problematic fossils
   provinciality
   range
   revision
   reworking
   sclerites
   septa
   shells
   skeletons
   skulls
   species diversity
   sutures
   typomorphism

foundations
A level 1 term as of 1978. Includes use on level 2 under dams(1). Used for geological studies on engineering foundations. If 1, term set options are:
bridges
   subtopic
buildings
   subtopic
construction
   subtopic
dams
   subtopic
design
   subtopic
earthworks
   subtopic
experimental studies
   subtopic
highways
   subtopic
instruments
   subtopic
land subsidence
   subtopic
materials, properties
   subtopic
piles
   subtopic
seepage
   subtopic
settlement
   subtopic
site exploration
   subtopic
stability
   subtopic
structures
   subtopic
theoretical studies
   subtopic
SA bearing capacity
   blasting
   bridges
   buildings
   construction
   dams
   design
   earthquakes
   earthworks
   engineering geology
   engineering properties
   explosions
   grouting
   highways
   land subsidence
   marine installations
   nuclear facilities

overconsolidated materials
piles
rock mechanics
seepage
settlement
site exploration
slope stability
soil mechanics
soils
stability
structures
tunnels
underground installations

Four Corners
Location where boundaries of four states meet. Index states as applicable.
BT United States
SA Arizona
     Colorado
     New Mexico
     Utah

Fourier analysis
Includes use on level 3 under automatic data processing(1). Also search harmonic analysis; Fourier transforms.
UF Fourier transformations
     Fourier transforms
     harmonic analysis
SA analysis
     automatic data processing
     mathematical geology

Fourier transformations
use Fourier analysis

Fourier transforms
use Fourier analysis

Foveaux Strait
Channel between S South Island and Stewart Island.
BT New Zealand

Fox Glacier
In the Southern Alps of Tasman National Park in E.
BT South Island
     New Zealand

Fox Hills Formation
Includes Milliken Sandstone Member, Colgate Member, Stoneville Member, Trail City Sandstone Member, Timber Lake Sandstone Member, Bullhead Member. Also search Fox Hills Sandstone.
BT Upper Cretaceous
     Cretaceous
SA Colorado
     Edmonton Formation
     Montana
     North Dakota
     South Dakota
     Wyoming

foyaite
Includes use on level 3 under igneous rocks(1) syenite family(2) and alkali gabbro family(2). See list H.
BA syenite family
BT igneous rocks
SA alkali gabbro family
     nepheline syenite

Fra Mauro
Includes use on level 3 under Moon(1). Also search Fra Mauro crater; Fra Mauro formation.
UF Fra Mauro crater
     Fra Mauro Formation
BT Moon

Fra Mauro crater
use Fra Mauro

Fra Mauro Formation
use Fra Mauro

fractional crystallization
Includes use on level 3 under magmas(1). Before 1978, search crystallization AND fractional.
SA crystallization
     differentiation
     magmas
     salt

fractionation
Includes use on level 2 under isotopes(1); on level 3 under geochemistry(1) processes(2).
SA geochemistry
     isotopes
     processes

fracture cleavage
Includes use on level 3 under foliation(1) style(2).
BA foliation
SA cleavage
     fractures
     joints

fracture strength
Includes use on level 3 under deformation(1) field studies(2).
UF breaking strength
     fracture stress
SA deformation
     elastic limit
     strength
     yield strength

fracture stress
use fracture strength

fracture zones
Includes use on level 3 under plate tectonics(1).
UF zones, fracture
SA faults
     mid-ocean ridges
     ocean floors
     plate tectonics
     sea-floor spreading

fractures
Includes use on level 1 (list A); on level 3 under deformation(1). Used for a break in a rock usually without displacement. If 1, term set options are:
distribution
    subtopic
experimental studies
    subtopic
genesis
    subtopic [release, shear, tension]
patterns
    subtopic
style
    subtopic [bedding, closed fractures, columnar joints, concentric fractures, conical fractures, cross fractures, dip fractures, extension fractures, feather fractures, joints, latent fractures, oblique fractures, polygonal fractures, radiating fractures, ring sheeting, strike fractures]
systems
    subtopic
theoretical studies
    subtopic
NA closed fractures
     columnar joints
     concentric fractures
     conical fractures
     cross fractures
     dip fractures
     extension fractures
     joints
     open fractures
     polygonal fractures
     release fractures
SA bedding
     cracks
     cross joints
     deformation
     engineering geology
     failure
     faults
     fissures
     folds
     foliation
     fracture cleavage
     lineation
     longitudinal orientation
     microfractures
     neotectonics
     oblique orientation
     shear zones
     strike
     structural analysis
     structural geology
     style
     systems
     tectonics
     tectonophysics
     tension
     thermal waters
     veins

fracturing
Includes use on level 3 under plate tectonics(1) effects(2).
SA hydraulic fracturing
     plate tectonics

fragmentation
Used as a general term.
SA fragments
     processes

fragments
SA clastic rocks
     clasts
     fragmentation
     lapilli
     particles
     regolith

framboidal texture
Term introduced in 1978. Before 1978, search framboidal. Also search framboidal pyrite; search framboids.
SA pyrite
     textures

framework silicates
Includes use on level 2 under minerals (1); in combination with feldspar group, nepheline group, scapolite group, silica group, sodalite group, and zeolite group (i.e. framework silicates, feldspar group) to form terms on level 2 under minerals(1). See list L. Also search tectosilicates.
UF tectosilicates
BA silicates
BT minerals
NA analcime
NZ cancrinite
NA chkalovite
     danburite
     feldspar group
NZ genthelvite
     hauyne
NA karpinskyite
NZ lazurite
     leifite
NA leucite
NZ meionite
NA nepheline group
     pollucite
     pseudoleucite
     reedmergnerite
     scapolite group
     silica minerals
     sodalite group
NZ thomsonite
     wenkite
NA zeolite group

France
Includes use on level 1 as an area term (list O). For term set options see list B.
CO N423000N510000
     E0083000W0050000
BA Europe
NT Adour Basin
     Aigoual Massif
     Aiguille Rouges
     Ain
     Aisne
     Allier
     Alpes-de-Haute Provence
     Alpes-Maritimes
     Alsace
     Anjou
     Aquitaine
     Aquitaine Basin
     Ardeche
     Ariege
     Arize Massif
     Armorican Massif
     Artois
     Aube
     Aubrac
     Aude
     Auvergne
     Aveyron
     Bas-Dauphine
     Bas-Rhin
     Bay of Bourgneuf
     Bay of Saint-Michel
     Bay of the Seine
     Bearn
     Beauce
     Belledonne Massif
     Bouches-du-Rhone
     Brittany
     Burgundy
     Calvados
     Cantal
     Causses
     Central Massif
     Cevennes
     Charente
     Charente-Maritime
     Charentes
     Cher
     Correze
NA Corsica
NT Cote-d'Or
     Cotes-du-Nord
     Creuse
     Dauphine
     Dauphine Alps
     Deux-Sevres
     Devoluy
     Dordogne
     Doubs
     Drac Valley
     Drome
     Durance Basin
     Eoulx Basin
     Esterel
     Eure
     Eure Valley
     Eure-et-Loir
     Finistere
     Franche-Comte
     French Coast
     Gard
     Garonne River
     Gers
     Gironde
     Gironde Estuary
     Haut-Rhin
     Haute-Garonne
     Haute-Loire
     Haute-Marne
     Haute-Saone
     Haute-Savoie
     Haute-Vienne
     Hautes-Alpes
     Hautes-Pyrenees
     Herault
     Ille-et-Vilaine
     Indre
     Indre-et-Loire
     Isere
     Isere Valley
     Jura
     Landes

Languedoc
Laval Basin
Lherz
Limagne
Limousin
Loir-et-Cher
Loire
Loire River
Loire Valley
Loire-Atlantique
Lorraine
Lot
Lot-et-Garonne
Lozere
Maine-et-Loire
Manche
Marche
Marne
Marne Valley
Mayenne
Meurthe-et-Moselle
Meuse
Montagne Noire
Monts du Lyonnais
Morbihan
Morvan
Moselle
Nievre
Nord
Nord-Pas-de-Calais Basin
Normandy
Oise
Oise River valley
Orleans
Orne
Paris
Paris Basin
Pas-de-Calais
Pelvoux Massif
Picardy
Poitou
Provence
Provence Alps
Puy-de-Dome
Pyrenees-Atlantiques
Pyrenees-Orientales
Quercy
Redon
Rennes
Rhone
Rodez Trough
Rouergue
Roussillon
Salat Valley
Saone Valley
Saone-et-Loire
Sarthe
Savoie
Seine
Seine Estuary
Seine River
Seine Valley
Seine-et-Marne
Seine-et-Oise
Seine-Maritime
Somme
Somme River valley
Tarn
Tarn-et-Garonne
Touraine
Val-d'Oise
Var
Vaucluse
Vendee
Vercors
Vienne
Vocontian Trough
Vosges
Vosges Mountains
Yonne
Yonne Valley
  SA Alps
      Annot Sandstone
      Ardennes
      Arve Valley
      Brest
      Brianconnais Zone
      Cottian Alps
      Jura Mountains
      Lake Geneva
      Maritime Alps
      Mediterranean region
      Meuse River
      Meuse Valley
      Moselle River
      Moselle Valley
      North Sea Coast
      Pyrenees
      Rhine Basin
      Rhine River
      Rhine Valley
      Rhone River
      Rhone Valley
      Savoy
      Scheldt River
      Upper Rhine Valley
      Versailles
      Voltzia Sandstone
      Western Alps

Franche-Comte
Historical region on the Swiss border S and SW of Alsace.
  BT France

Francis Creek Shale
W and central Illinois.
  BT Pennsylvanian
  SA Desmoinesian
      Illinois

Franciscan Formation
Includes Corral Hollow Shales, Oak Ridge Sandstone. W California. Also search Franciscan Complex; Franciscan Group.
  BT Mesozoic
  SA California
      Cretaceous
      Jurassic

francolite
  BA phosphates
  BT minerals
  SA apatite

Franconia
Former duchy of S central West Germany. Index states as applicable.
  UF Franken
  BT West Germany
  SA Baden-Wurttemberg
      Bavaria
      Hesse
      Upper Franconia

Franconia Formation
Includes Woodhill, Birkmose, Tomah, Reno and Mazomanie members. N Illinois, SE Minnesota, and SW Wisconsin. Index states as applicable.
  BT Upper Cambrian
      Cambrian
  SA Illinois
      Minnesota
      Wisconsin

Franconian Alb
use Franconian Jura

Franconian Forest
S outlier of the Thuringian Forest.
  BT West Germany

Franconian Jura
Plateau in central Bavaria. A northern continuation of the Swabian Jura. Also search Franconian Alb.
  CO N484500N500000
      E0120000E0104500
  UF Franconian Alb
  BT Bavaria
      West Germany

Franken
use Franconia

Frankenberg
City in SE part of country.
  BT Karl-Marx-Stadt
      East Germany

Frankfurt
City. Frankfurt am Main in West Germany and Frankfurt an der Oder in East Germany. Index countries as applicable.
  BT Europe
  SA East Germany
      West Germany

Frankfurt Stade
Europe. Term introduced in 1978. Before 1978, search Frankfurt.
  BA Weichselian
      upper Pleistocene
      Pleistocene
      Quaternary
  BT Cenozoic

Franklin County
Index states as applicable.
  BX Alabama
      Arkansas
      Florida
      Georgia
      Idaho
      Illinois
      Indiana
      Iowa
      Kansas
      Kentucky
      Maine
      Massachusetts
      Mississippi
      Missouri
      Nebraska
      New York
      North Carolina
      Ohio
      Pennsylvania
      Tennessee
      Texas
      Vermont
      Virginia
      Washington
  BT United States

Franklin District
use District of Franklin

Franklin Mountains
District of Mackenzie in Canada, and N of El Paso in the United States. Index Northwest Territories and/or Texas as applicable.
  SA Northwest Territories
      Rocky Mountains
      Texas

franklinite
  BA oxides
  BT minerals

Franz Josef Land
Archipelago in Arctic Ocean. Part of Arkhangelsk Oblast.
  BT Russian Republic
      USSR

Fraser Range
SE of Kalgoorlie in S central.
  BT Western Australia
      Australia

Fraser River
S central and SW British Columbia. Flows into Georgia Strait S of Vancouver.
  BT British Columbia
      Canada

Frasnian
Europe. Above Givetian, below Famennian. Includes use on level 3 under age terms(1). See list E.
  BA Upper Devonian
      Devonian
  BT Paleozoic
  SA Greenland Gap Group

Fredericksburg Group
Comprises Finlay Limestone, Kiamichi Formation, Edwards Limestone, Paluxy Sand, Walnut Formation, Comanche Peak Limestone, Goodland Formation. S Oklahoma and Texas.
  BT Comanchean
      Cretaceous
  SA Lower Cretaceous
      Oklahoma
      Texas

Fredericton
SW New Brunswick.
  BT New Brunswick
      Canada

Frederikshaab
use Frederikshab

Frederikshab
Settlement on Davis Strait in extreme SW Greenland. Also search Frederikshaab.
  UF Frederikshaab
  BT Greenland
      Arctic region

free energy
Includes use on level 3 under geochemistry(1) or phase equilibria(1). Also search Gibbs free energy.
  UF Gibbs free energy
      Helmholtz free energy
  SA energy
      geochemistry
      phase equilibria
      thermodynamic properties

free oscillations
Includes use on level 3 under Earth(1).
  UF free vibrations
  SA Earth
      oscillations

free vibrations
use free oscillations

free-air anomalies
Term introduced in 1978. Before 1978, search free-air.
  SA anomalies
      Bouguer anomalies
      gravity methods

freeze-and-thaw action
use frost action

freeze-thaw action
use frost action

Freiberg
City in SE part of country.
  BT Karl-Marx-Stadt
      East Germany

freibergite
Steel gray variety of tetrahedrite containing silver.
  BA sulfantimonites
      sulfosalts
  BT minerals
  SA tetrahedrite

Freiburg
City in SW part of country. Also search Freiburg im Breisgau.
  UF Freiburg im Breisgau
  BT Baden-Wurttemberg
      West Germany

Freiburg im Breisgau
use Freiburg

freieslebenite
  BA sulfantimonites
      sulfosalts
  BT minerals

Fremont County
Index states as applicable.
  BX Colorado
      Idaho
      Iowa
      Wyoming
  BT United States

Fremouw Formation
Beardmore Glacier area near the Ross Ice Shelf at N end of Queen Maud Range in Ross Dependency.
BT Triassic
SA Antarctica

French
Includes use on level 3 to indicate language of a document.

French Coast
Along British Channel and Bay of Biscay on the Atlantic Ocean side and the Mediterranean Sea on the S.
BT France

French Guiana
French overseas department on NE coast. Includes use on level 1 as an area term (list O). For term set options see list B.
BA South America
SA Guiana Basin
    Guianas
    Guyana Shield

French Somaliland
use Djibouti

frequency
Used as a general term.
SA periodicity

fresh water
Includes use on level 3 under ground water(1), water resources(1), ecology(1), paleoecology(1), and sedimentation(1). Also search fresh-water.
UF freshwater
SA ecology
    fresh-water environment
    ground water
    paleoecology
    salt water
    sedimentation
    water
    water quality
    water resources

fresh-water environment
Term introduced in 1978. Before 1978, search fresh-water or fresh water.
SA ecology
    environment
    fresh water
    paleoecology

freshwater
use fresh water

Freshwater
Town on Isle of Wight in the English Channel.
BT England
    Great Britain
    United Kingdom

Fresno
City in Fresno County in the San Joaquin Valley.
BT California
    United States

Fresno County
In the San Joaquin Valley of central.
CO N355500N373000
   W1182000W1205000
BT California
   United States

friction
Used as a general term.
SA mechanical properties
   physical properties

Friendly Islands
use Tonga

fringing reefs
UF shore reef
BT reefs
SA barrier reefs

Frio Formation
Consists of dark to very dark, varicolored shales and silty shales and massive to thin-bedded strata of sand and silty sand. Subsurface.
BT Oligocene
SA Louisiana
   Texas

Friuli-Venezia Giulia
Autonomous region in NE.
BT Italy
NT Tarvisio
   Trieste
SA Tagliamento Valley
   Venetia

Front Range
A range of Rocky Mountains in N central.
CO N383000N410000
   W1050000W1060000
BT Colorado
   United States
SA Rocky Mountains

Front Range urban corridor
BT Colorado
   United States

Frontenac County
Index provinces as applicable.
BT Canada
SA Ontario
   Quebec

Frontier Formation
In Colorado Group or Mancos Group. S Montana, and W Wyoming, Colorado, Idaho, and Utah.
BT Upper Cretaceous
   Cretaceous
SA Colorado
   Colorado Group
   Idaho
   Montana
   Utah
   Wyoming

froodite
BA bismuthides
   sulfides
BT minerals

Frosinone
Town and province in the Appenines SE of Rome.
BT Latium
   Italy

frost action
Includes use on level 2 under engineering geology(1) and geomorphology(1).
UF action, frost
   freeze-and-thaw action
   freeze-thaw action
SA congelifraction
   cryopedology
   cryoturbation
   engineering geology
   fossil ice wedges
   geomorphology
   glacial geology
   highways
   patterned ground
   periglacial environment
   permafrost
   polygons
   rock mechanics
   soil mechanics

frost bursting
use congelifraction

frost churning
use cryoturbation

frost features
Includes use on level 3 under sedimentary structures(1) bedding plane irregularities(2). See list K.
UF features, frost
BA bedding plane irregularities

sedimentary structures
SA fossil ice wedges

frost heaving
As of 1978, term is used on level 2 under permafrost(1).
UF heaving, frost
SA congelifraction
   cryoturbation
   permafrost

frost shattering
use congelifraction

frost splitting
use congelifraction

frost stirring
use cryoturbation

frost weathering
use congelifraction

frost wedging
use congelifraction

Frozen ground
UF gelisol
   ground, frozen
   merzlota
   taele
   tele
   tjaele
BA soils
SA ice
   periglacial features
   permafrost

fruits
Includes use on level 3 under soils(1) yields(2); includes use on level 3 as a fossil term.
SA soils
   yields

Frunze
City in N Kirghizia.
UF Pishpek
BT Kirghizia
   USSR

frustules
Refers to morphology of diatoms.
SA diatoms

Fuerteventura
Island in Spain's Las Palmas Province.
BT Canary Islands
   Atlantic Ocean

fugacity
Includes use on level 3 under geochemistry(1).
SA geochemistry
   partial pressure
   thermodynamic properties
   thermodynamics

Fujairah
Emirate. One of federation of 7 states at S end of Persian Gulf. Includes use on level 3 as an area term (list O).
BT United Arab Emirates
   Arabian Peninsula

Fuji
use Fujiyama

Fujiyama
Sacred mountain and quiescent volcano W of Yokohama. Also search Mount Fuji; Mt. Fuji; Mt. Fuji volcano.
UF Fuji
   Mount Fuji
   Mt. Fuji
   Mt. Fuji volcano
BT Honshu
   Japan

Fukui
City and prefecture on the Japan Sea in central.
BT Honshu
   Japan

Fukuoka
City and prefecture in N.
BT Kyushu
   Japan

Fukushima
City in Fukushima Prefecture N central.
BT Honshu
   Japan

Fukushima Prefecture
N central.
BT Honshu
   Japan

fuller's earth
Includes use as level 3 commodity term under clays(1). See list C.
SA clay
   clay mineralogy
   clays
   kaolin
   smectite
   soils

Fulton County
Index states as applicable.
BX Arkansas
   Georgia
   Illinois
   Indiana
   Kentucky
   New York
   Ohio
   Pennsylvania
BT United States

fulvic acids
Includes use on level 3 under organic materials(1).
BA organic materials
SA acids

fumaroles
As of 1978, term is used on level 2 under thermal waters(1).
NT solfataras
SA geothermal energy
   geysers
   heat flow
   springs
   sublimates
   thermal waters
   volcanism
   volcanology

functional morphology
BT morphology
SA fossils

functions
Used as a general term in a mathematical sense.
SA mathematical geology
   mathematical methods

fungi
Includes use on level 1 and 2 as a fossil term (list F); on level 3 under soils(1) biota(2).
BT Plantae
SA algae
   bacteria
   lichens
   microorganisms
   paleobotany
   parasites

Furnace Creek
Small creek which sinks into valley floor in Death Valley National Monument near Nevada border in SE.
BT California
   United States

fusinite
BA macerals
SA coal
   organic residues
   sedimentary rocks

fusion
Also search nuclear fusion.
SA fission

melting

Fusulinidae
Includes use on level 2 under foraminifera(1). See list F.
BA Fusulinina
  foraminifera
BT Invertebrata
NA Schwagerina
  Triticites

Fusulinina
Includes use on level 2 under foraminifera(1). See list F.
BA foraminifera
BT Invertebrata
NA Archaediscidae
  Endothyra
  Fusulinidae

Futaba Group
Underlain by the Paleozoic and granitic rocks with a distinct unconformity and covered disconformably by the Paleogene. N part of E border of Abukama Mountains in E central Honshu.
BT Senonian
  Upper Cretaceous
  Cretaceous
SA Honshu
  Japan

future
Used as a general term.

fyord
use fjords

Ga
use gallium

gabbro
Includes use on level 3 under igneous rocks(1) gabbro family(2). See list H.
BA gabbro family
BT igneous rocks
SA anorthositic gabbro
  gabbroic composition
  metagabbro
  microgabbro
  olivine gabbro

gabbro anorthosite
use gabbroic anorthosite

gabbro family
Includes use on level 2 under igneous rocks(1). See list H.
BT igneous rocks
NA anorthosite
  anorthositic gabbro
  eucrite
  gabbro
  gabbroic anorthosite
  labradoritite
  microgabbro
  norite
  olivine gabbro
  rodingite
  troctolite
SA alkali gabbro family
  gabbroic composition
  labradorite

gabbroic anorthosite
Also search gabbro anorthosite. See list H.
UF gabbro anorthosite
BA gabbro family
BT igneous rocks
SA anorthosite

gabbroic composition
Term introduced in 1978. Before 1978, search gabbroic; gabbroic rocks.
SA composition
  gabbro
  gabbro family

Gabilan Range
One of the Coast Ranges in W.
BT California
  United States
SA Coast Ranges

Gabon
Formerly part of French Equatorial Africa. Includes use on level 1 as an area term (list O). For term set options see list B.
CO S040000N023000
  E0143000E0083000
BA Africa
NT Oklo

Gabonella
Genus. Includes use on level 3 under foraminifera(1) Buliminacea(2).
BA Buliminacea
  Rotaliina
  foraminifera
BT Invertebrata

gadolinite
BA orthosilicates
  silicates
BT minerals

gadolinium
Includes use on level 1 and 2 as a chemical element (list D).
UF Gd
SA elements
  rare earths

gahnite
UF zinc spinel
BA oxides
BT minerals
SA spinel group

Gailtal Alps
S Austria.
BT Austria
SA Alps

Gairloch
Town and parish in NW.
BT Scotland
  Great Britain
  United Kingdom

gaize
BA clastic rocks
BT sedimentary rocks
SA clastic sediments
  sandstone

Galapagos Islands
A territory of Ecuador about 600 miles W of mainland. Includes use on level 1 as an area term (list O). For term set options see list B.
CO S013000N010000
  W0890000W0920000
UF Colon Archipelago
BA Pacific Ocean
SA Ecuador
  South America

galena
Also search galenite.
UF blue lead
  galenite
  lead glance
BA sulfides
BT minerals

galenite
use galena

galenobismutite
BA sulfobismuthites
  sulfosalts
BT minerals

Galezice
Syncline in the Swiety Krzyz Mountains between Vistula and Pilica rivers in SE central.
BT Poland
SA Swiety Krzyz Mountains

Galicia
Region and ancient kingdom in extreme NW Spain, and a former Austrian crownland now in SE Poland, and NW Ukraine. Index Ukraine and countries as applicable.
BT Europe
NT Arosa Bay
SA Poland
  Spain
  Ukraine

Galilean satellites
Includes Callisto, Europa, Ganymede, and Io.
SA Jupiter
  satellites

Galilee
Hilly region of N.
BT Israel
  Middle East

Gallatin County
SW Montana.
BT Montana
  United States

Gallatin Range
Index states as applicable.
BT United States
SA Montana
  Wyoming

gallium
Includes use on level 1 and 2 as a chemical element (list D). Also search Ga.
UF Ga
SA elements

Gallivare
Village N of the Arctic Circle.
UF Gellnare
BT Norrbotten
  Sweden

Galloway
District in SW.
BT Scotland
  Great Britain
  United Kingdom

Gallup Sandstone
In Mesaverde Group. NW New Mexico.
BT Upper Cretaceous
  Cretaceous
SA Mesaverde Group
  New Mexico

Galveston
City on Galveston Island on the Gulf of Mexico in Galveston County.
BT Texas
  United States

Galveston Bay
Inlet of Gulf of Mexico protected from gulf by Galveston Island and Bolivar Peninsula.
BT Texas
  United States

Galveston County
SE Texas.
BT Texas
  United States

Galveston Island
30 mile long island at the entrance of Galveston Bay.
BT Texas
  United States

Galway
County in W Ireland. Also a city.
CO N525500N534000
  W0080000W0101500
BT Ireland
NT Galway Granite

Galway Granite
Connemara District in W.
BT Galway
  Ireland
SA Devonian

Gambia
Includes use on level 1 as an area term (list O). For term set options see list B.
BA Africa
SA Georgetown
  West Africa

Gambier Embayment
Index states as applicable.
BT Australia
SA South Australia
  Victoria

gamma ray
A valid term through 1977. After 1977 use gamma-ray spectroscopy or gamma-ray methods.

gamma rays
Includes use on level 3 under spectroscopy(1), well-logging(1) and under chemical element(1).
UF rays, gamma
SA cosmic rays
  electromagnetic radiation
  gamma-gamma methods
  gamma-ray methods
  gamma-ray spectroscopy
  radioactivity
  scintillations
  X-rays

gamma-gamma methods
Term introduced in 1978. Includes use on level 3 under well-logging(1) and under geophysical methods(1). Before 1978, search gamma-gamma.
BA geophysical methods
SA gamma rays
  gamma-ray methods
  methods
  radioactivity
  well-logging

gamma-ray methods
Term introduced in 1978. Includes use on level 3 under well-logging. Before 1978, search gamma ray or gamma-ray.
BA geophysical methods
SA gamma rays
  gamma-gamma methods
  gamma-ray spectroscopy
  methods
  radioactivity
  well-logging

gamma-ray spectrometry
A valid term through 1978. After 1978, see spectroscopy(1).

gamma-ray spectroscopy
Term introduced in 1978. Includes use on level 3 under spectroscopy(1), under well-logging(1), and under chemical element(1) analysis(2). Before 1978, also search spectroscopy AND gamma ray.
BT spectroscopy
SA alpha-ray spectroscopy
  analysis
  gamma rays
  gamma-ray methods
  radioactivity
  well-logging

Gander Lake Group
UF Gander Lake Series
BT Ordovician
SA Canada
  Newfoundland

Gander Lake Series
use Gander Lake Group

Gandja
use Kirovabad

Gandzha
use Kirovabad

Ganga River
use Ganges River

Gangamopteris
Genus.
BA Glossopteris
gymnosperms
BT Plantae

Ganges River
Sacred river which rises in the Himalayas and flows into the Bay of Bengal. Index countries as applicable.
UF Ganga River
BT Asia
SA Bangladesh
India

Gangpur Group
use Gangpur Series

Gangpur Series
Subdivided into Ghoriajor Stage, Kumarmunda Stage, Birmitrapur Stage, and Laingar Stage.
UF Gangpur Group
BT Archean
SA India
Orissa

gangue
Includes use on level 3 under mineral deposits, genesis(1) or commodities (list C).
SA mineral deposits
ore deposits

Gard
Department in S.
CO N433000N443000
E0050000E0031500
BT France
NT Ales

Garfield County
Index states as applicable.
BX Colorado
Montana
Nebraska
Oklahoma
Utah
Washington
BT United States

Gargano
Mountain promontory extending into Adriatic Sea in NE Apulia. Also search Monte Gargano.
UF Monte Gargano
BT Apulia
Italy

Gargasian
Substage in Switzerland: upper Aptian; above Bedoulian Substage, Includes use on level 3 under age terms(1). See list E.
BA Lower Cretaceous
Cretaceous
BT Mesozoic
SA Aptian

Garhwal
District in N Uttar Pradesh.
UF Gurhwal
BT Uttar Pradesh
India

Garian
Town in extreme NW part of country.
UF Gharian
BT Tripolitania
Libya

Garlock Fault
Runs E-W in S central California. Also search Garlock fault zone.
BT California
United States

Garm
Town in N central.
BT Tadzhikistan
USSR

garnet
Includes use on level 3 under industrial minerals(1) and under minerals(1) orthosilicates, garnet group(2). See list C (commodities) and list L (minerals).
BA garnet group
orthosilicates
silicates
BT minerals
SA abrasives
gems
heavy minerals
spessartite

garnet group
Includes use in combination with orthosilicates (i.e. orthosilicates, garnet group) on level 2 under minerals(1). See list L.
BA orthosilicates
silicates
BT minerals
NA almandine
andradite
garnet
grossular
hydrogrossular
melanite
pyrope
schorlomite
spessartine
uvarovite
vanadium garnet

garnet lherzolite
BA ultramafic family
BT igneous rocks
SA lherzolite

garnet peridotite
BA ultramafic family
BT igneous rocks
SA peridotite

garnet pyroxenite
BA ultramafic family
BT igneous rocks
SA pyroxenite

Garnett
City in Anderson County in E.
BT Kansas
United States

garnierite
UF nepouite
BA sheet silicates
silicates
BT minerals
SA serpentine group

Garo Hills
In bend of Brahmaputra River in NE part of country.
BT India

Garonne River
Southwest.
BT France

garrelsite
BA orthosilicates
silicates
BT minerals

garronite
BA zeolite group
framework silicates
silicates
BT minerals

gas
A valid level 1 term through 1977 used in combination with natural (i.e. gas, natural). After 1977, use natural gas as level 1 term.

gas chromatography
Not a valid term from 1975 through 1977. After 1977, includes use as a valid index term on level 3.
SA chemical composition
chromatography
gases

gas fields, oil and
use oil and gas fields

Gas Hills
Fremont County in central.
BT Wyoming
United States

gas phase
use gaseous phase

gas storage
Includes use on level 3 under engineering geology(1) reservoirs(2).
SA engineering geology
natural gas
reservoirs
storage
underground installations

Gascoyne abyssal plain
Off western Australia.
BT Indian Ocean

gaseous phase
Term introduced in 1978. Includes use as general term under geochemistry(1). Also search gas phase.
UF gas phase
phase, gaseous
SA fluid phase
gases
gasification
geochemistry
liquid phase
solid phase

gases
Includes use on level 3 for other than natural gas.
SA Darcy's law
fluid inclusions
gas chromatography
gaseous phase
gasification
hydrogen sulfide
mining geology
natural gas
noble gases
oil-gas interface
phase equilibria
solfataras
sublimates

gasification
Includes use on level 3 under coal(1).
SA coal
gaseous phase
gases
natural gas

Gaspe
Village in E Quebec near E extremity of Gaspe Peninsula.
BT Quebec

Gaspe Peninsula
Extends into Gulf of Saint Lawrence. Also search Gaspe.
CO N475000N491500
W0641500W0680000
BT Quebec

gastaldite
use glaucophane

gastroliths
SA Pisces
Reptilia

Gastropoda
Includes use on level 2 under Mollusca(1). See list F.
BA Mollusca
BT Invertebrata
NA Archaeogastropoda
Cerithiidae
Harpidae
Muricacea
Planorbis
Prosobranchia
Pteropoda
Pupillidae
Turritellidae
Viviparus

Gatineau Valley
River valley in SW.
BT Quebec
Canada

gauges
use gauging

gauging
Term introduced in 1978. Also search gauges.
UF gauges
SA hydrology
water

Gault
Middle Europe. A lower Cretaceous clay formation in Great Britain. Includes use on level 3 under age terms (1). See list E. Also search Gault Clay and Gault Formation.
UF Gault Clay
BA Lower Cretaceous
Cretaceous
BT Mesozoic
SA Great Britain

Gault Clay
use Gault

Gaurdak
Town in Chardzhou Oblast in NE.
BT Turkmenia
USSR

Gauss epoch
Geomagnetic epoch.
UF Gauss Normal
BA Pliocene
Neogene
Tertiary
BT Cenozoic

Gauss Normal
use Gauss epoch

Gavarnie
Village in S part of country.
BT Hautes-Pyrenees
France

Gavrovo Zone
Central Peloponnesus. Also search Gavrovo.
UF Gavrovo-Tripolis Zone
BT Peloponnesus
Greece

Gavrovo-Tripolis Zone
use Gavrovo Zone

Gavur Mountains
use Amanos Mountains

Gay
Ore fields in the Southern Urals.
BT Russian Republic
USSR

Gaya
City in central.
BT Bihar
India

gaylussite
BA carbonates
BT minerals

Gd
use gadolinium

Gdansk
Province in N Poland. Also a city on the Gulf of Danzig.
UF Danzig
BT Poland
NT Leba
Puck Bay

Ge
use germanium

geanticlines
Includes use on level 3 under tectonics(1) structure(2).
SA anticlines
 anticlinoria
 geosynclines
 synclinoria
 tectonics

Geary County
NE central.
BT Kansas
 United States

Geauga County
NE Ohio.
BT Ohio
 United States

Gedinnian
Europe. Above Ludlovian (Silurian), below Siegenian. Includes use on level 3 under age terms(1). See list E.
BA Lower Devonian
 Devonian
BT Paleozoic

gedrite
BA amphibole group
 chain silicates
 silicates
BT minerals
SA amosite
 anthophyllite

gehlenite
BA melilite group
 orthosilicates
 silicates
BT minerals

Gehrden
Town in SE central.
BT Lower Saxony
 West Germany

geikielite
BA oxides
BT minerals

Geirud Formation
Central Elburz Mountains in N Iran.
BT Paleozoic
SA Carboniferous
 Devonian
 Iran
 Permian

Geisel River valley
use Geisel Valley

Geisel Valley
River valley in S central part of country. Also search Geisel River valley.
UF Geisel River valley
BT Halle
 East Germany

gelifraction
use congelifraction

gelisol
use Frozen ground

geliturbation
use cryoturbation

gelivation
use congelifraction

Gellnare
use Gallivare

gels
BT colloidal materials

Gemer
Village in S central.
BT Slovakia
 Czechoslovakia

gems
Includes use on level 1 as a commodity term (list C); on level 3 under economic geology(1). See also specific gems.
SA agate
 amethyst
 beryl
 chalcedony
 diamond
 diamonds
 economic geology
 emerald
 garnet
 minerals
 moonstone
 pearls
 peridot
 sapphire
 topaz
 zircon

general
Includes use as a level 2 or 3 term appropriate to a large number of topics, e.g. on level 2 under Mohorovicic discontinuity(1). As of 1976, the term is to be used on level 2 only when referring to geology. See list G.
SA education
 geology

generation
Includes use on level 2 under ocean waves(1).
SA ocean waves
 waves

Genesee County
BX Michigan
 New York
BT United States

Genesee Group
Comprises Geneseo Shale with Genundewa Limestone Lentil, Standish Flagstone, and West River Shale, Penn Yan Shale, Sherburne Flagstone, Renwick Shale, Ithaca Member. Maryland, New York, Pennsylvania, Western Virginia and northern West Virginia.
BT Devonian
SA Maryland
 Middle Devonian
 New York
 Pennsylvania
 Upper Devonian
 Virginia
 West Virginia

genesis
Includes use as level 1 in combination with mineral deposits (i.e. mineral deposits, genesis); as a level 2 or 3 term appropriate to a large number of topics, e.g. on level 2 under foliation(1), fractures(1), and organic materials(1). See list C (commodities) and list G (general terms).
SA crystallization
 mineral deposits

Geneva
Canton surrounding W end of Lake Geneva. Also a city.
BT Switzerland
SA Lake Geneva

Genoa
City on the Ligurian Sea. Also search Genova.
UF Genova
BT Liguria
 Italy

Genova
use Genoa

genthelvite
BZ framework silicates
 sulfides
BT minerals
SA helvite
 sodalite group

geobotanical methods
Includes use on level 2 under mineral exploration(1).
SA biogeochemical methods
 geochemical methods
 methods
 mineral exploration

geochemical
A valid term through 1977. After 1977, use geochemical maps, or geochemical methods, or geochemical processes.

geochemical controls
Includes use on level 3 under mineral deposits, genesis(1) controls(2). Also search geochemical.
SA controls
 mineral deposits
 sedimentation

geochemical exploration
A valid term through 1975. After 1975, use geochemical methods or geochemical prospecting.

geochemical maps
Term introduced in 1978. Before 1978, search maps AND geochemical.
BA maps

geochemical methods
Includes use on level 2 under mineral exploration(1). See geochemistry(1) methods(2). Before 1975, also search geochemical exploration. Also search geochemical.
NT biogeochemical methods
SA chemical analysis
 chemical methods
 dispersion patterns
 exploration
 geobotanical methods
 geochemical prospecting
 geochemistry
 haloes
 methods
 mineral exploration

Geochemical Ocean Sections Program
use GEOSECS

geochemical prospecting
Includes use on level 3 under mineral exploration(1) geochemical methods(2). Before 1976, also search geochemical exploration.
SA exploration
 geochemical methods
 mineral deposits
 mineral exploration
 prospecting

geochemical surveys
A valid term through 1978. After 1978, see surveys(2) under geochemistry(1).

geochemistry
Used for broad treatments as well as specific experiments. Includes use on level 1 (list A); on level 2 under major disciplines, area terms (list B), commodities (list C), intrusions(1), lava(1), organic materials(1), and rocks (list H, I, J). If 1, term set options are:
processes
 name of process [absorption, adsorption, chemical fractionation, chlorination, differentiation, diffusion, dissociation, electrolysis, endothermic reactions, fluorination, fractionation, hydration, hydrogenation, hydrolysis, ionization, ion exchange, nitrification, oxidation, photochemical reactions, photosynthesis, polymerization, pyrolysis, reduction, solution, substitution]
properties
 name of property [acidity, alkalinity, Eh, electrochemical properties, exchange capacity, pH, salinity, solubility, thermochemical properties, thermodynamic properties, etc.]
topic [e.g. bibliography, classification, concepts, cycles, education, experimental studies, instruments, methods, practice, properties, surveys, symposia, textbooks]
 subtopic
SA absolute age
 absorption
 activity
 adsorption
 alkalinity
 alumina
 ammonia compound
 ammonium
 aqueous solutions
 atmosphere
 biochemistry
 biogeochemical methods
 cadmium
 calcium carbonate
 carbon dioxide
 catalysis
 cation exchange capacity
 chemical analysis
 chemical properties
 chemistry
 combustion
 complexing
 condensation
 cosmochemistry
 crystal chemistry
 crystal growth
 crystallography
 cycles
 dehydration
 denitrification
 desorption
 devitrification
 diffusion
 dissociation
 Eh
 electrochemical properties
 electrolysis
 electrolytes
 elements
 enrichment
 enthalpy
 environmental geology
 equations of state
 exchange capacity
 fixation
 flocculation
 fluid inclusions
 fluid phase
 fractionation
 free energy
 fugacity
 gaseous phase
 geochemical methods
 geochronology
 geologic barometry
 geologic thermometry
 GEOSECS
 ground water
 halmyrolysis
 hydration
 hydrolysis
 inclusions
 ion exchange
 ions
 isotopes
 kinetics
 leaching
 luminescence
 metals
 metamorphic rocks
 metamorphism
 metasomatism
 meteorites

migration of elements
mineral deposits
mineral exploration
mobilization
organic materials
oxidation
paleomagnetism
paleosalinity
partitioning
pH
phase equilibria
photosynthesis
pollution
polymerization
precipitation
pyrolysis
reactions
reagents
reduction
saturation
sea water
sinks
sodium chloride
solid phase
solubility
solutes
solution
solutions
sorption
specific heat
spectra
spectroscopy
stability
standard materials
stoichiometry
sublimation
suspension
thermochemical properties
thermodynamic properties
trace metals
valency
volatiles
water
water vapor
weathering

geochronology
For methods and topics other than those treated under absolute age, mainly for non-isotope methods. Includes use on level 1 (list A); on level 2 under age sets (list E), and area terms (list B). If 1, term set options are:
methods
    subtopic
surveys (for several types)
    (method) or (topic)
time scales
    age (list E)
type of method [exposure age, fission-track dating, hydration of glass, lichenometry, optical mineralogy, paleomagnetism, particle-track dating (e.g. cosmic-ray track), racemization, radiation damage, tephrochronology, thermoluminescence, tree rings, varves]
    subtopic
UF geologic chronology
SA absolute age
    chemical analysis
    chronology
    dates
    declination
    exposure age
    fission-track dating
    geochemistry
    hydration of glass
    isochrons
    lichenometry
    metamorphic rocks
    metamorphism
    meteor craters
    optical mineralogy
    paleomagnetism
    paleontology
    palynology
    particle track
    racemization
    radiation damage
    radiometric properties
    relative age
    stratigraphy
    tephrochronology
    thermoluminescence
    time
    time scales
    tree rings
    varves

geocronite
BZ sulfantimonites
    sulfarsenites
BA sulfosalts
BT minerals

geodes
Includes use on level 3 under sedimentary structures(1) secondary structures(2). See list K.
BA secondary structures
    sedimentary structures
SA concretions
    nodules

geodesy
Used infrequently and for geologic applications only. Includes use on level 1 (list A); on level 3 under Earth(1) gravity field(2). If 1, term set options are:
topic [figure of Earth, geodetic coordinates, geoid, harmonics, methods, satellite measurements, surveys]
    subtopic
SA cartography
    changes of level
    coordinates
    deformation
    Earth
    Earth tides
    ellipticity
    engineering geology
    exploration
    figure of Earth
    geodetic coordinates
    geoid
    geomorphology
    geophysical methods
    geophysical surveys
    geophysics
    gravity field
    harmonics
    isostasy
    Laplace transformations
    leveling
    maps
    measurement
    neotectonics
    satellite measurements
    seismology
    selenodesy
    shape
    strainmeters
    surveys
    tilt
    topography
    triangulation

geodetic coordinates
Includes use on level 2 under geodesy(1).
SA coordinates
    geodesy
    geometry
    satellite measurements

geodynamics
SA Earth
    interior
    plate tectonics
    tectonics

geographic distribution
A valid term through 1978. For fossils, see distribution or biogeography.

geography
Also search physical geography.
UF physical geography
SA biogeography
    Earth
    geomorphology
    paleogeography

geoid
Includes use on level 2 under geodesy(1).
SA Earth
    figure of Earth
    geodesy

geologic
A valid term through 1977. After 1977, use geologic maps.

geologic barometry
As of 1978, term is used on level 2 under fluid inclusions(1).
UF barometry, geologic
SA fluid inclusions
    geochemistry
    geologic thermometry
    inclusions
    P-T conditions

geologic chronology
use geochronology

geologic engineering
use engineering geology

geologic hazards
A level 1 term as of 1978. Includes use on level 2 under impact statements(1), nuclear facilities(1), and land subsidence(1). Used for studies on the initiation, controls and effects of various geological phenomena which may be hazardous to human ecology. If 1, term set options are:
avalanches
    subtopic
catastrophies
    subtopic
causes
    subtopic
controls
    subtopic
earthquakes
    subtopic
effects
    subtopic
explosions
    subtopic
human ecology
    subtopic
faults
    subtopic
floods
    subtopic
land subsidence
    subtopic
landslides
    subtopic
mudflows
    subtopic
observations
    subtopic
prediction
    subtopic
rock bursts
    subtopic
site exploration
    subtopic
storms
    subtopic
sunspots
    subtopic
surveys
    subtopic
tsunamis
    subtopic
volcanoes
    subtopic
UF hazards, geologic
SA avalanches
    catastrophies
    destruction
    earthquakes
    engineering geology
    environmental geology
    erosion
    explosions
    faults
    floods
    ground motion
    human ecology
    hurricanes
    impact statements
    land subsidence
    land use
    landslides
    liquefaction
    marine installations
    mudflows
    nuclear facilities
    pollution
    reservoirs
    rock bursts
    rock mechanics
    seismic risk
    site exploration
    slope stability
    soil mechanics
    storms
    sunspots
    tsunamis
    tunnels
    underground installations
    volcanoes
    waste disposal

geologic mapping
A valid term through 1975. To search, see note under cartography. See maps(1).

geologic maps
Term introduced in 1978. Includes use on level 3 under maps(1). Before 1978, also search maps AND geologic.
BA maps

geologic surveys
use surveys

geologic thermometry
As of 1978, term is used on level 2 under fluid inclusions(1). Before 1974, also search paleotemperature; geothermometry.
UF geothermometry
SA C-13/C-12
    decrepitation
    fluid inclusions
    geochemistry
    geologic barometry
    inclusions
    O-18/O-16
    P-T conditions
    S-34/S-32
    temperature

geological exploration
A valid term through 1971. After 1971, use exploration under field of study. See maps(1) cartography(2).

geological methods
Includes use on level 2 under mineral exploration(1).
SA exploration
    geomorphological methods
    methods
    mineral exploration

geological oceanography
use marine geology

Geological Survey of Canada
SA Canada
    surveys

geologists
Before 1975, also search geology as a profession.
SA geology

geology ● geophysical methods

geology
For general treatments stressing the profession and the discipline. Includes use on level 1 (list A) and 2. If 1, term set options are:
topic [education, history, nomenclature, philosophy, practice, principles, research, textbooks]
subtopic
- SA areal geology
associations
automatic data processing
bibliography
catalogs
collections
current research
economic geology
education
elementary geology
engineering geology
environmental geology
extraterrestrial geology
general
geologists
geophysics
glossaries
historical geology
marine geology
mathematical geology
medical geology
mining geology
museums
physical geology
popular geology
research
structural geology
surficial geology
surveys
symposia
uniformitarianism

geology as a profession
A valid term through 1975. After 1975, use practice. See geology(1) and education(1).

geomagnetic secular variation
use secular variations

geomagnetism
A valid term through 1975. After 1975, see magnetic field under Earth (and other planets) and Moon; see paleomagnetism.

geometry
Includes use on level 2 under folds(1) and under plate tectonics(1).
- SA channel geometry
eccentricity
ellipticity
folds
geodetic coordinates
mathematical geology
obliquity
plate tectonics
quantitative geomorphology
statistical methods
structural analysis
tortuosity
waterways

geomorphic geology
use geomorphology

geomorphologic
A valid term through 1977. After 1977, use geomorphologic maps.

geomorphologic effects
Includes use on level 2 under isostasy(1).
- SA effects
isostasy

geomorphologic maps
Term introduced in 1978. Includes use on level 3 under maps(1). Before 1978, also search maps AND geomorphologic.
- BA maps

- SA geomorphology

geomorphological methods
Includes use on level 2 under mineral exploration(1).
- SA geological methods
geophysical methods
methods
mineral exploration

geomorphology
Includes use on level 1 (list A); on level 2 under area terms (list B), Moon (and other planets), and age terms (list E). Before 1976, also search physiography. If 1, term set options are:
features [cryptoexplosion features, eolian features, erosion features, fluvial features, frost action, impact features, lacustrine features, mass movements, shore features, solution features, volcanic features]
type of feature or topic [e.g. alluvial fans, beaches, caves, deltas, drainage basins, drainage patterns, estuaries, evolution, karst, lagoons, patterned ground, quantitative geomorphology, reefs, rivers, terraces]
topic [applications, bibliography, concepts, education, environment, history, interpretation, landform description, landform evolution, maps, methods, practice, processes, symposia, textbooks]
subtopic or area
- UF geomorphic geology
- SA aerial photography
aggradation
alluvial plains
arid environment
arroyos
astroblemes
barriers
bars
bays
beach ridges
beaches
benches
bogs
boulders
braided streams
bridges
buried channels
buried valleys
calderas
canyons
cartography
cascades
caverns
caves
changes of level
channel geometry
channels
cliffs
coastal plains
colluvium
corrosion
cryptoexplosion features
debris
degradation
deltas
denudation
depressions
deserts
domes
drainage basins
drainage patterns
dunes
ecology
engineering geology
environment
environmental geology
eolian features
erosion
erosion features
erosion surfaces
estuaries
etching
exfoliation
extinct lakes
fans
fjords
floodplains
fluvial features
frost action
geodesy
geography
geomorphologic maps
glacial geology
glacial lakes
glacis
gullies
highlands
hills
hummocks
hydrology
impact features
impacts
inselbergs
isostasy
karren
karst
lacustrine features
lagoons
landform description
landform evolution
landforms
landscapes
landslides
limnology
loess
maps
marshes
mass movements
meanders
meteor craters
morphometry
morphostructures
mountains
mud volcanoes
neotectonics
paleogeography
patterned ground
pediments
peneplains
permafrost
physiographic provinces
plains
plateaus
playas
quantitative geomorphology
reefs
relief
rivers
scarps
sedimentology
semi-arid environment
shoals
shore features
shorelines
sinkholes
slope stability
slopes
slumping
soils
solution cavities
solution features
speleology
spits
steppes
stream order
surface features
surficial geology
swamps
taiga environment
talus slopes
terrain classification
terrains
topography
tors
tundra
valleys
volcanic features
waterways
weathering
wetlands

geophysical
A valid term through 1977. After 1977, use geophysical maps, or geophysical methods, or geophysical surveys.

geophysical maps
Term introduced in 1978. Before 1978, search maps AND geophysical.
- BA maps

geophysical methods
Includes use on level 1 (list A); on level 2 under mineral exploration(1). Used for discussions which stress methodology of applied geophysics. If 1, term set options are:
type of method [acoustical methods, Earth-current methods, electrical methods, electromagnetic methods, gravity methods, infrared methods, magnetic methods, magnetotelluric methods, methods (for more than one method), radioactivity methods, remote sensing, seismic methods]
topic [applications, instruments, interpretation, techniques] or kind of platform [airborne methods, deep-tow methods, ground methods, marine methods, satellite methods], kind of method [induced polarization, resistivity, side-scanning methods, sonar methods]
- NA acoustical methods
airborne methods
deep-tow methods
Earth-current methods
echo sounding
electrical methods
electromagnetic methods
gamma-gamma methods
gamma-ray methods
gravity methods
ground methods
induced polarization
infrared methods
magnetic methods
magnetotelluric methods
marine methods
radioactivity methods
satellite methods
seismic methods
sonar methods
- SA arrays
autoradiography
Bouguer anomalies
deep sounding
electrical conductivity
electromagnetic field
electromagnetic logging
engineering geology
exploration
geodesy
geomorphological methods
geophysical surveys
geophysics
gravimeters
gravity anomalies
heat flow
homogeneity
induced polarization
induction
inverse problem
isostasy
Landsat
magnetic anomalies
magnetic field
maps
marine geology
methods
mineral exploration
mining geology

noise
paleomagnetism
remote sensing
resistivity
seismology
signals
sonobuoys
sounding
well-logging

geophysical surveys
Used for special bibliographies. for geophysical methods applied to specific areas. The methods are generally those of exploration and primarily concerned with the shallow structure of the Earth. Includes use on level 1 (list A); on level 2 under area terms (list B). If 1, term set options are:
type of survey [acoustical surveys, Earth-current surveys, electrical surveys, electromagnetic surveys, gravity surveys, infrared surveys, magnetic surveys, magnetotelluric surveys, radioactivity surveys, seismic surveys, surveys (for more than one survey)]
    area
BT surveys
NA acoustical surveys
    Earth-current surveys
    electrical surveys
    electromagnetic surveys
    gravity surveys
    infrared surveys
    magnetic surveys
    magnetotelluric surveys
    radar methods
    radioactivity surveys
    seismic surveys
SA acoustical properties
    airborne methods
    bathymetry
    Bouguer anomalies
    deep seismic sounding
    deep sounding
    echo sounding
    electromagnetic field
    electromagnetic logging
    engineering geology
    geodesy
    geophysical methods
    geophysics
    gravity anomalies
    gravity field
    gravity survey maps
    ground methods
    ground water
    heat flow
    imagery
    induced polarization
    inverse problem
    isostasy
    Landsat
    magnetic anomalies
    magnetic field
    magnetic survey maps
    magnetic susceptibility
    maps
    marine methods
    mineral exploration
    mining geology
    monitoring
    profiles
    reflection
    refraction
    remote sensing
    resistivity
    satellite methods
    self-potential methods
    sonar methods
    sonobuoys
    sounding
    thermal waters
    well-logging

geophysics
For general treatments stressing the profession and the discipline plus experimental studies on minerals and other materials. Includes use on level 1 (list A);on level 2 under bibliography(1), associations(1), education(1), symposia(1). If 1, term set options are:
topic [applications, bibliography, catalogs, classification, concepts, education, experimental studies, general, history, instruments, methods, nomenclature, objectives, observations, philosophy, practice, principles, symposia, textbooks, theoretical studies]
    subtopic
SA astrophysics and solar physics
    atmosphere
    automatic data processing
    core
    crust
    deep seismic sounding
    deformation
    demagnetization
    Earth
    Earth tides
    elastic constants
    elastic materials
    elastic properties
    engineering geology
    geodesy
    geology
    geophysical methods
    geophysical surveys
    heat flow
    inverse problem
    kinetics
    magnetosphere
    mathematical geology
    meteorology
    observatories
    paleomagnetism
    seismology
    structural analysis
    tectonophysics

geopressure
Related to reservoir rocks. Associated with areas of great thickness of sedimentation. Includes use on level 3 under inclusions(1), under commodity terms (list C), or material.
SA petroleum
    pressure
    reservoir rocks
    sedimentation

Georges Bank
E of Massachusetts and S of Nova Scotia. Also search Georges.
CO N400000N412000
    W0664500W0700000
BT Atlantic Ocean
SA Gulf of Maine

Georgetown
City in Georgetown County in South Carolina, a city on the Atlantic Ocean in Guyana, and a city on the Atlantic Ocean in Gambia. Index South Carolina and countries as applicable.
SA Gambia
    Guyana
    South Carolina

Georgetown County
On the Atlantic Ocean.
BT South Carolina
    United States

Georgia
Term is used on level 1 as an area term (list O) only when referring to the United States. Included use for the Georgian Soviet Socialist Republic in Transcaucasia through 1977, but after 1977 use Georgian Republic for that.
CO N302000N350000
    W0804500W0853500
BA United States
NT Altamaha River
    Atlanta
NX Berrien County
    Burke County
    Calhoun County
    Carroll County
NT Cartersville
NX Chatham County
    Cherokee County
    Clarke County
    Clay County
    Coffee County
    Columbia County
    Columbus
    Cook County
    Dade County
NT Doboy Sound
NX Douglas County
    Fayette County
    Floyd County
    Franklin County
    Fulton County
NT Glynn County
NX Greene County
    Hancock County
    Harris County
    Jackson County
    Jasper County
    Jefferson County
    Johnson County
    Lee County
    Lincoln County
NT Macon
NX Macon County
    Madison County
    Marion County
    McIntosh County
    Mitchell County
    Monroe County
    Montgomery County
    Morgan County
    Murray County
    Newton County
    Oconee County
    Pike County
    Polk County
    Pulaski County
    Putnam County
    Randolph County
NT Sapelo Island
    Stone Mountain
NX Taylor County
NT Towns County
    Twiggs County
NX Union County
    Walton County
    Warren County
    Washington County
    Wayne County
    Webster County
    Wheeler County
    White County
    Wilkinson County
SA Athens
    Atlantic Coastal Plain
    Bangor Limestone
    Blue Ridge Mountains
    Blue Ridge Province
    Brevard Zone
    Brunswick
    Chattanooga Shale
    Chickamauga Group
    Citronelle Formation
    Claiborne Group
    Clayton Formation
    Duplin Formation
    Eastern U.S.
    Fort Payne Formation
    Hawthorn Formation
    Jackson Group
    Knox Group
    Lake Chatuge
    Murphy Marble
    Ocala Group
    Okefenokee Swamp
    Pennington Formation
    Piedmont
    Red Mountain Formation
    Ripley Formation
    Roan Supergroup
    Rockwood Formation
    Rome Formation
    Saint Louis Limestone
    Sainte Genevieve Limestone
    Sand Hills
    Savannah River
    Shady Dolomite
    Suwannee Limestone
    Talladega Group
    Trenton Group
    Tuscaloosa Formation
    Twiggs Clay
    Valley and Ridge Province
    Wicomico Formation
    Wilcox Group

Georgian Bay
Inlet of E Lake Huron.
BT Ontario
    Canada

Georgian Republic
Term introduced in 1978 for the Georgian Soviet Socialist Republic in Transcancasia. Before 1978, also search USSR AND Georgia.
CO N420000N430000
    E0470000E0400000
BT USSR
NT Abkhazia
    Adzharistan
    Akhaltsikhe
    Alazani
    Aragvi
    Batumi
    Chiatura
    Colchis
    Dzirula Massif
    Inguri River
    Kakhetia
    Kartlia
    Kodor River basin
    Rioni Basin
    Shiraki Steppe
    Svanetia
    Tbilisi
    Tkibuli
    Trialet Range
SA Caucasus
    European USSR
    Greater Caucasus
    Kuban River
    Kuban Valley
    Kura River
    Lesser Caucasus
    Ossetia
    Terek River
    Transcaucasia

Georgina Basin
River drainage basin. Index Northern Territory and/or Queensland as applicable.
BT Australia
SA Northern Territory
    Queensland

GEOSECS
Acronym.
UF Geochemical Ocean Sections Program
SA geochemistry
    oceanography

geostrophic force
use Coriolis force

geosynclines
Includes use on level 1 (list A). Used for large, mobile downwarping of the crust, which subsides as rocks and sediments accumulate to thicknesses of thousands of meters. Term set options are:

geosynclines • Gilgai

evolution
    (name of geosyncline e.g. Alpine Geosyncline, Andean Geosyncline, Caledonian Geosyncline, Cordilleran Geosyncline, Hercynian Geosyncline, Mesogaea, Nevadan Geosyncline, Tethys) or subtopic
genesis
    (name of geosyncline) or subtopic
processes
    (name of geosyncline) or subtopic
structure
    (name of geosyncline) or subtopic
NT miogeosynclines
SA Adelaide Geosyncline
    Alpine Geosyncline
    Andean Geosyncline
    anticlinoria
    burial metamorphism
    Caledonian Geosyncline
    continental drift
    Cordilleran
    Cordilleran Geosyncline
    crust
    eugeosynclines
    geanticlines
    Hercynian Geosyncline
    igneous activity
    mantle
    metallogeny
    mobile belts
    ocean basins
    ophiolite
    orogeny
    paleogeography
    plate tectonics
    Propria Geosyncline
    sea-floor spreading
    synclinoria
    Tasman Geosyncline
    tectonics
    tectonophysics
    thermal history

geotectonics
use tectonics

geothermal energy
Includes use on level 1 as a commodity term (list C).
SA energy
    energy sources
    fumaroles
    geothermal processes
    geysers
    heat flow
    heat sources
    hot springs
    thermal waters

geothermal fields
Before 1978, search geothermal.
UF fields, geothermal
SA geothermal systems
    oil and gas fields

geothermal gradient
Includes use on level 2 under heat flow(1).
SA conductivity
    crust
    Earth
    heat flow
    mantle
    measurement
    regional patterns
    thermal conductivity
    thermal waters

geothermal phenomena
use geothermal systems

geothermal processes
Term introduced in 1978. Before 1978, search geothermal.
SA Earth
    geothermal energy
    heat sources
    processes

geothermal regimes
use geothermal systems

geothermal systems
Term introduced in 1978. Before 1978, also search geothermal; geothermal regions; geothermal phenomena.
UF geothermal phenomena
    geothermal regimes
SA geothermal fields
    systems

geothermometry
use geologic thermometry

Gera
District in SW East Germany. Also a city.
BT East Germany
NT Schwarzburg Anticlinorium
SA Thuringia
    Thuringian Basin
    Thuringian Forest

German
Includes use on level 3 to indicate language of a document.

germanates
BA oxides
BT minerals

germanite
BA sulfogermanates
    sulfosalts
BT minerals

germanium
Includes use on level 1 and 2 as a chemical element (list D). Also search Ge.
UF Ge
SA elements

Germany
Country in Central Europe. Divided in 1949 into West Germany and East Germany. Includes use on level 1 as an area term (list O). See list B for term set options.
BA Europe
NT Berlin
NA East Germany
NT Hesse
    Mainz Basin
    Odenwald
    Palatinate
NA West Germany
SA Allgau Alps
    Alps
    Baltic region
    Danube River
    Danube Valley
    European Platform
    Hanover
    Hunsruck Shale
    Lake Constance
    Molasse Basin
    Moselle Valley
    North Sea Coast
    Posidonia Shale
    Poznan Clays
    Rhine Graben
    Upper Rhine Valley

Gerona
Province in Catalonia. Also a city.
BT Spain
NT Ampurdan
    Osor
SA Catalonia
    Ter River basin

Gerrei
Region in SE.
BT Sardinia
    Italy

Gers
Department in SW France. Also a river.
CO N431500N440500
    E0011500W0001500
BT France

gersdorffite
UF nickel glance
BA sulfides
BT minerals

Getchell Mine
Humboldt County in NW.
BT Nevada
    United States

Getic Basin
use Getic Nappe

Getic Nappe
NW part of Transylvanian Alps in Hunedoara County in SW Transylvania and E Banat. Also search Getic Basin. Index regions as applicable.
UF Getic Basin
BT Romania
SA Banat
    Transylvania

Geula Cave
use Geula Caves

Geula Caves
Mount Carmel. Also search Geula Cave.
UF Geula Cave
BT Haifa
    Israel

geysers
As of 1978, term is used on level 2 under thermal waters(1).
UF gusher
    pulsating spring
SA fumaroles
    geothermal energy
    ground water
    hot springs
    springs
    thermal waters

Geysers, The
use The Geysers

Gezira
Region between Blue Nile and White Nile rivers in central part of country.
BT Blue Nile
    Sudan

Ghana
Britain's former Gold Coast colony combined with British trust territory of Togo. Includes use on level 1 as an area term (list O). For term set options see list B.
CO N043000N120000
    E0013000W0030000
BA Africa
NT Accra
    Bosumtwi Crater
SA Volta Basin
    West Africa

Gharian
use Garian

Ghats
Two mountain ranges in S India forming E and W edges of Deccan Plateau.
BT India
SA Eastern Ghats
    Western Ghats

Ghazni
Province. Also a city in E central.
BT Afghanistan

Ghost Ranch
Region in N central.
BT New Mexico
    United States

giant fields
UF fields, giant
BT oil and gas fields
SA natural gas
    petroleum

Giant's Causeway
Formation of prismatic basaltic columns extending into sea off N coast of County Antrim.
BT Northern Ireland
    United Kingdom

Gibbs fault zone
use Gibbs fracture zone

Gibbs fracture zone
S of Reykjanes Ridge which is SE of Iceland. Also search Gibbs Fault; Gibbs Fracture.
UF Gibbs fault zone
BT Atlantic Ocean

Gibbs free energy
use free energy

Gibbs phase rule
use phase rule

gibbsite
Also search hydrargillite.
UF hydrargillite
BA oxides
BT minerals
SA bayerite

Gibeon
Impact near village of Gibeon in S central South-West Africa. Also search Gibeon Meteorite.
UF Gibeon Meteorite
BT South-West Africa
    Africa
SA meteorites

Gibeon Meteorite
use Gibeon

Gibraltar
Peninsula and British colony at S tip of Spain.
BT Europe
SA Strait of Gibraltar

Gibraltar Point
Cape on N side of entrance to the Wash on the North Sea.
BT England
    Great Britain
    United Kingdom

Gifu
City and prefecture in central.
BT Honshu
    Japan

Gila County
E central.
BT Arizona
    United States

Gila River
Rises in SW New Mexico and empties into the Colorado River near Yuma, Arizona. Index states as applicable.
BT United States
SA Arizona
    New Mexico

Gilau
Village in Cluj County in central.
BT Transylvania
    Romania

Gilbert Islands
Island group SSE of Marshall Islands and NE of Solomon Islands. Forms main part of British Gilbert and Ellice Islands Colony.
BT Pacific Ocean
SA Micronesia

Giles Complex
Index states as applicable.
BT Australia
SA South Australia
    Western Australia

Gilgai
Settlement in NE.

BT New South Wales
   Australia

gillespite
BA sheet silicates
   silicates
BT minerals

Gilpin County
N central.
CO N394500N395000
   W1052000W1053500
BT Colorado
   United States

Ginkgo
Genus.
BA Ginkgoales
   gymnosperms
BT Plantae

Ginkgoales
Includes use on level 2 under gymnosperms(1). See list F.
BA gymnosperms
BT Plantae
NA Czekanowskiales
   Ginkgo
SA Dadoxylon

Gippsland
Region in SE Victoria. Includes part of the Gippsland Basin.
UF Gippsland region
BT Victoria
   Australia
SA Gippsland Basin

Gippsland Basin
Similar to area covered by Gippsland, however the Basin also includes an area offshore to S of Victoria.
CO S390000S380000
   E1490000E1470000
BT Victoria
   Australia
SA Gippsland

Gippsland region
use Gippsland

Girnar Hills
S central Kathiawar Peninsula in W Gujarat. Also search Girnar.
BT Gujarat
   India

Gironde
Department on Bay of Biscay.
CO N441500N453000
   E0003000W0013000
BT France
NT Arcachon Basin
   Bordeaux
   Bordelais
   Medoc
SA Gironde Estuary

Gironde Estuary
On Bay of Biscay. Formed by confluence of Garonne and Dordogne rivers.
BT France
SA Gironde

Girvan
Town on Firth of Clyde in Ayrshire in SW.
BT Scotland
   Great Britain
   United Kingdom

girvanella
BA biogenic structures
   sedimentary structures
SA algal biscuits

Gisborne
Seaport city in E.
BT North Island
   New Zealand

Gissar Range
use Hissar Range

Givet
Town in NE part of country.
BT Ardennes
   France

Givetian
Europe. Above Eifelian, below Frasnian. Includes use on level 3 under age terms(1). See list E.
BA Middle Devonian
   Devonian
BT Paleozoic
NT Horn Plateau Formation

glacial
A valid term through 1977. After 1977, use glacial environment.

glacial environment
Term introduced in 1978. Before 1978, search glacial.
SA ecology
   environment
   interglacial environment

glacial features
Includes use on level 2 under glacial geology(1).
UF features, glacial
NT cirques
   drumlins
   eskers
   kames
   outwash plains
SA extinct lakes
   firn
   fjords
   glacial geology
   glacial lakes
   glaciation
   glaciers
   ice
   ice caps
   icebergs
   lakes
   moraines
   outwash
   periglacial features
   striations
   till
   tillite
   varves

glacial geology
Includes use on level 1 (list A) for geologic features and effects resulting from the action of glaciers and ice sheets. Term set options are: ancient ice ages (older than Quaternary)
   age
glacial features
   name of feature [e.g. drumlins, eskers, fjords, fluting, glacial lakes, glacial polish, hanging valleys, kames, moraines, nunataks, sole marks, striations]
glaciation
   topic [e.g. deglaciation, deposition, erosion, extent, ice movement]
glaciers (for present-day glaciers)
   subtopic
periglacial features
   name of feature [e.g. extinct lakes, fans, ice wedges, nonglacial ice, patterned ground, permafrost, pingos, solifluction]
topic [applications, bibliography, catalogs, classification, concepts, experimental studies, history, instruments, methods, nomenclature, philosophy, practice, symposia, textbooks, theoretical studies]
   subtopic
SA ancient ice ages
   atmosphere
   boulders
   buried channels
   buried valleys
   changes of level
   cirques
   cryopedology
   deposition
   drumlins
   erosion
   erratics
   eskers
   extinct lakes
   fans
   firn
   fjords
   frost action
   geomorphology
   glacial features
   glacial lakes
   glacial transport
   glaciation
   glacier surges
   glaciers
   glaciofluvial environment
   hydrology
   ice movement
   ice sheets
   ice shelves
   ice wedges
   isostasy
   kames
   mass balance
   meltwater
   moraines
   non-glacial ice
   outwash
   paleoclimatology
   patterned ground
   periglacial environment
   periglacial features
   permafrost
   pingos
   sedimentary structures
   sedimentation
   sediments
   snow
   soils
   solifluction
   stratigraphy
   thawing
   till

glacial lakes
Includes use on level 3 under glacial geology(1) glacial features(2) and under geomorphology(1) lacustrine features(2).
SA cirques
   extinct lakes
   geomorphology
   glacial features
   glacial geology
   lacustrine features
   lakes
   limnology
   paleolimnology
   varves

glacial outwash
use outwash

glacial transport
includes use on level 3 under sedimentation(1) transport(2).
SA drift
   erratics
   glacial geology
   sedimentation
   till
   transport

glaciation
Includes use on level 2 under glacial geology(1); on level 3 under changes of level(1) and paleoclimatology(1).
SA abrasion
   changes of level
   deglaciation
   deposition
   erosion
   fjords
   glacial features
   glacial geology
   glaciers
   ice movement
   ice sheets
   ice shelves
   paleoclimatology

Glacier Bay
Narrow inlet of Pacific Ocean in N part of SE Alaska. In center of Glacier Bay National Monument.
BT Alaska
   United States

Glacier National Park
NW Montana.
BT Montana
   United States

glacier surges
Term introduced in 1978.
UF catastrophic advance
SA glacial geology
   glaciers
   surges

glaciers
Includes use on level 2 under glacial geology(1) for present day glaciers; on level 2 under hydrology(1); on level 3 under glacial geology(1) glacial features(2).
SA ablation
   accumulation
   deglaciation
   erratics
   firn
   glacial features
   glacial geology
   glaciation
   glacier surges
   glaciofluvial environment
   hydrology
   hydrosphere
   ice
   ice caps
   ice movement
   ice sheets
   ice shelves
   icebergs
   kames
   mass balance
   meltwater
   moraines
   nivation
   periglacial features
   postglacial environment
   rock glaciers
   snow

glaciofluvial environment
Term introduced in 1978. Before 1978 search glaciofluvial; fluvioglacial; fluvioglacial environment.
UF fluvioglacial environment
SA environment
   glacial geology
   glaciers

glacis
SA geomorphology
   slopes

Glamorgan
County on Bristol Channel in S Wales.
CO N513000N514500
   W0030500W0041500
UF Glamorganshire
BT Wales
   Great Britain
   United Kingdom

Glamorganshire
use Glamorgan

Glarus
Canton in E central.
BT Switzerland

Glarus Alps
N division of Central Alps. Chiefly in Glarus Canton. Also search Glarus.

Glarus Alps • Glossopteris

BT Switzerland
SA Alps
    Central Alps

Glasgow
City on Clyde River in S central.
BT Scotland
    Great Britain
    United Kingdom

glass
use glasses

glass, hydration of
use hydration of glass

Glass Mountains
Brewster County in SW.
BT Texas
    United States

glass, volcanic
use volcanic glass

glasses
Includes use on level 3, e.g. under tektites(1). Also search glass.
UF glass
SA Darwin glass
    devitrification
    pyroclastics
    pyroclastics and glasses
    tektites
    volcanic glass

glassy feldspar
use sanidine

glauberite
BA sulfates
BT minerals

glauconite
Includes use on level 1 as a commodity term (list C); on level 3 under minerals(1) sheet silicates, mica group(2). See list L (minerals).
BA sheet silicates
    silicates
BT minerals
SA celadonite
    clay mineralogy
    glauconitic composition
    glauconitization
    greensand
    mica group

glauconitic composition
Term introduced in 1978. Includes use on level 3 under sediments(1) and sedimentary rocks(1). Before 1978, also search glauconitic AND specific sediment type or sedimentary rock.
SA composition
    glauconite
    greensand
    sedimentary rocks
    sediments

glauconitic sand
A valid term through 1974. After 1974, use greensand.

glauconitic sandstone
use greensand

glauconitization
BT metamorphism
SA glauconite

glaucophane
Also search gastaldite.
UF gastaldite
BA amphibole group
    chain silicates
    silicates
BT minerals
SA aluminosilicates
    crossite

glaucophane schist
Includes use on level 3 under metamorphic rocks(1) schists(2). See list J.
BA schists

BT metamorphic rocks
SA blueschist
    schist

Glavatsi
Region in NW.
BT Bulgaria
    Europe

Gleicheniaceae
Family.
BA Filicopsida
    pteridophytes
BT Plantae
SA ferns

Glen Coe
Glen in Argyll County in W Scotland.
UF Glencoe
BT Scotland
    Great Britain
    United Kingdom

Glen Rose Formation
In Trinity Group. Includes Lower Glen Rose Formation, Glen Rose Anhydrite, upper Glen Rose Formation. Also search Glen Rose Limestone.
BT Lower Cretaceous
    Cretaceous
SA Texas
    Trinity Group

Glenarm Series
Includes Wakefield Marble, Silver Run Limestone, Ijamsville Phyllite, Urbana Phyllite, Marburg Schist, Setters Quartzite, Cockeysville Marble, Wissahickon Schist, Peters Creek Schist, Cardiff Conglomerate, Peach Bottom Slate. Lower Paleozoic (?).
BT lower Paleozoic
    Paleozoic
SA Maryland
    New Jersey
    Pennsylvania
    Virginia
    Wissahickon Formation

Glencoe
use Glen Coe

Glennie
Village in Alcona County in NE Lower Peninsula.
BT Michigan
    United States

Gley soils
use Gleys

Gleys
See list M. Also search Gley soils.
UF Gley soils
BA soils
SA Pseudogleys

gliding (tectonics)
use gravity sliding

Glinsk
Village in SW Suny Oblast in N.
BT Ukraine
    USSR

Gliridae
Family. Includes use on level 3 under Mammalia(1) Rodentia(2).
BA Rodentia
    Mammalia
BT Tetrapoda
    Vertebrata

global
General term used in a geographic sense. Also search world.
UF world
SA world ocean

Globigerina
Genus. Includes use on level 3 under foraminifera(1) Globigerinacea(2) or Rotaliina(2).
BA Globigerinidae
    Globigerinacea

    Rotaliina
    foraminifera
BT Invertebrata
NA Globigerina bulloides
    Globigerina pachyderma

Globigerina bulloides
Includes use on level 3 under foraminifera(1) Globigerinacea(2).
BA Globigerina
    Globigerinidae
    Globigerinacea
    Rotaliina
    foraminifera
BT Invertebrata

Globigerina pachyderma
Includes use on level 3 under foraminifera(1) Globigerinacea(2).
BA Globigerina
    Globigerinidae
    Globigerinacea
    Rotaliina
    foraminifera
BT Invertebrata

Globigerinacea
Includes use on level 2 under foraminifera(1). See list F.
BA Rotaliina
    foraminifera
BT Invertebrata
NA Globigerinidae
    Globorotaliidae
    Globotruncanidae
    Hantkenina
    Hedbergella
    Heterohelicidae
    Rotalipora

Globigerinidae
Family. Includes use on level 3 under foraminifera(1) Globigerinacea(2).
BA Globigerinacea
    Rotaliina
    foraminifera
BT Invertebrata
NA Globigerina
    Globigerinoides
    Orbulina
    Sphaeroidinella dehiscens

Globigerinoides
Genus. Includes use on level 3 under foraminifera(1) Globigerinacea(2).
BA Globigerinidae
    Globigerinacea
    Rotaliina
    foraminifera
BT Invertebrata
NA Globigerinoides ruber
    Globigerinoides trilobus

Globigerinoides ruber
Includes use on level 3 under foraminifera(1) Globigerinacea(2).
BA Globigerinoides
    Globigerinidae
    Globigerinacea
    Rotaliina
    foraminifera
BT Invertebrata

Globigerinoides trilobus
Includes use on level 3 under foraminifera(1) Globigerinacea(2).
BA Globigerinoides
    Globigerinidae
    Globigerinacea
    Rotaliina
    foraminifera
BT Invertebrata

Globorotalia
Genus. Includes use on level 3 under foraminifera(1) Globigerinacea(2).
BA Globorotaliidae
    Globigerinacea
    Rotaliina
    foraminifera

BT Invertebrata
NA Globorotalia menardii
    Globorotalia pachyderma
    Globorotalia truncatulinoides

Globorotalia menardii
Includes use on level 3 under foraminifera(1) Globigerinacea(2).
BA Globorotalia
    Globorotaliidae
    Globigerinacea
    Rotaliina
    foraminifera
BT Invertebrata

Globorotalia pachyderma
Includes use on level 3 under foraminifera(1) Globigerinacea(2).
BA Globorotalia
    Globorotaliidae
    Globigerinacea
    Rotaliina
    foraminifera
BT Invertebrata

Globorotalia truncatulinoides
Includes use on level 3 under foraminifera(1) Globigerinacea(2).
BA Globorotalia
    Globorotaliidae
    Globigerinacea
    Rotaliina
    foraminifera
BT Invertebrata

Globorotaliidae
Family. Includes use on level 3 under foraminifera(1) Globigerinacea(2).
BA Globigerinacea
    Rotaliina
    foraminifera
BT Invertebrata
NA Globorotalia

Globotruncana
Genus. Includes use on level 3 under foraminifera(1) Globigerinacea(2).
BA Globotruncanidae
    Globigerinacea
    Rotaliina
    foraminifera
BT Invertebrata

Globotruncanidae
Family. Includes use on level 3 under foraminifera(1) Globigerinacea(2).
BA Globigerinacea
    Rotaliina
    foraminifera
BT Invertebrata
NA Globotruncana

glossaries
Includes use on level 1 (list A) for glossaries and atlases. Term set options are:
subject (list B)
    subtopic (include language if not English)
SA atlas
    bibliography
    catalogs
    dictionaries
    education
    geology
    lexicons
    paleontology
    sedimentary petrology
    stratigraphy
    structural geology
    textbooks

Glossopteris
Genus. Includes use on level 2 under gymnosperms(1). See list F.
BA gymnosperms
BT Plantae
NA Gangamopteris
    Glossopteris flora

SA Dadoxylon
    Pecopteris
    Taeniopteris

Glossopteris flora
Includes use on level 3 under gymnosperms(1) Glossopteris(2).
BA Glossopteris
    gymnosperms
BT Plantae

Gloucestershire
County in W England. Also search Gloucester.
CO N513000N521000
    W0013000W0024000
BT England
    Great Britain
    United Kingdom

glowing avalanche
use ash flows

Glycymeris
Includes use on level 3 under Mollusca(1) Bivalvia(2).
BA Bivalvia
    Mollusca
BT Invertebrata

Glynn County
On the Atlantic Ocean in SE.
BT Georgia
    United States

gmelinite
BA zeolite group
    framework silicates
    silicates
BT minerals

gneiss
Includes use on level 3 under metamorphic rocks(1) gneisses(2). See list J. Also search hornblende gneiss.
BA gneisses
BT metamorphic rocks
SA augen gneiss
    banded gneiss
    biotite gneiss
    granite gneiss
    paragneiss
    sillimanite gneiss
    tonalite gneiss

gneisses
Includes use on level 2 under metamorphic rocks(1). See list J.
BT metamorphic rocks
NA augen gneiss
    banded gneiss
    biotite gneiss
    gneiss
    granite gneiss
    sillimanite gneiss
    tonalite gneiss

gneissic texture
Term introduced in 1978. Before 1978, search gneissic.
SA metamorphic rocks
    textures

Goa
Former Portuguese possession on the Arabian Sea annexed by India in 1962. Territory of Goa, Daman, and Diu.
BT India

Gobi Desert
Index Mongolia and/or Inner Mongolia as applicable. Also search Gobi.
CO N420000N460000
    E1170000E1020000
BT Asia
SA Inner Mongolia
    Mongolia

Goczalkowice
Borehole region in Upper Silesian coal basin in S part of country.
BT Katowice

    Poland

Godavari River
Rises in NW Maharashtra and flows SE into the Bay of Bengal. Index states as applicable. Also search Godavari.
BT India
SA Andhra Pradesh
    Maharashtra

Godavari Valley
River valley. Index states as applicable. Also search Godavari.
BT India
SA Andhra Pradesh
    Maharashtra
    Pranhita-Godavari Valley

Godthaab
Town on SW coast.
UF Godthab
BT Greenland
    Arctic region

Godthab
use Godthaab

goethite
UF gothite
    xanthosiderite
BA oxides
BT minerals

Goias
State in central Brazil. Also a town.
BT Brazil
NT Maranhao Basin
SA Brazilian Shield
    Estrada Nova Formation
    Irati Formation
    Tocantins River region

Golan Heights
Hilly region in SW.
BT Syria
    Middle East

gold
Includes use on level 1 and 2 as a commodity term (list C) and as a chemical element (list D); on level 2 under placers(1).
UF Au
SA calaverite
    elements
    heavy metals
    mineral deposits
    native elements and alloys
    placers

Golden
City in Jefferson County in N central.
BT Colorado
    United States

Golden Horn Batholith
Peak in San Juan and San Miguel counties in SW.
BT Colorado
    United States

Golden Valley Formation
Includes White Earth, South Ross, Lakeside, and East Tioga Clay Beds. SW North Dakota.
BT Eocene
SA North Dakota

golden-brown algae
use Chrysophyta

Goldfield
Village in Esmeralda County in SW.
BT Nevada
    United States

Golodnaya Step
use Bet-Pak-Dala

Golovanevsk
Village in Voronezh Oblast in S central European USSR.
BT Russian Republic
    USSR

Gomera
Island belonging to Spain.
BT Canary Islands
    Atlantic Ocean

gondite
BA metamorphic rocks

Gondwana
Theoretical ancient continent including India, Australia, Antarctica; and parts of southern Africa and South America. It is supposed to have fragmented and drifted apart in Post-Carboniferous time. Index countries and continents as applicable. Includes use on level 2 under continental drift (1). Before 1974, also search Gondwanaland. Also search lower Gondwana.
SA Africa
    Antarctica
    Australia
    continental drift
    India
    Laurasia
    Pangaea
    South America
    Southern Hemisphere

Gondwana System
Upper Carboniferous to Lower Cretaceous. Peninsular.
SA Carboniferous
    Cretaceous
    India
    Jurassic
    Lower Cretaceous
    Permian
    Triassic
    Upper Carboniferous

Gondwanaland
A valid term through 1973. After 1973, use Gondwana.

Goniatites
Genus. Includes use on level 3 under Mollusca(1) Cephalopoda(2).
BA Goniatitidae
    Ammonoidea
    Cephalopoda
    Mollusca
BT Invertebrata

Goniatitida
Order. Includes use on level 3 under Mollusca(1) Cephalopoda(2).
BA Ammonoidea
    Cephalopoda
    Mollusca
BT Invertebrata
NA Goniatitidae

Goniatitidae
Family. Includes use on level 3 under Mollusca(1) Cephalopoda(2).
BA Goniatitida
    Ammonoidea
    Cephalopoda
    Mollusca
BT Invertebrata
NA Goniatites

Good Hope
use Fort Good Hope

Gorda Ridge
use Gorda Rise

Gorda Ridges
use Gorda Rise

Gorda Rise
Off N California E of the Mendocino Escarpment. Also search Gorda Ridge and Gorda Ridges.
UF Gorda Ridge
    Gorda Ridges
BT Pacific Ocean

Gore Range
Part of Park Range in N central.
BT Colorado
    United States

Gorki
City and oblast. Also search Gorky.
UF Gorkiy
    Gorky
    Nizhni Novgorod
BT Russian Republic
    USSR

Gorkiy
use Gorki

Gorky
use Gorki

Gorlovka Basin
Industrial basin in the Donbas in E.
BT Ukraine
    USSR

Gornaya Shoriya
Mountainous region in Kemerovo Oblast just S of Kuznetsk Basin.
BT Russian Republic
    USSR

Gorny Altai
Mountain range in Altai Mountains of Gorno-Altai Autonomous Oblast.
CO N484500N523000
    E0900000E0840000
BT Russian Republic
    USSR
SA Altai Mountains

Goryachiy Plyazh
Beach on Kunashir Island of the Kuril Islands NE of Japan.
BT Russian Republic
    USSR
SA Kunashir Island

Gosau
Village in NW part of country.
BT Upper Austria
    Austria

Gosford
Town N of Sydney on the Tasman Sea.
BT New South Wales
    Australia

Goshen County
On the Nebraska border in SE.
BT Wyoming
    United States

Gosses Bluff
Meteor crater. Also search Gosses Bluff Meteor Crater.
UF Gosses Bluff Meteor Crater
BT Northern Territory
    Australia

Gosses Bluff Meteor Crater
use Gosses Bluff

Gota Valley
River valley between Lake Vanern and Kattegat in SW.
BT Sweden

Goteborg
City on the Kattegat in SW Sweden.
UF Gothenburg
BT Sweden

Gothard
use Gotthard Massif

Gothenburg
use Goteborg

gothite
use goethite

Gothland
use Gotland

Gothlandian
use Silurian

Gotland
Island and county in Baltic Sea off SE coast.
CO N570000N580000
    W0193000W0180000
UF Gothland
    Gottland

Gotland • granite-granodiorite family

BT Sweden
Gotlandian
 use Silurian
Goto Islands
 Island chain extending about 100 miles SW from NW Kyushu.
 BT Kyushu
   Japan
Gotthard Massif
 Mountain range of the Lepontine Alps in SE central Switzerland. Also search Gotthard; Saint Gotthard; St. Gotthard.
 UF Gothard
   Saint Gotthard
   St. Gotthard
 BT Switzerland
 SA Lepontine Alps
Gottingen
 City in SE.
 BT Lower Saxony
   West Germany
Gottland
 use Gotland
gouge
 Includes use on level 3 under faults(1) effects(2).
 SA breccia
   faults
Gour Oumelalen
 Region in NE Ahaggar in SE.
 BT Algeria
Gouverneur
 Village in Saint Lawrence County in N.
 BT New York
   United States
Gove County
 NW central.
 BT Kansas
   United States
Gower Peninsula
 Extends S into Bristol Channel from S.
 BT Wales
   Great Britain
   United Kingdom
Gowganda
 Village N of Georgian Bay.
 BT Ontario
   Canada
Gowganda Formation
 Consists of paraconglomerate, argillite, siltstone, subarkose, and greywacke. W Ontario.
 BT Huronian
   Proterozoic
   Precambrian
 SA Canada
   Ontario
grabens
 Includes use on level 3 under faults(1) systems(2).
 SA aulacogens
   faults
   horsts
   rift valleys
   systems
grade
 Includes use on level 2 under metamorphism(1).
 UF metamorphic grade
 SA facies
   high-grade metamorphism
   isograds
   low-grade metamorphism
   metamorphism
   P-T conditions
graded bedding
 Includes use on level 3 under sedimentary structures(1) turbidity current structures(2) and planar bedding structures(2). See list K.
 BA turbidity current structures
   sedimentary structures
 SA bedding
   planar bedding structures
   varves
gradient
 A valid term through 1978. After 1978 see geothermal gradient, or stream gradient for streams.
graduate-level education
 Term introduced in 1978. Before 1978, also search graduate level.
 SA education
Grafton County
 W New Hampshire.
 BT New Hampshire
   United States
graftonite
 BA phosphates
 BT minerals
Graham Coast
 use Graham Land
Graham County
 Index states as applicable.
 BX Arizona
   Kansas
   North Carolina
 BT United States
Graham Land
 Part of N Antarctic Peninsula S of Cape Horn.
 UF Graham Coast
 BT Antarctica
grahamite
 use mesosiderites
grain size
 Includes use on level 3, e.g. under sediments(1).
 SA dimensions
   fines
   granulometry
   sediments
   size
grains
 Includes use on level 3 under sedimentary rocks(1).
 SA dimensions
   granulometry
   particles
   recrystallization
   sand
   sedimentary rocks
   sediments
   shape analysis
   sorting
grainstone
 BA carbonate rocks
 BT sedimentary rocks
Graisivaudan
 use Gresivaudan
gramenite
 use nontronite
Gramineae
 Family.
 BA Monocotyledoneae
   angiosperms
 BT Plantae
Gran Canaria
 use Grand Canary
Gran Chaco
 use Chaco
Granada
 Province and ancient kingdom in S Spain. Also a city.
 BT Spain
 NT Granada Depression
 SA Andalusia
Granada Depression
 S part of country.
 BT Granada
   Spain
Grand Atlas Mountains
 use High Atlas
Grand Banks
 Shoal or banks E and S of Newfoundland.
 BT Atlantic Ocean
Grand Canary
 One of Spain's Canary Islands NW of Western Sahara. Also search Gran Canaria.
 UF Gran Canaria
 BT Canary Islands
   Atlantic Ocean
Grand Canyon
 Gorge in Colorado River in NW Arizona. Also search Grand Canyon region.
 UF Grand Canyon region
 BT Arizona
   United States
Grand Canyon region
 use Grand Canyon
Grand Cayman Island
 Largest of the Cayman Islands. About 200 miles NW of Jamaica. Also search Grand Cayman.
 BT Caribbean Sea
Grand County
 Index states as applicable.
 BX Colorado
   Utah
 BT United States
Grand Falls
 use Churchill Falls
Grand Forks
 City on the Red River of the North in Grand Forks County in E.
 BT North Dakota
   United States
Grand Isle
 Island in Jefferson Parish off SE coast.
 BT Louisiana
   United States
Grand Junction
 City in Mesa County in W.
 BT Colorado
   United States
Grand Manitoulin
 use Manitoulin Island
Grand Rapids
 City in Kent County in W Lower Peninsula.
 BT Michigan
   United States
Grand River
 River in numerous locations. Index Ontario and states as applicable.
 SA Louisiana
   Michigan
   Missouri
   Oklahoma
   Ontario
   South Dakota
Grand Traverse County
 NW Lower Peninsula.
 BT Michigan
   United States
Grandfather Mountain
 In the Blue Ridge Mountains in NW.
 BT North Carolina
   United States
 SA Blue Ridge Mountains
grandidierite
 BA orthosilicates
   silicates
 BT minerals
Graneros Shale
 In Colorado Group. E Colorado, Kansas, SE Montana, Nebraska, NE New Mexico, South Dakota, and E Wyoming.
 BT Cretaceous
 SA Benton Formation
   Colorado
   Colorado Group
   Kansas
   Lower Cretaceous
   Montana
   Nebraska
   New Mexico
   South Dakota
   Upper Cretaceous
   Wyoming
granite
 Includes use on level 1 as a commodity term (list C); on level 3 under igneous rocks(1) granite-granodiorite family(2). See list H. Also search granitic rocks.
 BA granite-granodiorite family
 BT igneous rocks
 SA albite
   alkalic granite
   apogranite
   biotite granite
   building stone
   construction materials
   dimension stone
   granitic composition
   granitization
   leucogranite
   metagranite
   microgranite
   muscovite granite
Granite County
 W Montana.
 BT Montana
   United States
granite gneiss
 Includes use on level 3 under metamorphic rocks(1) gneisses(2). See list J. Also search granitic gneiss.
 UF granitic gneiss
 BA gneisses
 BT metamorphic rocks
 SA gneiss
Granite Mountains
 Central.
 BT Wyoming
   United States
granite porphyry
 Includes use on level 3 under igneous rocks(1) granite-granodiorite family(2). See list H.
 BA granite-granodiorite family
 BT igneous rocks
 SA porphyry
granite-granodiorite family
 Includes use on level 2 under igneous rocks(1). See list H.
 UF granodiorite family
 BT igneous rocks
 NA adamellite
   alaskite
   alkalic granite
   aplite
   apogranite
   beresite
   biotite granite
   charnockite
   enderbite
   farsundite
   felsite
   granite
   granite porphyry
   granodiorite
   granodiorite porphyry
 NZ granosyenite
 NA graphic granite
   leucogranite
   microgranite
   micropegmatite

muscovite granite
pegmatite
quartz monzonite
rapakivi
silexite
SA orbicular texture

granitic
A valid term through 1977. After 1977, use granitic composition or granitic layer.

granitic composition
Term introduced in 1978. Before 1978, also search granitic; granitic rocks.
SA composition
granite
granitic layer

granitic gneiss
use granite gneiss

granitic layer
Term introduced in 1978. Before 1978, also search granitic; sial.
UF layer, granitic
sial
SA basaltic layer
granitic composition
granitic layer
lower crust

granitic rocks
Not a valid term. Use granite or granitic composition.

granitification
use granitization

granitization
Includes use on level 3 under metasomatism(1) processes(2) and under metamorphism(1).
UF granitification
BA metamorphism
SA crystallization
feldspathization
granite
metasomatism
processes

granitoid
A valid term through 1978. After 1978, use granite or granitic composition.

granodiorite
Includes use on level 3 under igneous rocks(1) granite-granodiorite family(2). See list H.
BA granite-granodiorite family
BT igneous rocks
SA plutonic rocks

granodiorite family
use granite-granodiorite family

granodiorite porphyry
See list H.
BA granite-granodiorite family
BT igneous rocks
SA porphyry

granophyre
BA volcanic rocks
BT igneous rocks

granosyenite
BZ granite-granodiorite family
syenite family
BT igneous rocks

Grant County
Index states as applicable.
BX Arkansas
Indiana
Kansas
Kentucky
Minnesota
Nebraska
New Mexico
North Dakota
Oklahoma
Oregon
South Dakota
Washington
West Virginia
Wisconsin
BT United States

Grant Range
NE Nye County in central.
BT Nevada
United States

Grants
Town in Valencia County in W.
BT New Mexico
United States

granular materials
Includes use on level 3 under soil mechanics(1). Also search granular.
SA engineering geology
granulometry
materials
soil mechanics

granulite
Includes use on level 3 under metamorphic rocks(1) granulites(2). See list J.
UF Erianfels
BA granulites
BT metamorphic rocks
SA leptite
pyroxene granulite

granulite facies
Includes use on level 3 under metamorphic rocks(1). See list J.
BT facies
SA metamorphic rocks

granulites
Includes use on level 2 under metamorphic rocks(1). See list J.
BT metamorphic rocks
NA eclogite
granulite
kinzigite
leptite
pyroxene granulite

granulometry
SA grain size
grains
granular materials
particles
shape analysis
size
size distribution

grapestone
Used to indicate a texture.
BA sedimentary rocks

graphic display
Includes use on level 3 under automatic data processing(1). Before 1978, also search graphic AND display.
UF display, graphic
SA automatic data processing

graphic granite
Includes use on level 3 under igneous rocks(1) granite-granodiorite family(2). See list H.
BA granite-granodiorite family
BT igneous rocks
SA micropegmatite
pegmatite

graphic methods
Before 1978, also search graphic.
SA histograms
methods

graphic texture
Term introduced in 1978. Before 1978, search graphic.
SA igneous rocks
textures

graphite
Includes use on level 1 as a commodity term (list C); on level 3 as a mineral.
UF black lead
BA native elements and alloys
BT minerals
SA diamond
diamonds
refractory materials

Graptolithina
Includes use on level 1 and 2 as a fossil term (list F).
BT Invertebrata
NA Dendroidea
Graptoloidea
Tuboidea
SA Hemichordata

Graptoloidea
Includes use on level 2 under Graptolithina(1). See list F.
BA Graptolithina
BT Invertebrata
NA Didymograptus
Diplograptina
Isograptus
Monograptina

grasslands
SA ecology
vegetation
wetlands

gratonite
BA sulfarsenites
sulfosalts
BT minerals

Gratz
use Graz

Graubunden
Canton in E Switzerland. Also search Grisons.
UF Grisons
BT Switzerland
NT Davos
Oberhalbstein

grauwacke
use graywacke

gravel
Includes sand when sand is used as a construction material. Includes use on level 1 as a commodity term (list C); on level 3 under sediments(1) clastic sediments(2) and carbonate sediments(2). See list N.
BA clastic sediments
BT sediments
SA aggregate
alluvium
boulders
breccia
carbonate sediments
cobbles
construction materials
pebbles
sand
shingle
terrigenous materials

gravimeters
Includes use on level 3 under geophysical methods(1). Also search gravity meters.
UF gravity meters
SA geophysical methods
gravity methods
instruments

gravitational gliding
use gravity sliding

gravitational sliding
use gravity sliding

gravity
A valid term through mid-1978. Use specific term, e.g. gravity field or gravity platforms. For gravity faults, use the term normal faults.

gravity anomalies
Includes use on level 3 under geophysical surveys(1) and geophysical methods(1). Also search gravity surveys AND anomalies; gravity methods AND anomalies.
NT Bouguer anomalies
SA anomalies
geophysical methods
geophysical surveys
gravity methods
gravity surveys
magnetic anomalies

gravity faults
use normal faults

gravity field
Includes use on level 2 under Earth(1) and Moon (1); on level 3 under geophysical surveys (1).
UF field, gravity
SA Earth
geodesy
geophysical surveys
Moon

gravity gliding
use gravity sliding

gravity meters
use gravimeters

gravity methods
Includes use on level 2 under geophysical methods(1).
BA geophysical methods
SA free-air anomalies
gravimeters
gravity anomalies
gravity survey maps
gravity surveys
methods

gravity platforms
Term introduced in 1978 used in engineering geology(1). Before 1978, search gravity AND platforms.
BT marine installations
SA marine platforms
platforms

gravity sliding
Includes use on level 2 under tectonics(1). Also search sliding.
UF ecoulement
gliding (tectonics)
gravitational gliding
gravitational sliding
gravity gliding
sliding, gravity
SA crustal shortening
tectonics

gravity survey maps
Term introduced in 1978. Also search maps AND gravity surveys.
BA maps
SA geophysical surveys
gravity methods
gravity surveys

gravity surveys
Includes use on level 2 under geophysical surveys(1).
BA geophysical surveys
BT surveys
SA gravity anomalies
gravity methods
gravity survey maps

gray antimony
use jamesonite

Gray desert soil
use Sierozems

Gray earth
use Sierozems

Gray forest soils
Also search Grey forest soils.
UF Grey forest soils
BA soils

gray hematite
use specularite

Grays Harbor
Inlet of Pacific Ocean on SW coast of Grays Harbor County.
 BT Washington
  United States

graywacke
Includes use on level 3 under sedimentary rocks(1) clastic rocks(2). See list I. Also search greywacke.
 UF grauwacke
  greywacke
 BA clastic rocks
 BT sedimentary rocks
 SA metagraywacke
  sandstone
  subgraywacke
  terrigenous materials

Graz
City in Styrian Alps in SE part of country.
 UF Gratz
 BT Styria
  Austria

Great Alfold
 use Alfold

Great Appalachian Valley
Term introduced in 1977. Before 1977, also search Great Valley. A longitudinal chain of lowlands of the Appalachians extending from Canada on NE to Alabama on SW.
 BT North America
 SA Appalachians

Great Artesian Basin
Index Northern Territory and states as applicable.
 CO S200000S180000
  E1490000E1370000
 BT Australia
 SA New South Wales
  Northern Territory
  Queensland
  South Australia

Great Australian Bight
Bay on S coast of Australia.
 BT Indian Ocean

Great Bahama Bank
Large shoal in the Bahamas SE of Miami and extending between Cuba and Andros Island.
 BT Atlantic Ocean

Great Barrier Reef
Largest coral reef in the world extending 1250 miles off NE Queensland, Australia.
 CO S230000S100000
  E1530000E1370000
 BT Coral Sea
  Pacific Ocean

Great Basin
Interior region between Sierra Nevada Mountains and S Cascade Range on W, and Wasatch Range and W face of Colorado Plateau on E. Introduced as a level 1 area term in 1978. Index states as applicable.
 CO N350000N430000
  W1110000W1200000
 BA United States
 SA California
  Idaho
  Nevada
  Oregon
  Utah

Great Bay
Inland tidal bay in SE.
 BT New Hampshire
  United States

Great Bear Lake
NW central District of Mackenzie.
 BT Northwest Territories
  Canada

Great Britain
Index political divisions as applicable. Includes use on level 1 as an area term (list O). For term set options see list B.
 CO N500000N580000
  E0013000W0080000
 BT United Kingdom
 BA Europe
 NT England
  Scotland
  Wales
 SA Caledonides
  Gault
  Upper Old Red Sandstone

Great Dyke
Part of great South African plateau.
 BT Rhodesia
  Africa

Great Glen Fault
Depression extending along ancient fault NE-SW across Scotland from Moray Firth to Loch Linnhe. Part of Caledonian Canal.
 BT Scotland
  Great Britain
  United Kingdom

Great Hungarian Plain
 use Alfold

Great Kabylia
W part of Kabylia which is a mountainous coastal region.
 BT Algeria
 SA Kabylia

Great Khingan Range
 use Khingan Range

Great Lake
Largest lake on the island in central.
 BT Tasmania
  Australia

Great Lakes
Largest group of fresh-water lakes in the world. Index lakes as applicable. Includes use on level 1 as an area term (list O). For term set options see list B.
 CO N414000N490000
  W0760000W0922000
 BA North America
 NA Lake Erie
  Lake Huron
  Lake Michigan
  Lake Ontario
  Lake Superior
 SA Canada
  Great Lakes region
  United States

Great Lakes region
Index Great Lakes, states, and provinces as applicable. Also search Great Lakes. Includes use on level 1 as an area term (list O). For term set options see list B.
 BA North America
 SA Canada
  Great Lakes
  Illinois
  Indiana
  Michigan
  Minnesota
  New York
  Ohio
  Ontario
  Pennsylvania
  United States
  Wisconsin

Great Meteor Seamount
About 800 miles S of Azores.
 BT Atlantic Ocean

Great Oolite Series
Consists of upper Estuarine Series, Great Oolite Limestone, Great Oolite Clays and Cornbrash. Oxfordshire and Gloucestershire in W central England.
 BT Bathonian
  Middle Jurassic
  Jurassic
 SA England

Great Plains
A level 1 area term as of 1978. Sloping plateau extending from Rocky Mountains in W to the margin of the Central Plains in the U. S. and to the margin of the Laurentian Highlands in Canada. Index states and provinces as applicable. Until 1978, High Plains was used for the Great Plains from Nebraska southward. Now use Great Plains, but also search High Plains.
 BA North America
 NA Northern Great Plains
  Southern Great Plains
 SA Alberta
  Canada
  Colorado
  Kansas
  Montana
  Nebraska
  New Mexico
  North Dakota
  Oklahoma
  Saskatchewan
  South Dakota
  Texas
  United States
  Wyoming

Great Rift Valley
A great fault system extending from the Sea of Galilee in SW Asia to Mozambique. Index East African Rift, and SW Asian countries, lakes, gulfs and seas. Also search Rift Valley.
 UF Rift Valley
 SA Afar Depression
  Dead Sea
  East African Rift
  Gulf of Aden
  Gulf of Aqaba
  Israel
  Jordan
  Jordan Valley
  Kenya Rift valley
  Red Sea
  Sea of Galilee
  Syria
  Wadi Araba

Great Salt Lake
In NW near Salt Lake City.
 CO N404000N414000
  W1121000W1125500
 BT Utah
  United States

Great Scar Limestone
Overlain by the Yoredale Series. Yorkshire in N England.
 BT Visean
  Lower Carboniferous
  Carboniferous
 SA England

Great Slave Lake
S Mackenzie District.
 CO N610000N630000
  W1100000W1170000
 BT Northwest Territories
  Canada

Great Smokies
 use Great Smoky Mountains

Great Smoky Mountains
Range of the Appalachians. Index states as applicable.
 UF Great Smokies
  Smokies
 BT United States
 SA Appalachians
  North Carolina
  Tennessee

Great South Bay
Long narrow inlet of Atlantic Ocean between Fire Island and S shore of Long Island.
 BT New York
  United States

Great Valley
A valid index term through 1977. After 1977, see entry for Central Valley in California, and see entry for Great Appalachian Valley.

Great Valley Sequence
In California.
 BT Mesozoic
 SA California
  Cretaceous
  Jurassic
  Upper Cretaceous
  Valley and Ridge Province

Greater Antilles
Includes the major islands of the Antilles. As of 1977, includes use as a level 1 area term (list O). See list B for term set options. Index islands as applicable.
 BA West Indies
 SA Antilles
  Caribbean region
  Cuba
  Hispaniola
  Jamaica
  Puerto Rico

Greater Caucasus
The main range of the Caucasus separating the Northern Caucasus from the Colchis and Kura Lowlands. Index Soviet republics as applicable.
 BT USSR
 SA Azerbaidzhan
  Caucasus
  Georgian Republic
  Russian Republic
  Sunzha

Greater Khingan Mountains
 use Khingan Range

Greater Khingan Range
 use Khingan Range

Greece
Includes use on level 1 as an area term (list O). For term set options see list B.
 CO N345500N414500
  E0284500E0193000
 BA Europe
 NT Aegean Islands
  Aegina
  Attica
  Boeotia
  Crete
  Epirus
  Euboea
  Hellenides
  Ionian Islands
  Kremasta
  Laurion
  Othrys
  Parnassus
  Peloponnesus
  Pindus Mountains
  Salonika
  Thessaly
  Vourinos
 SA Athens
  Balkan Peninsula
  Levant
  Macedonia
  Maritsa River
  Mediterranean region
  Near East
  Rhodope Mountains
  Struma River valley

Thrace
Vardar River

Greeley
City in Weld County in N.
BT Colorado
United States

green algae
use Chlorophyta

Green Bay
Inlet of NW Lake Michigan. Index Michigan and/or Wisconsin as applicable. Also a city in Brown County, Wisconsin.
BT United States
SA Michigan
Wisconsin

Green Gully
Near Keilor just NE of Melbourne.
BT Victoria
Australia

Green Lake
Green Lake County in S central.
BT Wisconsin
United States

Green Mountains
Range of the Appalachians. Index Quebec and U.S. states as applicable.
SA Appalachians
Massachusetts
Quebec
Vermont

Green River
Rises in the Wind River Range in W central Wyoming and flows S into the Colorado River N of Lake Powell. Index states as applicable.
BT United States
SA Colorado
Utah
Wyoming

Green River basin
Index states as applicable.
BT United States
SA Colorado
Utah
Wyoming

Green River Formation
Subdivided into Tipton Tongue, Laney Shale Member, and Morrow Creek Member. NW and central Colorado, E Utah, and SW Wyoming. Also search Green River Shale.
BT Eocene
SA Colorado
Laney Shale Member
lower Eocene
middle Eocene
Utah
Wyoming

green tuff
See list H.
BA pyroclastics and glasses
BT igneous rocks
SA tuff

Green Tuff Formation
Underlies Daijima Formation. SW Hokkaido.
BT Miocene
SA Hokkaido
Japan

Green tuff region
BT Japan

Greenbrier County
SE West Virginia.
BT West Virginia
United States

Greene County
Index states as applicable.
BX Alabama
Arkansas
Georgia
Illinois
Indiana
Iowa
Mississippi
Missouri
New York
North Carolina
Ohio
Pennsylvania
Tennessee
Virginia
BT United States

Greenhorn Limestone
Of Colorado Group. Includes Lincoln Limestone, Jetmore Chalk, Hartland Shale, Pfeifer Shale, Bridge Creek Limestone Member, Orman Lake Limestone Member.
BT Upper Cretaceous
Cretaceous
SA Benton Formation
Colorado
Colorado Group
Kansas
Montana
Nebraska
New Mexico
South Dakota
Wyoming

Greenland
Largest island in the world and a Danish province. Includes use on level 1 as an area term (list O). For term set options see list B.
CO N600000N840000
W0200000W0700000
BA Arctic region
NT Agto
Camp Century
Disko Island
NA East Greenland
NT Fiskenaesset
Frederikshab
Godthaab
Greenland Ice Sheet
Igaliko
Ivigtut
Jameson Land
Kuhn Island
Milne Land
Nugssuaq
Peary Land
NT Scoresby Land
NT Scoresby Sound
Skaergaard Intrusion
Sondre Strom Fjord
NA South Greenland
NT Thule
Umanak
NA West Greenland
SA Ilimaussaq

Greenland Gap Group
BT Upper Devonian
Devonian
SA Frasnian
Maryland
Virginia
West Virginia

Greenland Ice Sheet
Located in basin surrounded by coastal mountains covering roughly 85% of the island. Also search Greenland.
BT Greenland
Arctic region

Greenland Sea
Section of Arctic Ocean between NE coast of Greenland and Spitsbergen.
BT Arctic Ocean

greenockite
UF cadmium blende
cadmium ocher
xanthochroite
BA sulfides
BT minerals

greensand
Includes use on level 3 under sedimentary rocks(1) clastic rocks(2) and under sediments(1) carbonate sediments(2) and clastic sediments(2). Before 1975, also search glauconitic sand.
UF glauconitic sandstone
SA clastic rocks
clastic sediments
glauconite
glauconitic composition
sand
sandstone
sedimentary rocks

greenschist
Includes use on level 3 under metamorphic rocks(1) schists(2). See list J. Also search prasinite.
UF prasinite
BA schists
BT metamorphic rocks
SA greenschist facies
schist

greenschist facies
Includes use on level 3 under metamorphic rocks(1).
BT facies
SA blueschist facies
greenschist
metamorphic rocks

greenstone
Includes use on level 3 under metamorphic rocks(1) schists(2). See list J. Also used as mineral.
BA schists
BT metamorphic rocks
SA greenstone belts
nephrite

greenstone belts
UF belts, greenstone
SA greenstone

Greenwood County
Index states as applicable.
BX Kansas
South Carolina
BT United States

Gregory Rift
That part of the East African Rift extending SW from Lake Turkana into NW central Tanzania. Index countries as applicable. Also search Gregory Rift valley.
BT Africa
SA Kenya
Tanzania

greigite
Also search melnikovite.
UF melnikovite
BA sulfides
BT minerals

greisen
Compositional term. As of 1978, term is used on level 2 under metasomatic rocks(1).
BA metasomatic rocks
SA greisenization

greisenization
Includes use on level 3 under metasomatism(1).
BA metasomatism
SA greisen
hydrothermal alteration

Grenada
Island, southernmost of Windward Islands in British West Indies. Also a county and city in NW central Mississippi. Index West Indies or Mississippi as applicable.
SA Carriacou
Lesser Antilles
Mississippi
West Indies
Windward Islands

Grenoble
City in SE part of country.
BT Isere
France

Grenville
County in SE Ontario and town on E coast of Grenada Island in West Indies. Index Ontario or West Indies as applicable. Until 1977, used to indicate provincial series. Now see Grenvillian Orogeny for the orogeny.
SA Grenvillian Orogeny
Ontario
West Indies

Grenville Front
Index provinces as applicable.
BT Canada
SA Ontario
Quebec

Grenville Orogeny
use Grenvillian Orogeny

Grenville Province
Index New York, Labrador, and provinces as applicable.
CO N440000N560000
W0560000W0820000
SA Labrador
New York
Ontario
Quebec

Grenvillian
use Grenvillian Orogeny

Grenvillian Orogeny
Before 1978, also search Grenvillian; Grenville AND orogeny. Now Grenville is used as a place name.
UF Grenville Orogeny
Grenvillian
BT Precambrian
SA Grenville
orogeny

Gresivaudan
Alpine glacial trough in SE part of country.
UF Graisivaudan
BT Isere
France

Grey forest soils
use Gray forest soils

greywacke
use graywacke

Grimmertingen
Sands.
BT Oligocene
SA Belgium

grinding
BT erosion
SA abrasion

griquaite
Includes use on level 3 under igneous rocks(1) ultramafic family(2). See list H.
BA ultramafic family
BT igneous rocks

Grisons
use Graubunden

groins
BT marine installations
SA jetties

Groix
Groix Island was used through 1978. In Bay of Biscay off Brittany coast. Before 1978, also search Groix Island.
UF Groix Island
BT Morbihan
France

Groix Island
use Groix

Groningen
Province in NE Netherlands. Also a city.
BT Netherlands

groove casts
Includes use on level 3 under sedimentary structures(1) bedding plane irregularities(2). See list K.
UF casts, groove
BA bedding plane irregularities
    sedimentary structures
SA grooves

grooves
Includes use on level 3 under sedimentary structures(1) bedding plane irregularities(2). See list K.
BA bedding plane irregularities
    sedimentary structures
SA channels
    groove casts
    striations

Grosseto
City and province in S.
BT Tuscany
    Italy

grossular
Also search grossularite.
UF grossularite
BA garnet group
    orthosilicates
    silicates
BT minerals
SA hydrogrossular
    vanadium garnet

grossularite
use grossular

ground
A valid term through 1977. After 1977, use ground methods.

ground, frozen
use Frozen ground

ground ice
As of 1978, term is used on level 2 under permafrost(1).
SA ice
    permafrost
    thermokarst

ground mass
use matrix

ground methods
Term introduced in 1978. Includes use on level 3 under geophysical methods(1) and geophysical surveys(1). Before 1978, also search ground AND methods.
BA geophysical methods
SA geophysical surveys
    methods

ground motion
Includes use on level 3 under geologic hazards(1) earthquakes(2) or under earthquakes(1).
UF motion, ground
SA earthquakes
    geologic hazards
    seismology
    strong motion
    surface waves

ground, patterned
use patterned ground

ground truth
SA remote sensing

ground water
For economically oriented papers, see water resources. This term used for subsurface waters, especially for studies on specific areas. Includes use on level 1 (list A) and 2. If 1, term set options are:
topic [age, aquifers, artesian waters, composition, connate waters, contamination, genesis, geochemistry, hydrodynamics, levels, models, movement, recharge, salt-water intrusion, surveys, systems analogs] (area)
subtopic (if no area is given)
NT aquifers
    artesian waters
SA artificial recharge
    automatic data processing
    connate waters
    contamination
    Darcy's law
    desalinization
    development
    discharge
    drawdown
    engineering geology
    environmental geology
    flow regime
    fresh water
    geochemistry
    geophysical surveys
    geysers
    hot springs
    hydraulics
    hydrochemistry
    hydrodynamics
    hydrogeology
    hydrology
    hydrosphere
    impurities
    infiltration
    isotopes
    levels
    liquid waste
    mineral waters
    movement
    percolation
    permeability
    porosity
    pump tests
    purification
    recharge
    saline composition
    salt water
    salt-water intrusion
    seepage
    springs
    storage
    surface water
    surveys
    systems analogs
    thermal waters
    tracer experiments
    transmissivity
    unconfined aquifers
    waste disposal
    water
    water quality
    water recovery
    water resources

ground-water increment
use recharge

ground-water movement
A valid term through 1978. After 1978, see movement(2) under ground water(1).

ground-water recharge
use recharge

ground-water replenishment
use recharge

groundmass
use matrix

grout
use grouting

grouting
Includes use on level 3 under foundations(1). Also search grout.
UF grout
SA dams
    engineering geology
    foundations

groutite
BA oxides
BT minerals

growth
As of 1978, term is restricted to use for biological growth. For crystals, use crystal growth.
SA crystal growth

growth faults
Before 1978, also search flexure AND faults; contemporaneous faults; slump faults.
UF contemporaneous faults
    depositional fault
    flexure faults
    sedimentary fault
BA faults
SA flexure

growth spirals
Includes use on level 3 under crystal growth(1).
UF spirals, growth
SA crystal growth
    crystal structure
    crystallography

Grozny
City in Chechen-Ingush A.S.S.R. in Northern Caucasus.
BT Russian Republic
    USSR

Grudziadz
City in N central part of country.
BT Bydgoszcz
    Poland

Grundy County
Index states as applicable.
BX Illinois
    Iowa
    Missouri
    Tennessee
BT United States

grunerite
BA amphibole group
    chain silicates
    silicates
BT minerals

Gryphaea
Genus. Includes use on level 3 under Mollusca(1) Bivalvia(2).
BA Bivalvia
    Mollusca
BT Invertebrata

Grzybow
Sulfur mining region in S.
BT Poland

Guadalajara
Province in central Spain. Also a city.
BT Spain
SA Serrania de Cuenca

Guadalquivir
River in S Spain. Also search Guadalquivir Basin.
UF Guadalquivir River
BT Spain
SA Guadalquivir Basin

Guadalquivir Basin
River basin in S and SW Spain. Also search Guadalquivir; Gualdalquivir River; Guadalquivir River basin.
UF Guadalquivir River basin
BT Spain
SA Guadalquivir

Guadalquivir River
use Guadalquivir

Guadalquivir River basin
use Guadalquivir Basin

Guadalupe Island
Mexican island 180 miles off W central Baja California.
BT Pacific Ocean

Guadalupe Mountains
Index New Mexico and/or Texas as applicable.
BT United States
SA New Mexico
    Texas

Guadalupe River
SE Texas.
BT Texas
    United States

Guadalupian
Provincial series, North America: Lower and Upper Permian. Above Leonardian, below Ochoan. Includes use on level 3 under age terms(1). See list E.
BA Permian
BT Paleozoic
NT Bell Canyon Formation
    Capitan Formation
SA Lower Permian
    San Andres Formation
    Upper Permian

Guadeloupe
Combined islands of Basse-Terre and Grande-Terre which constitute an overseas department of France. Includes use on level 1 as an area term (list O). For term set options see list B.
BA West Indies
NX Soufriere

Guajira Peninsula
Northernmost part of country.
BT Colombia

Guam
Island and unincorporated U. S. territory.
BT Mariana Islands
    Pacific Ocean

Guanabara
State in SE.
BT Brazil
NT Sugarloaf Mountain

Guanajuato
State in central.
BT Mexico
NT Sierra Gorda

guano
BA organic sediments
BT sediments

Guapore
use Rondonia

Guarico
State in N central.
BT Venezuela

Guatemala
Includes use on level 1 as an area term (list O). For term set options see list B.
CO N135500N174500
    W0881500W0921500
BA Central America
NT Guatemala City
    Huehuetenango
    Pacaya

Guatemala City
City in S central.
BT Guatemala

Gudbrandsdal
use Gudbrandsdalen

Gudbrandsdalen
Valley in S central part of country.
UF Gudbrandsdal
BT Norway

gudmundite
BA sulfides
BT minerals

Guelma
Town in Constantine Mountains in NE part of country.
BT Constantine
    Algeria

Guelph
City in SE.
BT Ontario
Canada

Guerrero
State in SW.
BT Mexico
SA Sierra Madre del Sur

guest element
use trace elements

Guiana (Shield)
use Guyana Shield

Guiana Basin
Region between the Orinoco, Negro, and Amazon rivers and the Atlantic Ocean. Index countries as applicable. Also search Guiana.
BT South America
SA Brazil
French Guiana
Guyana
Surinam

Guiana Highland
use Guyana Shield

Guiana Massif
use Guyana Shield

Guianas
Index countries as applicable.
BT South America
SA French Guiana
Guyana
Surinam

Guichon Batholith
use Guichon Creek Batholith

Guichon Creek Batholith
S central British Columbia. Also search Guichon Batholith.
UF Guichon Batholith
BT British Columbia
Canada

guide fossils
use index fossils

guidebook
Includes use on level 3 under area terms(1) areal geology(2) and under associations(1).
SA areal geology
associations
directory
explanatory text
manuals
road log

guides, ore
use ore guides

Guinea
Formerly French Guinea. Also a term applied to coastal region of W Africa between 15°N and 15°S. Includes use on level 1 as an area term (list O). For term set options see list B.
BA Africa
NT Los Islands
SA Niger River
Niger Valley
Nimba Mountains
Senegal Basin
Senegal River
West Africa

Guinea-Bissau
Formerly Portuguese Guinea. Includes use on level 1 as an area term (list O). For term set options see list B. Before 1975, also search Portuguese Guinea.
CO N110000N130000
W0130000W0170000
BA Africa
SA West Africa

Guipuzcoa
One of the Basque Provinces in N.
BT Spain
NT San Sebastian
SA Basque Provinces

Gujarat
State in W.
BT India
NT Amba Dongar
Baroda
Bhuj
Broach
Cambay
Cambay Basin
Cutch
Girnar Hills
Kathiawar
Mount Girnar
Panch Mahals
Saurashtra
Wadia
SA Aravalli System
Bagh Beds
Bhuj Series
Nari Series
Narmada River
Narmada Valley

Gulbarga
Town in NE.
BT Mysore
India

Gulf Coast
use Gulf Coastal Plain

Gulf Coastal Plain
Index Mexican and U.S. states and Quintana Roo as applicable. Also search Gulf Coast. Includes use on level 1 or 2 as an area term (list O). If 1, see term set options under list B.
CO N250000N310000
W0790000W0980000
UF Gulf Coast
BA North America
SA Alabama
Atchafalaya Bay
Barataria Bay
Campeche
Florida
Louisiana
Mississippi
Quintana Roo
Tabasco
Tamaulipas
Texas
United States
Veracruz
Yucatan

Gulf of Aden
Between S coast of Arabian Peninsula and Somalia, E Africa. Includes use on level 1 or 2 as an area term (list O). If 1, see term set options under list B.
CO N103000N150000
E0520000E0430000
BA Arabian Sea
SA Great Rift Valley
Gulf of Tadjoura

Gulf of Alaska
Between Alaska Peninsula and Alaskan panhandle.
BT Pacific Ocean

Gulf of Aqaba
Between NW Saudi Arabia and the Sinai Peninsula. Also search Gulf of Eilat.
UF Gulf of Eilat
BT Red Sea
SA Great Rift Valley

Gulf of Bothnia
Between Finland and Sweden. Also search Bothnian Sea.
UF Bothnian Sea
BT Baltic Sea
SA Alno
Finland

Gulf of California
Between peninsula of Baja California and the Mexican states of Sonora and Sinaloa. Includes use on level 1 as an area term (list O). For term set options see list B.
BA Pacific Ocean

Gulf of Cambay
Inlet on W coast of India SE of Kathiawar Peninsula of Gujarat State.
BT Arabian Sea

Gulf of Cariaco
Inlet on NE coast of Venezuela S of Araya Peninsula in Sucre State.
BT Caribbean Sea

Gulf of Eilat
use Gulf of Aqaba

Gulf of Finland
Between Finland and Estonia.
BT Baltic Sea
SA Finland

Gulf of Gascony
French name for SE section of the Bay of Biscay between Gironde Department and Basque coast.
BT Bay of Biscay
Atlantic Ocean

Gulf of Guinea
Wide inlet on W coast of Africa just south of the continent's great bulge.
BT Atlantic Ocean

Gulf of Lion
Wide bay extending from French-Spanish border to Toulon. Also search Gulf of Lions.
UF Gulf of Lions
BT Mediterranean Sea

Gulf of Lions
use Gulf of Lion

Gulf of Maine
Off New England and S of Nova Scotia in Georges Bank area.
BT Atlantic Ocean
SA Georges Bank

Gulf of Mexico
Relatively shallow oceanic-type basin encircled by Cuba, Mexico, and the United States. Also search Gulf Basin. Includes use on level 1 or 2 as an area term (list O). If 1, see term set options under list B.
NT Alacran Reef
Alaminos Canyon
Campeche Bank
Challenger Knoll
Florida Bay
Sigsbee Deep
Yucatan Shelf
SA Gulf Stream

Gulf of Oman
Extends between N Oman on the Arabian Peninsula and SE Iran.
BT Arabian Sea

Gulf of Panama
On S coast of Panama.
BT Pacific Ocean

Gulf of Pozzuoli
NW inlet of Bay of Naples.
BT Mediterranean Sea

Gulf of Riga
Between Estonia and Latvia.
BT Baltic Sea

Gulf of Saint Lawrence
Off E coast of Canada encircled by Newfoundland, New Brunswick, Nova Scotia, and Quebec.
BT Atlantic Ocean

Gulf of Siam
Between S extension of Thailand on the Malay Peninsula on the W and Cambodia and Vietnam on the E. Also search Gulf of Thailand.
UF Gulf of Thailand
BT South China Sea

Gulf of Suez
Between the Sinai Peninsula and the Arabian Desert of Egypt. Joined to the Mediterranean Sea by the Suez Canal.
BT Red Sea

Gulf of Tadjoura
Inlet of the Gulf of Aden on coast of Afars and Issas Territory, E Africa. Also search Gulf of Tadjura.
UF Gulf of Tadjura
BT Arabian Sea
SA Gulf of Aden

Gulf of Tadjura
use Gulf of Tadjoura

Gulf of Thailand
use Gulf of Siam

Gulf Stream
Warm ocean current flowing out of Gulf of Mexico along E U.S. to mid-Atlantic Ocean where it merges with North Atlantic drift current and influences climate of Europe as far N as Norway.
BT Atlantic Ocean
SA Gulf of Mexico

Gulfian
North America, provincial series. Includes use on level 3 under age terms (list E).
BA Upper Cretaceous
Cretaceous
BT Mesozoic
NT Austin Group
Escondido Formation
Woodbine Formation

gullies
Includes use on level 3 under geomorphology(1) erosion features(2).
SA channels
cliffs
erosion features
geomorphology

Gumma
Prefecture in central.
BT Honshu
Japan

Gunflint Formation
use Gunflint Iron Formation

Gunflint Iron Formation
In Animikie Group. Gunflint Lake region and Vermilion District in NE Minnesota. Also search Gunflint Formation and Gunflint Iron-Formation.
UF Gunflint Formation
Gunflint Iron-Formation
BT Precambrian
SA Animikie Group
Minnesota

Gunflint Iron-Formation
use Gunflint Iron Formation

Gunnison County
W central.
BT Colorado
United States

Guntur
City and district in E central.
BT Andhra Pradesh
India

Gunz
Europe. Above Astian (Pliocene), below Mindel. Includes use on level 3 under age terms(1). See list E.
BA Pleistocene
Quaternary
BT Cenozoic

Gurev
 use Guryev

Gurghiu
 Village in Mures County in E central.
 BT Transylvania
  Romania

Gurghiu Mountains
 In SE Transylvania.
 UF Gurghiului Mountains
 BT Transylvania
  Romania

Gurghiului Mountains
 use Gurghiu Mountains

Gurhwal
 use Garhwal

Guryev
 City and oblast at mouth of Ural River at N end of Caspian Sea.
 UF Gurev
 BT Kazakhstan
  USSR

gusher
 use geysers

gustavite
 BA sulfosalts
 BT minerals

Gutenberg low-velocity zone
 use low-velocity layer

Gutii Mountains
 BT Moldavia
  Romania

Guyana
 Formerly British Guiana. Gained independence in 1966. Also search British Guiana. Includes use on level 1 as an area term (list O). For term set options see list B.
 UF British Guiana
 BA South America
 SA Georgetown
  Guiana Basin
  Guianas
  Guyana Shield
  Roraima Formation

Guyana Shield
 Highland area extending from E Venezuela across N Brazil and the Guianas. Index countries as applicable. Also search Guiana; Guiana Shield.
 UF Guiana (Shield)
  Guiana Highland
  Guiana Massif
 BT South America
 SA Brazil
  French Guiana
  Guyana
  Surinam

guyots
 use seamounts

gymnosperms
 Includes use on level 1 and 2 as a fossil term (list F).
 BT Plantae
 NA Bennettitales
  Caytoniales
  Coniferales
  Cordaitales
  Cycadales
  Dadoxylon
  Ginkgoales
  Glossopteris
  Pteridospermae
 SA angiosperms
  Callipteris
  Dicroidium
  Fagus
  fossil wood
  Juglans
  Pecopteris
  Pteropsida
  sporangia
  Taeniopteris

gypsite
 use gypsum

gypsum
 Includes use on level 1 as a commodity term (list C); on level 3 under sediments(1) chemically precipitated sediments(2). See list L (minerals) and list N (sediments) and list I (sed. rocks).
 UF gypsite
  plaster of paris
  plaster stone
 BA sulfates
 BT minerals
 SA alabaster
  anhydrite
  cap rocks
  chemically precipitated rocks
  chemically precipitated sediments
  evaporites
  halite
  sedimentary rocks
  sediments
  selenite

gyrolite
 UF centrallasite
 BA sheet silicates
  silicates
 BT minerals

gyttja
 BA organic sediments
 BT sediments
 SA sapropel
  soils

H

H
 use hydrogen

H-2
 use deuterium

H-3
 use tritium

Haast River
 W South Island.
 BT South Island
  New Zealand

habit
 Includes use on level 3 under crystal growth(1). Also search crystal habit.
 UF crystal habit
 SA crystal form
  crystal growth
  lattice
  minerals

habitat
 Includes use on level 2 and 3 under fossil group(1). See list F.
 SA ecology
  environment
  fossils

Hachijo-jima
 Second largest island in Izu-shichito group about 180 miles S of Tokyo.
 BT Pacific Ocean
 SA Izu-shichito

Hachinohe
 City in N.
 BT Honshu
  Japan

Hadley Rill
 use Hadley Rille

Hadley Rille
 UF Hadley Rill
 BT Moon

Hadrosauridae
 Family. Includes use on level 3 under Reptilia(1) Archosauria(2).
 BA Ornithischia
  Archosauria
  Reptilia
 BT Tetrapoda
  Vertebrata
 SA dinosaurs

Hadrynian
 BA upper Precambrian
  Precambrian

hafnium
 Includes use on level 1 and 2 as a chemical element (list D). Also search Hf.
 SA elements

Hagendorf
 Region in E.
 BT Bavaria
  West Germany

Haghimas Syncline
 BT Romania
 SA Eastern Carpathians

haidingerite
 BA arsenates
 BT minerals

Haifa
 District in NE Israel. Also a city on the Mediterranean Sea.
 BT Israel
  Middle East
 NT Geula Caves

Hail
 Town and oasis.
 UF Hayel
 BT Saudi Arabia

Hainaut
 Province in SW.
 BT Belgium
 NT Mons Basin
  Tournai

Haiti
 Occupies western third of island of Hispaniola. Includes use on level 1 as an area term (list O). For term set options see list B.
 BA Hispaniola
  West Indies

Hakodate
 City on Tsugaru Strait in S.
 BT Hokkaido
  Japan

Hakone
 Village and mountain resort in SE Honsu.
 BT Honshu
  Japan
 SA Mount Hakone

Hakone Volcano
 use Mount Hakone

Hakusan
 Extinct volcano in W Honshu. Also search Hakusan Volcano.
 UF Hakusan Volcano
 BT Honshu
  Japan

Hakusan Volcano
 use Hakusan

Halberstadt
 City in W part of country.
 BT Magdeburg
  East Germany

half space
 use half-space

half-space
 Includes use on level 3 under seismology(1). Also search half space.
 UF half space
 SA seismology
  velocity

halides
 Includes use on level 2 under minerals(1). See list L.
 UF halogenide
 BT minerals
 NA atacamite
 NZ bastnaesite
 NA bideauxite
 NZ boracite
 NA carnallite
  chlorides
 NZ clinohumite
 NA cryolite
 NZ eudialyte
  fluoborite
 NA fluorides
  fluorite
  halite
 NZ humite
  kainite
  leifite
  lepidolite
  microlite
  mimetite
  minyulite
 NA neighborite
 NZ norbergite
 NA paratacamite
 NZ phosgenite
 NA prosopite
 NZ pyrochlore
  pyromorphite
 NA rinneite
 NZ sarcopside
 NA sylvite
 NZ synchisite
 NA thomsenolite
 NZ triplite
  vanadinite
 NA villiaumite
  weberite
 NZ zinnwaldite
  zunyite
 SA sodium chloride

Halifax
 City and county on the Atlantic Ocean.
 BT Nova Scotia
  Canada

Halimeda
 Genus. Includes use on level 3 under algae(1) Chlorophyta(2).
 BA Codiaceae
  Chlorophyta
  algae
 BT Plantae

halite
 Includes use on level 3 under minerals(1) halides(2); on level 3 under sedimentary rocks(1) chemically precipitated rocks(2); on level 3 under sediments(1) chemically precipitated sediments(2). See list I (sed. rocks), list L (minerals), and list N (sediments).
 UF common salt
  rock salt
 BA halides
 BT minerals
 SA chemically precipitated rocks
  chemically precipitated sediments
  chlorides
  evaporites
  gypsum
  salt
  sediments

Halland
 County in SW.
 BT Sweden

Halle
 District in SW central part of country. Also a city.
 BT East Germany
 NT Aschersleben

Bitterfeld
Geisel Valley
Mansfeld Syncline
Merseburg
Rottleberode
Schlotheim
SA Saxony-Anhalt
Unstrut River

halloysite
Also search endellite; search metahalloysite.
UF endellite
metahalloysite
BA sheet silicates
silicates
BT minerals
SA clay minerals

Hallstatt Limestone
BT Triassic
SA Austria
Carnian
Ladinian
Upper Triassic

halmyrolysis
UF halmyrosis
submarine weathering
SA diagenesis
geochemistry
sea water

halmyrosis
use halmyrolysis

Halobia
Includes use on level 3 under Mollusca(1) Bivalvia(2).
BA Bivalvia
Mollusca
BT Invertebrata

haloes
Includes use on level 3 under mineral exploration(1). For intrusions and metamorphism, use aureoles. Also search halos.
UF halos
SA aureoles
geochemical methods
mineral exploration

Halog
Settlement in S.
BT Himachal Pradesh
India

halogenide
use halides

halokinesis
use salt tectonics

halos
use haloes

halotrichite
UF iron alum
BA sulfates
BT minerals

Halsingborg
City on Oresund in SW part of country. Also search Helsingborg.
UF Helsingborg
BT Malmohus
Sweden

Hamburg
City state on both sides of the Elbe River in N part of country.
BT West Germany

Hamersley Basin
BT Western Australia
Australia

Hamersley Group
BT Precambrian
SA Australia
Western Australia

Hamersley Range
In NW Western Australia.
BT Western Australia
Australia

Hamilton County
Index states as applicable.
BX Florida
Illinois
Indiana
Iowa
Kansas
Nebraska
New York
Ohio
Tennessee
Texas
BT United States

Hamilton Group
Divisions include Onondaga, Marcellus Formation, Mahantango Formation, Speeds Formation, Deputy Formation, Silver Creek Formation, Swanville Formation, Beachwood Formation, Bakoren Shale, Mount Marion Formation, Ashokan Formation, Kiskatom Formation, Skaneateles Shale, Ludlowville Shale, Moscow Shale, Montebello Formation, Sherman Ridge Formation.
BT Middle Devonian
Devonian
SA Mahantango Formation
Maryland
New York
Pennsylvania
West Virginia

hammarite
BA sulfobismuthites
sulfosalts
BT minerals

Hampshire Basin
Also search Hampshire.
BT England
Great Britain
United Kingdom

Hampshire County
Index states as applicable.
BX Massachusetts
West Virginia
BT United States

Han River basin
Also search Han River.
BT China
SA Shensi

Hanawa Mine
Akita Prefecture in N.
BT Honshu
Japan

Hancock County
Index states as applicable.
BX Georgia
Illinois
Indiana
Iowa
Kentucky
Maine
Mississippi
Ohio
Tennessee
West Virginia
BT United States

handling, data
use data handling

Hanford Reservation
Atomic energy facility. Before 1978, also search Hanford.
BT Washington
United States

Hangay
use Hangay Mountains

Hangay Mountains
W central Mongolia. Also search Hangay; Hangay Range; Khangai; Khangai Mountains.
CO N460000N481500
E1023000E0970000
UF Changai Mountains
Hangay
Hangay Range
Hangayn Mountains
Khangai
Khangai Mountains
BT Mongolia

Hangay Range
use Hangay Mountains

Hangayn Mountains
use Hangay Mountains

Hanna Basin
S central Wyoming. Also search Hanna.
BT Wyoming
United States

Hannover
use Hanover

Hanover
City in Lower Saxony and in New Hampshire. Index countries as applicable. Also search Hannover.
UF Hannover
SA Germany
United States

Hanover County
E central.
BT Virginia
United States

Hanson Lake
Near Manitoba border in E.
BT Saskatchewan
Canada

Hantkenina
Genus. Includes use on level 3 under foraminifera(1) Globigerinacea(2).
BA Globigerinacea
Rotaliina
foraminifera
BT Invertebrata

Haplophragmoides
Genus. Includes use on level 3 under foraminifera(1) Lituolacea(2).
BA Lituolidae
Lituolacea
Textulariina
foraminifera
BT Invertebrata

Harar
Province in E Ethiopia. Also a city. Also search Harrar.
UF Harer
Harrar
BT Ethiopia

harbors
As of 1978, term is used on level 2 under waterways(1).
SA waterways

Harbour Main Group
Overlain with angular unconformity by Lower Cambrian strata. The Conception Group conformably overlies the Harbour Main Group except near the Holyrood Granite where it is unconformable. E Newfoundland.
BT Precambrian
SA Canada
Newfoundland

Hardanger Plateau
use Hardangervidda

Hardangervidda
Plateau in Hordaland, SW Norway. Also search Hardanger Plateau.
UF Hardanger Plateau
Vidda
BT Norway

Hardeman County
Index states as applicable.
BX Tennessee
Texas
BT United States

Hardin County
Index states as applicable.
BX Illinois
Iowa
Kentucky
Ohio
Tennessee
Texas
BT United States

Harding County
Index states as applicable.
BX New Mexico
South Dakota
BT United States

Hardington County
use East Lothian

Hardingtonshire
use East Lothian

hardness
Includes use on level 3 under minerals(1).
SA mechanical properties
microhardness
minerals
physical properties
water

Hare Bay
Inlet near N tip of Newfoundland.
BT Atlantic Ocean

Harebell Formation
Unconformable contacts with overlying Pinyon Conglomerate and underlying Bacon Ridge Sandstone. Teton County in NW Wyoming.
BT Upper Cretaceous
Cretaceous
SA Wyoming

Harer
use Harar

Harford County
NE Maryland.
BT Maryland
United States

Harghita
County in E.
BT Transylvania
Romania

Harghita Mountains
E Transylvania.
UF Harghitei Mountains
BT Transylvania
Romania

Harghitei Mountains
use Harghita Mountains

Hariana
use Haryana

harkerite
BA orthosilicates
silicates
BT minerals

Harlech
Village in W.
BT Wales
Great Britain
United Kingdom

Harlech Stage
Term introduced in 1978. Before 1978, search Harlech.
BA Lower Cambrian
Cambrian
BT Paleozoic

harmonic analysis
use Fourier analysis

harmonic folds
Term introduced in 1978. Before 1978, search folds AND harmonic.
BA folds
SA disharmonic folds

harmonics
Includes use on level 2 under

geodesy(1).
SA geodesy
harmotome
BA zeolite group
framework silicates
silicates
BT minerals
Harney County
SE central.
CO N420000N440500
W1181000W1195500
BT Oregon
United States
Harper County
Index states as applicable.
BX Kansas
Oklahoma
BT United States
Harpidae
Family. Includes use on level 3 under Mollusca(1) Gastropoda(2).
BA Gastropoda
Mollusca
BT Invertebrata
Harrar
use Harar
Harrington Sound
BT Bermuda
Atlantic Ocean
Harris
Southern section of the island of Lewis in the Outer Hebrides off NW.
BT Scotland
Great Britain
United Kingdom
Harris County
Index states as applicable.
BX Georgia
Texas
BT United States
Harrisburg
City in Dauphin County in SE central.
BT Pennsylvania
United States
Harrison Lake
In S British Columbia.
BT British Columbia
Canada
Hartberg
Town in SE part of country.
BT Styria
Austria
Hartford County
N central.
BT Connecticut
United States
Haruna
Gumma Prefecture in N central Honshu. Also search Haruna Volcano.
UF Haruna Volcano
BT Honshu
Japan
Haruna Volcano
use Haruna
Harvey County
S central.
BT Kansas
United States
Haryana
State in N central India. Also search Hariana.
UF Hariana
BT India
NT Hissar
Narnaul
SA Delhi System
Harz Foreland
S Lower Saxony, West Germany, and W East Germany. Index countries as applicable. Also search Harz.
BT Europe

SA East Germany
West Germany
Harz Mountains
Mountain group between Elbe and Weser rivers in SE Lower Saxony, West Germany, and SW East Germany. Index countries as applicable. Also search Harz.
BT Europe
SA East Germany
West Germany
Harz region
Harz Foreland and Harz Mountain area of S Lower Saxony, West Germany, and W East Germany. Index countries as applicable. Also search Harz.
BT Europe
SA East Germany
West Germany
harzburgite
Includes use on level 3 under igneous rocks(1) ultramafic family(2). See list H.
BA ultramafic family
BT igneous rocks
SA peridotite
Hassan
Town in S.
BT Mysore
India
Hassi Messaoud
use Hassi Messaoud Field
Hassi Messaoud Field
Oil field near Great Eastern Erg in NE Algeria. Also search Hassi-Messaoud; Hassi Messaoud; Hassi Messaoud oil field.
UF Hassi Messaoud
Hassi Messaoud oil field
Hassi-Messaoud
BT Algeria
SA oil and gas fields
Hassi Messaoud oil field
use Hassi Messaoud Field
Hassi-Messaoud
use Hassi Messaoud Field
hastingsite
BA amphibole group
chain silicates
silicates
BT minerals
SA ferrohastingsite
Hat Creek Basalt
Recent. Hat Creek Valley 10 miles NNE of Lassen Peak. Also search Hat Creek.
BT Holocene
SA California
Hatay
Province in S Anatolia on the NW Syrian border.
BT Turkey
Middle East
hatchettolite
use betafite
Hateg
Town in Hunedoara county in SW.
BT Transylvania
Romania
Hatteras abyssal plain
Off SE coast of U. S.
BT Atlantic Ocean
Hattiesburg
City in Forrest County in S.
BT Mississippi
United States
hausmannite
BA oxides
BT minerals

Haut-Rhin
Department in NE.
CO N473000N481500
E0074500E0070000
BT France
NT Sainte-Marie-aux-Mines
SA Vosges Mountains
Haute-Garonne
Department in S.
BT France
NT Toulouse
SA Rodez Trough
Salat Valley
Haute-Loire
Department in S central.
CO N445500N451500
E0043000E0030000
BT France
NT Velay
SA Central Massif
Haute-Marne
Department in NE.
BT France
Haute-Saone
Department in E.
BT France
Haute-Savoie
Department in E.
CO N454500N463000
E0070000E0054500
BT France
NT Annecy
Argentiere Glacier
Chablais
Mont Blanc
Rumilly
Savoy Alps
Haute-Vienne
Department of W central.
BT France
NT Rochechouart
Saint-Sylvestre Massif
Hauterivian
Europe. Above Valanginian, below Barremian. Includes use on level 3 under age terms(1). See list E.
BA Lower Cretaceous
Cretaceous
BT Mesozoic
SA Zubair Formation
Hautes-Alpes
Department in SE.
BT France
NT Briancon
SA Eoulx Basin
Pelvoux Massif
Hautes-Pyrenees
Department in SW.
CO N423500N434000
E0005000W0001500
BT France
NT Gavarnie
Lourdes
hauyne
Also search hauynite.
UF hauynite
BZ framework silicates
sulfates
BT minerals
SA sodalite group
hauynite
use hauyne
Havana
City on Straits of Florida in NW part of country.
BT Cuba
Haveroe
Point of impact in Turku area in SW Finland. Also search Haveroe Ureilite.
UF Haveroe Ureilite
BT Finland

SA meteorites
Haveroe Ureilite
use Haveroe
Haviland
Village in Kiowa County in S central.
BT Kansas
United States
Hawaii
State including all of the Hawaiian Islands. Also southernmost and largest of the islands. Includes use on level 1 as an area term (list O). For term set options see list B.
CO N190000N283000
W1550000W1790000
BZ Pacific Ocean
United States
NT Alae Crater
Hawaii County
Hilo
Honolulu
Kaneohe Bay
Kauai
Kilauea
Kona
Koolau Range
Maui
Mauna Kea
Mauna Loa
Molokai
Oahu
Pearl Harbor
Red Hill
Waianae
Waimanalo
SA Polynesia
Western U.S.
Hawaii County
Entire island of Hawaii.
BT Hawaii
United States
Hawaii Ridge
use Hawaiian Ridge
Hawaiian Arch
use Hawaiian Ridge
Hawaiian Ridge
NW of Hawaiian Islands. Also search Hawaiian Arch and Hawaii Ridge.
UF Hawaii Ridge
Hawaiian Arch
BT Pacific Ocean
hawaiite
Includes use on level 3 under igneous rocks(1) alkali basalt family(2). See list H.
BA alkali basalt family
BT igneous rocks
Hawke's Bay
Provincial district on E coast.
BT North Island
New Zealand
Hawkesbury Sandstone
BT Triassic
SA Australia
New South Wales
Hawthorn Formation
In Alum Bluff Group. Central northern, N and S Florida; S and SE Georgia; South Carolina. Also, search Hawthorne Formation.
UF Hawthorne Formation
BT Miocene
SA Florida
Georgia
lower Miocene
middle Miocene
South Carolina
Hawthorne Formation
use Hawthorn Formation
Hayel
use Hail

Haymond Formation
Overlies Dimple Limestone; underlies Gaptank Formation. Marathon region in Brewster County in SW Texas.
BT Lower Pennsylvanian
Pennsylvanian
SA Texas

Hayward
City in Alameda County in W.
BT California
United States

Hayward Fault
NE, E, and SE of San Francisco Bay.
BT California
United States

Hazara
District.
BT North-West Frontier Province
Pakistan

hazards, geologic
use geologic hazards

Hazaribagh
Town and district in central.
BT Bihar
India

He
use helium

He-3
Includes use on level 3 under isotopes(1).
SA He-4/He-3
helium
isotopes

He-3/He-4
use He-4/He-3

He-4
Includes use on level 3 under isotopes(1).
SA He-4/He-3
helium
isotopes

He-4/He-3
Includes use on level 3 under isotopes(1). Also search He-3/He-4; search He-3 AND He-4.
UF He-3/He-4
SA absolute age
He-3
He-4
helium
isotopes

Healdiidae
Family. Includes use on level 3 under Ostracoda(1) Podocopida(2).
BA Podocopida
Ostracoda
BT Crustacea
Arthropoda
Invertebrata
SA Cavellinidae

health
use medical geology

Heart Mountain Fault
NW Wyoming.
BT Wyoming
United States

heat capacity
SA capacity
specific heat
thermal properties

heat content
use enthalpy

heat flow
Includes use on level 1 (list A) and 2. Used for the product of the thermal conductivity of a substance and the thermal gradient in the direction of the flow of heat. If 1, term set options are:
topic [anomalies, causes, changes, detection, distribution, experimental studies, genesis, geothermal gradient, heat sources, interpretation, measurement, observations, patterns, rates, regional patterns, temperature, theoretical studies, thermal conductivity]
subtopic (no area term)
NT heat flux
SA boreholes
conductivity
convection
core
crust
Curie point
Earth
energy sources
flow
fumaroles
geophysical methods
geophysical surveys
geophysics
geothermal energy
geothermal gradient
heat sources
heat transfer
hot springs
intrusions
mantle
measurement
plate tectonics
regional patterns
sea-floor spreading
tectonophysics
thermal conductivity
thermal diffusivity
thermal waters

heat flux
BT heat flow
SA flux

heat sources
Includes use on level 2 under heat flow(1).
UF sources, heat
SA core
crust
geothermal energy
geothermal processes
heat flow
mantle
regional patterns
thermal waters
thermodynamics
volcanism

heat, specific
use specific heat

heat transfer
UF transfer, heat
SA heat flow

heaving, frost
use frost heaving

heavy metals
Also search individual heavy metals.
BT metals
SA bismuth
cadmium
cobalt
copper
gold
iron
lead
manganese
nickel
platinum
silver
tantalum
tellurium
zinc

heavy minerals
Includes use on level 1 and 2 as a commodity term (list C) only when of economic value; on level 3 under sedimentary rocks(1), sedimentation(1), and sediments(1); on level 3 under rock type(1) topic(2). See also individual heavy minerals under minerals(1), list L.
BT minerals
SA accessory minerals
garnet
kyanite
light minerals
magnetite
monazite
placers
sedimentation
sediments
zircon

heavy oil
SA condensates
petroleum

heazlewoodite
BA sulfides
BT minerals

Hebgen Lake
W of Yellowstone National Park in Gallatin County.
BT Montana
United States

Hebrides
Islands in Atlantic Ocean W of Scotland.
CO N554000N583000
W0053000W0080000
UF Western Islands
BT Scotland
Great Britain
United Kingdom
SA Inner Hebrides
Outer Hebrides

hebronite
use amblygonite

Hecla
use Hekla

Hecla Hoek Formation
Spitsbergen Island.
BT Precambrian
SA Arctic region
Spitsbergen

Hector Formation
Hector and Corral Creek formations are lithologically similar to the Miette and Horsethief Creek Groups. SW Alberta.
BT Precambrian
SA Alberta

hectorite
BA sheet silicates
silicates
BT minerals
SA clay minerals

Hedbergella
Genus. Includes use on level 3 under foraminifera(1) Globigerinacea(1).
BA Globigerinacea
Rotaliina
foraminifera
BT Invertebrata

hedenbergite
BA pyroxene group
chain silicates
silicates
BT minerals

hedleyite
BA sulfides
BT minerals
SA native elements and alloys

Heemskirk Granite
W Tasmania.
BT Devonian
SA Australia
Tasmania

Hegau
Region in S.
BT Baden-Wurttemberg
West Germany

Heidelberg
City on Neckar River in N.
BT Baden-Wurttemberg
West Germany

Heilungkiang
Province in E Manchuria in NE.
BT China
NT Lesser Khingan Mountains

Heimaey
Island a few miles off S Iceland. Largest of Vestmannaeyjar group.
BT Iceland

heintzite
use kaliborite

Hekla
Volcano in SW Iceland. Also search Hecla.
UF Hecla
BT Iceland

Helderberg Group
Includes Coeymans Limestone, New Scotland Limestone, Mandata Shale, Becraft Limestone, Port Ewen Shale, Keyser Limestone, Port Jervis Limestone, Licking Creek Limestone, Alsen, Kalkberg, Manlius. New York, E Pennsylvania, W Maryland, Western Virginia, and N West Virginia.
BT Lower Devonian
Devonian
SA Maryland
New York
Pennsylvania
Virginia
West Virginia

Helderbergian
North America. Above upper Silurian, below Deerparkian. Includes use on level 3 under age terms(1). See list E.
BA Lower Devonian
Devonian
BT Paleozoic

Helena
City in Lewis and Clark County in W central.
BT Montana
United States

Helgoland
Island in North Sea 28 miles off mainland.
BT Schleswig-Holstein
West Germany

Helicopontosphaera
BA nannofossils
algae
BT Plantae

Helikian
North America.
BA Precambrian
SA Proterozoic

Heliolites
Genus. Includes use on level 3 under Coelenterata(1) Tabulata(2).
BA Heliolitidae
Tabulata
Anthozoa
Coelenterata
BT Invertebrata

Heliolitidae
Family. Includes use on level 3 under Coelenterata(1) Tabulata(1).
BA Tabulata
Anthozoa
Coelenterata
BT Invertebrata
NA Heliolites

helium
Includes use on level 1 and 2 as a commodity term (list C) and as a chemical element (list D). Also

search He.
UF He
BT noble gases
SA argon
elements
He-3
He-4
He-4/He-3
isotopes
krypton
neon
U/He

Hell Creek Formation
Includes Bull Creek Sand, Isabel-Firesteal Coal Member. E, N, and central S Montana; SW North Dakota; and NW and N South Dakota.
BT Upper Cretaceous
Cretaceous
SA Montana
North Dakota
South Dakota

Hellenides
Deep rooted mountain chain of Greece which is part of the Alpine orogenic belt. Bordered by Rhodope Mountains in N.
BT Greece
SA Alps
Crete
Peloponnesus
Pindus Mountains

Hell's Canyon
use Snake River canyon

Helmholtz free energy
use free energy

Helmstedt
City near East German border in W.
BT Lower Saxony
West Germany

Helsingborg
use Halsingborg

Helsingor
City on the Oresund N of Copenhagen.
BT Denmark

Helvetian
Europe. Above Burdigalian, below Tortonian. Includes use on level 3 under age terms(1). See list E.
BA Miocene
Neogene
Tertiary
BT Cenozoic
SA Tortonian

helvine
use helvite

helvite
Also search helvine.
UF helvine
BZ silicates
sulfides
BT minerals
SA genthelvite

hematite
Includes use as level 3 commodity term under iron(1). See list C.
BA oxides
BT minerals
SA iron
martite
specularite

Hemichordata
Includes use on level 1 and 2 as a fossil term (list F). Term is to be used only when more specific terms do not apply.
UF Stomachorda
SA Chordata
Dendroidea
Graptolithina
Monograpta
Tuboidea

hemihedrite
BA chromates
BT minerals

hemimorphite
BA orthosilicates
silicates
BT minerals

Hemingfordian
BA lower Miocene
Miocene
Neogene
Tertiary
BT Cenozoic

Hemphillian
BA Neogene
Tertiary
BT Cenozoic
SA Miocene
Pliocene

Henbury
Settlement in S central.
BT Northern Territory
Australia

Henderson
Village N of Denver in Adams County.
BT Colorado
United States

Hennessey Formation
Overlies Wichita Formation and underlies Duncan Sandstone, or Flowerpot Shale, where the Duncan is absent. Central and SW Oklahoma.
BT Permian
SA Oklahoma

Henry County
N Tennessee.
BT Tennessee
United States

Henry Mountains
Garfield County in S central.
BT Utah
United States

Hepaticae
Includes use on level 2 under bryophytes(1). See list F.
BA bryophytes
BT Plantae

Herault
Department in S. Also a river.
CO N431500N440000
E0041500E0023000
BT France
NT Bedarieux
Cabrieres
Lodeve
Lodeve Basin
Montpellier
Saint-Chinian
Sete

Hercegovina
use Herzegovina

Hercoglossa danica
Includes use on level 3 under Mollusca(1) Cephalopoda(2).
BA Nautiloidea
Cephalopoda
Mollusca
BT Invertebrata

Hercynian
A valid index term through 1977. After 1977, use Hercynian Orogeny.

Hercynian Geosyncline
Term introduced in 1978. Before 1978, search geosynclines AND Hercynian.
SA geosynclines
Hercynian Orogeny

Hercynian Orogeny
Term introduced in 1978. Late Paleozoic (Devonian-Lower Triassic) of Europe, extending through the Carboniferous and Permian. Also search orogeny AND Hercynian; Variscan; Variscan Orogeny.
UF Variscan Orogeny
BT Paleozoic
SA Devonian
Hercynian Geosyncline
Lower Triassic
Permian
Triassic
Variscides

hercynite
UF ferrospinel
iron spinel
BA oxides
BT minerals
SA spinel group

herderite
BA phosphates
BT minerals

Herja
Mining region in Maramures County in N.
BT Transylvania
Romania

Herkimer County
N central.
BT New York
United States

hermatypic taxa
Term introduced in 1978 on level 3 under fossil groups (list F). Before 1978, also search hermatypic.
UF taxa, hermatypic
SA ahermatypic taxa
Coelenterata
corals
reefs

Heron Island
Largest of Capricorn Islands in Coral Sea 40 miles off SE coast.
BT Queensland
Australia

Hertfordshire
SE England.
CO N513000N521000
E0001000W0004500
BT England
Great Britain
United Kingdom

hervidero
use mud volcanoes

Herzegovina
Region in N Bosnia and Herzegovina.
UF Hercegovina
BT Yugoslavia

herzenbergite
UF kolbeckine
BA sulfides
BT minerals

Hess Mountains
E central.
BT Yukon Territory
Canada

Hesse
Region in SW Germany, comprising the state of Hesse and the former Prussian province of Hesse-Nassau. Also search Hessen.
UF Hessen
BT Germany
NT Bundenbach
Darmstadt
Kassel
Marburg
Messel
Rheingau
Taunus
Vogelsberg
Waldeck
Wetterau
Wiesbaden
SA Franconia
Lahn River
Lahn River valley
Main River
Mainz Basin
Odenwald
Palatinate
Rhenish Schiefergebirge
Rhon Mountains
Steinheim
Swabia
Weser River
West Germany
Westerwald
Westphalia

Hessen
use Hesse

hessite
BA tellurides
sulfides
BT minerals

heterogeneity
use inhomogeneity

heterogeneous materials
Term introduced in 1978. Before 1978, search heterogeneous.

heterogenite
Also search stainierite.
UF stainierite
BA oxides
BT minerals

Heterohelicidae
Family. Includes use on level 3 under foraminifera(1) Globigerinacea(2).
BA Globigerinacea
Rotaliina
foraminifera
BT Invertebrata

heteromorphic taxa
Term introduced in 1978. Before 1978, search heteromorphs.
UF heteromorphs
taxa, heteromorphic

heteromorphite
BA sulfantimonites
sulfosalts
BT minerals

heteromorphs
use heteromorphic taxa

Heteroptera
Includes use on level 2 under Insecta(1). See list F.
BA Insecta
BT Arthropoda
Invertebrata

Heterostegina
Includes use on level 3 under foraminifera(1) Rotaliacea(2).
BA Rotaliacea
foraminifera
BT Invertebrata
NA Heterostegina depressa
SA Nummulitidae

Heterostegina depressa
Includes use on level 3 under foraminifera(1) Rotaliacea(2).
BA Heterostegina
Rotaliacea
foraminifera
BT Invertebrata

Heterostraci
Order. Includes use on level 3 under Pisces(1) Agnatha(2).
BA Agnatha
Pisces
BT Vertebrata

Hettangian
Europe. Above Rhaetian (Triassic), below Sinemurian. Includes use on

level 3 under age terms(1). See list E.
BA Lower Jurassic
   Jurassic
BT Mesozoic
SA Liassic

heulandite
BA zeolite group
   framework silicates
   silicates
BT minerals
SA clinoptilolite

hexahedrites
BA meteorites
SA iron meteorites

hexahydrite
BA sulfates
BT minerals

HF
Includes use on level 3 under ionosphere(1) wave propagation(2). Also search high frequency; high-frequency.
SA ELF
   ionosphere
   VLF
   wave propagation
   waves

Hg
use mercury

hibschite
use hydrogrossular

Hida
Former province in central Honshu. Now part of Gifu Prefecture.
BT Honshu
   Japan

Hida Belt
use Hida metamorphic belt

Hida metamorphic belt
Central Honshu. Also search Hida Belt; Hida metamorphic rocks.
UF Hida Belt
   Hida metamorphic rocks
BT Honshu
   Japan

Hida metamorphic rocks
use Hida metamorphic belt

Hida Mountains
Central.
BT Honshu
   Japan

Hidaka
Town in S.
BT Honshu
   Japan

Hidaka Belt
use Hidaka metamorphic belt

Hidaka metamorphic belt
In S Hokkaido. Also search Hidaka Belt.
UF Hidaka Belt
BT Hokkaido
   Japan

Hidaka Mountains
In S Hokkaido. Also search Hidaka.
BT Hokkaido
   Japan

Hidalgo
State in E central.
BT Mexico
NT Pachuca
SA Sierra Madre Oriental

Hidalgo County
Index states as applicable.
BX New Mexico
   Texas
BT United States

Hidas
Town in S. In Baranya.
BT Hungary

Hidden Valley Dolomite
Underlies Devonian Lost Burro Formation; overlies Ely Springs Dolomite, S California. Also search Hidden Valley.
BT Paleozoic
SA California
   Devonian
   Lower Devonian
   Silurian

Hierro
Spanish island off Western Sahara. Westernmost of the Canary Islands.
BT Canary Islands
   Atlantic Ocean

High Atlas
A range of the Atlas Mountains containing the highest peaks in the entire mountain system. In W and S part of country.
UF Grand Atlas Mountains
   High Atlas Mountains
BT Morocco
SA Atlas Mountains

High Atlas Mountains
use High Atlas

High Plains
A valid term through mid-1978 used to indicate the Great Plains from Nebraska southward. Now use Great Plains.

high pressure
Includes use on level 3 under metamorphism(1).
BT pressure
SA low pressure
   metamorphism
   pore pressure

high school
Includes use on level 3 under education(1).
SA college-level education
   education
   elementary school
   junior high school

High Tatra
use Tatra Mountains

High Tatra Mountains
use Tatra Mountains

High Tatras
use Tatra Mountains

high temperature
Includes use on level 3 under metamorphism(1). Also search high-temperature.
BT temperature
SA metamorphism

high-energy environment
Term introduced in 1978. Before 1978, also search high-energy or high energy.
SA environment
   low-energy environment
   sedimentation

high-grade metamorphism
Term introduced in 1978. Includes use on level 3 under metamorphism(1). Before 1978, also search high-grade AND metamorphism; high grade AND metamorphism.
BA metamorphism
SA grade

Highland County
Index states as applicable.
BX Ohio
   Virginia

BT United States

Highland Rim
BT Alabama
   United States

Highland Series
BT Precambrian
SA Sri Lanka

highlands
Includes use on level 3 under geomorphology(1) erosion features(2), or under Moon(1).
SA erosion features
   geomorphology
   maria
   Moon
   mountains
   plateaus

Highlands
That part of Scotland lying NW of a line from Loch Lomond to just S of Aberdeen.
BT Scotland
   Great Britain
   United Kingdom

highways
Used for geological studies on highways and materials. A level 1 term as of 1978. Includes use on level 2 under foundations(1). If 1, term set options are:
construction
   subtopic
embankments
   subtopic
feasibility studies
   subtopic
foundations
   subtopic
frost action
   subtopic
materials, properties
   subtopic
planning
   subtopic
site exploration
   subtopic
SA bridges
   construction
   embankments
   engineering geology
   engineering properties
   feasibility studies
   foundations
   frost action
   land subsidence
   planning
   rock mechanics
   site exploration
   slope stability
   soil mechanics
   soils
   tunnels

Hildesheim
City in S central.
BT Lower Saxony
   West Germany

Hildocerataceae
Superfamily. Includes use on level 3 under Mollusca(1) Cephalopoda(2).
BA Ammonoidea
   Cephalopoda
   Mollusca
BT Invertebrata

Hildoceratidae
Family. Includes use on level 3 under Mollusca(1) Cephalopoda(2).
BA Ammonoidea
   Cephalopoda
   Mollusca
BT Invertebrata

hills
Includes use on level 3 under geomorphology(1).
SA geomorphology
   landforms
   mountains
   ridges
   topography
   tors

Hilo
City in Hawaii county on E coast of island of Hawaii.
BT Hawaii
   United States

Himachal Pradesh
State in the Himalayas.
CO N310000N330000
   E0780000E0760000
BT India
NT Bilaspur
   Chamba
   Halog
   Kangra
   Mahasu
   Mandi
   Simla
   Simla Hills
SA Beas River
   Jumna River
   Kasauli Series
   Kumaun Himalayas

Himalaya
use Himalayas

Himalaya Mountains
use Himalayas

Himalayas
Mountain system extending from Jammu and Kashmir in the W to Assam in the E. Index Tibet and countries as applicable. Includes use on level 1 as an area term (list O). For term set options see list B. Also search Himalaya; Himalayan; Himalaya.
CO N270000N370000
   E0970000E0720000
UF Himalaya
   Himalaya Mountains
   The Himalaya
BA Asia
SA Bhutan
   India
   Kumaun Himalayas
   Lesser Himalayas
   Nepal
   Sikkim
   Tibet

Hindenburg
use Zabrze

Hindenburg in Oberschlesien
use Zabrze

Hindeodella
Genus.
BA conodonts

Hindu Kush
Mountain range in central Asia. Index Tadzhikistan and countries as applicable.
BT Asia
SA Afghanistan
   Pakistan
   Tadzhikistan

hinge faults
Term introduced in 1978. Before 1978, search faults AND hinge.
BA faults

hinsdalite
BZ phosphates
   sulfates
BT minerals

hintzeite
use kaliborite

Hipparion
Genus. Includes use on level 3 under Mammalia(1) Perissodactyla(2).
BA Equidae

Hipparion ● homogeneity

      Perissodactyla
    Mammalia
  BT Tetrapoda
      Vertebrata

Hippopotamus
Genus. Includes use on level 3 under Mammalia(1) Artiodactyla(2).
  BA Artiodactyla
      Mammalia
  BT Tetrapoda
      Vertebrata

Hippuritacea
Superfamily. Includes use on level 3 under Mollusca(1) Bivalvia(2).
  BA Bivalvia
      Mollusca
  BT Invertebrata

Hiroshima
City and prefecture on the Inland Sea in W.
  BT Honshu
      Japan

hisingerite
  BA sheet silicates
      silicates
  BT minerals
  SA clay minerals

Hispaniola
Island of the Greater Antilles between Cuba and Puerto Rico. Index countries as applicable.
  BT West Indies
  NA Dominican Republic
      Haiti
  SA Caribbean region
      Dominican Republic
      Greater Antilles

Hissar
City WNW of Delhi.
  BT Haryana
      India

Hissar Range
A branch of the Alai Range in NW Tadzhikistan. Also search Gissar Range and Hissar.
  CO N380000N381000
      E0700000E0673000
  UF Gissar Range
  BT Tadzhikistan
      USSR
  SA Alai Range
      Zeravshan-Hissar

histograms
  SA graphic methods
      kurtosis
      statistical analysis

histology
  SA fossils

historical geology
Includes use on level 3 under geology(1), paleontology(1) and stratigraphy (1). See list E (age terms).
  SA biostratigraphy
      education
      geology
      paleoenvironment
      paleontology
      physical geology
      stratigraphy

histories, case
  use case histories

history
Use term under various disciplines. Includes use as a level 2 or 3 term appropriate to a large number of topics, e.g. on level 2 under paleontology(1), geology(1), stratigraphy(1), and mineral exploration(1). See list G.
  SA uniformitarianism

Histosols
Includes use on level 3 under soils(1). See list M.
  BA soils
  SA Bog soils

Hitachi
City on the Pacific Ocean in E.
  BT Honshu
      Japan

Hiyoshi
An area of Mizunami city in Gifu Prefecture in central.
  BT Honshu
      Japan

Ho
  use holmium

Hobart
Coastal city in SE.
  BT Tasmania
      Australia

Hobdo
  use Kobdo

Hobsogol
  use Khubsugul

Hodgeman County
SW central.
  BT Kansas
      United States

Hodna Basin
Interior drainage basin in NE part of country.
  BT Constantine
      Algeria

hodographs
  use traveltime curves

Hodrusa
Ore bearing region.
  BT Slovakia
      Czechoslovakia

Hoggar Mountains
  use Ahaggar

Hoghiz
Village in Brasov County in SE.
  BT Transylvania
      Romania

Hohe Tauern
Range of Eastern Alps in SW.
  CO N470000N471500
      E0131500E0120000
  BT Austria
  SA Alps
      Eastern Alps
      Tauern Tunnel

Hohe Venn
Range of low mountains in E part of country.
  BT Liege
      Belgium

Hokkaido
Northernmost of the four main islands.
  BT Japan
  NT Abashiri
      Esan Cape
      Hakodate
      Hidaka metamorphic belt
      Hidaka Mountains
      Ikushumbetsu
      Ishikari Bay
      Ishikari Plain
      Kamikawa
      Kitami
      Koma-ga-take
      Kushiro coal field
      Nemuro Peninsula
      Okushiri Island
      Oshima Peninsula
      Shikotsu
      Sorachi
      Tarumae
      Tokachi
      Tokachi Plain
      Usu
  SA Green Tuff Formation
      Tsugaru Strait

Hoko Gunto
  use Penghu Islands

Hoko Shoto
  use Penghu Islands

Hokuriku
District.
  BT Honshu
      Japan

Holbrook
Town in Navajo County in E central.
  BT Arizona
      United States

Holderness
Peninsula in Yorkshire in NE.
  BT England
      Great Britain
      United Kingdom

Holland
  use Netherlands

hollandite
  BA oxides
  BT minerals
  SA coronadite

Hollister
City in San Benito County in W central.
  BT California
      United States

holmium
Includes use on level 1 and 2 as a chemical element (list D).
  UF Ho
  SA elements
      rare earths

holmquistite
  BA amphibole group
      chain silicates
      silicates
  BT minerals

Holocene
Recent. Late Quaternary, from Pleistocene to present time. Includes use on level 1 as an age term (list E); on level 2 under paleo- terms, e.g. paleoecology, paleogeography, paleomagnetism; on level 3 under age terms(1). Above Pleistocene. Also search Recent; Postglacial; Post-glacial.
  BA Quaternary
  BT Cenozoic
  NA Atlantic
      Boreal
      Flandrian
  NT Hat Creek Basalt
  NA lower Holocene
      Mesolithic
      middle Holocene
      Neolithic
      Preboreal
      Subboreal
      upper Holocene
      Versilian
  SA modern

holography
Used as a general term mostly in crystallography.
  SA crystallography

Holostei
Includes use on level 3 under Pisces(1) Osteichthyes(2). See list F.
  BA Osteichthyes
      Pisces
  BT Vertebrata
  SA fish

Holothuroidea
Includes use on level 2 under Echinodermata(1). See list F.
  BA Echinozoa
      Echinodermata
  BT Invertebrata
  SA sclerites

Holstein
Region in S.
  BT Schleswig-Holstein
      West Germany

Holsteinian
Europe.
  BA upper Pleistocene
      Pleistocene
      Quaternary
  BT Cenozoic

Holy Cross Mountains
  use Swiety Krzyz Mountains

Homalozoa
Subphylum. Includes use on level 2 under Echinodermata(1). See list F.
  BA Echinodermata
  BT Invertebrata

Homestake Mine
Largest gold mine in U. S. In Lead, Lawrence County, in W.
  BT South Dakota
      United States

Hominidae
Family. Includes use on level 3 under Mammalia(1) Primates(2).
  BA Primates
      Mammalia
  BT Tetrapoda
      Vertebrata
  NA Australopithecus
      Homo
      Neanderthal
      Paranthropus
      Pithecanthropus
      Ramapithecus

Homo
Genus. Includes use on level 3 under Mammalia(1) Primates(2).
  BA Hominidae
      Primates
      Mammalia
  BT Tetrapoda
      Vertebrata
  NA Homo erectus
      Homo habilis
      Homo sapiens

Homo erectus
Includes use on level 3 under Mammalia(1) Primates(2).
  BA Homo
      Hominidae
      Primates
      Mammalia
  BT Tetrapoda
      Vertebrata

Homo habilis
Includes use on level 3 under Mammalia(1) Primates(2).
  BA Homo
      Hominidae
      Primates
      Mammalia
  BT Tetrapoda
      Vertebrata

Homo neanderthalensis
  use Neanderthal

Homo sapiens
Includes use on level 3 under Mammalia(1) Primates(2).
  BA Homo
      Hominidae
      Primates
      Mammalia
  BT Tetrapoda
      Vertebrata

homogeneity
Includes use on level 3 under geophysical methods(1).
  SA electrical methods
      geophysical methods
      inhomogeneity

prospecting
homogenization
Used as a general term.
SA dispersion
petrology

homonymy
SA nomenclature
taxonomy

Honaker Trail Formation
In Hermosa Group. SE Utah.
BT Paleozoic
SA Pennsylvanian
Permian
Utah

Honduras
Includes use on level 1 as an area term (list O). For term set options see list B.
CO N130000N160000
W0831500W0891500
BA Central America

Hong Kong
British crown colony which includes the island of Hong Kong; and Kowloon Peninsula and New Territories on mainland. Includes use on level 1 as an area term (list O). For term set options see list B.
CO N220000N230000
E1150000E1130000
BA Asia

Honolulu
City and county on Oahu Island.
BT Hawaii
United States

Honshu
Largest of the Japanese islands.
BT Japan
NT Abukuma Mountains
Aichi
Aizu Basin
Akenobe Mine
Akita
Akiyoshi
Amagase
Aomori
Arita River
Asama
Ashio
Atera Fault
Atsumi Peninsula
Awashima
Chiba
Chiba Peninsula
Chichibu
Chichibu Belt
Chichibu Mine
Chugoku
Dogo
Fossa Magna
Fujiyama
Fukui
Fukushima
Fukushima Prefecture
Gifu
Gumma
Hachinohe
Hakone
Hakusan
Hanawa Mine
Haruna
Hida
Hida metamorphic belt
Hida Mountains
Hidaka
Hiroshima
Hitachi
Hiyoshi
Hokuriku
Ibaragi Complex
Ibaraki
Ikuno Mine
Ise Bay
Ishikawa
Ito
Iwaki
Iwate
Izu Peninsula
Izu-shichito
Izumi
Joban coal field
Kanagawa
Kanazawa
Kashiwazaki
Kii Peninsula
Kinki District
Kiso Mountains
Kitakami Mountains
Kosaka Mine
Kumamoto
Kwanto Plain
Kyoto
Lake Biwa
Lake Suwa
Maizuru
Matsukawa
Matsushiro
Mie
Miho Bay
Mikawa
Mitsuishi
Miura Peninsula
Miyagi
Miyake-Jima
Mizunami
Mizusawa
Mount Asama
Mount Hakone
Mount Mihara
Musashino
Myoko Mountain
Nagano
Nagano Prefecture
Naka-no-umi
Nanao
Nara
Narugo
Nasudake
Niigata
Nohi Rhyolite
Nojiri Lake
Noto Peninsula
Odate
Oga Peninsula
Ogasawara
Oguni
Ojika Peninsula
Okayama
Oki Islands
Omine Mine
Onikobe
Osaka
Otake
Rokko Mountains
Ryoke
Ryoke Belt
Ryujima
Sado Island
Sagami Bay
Saitama
Sendai
Shiga
Shimane
Shimokita Peninsula
Shinano River
Shinji Lake
Shiobara
Shirasu
Shizukuishi
Shizuoka
Shonai River
Suruga Bay
Takanuki
Tamagawa
Tamba Plateau
Tanzawa Mountains
Tochigi
Tohoku
Toki
Tokyo
Tottori
Towada
Toyama
Toyoma
Ube coal field
University of Tokyo
Wakayama
Yahagi River
Yake-Dake
Yamagata
Yamaguchi
Yamanashi
Yatsugatake
Yokohama
Yonezawa
Zao
SA Akiyoshi Limestone
Futaba Group
Izumi Group
Mikabu System
Motojuku Formation
Nankaido
Narita Formation
Osaka Group
Sambagawa Belt
Tono
Tsugaru Strait
Uonuma Group
Usuginu Conglomerate
Yamato Mountains

Hooke's law
Term introduced in 1978.
SA deformation
elastic limit
elastic properties
elastic strain
strain
stress

Hope
Town in SW.
BT British Columbia
Canada

Hopi Buttes
use Hopi Buttes Field

Hopi Buttes Field
In Hopi and Navajo Indian reservations in Navajo County. Also search Hopi Buttes volcanic field; Hopi Buttes.
UF Hopi Buttes
Hopi Buttes volcanic field
BT Arizona
United States

Hopi Buttes volcanic field
use Hopi Buttes Field

horizon differentiation
Includes use on level 3 under soils(1) genesis(2).
SA differentiation
horizons
morphology
profiles
soils

horizons
Includes use on level 3 under soils(1) morphology(2).
UF soil horizon
soil zone
SA horizon differentiation
morphology
parent materials
profiles
soils

horizontal
A valid term through 1977. After 1977, use horizontal orientation.

horizontal orientation
Term introduced in 1978. Reference to folds defined on the basis of orientation in relation to the geographic horizontal plane. Before 1978, also search folds AND horizontal.
SA folds
orientation

Horlick Mountains
SE of S tip of Ross Ice Shelf.
BT Antarctica

Horn Plateau Formation
S District of Mackenzie.
BT Givetian
Middle Devonian
Devonian
SA Northwest Territories

hornblende
Also search horn stone.
BA amphibole group
chain silicates
silicates
BT minerals
SA amphibole
kaersutite

hornblende gneiss
A valid metamorphic rock term through 1977. After 1977, use gneiss and hornblende.

hornblende schist
A valid metamorphic rock term through 1977. After 1977, use schist and hornblende.

hornblendite
Includes use on level 3 under igneous rocks(1) ultramafic family(2). See list H.
BA ultramafic family
BT igneous rocks

Horne Mine
Noranda area in SW.
BT Quebec
Canada

hornfels
Includes use on level 2 and 3 under metamorphic rocks(1). See list J.
BT metamorphic rocks

hornstein
use chert

Hornsund
Inlet on SW coast.
BT Spitsbergen
Arctic region

Horry County
Extreme E.
BT South Carolina
United States

horseflesh ore
use bornite

horsetail
use Sphenopsida

Horsham
City in W Sussex in S.
BT England
Great Britain
United Kingdom

horsts
Includes use on level 3 under faults(1) systems(2).
SA block structures
faults
grabens
systems

host materials
Includes use on level 3, e.g. under inclusions(1).
SA fluid inclusions
host rocks
inclusions
materials

host rocks
Includes use on level 3 under inclusions(1).
SA country rocks
fluid inclusions
host materials
inclusions
mineral deposits
mineral inclusions

rocks
veins
xenoliths

hot spots
Includes use on level 3 under plate tectonics(1).
SA magmas
    mantle
    plate tectonics
    plumes

hot springs
Includes use on level 2 under springs(1); on level 3 under thermal waters(1).
BT springs
SA geothermal energy
    geysers
    ground water
    heat flow
    hydrothermal processes
    thermal waters

Houghton County
NW Upper Peninsula.
BT Michigan
    United States

House Range
Millard County in W.
BT Utah
    United States

Houston
City in Harris County in E.
BT Texas
    United States

howardites
Includes use on level 3 under meteorites(1).
BA meteorites
SA achondrites
    eucrite
    stony irons

Howgill Fells
Westmoreland and Yorkshire counties in N.
BT England
    Great Britain
    United Kingdom

howieite
BA chain silicates
    silicates
BT minerals

Hoxnian
Europe.
BA upper Pleistocene
    Pleistocene
    Quaternary
BT Cenozoic

Hsinchu
Region in N.
BT Taiwan

Hubsugul
use Khubsugul

Hudson Bay
Large inland sea in E central. Administratively a part of District of Keewatin.
CO N550000N650000
    W0760000W0950000
BT Canada
SA Hudson Bay Lowlands
    Northwest Territories

Hudson Bay Lowland
use Hudson Bay Lowlands

Hudson Bay Lowlands
Index Northwest Territories and provinces as applicable. Also search Hudson Bay Lowland.
UF Hudson Bay Lowland
BT Canada
SA Hudson Bay
    Manitoba
    Northwest Territories
    Ontario
    Quebec

Hudson Highlands
Part of Appalachians lying along both banks of the Hudson River S of Newburgh.
BT New York
    United States
SA Appalachians

Hudson River
In New York and New Jersey.
BT United States
SA New Jersey
    New York

Hudson Valley
River valley primarily in New York. Index states as applicable.
BT United States
SA New Jersey
    New York

Hudsonian Orogeny
Name proposed for a time of plutonism, metamorphism, and deformation during the Precambrian in the Canadian Shield. Also search Hudsonian.
BT Precambrian
SA Canada
    orogeny

Hudspeth County
W Texas.
BT Texas
    United States

Huehuetenango
Department in Guatemala. Also a city.
BT Guatemala
    Central America

Huelva
Province in SW Spain. Also a city on the Gulf of Cadiz.
CO N364500N381500
    W0061500W0073000
BT Spain
NT Aracena
    Puebla de Guzman
    Rio Tinto
SA Andalusia

Huesca
Province in NE. Also a city.
BT Spain

Hughes County
E central.
BT Oklahoma
    United States

Hughes Creek Shale
Of Foraker Limestone. NE Kansas and SE Nebraska. Also search Hughes Creek Shale Member.
UF Hughes Creek Shale Member
BT Permian
SA Kansas
    Nebraska

Hughes Creek Shale Member
use Hughes Creek Shale

Hugoton
City in Stevens County in SW.
BT Kansas
    United States

Hugoton Embayment
SW Kansas and Oklahoma panhandle. Index states as applicable. Also search Hugoton.
BT United States
SA Kansas
    Oklahoma

Hugoton Field
SW Kansas. Also search Hugoton; Hugoton gas field; Hugoton-Panhandle Field.
UF Hugoton gas field
    Hugoton-Panhandle Field
BT Kansas
    United States
SA oil and gas fields

Hugoton gas field
use Hugoton Field

Hugoton-Panhandle Field
use Hugoton Field

Hull
City across Ottawa River from Ottawa, Ontario, in S.
BT Quebec
    Canada

hulsite
BA borates
BT minerals

human activities
use human activity

human activity
Also search anthropogenic; human activities; human impact; human influence; human interference.
UF anthropogenic (activity)
    human activities
    human impact
    human influence
    human interference
SA human ecology
    man

human ecology
As of 1978, term is used on level 2 under geologic hazards(1), land use(1), and pollution(1).
SA ecology
    environmental geology
    geologic hazards
    human activity
    human waste
    land use
    man
    medical geology
    pollution

human impact
use human activity

human influence
use human activity

human interference
use human activity

human waste
Term introduced in 1978.
UF waste, human
SA human ecology
    man
    waste disposal

humates
Includes use on level 2 under organic materials(1).
BA organic materials
SA humic acids

Humber Estuary
On E coast opening on North Sea.
BT England
    Great Britain
    United Kingdom

Humboldt County
Index states as applicable.
BX California
    Iowa
    Nevada
BT United States

Humboldt River valley
N central.
BT Nevada
    United States

humic acids
Includes use on level 2 under organic materials(1).
BA organic materials
SA acids
    humates
    humus

humic coal
Also search humite AND coal.
UF humite (coal)
BT coal

humid environment
Term introduced in 1978. Before 1978, search humid.
SA ecology
    environment
    humidity

humidity
Includes use on level 3 under meteorology(1) water(2).
SA atmosphere
    climate
    humid environment
    meteorology
    moisture
    raindrops
    water
    water vapor

humite
BZ halides
    orthosilicates
BT minerals
SA fluorides
    humite group

humite (coal)
use humic coal

humite group
Includes use in combination with orthosilicates (i.e. orthosilicates, humite group) on level 2 under minerals(1). See list L.
BA orthosilicates
    silicates
BT minerals
NA titanoclinohumite
SA clinohumite
    humite
    norbergite

hummocks
SA geomorphology
    periglacial features
    permafrost

Humphreys Peak
use San Francisco Mountain

humus
Includes use on level 3 under organic materials(1).
UF soil ulmin
BA organic materials
SA humic acids
    leaves
    Mor
    Mull
    peat
    soils

Hungarian Basin
use Alfold

Hungarian Great Plain
use Alfold

Hungarian Plain
use Alfold

Hungary
Includes use on level 1 as an area term (list O). For term set options see list B.
CO N454500N483000
    E0230000E0161000
BA Europe
NT Bakony Mountains
    Balaton region
    Borsod Basin
    Borzsony Mountains
    Budapest
    Bukk Mountains
    Debrecen
    Dorog
    Dorog Basin
    Eger
    Hidas
    Lake Balaton
    Matra Mountains
    Mecsek Mountains

Szolnok
Tokaj
Tokaj Mountains
Transdanubia
Vertes
Vertes Mountains
Veszprem
Villany Mountains
SA Alfold
Carpathians
Danube River
Danube Valley
Pannonia
Pannonian Basin
Somes Basin
Subcarpathians
Tisza River
Tokaj-Eperjes Mountains

Hunger Steppe
use Bet-Pak-Dala

Hunsruck
Mountain region S of the Mosel River between the Rhine and Saar rivers.
BT Rhineland-Palatinate
West Germany

Hunsruck Shale
Also search Hunsruck shales and Hunsrueck Shale.
UF Hunsruck shales
Hunsrueck Shale
BT Devonian
SA Germany
Rhineland-Palatinate

Hunsruck shales
use Hunsruck Shale

Hunsrueck Shale
use Hunsruck Shale

Hunter Valley
River valley in E central.
BT New South Wales
Australia

huntite
BA carbonates
BT minerals

Huntly
Town in Aberdeen County in E.
BT Scotland
Great Britain
United Kingdom

Hunton Group
Includes Chimneyhill Limestone, Henryhouse Shale, Haragan Shale, Bois d'Arc Limestone, Frisco Limestone, Kite Group with Haragan and Cravatt formations. SE Oklahoma. Also search Hunton Limestone.
BT Paleozoic
SA Devonian
Oklahoma
Silurian

Huon Peninsula
On E coast between Astrolabe Bay and Huon Gulf.
BT Papua New Guinea
Australasia

hureaulite
BA phosphates
BT minerals

Huronian
Great Lakes Region. Division of the Proterozoic of the Canadian Shield. Also search Huronian Supergroup.
UF Huronian Supergroup
BA Proterozoic
Precambrian
NT Gowganda Formation
Onaping Formation
Signal Hill Formation
SA Canada
United States

Huronian Supergroup
use Huronian

Hurricane Ridge Syncline
Southwest Virginia and SE West Virginia. Index states as applicable.
BT United States
SA Virginia
West Virginia

hurricanes
Includes use on level 3 under geologic hazards(1).
SA Agnes
geologic hazards
meteorology
monsoons
storms
winds

Hurwitz Group
Keewatin District.
BT Aphebian
lower Proterozoic
Proterozoic
Precambrian
SA Northwest Territories

Hut Point Peninsula
On Ross Island which is just off NW edge of Ross Ice Shelf in Ross Dependency.
BT Antarctica

Hutchinson
City in Reno County in S central.
BT Kansas
United States

Hutchinson Salt Member
Of Wellington Formation. Salt beds largely within Hutchinson city limits in Reno County in S central Kansas.
BT Permian
SA Kansas
Wellington Formation

Hyaenidae
Family. Includes use on level 3 under Mammalia(1) Carnivora(2).
BA Carnivora
Mammalia
BT Tetrapoda
Vertebrata

hyalite
UF water opal
BA silica minerals
framework silicates
silicates
BT minerals
SA opal

hyalophane
BA feldspar group
framework silicates
silicates
BT minerals

Hyalospongea
Includes use on level 2 under Porifera(1). See list F.
BA Porifera
BT Invertebrata

hybridization
SA magmas

Hyderabad
Princeley state prior to Indian independence. A city in Sind Province, Pakistan, and a city in Andhra Pradesh State in India. Index countries as applicable.
BT Asia
SA India
Pakistan

hydrargillite
use gibbsite

hydration
Includes use on level 3 under geochemistry(1) processes(2).
SA dehydration
geochemistry
processes
water

hydration of glass
Includes use on level 2 under geochronology(1).
UF glass, hydration of
SA geochronology

hydraulic conductivity
Also search permeability coefficient.
UF coefficient of permeability
permeability coefficient
SA conductivity
flow regime
hydraulics
permeability

hydraulic fracturing
Includes use on level 3.
SA fracturing
mining geology
rock mechanics
seismology

hydraulic maps
Term introduced in 1978. Before 1978, search maps AND hydraulic.
BA maps

hydraulic pressure
Term introduced in 1978. Before 1978, search hydraulic AND pressure.
BA pressure

hydraulics
Used when discussing channels, or engineering and practical applications. For mathematical aspects of movement, use hydrodynamics. As of 1978, term is used on level 2 under shorelines(1), and waterways(1).
SA competence
drawdown
engineering geology
flows
ground water
hydraulic conductivity
hydrodynamics
hydrographs
hydrostatic pressure
roughness
shorelines
soil mechanics
waterways

hydrobiotite
BA sheet silicates
silicates
BT minerals
SA aluminosilicates
mica group

hydroboracite
BA borates
BT minerals

hydrocarbons
Includes use on level 2 under organic materials(1).
BA organic materials
SA aromatic hydrocarbons
bitumens
carbon
ethane
methane
natural gas
petroleum
sapropel

hydrochemistry
SA chemistry
ground water
hydrology

hydrodynamics
Includes use on level 2 under ground water(1). Used when discussing mathematical aspects of ground water movement. For discussion of channels, engineering and practical applications, see hydraul-

ics.
SA Darcy's law
ground water
hydraulics

hydrogen
Includes use on level 1 and 2 as a chemical element (list D). Also search H.
UF H
SA D/H
deuterium
isotopes
stable isotopes
tritium

hydrogen sulfide
SA eutrophication
gases
lakes
natural gas
sulfides

hydrogeologic
A valid term through 1977. After 1977, use hydrogeologic maps.

hydrogeologic maps
Term introduced in 1978. Before 1978, search hydrogeologic AND maps.
BA maps
SA hydrogeology

hydrogeological controls
Includes use on level 3 under mineral deposits, genesis(1) controls(2).
SA controls
hydrogeology
mineral deposits

hydrogeology
Used for general studies. For the science that deals with subsurface waters and related geologic aspects of surface waters. To search for the hydrology of a particular river search for the name of the river in coordination with hydrology or hydrogeology. Includes use on level 1 (list A); on level 2 under area terms (list B). If 1, term set options are:
topic [applications, bibliography, catalogs, classification, concepts, education, experimental studies, general, history, instruments, methods, nomenclature, objectives, philosophy, practice, symposia, textbooks, theoretical studies]
subtopic
SA aquifers
artesian waters
contamination
dolines
ground water
hydrogeologic maps
hydrogeological controls
hydrologic cycle
hydrological methods
hydrology
hydrosphere
reservoirs
rivers
springs
thermal waters
transmissivity
water resources

hydrogoethite
BA oxides
BT minerals
SA limonite

hydrographs
SA flow regime
hydraulics
hydrology

hydrogrossular
Also search hibschite.
UF hibschite
BA garnet group
orthosilicates

silicates
BT minerals
SA grossular

hydrologic budget
use water balance

hydrologic cycle
Includes use on level 3 under hydrology(1). Also search water cycle.
UF water cycle
SA atmospheric precipitation
cycles
hydrogeology
hydrology
hydrosphere
meteorology
water

hydrologic maps
Term introduced in 1978. Before 1978, search maps AND hydrologic.
BA maps
SA hydrology

hydrological methods
Includes use on level 2 under mineral exploration(1).
SA hydrogeology
hydrology
methods
mineral exploration
water

hydrology
Used for continental surface water. In cases where the interchange between ground and surface water is intimate, this set may be used with cycles on level 2. This covers cases of modern sediment transport. To search for the hydrology of a particular river search for the name of the river in coordination with hydrology or hydrogeology. Includes use on level 1. Term set options are:
cycles
   subtopic
glaciers
   (area)
   subtopic
ice
   (area)
   subtopic (no area term)
instruments
   name of instrument
   subtopic
limnology
   (area)
   subtopic (no area term)
methods
   subtopic
precipitation
   (area)
   subtopic (no area term)
rivers and streams
   (area)
   subtopic (no area term)
seepage
   (area)
snow
   (area)
   subtopic (no area term)
surveys
   (area)
techniques
   name of technique or application
SA atmosphere
atmospheric precipitation
bedforms
cascades
channel geometry
cycles
discharge
ecology
engineering geology
environmental geology
eutrophication
evaporation
evapotranspiration
gauging
geomorphology
glacial geology
glaciers
ground water
hydrochemistry
hydrogeology
hydrographs
hydrologic cycle
hydrologic maps
hydrological methods
hydrosphere
hydrostatic pressure
ice
infiltration
lakes
limnology
meltwater
meteorology
percolation
pollution
precipitation
purification
rainfall
reservoirs
rivers
rivers and streams
runoff
sedimentation
sediments
snow
springs
surface water
suspension
transmissivity
water
water balance
water quality
water resources
water supply
watersheds
waterways

hydrolysis
Includes use on level 3 under geochemistry(1) processes(2); on level 3 under weathering(1).
SA geochemistry
oxidation
processes
weathering

hydromagnesite
BA carbonates
BT minerals

hydromica
BA sheet silicates
silicates
BT minerals
SA clay minerals
illite
mica group

hydromica schist
Not a valid term for GeoRef. Use schist and hydromica.

Hydromorphic soils
See list M.
BA soils
SA Bog soils
Intrazonal soils

hydromuscovite
BA sheet silicates
silicates
BT minerals
SA mica group
muscovite

hydrosodalite
Artificial mineral.
BA sodalite group
framework silicates
silicates
BT minerals
SA aluminosilicates

hydrosphere
Includes use on level 3 under hydrogeology(1).
SA atmosphere
biosphere
Earth
glaciers
ground water
hydrogeology
hydrologic cycle
hydrology
ice
lakes
rivers
rivers and streams
snow
streams
surface water
tectonosphere
water

hydrostatic pressure
BT pressure
SA flow regime
hydraulics
hydrology

hydrothermal
A valid term through 1977. After 1977, use hydrothermal alteration, or hydrothermal processes, or hydrothermal solutions.

hydrothermal alteration
Includes use on level 3 under metasomatism(1).
BA metasomatism
SA alteration
autometamorphism
greisenization
hydrothermal processes
propylitization
pyritization
pyrometasomatism
sericitization
serpentinization
thermal alteration
thermal waters
uralitization
wallrock alteration
zeolitization

hydrothermal fluids
use hydrothermal solutions

hydrothermal processes
Includes use on level 3 under mineral deposits, genesis(1) processes(2).
SA alteration
epithermal processes
hot springs
hydrothermal alteration
hydrothermal solutions
igneous processes
mineral deposits
processes
water

hydrothermal solutions
Also search hydrothermal fluids.
UF hydrothermal fluids
SA fluid inclusions
hydrothermal processes
metasomatism
mineral deposits
ore-forming fluids
solutions

hydroxides
BA oxides
BT minerals

hydroxyapatite
use hydroxylapatite

hydroxylapatite
Also search hydroxyapatite.
UF hydroxyapatite
BA phosphates
BT minerals
SA apatite

Hydrozoa
Includes use on level 2 under Coelenterata(1). See list F.
BA Coelenterata
BT Invertebrata

Hymenoptera
Includes use on level 2 under Insecta(1). See list F.
BA Insecta
BT Arthropoda
Invertebrata
SA Raphidiodea

Hyolithidae
Also search Hyolithes; Hyolithida; Hyolithimorpha.
BA problematic fossils
SA Mollusca

hypabyssal
A valid term through 1977. After 1977, use hypabyssal rocks on level 2 under igneous rocks(1).

hypabyssal rocks
Term introduced in 1978. Includes use on level 2 under igneous rocks(1). Before 1978, also search hypabyssal AND igneous rocks.
BT igneous rocks
SA beresite
intrusive rocks
plutonic rocks
rocks

hypersthene
BA pyroxene group
chain silicates
silicates
BT minerals

hypocenter
use focus

hypogene processes
SA enrichment
mineral deposits
processes
supergene processes

Hystrichosphaeridae
Includes use on level 3 under palynomorphs(1) acritarchs(2) or Dinoflagellata(2). Also search hystrichospheres; hystrichosphaerids.
UF hystrichosphaerids
hystrichospheres
BZ acritarchs
Dinoflagellata
BA palynomorphs

hystrichosphaerids
use Hystrichosphaeridae

hystrichospheres
use Hystrichosphaeridae

H2SO4
use sulfuric acid

I

I
use iodine

I-129
Includes use on level 3 under isotopes(1).
SA iodine
isotopes

I-131
Includes use on level 3 under isotopes(1).
SA iodine
isotopes

Iacobeni
Village in Suceava County in N.

Iapetus
Proto-Atlantic Ocean. For satellite of Saturn, use Iapetus Satellite.
SA Atlantic Ocean
Iapetus Satellite

Iapetus Satellite
Term introduced in 1978. Distinguish from Iapetus, proto-Atlantic Ocean.
SA Iapetus
Saturn

Ibadan
City in SW part of country. In Western Provinces.
BT Nigeria

Ibaragi Complex
Granitic complex in Osaka Prefecture.
UF Ibaragi Granitic Complex
BT Honshu
Japan

Ibaragi Granitic Complex
use Ibaragi Complex

Ibaraki
City in Osaka Prefecture. Also a prefecture in N central.
BT Honshu
Japan

Ibbenbueren
use Ibbenburen

Ibbenburen
Town in N North Rhine-Westphalia. Also search Ibbenbueren.
UF Ibbenbueren
BT North Rhine-Westphalia
West Germany

Iberia
use Iberian Peninsula

Iberian Cordillera
use Iberian Mountains

Iberian Mountains
Mountain system on E edge of great central plateau. Also search Iberian Cordillera.
UF Iberian Cordillera
BT Spain
Europe

Iberian Peninsula
Extreme SW Europe S of the Pyrenees and the Bay of Biscay. Also search Iberia. Index countries as applicable.
UF Iberia
BT Europe
SA Portugal
Spain

Ibiza
Third largest of Spain's Balearic Islands off E coast of Spanish mainland.
UF Iviza
BT Balearic Islands
Mediterranean Sea

ice
Includes use on level 2 under hydrology(1).
SA accumulation
Frozen ground
glacial features
glaciers
ground ice
hydrology
hydrosphere
ice rafting
meltwater
non-glacial ice
periglacial features
permafrost
sea ice
snow
thawing
water

ice ages
A valid term through 1977 used in combination with ancient (i.e. ice ages, ancient) on level 2 under glacial geology(1). After 1978, use ancient ice ages on level 2.

ice caps
SA glacial features
glaciers
ice sheets

ice mantle
use ice sheets

ice movement
Includes use on level 3 under glacial geology(1) glaciation(2). Before 1976, also search ice movements.
SA deposition
erosion
glacial geology
glaciation
glaciers
movement

ice rafting
Includes use on level 3 under sedimentation(1). Also search ice-rafting.
UF rafting, ice
SA ice
sedimentation
transport

ice sheets
Includes use on level 3 under glacial geology(1) glaciers(2).
UF ice mantle
SA glacial geology
glaciation
glaciers
ice caps
sheets

ice shelves
UF shelves, ice
SA glacial geology
glaciation
glaciers
marine environment

ice veins
use ice wedges

ice wedges
Includes use on level 3 under glacial geology(1) periglacial features(2) and under sedimentary structures(1) bedding plane irregularities(2). See list K.
UF ice veins
wedges, ice
SA bedding plane irregularities
fossil ice wedges
glacial geology
periglacial features
permafrost
sedimentary structures

ice wedges, fossil
use fossil ice wedges

ice-wedge cast
use fossil ice wedges

ice-wedge fill
use fossil ice wedges

ice-wedge pseudomorph
use fossil ice wedges

icebergs
SA glacial features
glaciers

Iceland
Independent country and island 155 miles SE of Greenland. Includes use on level 1 as an area term (list O). For term set options see list B.
CO N634000N663000
W0133000W0244500
BZ Atlantic Ocean
Europe
NT Breidamerkurjokull
Heimaey
Hekla
Kolbeinsey Island
Reykjanes Peninsula
Skeidararjokull
Snaefellsnes Peninsula
Surtsey
Vatnajokull
Vestmannaeyjar

Iceland crystal
use Iceland spar

Iceland spar
UF double-refracting spar
Iceland crystal
BA carbonates
BT minerals
SA calcite

Iceland-Faeroe Ridge
Between Iceland and Faeroe Islands. Also search Iceland-Faeroe Rise and Iceland-Faeroes Ridge.
UF Faeroe-Iceland Ridge
Iceland-Faeroe Rise
Iceland-Faeroes Ridge
BT Atlantic Ocean

Iceland-Faeroe Rise
use Iceland-Faeroe Ridge

Iceland-Faeroes Ridge
use Iceland-Faeroe Ridge

ichnofossils
Includes use on level 1 and 2 as a fossil term (list F); on level 3 under sedimentary structures(1) biogenic structures(2). See list K. Before 1973, also search trace fossils.
BA biogenic structures
sedimentary structures
NA Corophioides
Cruziana
Nereites
Planolites
Rusophycus
Skolithos
Thalassinoides
SA borings
burrows
chemical fossils
Collenia
fossilization
fossils
Invertebrata
lebensspuren
morphology
Ophiomorpha
Ophiomorpha nodosa
tracks
trails

Ichthyopterygia
Includes use on level 2 under Reptilia(1). See list F.
BA Reptilia
BT Tetrapoda
Vertebrata
NA Ichthyosaurus

Ichthyosaurus
Genus. Includes use on level 3 under Reptilia (1) Ichthyopterygia(2).
BA Ichthyopterygia
Reptilia
BT Tetrapoda
Vertebrata

Icriodus
Genus.
BA conodonts

Idaho
Includes use on level 1 as an area term (list O). For term set options see list B.
CO N420000N490000
W1110500W1171500
BA United States
NT Ada County
NX Adams County
NT Albion Range
Bannock County
Bingham County
NX Blaine County
NT Boise
NX Butte County
Clark County
Clearwater County
NT Coeur d'Alene
Coeur d'Alene River
NX Custer County
Franklin County
Fremont County
NT Idaho County
NX Jefferson County
Jerome
NT Kootenai County
Latah County
Lemhi County
Lemhi Range
NX Lincoln County
NT Mackay
NX Madison County
Oneida County
NT Owyhee County
Pioneer Mountains
Rainbow Mountain
Reynolds Creek
Salmon
Salmon River
Salmon River breaks
Sawtooth Range
Seven Devils Mountains
Shoshone County
Snake River plain
NX Teton County
Valley County
Washington County
NT Wilson Lake
SA Bald Mountain
Basin and Range Province
Bear Lake
Bear River basin
Bear River Range
Beaverhead Formation
Beaverhead Mountains
Belt Supergroup
Bitterroot Range
Blue Mountains
Cache Valley
Colorado Group
Columbia Plateau
Columbia River Basalt
Columbia River basin
Cordillera
Frontier Formation
Great Basin
Idaho Batholith
Lake Bonneville
Madison Group
Missoula Group
Overthrust Belt
Phosphoria Formation
Pocatello Formation
Prichard Formation
Purcell System
Raft River
Raft River basin
Ravalli Group
Snake River
Snake River basin
Snake River canyon
Stump Formation
Swan Peak Formation
Thaynes Formation
Wallace Formation
Wasatch Front
Wasatch Range
Western Interior
Yellowstone National Park

Idaho Batholith
In Bitterroot Range. Index states as applicable.
BT United States

SA Idaho
   Montana

Idaho County
N central.
BT Idaho
   United States

idaite
BA sulfides
BT minerals

ideal waves
Term introduced in 1978. Includes use on level 2 under ocean waves(1). Before 1978, also search ocean waves AND ideal.
BT ocean waves
SA waves

identification
Includes use as a level 2 or 3 term appropriate to a large number of topics, e.g. on level 2 under igneous rocks(1) and under sedimentary rocks(1). See list G.
SA classification
   definition
   nomenclature

Idjil
use Fort Gouraud

idocrase
use vesuvianite

Igaliko
Settlement 25 miles NE of Julianenhaab near S tip of island.
BT Greenland
   Arctic region

Igarka
City on the Yenisei River north of Arctic Circle in Krasnoyarsk Kray.
BT Russian Republic
   USSR

IGC
use International Geological Congress

Iglesiente
Region in SW.
BT Sardinia
   Italy

igneous activity
Includes use on level 3 under geosynclines(1) processes(2) and under orogeny(1).
UF activity, igneous
SA geosynclines
   igneous processes
   magmas
   orogeny
   volcanology

igneous intrusions
Not a valid index term for GeoRef. Use intrusions and igneous rocks.

igneous processes
Includes use on level 3 under mineral deposits, genesis(1) processes(2).
SA hydrothermal processes
   igneous activity
   magmas
   mineral deposits
   plutonic rocks
   processes

igneous rocks
Used to describe crystalline rocks of primary origin (list H). Includes use on level 1 (list A); on level 2 under weathering(1), phase equilibria(1), and paragenesis(1). If 1, term set options are:
   rock group [acidic composition, alkali basalt family, alkali gabbro family, alkalic composition, andesite-rhyolite family, basalt family, calc-alkalic composition, calcic composition, diorite family, gabbro family, granite-granodiorite family, hypabyssal rocks, lamprophyre and carbonatite family, mafic composition, plutonic rocks, pyroclastics and glasses, syenite family, trachyte-phonolite family, ultramafic family, volcanic rocks]
   rock name
   topic [alteration, classification, composition, distribution, experimental studies, genesis, geochemistry, identification, melting, nomenclature, occurrence, petrography, petrology, properties, textures]
      subtopic [e.g. chemical composition, mineral composition, physical properties]
NT alkali basalt family
   alkali gabbro family
   andesite-rhyolite family
   basalt family
   diorite family
   gabbro family
   granite-granodiorite family
   hypabyssal rocks
   lamprophyre and carbonatite family
   plutonic rocks
NA porphyry
NT pyroclastics and glasses
   syenite family
   trachyte-phonolite family
   ultramafic family
   volcanic rocks
SA acidic composition
   agglomerate
   alkalic composition
   alteration
   aphanitic texture
   autometamorphism
   banded structures
   basaltic composition
   basement
   batholiths
   calc-alkalic composition
   calcic composition
   chemical analysis
   complexes
   contact
   country rocks
   crystalline rocks
   cumulates
   differentiation
   dike swarms
   dikes
   emplacement
   extrusive rocks
   fabric
   fluid inclusions
   graphic texture
   inclusions
   intrusions
   intrusive rocks
   labradorite
   lava
   layered intrusions
   lineation
   lithology
   lopoliths
   mafic composition
   magmas
   matrix
   megacrysts
   metaigneous rocks
   metamorphism
   metasomatism
   paragenesis
   petrography
   petrology
   phase equilibria
   phenocrysts
   pipes
   plutons
   porphyritic texture
   protoliths
   pyroclastics
   schlieren
   spinifex
   abyssal rocks, lamprophyre and carbonatite family, mafic composition, plutonic rocks, pyroclastics and glasses, syenite family, trachyte-phonolite family, ultramafic family, volcanic rocks]
   standard materials
   stocks
   trap rock
   ultrabasic composition
   ultramafic composition
   volcanology

ignimbrite
Includes use on level 3 under igneous rocks(1) pyroclastics and glasses(2). See list H.
UF flood tuff
BA pyroclastics and glasses
BT igneous rocks
SA ash flows
   ash-flow tuff
   tuff
   volcanic glass
   welded tuff

ijolite
Includes use on level 3 under igneous rocks(2) alkali gabbro family(2). See list H.
BA alkali gabbro family
BT igneous rocks
SA melteigite
   plutonic rocks
   urtite

Ikh-Khayrkhan
Ore bearing region in central.
BT Mongolia

Iki Island
In Tsushima Strait off NW coast of Kyushu. Also search Iki-shima.
UF Iki-shima
BT Kyushu
   Japan

Iki-shima
use Iki Island

Ikuno Mine
Hyogo Prefecture in central.
BT Honshu
   Japan

Ikushumbetsu
Coal mine within town area of Mikasa in W central.
BT Hokkaido
   Japan

Ilerdian
Europe. See list E.
BA Paleogene
   Tertiary
BT Cenozoic

Iles de Loos
use Los Islands

Ili
Town on Ili River in E.
BT Kazakhstan
   USSR

Ili Basin
River basin. Index Kazakhstan and/or Sinkiang Uighur as applicable.
BT Asia
SA Kazakhstan
   Sinkiang Uighur

Ili Maussaq
use Ilimaussaq

Ilimaussaq
Alkaline intrusion in S Greenland, and Lovozero area of the Kola Peninsula in NW European USSR. Index Greenland and/or USSR. Also search Ilmaussaq and Ilimaussaq Intrusion.
UF Ili Maussaq
   Ilimaussaq Intrusion
   Ilmaussaq
SA Greenland
   USSR

Ilimaussaq Intrusion
use Ilimaussaq

Il'Intsa
Paleovolcano in Vinnitsa area of the Ukrainian Shield in W central.
BT Ukraine
   USSR

Ilion
use Troy

Ilium
use Troy

Illawarra Coal Measures
Top of the measures is marked by Bulli Coal Seam, which is conformably overlain by rocks of the Triassic Narrabeen Group. Illawarra District in SE New South Wales. Also search Illawarra.
BT Permian
SA Australia
   Bulli Seam
   New South Wales

Ille-et-Vilaine
Department in Brittany.
BT France

Illinoian
Pertaining to third glacial stage of Pleistocene Epoch in North America.
UF Illinoisan
BA upper Pleistocene
   Pleistocene
   Quaternary
BT Cenozoic

Illinois
Includes use on level 1 as an area term (list O). For term set options see list B.
CO N370000N423000
   W0873000W0913000
BA United States
NX Adams County
   Boone County
   Calhoun County
NT Carbondale
NX Carroll County
   Cass County
NT Chicago
NX Clark County
   Clay County
   Columbia
   Cook County
   Cumberland County
   Douglas County
   Edwards County
   Fayette County
   Franklin County
   Fulton County
   Greene County
   Grundy County
   Hamilton County
   Hancock County
   Hardin County
NT Iola Field
   Iroquois County
NX Jackson County
   Jasper County
   Jefferson County
   Johnson County
   Kane County
   Lake County
   Lawrence County
   Lee County
   Livingston County
   Macon County
   Madison County
   Marion County
   Marshall County
   Mason County
NT Mazon Creek
NX McHenry County
   McLean County
   Mercer County
   Monroe County
   Montgomery County
   Morgan County
   Perry County

Pike County
Pope County
Pulaski County
Putnam County
Randolph County
NT Rock Island
NX Saint Clair County
NT Sangamon
NX Scott County
Stark County
Union County
Urbana
Warren County
Washington County
Wayne County
White County
NT Will County
NX Williamson County
Winnebago County
NT Wood River
SA Bainbridge Formation
Brassfield Formation
Carbondale Formation
Caseyville Formation
Cedar Valley Formation
Chattanooga Shale
Chesterian
Fernvale Formation
Francis Creek Shale
Franconia Formation
Great Lakes region
Illinois Basin
Illinois State Geological Survey
Keokuk Limestone
La Salle Limestone
Lake Michigan
Maquoketa Formation
Mattoon Formation
Midcontinent
Midwest
Mississippi Embayment
Mississippi River
Mississippi Valley
Ohio River
Ohio River valley
Peoria Loess
Platteville Formation
Pottsville Group
Richmond Group
Ripley Formation
Saint Laurent Limestone
Saint Louis Limestone
Saint Peter Sandstone
Sainte Genevieve Limestone
Salem Limestone
Shakopee Formation
Springfield
Springfield Coal Member
Upper Mississippi Valley
Versailles
Warsaw Formation
Wedron Formation
Wilcox Group
Woodfordian

Illinois Basin
Index states as applicable.
BT United States
SA Illinois
Indiana
Kentucky

Illinois State Geological Survey
SA Illinois
surveys

Illinoisan
use Illinoian

illite
BA sheet silicates
silicates
BT minerals
SA clay minerals
crystallinity
hydromica
muscovite

Ilmaussaq
use Ilimaussaq

Ilmen Mountains
Near Chelyabinsk in the Central Urals.
BT Russian Republic
USSR

ilmenite
Also search menaccanite.
UF menaccanite
BA oxides
BT minerals
SA magnetic minerals

Iloilo Basin
SW Panay.
BT Philippine Islands

ilvaite
BA orthosilicates
silicates
BT minerals

imagery
As of 1978, term is used on level 2 under remote sensing(1).
SA geophysical surveys
Landsat
remote sensing

Imandra
Lake on W Kola Peninsula, Murmansk Oblast.
BT Russian Republic
USSR

Imataca Complex
In Serrania de Imataca Mountains S of Orinoco Delta.
BT Bolivar
Venezuela

imbrication
Includes use on level 3 under sedimentary structures(1) planar bedding structures(2). See list K.
BA planar bedding structures
sedimentary structures
SA bedding plane irregularities
tectonics

immiscibility
Includes use on level 3 under magmas(1). Also search miscibility.
UF miscibility
SA magmas
phase equilibria
solubility

imogolite
BA sheet silicates
silicates
BT minerals
SA clay minerals

impact features
Includes use on level 2 under geomorphology(1).
UF features, impact
SA astroblemes
cratering
craters
cryptoexplosion features
geomorphology
impactite
impacts
meteor craters
meteorites
suevite

impact phenomena
A valid term through 1972. From 1972 through 1978 impact was used. After 1978, use impacts.

impact statements
A level 1 term as of 1978. For official documents detailing the impact of contemplated works or land use on the environment. Includes use on level 2 under land use(1), nuclear facilities(1), and pollution(1). If 1, term set options are:
conservation
subtopic
geologic hazards
subtopic
land use
subtopic
pollution
subtopic
reclamation
subtopic
waste disposal
subtopic
SA conservation
ecology
engineering geology
environmental geology
geologic hazards
land use
nuclear facilities
pollution
reclamation
waste disposal

impactite
SA cryptoexplosion features
Darwin glass
impact features
metamorphic rocks
meteorites
suevite

impacts
Includes use on level 3 under seismology(1) and volcanology(1). Before 1973, also search impact phenomena; impact.
SA craters
cryptoexplosion features
geomorphology
impact features
meteor craters
meteorites
planetology

Imperial County
BT California
United States

Imperial Valley
Imperial County in extreme SE.
CO N324000N333000
W1151500W1161500
BT California
United States

impurities
A general term used mostly in reference to water.
SA contamination
ground water
minerals
purification
water

In
use indium

in situ
Applicable to many fields, especially rocks, soils, or fossils when in the situation in which they were originally formed or deposited.

Inadunata
Subclass. Includes use on level 3 under Echinodermata(1) Crinoidea(2).
BA Crinoidea
Crinozoa
Echinodermata
BT Invertebrata

Inangahua
Village in NW.
BT South Island
New Zealand

Inari
Town on Lake Inari in N part of country.
UF Enari
BT Finland

Inarticulata
Includes use on level 2 under Brachiopoda(1). See list F.
BA Brachiopoda
BT Invertebrata
NA Lingula
SA Articulata

incarbonization
use coalification

Inceptisols
Includes use on level 3 under soils(1). See list M.
BA soils

inclined folds
Term introduced in 1978. Includes use on level 3 under folds(1) orientation(2). Before 1978, also search folds AND inclined.
BA folds
SA orientation

inclusions
Includes use on level 1 (list A) and 2 for studies on enclosures or enclaves in rocks or minerals. If 1, term set options are:
mineral inclusions
type of inclusion or host materials [e.g. name of mineral, name of rock, minerals, ore deposits, rocks] or topic or area
topic [analysis, changes, composition, detection, experimental studies, genesis, geologic barometry, geologic thermometry, observations, theoretical studies]
subtopic
xenoliths
type of xenolith or host materials or topic
NT mineral inclusions
xenoliths
SA chemical analysis
crystal growth
crystal structure
fluid inclusions
geochemistry
geologic barometry
geologic thermometry
host materials
host rocks
igneous rocks
isotopes
lava
magmas
mineral deposits
minerals
P-T conditions
petrology
phase equilibria
rocks

incoalation
use coalification

incompressibility modulus
use bulk modulus

increment
use recharge

index beds
use key beds

index fossils
Includes use on level 3 under paleontology(1). Also search guide fossils.
UF guide fossils
SA assemblages
biostratigraphy
fossils
microfossils
nannofossils
stratigraphy

index maps
BA maps

index of refraction
use refractive index

India
Includes use on level 1 as an area term (list O). For term set options see list B.
CO N070000N370000

India

```
        E0970000E0680000
BA   Asia
NT   Ajay River
     Andaman Islands
     Andhra Pradesh
     Assam
     Beas River
     Bihar
     Bundelkhand
     Cauvery Basin
     Coromandel Coast
     Damodar Valley
     Deccan Plateau
     Delhi
     Eastern Ghats
     Garo Hills
     Ghats
     Goa
     Godavari River
     Godavari Valley
     Gujarat
     Haryana
     Himachal Pradesh
     Jumna River
     Kerala
     Krishna
     Kunavaram Series
     Madhya Pradesh
     Mahanadi Valley
     Maharashtra
     Mysore
     Narmada River
     Narmada Valley
     Neill Island
     Nicobar Islands
     Orissa
     Pondicherry
     Pranhita-Godavari Valley
     Rajasthan
     Satpura Range
     Sikkim
     Singhbhum shear zone
     Son Valley
     Tamil Nadu
     Trans-Aravalli Vindhyan Basin
     Tripura
     Uttar Pradesh
     West Bengal
     Western Ghats
SA   Aravalli System
     Ariyalur Stage
     Athgarh Sandstone
     Badami Series
     Bagh Beds
     Barakar Stage
     Bhander Group
     Bhuj Series
     Bijawar System
     Brahmaputra River
     Cuddalore Series
     Cuddapah System
     Damuda Series
     Deccan Traps
     Delhi System
     Dharwars
     Ganges River
     Gangpur Series
     Gondwana
     Gondwana System
     Himalayas
     Hyderabad
     Indian Plate
     Indian Shield
     Intertrappean Beds
     Jabalpur Series
     Jammu and Kashmir
     Kaimur Sandstone
     Kaladgi System
     Karakoram
     Karharbari Stage
     Kasauli Series
     Kashmir
     Kurnool System
     Ladakh
     Lathi Formation
     Lesser Himalayas
     Muth Quartzite
     Nari Series
     Neyveli Lignite
     Niniyur Group
     Panchet Series
     Punjab
     Raghavapuram Shales
     Ramnagar
     Raniganj Stage
     Sausar Series
     Semri Series
     Singhbhum Granite
     Siwalik Range
     Siwalik System
     Tal Formation
     Talchir Formation
     Talchir Series
     Tons Member
     Vindhyan
     Vindhyan Basin
     Warkalli Formation
```

Indian Ocean
Includes use on level 1 or 2 as an area term (list O). If 1, see term set options under list B.
```
CO   S700000N250000
     E1200000E0200000
NT   Agulhas Bank
     Amsterdam Island
     Anjouan Island
NA   Arabian Sea
NT   Argo abyssal plain
     Bay of Bengal
     Carlsberg Ridge
     Christmas Island
NA   Comoro Islands
NT   Crozet Islands
     Cuvier abyssal plain
NA   East Indian Ocean
NT   Gascoyne abyssal plain
     Great Australian Bight
     Kerguelen Islands
     Madagascar Basin
     Mahe Island
NA   Maldive Islands
NT   Mascarene Basin
     Mascarene Islands
NA   Mauritius
NT   Mid-Indian Ridge
     Mozambique Channel
     Ninetyeast Ridge
NA   Persian Gulf
NT   Perth abyssal plain
NA   Reunion
NT   Rodriguez Island
NA   Seychelles
NT   Somali Basin
     Timor Sea
NA   West Indian Ocean
NT   Wharton Basin
SA   Antarctic Ocean
     Indian Plate
```

Indian Ocean Ridge
use Mid-Indian Ridge

Indian Plate
Tectonic plate including the India-Himalaya-Burma area. Index Tibet and countries as applicable.
```
BT   Asia
SA   Bangladesh
     India
     Indian Ocean
     Pakistan
     plate tectonics
     Tibet
```

Indian Ridge
use Mid-Indian Ridge

Indian Shield
Plateau region in S and central India plus Sri Lanka. Index countries as applicable. Also search Peninsular Shield.
```
UF   Peninsular Shield
BT   Asia
SA   Deccan Plateau
     India
     Sri Lanka
```

Indiana
Includes use on level 1 as an area term (list O). For term set options see list B.
```
CO   N374500N414500
     W0844500W0881000
BA   United States
NX   Adams County
     Benton County
NT   Blackford County
NX   Boone County
     Brown County
     Carroll County
     Cass County
     Clark County
     Delaware County
     Fayette County
     Floyd County
     Franklin County
     Fulton County
     Grant County
     Greene County
     Hamilton County
     Hancock County
     Jackson County
     Jasper County
     Jefferson County
     Johnson County
NT   Kentland
NX   Lake County
     Lawrence County
     Madison County
     Marion County
     Marshall County
     Monroe County
     Montgomery County
     Morgan County
     Newton County
     Orange County
     Perry County
     Pike County
     Pulaski County
     Putnam County
     Randolph County
     Rush County
     Scott County
     Sullivan County
NT   Tippecanoe County
NX   Union County
     Warren County
     Washington County
     Wayne County
     White County
SA   Borden Group
     Brassfield Formation
     Chesterian
     Cincinnati Arch
     Eden Shale
     Fairview Formation
     Great Lakes region
     Illinois Basin
     Lake Michigan
     Mansfield Formation
     Maumee River valley
     Midcontinent
     Midwest
     New Albany Shale
     Ohio River
     Ohio River valley
     Richmond Group
     Saint Louis Limestone
     Saint Peter Sandstone
     Sainte Genevieve Limestone
     Salem Limestone
     Staunton Formation
     Versailles
     Wabash Formation
     Waldron Shale
     Warsaw Formation
     Wedron Formation
     White River
     Whitewater River valley
```

indicators
Includes use on level 2 under paleoclimatology(1), under paleoecology(1), and under plate tectonics(1).
```
SA   diagenesis
     paleoclimatology
     paleoecology
     plate tectonics
```

Indigirka River
NE Yakutia. Flows into East Siberian Sea. Also search Indigirka.
```
BT   Russian Republic
     USSR
```

indium
Includes use on level 1 and 2 as a chemical element (list D).
```
UF   In
SA   elements
```

Indochina
The SE peninsula of Asia including the Malay Peninsula. (French Indochina consisted of current Cambodia, Laos, and Vietnam). Index countries as applicable. Includes use on level 1 as an area term (list O). For term set options see list B.
```
UF   Farther India
BA   Asia
SA   Burma
     Cambodia
     Far East
     Laos
     Malaysia
     Thailand
     Vietnam
```

indochinite
BA tektites

Indonesia
Includes use on level 1 as an area term (list O). For term set options see list B.
```
CO   S113000N063000
     E1413000E0950000
BA   Asia
NT   Bali
     Bangka
     Billiton
     Celebes
     Irian Jaya
     Java
     Kalimantan
     Krakatoa
     Lesser Sunda Islands
     Moluccas
     Sumatra
SA   Arafura Sea
     Far East
     Malay Archipelago
     Timor
```

Indre
River which flows NW into the Loire River, and a department in W central.
```
BT   France
SA   Touraine
```

Indre-et-Loire
Department in N central.
```
BT   France
NT   Tours
SA   Touraine
```

induced polarization
Includes use on level 3 under geophysical methods(1) and under geophysical surveys(1).
```
BA   geophysical methods
SA   electrical field
     electrical methods
     electrical surveys
     geophysical methods
     geophysical surveys
     polarization
     well-logging
```

induction
Used for geophysical properties.
```
SA   electromagnetic methods
     geophysical methods
```

magnetic field
paleomagnetism
well-logging

Indus River
Rises in SW Tibet and flows into Arabian Sea. Index Jammu and Kashmir, Pakistan, and Tibet as applicable.
BT   Asia
SA   Jammu and Kashmir
     Pakistan
     Tibet

industrial ash
use ash

industrial minerals
Includes use on level 1 as a commodity term (list C).
BT   minerals
SA   abrasives
     carbonado
     economic geology
     nonmetals
     refractory materials

industrial waste
As of 1978, term is used on level 2 under waste disposal(1).
UF   waste, industrial
SA   environmental geology
     liquid waste
     pollution
     radioactive waste
     solid waste
     waste disposal
     waste water

industry
Used as a general term.
SA   annual report
     economic geology
     engineering geology
     environmental geology

inert gases
use noble gases

inertinite
BA   macerals
SA   coal
     organic residues
     sedimentary rocks

infiltration
Includes use on level 3 under hydrology(1).
SA   ground water
     hydrology
     percolation
     recharge

information systems
Includes use on level 3 under automatic data processing(1).
SA   automatic data processing
     data bases
     data handling
     data processing
     data retrieval
     data storage
     systems

Infracambrian
BA   upper Precambrian
     Precambrian
SA   Eocambrian
     Vendian

infrared
A valid term through 1977. After 1977, use infrared spectroscopy.

infrared methods
Includes use on level 2 under geophysical methods(1). See spectroscopy(1) methods(2).
BA   geophysical methods
SA   infrared surveys
     methods
     remote sensing
     thermal emission

infrared spectra
Includes use on level 3 under chemical analysis(1) or spectroscopy(1). Before 1978, also search infrared.
SA   chemical analysis
     infrared spectroscopy
     spectra
     spectroscopy

infrared spectroscopy
Not valid term through 1977. After 1977, includes use on level 3 under spectroscopy(1) and under chemical element(1). Before 1978, also search spectroscopy AND infrared.
BT   spectroscopy
SA   analysis
     infrared spectra

infrared surveys
Includes use on level 2 under geophysical surveys(1).
BA   geophysical surveys
BT   surveys
SA   infrared methods

Ingaladhal
use Ingaldhal

Ingaldhal
Ore bearing region in Chitradurga District in central Mysore.
UF   Ingaladhal
BT   Mysore
     India

Ingham County
S central Lower Peninsula. Also search Ingham.
BT   Michigan
     United States

Ingleborough
Mountain in W Yorkshire in N.
BT   England
     Great Britain
     United Kingdom

Ingleton
Village and parish in W Yorkshire in N.
BT   England
     Great Britain
     United Kingdom

Inglewood
City in Los Angeles County in S.
BT   California
     United States

Ingul River
S Ukraine. Also search Ingul.
BT   Ukraine
     USSR

Ingulets River
S Ukraine. Also search Ingulets.
BT   Ukraine
     USSR

Ingur (River)
use Inguri River

Inguri River
Flows into Black Sea. Also search Ingur; Inguri.
UF   Ingur (River)
BT   Georgian Republic
     USSR

inhomogeneity
Also search heterogeneity.
UF   heterogeneity
SA   electrical methods
     homogeneity
     prospecting

injection
Includes use on level 3 under environmental geology(1) waste disposal(2).
SA   engineering geology
     environmental geology
     fluid injection
     waste disposal

Inland Sea
Extends E-W with W Honshu on the N, Shikoku on the S, and Kyushu on the W. Connected with Pacific Ocean by 4 channels.
UF   Seto Inland Sea
     Seto Sea
     Seto-chi-umi
     Seto-Naikai
     Setouchi
BT   Japan
     Asia

inlets
SA   bays
     coastal environment
     shore features
     shorelines
     tidal inlets

Inn Valley
River valley. Index Bavaria and countries as applicable.
BT   Europe
SA   Austria
     Bavaria
     Switzerland

innelite
BA   orthosilicates
     silicates
BT   minerals

inner core
BA   core
SA   outer core

Inner Hebrides
Islands immediately off W coast of Scotland. Separated from Outer Hebrides by the Straits of The Minch, the Little Minch, and the Sea of Hebrides.
BT   Scotland
     Great Britain
     United Kingdom
NT   Mull Island
SA   Hebrides
     Islay
     Isle of Skye
     Outer Hebrides
     Raasay
     Rhum

Inner Mongolia
Autonomous region S, SE, and E of Mongolia.
BT   China
NT   Khingan Range
SA   Gobi Desert
     Kerulen River

inner shelf
SA   continental shelf
     outer shelf

inner slope
Term introduced in 1978.
SA   continental slope
     outer slope

inner transition elements
use rare earths

Innsbruck
City in central.
BT   Tyrol
     Austria

Inoceramidae
Family. Includes use on level 3 under Mollusca(1) Bivalvia(2).
BA   Bivalvia
     Mollusca
BT   Invertebrata

Inoceramus
Genus. Includes use on level 3 under Mollusca(1) Bivalvia(2).
BA   Bivalvia
     Mollusca
BT   Invertebrata

inorganic
A valid term through 1977. After 1977, use inorganic materials.

inorganic materials
Before 1978, also search inorganic AND materials.
SA   materials
     organic materials

inosilicates
use chain silicates

Inowroclaw
City in NW central part of country.
BT   Bydgoszcz
     Poland

INQUA
Acronym. Includes use on level 3 under associations(1).
UF   International Association for Quaternary Research
SA   associations

Insecta
Class. Includes use on level 1 and 2 as a fossil term (list F).
BT   Arthropoda
     Invertebrata
NA   Archostemata
     Coleoptera
     Diptera
     Heteroptera
     Hymenoptera
     Lepidoptera
     Odonata
     Plecoptera
     Plectoptera
     Protorthoptera
     Raphidiodea

Insectivora
Includes use on level 2 under Mammalia(1). See list F.
BA   Mammalia
BT   Tetrapoda
     Vertebrata

inselbergs
SA   erosion features
     geomorphology

insolation
SA   mechanical weathering
     weathering

insoluble residues
Includes use on level 3 under sedimentary rocks(1). Also search IR.
UF   IR
     residues, insoluble
SA   sedimentary rocks
     siliceous composition
     solubility

instabilities, plasma
use plasma instabilities

installations, marine
use marine installations

installations, submarine
use submarine installations

installations, underground
use underground installations

instruments
Includes use as a level 2 or 3 term appropriate to a large number of topics, e.g. on level 2 under meteorology(1), geochemistry(1), heat flow(1), and hydrology(1). See list G.
SA   design
     extensometers
     gravimeters
     magnetometers
     ocean bottom seismographs
     Seislog
     seismographs
     seismometers
     techniques
     tiltmeters

intensity
Used as general term. Includes use

on level 2 under earthquakes(1); on level 3 under magnetosphere(1) cosmic rays(2) and under interplanetary space(1). Also search intensities.
SA amplitude
aurora
bow shock waves
cosmic rays
disturbances
earthquakes
interplanetary space
magnetic field
magnetic storms
magnetosphere
magnitude
modified Mercalli scale
paleomagnetism
particle precipitation
particles
solar cycles
sudden commencements

interaction
A valid general term through mid-1978. Use layered media for the geophysical meaning.

interactions, boundary
use boundary interactions

interface, air-sea
use air-sea interface

interface, oil-gas
use oil-gas interface

interface, oil-water
use oil-water interface

interface, sediment-water
use sediment-water interface

interferometry
SA optical methods

interglacial environment
Term introduced in 1978. Before 1978, search interglacial.
SA environment
glacial environment
paleoclimatology

intergrowths
Includes use on level 3 under crystal growth(1).
SA crystal growth
crystallization
crystallography
crystals

interior
Includes use on level 2 under Earth(1), Moon(1), and seismology(1).
SA core
Earth
geodynamics
Moon
seismology

intermediate earthquake
use intermediate-focus earthquakes

intermediate-focus earthquakes
Term introduced in 1978. Includes use on level 3 under earthquakes(1). Before 1978, search intermediate-focus.
UF intermediate earthquake
SA deep-focus earthquakes
earthquakes
focus
shallow-focus earthquakes

intermontane basins
SA basins

internal waves
Includes use on level 2 under ocean waves(1).
SA ocean waves
waves

International Association for Quaternary Research
use INQUA

international cooperation
Includes use on level 3 under associations(1).
UF cooperation, international
SA associations
policy

International Decade of Ocean Exploration
SA oceanography

International Geological Congress
Includes use on level 3 under associations(1).
UF IGC
SA associations

International Union of Geological Sciences
use IUGS

interplanetary dust
Includes use on level 2 under interplanetary space(1).
SA cosmic dust
dust
interplanetary space
meteorites
particles

interplanetary space
Usually out-of-scope for GeoRef. Includes use on level 1 for special bibliographies. Also search space. Term set options are:
comets
  subtopic
cosmic rays
  albedo
  cut-off rigidities
  intensity
  ionization
  subtopic (flares, measuring platform, particles, solar cycles)
electromagnetic radiation
  subtopic
instruments
  name of instrument or platform or subtopic
interplanetary dust
  subtopic
shock waves
  subtopic
solar wind
  subtopic
UF outer space
  space
SA aeronomy
albedo
asteroids
astrophysics and solar physics
comets
cosmic dust
cosmic rays
cutoff rigidities
electromagnetic radiation
intensity
interplanetary dust
ionization
magnetosphere
meteoroids
meteors
particles
planetology
plasma
shock waves
solar cycles
solar flares
solar wind

interpretation
Includes use as a level 2 or 3 term appropriate to a large number of topics, e.g. on level 2 under foliation(1) and heat flow(1). See list G.

interstitial water
use pore water

intertidal environment
Term introduced in 1978. Before 1978, search intertidal.
SA environment
littoral environment

Intertrappean Beds
Upper Cretaceous to lower Eocene. Sedimentary deposits of fossiliferous cherts freshwater limestone or of bole intercalated between the flows of Deccan Trap Basalt. Also search Deccan Intertrappean; Deccan Intertrappean Beds; Deccan Intertrappean Series; Intertrappean.
UF Deccan Intertrappean
Deccan Intertrappean Beds
Deccan Intertrappean Series
SA Cretaceous
Eocene
India
lower Eocene
Upper Cretaceous

intraclastic texture
use intraclasts

intraclasts
UF intraclastic texture
SA carbonate rocks
textures

intrafolial folds
Term introduced in 1978. Before 1978, search folds AND intrafolial.
BA folds

intraplate
Includes use on level 3 under plate tectonics(1) geometry(2).
SA microplates
plate tectonics
plates

Intrazonal soils
See list M.
BA soils
SA Bog soils
Brown forest soils
Hydromorphic soils
Meadow soils
Planosols
Solonchak soils
Solonetz soils

intrusion (ground water)
use salt-water intrusion

intrusions
Includes use on level 1 (list A). Used for the igneous rocks mass formed by emplacement of magma in pre-existing rock. It is also used for the process of emplacement. Term set options are:
kind of intrusion [batholiths, diatremes, dikes, laccoliths, layered intrusions, lopoliths, pipes, plugs, plutons, ring complexes, sheets, sills, stocks]
  topic or area
topic [age, classification, composition, contact, distribution, evolution, extent, genesis, geochemistry, mechanism, occurrence, petrology, structure]
  subtopic
UF invasion (intrusion)
  irruption (intrusion)
NA batholiths
diatremes
dikes
laccoliths
layered intrusions
lopoliths
pipes
plugs
plutons
ring complexes
sills
stocks
SA aureoles
breccia pipes
contact
country rocks
crystallization
diapirism
differentiation
dike swarms
domes
emplacement
foliation
heat flow
igneous rocks
intrusive rocks
lineation
magma chambers
magmas
metamorphic rocks
metamorphism
metasomatism
petrology
ring structures
sheets
swarms
veins
volcanology
zoning

intrusive
A valid term through 1977. After 1977, use intrusive rocks.

intrusive mountain
use batholiths

intrusive rocks
Term introduced in 1978. Includes use on level 3 under igneous rocks(1). Before 1978, also search igneous rocks AND intrusive.
SA hypabyssal rocks
igneous rocks
intrusions
rocks

Inuvik
Village in NW District of Mackenzie.
BT Northwest Territories
Canada

invasion (intrusion)
use intrusions

inventory
Includes use on level 2 and 3 under commodity terms (list C); on level 3 under environmental geology and under engineering geology(1).
SA engineering geology
environmental geology

Inverness
City in Inverness-shire in N.
CO N564000N580000
W0033000W0073000
BT Scotland
Great Britain
United Kingdom

Inverness-shire
County in NW.
BT Scotland
Great Britain
United Kingdom

inverse problem
Includes use on level 3 under geophysics(1), geophysical methods(1), geophysical surveys(1), and seismology(1). Also search inversion.
UF inversion (seismology)
SA geophysical methods
geophysical surveys
geophysics
seismology

inversion (seismology)
use inverse problem

Invertebrata
Includes use on level 1 and 2 as a fossil term (list F); on level 3 under soils(1) biota(2). Term is to be used

on level 1 only when more specific terms do not apply or are too numerous to be recorded.
NT Annelida
Archaeocyatha
Arthropoda
Brachiopoda
Bryozoa
Coelenterata
Echinodermata
foraminifera
Graptolithina
Mollusca
Porifera
Pterobranchia
Radiolaria
SA biota
coprolites
ichnofossils
shells
tests

Inyo County
E California.
CO N354500N373000
W1154000W1184500
BT California
United States

Inyo Mountains
Range in W central Inyo County in E California. Also search White-Inyo Mountains and White-Inyo Range.
BT California
United States

Io
use ionium

Io/Th
Includes use on level 3 under absolute age(1) methods(2).
SA absolute age
ionium
thorium

Io/U
Includes use on level 3 under absolute age(1) methods(2).
SA absolute age
ionium
uranium

iodates
Includes use on level 2 under minerals(1). See list L.
BT minerals

iodine
Includes use on level 1 and 2 as a chemical element (list D) and as a commodity term (list C); on level 3 under brines(1) and salt(1).
UF I
SA brines
elements
I-129
I-131
salt

Iola Field
Term introduced in 1978. Gas field in Clay County in S central Illinois. Before 1978, search Iola; Iola gas field.
UF Iola gas field
BT Illinois
United States

Iola gas field
use Iola Field

ion densities and temperatures
Includes use on level 2 under ionosphere(1). Also search ion densities.
SA densities and temperatures
density
electron content
ionosphere
ions
temperature

ion exchange
As of 1978, term is used on level 2 under crystal chemistry(1). Also search cation exchange.
UF base exchange
SA anions
cation exchange capacity
clay mineralogy
crystal chemistry
crystal structure
diffusion
electrolytes
exchange capacity
geochemistry
ions
sea water
soils

ion probe
Includes use on level 3 under chemical analysis(1) methods(2) and under spectroscopy(1) methods(2). For methodology, not data.
UF probe, ion
SA analysis
chemical analysis
electron probe
ion probe data
ions
spectroscopy

ion probe data
Term introduced in 1978. Before 1978, search ion probe AND data.
SA chemical analysis
data
ion probe
spectroscopy

Ionian Islands
Administrative region comprising 7 islands in Ionian Sea off W coast: Corfu, Paxos, Leukas, Ithaca, Cephalonia, Zante, and Cerigo.
BT Greece
NT Cephalonia
SA Ithaca

Ionian Sea
Between SE coast of Italy and W Greece. Includes use on level 1 as an area term (list O). For term set options see list B.
BA Mediterranean Sea

Ionian Zone
S, SW, and W.
BT Albania
SA Cretaceous
Jurassic
Mesozoic
Tertiary

ionium
Radioactive isotope of radon of uranium-radium series. Also search Io.
UF Io
SA Io/Th
Io/U
isotopes
radon

ionization
Includes use on level 2 under aeronomy(1); on level 3 under interplanetary space(1) cosmic rays(2).
SA aeronomy
cosmic rays
interplanetary space
ionosphere
ions
particle radiation
solar wind

ionosphere
Usually out-of-scope for GeoRef. Includes use on level 1 for special bibliographies. Term set options are:
airglow
type of emission or latitude or mechanism
albedo
[magnetic albedo, neutron albedo, also albedo of electromagnetic waves]
aurora
subtopic
currents
topic [e.g. field-aligned, ring, etc.]
disturbances
subtopic
D-region
subtopic
electrical field
subtopic
electrojet
equatorial regions
polar regions
subtopic
electron content
topic [e.g. density, region of atmosphere, temperature]
E-region
topic [e.g. density, sporadic E, temperature, variations]
F-region
topic [airglow, density, spread-F, temperature]
instruments
name of instrument or phenomena or platform
ion densities and temperatures
name of ion or phenomenon or platform
name of phenomena or platform
name of region
subtopic
particle precipitation
name of particle or phenomenon or platform
phenomenon or platform
scintillations
subtopic
wave propagation
name of phenomenon or subtopic
type of wave [e.g. ELF, HF, VHF, VLF] or platform
SA aeronomy
airglow
albedo
atmosphere
aurora
currents
D-region
disturbances
E-region
electrical field
electrojet
electron content
ELF
equatorial region
F-region
HF
ion densities and temperatures
ionization
magnetosphere
meteorology
particle precipitation
particle radiation
polar regions
scintillations
solar cycles
stratosphere
VLF
wave propagation

ions
Includes use on level 2 under meteorology(1); on level 3 under geochemistry(1). Also search cations.
SA anions
cation exchange capacity
chloride ion
electrolytes
exchange capacity
fixation
fluoride ion
geochemistry
ion densities and temperatures
ion exchange
ion probe
ionization
meteorology
particle precipitation
particles
trapped particles

Iowa
Includes use on level 1 as an area term (list O). For term set options see list B.
CO N402500N433000
W0901000W0963500
BA United States
NX Adams County
Benton County
Boone County
Buchanan County
Butler County
Calhoun County
Carroll County
Cass County
Cherokee County
Clarke County
Clay County
Dallas County
NT Davenport
NX Delaware County
Dickinson County
Fayette County
Floyd County
Franklin County
Fremont County
Greene County
Grundy County
Hamilton County
Hancock County
Hardin County
Humboldt County
Iowa County
Jackson County
Jasper County
Jefferson County
Johnson County
Lee County
Madison County
NT Mahaska County
NX Marion County
Marshall County
Mitchell County
Monroe County
Montgomery County
Pocahontas County
Polk County
Scott County
Sioux County
Taylor County
Union County
Warren County
Washington County
Wayne County
Webster County
Winnebago County
SA Beil Limestone Member
Cedar Valley Formation
Cherokee Group
Colorado Group
Desmoinesian
Douglas Group
Ervine Creek Limestone
Forest City Basin
Kansas City Group
Keokuk Limestone
Labette Shale
Lawrence Formation
Lecompton Limestone
Lime Creek Formation
Maquoketa Formation
Marmaton Group
Midcontinent
Midwest
Mississippi River
Mississippi Valley
Missouri River
Missouri River basin

Missouri River valley
Oread Limestone
Peoria Loess
Platteville Formation
Plattsburg Limestone
Plattsmouth Limestone Member
Pleasanton Group
Saint Louis Limestone
Saint Peter Sandstone
Sainte Genevieve Limestone
Salem Limestone
Shakopee Formation
Sioux Quartzite
Stanton Formation
Upper Mississippi Valley
Wabaunsee Group
Warsaw Formation
Waterloo
Wyandotte Limestone

Iowa County
Index states as applicable.
BX Iowa
Wisconsin
BT United States

iozite
use wustite

Ipswich
City in SE.
BT Queensland
Australia

Ipswich Coal Measures
SE Queensland.
BT Triassic
SA Australia
Queensland

Ipswichian
Europe.
BA upper Pleistocene
Pleistocene
Quaternary
BT Cenozoic

Ir
use iridium

IR
use insoluble residues

Iran
Includes use on level 1 as an area term (list O). For term set options see list B.
CO N250000N393000
E0632000E0440000
UF Persia
BA Asia
NT Anarak
Azerbaijan
Dasht-e-Bayaz
Djadjerud Valley
Elburz
Fars
Kerman
Lar
Lut Desert
Sangun
Songhor
Talysh Mountains
Teheran
SA Bakhtiari Formation
Caspian Basin
Caspian Sea
Geirud Formation
Kopet-Dag Range
Kura Lowland
Middle East
Zagros

Iraq
Includes use on level 1 as an area term (list O). For term set options see list B.
CO N290000N371500
E0483000E0390000
BA Middle East
NT Kirkuk
SA Asia

Bakhtiari Formation
Euphrates River
Mesopotamia
Near East
Shiranish Formation
Sinjar Formation
Zagros
Zubair Formation

Irati Formation
BT Permian
SA Brazil
Goias
Mato Grosso
Minas Gerais
Parana
Passa Dois Group
Sao Paulo

Ireland
The Republic of Ireland, also called Eire, which occupies the 26 counties in the S, central, and NW of the island of Ireland. Includes use on level 1 as an area term (list O). For term set options see list B.
CO N513000N552000
W0063000W0103000
UF Eire
BA Europe
NT Clare
Connemara
Cork
Dingle Peninsula
Donegal
Dublin
Galway
Kerry
Kildare
Leinster
Limerick
Mayo
Meath
Sligo
Tipperary
Tynagh
Wexford
Wicklow
SA Northern Ireland

Irgiz
Village on Irgiz River in Aktyubinsk Oblast in E.
BT Kazakhstan
USSR

Irian Barat
use Irian Jaya

Irian Jaya
W one-half of island of New Guinea. Also search West Irian.
UF Dutch New Guinea
Irian Barat
Netherlands New Guinea
West Irian
West New Guinea
BT Indonesia
NT False Cape
SA New Guinea

iridium
Includes use on level 1 and 2 as a chemical element (list D). Also search Ir.
UF Ir
SA elements

Irish Sea
Between England and Ireland. Includes use on level 1 as an area term (list O). For term set options see list B.
CO N530000N552000
W0030000W0063000
BA Atlantic Ocean
SA Cardigan Bay

Irkutsk
City on Angara River near SW shore of Lake Baikal. Also an oblast.
CO N520000N530000

E1060000E1070000
BT Russian Republic
USSR

Irkutsk Amphitheater
use Irkutsk Basin

Irkutsk Basin
Coal Basin extending 150 miles NW along Trans-Siberian RR from Lake Baikal. Also search Cheremkhovo Basin and Irkutsk Amphitheater.
UF Cheremkhovo Basin
Cheremkovo coal basin
Irkutsk Amphitheater
BT Russian Republic
USSR

iron
Includes use on level 1 and 2 as a commodity term (list C) and as a chemical element (list D); on level 2 under mineral deposits, genesis(2) and paragenesis(1). Also search Fe.
UF Fe
SA elements
Fe-55
Fe-57
ferric iron
ferromanganese composition
ferrous iron
ferruginous composition
heavy metals
hematite
iron meteorites
native elements and alloys

iron alum
use halotrichite

Iron County
Index states as applicable.
BX Michigan
Missouri
Utah
Wisconsin
BT United States

iron formations
Includes use on level 3 with iron-rich composition under sedimentary rocks(1) chemically precipitated rocks(1). See list I. Also search iron formation.
UF formations, iron
iron-formations
BA chemically precipitated rocks
BT sedimentary rocks
SA iron-rich composition
itabirite
jaspilite
taconite

iron meteorites
Includes use on level 3 under meteorites(1). Before 1974, also search octahedrites.
UF meteoric iron
siderite (meteorite)
BA meteorites
SA hexahedrites
iron
kamacite

iron oxides
BA oxides
BT minerals

iron spinel
use hercynite

iron sulfides
BA sulfides
BT minerals

iron-formations
use iron formations

iron-rich composition
Term introduced in 1978 on level 3 under sedimentary rocks(1) chemically precipitated rocks(2) or under sediments(1) chemically precipitated sediments(2). Before 1978, search iron-rich rocks; iron-rich sediments.
SA chemically precipitated rocks
chemically precipitated sediments
composition
iron formations
sedimentary rocks
sediments

iron-rich rocks
A valid term through mid-1978. Use iron-rich composition under sedimentary rocks(1) chemically precipitated rocks(1).

iron-rich sediments
A valid term through mid-1978. Use iron-rich composition under sediments(1) chemically precipitated sediments(2).
SA ferruginous composition

iron-stony meteorite
use stony irons

ironstone
Includes use on level 3 under sedimentary rocks(1) chemically precipitated rocks(2). See list I.
BA chemically precipitated rocks
BT sedimentary rocks

Iroquois County
E Illinois.
BT Illinois
United States

irrigation
As of 1978, term is used on level 2 under waterways(1).
SA ponds
soils
surface water
treatment
utilization
water
water quality
water regimes
water supply
waterways

irrotational wave
use P-waves

irruption (intrusion)
use intrusions

Irtish River
use Irtysh River

Irtysh River
Largest tributary of the Ob River. Index Sinkiang Uighur and Soviet republics as applicable. Also search Irtysh.
UF Irtish River
BT Asia
SA Kazakhstan
Ob River
Ob-Irtysh Interfluve
Russian Republic
Sinkiang Uighur

Isachsen Formation
Conformably overlies Deer Bay Formation on Axel Heiberg and Ellef Ringnes Islands. On Isachsen Peninsula of Ellef Rignes Island in Franklin District.
BT Lower Cretaceous
Cretaceous
SA Northwest Territories

Isar Valley
River valley. Index Austrian and German states as applicable. Also search Isar River.
BT Europe
SA Bavaria
Tyrol

Ischia
Island in Tyrrhenian Sea outside Bay of Naples.
BT Campania

Italy

Ise Bay
Inlet of Pacific Ocean on S coast.
BT Honshu
    Japan

Isera Mountains
use Izera Mountains

Isere
Department in SE.
CO N444500N455000
    E0063000E0044500
BT France
NT Grenoble
    Gresivaudan
SA Pelvoux Massif
    Vercors

Isere Valley
River valley in SE. Also search Isere.
BT France

Isergebirge
Mountain range of the W Sudeten Mountains. Index countries as applicable.
BT Europe
SA Czechoslovakia
    Poland
    Sudeten Mountains

Ishikari Bay
Inlet of Sea of Japan on W Hokkaido. Also search Ishikari.
BT Hokkaido
    Japan
SA Ishikari Plain

Ishikari Plain
S of Ishikari Bay in SW Hokkaido. Also search Ishikari.
BT Hokkaido
    Japan
SA Ishikari Bay

Ishikawa
Prefecture on the Sea of Japan N of Nagoya in central.
BT Honshu
    Japan

Ishim
River which flows into the Irtysh River. Index Soviet republics as applicable.
BT USSR
SA Kazakhstan
    Russian Republic

Iskar River
use Isker River

Isker River
NW central Bulgaria. Also search Isker.
UF Iskar River
    Iskur River
BT Bulgaria

Iskur River
use Isker River

island arcs
Includes use on level 2 under plate tectonics(1).
UF arcs, volcanic
    volcanic arcs
SA continental margin
    continents
    islands
    plate tectonics
    tectonophysics
    trenches

islands
Includes use on level 3, e.g. under tectonophysics(1).
SA atolls
    barrier islands
    continents
    island arcs

Islay
Most southerly island of Inner Hebrides off W coast.
CO N553500N555500
    W0060000W0063000
BT Scotland
    Great Britain
    United Kingdom
SA Inner Hebrides

Isle of Man
Island in Irish Sea.
CO N541000N543000
    W0041500W0044500
BT United Kingdom

Isle of Pines
Island in Caribbean S of W Cuba.
BT Cuba
    West Indies

Isle of Skye
Island of Inner Hebrides off NW coast. Also search Skye.
CO N570000N574000
    W0054000W0064500
UF Skye
BT Scotland
    Great Britain
    United Kingdom
SA Inner Hebrides

Isle of Wight
Island in English Channel just S of Southampton.
CO N504000N504500
    W0010000W0013000
BT England
    Great Britain
    United Kingdom

Isle Royale National Park
In NW Lake Superior. Includes Isle Royale and neighboring islands.
BT Michigan
    United States

isochrons
Includes use on level 3 under geochronology(1) and absolute age(1).
SA absolute age
    geochronology
    seismology

isoclinal
A valid term through 1977. After 1977, use isoclinal folds.

isoclinal folds
Term introduced in 1978. Includes use on level 3 under folds(1) style(2). Before 1978, also search folds AND isoclinal; isoclines.
UF isoclines
BA folds
SA deformation

isoclines
use isoclinal folds

isograd maps
Term introduced in 1978. Before 1978, search maps AND isograds.
BA maps
SA isograds

isograds
SA facies
    grade
    isograd maps
    maps
    metamorphism

Isograptus
Genus. Includes use on level 3 under Graptolithina(1) Graptoloidea(2).
BA Graptoloidea
    Graptolithina
BT Invertebrata

isomorphism
Used as a general term.
SA crystal chemistry
    paleontology

isopach maps
Before 1978, also search maps AND isopach.
BA maps
SA isopleth maps
    stratigraphy

isopleth maps
BA maps
SA isopach maps

isoprenoids
BA organic materials

isoseismic maps
BA maps
SA earthquakes
    seismology

isostasy
To be used for the condition of equilibrium of the crust above the mantle, and related features. Includes use on level 1 (list A). Term set options are:
topic [anomalies, causes, changes, compensation, concepts, detection, genesis, geomorphologic effects, interpretation, mechanism, observations]
    subtopic (no area term)
SA asthenosphere
    changes of level
    compensation
    continents
    crust
    density
    Earth
    epeirogeny
    eustacy
    geodesy
    geomorphologic effects
    geomorphology
    geophysical methods
    geophysical surveys
    glacial geology
    mantle
    Mohorovicic discontinuity
    neotectonics
    tectonics
    tectonophysics
    vertical tectonics

isothermal remanent magnetization
BA remanent magnetization
SA paleomagnetism

isotope ratios
A valid term through 1972. After 1972, use ratios. See isotopes, absolute age and name of chemical element(1) isotopes(2).

isotopes
Includes use on level 1 (list A); on level 2 under chemical elements (list D). Used for the application of the study of radioactive and stable isotopes, especially their abundances, to geology. Also search isotopic. If 1, term set options are:
element
    specific isotope
material (first-level terms only)
    type of material
topic [abundance, analysis, fractionation, methods, ratios, tracer experiments, tracers]
    subtopic
NA radioactive isotopes
    stable isotopes
SA absolute age
    activation analysis
    Al-20
    analysis
    Ar-36
    Ar-37
    Ar-38
    Ar-39
    Ar-40
    argon
    Be-10
    Be-7
    C-12
    C-13
    C-13/C-12
    C-14
    carbon
    chemical analysis
    cosmogenic elements
    Cs-137
    D/H
    deuterium
    elements
    fallout
    Fe-55
    Fe-57
    fractionation
    geochemistry
    ground water
    He-3
    He-4
    He-4/He-3
    helium
    hydrogen
    I-129
    I-131
    inclusions
    ionium
    K-40
    krypton
    lead
    Li-6
    Li-7
    metals
    meteorites
    mineral deposits
    Mn-53
    Mn-54
    N-15/N-14
    Na-22
    Na-24
    Ne-20
    Ne-21
    Ne-22
    neon
    noble gases
    O-16
    O-18
    O-18/O-16
    organic materials
    oxygen
    Pa-231
    paleoclimatology
    Pb-204
    Pb-206
    Pb-206/Pb-204
    Pb-207
    Pb-207/Pb-204
    Pb-207/Pb-206
    Pb-208
    Pb-208/Pb-204
    Pb-210
    Po-210
    potassium
    Pu-239
    Pu-244
    Ra-226
    radioactive decay
    radioactive tracers
    radioactivity
    radiometric properties
    radium
    radon
    ratios
    Rn-222
    rubidium
    S-32
    S-34
    S-34/S-32
    sea water
    Si-32
    spallation
    Sr-86
    Sr-87
    Sr-87/Sr-86
    Sr-90
    standard materials
    strontium

sulfur
tektites
Th-228
Th-230
Th-232
Th-232/Th-230
Th-234
thorium
tracer experiments
tracers
tritium
U-234
U-235
U-238
U-238/U-234
uranium
uranium disequilibrium
Xe-124
Xe-126
Xe-128
Xe-129
Xe-131
Xe-132
xenon
Zr-95

isotropic materials
Term introduced in 1978. Before 1978, search isotropic.
SA   materials

Israel
Includes use on level 1 as an area term (list O). For term set options see list B.
CO   N293000N333000
     E0353000E0342000
BA   Middle East
NT   Elath
     Galilee
     Haifa
     Makhtesh Ramon
     Mount Carmel
     Negev
     Sea of Galilee
SA   Asia
     Dead Sea
     Dead Sea Rift
     Great Rift Valley
     Jerusalem
     Jordan River
     Jordan Valley
     Levant
     Mediterranean region
     Mishash Formation
     Palestine
     Red Sea Basin
     Samaria
     Sinai

Issyk-kul Lake
NE Kirghizia. Also search Issyk-Kul.
BT   Kirghizia
     USSR

Istanbul
Province on both sides of Bosporus. Also a city on the Bosporus in Europe.
BT   Turkey
     Middle East

Istranca Mountains
Range along Black Sea coast in SE Bulgaria, and Turkey in Europe. Index countries as applicable.
UF   Strandzha Mountains
SA   Bulgaria
     Turkey

Istria
Peninsula in NW jutting into Adriatic Sea. Index republics as applicable.
BT   Yugoslavia
SA   Croatia
     Slovenia

Itabira
City in E central Minás Gerais.
UF   Presidente Vargas
BT   Minas Gerais
     Brazil

itabirite
BA   chemically precipitated rocks
BT   sedimentary rocks
SA   iron formations

Itaborai
City in S central.
BT   Rio de Janeiro
     Brazil

Itaka
Town in central Chita Oblast in Transbaikalia.
BT   Russian Republic
     USSR

Italy
Includes use on level 1 as an area term (list O). For term set options see list B.
CO   N363000N473000
     E0190000E0063000
BA   Europe
NT   Abruzzi
     Acceglio
     Acqui
     Adamello Massif
     Alessandria
     Alto Adige
     Apennine Front
     Apulia
     Argentera
     Argentera Massif
     Asti
     Basilicata
     Biella
     Calabria
     Campania
     Campobasso
     Canavese Zone
     Dolomites
     Emilia-Romagna
     Friuli-Venezia Giulia
     Ivrea
     Ivrea Zone
     Ivrea-Verbano Zone
     Langhe
     Lanzo Massif
     Latium
     Lessini Mountains
     Liguria
     Lombardy
     Lucania
     Marches
     Marmolada Glacier
     Novara
NX   Piedmont
NT   Piedmont Alps
     Po River
     Po Valley
NA   Sardinia
NT   Sesia Valley
     Sesia Zone
     Sesia-Lanzo Zone
     Sicily
     Strona Valley
     Tagliamento Valley
     Tiber Valley
     Trentino-Alto Adige
     Turin
     Tuscany
     Umbria
     Valle d'Aosta
     Venetia
     Veneto
     Vercelli
     Villafranca d'Asti
SA   Adriatic region
     Alps
     Apennines
     Brianconnais Zone
     Carnic Alps
     Ceneri Zone
     Central Alps
     Cottian Alps
     Julian Alps
     Karst region
     Lago Maggiore
     Lepontine Alps
     Maiolica Limestone
     Maritime Alps
     Mediterranean region
     Monte Rosa
     Otztal Alps
     Pennine Alps
     Piedmont
     Prealps
     Rhaetian Alps
     Savoy
     Scaglia Formation
     Simplon region
     Val Gardena Sandstone
     Western Alps
     Zillertal Alps

Itasca County
N central.
BT   Minnesota
     United States

Ithaca
City in Tompkins County in S central; one of Ionian Islands in Greece.
SA   Ionian Islands
     New York

Ito
City on E Izu Peninsula on Sagami Sea in central.
BT   Honshu
     Japan

Iturup Island
Largest of the Kuril Islands, SW of Kamchatka. Also search Iturup.
BT   Russian Republic
     USSR
SA   Kuril Islands

IUGS
Acronym. Includes use on level 3 under associations(1).
UF   International Union of Geological Sciences
SA   associations

Ivano-Frankovsk
City and oblast in W Ukraine. Formerly part of Poland. Also search Stanislawow.
UF   Stanislav
     Stanislawow
BT   Ukraine
     USSR

Ivigtut
Settlement on SW coast.
BT   Greenland
     Arctic region

Iviza
use Ibiza

Ivory Coast
Includes use on level 1 as an area term (list O). For term set options see list B.
BA   Africa
NT   Bandama River
     Toumodi
SA   Nimba Mountains
     Trou Sans Fond
     Volta Basin
     West Africa

Ivrea
City NE of Turin in central.
BT   Italy
SA   Piedmont

Ivrea Zone
Central.
BT   Italy
SA   Piedmont

Ivrea-Verbano Zone
Central and NW.
BT   Italy
SA   Piedmont

Iwaki
Mountain peak in Aomori Prefecture in N Honshu. Also a river and city.
BT   Honshu
     Japan

Iwate
Prefecture in NW Honshu. Also a dormant volcano.
BT   Honshu
     Japan

Iwo-Jima
Center island of volcano group located 660 nautical miles S of Tokyo. Returned to Japan in 1968.
BT   Pacific Ocean

ixiolite
BA   oxides
BT   minerals
SA   niobates

Iya River
SW Irkutsk Oblast. Also search Iya.
BT   Russian Republic
     USSR

Izera Mountains
Range of the Sudeten Mountains in N Bohemia and Lower Silesia. Also search Isera Mountains. Index countries as applicable.
UF   Isera Mountains
BT   Europe
SA   Czechoslovakia
     Poland
     Sudeten Mountains

Izu Islands
use Izu-shichito

Izu Peninsula
Extends into Pacific Ocean between Suruga Bay and Sagami Sea in S central Honshu. Also search Izu.
BT   Honshu
     Japan

Izu-shichito
Group of volcanic islands in Pacific Ocean off Izu Peninsula. Also search Izu Islands.
UF   Izu Islands
BT   Honshu
     Japan
SA   Hachijo-jima
     Miyake-Jima
     O-shima

Izumi
City in Osaka Prefecture.
BT   Honshu
     Japan

Izumi Group
Kii Peninsula in S central Honshu, and Shikoku.
BT   Upper Cretaceous
     Cretaceous
SA   Honshu
     Japan
     Senonian
     Shikoku

J

Jabalpur
City and district in central Madhya Pradesh. Also search Jubbulpore.
UF   Jubbulpore
BT   Madhya Pradesh
     India

Jabalpur Series
Consists of clays, shales, earthy sandstones, and thin coal seams. Near city of Jabalpur.
BT   Upper Jurassic
     Jurassic
SA   India

Madhya Pradesh

Jachymov
Town in Erzgebirge in NW.
BT   Bohemia
     Czechoslovakia

Jackfork Group
Divided into Wildhorse Mountain Formation, Prairie Mountain Formation, Markham Mill Formation, Wesley Formation, Game Refuge Formation. SE and central S Oklahoma; and SW Arkansas. Also search Jackfork Sandstone.
BT   Mississippian
SA   Arkansas
     Oklahoma

Jackson County
BX   Alabama
     Arkansas
     Colorado
     Florida
     Georgia
     Illinois
     Indiana
     Iowa
     Kansas
     Kentucky
     Louisiana
     Michigan
     Minnesota
     Mississippi
     Missouri
     North Carolina
     Ohio
     Oklahoma
     Oregon
     South Dakota
     Tennessee
     Texas
     West Virginia
     Wisconsin
BT   United States

Jackson Group
Includes Caddell (Moodys Marl) Formation, Wellborn Formation, Manning Formation, and Whitsett Formation, Gosport Sand, Yazoo Clay, Ocata Limestone, Barnwell Formation, McElroy Formation, Mosley Hill Formation. Gulf Coastal Plain from Georgia to S Texas.
BT   upper Eocene
     Eocene
SA   Alabama
     Florida
     Georgia
     Louisiana
     Mississippi
     Texas
     Yazoo Clay

Jackson Hole
Valley in Teton County in NW.
BT   Wyoming
     United States

jacobsite
BA   oxides
BT   minerals
SA   spinel group

Jacobsville Sandstone
Precambrian or Cambrian. Underlies Munising Formation. N Michigan.
SA   Cambrian
     Michigan
     Precambrian

jade
BA   pyroxene group
     chain silicates
     silicates
BT   minerals

Jade Bay
Inlet of North Sea on N coast of Oldenburg region.
BT   Lower Saxony
     West Germany

jadeite
BA   pyroxene group
     chain silicates
     silicates
BT   minerals

Jaen
Province in S central Spain. Also a city.
BT   Spain
SA   Andalusia

Jaguaribe
Town and river in NE part of country.
BT   Ceara
     Brazil

Jaipur
City and former princely state in E central.
BT   Rajasthan
     India

Jaisalmer
Town and former princely state in W.
BT   Rajasthan
     India

Jalisco
State in W.
BT   Mexico

Jamaica
Includes use on level 1 as an area term (list O). For term set options see list B.
BA   West Indies
NT   Discovery Bay
     Maroon Town
SA   Blue Mountains
     Greater Antilles

James Bay
Southern extension of Hudson Bay between NE Ontario and W Quebec.
BT   Canada

James River
E central.
BT   Virginia
     United States

James W. Ellsworth Land
use   Ellsworth Land

Jameson Land
Central E coast.
BT   Greenland
     Arctic region

jamesonite
UF   gray antimony
BA   sulfantimonites
     sulfosalts
BT   minerals

Jamkhandi
Town and former princely state in S.
BT   Maharashtra
     India

Jammu
City and district in S Jammu and Kashmir.
UF   Jummoo
BT   Jammu and Kashmir
     India

Jammu and Kashmir
Official name of state. Formerly princely state of Kashmir and Jammu. Partitioned in 1949, with Pakistan controlled area in NW called Azad Kashmir, and the remaining, larger part, known as the Indian state of Jammu and Kashmir. Index countries as applicable. State comprises Jammu province, Kashmir province and frontier districts of Ladakh, Astor, and Gilgit. Also search Kashmir.
UF   Kashmir and Jammu
BT   Asia
NT   Anantnag
     Doda
     Jammu
     Kashmir
     Kashmir Valley
     Kishtwar
     Naubug
     Riasi
     Srinagar
     Tawi Valley
SA   India
     Indus River
     Muth Quartzite
     Pakistan
     Pamirs

Jamtland
County in W.
BT   Sweden

Jan Mayen
Norwegian volcanic island between Greenland and Norway. Includes use on level 1 as an area term (list O). For term set options see list B.
CO   N700000N720000
     W0073000W0090000
BZ   Arctic Ocean
     Atlantic Ocean
SA   Arctic region
     Beerenberg

Japan
Includes use on level 1 as an area term (list O). For term set options see list B.
CO   N300000N450000
     E1470000E1290000
BA   Asia
NT   Green tuff region
     Hokkaido
     Honshu
     Inland Sea
     Kuroko
     Kyushu
     Nankaido
     O-shima
     Ryukyu Islands
     Sakura-jima
     Shikoku
     Tsugaru Strait
SA   Akiyoshi Limestone
     Far East
     Futaba Group
     Green Tuff Formation
     Izumi Group
     Median Tectonic Line
     Mikabu System
     Motojuku Formation
     Muro Group
     Narita Formation
     Osaka Group
     Semata Formation
     Shimanto Group
     Shimosa Group
     Taishu Group
     Uonuma Group
     Usuginu Conglomerate
     Yamato Mountains

Japan Current
use   Kuroshio

Japan Sea
Between Japan in the E; and Korea and Primorye Kray in the Soviet Far East on the W. Includes use on level 1 as an area term (list O). For term set options see list B. Before 1974, also search Sea of Japan.
CO   N340000N450000
     E1420000E1271500
BA   Pacific Ocean

Japan Trench
Submarine depression extending from Kuril Islands along E coast of Hokkaido and E coast of N and central Honshu.
BT   Pacific Ocean

Jaramillo Event
Geomagnetic event.
BA   Quaternary
BT   Cenozoic

Jarlsberg
use   Vestfold

jarosite
UF   utahite
BA   sulfates
BT   minerals
SA   alunite

jasper
BA   silica minerals
     framework silicates
     silicates
BT   minerals
SA   chert
     jasperoid

Jasper County
Index states as applicable.
BX   Georgia
     Illinois
     Indiana
     Iowa
     Mississippi
     Missouri
     South Carolina
     Texas
BT   United States

Jasper National Park
On British Columbia border in W Alberta. Also search Jasper.
BT   Alberta
     Canada

jasperoid
See list I.
BA   chemically precipitated rocks
BT   sedimentary rocks
SA   jasper
     limestone
     siliceous composition

jaspilite
Includes use on level 3 under sedimentary rocks(1) chemically precipitated rocks(2). See list I. Also search jaspillite.
BA   chemically precipitated rocks
BT   sedimentary rocks
SA   iron formations

Java
Island in the Greater Sunda group. Most populous and economically most important.
BT   Indonesia

Java Sea
N of Java and S of Borneo.
BT   Pacific Ocean
NT   Sunda Shelf

jaws
Includes use on level 3 under fossil groups(1). See list F.
SA   bones
     fossils
     paleontology
     skulls
     teeth
     Vertebrata

Jaxartes River
use   Syr Darya

Jefferson County
BX   Alabama
     Arkansas
     Colorado
     Florida
     Georgia
     Idaho
     Illinois
     Indiana
     Iowa
     Kansas
     Kentucky
     Louisiana
     Mississippi
     Missouri
     Montana

Nebraska
New York
Ohio
Oklahoma
Oregon
Pennsylvania
Tennessee
Texas
Washington
West Virginia
Wisconsin
BT   United States

Jefferson Parish
W and S of New Orleans.
BT   Louisiana
       United States

Jemez Mountains
use Valle Grande Mountains

Jerome
Town in central Arizona and city in S Idaho. Index states as applicable.
BX Arizona
       Idaho
BT   United States

Jerusalem
District in E central Israel. Also a city. Index countries as applicable.
BT   Middle East
SA   Israel
       Jordan

jetties
BT   marine installations
SA   groins

Jhabua
Village, district and former princely state in W.
BT   Madhya Pradesh
       India

Jhansi
City and division in S.
BT   Uttar Pradesh
       India

Jharia
Town in SE Bihar.
UF   Jherria
BT   Bihar
       India
SA   Jharia coal field

Jharia coal field
In SE Bihar. Also search Jharia Coalfield; Jharia.
UF   Jharia Coalfield
BT   Bihar
       India
SA   Jharia

Jharia Coalfield
use Jharia coal field

Jherria
use Jharia

Jhunjhunu
Town and district in NE.
BT   Rajasthan
       India

Jiu River valley
Index regions as applicable. Also search Jiu; Jiu River.
BT   Romania
SA   Transylvania
       Walachia

joaquinite
BA   orthosilicates
       silicates
BT   minerals

Joban coal field
A major coal field on Pacific Ocean in central Honshu. Also search Joban; Joban Coalfield.
UF   Joban Coalfield
BT   Honshu
       Japan

Joban Coalfield
use Joban coal field

Jodhpur
City and region in central Rajasthan. Former princely state.
BT   Rajasthan
       India

Johanna Island
use Anjouan Island

Johannesburg
City in S central.
BT   Transvaal
       South Africa

johannsenite
BA   pyroxene group
       chain silicates
       silicates
BT   minerals

John Day Formation
Tentatively divided into three unnamed members. N central Oregon.
BT   Tertiary
       Cenozoic
SA   lower Miocene
       Miocene
       Oligocene
       Oregon
       upper Oligocene

Johnnie Formation
Underlies Stirling Quartzite and overlies Noonday Dolomite. E California and SE Nevada.
BT   Precambrian
SA   California
       Nevada

Johns Valley Formation
Scott County Arkansas, and Ouachita region of SE Oklahoma.
BT   Carboniferous
SA   Arkansas
       Mississippian
       Oklahoma
       Pennsylvanian

Johnson County
BX Arkansas
       Georgia
       Illinois
       Indiana
       Iowa
       Kansas
       Kentucky
       Missouri
       Nebraska
       Tennessee
       Texas
       Wyoming
BT   United States

Johore
State just N of Singapore on the Malay Peninsula.
BT   Malaysia
SA   West Malaysia

joints
Includes use on level 3 under fractures(1) style(2) and under structural analysis(1) theoretical studies(2).
BA   fractures
NT   cross joints
SA   columnar joints
       conjugate folds
       cross fractures
       fissures
       fracture cleavage
       longitudinal orientation
       oblique orientation
       release fractures
       strike
       structural analysis

Joliba River
use Niger River

Joliba Valley
use Niger Valley

Jones Mountains
On Eights Coast in Ellsworth Land on Pacific Ocean side.
BT   Antarctica

Joplin
City in Jasper and Newton counties in SW.
BT   Missouri
       United States

Jordan
Includes use on level 1 as an area term (list O). For term set options see list B.
CO   N290000N333000
       E0390000E0343000
BA   Middle East
SA   Asia
       Dead Sea
       Dead Sea Rift
       Great Rift Valley
       Jerusalem
       Jordan River
       Jordan Valley
       Near East
       Palestine
       Red Sea Basin
       Samaria
       Wadi Araba

Jordan Rift valley
use Jordan Valley

Jordan River
Rises in Syria and flows S through the Sea of Galilee to N end of the Dead Sea. Index countries as applicable.
BT   Middle East
SA   Israel
       Jordan
       Syria

Jordan River valley
use Jordan Valley

Jordan Valley
River valley between Sea of Galilee and Dead Sea which is part of the Great Rift Valley. Also search Jordan Rift valley; Jordan River valley; Tiberias-Dead Sea Rift valley. Index countries as applicable.
UF   Jordan Rift valley
       Jordan River valley
BT   Middle East
SA   Great Rift Valley
       Israel
       Jordan
       Syria

jordanite
BA   sulfarsenites
       sulfosalts
BT   minerals

Jordanow
Town in S part of country.
BT   Krakow
       Poland

jordisite
BA   sulfides
BT   minerals

Jos Plateau
Central Nigeria. Also search Bauchi Plateau.
UF   Bauchi Plateau
BT   Nigeria

joseite
BA   tellurides
       sulfides
BT   minerals

Joseph Bonaparte Gulf
Inlet of Timor Sea. Index Northern Territory and/or Western Australia as applicable.
BT   Australia
SA   Bonaparte Gulf basin
       Northern Territory
       Western Australia

josephinite
BA   native elements and alloys
BT   minerals

Jotnian
BA   upper Precambrian
       Precambrian

Jotunheim Mountains
Range in S central Norway. Also search Jotunheimen; Jotunheimen Mountains.
UF   Jotunheimen Mountains
BT   Norway

Jotunheimen Mountains
use Jotunheim Mountains

jouravskite
BA   sulfates
BT   minerals

Juab County
W central.
BT   Utah
       United States

Juan de Fuca Ridge
Off coast of Washington State. Also search Juan de Fuca Rise.
UF   Juan de Fuca Rise
BT   Pacific Ocean

Juan de Fuca Rise
use Juan de Fuca Ridge

Juan de Fuca Strait
Between Vancouver Island on the N and mainland of Washington on the S. Index British Columbia and/or Washington as applicable. Also search Strait of Juan de Fuca.
UF   Strait of Juan de Fuca
SA   British Columbia
       Washington

Jubbulpore
use Jabalpur

Judith River Formation
In Montana Group. Comprises two members: Parkman Sandstone and an upper unnamed member. N, Central, S, and SE Montana.
BT   Upper Cretaceous
       Cretaceous
SA   Montana
       Montana Group

Juglandaceae
BA   Dicotyledoneae
       angiosperms
BT   Plantae
SA   Juglans

Juglans
Genus.
BA   Dicotyledoneae
       angiosperms
BT   Plantae
SA   Filicopsida
       gymnosperms
       Juglandaceae
       pteridophytes

Jujuy
Province in extreme NW Argentina. Also a city.
CO   S250000S220000
       W0640000W0673000
BT   Argentina
SA   Yacoraite Formation

julgoldite
BA   orthosilicates
       silicates
BT   minerals

Julian Alps
Division of Eastern Alps. Index countries as applicable.
BT   Alps
       Europe
SA   Eastern Alps
       Italy
       Yugoslavia

Jumilla
City in SE part of country.
BT Murcia
 Spain

Jummoo
use Jammu

Jumna River
N central India. Also search Yamuna River. Index states as applicable.
UF Yamuna River
BT India
SA Himachal Pradesh
 Punjab
 Uttar Pradesh

Juneau
City in panhandle near border of Yukon Territory.
BT Alaska
 United States

Juneau Glacier
use Juneau ice field

Juneau ice field
N of Juneau in panhandle. Also search Juneau Icefield; Juneau Glacier.
UF Juneau Glacier
 Juneau Icefield
BT Alaska
 United States

Juneau Icefield
use Juneau ice field

Juniata Formation
Includes East Waterford Red Sandstone Member, Plummer Hollow Red Mudstone, Run Gap Red Sandstone Member. S central and E Pennsylvania, W Maryland, E Tennessee, W Virginia, and E West Virginia.
BT Upper Ordovician
 Ordovician
SA Maryland
 Pennsylvania
 Richmond Group
 Tennessee
 Virginia
 West Virginia

Junin
Department in W central.
BT Peru
SA Pucara Group

junior high school
Term introduced in 1978. Includes use on level 3 under education(1). Before 1978, search junior high.
SA college-level education
 education
 elementary school
 high school

Jupiter
Includes use on level 1 and 2. See term set options under Moon(1).
SA Galilean satellites
 outer planets
 Pioneer 10
 planetology
 planets
 satellites
 solar system

Jura
Department on the Swiss border.
BT France
NT Morez
SA Jura Mountains

Jura Mountains
Index countries as applicable. Also search Jura.
BT Europe
SA France
 Jura
 Switzerland

Jurassic
Includes use on level 1 as an age term (list E); on level 2 under paleoterms, e.g. paleoecology, paleogeography, paleomagnetism. Above Triassic, below Cretaceous.
BA Mesozoic
NA Aalenian
NT Athgarh Sandstone
 Bowser Formation
 Ferrar Group
 Lathi Formation
NA Lower Jurassic
NT Mansalay Formation
NA Middle Jurassic
NT Norphlet Formation
 Posidonia Shale
 Shaunavon Formation
 Solnhofen Limestone
NA Upper Jurassic
NT Younger Granites
SA Beacon Supergroup
 Brianconnais Zone
 Franciscan Formation
 Gondwana System
 Great Valley Sequence
 Ionian Zone
 Karroo System
 Khorat Group
 Kootenay Formation
 Maiolica Limestone
 Navajo Sandstone
 Novorayskoe Formation
 Pucara Group
 Serra Gerral Formation
 Tal Formation

Jutland
Peninsula including Schleswig in Germany and the Danish mainland. Politically it applies only to the mainland of Denmark. Index Denmark and/or Schleswig-Holstein as applicable. Also search Jylland.
UF Jylland
BT Europe
NT Djursland
SA Denmark
 Schleswig-Holstein

Jylland
use Jutland

K
use potassium

K-feldspar
Also search potassium feldspar; potash feldspar.
UF potash feldspar
 potassium feldspar
BA feldspar group
 framework silicates
 silicates
BT minerals
SA adularia
 alkali feldspar
 microcline
 orthoclase
 sanidine

K-40
Includes use on level 3 under isotopes(1).
SA isotopes
 potassium

K/Ar
Includes use on level 3 under absolute age(1) methods(2). Also search potassium-argon.
SA absolute age
 argon
 potassium

Kaap Valsch
use False Cape

Kabardia
N part of Kabardin-Balkar A.S.S.R. which is on N slopes of Caucasus.
BT Russian Republic
 USSR

Kabardin-Balkar
Autonomous Soviet Socialist Republic. Also search Kabardinia-Balkar, and Kabardino-Balkar.
UF Kabardinia-Balkar
 Kabarnino-Balkar
BT Russian Republic
 USSR

Kabardinia-Balkar
use Kabardin-Balkar

Kabarnino-Balkar
use Kabardin-Balkar

Kabinda
use Cabinda

Kabul
Province in NE. Also a city.
BT Afghanistan

Kabylia
Montainous coastal region in N.
BT Algeria
SA Great Kabylia

Kadah
Northernmost state in West Malaysia on the Malay Peninsula.
BT Malaysia

Kadzharan
Town in S.
BT Armenia
 USSR

kaersutite
BA amphibole group
 chain silicates
 silicates
BT minerals
SA hornblende

Kafan
City in SE.
BT Armenia
 USSR

Kafue
Township and river in N.
BT Zambia

Kaganovich Peak
use Darvaz

Kagawa
Prefecture on the Inland Sea in NE.
BT Shikoku
 Japan

Kagoshima
City and prefecture in S.
BT Kyushu
 Japan

Kaibab Formation
Divided vertically into three members. N Arizona, SE California, SE Nevada and W Utah.
BT Permian
SA Arizona
 California
 Nevada
 Utah

Kaikoura
Mountain range in NE.
BT South Island
 New Zealand

Kaimur Sandstone
Constitutes part of the Semri Group in the Gwalior and Mewar areas. Upper Kaimur Sandstone and lower Kaimur Sandstone are included in Kaimur Series.
SA Cambrian
 India
 Madhya Pradesh
 Precambrian
 Rajasthan
 Uttar Pradesh

kainite
BZ halides
 sulfates
BT minerals
SA chlorides

Kaipara Harbor
Inlet of Pacific Ocean on W coast of N extension of North Island. Also search Kaipara.
BT North Island
 New Zealand

Kaiparowits Plateau
W of Colorado River between Escalante and Paria rivers in S central.
BT Utah
 United States

Kaiserstuhl
Mountain group in SW.
BT Baden-Wurttemberg
 West Germany

Kakagi Lake
East of Lake of the Woods in W.
BT Ontario
 Canada

Kakanui
Mountains in SE, north of Dunedin.
BT South Island
 New Zealand

Kakhetia
Region in E.
BT Georgian Republic
 USSR

kakortokite
Includes use on level 3 under igneous rocks(1) syenite family(2). See list H.
BA syenite family
BT igneous rocks
SA nepheline syenite

Kakuto
Village in SE on Lake Victoria.
BT Uganda

Kaladgi
Village in W central.
BT Mysore
 India

Kaladgi System
Precambrian? Divided into upper Kaladgi and lower Kaladgi series.
BT Precambrian
SA India
 Maharashtra
 Mysore

Kalahari Desert
Plateau and desert region. Index South-West Africa and countries as applicable.
BT Africa
SA Botswana
 South Africa
 South-West Africa

Kalamazoo County
SW Lower Peninsula.
BT Michigan
 United States

Kalba Range
W branch of Altai Mountains in E Kazakhstan. Also search Kalba.
CO N490000N492000
 E0830000E0823000
BT Kazakhstan
 USSR
SA Altai Mountains

Kalgoorlie
Municipality in S.
CO S293000S293000

Kalgoorlie ● Kansas

    E1220000E1220000
BT Western Australia
    Australia

Kalgoorlie System
Includes "Older Greenstones", Black Flag Group, Yindarlgooda Group, Kundana Group, and "Younger Greenstones". S part of state.
BT Archean
SA Western Australia

Kali Gandaki Valley
River valley in central.
BT Nepal

kaliborite
Also search paternoite.
UF heintzite
    hintzeite
    paternoite
BA borates
BT minerals

Kalimantan
All of island of Borneo excepting Sarawak, Sabah and Brunei.
BT Indonesia
SA Borneo

Kalinin
City and oblast.
BT Russian Republic
    USSR

Kaliningrad
City and oblast.
BT Russian Republic
    USSR

kaliophilite
BA nepheline group
    framework silicates
    silicates
BT minerals
SA aluminosilicates

Kalmar
County in S.
BT Sweden
NT Oland
    Vastervik
SA Smaland

Kalmyk
Autonomous Soviet Socialist Republic on NW shore of Caspian Sea.
BT Russian Republic
    USSR

kalsilite
BA nepheline group
    framework silicates
    silicates
BT minerals
SA aluminosilicates

Kama
Village in Kivu, Zaire; village in Thayetmyo district, Upper Burma. Use Kama River for the river.
SA Burma
    Kama River
    Zaire

Kama River
Rises in central Urals and flows into Volga River. Also search Kama AND USSR.
CO N610000N550000
    E0560000E0520000
BT Russian Republic
    USSR
SA Kama

kamacite
Meteorite mineral.
BA native elements and alloys
BT minerals
SA iron meteorites
    meteorites
    plessite
    taenite

Kambalda
Settlement S of Kalgoorlie in S.
BT Western Australia
    Australia

Kamchatka
Use for the oblast. Includes Kamchatka Peninsula.
CO N510000N620000
    E1750000E1550000
UF Kamchatka Oblast
BT Russian Republic
    USSR

Kamchatka Oblast
use Kamchatka

Kamchatka Peninsula
Between Okhotsk Sea and Bering Sea.
BT Russian Republic
    USSR
SA Soviet Far East

Kamen
City in the Ruhr in W.
BT North Rhine-Westphalia
    West Germany

Kamenka River
Flows into the Angara River in S Krasnoyarsk Kray.
BT Russian Republic
    USSR

kames
Includes use on level 3 under glacial geology(1) glacial features(2).
BT glacial features
SA cirques
    eskers
    glacial geology
    glaciers

Kamikawa
Sub-prefect in central.
BT Hokkaido
    Japan

Kaminak Lake
SW District of Keewatin.
BT Northwest Territories
    Canada

Kamloops
City in S.
BT British Columbia
    Canada

kammererite
BA sheet silicates
    silicates
BT minerals
SA chlorite group

Kamoto
Mine five miles W of village of Kolwezi in S part of country.
BT Shaba
    Zaire

Kamp Valley
River valley in NE part of country.
BT Lower Austria
    Austria

Kampinos Forest
E central.
BT Poland

Kampuchea
use Cambodia

Kamyshin
City on the Volga River in Volgograd Oblast.
BT Russian Republic
    USSR

Kanagawa
Prefecture whose capital is Yokohama.
BT Honshu
    Japan

Kanara
Region. Former district of North Kanara in W.
BT Mysore
    India

Kanawha County
W central.
BT West Virginia
    United States

Kanazawa
City in W.
BT Honshu
    Japan

Kane County
BX Illinois
    Utah
BT United States

Kanem
Prefecture NE of Lake Chad in W.
BT Chad

Kaneohe Bay
Wide inlet on E coast of Oahu Island.
BT Hawaii
    United States

Kanev
City in Kiev Oblast in N central.
BT Ukraine
    USSR

Kangaroo Island
In Indian Ocean S of Yorke Peninsula.
BT South Australia
    Australia

Kangra
Town in W.
BT Himachal Pradesh
    India

Kangwon
Province in SE North Korea, and in NE South Korea. Index countries as applicable.
BT Asia
SA North Korea
    South Korea

Kanhan Valley
River valley in central Madhya Pradesh. Also search Kanhan.
BT Madhya Pradesh
    India

Kanin Peninsula
Projects into Barents Sea on N coast of Nenets National Okrug in Arkhangelsk Oblast.
BT Russian Republic
    USSR

Kansan
North America. Pertains to second glacial stage of Pleistocene Epoch.
BA lower Pleistocene
    Pleistocene
    Quaternary
BT Cenozoic

Kansas
Includes use on level 1 as an area term (list O). For term set options see list B.
CO N370000N383000
    W0943500W0980000
BA United States
NT Abilene
    Admire
NX Allen County
    Anderson County
NT Arkansas City
    Atchison
    Atchison County
NX Barton County
    Brown County
    Butler County
    Chase County
    Chautauqua County
    Cherokee County
    Cheyenne County
    Clark County
    Clay County
    Cloud County
    Coldwater
NX Comanche County
NT Cowley County
    Decatur County
NX Dickinson County
NT Doniphan County
NX Douglas County
    Edwards County
    Ellis County
NT Ellsworth County
    Emporia
NX Eureka
NT Finney County
NX Franklin County
NT Garnett
    Geary County
    Gove County
NX Graham County
    Grant County
    Greenwood County
    Hamilton County
    Harper County
NT Harvey County
    Haviland
    Hodgeman County
    Hugoton
    Hugoton Field
    Hutchinson
NX Jackson County
    Jefferson County
    Johnson County
NT Kansas River
    Kansas River valley
    Kingman County
NX Kiowa County
NT Kirwin
    Labette County
NX Lane County
NT Leavenworth County
NX Lincoln County
    Manhattan
    Marion County
    Marshall County
    McPherson County
    Meade County
    Mitchell County
    Montgomery County
    Morton County
    Nemaha County
NT Ness County
    Norton County
    Norton County Meteorite
NX Osage County
NT Osborne County
NX Ottawa County
    Pawnee County
    Phillips County
    Pottawatomie County
NT Pratt County
    Rawlins County
    Reno County
    Republic County
NX Rice County
NT Riley County
    Rooks County
NX Rush County
    Russell County
    Scott County
NT Sedgwick Basin
NX Sedgwick County
    Seward County
NT Shawnee County
NX Sheridan County
    Sherman County
    Smith County
NT Smoky Hill River basin
NX Stafford County
    Stanton County
    Stevens County
    Sumner County
NT Topeka
    Trego County
    Tuttle Creek Dam
    Wabaunsee County
    Wallace County
NX Washington County

NT Wichita
NX Wichita County
   Wilson County
NT Woodson County
   Wyandotte County
SA Americus Limestone Member
   Anadarko Basin
   Arkansas River
   Arkansas River valley
   Ash Hollow Formation
   Beil Limestone Member
   Benton Formation
   Bevier Coal
   Burbank Sand
   Cabaniss Formation
   Cambridge Arch
   Carlile Shale
   Chase Group
   Cherokee Group
   Cheyenne Sandstone
   Colorado Group
   Cottonwood Limestone
   Crooked Creek Formation
   Dakota Formation
   Desmoinesian
   Dockum Group
   Douglas Group
   Ervine Creek Limestone
   Eskridge Shale
   Excello Shale
   Flint Hills
   Forest City Basin
   Fort Hays Limestone Member
   Fort Riley Limestone
   Graneros Shale
   Great Plains
   Greenhorn Limestone
   Hughes Creek Shale
   Hugoton Embayment
   Hutchinson Salt Member
   Kansas City Group
   Kingsdown Formation
   Kiowa Formation
   Labette Shale
   Lansing Group
   Lawrence Formation
   Lecompton Limestone
   Loup Fork Group
   Marmaton Group
   McPherson Formation
   Mentor Beds
   Midcontinent
   Midwest
   Missouri River
   Missouri River basin
   Missouri River valley
   Montana Group
   Morrison Formation
   Nemaha Ridge
   Neosho River valley
   Niobrara Formation
   Ogallala Formation
   Oread Limestone
   Peoria Loess
   Plattsburg Limestone
   Plattsmouth Limestone Member
   Pleasanton Group
   Red Eagle Limestone
   Rexroad Formation
   Roca Formation
   Rock Lake Shale Member
   Saint Peter Sandstone
   Shawnee Group
   Smoky Hill Chalk Member
   Stanton Formation
   Stearns Shale
   Stone Corral Formation
   Tonganoxie Sandstone
   Valentine Formation
   Verdigris River valley
   Wabaunsee Group
   Wellington Formation
   Wreford Limestone
   Wyandotte Limestone

Kansas City Group
Subdivided into Bronson, Linn, and Zarah subgroups. Includes Hertha, Ladore, Swope, Galesburg, Dennis, Fontana, Sarpy, Drum, Chanute, Iola, Lane, Wyandotte and Bonner Springs formations.
BT Missourian
   Upper Pennsylvanian
   Pennsylvanian
SA Iowa
   Kansas
   Missouri
   Nebraska
   Wyandotte Limestone

Kansas River
Flows into Missouri River at Kansas City.
BT Kansas
   United States

Kansas River basin
use Kansas River valley

Kansas River valley
NE central Kansas. Also search Kansas River basin.
UF Kansas River basin
BT Kansas
   United States

kansite
use mackinawite

Kansk
City E of Krasnoyarsk on the Trans-Siberian R.R. in S Krasnoyarsk Kray.
BT Russian Republic
   USSR

Kansk-Achinsk Basin
S and SW Krasnoyarsk Kray along Trans-Siberian R.R.
BT Russian Republic
   USSR

Kanto
use Kwanto Plain

kaolin
Includes use on level 1 as a commodity term (list C).
SA clay
   clay mineralogy
   clay minerals
   clays
   fuller's earth
   kaolinite
   kaolinization
   metakaolin
   refractory materials

kaolinisation
use kaolinization

kaolinite
BA sheet silicates
   silicates
BT minerals
SA clay minerals
   kaolin

kaolinitization
use kaolinization

kaolinization
Includes use on level 3 under metasomatism(1) processes(2). Also search kaolinitization.
UF kaolinisation
   kaolinitization
   kaolisation
BA metasomatism
SA alteration
   clay minerals
   kaolin
   processes
   weathering

kaolisation
use kaolinization

Kapoeta
Point of impact near Kapoeta in S. Also search Kapoeta Meteorite.
UF Kapoeta Meteorite
BT Sudan
SA meteorites

Kapoeta Meteorite
use Kapoeta

Kara Sea
Between Novaya Zemlya Island on W; and Yamal and Tamyr peninsulas on the E.
BT Arctic Ocean

Kara-Bogaz Gulf
Large shallow gulf of the Caspian Sea in NW Krasnovodsk Kray. Also search Kara-Bogaz-Gol Gulf; Kara-Bogaz-Gol region.
UF Kara-Bogaz-Gol Gulf
   Kara-Bogaz-Gol region
BT Turkmenia
   USSR

Kara-Bogaz-Gol
Town on the Caspian Sea at entrance to Kara-Bogaz Gulf.
BT Turkmenia
   USSR
SA Caspian Sea

Kara-Bogaz-Gol Gulf
use Kara-Bogaz Gulf

Kara-Bogaz-Gol region
use Kara-Bogaz Gulf

Kara-Kum
use Karakum

Kara-Mazar
Village in Leninabad Oblast in N.
BT Tadzhikistan
   USSR

Kara-Tau (Range)
use Karatau Range

Karabakh
Mountain area. Now Nagorno-Karabakh Autonomous Oblast in S.
BT Azerbaidzhan
   USSR

Karafuto
use Sakhalin

Karaganda
City and oblast in NE central.
BT Kazakhstan
   USSR

Karaganda Basin
Coal basin in Karaganda Oblast in NE central. Also search Karaganda.
BT Kazakhstan
   USSR

Karaginski Island
use Karaginskiy Island

Karaginskiy Island
In Bering Sea just off coast of N Kamchatka Peninsula. Also search Karaginski Island.
UF Karaginski Island
BT Russian Republic
   USSR

Karakoram
Mountain range in N Jammu and Kashmir. Index countries as applicable. Also search Karakorum.
UF Karakorum
BT Asia
SA India
   Pakistan

Karakorum
use Karakoram

Karakum
Desert area S of the Aral Sea stretching from the Caspian Sea to the Amu Darya River. Also search Kara-Kum; Karakumy.
UF Kara-Kum
   Karakumy
   Qaraqum
BT Turkmenia
   USSR

Karakumy
use Karakum

Karamazar
Village in Leninabad Oblast in N.
BT Tadzhikistan
   USSR

Karanpura
Village in central.
BT Bihar
   India

Karanpura Basin
use Karanpura coal field

Karanpura coal field
Central Bihar. Also search Karanpura; Karanpura Basin; Karanpura Coalfield.
UF Karanpura Basin
   Karanpura Coalfield
BT Bihar
   India

Karanpura Coalfield
use Karanpura coal field

Karatau
use Karatau Range

Karatau Range
W branch of the Tien Shan in SE Kazakhstan. Also search Karatau; Kara-Tau.
CO N420000N443000
   E0710000E0670000
UF Kara-Tau (Range)
   Karatau
BT Kazakhstan
   USSR

Karategin Range
SW Tadzhikistan.
BT Tadzhikistan
   USSR

Karawanken
Range of Eastern Alps. A continuation of Carnic Alps. Index countries as applicable. Also search Karawanken Mountains; Karawanken Alps.
UF Karawanken Alps
   Karawanken Mountains
BT Alps
   Europe
SA Austria
   Carnic Alps
   Eastern Alps
   Yugoslavia

Karawanken Alps
use Karawanken

Karawanken Mountains
use Karawanken

Karelia
Karelian Autonomous Soviet Socialist Republic in NW European USSR.
CO N600000N680000
   E0400000E0300000
BT Russian Republic
   USSR

Karelian
Includes use as an age term (list E).
BA middle Precambrian
   Precambrian
SA Karelian Orogeny

Karelian Orogeny
Term introduced in 1978. Before 1978, also search Karelian AND orogeny.
BT Precambrian
SA Karelian
   orogony

Karharbari
Town in E central.
BT Bihar
   India

Karharbari Stage
Basal member of the Damuda Series. Named after Karharbari in E cen-

tral Bihar.
  BT Upper Carboniferous
       Carboniferous
  SA Bihar
       Damuda Series
       India

Kariba
Town on Lake Kariba in NW.
  BT Rhodesia

Kariba Lake
use Lake Kariba

Karibib
Town in W central.
  BT South-West Africa

Karkonosze
use Karkonosze Mountains

Karkonosze Mountains
Highest range of the Sudeten Mountains. Index Bohemia and/or Poland as applicable. Also search Karkonosze.
  UF Karkonosze
  BT Europe
  SA Bohemia
       Poland
       Sudeten Mountains

Karl-Marx-Stadt
District in SE. Also a city.
  BT East Germany
  NT Altenberg
       Frankenberg
       Freiberg
       Schneeberg
       Schwarzenberg
       Vogtland
       Wildenfels
       Zwickau
  SA Saxony

Karlovy Vary
Town and health resort in W Bohemia. Used for Carlsbad when referring to Czechoslovakia. Before 1977, also search Carlsbad AND Czechoslavakia.
  BT Bohemia
       Czechoslovakia

Karlsruhe
City in NW central.
  BT Baden-Wurttemberg
       West Germany

Karnes County
S central.
  BT Texas
       United States

Karnul
use Kurnool

Karoo
use Karroo Basin

Karpathos
use Karpathos Island

Karpathos Island
In the Dodecanese. Also search Karpathos.
  UF Karpathos
  BT Aegean Islands
       Greece

karpinskyite
  BA framework silicates
       silicates
  BT minerals

karren
  BT solution features
  SA geomorphology
       karst

Karroo Basalts
use Karroo System

Karroo Basin
An arid tableland region consisting of North Karroo along the Orange River in N, Great Karroo or Central Karroo in S central, and Southern Karroo or Little Karroo along S coast. Also search Karroo.
  UF Karoo
  BT Cape Province
       South Africa

Karroo System
Carboniferous to Jurassic of South and South-West Africa. Divided into Dwyka Series, Ecca Series, Beaufort Series, and Stormberg Series. Also search Karroo Basalts; Karroo.
  UF Karroo Basalts
  SA Africa
       Botswana
       Carboniferous
       Dwyka Series
       Jurassic
       Permian
       South Africa
       South-West Africa
       Triassic

Karshi
Town in Kashkadarya Oblast in S Uzbekistan.
  UF Bek-Budi
  BT Uzbekistan
       USSR

Karshi Steppe
In Kashkadarya Oblast in S Uzbekistan. Also search Karshi.
  BT Uzbekistan
       USSR

karst
Includes use on level 3 under geomorphology(1) solution features(2). Before 1977, Karst was infrequently used to indicate the region. Now use Karst region.
  UF carst
  BT solution features
  SA caves
       dolines
       geomorphology
       karren
       sinkholes
       solution cavities
       thermokarst

Karst region
Limestone plateau N of Trieste and E of Isonzo River. Index countries as applicable. Also search Karst regions. Before 1977, Karst was used to indicate the region.
  BT Europe
  SA Italy
       Yugoslavia

Kartalinia
use Kartlia

Kartlia
Hilly central region.
  UF Kartalinia
  BT Georgian Republic
       USSR

Karvina
City in NE Moravia.
  UF Karvinna
  BT Moravia
       Czechoslovakia

Karvinna
use Karvina

Kasai River
Flows into the Congo. Also search Kasai. Index countries as applicable.
  BT Africa
  SA Angola
       Zaire

Kasauli Series
  BT lower Miocene
       Miocene
  SA Himachal Pradesh
       India
       Punjab

Kashiwazaki
City on Japan Sea in central.
  BT Honshu
       Japan

Kashmir
Part of Jammu and Kashmir.
  BT Jammu and Kashmir
  SA India

Kashmir and Jammu
use Jammu and Kashmir

Kashmir Valley
Intermontane valley in the Himalayas of W Jammu and Kashmir.
  UF Vale of Kashmir
  BT Jammu and Kashmir
       India

Kaskawulsh Glacier
In the Saint Elias Mountains in SW.
  BT Yukon Territory
       Canada

kasolite
  BA orthosilicates
       silicates
  BT minerals

Kassel
City in N.
  BT Hesse
       West Germany

katagenesis
use catagenesis

Katanga
use Shaba

Katangan Orogeny
Term introduced in 1978. Before 1978, search Katangan AND orogeny.
  BT Precambrian
  SA Africa
       orogeny
       Zaire

Katherine
River and settlement in N central.
  BT Northern Territory
       Australia

Kathiawar
Peninsula jutting into Arabian Sea.
  BT Gujarat
       India

Katmai National Monument
At N end of Alaska Peninsula in S.
  BT Alaska
       United States

Katowice
Province in S central. Also a city.
  BT Poland
  NT Bielsko
       Bytom
       Chorzow
       Cieszyn
       Czestochowa
       Czestochowa-Zawierce Basin
       Goczalkowice
       Ruda
       Rybnik
       Sosnowiec
       Upper Silesian coal basin
       Zabrze
       Zawiercie
  SA Silesian coal basin

Kattegat
Strait between Sweden and Jutland, Denmark. Also search Kattegatt.
  UF Kattegatt
  BT North Sea

Kattegatt
use Kattegat

Katzenbuckel
Mountain. Highest point in the Odenwald in N.
  BT Baden-Wurttemberg
       West Germany

Kauai
One of the major Hawaiian Islands.
  BT Hawaii
       United States

Kavalerovo
Village in SE Primorye Kray in Soviet Far East.
  BT Russian Republic
       USSR

Kaveri Basin
use Cauvery Basin

Kayenta Formation
In Glen Canyon Group. Upper Triassic(?). NE Arizona, SW Colorado, and S and SE Utah.
  BT Upper Triassic
       Triassic
  SA Arizona
       Colorado
       Utah

Kazakhstan
Kazakh Soviet Socialist Republic. In Soviet Central Asia.
  CO N410000N550000
       E0860000E0470000
  BT USSR
  NT Akbastau
       Akchatau
       Aksu
       Aktyubinsk
       Akzhal
       Alakol
       Alma-Ata
       Atasu
       Balkhash
       Balkhash region
       Barsakelmes
       Bet-Pak-Dala
       Biikzhal
       Boshchekul
       Char
       Chingis-Tau
       Chu
       Chu-Ili Mountains
       Dzhezkazgan
       Ekibastuz
       Emba
       Emba River
       Guryev
       Ili
       Irgiz
       Kalba Range
       Karaganda
       Karaganda Basin
       Karatau Range
       Kokchetav
       Kustanay
       Lake Balkhash
       Leninogorsk
       Mangyshlak Peninsula
       Mirgalimsay
       Mugodzhar Hills
       Muyunkum
       Pavlodar
       Sarysu
       Sayak
       Semipalatinsk
       Stepnyak
       Tekeli
       Tengiz
       Tokrau Synclinorium
       Turgay
       Turgay Basin
       Ulu-Tau
       Uspenskiy
       Zaisan
       Zaisan Basin
       Zerenda
       Zhetybay
       Zyryanovsk
  SA Aral region
       Asiatic USSR
       Caspian Basin
       Caspian Depression
       Central Asia
       Chirchik River

Dzhungarian Alatau
Ili Basin
Irtysh River
Ishim
Kulunda Steppe
Kyzylkum
Ob-Irtysh Interfluve
Pskem Range
Rudny Altai
Siberia
Steppes
Syr Darya
Tarbagatay Range
Turan
Turanian Platform
Ural River
Ustyurt
Uzen
Volga-Ural region

Kazan
City on left bank of Volga in Tatar A.S.S.R.
BT   Russian Republic
      USSR

Kazanian
Europe. Above Kungurian, below Tatarian. Includes use on level 3 under age terms(1). See list E.
BA   Upper Permian
      Permian
BT   Paleozoic

keatite
BA   silica minerals
      framework silicates
      silicates
BT   minerals

Keban Mine
In E central. Also search Keban.
BT   Turkey

Kebnekaise
Peak in Kjolen Mountains in NW Norrbotten. Highest in country.
BT   Norrbotten
      Sweden

Kedabek
Town in W.
BT   Azerbaidzhan
      USSR

Keel
Village on Achill Island in Atlantic Ocean.
BT   Mayo
      Ireland

Keewatin
Town in W. Ontario and village in NE Minnesota. Index Ontario or Minnesota as applicable. Before 1978, also used for District of Keewatin.
SA   District of Keewatin
      Minnesota
      Ontario

Keewatin District
use District of Keewatin

Kefallinia
use Cephalonia

Keg River Formation
Subsurface. Includes the Rainbow (reef) Member. Eastward in Saskatchewan, the Keg River Formation is termed the Winnipegosis Formation. NW Alberta.
BT   Devonian
SA   Alberta
      Canada

Kelantan
State on Thailand border on the Malay Peninsula in West Malaysia.
BT   Malaysia
SA   West Malaysia

Kelheim
Town in central.
BT   Bavaria
      West Germany

Kemerovo
City in the Kuznetsk Basin of Kemerovo Oblast in S central Siberia.
BT   Russian Republic
      USSR

Kenai Formation
use Kenai Group

Kenai Group
Eocene(?) and Oligocene, Central S Alaska. Also search Kenai Formation.
UF   Kenai Formation
BT   Paleogene
      Tertiary
      Cenozoic
SA   Alaska

Kenai Peninsula
Between Cook Inlet on W and Prince William Sound on E in S.
BT   Alaska
      United States

Keno Hill
Village in central.
BT   Yukon Territory
      Canada

Kenora
City and district in W.
BT   Ontario
      Canada

Kenoran Orogeny
A time of plutonism, metamorphism, and deformation during the Precambrian of the Canadian Shield. Before 1978, also search Kenoran AND orogeny.
BT   Precambrian
SA   Canada
      orogeny

Kentland
Town in Newton County in NW.
BT   Indiana
      United States

Kentucky
Includes use on level 1 as an area term (list O). For term set options see list B.
CO   N363000N391000
      W0820000W0894000
BA   United States
NX   Allen County
      Anderson County
NT   Benham
NX   Boone County
      Butler County
      Campbell County
      Carroll County
      Carter County
      Clark County
      Clay County
      Columbia
      Crittenden County
      Cumberland County
NT   Elliott County
NX   Fayette County
      Floyd County
      Franklin County
      Fulton County
      Grant County
      Hancock County
      Hardin County
      Jackson County
      Jefferson County
      Johnson County
NT   Kentucky River
      Kentucky River valley
NX   Lawrence County
      Lee County
      Lincoln County
      Livingston County
NT   Louisville
NX   Madison County
NT   Mammoth Cave
NX   Marion County
      Marshall County
      Mason County
      McLean County
      Meade County
      Mercer County
      Monroe County
      Montgomery County
      Morgan County
      Nelson County
      Pendleton County
      Perry County
      Pike County
      Powell County
      Pulaski County
NT   Rough Creek fault zone
NX   Rowan County
      Russell County
      Scott County
      Taylor County
      Union County
      Warren County
      Washington County
      Wayne County
      Webster County
SA   Allegheny Group
      Allegheny Plateau
      Ames Limestone
      Appalachian Basin
      Appalachian Plateau
      Bedford Shale
      Berea Sandstone
      Borden Group
      Brassfield Formation
      Breathitt Formation
      Carbondale Formation
      Caseyville Formation
      Chattanooga Shale
      Chesterian
      Cincinnati Arch
      Cumberland
      Cumberland Plateau
      Eden Shale
      Fairview Formation
      Fort Payne Formation
      Illinois Basin
      Keokuk Limestone
      Lee Formation
      Lexington Limestone
      Logan Formation
      Midcontinent
      Midwest
      Mississippi Embayment
      Mississippi River
      Mississippi Valley
      New Albany Shale
      Newman Limestone
      Ohio River
      Ohio River valley
      Ohio Shale
      Pennington Formation
      Pottsville Group
      Richmond Group
      Ripley Formation
      Saint Louis Limestone
      Saint Peter Sandstone
      Sainte Genevieve Limestone
      Salem Limestone
      Somerset
      Tennessee River
      Tennessee Valley
      Versailles
      Waldron Shale
      Warsaw Formation
      Wilcox Group

Kentucky River
Formed by junction of North and Middle forks 4 miles ENE of Beattyville.
BT   Kentucky
      United States
SA   Kentucky River valley

Kentucky River valley
NE central.
BT   Kentucky
      United States

Kenya
Includes use on level 1 as an area term (list O). For term set options see list B.
CO   S043000N043000
      E0420000E0340000
BA   Africa
NT   East Rudolf
      Fort Ternan
      Kenya Rift valley
      Lake Magadi
      Mount Kenya
      Nairobi
      Nakuru Basin
      Turkana District
SA   East African Rift
      Gregory Rift
      Lake Natron
      Lake Turkana
      Lake Victoria
      Ngorora Formation
      Umba

Kenya Rift
use Kenya Rift valley

Kenya Rift valley
That segment of the Great Rift Valley running N and S through W central Kenya. Also search Kenya Rift; Kenya Rift zone.
UF   Kenya Rift
      Kenya Rift zone
BT   Kenya
SA   Great Rift Valley

Kenya Rift zone
use Kenya Rift valley

kenyaite
BA   scapolite group
      framework silicates
      silicates
BT   minerals

Keokuk Limestone
Uppermost formation of Osagian Series. Iowa, Illinois, W Kentucky, and E Missouri.
BT   Osagian
      Lower Mississippian
      Mississippian
SA   Illinois
      Iowa
      Kentucky
      Missouri

Keonjhar
District in N.
BT   Orissa
      India

kerabitumen
use kerogen

Kerala
State on Arabian Sea in SW.
BT   India
SA   Warkalli Formation

keratophyre
Includes use on level 3 under igneous rocks(1) trachyte-phonolite family(2). See list H.
BA   trachyte-phonolite family
BT   igneous rocks
SA   quartz keratophyre

Kerch
City on the Kerch Peninsula in E Crimea Oblast.
BT   Ukraine
      USSR

Kerch Peninsula
E section of the Crimea between the Azov Sea on the N and the Black Sea on the S. Also search Kerch.
BT   Ukraine
      USSR

Kerguelen Islands
French island group, 1,400 miles off

Antarctic mainland.
UF Desolation Islands
BT Indian Ocean

Kermadec Islands
Volcanic island group 600 miles NNW of New Zealand. Annexed to New Zealand in 1887. Also search Kermadec.
BT Pacific Ocean
SA Raoul Island

Kerman
Province in SE central. Also a city.
BT Iran
NT Zarand Basin

Kermine
Town in Bukhara Oblast in central.
BT Uzbekistan
    USSR

Kern County
S central.
BT California
    United States

kernite
UF rasorite
BA borates
BT minerals

kerogen
Includes use on level 2 under organic materials(1).
UF kerabitumen
    petrologen
BA organic materials
SA bitumens
    oil shale
    petroleum
    sedimentary rocks

Kerry
County in SW.
BT Ireland

kersantite
Includes use on level 3 under igneous rocks(1) lamprophyre and carbonatite family(2). See list H.
BA lamprophyre and carbonatite family
BT igneous rocks

Kerulen
use Choybalsan

Kerulen River
A headstream of the Amur River. Index Inner Mongolia and/or Mongolia as applicable. Also search Kerulen.
BT Asia
SA Inner Mongolia
    Mongolia

kesterite
BA sulfides
BT minerals

Kettleman Hills
Along W side of San Joaquin Valley.
BT California
    United States

Keuper
Europe. Above Muschelkalk, below Jurassic. Includes use on level 3 under age terms(1). See list E.
BA Upper Triassic
    Triassic
BT Mesozoic

Keweenaw County
NW Upper Peninsula.
BT Michigan
    United States

Keweenaw Peninsula
Juts into Lake Superior in NW Upper Peninsula.
BT Michigan
    United States

Keweenawan
Provincial series, Michigan and Wisconsin. Includes use on level 3 under age terms(1). See list E.
BA Precambrian
NT Portage Lake Lava Series
SA Michigan
    Wisconsin

key beds
Term introduced in 1978 for use in stratigraphy(1). Before 1978, search marker beds.
UF beds, key
    index beds
SA marker beds
    stratigraphy

Key Largo
Island off extreme SE coast. Largest of the Florida Keys and the first island link of the Overseas Highway.
BT Florida
    United States

Key Largo Limestone
Pleistocene deposit older than Pamlico Sand. S Florida.
BT Pleistocene
SA Florida

Khabarovsk
City in S Khabarovsk Kray in Soviet Far East.
BT Russian Republic
    USSR

Khamar-Daban Range
S of Lake Baikal in Buryat A.S.S.R. Also search Khamar-Daban.
BT Russian Republic
    USSR

Khammam
Town and district in E central.
BT Andhra Pradesh
    India

Khangai
use Hangay Mountains

Khangai Mountains
use Hangay Mountains

Khanka Lake
N of Vladivostok. Index China and/or Russian Republic as applicable. Also search Khanka.
BT Asia
SA China
    Manchuria
    Russian Republic

Khapcheranga
Town in SW Chita Oblast in Transbaikalia.
BT Russian Republic
    USSR

Kharga
use Kharga Oasis

Kharga Oasis
Valley and oasis in S central Egypt. Also search El Kharga; Kharga.
UF El Kharga
    Kharga
    The Great Oasis
BT Egypt

Kharkov
City and oblast in NE.
BT Ukraine
    USSR

Khaskovo
Province in S. Also a city.
CO N411000N421000
    E0264000E0244000
BT Bulgaria

Khatanga
Town in Taymyr National Okrug about 150 miles from mouth of Khatanga River in N central Siberia.
BT Russian Republic
    USSR

Khatanga Basin
River basin in E Taymyr National Okrug in N central Siberia. Also search Khatanga.
CO N713000N740000
    E1080000E1013000
BT Russian Republic
    USSR
SA Kotui
    Yenisei-Khatanga basin

Khatanga River
Formed by the union of the Kheta and Kotui rivers in N central Siberia. Flows into Khatanga Gulf an inlet of the Laptev Sea. Also search Khatanga.
BT Russian Republic
    USSR
SA Kheta River
    Kotui

Khawr al Bazam
use Khor al Bazam

Kheta River
Chief tributary of the Khatanga River in Taymr National Okrug in N central Siberia.
BT Russian Republic
    USSR
SA Khatanga River

Khetri
Town in E.
BT Rajasthan
    India

Khetri copper belt
In N Aravalli Range in central. Also search Khetri.
BT Rajasthan
    India

Khewra
Village in N.
BT Punjab
    Pakistan

Khibiny Massif
use Khibiny Mountains

Khibiny Mountains
Central Kola Peninsula in extreme NW European USSR. Also search Khibiny; Khibiny Massif.
CO N670000N680000
    E0340000E0330000
UF Khibiny Massif
BT Russian Republic
    USSR

Khingan Range
At E end of Mongolian Plateau in N Inner Mongolia. Also search Khingan; Greater Khingan Mountains.
UF Great Khingan Range
    Greater Khingan Mountains
    Greater Khingan Range
BT Inner Mongolia
    China

Khios
use Chios

Khiva
Town in oasis region on left bank of the lower Amu Darya in Khorezm Oblast in NW Uzbekistan. The former Khanate of Khiva is coextensive with current Khorezm Oblast.
BT Uzbekistan
    USSR

Khmer Republic
use Cambodia

Khobdo
use Kobdo

khondalite
Includes use on level 3 under metamorphic rocks(1). See list J.
BA metasedimentary rocks
BT metamorphic rocks

Khor al Bazam
A strait off coast of central United Arab Emirates W of Abu Dhabi.
UF Khawr al Bazam
BT Persian Gulf

Khorat Group
Includes the Kamawkala Limestone. May also include beds of Permian as well as Cretaceous age or younger. Also search Korat series. Widespread throughout the country.
UF Korat Series
BT Mesozoic
SA Jurassic
    Thailand
    Triassic

Khorat Plateau
Tableland between the Mekong River on N and E and Cambodia on the S.
UF Korat Plateau
BT Thailand

Khovu-Aksy
Village in Tuva A.S.S.R. just N of W Mongolia.
BT Russian Republic
    USSR

Khubsugul
Province in N Mongolia.
UF Hobsogol
    Hubsugul
BT Mongolia

Kicking Horse River valley
In Banff National Park area. Index provinces as applicable. Also search Kicking Horse River.
BT Canada
SA Alberta
    British Columbia

Kiel
City on the Baltic Sea in N.
BT Schleswig-Holstein
    West Germany

Kiel Bay
Off NE coast of Schleswig-Holstein.
BT Baltic Sea
    Atlantic Ocean

Kielce
Province in SE central. Also a city.
BT Poland
NT Nida Basin
    Ostrowiec Swietokrzyski
    Sandomierz
    Swiety Krzyz Mountains

kieserite
BA sulfates
BT minerals

Kiev
City and oblast in N central.
BT Ukraine
    USSR

Kiev Member
Also search Kiev Series.
BT lower Tertiary
    Tertiary
SA Ukraine
    USSR

Kiglapait
Cape in N.
BT Labrador
    Canada

Kii Peninsula
Between Kii Channel on W and Kumano Sea on E in S Honshu. Also search Kii.
BT Honshu
    Japan

Kilauea
Active crater on E side of Mauna Loa in Volcanoes National Park. Also search Kilauea Volcano.
CO N191500N192500

W1550500W1552000
UF Kilauea Volcano
BT Hawaii
United States
SA Alae Crater

Kilauea Volcano
use Kilauea

kilchoanite
BA orthosilicates
silicates
BT minerals

Kildare
County in E central. Also a town.
BT Ireland

Kilimanjaro
Highest mountain in Africa. Near Kenya border in NE.
BT Tanzania

Kilo
use Bambu

Kilo-Mines
use Bambu

Kimberley
City. World's diamond center with mines nearby. Near Orange Free State border in N.
BT Cape Province
South Africa

kimberlite
Includes use on level 3 under igneous rocks(1) ultramafic family(2). See list H.
BA ultramafic family
BT igneous rocks
SA diamond
diamonds

Kimmerian
use Cimmerian

Kimmeridgian
Europe. Above Oxfordian, below Portlandian. Includes use on level 3 under age terms(1). See list E.
BA Upper Jurassic
Jurassic
BT Mesozoic
SA Malm

Kinderhookian
Provincial series, North America. Above Chautauquan (Devonian), below Osagian. Includes use on level 3 under age terms(1). See list E.
BA Lower Mississippian
Mississippian
BT Paleozoic
SA Carboniferous

kinematics
Branch of dynamics that deals with aspects of motion apart from considerations of mass and force.
SA kinetics
velocity

kinetics
Used as a general term. Branch of dynamics that deals with the effects of forces upon the motions of material bodies.
SA energy
geochemistry
geophysics
kinematics
velocity

King County
BX Texas
Washington
BT United States

King George Island
Largest of the South Shetland Islands in British Antarctic Territory N of the Antarctic Peninsula.
BT Antarctica
SA South Shetland Islands

Kingfisher County
Central.
BT Oklahoma
United States

Kingman County
S central.
BT Kansas
United States

Kings Bay
Inlet on NW coast of Norwegian island of Spitsbergen.
BT Spitsbergen
Arctic region

Kings Mountain
Isolated ridge. Index states as applicable.
BT United States
SA North Carolina
South Carolina

Kingsdown Formation
In Sanborn Group. Overlies Crooked Creek Formation; underlies Vanhem Formation. SW Kansas.
BT Pleistocene
SA Kansas

Kingsport Formation
In Knox Group. E Tennessee and SW Virginia.
BT Lower Ordovician
Ordovician
SA Knox Group
Tennessee
Virginia

kink folds
Term introduced in 1978. Includes use on level 3 under folds(1) style(2). Before 1978, also search folds AND kink.
BA folds
SA chevron folds
kink-band structures

kink-band structures
Term introduced in 1978. Includes use on level 3 under deformation(1) experimental studies(2). Also search kink bands; kink-band.
SA deformation
folds
kink folds
structures

Kinki District
The Osaka, Kobe and Kyoto region in S central.
BT Honshu
Japan

Kinnashih
use Chinkuashih

Kinshasa
City and Federal District on left bank of the Congo River in W Zaire.
UF Leopoldville
BT Zaire

Kinta Valley
River valley in NW West Malaysia.
BT Perak
Malaysia

Kinwashih Mine
use Chinkuashih Mine

Kinzers Formation
Overlies Vintage Limestone; underlies Ledger Dolomite. Maryland, SE Pennsylvania, and Virginia.
BT Lower Cambrian
Cambrian
SA Maryland
Pennsylvania
Virginia

kinzigite
Includes use on level 3 under metamorphic rocks(1) granulites(2). See list J.
BA granulites

BT metamorphic rocks

Kiowa County
Index states as applicable.
BX Colorado
Kansas
Oklahoma
BT United States

Kiowa Formation
Overlies Cheyenne Sandstone Member; underlies Dakota Sandstone. Central S Kansas, E Colorado, W Oklahoma. Also search Kiowa Shale.
UF Kiowa Shale
BT Lower Cretaceous
Cretaceous
SA Colorado
Kansas
Oklahoma

Kiowa Shale
use Kiowa Formation

Kipawa Lake
SW Quebec. Also search Kipawa.
BT Quebec
Canada

Kirghiz (Soviet Socialist Republic)
use Kirghizia

Kirghizia
Kirghiz Soviet Socialist Republic in Soviet Central Asia. Also search Kirghiz; Kirgizia.
CO N390000N420000
E0800000E0690000
UF Kirghiz (Soviet Socialist Republic)
Kirgizia
BT USSR
NT Alai Range
Chatkal Range
Frunze
Issyk-kul Lake
Naryn
Suzak
Talas Range
SA Asiatic USSR
Central Asia
Fergana Basin
Pskem Range
Syr Darya
Tien Shan

Kirgizia
use Kirghizia

Kirishima
Mountain in Kirishima Range in S Kyushu. Also search Kirishima-Yama.
UF Kirishima-Yama
BT Kyushu
Japan

Kirishima-Yama
use Kirishima

Kirkcudbrightshire
County in S.
CO N544500N551500
W0033000W0043000
BT Scotland
Great Britain
United Kingdom

Kirkland Lake
Town in E Ontario near Quebec border.
BT Ontario
Canada

Kirkuk
Town near oil fields in NE.
BT Iraq
Middle East

Kirkwood Formation
Arikareean and Hemingfordian. E New Jersey.
BT middle Miocene
Miocene

SA New Jersey

Kirov
City and oblast in E central European USSR.
UF Vyatka
BT Russian Republic
USSR

Kirovabad
City in W Azerbaidzhan.
UF Gandja
Gandzha
BT Azerbaidzhan
USSR

Kirovograd
City and oblast in central S.
BT Ukraine
USSR

Kiruna
City in N central.
BT Norrbotten
Sweden

Kirwin
Town in Phillips County in N.
BT Kansas
United States

Kishtwar
Town in SW.
BT Jammu and Kashmir
India

Kiso Mountains
In central. Also search Kiso.
BT Honshu
Japan

Kistna River
use Krishna

Kitakami Massif
use Kitakami Mountains

Kitakami Mountains
N Honshu. Also search Kitakami; Kitakami Massif.
UF Kitakami Massif
BT Honshu
Japan

Kitami
City in NE.
BT Hokkaido
Japan

Kitsap Peninsula
Bounded on W by Hood Canal and on E by Puget Sound. W of Seattle in NW.
BT Washington
United States

Kivu
Province in E.
BT Zaire
NT Nyiragongo

Kiya River
In Kemerovo and Tomsk oblasts E and NE of Novosibirsk.
BT Russian Republic
USSR

Kizel coal basin
In Central Urals of European USSR in Perm Oblast. Also search Kizel.
BT Russian Republic
USSR

Kizil Kum
use Kyzylkum

Kizil-Dere
use Kizyl-Dere

Kizildere
use Kizyl-Dere

Kizyl-Dere
Gorge near Iranian border in S Turkmenia. Also search Kizil-Dere; Kizildere.
UF Kizil-Dere
Kizildere
BT Turkmenia

Kladno
City NW of Prague.
BT Bohemia
Czechoslovakia

Klamath Mountains
Part of the Coast Ranges. Index states as applicable.
CO N410000N430000 W1230000W1243000
BT United States
SA California
Coast Ranges
Oregon

Kletno
Village in central.
BT Byelorussia
USSR

Klichka
Village in S Chita Oblast near Inner Mongolian border.
BT Russian Republic
USSR

klippen
Also search klippe.
SA erosion features
faults
nappes
tectonics

Klodawa
Town in central part of country.
BT Poznan
Poland

Klondike
Region in Yukon River basin in SW central.
BT Yukon Territory
Canada

Kluane Lake
Along N slope of Saint Elias Mountains in SW.
BT Yukon Territory
Canada

Klyuchevskaya Sopka
Active volcano in Eastern Range of E central Kamchatka Peninsula. Also search Klyuchevskaya. Highest mountain in Siberia.
BT Russian Republic
USSR

Knee Lake
NE Manitoba.
BT Manitoba
Canada

Knife Lake Group
Comprises rocks formerly called Ogishke Conglomerate, Agawa Iron-formation, and Knife Lake Slates. Includes many members which cannot be separated into the three former divisions.
BT Precambrian
SA Minnesota

Knik Arm
Extension of Cook Inlet N of Anchorage. Also search Knik River.
UF Knik River
BT Alaska
United States

Knik River
use Knik Arm

Knox Group
Includes Copper Ridge Dolomite, Chepultepec Dolomite or Limestone, Longview or Nittany Dolomite, Kingsport Limestone, Mascot Dolomite, Newala Formation, Conococheague Limestone, Jonesboro Limestone. NW Georgia, W North Carolina, E Tennessee, and SW Virginia.
BT Paleozoic

SA Cambrian
Georgia
Kingsport Formation
Lower Ordovician
Mascot Dolomite
North Carolina
Ordovician
Tennessee
Upper Cambrian
Virginia

Knysna
Town in S.
BT Cape Province
South Africa

Kobdo
Town and river in W Mongolia.
UF Hobdo
Khobdo
BT Mongolia

kobellite
BA sulfantimonites
sulfosalts
BT minerals

Kobenhavn
use Copenhagen

Kobystan
Region near Baku in E.
BT Azerbaidzhan
USSR

Kocaeli
Province in NW Anatolia.
BT Turkey
Middle East

Kochi
City and prefecture in S.
BT Shikoku
Japan

Kochkar
Village in central Chelyabinsk Oblast in W Siberia.
BT Russian Republic
USSR

Kodaikanal
City in S.
BT Madras
India

Kodar Mountains
use Kodar Range

Kodar Range
NE of Lake Baikal in N Chita Oblast. Also search Kodar; Kodar Mountains.
UF Kodar Mountains
BT Russian Republic
USSR

Kodiak Island
In Gulf of Alaska SE of Alaska Peninsula.
BT Alaska
United States

Kodor River basin
In Abkhaz A.S.S.R. in W.
BT Georgian Republic
USSR

Kokchetav
City and oblast in N.
BT Kazakhstan
USSR

Kola
Town and river in Murmansk Oblast on the Kola Peninsula in extreme NW European USSR.
BT Russian Republic
USSR

Kola Peninsula
Between White Sea and Barents Sea in extreme NW European USSR.
CO N660000N700000 E0420000E0320000
BT Russian Republic

SA Allarechenskiy

Kolar
Town and district in E.
BT Mysore
India

Kolar Gold Field
use Kolar Gold Fields

Kolar Gold Fields
City in E Mysore. Center of gold-mining industry. Also search Kolar; Kolar Gold Field; Kolar Goldfields.
UF Kolar Gold Field
Kolar Goldfields
BT Mysore
India

Kolar Goldfields
use Kolar Gold Fields

kolbeckine
use herzenbergite

Kolbeinsey Island
Off N coast in Norwegian Sea.
BT Iceland
SA Norwegian Sea

Kolima Mountains
use Kolyma Uplift

Kolima River
use Kolyma River

Kolima River basin
use Kolyma River basin

Kolyma Massif
use Kolyma Uplift

Kolyma River
Rises in SE Cherskiy Range and flows generally N into the East Siberian Sea. Also search Kolyma.
UF Kolima River
BT Russian Republic
USSR

Kolyma River basin
N Khabarovsk Kray and E Yakutia. Also search Kolyma.
UF Kolima River basin
BT Russian Republic
USSR

Kolyma Uplift
Mountain range N of Okhotsk Sea in NE Khabarovsk Kray. Also search Kolyma and Kolyma Massif.
UF Kolima Mountains
Kolyma Massif
BT Russian Republic
USSR

Kom Ombo
Village in S part of country.
UF Kum Umbu
BT Aswan
Egypt

Koma-ga-take
Volcanic peak near SW entrance to Uchiura Bay in SW.
BT Hokkaido
Japan

komatiite
Includes use on level 3 under igneous rocks(1) ultramafic family(2). See list H.
BA ultramafic family
BT igneous rocks

Komi
Autonomous Soviet Socialist Republic in NE, west of the Northern Urals.
BT Russian Republic
USSR

Kommunar
Town in Khakass Autonomous Oblast in SW Krasnoyarsk Kray.
BT Russian Republic
USSR

Komsomolsk
City in S Khabarovsk Kray in Soviet Far East. Also search Komsomolsk-on-Amur.
UF Komsomolsk-na-Amure
Komsomolsk-on-Amure
BT Russian Republic
USSR

Komsomolsk-na-Amure
use Komsomolsk

Komsomolsk-on-Amure
use Komsomolsk

Kona
Divisions, both North and South, of Hawaii County on W and SW side of island of Hawaii.
BT Hawaii
United States

Kondapalli
Town in NE.
BT Tamil Nadu
India

Kongo River
use Congo River

Kongsberg
Town in SE.
BT Norway

Konia
use Konya

Konieh
use Konya

Konin
Town in central part of country.
BT Poznan
Poland

Konstantinovka
City in Donetsk Oblast in SE.
BT Ukraine
USSR

Konya
Province in SW central Anatolia. Also a city.
UF Konia
Konieh
BT Turkey
Middle East

Koolau Range
Extends along E side of Oahu Island.
BT Hawaii
United States

Kootenai County
In northern panhandle.
BT Idaho
United States

Kootenai Formation
Includes Moose Mountain Member, Adanac, Hillcrest, and Mutz members. Also search Kootenai Formation.
BT Mesozoic
SA British Columbia
Canada
Cretaceous
Jurassic

Kootenay Lake
SE British Columbia.
UF Kutenai Lake
BT British Columbia
Canada

Kopaonik
Central Serbia. Also search Kopaonik Mountains.
UF Kopaonik Mountains
BT Serbia
Yugoslavia

Kopaonik Mountains
use Kopaonik

Kopet Dag (Range)
use Kopet-Dag Range

Kopet Dagh (Range)
use Kopet-Dag Range

Kopet-Dag Range
Index Iran and/or Turkmenia as applicable. Also search Kopet Dag; Kopet Dagh; Kopet-Dag.
CO N360000N393000
E0610000E0550000
UF Kopet Dag (Range)
Kopet Dagh (Range)
BT Asia
SA Iran
Turkmenia

Koralpe Range
Small range of Noric Alps in S Austria. Also search Koralpe.
BT Austria

Koraput
Village and district in SW.
BT Orissa
India

Korat Plateau
use Khorat Plateau

Korat Series
use Khorat Group

Kordofan
Province in central.
BT Sudan

Korea
A peninsula on E coast of Asia, since 1948 partitioned into two republics. Includes use on level 1 as an area term (list O). For term set options see list B. Use North Korea and/or South Korea on level 3.
CO N330000N430000
E1300000E1241500
BA Asia
NT Cheju Island
North Korea
South Korea
SA Far East

Korenevskaya Formation
Dnieper-Donets region.
SA Permian
Triassic
Ukraine
USSR

kornerupine
UF prismatine
BA orthosilicates
silicates
BT minerals

Korosten
City in Zhitomir Oblast in NW central.
BT Ukraine
USSR

Korsun
Old name of city in Cherkassy Oblast in central Ukraine. Now called Korsun-Shevchenkovski.
BT Ukraine
USSR

Koryak Range
Extends from neck of Kamchatka Peninsula in Khabarovsk Kray into the Chukchi National Okrug. Also search Koryak.
BT Russian Republic
USSR

Korytnica
Health resort on E slope of the Tatra Mountains.
BT Slovakia
Czechoslovakia

Kos
One of the islands of the Dodeconese off SW Turkey. Also search Coo.
UF Coo
Cos
BT Aegean Islands
Greece

Kosaka Mine
Near town of Kosaka in N.
BT Honshu
Japan

Kosi Basin
River basin. Index Bihar and/or Nepal as applicable. Also search Kosi River.
BT Asia
SA Bihar
Nepal

Kosov
Town in Ivano-Frankovsk Oblast in SE.
BT Ukraine
USSR

Kosseir
E Egypt, port on Red Sea. Also search Quseir.
UF Al-Qusayr
Al-Quseir
Quseir
BT Egypt

Kostolac
Village and arm of the Danube River in E Serbia.
UF Kostolats
BT Serbia
Yugoslavia

Kostolats
use Kostolac

Kostroma
City on left Bank of the Volga River. Also an oblast.
BT Russian Republic
USSR

Koszalin
Province in NW. Also a city.
BT Poland

Kota
Town in E central.
BT Madhya Pradesh
India

Koto
River in E and S which flows into the Ubangi River.
UF Kotto
BT Central African Republic

kotoite
BA borates
BT minerals

Kotto
use Koto

Kotui
River in Evenk and Taymyr National okrugs in N central Siberia. Along with the Kheta River, it forms Khatanga River. Also search Kotuy; Kotuy River.
UF Kotuy River
BT Russian Republic
USSR
SA Khatanga Basin
Khatanga River

Kotuy River
use Kotui

Kouchibouguac Bay
On N North Umberland Strait in E.
BT New Brunswick
Canada

Koyna
River in W.
BT Maharashtra
India

Kozhim
Village and river in Komi A.S.S.R. W of the Northern Urals.
BT Russian Republic
USSR

Kr
use krypton

Kraishte
use Krajiste

Krajiste
Highland. Index countries as applicable. Also search Kraishte; Krayishte.
UF Kraishte
Krayishte
BT Europe
SA Bulgaria
Yugoslavia

Krakatao
use Krakatoa

Krakatau
use Krakatoa

Krakatoa
Island volcano in center of Sunda Strait between Sumatra and Java. Also search Krakatao; Krakatau.
UF Krakatao
Krakatau
BT Indonesia

Krakow
Province in S Poland. Also a City province and a city. Also search Crakow.
UF Cracow
BT Poland
NT Andrychow
Bochnia
Jordanow
Krzeszowice
Miechow
Nowy Sacz
Nowy Targ
Olkusz
Pieniny Klippen Belt
Pieniny Mountains
Pilica
Podhale
Tarnow
Wadowice
Wieliczka
Zakopane

Krakow-Czestochowa Jura
Mountain range in S between Krakow and Czestochowa.
BT Poland

Krappfeld
Region in S part of country.
BT Carinthia
Austria

Krasnodar
City in Krasnodar Kray in the Northern Caucasus.
BT Russian Republic
USSR

Krasnoyarsk
City in Krasnoyarsk Kray in S central Siberia.
BT Russian Republic
USSR

Krayishte
use Krajiste

KREEP
Acronym for a basaltic lunar rock type first found in Apollo 12 fines and breccias and characterized by unusually high contents of potassium (K), rare-earth elements (REE), phosphorus (P), and other trace elements in comparison to other lunar rock types.
UF nonmare basalt
SA Moon
phosphorus
potassium
rare earths

Krefeld
City in W.
BT North Rhine-Westphalia
West Germany

Kremasta
Village in NW. Central Greece and Euboea.
BT Greece

Kremenchug
City in Poltava Oblast in NE central.
BT Ukraine
USSR

Kremnica
Town in W central.
BT Slovakia
Czechoslovakia

Kremnica Mountains
In W central.
BT Slovakia
Czechoslovakia

Krems
City on the Danube River at the mouth of the Krems River.
UF Krems an dar Donau
BT Lower Austria
Austria

Krems an dar Donau
use Krems

Kreuth
Resort in Bavarian Alps in S.
BT Bavaria
West Germany

Krishna
River which flows into the Bay of Bengal. Index states as applicable. Also search Kistna River.
UF Kistna River
BT India
SA Andhra Pradesh
Maharashtra
Mysore

Kristianstad
County in S. Also a city.
BT Sweden
SA Skane

Kristiansund
Seaport city, W Norway.
UF Christiansund
BT Norway

Krivoi Rog
use Krivoy Rog

Krivoi Rog Basin
use Krivoy Rog Basin

Krivoi Rog region
use Krivoy Rog Basin

Krivoi Rog Series
use Krivoy Rog Series

Krivoy Rog
City in W Dnepropetrovsk Oblast in S central Ukraine. Also search Krivoi Rog.
UF Krivoi Rog
BT Ukraine
USSR

Krivoy Rog Basin
Basin of the Ingulets River in the Krivoy Rog region. An industrial basin in W Dnepropetrovsk Oblast in S central Ukraine. Also search Krivoi Rog Basin; Krivoy Rog region.
CO N473000N480000
E0353000E0323000
UF Krivoi Rog Basin
Krivoi Rog region
BT Ukraine
USSR

Krivoy Rog Series
Also search Krivoi Rog Series.
UF Krivoi Rog Series
BT Archean
SA Ukraine
USSR

Krosno Beds
Lattorfian or Rupelian of some authors. In the Polish Carpathians.

BT Oligocene
SA Poland
   Rupelian
   Tongrian

Krusne Hory
use Erzgebirge

Krusnehory Mountains
use Erzgebirge

Krusny Hory Mountains
use Erzgebirge

krypton
Includes use on level 1 and 2 as a chemical element (list D).
UF Kr
BT noble gases
SA argon
   elements
   helium
   isotopes
   neon
   radon
   xenon

Krzeszowice
Town in S part of country.
BT Krakow
   Poland

Ksiaz
Town in W central part of country.
UF Ksiaz Wielkopolski
BT Poznan
   Poland

Ksiaz Wielkopolski
use Ksiaz

Kuala Lumpur
City in SE West Malaysia on the Malay Peninsula.
BT Selangor
   Malaysia

Kuban
Steppe region in the W Northern Caucasus. Formerly an oblast.
BT Russian Republic
   USSR

Kuban River
Rises in Georgian Republic in the Caucasus and flows N and NW into the Sea of Azov. Index Soviet republics as applicable. Also search Kuban.
BT USSR
SA Georgian Republic
   Russian Republic

Kuban Valley
River valley. Index Soviet republics as applicable. Also search Kuban.
BT USSR
SA Georgian Republic
   Russian Republic

Kuchinoerabu-Jima
Small island off S coast.
BT Kyushu
   Japan

Kufara
use Kufra Basin

Kufra Basin
Group of 5 oases in central Libyan Desert.
UF Al-Kufrah
   Kufara
BT Libya

Kugitang (-Tau)
use Kugitang-Tau

Kugitang-Tau
Mountain range in E Chardzou Oblast in E Turkmenia. S spur of Baisun-Tau. Also search Kugitang.
UF Kugitang (-Tau)
BT Turkmenia
   USSR

Kuhn Island
Off NE coast.
BT Greenland
   Arctic region

Kuibyshev
City and oblast. Also search Kuybyshev.
UF Kuybyshev
BT Russian Republic
   USSR

Kujawy
Region in N central.
BT Poland

Kuju
Mountain in Oita Prefecture in central.
BT Kyushu
   Japan

Kular Range
N central Yakutia. Also search Kular.
BT Russian Republic
   USSR

Kul'dzhuktau
Mountains in central.
BT Uzbekistan
   USSR

Kulunda Steppe
Between Ob River in Novosibirsk Oblast and Altay Kray on the E, and E Pavlodar Oblast in Kazakhstan on the W. Index Soviet republics as applicable.
BT USSR
SA Kazakhstan
   Russian Republic

Kulyab
Town in SW.
BT Tadzhikistan
   USSR

Kum Umbu
use Kom Ombo

Kuma Basin
River basin in Northern Caucasus. Also search Kuma; Kuma River.
BT Russian Republic
   USSR

Kumamoto
City and prefecture in W central.
BT Honshu
   Japan

Kumaon Himalaya
use Kumaun Himalayas

Kumaon Himalayas
use Kumaun Himalayas

Kumaun Himalayas
W central subdivision of the Himalayas. Index Nepal, Tibet, and Indian states as applicable. Also search Kumaon Himalaya; Kumaon Himalayas; Kumuan Himalaya.
UF Kumaon Himalaya
   Kumaon Himalayas
   Kumuan Himalaya
BT Asia
SA Himachal Pradesh
   Himalayas
   Nepal
   Tibet
   Uttar Pradesh

Kumuan Himalaya
use Kumaun Himalayas

Kunashir Island
Second largest of the Kuril Islands which are administratively part of Sakhalin Oblast. Also search Kunashir.
BT Russian Republic
   USSR
SA Goryachiy Plyazh
   Kuril Islands

Kunavaram Series
BT India

Kundelungu Plateau
SE part of country.
BT Zaire

Kungurian
Europe: Lower Permian or Middle Permian. Above Artinskian, below Kazanian. Includes use on level 3 under age terms(1). See list E.
BA Permian
BT Paleozoic
SA Lower Permian
   Middle Permian

Kuno
Danish island in the Faeroe Islands.
BT Atlantic Ocean
SA Faeroe Islands

kunzite
Transparent gem variety of spodumene.
BA pyroxene group
   chain silicates
   silicates
BT minerals
SA spodumene

Kura Depression
use Kura Lowland

Kura Lowland
Extensive plain along SW shores of Caspian Sea. Index Azerbaidzhan and/or Iran as applicable. Also search Kura and Kura Depression.
UF Kura Depression
SA Azerbaidzhan
   Iran

Kura River
Flows into Caspian Sea S of Baku. Index Turkey and Soviet republics as applicable. Also search Kura.
SA Azerbaidzhan
   Georgian Republic
   Turkey

Kurama Range
Branch of the Tien Shan. Index Soviet republics as applicable. Also search Kurama.
BT USSR
SA Tadzhikistan
   Tien Shan
   Uzbekistan

kurchatovite
BA borates
BT minerals

Kureika River (region)
use Kureyka River region

Kureyka River region
N Krasnoyarsk Kray in NW central Siberia. Also search Kureika River; Kureyka.
UF Kureika River (region)
BT Russian Republic
   USSR

Kurgan
City and oblast E of the Southern Urals.
BT Russian Republic
   USSR

Kuril Islands
Group of 56 islands extending from Kamchatka Peninsula to Hokkaido. Administratively part of Sakhalin Oblast. Also search Kurile Islands; Kurils.
CO N450000N510000
   E1560000E1450000
UF Kurile Islands
   Kurils
BT Russian Republic
   USSR
SA Iturup Island
   Kunashir Island
   Kuril Trench
   Paramushir
   Soviet Far East
   Urup

Kuril Trench
E of the Kuril Islands and along E coast of SE Kamchatka Peninsula. Also search Kuril-Kamchatka; Kuril-Kamchatka Trench.
UF Kuril-Kamchatka Trench
BT Pacific Ocean
SA Kuril Islands

Kuril-Kamchatka Trench
use Kuril Trench

Kurile Islands
use Kuril Islands

Kurils
use Kuril Islands

Kurland Spit
use Courland Spit

kurnakovite
BA borates
BT minerals

Kurnool
Town in W Andra Pradesh.
UF Karnul
BT Andhra Pradesh
   India

Kurnool System
Included in the Purana Group. Subdivided into Kundair Stage, Paniam Stage, Jammalamadugu Stage, Banganapalli Stage. Named after town of Kurnool.
UF Kurnul System
SA Andhra Pradesh
   Cambrian
   India
   Precambrian

Kurnul System
use Kurnool System

Kuroko
As of 1976, usage restricted to geographic place name. Before 1976, term was also used to indicate type of ore deposit.
BT Japan

kuroko-type
Metallic sulfide ore. Before 1977, also search kuroko.
SA mineral deposits

Kuroshio
A branch of the equatorial current flowing along E coast of Taiwan, thence NE along E coast of Honshu, past the Aleutians, and S along North American coast. More commonly known as Japan Current. Also search Kuroshio Current.
UF Japan Current
   Kuroshio Current

Kuroshio Current
use Kuroshio

Kursk
City and Oblast in SW European USSR.
CO N510000N521500
   E0380000E0341500
BT Russian Republic
   USSR

Kursk magnetic anomaly
SW European USSR. Extensive iron-ore region, 90 miles long and 10 miles wide, in central SE Kursk Oblast in SW European USSR. Also search Kursk.
BT Russian Republic
   USSR

Kursk Series
Named after city in Kursk Oblast in SW Russian Republic.
BT Proterozoic
   Precambrian
SA Russian Republic

USSR
Kurt
Village S of Aral Sea.
BT Turkmenia
USSR
kurtosis
SA histograms
skewness
statistical analysis
Kushiro coal field
SE Hokkaido. Also search Kushiro.
BT Hokkaido
Japan
Kushva
City in Ural Mountains in W Sverdlovsk Oblast in West Siberia.
BT Russian Republic
USSR
Kustanai
use Kustanay
Kustanay
City and oblast in NW Kazakhstan.
UF Kustanai
BT Kazakhstan
USSR
Kutch
use Cutch
Kutenai Lake
use Kootenay Lake
Kutna Hora
Town in central.
BT Bohemia
Czechoslovakia
Kuwait
Includes use on level 1 as an area term (list O). For term set options see list B.
CO N283000N301500
E0484500E0463000
BA Arabian Peninsula
SA Asia
Middle East
Near East
Kuybyshev
use Kuibyshev
Kuzbas
use Kuznetsk Basin
Kuznetsk Alatau
Mountain system along borders of Khakass Autonomous Oblast of Krasnoyarsk Kray, and Kemerovo Oblast in SSW Siberia. An outlier of the Altai Mountains.
CO N530000N553000
E0890000E0860000
BT Russian Republic
USSR
SA Altai Mountains
Kuznetsk Basin
Coal basin, 150 miles long and 65 miles wide, in Kemerovo Oblast extending from Tomsk to Novokuznetsk in SSW Siberia.
CO N520000N561500
E0890000E0840000
UF Kuzbas
BT Russian Republic
USSR
Kwanto Plain
Largest and most densely populated plain in country. In central part of island and includes Tokyo-Yokohama industrial region. Also search Kwanto.
UF Kanto
BT Honshu
Japan
Kwanza Basin
use Cuanza Basin
Kworra River
use Niger River

Kworra Valley
use Niger Valley
kyanite
Includes use as level 3 commodity term under ceramic materials(1). See list C.
BA orthosilicates
silicates
BT minerals
SA andalusite
ceramic materials
heavy minerals
sillimanite
Kyffhauser Range
In SW. Also search Kyffhauser.
BT East Germany
Kyongsang Basin
In North Kyongsang and South Kyongsang provinces in SE.
BT South Korea
Kyoto
City and prefecture in central.
BT Honshu
Japan
Kyushu
Southernmost of the four main islands.
BT Japan
NT Aira Caldera
Ariake Bay
Aso
Aso Caldera
Beppu
Ebino
Fukuoka
Goto Islands
Iki Island
Kagoshima
Kirishima
Kuchinoerabu-Jima
Kuju
Miyazaki
Nagasaki
Oita
Ontake
Osumi Peninsula
Saga
Sasebo
Satsuma Peninsula
Tanega-shima
Tsushima
Unzen
Usuki
SA Sambagawa Belt
Shimanto Group
Taishu Group
Kyzyl Kum
use Kyzylkum
Kyzyl Kum Desert
use Kyzylkum
Kyzyl-Kum
use Kyzylkum
Kyzylkum
Desert SE of Aral Sea between the Amu Darya and the Syr Darya. Also search Kyzyl Kum; Kyzyl-Kum; Kyzyl Kum Desert. Index Soviet republics as applicable.
CO N400000N440000
E0690000E0610000
UF Kizil Kum
Kyzyl Kum
Kyzyl Kum Desert
Kyzyl-Kum
Qizil Qum
BT USSR
SA Kazakhstan
Uzbekistan

L

L waves
use surface waves
La
use lanthanum
La Caridad
Settlement in NW central part of country.
BT Oviedo
Spain
La Coruna
Province in extreme NW. Also a city.
BT Spain
NT Cabo Ortegal
Santiago de Compostela
La Jolla
NW section of San Diego in San Diego County.
BT California
United States
La Ligua
Town and river in central part of country.
BT Aconcagua Province
Chile
La Pampa
Province in central.
BT Argentina
SA Salado Basin
La Paz
Department in W Bolivia. Also a city.
CO S180000S115000
W0670000W0693000
BT Bolivia
La Rioja
Province in NW.
CO S330000S270000
W0650000W0700000
BT Argentina
NT Paganzo
SA Chanares Formation
Salado Basin
La Ronge
Village and lake in N central.
BT Saskatchewan
Canada
La Salle Limestone
In McLeansboro Group. NE Illinois.
BT Pennsylvanian
SA Illinois
Middle Pennsylvanian
Upper Pennsylvanian
La Ventana Sandstone
Of Mesaverde Formation. NW New Mexico.
BT Upper Cretaceous
Cretaceous
SA Mesaverde Group
New Mexico
Laba Basin
River basin in Krasnodar Kray of the Northern Caucasus. Also search Laba.
BT Russian Republic
USSR
Labette County
SE Kansas.
BT Kansas
United States
Labette Shale
In Marmaton Group.
BT Desmoinesian
Middle Pennsylvanian
Pennsylvanian
SA Iowa
Kansas

Marmaton Group
Missouri
Oklahoma
laboratory studies
UF studies, laboratory
SA experimental studies
theoretical studies
Labrador
Now part of Newfoundland. Includes use on level 1 as an area term (list O). For term set options see list B.
CO N513000N610000
W0553000W0673000
BA Canada
NT Churchill Falls
Kiglapait
Mistastin Lake
Nain
Nain Massif
Wabush
SA Churchill Province
Grenville Province
Newfoundland
Sokoman Formation
Ungava
Labrador Basin
E of Labrador and S of Greenland. Also search Labrador Trough.
UF Labrador Trough
BT Atlantic Ocean
Labrador Sea
Between Labrador and SW Greenland.
BT Atlantic Ocean
Labrador Trough
use Labrador Basin
labradorite
BA feldspar group
framework silicates
silicates
BT minerals
SA gabbro family
igneous rocks
plagioclase
labradoritite
Includes use on level 3 under igneous rocks(1) gabbro family(2). See list H.
BA gabbro family
BT igneous rocks
labuntsovite
BA orthosilicates
silicates
BT minerals
Labyrinthodontia
Includes use on level 2 under Amphibia(1). See list F.
BA Amphibia
BT Tetrapoda
Vertebrata
Lac La Ronge
Lake in N central.
BT Saskatchewan
Canada
laccoliths
Includes use on level 2 under intrusions(1).
UF cistern rock
BA intrusions
SA batholiths
differentiation
dikes
domes
lopoliths
Lacertilia
Suborder.
BA Lepidosauria
Reptilia
BT Tetrapoda
Vertebrata
lacustrine
A valid term through 1977. After

lacustrine ● **Lake Michigan** 158

1977, use lacustrine features.
lacustrine features
Includes use on level 2 under geomorphology(1). Also search lacustrine.
UF features, lacustrine
SA beaches
extinct lakes
geomorphology
glacial lakes
lakes
limnology
shoals
terraces
Ladak
use Ladakh
Ladakh
Region in Jammu and Kashmir. S Ladakh divided between India and Pakistan. Index countries as applicable.
UF Ladak
BT Asia
SA India
Pakistan
Ladinian
Europe: upper Middle Triassic. Above Anisian, below Carnian. Includes use on level 3 under age terms(1). See list E.
BA Middle Triassic
Triassic
BT Mesozoic
SA Hallstatt Limestone
Lafayette County
BX Arkansas
Florida
Louisiana
Mississippi
Missouri
Wisconsin
BT United States
Lagenidae
Includes use on level 3 under foraminifera(1) Rotaliina(2).
BA Rotaliina
foraminifera
BT Invertebrata
Laghouat
Town and oasis in N central.
BT Algeria
Lago Maggiore
Lake. Index countries as applicable.
BT Europe
SA Italy
Switzerland
Lagomorpha
Includes use on level 2 under Mammalia(1). See list F.
BA Mammalia
BT Tetrapoda
Vertebrata
lagoonal
A valid term through 1977. After 1977, use lagoons.
lagoons
Includes use on level 3 under ecology(1), geomorphology(1), and paleoecology(1); on level 3 under sedimentation(1) for type of environment. Also search lagoonal.
SA atolls
barrier reefs
beaches
coastal environment
ecology
geomorphology
lakes
paleoecology
reefs
sedimentation
Lagos Lagoon
Between Lagos Island just off the Gulf of Guinea and the mainland. City of Lagos at W end of lagoon. Also search Lagos.
BT Nigeria
lahars
SA lava flows
mass movements
mudflows
volcanism
Lahn River
Flows S and SW into the Rhine River SE of Koblenz. Index states as applicable. Also search Lahn.
BT West Germany
SA Hesse
Rhineland-Palatinate
Lahn River valley
Index states as applicable. Also search Lahn.
BT West Germany
SA Hesse
Rhineland-Palatinate
Laisvall
Village in N part of country.
BT Norrbotten
Sweden
Lake Agassiz
Glacial lake which existed in Pleistocene Epoch. Covered parts of N central U. S. and S central Canada. Index Manitoba and states as applicable.
BT North America
SA Manitoba
Minnesota
North Dakota
Lake Albert
Index Uganda and/or Zaire as applicable.
BT Africa
SA East African Rift
Uganda
Zaire
Lake Baikal
Largest freshwater lake in Eurasia. Bounded on N, E, and S by the Buryat A.S.S.R., and on the W by Irkutsk Oblast. Also search Baikal AND lake or lakes.
CO N520000N560000
E1100000E1040000
UF Baikal (Lake)
BT Russian Republic
USSR
SA Baikal region
Lake Balaton
W Hungary. Also search Balaton.
UF Balaton (Lake)
BT Hungary
SA Balaton region
Lake Balkhash
E Kazakhstan. Also search Balkhash.
CO N450000N470000
E0790000E0730000
UF Balkhash (Lake)
BT Kazakhstan
USSR
SA Balkhash region
lake biscuits
use algal biscuits
Lake Biwa
W central Honshu. Also search Biwa; Biwa Lake.
UF Biwa Lake
Omi (Lake)
BT Honshu
Japan
Lake Bonneville
Large prehistoric body of water in present Great Salt Lake area. Index states as applicable.
BT United States
SA Idaho
Nevada
Utah
Lake Bonney
Southeast.
BT South Australia
Australia
Lake Chad
Index countries as applicable.
CO N123000N143000
E0160000E0123000
BT Africa
SA Cameroon
Chad
Chad Basin
Niger
Nigeria
Lake Champlain
Between New York and Vermont extending into Quebec. Index Quebec and states as applicable.
SA Champlain Valley
New York
Quebec
Vermont
Lake Chatuge
Index states as applicable.
BT United States
SA Georgia
North Carolina
Lake Constance
Index countries as applicable. Also search Bodensee; Constance Lake.
UF Bodensee
Constance Lake
BT Europe
SA Austria
Germany
Rhine Basin
Rhine River
Switzerland
Lake County
BX California
Colorado
Florida
Illinois
Indiana
Michigan
Minnesota
Montana
Ohio
Oregon
South Dakota
Tennessee
BT United States
Lake District
Mountain and lake region in NW.
BT England
Great Britain
United Kingdom
Lake Edward
S of Lake Albert. Index countries as applicable.
BT Africa
SA East African Rift
Uganda
Zaire
Lake Erie
Index Ontario and states as applicable.
BA Great Lakes
BT North America
SA Michigan
New York
Ohio
Ontario
Pennsylvania
Lake Eyre
Northeast.
BT South Australia
Australia
Lake Geneva
Index countries as applicable. Also search Lake Leman; Leman Lake.
UF Lake Leman
Leman Lake
BT Europe
SA France
Geneva
Rhone River
Rhone Valley
Switzerland
Lake George
NE New York.
BT New York
United States
Lake Huron
Index Michigan and/or Ontario.
BA Great Lakes
BT North America
SA Michigan
Ontario
Lake Kariba
Formed by dam in Kariba Gorge of Zambezi River. Index countries as applicable. Also search Kariba; Kariba Lake.
UF Kariba Lake
BT Africa
SA Rhodesia
Zambia
Lake Kinneret
use Sea of Galilee
Lake Kivu
Index countries as applicable.
BT Africa
SA East African Rift
Rwanda
Zaire
Lake Ladoga
E and NE of Leningrad.
BT Russian Republic
USSR
Lake Lappajarvi
SW central. Vaasa.
BT Finland
Lake Leman
use Lake Geneva
Lake Louise
Village and small lake in Banff National Park in SW.
BT Alberta
Canada
Lake Magadi
Soda lake in S central.
BT Kenya
Lake Malawi
Index countries as applicable. Also search Lake Nyasa.
UF Lake Nyasa
BT Africa
SA East African Rift
Malawi
Mozambique
Tanzania
Lake Maracaibo
NW part of country.
BT Zulia
Venezuela
SA Maracaibo Basin
Lake Mead
Within Lake Mead National Recreation Area. Index states as applicable.
BT United States
SA Arizona
Nevada
Lake Mendota
In Dane County in S central.
BT Wisconsin
United States
Lake Michigan
Index states as applicable.
CO N414500N460000
W0844500W0875000
BA Great Lakes
BT North America
SA Illinois

Indiana
Michigan
United States
Wedron Formation
Wisconsin
Lake Natron
Index countries as applicable.
BT Africa
SA Kenya
Tanzania
Lake Nyasa
use Lake Malawi
Lake of Bienne
use Biel Lake
Lake of Zurich
N Switzerland.
BT Switzerland
Lake Okeechobee
S central.
BT Florida
United States
Lake Onega
In Karelian A.S.S.R. Also search Onega.
BT Russian Republic
USSR
SA Onega
Lake Ontario
Index New York and/or Ontario as applicable.
BA Great Lakes
BT North America
SA New York
Ontario
Lake Pontchartrain
S Louisiana.
BT Louisiana
United States
Lake Powell
BT United States
SA Arizona
Utah
Lake Rudolf
use Lake Turkana
Lake Rudolph
use Lake Turkana
Lake Sevan
N Armenia.
BT Armenia
USSR
SA Sevan
Lake Superior
Index Ontario and states as applicable.
BA Great Lakes
BT North America
SA Michigan
Minnesota
Ontario
Wisconsin
Lake Suwa
W central.
BT Honshu
Japan
Lake Tahoe
Index states as applicable. Also search Lake Tahoe Basin.
UF Lake Tahoe Basin
BT United States
SA California
Nevada
Lake Tahoe Basin
use Lake Tahoe
Lake Tanganyika
Index countries as applicable.
BT Africa
SA Burundi
East African Rift
Tanzania
Zaire
Zambia

Lake Tiberias
use Sea of Galilee
Lake Timiskaming
Index provinces as applicable.
BT Canada
SA Ontario
Quebec
Lake Titicaca
Index countries as applicable.
BT South America
SA Bolivia
Peru
Lake Torrens
E central.
BT South Australia
Australia
Lake Turkana
New name for Lake Rudolf. Primarily in Kenya. Index countries as applicable. Also search Lake Rudolf; Lake Rudolph; Rudolf.
UF Lake Rudolf
Lake Rudolph
Rudolf (Lake)
BT Africa
SA East African Rift
Ethiopia
Kenya
Sudan
Lake Valley Formation
Includes Alamogordo, Arcente, Dona Ana, and Nunn members.
BT Lower Mississippian
Mississippian
SA New Mexico
Lake Victoria
Index countries as applicable.
BT Africa
SA Kenya
Tanzania
Uganda
Lake Winnipeg
S central.
BT Manitoba
Canada
lakes
Includes use on level 3 under geomorphology(1) lacustrine features(2) and under glacial geology(1) glacial features(2); on level 3 under ecology(1) and paleoecology(1); under sedimentation(1) for type of environment.
SA drainage
ecology
eutrophication
extinct lakes
glacial features
glacial lakes
hydrogen sulfide
hydrology
hydrosphere
lacustrine features
lagoons
lava lakes
limnology
paleoecology
periglacial features
ponds
sedimentation
surface water
water balance
water supply
watersheds
lamellae
Restricted to use with minerals. Not used in biological sense.
SA deformation
minerals
laminations
Includes use on level 3 under sedimentary structures(1) planar bedding structures(2); on level 3 under

foliation(1) style(2) and under structural anaylsis(1) theoretical studies(2). See list K. Before 1976, also search lamination.
BA planar bedding structures
sedimentary structures
SA bedding
cross-bedding
cross-laminations
foliation
ripple drift-cross laminations
stratification
lamproite
Includes use on level 3 under igneous rocks(1) trachyte-phonolite family(2). See list H.
BA trachyte-phonolite family
BT igneous rocks
lamprophyre
Includes use on level 3 under igneous rocks(1) lamprophyre and carbonatite family(2). See list H.
BA lamprophyre and carbonatite family
BT igneous rocks
lamprophyre and carbonatite family
Includes use on level 2 under igneous rocks(1). See list H.
UF carbonatite family, lamprophyre
BT igneous rocks
NA camptonite
carbonatite
kersantite
lamprophyre
minette
monchiquite
sovite
spessartite
Lancara Formation
Lower Acadian. Overlies The Herreira Sandstone. NW part of country.
BT Acadian
Middle Cambrian
Cambrian
SA Spain
Lancashire
County in NW.
CO N532000N543000
W0020000W0031500
BT England
Great Britain
United Kingdom
Lancaster Sound
Channel between Devon Island and N Baffin Island in E District of Franklin.
BT Northwest Territories
Canada
Lance Formation
Includes Colgate Sandstone Member, Cannonball Marine Member, Ludlow Lignitic Member, Tullock Member, Hell Creek Member, Torrington Member, and Ilo Ridge Member. Montana, W North Dakota, W South Dakota, and Wyoming.
BT Upper Cretaceous
Cretaceous
SA Colorado
Montana
North Dakota
South Dakota
Wyoming
land canyons
use canyons
land forms
use landforms
land leases
Term introduced in 1978.
SA conservation
reclamation

land subsidence
A level 1 term as of 1978. Includes use on level 2 under foundations(1) and geologic hazards(1). Used for geotechnical studies on land subsidence. If 1, term set options are:
causes
   subtopic
controls
   subtopic
foundations
   subtopic
geologic hazards
   subtopic
mines
   subtopic
site exploration
   subtopic
settlement
   subtopic
stability
   subtopic
solution features
   subtopic
SA collapse structures
dams
engineering geology
explosions
foundations
geologic hazards
highways
mechanics
mines
oil and gas fields
rock mechanics
settlement
site exploration
slope stability
soil mechanics
solution features
stability
subsidence
tunnels
underground installations

land use
A level 1 term as of 1978. Includes use on level 2 under impact statements(1). If 1, term set options are:
agriculture
   subtopic
arid environment
   subtopic
changes
   subtopic
classification
   subtopic
conservation
   subtopic
controls
   subtopic
effects
   subtopic
human ecology
   subtopic
impact statements
   subtopic
inventory
   subtopic
legislation
   subtopic
management
   subtopic
maps
   cartography
mines
   subtopic
natural resources
   subtopic
planning
   subtopic
preservation
   subtopic
reclamation
   subtopic
recreation
   subtopic
regional planning

  subtopic
 remote sensing
  subtopic
 soils
  subtopic
 surveys
  subtopic
 underground space
  subtopic
 urban planning
  subtopic
SA agriculture
  arid environment
  conservation
  engineering geology
  environmental geology
  geologic hazards
  human ecology
  impact statements
  legislation
  management
  maps
  mines
  natural resources
  planning
  pollution
  preservation
  reclamation
  recreation
  regional planning
  remote sensing
  soils
  strip mining
  underground space
  urban planning
  urbanization
  waste disposal

Landenian
 Belgium. Includes use on level 3 under age terms(1). See list E.
 BA upper Paleocene
  Paleocene
  Paleogene
  Tertiary
 BT Cenozoic

Lander County
 N central.
 CO N391000N410000
  W1163500W1174500
 BT Nevada
  United States

Landes
 Department in SW. Also a region.
 BT France
 NT Capbreton

landfills
 Includes use on level 3 under environmental geology(1) waste disposal(2); on level 3 under engineering geology(1).
 NT sanitary landfills
 SA engineering geology
  environmental geology
  waste disposal

landform description
 Includes use on level 2 under geomorphology(1). Before 1975, also search surficial geology.
 SA aerial photography
  description
  geomorphology
  landform evolution
  landforms
  landscapes
  morphometry
  mountains
  relief

landform evolution
 Includes use on level 2 under geomorphology(1).
 SA beaches
  erosion
  etching
  evolution
  geomorphology
  landform description
  paleorelief
  relief

landforms
 Also search landforms; relief features.
 UF land forms
  relief features
 SA geomorphology
  hills
  landform description
  landscapes
  mountains
  plains
  plateaus
  valleys

landing sites
 UF sites, landing
 SA Apollo
  extraterrestrial geology
  Mars
  Moon
  Venus
  Viking

Land's End
 Cape. SW end of Cornwall. Use for Lands End.
 BT England
  Great Britain
  United Kingdom

Landsat
 Term introduced in 1978. Includes use on level 3 under geophysical methods(1) and geophysical surveys(1), and remote sensing(1). Before 1978, also search ERTS; ERTS-1.
 SA geophysical methods
  geophysical surveys
  imagery
  remote sensing
  satellite methods

landscapes
 SA geomorphology
  landform description
  landforms

landslides
 As of 1978, term is used on level 2 under geologic hazards(1), and slope stability(1).
 SA avalanches
  creep
  engineering geology
  erosion
  geologic hazards
  geomorphology
  mass movements
  mass wasting
  mudflows
  rockslides
  slope stability
  slopes
  slumping
  soils
  talus slopes

Lane County
 Index states as applicable.
 BX Kansas
  Oregon
 BT United States

Lanersbach
 Village SE of Innsbruck.
 BT Tyrol
  Austria

Laneuville-devant-Nancy Boring
 Boring at or near outer SE suburb of Nancy.
 BT Meurthe-et-Moselle
  France

Laney Shale
 use Laney Shale Member

Laney Shale Member
 Of Green River Formation. SW Wyoming. Also search Laney Shale.
 UF Laney Shale
 BT middle Eocene
  Eocene
 SA Colorado
  Green River Formation
  Wyoming

langbeinite
 BA sulfates
 BT minerals

Langebaanweg
 Village in SW.
 BT Cape Province
  South Africa

Langesund Fjord
 Inlet of the Skaggerak.
 BT Telemark
  Norway

Langhe
 Region NW of Genoa.
 BT Italy
 SA Piedmont

Langhian
 Europe. See list E.
 BA Miocene
  Neogene
  Tertiary
 BT Cenozoic

Langkawi Islands
 Group in Andaman Sea near entrance to Strait of Malacca. Administrated by Kedah state in West Malaysia. Also search Langkawi.
 BT Malaysia

Languedoc
 Historical region of S.
 BT France

Lansing
 City in Clinton, Eaton, and Ingham counties in S central Lower Peninsula.
 BT Michigan
  United States

Lansing Group
 Includes Lane Shale, Plattsburg Limestone, Vilas Shale, Stanton Limestone.
 BT Missourian
  Upper Pennsylvanian
  Pennsylvanian
 SA Kansas
  Plattsburg Limestone
  Stanton Formation

lanthanide series
 use rare earths

lanthanides
 use rare earths

lanthanoans
 use rare earths

lanthanum
 Includes use on level 1 and 2 as a chemical element (list D).
 UF La
 SA cerium
  elements
  rare earths

Lanzarote
 Easternmost of the Spanish controlled island group.
 BT Canary Islands
  Atlantic Ocean

Lanzo Massif
 NW of Turin in Piedmont.
 BT Italy
 SA Piedmont

Laos
 Includes use on level 1 as an area term (list O). For term set options see list B.
 BA Asia
 SA Indochina

lapilli
 SA ejecta
  fragments
  pyroclastics
  pyroclastics and glasses

Laplace azimuths
 use Laplace transformations

Laplace points
 use Laplace transformations

Laplace tidal equation
 use Laplace transformations

Laplace transformations
 Term introduced in 1978. Search Laplace azimuths; Laplace points; Laplace tidal equation; Laplace transforms.
 UF Laplace azimuths
  Laplace points
  Laplace tidal equation
  Laplace transforms
  transformations, Laplace
 SA geodesy

Laplace transforms
 use Laplace transformations

Lapland
 Region N of the Arctic Circle including the Kola Peninsula of the Russian Republic. Index countries as applicable.
 BT Europe
 SA Finland
  Norway
  Sweden
  USSR

Laptev Sea
 Along N coast of Siberia between Taymyr Peninsula and New Siberian Islands.
 BT Arctic Ocean

Lar
 River valley in N, and city in Laristan region in S.
 BT Iran

Lara
 State in NW.
 BT Venezuela

Laramian Orogenic Phase
 use Laramide Orogeny

Laramian Orogeny
 use Laramide Orogeny

Laramic Orogeny
 use Laramide Orogeny

Laramide Orogeny
 A time of deformation typically developed in the eastern Rocky Mountains of the United States, whose several phases extended from Upper Cretaceous until the end of the Paleocene. Term introduced in 1978. Before 1978, search orogeny AND Laramide; Laramian.
 UF Laramian Orogenic Phase
  Laramian Orogeny
  Laramic Orogeny
  Laramide Revolution
 SA Cordilleran Orogeny
  Cretaceous
  orogeny
  Tertiary

Laramide Revolution
 use Laramide Orogeny

Laramie Basin
 High plateau in SE between Laramie Mountains on E and Medicine Bow Mountains on the W.
 BT Wyoming
  United States

Laramie County
 Extreme SE.
 BT Wyoming
  United States

Laramie Formation
Overlies Fox Hills Formation; unconformably underlies Denver Formation. Denver Basin region of E Colorado.
BT Upper Cretaceous
Cretaceous
SA Colorado

Laramie Mountains
A range of the Rocky Mountains. Index states as applicable. Also search Laramie Range.
UF Laramie Range
BT United States
SA Colorado
Rocky Mountains
Wyoming

Laramie Range
use Laramie Mountains

Lardeau Group
Includes Triune and Sharon Creek Formations.
UF Lardeau Series
BT Cambrian
SA British Columbia
Canada
Precambrian

Lardeau Series
use Lardeau Group

Larderello
Village in central part of country.
BT Tuscany
Italy

Largentiere
Village in S part of country.
BT Ardeche
France

Larimer County
On the Wyoming border in N.
CO N401500N410000
W1045000W1062000
BT Colorado
United States

larnite
BA orthosilicates
silicates
BT minerals

Larvik
Town at head of Larvik Fjord near entrance to Oslofjord.
BT Vestfold
Norway

Las Animas County
SE central.
CO N370000N375000
W1030500W1051500
BT Colorado
United States

Las Vegas
City in Clark County in extreme S.
BT Nevada
United States

Las Villas
Province in W central Cuba.
BT Cuba
West Indies

LASA
Array of seismographs. Includes use on level 3 under seismology(1).
SA arrays
Montana

Lassen Peak
Active volcano in Lassen Volcanic National Park in NE.
BT California
United States

Lassen Volcanic National Park
NE California.
BT California
United States

Latah County
On the Washington border in northern panhandle.
BT Idaho
United States

Latdorfian
use Lattorfian

late diagenesis
Term introduced in 1978.
BA diagenesis
SA early diagenesis

Latemar Massif
In Dolomites.
BT Trentino-Alto Adige
Italy
SA Dolomites

Laterite soils
use laterites

laterites
Includes use on level 3 under soils(1) soil group(2). See list M. Also search laterite; laterite soils; lateritic soils.
UF Laterite soils
BA soils
SA bauxite
bauxitization
laterization
Latosols

laterization
SA bauxitization
laterites
soils

Lathi Beds
use Lathi Formation

Lathi Formation
UF Lathi Beds
BT Jurassic
SA India
Rajasthan

latite
Includes use on level 3 under igneous rocks(1) andesite-rhyolite family(2). See list H.
BA andesite-rhyolite family
BT igneous rocks
SA quartz latite
trachybasalt

Latium
Autonomous region in W central.
BT Italy
NT Alban Hills
Frosinone
Rome
Sabatini Mountains
Tolfa Hills
University of Rome
Viterbo
SA Tiber Valley

Latosols
See list M.
BA soils
SA laterites

lattice
Includes use on level 3 under crystal structure(1). Also search Bravais lattice; crystal lattice; space lattice.
UF Bravais lattice
crystal lattice
space lattice
translation lattice
SA cell dimensions
crystal structure
habit
lattice
lattice parameters
substitution
twinning
unit cell

lattice constants
use lattice parameters

lattice parameters
Also search lattice constants.
UF lattice constants
SA crystal chemistry
crystal growth
crystal structure
crystallography
lattice
unit cell

Lattorfian
Europe.
UF Latdorfian
BA upper Eocene
Eocene
Paleogene
Tertiary
BT Cenozoic
SA Tongrian

Latvia
Latvian Soviet Socialist Republic.
CO N560000N580000
E0280000E0210000
BT USSR
NT Riga
SA Baltic region
Baltic Shield
Dvina River
European USSR
Russian Plain
Russian Platform

Lau Basin
West of the Tonga Islands and S of the Lau group of the Fiji Islands
BT Pacific Ocean

laumonite
use laumontite

laumontite
UF laumonite
lomonite
lomontite
BA zeolite group
framework silicates
silicates
BT minerals
SA leonhardite

Lauraceae
Family.
BA Dicotyledoneae
angiosperms
BT Plantae

Laurasia
The protocontinent of the Northern Hemisphere which is a combination of Laurentia and Eurasia. It included most of North America, Greenland, and much of Eurasia excluding India. Includes use on level 2 under continental drift(1).
SA continental drift
Eurasia
Gondwana
Northern Hemisphere
Pangaea

Laurentian Highlands
use Canadian Shield

Laurentian Plateau
use Canadian Shield

Laurentide Ice Sheet
SA Saint Lawrence Lowlands
Saint Lawrence Valley

Laurion
City SE of Athens. Central Greece and Euboea. Also search Laurium.
UF Laurium
Lavrion
BT Greece

Laurium
use Laurion

Lausitz
use Lusatia

lava
This set is used for all molten extrusives and the rocks that solidify from them. Includes use on level 1 (list A); on level 2 under isotopes(1). If 1, term set options are:
topic [age, alteration, analysis, classification, composition, distribution, flow mechanism, genesis, geochemistry, mechanism, nomenclature, observations, occurrence, petrology, properties, structure, temperature, viscosity]
subtopic [e.g. electrical properties, magnetic properties, pillow lava, pillow structure]
NT aa lava
pahoehoe
SA basalt
eruptions
extrusive rocks
flow mechanism
igneous rocks
inclusions
lava channels
lava fields
lava flows
lava lakes
lava tubes
lava tunnels
magmas
mud volcanoes
paleomagnetism
perlite
petrology
pillow lava
pillow structure
plugs
viscosity
volcanic features
volcanic rocks
volcanism
volcanoes
volcanology

lava channels
BT volcanic features
SA channels
lava

lava domes
use shield volcanoes

lava fields
UF fields, lava
BT volcanic features
SA lava

lava flows
Includes use on level 3 under lava(1) and under volcanology(1).
UF flow (volcanic)
nappe (volcanic)
SA flow mechanism
flows
lahars
lava
lava tubes
volcanic features
volcanology

lava lakes
BT volcanic features
SA lakes
lava

lava tubes
Includes use on level 3 under lava(1).
UF tubes, lava
SA lava
lava flows
lava tunnels
volcanic features

lava tunnels
BT volcanic features
SA lava
lava tubes
tunnels

Laval Basin
BT France

Lavrion
use Laurion

Lawrence County
BX Alabama
　　Arkansas
　　Illinois
　　Indiana
　　Kentucky
　　Mississippi
　　Missouri
　　Ohio
　　Pennsylvania
　　South Dakota
　　Tennessee
BT United States

Lawrence Formation
Includes two unnamed shale members separated by Amazonia Limestone Member, Ireland Limestone Member. SW Iowa, E Kansas, NW Missouri, and SE Nebraska.
UF Lawrence Shale
BT Virgilian
　　Upper Pennsylvanian
　　Pennsylvanian
SA Douglas Group
　　Iowa
　　Kansas
　　Missouri
　　Nebraska

Lawrence Shale
use Lawrence Formation

lawsonite
BA orthosilicates
　　silicates
BT minerals
SA aluminosilicates

Laxfordian
Europe. See list E.
BA upper Precambrian
　　Precambrian

layer, active
use active layer

layer, basaltic
use basaltic layer

layer, boundary
use boundary layer

layer, granitic
use granitic layer

layer, low-velocity
use low-velocity layer

layer, nepheloid
use nepheloid layer

layered
A valid term on level 2 through 1977. After 1977, use layered intrusions on level 2. For engineering geology, use layered media.

layered intrusions
Not a valid term from 1975 through 1977. After 1977, includes use on level 2. Before 1978, also search intrusions AND layered.
BA intrusions
SA igneous rocks

layered media
Term introduced in 1978. Includes use on level 3, especially under engineering geology(1). Before 1978, also search layered; layers.
SA engineering geology
　　media

layers
A valid term through 1978. After 1978, use layered media.

lazurite
BZ framework silicates
　　sulfates
　　sulfides
BT minerals
SA sodalite group

Le Havre
City on the English Channel.
BT Seine-Maritime
　　France

Le Mans
City in NE central part of country.
BT Sarthe
　　France

Le Mont-Dore
use Mont-Dore

Le Trou Sans Fond
use Trou Sans Fond

leaching
Includes use on level 3 under soils(1) genesis(2); on level 3 under geochemistry(1) and under weathering(1).
UF lixiviation
SA alteration
　　degradation
　　desalinization
　　geochemistry
　　mining geology
　　percolation
　　soils
　　solution
　　solution mining
　　weathering

lead
Includes use on level 1 and 2 as a commodity term (list C) and as a chemical element (list D); on level 2 under mineral deposits, genesis(1) and paragenesis(1). Also search Pb.
UF Pb
SA elements
　　heavy metals
　　isotopes
　　lead-zinc deposits
　　mississippi valley-type
　　Pb-204
　　Pb-206
　　Pb-206/Pb-204
　　Pb-207
　　Pb-207/Pb-204
　　Pb-207/Pb-206
　　Pb-208
　　Pb-208/Pb-204
　　Pb-210
　　Pb/Pb
　　Pb/Th
　　U-238/Pb-206
　　U/Pb
　　U/Th/Pb

lead glance
use galena

lead-zinc deposits
As of 1978, term includes use on level 1 and 2 as a commodity (list C). Before 1978, also search lead-zinc.
SA deposits
　　lead
　　mississippi valley-type
　　zinc

Leadville
City in Lake County in central.
BT Colorado
　　United States

Leadville Formation
Includes Gilman Sandstone Member.
UF Leadville Limestone
BT Mississippian
SA Colorado
　　Lower Mississippian
　　Upper Mississippian
　　Yule Marble

Leadville Limestone
use Leadville Formation

leaf
use leaves

least-squares analysis
Also search least-squares method.
UF least-squares method
BA statistical methods
SA analysis
　　mathematical methods
　　regression analysis

least-squares method
use least-squares analysis

Leavenworth County
NE Kansas.
BT Kansas
　　United States

leaves
Also search leaf.
UF leaf
SA humus
　　paleobotany
　　Plantae
　　vegetation

Leba
Town and river in N part of country.
BT Gdansk
　　Poland

Lebanon
Includes use on level 1 as an area term (list O). For term set options see list B.
CO N330000N344500
　　E0364500E0350000
BA Middle East
NT Beirut
SA Asia
　　Levant
　　Mediterranean region

Lebanon County
SE central.
BT Pennsylvania
　　United States

Lebedin
City in Sumy Oblast in NE central.
BT Ukraine
　　USSR

lebensspuren
Includes use on level 3 under sedimentary structures(1) biogenic structures(2). See list K.
BA biogenic structures
　　sedimentary structures
SA burrows
　　ichnofossils
　　tracks

Lebombo Mountains
Index Mozambique and/or Transvaal as applicable. Also search Lebombo.
BT Africa
SA Mozambique
　　Transvaal

Lecce
City and province in heel of Italian boot.
BT Apulia
　　Italy

lechatelierite
BA silica minerals
　　framework silicates
　　silicates
BT minerals

Lechtal Alps
W Tyrol.
BT Tyrol
　　Austria

Lecompton Formation
use Lecompton Limestone

Lecompton Limestone
In Shawnee Group. Comprises Spring Branch Limestone, Doniphan Shale, Big Springs Limestone, Queen Hill Shale, Beil Limestone, King Hill Shale, Avoca Limestone members. E Kansas, NW Missouri, SE Nebraska, SW Iowa, central N Oklahoma. Also search Lecompton Formation.
UF Lecompton Formation
BT Virgilian
　　Upper Pennsylvanian
　　Pennsylvanian
SA Beil Limestone Member
　　Iowa
　　Kansas
　　Missouri
　　Nebraska
　　Oklahoma
　　Shawnee Group

Leczyca
Town in central.
BT Lodz
　　Poland

Leda Clay
BT Cenozoic
SA Canada
　　Ontario
　　Ottawa

Leduc
Town S of Edmonton in central.
BT Alberta
　　Canada

Lee County
BX Alabama
　　Arkansas
　　Florida
　　Georgia
　　Illinois
　　Iowa
　　Kentucky
　　Mississippi
　　North Carolina
　　South Carolina
　　Texas
　　Virginia
BT United States

Lee Formation
Includes Rockcastle Sandstone. Defined as all of lower Pennsylvanian rocks from Mississippian-Pennsylvanian contact to top of Zachariah Coal Bed. E Kentucky, E Tennessee, and SW Virginia.
BT Lower Pennsylvanian
　　Pennsylvanian
SA Kentucky
　　Tennessee
　　Virginia

Leeward Islands
The northern chain of islands in the Lesser Antilles extending from the Virgin Islands in N to Dominica in S.
BT West Indies
SA Lesser Antilles

Leg 18
Includes use on level 3 under ocean floors(1).
SA Deep Sea Drilling Project
　　marine geology
　　ocean floors

Leg 25
Includes use on level 3 under ocean floors(1).
SA Deep Sea Drilling Project
　　ocean floors

Leg 27
Includes use on level 3 under ocean floors(1).
SA Deep Sea Drilling Project
　　ocean floors

Leg 45
Term introduced in 1978. Includes use on level 3 under ocean floors(1).
SA Deep Sea Drilling Project
　　marine geology
　　ocean floors

Leg 46
Term introduced in 1978. Includes

use on level 3 under ocean floors(1).
SA Deep Sea Drilling Project
   marine geology
   ocean floors

Leg 48
Term introduced in 1978. Includes use on level 3 under ocean floors(1).
SA Deep Sea Drilling Project
   marine geology
   ocean floors

legend
SA cartography
   diagrams
   maps

legislation
As of 1978, term is used on level 2 under land use(1).
SA economic geology
   engineering geology
   environmental geology
   land use
   policy

legrandite
BA arsenates
BT minerals

Leguminosae
Family.
BA Dicotyledoneae
   angiosperms
BT Plantae

lehm
use loess

Leicestershire
County in central.
CO N523000N530000
   W0004000W0014000
BT England
   Great Britain
   United Kingdom

leifite
BZ framework silicates
   halides
BT minerals
SA aluminosilicates
   fluorides

Leine Valley
River valley in SE and central Lower Saxony. Also search Leine.
BT Lower Saxony
   Germany

Leinster
Province in E comprising 12 counties.
BT Ireland

Leipzig
District in S central. Also a city.
BT East Germany
SA Saxony

Leiria
District in W. Also a city.
BT Portugal

Leman Lake
use Lake Geneva

Lemhi County
E central.
BT Idaho
   United States

Lemhi Range
E central Idaho. Primarily in Lemhi and Butte counties.
BT Idaho
   United States

Lena Basin
River basin. E central and N Siberia including areas in Irkutsk Oblast, and Yakutia. Also search Lena; Lena region; Lena River basin.
UF Lena region
   Lena River basin
BT Russian Republic
   USSR
SA Angara-Lena Basin

Lena region
use Lena Basin

Lena River
Rises in the Baikal Mountains W of Lake Baikal in Irkutsk Oblast and flows NE into a great bend in Yakutia, and then NW through a large delta into the Laptev Sea. Also search Lena.
BT Russian Republic
   USSR

Lena River basin
use Lena Basin

Leninabad
City in NW.
BT Tadzhikistan
   USSR

Leningrad
City on the Gulf of Finland. Also an oblast.
BT Russian Republic
   USSR

Leningrad Mining Institute
In Leningrad.
BT Russian Republic
   USSR

Leninogorsk
City in Vostochno Kazakhstan Oblast in extreme E Kazakhstan.
UF Ridder
BT Kazakhstan
   USSR

Lenkoran
City on Caspian Sea near Iranian border.
BT Azerbaidzhan
   USSR

lenses
Any lense-shaped deposit.
SA deposits
   ore deposits
   sedimentary rocks

Lenticulina
Genus. Includes use on level 3 under foraminifera(1) Nodosariacea(2).
BA Nodosariidae
   Nodosariacea
   Rotaliina
   foraminifera
BT Invertebrata

Leon
Province, city, region and former kingdom in Spain. City and department in Nicaragua. Counties in Florida and Texas. City in central Mexico. Index states and countries as applicable.
SA Florida
   Mexico
   Nicaragua
   Spain
   Texas

Leonardian
Provincial series, North America. Above Wolfcampian, below Guadalupian. Includes use on level 3 under age terms(1). See list E.
BA Lower Permian
   Permian
BT Paleozoic
SA San Andres Formation

leonardite
use leonhardite

leonhardite
Also search leonardite.
UF leonardite
BA zeolite group
   framework silicates
   silicates
BT minerals

SA laumontite

leopoldite
use sylvite

Leopoldville
use Kinshasa

Leoville
Village in W central.
BT Saskatchewan
   Canada

Leperditicopida
Includes use on level 2 under Ostracoda(1). See list F.
BA Ostracoda
BT Crustacea
   Arthropoda
   Invertebrata

lepidocrocite
BA oxides
BT minerals

Lepidocyclina
Includes use on level 3 under foraminifera(1) Orbitoididae(2).
BA Orbitoididae
   Orbitoidacea
   Rotaliina
   foraminifera
BT Invertebrata

lepidolite
BZ halides
   sheet silicates
BT minerals
SA cookeite
   fluorides
   mica group

lepidomelane
BA sheet silicates
   silicates
BT minerals
SA mica group

Lepidoptera
Includes use on level 2 under Insecta(1). See list F.
BA Insecta
BT Arthropoda
   Invertebrata
SA Raphidiodea

Lepidosauria
Includes use on level 2 under Reptilia(1). See list F.
BA Reptilia
BT Tetrapoda
   Vertebrata
NA Lacertilia
   Mosasauridae
   Squamata

Lepontine Alps
Division of Central Alps. Index countries as applicable.
CO N460000N464500
   E0091500E0081500
BT Alps
   Europe
SA Central Alps
   Gotthard Massif
   Italy
   Piedmont Alps
   Switzerland

Lepospondyli
Includes use on level 2 under Amphibia(1). See list F.
BA Amphibia
BT Tetrapoda
   Vertebrata

leptite
Includes use on level 3 under metamorphic rocks(1) granulites(2). See list J. Before 1978, also search leptynite.
UF leptynite
BA granulites
BT metamorphic rocks
SA granulite

Leptocythere
Genus. Includes use on level 3 under Ostracoda(1) Podocopida(2).
BA Cytheracea
   Podocopida
   Ostracoda
BT Crustacea
   Arthropoda
   Invertebrata

leptynite
use leptite

Lerida
Province in Catalonia in NE.
BT Spain
NT Montsech
   Pobla de Segur
   Segre Valley
   Tremp
SA Catalonia

Lesbos
Island in E Aegean Sea off NW Anatolia.
BT Aegean Islands
   Greece

Lesna
Town in Lower Silesia in SW part of the country.
BT Wroclaw
   Poland

Lesotho
Former British colony of Basutoland. Includes use on level 1 as an area term (list O). For term set options see list B. Also search Basutoland.
UF Basutoland
BA Africa
SA Orange River

Lesser Antilles
The smaller islands of the Antilles extending in an arc from Puerto Rico to the islands N of Venezuela. Index Barbados, Trinidad and island groups as applicable. As of 1977, includes use as a level 1 area term (list O). See list B for term set options.
BA West Indies
NT Martinique
SA Antilles
   Barbados
   British Virgin Islands
   Caribbean region
   Grenada
   Leeward Islands
   Netherlands Antilles
   Trinidad
   Windward Islands

Lesser Caucasus
Mountain system formed by the N frontal ranges of the Armenian Highland and separated from the Greater Caucasus by the Colchis and Kura lowlands. Index Soviet republics as applicable.
CO N420000N450000
   E0464000E0360000
BT USSR
SA Armenia
   Azerbaidzhan
   Caucasus
   Georgian Republic
   Shakhdag Range
   Trialet Range

Lesser Himalayas
The central range of the Himalayas running parallel to the Greater Himalayas to the N. Index countries as applicable.
BT Asia
SA Bhutan
   Himalayas
   India
   Nepal
   Sikkim

Lesser Khingan Mountains
E Heilungkiang. Separates Amur River from Sungari Valley. Also search Lesser Khingan Range.
UF Lesser Khingan Range
Little Khingan Mountains
BT Heilungkiang
China

Lesser Khingan Range
use Lesser Khingan Mountains

Lesser Sunda Islands
Chain of islands E from Bali to and including Alor and Timor, but not Wetar.
BT Indonesia
NT Mount Agung
Pantar
SA Timor

Lesser Walachia
use Oltenia

Lessini Mountains
In SW Dolomites in NE.
BT Italy
SA Dolomites

letter
A valid term through mid-1978, used in title annotation when document is correspondence. No longer a valid index term.

leucite
UF Vesuvian garnet
BA framework silicates
silicates
BT minerals

leucitite
Includes use on level 3 under igneous rocks(1) alkali basalt family(2) and ultramafic family(2). See list H.
BA alkali basalt family
BT igneous rocks
SA ultramafic family

leucogranite
See list H.
BA granite-granodiorite family
BT igneous rocks
SA granite

leucosphenite
BA sheet silicates
silicates
BT minerals

Leuven
use Louvain

Levant
Name applied to countries along the E shore of the Mediterranean including Egypt and Greece. Index countries as applicable.
BT Middle East
SA Egypt
Greece
Israel
Lebanon
Syria
Turkey

Levantinian
Austria. Includes use on level 3 under age terms(1). See list E.
BA upper Pliocene
Pliocene
Neogene
Tertiary
BT Cenozoic

levees
SA channels
embankments
marine geology
streams

level, changes of
use changes of level

leveling
SA cartography
geodesy

levels
Includes use on level 2 under ground water(1).
SA drawdown
ground water
reservoirs

Levis Shale
BT Devonian
SA Quebec

Lewis and Clark County
W central.
BT Montana
United States

Lewis Shale
Lewis and Meeteetse formations throughout N central Wyoming considered stratigraphic equivalents with Lewis being applied generally in areas where marine strata are present and Meeteetse in areas where rocks are nonmarine in character. W Colorado, NW New Mexico, and S and central Wyoming.
BT Upper Cretaceous
Cretaceous
SA Colorado
New Mexico
Wyoming

Lewisian
Europe. See list E.
BA Precambrian

lexicons
Includes use on level 1 (list A) for systematic arrangement of words in a particular language, or of a considerable number of them, and their definition. Term set options are: topic (List B or extraterrestrial geology or geophysics)
language (if not English)
SA bibliography
catalogs
dictionaries
glossaries
mineralogy
paleontology
petrology
sedimentary petrology
stratigraphy
structural geology

Lexington Limestone
Comprises Curdsville Limestone, Logana Formation, Jessamine Limestone, Benson Limestone, Brannon Limestone with Woodburn Phosphatic Member. Central Kentucky.
BT Middle Ordovician
Ordovician
SA Kentucky

Lherz
Lake in the Pyrenees in S.
BT France

lherzolite
Includes use on level 3 under igneous rocks(1) ultramafic family(2). See list H.
BA ultramafic family
BT igneous rocks
SA garnet lherzolite
peridotite
spinel lherzolite

Li
use lithium

Li-6
Includes use on level 3 under isotopes(1).
SA isotopes
lithium

Li-7
Includes use on level 3 under isotopes(1).
SA isotopes
lithium

Liaoning
Province in S.
BT Manchuria
China

Liard River
Index British Columbia and territories as applicable.
BT Canada
SA British Columbia
Northwest Territories
Yukon Territory

Liassic
Middle Europe. use on level 3 under age terms(1). See list E. Also search Lias.
BA Lower Jurassic
Jurassic
BT Mesozoic
NT Luxembourg Sandstone
SA Hettangian
Pliensbachian
Sinemurian
Toarcian

Liberia
Includes use on level 1 as an area term (list O). For term set options see list B.
CO N040000N080000
W0070000W0120000
BA Africa
NT Bong Range
Liberian Shield
SA Nimba Mountains
West Africa

Liberian Shield
Inland plateau.
BT Liberia

Libya
Includes use on level 1 as an area term (list O). For term set options see list B.
CO N193000N330000
E0250000E0093000
BA Africa
NT Cyrenaica
Fezzan
Kufra Basin
Murzuk Basin
Sirte Basin
Tibesti
Tripoli
Tripolitania
Zelten
SA Libyan Desert
Mediterranean region
Near East
North Africa
Sahara

Libyan Desert
Desert area of the E Sahara, W of the Nile. Index countries as applicable.
BT Africa
SA Egypt
Libya
Sudan
Western Desert

lichenometry
Includes use on level 2 under geochronology(1).
SA geochronology
lichens

lichens
Includes use on level 1 and 2 as a fossil term (list F).
BT Plantae
SA algae
fungi
lichenometry
thallophytes

Lichida
Includes use on level 2 under Trilobita(1). See list F.
BA Trilobita
BT Trilobitomorpha
Arthropoda
Invertebrata

Liechtenstein
Independent principality between NE Switzerland and W Austria.
BT Europe
SA Rhine Basin
Rhine River

Liege
Province in E Belgium. Also a city.
CO N501400N504800
E0063000E0045800
BT Belgium
NT Hohe Venn
Stavelot
Vesdre Valley

life
A valid level 2 term through 1977 used in combination with origin. After 1977, use life origin on level 2 under paleontology(1).

life assemblage
use biocenoses

life origin
Term introduced in 1978. Includes use on level 2 under paleontology(1). Before 1978, also search origin AND life.
SA paleontology

light minerals
BT minerals
SA calcite
dolomite
feldspar
heavy minerals
muscovite
opaque minerals
quartz

lightning
Includes use on level 3 under meteorology(1) electrical phenomena(2).
SA currents
electrical phenomena
meteorology
thunderstorms
whistlers

lignite
Includes use on level 1 and 2 as a commodity term (list C); on level 3 under sedimentary rocks(1) organic residues(2). See list I.
BA organic residues
BT sedimentary rocks
SA anthracite
brown coal
coal
organic materials
peat

Liguria
Autonomous region in NW.
CO N434500N444000
E0100500E0074500
BT Italy
NT Genoa
Ligurian Alps
Ligurian Apennines
Monte Antola
Savona

Ligurian Alps
E extention of Maritime Alps along coast of Ligurian Sea.
BT Liguria
Italy
SA Maritime Alps

Ligurian Apennines
Extend from near Savona SE to just N of La Spezia along coast of Ligurian Sea.
BT Liguria

Italy
SA Apennines
Monte Antola

Ligurian Sea
Enclosed by Italian autonomous regions of Liguria and Tuscany on the N and E, and by Corsica on the S.
BT Mediterranean Sea

Likhvin
use Chekalin

Lille
City near Belgian frontier in N part of country.
BT Nord
France

lillianite
BA sulfobismuthites
sulfosalts
BT minerals

Lilydale Limestone
Upper Yarra District in S central Victoria.
BT Devonian
SA Australia
Victoria

Lima
Department in W. Also a city.
BT Peru
NT Casapalca
Santa Eulalia
SA Pucara Group

Limagne
Fertile lowland in central part of country. Index departments as applicable.
BT France
SA Allier
Puy-de-Dome

Limburg
Name of Town and province in Belgium. Also province in SE Netherlands, and a region in W Europe. Index countries as applicable.
BT Europe
SA Belgium
Maastricht
Netherlands

limburgite
Includes use on level 3 under igneous rocks(1) ultramafic family(2). See list H.
BA ultramafic family
BT igneous rocks

lime
Used as a level 3 commodity term (list C).
SA calcium carbonate
limestone

Lime Creek Formation
Includes Juniper Hill, Cerro Gordo, and Owen members. Central N Iowa.
BT Upper Devonian
Devonian
SA Iowa

Limerick
County in W central.
BT Ireland

limestone
Includes use on level 1 as a commodity term (list C); on level 3 under sedimentary rocks(1) carbonate rocks(2). See list I.
BA carbonate rocks
BT sedimentary rocks
NA algal limestone
magnesian limestone
oolitic limestone
SA algal mounds
biosparite
building stone
calcarenite
calcilutite
calcite
carbonate sediments
cement materials
chalk
construction materials
dimension stone
dolomitization
jasperoid
lime
metalimestone
micrite
oolite
oolitic texture
ophicalcite
pisolites
travertine

Limestone Alps
BT Austria
SA Alps

Limestone County
Index states as applicable.
BX Alabama
Texas
BT United States

limit, elastic
use elastic limit

limits, Atterberg
use Atterberg limits

Limnocardium
Genus. Includes use on level 3 under Mollusca(1) Bivalvia(2).
BA Bivalvia
Mollusca
BT Invertebrata

limnology
Includes use on level 2 under hydrology(1); on level 3 under geomorphology(1) lacustrine features(2).
SA geomorphology
glacial lakes
hydrology
lacustrine features
lakes
paleolimnology
ponds
seiches

limonite
Includes use as level 3 commodity term under iron(1). See list C.
BA oxides
BT minerals
SA hydrogoethite

Limousin
Historical region in S central.
CO N450000N461000
E0030000E0003000
BT France

Limpopo Basin
River basin. Also search Limpopo. Index Transvaal and countries as applicable.
UF Crocodile River basin
BT Africa
SA Botswana
Mozambique
Rhodesia
Transvaal

Limski Channel
On Istrian Peninsula SW of Rijeka.
BT Croatia
Yugoslavia

Lincoln County
BX Arkansas
Colorado
Georgia
Idaho
Kansas
Kentucky
Louisiana
Maine
Minnesota
Mississippi
Missouri
Montana
Nebraska
Nevada
New Mexico
North Carolina
Oklahoma
Oregon
Oklahoma
Oregon
South Dakota
Tennessee
Washington
West Virginia
Wisconsin
Wyoming
BT United States

Lincoln Creek Formation
Vaguely defined. Also search Lincoln Formation. SW Washington.
BT Oligocene
SA Washington

Lincolnshire
County on the North Sea in E.
CO N524000N534500
E0002000W0010000
BT England
Great Britain
United Kingdom

Lincolnshire Limestone
Includes Edison Cherty Limestone Member and Hogskin Member. SW Virginia and NE Tennessee.
BT Middle Ordovician
Ordovician
SA Tennessee
Virginia

Lindlar
Village in S.
BT North Rhine-Westphalia
West Germany

Line Islands
Group S of Hawaiian Islands, N and S of Equator. Some belong to U.S. and some to U.K.
BT Pacific Ocean
NT Vostok
SA Fanning Island
Polynesia

line of strike
use strike

lineaments
Includes use on level 3 under tectonics(1) structure(2).
SA fault zones
faults
remote sensing
tectonics

linear orientation
Term introduced in 1978. Before 1978, search linear.
SA orientation

lineation
Small-scale features. Includes use on level 1 (list A); on level 2 and 3 under structural analysis(1). Also search lineations. If level 1, term set options are:
experimental studies
  subtopic
genesis
  subtopic
interpretation
  subtopic
style
  subtopic [e.g. boudinage, elongate minerals, flow lines, mullions, slickensides]
SA bedding
boudinage
boudins
cleavage
deformation
elongate minerals
fabric
flow lines
folds
foliation
fractures
igneous rocks
intrusions
metamorphic rocks
metamorphism
mullions
orientation
parting lineation
preferred orientation
slickensides
structural analysis
structural geology
style
tectonics

lines, flow
use flow lines

Lingula
Genus. Includes use on level 3 under Brachiopoda(1) Inarticulata(2).
BA Inarticulata
Brachiopoda
BT Invertebrata

linnaeite
Also search linneite.
UF cobalt pyrites
linneite
BA sulfides
BT minerals
SA carrollite
polydymite

linneite
use linnaeite

Linz
City in NW part of country.
BT Upper Austria
Austria

Lipari Island
BT Sicily
Italy
SA Lipari Islands

Lipari Islands
Group of small volcanic islands in SE Tyrrhenian Sea off N coast of Sicily. Also search Lipari.
UF Aeolian Islands
BT Sicily
Italy
SA Lipari Island
Vulcano

liparite
Includes use on level 3 under igneous rocks(1) andesite-rhyolite family(2). See list H.
BA andesite-rhyolite family
BT igneous rocks

liparite porphyry
See list H.
BA andesite-rhyolite family
BT igneous rocks
SA porphyry

lipids
BA organic materials

Lippe
Former German state in NW part of country. Also a river.
BT North Rhine-Westphalia
West Germany

liquefaction
Not to be used under phase equilibria.
SA geologic hazards
mass movements
sediments
slumping
soil mechanics

liquid
use liquid phase

liquid phase
Term introduced in 1978. Before 1978, also search liquid.
UF liquid
SA fluid phase
gaseous phase
melts

liquid waste
As of 1978, term is used on level 2 under waste disposal(1).
UF waste, liquid
SA detergents
engineering geology
environmental geology
ground water
industrial waste
radioactive waste
solid waste
storage
waste disposal
waste water

Lisburne Group
Includes Wachsmuth Limestone, Alapah Limestone, Kayak Shale, Kanayut Conglomerate. N Alaska.
BT Paleozoic
SA Alaska
Lower Mississippian
Mississippian
Pennsylvanian
Permian
Upper Mississippian

Lissamphibia
Includes use on level 2 under Amphibia(1). See list F.
BA Amphibia
BT Tetrapoda
Vertebrata
NA Anura

list, faunal
use faunal list

listvenite
BA sheet silicates
silicates
BT minerals
SA mica group

listwanite
Compositional term. Includes use on level 3 under metamorphic rocks(1). See list J.
BA schists
BT metamorphic rocks

Litchfield County
NW Connecticut.
BT Connecticut
United States

lithic texture
Term introduced in 1978. Includes use on level 3 under sediments(1) and sedimentary rocks(1). Before 1978, also search lithic or lithologic AND specific sediment type or sedimentary rock.
SA sedimentary rocks
sediments
textures

lithification
Includes use on level 3 under sedimentation(1) or diagenesis(1).
SA cementation
coal
compaction
compression
consolidation
crystallization
diagenesis
recrystallization
sedimentation
sediments

lithiophilite
BA phosphates
BT minerals

lithiophorite
BA oxides
BT minerals

lithium
Includes use on level 1 and 2 as a commodity term (list C) and as a chemical element (list D). Also search Li.
UF Li
SA alkali metals
amblygonite
elements
Li-6
Li-7

lithofacies
Includes use on level 2 under sedimentary rocks(1), sediments(1), and reefs(1).
SA facies
reefs
sedimentary rocks
sediments

lithologic controls
Includes use on level 3 under mineral deposits, genesis(1) controls(2). Also search lithologic.
SA controls
lithology
mineral deposits
stratigraphic controls

lithologic maps
Term introduced in 1978. Before 1978, search maps AND lithologic.
BA maps

lithology
Includes use on level 3 under sedimentary rocks(1), metamorphic rocks(1), igneous rocks(1), and under age terms(1). See list E.
SA domains
igneous rocks
lithologic controls
metamorphic rocks
petrography
sedimentary rocks

lithophile elements
Term introduced in 1978. Before 1978, also search lithophile.
SA elements

Lithophyllum
Includes use on level 3 under algae(1) Rhodophyta(2).
BA Corallinaceae
Rhodophyta
algae
BT Plantae

lithosiderite
use stony irons

lithosphere
Includes use on level 3 under tectonophysics(1) and under plate tectonics(1).
SA asthenosphere
Benioff zone
biosphere
core
crust
Earth
lower crust
mantle
Mohorovicic discontinuity
plate tectonics
tectonophysics
tectonosphere

lithostratigraphy
For general description of the sedimentary sequence of an area. Includes use on level 2 under sedimentary rocks(1), metamorphic rocks, and under sediments(1).
UF petrostratigraphy
rock-stratigraphy
SA metamorphic rocks
sedimentary rocks
sediments
stratigraphy

lithothamnion
use Lithothamnium

Lithothamnium
Genus. Includes use on level 3 under algae(1) Rhodophyta(2). Also search lithothamnion.
UF lithothamnion
BA Corallinaceae
Rhodophyta
algae
BT Plantae

Lithuania
Lithuanian Soviet Socialist Republic.
CO N530000N560000
E0270000E0210000
BT USSR
NT Sventoji River
Vilna
SA Baltic region
Courland Spit
European USSR
Neman River basin
Russian Plain
Russian Platform

Little Belt Mountains
Range of the Rocky Mountains in central.
BT Montana
United States
SA Rocky Mountains

Little Falls Dolomite
use Little Falls Formation

Little Falls Formation
Represents an offshore, more carbonate phase of Potsdam Sandstone, Galway Formation, and perhaps even part of Hoyt Limestone. E central and E New York.
UF Little Falls Dolomite
BT Upper Cambrian
Cambrian
SA New York

Little Khingan Mountains
use Lesser Khingan Mountains

Little Missouri River basin
Index states as applicable. Also search Little Missouri River.
BT United States
SA Montana
North Dakota
South Dakota
Wyoming

Little Walachia
use Oltenia

Littleton Formation
Includes Gove Member, Hubbard Hill Member, May Pond Member, Dakin Hill Member, Pittsfield Member, Jenness Pond Member, Durgin Brook Member. Central and S New Hampshire, N central Massachusetts, and SE Vermont.
BT Lower Devonian
Devonian
SA Massachusetts
New Hampshire
Vermont

littoral
A valid term through 1977. After 1977, use littoral environment.

littoral drift
Also search longshore drift.
UF longshore drift
shore drift
SA drift
longshore currents
progradation
sediments

littoral environment
Term introduced in 1978. Before 1978, search littoral.
SA ecology
environment
intertidal environment
sedimentation
sublittoral environment
tidal flats

Lituolacea
Includes use on level 2 under foraminifera(1). See list F.
BA Textulariina
foraminifera
BT Invertebrata
NA Ammobaculites
Ataxophragmiidae
Lituolidae
Orbitolinidae

Lituolidae
Family. Includes use on level 3 under foraminifera(1) Lituolacea(2).
BA Lituolacea
Textulariina
foraminifera
BT Invertebrata
NA Haplophragmoides

Liverpool Bay
Inlet of Irish Sea off Liverpool. Also search Liverpool.
BT England
Great Britain
United Kingdom

living culture
UF culture, living
SA algae
bacteria

living materials
Term introduced in 1978. Distinguish from fossiliferous materials. Before 1978, search living.
SA fossiliferous materials
materials

Livingston County
BX Illinois
Kentucky
Louisiana
Michigan
Missouri
New York
BT United States

Livingston Island
One of South Shetland Islands in British Antarctic Territory off the Antarctic Peninsula.
BT Antarctica
SA South Shetland Islands

Livorno
Province in NW Tuscany. Also a city on the Ligurian Sea now known as Leghorn.
BT Tuscany
Italy

lixiviation
use leaching

lizardite
BA sheet silicates
silicates
BT minerals
SA serpentine group

Ljubljana
City in NW part of country.
UF Lyublyana
BT Slovenia
Yugoslavia

Llandeilian
Europe. Above Llanvirnian, below lower Caradocian. Includes use on level 3 under age terms(1). See list E.
BA Middle Ordovician
Ordovician
BT Paleozoic

Llandeilian ● London Clay

SA Borrowdale Volcanic Series

Llandoverian
Valentian. Europe. Above Ashgillian (Ordovician), below Tarannon. Includes use on level 3 under age terms(1). See list E.
BA Lower Silurian
      Silurian
BT Paleozoic

Llandovery
Town in S.
BT Wales
      Great Britain
      United Kingdom

Llano
Town in Llano County in central.
BT Texas
      United States

Llano County
Central.
BT Texas
      United States

Llano Estacado
Vast plateau. Index states as applicable.
UF Staked Plain
BT United States
SA New Mexico
      Oklahoma
      Texas

Llano Uplift
BT Texas

Llanos
Vast plains drained by the Orinoco River. Index countries as applicable.
BT South America
SA Colombia
      Venezuela

Llanvirnian
Europe. Above Arenigian, below Llandeilian. Includes use on level 3 under age terms(1). See list E.
BA Middle Ordovician
      Ordovician
BT Paleozoic
SA Borrowdale Volcanic Series

Llewn Promontory
use Lleyn Peninsula

Lleyn Peninsula
Headland extending SW into Saint Georges channel from NW Wales.
UF Llewn Promontory
BT Wales
      Great Britain
      United Kingdom

Llogrebat River basin
In Catalonia in NE part of country. Also search Llogrebat River.
BT Barcelona
      Spain

load casts
Includes use on level 3 under sedimentary structures(1) soft sediment deformation(2) turbidity current structures(2), and bedding plane irregularities(2). See list K.
UF casts, load
BA turbidity current structures
      sedimentary structures
SA bedding plane irregularities
      flame structures
      flute casts
      soft sediment deformation
      sole marks

load pressure
Term introduced in 1978.
SA engineering geology
      pressure

loading
Includes use on level 3 under soil mechanics(1).

SA cyclic loading
      soil mechanics

loam
Includes use on level 3 under soils(1). See list M.
BA soils
SA clay
      loess
      sand
      silt

Loanda
use Luanda

localization
Used for concentration of minerals, e.g. under sedimentation(1), mineral deposits, genesis(1), or metasomatism(1).
SA concentration
      mineral deposits
      minerals
      sedimentation

Loch Lomond
S central.
BT Scotland
      Great Britain
      United Kingdom

Lochkovian
Europe. See list E.
BA Lower Devonian
      Devonian
BT Paleozoic

Lockport Formation
Includes Gasport Dolomite, Suspension Bridge Dolomite, Eramosa Dolomite members, and DeCew Waterlime, Gasport Limestone, Goat Island, Oak Orchard members, Devils Hole Dolomite, and Oakfield Limestone.
BT Niagaran
      Middle Silurian
      Silurian
SA Michigan
      New York
      Ontario

locomotion
Includes use on level 3 under appropriate fossil group (list F).
SA fossils

Lodeve
Town in S part of country.
BT Herault
      France

Lodeve Basin
S part of country. Also search Lodeve.
BT Herault
      France

Lodgepole Formation
In Madison Group. Includes Paine member, Woodhurst Member, Little Chief Canyon Member. SW Montana, NE Utah, and E Wyoming. Also search Lodgepole Limestone.
UF Lodgepole Limestone
BT Lower Mississippian
      Mississippian
SA Madison Group
      Montana
      Utah
      Wyoming

Lodgepole Limestone
use Lodgepole Formation

Lodz
Province in central part of country. Also a city.
BT Poland
NT Belchatow
      Leczyca
      Lowicz
      Tomaszow Mazowiecki
      Widawka Basin
      Wielun

loellingite
use lollingite

loess
Includes use on level 3 under sediments(1) clastic sediments(2); on level 3 under geomorphology(1) eolian features(2). See list N.
UF bluff formation
      lehm
BA clastic sediments
BT sediments
SA drift
      eolian features
      geomorphology
      loam
      marl
      silt
      soil mechanics
      terrigenous materials

Lofoten Islands
Island group in Norwegian Sea off NW mainland. Also search Lofoten.
BT Nordland
      Norway

log, road
use road log

Logan Formation
Comprises Beyer, Allensville, and Vinton members. NE Kentucky, and Ohio.
BT Mississippian
SA Kentucky
      Ohio

Logar
Province in E.
BT Afghanistan

logging, acoustical
use acoustical logging

logging, caliper
use caliper logging

logging, dipmeter
use dipmeter logging

logging, electrical
use electrical logging

logging, electromagnetic
use electromagnetic logging

Logone River
Flows into the Shari River at Fort Lamy in Chad S of Lake Chad. Index countries as applicable. Also search Logone.
BT Africa
SA Cameroon
      Chad

Logrono
Province in N central. Also a city.
BT Spain

Logudoro
Region in NW.
BT Sardinia
      Italy

Loir-et-Cher
Department in W central.
BT France
SA Beauce

Loire
Department in S central.
CO N452000N462000
      E0050000E0034000
BT France
NT Saint-Etienne coal basin
SA Monts du Lyonnais

Loire River
Longest river in France. Rises in Ardeche Department in SE and flows N and NE into Loiret Department where it turns E emptying into the Bay of Biscay at Saint-Nazaire.
BT France

Loire Valley
River valley in S central, central, and E central part of country. Also search Loire.
BT France

Loire-Atlantique
Department in W France. Also search Loire-Inferieure.
CO N470000N475000
      W0005000W0023000
BT France
NT Ancenis
      Nantes

Loire-Inferieure
A valid term through 1976. After 1976, use Loire-Atlantique.

lollingite
Also search loellingite.
UF loellingite
BA arsenides
      sulfides
BT minerals
SA arsenopyrite

Lom Depression
In vicinity of Lom River and city of Lom in NW part of country.
BT Vidin
      Bulgaria

Lombardy
Autonomous region in N.
BT Italy
NT Bergamo
      Brescia
      Como
      Milan
      Pavia
      Sondrio
      Trompia Valley
      Valtellina
      Varese
SA Po River
      Po Valley
      Sesia-Lanzo Zone

lomonite
use laumontite

Lomonsov Range
use Lomonsov Ridge

Lomonsov Ridge
N of New Siberian Islands across top of world to N of Greenland.
UF Lomonsov Range
BT Arctic Ocean

lomontite
use laumontite

Lompoc
City in Santa Barbara County in SW.
BT California
      United States

London
City on the Thames River in SE.
BT England
      Great Britain
      United Kingdom

London Basin
SE England.
BT England
      Great Britain
      United Kingdom

London Clay
Rests on Reading, Woolwich, Blackheath or Oldhaven Beds except in Norfolk to the west of Great Yarmouth. Overlain by the Claygate or Bagshot Beds, except in East Anglia. Occupies greater part of a large triangular area in London Basin bounded roughly by line from Herne Bay in Kent to Croyden, Farnham, and Newbury and from there NE to Aldeburgh in Suffolk.
BT Ypresian
      lower Eocene
      Eocene
SA England

United Kingdom
Londonderry
County in NW Northern Ireland. Also a city.
  BT Northern Ireland
       United Kingdom
Long Beach
City in Los Angeles County in S.
  BT California
       United States
Long Island
Lying between Long Island Sound on N and Atlantic Ocean on S.
  CO N403000N411500
       W0715000W0741500
  BT New York
       United States
Long Island Sound
Body of water between S shore of Connecticut and N shore of Long Island, New York.
  BT Atlantic Ocean
long period
use long-period waves
long-period waves
Includes use on level 3 under seismology(1). Before 1978, search long period or long-period AND waves.
  UF long period
  SA elastic waves
       seismology
       waves
longitudinal
A valid term through 1977 used in combination with folds, fractures, or faults. Now use longitudinal orientation in combination with folds or fractures, and use strike faults for longitudinal faults.
longitudinal faults
use strike faults
longitudinal orientation
Term introduced in 1978. Includes use as orientation of folds relative to spatially associated macroscopic structures such as large folds, fold systems, and orogenic zones. Before 1978, also search folds or fractures AND longitudinal.
  SA discordant folds
       folds
       fractures
       joints
       orientation
       strike faults
longitudinal wave
use P-waves
longshore bars
Term introduced in 1978.
  UF barrier bars
  BT bars
longshore currents
  BT currents
  SA littoral drift
       ocean circulation
       ocean waves
longshore drift
use littoral drift
lonsdaleite
  BA native elements and alloys
  BT minerals
  SA meteorites
loparite
  BA oxides
  BT minerals
  SA niobates
lopoliths
Includes use on level 2 under intrusions(1).
  BA intrusions

  SA batholiths
       igneous rocks
       laccoliths
Lord Howe Island
Volcanic island 435 miles NE of Sydney. Dependency of New South Wales.
  BT Pacific Ocean
Lord Howe Ridge
use Lord Howe Rise
Lord Howe Rise
Extends from SW of New Caledonia to W of New Zealand. Also search Lord Howe Ridge.
  UF Lord Howe Ridge
       Lord Howe-Chesterfield Ridge
       Lord Howe-New Zealand Ridge
       Lord Howe-New Zealand Rise
  BT Pacific Ocean
Lord Howe-Chesterfield Ridge
use Lord Howe Rise
Lord Howe-New Zealand Ridge
use Lord Howe Rise
Lord Howe-New Zealand Rise
use Lord Howe Rise
Lorrain Formation
Huronian Supergroup of the Cobalt Group. Overlies The Gowganda conformably and gradationally.
  UF Lorrain Series
  BT Proterozoic
       Precambrian
  SA Ontario
       Quebec
Lorrain Series
use Lorrain Formation
Lorraine
Region and former province on Belgium and Luxembourg borders in NE.
  BT France
Los Alamos Scientific Laboratory
Los Alamos County, 35 miles NW of Santa Fe, in N central.
  BT New Mexico
       United States
  SA University of California
Los Angeles
City in Los Angeles County in S.
  BT California
       United States
Los Angeles Basin
Greater Los Angeles area.
  BT California
       United States
Los Angeles County
S California.
  CO N334500N344500
       W1174000W1185000
  BT California
       United States
Los Islands
Group of small islands in Atlantic Ocean off Conakry.
  UF Iles de Loos
  BT Guinea
Los Pedroches
Region N of city of Cordoba in SW central part of country.
  BT Cordoba
       Spain
Los Testigos
Small group of islands SW of Grenada and N of Sucre, Venezuela.
  BT Caribbean Sea
Lost Burro Formation
Includes Lippincott Member, Quartz Spring Sandstone Member. S California.
  BT Devonian
  SA California

       Middle Devonian
       Upper Devonian
Lost City
Ancient Indian city now covered by Lake Mead. Relics now housed in Overton in Clark County 5 miles N of original location in S.
  BT Nevada
       United States
Lot
Department in S central.
  BT France
  SA Quercy
Lot-et-Garonne
Department in SW.
  BT France
  NT Durance
Lotharingian
Europe. Above Sinemurian, below Pliensbachian. Includes use as level 3 term under age terms(1). See list E.
  BA Lower Jurassic
       Jurassic
  BT Mesozoic
Louisiade Archipelago
Island group in Solomon Sea SE of E tip of New Guinea.
  BT Papua New Guinea
       Australasia
Louisiana
Includes use on level 1 as an area term (list O). For term set options see list B.
  CO N290000N330000
       W0890000W0940500
  BA United States
  NT Atchafalaya Bay
       Barataria Bay
       Baton Rouge
       Belle Isle
       Calcasieu Parish
       Cameron Parish
       Claiborne Parish
  NX Columbia
  NT Evangeline Parish
       Grand Isle
  NX Jackson County
       Jefferson County
  NT Jefferson Parish
  NX Lafayette County
  NT Lake Pontchartrain
  NX Lincoln County
       Livingston County
  NT Mississippi Delta
       New Orleans
       Rapides Parish
       Saint Mary Parish
       South Pass
       Vermilion Parish
       Winnfield salt dome
  SA Amite River
       Bossier Formation
       Buckner Formation
       Citronelle Formation
       Claiborne Group
       Cook Mountain Formation
       Cotton Valley Group
       Fleming Formation
       Frio Formation
       Grand River
       Gulf Coastal Plain
       Jackson Group
       Mississippi River
       Mississippi Valley
       Norphlet Formation
       Queen City Formation
       Red River
       Red River valley
       Sabine Lake
       Schuler Formation
       Smackover Formation
       Trinity Group
       Vicksburg Group
       Washita Group

       Wilcox Group
       Woodbine Formation
       Yazoo Clay
Louisville
City on the Ohio River in Jefferson County.
  BT Kentucky
       United States
Loup Fork Group
Miocene, Pliocene, and Pleistocene(?). Also search Loup Fork Formation.
  BT Cenozoic
  SA Colorado
       Kansas
       Miocene
       Nebraska
       New Mexico
       Pleistocene
       Pliocene
       South Dakota
       Texas
       Wyoming
Lourdes
Town in SW part of country.
  BT Hautes-Pyrenees
       France
Louvain
City E of Brussels in central part of country.
  UF Leuven
  BT Brabant
       Belgium
Love waves
Includes use on level 3 under seismology(1).
  UF Q waves
  BT surface waves
  SA elastic waves
       Q
       seismology
       waves
Loveland
City in Larimer County in N.
  BT Colorado
       United States
Lovelock
City in Pershing County in NW.
  BT Nevada
       United States
Lovozero
Village in Murmansk Oblast on Kola Peninsula in NW European USSR.
  CO N670000N680000
       E0350000E0340000
  BT Russian Republic
       USSR
Lovozero Massif
Murmansk Oblast on the Kola Peninsula in NW European USSR.
  BT Russian Republic
       USSR
Low Archipelago
use Tuamotu Islands
Low Jesenik Mountains
use Nizky Jezenik Mountains
low pressure
Includes use on level 3 under phase equilibria(1).
  BT pressure
  SA high pressure
       metamorphism
       phase equilibria
Low Tatra Mountains
Section of the Carpathians parallel to and S of the Tatra Mountains in N central Slovakia. Also search Low Tatra; Low Tatras.
  UF Low Tatras
  BT Slovakia
       Czechoslovakia
  SA Carpathians

Tatra Mountains
Low Tatras
use Low Tatra Mountains
low temperature
Includes use on level 3 under phase equilibria(1).
BT temperature
SA metamorphism
phase equilibria
low velocity layer
use low-velocity layer
low velocity zone
use low-velocity zones
low-angle faults
Term introduced in 1978. Before 1978, search faults AND low-angle.
BA faults
low-energy environment
Term introduced in 1978. Before 1978, search low-energy.
SA environment
high-energy environment
sedimentation
low-grade
A valid term through 1977. After 1977, use low-grade metamorphism.
low-grade metamorphism
Term introduced in 1978. Includes use on level 3 under metamorphism(1). Before 1978, search low-grade AND metamorphism.
BA metamorphism
SA grade
low-velocity layer
Applies only to mantle. Includes use on level 2 under mantle(1). Also search low velocity layer.
UF B layer
Gutenberg low-velocity zone
layer, low-velocity
low velocity layer
SA Earth
low-velocity zones
mantle
low-velocity zones
Term introduced in 1978. Used as a general term, i.e. applicable to crust, mantle, etc. Also search low velocity zone; low velocity zones.
UF low velocity zone
zones, low-velocity
SA crust
low-velocity layer
mantle
Lower Austria
State in NE.
BT Austria
NT Bad Deutsch Altenburg
Eichkogel
Kamp Valley
Krems
Schwechat Valley
Spitz
Vienna
Wiener Wald
SA Enns Valley
Semmering
Wechsel
lower boundary
A valid general term through 1976. For stratigraphic meaning use boundary.
Lower Cambrian
BA Cambrian
BT Paleozoic
NA Aldanian
NT Antietam Formation
Chilhowee Group
NA Harlech Stage
NT Kinzers Formation
Murphy Marble
Rome Formation
Shady Dolomite
Usa Series
Yudoma Series
SA Middle Cambrian
Tintic Quartzite
Upper Cambrian
Wood Canyon Formation
Lower Carboniferous
BA Carboniferous
BT Paleozoic
NA Avonian
Culm
Dinantian
Tournaisian
Visean
SA Mississippian
Namurian
Upper Carboniferous
lower Cenozoic
BA Cenozoic
Lower Cretaceous
BA Cretaceous
BT Mesozoic
NT Antlers Sands
Areado Formation
Baquero Formation
NA Barremian
Berriasian
NT Cedar Mountain Formation
Cheyenne Sandstone
Christopher Formation
Fall River Formation
NA Gargasian
Gault
NT Glen Rose Formation
NA Hauterivian
NT Isachsen Formation
Kiowa Formation
Mannville Formation
Muddy Sandstone
NA Neocomian
NT Pasayten Group
NA Purbeckian
NT Speeton Clay
NA Urgonian
Valanginian
Wealden
NT Zubair Formation
SA Albian
Aptian
Benton Formation
Blairmore Group
Colorado Group
Comanchean
Dakota Formation
Edwards Formation
Fredericksburg Group
Gondwana System
Graneros Shale
Mancos Shale
Mentor Beds
Middle Cretaceous
Paluxy Formation
Potomac Group
Upper Cretaceous
Vraconian
Washita Group
lower crust
BA crust
SA asthenosphere
basaltic layer
granitic layer
lithosphere
upper crust
upper mantle
Lower Devonian
BA Devonian
BT Paleozoic
NT Coeymans Formation
NA Dittonian
Emsian
Gedinnian
NT Helderberg Group
NA Helderbergian
NT Littleton Formation
NA Lochkovian
NT Manlius Formation
Matagamon Sandstone
Oriskany Sandstone
NA Siegenian
SA Downtonian
Hidden Valley Dolomite
Middle Devonian
Read Bay Formation
Rondout Formation
Upper Devonian
lower Eocene
BA Eocene
Paleogene
Tertiary
BT Cenozoic
NT Ager Formation
Aquia Formation
NA Cuisian
NT San Jose Formation
NA Sparnacian
NT Wilcox Group
Willwood Formation
Wind River Formation
NA Ypresian
SA Deccan Traps
Green River Formation
Intertrappean Beds
middle Eocene
Nanjemoy Formation
Rose Canyon Formation
Sinjar Formation
Umpqua Formation
upper Eocene
Lower Greensand
Aptian and lower Albian.
BT Cretaceous
SA England
United Kingdom
lower Holocene
BA Holocene
Quaternary
BT Cenozoic
Lower Jurassic
BA Jurassic
BT Mesozoic
NA Carixian
Domerian
Hettangian
Liassic
Lotharingian
Pliensbachian
Sinemurian
NT Sunrise Formation
NA Toarcian
SA Aalenian
Middle Jurassic
Novorayskoe Formation
Upper Jurassic
Lower Lusatia
That part of Lusatia between the Neisse and the Bober rivers in SW.
BT Poland
SA Lusatia
Upper Lusatia
lower mantle
Includes use on level 3 under mantle(1).
BA mantle
SA upper mantle
lower Mesozoic
BA Mesozoic
lower Miocene
BA Miocene
Neogene
Tertiary
BT Cenozoic
NA Altonian
Aquitanian
Awamoan
Burdigalian
NT Chipola Formation
NA Eggenburgian
Hemingfordian
NT Kasauli Series
NA Saucesian
NT Waitemata Group
SA Arikareean
Badenian
Hawthorn Formation
John Day Formation
middle Miocene
upper Miocene
Lower Mississippi Valley
Term introduced in 1978.
BA Mississippi Valley
Lower Mississippian
BA Mississippian
BT Paleozoic
NT Fort Payne Formation
NA Kinderhookian
NT Lake Valley Formation
Lodgepole Formation
NA Osagian
NT Pocono Formation
SA Borden Group
Catskill Formation
Leadville Formation
Lisburne Group
Madison Group
Monte Cristo Limestone
Rampart Group
Upper Mississippian
Valmeyeran
lower Oligocene
BA Oligocene
Paleogene
Tertiary
BT Cenozoic
NT Chadron Formation
NA Chadronian
Sannoisian
Stampian
SA middle Oligocene
upper Oligocene
Lower Ordovician
BA Ordovician
BT Paleozoic
NA Arenigian
Canadian
NT Fillmore Formation
Kingsport Formation
Manitou Formation
Mascot Dolomite
Shakopee Formation
Smithville Formation
NA Tremadocian
NT White Limestone
SA Antelope Valley Limestone
Arbuckle Group
Deadwood Formation
Knox Group
Middle Ordovician
Pogonip Group
Upper Ordovician
lower Paleocene
BA Paleocene
Paleogene
Tertiary
BT Cenozoic
NA Danian
SA Difunta Group
upper Paleocene
lower Paleozoic
BA Paleozoic
NT Ashe Formation
Glenarm Series
SA Wissahickon Formation
Lower Peninsula
S part of state S of Straits of Mackinack. Also search Southern Peninsula.
UF Southern Peninsula
BT Michigan
United States
SA Upper Peninsula

Lower Pennsylvanian
- BA Pennsylvanian
- BT Paleozoic
- NT Caseyville Formation
  - Haymond Formation
  - Lee Formation
- NA Morrowan
- NT Pocahontas Formation
- SA Mansfield Formation
  - Marble Falls Group
  - Middle Pennsylvanian
  - Pottsville Group
  - Upper Pennsylvanian

Lower Permian
- BA Permian
- BT Paleozoic
- NA Artinskian
  - Autunian
- NT Barakar Stage
- NA Leonardian
  - Sakmarian
  - Wolfcampian
- SA Bird Spring Formation
  - Guadalupian
  - Kungurian
  - Middle Permian
  - Rotliegendes
  - San Andres Formation
  - Tensleep Sandstone
  - Upper Permian

lower Pleistocene
- BA Pleistocene
  - Quaternary
- BT Cenozoic
- NA Aftonian
  - Calabrian
  - Danube Stage
  - Kansan
  - Nebraskan
- SA Matuyama Epoch
  - Uonuma Group
  - upper Pleistocene

lower Pliocene
- BA Pliocene
  - Neogene
  - Tertiary
- BT Cenozoic
- NA Plaisancian
  - Tabianian
- NT Waccamaw Formation
- SA Capistrano Formation
  - Meotian
  - middle Pliocene
  - upper Pliocene
  - Yakima Basalt

lower Precambrian
- BA Precambrian
- NA Archean
- NT Bulawayan Group
- SA upper Precambrian

lower Proterozoic
- BA Proterozoic
  - Precambrian
- NA Aphebian

lower Quaternary
- BA Quaternary
- BT Cenozoic

Lower Rhine Basin
That section of the Rhine Basin between Bonn and the North Sea; in the Netherlands it includes the basins of the Lower Rhine, Lek, and Waal rivers. Index countries as applicable. Also search Lower Rhine.
- BT Europe
- SA Netherlands
  - Rhine Basin
  - Rhine Valley
  - West Germany

Lower Rhine Graben
- BT Rhine Graben
  - Europe

Lower Saxony
State in N.
- CO N511500N535000 E0113000E0070000
- BT West Germany
- NT Bad Pyrmont
  - Bramsche
  - East Frisian Islands
  - Eilenriede
  - Emsland
  - Gehrden
  - Gottingen
  - Helmstedt
  - Hildesheim
  - Jade Bay
  - Leine Valley
  - Osnabruck
  - Rammelsberg
  - Weser-Ems
- SA Ems River
  - Teutoburg Forest
  - Weser River
  - Westphalia

Lower Silesia
Most of former German Silesia which became part of Poland following World War II. It lies on both sides of the upper Oder River centering on Wroclaw with Czechoslovakia and the Neisse River constituting the SE and E borders respectively.
- BT Poland
- SA Silesia
  - Silesian coal basin
  - Upper Silesia

Lower Silurian
- BA Silurian
- BT Paleozoic
- NA Alexandrian
- NT Brassfield Formation
- NA Llandoverian
- NT Medina Formation
  - Tuscarora Formation
- SA Middle Silurian
  - Rockwood Formation
  - Upper Silurian

lower Tertiary
- BA Tertiary
- BT Cenozoic
- NT Kiev Member
  - Taishu Group
- SA Bauru Formation
  - Ewekoro Formation
  - middle Tertiary
  - Paleogene
  - upper Tertiary

Lower Triassic
- BA Triassic
- BT Mesozoic
- NA Bunter
- NT Panchet Series
- NA Scythian
- NT Serebryanka Formation
  - Thaynes Formation
- NA Werfenian
- SA Hercynian Orogeny
  - Middle Triassic
  - Moenkopi Formation
  - Sadlerochit Formation
  - Upper Triassic

Lower Tunguska River
Rises in N central Irkutsk Oblast and flows N crossing into Evenk National Okrug then W into the Yenisei River at Turukhansk.
- BT Russian Republic
  - USSR
- SA Tunguska
  - Tunguska River

Lowicz
Town in central part of country.
- BT Lodz
  - Poland

Lowlands
Region S of Dumbarton-Stonehaven line to the Southern Uplands.
- BT Scotland
  - Great Britain
  - United Kingdom

Loxoconcha
Includes use on level 3 under Ostracoda(1).
- BA Podocopida
  - Ostracoda
- BT Crustacea
  - Arthropoda
  - Invertebrata

Loyalty Islands
Island group 60 miles E of New Caledonia belonging to France.
- BT Pacific Ocean

Lozere
Department in S France. Also a mountain range.
- CO N441500N451000 E0040000E0030000
- BT France

Lu
use lutetium

Luanda
District in NW part of country. Also a city.
- UF Loanda
- BT Angola

Lubbock
City in Lubbock County in NW.
- BT Texas
  - United States

Luben
use Lubin

Lubin
City in SW part of country.
- UF Luben
- BT Wroclaw
  - Poland

Lubin Legnicki
Region in and around cities of Lubin and Legnica in SW part of country.
- BT Wroclaw
  - Poland

Lublin
Province in SE part of country. Also a city.
- BT Poland
- NT Lublin Upland
  - Lukow
  - Tomaszow Lubelski

Lublin Upland
SW Lublin. Also search Lublin.
- BT Lublin
  - Poland

Lucania
Ancient district in S including Basilicata autonomous region and part of Salerno Province of Campania. Index autonomous regions as applicable.
- BT Italy
- SA Basilicata
  - Campania

Lucca
City and province in central part of country.
- BT Tuscany
  - Italy

Lucerne
Canton in N central Switzerland. Also a city.
- UF Luzern
- BT Switzerland

Lucinidae
Family.
- BA Bivalvia
  - Mollusca
- BT Invertebrata

Luderitz
Town in SW South-West Africa. Formerly Angra Pequena.
- BT South-West Africa

Ludian
Europe. Above Bartonian, below Tongrian (Oligocene). Includes use on level 3 under age terms(1). See list E.
- BA upper Eocene
  - Eocene
  - Paleogene
  - Tertiary
- BT Cenozoic
- SA Priabonian

Ludlovian
Europe. Above Wenlockian, below Gedinnian (Devonian). Includes use on level 3 under age terms(1). See list E.
- BA Upper Silurian
  - Silurian
- BT Paleozoic

ludwigite
- BA borates
- BT minerals

lujavrite
Includes use on level 3 under igneous rocks(1) syenite family(2). See list H.
- BA syenite family
- BT igneous rocks
- SA nepheline syenite

Lukov
use Lukow

Lukow
Town in E part of country.
- UF Lukov
- BT Lublin
  - Poland

Lukuga River valley
Eastern.
- BT Zaire

luminescence
Includes use on level 3 under discipline or material, e.g. under minerals(1), Moon(1), and under geochemistry(1).
- SA airglow
  - cathodoluminescence
  - electromagnetic radiation
  - fluorescence
  - geochemistry
  - minerals
  - Moon
  - thermoluminescence
  - X-ray fluorescence

Luna County
SW New Mexico.
- BT New Mexico
  - United States

Luna 16
Includes use on level 3 under Moon(1).
- SA Moon
  - satellite methods

Luna 20
Includes use on level 3 under Moon(1).
- SA Moon
  - satellite methods

lunar materials
Term introduced in 1978. Before 1978, search lunar.
- SA lunar samples
  - materials
  - Moon

lunar samples
Distinguish from terrestrial samples. Also search lunar rocks; lunar soils.
- SA lunar materials
  - Moon
  - samples

lunar soils
A valid term through mid-1978. Use lunar samples or lunar materials, or search soils AND Moon.

Lund
City in extreme S part of country.
BT Malmohus
Sweden

Lusaka
City in central.
BT Zambia

Lusatia
Region in SE East Germany, and in SW Poland. It comprises both Lower and Upper Lusatia. Also search Lausitz. Index countries as applicable.
UF Lausitz
BT Europe
SA East Germany
Lower Lusatia
Poland
Upper Lusatia

Lushs Bight Group
Also search Lush's Bight Group.
BT Ordovician
SA Newfoundland

Lusitanian
Europe. Above Oxfordian, below Kimmeridgian. Includes use on level 3 under age terms(1). See list E.
BA Upper Jurassic
Jurassic
BT Mesozoic
SA Malm
Rauracian

Lut Desert
Great sandy and stony desert in E central Iran. Also search Dash-e-Lut Basin; Dash-i-Lut.
UF Dasht-e-Lut Basin
Dasht-i-Lut
BT Iran

Lutetian
Europe. Above Cuisian, below Auversian. Includes use on level 3 under age terms (1). See list E.
BA Eocene
Paleogene
Tertiary
BT Cenozoic

lutetium
Includes use on level 1 and 2 as a chemical element (list D).
UF Lu
SA elements
rare earths

Lutzow-Holm Bay
Between Prince Olav Coast and Prince Harald Coast in Queen Maud Land of the Norwegian Sector. Also search Lutzow-Holmbukta.
UF Lutzow-Holmbukta
BT Antarctica

Lutzow-Holmbukta
use Lutzow-Holm Bay

Luvisols
BA soils

Luxembourg
Includes use on level 1 as an area term (list O). For term set options see list B.
CO N493000N501500
E0063000E0054500
BA Europe
SA Ardennes
Luxembourg Sandstone
Moselle River
Moselle Valley

Luxembourg Sandstone
Hettangian and Sinemurian. Also called Hettange Sandstone or Orval Sandstone.

BT Liassic
SA Belgium
Luxembourg

Luzern
use Lucerne

Luzerne County
NE Pennsylvania.
BT Pennsylvania
United States

Luzon
Northernmost and most important island.
BT Philippine Islands
Asia
NT Bondoc Peninsula
Camarines Norte
Taal
Wawa

luzonite
BA sulfosalts
BT minerals
SA famatinite

Lvov
City and oblast in SW European USSR. Also search L'vov.
UF Lwow
BT Ukraine
USSR

Lvov Basin
In Lvov Oblast on the Polish border. Also search Lvov.
BT Ukraine
USSR

L'vov Volyn Basin
use Lvov-Volyn Basin

Lvov-Volyn Basin
Includes parts of Lvov Oblast and adjoining Volyn Oblast along the Polish border. Also search L'vov Volyn Basin.
UF L'vov Volyn Basin
BT Ukraine
USSR

Lwow
use Lvov

Lyangar
Town in Samarkand Oblast in S central.
BT Uzbekistan
USSR

Lycian Taurus
Mountains in ancient district of SW Anatolia called Lycia.
BT Turkey
Middle East

Lycopoda
use Lycopsida

Lycopodiales
use Lycopsida

lycopods
use Lycopsida

Lycopsida
Including Lycopodiales. Includes use on level 2 under pteridophytes(1). See list F. Also search Lycopoda; lycopods.
UF Lycopoda
Lycopodiales
lycopods
BA pteridophytes
BT Plantae
NA Sigillaria
Stigmaria

Lyngen Peninsula
Along Lyngen Fjord 30 miles E of Tromso in N.
BT Troms
Norway

Lynn Lake
Town in NW.
BT Manitoba

Canada

Lyon
use Lyons

Lyons
City in SE central part of country. Also search Lyon.
UF Lyon
BT Rhone
France

lysoclines
SA carbonate compensation depth
solution

Lystrosaurus
Genus. Includes use on level 3 under Reptilia(1) Synapsida(2).
BA Synapsida
Reptilia
BT Tetrapoda
Vertebrata

Lyublyana
use Ljubljana

M

M-discontinuity
use Mohorovicic discontinuity

maars
BT volcanic features
SA craters
volcanism

Maas River
use Meuse River

Maas Valley
use Meuse Valley

Maastricht
City near the German border in extreme SE part of country.
UF Maestricht
BT Netherlands
SA Limburg

Maastrichtian
use Maestrichtian

Macacu River
Central Rio de Janeiro.
BT Rio de Janeiro
Brazil

macaluba
use mud volcanoes

Macanao Peninsula
On Margarita Island 15 miles off NE coast of Venezuela.
BT Nueva Esparta
Venezuela

Macedonia
Region in Bulgaria, department in Greece, and republic in Yugoslavia. Index countries as applicable.
BT Europe
NT Skopje
SA Bulgaria
Greece
Salonika
Serbo-Macedonian Massif
Vourinos
Yugoslavia

macerals
A textural term. Includes use on level 3 under sedimentary rocks(1) organic residues(2). See list I.
UF micropetrological unit
NA exinite
fusinite
inertinite
micrinite
resinite
sporinite

vitrinite
SA coal
organic residues
petrography
sedimentary rocks

Mackay
Town in Custer County in S central.
BT Idaho
United States

Mackenzie Delta
On Beaufort Sea in District of Mackenzie. Also search Mackenzie and Mackenzie River delta.
UF Mackenzie River delta
BT Northwest Territories
Canada

Mackenzie District
use District of Mackenzie

Mackenzie Mountains
Index territories as applicable. Also search Mackenzie.
BT Canada
SA Northwest Territories
Rocky Mountains
Yukon Territory

Mackenzie River delta
use Mackenzie Delta

Mackenzie River valley
W District of Mackenzie. Also search Mackenzie, Mackenzie River, and Mackenzie River delta.
UF Mackenzie Valley
BT Northwest Territories
Canada

Mackenzie Valley
use Mackenzie River valley

mackinawite
UF kansite
BA sulfides
BT minerals

Macleay River
E New South Wales.
BT New South Wales
Australia

Macomb County
On Lake St. Clair just N of Detroit.
BT Michigan
United States

Macon
City in Bibb County in central.
BT Georgia
United States

Macon County
Index states as applicable.
BX Alabama
Georgia
Illinois
Missouri
North Carolina
Tennessee
BT United States

Macquarie Island
850 miles SE of Tasmania by whom it is administered.
BT Pacific Ocean

Macquarie Ridge
Extends between Macquarie Island on S and South Island of New Zealand on N.
BT Pacific Ocean

Macropodidae
Family. Includes use on level 3 under Mammalia(1) Marsupialia(2).
BA Marsupialia
Mammalia
BT Tetrapoda
Vertebrata

Mactra
Genus. Includes use on level 3 under Mollusca(1) Bivalvia(2).
BA Bivalvia

Mactra ● magnesium

        Mollusca
   BT Invertebrata

Madagascar
A valid term through 1974. After 1974, use Malagasy Republic.

Madagascar Basin
SE of Madagascar. Also search Madagascar.
   BT Indian Ocean

Madan
Village in SE Rhodope Mountains in S part of country.
   BT Plovdiv
        Bulgaria

Madeira
Island group belonging to Portugal 600 miles SW of Lisbon and 400 miles W of Morocco. Includes use on level 1 as an area term (list O). For term set options see list B.
   UF Madeira Archipelago
        Madeira Island
        Madeira Islands
        Madeiras
   BA Atlantic Ocean
   NT Porto Santo Island

Madeira Archipelago
use Madeira

Madeira Island
use Madeira

Madeira Islands
use Madeira

Madeiras
use Madeira

Madera County
Central California.
   BT California
        United States

Madhya Pradesh
State in central.
   CO N180000N270000
        E0840000E0740000
   BT India
   NT Balaghat
        Bastar
        Chhindwara
        Chhindwara District
        Jabalpur
        Jhabua
        Kanhan Valley
        Kota
        Pauni
        Pench Valley
        Raipur
        Rewa
        Satna
        Sidhi
        Umaria
        Umrer
   SA Bagh Beds
        Bhander Group
        Bijawar System
        Bundelkhand
        Damuda Series
        Deccan Plateau
        Jabalpur Series
        Kaimur Sandstone
        Mahanadi Valley
        Narmada River
        Narmada Valley
        Satpura Range
        Semri Series
        Son Valley
        Trans-Aravalli Vindhyan Basin

Madison
City in Dane County in S central.
   BT Wisconsin
        United States

Madison County
Index states as applicable.
   BX Alabama
        Arkansas
        Florida
        Georgia
        Idaho
        Illinois
        Indiana
        Iowa
        Kentucky
        Mississippi
        Missouri
        Montana
        Nebraska
        New York
        North Carolina
        Tennessee
        Texas
        Virginia
   BT United States

Madison Group
Lower and upper Mississippian. Consists of Lodgepole Limestone, Mission Canyon Limestone, and Charles Formation. Also search Madison Formation.
   BT Mississippian
   SA Colorado
        Idaho
        Lodgepole Formation
        Lower Mississippian
        Montana
        Upper Mississippian
        Utah
        Wyoming

Madoc
Village in SE.
   BT Ontario
        Canada

Madonie Mountains
NW central.
   BT Sicily
        Italy

Madras
City on the Bay of Bengal in SE part of country.
   BT Tamil Nadu
        India
   NT Kodaikanal

Madreporaria
use Scleractinia

Madrid
Province in central part of country. Also a city.
   BT Spain

Maestrazgo
Mountainous district in E Spain. Index provinces as applicable.
   BT Spain
   SA Castellon de la Plana
        Teruel

Maestricht
use Maastricht

Maestrichtian
Europe. Above Campanian, below Danian (Tertiary). Includes use on level 3 under age terms(1). See list E.
   UF Maastrichtian
   BA Upper Cretaceous
        Cretaceous
   BT Mesozoic
   SA Rosario Formation
        Senonian

mafic
A valid term through 1977. After 1977, use mafic composition on level 2 under igneous rocks(1).

mafic composition
Term introduced in 1978. Includes use on level 2 under igneous rocks(1). Before 1978, also search igneous rocks AND mafic. Also search basic rocks.
   SA composition
        igneous rocks
        ultramafic composition

Magadan
City on N shore of Okhotsk Sea. Also an oblast.
   BT Russian Republic
        USSR

magadiite
   BA scapolite group
        framework silicates
        silicates
   BT minerals

Magallanes
Southernmost province.
   BT Chile

Magdalena
Department in N Colombia. Village in W central New Mexico. Towns in Argentina, Bolivia and Peru. Index countries as applicable.
   SA Argentina
        Bolivia
        Colombia
        Mexico
        New Mexico
        Peru

Magdalena Basin
use Magdalena Valley

Magdalena Delta
At the mouth of the Magdalena River on the Caribbean Sea near Barranquilla. Also search Magdalena.
   BT Bolivar
        Colombia

Magdalena Mountains
In Socorro County in W central.
   BT New Mexico
        United States

Magdalena River
Rises on E slope of Andes in S Colombia and flows N into the Caribbean Sea near Barranquilla. Also search Magdalena AND appropriate area.
   BT Colombia

Magdalena Valley
River valley extending from NW to SW. Also search Magdalena Basin.
   UF Magdalena Basin
   BT Colombia
        South America

Magdalenian
Archaeologic classification. Europe, North Africa. Upper Paleolithic.
   BA Paleolithic
   BT Cenozoic
   SA Pleistocene
        Quaternary

Magdeburg
District in W central East Germany. Also a city.
   BT East Germany
   NT Elbingerode
        Halberstadt
        Stassfurt
   SA Saxony-Anhalt

maghemite
   UF oxymagnite
   BA oxides
   BT minerals
   SA spinel group
        titanomaghemite

magma chambers
Includes use on level 3 under magmas(1) and under intrusions(1).
   UF chambers, magma
        magma reservoir
   SA intrusions
        magmas

magma reservoir
use magma chambers

magmas
Includes use on level 1 (list A); on level 2 under phase equilibria(1). Used for naturally occurring mobile rock material, generated within the Earth and capable of intrusion and extrusion. Also search magma. If 1, term set options are:
topic [age, classification, composition, differentiation, evolution, genesis, geochemistry, properties, temperature, viscosity]
        subtopic
   SA anatexis
        asthenosphere
        contamination
        crust
        crystallites
        crystallization
        cumulates
        differentiation
        emplacement
        fractional crystallization
        hot spots
        hybridization
        igneous activity
        igneous processes
        igneous rocks
        immiscibility
        inclusions
        intrusions
        lava
        magma chambers
        palingenesis
        partial melting
        petrology
        phase equilibria
        viscosity
        volcanology

magmatism
A valid term through 1974. Use igneous activity or magmas.

magnesian calcite
Also search magnesium calcite.
   UF magnesium calcite
   BA carbonates
   BT minerals
   SA calcite

magnesian limestone
   BA limestone
        carbonate rocks
   BT sedimentary rocks

magnesian spar
use dolomite

magnesioferrite
   UF magnoferrite
   BA oxides
   BT minerals
   SA spinel group

magnesioriebeckite
   BA amphibole group
        chain silicates
        silicates
   BT minerals

magnesite
Includes use on level 1 as a commodity term (list C).
   BA carbonates
   BT minerals
   SA refractory materials

magnesium
Includes use on level 1 and 2 as a chemical element (list D). Also search Mg.
   UF Mg
   SA elements

magnesium calcite
use magnesian calcite

magnetic anomalies
Includes use on level 3 under geophysical methods(1) and geophysical surveys(1).
SA anomalies
geophysical methods
geophysical surveys
gravity anomalies
magnetic field
magnetic methods
magnetic surveys
paleomagnetism
sea-floor spreading

magnetic domains
Level 2 term introduced in 1978 under paleomagnetism(1).
SA domain structure
domains
magnetic properties
paleomagnetism

magnetic field
Includes use on level 2 under astrophysics and solar physics(1), under aurora(1), Earth(1), Mars(1), Moon(1) and any of the planets; on level 3 under geophysics(1) under geophysical methods(1), and geophysical surveys(1). Before 1974, also search magnetic fields.
UF field, magnetic
SA astrophysics and solar physics
aurora
coercivity
core
cosmic rays
currents
declination
dipole moment
diurnal variations
Earth
Earth-current methods
electrical field
electrojet
electromagnetic field
electromagnetic radiation
electromagnetic surveys
electromagnetic waves
geophysical methods
geophysical surveys
induction
intensity
magnetic anomalies
magnetic methods
magnetic storms
magnetic surveys
magnetic susceptibility
magnetic tail
magnetization
magnetohydrodynamics
magnetometers
magnetosphere
magnetotelluric methods
magnetotelluric surveys
micropulsations
Moon
natural remanent magnetization
paleomagnetism
plasma instabilities
pole positions
pulsations
remanent magnetization
reversals
secular variations
solar wind
spherical harmonic analysis
thermoremanent magnetization
trapped particles
variations
VLF
whistlers

magnetic iron ore
use magnetite

magnetic methods
Includes use on level 2 under geophysical methods(1).
BA geophysical methods
SA airborne methods
magnetic anomalies
magnetic field
magnetic properties
magnetic surveys
magnetometers
methods

magnetic minerals
BT minerals
SA ilmenite
magnetic properties
magnetite
titanomagnetite

magnetic polarization
use magnetization

magnetic properties
Includes use on level 3 under appropriate material name, e.g. under lava(1).
SA Curie point
magnetic domains
magnetic methods
magnetic minerals
magnetic susceptibility
properties
remanent magnetization

magnetic storms
Includes use on level 2 under magnetosphere(1); on level 3 under aurora(1) magnetic field(2).
SA aurora
disturbances
intensity
magnetic field
magnetosphere
main phase
plasmasphere
polar regions
solar cycles
solar flares
solar wind
storms
substorms
sudden commencements
trapped particles

magnetic survey maps
Term introduced in 1978. Before 1978, also search maps AND magnetic surveys.
BA maps
SA geophysical surveys
magnetic surveys

magnetic surveys
Includes use on level 2 under geophysical surveys(1). Before 1974, Also search aeromagnetic surveys.
BA geophysical surveys
BT surveys
SA airborne methods
magnetic anomalies
magnetic field
magnetic methods
magnetic survey maps
magnetometers

magnetic susceptibility
Includes use on level 3 under geophysical surveys(1) or paleomagnetism(1).
UF magnetic susceptibility anisotropy
volume susceptibility (magnetic)
SA anisotropy
Curie point
geophysical surveys
magnetic field
magnetic properties
paleomagnetism

magnetic susceptibility anisotropy
use magnetic susceptibility

magnetic tail
Includes use on level 2 under magnetosphere(1).
SA magnetic field
magnetopause
magnetosphere
solar wind

magnetism, paleo-
use paleomagnetism

magnetite
Includes use on level 3 under minerals(1) as a commodity term on level 3 under iron(1). See list C (commodities) and list L (minerals).
UF magnetic iron ore
octahedral iron ore
BA oxides
BT minerals
SA heavy minerals
magnetic minerals
spinel group
titanomagnetite

magnetization
The magnetic moment per unit volume.
UF magnetic polarization
SA chemical remanent magnetization
demagnetization
magnetic field
natural remanent magnetization
paleomagnetism
remanent magnetization
thermoremanent magnetization

magnetohydrodynamics
SA core
magnetic field

magnetometers
SA instruments
magnetic field
magnetic methods
magnetic surveys

magnetopause
Includes use on level 2 under magnetosphere(1).
SA bow shock waves
magnetic tail
magnetosheath
magnetosphere
solar wind

magnetosheath
Includes use on level 2 under magnetosphere(1).
SA bow shock waves
magnetopause
magnetosphere

magnetosphere
Includes use on level 1 for special indexes. Term set options are:
bow shock waves
topic
configuration
topic
coordinate systems
topic
cosmic rays
albedo
cutoff rigidities
intensity
electrical field
topic or platform
general
subtopic
instruments
name of instrument
magnetic tail
topic or platform
magnetic storms
topic [e.g. initial phase, main phase, plasmasphere, recovery phase, substorms, sudden-commencements, trapped particles]
magnetopause
topic or platform
magnetosheath
topic or platform
plasma instabilities
topic or platform
plasma motion
topic [circulation, convection]
plasmapause
topic or platform
solar wind
subtopic
techniques
name of technique
type of phenomena
trapped particles
type of particle
type of phenomena
variations
topic [e.g. micropulsations, solar eclipses]
wave propagation
subtopic
type of wave
whistlers
(platform if satellite)
topic
SA aeronomy
albedo
astrophysics and solar physics
atmosphere
aurora
bow shock waves
configuration
coordinate systems
cosmic rays
cutoff rigidities
disturbances
Earth
electrical field
geophysics
intensity
interplanetary space
ionosphere
magnetic field
magnetic storms
magnetic tail
magnetopause
magnetosheath
main phase
meteorology
plasma instabilities
plasma motion
plasmapause
plasmasphere
pulsations
solar wind
substorms
sudden commencements
trapped particles
variations
wave propagation
whistlers

magnetostratigraphy
Term introduced in 1978.
BT stratigraphy

magnetotelluric methods
Includes use on level 2 under geophysical methods(1).
BA geophysical methods
SA Earth-current methods
electrical field
electromagnetic field
electromagnetic methods
magnetic field
magnetotelluric surveys
methods

magnetotelluric surveys
Includes use on level 2 under geophysical surveys(1).
BA geophysical surveys
BT surveys

magnetotelluric surveys ● **Malaya**

  SA Earth-current surveys
      electromagnetic methods
      electromagnetic surveys
      magnetic field
      magnetotelluric methods

Magnitogorsk
City just E of the South Urals in Chelyasinsk Oblast.
  BT Russian Republic
      USSR

magnitude
Includes use on level 2 under earthquakes(1).
  UF earthquake magnitude
  SA earthquakes
      intensity

magnoferrite
  use magnesioferrite

Magothy Formation
Consists essentially of light-gray crossbedded coarse sand containing small amounts of glauconite and pyrite; particles of carbonaceous matter or lignite common throughout.
  BT Upper Cretaceous
      Cretaceous
  SA Delaware
      Maryland
      New Jersey
      New York

Magura
Village W of Dacca.
  BT Bangladesh

Mahanadi Valley
River valley. Index states as applicable. Also search Mahanadi River.
  BT India
  SA Madhya Pradesh
      Orissa

Mahantango Formation
Central Pennsylvania. In Hamilton Group. Consists principally of greenish-gray, thin - to medium bedded slightly carbonaceous shale. Sporadic sandstone zones appear throughout area and thicken eastward. In central Pennsylvania.
  BT Middle Devonian
      Devonian
  SA Hamilton Group
      Pennsylvania

Maharashtra
State in W central.
  BT India
  NT Bhandara
      Bombay
      Jamkhandi
      Koyna
      Nagpur
      Poona
      Wardha River valley
  SA Deccan Plateau
      Godavari River
      Godavari Valley
      Kaladgi System
      Krishna
      Pranhita-Godavari Valley
      Satpura Range

Mahaska County
S central.
  BT Iowa
      United States

Mahasu
District in NW part of country.
  BT Himachal Pradesh
      India

Mahe Island
Most important of Seychelles, a British Colony, 1100 miles E of Kenya and 700 miles N of the Malagasy Republic. Also search Mahe.
  BT Indian Ocean
  SA Seychelles

Mahoning County
NE Ohio.
  BT Ohio
      United States

Maikop
City in Adygey Autonomous Oblast in the Northern Caucusus. Also search Maykop.
  UF Maykop
  BT Russian Republic
      USSR

Maikop Series
Oligocene-Miocene. The following horizons have been established: Zuramakent, Riki, Mutsidakal, Miatly, and Khadum. Northern Caucusus and Crimea.
  BT Tertiary
  SA Russian Republic
      Ukraine
      USSR

Maimecha
  use Maymecha

Maimecha-Kotui
  use Maymecha-Kotuy

main phase
Includes use on level 3 under magnetosphere(1) magnetic storms(2).
  SA magnetic storms
      magnetosphere

Main River
Also search Main. Index states as applicable.
  BT West Germany
  SA Baden-Wurttemberg
      Bavaria
      Hesse

Maine
Includes use on level 1 as an area term (list O). For term set options see list B.
  CO N430000N473000
      W0670000W0710500
  BA United States
  NT Augusta
  NX Cumberland County
      Franklin County
      Hancock County
      Lincoln County
  NT Oxford County
      Penobscot Bay
  NX Presque Isle
      Somerset County
      Washington County
      York County
  SA Atlantic Coastal Plain
      Eastern U.S.
      Matagamon Sandstone
      Merrimack Synclinorium
      New England
      Rangeley Lakes
      Saco River
      Waterville Formation

Maine-et-Loire
Department in NW central.
  CO N470000N474500
      E0001500W0012000
  BT France

maintenance
As of 1978, term is used on level 2 under reservoirs(1).
  SA reservoirs

Mainz
City of the Rhine River in W central part of country.
  BT Rhineland-Palatinate
      Germany
  SA Mainz Basin

Mainz Basin
Area around confluence of the Main and Rhine Rivers. Index states as applicable.
  BT Germany

  SA Hesse
      Mainz
      Rhineland-Palatinate

Maiolica Limestone
Jurassic-Cretaceous. White, marly limestone. In Lombardy, Italy. Also search Maiolica Formation.
  BT Mesozoic
  SA Cretaceous
      Italy
      Jurassic

Maizuru
City in Kyoto Prefecture in central.
  BT Honshu
      Japan

Majdan Pek
  use Majdanpek

Majdanpek
City in E Serbia.
  UF Majdan Pek
  BT Serbia
      Yugoslavia

Major County
NW central.
  BT Oklahoma
      United States

major elements
Includes use on level 3 under soils(1) nutrients(2).
  SA elements
      major-element analyses
      minor elements
      nutrients
      trace elements

major-element analyses
For analytical methods. Includes use on level 3 under chemical analysis(1) and spectroscopy(1). For data, see material name.
  SA analysis
      chemical analysis
      elements
      major elements
      minor-element analyses
      spectroscopy
      trace-element analyses

Majorca
Largest of Spain's Balearic Islands off E coast of Spanish mainland.
  CO N391000N400000
      E0033000E0021500
  UF Mallorca
  BT Balearic Islands
      Mediterranean Sea

Majunga Basin
  use Betsiboka Basin

Makhachkala
City in Dagestan Autonomous Soviet Socialist Republic on Caspian Sea.
  UF Petrovsk
  BT Russian Republic
      USSR

Makhtesh Ramon
Canyon in the Negev.
  BT Israel

Malacca Strait
  use Strait of Malacca

Malacca Straits
  use Strait of Malacca

malachite
  BA carbonates
  BT minerals

Malacostraca
Includes use on level 2 under Arthropoda(1). See list F.
  BA Crustacea
      Arthropoda
  BT Invertebrata
  NA Ophiomorpha

Malaga
Privince on the Mediterranean Sea in S central Spain. Also a city.
  BT Spain
  SA Andalusia
      Serrania de Ronda

Malagasy Republic
Formerly Madagascar. Became Malagasy Republic in 1958. Also search Madagascar. Includes use on level 1 as an area term (list O). For term set options see list B. Before 1975, also search Madagascar.
  CO S254000S115200
      E0503000E0430000
  BA Africa
  NT Ambre Mountain
      Andriamena
      Betsiboka Basin
      Sakoa Basin
      Tananarive
      Tulear Basin

Malaita
Long, narrow island in SE Solomon Islands.
  BT Pacific Ocean
  SA Solomon Islands

Malakal
Town in S part of country.
  BT Sudan

Malatya
Province in E central Anatolia. Also a city.
  BT Turkey
      Middle East

Malawi
Formerly Nyasaland. At one time it was part of British Central African Protectorate and later part of the Federation of Rhodesia & Nyasaland. Includes use on level 1 as an area term (list O). For term set options see list B.
  CO S173000S090000
      E0363000E0330000
  UF Nyasaland
  BA Africa
  SA East African Rift
      Lake Malawi

Malay Archipelago
Largest island group in world off SE coast of Asia between Pacific and Indian oceans. Index countries as applicable. Includes use on level 1 or 2 as an area term (list O). If 1, see term set options under list B.
  BA Asia
  NT Borneo
  SA Celebes Sea
      Far East
      Indonesia
      Malaysia
      New Guinea
      Oceania
      Papua New Guinea
      Sarawak
      Singapore
      Sumatra

Malay Peninsula
Comprises West Malaysia and SW part of Thailand. Index countries as applicable.
  BT Asia
  SA Malaysia
      Pahang
      Thailand
      West Malaysia

Malaya
A valid index term through 1976. After 1976, see Malaysia. Former Federation of Malaya which was a federation of 9 Malay states of the Malay Peninsula, plus 2 of the Straits settlements, Malacca and Penang. It constituted what is now West Malaysia of the present Federation

of Malaysia.

malayaite
BA ring silicates
silicates
BT minerals

Malaysia
Or officially known as Federation of Malaysia. Independent federation, SE Asia, consisting of eleven states (West Malaysia) on the Malay Peninsula and two states (East Malaysia) on the island of Borneo. Includes use on level 1 as an area term (list O). For term set options see list B.
CO N013000N073000
E1190000E1000000
BA Asia
NT Darvel Bay
NA East Malaysia
NT Johore
Kadah
Kelantan
Langkawi Islands
Pahang
Perak
Sabah
Sandakan
Sarawak
Selangor
NA West Malaysia
SA Far East
Indochina
Malay Archipelago
Malay Peninsula

Maldive Islands
Group of 19 atolls in the Indian Ocean 300 miles SW of southern tip of India. Includes use on level 1 as an area term (list O). For term set options see list B.
BA Indian Ocean
SA Asia

Malekula
An island of the New Hebrides which lies E of Queensland, Australia, between the Solomon Islands and New Caledonia.
BT Pacific Ocean
SA New Hebrides

Malgobek
City in North Ossetian Autonomous Soviet Socialist Republic in the Northern Caucasus. A petroleum producing center.
BT Russian Republic
USSR

Malheur County
Extreme E and SE.
CO N420000N442500
W1170500W1181500
BT Oregon
United States

Mali
Includes use on level 1 as an area term (list O). For term set options see list B.
CO N100500N250000
E0040000W0130000
BA Africa
NT Taoudenni
SA Mali-Niger Syneclise
Niger River
Niger Valley
Sahara
Sahel
Senegal Basin
Senegal River
Tanezrouft
West Africa
West African Shield

Mali-Niger Syneclise
Index countries as applicable.
UF Mali-Nigeria Syneclise
BT Africa
SA Benin
Mali
Niger
Nigeria

Mali-Nigeria Syneclise
use Mali-Niger Syneclise

Malines
City N of Brussels.
UF Mechelen
Mechlin
BT Antwerp
Belgium

Mallorca
use Majorca

Malm
Middle Europe. Above Dogger, below Neocomian (Cretaceous). Includes use on level 3 under age terms (1). See list E.
BA Upper Jurassic
Jurassic
BT Mesozoic
SA Kimmeridgian
Lusitanian
Oxfordian
Portlandian
Volgian

Malmohus
County in S.
BT Sweden
NT Halsingborg
Lund
University of Lund
SA Skane

Malta
Comprises 3 islands. Includes use on level 1 as an area term (list O). For term set options see list B.
BA Mediterranean Sea
SA Mediterranean region

Malvern
City in Worcestershire in W central.
BT England
Great Britain
United Kingdom

Malvern Hills
Between Worcestershire and Herefordshire in W England.
BT England
Great Britain
United Kingdom

Mama
Town on the Vitim River at the mouth of the Mama River in NE Irkutsk Oblast.
BT Russian Republic
USSR

Mama River
Rises NE of Lake Baikal and flows NE into Vitim River at Mama. Also search Mama.
BT Russian Republic
USSR

Mamainse Point
Village on Lake Superior N of Sault Ste. Marie.
BT Ontario
Canada

Mammalia
Includes use on level 1 and 2 as a fossil term (list F).
BT Tetrapoda
Vertebrata
NA Amblypoda
Artiodactyla
Carnivora
Cetacea
Chiroptera
Condylarthra
Creodonta
Desmostylia
Docodonta
Edentata
Insectivora
Lagomorpha
Marsupialia
Multituberculata
Notoungulata
Perissodactyla
Primates
Proboscidea
Rodentia
Sirenia
Theria
Tillodontia
Ungulata
SA fossil man
man

Mammoth Cave
Cave and national park in Edmonson County in SW central.
BT Kentucky
United States

mammoths
use Mammuthus

Mammut
use Mastodon

Mammuthus
Genus. Includes use on level 3 under Mammalia(1) Proboscidea(2). Also search mammoths.
UF mammoths
BA Elephantidae
Proboscidea
Mammalia
BT Tetrapoda
Vertebrata
NA Mammuthus primigenius

Mammuthus primigenius
Includes use on level 3 under Mammalia(1) Proboscidea(2).
BA Elephantidae
Proboscidea
Mammalia
BT Tetrapoda
Vertebrata

Mammutidae
use Mastodontidae

man
Used through 1977 in combination with fossil (i.e., man, fossil) on level 1 and 2. After 1977, term is used on level 3 only. Use fossil man on level 1 and 2.
SA anthropology
archaeology
artifacts
fossil man
human activity
human ecology
human waste
Mammalia
Primates
Vertebrata

management
Used as a general term. As of 1978, term is used on level 2 under land use(1) and shorelines(1).
SA land use
planning
shorelines
soil management

Managua
Department in W part of country. Also a city.
BT Nicaragua

Manawatu River valley
SW North Island. Also search Manawatu.
BT North Island
New Zealand

Manche
Department on the English Channel in NW.
CO N483000N494500
W0004500W0020000
BT France
NT Cotentin Peninsula
SA Normandy

Manchester
City in Lancashire in NW England, and city in Hillsborough County in S New Hampshire. Index countries as applicable.
SA New Hampshire
United Kingdom
United States

Manchuria
Region in NE.
BT China
NT Liaoning
SA Amur Basin
Amur River
Argun River
Khanka Lake

Mancos Shale
Lower and upper Cretaceous. In Colorado, it is restricted to thick succession of shale, sandy shale, and thin-bedded sandstone overlying Niobrara Formation and underlying Mesaverde Group. NE Arizona, W Colorado, NW New Mexico, E Utah, and S and central Wyoming.
BT Cretaceous
SA Arizona
Colorado
Ferron Sandstone Member
Lower Cretaceous
New Mexico
Upper Cretaceous
Utah
Wyoming

Mandi
City in Mandi District in NW part of country.
BT Himachal Pradesh
India

Mangalia
Town on the Black Sea in Constanta County.
BT Dobruja
Romania

Mangalore
City in SW.
BT Mysore
India

manganblende
use alabandite

manganese
Includes use on level 1 and 2 as a commodity term (list C) and as a chemical element (list D); on level 2 under mineral deposits, genesis(1). Also search Mn.
UF Mn
SA elements
ferromanganese composition
heavy metals
manganese oxide
Mn-53
Mn-54

manganese oxide
BA oxides
BT minerals
SA manganese
psilomelane

manganite
BA oxides
BT minerals

manganosite
BA oxides
BT minerals

mangerite
Includes use on level 3 under igneous rocks(1) diorite family(2). See list H.
BA diorite family

mangerite ● Marajo

BT igneous rocks
SA plutonic rocks

Mangishlak Peninsula
use Mangyshlak Peninsula

Mangla Dam
SE of Rawalpindi in N part of country. Also search Mangla.
BT Punjab
    Pakistan

mangrove swamps
Term introduced in 1978.
SA ecology
    swamps

Mangyshlak Peninsula
On E coast of N Caspian Sea. Also search Mangyshlak.
CO N423000N450000
    E0543000E0500000
UF Mangishlak Peninsula
BT Kazakhstan
    USSR

Manhattan
Island and borough of New York City at N end of New York Bay; and city in Riley County in NE central Kansas. Index states as applicable.
BX Kansas
    New York
BT United States

Manhattan Formation
Dominantly a garnetiferous quartz-biotite-plagioclase gneiss characterized by sillimimanite and locally much muscovite. SE New York, and W Connecticut.
BT Precambrian
SA Connecticut
    New York
    New York City Group

Manicouagan
Lake and river in Saguenay County, N of St. Lawrence River.
BT Quebec
    Canada

Manihiki Plateau
Undersea feature in the N Cook Islands area E of the Samoa Islands.
BT Pacific Ocean

Manila Trench
Just W of Luzon.
BT South China Sea
    Pacific Ocean

Manildra
Village in E central.
BT New South Wales
    Australia

Manistee County
On Lake Michigan in the Lower Peninsula.
BT Michigan
    United States

Manitoba
Includes use on level 1 as an area term (list O). For term set options see list B.
CO N490000N600000
    W0893000W1020000
BA Canada
NT Beresford Lake
    Bernic Lake
    Bird River
    Churchill
    Flin Flon
    Fort Churchill
    Knee Lake
    Lake Winnipeg
    Lynn Lake
    Nelson River
    Nelson River basin
    Riding Mountain National Park
    Setting Lake
    Snow Lake
    Steinbach
    Tanco Pegmatite
    Two Creeks
    Winnipeg
SA Amisk Group
    Assiniboine River
    Canadian Shield
    Churchill Province
    Elk Point Basin
    Elk Point Group
    Hudson Bay Lowlands
    Lake Agassiz
    Missi Group
    Prairie Evaporite
    Red River Formation
    Rice Lake
    Rice Lake Group
    Saskatchewan River
    Williston Basin
    Winnipeg Formation
    Winnipegosis Formation

Manitou Formation
Consists of finely to coarsely crystalline limestone and minor amounts of dolomite. E Colorado.
BT Lower Ordovician
    Ordovician
SA Colorado

Manitoulin Island
Largest lake island in world. In northern Lake Huron at NW end of Georgian Bay. Also search Manitoulin.
UF Grand Manitoulin
BT Ontario
    Canada

Manitouwadge
Town N of Lake Superior.
BT Ontario
    Canada

Manlius Formation
According to the latest published lexique, the age of Manlius Limestone (in Helderberg Group) is Lower Devonian.
BT Lower Devonian
    Devonian
SA New York

Manning Park
Provincial park in Cascade Mountains E of Vancouver.
BT British Columbia
    Canada

Mannville Formation
In the lower Mannville Formation, basal well-sorted quartzose sandstones were deposited as shoreline sediments or as channel-fills. In the upper Mannville, stratigraphic traps formed in part by intertonguing of sandstones and shales. Also search Manville Group.
UF Manville Formation
BT Lower Cretaceous
    Cretaceous
SA Alberta
    Canada

Mansalay Formation
Composed of hard calcareous mudstones, and calcareous and siliceous carbonaceous sandstone. S Mindoro Island.
BT Jurassic
SA Mindoro
    Philippine Islands

Mansehra
Village in N part of country.
BT Punjab
    Pakistan

Mansfeld Syncline
Also search Mansfeld.
BT Halle
    East Germany

Mansfield Formation
Lower and Middle Pennsylvanian. Contains large amounts of shale, thin beds of coal under clay, and limestone, and is only locally predominantly sandstone.
BT Pennsylvanian
SA Indiana
    Lower Pennsylvanian
    Middle Pennsylvanian

mantle
Includes use on level 1 (list A); on level 2 under seismology(1). Used for the zone of Earth below the crust and above the core (to a depth of 3480 km). If 1, term set options are: topic [age, composition, earthquakes, elastic waves, evolution, genesis, geochemistry, interpretation, low-velocity layer, processes, properties, structure, temperature]
    subtopic (no area term)
NA lower mantle
    upper mantle
SA asthenosphere
    continental drift
    convection
    convection cells
    convection currents
    core
    crust
    degassing
    discontinuities
    Earth
    earthquakes
    elastic waves
    geosynclines
    geothermal gradient
    heat flow
    heat sources
    hot spots
    isostasy
    lithosphere
    low-velocity layer
    low-velocity zones
    Mohorovicic discontinuity
    partial melting
    plate tectonics
    plates
    sea-floor spreading
    seismology
    tectonophysics
    transition zones

manuals
Includes use on level 3 as a term appropriate to a large number of topics, e.g. under education(1). See list G. Also search manual.
SA guidebook
    textbooks

Manville Formation
use Mannville Formation

Mapimi
Town in NE.
BT Durango
    Mexico

mapping
A valid term through 1977. From 1978 on use cartography. To search, see note under cartography.

maps
Includes use on level 1 (list A); on level 2 under geomorphology(1), land use(1), and environmental geology(1). Also search map. If 1, term set options are:
    cartography [for method, instrument, program]
    subtopic
    topic (for global maps; List B except for areal geology)
        subtopic (no area term) e.g. type of map [e.g. economic geology, environmental geology, geologic maps, geomorphologic maps, hydrogeologic maps, photogeologic maps, soils maps, surficial geology, tectonic maps]
NA bathymetric maps
    contour maps
    economic geology maps
    geochemical maps
    geologic maps
    geomorphologic maps
    geophysical maps
    gravity survey maps
    hydraulic maps
    hydrogeologic maps
    hydrologic maps
    index maps
    isograd maps
    isopach maps
    isopleth maps
    isoseismic maps
    lithologic maps
    magnetic survey maps
    paleogeographic maps
    photogeologic maps
    stratigraphic maps
    structural maps
    structure contour maps
    tectonic maps
    topographic maps
SA aerial photography
    areal geology
    atlas
    cartography
    catalogs
    coordinates
    diagrams
    engineering geology
    environmental geology
    explanatory text
    geodesy
    geomorphology
    geophysical methods
    geophysical surveys
    isograds
    land use
    legend
    mineral exploration
    Moon
    photogeology
    road log
    structural geology
    surficial geology
    surveys

Maputo
New name for city of Lourenco Marques located on Delagoa Bay in S part of country.
BT Mozambique

Maquoketa Formation
In Missouri, it is typically thin laminated shale interbedded with shaly limestone members. W Illinois, E Iowa, S Minnesota, E Missouri, and SW Wisconsin.
BT Upper Ordovician
    Ordovician
SA Illinois
    Iowa
    Minnesota
    Missouri
    Wisconsin

Maracaibo Basin
NW part of country. Index Lake Maracaibo and states as applicable.
BT Venezuela
SA Lake Maracaibo
    Merida
    Trujillo
    Zulia

Marajo
Largest island in Amazon Delta.
BT Para
    Brazil

Maramures
County in N.
BT Transylvania
    Rumania

Maranhao
State in NE.
BT Brazil
NT Barreirinhas Basin
SA Brazilian Shield
    Parnaiba Basin
    Tocantins River region

Maranhao Basin
River basin in central.
BT Goias
    Brazil

Marathon Basin
North of Big Bend National Park in W.
BT Texas
    United States

Marathon County
Central Wisconsin.
BT Wisconsin
    United States

marble
Includes use on level 1 as a commodity term (list C); on level 3 under metamorphic rocks(1). See list J. Before 1978, also search crystalline limestone. Also search dolomite marble.
UF bardiglio
    crystalline limestone
BA marbles
BT metamorphic rocks
SA building stone
    calcite
    construction materials
    dimension stone
    dolomite

Marble Canyon
Gorge along the Colorado River. Often considered the upper part of Grand Canyon. Established as National Monument in 1969.
BT Arizona
    United States

Marble Falls Group
Lower and middle Pennsylvanian. Comprises Sloan Formation below and Big Saline Formation above. Central Texas. Also search Marble Falls Formation.
BT Pennsylvanian
SA Lower Pennsylvanian
    Middle Pennsylvanian
    Texas

marbles
Includes use on level 2 under metamorphic rocks(1). See list J.
BT metamorphic rocks
NA brucite marble
    calciphyre
    marble
    ophicalcite
    ophite
SA dolomite

Marburg
City on Lahn River N of Frankfurt.
UF Marburg an der Lahn
BT Hesse
    West Germany

Marburg an der Lahn
use Marburg

Marca Shale Member
Of Moreno Formation. Underlies Dos Palos Shale Member; overlies Tierra Loma Shale Member. S California.
BT Upper Cretaceous
    Cretaceous
SA California
    Moreno Formation

marcasite
BA sulfides
BT minerals

Marche
Historical region in central part of country.
BT France

Marches
Autonomous region in E central Italy. Also search The Marches.
UF The Marches
BT Italy
NT Ancona
    Ascoli Piceno
    Pesaro

Marcy Massif
Mt. Marcy. Highest peak in the Adirondack Mountains.
BT New York
    United States

Mardin Formation
Cretaceous-Paleocene. Composed of clayey schist and marly sandstone. In Mardin in SE Anatolia.
BT Mesozoic
SA Cretaceous
    Paleocene
    Turkey

mare
use maria

Mare Crisium
use Sea of Crises

Mare Imbrium
use Sea of Rains

Mare Nectaris
use Sea of Nectar

Mare Serenitatis
use Sea of Serenity

Mare Spumans
use Foaming Sea

Mare Tranquillitatis
use Sea of Tranquillity

Mare Undarum
use Sea of Waves

Margarita Island
In the Caribbean Sea 15 miles off NE coast.
BT Nueva Esparta
    Venezuela

margarite
BA sheet silicates
    silicates
BT minerals
SA mica group

margin, continental
use continental margin

marginal basins
Includes use on level 3 under plate tectonics(1) evolution(2), structure(2), and subduction(2).
SA basins
    bottom features
    continental slope
    marginal seas
    ocean basins
    ocean floors
    plate tectonics
    subduction

marginal seas
Includes use on level 3 under plate tectonics(1) evolution(2), structure(2), and subduction(2).
UF seas, marginal
SA continents
    marginal basins
    plate tectonics
    subduction

marginal trench
use trenches

margins
Use (instead of boundaries) when referring to plates. Includes use on level 3 under plate tectonics(1).
SA plate tectonics
    plates

Marguerite Bay
Inlet on W coast of the Antarctic Peninsula between Adelaide and Alexander I Islands.
BT Antarctica

maria
Plural of mare. Includes use on level 3 under Moon(1). Also search mare.
UF mare
SA highlands
    mascons
    Moon

Mariana Islands
Group of islands 1500 miles E of Philippine Islands. The islands, not including Guam, formerly were part of the U.S. Trust Territory of the Pacific Islands. Commonwealth status within the U.S. was achieved on March 24, 1976. Includes use on level 1 as an area term (list O). For term set options see list B.
UF Marianas
    Marianas Islands
BA Micronesia
NT Guam
    Saipan

Marianas
use Mariana Islands

Marianas Islands
use Mariana Islands

Marianna Limestone
In Vicksburg Group. Consists of soft chalky limestone, locally called chimney rock; some local, hard limestone; tough to hard ledges in the chimney rock; sandy glauconitic limestone; marl; calcareous sand; and lignitic clay. S Alabama, W Florida, and S Mississippi.
BT middle Oligocene
    Oligocene
SA Alabama
    Florida
    Mississippi
    Vicksburg Group

Maricopa County
SW central Arizona.
BT Arizona
    United States

Marie Byrd Land
Large section E of Ross Ice Shelf and Ross Sea and extending E to Ellsworth Land. Claimed for U.S. by Richard E. Byrd in 1924.
BT Antarctica

Mariinsk
City in Kemerovo Oblast on Trans-Siberian R.R.
BT Russian Republic
    USSR

Marilia
City in W central.
BT Sao Paulo
    Brazil

Marin County
Just N of San Francisco.
BT California
    United States

Marinduque
Island and province just S of Luzon.
BT Philippine Islands
    Malay Archipelago

marine
A valid index term through 1977. After 1977, use marine environment for type of environment or marine methods for geophysical method, or use marine platforms for type of platform.

marine environment
Term introduced in 1978. Includes use on level 3 under ecology(1) and under sedimentation(1). Before 1978, also search marine.
SA abyssal environment
    ecology
    environment
    ice shelves
    paleoecology
    sedimentation
    spits
    tidal flats

marine features
A valid term through 1975. Use shore features or ocean floors, or use the specific feature.

marine geology
Includes use on level 1 (list A). Used for that aspect of the study of the ocean which deals specifically with the ocean floor and the ocean-continent border. Term set options are: topic [applications, bibliography, bottom features, catalogs, concepts, education, experimental studies, geochemistry, history, instruments, methods, nomenclature, objectives, observations, practice, principles, textbooks, theoretical studies]
    subtopic (no area term)
UF geological oceanography
SA bathymetry
    benthonic taxa
    bottom features
    continental shelf
    continental slope
    environmental geology
    geology
    geophysical methods
    Leg 18
    Leg 45
    Leg 46
    Leg 48
    levees
    nodules
    ocean basins
    ocean circulation
    ocean floors
    ocean waves
    oceanography
    reefs
    sea water
    sea-floor spreading
    sedimentary petrology
    sedimentation
    sedimentology
    sediments
    submarine canyons

marine installations
Term introduced in 1976 and used on level 1 beginning 1978. Before 1978, included use on level 2 under engineering geology(1). Term set options are:
construction
    subtopic
design
    subtopic
experimental studies
    subtopic
feasibility studies
    subtopic
foundations
    subtopic
instruments
    subtopic
methods
    subtopic
pipelines
    subtopic
site exploration

marine installations ● Martinsburg Formation 178

    subtopic
submarine installations
    subtopic
theoretical studies
    subtopic
UF installations, marine
NT breakwaters
    gravity platforms
    groins
    jetties
    marine platforms
    piers
    submarine installations
SA construction
    design
    engineering geology
    explosions
    feasibility studies
    foundations
    geologic hazards
    nuclear facilities
    pipelines
    rock mechanics
    shorelines
    site exploration
    soil mechanics
    subsidence
    uplifts
    waste disposal
    waterways

marine methods
Term introduced in 1978. Includes use on level 3 under geophysical methods(1) and geophysical surveys(1). Before 1978, also search marine.
BA geophysical methods
SA geophysical surveys
    methods
    sonobuoys

marine platforms
Term introduced in 1978. Before 1978, search platforms AND marine.
UF marine-cut platform
BT marine installations
SA gravity platforms
    platforms

marine sediments
Not a valid term for GeoRef. Use sediments and marine environment.

marine transport
Includes use on level 3 under sedimentation(1).
SA sedimentation
    transport

marine-cut platform
use marine platforms

Mariner 6
Includes use on level 3 under Mars(1).
SA Mars
    satellite methods

Mariner 7
Includes use on level 3 under Mars(1).
SA Mars
    satellite methods

Mariner 9
Includes use on level 3 under Mars(1).
SA Mars
    satellite methods

Mariner 10
Includes use on level 3 under Mars(1).
SA Mars
    satellite methods

Marion County
Index states as applicable.
BX Alabama
    Arkansas
    Florida
    Georgia
    Illinois
    Indiana
    Iowa
    Kansas
    Kentucky
    Mississippi
    Missouri
    Ohio
    Oregon
    South Carolina
    Tennessee
    Texas
    West Virginia
BT United States

Marion Island
Subantarctic island 1200 miles SE of Capetown and just SW of Prince Edward Islands. Annexed by South Africa.
BT Pacific Ocean

Mariposa County
Central California.
BT California
    United States

Maritime Alps
S division of Western Alps. Index countries as applicable.
BT Alps
    Europe
SA France
    Italy
    Ligurian Alps
    Piedmont Alps
    Western Alps

Maritime Kray
use Primorye

Maritime Provinces
Often called the Maritimes. Index provinces as applicable. Introduced as level 1 area term in 1978.
UF Maritimes
BA Canada
SA New Brunswick
    Nova Scotia
    Prince Edward Island

Maritime Territory
use Primorye

Maritimes
use Maritime Provinces

Maritsa River
Rises in Bulgaria and flows into Aegean Sea. Also search Maritsa. Index countries as applicable.
UF Evros River
    Meric River
BT Europe
SA Bulgaria
    Greece
    Turkey

Mariupol
use Zhdanov

marker beds
As of 1978, term is restricted for use in seismology(1). Before 1978, term also included use for key beds (stratigraphy).
UF beds, marker
SA key beds
    seismology

markings, current
use current markings

marl
Includes use on level 3 under sedimentary rocks(1) clastic rocks(2) and under carbonate rocks(2). See list I. Also search marlstone.
UF calcareous clay
    marlstone
BZ clastic rocks
    carbonate rocks
BT sedimentary rocks
SA argillaceous texture
    clay
    loess
    terrigenous materials

marlstone
use marl

Marmaton Group
Includes (ascending) Fort Scott, Labette, Pawnee, Bandera, Altamont, Nowata, and Lenapan formations. SW Iowa, E Kansas, W Missouri, and NE Oklahoma.
BT Desmoinesian
    Middle Pennsylvanian
    Pennsylvanian
SA Iowa
    Kansas
    Labette Shale
    Missouri
    Oklahoma
    Wewoka Formation

Marmolada Glacier
At Marmolada Peak, the highest in the Dolomites. Index Autonomous regions as applicable.
BT Italy
SA Trentino-Alto Adige
    Veneto

Marne
Department in NE.
CO N483000N493000
    E0051000E0031500
BT France

Marne Valley
River valley in NE central France. Also search Marne River.
BT France

Maroon Town
Town in NW.
BT Jamaica
    West Indies

Marquette
City on Lake Superior in Marquette County, Upper Peninsula.
BT Michigan
    United States

Marquette Iron Range
use Marquette Range

Marquette Range
Iron range in Marquette County in Upper Peninsula.
UF Marquette Iron Range
BT Michigan
    United States

Mars
Includes use on level 1 and 2. See entry under Moon(1) for term set options.
NT Tharsis
SA Deimos
    landing sites
    Mariner 10
    Mariner 6
    Mariner 7
    Mariner 9
    Moon
    Phobos
    planetology
    planets
    satellites
    solar system
    terrestrial planets
    Viking

Marseilles
City on Gulf of Lion.
BT Bouches-du-Rhone
    France

Marshall County
Index states as applicable.
BX Alabama
    Illinois
    Indiana
    Iowa
    Kansas
    Kentucky
    Minnesota
    Mississippi
    Oklahoma
    South Dakota
    Tennessee
    West Virginia
BT United States

Marshall Islands
Group of 34 atolls and coral islands, including Eniwetok and Kwajalein, SW of Hawaii and E of Guam. Part of the U.S. Trust Territory of the Pacific Islands. Includes use on level 1 as an area term (list O). For term set options see list B.
BA Micronesia
NT Eniwetok Atoll

Marshalltown Formation
In Matawan Group. Overlies Englishtown Formation; underlies Wenonah Sand.
BT Upper Cretaceous
    Cretaceous
SA Delaware
    New Jersey

marshes
Includes use on level 3 under geomorphology(1).
SA bogs
    geomorphology
    salt marshes
    swamps

Marsica
Mountain range in E central part of country.
BT Abruzzi
    Italy

Marsupialia
Includes use on level 2 under Mammalia(1). See list F.
BA Mammalia
BT Tetrapoda
    Vertebrata
NA Macropodidae

Martha's Vineyard
Island in Atlantic Ocean off SW coast of Cape Cod.
BT Massachusetts
    United States

Martin Formation
Composed of three unnamed members (ascending): conglomerate sandstone and dolomite limestone member; sandstone and limestone member; and calcareous sandstone, sandy limestone and shale member. Central and E Arizona.
BT Devonian
SA Arizona
    Middle Devonian
    Upper Devonian

Martin Lake
Formed by dam across Tallapoosa River in E central.
BT Alabama
    United States

Martin River Glacier
NE of Cordova in S.
BT Alaska
    United States

Martinique
Island and department of France in Windward Islands.
BT Lesser Antilles
    West Indies
SA Windward Islands

Martinsburg Formation
According to the latest published Lexique, the age of Martinsburg Shale is middle and upper Ordovician. In Pennsylvania, comprises (ascending) Cocalio Shale, unnamed

and undifferentiated dark shales, Jonestown Beds (new), and Fairview Shochary Sandstones. Maryland, New Jersey, SE Pennsylvania, Tennessee, W Virginia, and West Virginia. Also search Martinsburg Shale.
- BT Ordovician
- SA Maryland
  - Middle Ordovician
  - New Jersey
  - Pennsylvania
  - Upper Ordovician
  - Virginia
  - West Virginia

martite
- BA oxides
- BT minerals
- SA hematite

Maryborough Basin
SE Queensland. Also search Maryborough.
- BT Queensland
  - Australia

Maryland
Includes use on level 1 as an area term (list O). For term set options see list B.
- CO N375500N394300 W0750500W0793000
- BA United States
- NX Allegany County
- NT Baltimore
  - Baltimore County
- NX Carroll County
- NT Harford County
- NX Montgomery County
- NT Patuxent River
  - Prince Georges County
  - Savage River
- NX Somerset County
  - Washington County
  - Worcester County
- SA Allegheny Front
  - Allegheny Group
  - Allegheny Mountains
  - Ames Limestone
  - Anacostia River basin
  - Antietam Formation
  - Appalachian Basin
  - Appalachian Plateau
  - Aquia Formation
  - Atlantic Coastal Plain
  - Baltimore Gneiss
  - Bloomsburg Formation
  - Blue Ridge Province
  - Calvert Formation
  - Catoctin Formation
  - Catskill Delta
  - Catskill Formation
  - Chesapeake Bay
  - Chilhowee Group
  - Choptank Formation
  - Choptank River
  - Clinton Group
  - Coeymans Formation
  - Conemaugh Group
  - Cumberland
  - Delmarva Peninsula
  - Dunkard Group
  - Eastern U.S.
  - Genesee Group
  - Glenarm Series
  - Greenland Gap Group
  - Hamilton Group
  - Helderberg Group
  - Juniata Formation
  - Kinzers Formation
  - Magothy Formation
  - Martinsburg Formation
  - McKenzie Formation
  - Monmouth Group
  - Monongahela Group
  - Nanjemoy Formation
  - Newark Group
  - Newark-Gettysburg Basin
  - Onondaga Limestone
  - Oriskany Sandstone
  - Piedmont
  - Pocono Formation
  - Potomac Group
  - Potomac River
  - Potomac River basin
  - Pottsville Group
  - Raritan Formation
  - Rochester Formation
  - Saint Marys Formation
  - Sharon Conglomerate
  - South Mountain
  - Susquehanna River
  - Susquehanna River basin
  - Tonoloway Limestone
  - Tuscarora Formation
  - Valley and Ridge Province
  - Wicomico Formation
  - Williamsport Sandstone
  - Wissahickon Formation
  - Yorktown Formation

Marysvale
City in Piute County in S central.
- BT Utah
  - United States

Mascarene Basin
E of N Malagasy Republic.
- BT Indian Ocean

Mascarene Islands
Group between 400 and 500 miles E of the Malagasy Republic. Index islands as applicable.
- BT Indian Ocean
- NT Piton de la Fournaise
  - Piton des Neiges
  - Saint-Louis
- SA Mauritius
  - Reunion
  - Rodriguez Island

mascons
Includes use on level 3 under Moon(1).
- SA maria
  - Moon

Mascot Dolomite
In Knox Group. Consists of light- and dark-gray dolomite and limestone; moderately cherty; base marked by chert matrix sandstone. E Tennessee and Virginia. Index states as applicable.
- BT Lower Ordovician
  - Ordovician
- SA Knox Group
  - Tennessee
  - Virginia

maskelynite
- BA feldspar group
  - framework silicates
  - silicates
- BT minerals
- SA aluminosilicates

Mason County
Index states as applicable.
- BX Illinois
  - Kentucky
  - Michigan
  - Texas
  - Washington
  - West Virginia
- BT United States

mass
Includes use on level 3 under Earth.
- SA mass balance
  - mass spectroscopy

mass balance
Includes use on level 3 under glacial geology(1) glaciers(2). Also search balance.
- UF balance, mass
  - mass budget
- SA ablation
  - glacial geology
  - glaciers
  - mass

mass budget
- use mass balance

mass movements
Includes use on level 2 under geomorphology(1). Also search mass transport.
- UF mass transport
  - movements, mass
- SA creep
  - flows
  - geomorphology
  - lahars
  - landslides
  - liquefaction
  - mass wasting
  - mudflows
  - rockslides
  - slope stability
  - slumping

mass spectrometry
A valid term through 1973. After 1973, see methods(2) under chemical analysis(1), and spectroscopy(1).

mass spectroscopy
Not a valid term from 1972 through 1977. After 1977, includes use on level 3 under spectroscopy(1) methods(2) and under chemical element (1). Before 1978, also search mass spectrometry; spectroscopy AND mass.
- BT spectroscopy
- SA mass

mass transfer
Used as a general term.
- UF transfer, mass

mass transport
- use mass movements

mass wasting
Includes use on level 3 under engineering geology(1) landslides(2).
- UF wasting, mass
- SA creep
  - engineering geology
  - erosion
  - etching
  - landslides
  - mass movements
  - solifluction

Massachusetts
Includes use on level 1 as an area term (list O). For term set options see list B.
- CO N411500N425500 W0695500W0733000
- BA United States
- NT Barnstable County
  - Berkshire Hills
  - Boston
  - Buzzards Bay
- NX Cambridge
- NT Cape Cod
- NX Essex County
  - Franklin County
  - Hampshire County
- NT Martha's Vineyard
- NX Middlesex County
- NT Monomoy Island
  - Nantucket Island
- NX Suffolk County
- NT Wachusett-Marlborough Tunnel
  - Ware
  - Woods Hole
  - Woods Hole Oceanographic Institution
  - Worcester
- NX Worcester County
- SA Atlantic Coastal Plain
  - Connecticut River
  - Connecticut Valley
  - Eastern U.S.
  - Green Mountains
  - Littleton Formation
  - Merrimack River valley
  - Merrimack Synclinorium
  - New England
  - Normanskill Formation
  - Springfield

masses, median
- use median masses

Massiac
Village in S central part of country.
- BT Cantal
  - France

Massif Central
- use Central Massif

massifs
A massive topographic and structural feature in an orogenic belt. Not to be used for complexes.
- SA complexes
  - shields
  - tectonics

massive
A valid index term through 1977. After 1977, use massive bedding for the sedimentary structure, and use massive deposits for ore deposits.

massive bedding
Term introduced in 1978. Includes use on level 3 under sedimentary structures(1) planar bedding structures(2). Before 1978, also search massive AND bedding.
- BA planar bedding structures
  - sedimentary structures
- SA bedding

massive deposits
Term introduced in 1978. Includes use on level 3 for ore deposits. Before 1978, also search massive AND ore deposits; massive AND deposits.
- SA deposits
  - mineral deposits
  - ore deposits

Mastodon
Genus. Includes use on level 3 under Mammalia(1) Proboscidea(2). Also search Mammut.
- UF Mammut
- BA Mastodontidae
  - Proboscidea
  - Mammalia
- BT Tetrapoda
  - Vertebrata

Mastodontidae
Family. Includes use on level 3 under Mammalia(1) Proboscidea(2). Also search Mammutidae.
- UF Mammutidae
- BA Proboscidea
  - Mammalia
- BT Tetrapoda
  - Vertebrata
- NA Mastodon

Matadi
Port city on Congo River in W part of county. Bas-Zaire.
- BT Zaire

Matagami
Lake SE of James Bay.
- UF Mattagami
- BT Quebec
  - Canada

Matagamon Sandstone
- BT Lower Devonian
  - Devonian
- SA Maine

Matagorda Bay
Inlet of the Gulf of Mexico in SE.
- BT Texas

United States
Matanuska Valley
River valley NE of Anchorage in S Alaska. Also search Matanuska River.
BT Alaska
United States
Matera
City and province in S part of country.
BT Basilicata
Italy
materials
Includes use in combination with properties (i.e. materials, properties) on level 2 under engineering geology(1); on level 3 under education(1) and engineering geology(1).
SA amorphous materials
anisotropic materials
banded materials
cement materials
ceramic materials
coexisting materials
cohesive materials
colloidal materials
competent materials
construction materials
dissolved materials
education
elastic materials
engineering geology
expansive materials
fossiliferous materials
granular materials
host materials
inorganic materials
isotropic materials
living materials
lunar materials
natural materials
nonterrigenous materials
organic materials
overconsolidated materials
parent materials
plastic materials
properties
refractory materials
rocks
saturated materials
standard materials
suspended materials
synthetic materials
terrestrial materials
terrigenous materials
toxic materials
unconsolidated materials
viscous materials
mathematical geology
For mathematics as applied to geology. Includes use on level 1 (list A). Term set options are:
topic [bibliography, concepts, education, history, interpretation, methods, nomenclature, objectives, philosophy, principles, symposia, textbooks, theoretical studies]
subtopic
SA algorithms
analog simulation
automatic data processing
biometry
canonical analysis
cluster analysis
discriminant analysis
equations
factor analysis
finite difference analysis
finite element analysis
Fortran IV
Fourier analysis
functions
geology
geometry
geophysics
mathematical methods

mathematical models
multivariate analysis
regression analysis
simulation
statistical analysis
statistical methods
trend-surface analysis
mathematical methods
Used as a general term.
SA autocorrelation
eigenvalues
functions
least-squares analysis
mathematical geology
methods
numerical analysis
probability
statistical methods
mathematical models
Includes use on level 3 under mathematical geology(1) or automatic data processing(1).
BA models
SA automatic data processing
mathematical geology
matildite
Also search schapbachite.
UF plenargyrite
schapbacite
BA sulfosalts
BT minerals
Mato Grosso
State in SW central Brazil.
UF Matto Grosso
BT Brazil
SA Botucatu Formation
Brazilian Shield
Irati Formation
Matra Mountains
S spur of the Carpathians. Also search Matra.
BT Hungary
SA Carpathians
matrix
The fine-grained interstitial material of an igneous rock. Also search ground mass; groundmass.
UF ground mass
groundmass
SA igneous rocks
megacrysts
Matsukawa
Town in N central.
BT Honshu
Japan
Matsushiro
Town in central.
BT Honshu
Japan
Matsuyama
City in W near Inland Sea.
BT Shikoku
Japan
Mattagami
use Matagami
Matto Grosso
use Mato Grosso
Mattoon Formation
McLeansboro Group. Predominantly shale and sandstone.
BT Pennsylvanian
SA Illinois
maturity
Used as a general term.
Matuyama Epoch
Geomagnetic epoch.
UF Matuyama Reversed
BA Cenozoic
SA lower Pleistocene
Quaternary
Tertiary
upper Pliocene

Matuyama Reversed
use Matuyama Epoch
maucherite
BA arsenides
sulfides
BT minerals
Maui
Island in the Hawaiian Islands NW of the island of Hawaii.
BT Hawaii
United States
Maumee River valley
NE Indiana and NW Ohio. Index states as applicable. Also search Maumee River.
BT United States
SA Indiana
Ohio
Mauna Kea
Extinct volcano on N central island of Hawaii.
CO N193000N194000
W1552000W1553500
UF Mauna Kea Volcano
BT Hawaii
United States
Mauna Kea Volcano
use Mauna Kea
Mauna Loa
Volcano on S central island of Hawaii.
CO N192000N193500
W1552000W1554000
UF Mauna Loa Volcano
BT Hawaii
United States
Mauna Loa Volcano
use Mauna Loa
Maures Massif
On the Mediterranean coast at W end of the Riviera. Also search Maures; Monts des Maures.
UF Monts des Maures
BT Var
France
Mauritania
Includes use on level 1 as an area term (list O). For term set options see list B.
CO N145500N273000
W0045500W0173000
BA Africa
NT Adrar
Akjoujt
Aouelloul
Fort Gouraud
Nouakchott
Richat Mountain
SA Cap Blanc
Reguibat Ridge
Sahara
Sahel
Senegal Basin
Senegal River
West Africa
Mauritius
An independent state in the Mascarene Islands about 450 miles E of the Malagasy Republic. Includes use on level 1 as an area term (list O). For term set options see list B.
BA Indian Ocean
SA Mascarene Islands
Maury Channel
Sea channel E of central Reykjanes Ridge and S of Iceland.
BT Atlantic Ocean
Maverick County
On the Rio Grande in SW.
BT Texas
United States

mawsonite
BA sulfides
BT minerals
May-sur-Orne
Village on the Orne River in NW part of country.
BT Calvados
France
Maya Mountains
Range in S Belize. Also search Maya AND appropriate area term.
BT Belize
Central America
Maya River basin
Chiefly in central Khabarovsk Kray but also in Yakutia. Also search Maya River.
BT Russian Republic
USSR
Mayaguez Bay
Off city of Mayaguez in W.
BT Puerto Rico
West Indies
Mayenne
Department in NW.
BT France
Maykop
use Maikop
Maymecha
River. Rises in Evenk National Okrug and flows N into the Kheta River in the Taymyr National Okrug. Also search Maimecha.
UF Maimecha
BT Russian Republic
United States
Maymecha Kotuy
use Maymecha-Kotuy
Maymecha-Kotui
use Maymecha-Kotuy
Maymecha-Kotuy
Region in valleys of the Maymecha, Kotuy, and Kheta rivers in the Taymyr and Evenki national okrugs. Also search Maimecha-Kotui.
CO N700000N711500
E1033000E0993000
UF Maimecha-Kotui
Maymecha Kotuy
Maymecha-Kotui
BT Russian Republic
USSR
Mayo
County in NW.
CO N533000N542000
W0084000W0101500
BT Ireland
NT Keel
Mayurbhanj
District in E part of country.
BT Orissa
India
Mazama Ash
At site of Mount Mazama, a prehistoric volcanic mountain in the Cascade Mountains of Klamath County in S Oregon. Caldera now occupied by Crater Lake.
BT Oregon
United States
Mazatlan
City on the Pacific Ocean.
BT Sinaloa
Mexico
Mazon Creek
NE Illinois.
BT Illinois
United States
McHenry County
Index states as applicable.
BX Illinois

North Dakota
BT United States

McIntosh County
Index states as applicable.
BX Georgia
   North Dakota
   Oklahoma
BT United States

McKenzie Formation
In Maryland, it includes Rabble Run Sandstone Member. Underlies Bloomsburg Formation and overlies Rochester Formation. W Maryland, central Pennsylvania, N Virginia, and NE West Virginia.
BT Middle Silurian
   Silurian
SA Maryland
   Pennsylvania
   Virginia
   West Virginia

McKinley County
NW New Mexico.
BT New Mexico
   United States

McLean County
Index states as applicable.
BX Illinois
   Kentucky
   North Dakota
BT United States

McMurdo Ice Shelf
Part of Ross Ice Shelf in McMurdo Sound area in Ross Dependency on the Pacific Ocean side.
BT Antarctica
SA Ross Ice Shelf

McMurdo Sound
Inlet of SW Ross Sea at the edge of the Ross Ice Shelf beween Ross Island and coast of Victoria Land. Site of major U.S. research and exploration base in Ross Dependency on the Pacific Ocean side.
BT Antarctic Ocean

McPherson County
Index states as applicable.
BX Kansas
   Nebraska
   South Dakota
BT United States

McPherson Formation
Consists of early Pleistocene stream deposits, later and coarser Pleistocene stream channel deposits, and still later Pleistocene silt, clay, and fine sand. Central Kansas.
BT Pleistocene
SA Kansas

Meade Basin
River basin S of Point Barrow in N. Coal mining in area.
BT Alaska
   United States

Meade County
Index states as applicable.
BX Kansas
   Kentucky
   South Dakota
BT United States

Meadow soils
BA soils
SA Intrazonal soils

Meagher County
W central Montana.
BT Montana
   United States

meanders
Includes use on level 3 under geomorphology(1) fluvial features(2).
BT fluvial features
SA floodplains
   geomorphology
   rivers
   rivers and streams
   sinuosity
   streams

measurement
Includes use on level 2 under heat flow(1).
SA conductivity
   geodesy
   geothermal gradient
   heat flow
   satellite measurements
   thermal conductivity

Meath
County on the Irish Sea in NE.
BT Ireland

mechanical controls
Includes use on level 3 under mineral deposits, genesis(1) controls(2).
SA controls
   mineral deposits

mechanical erosion
use abrasion

mechanical properties
SA anisotropy
   creep
   deformation
   elastic limit
   elastic properties
   elasticity
   engineering geology
   friction
   hardness
   physical properties
   plasticity
   Poisson's ratio
   properties
   rigidity
   shear strength
   soil mechanics
   strain
   strength
   tensile strength
   weathering
   yield strength

mechanical weathering
Term introduced in 1978. Before 1978, also search weathering AND mechanical.
BT weathering
SA insolation
   physical weathering

mechanics
Includes use on level 2 under faults(1) and folds(1); on level 3 under plate tectonics(1).
SA decollement
   faults
   flexural-slip
   folds
   land subsidence
   plate tectonics
   rock mechanics
   shear
   soil mechanics
   stick-slip

mechanism
Includes use as a level 2 or 3 term appropriate to a large number of topics, e.g. on level 2 under orogeny(1), earthquakes(1), under salt tectonics(1), and under plate tectonics(1). See list G.
SA flow mechanism
   focal mechanism

Mechelen
use Malines

Mechlin
use Malines

Mecklenburg
Region on the Baltic Sea in N. A former German state.
BT East Germany

Mecsek Mountains
Near Yugoslav border in S part of country. In Baranya. Also search Mecsek.
UF Baranya Mountains
BT Hungary

media
Used as a general term.
SA anelastic media
   layered media
   porous media
   seismology

median masses
A structural unit. Includes use on level 3 under tectonics(1). Also search Zwischengebirge.
UF betwixt mountains
   masses, median
   Zwischengebirge
SA tectonics

Median Tectonic Line
Southwest Japan.
SA Japan
   plate tectonics
   tectonics

median valley
UF valley, median
SA Mid-Atlantic Ridge
   mid-ocean ridges
   rift zones
   rifting

medical geology
Term introduced in 1978. Before 1978, also search health.
UF health
SA environmental geology
   geology
   human ecology
   pollution

Medicine Bow Mountains
Range of the Rocky Mountains. Index states as applicable.
BT United States
SA Colorado
   Rocky Mountains
   Wyoming

Medicine Hat
City in SE.
BT Alberta
   Canada

Medicine Lake
In Sheridan County in NE.
BT Montana
   United States

Medina Formation
According to the latest published Lexique, the age of Medina Group is lower Silurian; however, the group has either been abandoned by author or restricted for use by the U. S. Geological Survey.
BT Lower Silurian
   Silurian
SA Michigan
   New York

Mediterranean (region)
use Mediterranean region

Mediterranean region
The Mediterranean Sea and its islands, the Adriatic Sea, and parts of those countries along their shores in S Europe, N Africa, and the Middle East. Index Adriatic Sea, Mediterranean Sea, countries, and islands as applicable. Also search Mediterranean. Includes use on level 1 or 2 as an area term (list O). If 1, see term set options under list B.
CO N300000N473000
   E0380000W0050000
UF Mediterranean (region)
SA Adriatic Sea
   Albania
   Algeria
   Corsica
   Cyprus
   Egypt
   France
   Greece
   Israel
   Italy
   Lebanon
   Libya
   Malta
   Mediterranean Sea
   Mesogaea
   Morocco
   Sardinia
   Sicily
   Spain
   Syria
   Tunisia
   Turkey
   Yugoslavia

Mediterranean Ridge
Between Crete and the coast of E Libya.
BT Mediterranean Sea

Mediterranean Sea
Enclosed by Europe on the W and N, Asia on the E, and Africa on the S. Includes use on level 1 or 2 as an area term (list O). If 1, see term set options under list B.
CO N300000N450000
   E0370000W0060000
NA Adriatic Sea
   Aegean Sea
NT Alboran Sea
   Balearic Basin
NA Balearic Islands
   East Mediterranean
NT El Alboran Island
   Gulf of Lion
   Gulf of Pozzuoli
NA Ionian Sea
NT Ligurian Sea
NA Malta
NT Mediterranean Ridge
   Strait of Sicily
NA Tyrrhenian Sea
NT West Mediterranean
SA Mediterranean region
   Sardinia
   Turkey

Medo
use Tokyo

Medoc
District between the Gironde River and the Bay of Biscay in NW.
BT Gironde
   France

meerschaum
use sepiolite

Meerut
City, district, and division in N part of country.
BT Uttar Pradesh
   India

meetings
A valid term through 1975. After 1975, use associations or symposia. Also search meeting.

megacrysts
SA igneous rocks
   matrix
   metamorphic rocks
   textures

megacyclothems
BA cyclothems
   planar bedding structures
   sedimentary structures

Megalopolis Basin
In Arcadia Department in central Peloponnesus. Also search

Megalopolis.
BT Peloponnesus
Greece

megaripples
Includes use on level 3 under sedimentary structures(1) bedding plane irregularities(2). See list K.
BA bedding plane irregularities
sedimentary structures
SA sand waves

megaspores
Includes use on level 2 under palynomorphs(1). See list F.
BA palynomorphs
SA miospores
spores

Megion Field
Oil field in Khanty-Mansino National Okrug in West Siberian Plain. Also search Megion.
UF Megion Oil Field
BT Russian Republic
USSR

Megion Oil Field
use Megion Field

Megri
Town in S near Iranian border.
BT Armenia
USSR

Megrinskiy Pluton
In Megrinskiy Mountains in S.
BT Armenia
USSR

Meguma Group
Composed of Goldenville and Halifax formations. In Nova Scotia.
BT Ordovician
SA Canada
Nova Scotia

Mehadia
Village at W edge of the Transylvanian Alps.
BT Banat
Romania

Mehedinti Plateau
SW part of country. Index regions as applicable.
BT Romania
SA Banat
Walachia

Meighen Island
W of Axel Heiberg Island in District of Franklin.
BT Northwest Territories
Canada

meimechite
Includes use on level 3 under igneous rocks(1) ultramafic family(2). See list H. Before 1975, also search meymechite.
UF meymechite
BA ultramafic family
BT igneous rocks

meionite
BZ carbonates
framework silicates
BT minerals
SA chlorides
scapolite group

Mekong Delta
On the South China Sea S of Saigon.
BT Vietnam

Melanesia
Collective name for islands NE of Australia. Index island groups as applicable. Includes use on level 1 and 2 as an area term (list O). For term set options see list B.
BA Pacific Ocean
NA Fiji
New Hebrides
Solomon Islands
SA Bismarck Archipelago
New Caledonia
Oceania
Santa Isabel
Solomon Islands

melange
Includes use on level 3 under deformation(1), and under structural analysis(1); on level 3 under plate tectonics(1).
UF block clay
SA boudinage
boudins
deformation
plate tectonics
structural analysis

melanite
BA garnet group
orthosilicates
silicates
BT minerals

melanocerite
BA orthosilicates
silicates
BT minerals

melanophlogite
BA silica minerals
framework silicates
silicates
BT minerals

melanterite
BA sulfates
BT minerals

melaphyre
See list H.
BA basalt family
BT igneous rocks
SA basalt
porphyry

Melbourne
City on the Tasman Sea in SE part of country.
BT Victoria
Australia

Meldon Aplite
Nothumberland County in extreme NE.
BT England
Great Britain
United Kingdom

melilite
UF mellilite
BA melilite group
orthosilicates
silicates
BT minerals

melilite group
Includes use in combination with orthosilicates (i.e. orthosilicates, melilite group) on level 2 under minerals(1). See list L.
BA orthosilicates
silicates
BT minerals
NA akermanite
gehlenite
melilite

melilitite
Includes use on level 3 under igneous rocks(1) alkali basalt family(2) and ultramafic family(2). See list H.
BA alkali basalt family
BT igneous rocks
SA ultramafic family

mellilite
use melilite

melnikovite
use greigite

Melosira
BA diatoms
algae
BT Plantae

melt water
use meltwater

melteigite
Includes use on level 3 under igneous rocks(1) alkali gabbro family(2). See list H.
BA alkali gabbro family
BT igneous rocks
SA ijolite
urtite

melting
Includes use on level 2 under phase equilibria(1).
SA fusion
partial melting
phase equilibria

melting relations
A valid term through mid-1978. Use melting or phase equilibria.

melts
Includes use on level 3 under phase equilibria(1).
SA liquid phase
partial melting
petrology
phase equilibria

meltwater
Includes use on level 3 under glacial geology(1) glaciers(2). Also search melt water.
UF melt water
SA glacial geology
glaciers
hydrology
ice
snow
water

Melville Peninsula
N projection of Canadian mainland in District of Franklin between Committee Bay on the W and Foxe Basin on the E.
BT Northwest Territories
Canada

Memel River basin
use Neman River basin

menaccanite
use ilmenite

Mendeleyev Ridge
N of Alaska and E of Siberia.
BT Arctic Ocean

Mendeleyev Volcano
On Kunashir, southernmost of the Kuril Islands. Also search Medeleyeva Volcano.
UF Mendeleyeva Volcano
BT Russian Republic
USSR

Mendeleyeva Volcano
use Mendeleyev Volcano

Mendip Hills
SW England. Also search Mendip.
BT England
Great Britain
United Kingdom

Mendocino (Cape)
use Cape Mendocino

Mendocino County
Along Pacific Ocean in NW. Also search Mendocino.
BT California
United States

Mendocino fracture zone
Running in an E-W direction approximately 1250 miles W of N California.
BT Pacific Ocean

Mendoza
Province in W.
CO S380000S330000
W0660000W0710000
BT Argentina
NT Payun Matru
SA Salado Basin

Menefee Formation
In Mesaverde Group. Includes Cleary Coal Member and above that a bed formerly called Allison Barren Member. SW Colorado, and NW New Mexico.
BT Upper Cretaceous
Cretaceous
SA Colorado
Mesaverde Group
New Mexico

meneghinite
BA sulfantimonites
sulfosalts
BT minerals

Menorca
use Minorca

Mentor Beds
According to the latest published Lexique, the age of Mentor Sandstone Member (or Belvedere Formation) is Lower Cretaceous. In central Kansas.
BT Comanchean
Cretaceous
SA Kansas
Lower Cretaceous

Meotian
European stage, Black Sea area. Above Sarmatian, below Cimmerian. Includes use on level 3 under age terms(1). See list E.
BA Neogene
Tertiary
BT Cenozoic
SA lower Pliocene
Miocene
Pliocene
upper Miocene

Meramecian
Provincial series, North America. Above Osagian, below Chesterian. Includes use on level 3 under age terms(1). See list E.
BA Upper Mississippian
Mississippian
BT Paleozoic
NT Saint Louis Limestone
Sainte Genevieve Limestone
Salem Limestone
Warsaw Formation
SA Chesterian
Valmeyeran

Merano
City in Bolzano Province in NE part of country.
BT Trentino-Alto Adige
Italy

Mercalli scale, modified
use modified Mercalli scale

Merced County
W center.
BT California
United States

Mercenaria
Genus. Includes use on level 3 under Mollusca(1) Bivalvia(2).
BA Bivalvia
Mollusca
BT Invertebrata

Mercer County
Index states as applicable.
BX Illinois
Kentucky
Missouri
New Jersey
North Dakota
Ohio
Pennsylvania
West Virginia

BT United States

mercury
Includes use on level 1 and 2 as a chemical element (list D) and as a commodity term (list C); on level 2 under paragenesis(1) and mineral deposits, genesis(1). For documents about the planet see Mercury Planet. Also search quicksilver.
UF Hg
quicksilver
SA cinnabar
elements

Mercury Planet
Includes use on level 1 and 2. See entry under Moon(1) for term set options. Before 1972, also search Mercury in combination with the category (extraterrestrial geology). For the element or commodity, see mercury.
SA planetology
planets
solar system
terrestrial planets

Meric River
use Maritsa River

Merida
State in W Venezuela, and city in Yucatan in N part of Yucatan Peninsula in Mexico. Index countries as applicable.
SA Maracaibo Basin
Mexico
Venezuela

Merionethshire
County in W.
CO N523000N530000
W0031500W0041000
BT Wales
Great Britain
United Kingdom

Merostomata
Includes use on level 2 under Arthropoda(1). See list F.
BA Chelicerata
Arthropoda
BT Invertebrata
NA Eurypterida
Xiphosura

Merrimack River valley
Index states as applicable.
BT United States
SA Massachusetts
New Hampshire

Merrimack Synclinorium
Index states as applicable.
BT United States
SA Maine
Massachusetts
New Hampshire

Merseburg
City in SW central.
BT Halle
East Germany

Mersey River
Name of rivers in N Tasmania and in NW England. Index England or Australia as applicable.
SA Australia
England
Mersey Valley

Mersey Valley
River valley in NW England and in Tasmania.
SA Australia
England
Mersey River

merwinite
BA orthosilicates
silicates
BT minerals

merzlota
use Frozen ground

Mesa County
W central Colorado.
BT Colorado
United States

Mesabi Range
Iron-ore bearing range of low hills in NE.
BT Minnesota
United States

Mesaverde Group
Throughout San Juan Basin of SW Colorado and NW New Mexico, composed of Menefee Formation, Point Lookout Sandstone, Cliff House Sandstone, Gallup Sandstone, and Crevasse Canyon Formation. Arizona, W Colorado, NW New Mexico, E Utah, and Wyoming. Also search Mesaverde Formation.
BT Upper Cretaceous
Cretaceous
SA Almond Formation
Arizona
Blackhawk Formation
Colorado
Gallup Sandstone
La Ventana Sandstone
Menefee Formation
New Mexico
Utah
Wyoming

Meseta
Geographic term for the entire interior of Spain covering almost 3/4 of country and consisting of an immense plateau with Madrid at its center.
BT Spain

Mesogaea
Term introduced in 1978. Includes use on level 3 under paleogeography(1).
SA Mediterranean region
paleogeography
Tethys

mesolite
BA zeolite group
framework silicates
silicates
BT minerals

Mesolithic
Archaeologic classification.
UF Middle Stone Age
BA Holocene
Quaternary
BT Cenozoic

Mesopotamia
Region in SW Asia between Tigris and Euphrates rivers. Primarily in Iraq but partly in NE Syria. Index countries as applicable.
BT Middle East
SA Iraq
Syria

mesosiderites
UF grahamite
BA meteorites
SA stony irons

Mesozoic
Includes use on level 1 as an age term (list E); on level 2 under paleoterms, e.g. paleoecology, paleogeography, paleomagnetism. From the end of the Paleozoic to the beginning of the Cenozoic.
BA Phanerozoic
NT Caracas Group
NA Cretaceous
NT Franciscan Formation
Great Valley Sequence
NA Jurassic
NT Khorat Group
Kootenay Formation
NA lower Mesozoic
NT Maiolica Limestone
Mardin Formation
NA middle Mesozoic
NT Navajo Sandstone
Pucara Group
Serra Gerral Formation
Shimanto Group
Tal Formation
NA Triassic
upper Mesozoic
SA Alpine Orogeny
Beacon Supergroup
Botucatu Formation
Brianconnais Zone
Ferrar Group
Ionian Zone
Shuswap Complex

Messel
Village near Darmstadt in S.
BT Hesse
West Germany

Messina
City and province in NE.
BT Sicily
Italy

Messinian
BA upper Miocene
Miocene
Neogene
Tertiary
BT Cenozoic

meta-andesite
Also search metaandesite.
UF metaandesite
BA metaigneous rocks
BT metamorphic rocks
SA andesite

meta-arkose
Also search metaarkose.
UF metaarkose
BA metasedimentary rocks
BT metamorphic rocks
SA arkose

meta-limestone
use metalimestone

metaandesite
use meta-andesite

metaarkose
use meta-arkose

metabasalt
BA metaigneous rocks
BT metamorphic rocks
SA basalt

metabasite
BA metaigneous rocks
BT metamorphic rocks

metabentonite
Includes use as a rock and a mineral.
BA metasedimentary rocks
BT metamorphic rocks
SA bentonite
clay minerals

metabolism
Includes use under fossil groups(1) or paleontology(1).
SA nutrition

metacinnabar
Also search metacinnabarite.
UF metacinnabarite
BA sulfides
BT minerals
SA onofrite

metacinnabarite
use metacinnabar

metaconglomerate
BA metasedimentary rocks
BT metamorphic rocks
SA conglomerate

metadiabase
BA metaigneous rocks
BT metamorphic rocks
SA diabase

metadiorite
BA metaigneous rocks
BT metamorphic rocks
SA diorite

metadolerite
See list J.
BA metaigneous rocks
BT metamorphic rocks
SA dolerite

metagabbro
BA metaigneous rocks
BT metamorphic rocks
SA gabbro

metagranite
BA metaigneous rocks
BT metamorphic rocks
SA granite

metagraywacke
Also search metagreywacke.
UF metagreywacke
BA metasedimentary rocks
BT metamorphic rocks
SA graywacke

metagreywacke
use metagraywacke

metahalloysite
use halloysite

metaigneous
A valid index term on level 2 through 1977. After 1977, use metaigneous rocks.

metaigneous rocks
Includes use on level 2 under metamorphic rocks(1). Before 1978, also search metaigneous.
BT metamorphic rocks
NA meta-andesite
metabasalt
metabasite
metadiabase
metadiorite
metadolerite
metagabbro
metagranite
metaperidotite
metapyroxenite
metarhyolite
metatuff
NZ serpentinite
SA igneous rocks
rocks

metakaolin
Also search metakaolinite.
UF metakaolinite
SA clay minerals
kaolin

metakaolinite
use metakaolin

metalimestone
Also search meta-limestone.
UF meta-limestone
BA metasedimentary rocks
BT metamorphic rocks
SA limestone

Metaline Falls
Town in Pend Oreille County in extreme NE.
BT Washington
United States

metallogenic provinces
Includes use on level 3 under metallic commodities (list C) or economic geology(1).
UF provinces, metallogenic
SA economic geology
mineral deposits
mineralization
ore deposits

metallogeny ● metasomatism 184

metallogeny
Includes use on level 3 under geosynclines(1) processes (2) and under economic geology (1).
SA economic geology
　　geosynclines
　　mineral deposits
　　ore deposits

metallography
Not a valid index term for GeoRef. See mineralogy(1) methods(2) and petrology(1) methods(2).

metallurgy
SA alloys
　　metals
　　ore deposits

metals
Use mainly when of economic value. Includes use on level 1 and 2 as a commodity term (list C).
NT alkali metals
　　heavy metals
　　rare metals
SA alloys
　　base metals
　　complexing
　　economic geology
　　elements
　　geochemistry
　　isotopes
　　metallurgy
　　native elements and alloys
　　nonmetals
　　ore minerals
　　pollution
　　trace metals

metamict
Includes use on level 3 under minerals(1) and mineralogy(1).
SA crystal structure
　　metamictization
　　minerals
　　radiation damage

metamictization
Includes use on level 3 under minerals(1) and mineralogy(1).
SA crystal structure
　　metamict
　　minerals
　　radiation damage

metamorphic aureoles
use aureoles

metamorphic belts
Includes use on level 3 under plate tectonics(1).
UF belts, metamorphic
SA plate tectonics

metamorphic differentiation
Not a valid index term for GeoRef. See metamorphism(1) migration of elements(2) and metamorphism(1) differentiation(2).

metamorphic grade
use grade

metamorphic rocks
See list J for rock groups and names. Includes use on level 1 (list A); on level 2 under paragenesis(1), phase equilibria(1), and weathering(1). If 1, term set options are:
rock group [amphibolites, cataclasites, gneisses, granulites, hornfels, marbles, metaigneous rocks, metaplutonic rocks, metasedimentary rocks, metavolcanic rocks, migmatites, mylonites, phyllites, phyllonites, schists, slates]
　　rock name
topic [Used for general studies where several rock types are discussed or where rock is not assignable to one of the groups above; age, classification, composition, correlation, distribution, evolution, experimental studies, facies, genesis, geochemistry, lithostratigraphy, mineral assemblages, occurrence, petrography, petrology, properties, textures]
　　subtopic
NT amphibolites
NA anatexite
NT cataclasites
NA ferruginous quartzite
NT gneisses
NA gondite
NT granulites
　　hornfels
　　marbles
　　metaigneous rocks
　　metaplutonic rocks
　　metasedimentary rocks
　　metavolcanic rocks
　　migmatites
　　mylonites
　　phyllites
　　phyllonites
NA quartzite
NT schists
　　slates
SA actinolite facies
　　amphibolite facies
　　banded structures
　　basement
　　blueschist facies
　　complexes
　　crystalline rocks
　　deformation
　　epidote-amphibolite facies
　　fabric
　　facies
　　folds
　　foliation
　　geochemistry
　　geochronology
　　gneissic texture
　　granulite facies
　　greenschist facies
　　impactite
　　intrusions
　　lineation
　　lithology
　　lithostratigraphy
　　megacrysts
　　metamorphism
　　metasomatism
　　mineral assemblages
　　mullions
　　paragenesis
　　pelitic texture
　　petrography
　　petrology
　　phase equilibria
　　porphyroblastic texture
　　prehnite-pumpellyite facies
　　protoliths
　　rocks
　　schistosity
　　skarn
　　suevite
　　zeolite facies

metamorphic zone
use aureoles

metamorphism
Includes use on level 1 (list A). Term set options are:
kind of metamorphism [anchimetamorphism, autometamorphism, burial metamorphism, contact metamorphism, dynamic metamorphism, grade, polymetamorphism, prograde metamorphism, regional metamorphism, retrograde metamorphism, shock metamorphism, thermal metamorphism, ultrametamorphism]
　　topic (high-grade metamorphism, low-grade metamorphism)
topic [For more than one kind or for a kind not assignable to one of the categories above; age, causes, classification, concepts, environment, evolution, extent, genesis, mechanism, migration of elements, processes, P-T conditions, rates, rheomorphism, temperature, theoretical studies, zoning]
　　subtopic
NA autometamorphism
　　burial metamorphism
　　contact metamorphism
　　dynamic metamorphism
NT glauconitization
NA granitization
　　high-grade metamorphism
　　low-grade metamorphism
　　migmatization
　　phyllonitization
　　polymetamorphism
　　prograde metamorphism
　　regional metamorphism
　　retrograde metamorphism
　　shock metamorphism
　　thermal metamorphism
　　ultrametamorphism
SA anatexis
　　aureoles
　　complexes
　　contact
　　deformation
　　diagenesis
　　feldspathization
　　folds
　　foliation
　　geochemistry
　　geochronology
　　grade
　　high pressure
　　high temperature
　　igneous rocks
　　intrusions
　　isograds
　　lineation
　　low pressure
　　low temperature
　　metamorphic rocks
　　metaplutonic rocks
　　metasomatism
　　meteor craters
　　migration of elements
　　orogeny
　　P-T conditions
　　paragenesis
　　petrology
　　phase equilibria
　　protoliths
　　recrystallization
　　rheomorphism
　　shatter cones
　　shock waves
　　slaty cleavage
　　uralitization
　　zoning

metapelite
BA metasedimentary rocks
BT metamorphic rocks

metaperidotite
BA metaigneous rocks
BT metamorphic rocks
SA peridotite

metaplutonic rocks
Term introduced in 1978. Includes use on level 2 under metamorphic rocks(1). Before 1978, also search metamorphic rocks AND metaplutonic.
BT metamorphic rocks
SA metamorphism
　　plutonic rocks
　　rocks

metapyroxenite
BA metaigneous rocks
BT metamorphic rocks
SA pyroxenite

metarhyolite
BA metaigneous rocks
BT metamorphic rocks
SA rhyolite

metasandstone
BA metasedimentary rocks
BT metamorphic rocks
SA sandstone

metasedimentary
A valid index term on level 2 through 1977. After 1977, use metasedimentary rocks.

metasedimentary rocks
Includes use on level 2 under metamorphic rocks(1). Before 1978, also search metasedimentary.
BT metamorphic rocks
NA khondalite
　　meta-arkose
　　metabentonite
　　metaconglomerate
　　metagraywacke
　　metalimestone
　　metapelite
　　metasandstone
　　metasiltstone
　　paragneiss
SA rocks
　　sedimentary rocks

metasediments
A valid term through 1974. Use metasedimentary rocks.

metasiltstone
BA metasedimentary rocks
BT metamorphic rocks
SA siltstone

metasomatic rocks
As of 1978, term is used on level 1 for any rock produced by replacement processes at constant volume with little disturbance of textural or structural features. Includes use on level 3 under petrology(1) and under metasomatism(1). Also search metasomatite and metasomatites. Term set options are:
rock group [greisen, propylite, skarn]
　　rock name or process or subtopic
topic [age, classification, composition, distribution, evolution, facies, genesis, geochemistry, mineral assemblages, occurrence, petrography, petrology, textures]
　　subtopic
UF metasomatite
　　metasomatites
NA fenite
　　greisen
　　propylite
NZ serpentinite
NA skarn
　　tactite
　　talc rock
SA alteration
　　metasomatism
　　rocks

metasomatism
Includes use on level 1 (list A). For documents before 1976, also search replacement. Term set options are:
materials
　　name of material [e.g. carbonate rocks, granite, igneous rocks, ore deposits, sedimentary rocks]
processes
　　name of process [e.g. albitization, alunitization, analcimization, argillization, dolomitization, granitization, greisenization, hydrothermal alteration,

metasomatism • methods

kaolinization, laumontization, microclinization, muscovitization, palagonitization, propylitization, pyrometasomatism, scapolitization, serpentinization, silicification, zeolitization]
topic [age, causes, environment, experimental studies, geochemistry, interpretation, mechanism, rates, theoretical studies]
    subtopic
NA albitization
    autometasomatism
    chloritization
    feldspathization
    fenitization
    greisenization
    hydrothermal alteration
    kaolinization
    propylitization
    pyritization
    pyrometasomatism
    sericitization
    serpentinization
    spilitization
    uralitization
    wallrock alteration
    zeolitization
SA alteration
    diagenesis
    dolomitization
    enrichment
    fenite
    geochemistry
    granitization
    hydrothermal solutions
    igneous rocks
    intrusions
    metamorphic rocks
    metamorphism
    metasomatic rocks
    mineral deposits
    ore deposits
    P-T conditions
    palingenesis
    paragenesis
    petrology
    phase equilibria
    plutons
    silicification
    zoning

metasomatite
use metasomatic rocks

metasomatites
use metasomatic rocks

metastibnite
BA sulfides
BT minerals
SA stibnite

metatuff
BA metaigneous rocks
BT metamorphic rocks
SA tuff

metavolcanic
A valid term through 1977. After 1977, use metavolcanic rocks on level 2 under metamorphic rocks(1).

metavolcanic rocks
Term introduced in 1978. Includes use on level 2 under metamorphic rocks(1). Before 1978, also search metamorphic rocks AND metavolcanic.
BT metamorphic rocks
SA rocks
    volcanic rocks

Metazoa
Used as a general fossil term. Also search individual phyla and groups.
SA Protista

meteor craters
Includes use on level 1 (list A). Used for impact craters formed by the falling of a large meteorite onto a surface. Also search meteor crater. If 1, term set options are:
topic [age, causes, classification, concepts, distribution, evolution, genesis, interpretation, observations, occurrence, patterns, structure]
    subtopic
UF meteorite crater
SA astroblemes
    breccia
    craters
    cryptoexplosion features
    deformation
    geochronology
    geomorphology
    impact features
    impacts
    metamorphism
    meteorites
    shatter cones
    shock metamorphism
    tectonics

meteoric iron
use iron meteorites

meteorite crater
use meteor craters

meteorites
Includes use on level 1 (list A) and 2. Used for any meteoroid that has fallen to the Earth's surface in one piece or in fragments without being completely vaporized. Term set options are:
topic [age, bibliography, catalogs, collections, composition, detection, distribution, experimental studies, flux, genesis, geochemistry, interpretation, isotopes, organic materials, phase equilibria, properties, radioactivity, textures]
    name of meteorite or type [e.g. achondrites, chondrites, howardites, iron meteorites, stony irons] or subtopic
UF cosmolites
    skystones
NA achondrites
    chondrites
    hexahedrites
    howardites
    iron meteorites
    mesosiderites
    micrometeorites
    pallasites
    stony irons
    ureilites
SA Allende
    Ashmore
    asteroids
    australite
    Barwell
    Bjurbole
    Campo del Cielo
    Canyon Diablo
    Chainpur
    chaoite
    chondrules
    cohenite
    Coldwater
    comets
    cosmic dust
    extraterrestrial geology
    flux
    geochemistry
    Gibeon
    Haveroe
    impact features
    impactite
    impacts
    interplanetary dust
    isotopes
    kamacite
    Kapoeta
    lonsdaleite
    meteor craters
    meteoroids
    meteors
    Moon
    Norton County Meteorite
    Orgueil
    Pasamonte
    petrology
    planetology
    plessite
    Sharps
    shatter cones
    solar system
    suevite
    taenite
    tektites
    Toluca
    troilite
    Vigarano
    Warburton
    Weekeroo Station
    Wolf Creek

meteoroids
SA asteroids
    interplanetary space
    meteorites
    meteors
    planets

meteorology
Primarily used for special indexes. Includes use on level 1. Term set options are:
aerosols
    topic [composition, convection, particles, radioactive tracers, turbidity]
boundary interactions
    topic
circulation
    topic [meridional circulation, zonal circulation, etc.]
composition
    elements or compound or region of atmosphere
convection
    topic
diffusion
    models
    property diffused
    type of diffusion
electromagnetic waves
    topic [currents, lightning, raindrops, space charge, thunderstorms]
experimental studies
    topic
instruments
    name of instrument, free
    type of phenomenon
ions
    name of ion
particles
    name of particle or topic or subtopic
    models
storms
    models
    type of storm
techniques
    type of phenomenon
    technique
temperature
    topic
theoretical studies
    topic
turbulence
    area
    subtopic
    topic
water
    topic [clouds, droplets, humidity, precipitation, raindrops, name of isotope(s)], isotope(s)]
waves
    type of wave
winds
    type of wind
SA aeronomy
    aerosols
    atmosphere
    atmospheric precipitation
    aurora
    boundary interactions
    circulation
    climate
    clouds
    convection
    currents
    diffusion
    droplets
    electrical phenomena
    electromagnetic waves
    geophysics
    humidity
    hurricanes
    hydrologic cycle
    hydrology
    ionosphere
    ions
    lightning
    magnetosphere
    particles
    raindrops
    storms
    thunderstorms
    turbidity
    turbulence
    water
    waves
    winds

meteors
Includes use on level 3 under meteorites(1) and under planetology(1).
UF shooting star
SA comets
    interplanetary space
    meteorites
    meteoroids
    planetology

methane
Includes use on level 3 under organic materials(1) hydrocarbons(2).
BA organic materials
SA ethane
    hydrocarbons
    natural gas

method, transmission
use transmission method

methods
This term applies to theoretical and experimental studies, while the term techniques is used for discussion of samples. Includes use as a level 2 or 3 term appropriate to a large number of topics, e.g. on level 2 under absolute age(1), geochemistry(1), geophysical methods(1) organic materials(1), and soils(1) and mineral exploration(1). See list G.
SA acoustical methods
    airborne methods
    analytical methods
    Ar/Ar
    autoradiography
    biogeochemical methods
    chemical methods
    common-depth-point method
    deep-tow methods
    Earth-current methods
    electrical methods
    electromagnetic methods
    electron paramagnetic resonance
    electron probe
    fission-track dating
    gamma-gamma methods
    gamma-ray methods
    geobotanical methods

geochemical methods
geological methods
geomorphological methods
geophysical methods
graphic methods
gravity methods
ground methods
hydrological methods
infrared methods
magnetic methods
magnetotelluric methods
marine methods
mathematical methods
microscope methods
microwave methods
new methods
optical methods
photogeologic methods
physical methods
quantitative analysis
quantitative methods
radar methods
radio-frequency spectroscopy
radio-wave methods
radioactivity methods
satellite methods
scanning method
seismic methods
self-potential methods
single-crystal method
sonar methods
statistical methods
techniques
U-238/Pb-206
U/He
U/Pb
U/Th/Pb
wet methods

Methow River valley
Okanogan County on British Columbia border. Also search Methow River.
BT Washington
    United States

Metohija
District in SW section of country forming part of autonomous province of Kosovo-Metohija.
BT Serbia
    Yugoslavia

Meurthe-et-Moselle
Department in NE.
CO N483000N493000
   E0071500E0053500
BT France
NT Laneuville-devant-Nancy Boring

Meuse
Department in NE.
BT France
NT Verdun

Meuse River
Rises in NE France, flows N across E Belgium and, as the Maas River, flows through the Netherlands centering the North Sea through its estuary, the Hollandsch Diep. Index countries as applicable. Also search Maas River; Meuse.
UF Maas River
BT Europe
SA Belgium
    France
    Netherlands

Meuse Valley
River valley. Known as the Maas Valley in the Netherlands. Index countries as applicable.
UF Maas Valley
BT Europe
SA Belgium
    France
    Netherlands

Mexicali
City on California border.
BT Baja California
    Mexico

Mexico
Includes use on level 1 or 2 as an area term (list O). If 1, see term set options under list B.
CO N143000N330000
   W0864500W1170000
BA North America
NT Baja California
   Campeche
   Chiapas
   Chihuahua
   Coahuila
   Colorado River delta
   Durango
   Federal District
   Guanajuato
   Guerrero
   Hidalgo
   Jalisco
   Mexico state
   Michoacan
   Nuevo Leon
   Oaxaca
   Puebla
   Quintana Roo
   San Luis Potosi
   Sierra Madre
   Sierra Madre del Sur
   Sierra Madre Occidental
   Sierra Madre Oriental
   Sinaloa
   Sonora
   Tabasco
   Tamaulipas
   Toluca
   Valley of Mexico
   Veracruz
   Yucatan
   Yucatan Peninsula
   Zacatecas
   Zaragoza
SA Difunta Group
   Leon
   Magdalena
   Merida
   Rosario Formation
   San Felipe Formation
   Zuloaga Limestone

Mexico Basin
use Sigsbee Deep

Mexico City
City.
UF Mexico D.F.
BT Federal District
    Mexico

Mexico D.F.
use Mexico City

Mexico state
State N, E and W of Federal District.
BT Mexico
SA Valley of Mexico

meymechite
use meimechite

Mezen
Town on the Mezen River in Arkhangelsk Oblast in NW European USSR.
BT Russian Republic
    USSR

Mezen River basin
N central Arkhangelsk Oblast in NW European USSR. Also search Mezen River.
BT Russian Republic
    USSR

Mezhrechye
Village in W.
BT Byelorussia
    USSR

Mg
use magnesium

Miami
City in Dade County in SE.
BT Florida
    United States

Miami Limestone
According to the latest published Lexique, the age of Miami Oolite is Pleistocene. It is a soft white oolitic limestone, containing streaks of thin irregular layers of calcite separating less crystalline streaks. In S Florida. Also search Miami Oolite.
BT Pleistocene
SA Florida

Mianwali
Town and district in NE part of country.
BT Punjab
    Pakistan

Miaoli
Town in NW.
BT Taiwan

miargyrite
BA sulfantimonites
   sulfosalts
BT minerals

miaskite
Includes use on level 3 under igneous rocks(1) syenite family(2). See list H.
BA syenite family
BT igneous rocks

mica
Includes use on level 1 as a commodity term (list C).
BA sheet silicates
BT minerals
SA mica group
   vermiculite

mica group
Includes use in combination with sheet silicates (i.e. sheet silicates, mica group) on level 2 under minerals(1). See list L.
BA sheet silicates
   silicates
BT minerals
SA biotite
   celadonite
   clintonite
   fluor-phlogopite
   glauconite
   hydrobiotite
   hydromica
   hydromuscovite
   lepidolite
   lepidomelane
   listvenite
   margarite
   mica
   muscovite
   paragonite
   phengite
   phlogopite
   polylithionite
   sericite
   vermiculite
   zinnwaldite
   zussmanite

mica schist
Includes use on level 3 under metamorphic rocks(1) schists(2). See list J.
BA schists
BT metamorphic rocks
SA schist

Michelle Formation
Composed of black calcareous shale and aphanitic silty limestone. In W Northwest Territories and Yukon Territory.
BT Middle Devonian
   Devonian
SA Canada
   Northwest Territories
   Yukon Territory

michenerite
BA sulfides
BT minerals

Michigan
Includes use on level 1 as an area term (list O). For term set options see list B.
CO N414500N473000
   W0823000W0901500
BA United States
NT Alpena County
   Ann Arbor
NX Berrien County
   Calhoun County
   Cass County
NT Charlevoix County
NX Dickinson County
   Genesee County
NT Glennie
   Grand Rapids
   Grand Traverse County
   Houghton County
   Ingham County
NX Iron County
NT Isle Royale National Park
NX Jackson County
NT Kalamazoo County
   Keweenaw County
   Keweenaw Peninsula
NX Lake County
NT Lansing
NX Livingston County
NT Lower Peninsula
   Macomb County
   Manistee County
   Marquette
   Marquette Range
NX Mason County
NT Michigan Basin
NX Midland County
   Monroe County
NT Negaunee Iron Formation
   Ontonagon County
NX Ottawa County
NT Porcupine Mountains
NX Presque Isle
NT Roscommon County
NX Saint Clair County
NT University of Michigan
   Upper Peninsula
NX Utica
   Wayne County
NT White Pine
   White Pine Mine
SA Animikie Group
   Berea Sandstone
   Canadian Shield
   Charlevoix
   Clinton Group
   Copper Harbor Conglomerate
   Detroit River Group
   Grand River
   Great Lakes region
   Green Bay
   Jacobsville Sandstone
   Keweenawan
   Lake Erie
   Lake Huron
   Lake Michigan
   Lake Superior
   Lockport Formation
   Medina Formation
   Midcontinent
   Midwest
   Nonesuch Shale
   Penokean Orogeny
   Portage Lake Lava Series
   Saginaw Formation
   Saint Clair River
   Saint Clair River delta
   Saint Peter Sandstone

Salina Group
Sault Sainte Marie
Sylvania Formation
Traverse Group
Trenton Group
Utica Shale

Michigan Basin
Lower Peninsula.
BT Michigan
    United States

Michipicoten Island
NE Lake Superior.
BT Ontario
    Canada

Michoacan
State in SW..
BT Mexico
NT Paricutin

micrinite
BA macerals
SA coal
    organic residues
    sedimentary rocks

micrite
Includes use on level 3 under sedimentary rocks(1) carbonate rocks(2). See list I.
BA carbonate rocks
BT sedimentary rocks
SA biomicrite
    limestone

microbreccia
BA clastic rocks
BT sedimentary rocks
SA breccia
    sandstone
    terrigenous materials

microcline
BA feldspar group
    framework silicates
    silicates
BT minerals
SA amazonite
    K-feldspar

Microcodium
Includes use on level 3 under algae(1).
BA algae
BT Plantae

microcomputers
Term introduced in 1978. Includes use on level 3 under automatic data processing(1).
SA automatic data processing
    computers

microcracks
BT cracks

microcraters
Includes use on level 2 under Moon(1).
SA craters
    Moon

microdiorite
BA diorite family
BT igneous rocks
SA diorite

microearthquakes
Includes use on level 2 under seismology(1).
SA earthquakes
    seismology

microelement
use trace elements

microfacies
Includes use on level 3 under micropaleontology(1).
SA facies
    micropaleontology
    petrography
    thin sections

microfauna
Includes use on level 3 under fossil groups (list F) or micropaleontology(1).
SA ecology
    fauna
    micropaleontology
    paleontology

microflora
use microfossils

microfossils
Includes use on level 3 under fossil groups (list F). Also search microflora.
UF microflora
SA fauna
    flora
    fossils
    index fossils
    micropaleontology
    otoliths
    pollen

microfractures
Term used mostly in relation to deformation(1) and engineering geology(1).
SA deformation
    engineering geology
    fissures
    fractures
    structural analysis

microgabbro
BA gabbro family
BT igneous rocks
SA gabbro

microgranite
BA granite-granodiorite family
BT igneous rocks
SA granite

microhardness
Includes use on level 3 under minerals(1).
SA hardness
    minerals
    physical properties

microlite
BZ halides
    oxides
BT minerals
SA niobates
    pyrochlore
    tantalates

micrometeorites
Very small meteorite or meteoritic particle with diameter generally less than a millimeter.
BA meteorites

micromorphology
Includes use on level 3 under soils(1) morphology(2).
SA morphology
    profiles
    soils

Micronesia
Collective name for islands E of the Philippine Islands and S of Japan. Index island groups as applicable. Includes use on level 1 or 2 as an area term (list O). If 1, see term set options under list B.
BA Pacific Ocean
NA Mariana Islands
    Marshall Islands
SA Caroline Islands
    Gilbert Islands
    Oceania

microorganisms
For several types. Includes use on level 3 under soils(1) biota(2); on level 3 under geochemistry(1).
SA algae
    bacteria
    biota
    fauna
    faunal studies
    flora
    floral studies
    fungi
    organisms
    pollen
    Protista
    soils

micropaleontology
Used for the discipline as a whole, not for specific applications. Includes use on level 1 (list A). See list F (microfossil groups). Term set options are:
topic [applications, bibliography, catalogs, concepts, experimental studies, history, instruments, methods, photography, practice, techniques, theoretical studies]
    subtopic [e.g. sample preparation]
SA automatic data processing
    foraminifera
    fossils
    microfacies
    microfauna
    microfossils
    paleobotany
    paleontology
    palynology
    sample preparation
    stratigraphy

micropegmatite
BA granite-granodiorite family
BT igneous rocks
SA graphic granite
    pegmatite

microperthite
BA feldspar group
    framework silicates
    silicates
BT minerals
SA perthite

micropetrological unit
use macerals

microplankton
use plankton

microplates
Includes use on level 3 under plate tectonics(1) geometry(2).
SA intraplate
    plate tectonics
    plates

microprocessors
Term introduced in 1978. Includes use on level 3 under automatic data processing(1).
SA automatic data processing

micropulsations
Includes use on level 3 under aurora(1) magnetic field(2).
SA aurora
    magnetic field
    pulsations

microscope methods
Includes use on level 3 under petrology(1) and mineralogy(1).
SA methods
    mineralogy
    ore microscopy
    petrography
    petrology
    polished sections
    reflectance
    reflectivity
    thin sections
    universal stage

microscopy, electron
use electron microscopy

microscopy, ore
use ore microscopy

microseisms
Includes use on level 2 under seismology(1); on level 3 under earthquakes(1).
SA earthquakes
    noise
    seismology

microstructure
A valid term through 1978. After 1978, use ultrastructure.

microstylolites
See list K.
BA secondary structures
    sedimentary structures
SA stylolites

microsyenite
BA syenite family
BT igneous rocks
SA syenite

microtectonics
A valid term through 1975. After 1975, use structural analysis.

microtektites
BA tektites

microwave methods
Term introduced in 1978. Includes use on level 3 under spectroscopy(1) methods (2). Before 1978, also search microwave.
SA analysis
    methods
    microwave spectroscopy
    spectroscopy

microwave spectroscopy
Term introduced in 1978. Includes use on level 3 under spectroscopy(1), and under chemical element(1) analysis(2). Before 1978, also search spectroscopy AND microwave.
BT spectroscopy
SA analysis
    microwave methods

Mid-Atlantic Ridge
Extends parallel to the continental margins in mid ocean in both the North and South Atlantic Ocean. Rises 6000 ft. above the ocean floor and surfaces at the Azores, Ascension, St. Helena, and Tristan de Cunha islands.
CO S570000N600000
    E0030000W0620000
BT Atlantic Ocean
SA Atlantis fracture zone
    FAMOUS
    median valley

Mid-Continent
use Midcontinent

Mid-Indian Ocean Ridge
use Mid-Indian Ridge

Mid-Indian Ridge
Extends from S of the Maldives, S and SE into Antarctic waters. Also search Indian Ocean Ridge; Indian Ridge; Mid-Indian Ocean Ridge.
UF Indian Ocean Ridge
    Indian Ridge
    Mid-Indian Ocean Ridge
BT Indian Ocean
NT East Indian Ridge

mid-ocean ridges
Includes use on level 2 under ocean floors(1). Also search mid-oceanic ridges; mid-oceanic ridge; midoceanic ridges.
UF mid-oceanic ridge
    mid-oceanic ridges
    midoceanic ridges
SA aseismic ridges
    bottom features
    fracture zones
    median valley

mid-ocean ridges • Middle Triassic

    ocean floors
    ridges
    rift valleys
    sea-floor spreading

mid-oceanic ridge
  use mid-ocean ridges

mid-oceanic ridges
  use mid-ocean ridges

Mid-Pacific Rise
  Term introduced in 1978.
  SA Pacific Ocean

Midcontinent
Mid-America between the Appalachians and the Rocky Mountains. Index states as applicable. Also search Mid-continent.
- UF Mid-Continent
- BT United States
- SA Arkansas
    Colorado
    Illinois
    Indiana
    Iowa
    Kansas
    Kentucky
    Michigan
    Minnesota
    Missouri
    Montana
    Nebraska
    New Mexico
    North Dakota
    Ohio
    Oklahoma
    South Dakota
    Tennessee
    Texas
    Wisconsin
    Wyoming

Middle America Trench
Off W coast of Mexico and Central America.
- BT Pacific Ocean

Middle Atlas
Range of the Atlas Mountains N and E of the High Atlas Mountains in North Central.
- BT Morocco
- SA Atlas Mountains

Middle Cambrian
- BA Cambrian
- BT Paleozoic
- NA Acadian
    Barrandian
- NT Burgess Shale
    Flathead Sandstone
    Pagoda Formation
- SA Lower Cambrian
    Tintic Quartzite
    Upper Cambrian

Middle Carboniferous
- BA Carboniferous
- BT Paleozoic
- NA Westphalian

Middle Cretaceous
- BA Cretaceous
- BT Mesozoic
- NT Bhuj Series
- SA Albian
    Aptian
    Cenomanian
    Lower Cretaceous
    Middle Cretaceous
    Turonian
    Upper Cretaceous

Middle Devonian
- BA Devonian
- BT Paleozoic
- NT Cedar Valley Formation
    Columbus Limestone
- NA Couvinian
- NT Delaware Limestone
    Detroit River Group
- NA Eifelian
- NT Elk Point Group
- NA Givetian
- NT Hamilton Group
    Mahantango Formation
    Michelle Formation
    Nahanni Formation
    Onondaga Limestone
    Prairie Evaporite
    Saint Laurent Limestone
    Sylvania Formation
    Tioga Bentonite
    Tully Limestone
    Winnipegosis Formation
- SA Catskill Formation
    Genesee Group
    Lost Burro Formation
    Lower Devonian
    Martin Formation
    Muth Quartzite
    Traverse Group
    Upper Devonian

Middle East
An indefinite and unofficial term comprising a region including Cyprus, Egypt, and the countries of SW Asia. Index countries as applicable. Includes use on level 1 or 2 as an area term (list O). If 1, see term set options under list B.
- CO N290000N370000
    E0480000E0323000
- NA Cyprus
- NT Dead Sea
    Dead Sea Rift
- NA Iraq
    Israel
- NT Jerusalem
- NA Jordan
- NT Jordan River
    Jordan Valley
- NA Lebanon
- NT Levant
    Mesopotamia
    Red Sea Basin
    Samaria
- NA Syria
    Turkey
- NT Wadi Araba
- SA Arabian Peninsula
    Egypt
    Iran
    Kuwait
    Oman
    Qatar
    Sinjar Formation
    Southern Yemen
    United Arab Emirates
    Yemen
    Zubair Formation

middle Eocene
- BA Eocene
    Paleogene
    Tertiary
- BT Cenozoic
- NT Claiborne Group
    Cook Mountain Formation
    Laney Shale Member
    Queen City Formation
    Tyee Formation
- NA Ulatisian
- SA Biarritzian
    Castle Hayne Limestone
    Green River Formation
    lower Eocene
    Nanjemoy Formation
    Narizian
    Rose Canyon Formation
    Umpqua Formation
    upper Eocene

Middle Franconia
Part of old historical region of Franconia. A hilly region in Franconian Jura in N central.
- BT Bavaria
    West Germany

middle Holocene
Term introduced in 1978.
- BA Holocene
    Quaternary
- BT Cenozoic

Middle Jurassic
- BA Jurassic
- BT Mesozoic
- NA Bajocian
    Bathonian
    Dogger
- SA Aalenian
    Bowser Formation
    Lower Jurassic
    Upper Jurassic

middle Mesozoic
- BA Mesozoic
- SA Caracas Group
    upper Mesozoic

middle Miocene
- BA Miocene
    Neogene
    Tertiary
- BT Cenozoic
- NT Calvert Formation
    Choptank Formation
    Kirkwood Formation
    San Onofre Breccia
- NA Serravallian
- SA Badenian
    Hawthorn Formation
    lower Miocene
    Monterey Formation
    Saint Marys Formation
    Siwalik System
    upper Miocene

Middle Mississippian
- BA Mississippian
- BT Paleozoic

middle Oligocene
- BA Oligocene
    Paleogene
    Tertiary
- BT Cenozoic
- NT Marianna Limestone
    Vicksburg Group
- SA Brule Formation
    lower Oligocene
    upper Oligocene

Middle Ordovician
- BA Ordovician
- BT Paleozoic
- NT Bigby-Cannon Limestone
    Black River Group
- NA Blackriverian
    Champlainian
    Chazyan
- NT Cloridorme Formation
    Everton Formation
    Lexington Limestone
    Lincolnshire Limestone
- NA Llandeilian
    Llanvirnian
- NT Normanskill Formation
    Platteville Formation
    Saint Peter Sandstone
    Schenectady Formation
    Simpson Group
    Swan Peak Formation
    Trenton Group
- NA Trentonian
- NT Winnipeg Formation
- SA Antelope Valley Limestone
    Bala
    Chickamauga Group
    Lower Ordovician
    Martinsburg Formation
    Pogonip Group
    Upper Ordovician
    Viola Limestone

middle Paleocene
- BA Paleocene
    Paleogene
    Tertiary
- BT Cenozoic

middle Paleozoic
- BA Paleozoic

Middle Pennsylvanian
- BA Pennsylvanian
- BT Paleozoic
- NT Allegheny Group
- NA Atokan
- NT Breathitt Formation
    Carbondale Formation
- NA Desmoinesian
- NT Paradox Member
    Staunton Formation
    Strawn Series
- SA La Salle Limestone
    Lower Pennsylvanian
    Mansfield Formation
    Marble Falls Group
    Pottsville Group
    Upper Pennsylvanian

Middle Permian
- BA Permian
- BT Paleozoic
- NA Saxonian
- SA Kungurian
    Lower Permian
    Rotliegendes
    Upper Permian

middle Pleistocene
- BA Pleistocene
    Quaternary
- BT Cenozoic

middle Pliocene
- BA Pliocene
    Neogene
    Tertiary
- BT Cenozoic
- NT Bone Valley Formation
- SA lower Pliocene
    upper Pliocene

middle Precambrian
- BA Precambrian
- NA Carpentarian
    Karelian
    Svecofennian
- SA Aphebian

middle Proterozoic
- BA Proterozoic
    Precambrian

middle Quaternary
- BA Quaternary
- BT Cenozoic

Middle Silurian
- BA Silurian
- BT Paleozoic
- NT Clinton Group
    McKenzie Formation
- NA Niagaran
- NT Roberts Mountains Formation
    Rochester Formation
    Waldron Shale
- NA Wenlockian
- SA Lower Silurian
    Read Bay Formation
    Rockwood Formation
    Upper Silurian

Middle Stone Age
  use Mesolithic

middle Tertiary
- BA Tertiary
- BT Cenozoic
- SA lower Tertiary
    San Juan Formation
    upper Tertiary

Middle Triassic
- BA Triassic
- BT Mesozoic
- NA Anisian
    Ladinian
    Muschelkalk
- SA Lower Triassic
    Moenkopi Formation
    Serebryanka Formation

Upper Triassic

Middle Tunguska River
use Stony Tunguska River

Middle West (United States)
use Midwest

Middlesex County
Index states as applicable.
BX Connecticut
Massachusetts
New Jersey
Virginia
BT United States

Midland
City in Midland County in W.
BT Texas
United States

Midland Basin
W Texas. Also search Midland.
BT Texas
United States

Midland County
Index states as applicable.
BX Michigan
Texas
BT United States

Midlands
Region including the highly industrialized central counties.
CO N513000N533000
W0013000W0023000
BT England
Great Britain
United Kingdom

Midlothian
Formerly Edinburghshire, county. SE Scotland on S shore of Firth of Forth.
CO N554000N560000
W0024500W0034500
BT Scotland
Great Britain
United Kingdom

midoceanic ridges
use mid-ocean ridges

Midway
Two small islands (Eastern and Sand), parts of a low coral atoll under U.S. Navy control 1300 miles WNW of Honolulu. Also search Midway Atoll.
UF Midway Atoll
BT Pacific Ocean

Midway Atoll
use Midway

Midwest
A level 1 area term as of 1978. N part of central U.S. Region comprising states N of the Ohio and Missouri Rivers plus the E edge of the Great Plains including Kansas and Nebraska. Also search United States AND north-central or north central.
UF Middle West (United States)
North Central (United States)
North-Central (United States)
BA United States
SA Illinois
Indiana
Iowa
Kansas
Kentucky
Michigan
Minnesota
Missouri
Nebraska
North Dakota
Ohio
South Dakota
Wisconsin

Mie
Prefecture in S central Honshu.
UF Miye

BT Honshu
Japan

Miechow
Town in S part of country.
BT Krakow
Poland

Miette Complex
Devonian reef complex. Distinguish from Miette Group.
BT Devonian
SA Canada
Miette Group

Miette Group
Composed of a basal unit of argillite and argillaceous sandstone, succeeded by 2,000 feet of sandstone, grit, conglomerate, and argillite, which is overlain by 3,000 feet of argillite, and at the top by a carbonate. Distinguish from the Devonian Miette Complex.
BT Proterozoic
Precambrian
SA Alberta
British Columbia
Canada
Miette Complex

migmatite
BA migmatites
BT metamorphic rocks
SA migmatization

migmatites
Includes use on level 2 under metamorphic rocks(1). See list J.
BT metamorphic rocks
NA agmatite
ectinite
migmatite
SA migmatization
ptygmatic folds

migmatitization
use migmatization

migmatization
Includes use on level 3 under metamorphism(1).
UF migmatitization
BA metamorphism
SA migmatite
migmatites

migration
Includes use on level 3 under soils(1) genesis(2) and under petroleum(1) genesis(2).
SA petroleum
soils

migration of elements
Includes use on level 2 under metamorphism(1); on level 3 under geochemistry(1) and under weathering(1).
SA depletion
elements
enrichment
geochemistry
metamorphism
mobilization
thermal waters
weathering

Mihara (Mount)
use Mount Mihara

Mihara Volcano
use Mount Mihara

Mihara-Yama
use Mount Mihara

Miho Bay
Inlet of Sea of Japan in S.
BT Honshu
Japan

Mikabu System
According to the Japan Lexique, the age of Mikabu Series is Paleozoic. Consists of amphibole, pyroxene, semi-schists, and phyllites with limestones and quartzites.
BT Paleozoic
SA Honshu
Japan

Mikawa
Former province in central Honshu. Now part of Aichi Prefecture.
BT Honshu
Japan

Mikhailovka
City in Volgograd Oblast 110 miles NW of Volgograd in European USSR.
BT Russian Republic
USSR

Milam County
NE of Austin in central.
BT Texas
United States

Milan
City and province in NW part of country.
UF Milano
BT Lombardy
Italy

Milano
use Milan

milarite
BA ring silicates
silicates
BT minerals
SA aluminosilicates

Mildura
Town on the Murray River 300 miles NW of Melbourne in NW.
BT Victoria
Australia

Miliolacea
Includes use on level 2 under foraminifera(1). See list F.
BA Miliolina
foraminifera
BT Invertebrata
NA Miliolidae
Ophthalmidium
Quinqueloculina

Miliolidae
Family. Includes use on level 3 under foraminifera(1) Miliolacea(2).
BA Miliolacea
Miliolina
foraminifera
BT Invertebrata

Miliolina
Includes use on level 2 under foraminifera(1). See list F.
BA foraminifera
BT Invertebrata
NA Alveolinellidae
Miliolacea

Milk River Formation
Composed of well-sorted sandstones. In S Alberta.
BT Upper Cretaceous
Cretaceous
SA Alberta
Canada

Mill Creek
Rises in Lassen Volcanic National Park and flows into Sacramento River.
BT California
United States

Millard County
W central Utah.
BT Utah
United States

millerite
UF nickel pyrites
BA sulfides
BT minerals

Millstone Grit
Provincial Series, Europe.
BA Upper Carboniferous
Carboniferous
BT Paleozoic

Milne Land
An island in Scoresby Sound which is a deep inlet of Greenland Sea on E coast.
BT Greenland
Arctic region

Milwaukee
City on Lake Michigan in Milwaukee County in SE.
BT Wisconsin
United States

mimetite
BZ arsenates
halides
BT minerals
SA chlorides

Minas Basin
NE extension of Bay of Fundy in central Nova Scotia. Also search Minas AND appropriate area term.
BT Nova Scotia
Canada
SA Bay of Fundy
Cobequid Bay

Minas Gerais
State in E central.
CO S233000S140000
W0394500W0510000
BT Brazil
NT Araxa
Itabira
Ouro Preto
Pocos de Caldas
SA Areado Formation
Bambui Group
Bauru Formation
Brazilian Shield
Irati Formation
Sao Francisco Basin

Minas Series
BT Precambrian
SA Brazil

Mindanao
Second largest and southernmost major island.
BT Philippine Islands
Malay Archipelago

Mindel
Europe. Above Gunz, below Riss. Includes use on level 3 under age terms(1). See list E.
BA Pleistocene
Quaternary
BT Cenozoic

Mindel-Riss
use Mindel/Riss Interglacial

Mindel-Riss Interglacial
use Mindel/Riss Interglacial

Mindel/Riss Interglacial
Europe. Term applied in the Alps to the second interglacial stage of the Pleistocene Epoch, after the Mindel glacial stage and before the Riss. Also search Mindel-Riss; Mindel-Riss Interglacial.
UF Mindel-Riss
Mindel-Riss Interglacial
BA Pleistocene
Quaternary
BT Cenozoic

Mindoro
Island in central Philippines SW of Luzon.
BT Philippine Islands
SA Mansalay Formation

mineragraphy
A valid term through 1976. After

1976, use ore microscopy.
mineral assemblages
Includes use on level 2 under metamorphic rocks(1).
SA   assemblages
      autometamorphism
      metamorphic rocks

mineral associations
A valid term through 1977. Use mineral assemblages.

mineral collecting
Includes use on level 3 under mineralogy(1) and minerals(1). Also use collecting on level 3 under minerals(1).
SA   collecting
      collections
      mineralogy
      minerals

mineral composition
Includes use on level 3 under igneous rocks(1), sedimentary rocks(1), metamorphic rocks(1), and under soils(1) composition(2).
SA   composition

Mineral County
Index states as applicable.
BX   Colorado
      Montana
      Nevada
      West Virginia
BT   United States

mineral data
Includes use on level 2 under clay mineralogy(1); on level 3 under minerals(1).
SA   areal studies
      clay mineralogy
      data
      mineralogy
      minerals

mineral deposits
Do not include water resources or fuels. Used for substantial discussions of the genesis of ore deposits. Includes use on level 1 (list A) in combination with genesis (i.e. mineral deposits, genesis). Term set options are:
commodity (List C)
    topic
    controls
        type of control [geochemical controls, hydrogeological controls, lithologic controls, mechanical controls, paleogeographic controls, stratigraphic controls, structural controls]
    processes
        type of process [endogene processes, epigene processes, exhalative processes, exogene processes, hydrothermal processes, igneous processes, metamorphism, plate tectonics, sedimentary processes, supergene processes, syngenesis, volcanism, weathering]
topic [age, cause, concepts, environment, experimental studies, interpretation, mechanism, patterns, theoretical studies]
    subtopic
SA   alpine-type
      bauxitization
      controls
      deposits
      discoveries
      disseminated deposits
      economic geology
      endogene processes
      enrichment
      epigene processes
      epithermal processes
      exhalative processes
      exogene processes
      fluid inclusions
      gangue
      genesis
      geochemical controls
      geochemical prospecting
      geochemistry
      gold
      host rocks
      hydrogeological controls
      hydrothermal processes
      hydrothermal solutions
      hypogene processes
      igneous processes
      inclusions
      isotopes
      kuroko-type
      lithologic controls
      localization
      massive deposits
      mechanical controls
      metallogenic provinces
      metallogeny
      metasomatism
      mineral exploration
      mineral resources
      mineralization
      mining
      mississippi valley-type
      ore deposits
      ore sources
      ore transport
      ore-forming fluids
      oxidation zone
      paleogeographic controls
      paragenesis
      phase equilibria
      placers
      porphyry copper
      sedimentary processes
      stockwork deposits
      stratabound deposits
      stratiform deposits
      stratigraphic controls
      structural controls
      sublimation
      supergene processes
      syngenesis
      veins
      wallrock alteration
      weathering

mineral economics
Includes use on level 2 under commodities (list C); on level 3 under economic geology (1).
SA   economic geology
      economics
      mineral resources

mineral exploration
Used for substantial discussions of exploration for all "commodities" (list C) excluding fuel deposits and water resources. Includes use on level 1 (list A). Term set options are:
methods [biogeochemical methods, geobotanical methods, geochemical methods, geological methods, geomorphological methods, geophysical methods, hydrological methods, methods, photogeologic methods, remote sensing, statistical methods]
    topic
topic [applications, bibliography, concepts, history, instruments, objectives, ore guides, programs, techniques]
    subtopic
SA   biogeochemical methods
      dispersion patterns
      economic geology
      electromagnetic methods
      exploration
      geobotanical methods
      geochemical methods
      geochemical prospecting
      geochemistry
      geological methods
      geomorphological methods
      geophysical methods
      geophysical surveys
      haloes
      hydrological methods
      maps
      mineral deposits
      mineral resources
      minerals
      networks
      ore guides
      photogeologic methods
      placers
      primary dispersion
      programs
      remote sensing
      secondary dispersion
      sediment sampling
      statistical methods
      stream sediments

mineral facies
A valid term through 1976. Use facies.

mineral inclusions
Includes use on level 2 under inclusions(1).
BT   inclusions
SA   fluid inclusions
      host rocks
      xenoliths

mineral resources
For very general treatments. Includes use on level 1 as a commodity term (list C); on level 2 under ocean floors(1).
SA   economic geology
      environmental geology
      exploration
      mineral deposits
      mineral economics
      mineral exploration
      minerals
      ocean floors
      resources

mineral sequence
use paragenesis

mineral soap
use bentonite

mineral synthesis
A valid term through 1970. After 1970, see synthesis under crystal growth(1) and under phase equilibria(1).

mineral waters
Includes use on level 2 and 3 under springs(1).
SA   ground water
      springs
      thermal waters
      water

mineral zoning
A valid term through 1975. After 1975, use zoning.

mineralization
Restricted to ore deposits. Not used for paleontology.
SA   metallogenic provinces
      mineral deposits
      ore deposits
      ore-forming fluids

mineralogy
Used for the discipline as a whole. Includes use on level 1 (list A); on level 2 under areas (list B), under symposia(1), and under education(1). If 1, term set options are:
topic [applications, bibliography, catalogs, classification, concepts, education, experimental studies, history, instruments, methods, nomenclature, objectives, philosophy, practice, principles, symposia, textbooks, theoretical studies]
    subtopic
SA   atomic packing
      bibliography
      catalogs
      clay mineralogy
      collecting
      color centers
      crystal chemistry
      crystal growth
      crystal structure
      crystallography
      decrepitation
      lexicons
      microscope methods
      mineral collecting
      mineral data
      minerals
      optical methods
      optical mineralogy
      optical properties
      ore microscopy
      petrology
      phase equilibria
      reflectance
      reflectivity

minerals
Includes use on level 1 (list A); on level 2 under phase equilibria(1) and weathering(1). Used for descriptions of mineral occurrence or of minerals themselves. Also search mineral. If 1, term set options are:
mineral group (List L)
    mineral species
topic [chemical analysis, crystal chemistry, crystal structure, mineral data, occurrence, optical properties, spectra] (more than one single mineral)
    subtopic
NT   accessory minerals
      antimonates and antimonites
      arsenates
      arsenites
      authigenic minerals
      borates
      carbonates
      chromates
      coexisting minerals
      elongate minerals
      halides
      heavy minerals
      industrial minerals
      iodates
      light minerals
      magnetic minerals
      miscellaneous minerals
      mixed-layer minerals
      molybdates
      native elements and alloys
      new minerals
      nitrates
      opaque minerals
      ore minerals
      organic compounds
      oxides
      phosphates
      secondary minerals
      selenates and selenites
      silicates
      sulfates
      sulfides
      sulfosalts
      tellurates and tellurites
      tungstates
      uranium minerals
      vanadates
SA   authigenesis
      birefringence
      cell dimensions
      chemical analysis
      clay mineralogy
      cleavage
      crystal chemistry
      crystal field

crystal growth
crystal structure
crystallography
crystals
dehydration
diamonds
dimorphism
extinction
fluid inclusions
formula
gems
habit
hardness
impurities
inclusions
lamellae
localization
luminescence
metamict
metamictization
microhardness
mineral collecting
mineral data
mineral exploration
mineral resources
mineralogy
optical methods
optical properties
overgrowths
paragenesis
phenocrysts
pleochroism
polymorphism
polytypism
pseudomorphism
reflectance
reflectivity
refractive index
single-crystal method
spectra
standard materials
synthesis
synthetic materials
thermodynamic properties
transformations
typomorphism
unit cell

minerals in meteorites
Not a valid index term. Place minerals in appropriate group. See list L.

mines
As of 1978, term is used on level 2 under land subsidence(1), land use(1), and underground installations(1).
SA acid mine drainage
      land subsidence
      land use
      mining
      mining geology
      open-pit mining
      quarries
      solution mining
      underground installations

minette
Includes use on level 3 under igneous rocks(1) lamprophyre and carbonatite family(2). See list H.
BA lamprophyre and carbonatite family
BT igneous rocks

minicomputers
Includes use on level 3 under automatic data processing(1).
SA automatic data processing
      computers

mining
SA fines
      mineral deposits
      mines
      mining geology
      open-pit mining
      recovery
      solution mining
      strip mining
      tailings

mining geology
Geology applied to mining operations. Includes use on level 1 (list A). Term set options are:
topic [applications, bibliography, catalogs, classification, concepts, evaluation, history, instruments, methods, nomenclature, objectives, practice, production control, symposia, technology, textbooks]
    subtopic
SA dredging
      economic geology
      engineering geology
      environmental geology
      evaluation
      expeditions
      explosions
      gases
      geology
      geophysical methods
      geophysical surveys
      hydraulic fracturing
      leaching
      mines
      mining
      open-pit mining
      ore deposits
      production control
      quarries
      reclamation
      recovery
      solution mining
      strip mining
      technology

mining, open-pit
use open-pit mining

Minneapolis
City in Hennepin County in SE central.
BT Minnesota
      United States

Minnelusa Formation
Pennsylvanian and Permian. Composed of sandstones and evaporite deposits. W South Dakota, and NE Wyoming.
BT Paleozoic
SA Pennsylvanian
      Permian
      South Dakota
      Wyoming

Minnesota
Includes use on level 1 as an area term (list O). For term set options see list B.
CO N433000N490000
      W0894500W0971000
BA United States
NX Benton County
      Cass County
      Clay County
      Clearwater County
      Cook County
      Douglas County
NT Duluth
      Duluth Complex
NX Ely
      Grant County
NT Itasca County
NX Jackson County
      Lake County
      Lincoln County
      Marshall County
NT Mesabi Range
      Minneapolis
      Minnesota River valley
NX Murray County
NT New Ulm
NX Pennington County
      Polk County
      Pope County
NT Red Lake
NX Rice County
      Saint Louis County
NT Saint Paul
NX Scott County
      Stevens County
NT Two Harbors
      Vermilion Range
NX Washington County
SA Animikie Group
      Benton Formation
      Biwabik Iron Formation
      Cedar Valley Formation
      Dakota Formation
      Franconia Formation
      Great Lakes region
      Gunflint Iron Formation
      Keewatin
      Knife Lake Group
      Lake Agassiz
      Lake Superior
      Maquoketa Formation
      Midcontinent
      Midwest
      Mississippi River
      Mississippi Valley
      Niobrara Formation
      North Shore Volcanics
      Penokean Orogeny
      Pierre Shale
      Platteville Formation
      Rainy Lake
      Rainy River
      Saganaga Lake
      Saint Peter Sandstone
      Shakopee Formation
      Sioux Quartzite
      Thomson Slate
      Upper Mississippi Valley

Minnesota River valley
S Minnesota.
BT Minnesota
      United States

minor elements
Includes use on level 3. See list D.
SA elements
      major elements
      nutrients
      trace elements

minor planets
use asteroids

minor-element analyses
Includes use on level 3 under chemical analysis(1) methods (2). Before 1974, also search minor element analyses.
SA analysis
      chemical analysis
      elements
      major-element analyses
      trace-element analyses

Minorca
Second largest of Spain's Balearic Islands off E coast of Spanish mainland.
UF Menorca
BT Balearic Islands
      Mediterranean Sea.

Minsk
City and oblast in central.
BT Byelorussia
      USSR

Minturn Formation
Consists chiefly of grayish to greenish sandstones, conglomerates, and shales or siltstones. Central Colorado.
BT Pennsylvanian
SA Colorado

Minusinsk
Town in SW Krasnoyarsk Kray on the Yenisei River in SSW Siberia.
BT Russian Republic
      USSR

Minusinsk Basin
Coal basin in extreme S Krasnoyarsk Kray in SSW Siberia. A good agricultural region. Also search Minusinsk.
BT Russian Republic
      USSR

minyulite
BZ halides
      phosphates
BT minerals
SA fluorides

Miocene
World. Above Oligocene, below Pliocene. Includes use on level 1 as an age term (list E); on level 2 under paleo- terms, e.g. paleoecology, paleogeography, paleomagnetism; on level 3 under age terms(1).
BA Neogene
      Tertiary
BT Cenozoic
NT Arikaree Group
NA Badenian
      Barstovian
NT Fleming Formation
      Green Tuff Formation
      Hawthorn Formation
NA Helvetian
      Langhian
      lower Miocene
      middle Miocene
NT Monterey Formation
      Motojuku Formation
      Ngorora Formation
      Oktemberyan Series
NA Ottnangian
NT Pirabas Formation
      Saint Marys Formation
NA Sarmatian
      upper Miocene
      Vallesian
      Vindobonian
NT Wood Mountain Formation
SA Columbia River Basalt
      Cuddalore Series
      Hemphillian
      John Day Formation
      Loup Fork Group
      Meotian
      Neyveli Lignite
      Pannonian
      Poltava Series
      Siwalik System
      Snoqualmie Batholith
      Warkalli Formation

miogeosynclines
Includes use on level 3 under geosynclines(1) or tectonics(1).
UF miomagmatic zone
BT geosynclines
SA eugeosynclines
      tectonics

Miogypsina
Genus. Includes use on level 3 under foraminifera(1) Rotaliacea(2).
BA Miogypsinidae
      Rotaliacea
      Rotaliina
      foraminifera
BT Invertebrata

Miogypsinidae
Family. Includes use on level 3 under foraminifera(1) Rotaliacea(2).
BA Rotaliacea
      Rotaliina
      foraminifera
BT Invertebrata
NA Miogypsina

miomagmatic zone
use miogeosynclines

miospores
Includes use on level 2 under palynomorphs(1). See list F.
BA palynomorphs
NA Cicatricosisporites
      Cladophlebis

Classopollis
Dryas
Florinites
SA Artemisia
exine
megaspores
monolete taxa
spores
sporopollenin

mirabilite
BA sulfates
BT minerals

Mirador Formation
BT Eocene
SA Venezuela

Miramichi Bay
Inlet of Gulf of Saint Lawrence. Also search Miramichi.
BT New Brunswick
Canada

Miranda
State on the Caribbean Sea.
BT Venezuela

Mirgalimsay
Village in S near Syr Darya River.
BT Kazakhstan
USSR

Mirny
use Mirnyy

Mirnyy
USSR IGY station on Davis Sea near Shackleton Ice Shelf on the Indian Ocean side. Also search Mirny.
UF Mirny
BT Antarctica

mirrorstone
use muscovite

Mirsk
Town near Czechoslovak border.
BT Wroclaw
Poland

Mirzapur
City in SE.
BT Uttar Pradesh
India

miscellanea
Includes use on level 2 under fossil group(1) if fossil is not yet classified. See list F.

miscellaneous minerals
Includes use on level 2 under minerals(1) for the case of several mineral groups when the topic is not emphasized or for a mineral of unknown affinity. See list L.
BT minerals
SA uranium minerals

miscibility
use immiscibility

miscibility gap
SA phase equilibria

Mishash Formation
BT Cretaceous
SA Israel

Missi Group
Its basal unit is normally composed of a conglomerate consisting of angular to subrounded boulders or pebbles along with minor amounts of chert, jasper and quartz. The unit grades upwards to graywacke and subgraywacke with thin beds of conglomerate. In the Flin Flon area of W Manitoba and E Saskatchewan.
BT Precambrian
SA Canada
Manitoba
Saskatchewan

Mississippi
Includes use on level 1 as an area term (list O). For term set options see list B.
CO N301500N350000
W0880500W0914000
BA United States
NX Adams County
Benton County
Calhoun County
Carroll County
Clarke County
Clay County
Franklin County
Greene County
Hancock County
NT Hattiesburg
NX Jackson County
Jasper County
Jefferson County
Lafayette County
Lawrence County
Lee County
Lincoln County
Madison County
Marion County
Marshall County
Monroe County
Montgomery County
Newton County
NT Pascagoula River basin
NX Perry County
Pike County
Pontotoc County
NT Rankin County
NX Scott County
Smith County
NT Tishomingo
NX Union County
Warren County
Washington County
Wayne County
Webster County
Wilkinson County
SA Amite River
Arkansas River
Black Warrior Basin
Buckner Formation
Chattanooga Shale
Chickasawhay Formation
Citronelle Formation
Claiborne Group
Clayton Formation
Cotton Valley Group
Fort Payne Formation
Grenada
Gulf Coastal Plain
Jackson Group
Marianna Limestone
Mississippi Embayment
Mississippi River
Mississippi Valley
Pottsville Group
Ripley Formation
Smackover Formation
Talladega Group
Tuscaloosa Formation
Vicksburg Group
Warsaw Formation
Wilcox Group
Yazoo Clay

Mississippi Delta
On Gulf of Mexico 100 miles SE of New Orleans. Also search Mississippi River delta.
UF Mississippi River delta
BT Louisiana
United States
SA Mississippi River

Mississippi Embayment
A geosynclinal area including a section of the US Gulf Coast and extending northward up the Mississippi Valley. Index states as applicable.
BT United States
SA Alabama
Arkansas
Illinois
Kentucky
Mississippi
Mississippi Valley
Missouri
Tennessee

Mississippi River
Rises in Lake Itasca in NW Minnesota and flows SE and S into the Gulf of Mexico. Index states as applicable.
BT United States
SA Arkansas
Illinois
Iowa
Kentucky
Louisiana
Minnesota
Mississippi
Mississippi Delta
Mississippi River basin
Mississippi Valley
Missouri
Tennessee
Wisconsin

Mississippi River basin
Drainage basin including the Ohio and Missouri river systems comprises 2/5 of the total U.S. area covering all or parts of 31 states plus S Alberta and Saskatchewan. It drains much of the interior lowlands, Great Plains, central Gulf coastal plain, a portion of the Appalachians, and the Rocky Mountains W to the Continental Divide. Index countries as applicable.
SA Canada
Mississippi River
United States

Mississippi River delta
use Mississippi Delta

Mississippi River valley
use Mississippi Valley

Mississippi Valley
A level 1 area term as of 1978. River valley. Also search Mississippi River valley. Index states as applicable.
UF Mississippi River valley
BA United States
NA Lower Mississippi Valley
Upper Mississippi Valley
SA Arkansas
Illinois
Iowa
Kentucky
Louisiana
Minnesota
Mississippi
Mississippi Embayment
Mississippi River
Missouri
Tennessee
Wisconsin

mississippi valley type
use mississippi valley-type

mississippi valley-type
Also search mississippi valley type.
UF mississippi valley type
SA lead
lead-zinc deposits
mineral deposits
zinc

Mississippian
Includes use on level 1 as an age term (list E); on level 2 under paleoterms, e.g. paleoecology, paleogeography, paleomagnetism. After the Devonian and before the Pennsylvanian. Approximate equivalent of Lower Carboniferous of European usage.
BA Paleozoic
NT Bear Gulch Limestone Member
Borden Group
Jackfork Group
Leadville Formation
Logan Formation
NA Lower Mississippian
NT Madison Group
NA Middle Mississippian
NT Monte Cristo Limestone
Newman Limestone
Price Formation
Rampart Group
Redwall Limestone
Stanley Group
NA Upper Mississippian
Valmeyeran
NT Windsor Group
Yule Marble
SA Antler Orogeny
Bedford Shale
Berea Sandstone
Bird Spring Formation
Carboniferous
Catskill Formation
Chattanooga Shale
Johns Valley Formation
Lisburne Group
Lower Carboniferous
Pennsylvanian
Springer Formation
Woodford Shale

Missoula Group
Includes numerous named formations: among these are Marsh Shale in Helena region; Striped Peak and Libby formations in NW Montana, five formations near Missoula, and others in and S of Glacier National Park. In N Idaho and NW Montana.
BT Precambrian
SA Belt Supergroup
Bonner Formation
Idaho
Montana

Missouri
Includes use on level 1 as an area term (list O). For term set options see list B.
CO N360000N403500
W0890500W0954500
BA United States
NX Barton County
Benton County
Boone County
Buchanan County
Butler County
Carroll County
Carter County
Cass County
Clark County
Clay County
Dade County
Dallas County
NT Decaturville
Devils Icebox
NX Douglas County
Franklin County
Greene County
Grundy County
Iron County
Jackson County
Jasper County
Jefferson County
Johnson County
NT Joplin
NX Lafayette County
Lawrence County
Lincoln County
Livingston County
Macon County
Madison County
Marion County
Mercer County
Monroe County
Montgomery County
Morgan County
NT New Madrid
NX Newton County
Osage County

Missouri ● modulus of rigidity

Perry County
Pike County
Polk County
Pulaski County
Putnam County
Randolph County
Saint Clair County
- NT Saint Francois Mountains
  Saint Louis
- NX Saint Louis County
  Scott County
  Shannon County
  Sullivan County
  Texas County
  Warren County
  Washington County
  Wayne County
  Webster County
- SA Bainbridge Formation
  Bonneterre Formation
  Cabaniss Formation
  Cedar City Formation
  Chattanooga Shale
  Cherokee Group
  Chesterian
  Clayton Formation
  Desmoinesian
  Douglas Group
  Ervine Creek Limestone
  Everton Formation
  Excello Shale
  Fayetteville Formation
  Fernvale Formation
  Forest City Basin
  Grand River
  Kansas City Group
  Keokuk Limestone
  Labette Shale
  Lawrence Formation
  Lecompton Limestone
  Maquoketa Formation
  Marmaton Group
  Midcontinent
  Midwest
  Mississippi Embayment
  Mississippi River
  Mississippi Valley
  Missouri River
  Missouri River basin
  Missouri River valley
  Oread Limestone
  Ozark Mountains
  Plattsburg Limestone
  Plattsmouth Limestone Member
  Pleasanton Group
  Red Eagle Limestone
  Richmond Group
  Ripley Formation
  Rock Lake Shale Member
  Saint Laurent Limestone
  Saint Louis Limestone
  Saint Peter Sandstone
  Sainte Genevieve Limestone
  Salem Limestone
  Smithville Formation
  Springfield
  Stanton Formation
  Tonganoxie Sandstone
  Upper Mississippi Valley
  Versailles
  Wabaunsee Group
  Warsaw Formation
  White River
  Wilcox Group
  Wyandotte Limestone

Missouri River
Rises in S Montana and flows E and SE to join the Mississippi River N of St. Louis. Index states as applicable.
- BT United States
- SA Iowa
  Kansas
  Missouri
  Missouri River basin
  Missouri River valley
  Montana

  Nebraska
  North Dakota
  South Dakota

Missouri River basin
Drainage basin. Index states and provinces as applicable.
- SA Alberta
  Colorado
  Iowa
  Kansas
  Missouri
  Missouri River
  Montana
  Nebraska
  North Dakota
  Saskatchewan
  South Dakota
  Wyoming

Missouri River valley
Index states as applicable.
- BT United States
- SA Iowa
  Kansas
  Missouri
  Missouri River
  Montana
  Nebraska
  North Dakota
  South Dakota

Missourian
Provincial series, North America: lower Upper Pennsylvanian, above Desmoinesian, below Virgilian. Includes use on level 3 under age terms(1). See list E.
- UF Missourian Series
- BA Upper Pennsylvanian
  Pennsylvanian
- BT Paleozoic
- NT Kansas City Group
  Lansing Group
  Plattsburg Limestone
  Pleasanton Group
  Rock Lake Shale Member
  Seminole Formation
  Stanton Formation
  Wann Formation
  Wyandotte Limestone

Missourian Series
use Missourian

Mistastin Lake
N Labrador.
- BT Labrador
  Canada

Mitchell County
Index states as applicable.
- BX Georgia
  Iowa
  Kansas
  North Carolina
  Texas
- BT United States

Mitsuishi
Town in SW Honshu. Sometimes spelled Mituisi.
- UF Mituisi
- BT Honshu
  Japan

Mitterberg
Village in W central part of country.
- BT Salzburg
  Austria

Mituisi
use Mitsuishi

Miura Peninsula
SE Honshu. Extends into Sagami Sea S of Yokohama and Tokyo Bay.
- BT Honshu
  Japan

mixed crystals
use solid solution

mixed-layer minerals
Term introduced in 1978. Before 1978, search clay minerals AND mixed-layer.
- BT minerals
- SA clay minerals

mixing
Used as a general term.
- SA processes
  separation

mixite
- BA arsenites
- BT minerals
- SA mixtite

mixtite
- BA clastic rocks
- BT sedimentary rocks
- SA diamictite
  mixite

Miyagi
Prefecture in N.
- BT Honshu
  Japan

Miyake-Jima
Island of Izu-shichito group in Greater Tokyo, S of O-shima Island and Sagami Sea. Active volcano on island.
- UF Miyaki Island
  Miyaki-Jima Island
- BT Honshu
  Japan
- SA Izu-shichito

Miyaki Island
use Miyake-Jima

Miyaki-Jima Island
use Miyake-Jima

Miyazaki
City and Prefecture in SE.
- BT Kyushu
  Japan

Miye
use Mie

Mizunami
Town in central.
- BT Honshu
  Japan

Mizusawa
Town in N.
- BT Honshu
  Japan

MM scale
use modified Mercalli scale

Mn
use manganese

Mn-53
Includes use on level 3 under isotopes(1).
- SA isotopes
  manganese

Mn-54
Includes use on level 3 under isotopes(1).
- SA isotopes
  manganese

Mo
use molybdenum

Moab
City in Grand County in E.
- BT Utah
  United States

Moapa Valley
NE of Las Vegas in Clark County in S.
- BT Nevada
  United States

Mobile
City on Mobile Bay in Mobile County in S.
- BT Alabama

  United States

mobile belts
Includes use on level 3 under geosynclines(1).
- UF belts, mobile
- SA crust
  geosynclines
  tectonics

Mobile County
On Mobile Bay in S.
- BT Alabama
  United States

mobility
Used as a general term.
- SA erosion
  tectonics

mobilization
Restricted to geochemical meaning.
- SA deformation
  geochemistry
  migration of elements

Mocamedes
use Mossamedes

Moctezuma
Town in E central.
- BT Sonora
  Mexico

modal analysis
Includes use on level 3 under mineralogy(1) methods(2), under petrology(1) methods(2), and under sedimentary petrology(1) methods(2).
- SA analysis
  petrology

models
Includes use on level 2 under ground water(1).
- NA mathematical models
  one-dimensional models
  physical models
  three-dimensional models
  two-dimensional models
- SA automatic data processing
  simulation

Modena
City and province in central.
- BT Emilia-Romagna
  Italy

modern
Used for present-day studies in conjunction with Holocene.
- SA Holocene

modified Mercalli scale
Term introduced in 1978. Includes use on level 3 under earthquakes(1).
- UF Mercalli scale, modified
  MM scale
- SA earthquakes
  intensity
  seismology

Modoc County
In extreme NE. Also search Modoc.
- BT California
  United States

Modoc Plateau
High, semiarid, and volcanic plateau in NE California. Also search Modoc.
- BT California
  United States

modulus, bulk
use bulk modulus

modulus of compression
use compressibility

modulus of elasticity
use elastic constants

modulus of incompressibility
use bulk modulus

modulus of rigidity
use shear modulus

modulus, shear
use shear modulus

Moenkopi Formation
Lower and middle (?) Triassic or Triassic (?). In SW Utah, subdivided into (ascending) Timpoweap, Lower Red, Virgin Limestone, Middle Red, Shnabkaib, and Upper Red members. Arizona, California, Colorado, Nevada, and S Utah.
BT Triassic
SA Arizona
California
Colorado
Lower Triassic
Middle Triassic
Nevada
Utah

Moesia
Ancient region S of the lower Danube. Includes Bulgaria, SE Romania, and Serbia. Index Serbia and countries as applicable.
BT Europe
SA Bulgaria
Romania
Serbia

Moesian Platform
Index countries as applicable.
BT Europe
SA Bulgaria
Romania

Mogilev
City on the Dnieper River. Also an oblast.
UF Mogilev on the Dnieper
BT Byelorussia
USSR

Mogilev on the Dnieper
use Mogilev

Mogilno
Town in NW central.
BT Bydgoszcz
Poland

Mogollon Plateau
Tableland S of Winslow in E central.
BT Arizona
United States

Mohave County
S of Lake Mead in NW.
BT Arizona
United States

Mohave Desert
use Mojave Desert

Mohawk River Valley
use Mohawk Valley

Mohawk Valley
River valley in E central New York.
UF Mohawk River Valley
BT New York
United States

Moho
use Mohorovicic discontinuity

Mohorovicic discontinuity
Includes use on level 1 (list A) for discussions which stress the crust-mantle transition. Term set options are:
topic [causes, composition, depth, detection, extent, geometry, identification, interpretation, Mohole project, observations, patterns, properties, structure, surveys, temperature]
subtopic (no area term)
UF M-discontinuity
Moho
SA asthenosphere
basement
continental drift
core
crust

depth
discontinuities
Earth
earthquakes
elastic waves
isostasy
lithosphere
mantle
plate tectonics
sea-floor spreading
seismology
tectonophysics

Moinian
Europe.
BA upper Precambrian
Precambrian

moissanite
Meteorite mineral.
BA native elements and alloys
BT minerals
SA carbides

moisture
Includes use on level 3 under soils(1). Also search moisture content.
UF moisture content
SA atmospheric precipitation
desiccation
humidity
soils
water
water vapor

moisture content
use moisture

Mojave Desert
Arid basin in S California. Also search Mojave.
UF Mohave Desert
BT California
United States

Mokoia
Village in SW.
BT North Island
New Zealand

molasse
Includes use on level 3 under sedimentary rocks(1) clastic rocks(2). See list I.
BA clastic rocks
BT sedimentary rocks
SA terrigenous materials

Molasse Basin
A geological and geographic term. Index countries as applicable as an example of such a basin. Also search molasse in combination with country name.
BT Europe
SA Austria
Germany
Switzerland

Moldanubian
Also search Moldanubian Complex; Moldanubian Zone; Moldanubicum; Moldanubicum Complex; Moldanubikum.
UF Moldanubian Complex
Moldanubian Zone
Moldanubicum
Moldanubicum Complex
Moldanubikum
BT Precambrian
SA Archean
Czechoslovakia

Moldanubian Complex
use Moldanubian

Moldanubian Zone
use Moldanubian

Moldanubicum
use Moldanubian

Moldanubicum Complex
use Moldanubian

Moldanubikum
use Moldanubian

Moldava Valley
River valley. Also search Moldava River and Moldova Valley.
UF Moldova Valley
BT Moldavia
Romania

Moldavia
Region E of Transylvania and N of E Walachia in Romania, and the Moldavian Soviet Socialist Republic in the USSR. Index countries as applicable.
CO N450000N480000 E0300000E0350000
BT Europe
NT Bacau
Bessarabian
Bicaz Valley
Bistrita Valley
Gutii Mountains
Iacobeni
Moldava Valley
Moldavian Platform
Piatra-Neamt
Rarau Massif
Suceava
Teleajen Valley
Tiraspol
Trotus Valley
SA Dniester River
Dniester-Prut Interfluve
European USSR
Prut River
Romania
Romanian Plain
Russian Plain
Russian Platform
Siret River
USSR

Moldavian Plateau
use Moldavian Platform

Moldavian Platform
Also search Moldavian Plateau.
UF Moldavian Plateau
BT Moldavia
Romania

moldavite
BA tektites

Moldova Noua
Town in E near the Serbian border. Also search Moldova-Noua.
UF Moldova-Noua
BT Banat
Romania

Moldova Valley
use Moldava Valley

Moldova-Noua
use Moldova Noua

molecular structure
SA crystal structure
structure

mollisol
use active layer

Mollisols
Includes use on level 3 under soils(1). See list M.
BA soils

Mollusca
Includes use on level 1 and 2 as a fossil term (list F).
BT Invertebrata
NA Bivalvia
Cephalopoda
Gastropoda
Monoplacophora
Polyplacophora
Scaphopoda
SA Dacryoconarida
Hyolithidae
mollusks
pearls

Tentaculites
Tentaculitida
Tentaculitidae

mollusks
SA fossiliferous materials
Mollusca

Molodezhnaya Station
USSR station on Alasheyev Bight in Enderby Land between Lutzow-Holm Bay and Cape Ann on the Indian Ocean side. Also search Molodezhnaya.
BT Antarctica

Molokai
Island in the Hawaiian Islands between Oahu and Maui.
BT Hawaii
United States

Molotov
City W of the Urals in Perm Oblast.
BT Russian Republic
USSR
SA Perm

Molteno
Town in E.
BT Cape Province
South Africa

Moluccas
Group of islands between Celebes and New Guinea.
UF Spice Islands
BT Indonesia

Moluya
use Moulouya River

molybdates
Includes use on level 2 under minerals(1). See list L.
BT minerals
NA ferrimolybdite
powellite
umohoite
wulfenite
SA tungstates

molybdenite
BA sulfides
BT minerals
SA molybdenum

molybdenum
Includes use on level 1 and 2 as a commodity term (list C) and as a chemical element (list D); on level 2 under mineral deposits, genesis(1) and paragenesis(1).
UF Mo
SA elements
molybdenite

Monaco
On the Mediterranean Sea near the French-Italian border. Includes use on level 1 as an area term (list O). For term set options see list B.
BA Europe
NT Monte Carlo

monazite
Includes use on level 1 as a commodity term (list C); on level 2 under placers(1).
BA phosphates
BT minerals
SA heavy minerals
placers

Monchegorsk
City in Murmansk Oblast about 70 miles S of Murmansk in extreme NW European USSR.
BT Russian Republic
USSR

monchiquite
Includes use on level 3 under igneous rocks(1) lamprophyre and carbonatite family. See list H.
BA lamprophyre and carbonatite

family
BT igneous rocks

Moneron Island
Soviet island in Japan Sea 30 miles off SW Sakhalin.
BT Russian Republic
USSR

monetite
BA phosphates
BT minerals

Monghyr
City in NE central.
BT Bihar
India

Mongolia
Includes use on level 1 as an area term (list O). For term set options see list B.
CO N420000N520000
E1200000E0870000
UF Outer Mongolia
BA Asia
NT Altan Teeli
Bayn Dzak
Choybalsan
Hangay Mountains
Ikh-Khayrkhan
Khubsugul
Kobdo
Mongolian Altai
Selenga
Ulan Bator
SA Altai Mountains
Altai-Sayan region
Amur Basin
Djadokhta Formation
Far East
Gobi Desert
Kerulen River
Selenga River valley
Tannu-Ola Range
Yenisei Basin
Yenisei-Khatanga basin

Mongolian Altai
Southeastern extension of the Altai Mountains.
BT Mongolia
SA Altai Mountains

monitoring
SA geophysical surveys

Monmouth County
On the Atlantic Ocean in E.
BT New Jersey
United States

Monmouth Group
Includes (ascending) Mount Laurel, Navesink, Red Bank, and Tinton formations. Delaware, NE Maryland, and New Jersey.
BT Upper Cretaceous
Cretaceous
SA Delaware
Maryland
Navesink Formation
New Jersey

Monmouthshire
County in SE.
CO N513000N520000
W0024000W0032000
BT Wales
Great Britain
United Kingdom

Mono Basin
Mono County in E.
BT California
United States

Mono County
Along the Nevada border in E.
CO N373000N384000
W1175000W1193000
BT California
United States

Mono Craters
Range of about 20 geologically recent volcanic cones just S of Mono Lake, Mono County in E.
BT California
United States

Mono Lake
In central Mono County in E.
BT California
United States

monoclines
Not valid term from 1975 through 1977. After 1977, includes use on level 3 under folds(1) style(2). Before 1978, also search folds AND monoclinal.
BA folds
SA flexure folds

monoclinic system
Term introduced in 1978. Before 1978, search monoclinic.
SA crystallography

Monocotyledoneae
Includes use on level 2 under angiosperms(1). See list F.
BA angiosperms
BT Plantae
NA Gramineae
Palmae
SA Dicotyledoneae

Monograptina
Includes use on level 2 under Graptolithina(1). See list F.
BA Graptoloidea
Graptolithina
BT Invertebrata
NA Monograptus
SA Hemichordata

Monograptus
Genus. Includes use on level 3 under Graptolithina(1) Monograptina(2).
BA Monograptina
Graptoloidea
Graptolithina
BT Invertebrata
NA Monograptus uniformis

Monograptus uniformis
Includes use on level 3 under Graptolithina(1) Monograptina(2).
BA Monograptus
Monograptina
Graptoloidea
Graptolithina
BT Invertebrata

monohydrocalcite
BA carbonates
BT minerals

monolete taxa
Term introduced in 1978. Before 1978, also search monoletes.
UF monoletes
SA miospores
palynomorphs

monoletes
use monolete taxa

Monomoy Island
A 10 mile sandspit S of Chantham on Cape Cod. Sometimes it is connected to land and is not an island.
BT Massachusetts
United States

Monongahela Group
In Pennsylvania, contains Pittsburgh Coal Seam at base; top of Waynesburg Coal marks upper boundary. W Maryland, E Ohio, W Pennsylvania, W Virginia, and West Virginia.
BT Pennsylvanian
SA Maryland
Ohio
Pennsylvania
Virginia
West Virginia

Monoplacophora
Includes use on level 2 under Mollusca(1). See list F.
BA Mollusca
BT Invertebrata

Monotis
Genus. Includes use on level 3 under Mollusca(1) Bivalvia(2).
BA Bivalvia
Mollusca
BT Invertebrata

Monroe County
Index states as applicable.
BX Alabama
Arkansas
Florida
Georgia
Illinois
Indiana
Iowa
Kentucky
Michigan
Mississippi
Missouri
New York
Ohio
Pennsylvania
Tennessee
West Virginia
Wisconsin
BT United States

Mons Basin
Coal mining area near the French border. Also search Mons.
BT Hainaut
Belgium

monsoons
SA hurricanes
storms

Mont Blanc
In Savoy Alps near Italian border. Highest mountains in Alps.
BT Haute-Savoie
France
SA Savoy Alps

Mont Dore
use Mont-Dore

Mont-Dore
Town and thermal station in S central part of country. Also search Mont Dore.
UF Le Mont-Dore
Mont Dore
Mont-Dore-les-Bains
BT Puy-de-Dome
France

Mont-Dore-les-Bains
use Mont-Dore

Mont-Saint-Michel Bay
use Bay of Saint-Michel

Montagne Noire
Southernmost range of Central Massif. Index departments as applicable.
BT France
SA Aude
Central Massif
Tarn

Montague Island
At entrance of Prince William Sound, SE of Anchorage in S.
BT Alaska
United States

Montana
Includes use on level 1 as an area term (list O). For term set options see list B.
CO N443000N490000
W1040200W1160200
BA United States
NT Beaverhead County
Bell Creek Field
Belt Basin
Big Belt Mountains
NX Big Horn County
NT Big Snowy Mountains
Billings
NX Blaine County
NT Boulder Batholith
Broadwater County
Butte
Butte District
NX Carbon County
Carter County
Custer County
NT Fergus County
Flathead Lake
Flint Creek Range
Gallatin County
NX Garfield County
NT Glacier National Park
Granite County
Hebgen Lake
Helena
NX Jefferson County
Lake County
NT Lewis and Clark County
NX Lincoln County
NT Little Belt Mountains
NX Madison County
NT Meagher County
Medicine Lake
NX Mineral County
NT Mount Morgan
NX Park County
Phillips County
Powell County
NT Rattlesnake Mountain
NX Roosevelt County
NT Ruby Range
NX Sheridan County
NT Shonkin Sag Laccolith
Silver Bow County
NX Teton County
NT Three Forks
Tobacco Root Mountains
NX Valley County
NT Vulture Mountain
SA Absaroka Range
Altyn Limestone
Arikaree Group
Bear Gulch Limestone Member
Bearpaw Formation
Beartooth Mountains
Beaverhead Formation
Beaverhead Mountains
Belt Supergroup
Benton Formation
Bighorn Basin
Bighorn Mountains
Bitterroot Range
Bonner Formation
Carlile Shale
Colorado Group
Columbia River basin
Cordillera
Dakota Formation
Deadwood Formation
Elk Point Group
Flathead Sandstone
Fort Union Formation
Fox Hills Formation
Frontier Formation
Gallatin Range
Graneros Shale
Great Plains
Greenhorn Limestone
Hell Creek Formation
Idaho Batholith
Judith River Formation
Lance Formation
LASA
Little Missouri River basin
Lodgepole Formation
Madison Group
Midcontinent
Missoula Group
Missouri River
Missouri River basin

Missouri River valley
Montana Group
Morrison Formation
Mowry Shale
Muddy Sandstone
Niobrara Formation
Overthrust Belt
Pagoda Formation
Parkman Sandstone
Phosphoria Formation
Pierre Shale
Pilgrim Formation
Powder River basin
Prichard Formation
Purcell System
Ravalli Group
Red River Formation
Rocky Mountain Trench
Sentinel Butte Formation
Stillwater Complex
Sundance Formation
Sweetgrass Arch
Tensleep Sandstone
Thaynes Formation
Tongue River Formation
Wallace Formation
Wasatch Formation
Western Interior
White River Group
Williston Basin
Winnipeg Formation
Winnipegosis Formation
Yellowstone National Park

Montana Group
Includes Bearpaw Shale, Claggett Shale, Cody Shale (upper part), Eagle Sandstone, Fox Hills Sandstone, Horsethief Sandstone, Judith River Formation, Lennep Sandstone, Parkman Sandstone, Pierre Shale, Trinidad Sandstone, Two Medicine Formation, and Virgelle Sandstone.
 BT Upper Cretaceous
  Cretaceous
 SA Bearpaw Formation
  Cody Shale
  Colorado
  Judith River Formation
  Kansas
  Montana
  New Mexico
  North Dakota
  Parkman Sandstone
  Pierre Shale
  South Dakota
  Utah
  Wyoming

Montastrea
Genus. Includes use on level 3 under Coelenterata(1) Scleractinia(2).
 BA Scleractinia
  Anthozoa
  Coelenterata
 BT Invertebrata

Montauban
 use Orgueil

Monte Amiata
Extinct volcano in the Apennines.
 BT Tuscany
  Italy

Monte Antola
Peak in Ligurian Apennines in NW part of country.
 BT Liguria
  Italy
 SA Ligurian Apennines

Monte Baldo
Mountain range between Lake Garda and Adige River in NE part of country.
 BT Veneto
  Italy

Monte Capanne
W end of island of Elba in Mediterranean Sea off Tuscany.
 BT Tuscany
  Italy

Monte Carlo
Resort on the Ligurian Sea in the Riviera.
 BT Monaco

Monte Cristo Limestone
Lower and Upper Mississippian. Consists of (ascending) Dawn Limestone, Anchor Limestone, Bullion Limestone, and Yellowpine Limestone members. SE California and SE Nevada. Index states as applicable.
 BT Mississippian
 SA California
  Lower Mississippian
  Nevada
  Upper Mississippian

Monte Gargano
 use Gargano

Monte Pisano
Mountain group near Pisa in NW.
 BT Tuscany
  Italy

Monte Rosa
Highest mountain group in the Pennine Alps. Index countries as applicable.
 BT Europe
 SA Italy
  Pennine Alps
  Switzerland

Monte Somma
Simicircular ridge on N and E sides of Vesuvius in E.
 BT Campania
  Italy

Monteagle Limestone
 BT Upper Mississippian
  Mississippian
 SA Alabama
  Tennessee

montebrasite
 BA phosphates
 BT minerals
 SA amblygonite

Montenegro
Consituent republic in SW.
 BT Yugoslavia
 NT Niksic

Monterey
City in Monterey County on the Pacific Ocean.
 BT California
  United States

Monterey Bay
Inlet of Pacific Ocean in Santa Cruz and Monterey Counties.
 BT California
  United States

Monterey Canyon
Submarine canyon just off coast of central California.
 UF Monterey Deep-Sea Channel
  Monterey Gorge
  Monterey Seavalley
  Monterey Submarine Canyon
  Monterey Trough
 BT Pacific Ocean

Monterey County
On Pacific Ocean in W.
 BT California
  United States

Monterey Deep-Sea Channel
 use Monterey Canyon

Monterey Deep-Sea Fan
 use Monterey Fan

Monterey Fan
Off coast of central California just to the W of the Monterey Canyon.
 UF Monterey Deep-Sea Fan
 BT Pacific Ocean

Monterey Formation
Middle and upper Miocene. Comprises (ascending) Gould Shale, Devilwater Silt, McDonald Shale, Antelope Shale, and Chico-Martinez Chert members. W California.
 BT Miocene
 SA California
  middle Miocene
  upper Miocene

Monterey Gorge
 use Monterey Canyon

Monterey Seavalley
 use Monterey Canyon

Monterey Submarine Canyon
 use Monterey Canyon

Monterey Trough
 use Monterey Canyon

Montes de Toledo
Mountain range in S central part of country. Index provinces as applicable.
 BT Spain
 SA Ciudad Real
  Toledo

Montesano Formation
Largely massive coarse-grained light-brown sandstones, with many intercalated lenses of conglomerate and grit; shales subordinate in lower part but common in upper part. SW and NW Washington.
 BT upper Miocene
  Miocene
 SA Washington

Montevideo
Department in S Uruguay. Also a city on the Rio de la Plata.
 BT Uruguay

Montgomery County
Index states as applicable.
 BX Alabama
  Arkansas
  Georgia
  Illinois
  Indiana
  Iowa
  Kansas
  Kentucky
  Maryland
  Mississippi
  Missouri
  New York
  North Carolina
  Ohio
  Pennsylvania
  Tennessee
  Texas
  Virginia
 BT United States

Monti Berici
Range of volcanic hills SW of Vicenza in NE part of country.
 BT Veneto
  Italy

Monti Caronie
 use Nebrodi Mountains

Monti Lessini
Mountain group separated from Monte Baldo group by Adige River in NE part of country.
 BT Veneto
  Italy

Montian
Europe. Above Danian, below Thanetian. Includes use on level 3 under age terms(1). See list E.
 BA Paleocene
  Paleogene
  Tertiary
 BT Cenozoic

monticellite
 BA orthosilicates
  silicates
 BT minerals
 SA olivine group

Monticello
City in San Juan County in SW.
 BT Utah
  United States

montmorillonite
 BA sheet silicates
  silicates
 BT minerals
 SA beidellite
  clay minerals
  nontronite
  smectite

Montpellier
City in S part of country.
 BT Herault
  France

Montreal
City on an island in the Saint Lawrence River.
 BT Quebec
  Canada

Monts des Maures
 use Maures Massif

Monts Dome
Division of Auvergne Mountains in S central part of country.
 UF Chaine des Puys
 BT Puy-de-Dome
  France

Monts du Lyonnais
Index departments as applicable.
 BT France
 SA Loire
  Rhone

Montsech
Foothills of the Pyrenees.
 BT Lerida
  Spain

Montserrat
Volcanic island in the Leeward Islands belonging to the United Kingdom.
 BT West Indies
 NX Soufriere

monzodiorite
 BA diorite family
 BT igneous rocks
 SA plutonic rocks

monzonite
Includes use on level 3 under igneous rocks(1) syenite family(2). See list H.
 BA syenite family
 BT igneous rocks

Moon
See also other planets, i.e. Mars, Venus, etc. Includes use on level 1 and 2, e.g. under maps(1). Term set options are:
 name of discipline [List B except for oceanography, paleobotany and paleontology]
  topic (similar to terms called for on 3rd level in List B)
 theoretical studies
  subtopic
 topic [age, anomalies, atmosphere, bibliography, catalogs, composition, concepts, environment, evolution, exploration, genesis, gravity field, interior, magnetic field, microcraters, motions, observations, paleomagnetism, surface properties]
  subtopic
 NT Foaming Sea

Fra Mauro
Hadley Rille
North Ray Crater
Ocean of Storms
Sculptured Hills
Sea of Crises
Sea of Fertility
Sea of Nectar
Sea of Rains
Sea of Serenity
Sea of Tranquillity
Sea of Waves
Shorty Crater
South Ray Crater
Taurus-Littrow
SA accretion
   agglutinates
   albedo
   Apollo
   Apollo 9
   Apollo 11
   Apollo 12
   Apollo 14
   Apollo 15
   Apollo 16
   Apollo 17
   asteroids
   atmosphere
   breccia
   condensation
   coordinates
   cosmic dust
   cosmic rays
   craters
   Earth
   Eratosthenian
   exploration
   Explorer 35
   farside
   fines
   gravity field
   highlands
   interior
   KREEP
   landing sites
   luminescence
   Luna 16
   Luna 20
   lunar materials
   lunar samples
   magnetic field
   maps
   maria
   Mars
   mascons
   meteorites
   microcraters
   moonquakes
   motions
   orange material
   particles
   planetology
   planets
   rilles
   selenodesy
   solar flares
   solar system
   solar wind
   surface features
   surface properties
   Surveyor 3
   temperature
   terrestrial materials

moon rocks
A valid term through mid-1978. Use lunar samples.

moonquakes
Includes use on level 3 under Moon(1).
SA earthquakes
   Moon

moonstone
BA feldspar group
      framework silicates
      silicates
BT minerals

SA gems

Moore County
Index states as applicable.
BX North Carolina
   Tennessee
   Texas
BT United States

Moose River basin
SW of lower James Bay.
BT Ontario
   Canada

Mor
See list M.
UF raw humus
BA soils
SA humus
   Mull

morainal plains
use outwash plains

moraines
Includes use on level 3 under glacial geology(1) glacial features(2).
SA drift
   drumlins
   glacial features
   glacial geology
   glaciers
   outwash plains
   till

Morava
use Moravia

Morava River valley
River valley in NE central Serbia; and a river valley in NE Austria and central Czechoslovakia. Index Austria, Czechoslovakian regions and Serbia. Also search Morava; Morava River.
BT Europe
SA Austria
   Moravia
   Serbia
   Slovakia

Moravia
Region between Bohemia and Slovakia.
UF Morava
BT Czechoslovakia
NT Brno
   Karvina
   Moravian Karst
   Nizky Jezenik Mountains
   Ostrava
   Ostrava-Karvina
   Stramberk
   Svratka
SA Morava River valley
   Sudeten
   Sudeten Mountains
   Sudetic Basin

Moravian Karst
BT Moravia
   Czechoslovakia

Moravska Ostrava
use Ostrava

Morbihan
Department in NW.
BT France
NT Groix
   Vilaine Bay

mordenite
Also search ashtonite.
UF arduinite
   ashtonite
   flokite
   ptilolite
BA zeolite group
      framework silicates
      silicates
BT minerals
SA aluminosilicates

Morecambe Bay
Inlet of the Irish Sea in NW.

BT England
   Great Britain
   United Kingdom

morencite
use nontronite

Moreno Formation
In Chico Group. Upper Cretaceous and Paleocene (?). Includes four members (ascending): Dosados Sand and Shale, Tierra Loma Shale (including Mercy Sandstone lens), Marca Shale, and Dos Palos Shale (includes Cima Sandstone lens). S California.
BT Upper Cretaceous
   Cretaceous
SA California
   Marca Shale Member
   Paleocene

Moreton
Village on Cape York Peninsula in N.
BT Queensland
   Australia

Moreton Bay
In SE just N of Brisbane.
BT Queensland
   Australia

Morez
Town in in E part of country.
BT Jura
   France

Morgan County
Index states as applicable.
BX Alabama
   Colorado
   Georgia
   Illinois
   Indiana
   Kentucky
   Missouri
   Ohio
   Tennessee
   Utah
   West Virginia
BT United States

Morin Anorthosite
use Morin Complex

Morin Complex
Also search Morin Anorthosite.
CO N454500N463000
   W0733000W0743000
UF Morin Anorthosite
   Morin Plutonic Complex
BT Quebec
   Canada

Morin Plutonic Complex
use Morin Complex

Morlaix
Town on inlet of English Channel.
BT Finistere
   France

Mormoiron
Village in SE part of country.
BT Vaucluse
   France

Mormon Mountains
SE Nevada.
BT Nevada
   United States

Morocco
Includes use on level 1 as an area term (list O). For term set options see list B.
CO N230000N360000
   W0020000W0153000
BA Africa
NT Anti-Atlas
   Beni Bouchera
   Bou Azzer
   Casablanca
   High Atlas
   Middle Atlas

   Moulouya River
   Rabat
   Rehamna
   Rif
   Tafilalt
   Tangier
   Tarfaya
   Taza
SA Atlas Mountains
   Mediterranean region
   North Africa
   Sahara
   Taourirt
   West Africa

Moroto Mountain
Just E of town of Moroto near Kenya border.
BT Uganda

morphology
Includes use on level 2 under fossil terms (list F), and under aurora(1) and soils(1).
NT functional morphology
SA aurora
   coiling
   consistency
   description
   horizon differentiation
   horizons
   ichnofossils
   micromorphology
   physiology
   profiles
   septa
   soils
   sutures
   valves

morphometry
General term used under geomorphology(1).
SA geomorphology
   landform description

morphostructures
SA geomorphology

Morrison Formation
In Plata Mountains of Colorado, comprises (ascending) Pony Express Limestone, Bilk Creek Sandstone, Wanakah Marl (restricted), and Junction Creek Sandstone.
BT Upper Jurassic
   Jurassic
SA Arizona
   Colorado
   Kansas
   Montana
   New Mexico
   Oklahoma
   South Dakota
   Utah
   Wyoming

Morrow Formation
Composed of shales with some thin limestones and occassional sandstones. SE Arkansas, and central E and NE Oklahoma.
BT Pennsylvanian
SA Arkansas
   Morrowan
   Oklahoma

Morrow Series
use Morrowan

Morrowan
Provincial series, North America. Above Chesterian (Mississippian), below Atokan. Includes use on level 3 under age terms(1). See list E.
UF Morrow Series
   Morrowan Series
BA Lower Pennsylvanian
      Pennsylvanian
BT Paleozoic
NT Bloyd Formation
SA Chesterian

Morrow Formation
Springer Formation

Morrowan Series
use Morrowan

Mortagne
Town in NW Orne.
UF Mortagne-au-Perchey
BT Orne
France

Mortagne-au-Perchey
use Mortagne

Morton County
Index states as applicable.
BX Kansas
North Dakota
BT United States

Morvan
Northernmost spur of the Central Massif. Index departments as applicable.
UF Morvan Massif
Morvan Mountains
BT France
SA Central Massif
Cote-d'Or
Nievre
Saone-et-Loire
Yonne

Morvan Massif
use Morvan

Morvan Mountains
use Morvan

Mosabani Mine
use Mosaboni copper mines

Mosaboni copper mines
SE Bihar, India. Also search Mosaboni.
UF Mosabani Mine
Mosaboni Mine
Mosaboni Mines
Musabani copper mines
Mushabani copper mines
BT Bihar
India

Mosaboni Mine
use Mosaboni copper mines

Mosaboni Mines
use Mosaboni copper mines

Mosasauridae
Family. Includes use on level 3 under Reptilia(1) Lepidosauria(2). Also search Mosasaurus; mosasaurs.
UF mosasaurs
BA Lepidosauria
Reptilia
BT Tetrapoda
Vertebrata

mosasaurs
use Mosasauridae

Mosbas
use Moscow Basin

Moscovian
Russia. Includes use on level 3 under age terms(1). See list E.
BA Upper Carboniferous
Carboniferous
BT Paleozoic

moscovite
use muscovite

Moscow
City and oblast.
UF Moskva
BT Russian Republic
USSR

Moscow Basin
Lignite basin extending over 600 miles in an arc from Borovichi in Leningrad Oblast to Skopin, SE of Moscow, in Ryazan Oblast. Also search Moscow.
UF Mosbas
Moscow coal basin
BT Russian Republic
USSR

Moscow coal basin
use Moscow Basin

Moscow River
use Moskva River

Moscow State University
use Moscow University

Moscow Syneclise
Central European USSR. Also search Moscow.
CO N540000N573000
E0400000E0350000
BT Russian Republic
USSR

Moscow University
In Moscow. Also search Moscow State University.
UF Moscow State University
BT Russian Republic
USSR

Mosel Valley
use Moselle Valley

Moselle
Department in NE.
BT France
SA Saar Basin
Saar-Nahe Basin

Moselle River
Rises in NE France and flows N and NE entering the Rhine River at Koblenz. Index countries as applicable. Also search Moselle.
BT Europe
SA France
Luxembourg
West Germany

Moselle Valley
River valley. Index countries as applicable.
UF Mosel Valley
BT Europe
SA France
Germany
Luxembourg

Moskva
use Moscow

Moskva River
Flows through Moscow and then SE into the Oka River.
UF Moscow River
BT Russian Republic
USSR

Mosquito Range
S part of Park Range near Leadville.
BT Colorado
United States

Mossamedes
District and desert basin in extreme SW part of country. Also a city on the Atlantic Ocean.
UF Mocamedes
BT Angola

Mossbauer
A valid index term through 1977. After 1977, use Mossbauer spectroscopy.

Mossbauer spectra
Term introduced in 1978. Before 1978, also search Mossbauer.
UF Mossbauer spectrum
SA Mossbauer spectroscopy
spectra

Mossbauer spectroscopy
Term introduced in 1978. Includes use on level 3 under spectroscopy(1) and under chemical element(1). Before 1978, also search Mossbauer AND spectroscopy.
BT spectroscopy
SA analysis
Mossbauer spectra

Mossbauer spectrum
use Mossbauer spectra

mosses
use bryophytes

motion, ground
use ground motion

motion, plasma
use plasma motion

motion, strong
use strong motion

motions
Includes use on level 2 under Earth(1) and Moon(1); on level 3 under aurora(1) morphology(2).
SA aurora
Chandler wobble
Earth
Moon
oscillations
seismometers

Motojuku Formation
BT Miocene
Neogene
Tertiary
SA Honshu
Japan

Moulouya River
Rises in central part of country and flows NE into Mediterranean Sea. Also search Moulouya.
UF Moluya
Mulwiya
BT Morocco

mounds
Includes use on level 3 under sedimentary structures(1) bedding plane irregularities(2). See list K.
BA bedding plane irregularities
sedimentary structures

mounds, algal
use algal mounds

Mount Agoeng
use Mount Agung

Mount Agung
Highest volcanic peak on Bali.
UF Mount Agoeng
BT Lesser Sunda Islands
Indonesia

Mount Asama
Active volcano in central.
BT Honshu
Japan

Mount Carmel
Mountain in N near Mediterranean coast.
BT Israel
Middle East

Mount Elbrus
use Elbrus

Mount Etna
Active volcano in NE Sicily. Also search Etna and Mt. Etna.
CO N374500N374500
E0150100E0150100
UF Etna
Mt. Etna
BT Sicily
Italy

Mount Fuji
use Fujiyama

Mount Girnar
Village in S central Kathiawar Peninsula. Also search Mount Girnar Massif.
UF Mount Girnar Massif
BT Gujarat
India

Mount Girnar Massif
use Mount Girnar

Mount Hakone
Extinct volcano in central Honshu. Also search Hakone; Hakone Volcano.
UF Hakone Volcano
BT Honshu
Japan
SA Hakone

Mount Hood
Peak in Cascade Range in Clackmas and Hood River counties. Also search Mt. Hood.
CO N452500N452500
W1214000W1214000
UF Mt. Hood
BT Oregon
United States

Mount Isa
Town in W.
CO S204500S204500
E1393000E1393000
BT Queensland
Australia

Mount Kenya
Extinct volcano in central.
BT Kenya

Mount Lofty Ranges
SE South Australia.
UF Mt. Lofty Ranges
BT South Australia
Australia

Mount Lyell
Peak in Sierra Nevada near junction of Madera, Mariposa, and Tuolumne counties in E central.
UF Mt. Lyell
BT California
United States

Mount Mihara
Active volcano on central O-shima of Izu-shichito Island Group S of Sagami Bay. Also search Mihara.
UF Mihara (Mount)
Mihara Volcano
Mihara-Yama
Mt. Mihara
BT Honshu
Japan

Mount Morgan
Peak in Glacier National Park in NW.
BT Montana
United States

Mount Rainier
Volcanic peak of Cascade Range in Mount Rainier National Park in W central Washington. Also search Mt. Rainier.
UF Mt. Rainier
BT Washington
United States
SA Cascade Range

Mount Rainier National Park
W central part of state. Mt. Rainier occupies one fourth of park area in W central Washington.
UF Mt. Rainier National Park
BT Washington
United States

Mount Saint Helens
Peak in NW Skamania County in S Washington.
UF Mount St. Helens
Mt. St. Helens
BT Washington
United States

Mount St. Helens
use Mount Saint Helens

Mount Wood
Peak of St. Elias Mountains in SW.
BT Yukon Territory

Canada
mountain building
A valid term through 1973. After 1973, use orogeny.
mountain crystal
use quartz crystal
mountains
Includes use on level 3 under geomorphology(1) landform description(2).
SA geomorphology
  highlands
  hills
  landform description
  landforms
  ridges
Mousterian
Archeologic classification.
BA Paleolithic
BT Cenozoic
SA Pleistocene
  Quaternary
Mouthoumet Massif
S part of country. Also search Mouthoumet.
BT Aude
  France
Mouydir
Mountains in S central.
BT Algeria
movement
Includes use on level 2 under ground water(1) and plate tectonics(1); on level 3 under soils(1) water regimes(2).
SA ground water
  ice movement
  plate tectonics
  soils
  water regimes
movements, mass
use mass movements
movements, vertical
use vertical movements
Mowry Shale
In Colorado Group. Consists of hard lighter-gray shales and thin bedded sandstones that weather light gray and form ridges. Montana, W South Dakota, and Wyoming. Index states as applicable.
BT Upper Cretaceous
  Cretaceous
SA Colorado Group
  Montana
  South Dakota
  Wyoming
Mozambique
Includes use on level 1 as an area term (list O). For term set options see list B.
CO S270000S100000
  E0410000E0300000
BA Africa
NT Maputo
  Mozambique Belt
  Tete
  Zambezia
SA East African Rift
  Lake Malawi
  Lebombo Mountains
  Limpopo Basin
  Zambezi Valley
Mozambique Belt
Orogenic belt. Mountains in N extend to the rim of the Great Rift Valley.
BT Mozambique
Mozambique Channel
Strait between the Malagasy Republic and Mozambique.
BT Indian Ocean

mroseite
BA tellurates and tellurites
BT minerals
Mt. Elbrus
use Elbrus
Mt. Etna
use Mount Etna
Mt. Fuji
use Fujiyama
Mt. Fuji volcano
use Fujiyama
Mt. Hood
use Mount Hood
Mt. Lofty Ranges
use Mount Lofty Ranges
Mt. Lyell
use Mount Lyell
Mt. Mihara
use Mount Mihara
Mt. Rainier
use Mount Rainier
Mt. Rainier National Park
use Mount Rainier National Park
Mt. St. Helens
use Mount Saint Helens
Mubarek
Village SW of Samarkand.
BT Uzbekistan
  USSR
mud
Includes use on level 3 under sediments(1) carbonate sediments(2) and clastic sediments(2) See list N.
BA clastic sediments
BT sediments
SA clay
  mud banks
  mud lumps
  mud volcanoes
  ooze
  silt
  terrigenous materials
mud balls, armored
use armored mud balls
mud banks
Also search mudbanks.
UF mudbanks
SA fluvial features
  mud
  shore features
mud flats
UF mudflats
BT shore features
SA tidal flats
mud flows
use mudflows
mud lumps
Includes use on level 3 under sedimentary structures(1) bedding plane irregularities(2); on level 3 under geomorphology(1) shore features(2) and fluvial features(2). See list K.
UF mudlumps
BA bedding plane irregularities
  sedimentary structures
SA clay
  mud
mud volcanoes
Includes use on level 1 (list A). Used for an accumulation, usually conical, of mud and rock ejected by volcanic gases; also for a similar accumulation formed by escaping proliferous gases; also, for mud cones not of eruptive origin. Term set options are:
topic [age, causes, classification, detection, distribution, evolution, extent, genesis, interpretation, mechanism, occurrence, structure, tem-

perature]
subtopic (no area term)
UF hervidero
  macaluba
SA ejecta
  geomorphology
  lava
  mud
  volcanoes
  volcanology
mudbanks
use mud banks
mudcracks
Includes use on level 3 under sedimentary structures(1) bedding plane irregularities(2). See list K.
BA bedding plane irregularities
  sedimentary structures
SA shrinkage cracks
Muddy Sandstone
Of Thermopolis Shale. Overlies a lower black shale member and underlies a black shale member. Surface and subsurface in central N Wyoming and subsurface in central S Montana. Also search Muddy Formation.
UF Muddy Sandstone Member
BT Lower Cretaceous
  Cretaceous
SA Montana
  Wyoming
Muddy Sandstone Member
use Muddy Sandstone
mudflats
use mud flats
mudflows
As of 1978, term is used on level 2 under geologic hazards(1), and slope stability(1). Also search mud flows.
UF mud flows
SA erosion
  geologic hazards
  lahars
  landslides
  mass movements
  mudslides
  slope stability
mudlumps
use mud lumps
mudslides
SA earthflows
  mudflows
mudstone
Includes use on level 3 under sedimentary rocks(1) clastic rocks(2). See list I.
BA clastic rocks
BT sedimentary rocks
SA claystone
  siltstone
  terrigenous materials
Mudurnu
Village in NW part of Anatolia, SW of Bolu.
BT Turkey
Muenster
use Munster
Muenster in Westfalen
use Munster
Muensterland
use Munsterland
Mugan Steppe
Part of Kura Lowland S of Aras and Kura rivers in SE Azerbaidzhan. Also search Mugan.
BT Azerbaidzhan
  USSR
mugearite
Includes use on level 3 under igne-

ous rocks(1) basalt family(2). See list H.
BA basalt family
BT igneous rocks
Mugodzhar Hills
Southernmost extension of the Urals in Aktyubinsk Oblast. Also search Mugodzhar, Mugodzhar Mountains, Mugodzhars, and Mugodzhry.
UF Mugodzhar Mountains
  Mugodzhar Range
  Mugodzhars
  Mugodzhary
BT Kazakhstan
  USSR
Mugodzhar Mountains
use Mugodzhar Hills
Mugodzhar Range
use Mugodzhar Hills
Mugodzhars
use Mugodzhar Hills
Mugodzhary
use Mugodzhar Hills
Mukhor-Tala
Village E of Lake Baikal in Buryat Autonomous Soviet Socialist Republic.
BT Russian Republic
  USSR
Mule Ear Diatreme
In Mule Ear Peaks, a mountain in S Brewster County, in W. Also search Mule Ear.
BT Texas
  United States
Mull
See list M.
BA soils
SA humus
  Mor
  Mull Island
Mull Island
Term introduced in 1978. Island of the Inner Hebrides, Argyll, Scotland. Before 1978, also search Mull. After 1978, Mull is restricted for use as the soil.
CO N561500N564000
  W0054000W0063000
BT Inner Hebrides
  Scotland
SA Mull
mullions
Includes use on level 3 under lineation(1) style(2) and under structural analysis(1).
SA lineation
  metamorphic rocks
  structural analysis
mullite
BA orthosilicates
  silicates
BT minerals
multispectral analysis
Term introduced in 1978. Before 1978, search multispectral AND analysis.
SA analysis
  remote sensing
Multituberculata
Includes use on level 2 under Mammalia(1). See list F.
BA Mammalia
BT Tetrapoda
  Vertebrata
multivariate analysis
Includes use on level 3 under mathematical geology(1).
SA analysis
  mathematical geology
  statistical analysis

Mulwiya
   use Moulouya River

Munchberg Gneiss Massif
   In Fichtelgebirge area in NE Bavaria. Also search Munchberg Gneiss.
   UF   Munchberg Massif
   BT   Bavaria
          West Germany

Munchberg Massif
   use Munchberg Gneiss Massif

Munchen
   use Munich

Mundrabilla
   Village in SE.
   BT   Western Australia
          Australia

Munich
   City in S Bavaria.
   UF   Munchen
   BT   Bavaria
          West Germany

Munster
   City in N North Rhine-Westphalia.
   UF   Muenster
          Muenster in Westfalen
   BT   North Rhine-Westphalia
          West Germany

Munsterland
   Area around Munster in N North Rhine-Westphalia. Also search Munsterland.
   UF   Muensterland
   BT   North Rhine-Westphalia
          West Germany

Muntenia
   Region and former province in E.
   BT   Walachia
          Romania

Muntii Metalici
   Metal Mountains. S part of Apuseni Mountains in W central part of country.
   BT   Transylvania
          Romania
   SA   Apuseni Mountains

muong nong type
   Also search muong nong.
   UF   muong nong-type
   BA   tektites

muong nong-type
   use muong nong type

Murcia
   Province in SE Spain. Formerly in ancient kingdom which included a larger area in SE.
   BT   Spain
   NT   Caravaca
          Jumilla

Mures
   Country in N central part of country.
   UF   Mures Province
   BT   Transylvania
          Romania

Mures Province
   use Mures

Murgab Basin
   River basin. Index Afghanistan and/or Turkmenia as applicable. Also search Murgab.
   UF   Murghab (Basin)
   BT   Asia
   SA   Afghanistan
          Turkmenia

Murge
   Plateau region SW of Bari near the heel of the Italian boot.
   BT   Apulia
          Italy

Murghab (Basin)
   use Murgab Basin

Muricacea
   Superfamily. Includes use on level 3 under Mollusca(1) Gastropoda(2).
   BA   Gastropoda
          Mollusca
   BT   Invertebrata

Muridae
   Family. Includes use on level 3 under Mammalia(1) Rodentia(2).
   BA   Rodentia
   BT   Tetrapoda
          Vertebrata

Murmansk
   City on the Arctic Ocean in extreme NW European USSR. Also an oblast.
   BT   Russian Republic
          USSR

Muro Group
   SA   Cretaceous
          Japan
          Tertiary

Murphy Marble
   Sequence (ascending) is Valleytown Formation, Murphy Marble, Andrews Schist, and Nottely Quartzite. N Georgia, W North Carolina and E Tennessee.
   BT   Lower Cambrian
          Cambrian
   SA   Georgia
          North Carolina
          Tennessee

Murray Basin
   River basin. Index states as applicable. Also search Murray AND basin.
   CO   S360000S320000
          E1450000E1400000
   BT   Australia
   SA   New South Wales
          South Australia
          Victoria

Murray County
   Index states as applicable.
   BX   Georgia
          Minnesota
          Oklahoma
   BT   United States

Murray fracture zone
   NE of Hawaii. Also search Murray AND fracture zone.
   UF   Murray Seascarp
   BT   Pacific Ocean

Murray River
   The major river of Australia. It rises in E Victoria and flows NW and then S into Encounter Bay on the Indian Ocean. Index states as applicable. Also search Murray.
   BT   Australia
   SA   New South Wales
          South Australia
          Victoria

Murray Seascarp
   use Murray fracture zone

Murrumbidgee River
   Flows W from Great Dividing Range near Canberra to join Murray River in S.
   BT   New South Wales
          Australia

Murrumbidgee Valley
   River valley in S.
   BT   New South Wales
          Australia

Mururoa Atoll
   In Tuamotu Archipelago.
   BT   Polynesia
          Pacific Ocean
   SA   Tuamotu Islands

Murzuk Basin
   Dune region in SW Libya. Also search Murzuk.
   UF   Murzuq
   BT   Libya
          Africa

Murzuq
   use Murzuk Basin

Musabani copper mines
   use Mosaboni copper mines

Musashino
   City in Tokyo Prefecture.
   BT   Honshu
          Japan

Muschelkalk
   Europe. Includes use on level 3 under age terms(1). See list E.
   BA   Middle Triassic
          Triassic
   BT   Mesozoic

Musci
   Includes use on level 2 under bryophytes(1). See list F.
   BA   bryophytes
   BT   Plantae

muscovite
   Also search white mica.
   UF   common mica
          mirrorstone
          moscovite
          potash mica
          white mica
   BA   sheet silicates
          silicates
   BT   minerals
   SA   hydromuscovite
          illite
          light minerals
          mica group
          phengite
          sericite

muscovite granite
   See list H.
   BA   granite-granodiorite family
   BT   igneous rocks
   SA   granite

museums
   Includes use on level 1 (list A). Used for papers discussing the functioning of a geology museum and/or its history. Term set options are:
   topic (list B)
         name of museum (in original language)
   SA   associations
          catalogs
          collections
          education
          geology
          organization

Musgrave Range
   use Musgrave Ranges

Musgrave Ranges
   Index Northern Territory and/or South Australia as applicable. Also search Musgrave.
   UF   Musgrave Range
   BT   Australia
   SA   Northern Territory
          South Australia

Mushabani copper mines
   use Mosaboni copper mines

Muskogee County
   E Oklahoma.
   BT   Oklahoma
          United States

Muskox Intrusion
   In District of Mackenzie.
   BT   Northwest Territories
          Canada

Mussoorie
   Hill station and sanitarium in N.
   BT   Uttar Pradesh
          India

Mustang Island
   In Neuces County between Corpus Christi Bay and the Gulf of Mexico.
   BT   Texas
          United States

Mustelidae
   Family of Mammalia. Includes use on level 3 under Mammalia(1) Carnivora(2). Also search mustelids.
   UF   mustelids
   BA   Carnivora
          Mammalia
   BT   Tetrapoda
          Vertebrata

mustelids
   use Mustelidae

Muth Quartzite
   Middle to Upper Devonian. Includes a thick succession of snow-white to greenish quartzites.
   BT   Devonian
   SA   India
          Jammu and Kashmir
          Middle Devonian
          Punjab
          Upper Devonian
          Uttar Pradesh

Muyun-Kum
   use Muyunkum

Muyunkum
   Sandy desert in Dzhambul Oblast, S of the Chu River near Kirghizia border.
   UF   Muyun-Kum
   BT   Kazakhstan
          USSR

mylonite
   Includes use on level 3 under metamorphic rocks(1) mylonites(2); on level 3 under faults(1) effects(2). See list J.
   BA   mylonites
   BT   metamorphic rocks
   SA   blastomylonite
          faults
          mylonitization

mylonites
   Includes use on level 2 under metamorphic rocks(1). See list J.
   BT   metamorphic rocks
   NA   blastomylonite
          mylonite
          pseudotachylite

mylonitization
   SA   deformation
          faults
          mylonite
          phyllonitization

Myodocopida
   Includes use on level 2 under Ostracoda(1). See list F.
   BA   Ostracoda
   BT   Crustacea
          Arthropoda
          Invertebrata

Myoko Mountain
   W central Honshu. Also search Myoko Volcano.
   UF   Myoko Volcano
          Myoko-san
          Myoko-zan
   BT   Honshu
          Japan

Myoko Volcano
   use Myoko Mountain

Myoko-san
   use Myoko Mountain

Myoko-zan
   use Myoko Mountain

Myophoria
   Genus. Includes use on level 3 under Mollusca(1) Bivalvia(2).

BA Bivalvia
   Mollusca
BT Invertebrata

Myriapoda
Superclass. Use classes on level 3: Archipolypoda, Chilopoda, Diploda, Pauropoda, Symphyla. Includes use on level 2 under Arthropoda(1).
BA Arthropoda
BT Invertebrata

myrmekite
BA feldspar group
   framework silicates
   silicates
BT minerals

Mysore
State on the Arabian Sea in SW.
BT India
NT Bagalkot
   Bangalore
   Belgaum
   Bellary
   Bijapur
   Chickmagalur
   Chitaldrug schist belt
   Chitradurga
   Dharwar
   Gulbarga
   Hassan
   Ingaldhal
   Kaladgi
   Kanara
   Kolar
   Kolar Gold Fields
   Mangalore
   Sagar
   Sandur
SA Badami Series
   Cauvery Basin
   Deccan Plateau
   Kaladgi System
   Krishna

Mytilus
Genus. Includes use on level 3 under Mollusca(1) Bivalvia(2).
BA Bivalvia
   Mollusca
BT Invertebrata
NA Mytilus edulis

Mytilus edulis
Includes use on level 3 under Mollusca(1) Bivalvia(2).
BA Mytilus
   Bivalvia
   Mollusca
BT Invertebrata

N

N
use nitrogen

N-15/N-14
SA isotopes
   nitrogen

Na
use sodium

Na-22
Includes use on level 3 under isotopes(1).
SA isotopes
   sodium

Na-24
Includes use on level 3 under isotopes(1).
SA isotopes
   sodium

Nabburg
Town in E.
BT Bavaria
   West Germany

Nacimiento Mountains
Range in NW New Mexico. Also search Nacimiento Uplift.
UF Nacimiento Range
   Nacimiento Uplift
BT New Mexico
   United States

Nacimiento Range
use Nacimiento Mountains

Nacimiento Uplift
use Nacimiento Mountains

Nacozari de Garcia
Town in NE Sonora. Also search Nacozari.
BT Sonora
   Mexico

nacrite
BA sheet silicates
   silicates
BT minerals
SA clay minerals

Nagano
City in central.
BT Honshu
   Japan

Nagano Prefecture
Central Honshu.
BT Honshu
   Japan

Nagasaki
City and prefecture in W.
BT Kyushu
   Japan

Nagpur
City, district, and division in NE.
BT Maharashtra
   India

Naha test well
In Navajo Indian Reservation in NE.
BT Arizona
   United States

Nahanni Formation
British Columbia, Mackenzie District of Northwest Territories, and Yukon Territory.
BT Middle Devonian
   Devonian
SA British Columbia
   Canada
   Northwest Territories
   Yukon Territory

nahcolite
BA carbonates
BT minerals

Nahe
River which flows into Rhine at Bingen. Index states as applicable.
BT West Germany
SA Rhineland-Palatinate
   Saar-Nahe Basin
   Saarland

Naiba River basin
use Nayba River basin

Nain
Village on central coast.
BT Labrador
   Canada

Nain Anorthosite
use Nain Massif

Nain Anorthosite Massif
use Nain Massif

Nain Massif
Anorthosite massif. Also search Nain Anorthosite; Nain Anorthosite Massif.
UF Nain Anorthosite
   Nain Anorthosite Massif
BT Labrador
   Canada

Naini Tal
Town, hill station and district in NE Uttar Pradesh.
CO N292000N292500
   E0793000E0792200
UF Naini-Tal
   Nainital
BT Uttar Pradesh
   India

Naini-Tal
use Naini Tal

Nainital
use Naini Tal

Nairobi
City in S part of country.
BT Kenya

Naka-no-umi
Inlet of Sea of Japan in SW Honshu. Search Nakanoumi Lake.
UF Nakano umi
   Nakanoumi Lake
BT Honshu
   Japan

Nakano umi
use Naka-no-umi

Nakanoumi Lake
use Naka-no-umi

Nakhichevan
Town and Autonomous Soviet Socialist Republic on the Iranian border separated from Azerbaidzhan proper by a narrow strip of Armenia.
BT Azerbaidzhan
   USSR

Nakhla
Village.
BT Qatar
   Arabian Peninsula

Nakuru Basin
In Great Rift Valley in Lake Nakuru area of W central Rift Valley Province. Also search Nakuru.
BT Kenya

Nama Beds
use Nama System

Nama Group
use Nama System

Nama Series
use Nama System

Nama Supergroup
use Nama System

Nama System
In SW and S Africa. Also search Nama Group; Nama Beds; Nama Series; Nama Supergroup.
UF Nama Beds
   Nama Group
   Nama Series
   Nama Supergroup
   Nama Tillite
BT Cambrian
SA South Africa
   South-West Africa

Nama Tillite
use Nama System

Namaland
use Namaqualand

Namaqualand
Coastal region of sandy plains and bare hills. Index Cape Province and/or South-West Africa as applicable.
UF Namaland
BT Africa
SA Cape Province
   South-West Africa

Namib Desert
Arid region extending along entire coast.
BT South-West Africa
   Canada

Namur
Province in S.
BT Belgium
NT Dinant
   Dinant Basin

Namurian
Europe. Above Visean, below Westphalian. Includes use on level 3 under age terms(1). See list E.
BA Carboniferous
BT Paleozoic
SA Lower Carboniferous
   Upper Carboniferous

Nanao
City on W coast on E side of Noto Peninsula.
BT Honshu
   Japan

Nandewar Mountains
Range in NE.
BT New South Wales
   Australia

Nanjemoy Formation
In Pamunkey Group. Includes Potapaco Clay Member, Woodstock Greensand, Marlboro Clay Member. E Maryland and E Virginia.
BT Eocene
SA lower Eocene
   Maryland
   middle Eocene
   Virginia

Nankaido
Former division including Awaji Islands, S central part of S coast of Honshu, and all of Shikoku. Index major islands as applicable.
BT Japan
SA Honshu
   Shikoku

nannoconids
Includes use on level 3 under algae(1) nannofossils(2). See list F. Before 1976, also search Nannoconus.
BA nannofossils
   algae
BT Plantae
SA Nannoconus

Nannoconus
Genus. Includes use on level 3 under algae(1) nannofossils(2).
BA nannofossils
   algae
BT Plantae
SA nannoconids

nannofossils
Discoasters and nannoconids are included here. Includes use on level 2 under algae(1). Also search calcareous nannofossils. See list F.
UF calcareous nannofossils
BA algae
BT Plantae
NA Braarudosphaeridae
   discoasters
   Helicopontosphaera
   nannoconids
   Nannoconus
   Sphenolithus
SA Coccolithophoraceae
   fossils
   index fossils
   problematic fossils

Nantes
City on the Loire River.
BT Loire-Atlantique
   France

Nantucket (Island)
use Nantucket Island

Nantucket Island
In Atlantic Ocean, S of Cape Cod. Also search Nantucket.

UF Nantucket (Island)
BT Massachusetts
United States

Napa County
W central.
BT California
United States

Naples
City on the Tyrrhenian Sea.
UF Napoli
BT Campania
Italy

Napoli
use Naples

nappe (volcanic)
use lava flows

nappes
Folds defined on the basis of orientation in relation to the geographic horizontal plane. Includes use on level 3 under folds(1) orientation(2) and under faults(1) displacements(2) overthrust(3); on 3 under tectonics(1).
SA displacements
faults
folds
klippen
orientation
overthrust faults
tectonics

Nara
City and prefecture E of Osaka.
BT Honshu
Japan

Narbada River
use Narmada River

Narbada Valley
use Narmada Valley

Narbonne
City near the Mediterranean Sea.
BT Aude
France

Nari Series
Underlies the Gaj and overlies the Khirthar Series.
BT Oligocene
SA Gujarat
India
Pakistan
Sind

Narita Formation
Tokyo Bay area.
BT Pliocene
SA Honshu
Japan

Narizian
North America. Above Ulatisian, below Fresnian. Includes use on level 3 under age terms(1). See list E.
BA Eocene
Paleogene
Tertiary
BT Cenozoic
SA middle Eocene
upper Eocene

Narmada River
Rises in the Maikala Range in Madhya Pradesh and flows W into Gulf of Cambay. Index states as applicable. Also search Narmada.
UF Narbada River
Nerbudda River
BT India
SA Gujarat
Madhya Pradesh

Narmada Valley
River valley. Index states as applicable.
UF Narbada Valley
Nerbuda Valley
BT India

SA Gujarat
Madhya Pradesh

Narnaul
Town 80 miles WSW of Delhi.
BT Haryana
India

Narpaign Fjord
use Narpaing Fjord

Narpaing Fiord
use Narpaing Fjord

Narpaing Fjord
Narrow inlet of Davis Strait on Baffin Island, District of Keewatin.
UF Narpaign Fjord
Narpaing Fiord
BT Northwest Territories
Canada

Narrabeen Group
Includes the Gosford Formation and Clifton Sub-Group. N of Sydney.
BT Triassic
SA Australia
New South Wales

Narragansett Bay
Inlet of Atlantic Ocean in SE.
BT Rhode Island
United States

Narrows
Strait between W end of Long Island and Staten Island separating Upper New York Bay from Lower New York Bay.
BT New York
United States

Narugo
Town in N.
BT Honshu
Japan

Narvik
City on Ofot Fjord in N part of country.
BT Nordland
Norway

Narym
Village in Ob River in central Tomsk Oblast in W central Siberia.
BT Russian Republic
USSR

Naryn
Town on Naryn River in Naryn Oblast in SE.
BT Kirghizia
USSR

Nasca Plate
use Nazca Plate

Nassau County
Index states as applicable.
BX Florida
New York
BT United States

Nassellaria
use Nassellina

Nassellina
Suborder. Includes use on level 2 under Radiolaria(1). See list F. Also search Nassellaria.
UF Nassellaria
BA Osculosida
Radiolaria
BT Invertebrata

nasturan
use pitchblende

Nasu (dake)
use Nasudake

Nasu Mountain
use Nasudake

Nasudake
Volcanic peak NE of Nikko in N central Honshu. Also search Nasu Mountain.

UF Nasu (dake)
Nasu Mountain
BT Honshu
Japan

Natal
South African province on the Indian Ocean, and city in Rio Grande do Norte, Brazil. Index countries as applicable.
NT Durban
Tugela Basin
Zululand
SA Brazil
South Africa

native elements and alloys
Includes use on level 2 under minerals(1). See list L.
BT minerals
NA awaruite
carbides
carbonado
chaoite
cohenite
diamond
ferrosilicon
graphite
josephinite
kamacite
lonsdaleite
moissanite
NZ niggliite
NA nitrides
phosphides
plessite
quicksilver
NZ schreibersite
NA silicides
taenite
SA alloys
antimony
arsenic
bismuth
copper
gold
hedleyite
iron
metals
silver
sulfur

native water
use connate waters

natroborocalcite
use ulexite

natrolite
BA zeolite group
framework silicates
silicates
BT minerals

Natrona County
Central Wyoming.
BT Wyoming
United States

natural
A valid level 1 term through 1977 used in combination with gas (i.e. gas, natural). After 1977, use natural gas as level 1 term.

natural gas
Not a valid term through 1977. After 1977, includes use on level 1 as a commodity term. Before 1978, also search gas AND natural; petroleum-gas; petroleum-natural gas.
SA condensates
energy sources
gas storage
gases
gasification
giant fields
hydrocarbons
hydrogen sulfide
methane
oil and gas fields
oil-gas interface

petroleum
pipelines
reservoir properties
reservoir rocks
source rocks
stratigraphic traps
structural traps
traps

natural materials
Term introduced in 1968. To distinguish from artificial materials.
SA materials
synthetic materials

natural remanence
use natural remanent magnetization

natural remanent magnetism
use natural remanent magnetization

natural remanent magnetization
Includes use on level 3 under paleomagnetism(1).
UF natural remanence
natural remanent magnetism
NRM
BA remanent magnetization
SA magnetic field
magnetization
paleomagnetism
remagnetization

natural resources
As of 1978, term is used on level 2 under conservation(1), land use(1), and reclamation(1).
SA conservation
depletion
environmental geology
land use
reclamation
resources

natural selection
UF selection, natural
SA evolution

Naubug
Village SE of Srinager.
BT Jammu and Kashmir
India

naujaite
Includes use on level 3 under igneous rocks(1) syenite family(2). See list H.
BA syenite family
BT igneous rocks

Nautiloidea
Order or other subdivision. Includes use on level 3 under Mollusca(1) Cephalopoda(2).
BA Cephalopoda
Mollusca
BT Invertebrata
NA Aturia
Hercoglossa danica
Nautilus

Nautilus
Genus. Includes use on level 3 under Mollusca(1) Cephalopoda(2).
BA Nautiloidea
Cephalopoda
Mollusca
BT Invertebrata

Navajo County
NE Arizona.
BT Arizona
United States

Navajo Indian Reservation
NE Arizona, NW New Mexico, and SE Utah. Index states as applicable.
BT United States
SA Arizona
New Mexico
Utah

Navajo Sandstone
In Glen Canyon Group. Upper Triassic(?) and Jurassic. N Arizona, W Colorado, NW New Mexico, and SE Utah.
BT Mesozoic
SA Arizona
Colorado
Jurassic
New Mexico
Upper Triassic
Utah

Navarre
Region and former kingdom of N Spain and SW France.
BT Europe
NT Asturreta
Pamplona
SA Spain

Navesink Formation
In Monmouth Group.
BT Upper Cretaceous
Cretaceous
SA Delaware
Monmouth Group
New Jersey

Navicula
BA diatoms
algae
BT Plantae

Naxos
Largest island of the Cyclades in the Aegean Sea.
BT Aegean Islands
Greece

Nayba River basin
S Sakhalin Island. Also search Nayba.
UF Naiba River basin
BT Russian Republic
USSR

Nazca Plate
S of the Galapagos Islands. Also search Nasca Plate.
UF Nasca Plate
BT Pacific Ocean
SA plate tectonics

Nb
use niobium

Nchanga
Mining township in N near Zaire border.
BT Zambia

Nd
use neodymium

Ne
use neon

Ne-20
Includes use on level 3 under isotopes(1).
SA isotopes
neon

Ne-21
Includes use on level 3 under isotopes(1).
SA isotopes
neon

Ne-22
Includes use on level 3 under isotopes(1).
SA isotopes
neon

Neanderthal
Includes use on level 3 under fossil man(1) or Mammalia(1) Primates(2). Also search Neanderthal man; Homo neanderthalensis.
UF Homo neanderthalensis
Neanderthal man
BA Hominidae
Primates
Mammalia
BT Tetrapoda
Vertebrata
SA fossil man

Neanderthal man
use Neanderthal

Near East
An indefinite and unofficial term including the countries of the Middle East plus Libya, Sudan, and occasionally Greece. Index countries as applicable.
SA Cyprus
Egypt
Greece
Iraq
Jordan
Kuwait
Libya
Oman
Qatar
Southern Yemen
Sudan
Syria
Turkey
United Arab Emirates
Yemen

nearshore
A valid term through 1977. After 1977, use nearshore environment.

nearshore environment
Term introduced in 1978. Before 1978, search nearshore.
SA environment

Nebraska
Includes use on level 1 as an area term (list O). For term set options see list B.
CO N400000N430000
W0952000W1040500
BA United States
NX Adams County
NT Beaver Lake
NX Blaine County
Boone County
Butler County
Cass County
Chase County
Cheyenne County
Clay County
Colfax County
Custer County
Douglas County
Franklin County
Garfield County
Grant County
Hamilton County
Jefferson County
Johnson County
Lincoln County
Madison County
McPherson County
Nemaha County
Pawnee County
NT Platte River
NX Polk County
Seward County
Sheridan County
Sherman County
Sioux County
Stanton County
Valley County
Washington County
Wayne County
Webster County
Wheeler County
York County
SA Americus Limestone Member
Ash Hollow Formation
Beil Limestone Member
Benton Formation
Brule Formation
Cambridge Arch
Carlile Shale
Chadron Arch
Chadron Formation
Chase Group
Cherokee Group
Colorado Group
Cottonwood Limestone
Dakota Formation
Denver Basin
Desmoinesian
Douglas Group
Ervine Creek Limestone
Eskridge Shale
Forest City Basin
Graneros Shale
Great Plains
Greenhorn Limestone
Hughes Creek Shale
Kansas City Group
Lawrence Formation
Lecompton Limestone
Loup Fork Group
Midcontinent
Midwest
Missouri River
Missouri River basin
Missouri River valley
Nemaha Ridge
Niobrara Formation
Ogallala Formation
Oread Limestone
Peoria Loess
Pierre Shale
Plattsburg Limestone
Plattsmouth Limestone Member
Red Eagle Limestone
Roca Formation
Rock Lake Shale Member
Sand Hills
Shawnee Group
Sioux Quartzite
South Platte River valley
Spearfish Formation
Stanton Formation
Stearns Shale
Sundance Formation
Valentine Formation
Wabaunsee Group
White River Group
Wreford Limestone
Wyandotte Limestone

Nebraskan
Pertaining to first glacial stage of Pleistocene Epoch in North America.
BA lower Pleistocene
Pleistocene
Quaternary
BT Cenozoic

Nebrodi Mountains
N Sicily.
UF Monti Caronie
BT Sicily
Italy

nebula, solar
use solar nebula

Neckar River
Rises in the Black Forest and flows N past Heidelberg into the Rhine River. Also search Neckar.
BT Baden-Wurttemberg
West Germany

Needles
Town on the Colorado River in San Bernardino County in SE.
BT California
United States

Neftechala
Town in SE.
BT Azerbaidzhan
USSR

Neftyanyye Kamni
Oil well in Caspian Sea off Apsheron Peninsula.
BT Azerbaidzhan
USSR
SA Caspian Sea

Negaunee Formation
use Negaunee Iron Formation

Negaunee Iron Formation
Upper Peninsula.
UF Negaunee Formation
BT Michigan
United States

Negev
Desert region in S Israel. Also search Negev Desert.
UF Negev Desert
BT Israel
Middle East

Negev Desert
use Negev

neighborite
BA halides
BT minerals
SA fluorides

Neill Island
Small island E of South Andaman Island in the Andaman Islands in E Bay of Bengal. Also search Neill AND appropriate area term.
BT India

Neiveli
use Neyveli

Nellore
City in S.
BT Andhra Pradesh
India

Nellore mica belt
In Nellore area. Also search Nellore.
BT Andhra Pradesh
India

Nelson County
Index states as applicable.
BX Kentucky
North Dakota
Virginia
BT United States

Nelson River
Flows NE out of N Lake Winnipeg into Hudson Bay at Fort Nelson.
BT Manitoba
Canada

Nelson River basin
N central.
BT Manitoba
Canada

Nemaha County
Index states as applicable.
BX Kansas
Nebraska
BT United States

Nemaha Ridge
Index states as applicable.
BT United States
SA Kansas
Nebraska
Oklahoma

Neman River basin
Index Soviet republics as applicable. Also search Neman AND appropriate area term.
UF Memel River basin
Niemen River basin
Nyeman River basin
BT USSR
SA Byelorussia
Lithuania

Nemuro Peninsula
E Hokkaido. Also search Nemuro.
BT Hokkaido
Japan

nenadkevichite
BZ orthosilicates
oxides
BT minerals
SA tantalates

Nenets
National okrug of NE Arkhangelsk Oblast in N European USSR.
BT Russian Republic
    USSR

Neocomian
Europe. Includes Berriasian (oldest), Valanginian, Hauterivian, Barremian. Includes use on level 3 under age terms(1). See list E.
BA Lower Cretaceous
    Cretaceous
BT Mesozoic
SA Berriasian
    Valanginian

neodymium
Includes use on level 1 and 2 as a chemical element (list D).
UF Nd
SA elements
    rare earths

Neogene
As of 1978, term includes use on level 1 and 2 (list E). An interval of geologic time incorporating the Miocene and Pliocene of the Tertiary period; the upper Tertiary.
BA Tertiary
    Cenozoic
NT Capistrano Formation
NA Hemphillian
    Meotian
    Miocene
    Pannonian
    Pliocene
SA Arikareean
    Siwalik System
    upper Tertiary

Neolithic
Archaeologic classification.
BA Holocene
    Quaternary
BT Cenozoic

neon
Includes use on level 1 and 2 as a chemical element (list D). Also search Ne.
UF Ne
BT noble gases
SA argon
    elements
    helium
    isotopes
    krypton
    Ne-20
    Ne-21
    Ne-22
    radon
    xenon

Neornithes
Includes use on level 2 under Aves(1). See list F.
BA Aves
BT Tetrapoda
    Vertebrata
NA Odontornithes
SA Archaeornithes

Neosho River valley
Index states as applicable. The lower course of the Neosho is called Grand River in Oklahoma. Also search Neosho River.
SA Kansas
    Oklahoma

neotectonics
A level 1 term as of 1978. Used for the study of the last structures and structural history of the Earth's crust, after the Miocene and during the later Tertiary and the Quaternary. Before 1976, also search paleotectonics. Term set options are:
concepts
    subtopic

effects
    topic [e.g. changes of level, geomorphologic effects, seismicity]
measurement
    subtopic
observations
    subtopic
rates
    subtopic [e.g. absolute age, geodetic coordinates, remote sensing, satellite measurements]
subsidence
    subtopic
uplifts
    subtopic
SA changes of level
    crust
    epeirogeny
    faults
    folds
    foliation
    fractures
    geodesy
    geomorphology
    isostasy
    seismology
    subsidence
    tectonics
    tectonophysics
    triangulation
    uplifts
    vertical movements

neotypes
SA fossils
    type specimens

Nepal
Includes use on level 1 as an area term (list O). For term set options see list B.
CO N270000N301500
    E0880000E0800000
BA Asia
NT Kali Gandaki Valley
SA Himalayas
    Kosi Basin
    Kumaun Himalayas
    Lesser Himalayas
    Siwalik Range

nepheline
BA nepheline group
    framework silicates
    silicates
BT minerals

nepheline basalt
Includes use on level 3 under igneous rocks(1) alkali basalt family(2). See list H.
BA alkali basalt family
BT igneous rocks
SA ankaratrite
    basalt
    olivine nephelinite

nepheline group
Includes use in combination with framework silicates (i.e. framework silicates, nepheline group) on level 2 under minerals(1). See list L.
BA framework silicates
    silicates
BT minerals
NA kaliophilite
    kalsilite
    nepheline
    petalite

nepheline syenite
Includes use on level 3 under igneous rocks(1) syenite family(2). See list H.
BA syenite family
BT igneous rocks
SA foyaite
    kakortokite
    lujavrite
    syenite

nephelinite
Includes use on level 3 under igneous rocks(1) alkali basalt family(2) and ultramafic family. See list H.
BA alkali basalt family
BT igneous rocks
SA olivine nephelinite
    ultramafic family

nepheloid layer
Also search nepheloid.
UF layer, nepheloid
    nepheloid zone
SA continental rise
    sea water

nepheloid zone
use nepheloid layer

nephrite
BA amphibole group
    chain silicates
    silicates
BT minerals
SA greenstone

Nephrolepidina
Includes use on level 3 under foraminifera(1) Orbitoidacea(2).
BA Orbitoidacea
    Rotaliina
    foraminifera
BT Invertebrata

nepouite
use garnierite

Neptune
Includes use on level 1 and 2. See entry under Moon(1) for term set options.
SA outer planets
    planetology
    planets
    solar system

neptunite
BA chain silicates
    silicates
BT minerals

neptunium
Includes use on level 1 and 2 as chemical element (list D).
UF Np
SA elements

Nerbuda Valley
use Narmada Valley

Nerbudda River
use Narmada River

Nereites
Genus.
BA ichnofossils

Neretva Valley
River valley in Bosnia and Herzegovina.
BT Yugoslavia

nesosilicates
use orthosilicates

nesquehonite
BA carbonates
BT minerals

Ness County
W central Kansas.
BT Kansas
    United States

Netherlands
Also search Holland. Includes use on level 1 as an area term (list O). For term set options see list B.
CO N504500N533000
    E0071500E0031500
UF Holland
BA Europe
NT Amsterdam
    Delta Area
    Groningen
    Maastricht
    Overijssel
    Utrecht
    Wadden Zee
    Zeeland
SA Campine
    European Platform
    Limburg
    Lower Rhine Basin
    Meuse River
    Meuse Valley
    North Sea Coast
    Rhine Basin
    Rhine River
    Rhine Valley
    Scheldt River

Netherlands Antilles
Formerly known as Curacao territory. Islands of Aruba, Bonaire, and Curacao in the Caribbean Sea off the coast of Venezuela, plus the Dutch section of St. Martin at N end of Leeward Islands. Includes use on level 1 as an area term (list O). For term set options see list B.
BA Caribbean region
NT Bonaire
    Curacao
SA Lesser Antilles
    West Indies

Netherlands Guiana
use Surinam

Netherlands New Guinea
use Irian Jaya

network deposits
use stockwork deposits

networks
Includes use on level 3 under seismology(1) or mineral exploration(1).
SA mineral exploration
    prospecting
    seismology
    surveys
    triangulation

Neuburg
City in W central Bavaria. Also Neuburg an der Donau.
UF Neuburg au der Donau
BT Bavaria
    West Germany

Neuburg au der Donau
use Neuburg

Neuchatel
Canton in W Switzerland. Also a city on Lake Neuchatel.
BT Switzerland

Neuquen
Province in W central.
CO S410000S360000
    W0670000W0730000
BT Argentina
SA Neuquen Basin

Neuquen Basin
River basin W central part of country. Index provinces as applicable.
BT Argentina
SA Neuquen
    Rio Negro

Neuropteris
BA Pteridospermae
    gymnosperms
BT Plantae
SA pteridophytes

Neuse River
Rises in the Piedmont and flows SE into Pamlico Sound.
BT North Carolina
    United States

neutral stress
use pore pressure

neutron
A valid term through mid-1978 used as a modifier for activation

analysis. Use *neutron activation analysis.*

neutron activation analysis
Includes use on level 3 under chemical analysis(1) methods(2). Before 1978, also search activation analysis AND neutron.
SA activation analysis
      analysis
      chemical analysis
      quantitative analysis

neutron diffraction analysis
Includes use on level 3 under chemical element(1) analysis(2). See list D.
SA analysis
      electron diffraction analysis
      X-ray diffraction analysis

neutrons
Used as a general term.
SA electrons
      particles

Nevada
Includes use on level 1 as an area term (list O). For term set options see list B.
CO N350000N420000
      W1140500W1200000
BA United States
NT Arrow Canyon Range
      Beatty
      Black Rock Desert
      Carlin
      Carlin Mine
      Churchill County
NX Clark County
NT Copper Canyon
      Cortez Mountains
      Dixie Valley
NX Douglas County
NT Elko County
NX Ely
NT Esmeralda County
NX Eureka
NT Eureka County
      Fairview Peak
      Getchell Mine
      Goldfield
      Grant Range
NX Humboldt County
NT Humboldt River valley
      Lander County
      Las Vegas
NX Lincoln County
NT Lost City
      Lovelock
NX Mineral County
NT Moapa Valley
      Mormon Mountains
      Nevada Test Site
      Nye County
      Pahute Mesa
      Pequop Mountains
      Pershing County
      Pinon Range
      Pyramid Lake
      Reno
      Roberts Mountains
      Ruby Mountains
      Santa Rosa Range
      Shoshone Mountains
      Silver Peak Mountains
      Snake Range
      Spring Mountains
      Tonopah
      Walker River
      Washoe County
      White Pine County
      Winnemucca
      Yerington
      Yucca Flat
SA Amargosa Desert
      Antelope Valley Limestone
      Antler Orogeny
      Basin and Range Province
      Bird Spring Formation
      Chinle Formation
      Coconino Sandstone
      Colorado River
      Columbia River basin
      Cordillera
      Fillmore Formation
      Great Basin
      Johnnie Formation
      Kaibab Formation
      Lake Bonneville
      Lake Mead
      Lake Tahoe
      Moenkopi Formation
      Monte Cristo Limestone
      Pogonip Group
      Roberts Mountains Formation
      Shadow Mountains
      Snake River basin
      Stirling Quartzite
      Sunrise Formation
      Supai Formation
      Tor Formation
      Toroweap Formation
      Truckee River
      Virgin River valley
      Western U.S.
      White Mountains
      Wood Canyon Formation

Nevada County
Index states as applicable.
BX Arkansas
      California
BT United States

Nevada Test Site
NW of Las Vegas in S.
BT Nevada
      United States

New Albany Shale
Includes Devonian Blocher and Blackiston formations, and Mississippian Sanderson, Underwood, and Henryville formations. Indiana and N central Kentucky.
BT Upper Devonian
      Devonian
SA Indiana
      Kentucky

New Britain
Largest island in the Bismarck Archipelago NE of New Guinea.
BT Papua New Guinea
      Australasia
SA Bismarck Archipelago

New Brunswick
Includes use on level 1 as an area term (list O). For term set options see list B.
CO N450000N480500
      W0634500W0690000
BA Canada
NT Bathurst
      Fredericton
      Kouchibouguac Bay
      Miramichi Bay
      Saint John
      Woodstock
SA Atlantic Coastal Plain
      Chaleur Bay
      Charlotte County
      Maritime Provinces
      Restigouche Estuary
      Windsor Group

New Caledonia
French overseas territory E of Queensland, Australia. Also its main island. Includes use on level 1 as an area term (list O). For term set options see list B.
CO S223000S200000
      E1670000E1640000
BA Pacific Ocean
NT Noumea
      Ouegoa
SA Melanesia

New Caledonia Basin
Between central Lord Howe Rise and Norfolk Island in SW.
BT Pacific Ocean

new data
Used as a general term.
SA data

New Delhi
Capital city SSW of Delhi.
BT Delhi
      India

New England
Index states as applicable. Includes use on level 1 as an area term (list O). For term set options see list B.
CO N420000N470000
      W0670000W0730000
BA United States
SA Connecticut
      Maine
      Massachusetts
      New Hampshire
      Rhode Island
      Vermont

New England Batholith
BT Australia
SA New South Wales
      Queensland

new genera
A valid term through 1973. Use new taxa.

new genus
A valid term through 1973. Use new taxa.

New Guinea
Refers to whole island. Index Irian Jaya and/or Papua New Guinea. Includes use on level 1 as an area term (list O). For term set options see list B.
CO S101500S023000
      E1510000E1410000
BA Australasia
SA Irian Jaya
      Malay Archipelago
      Oceania
      Pacific mobile belt
      Papua New Guinea

New Hampshire
Includes use on level 1 as an area term (list O). For term set options see list B.
CO N424500N452000
      W0704500W0723500
BA United States
NT Bronson Hill Anticlinorium
NX Carroll County
NT Grafton County
      Great Bay
      Ossipee Mountains
NX Portland
      Rockingham County
      Sullivan County
SA Atlantic Coastal Plain
      Connecticut River
      Connecticut Valley
      Eastern U.S.
      Littleton Formation
      Manchester
      Merrimack River valley
      Merrimack Synclinorium
      New England
      Rangeley Lakes
      Saco River
      White Mountains

New Hanover County
Extreme SE.
BT North Carolina
      United States

New Hebrides
Group of islands NE of New Caledonia and W of Fiji. Under joint British and French administration. Includes use on level 1 as an area term (list O). For term set options see list B.
BA Melanesia
SA Malekula
      Pacific Ocean

New Ireland
Island in the Bismarck Archipelago NNE of New Britain in SW Pacific.
BT Papua New Guinea
      Australasia
SA Bismarck Archipelago

New Jersey
Includes use on level 1 as an area term (list O). For term set options see list B.
CO N385500N412000
      W0735500W0753500
BA United States
NT Beemerville
NX Cumberland County
      Essex County
      Mercer County
      Middlesex County
NT Monmouth County
      Newark
      Pine Barrens
      Sandy Hook
NX Somerset County
      Sussex County
NT Trenton
NX Union County
      Warren County
      Woodbury
SA Atlantic Coastal Plain
      Delaware Bay
      Delaware River
      Delaware River basin
      Eastern U.S.
      Glenarm Series
      Hudson River
      Hudson Valley
      Kirkwood Formation
      Magothy Formation
      Marshalltown Formation
      Martinsburg Formation
      Monmouth Group
      Navesink Formation
      Newark Basin
      Newark Group
      Newark-Gettysburg Basin
      Piedmont
      Raritan Formation
      Valley and Ridge Province
      Woodbury Clay

New Liskeard
Town in SE.
BT Ontario
      Canada

New Madrid
Town on Mississippi River in New Madrid County.
BT Missouri
      United States

new methods
Used as a general term, e.g. on level 3 under absolute age(1).
SA absolute age
      methods

New Mexico
Includes use on level 1 as an area term (list O). For term set options see list B.
CO N313000N370000
      W1030000W1090500
BA United States
NT Albuquerque
      Ambrosia Lake
      Animas Mountains
      Bernalillo County
      Black Range
      Carlsbad
      Carlsbad Caverns
      Catron County
      Chama Basin

New Mexico ● New York Bight

Chaves County
Clovis
NX Colfax County
Curry County
NT Dona Ana County
NX Eddy County
NT Florida Mountains
Ghost Ranch
NX Grant County
NT Grants
NX Harding County
Hidalgo County
Lincoln County
NT Los Alamos Scientific Laboratory
Luna County
Magdalena Mountains
McKinley County
Nacimiento Mountains
NX Otero County
NT Pasamonte
Picuris Range
Questa Mine
Rio Arriba County
NX Roosevelt County
NT Sacramento Mountains
NX San Juan County
NT San Mateo Mountains
Sandia Mountains
Sandoval County
Santa Fe County
Santa Rita
Sierra Blanca
NX Sierra County
NT Silver City
Socorro
Socorro County
Taos County
Tatum Basin
Tierra Amarilla
Tucumcari
Tusas Mountains
NX Union County
NT Ute Creek
Valencia County
Valle Grande Mountains
Valles Caldera
White Sands
Zuni Mountains
SA Basin and Range Province
Bell Canyon Formation
Benton Formation
Bidahochi Formation
Canadian River
Capitan Formation
Carlile Shale
Castile Formation
Chihuahua tectonic belt
Chinle Formation
Colorado Group
Colorado Plateau
Cutler Formation
Dakota Formation
Delaware Basin
Dockum Group
Fort Hays Limestone Member
Four Corners
Gallup Sandstone
Gila River
Graneros Shale
Great Plains
Greenhorn Limestone
Guadalupe Mountains
La Ventana Sandstone
Lake Valley Formation
Lewis Shale
Llano Estacado
Loup Fork Group
Magdalena
Mancos Shale
Menefee Formation
Mesaverde Group
Midcontinent
Montana Group
Morrison Formation
Navajo Indian Reservation
Navajo Sandstone

Niobrara Formation
Ogallala Formation
Ojo Alamo Sandstone
Paradox Basin
Pecos River valley
Pedregosa Basin
Raton Basin
Red River
Red River valley
Redonda Formation
Rio Grande
Rio Grande Rift
Rio Grande Valley
San Andres Formation
San Francisco Mountains
San Jose Formation
San Juan Basin
San Juan River
San Luis Valley
Sangre de Cristo Mountains
Santa Fe
Smoky Hill Chalk Member
Southwestern U.S.
Supai Formation
Todilto Formation
Trans-Pecos
Wasatch Formation
Yates Formation

new minerals
Used to indicate new mineral names.
BT minerals
SA authigenesis

new names
Used as a general term.
SA nomenclature
taxonomy

New Orleans
City on the Mississippi River in Orleans Parish.
BT Louisiana
United States

New Quebec
use Ungava

New Red Sandstone
England. Permian and Triassic. Includes use on level 3 under age terms(1). See list E.
SA Permian
Triassic

New River
Rises in NW North Carolina and joins with Gauley River in West Virginia to form the Kanawha River. Index states as applicable.
BT United States
SA North Carolina
Virginia
West Virginia

New South Wales
Includes use on level 1 as an area term (list O). For term set options see list B.
CO S373000S281500
E1533000E1410000
BA Australia
NT Armidale
Australian Capital Territory
Broken Bay
Broken Hill
Bungonia Caves
Cooma
Gilgai
Gosford
Hunter Valley
Macleay River
Manildra
Murrumbidgee River
Murrumbidgee Valley
Nandewar Mountains
Round Mountain
Sydney
Sydney Basin
Tamworth

Wagga Wagga
Yass
Yeoval
SA Bulli Seam
Great Artesian Basin
Hawkesbury Sandstone
Illawarra Coal Measures
Murray Basin
Murray River
Narrabeen Group
New England Batholith
Singleton Coal Measures
Snowy Mountains
Tasman Geosyncline
Willyama Complex

new species
A valid term through 1973. Use new taxa.

new taxa
Includes use on level 3 under fossil group(1) whenever a new taxonomic group is discussed. See list F.
UF taxa, new
SA revision
taxonomy

New Ulm
City in Brown County in S central.
BT Minnesota
United States

New World Island
In Notre Dame Bay off NE.
BT Newfoundland
Canada

New York
Includes use on level 1 as an area term (list O). For term set options see list B.
CO N403000N450000
W0715500W0794500
BA United States
NT Adirondack Mountains
Albany
NX Albany County
Allegany County
NT Balmat
Bay Park
Becrfact Mountain
Benson Mines
Binghamton
Blue Mountain Lake
Catskill Mountains
NX Chautauqua County
NT Chenango River
NX Columbia County
Delaware County
NT Dutchess County
NX Erie County
Essex County
NT Finger Lakes
Fire Island
NX Franklin County
Fulton County
Genesee County
NT Gouverneur
Great South Bay
NX Greene County
Hamilton County
NT Herkimer County
Hudson Highlands
NX Jefferson County
NT Lake George
NX Livingston County
NT Long Island
NX Madison County
Manhattan
NT Marcy Massif
Mohawk Valley
NX Monroe County
Montgomery County
NT Narrows
NX Nassau County
NT New York Bight
New York City
Ogdensburg
NX Oneida County

NT Oneida Lake
Onondaga County
NX Orange County
NT Oswego County
Peekskill
Port Jervis
NX Putnam County
NT Rensselaer County
Rochester
Saint Lawrence County
Schoharie County
Seneca Lake
Shawangunk Mountains
Springwater
Staten Island
NX Suffolk County
Sullivan County
NT Ticonderoga
Ulster County
NX Utica
Warren County
Washington County
Wayne County
NT Westchester County
SA Adirondack Anorthosite
Allegheny Plateau
Appalachian Basin
Appalachian Plateau
Atlantic Coastal Plain
Black River Group
Canadian Shield
Catskill Delta
Catskill Formation
Champlain Valley
Clinton Group
Coeymans Formation
Delaware River
Delaware River basin
Eastern U.S.
Genesee Group
Great Lakes region
Grenville Province
Hamilton Group
Helderberg Group
Hudson River
Hudson Valley
Ithaca
Lake Champlain
Lake Erie
Lake Ontario
Little Falls Formation
Lockport Formation
Magothy Formation
Manhattan Formation
Manlius Formation
Medina Formation
New York City Group
Newark Basin
Newark Group
Newark-Gettysburg Basin
Niagara Escarpment
Niagara Falls
Niagara River
Normanskill Formation
Onondaga Limestone
Oriskany Sandstone
Piedmont
Potsdam Sandstone
Rochester Formation
Rondout Formation
Saint Lawrence Lowlands
Saint Lawrence River
Saint Lawrence Valley
Schenectady Formation
Susquehanna River
Susquehanna River basin
Syracuse
Tioga Bentonite
Trenton Group
Tully Limestone
Utica Shale
Valley and Ridge Province

New York Bight
CO N390000N410000
W0720000W0750000
BT New York
United States

SA Atlantic Coastal Plain
New York City
City on W tip of Long Island, Manhattan Island, Staten Island, and S tip of mainland.
BT New York
United States
New York City Group
Precambrian or Paleozoic (pre-Upper Devonian). Includes Fordham Gneiss, Inwood Marble, Manhattan Formation. W Connecticut and SE New York.
SA Connecticut
Manhattan Formation
New York
Paleozoic
Precambrian
New Zealand
Includes use on level 1 as an area term (list O). For term set options see list B.
CO S473000S343000
E1783000E1663000
BA Australasia
NT Alpine Fault
Broadlands
Foveaux Strait
North Island
South Island
Tasman orogenic zone
SA Oceania
Pacific mobile belt
Torlesse Supergroup
Waitemata Group
Newark
City on Newark Bay just W of Jersey City and New York City.
BT New Jersey
United States
Newark Basin
Index states as applicable.
BT United States
SA New Jersey
New York
Pennsylvania
Newark Group
Includes New Oxford Formation, Gettysburg Shale, Stockton Formation, Lockatong Formation, Brunswick Formation, New Haven Arkose, Meriden Formation, Portland Arkose, Pekin Formation, Cumnock Formation, Sanford Formation, Talcott Basalt, Shuttle Meadow Formation, Holyoke Basalt, East Berlin Formation, Hampden Basalt, Otterdale Sandstone.
BT Upper Triassic
Triassic
SA Connecticut
Delaware
Maryland
New Jersey
New York
North Carolina
Virginia
Newark-Gettysburg Basin
Newark Basin plus southwestward extension into SE Pennsylvania and N central Maryland. Index states as applicable.
BT United States
SA Maryland
New Jersey
New York
Pennsylvania
newberyite
BA phosphates
BT minerals
Newcastle
City on the Tyne River in Northumberland County in N England. Also search Newcastle upon Tyne.
UF Newcastle upon Tyne
BT England
Great Britain
United Kingdom
Newcastle upon Tyne
use Newcastle
Newfoundland
Comprises island of Newfoundland plus Labrador. Includes use on level 1 as an area term (list O). For term set options see list B.
CO N463000N604500
W0523000W0673000
BA Canada
NT Avalon Peninsula
Bay of Islands
Bell Island
Betts Cove
Daniel's Harbour
New World Island
Notre Dame Bay
Port au Port Peninsula
Saint John's
White Bay
SA Davidsville Group
Gander Lake Group
Harbour Main Group
Labrador
Lushs Bight Group
Signal Hill Formation
Newman Limestone
Overlies Grainger Formation; underlies Pennington Formation. E Kentucky, E Tennessee, and SW Virginia.
BT Mississippian
SA Kentucky
Tennessee
Virginia
Newmarkt
use Nowy Targ
Newport
City at mouth of Narragansett Bay in S Rhode Island, and city on the Pacific Ocean in W Oregon. Index states as applicable.
BX Oregon
Rhode Island
BT United States
Newport Bay
Dredged harbor in Orange County in S.
BT California
United States
Newport-Inglewood Fault
In Los Angeles area in S California. Also search Newport-Inglewood fault zone.
UF Newport-Inglewood fault zone
BT California
United States
Newport-Inglewood fault zone
use Newport-Inglewood Fault
Newton County
Index states as applicable.
BX Arkansas
Georgia
Indiana
Mississippi
Missouri
Texas
BT United States
Neyveli
Town in E Tamil Nadu.
UF Neivelli
BT Tamil Nadu
India
Neyveli Lignite
Associated with sands, conglomerates and clays of Cuddalore Sandstone Formation. E and NE of Tamil Nadu. Also search Neyveli.
BT Tertiary

SA India
Miocene
Pliocene
Tamil Nadu
upper Miocene
Ngorora Formation
BT Miocene
Neogene
Tertiary
SA Kenya
Ni
use nickel
Niagara Escarpment
Site of the great falls of the Niagara River. Index state and province as applicable. Also search Niagara.
BT North America
SA New York
Ontario
Niagara Falls
Great falls of the Niagara River. Index state and province as applicable. Also search Niagara.
CO N430600N430600
W0790300W0790300
BT North America
SA New York
Ontario
Niagara River
Flows between Lake Erie and Lake Ontario going over the Niagara Escarpment. Index province and state as applicable. Also search Niagara.
BT North America
SA New York
Ontario
Niagaran
Provincial series, North America. Above Alexandrian, below Cayugan. Includes use on level 3 under age terms(1). See list E.
BA Middle Silurian
Silurian
BT Paleozoic
NT Bainbridge Formation
Lockport Formation
Nicaragua
Includes use on level 1 as an area term (list O). For term set options see list B.
BA Central America
NT Managua
SA Leon
Nicaragua Rise
Between Jamaica and Nicaragua.
UF Nicaraguan Rise
BT Caribbean Sea
Nicaraguan Rise
use Nicaragua Rise
niccolite
UF nickeline
nicolite
BA arsenides
sulfides
BT minerals
SA nickel
Nice
City on the Mediterranean Sea on the Riviera.
BT Alpes-Maritimes
France
nickel
Includes use on level 1 and 2 as a commodity term (list C) and as a chemical element (list D); on level 2 under mineral deposits, genesis(1). Also search Ni.
UF Ni
SA elements
heavy metals
niccolite
pentlandite

nickel glance
use gersdorffite
nickel pyrites
use millerite
nickel-antimony glance
use ullmannite
nickeline
use niccolite
Nicobar Island
use Nicobar Islands
Nicobar Islands
Island group in Bay of Bengal NW of Sumatra, forming S part of Andaman and Nicobar Islands. Also search Nicobar Island.
UF Nicobar Island
Nicobars
BT India
Nicobars
use Nicobar Islands
Nicola Group
S British Columbia.
BT Triassic
SA British Columbia
Canada
Nicolaus
Village N of Sacramento in Sutter County in N central.
BT California
United States
nicolite
use niccolite
nicopyrite
use pentlandite
Nida Basin
River basin in S part of country. Also search Nida; Nida River basin.
UF Nida River basin
BT Kielce
Poland
Nida River basin
use Nida Basin
Niemen River basin
use Neman River basin
Nievre
Department in central part of country.
CO N463000N473000
E0041000E0024500
BT France
NT Nivernais Plateau
SA Morvan
Yonne Valley
Niger
Includes use on level 1 as an area term (list O). For term set options see list B.
CO N120000N233000
E0160000E0001500
BA Africa
NT Agades
SA Chad Basin
Lake Chad
Mali-Niger Syneclise
Niger River
Niger Valley
Sahara
Sahel
West Africa
West African Shield
Younger Granites
Niger Delta
On the Gulf of Guinea.
BT Nigeria
Niger River
Rises in Guinea and flows NE into a great curve and then SE into the Gulf of Guinea. Index countries as applicable.
UF Joliba River
Kworra River

Niger River

- BT Africa
- SA Benin
  - Guinea
  - Mali
  - Niger
  - Nigeria

Niger Valley
River valley. Index countries as applicable. Also search Niger.
- UF Joliba Valley
  - Kworra Valley
- BT Africa
- SA Benin
  - Guinea
  - Mali
  - Niger
  - Nigeria

Nigeria
Includes use on level 1 as an area term (list O). For term set options see list B.
- CO N040000N140000
  E0143000E0023000
- BA Africa
- NT Bida
  - Ewekoro
  - Ibadan
  - Jos Plateau
  - Lagos Lagoon
  - Niger Delta
  - Sokoto Basin
  - Zaria
- SA Benue Valley
  - Chad Basin
  - Ewekoro Formation
  - Lake Chad
  - Mali-Niger Syncline
  - Niger River
  - Niger Valley
  - West Africa
  - Younger Granites

nigerite
- BA oxides
- BT minerals

niggliite
- BZ native elements and alloys
  - sulfides
- BT minerals
- SA tellurides

Niigata
City and prefecture in NW.
- BT Honshu
  - Japan

Nikitovka
City in central Stalino Oblast in Donbas.
- BT Ukraine
  - USSR

Nikopol
City in Dnepropetrovsk Oblast in E central.
- BT Ukraine
  - USSR

Niksic
Town in SW part of country.
- BT Montenegro
  - Yugoslavia

Nile Delta
On Mediterranean Sea E of Alexandria.
- UF Nile River delta
- BT Egypt
- SA Nile River

Nile River
The longest river in the world. Its remotest headstream is the Luvironza River in Burundi. The flow is through Lake Victoria and Lake Albert, and then, as the White Nile, into Sudan where it unites with the Blue Nile from Ethiopia as the Nile proper which flows into the Mediterranean Sea. Index countries as applicable.
- CO N220000N310000
  E0370000E0310000
- BT Africa
- SA Blue Nile
  - Burundi
  - Egypt
  - Ethiopia
  - Nile Delta
  - Sudan
  - Uganda

Nile River delta
use Nile Delta

Nile Valley
River valley of the Nile River proper. Index countries as applicable.
- BT Africa
- SA Egypt
  - Sudan

Nilgiri
Former Indian state near coast of NW Bay of Bengal.
- BT Orissa
  - India

Nimba Mountains
West African bulge. Index countries as applicable. Also search Nimba.
- BT Africa
- SA Guinea
  - Ivory Coast
  - Liberia

Ninetyeast Ridge
Undersea ridge extending N and S along line of 90° E longitude from just S of Nicobar Islands to approximately 30° S latitude.
- BT Indian Ocean

niningerite
Meteorite mineral.
- BA sulfides
- BT minerals

Niniyur Group
S part of country.
- BT Danian
  - lower Paleocene
  - Paleocene
  - Paleogene
  - Tertiary
- SA Cretaceous
  - India
  - Tamil Nadu
  - Upper Cretaceous

niobates
Includes use on level 3 under minerals(1) oxides(2). See list L.
- BA oxides
- BT minerals
- SA aeschynite
  - betafite
  - euxenite
  - fergusonite
  - ixiolite
  - loparite
  - microlite
  - pyrochlore
  - samarskite
  - stibiotantalite
  - tantalite
  - tapiolite

niobium
Includes use on level 1 and 2 as a commodity term (list C) and as a chemical element (list D). Also search Nb.
- UF columbium
  - Nb
- SA elements
  - tantalum

Niobrara Chalk
use Niobrara Formation

Niobrara Formation
In Colorado Group. Includes Fort Hays Limestone, Smoky Hill Chalk members. Also search Niobrara Chalk.
- UF Niobrara Chalk
- BT Upper Cretaceous
  - Cretaceous
- SA Colorado
  - Colorado Group
  - Fort Hays Limestone Member
  - Kansas
  - Minnesota
  - Montana
  - Nebraska
  - New Mexico
  - North Dakota
  - Smoky Hill Chalk Member
  - South Dakota
  - Wyoming

Nipissing Diabase
- BT Precambrian
- SA Canada
  - Ontario
  - Quebec

Nisqually Glacier
On S slope of Mt Rainier in Mt. Rainier National Park in W central.
- BT Washington
  - United States

nitrates
Includes use on level 1 as a commodity term (list C); on level 2 under minerals(1). See list L.
- BT minerals

nitrides
Includes use on level 3 under minerals(1) native elements and alloys(2). See list L.
- BA native elements and alloys
- BT minerals

nitrogen
Includes use on level 1 and 2 as a chemical element (list D); on level 3 under soils(1) chemistry(2). Also search N.
- UF N
- SA elements
  - fertilizers
  - N-15/N-14
  - soils

Nittany Valley
In Centre and Clinton counties in central.
- BT Pennsylvania
  - United States

Nitzschia
Genus.
- BA diatoms
  - algae
- BT Plantae

nivation
- UF snow-patch erosion
- BT erosion
- SA glaciers

Nivernais Hills
use Nivernais Plateau

Nivernais Plateau
In central part of country.
- UF Nivernais Hills
- BT Nievre
  - France

Nizhni Novgorod
use Gorki

Nizhniy Ufaley
use Ufaley

Nizke Jesenik Mountains
use Nizky Jezenik Mountains

Nizky Jezenik Mountains
In NW near the Polish border. Also search Nizky Jezenik.
- UF Low Jesenik Mountains
  - Nizke Jesenik Mountains
- BT Moravia
  - Czechoslovakia

noble gases
Includes use on level 1 and 2. See list D for term set options. Also see list C (commodities).
- UF inert gases
  - rare gases
- NT argon
  - helium
  - krypton
  - neon
  - radon
  - xenon
- SA gases
  - isotopes

Nodosaria
Genus. Includes use on level 3 under foraminifera(1) Nodosariacea(2).
- BA Nodosariidae
  - Nodosariacea
  - Rotaliina
  - foraminifera
- BT Invertebrata

Nodosariacea
Includes use on level 2 under foraminifera(1). See list F.
- BA Rotaliina
  - foraminifera
- BT Invertebrata
- NA Nodosariidae

Nodosariidae
Family. Includes use on level 3 under foraminifera(1) Nodosariacea(1).
- BA Nodosariacea
  - Rotaliina
  - foraminifera
- BT Invertebrata
- NA Lenticulina
  - Nodosaria

nodules
Includes use on level 1 (list A). See list K. If level 1, term set options include:
kind of nodule [carbonate composition, ferromanganese composition, manganese composition, phosphate composition]
topic
topic [age, classification, composition, distribution, genesis, observations, properties, structure]
subtopic
- BA secondary structures
  - sedimentary structures
- SA carbonate composition
  - concretions
  - ferromanganese composition
  - geodes
  - marine geology
  - ocean floors
  - phosphate composition
  - sedimentary rocks
  - sedimentation

Nohi Rhyolite
Complex in central Japan.
- BT Honshu
  - Japan

noise
Includes use on level 3 under seismology(1) or geophysical methods(1). Before 1978, also search seismic noise.
- SA geophysical methods
  - microseisms
  - seismology
  - signals

Nojiri Lake
Central Honshu.
- BT Honshu
  - Japan

Nome
City on S side of Seward Peninsula in W.
- BT Alaska
  - United States

nomenclature
Includes use as a level 2 or 3 term appropriate to a large number of topics, e.g. on level 2 under sedimentary petrology(1). See list G.
- SA classification
  - definition
  - homonymy
  - identification
  - new names

nomograms
- UF nomographs
- SA automatic data processing

nomographs
use nomograms

non-glacial
A valid index term through 1976. Now use non-glacial ice. Before 1976 also search ice AND non-glacial.

non-glacial ice
Term introduced in 1976. Includes use on level 3 under glacial geology(1). Before 1976 also search ice AND non-glacial.
- UF nonglacial (ice)
- SA glacial geology
  - ice
  - periglacial features

Noncalcarea
use Silicispongiae

Nonesuch Shale
In Oronto Group. N Michigan and NE Wisconsin.
- BT Precambrian
- SA Michigan
  - Wisconsin

Nong-Son
Village in Quang Nam Province in N South Vietnam.
- UF Nongson
- BT Vietnam
  - Asia

nonglacial (ice)
use non-glacial ice

Nongson
use Nong-Son

nonmare basalt
use KREEP

nonmetals
Use mainly when of economic value. Includes use on level 1 and 2 as a commodity term (list C).
- SA economic geology
  - elements
  - industrial minerals
  - metals

nonterrigenous
A valid level 2 term through 1977 used in combination with clastic rocks or clastic sediments. After 1977, use nonterrigenous materials only on level 3, i.e. under sediments(1) or sedimentary rocks(1).

nonterrigenous materials
Term introduced in 1978. Includes use on level 3 under sediments(1) and sedimentary rocks(1). Before 1978, also search nonterrigenous AND clastic rocks or clastic sediments.
- SA clastic rocks
  - clastic sediments
  - diatomaceous earth
  - diatomite
  - materials
  - novaculite
  - ooze
  - radiolarite
  - sedimentary rocks
  - sediments
  - spongolite
  - terrigenous materials
  - turbidite

nontronite
Also search chloropal.
- UF chloropal
  - gramenite
  - morencite
  - pinquite
- BA sheet silicates
  - silicates
- BT minerals
- SA clay minerals
  - montmorillonite

Noonday Dolomite
Underlies Johnnie Formation; overlies Kingston Peale Formation with disconformity or slight angular unconformity.
- BT Precambrian
- SA California

Noranda
Mining city in SW.
- BT Quebec
  - Canada

norbergite
- BZ halides
  - orthosilicates
- BT minerals
- SA fluorides
  - humite group

Nord
Department in extreme N.
- BT France
- NT Dunkirk
  - Lille
  - Valenciennes
- SA Nord-Pas-de-Calais Basin

Nord Fjord
Inlet of Norwegian Sea.
- BT Norway

Nord-Pas-de-Calais Basin
In extreme N part of country. Index departments as applicable.
- BT France
- SA Nord
  - Pas-de-Calais

Nord-Trondelag
County N of Trondheim Fjord.
- BT Norway
- SA Trondelag

Nordauslandet
Norwegian for North East Land. One of the islands of the Spitsbergen group in Barents Sea NE of island of Spitsbergen.
- BT Spitsbergen
  - Arctic region

Nordland
County in W.
- BT Norway
- NT Bleikvassli
  - Lofoten Islands
  - Narvik
  - Ofoten
  - Rana Fjord
- SA Vesteralen

Nordlingen
use Ries Crater

Nordlinger Ries
use Ries Crater

Nordlinger Ries Crater
use Ries Crater

nordstrandite
- BA oxides
- BT minerals

Norfolk
City on Hampton Roads in SE.
- CO N523000N530000
  - E0014500E0000500
- BT Virginia
  - United States

Norfolk Island
Midway between New Caledonia and N New Zealand. An external territory of Australia.
- BT Pacific Ocean

Norfolk Ridge
Between New Caledonia and Norfolk Island in SW.
- BT Pacific Ocean

Norian
Europe. Above Carnian, below Rhaetian. Includes use on level 3 under age terms(1). See list E.
- BA Upper Triassic
  - Triassic
- BT Mesozoic

Norilsk
Town in Taymyr National Okrug in N Siberia. Also search Noril'sk.
- CO N692000N692200
  - E0880300E0880100
- BT Russian Republic
  - USSR

Norilsk region
A mining area in the Taymyr National Okrug connected to the town of Dudinka on the Yenisei River by a 60 mile narrow gage rail line. Also search Norilsk.
- CO N690000N694500
  - E0890000E0870000
- BT Russian Republic
  - USSR

norite
Includes use on level 3 under igneous rocks(1) gabbro family(2). See list H.
- BA gabbro family
- BT igneous rocks

normal
A valid term through 1977. After 1977, use normal faults or normal folds.

normal earthquake
use shallow-focus earthquakes

normal faults
Term introduced in 1978. Includes use on level 3 under faults(1) displacements(2). Before 1978, also search faults AND normal; slump faults.
- UF gravity faults
  - normal slip faults
- BA faults
- SA block structures

normal folds
Term introduced in 1978. Before 1978 search folds AND normal.
- BA folds
- SA symmetric folds

normal slip faults
use normal faults

Norman Wells
Village on the Mackenzie River in W District of Mackenzie.
- BT Northwest Territories
  - Canada

Normandie
use Normandy

Normandy
Historical region of NW part of country. Index departments as applicable.
- UF Normandie
- BT France
- SA Calvados
  - Eure
  - Manche
  - Orne
  - Seine-Maritime

Normanskill Chalk
use Normanskill Formation

Normanskill Formation
Comprises Mount Merino Chert and Shale, Austin Glen Grit and Shale. NW Massachusetts, E New York, and SW Vermont.
- UF Normanskill Chalk
- BT Middle Ordovician
  - Ordovician
- SA Massachusetts
  - New York
  - Vermont

Norphlet Formation
Underlies Smackover Formation; overlies Louann Salt. Subsurface.
- BT Jurassic
- SA Arkansas
  - Louisiana
  - Texas

Norrbotten
Northernmost county.
- BT Sweden
- NT Gallivare
  - Kebnekaise
  - Kiruna
  - Laisvall
- SA Skellefte

Norseman
Town in S central.
- BT Western Australia
  - Australia

norsethite
- BA carbonates
- BT minerals

north
General term used after any of the main geographic terms (list O).
- UF northern
- SA north-central
  - northeast
  - northwest

North Africa
A region. Index countries as applicable.
- BA Africa
- SA Algeria
  - Central Africa
  - East Africa
  - Libya
  - Morocco
  - Southern Africa
  - Tunisia
  - West Africa

North America
Includes use on level 1 or 2 as an area term (list O). If 1, see term set options under list B. To retrieve all documents, individual countries and physiographic regions should also be searched (see list O).
- CO N080000N840000
  - W0100000W1730000
- NA Appalachians
  - Atlantic Coastal Plain
- NT Canada
  - Cordillera
  - Great Appalachian Valley
- NA Great Lakes
  - Great Lakes region
  - Great Plains
  - Gulf Coastal Plain
- NT Lake Agassiz
- NA Mexico
- NT Niagara Escarpment
  - Niagara Falls
  - Niagara River
  - Okanagan Valley
  - Rio Grande Rift
  - Rocky Mountain Trench
- NA Rocky Mountains
- NT Saint Clair River
  - Saint Clair River delta
  - Saint Elias Mountains
  - Saint Lawrence Lowlands

Saint Lawrence River
United States
NA　Western Interior
NT　White River
SA　North American Plate
　　Pacific mobile belt

North American Plate
Includes North America N of central America as well as the North Atlantic Ocean W of the Mid-Ocean Ridge.
SA　North America
　　plate tectonics

North Arcot
District in E central.
BT　Tamil Nadu
　　India

North Atlantic
Term introduced in 1978. Before 1978, search Atlantic Ocean AND north.
BA　Atlantic Ocean
SA　Northeast Atlantic
　　Northwest Atlantic
　　South Atlantic
　　Southeast Atlantic
　　Southwest Atlantic
　　West Atlantic

North Bay
City on Lake Nipissing in SE.
BT　Ontario
　　Canada

North Borneo
use　Sabah

North Carolina
Includes use on level 1 as an area term (list O). For term set options see list B.
CO　N335000N363500
　　W0753000W0841500
BA　United States
NX　Alleghany County
NT　Beaufort
　　Beaufort County
　　Buncombe County
NX　Burke County
NT　Cape Fear Arch
　　Cape Hatteras
　　Cape Lookout
　　Castle Hayne
NX　Chatham County
　　Cherokee County
　　Clay County
　　Cumberland County
NT　Dare County
　　Davidson
NX　Davidson County
　　Franklin County
　　Graham County
NT　Grandfather Mountain
NX　Greene County
　　Jackson County
　　Lee County
　　Lincoln County
　　Macon County
　　Madison County
　　Mitchell County
　　Montgomery County
　　Moore County
NT　Neuse River
　　New Hanover County
　　Onslow Bay
NX　Orange County
NT　Outer Banks
　　Pamlico River
　　Pamlico Sound
NX　Polk County
NT　Pungo River
NX　Randolph County
　　Rockingham County
　　Rowan County
NT　Stokes County
NX　Surry County
　　Union County
　　Warren County
　　Washington County
　　Wayne County
　　Wilson County
SA　Appalachian Basin
　　Appalachian Plateau
　　Ashe Formation
　　Atlantic Coastal Plain
　　Beaufort Formation
　　Blue Ridge Mountains
　　Blue Ridge Province
　　Brevard Zone
　　Carolina slate belt
　　Castle Hayne Limestone
　　Chilhowee Group
　　Dan River basin
　　Duplin Formation
　　Eastern U.S.
　　Great Smoky Mountains
　　Kings Mountain
　　Knox Group
　　Lake Chatuge
　　Murphy Marble
　　New River
　　Newark Group
　　Peedee Formation
　　Piedmont
　　Roan Supergroup
　　Rome Formation
　　Sand Hills
　　Shady Dolomite
　　Talladega Group
　　Trent Valley
　　Tuscaloosa Formation
　　Valley and Ridge Province
　　Waccamaw Formation
　　Wicomico Formation
　　Wilmington
　　Yorktown Formation

North Caucasus
use　Northern Caucasus

north central
use　north-central

North Central (United States)
use　Midwest

North Dakota
Includes use on level 1 as an area term (list O). For term set options see list B.
CO　N455500N490000
　　W0963500W1040500
BA　United States
NX　Adams County
NT　Billings County
NX　Burke County
　　Cass County
　　Eddy County
NT　Grand Forks
NX　Grant County
　　McHenry County
　　McIntosh County
　　McLean County
　　Mercer County
　　Morton County
　　Nelson County
NT　Oliver County
NX　Sheridan County
　　Sioux County
　　Stark County
NT　Stutsman County
　　Walsh County
NX　Ward County
SA　Badlands
　　Colorado Group
　　Dakota Formation
　　Elk Point Group
　　Fort Union Formation
　　Fox Hills Formation
　　Golden Valley Formation
　　Great Plains
　　Hell Creek Formation
　　Lake Agassiz
　　Lance Formation
　　Little Missouri River basin
　　Midcontinent
　　Midwest
　　Missouri River
　　Missouri River basin
　　Missouri River valley
　　Montana Group
　　Niobrara Formation
　　Pierre Shale
　　Prairie Evaporite
　　Red River Formation
　　Sentinel Butte Formation
　　Souris River basin
　　Tongue River Formation
　　Wasatch Formation
　　White River Group
　　Williston Basin
　　Winnipeg Formation
　　Winnipegosis Formation

North Devon Island
use　Devon Island

North German Plain
Roughly the northern one-third of West Germany and the northern half of East Germany. Index countries as applicable.
BT　Europe
SA　East Germany
　　West Germany

North Island
The northernmost of 3 main islands of New Zealand.
CO　S413000S343000
　　E1783000E1720000
BT　New Zealand
NT　Auckland
　　Coromandel Peninsula
　　Gisborne
　　Hawke's Bay
　　Kaipara Harbor
　　Manawatu River valley
　　Mokoia
　　Raukumara Peninsula
　　Ruapehu
　　Taranaki
　　Tarawera volcanic complex
　　Taupo
　　Taupo volcanic zone
　　Te Aroha
　　Waikato Basin
　　Wairakei
　　Wairarapa
　　Wanganui
　　Wanganui Valley
　　Wellington
　　White Island
SA　Tasman orogenic zone
　　Waitemata Group

North Korea
Officially Democratic People's Republic of Korea. Republic, on E coast of Asia, bounded on N by China, on NE by USSR, on E by Japan Sea, on S by South Korea, and on W by the Yellow Sea and Korea Bay.
CO　N374500N430000
　　E1303000E1241500
BT　Korea
SA　Kangwon
　　South Korea

North Ossetia
North Ossetian Autonomous Soviet Socialist Republic. On N slopes of central Caucasus Mountains.
BT　Russian Republic
　　USSR

North Pacific
Term introduced in 1978. Before 1978, search Pacific Ocean AND north.
BA　Pacific Ocean
SA　East Pacific
　　Equatorial Pacific
　　Northeast Pacific
　　Northwest Pacific
　　Pacific Basin
　　South Pacific
　　Southeast Pacific
　　Southwest Pacific
　　West Pacific

North Polar Sea
use　Arctic Ocean

North Pole
The N extremity of the Earth's axis at 90° N latitude and the point from which all directions are S.
SA　Polar regions

North Ray Crater
BT　Moon

North Rhine
North portion of former Prussian Rhine Province which included Aachen, Cologne and Dusseldorf.
BT　North Rhine-Westphalia
　　West Germany

North Rhine-Westphalia
State.
CO　N502000N523000
　　E0093000E0060000
BT　West Germany
NT　Aachen
　　Bergisch Gladbach
　　Bochum
　　Cologne
　　Ibbenburen
　　Kamen
　　Krefeld
　　Lindlar
　　Lippe
　　Munster
　　Munsterland
　　North Rhine
　　Paderborn
　　Plettenberg
　　Ruhr
　　Sauerland
　　Siebengebirge
　　University of Bonn
　　Wiehen Mountains
SA　Ems River
　　Essen Beds
　　Rhenish Schiefergebirge
　　Rhineland
　　Steinheim
　　Teutoburg Forest
　　Weser River
　　Westphalia

North Saskatchewan River
Rises in Columbia Ice Field at foot of Mt. Saskatchewan in Alberta. Index provinces as applicable.
BT　Canada
SA　Alberta
　　Saskatchewan
　　Saskatchewan River
　　South Saskatchewan River

North Sea
Between the European continent on the S and E, and Great Britain on the W. Also search North Sea Basin. Includes use on level 1 as an area term (list O). For term set options see list B.
CO　N510000N610000
　　E0080000W0020000
UF　North Sea Basin
BA　Atlantic Ocean
NT　Dogger Bank
　　Kattegat
　　Norwegian Channel
　　Skagerrak
SA　Forties Field
　　North Sea region

North Sea Basin
use　North Sea

North Sea Coast
Index England, Scotland and European countries as applicable.
BT　Europe
SA　Belgium
　　Denmark
　　England

France
Germany
Netherlands
Norway
Scotland
Sweden

North Sea region
SA   North Sea

North Shore Volcanic Group
use North Shore Volcanics

North Shore Volcanics
In Keweenawan Group. NE Minnesota. Also search North Shore Volcanic Group.
UF   North Shore Volcanic Group
BT   Precambrian
SA   Minnesota

North Slope
Arctic plains N of Brooks Range to Arctic Sea.
BT   Alaska
     United States
SA   Prudhoe Bay

North Sudetic Basin
North part of the Sudetic Basin in the Sudeten Mountains. Index countries as applicable.
BT   Europe
SA   Czechoslovakia
     Poland
     Sudeten Mountains
     Sudetic Basin

north west
use northwest

north-central
General term used after any of the main geographic terms (list O).
UF   north central
     northcentral
SA   central
     north

North-Central (United States)
use Midwest

north-east
use northeast

north-west
use northwest

North-West Frontier Province
W Pakistan.
BT   Pakistan
NT   Chitral
     Hazara
     Peshawar
     Swat
SA   Sulaiman Range

Northamptonshire
County in central.
CO   N520000N524000
     W0002000W0012000
BT   England
     Great Britain
     United Kingdom

northcentral
use north-central

northeast
General term used after any of the main geographic terms (list O).
UF   north-east
     northeastern
SA   east
     north

Northeast Atlantic
Term introduced in 1978. Before 1978, search Atlantic Ocean AND northeast.
BA   Atlantic Ocean
SA   East Atlantic
     North Atlantic
     Northwest Atlantic
     South Atlantic
     Southeast Atlantic
     Southwest Atlantic
     West Atlantic

Northeast Pacific
Term introduced in 1978. Before 1978, search Pacific Ocean AND northeast.
BA   Pacific Ocean
SA   East Pacific
     Equatorial Pacific
     North Pacific
     Northwest Pacific
     South Pacific
     Southeast Pacific
     Southwest Pacific
     West Pacific

Northeast Providence Channel
Strait NE of Nassau between Great Abaco and Eleuthera islands.
BT   Bahamas
     West Indies

northeastern
use northeast

Northeastern USSR
Term introduced in 1978. Before 1978, search USSR AND northeast.
BA   USSR

northern
use north

Northern Andes
Term introduced in 1978. Before 1978, search Andes AND north or northern.
BA   Andes
BT   South America
SA   Central Andes
     Southern Andes

Northern Appalachians
Term introduced in 1978. Before 1978, search Appalachians AND north or northern.
BA   Appalachians
BT   North America
SA   Central Appalachians
     Southern Appalachians

Northern Caucasia
use Northern Caucasus

Northern Caucasus
Region N of the Caucasus comprising Chechen-Ingush A.S.S.R., S half of Krasnodar Kray, and the Daghestan, Kabardin-Balkar, and North Ossetian A.S.S.Rs. Also search Ciscaucasia.
CO   N440000N460000
     E0450000E0380000
UF   Ciscaucasia
     North Caucasus
     Northern Caucasia
BT   Russian Republic
     USSR
SA   Caucasus Foreland
     Transcaucasia

Northern Dvina River
Chief river of the White Sea Basin in NW European USSR.
BT   Russian Republic
     USSR

Northern Great Plains
Term introduced in 1978. Before 1978, search Great Plains AND north or northern.
BA   Great Plains
BT   North America
SA   Southern Great Plains

Northern Hemisphere
Used when discussing many large areas too numerous to mention. Includes use on level 1 and 2 as an area term (list O). If 1, see term set options under list B.
SA   Eastern Hemisphere
     Laurasia
     Pangaea
     Southern Hemisphere
     Western Hemisphere

Northern Ireland
Six counties comprising the NE part of island of Ireland. Also search Ulster. Includes use on level 1 as an area term (list O). For term set options see list B.
CO   N540000N553000
     W0053000W0080000
UF   Ulster
BT   United Kingdom
BA   Europe
NT   Antrim
     Belfast
     Fermanagh
     Giant's Causeway
     Londonderry
     Tyrone
SA   Ireland

Northern Light Lake
In SW just N of Minnesota border.
BT   Ontario
     Canada

Northern Limestone Alps
The northern part of the Dinaric Alps which are a great belt of limestone ranges and plateaus along the Dalmatian coast of the Adriatic Sea.
BT   Croatia
     Yugoslavia
SA   Alps
     Dinaric Alps

Northern Range
N Trinidad.
BT   Trinidad and Tobago
     West Indies

Northern Rhodesia
use Zambia

Northern Rocky Mountains
Term introduced in 1978. Before 1978, also search Rocky Mountains AND north or northern.
BA   Rocky Mountains
BT   North America
SA   Central Rocky Mountains
     Southern Rocky Mountains

Northern Territory
Includes use on level 1 as an area term (list O). For term set options see list B.
CO   S260000S110000
     E1380000E1290000
BA   Australia
NT   Alice Springs
     Darwin
     Gosses Bluff
     Henbury
     Katherine
     Tennant Creek
     Victoria Valley
SA   Amadeus Basin
     Arunta Complex
     Bitter Springs Formation
     Carpentaria Basin
     Georgina Basin
     Great Artesian Basin
     Joseph Bonaparte Gulf
     Musgrave Ranges
     Simpson Desert

Northern Urals
Term introduced in 1978. Before 1978, search Urals AND north or northern.
CO   N620000N680000
     E0680000E0590000
BA   Urals
BT   Russian Republic
     USSR
SA   Polar Urals

Northumberland
County in N on border of Scotland.
CO   N544500N554500
     W0013000W0023000
BT   England
     Great Britain
     United Kingdom

Northumberland Strait
Channel of the Gulf of Saint Lawrence between Prince Edward Island on the N, and New Brunswick and Nova Scotia on the SE and E respectively.
BT   Canada

northwest
General term used after any of the main geographic terms (list O).
UF   north west
     north-west
     northwestern
SA   north
     west

Northwest Atlantic
Term introduced in 1978. Before 1978, search Atlantic Ocean AND northwest.
BA   Atlantic Ocean
SA   East Atlantic
     North Atlantic
     Northeast Atlantic
     South Atlantic
     Southeast Atlantic
     Southwest Atlantic
     West Atlantic

Northwest Pacific
Term introduced in 1978. Before 1978, search Pacific Ocean AND northwest.
BA   Pacific Ocean
SA   East Pacific
     Equatorial Pacific
     North Pacific
     Northeast Pacific
     South Pacific
     Southeast Pacific
     Southwest Pacific
     West Pacific

Northwest Territories
Index districts as applicable. Includes use on level 1 as an area term (list O). For term set options see list B.
CO   N600000N840000
     W0600000W1360000
BA   Canada
NT   Agricola Lake
     Amund Ringnes Island
     Arctic Archipelago
     Axel Heiberg Island
     Baffin Island
     Banks Island
     Barnes ice cap
     Bathurst Island
     Bear Province
     Bear-Slave Operation
     Belcher Islands
     Boothia Peninsula
     Cape Dyer
     Coppermine River
     Cornwallis Island
     Cumberland Peninsula
     Devon Island
     District of Franklin
     District of Keewatin
     District of Mackenzie
     Ellef Ringnes Island
     Ellesmere Island
     Fort Good Hope
     Fort Ross
     Great Bear Lake
     Great Slave Lake
     Inuvik
     Kaminak Lake
     Lancaster Sound
     Mackenzie Delta
     Mackenzie River valley
     Meighen Island
     Melville Peninsula
     Muskox Intrusion

Narpaing Fjord
Norman Wells
Otto Fjord
Queen Elizabeth Islands
Radstock Bay
Richards Island
Somerset Island
South Nahanni River
Sverdrup Basin
Tanquary Fiord
Tuktoyaktuk Peninsula
Wager
Yellowknife
SA Canadian Shield
Carswell Structure
Christopher Formation
Churchill Province
Edmonton Formation
Eureka Sound Formation
Franklin Mountains
Horn Plateau Formation
Hudson Bay
Hudson Bay Lowlands
Hurwitz Group
Isachsen Formation
Liard River
Mackenzie Mountains
Michelle Formation
Nahanni Formation
Peel River
Peel Sound Formation
Prince of Wales Island
Ramparts Formation
Richardson Mountains
Slave Province
Yellowknife Group

northwestern
use northwest

Northwestern USSR
Term introduced in 1978. Before 1978, search USSR AND northwest.
BA USSR

Norton County
NW Kansas.
BT Kansas
United States

Norton County Meteorite
Term introduced in 1978 to distinguish from place name, Norton County. Before 1978, also search Norton County for the meteorite.
BT Kansas
United States
SA meteorites

Norway
Includes use on level 1 as an area term (list O). For term set options see list B.
CO N580000N710000
E0310000W0040000
BA Europe
NT Arendal
Bergen
Bjerkrem-Sogndal Massif
Drammen
Finnmark
Folldal
Gudbrandsdalen
Hardangervidda
Jotunheim Mountains
Kongsberg
Kristiansund
Nord Fjord
Nord-Trondelag
Nordland
Oslo
Ringerike
Rogaland
Sogn
Solund Islands
Sor-Trondelag
Soroy
Telemark
Troms
Trondelag
Tunsbergdalsbreen
Valdres
Vest-Agder
Vesteralen
Vestfold
SA Arctic region
Baltic Shield
Fennoscandia
Lapland
North Sea Coast
Porsanger Dolomite Formation
Ringerike Sandstone
Scandinavia
Sparagmite Group
Sulitjelma

Norwegian Channel
Undersea feature just off SW Norway.
UF Norwegian Deep
Norwegian Trough
BT North Sea
Atlantic Ocean

Norwegian Deep
use Norwegian Channel

Norwegian Sea
Part of Atlantic Ocean, off coast of Norway, opening N on Greenland Sea, NE on Barents Sea, S on North Sea, and SW on the open Atlantic.
BT Atlantic Ocean
NT Voring Plateau
SA Kolbeinsey Island

Norwegian Trough
use Norwegian Channel

Norwich
City in E.
BT England
Great Britain
United Kingdom

nosean
Also search noselite.
UF noselite
BA sodalite group
framework silicates
silicates
BT minerals

noselite
use nosean

Noto Peninsula
Large headland projecting N into Japan Sea in central.
BT Honshu
Japan

Notoungulata
Order. Includes use on level 2 under Mammalia(1). See list F.
UF Notungulata
BA Mammalia
BT Tetrapoda
Vertebrata

Notre Dame Bay
Inlet of Atlantic Ocean on N coast of island of Newfoundland.
BT Newfoundland
Canada

Nottingham
City in N central.
BT England
Great Britain
United Kingdom

Nottinghamshire
County in N central.
CO N524500N533000
W0004000W0013000
BT England
Great Britain
United Kingdom

Notungulata
use Notoungulata

Nouakchott
City in SW near coast.
BT Mauritania

Noumea
Town on SW coast of French island of New Caledonia E of Queensland, Australia.
BT New Caledonia
Pacific Ocean

Nova Scotia
Includes use on level 1 as an area term (list O). For term set options see list B.
CO N433000N470000
W0594500W0661500
BA Canada
NT Annapolis Valley
Antigonish
Cape Breton Island
Cape Sable
Chedabucto Bay
Cobequid Bay
Halifax
Minas Basin
Sable Island
Yarmouth
SA Atlantic Coastal Plain
Maritime Provinces
Meguma Group
Windsor Group

Nova Scotia Shelf
use Scotian Shelf

Nova Scotian Shelf
use Scotian Shelf

novaculite
Includes use on level 3 under sedimentary rocks(1) clastic rocks(2). See list I.
UF razor stone
BA clastic rocks
BT sedimentary rocks
SA chemically precipitated rocks
nonterrigenous materials

Novaky
Village in W central.
BT Slovakia
Czechoslovakia

Novara
City and province in NW part of country.
BT Italy
SA Piedmont

Novaya Zemlya
Two large islands of Arkhangelsk Oblast between Barents and Kara seas.
CO N700000N770000
E0700000E0500000
BT Russian Republic
USSR

Novgorod
City S of Leningrad.
BT Russian Republic
USSR

Novoraskoe Series
use Novorayskoe Formation

Novorayskoe Formation
In Donbas in E Ukraine.
UF Novoraskoe Series
SA Donets Basin
Jurassic
Lower Jurassic
Triassic
Ukraine
Upper Triassic
USSR

Novosibirsk
City and oblast N of NE Kazakhstan in W Siberia.
BT Russian Republic
USSR

Nowa Ruda
Town near the border of Czechoslovakia in SW part of country.
BT Wroclaw
Poland

Nowa Sol
City on the Odor River in W part of country.
BT Zielona Gora
Poland

Nowy Sacz
City in N foothills of Carpathians in S part of country.
BT Krakow
Poland

Nowy Targ
City at foot of Tatra Mountains in S part of country.
UF Newmarkt
BT Krakow
Poland

Np
use neptunium

NRM
use natural remanent magnetization

nsutite
BA oxides
BT minerals

Nubia
Region in Nile Valley extending from about 16° N to include Aswan and First Cataract. Index countries as applicable.
BT Africa
SA Egypt
Sudan

Nubian Sandstone
Cretaceous?
BT Cretaceous
SA Africa
Egypt
Sudan

nuclear energy
Also search nuclear AND energy.
SA energy
energy sources
nuclear facilities

nuclear explosions
Includes use on level 2 under seismology(1); on level 3 under engineering geology(1) explosions(2).
BT explosions
SA engineering geology
seismology
shatter cones

nuclear facilities
A level 1 term as of 1978. Before 1978, included use on level 2 under engineering geology(1). Also search nuclear AND facilities. If 1, term set options are:
design
  subtopic
earthquakes
  subtopic
faults
  subtopic
feasibility studies
  subtopic
foundations
  subtopic
geologic hazards
  subtopic
impact statements
  subtopic
pollution
  subtopic
rock mechanics
  subtopic
seepage
  subtopic
site exploration
  subtopic
soil mechanics
  subtopic

UF facilities, nuclear
SA design
    earthquakes
    engineering geology
    faults
    feasibility studies
    foundations
    geologic hazards
    impact statements
    marine installations
    nuclear energy
    power plants
    rock mechanics
    seepage
    seismic risk
    site exploration
    soil mechanics

nuclear fission
use fission

nuclear magnetic resonance
Includes use on level 3 under spectroscopy(1) methods(2); in combination with spectroscopy (i.e. spectroscopy, nuclear magnetic resonance) on level 3 under chemical element(1) analysis(2). See list D.
SA analysis
    electron paramagnetic resonance
    resonance
    spectroscopy

nucleation
Restricted to use under crystal growth.
SA crystal growth

Nuculanidae
Family. Includes use on level 3 under Mollusca(1) Bivalvia(2).
BA Bivalvia
    Mollusca
BT Invertebrata
SA Nuculidae

Nuculidae
Family. Includes use on level 3 under Mollusca(1) Bivalvia(2).
BA Bivalvia
    Mollusca
BT Invertebrata
SA Nuculanidae

Nueces River
Flows into Nueces Bay at head of Corpus Christi Bay in S.
BT Texas
    United States

Nueva Esparta
State comprising an island group in the Caribbean Sea off N coast.
BT Venezuela
NT Macanao Peninsula
    Margarita Island

Nuevo Leon
State in NE.
BT Mexico
SA Rio Grande
    Rio Grande Valley
    Sierra Madre Oriental

Nugssuak
use Nugssuaq

Nugssuaq
Peninsula in central part of W coast. Also search Nugssuak.
UF Nugssuak
BT Greenland
    Arctic region

Nullarbor Plain
Extends inland along entire Great Australian Bight in S part of country. Index states as applicable.
BT Australia
SA South Australia
    Western Australia

numerical analysis
SA analysis
    mathematical methods
    statistical analysis

Nummulites
Genus. Includes use on level 3 under foraminifera(1) Nummulitidae(2).
BA Nummulitidae
    Rotaliacea
    Rotaliina
    foraminifera
BT Invertebrata
NA Operculina

Nummulitidae
Includes use on level 2 under foraminifera(1). See list F.
BA Rotaliacea
    Rotaliina
    foraminifera
BT Invertebrata
NA Nummulites
SA Heterostegina

Nunivak Island
Second largest island in Bering Sea. Off W coast.
BT Alaska
    United States

Nura-Tau
Range in N Samarkand Oblast. Also search Nuratau.
CO N380000N400000 E0660000E0640000
UF Nuratau
BT Uzbekistan
    USSR

Nuratau
use Nura-Tau

Nurek
Village in NE Dushanbe Oblast in W.
BT Tadzhikistan
    USSR

Nuremberg
City in N central Bavaria.
UF Nurnberg
BT Bavaria
    West Germany

Nurnberg
use Nuremberg

nutrients
Includes use on level 2 under soils(1). See list M.
SA fertilization
    fertilizers
    major elements
    minor elements
    nutrition
    soils
    trace elements

nutrition
Includes use on level 3 under fossil groups (list F) or paleontology(1).
SA metabolism
    nutrients
    paleontology

Nuwuk
use Point Barrow

Ny Friesland
Region on NE island of Spitsbergen.
BT Spitsbergen
    Arctic region

Nyasaland
use Malawi

Nye County
Central and S.
BT Nevada
    United States

Nyeman River basin
use Neman River basin

Nyiragongo
Volcano at N end of Lake Kivu in E part of country.
UF Nyiragongo Volcano
BT Kivu
    Zaire

Nyiragongo Volcano
use Nyiragongo

Nykoeping
use Nykoping

Nykoping
City on Baltic Sea.
UF Nykoeping
BT Sweden

O
use oxygen

O-shima
Island or groups of islands in Japan. Index as applicable.
UF Vries Island
BT Japan
SA Amami-O-shima
    Izu-shichito

O-16
Includes use on level 3 under isotopes(1).
SA isotopes
    O-18/O-16
    oxygen

O-16/O-18
use O-18/O-16

O-18
Includes use on level 3 under isotopes(1).
SA isotopes
    O-18/O-16
    oxygen

O-18/O-16
Includes use on level 3 under isotopes(1). Also search O-16/O-18; O-16 AND O-18.
UF O-16/O-18
SA geologic thermometry
    isotopes
    O-16
    O-18
    oxygen
    paleoclimatology
    stable isotopes

Oahu
Third largest of the Hawaiian islands. Island on which Honolulu is located.
BT Hawaii
    United States

Oamaru
Town on Pacific Ocean in SE.
BT South Island
    New Zealand

Oaxaca
State in S Mexico. Also a city.
BT Mexico
SA Sierra Madre del Sur

Ob River
Rises in Altai Mountains and flows NE and N through West Siberian Plain into Gulf of Ob on Kara Sea. Also search Ob.
BT Russian Republic
    USSR
SA Irtysh River
    Ob-Irtysh Interfluve

Ob-Irtysh Interfluve
Extends from source of both rivers in Altai Mountains to their conjunction in West Siberian Plain. Index Soviet Republics as applicable.
BT USSR
SA Irtysh River
    Kazakhstan
    Ob River
    Russian Republic

obduction
Includes use on level 2 under plate tectonics(1).
SA plate tectonics
    subduction

Oberhalbstein
Valley NW of Saint Moritz in E part of country.
BT Graubunden
    Switzerland

objectives
Includes use as a level 2 or 3 term appropriate to a large number of topics, e.g. on level 2 under geology(1). See list G.

oblique orientation
Term introduced in 1978. Includes use as orientation of folds relative to spatially associated macroscopic structures such as large folds, fold systems and orogenic zones. Includes use on level 3 under folds(1) orientation(2); on level 3 under faults(1) orientation(2) and fractures(1) style(2). Before 1978, also search folds or faults AND oblique.
SA faults
    folds
    fractures
    joints
    orientation

obliquity
Used as a general term.
SA geometry

observations
Includes use as a level 2 or 3 term appropriate to a large number of topics, e.g. on level 2 under seismology(1). See list G.
SA orbital observations

observatories
Includes use on level 2 under geophysics(1) and seismology(1).
SA geophysics
    seismology

obsidian
Includes use on level 3 under igneous rocks(1) pyroclastics and glasses(2). See list H.
BA pyroclastics and glasses
BT igneous rocks
SA volcanic glass

Ocala Group
Includes Tivola Tongue, Inglis Formation, Williston Formation, Crystal River Formation.
BT upper Eocene
    Eocene
SA Alabama
    Crystal River Formation
    Florida
    Floridan Aquifer
    Georgia

occurrence
Includes use as a level 2 or 3 term appropriate to a large number of topics, e.g. on level 2 under metamorphic rocks(1). See list G. Also search occurrences.

ocean basins
Used for studies on the major ocean basins, their origin, evolution and present configuration. For basins found within the ocean and for sedimentation studies, see ocean floors. Includes use on level 1 (list A). Term set options are:
topic [age, evolution, genesis, pat-

terns, structure]
    subtopic (no area term)
  SA abyssal plains
      basins
      continental drift
      crust
      Deep Sea Drilling Project
      geosynclines
      marginal basins
      marine geology
      ocean floors
      oceanography
      paleo-oceanography
      paleogeography
      plate tectonics
      sea-floor spreading
      tectonophysics

ocean bottom seismographs
Term introduced in 1978.
  UF ocean-bottom seismometers
  SA instruments
      seismic methods
      seismographs
      seismology

ocean circulation
Level 1 term introduced in 1978. Primarily used for special indexes. Before 1978, also search circulation AND oceans. Term set options are:
currents
  (name of current)
topic [anomalies, biocirculation, boundary layer, causes, climate-induced circulation, Coriolis force, detection, diffusion, distribution, Ekman spiral, genesis, patterns, thermal circulation, thermohaline circulation, tides, turbulence]
    subtopic
  SA biocirculation
      bottom currents
      boundary layer
      circulation
      climate-induced circulation
      continental shelf
      continental slope
      Coriolis force
      current directions
      currents
      diffusion
      Earth tides
      Ekman spiral
      longshore currents
      marine geology
      ocean floors
      ocean waves
      oceanography
      paleocirculation
      sedimentation
      shorelines
      thermal circulation
      thermohaline circulation
      tides
      turbulence
      upwelling
      waterways

ocean floors
Used for discussions of processes taking place on the ocean floor as well as features thereof. For tectonics, see ocean basins. Includes use on level 1 (list A). Term set options are:
topic [anomalies, bottom features, exploration, mid-ocean ridges, mineral resources, rift valleys, seamounts, sedimentation, submarine canyons, trenches, troughs]
    subtopic (no area term)
  SA abyssal plains
      bathymetric maps
      bathymetry
      bottom features
      continental margin
      continental rise
      continental shelf
      continental slope
      Deep Sea Drilling Project
      dredging
      fracture zones
      Leg 18
      Leg 25
      Leg 27
      Leg 45
      Leg 46
      Leg 48
      marginal basins
      marine geology
      mid-ocean ridges
      mineral resources
      nodules
      ocean basins
      ocean circulation
      oceanography
      paleo-oceanography
      paleobathymetry
      relief
      rift valleys
      sea-floor spreading
      seamounts
      sedimentation
      sediments
      submarine canyons
      submarine fans
      topography
      trenches
      troughs

Ocean of Storms
Also search Oceanus Procellarum.
  UF Oceanus Procellarum
  BT Moon

ocean waves
Includes use on level 1 (list A) for special indexes or for papers relating physical oceanography to marine geology. Term set options are:
topic [airy waves, anomalies, breaking waves, catastrophic waves, causes, diffraction, effects, generation, ideal waves, internal waves, reflection waves, refraction waves, shoaling, solitary waves, transformations]
    subtopic
  NT breaking waves
      catastrophic waves
      ideal waves
  SA airy waves
      circulation
      diffraction
      generation
      internal waves
      longshore currents
      marine geology
      ocean circulation
      oceanography
      reflection waves
      refraction waves
      shoaling
      shorelines
      solitary waves
      surges
      transformations
      tsunamis
      waves

ocean, world
use world ocean

ocean-bottom seismometers
use ocean bottom seismographs

ocean-floor spreading
use sea-floor spreading

Oceania
Collective name for the islands and island groups of the central and south Pacific Ocean. Index island divisions as applicable. Australia, New Zealand, and Malay Archipelago are sometimes considered within the term. Includes use on level 1 or 2 as an area term (list O). If 1, see term set options under list B.
  UF Oceanica
  BA Pacific Ocean
  SA Australasia
      Australia
      Bismarck Archipelago
      Malay Archipelago
      Melanesia
      Micronesia
      New Guinea
      New Zealand
      Polynesia

oceanic
A valid term through 1977. After 1977, use oceanic type (for the crust).

oceanic crust
use oceanic type

oceanic genesis
use oceanic type

oceanic ridges
A valid term through 1971. After 1971, use mid-ocean ridges.

oceanic trench
use trenches

oceanic type
Term introduced in 1978. Also search oceanic crust; oceanic AND crust; oceanic AND genesis.
  UF oceanic crust
      oceanic genesis
  SA crust
      plate tectonics

Oceanica
use Oceania

oceanite
Includes use on level 3 under igneous rocks(1) basalt family(2). See list H.
  BA basalt family
  BT igneous rocks

oceanography
Includes use on level 1 (list A) for special indexes and for treating the discipline as a whole; on level 2 under area terms (list B), bibliography(1), education(1), continental shelf(1), and continental slope(1). In January 1976, the following terms were deleted from level 2: boundary layer, diffusion, distribution, and turbulence. These terms are now used on level 2 under ocean circulation(1). When oceanography is a level 1 term, set options are:
experimental studies
  topic
instruments
  name of instrument or platform
phenomenon
sea ice
  topic
techniques
  topic
theoretical studies
  topic
topic [applications, bibliography, catalogs, classification, concepts, education, general, history, methods, nomenclature, objectives, practice, principles, research, symposia, textbooks]
    subtopic
  UF oceanology
  SA abyssal environment
      abyssal plains
      air-sea interface
      bathymetry
      bibliography
      bottom water
      boundary layer
      catalogs
      continental shelf
      continental slope
      convection
      Deep Sea Drilling Project
      diffusion
      echo sounding
      education
      GEOSECS
      International Decade of Ocean Exploration
      marine geology
      ocean basins
      ocean circulation
      ocean floors
      ocean waves
      paleo-oceanography
      pelagic environment
      sea ice
      sea water
      sedimentary petrology
      sedimentation
      shoals
      submersibles
      turbulence
      upwelling

oceanology
use oceanography

oceans
A valid level 1 term through 1977 used in combination with circulation (i.e. oceans, circulation). After 1977, use ocean circulation on level.

Oceanus Procellarum
use Ocean of Storms

Ocoee Series
Provincial series, Virginia, Tennessee, North Carolina, and Georgia. Includes use on level 3 under age terms(1). See list E.
  BA Precambrian

Oconee County
Index states as applicable.
  BX Georgia
      South Carolina
  BT United States

octahedral iron ore
use magnetite

octahedrite (mineral)
use anatase

octahedrites
A valid term through 1975. After 1975, use iron meteorites.

Octocorallia
Includes use on level 2 under Coelenterata(1). See list F.
  BA Coelenterata
  BT Invertebrata

Odate
City in N.
  BT Honshu
      Japan

Odenwald
Mountainous region in S central part of country. Index states as applicable.
  UF Odenwald Massif
      Odenwald Mountains
  BT Germany
  SA Baden-Wurttemberg
      Bavaria
      Hesse

Odenwald Massif
use Odenwald

Odenwald Mountains
use Odenwald

Oder Valley
River valley. Index countries as applicable. Also search Odra River valley.
  UF Odra River valley
  BT Europe
  SA Czechoslovakia
      East Germany

Odessa
City and oblast on the Black Sea.
BT Ukraine
USSR

Odonata
Includes use on level 2 under Insecta(1). See list F.
BA Insecta
BT Arthropoda
Invertebrata

Odontopleurida
Includes use on level 2 under Trilobita(1). See list F.
BA Trilobita
BT Trilobitomorpha
Arthropoda
Invertebrata

Odontornithes
Subclass. Includes use on level 3 under Aves(1) Neornithes(2).
BA Neornithes
Aves
BT Tetrapoda
Vertebrata

Odra River valley
use Oder Valley

Oerebro
use Orebro

Oetztal Alps
use Otztal Alps

Oetztal Massif
use Otztal Alps

Oetztaler
use Otztal Alps

off-lying
use offshore

Officer Basin
Sedimentary basin.
CO S300000S240000
E1280000E1240000
BT Australia
SA South Australia
Western Australia

offretite
BA zeolite group
framework silicates
silicates
BT minerals
SA aluminosilicates

offshore
Used to indicate area situated off or at a distance from the shore. Not used as an adjective.
UF off-lying
SA continental shelf
continental slope
onshore
shorelines

Oficina
Oil field near town of El Tigre in NE central part of country.
BT Anzoategui
Venezuela

Ofot Fjord
use Ofoten

Ofoten
Fjord. NE extension of Vest Fjord on which Narvik is located.
UF Ofot Fjord
BT Nordland
Norway

Oga Peninsula
Extends into Japan Sea in N.
BT Honshu
Japan

Ogallala Formation
Includes Burge Sands, Valentine Beds, Ash Hollow Member, Kimball Member. NE Colorado, W and central Kansas, W Nebraska, E New Mexico, W Oklahoma, NW Texas and SE Wyoming. Also search Ogallala Aquifer.
BT Pliocene
SA Ash Hollow Formation
Colorado
Kansas
Nebraska
New Mexico
Oklahoma
Texas
Valentine Formation
Wyoming

Ogasawara
Town in Yamanashi Prefecture in central.
BT Honshu
Japan

Ogasawara Islands
use Bonin Islands

Ogdensburg
City on the Saint Lawrence River in Saint Lawrence County.
BT New York
United States

Ogilvie Mountains
Range in central.
BT Yukon Territory
Canada

Oguni
Village in N central.
BT Honshu
Japan

Ohio
Includes use on level 1 as an area term (list O). For term set options see list B.
CO N382500N420000
W0803030W0845000
BA United States
NX Adams County
NT Akron
NX Allen County
Butler County
Carroll County
NT Cincinnati
NX Clark County
NT Cleveland
NX Columbus
Delaware County
Erie County
Fayette County
Franklin County
Fulton County
NT Geauga County
NX Greene County
Hamilton County
Hancock County
Hardin County
Highland County
Jackson County
Jefferson County
Lake County
Lawrence County
NT Mahoning County
NX Marion County
Mercer County
Monroe County
Montgomery County
Morgan County
NT Ohio State University
NX Ottawa County
Perry County
Pike County
Portage County
Putnam County
NT Scioto River basin
NX Stark County
Summit County
NT Tuscarawas County
NX Union County
Urbana
Warren County
Washington County
Wayne County
Wood County
SA Allegheny Group
Allegheny Plateau
Appalachian Basin
Appalachian Plateau
Athens
Bedford Shale
Berea Sandstone
Brassfield Formation
Cincinnati Arch
Columbus Limestone
Conemaugh Group
Delaware Limestone
Detroit River Group
Dunkard Group
Eden Shale
Fairview Formation
Great Lakes region
Lake Erie
Logan Formation
Maumee River valley
Midcontinent
Midwest
Monongahela Group
Ohio River
Ohio River valley
Ohio Shale
Olentangy Shale
Pottsville Group
Richmond Group
Saint Peter Sandstone
Salina Group
Sharon Conglomerate
Springfield
Sylvania Formation
Tioga Bentonite
Toledo
Trenton Group
Whitewater River valley

Ohio Range
N of the Horlick Mountains and S of Marie Byrd Land.
BT Antarctica

Ohio River
Formed by confluence of the Allegheny and Monongahela rivers at Pittsburgh. It flows W and SW into the Mississippi River at Cairo, Illinois. Index states as applicable.
BT United States
SA Illinois
Indiana
Kentucky
Ohio
Pennsylvania
West Virginia

Ohio River valley
Index states as applicable. Also search Ohio Valley.
UF Ohio Valley
BT United States
SA Illinois
Indiana
Kentucky
Ohio
Pennsylvania
West Virginia

Ohio Shale
Includes Cleveland Member, Huron Member, Chagrin Member. Ohio, N central Kentucky, and West Virginia.
BT Upper Devonian
Devonian
SA Kentucky
Ohio
West Virginia

Ohio State University
In Columbus, Franklin County.
BT Ohio
United States

Ohio Valley
use Ohio River valley

oil (petroleum)
use petroleum

oil and gas fields
For detailed descriptions of individual fields or for discussions of the origin of several fields. Includes use on level 1 (list A). See also list C (commodities). Term set options are: topic [classification, distribution, genesis, structure]
name of oil field
subtopic (no area term)
UF fields, oil and gas
gas fields, oil and
NT giant fields
SA economic geology
Forties Field
geothermal fields
Hassi Messaoud Field
Hugoton Field
land subsidence
natural gas
petroleum
petroleum engineering
reservoir properties
reservoir rocks
stratigraphic traps
traps

oil sands
Includes use on level 1 as a commodity term (list C).
UF tar sands
SA oil shale
organic materials
petroleum

oil seeps
Also search petroleum seepage.
UF petroleum seepage
seeps, oil
SA bitumens
petroleum
seepage

oil shale
Includes use on level 1 as a commodity term (list C).
UF shale oil
SA anthraxolite
kerogen
oil sands
organic materials
petroleum
shale
torbanite

oil spills
As of 1978, term is used on level 2 under pollution(1).
UF spills, oil
SA petroleum
pollution

oil-gas interface
Term introduced in 1978. Before 1978, search interface AND oil-gas.
UF interface, oil-gas
SA gases
natural gas
oil-water interface
petroleum

oil-water contact
use oil-water interface

oil-water interface
Term introduced in 1978. Before 1978, search oil-water AND interface; oil-water contact.
UF interface, oil-water
oil-water contact
SA oil-gas interface
petroleum
water

Oise
Department in N.
BT France
NT Clermont
Creil
SA Oise River valley

Oise River valley
Index departments as applicable. Also search Oise Valley.
UF Oise Valley
BT France
SA Aisne
Oise
Val-d'Oise

Oise Valley
use Oise River valley

Oita
City and prefecture in NE.
BT Kyushu
Japan

Ojika Peninsula
Between Ishinomaki Bay and the Pacific Ocean in NE.
BT Honshu
Japan

Ojo Alamo Sandstone
NW New Mexico.
BT Upper Cretaceous
Cretaceous
SA New Mexico

Ojo de Liebre Lagoon
Inlet of Sebastian Vizcaino Bay on NW coast.
BT Baja California
Mexico

Oka
River W of Lake Baikal in central Irkutsk Oblast which flows into the Angara River, and a river in central European USSR which flows into the Volga at Gorkiy.
BT Russian Republic
USSR

Okanagan Valley
River valley. Index British Columbia and/or Washington as applicable. Called Okanogan Valley in the U.S.
UF Okanagen Valley
Okanogan Valley
BT North America
SA British Columbia
Washington

Okanagen Valley
use Okanagan Valley

Okanogan County
N Washington.
BT Washington
United States

Okanogan Valley
use Okanagan Valley

Okayama
City and prefecture on N side of Inland Sea W of Kobe.
BT Honshu
Japan

Okefenokee Swamp
Mostly in SE Georgia. Index states as applicable.
BT United States
SA Florida
Georgia

Okehampton
Town in Devonshire in SW.
BT England
Great Britain
United Kingdom

Okhotsk
Town on NW coast of Okhotsk Sea in Khabarovsk Kray in E Siberia.
CO N580000N600000
E1450000E1400000
BT Russian Republic
USSR

Okhotsk Massif
Just N of Magadan in S Magadan Oblast in E Siberia.
UF Okhotskoye Ploskogor'ye
BT Russian Republic
USSR

Okhotsk Sea
West of Kamchatka Peninsula and the Kuril Islands. Includes use on level 1 as an area term (list O). For term set options see list B. Before 1974, also search Sea of Okhotsk.
CO N440000N463000
E1650000E1350000
BA Pacific Ocean

Okhotsk-Chukchi
Large region extending NE of Okhotsk Sea, including area of Chukchi Peninsula.
BT Russian Republic
USSR
SA Chukchi Peninsula
Soviet Far East

Okhotsk-Chukchi volcanic belt
Includes numerous ranges extending S from the Chukchi Peninsula around both sides of Okhotsk Sea. Many volcanoes are still active in Kamchatka Peninsula. Also search Okhotsk-Chukchi.
UF Okhotsk-Chukot volcanic belt
BT Russian Republic
USSR

Okhotsk-Chukot volcanic belt
use Okhotsk-Chukchi volcanic belt

Okhotskoye Ploskogor'ye
use Okhotsk Massif

Oki Archipelago
use Oki Islands

Oki Islands
In SE Japan Sea 44 miles off W coast.
CO N360000N370000
E1340000E1323000
UF Oki Archipelago
Oki Retto
BT Honshu
Japan
SA Dogo

Oki Retto
use Oki Islands

Okinawa
Major island in Okinawa Island group midway between the main islands of Japan and Taiwan.
BT Ryukyu Islands
Japan

Oklahoma
Includes use on level 1 as an area term (list O). For term set options see list B.
CO N333500N370000
W0942500W1030000
BA United States
NT Arbuckle Mountains
NX Beaver County
Blaine County
NT Caddo County
NX Carter County
Cherokee County
NT Cimmaron County
Coal County
NX Comanche County
NT Cotton
NX Custer County
Delaware County
Dewey County
Ellis County
Garfield County
Grant County
Harper County
NT Hughes County
NX Jackson County
Jefferson County
NT Kingfisher County
NX Kiowa County
Lincoln County
NT Major County
NX Marshall County
McIntosh County
Murray County
NT Muskogee County
NX Osage County
Ottawa County
NT Pawhuska Rock Plain
NX Pawnee County
NT Payne County
Picher
NX Pontotoc County
Pottawatomie County
NT Stillwater
NX Texas County
NT Tulsa
Tulsa County
NX Washington County
NT Wichita Mountains
Woods County
Woodward
SA Americus Limestone Member
Anadarko Basin
Antlers Sands
Arbuckle Group
Arkansas River
Arkansas River valley
Arkoma Basin
Atoka Formation
Bartlesville Sand
Blaine Formation
Bloyd Formation
Burbank Sand
Cabaniss Formation
Canadian River
Chase Group
Chattanooga Shale
Cheyenne Sandstone
Cottonwood Limestone
Crooked Creek Formation
Dakota Formation
Desmoinesian
Dockum Group
Eskridge Shale
Excello Shale
Fayetteville Formation
Fernvale Formation
Flint Hills
Fort Riley Limestone
Fredericksburg Group
Grand River
Great Plains
Hennessey Formation
Hugoton Embayment
Hunton Group
Jackfork Group
Johns Valley Formation
Kiowa Formation
Labette Shale
Lecompton Limestone
Llano Estacado
Marmaton Group
Midcontinent
Morrison Formation
Morrow Formation
Nemaha Ridge
Neosho River valley
Ogallala Formation
Oread Limestone
Ouachita Mountains
Ozark Mountains
Paluxy Formation
Panhandle
Pitkin Limestone
Pleasanton Group
Reagan Sandstone
Red Eagle Limestone
Red Fork Sandstone
Red River
Red River valley
Rexroad Formation
Roca Formation
Saint Peter Sandstone
Savanna Formation
Seminole Formation
Simpson Group
Southwestern U.S.
Springer Formation
Stanley Group
Stanton Formation
Trinity Group
Verdigris River valley
Viola Limestone
Wabaunsee Group
Wann Formation
Wapanucka Limestone
Washita Group
Washita River valley
Wellington Formation
Wewoka Formation
Woodbine Formation
Woodford Shale
Wreford Limestone

Oklo
BT Gabon

Oktemberyan Series
BT Miocene
Neogene
Tertiary
SA Armenia
USSR

Oktyabr
Town in SE Uzbekistan.
BT Uzbekistan
USSR

Okujiri
use Okushiri Island

Okushiri Island
In Japan Sea off SW coast.
UF Okujiri
Okushiri-shima
BT Hokkaido
Japan

Okushiri-shima
use Okushiri Island

Oland
Term introduced in 1978 to refer to island and province in Baltic Sea off SE coast of Sweden. Before 1978, also search Oland Island.
UF Oland Island
BT Kalmar
Sweden

Oland Island
use Oland

Old Crow
Village in N.
BT Yukon Territory
Canada

Old Faithful Geyser
In Yellowstone National Park.
BT Wyoming
United States

Old Red Sandstone
Great Britain. Includes use on level 3 under age terms(1). See list E.
BA Devonian
BT Paleozoic
NA Downtonian
SA Dittonian

Old Stone Age
use Paleolithic

Older Dryas
Term used primarily in Europe for an interval of late glacial time following the Bolling and preceding the Allerod. Before 1978, search Dryas.
BA Weichselian
upper Pleistocene
Pleistocene
Quaternary
BT Cenozoic
SA Dryas

Oldest Dryas
Term introduced in 1978. Used primarily in Europe for an interval of late-glacial time preceding the Bolling. Before 1978, search Dryas.
BA Weichselian

upper Pleistocene
Pleistocene
Quaternary
BT Cenozoic
SA Dryas

Oldman Formation
Of the Belly River Group.
BT Upper Cretaceous
Cretaceous
SA Alberta

Oldoinyo Lengai
Village in N part of country.
BT Tanzania

Olduvai Gorge
Site of rich fossil beds in N part of country. Also search Olduvai.
CO S031500S031500
E0353000E0353000
BT Tanzania

O'Leary Peak
Coconino County in N central.
BT Arizona
United States

Olekma
River in N Chita Oblast and S Yakutia in SE central Siberia.
BT Russian Republic
USSR

Olekma-Vitim Highlands
In NE Buryat ASSR and NW Chita Oblast in N Transbaikalia.
UF Olekma-Vitim mountain country
Olekma-Vitim Mountains
BT Russian Republic
USSR

Olekma-Vitim mountain country
use Olekma-Vitim Highlands

Olekma-Vitim Mountains
use Olekma-Vitim Highlands

Olenek River
Rises in Central Siberian Plateau and flows NE through Yakutia into Laptev Sea. Also search Olenek.
BT Russian Republic
USSR

Olenidae
Family. Includes use on level 3 under Trilobita(1) Ptychopariida(2).
BA Ptychopariida
Trilobita
BT Trilobitomorpha
Arthropoda
Invertebrata

Olentangy Shale
Plum Brook Shale has been correlated with Olentangy, and so-called Prout Limestone has been regarded as member of Olentangy Shale. Central Ohio.
BT Upper Devonian
Devonian
SA Ohio

Oligocene
World. Above Eocene, below Miocene. Includes use on level 1 as an age term (list E); on level 2 under paleo- terms, e.g. paleoecology, paleogeography, paleomagnetism.
BA Paleogene
Tertiary
Cenozoic
NT Bembridge Marls
Brule Formation
Cypress Hills Formation
Frio Formation
Grimmertingen
Krosno Beds
Lincoln Creek Formation
NA lower Oligocene
middle Oligocene
NT Nari Series
NA Rupelian

upper Oligocene
NT White River Group
SA Duchesne River Formation
John Day Formation
Poltava Series
Refugian
San Lorenzo Formation
Saucesian
Sespe Formation
Twin River Formation

oligoclase
BA feldspar group
framework silicates
silicates
BT minerals
SA plagioclase

olistoliths
Includes use on level 3 under sedimentary structures(1) turbidity current structures(2) and soft sediment deformation(2). See list K.
BA soft sediment deformation
sedimentary structures
SA olistostromes
turbidity current structures

olistostromes
Includes use on level 3 under sedimentary structures(1) turbidity current structures(2) and soft sediment deformation(2). See list K.
BA soft sediment deformation
sedimentary structures
SA olistoliths
turbidity current structures

Oliver County
W central North Dakota.
BT North Dakota
United States

olivine
BA olivine group
orthosilicates
silicates
BT minerals

olivine basalt
Includes use on level 3 under igneous rocks(1) basalt family(2). See list H.
BA basalt family
BT igneous rocks
SA alkali olivine basalt

olivine diabase
See list H.
BA basalt family
BT igneous rocks
SA diabase

olivine dolerite
See list H.
BA basalt family
BT igneous rocks
SA dolerite

olivine gabbro
See list H.
BA gabbro family
BT igneous rocks
SA gabbro

olivine group
Includes use in combination with orthosilicates (i.e. orthosilicates, olivine group) on level 2 under minerals(1). See list L.
BA orthosilicates
silicates
BT minerals
NA fayalite
forsterite
olivine
peridot
tephroite
SA monticellite

olivine nephelinite
Includes use on level 3 under igneous rocks(1) alkali basalt family(2). See list H.

BA alkali basalt family
BT igneous rocks
SA ankaratrite
nepheline basalt
nephelinite

olivine tholeiite
BA basalt family
BT igneous rocks
SA tholeiite

olivinite
Includes use on level 3 under igneous rocks(1) ultramafic family(2). See list H.
BA ultramafic family
BT igneous rocks

Olkusz
Town in S part of country.
BT Krakow
Poland

Olsztyn
Province in N part of country. Also city SE of Gdansk.
BT Poland

Olt River
Central and S Romania, in Transylvania and Walachia. Also search Olt and Olt River valley.
UF Oltul
BT Romania
SA Transylvania
Walachia

Olt River valley
Index regions as applicable. Also search Olt, Olt River, and Olt Valley.
UF Olt Valley
Oltul Valley
BT Romania
SA Transylvania
Walachia

Olt Valley
use Olt River valley

Oltenia
A sub-region. W part of Walachia.
UF Lesser Walachia
Little Walachia
BT Walachia
Romania

Oltul
use Olt River

Oltul Valley
use Olt River valley

Olympic Mountains
Part of the Coast Ranges in Olympic National Park on the Olympic Peninsula in NW.
BT Washington
United States

Olympic Peninsula
Between the Pacific Ocean and Puget Sound in NW.
BT Washington
United States

Olympus
Highest mountain in Greece in NE part of country.
BT Thessaly
Greece

Oman
Formerly Muscat and Oman. Includes use on level 1 as an area term (list O). For term set options see list B.
CO N170000N270000
E0600000E0523000
BA Arabian Peninsula
Asia
NT Oman Mountains
SA Middle East
Near East

Oman Mountains
BT Oman

Arabian Peninsula

Omi (Lake)
use Lake Biwa

Omine Mine
Near town of Omine in SW.
BT Honshu
Japan

Omineca Mountains
NW British Columbia.
BT British Columbia
Canada

Omo
Village in SW part of country.
BT Ethiopia

Omo Basin
use Omo Valley

Omo River
Flows through SW Ethiopia into Lake Turkana (Lake Rudolph). Also search Omo.
BT Ethiopia

Omo Valley
River valley in SW Ethiopia. Also search Omo, and Omo Basin.
UF Omo Basin
BT Ethiopia

Omolon
River in N Khabarovsk Kray, W Chukchi National Okrug, and NE Yakutia which flows into the Kolyma River in NE Siberia.
BT Russian Republic
USSR

Omolon Block
NE Siberia in the Omolon River basin. Also search Omolon.
BT Russian Republic
USSR

omphacite
BA pyroxene group
chain silicates
silicates
BT minerals
SA augite

Omsk
City and oblast in SW Siberia.
BT Russian Republic
USSR

Omulevka
River in N Magadan Oblast in NE Siberia.
BT Russian Republic
USSR

Onaping Formation
Of the Whitewater Group.
BT Huronian
Proterozoic
Precambrian
SA Canada
Ontario

oncolites
Includes use on level 3 under sedimentary structures(1) biogenic structures(2). See list K.
BA biogenic structures
sedimentary structures
SA stromatolites

one-dimensional models
Term introduced in 1978. Before 1978, search one dimensional; one-dimensional AND models.
BA models
SA three-dimensional models
two-dimensional models

Onega
City on Onega Bay at mouth of Onega River in NW Arhangelsk Oblast in NW European USSR.
BT Russian Republic
USSR
SA Lake Onega

Oneida County
Index states as applicable.
- BX Idaho
  New York
  Wisconsin
- BT United States

Oneida Lake
SE of Lake Ontario in central.
- BT New York
  United States

Onikobe
Region.
- BT Honshu
  Japan

onofrite
- BA sulfides
- BT minerals
- SA metacinnabar
  selenium

Onondaga County
W central New York.
- BT New York
  United States

Onondaga Limestone
Includes Springfield Center Member, Babcock Hill Member, Edgecliff Member, Nedrow Member, Moorehouse Member, Seneca Member, Needmore Shale, Selinsgrove Limestone. W Maryland, New York, Pennsylvania, Western Virginia, and N West Virginia.
- BT Middle Devonian
  Devonian
- SA Maryland
  New York
  Pennsylvania
  Virginia
  West Virginia

onshore
- SA offshore
  shorelines

Onslow Bay
Between Cape Lookout and Cape Fear.
- BT North Carolina
  United States

Ontake
On a peninsula in Kagoshima Bay in S Kyushu. Also search Ontake Volcano.
- UF Ontake Volcano
- BT Kyushu
  Japan

Ontake Volcano
use Ontake

Ontario
Includes use on level 1 as an area term (list O). For term set options see list B.
- CO N420000N570000
  W0740000W0950000
- BA Canada
- NT Algoma
  Bancroft
  Batchawana Bay
  Blind River
  Brent Crater
  Callander Bay
  Cobalt
  Cochrane
  District of Kenora
  Echo Bay
  Elliot Lake
  Flack Lake
  Georgian Bay
  Gowganda
  Guelph
  Kakagi Lake
  Kenora
  Kirkland Lake
  Madoc
  Mamainse Point
  Manitoulin Island
  Manitouwadge
  Michipicoten Island
  Moose River basin
  New Liskeard
  North Bay
  Northern Light Lake
  Ottawa
  Parry Sound
  Quirke Lake
  Saint Catharines
  Sudbury
  Sudbury Basin
  Sudbury Irruptive
  Superior Province
  Temagami Mine
  Thunder Bay
  Timmins
  Toronto
  Uchi Lake
  Welland Canal
  Whetstone Lake
  Windsor
- SA Animikie Group
  Canadian Shield
  Churchill Province
  Detroit River Group
  Espanola Formation
  Flinton Group
  Frontenac County
  Gowganda Formation
  Grand River
  Great Lakes region
  Grenville
  Grenville Front
  Grenville Province
  Hudson Bay Lowlands
  Keewatin
  Lake Erie
  Lake Huron
  Lake Ontario
  Lake Superior
  Lake Timiskaming
  Leda Clay
  Lockport Formation
  Lorrain Formation
  Niagara Escarpment
  Niagara Falls
  Niagara River
  Nipissing Diabase
  Onaping Formation
  Osler Series
  Ottawa Valley
  Rainy Lake
  Rainy River
  Rice Lake
  Rochester Formation
  Saganaga Lake
  Saint Clair River
  Saint Clair River delta
  Saint Lawrence Lowlands
  Saint Lawrence River
  Saint Lawrence Valley
  Sault Sainte Marie
  Sylvania Formation
  Timiskaming
  Trent Valley
  Utica Shale
  Waterloo

ontogeny
Includes use on level 2 and 3 under fossil group(1). See list F.
- SA fossils
  paleontology
  phylogeny

Ontonagon County
In Upper Peninsula.
- BT Michigan
  United States

Ontong Java Plateau
Between Solomon Islands on the W and Lord Howe Islands (Ontong Java Islands) on the E.
- UF Antong Java Rise
  Ontong Java Rise
- BT Pacific Ocean

Ontong Java Rise
use Ontong Java Plateau

onvarovite
use uvarovite

Onverwacht Group
- BT Precambrian
- SA South Africa

onyx marble
use alabaster

oolite
Includes use on level 3 under sedimentary rocks(1) carbonate rocks(2); on level 3 under sediments(1) carbonate sediments(2). See list I (sed. rocks), list N (sediments). Also search oolites.
- UF eggstone
  ooliths
  roestone
- SA carbonate rocks
  carbonate sediments
  limestone
  oolitic texture
  pellets
  sedimentary rocks

oolites
A valid term through 1976. After 1976, use oolite.

ooliths
use oolite

oolitic limestone
- BA limestone
  carbonate rocks
- BT sedimentary rocks

oolitic texture
Term introduced in 1978. Includes use on level 3 under sediments(1) and sedimentary rocks(1). Before 1978, also search oolitic.
- SA limestone
  oolite
  sedimentary rocks
  sediments
  textures

ooze
Includes use on level 3 under sediments(1) clastic sediments(2).
- BA clastic sediments
- BT sediments
- SA calcareous composition
  coccoliths
  mud
  nonterrigenous materials
  sapropel

opal
Also search opaline.
- UF opaline
- BA silica minerals
  framework silicates
  silicates
- BT minerals
- SA hyalite

opaline
use opal

opaque minerals
Includes use on level 3 under minerals(1).
- BT minerals
- SA light minerals

open fractures
Term introduced in 1978. Includes use on level 3 under fractures(1) style(2). Before 1978, also search fractures AND open.
- BA fractures
- SA cracks

open pit mines
use open-pit mining

Ontong Java Rise
use Ontong Java Plateau

open pit mining
use open-pit mining

open systems
Used as a general term.
- SA closed systems
  systems

open-pit mining
Includes use on level 3 under mining geology(1). Also search open pit mines; open pit mining.
- UF mining, open-pit
  open pit mines
  open pit mining
  opencast mining
  opencut mining
- SA coal
  mines
  mining
  mining geology
  strip mining

opencast mining
use open-pit mining

opencut mining
use open-pit mining

Operculina
Includes use on level 3 under foraminifera(1) Nummulitidae(2).
- BA Nummulites
  Nummulitidae
  Rotaliacea
  Rotaliina
  foraminifera
- BT Invertebrata

ophicalcite
See list J.
- BA marbles
- BT metamorphic rocks
- SA limestone

ophiolite
Includes use on level 3 under igneous rocks(1) ultramafic family(2); on level 3 under plate tectonics(1). See list H.
- BA ultramafic family
- BT igneous rocks
- SA geosynclines
  plate tectonics

Ophiomorpha
Order. Includes use on level 3 under Arthropoda(1) Malacostraca(2).
- BA Malacostraca
  Crustacea
  Arthropoda
- BT Invertebrata
- NA Ophiomorpha nodosa
- SA Decapoda
  ichnofossils

Ophiomorpha nodosa
Includes use on level 3 under Arthropoda(1) Malacostraca(2).
- BA Ophiomorpha
  Malacostraca
  Crustacea
  Arthropoda
- BT Invertebrata
- SA Decapoda
  ichnofossils

ophite
Includes use on level 3 under metamorphic rocks(1) marbles(2). See list J.
- BA marbles
- BT metamorphic rocks

Ophiuroidea
Includes use on level 3 under Echinodermata(1) Stelleroidea(2). See list F.
- BA Stelleroidea
  Asterozoa
  Echinodermata
- BT Invertebrata

Ophthalmidium
Genus. Includes use on level 3 under foraminifera(1) Miliolacea(2).
- BA Miliolacea
    Miliolina
    foraminifera
- BT Invertebrata

Opole
Province in SW part of country. Also a city.
- BT Poland
- NT Raciborz

optical
A valid term through 1977. After 1977, use optical spectroscopy.

optical methods
- SA interferometry
    methods
    mineralogy
    minerals
    optical properties

optical mineralogy
Includes use on level 2 under geochronology(1); on level 3 under minerals(1) methods(2).
- SA geochronology
    mineralogy
    ore microscopy

optical properties
As of 1978, term is used on level 2 under minerals(1).
- SA birefringence
    color centers
    extinction
    fluorescence
    mineralogy
    minerals
    optical methods
    pleochroism
    polarization
    properties
    refractive index
    scattering

optical spectra
Used for data, to distinguish from methodology. Before 1978, search optical AND spectroscopy.
- SA spectra
    spectroscopy

optical spectroscopy
Term introduced in 1978. Includes use on level 3 under spectroscopy(1) methods (2) and under chemical element(1) analysis(2). Before 1978, also search spectroscopy AND optical.
- BT spectroscopy
- SA analysis

optimization
Used as a general term.

Oquirrh Mountains
S of Great Salt Lake.
- UF Oquirrh Range
- BT Utah
    United States

Oquirrh Range
use Oquirrh Mountains

Oran
Department in NW Algeria. Also a city on the Mediterranean Sea.
- BT Algeria
- NT Descartes

Orange County
Index states as applicable.
- BX California
    Florida
    Indiana
    New York
    North Carolina
    Texas
    Vermont
    Virginia
- BT United States

Orange Free State
Province in E central.
- BT South Africa
- NT Swartkrans
    Vredefort Dome
- SA Orange River
    Vaal River

orange material
- SA fines
    Moon

Orange River
Rises in the Drakensberg Range in Lesotho and flows W into the Pacific Ocean. Index South-West Africa, Lesotho and provinces of South Africa.
- BT Africa
- SA Cape Province
    Lesotho
    Orange Free State
    South-West Africa

Orava Valley
River valley in N Slovakia. Also search Orava.
- BT Slovakia
    Czechoslovakia

orbicular texture
Term introduced in 1978. Before 1978, search orbicular AND texture.
- SA diorite family
    granite-granodiorite family
    spherulites
    textures

orbital observations
Term introduced in 1978. Before 1978, search observations AND orbital.
- SA observations

Orbitoidacea
Includes use on level 2 under foraminifera(1). See list F.
- BA Rotaliina
    foraminifera
- BT Invertebrata
- NA Amphistegina
    Asterocyclina
    Cibicides
    Discocyclina
    Nephrolepidina
    Orbitoididae

Orbitoidae
use Orbitoididae

Orbitoididae
Includes use on level 2 under foraminifera(1). See list F. Also search Orbitoidae.
- UF Orbitoidae
- BA Orbitoidacea
    Rotaliina
    foraminifera
- BT Invertebrata
- NA Lepidocyclina

Orbitolina
Genus. Includes use on level 3 under foraminifera(1) Lituolacea(2).
- BA Orbitolinidae
    Lituolacea
    Textulariina
    foraminifera
- BT Invertebrata

Orbitolinidae
Family. Includes use on level 3 under foraminifera(1) Lituolacea(2).
- BA Lituolacea
    Textulariina
    foraminifera
- BT Invertebrata
- NA Orbitolina

Orbulina
Genus. Includes use on level 3 under foraminifera(1) Globigerinacea(2).
- BA Globigerinidae
    Globigerinacea
    Rotaliina
    foraminifera
- BT Invertebrata
- NA Orbulina universa

Orbulina universa
Includes use on level 3 under foraminifera(1) Globigerinacea(2).
- BA Orbulina
    Globigerinidae
    Globigerinacea
    Rotaliina
    foraminifera
- BT Invertebrata

order-disorder
As of 1978, term is used on level 2 under crystal chemistry(1).
- SA bonding
    crystal chemistry
    substitution

Ordovician
Includes use on level 1 as an age term (list E); on level 2 under paleo- terms, e.g. paleoecology, paleogeography, paleomagnetism. After the Cambrian, before the Silurian.
- BA Paleozoic
- NT Antelope Valley Limestone
- NA Bala
- NT Borrowdale Volcanic Series
    Chickamauga Group
    Davidsville Group
    Gander Lake Group
- NA Lower Ordovician
- NT Lushs Bight Group
    Martinsburg Formation
    Meguma Group
- NA Middle Ordovician
- NT Pogonip Group
    Shap Granite
    Skiddaw Slates
- NA Upper Ordovician
- NT Viola Limestone
- NA Viruan
- SA Arbuckle Group
    Knox Group
    Taconic Orogeny
    Talladega Group

Ordubad
City in Nakhichevan Autonomous Soviet Socialist Republic (W enclave of Azerbaidzhan).
- BT Azerbaidzhan
    USSR

ore bodies
- UF bodies, ore
- SA ore deposits

ore deposits
Includes use on level 2 or 3 under metallic commodity terms (list C); on level 3 under mineral deposits, genesis(1) and under metasomatism(1) materials(2).
- SA deposits
    disseminated deposits
    exploration
    gangue
    lenses
    massive deposits
    metallogenic provinces
    metallogeny
    metallurgy
    metasomatism
    mineral deposits
    mineralization
    mining geology
    ore bodies
    ore minerals
    ore sources
    ore transport
    placers
    porphyry copper
    reflectivity
    reserves
    roll-type deposits
    stockwork deposits
    stratabound deposits
    stratiform deposits
    talc
    veins
    wallrock alteration

ore finding
Not a valid index term for GeoRef. See mineral exploration.

ore guides
Includes use on level 2 under mineral exploration(1).
- UF guides, ore
- SA mineral exploration

ore microscopy
Includes use on level 3 under mineralogy(1) methods(2) and under petrology(1) methods(2). See list C (commodities). Before 1976, also search mineragraphy.
- UF microscopy, ore
- SA microscope methods
    mineralogy
    optical mineralogy
    universal stage

ore minerals
Includes use on level 3 under economic geology(1).
- BT minerals
- SA economic geology
    metals
    ore deposits
    ore-forming fluids

ore of sedimentation
use placers

ore sources
Before 1978, search ore deposits AND sources.
- UF sources, ore
- SA mineral deposits
    ore deposits
    ore transport

ore transport
Includes use on level 3 under mineral deposits, genesis(1). Before 1978, search ore deposits AND transport.
- SA mineral deposits
    ore deposits
    ore sources
    transport

ore-forming fluids
Includes use on level 3 under mineral deposits, genesis(1) or economic geology(1). Also search mineralizers.
- UF fluids, ore-forming
- SA hydrothermal solutions
    mineral deposits
    mineralization
    ore minerals

Oread Limestone
In Shawnee Group. Includes Toronto Limestone, Snyderville Shale, Leavenworth Limestone, Heebner Shale, Plattsmouth Limestone, Heumader Shale, Kereford Limestone members. SW Iowa, E Kansas, NW Missouri, SE Nebraska, and N Oklahoma.
- BT Virgilian
    Upper Pennsylvanian
    Pennsylvanian
- SA Iowa
    Kansas
    Missouri
    Nebraska
    Oklahoma
    Plattsmouth Limestone Member
    Shawnee Group

Orebro
County in S central part of country. Also a city. Search Oerebro.
- UF Oerebro
- BT Sweden

Oregon
Includes use on level 1 as an area term (list O). For term set options see list B.
- CO N420000N462000
      W1163500W1243500
- BA United States
- NX Benton County
- NT Clarno
      Clatsop County
- NX Columbia County
- NT Coos Bay
      Coos County
      Crater Lake
- NX Crook County
      Curry County
      Douglas County
      Grant County
- NT Harney County
- NX Jackson County
      Jefferson County
      Lake County
      Lane County
      Lincoln County
- NT Malheur County
- NX Marion County
- NT Mazama Ash
      Mount Hood
- NX Newport
      Polk County
      Portland
- NT Riddle
      Rogue River
- NX Sherman County
- NT Sixes River
      Steens Mountain
      Tillamook County
- NX Union County
- NT Wallowa Mountains
- NX Washington County
      Wheeler County
- NT Willamette Valley
      Willow Creek
      Yaquina Bay
- SA Bald Mountain
      Basin and Range Province
      Blue Mountains
      Cascade Range
      Clarno Formation
      Coast Ranges
      Columbia Plateau
      Columbia River
      Columbia River Basalt
      Columbia River basin
      Columbia River estuary
      Cordillera
      Great Basin
      John Day Formation
      Klamath Mountains
      Pacific Coast
      Snake River
      Snake River basin
      Snake River canyon
      Tyee Formation
      Umpqua Formation
      Western U.S.
      Yamhill Formation

Orenburg
City on Ural River in the Southern Urals in European USSR. Also an oblast.
- UF Chkalov
- BT Russian Republic
      USSR

Orense
Province in NW part of country. Also a city.
- BT Spain

ores in sedimentary rocks
A valid term through 1973. Use sedimentary rocks or sedimentation in conjunction with the individual ore.

ores, polymetallic
use polymetallic ores

Oresund
Strait connecting the Kattegat with the Baltic Sea. Index countries as applicable.
- BT Europe
- SA Denmark
      Sweden

organic
A valid term through 1977. After 1977, use organic materials.

organic carbon
Includes use on level 3 under organic materials(1).
- BA organic materials
- SA carbon

organic compounds
Includes use on level 2 under minerals(1). See list L.
- BT minerals
- NA amber
      ozocerite
- SA compounds
      porphyrins

organic materials
Includes use on level 1 (list A); on level 2 under soils(1). Used for discussions of mostly very small concentrations of organic materials in rocks. Also search organic matter. If 1, term set options are:
  kind of material [amino acids, bitumens, carbohydrates, fatty acids, humates, humic acids, hydrocarbons, kerogen, phenols]
  specific kind of material or topic
  topic [abundance, age, alteration, analysis, classification, composition, detection, distribution, experimental studies, genesis, geochemistry, identification, nomenclature, observations, occurrence, properties, varieties]
      subtopic
- NA amino acids
      aromatic hydrocarbons
      bitumens
      carbohydrates
      carotenoids
      chlorophyll
      collagen
      conchiolin
      enzymes
      ethane
      fatty acids
      fulvic acids
      humates
      humic acids
      humus
      hydrocarbons
      isoprenoids
      kerogen
      lipids
      methane
      organic carbon
      phenols
      pigments
      porphyrins
      proteins
      purines
      resins
      sapropel
      sporopollenin
      sugars
- SA abundance
      anthracite
      anthraxolite
      carbon
      carbonaceous composition
      caustobiolith
      chemical analysis
      coal
      complexing
      geochemistry
      inorganic materials
      isotopes
      lignite
      materials
      oil sands
      oil shale
      organic residues
      organic sediments
      peat
      petroleum
      polymerization
      reflectance
      sedimentary rocks
      sedimentation
      sediments
      soils

organic mound
use bioherms

organic residues
Includes use on level 2 under sedimentary rocks(1). See list I (sed. rocks). Until 1976, term was also used on level 2 under sediments.
- UF residues, organic
- BT sedimentary rocks
- NA anthracite
      bituminous coal
      brown coal
      caustobiolith
      coal
      durain
      lignite
      sapropelite
      torbanite
      vitrain
- SA exinite
      fusinite
      inertinite
      macerals
      micrinite
      organic materials
      organic sediments
      resinite
      sediments
      sporinite
      vitrinite

organic sediments
Term introduced in 1976. Includes use on level 2 under sediments(1). See list N.
- BA sediments
- NA guano
      gyttja
      peat
- SA organic materials
      organic residues

organisms
Includes use on level 3 under soils(1) genesis(2) as type of factor.
- SA biogeography
      biology
      biomass
      biota
      bioturbation
      factors
      microorganisms
      paleobotany
      soils

organization
Includes use on level 3 under surveys(1) and associations(1).
- SA associations
      museums
      surveys
      symposia

Orgueil
Also search Montauban.
- UF Montauban
      Orgueil Meteorite
- BT Tarn-et-Garonne
      France
- SA meteorites

Orgueil Meteorite
use Orgueil

orientation
Includes use as attitude of fold elements with respect to external coordinates. Includes use on level 2 under faults(1) and folds(1); on level 3 to index orientation grains under structural analysis(1).
- SA dip
      discordant folds
      drag folds
      fabric
      faults
      folds
      horizontal orientation
      inclined folds
      linear orientation
      lineation
      longitudinal orientation
      nappes
      oblique orientation
      overturned folds
      petrofabrics
      plunging folds
      preferred orientation
      recumbent folds
      schistosity
      strike
      structural analysis
      style
      superposed folds
      transverse faults
      vertical orientation

Oriente
Province in E.
- BT Cuba
      West Indies

origin
A valid level 2 term through 1977 used in combination with life. After 1977, use life origin on level 2 under paleontology(1).

Orinoco River
Rises in Serra Parima Mountains in S Venezuela, turns E in central Venezuela and empties through a wide delta into the Atlantic Ocean. Index countries as applicable. Also search Orinoco.
- BT South America
- SA Colombia
      Venezuela

Oriskany Sandstone
Comprises Shriver Chert, and Ridgeley Sandstone Member. W Maryland, New York, Pennsylvania, Western Virginia and E West Virginia.
- BT Lower Devonian
- SA Maryland
      New York
      Pennsylvania
      Virginia
      West Virginia

Orissa
State on the Bay of Bengal in E.
- BT India
- NT Balasore
      Bolangir
      Cuttack
      Dhenkanal
      Keonjhar
      Koraput
      Mayurbhanj
      Nilgiri
      Rampur coal field
      Sambalpur
      Talchir
      Talchir coal field
- SA Deccan Plateau
      Gangpur Series
      Mahanadi Valley
      Singhbhum Granite
      Singhbhum shear zone

Talchir Series

Orkney Islands
Archipelago off NE coast.
CO N584500N593000
W0061000W0080000
BT Scotland
Great Britain
United Kingdom

Orleans
City in N central part of country.
BT France

Orlov
Old name for city of Khalturin just W of Kirov in Kirov Oblast of European USSR.
BT Russian Republic
USSR

ornamentation
SA paleontology
shells

Orne
Department in NW.
BT France
NT Mortagne
SA Eure Valley
Normandy

Ornithischia
Order. Includes use on level 3 under Reptilia(1) Archosauria(2).
BA Archosauria
Reptilia
BT Tetrapoda
Vertebrata
NA Hadrosauridae

orogenesis
A valid term through 1974. After 1974, use orogeny.

orogeny
Includes use on level 1 (list A). Used for discussions of either individual orogenies or detailed general treatments on several orogenies. Before 1974, also search mountain building; before 1975, also search orogenesis. Term set options are: topic [absolute age, causes, evolution, extent, mechanism, periodicity] name of orogeny
subtopic
SA absolute age
Acadian Phase
Allegheny Orogeny
Alpine Orogeny
Andean Orogeny
Antler Orogeny
Appalachian Phase
Assyntic Orogeny
Asturian Orogeny
Avalonian Orogeny
Baikalian Phase
Cadomian Orogeny
Caledonian Orogeny
Cordilleran Orogeny
deformation
diastrophism
epeirogeny
faults
folds
geosynclines
Grenvillian Orogeny
Hudsonian Orogeny
igneous activity
Karelian Orogeny
Katangan Orogeny
Kenoran Orogeny
Laramide Orogeny
metamorphism
Pan-African Orogeny
Penokean Orogeny
periodicity
plate tectonics
Pyrenean Orogeny
Saalian Phase
structural geology

Taconic Orogeny
taphrogeny
tectonics
volcanology

Orosei
Village and gulf of Tyrrhenian Sea.
BT Sardinia
Italy

Oroville Dam
Earth fill dam on Feather River which forms Oroville Reservoir in Butte County of N central California. Also search Oroville Reservoir; reservoirs AND Oroville; surface reservoirs AND Oroville.
BT California
United States
SA dams
reservoirs
surface reservoirs

orpiment
UF yellow arsenic
BA sulfides
BT minerals

Orsha
City in Vitebsk Oblast in N.
BT Byelorussia
USSR

Orsk
City in Orenburg Oblast in Southern Urals near the Kazakh border.
BT Russian Republic
USSR

Orthida
Includes use on level 2 under Brachiopoda(1). See list F.
BA Articulata
Brachiopoda
BT Invertebrata
NA Enteletacea

orthite
use allanite

ortho-amphibolite
use orthoamphibolite

orthoamphibolite
See list J.
UF ortho-amphibolite
BA amphibolites
BT metamorphic rocks

orthoclase
BA feldspar group
framework silicates
silicates
BT minerals
SA adularia
anorthoclase
K-feldspar
sanidine

orthoenstatite
BA pyroxene group
chain silicates
silicates
BT minerals

orthopyroxene
BA pyroxene group
chain silicates
silicates
BT minerals

orthoquartzite
Includes use on level 3 under sedimentary rocks(1) clastic rocks(2). See list I. Also search sedimentary quartzite.
UF orthoquartzitic sandstone
sedimentary quartzite
BA clastic rocks
BT sedimentary rocks
SA sandstone
terrigenous materials

orthoquartzitic sandstone
use orthoquartzite

orthosilicates
Sorosilicates and neosilicates. Includes use on level 2 under minerals(1); in combination with epidote group, garnet group, humite group, melilite group, and olivine group (i.e. orthosilicates, epidote group) to form terms on level 2 under minerals(1). See list L. Also search sorosilicates.
UF nesosilicates
sorosilicates
BA silicates
BT minerals
NA andalusite
braunite
chevkinite
chloritoid
NZ clinohumite
NA coffinite
cyrtolite
datolite
dumortierite
epidote group
euclase
eucryptite
gadolinite
garnet group
garrelsite
grandidierite
harkerite
hemimorphite
NZ humite
NA humite group
ilvaite
innelite
joaquinite
julgoldite
kasolite
kilchoanite
kornerupine
kyanite
labuntsovite
larnite
lawsonite
melanocerite
melilite group
merwinite
monticellite
mullite
NZ nenadkevichite
norbergite
NA olivine group
perrierite
phenakite
pumpellyite
roggianite
sapphirine
sillimanite
NZ spurrite
NA staurolite
stillwellite
thalenite
NZ thaumasite
NA thorite
thortveitite
titanite
topaz
uranophane
vesuvianite
viridine
willemite
yttrialite
zircon
NZ zunyite

Orulgan Mountains
Extend N and S, E of Lena River in N central Russian Republic. Also search Orulgan; Orulgan Range.
UF Orulgan Range
BT Russian Republic
USSR

Orulgan Range
use Orulgan Mountains

Oruro
Department in W part of the country on the Chilean border. Also a city.
BT Bolivia

os
use eskers

Os
use osmium

Osage County
Index states as applicable.
BX Kansas
Missouri
Oklahoma
BT United States

Osagean
use Osagian

Osagian
Provincial series, North America. Above Kinderhookian, below Meramecian. Includes use on level 3 under age terms(1). See list E. Before 1976, also search Osagean.
UF Osagean
BA Lower Mississippian
Mississippian
BT Paleozoic
NT Keokuk Limestone
SA Valmeyeran

Osaka
City and prefecture in S.
BT Honshu
Japan

Osaka Group
BT Cenozoic
SA Honshu
Japan
Pleistocene
Pliocene

Osborne County
N Kansas.
BT Kansas
United States

oscillations
As of 1978 term is restricted to refer to motions under Earth(1).
SA Earth
free oscillations
motions
resonance
vibration

Osculosida
Order. Includes use on level 3 under Radiolaria(1).
BA Radiolaria
BT Invertebrata
NA Nassellina

Osetia
use Ossetia

Oshima
Fishing town in Kumamoto Prefecture W Kyushu, Japan.
UF Amami
BT Ryukyu Islands
Japan

Oshima Peninsula
N of Hakodate in Oshima sub-prefecture in SW.
BT Hokkaido
Japan

Osler Series
SW Ontario.
BT Precambrian
SA Canada
Ontario

Oslo
County in SE Norway. Also a city on Oslo Fjord an inlet of the Skagerrak.
BT Norway

osmium
Includes use on level 1 and 2 as a chemical element (list D).
UF Os
SA elements

Osnabruck
City in NW part of country. Also search Osnabrueck.
UF Osnabrueck
BT Lower Saxony
    West Germany

Osnabrueck
use Osnabruck

Osogovo Mountains
Index countries as applicable. Also search Osogov; Osogovo.
BT Europe
SA Bulgaria
    Yugoslavia

Osor
Village in NE part of country.
BT Gerona
    Spain

Ossetia
Region of the central Caucasus. Divided into North Ossetian A.S.S.R., and the South Ossetian Autonomous Oblast. Index Soviet Republics as applicable.
UF Osetia
BT USSR
SA Georgian Republic
    Russian Republic

Ossipee Mountains
In Carrol County in E.
BT New Hampshire
    United States

Ostashkovichi
Village in S.
BT Byelorussia
    USSR

Osteichthyes
For Actinopterygii, Holostei, Teleostei. Includes use on level 2 under Pisces(1). See list F.
BA Pisces
BT Vertebrata
NA Acanthodes
    Actinopterygii
    Crossopterygii
    Cyprinidae
    Holostei
    Teleostei
SA fish

osteology
SA bones
    paleontology

Ostracoda
Includes use on level 1 and 2 as a fossil term (list F).
BT Crustacea
    Arthropoda
    Invertebrata
NA Archeocopida
    Cyprididae
    Leperditicopida
    Myodocopida
    Paleocopida
    Podocopida
SA plankton

ostracoderms
Included in Agnatha.
BA Agnatha
    Pisces
BT Vertebrata

Ostrava
City in N Moravia.
UF Moravska Ostrava
BT Moravia
    Czechoslovakia

Ostrava-Karvina
Coal mining and steel producing region near the Polish border in N.
BT Moravia
    Czechoslovakia

Ostrea
Genus. Includes use on level 3 under Mollusca(1) Bivalvia(2).
BA Ostreidae
    Bivalvia
    Mollusca
BT Invertebrata

Ostreidae
Family. Includes use on level 3 under Mollusca(1) Bivalvia(2).
BA Bivalvia
    Mollusca
BT Invertebrata
NA Ostrea

Ostrowiec
use Ostrowiec Swietokrzyski

Ostrowiec Swietokrzyski
City in SW central part of country. Also search Ostrowiec.
UF Ostrowiec
BT Kielce
    Poland

Osumi Peninsula
Between Kagoshima Bay and Osumi Strait in S.
BT Kyushu
    Japan

osumilite
BA ring silicates
    silicates
BT minerals
SA aluminosilicates

Oswego County
On SE Lake Ontario.
BT New York
    United States

Otago
Provincial district in S.
BT South Island
    New Zealand

Otago Peninsula
On E side of Otago Harbor in Dunedin area on SE.
BT South Island
    New Zealand

Otake
Town on Hiroshima Bay in SW.
BT Honshu
    Japan

Otavi
Town in N.
BT South-West Africa

Otero County
Index states as applicable.
BX Colorado
    New Mexico
BT United States

Othris
use Othrys

Othrys
Mountain range in central part of country. Index administrative regions as applicable.
UF Othris
    Othrys Massif
BT Greece
SA Thessaly

Othrys Massif
use Othrys

otoliths
Includes use on level 3 under Pisces(1).
SA fossils
    microfossils
    Pisces

Otranto
Town on Strait of Otranto in heel of Italian boot.
BT Apulia
    Italy

Ottawa
City in SE.
BT Ontario
    Canada
SA Leda Clay

Ottawa County
Index states as applicable.
BX Kansas
    Michigan
    Ohio
    Oklahoma
BT United States

Ottawa Valley
River valley. Index provinces as applicable.
BT Canada
SA Ontario
    Quebec

Ottnangian
Europe.
BA Miocene
    Neogene
    Tertiary
BT Cenozoic

Otto Fiord
use Otto Fjord

Otto Fjord
NW Ellesmere Island, District of Franklin. Also search Otto.
UF Otto Fiord
BT Northwest Territories
    Canada

Otway Basin
Primarily in W and SW Victoria. Index states as applicable.
BT Australia
SA South Australia
    Victoria

Otztal Alps
Mountain range of the Eastern Alps. Index countries as applicable. Also search Oetztal Alps.
UF Oetztal Alps
    Oetztal Massif
    Oetztaler
    Otztaler Alps
BT Alps
    Europe
SA Austria
    Eastern Alps
    Italy
    Stubai Alps

Otztaler Alps
use Otztal Alps

Ouachita Mountains
Index states as applicable.
CO N340000N344000
    W0943000W0960000
BT United States
SA Arkansas
    Oklahoma

Ouegoa
Village.
BT New Caledonia
    Pacific Ocean

Ougarta
Village W of the Great Western Erg in W central.
BT Algeria

Ouray County
SW Colorado.
BT Colorado
    United States

Ouro Preto
City in S.
BT Minas Gerais
    Brazil

outcropping
use outcrops

outcrops
Used as a general term. Also search outcropping.
UF outcropping
SA bedrock

Outer Banks
Chain of sandy barrier islands stretching length of coast.
UF The Banks
BT North Carolina
    United States

outer core
BA core
SA inner core

Outer Hebrides
Outer group of the Hebrides, W of Little Minch. Sometimes called Long Island.
CO N564500N583000
    W0061000W0080000
BT Scotland
    Great Britain
    United Kingdom
SA Hebrides
    Inner Hebrides

outer mantle
use upper mantle

Outer Mongolia
use Mongolia

outer planets
SA Jupiter
    Neptune
    planetology
    planets
    Pluto
    Saturn
    terrestrial planets
    Uranus

outer shelf
SA continental shelf
    inner shelf

outer slope
Term introduced in 1978.
UF slope, outer
SA continental slope
    inner slope

outer space
use interplanetary space

Outokumpu
Village in SE part of country.
BT Finland

outwash
Includes use on level 3 under sediments(1) clastic sediments(2); on level 3 under glacial geology(1) glacial features(2). See list N.
UF glacial outwash
    outwash drift
BA clastic sediments
BT sediments
SA glacial features
    glacial geology
    terrigenous materials

outwash aprons
use outwash plains

outwash drift
use outwash

outwash plains
Also search sandurs.
UF morainal plains
    outwash aprons
    overwash plain
    sandurs
BT glacial features
SA moraines
    plains

overburden
SA regolith

overconsolidated materials
Term introduced in 1978. Includes use on level 3 under engineering geology(1). Before 1978, search overconsolidated AND materials.
SA consolidation
    engineering geology
    foundations

overgrowths
SA crystal growth
   minerals

Overijssel
Province in E.
BT Netherlands
NT Twente

overthrust
A valid term through 1977. After 1977, use overthrust faults.

Overthrust Belt
Also search Thrust Belt.
UF Thrust Belt
BT United States
SA Idaho
   Montana
   Utah
   Wyoming

overthrust faults
Term introduced in 1978. Includes use on level 3 under faults(1) displacements(2). Before 1978, also search faults AND overthrust.
BA faults
SA displacements
   nappes
   thrust faults

Overton County
N Tennessee.
BT Tennessee
   United States

overturned folds
Term introduced in 1978. Includes use as folds defined on the basis of orientation in relation to the geographic horizontal plane. Includes use on level 3 under folds(1) orientation(2). Before 1978, also search folds AND overturned.
BA folds
SA orientation

overwash
Term introduced in 1978. Also search washover.
UF washover
SA fluvial features

overwash plain
use outwash plains

Oviedo
Province in NW Spain. Also a city.
BT Spain
NT Asturian Massif
   Aviles
   La Caridad

Ovruch Series
Divided into two subseries.
BT Proterozoic
   Precambrian
SA Ukraine
   USSR

Owens Valley
River valley between Sierra Nevada on the W, and White and Inyo Mountains on E.
BT California
   United States

Owyhee County
Extreme SW Idaho.
BT Idaho
   United States

Oxford
City 52 miles WNW of London.
BT England
   Great Britain
   United Kingdom

Oxford Clay
BT Oxfordian
   Jurassic
SA England
   United Kingdom

Oxford County
W Maine.
BT Maine
   United States

Oxfordian
Europe. Above Callovian, below Kimmeridgian. Includes use on level 3 under age terms(1). See list E.
BA Upper Jurassic
   Jurassic
BT Mesozoic
NT Oxford Clay
SA Malm

Oxfordshire
County in central.
CO N513000N521000
   W0005000W0014500
BT England
   Great Britain
   United Kingdom

oxidation
Includes use on level 3 under geochemistry(1) processes(2).
SA Eh
   geochemistry
   hydrolysis
   oxidation zone
   oxygen
   processes
   reduction

oxidation zone
UF zone, oxidation
SA Eh
   mineral deposits
   oxidation

oxidation-reduction potential
use Eh

oxides
Includes niobates and tantalates. Includes use on level 2 under minerals(1). See list L.
BT minerals
NA aeschynite
   akaganeite
   alexandrite
   anatase
   armalcolite
   baddeleyite
   bayerite
   betafite
   birnessite
   bixbyite
   boehmite
   brannerite
   bromellite
   brookite
   brucite
   cassiterite
   chrome spinel
   chromite
   chrysoberyl
   columbite
   coronadite
   corundum
   cryptomelane
   cuprite
   delafossite
   diaspore
   euxenite
   fergusonite
   ferropseudobrookite
   franklinite
   gahnite
   geikielite
   germanates
   gibbsite
   goethite
   groutite
   hausmannite
   hematite
   hercynite
   heterogenite
   hollandite
   hydrogoethite
   hydroxides
   ilmenite
   iron oxides
   ixiolite
   jacobsite
   lepidocrocite
   limonite
   lithiophorite
   loparite
   maghemite
   magnesioferrite
   magnetite
   manganese oxide
   manganite
   manganosite
   martite
NZ microlite
   nenadkevichite
NA nigerite
   niobates
   nordstrandite
   nsutite
   periclase
   perovskite
   pitchblende
   pseudobrookite
   psilomelane
NZ pyrochlore
NA pyrolusite
   ramsdellite
   rancieite
   rutile
   samarskite
   sapphire
   senarmontite
   specularite
   spinel
   spinel group
   stibiotantalite
   taaffeite
   tantalates
   tantalite
   tapiolite
   thorianite
   titanomaghemite
   titanomagnetite
   todorokite
   trevorite
   ulvospinel
   uraninite
   valentinite
   wodginite
   wustite
   zincite
   zirconolite
NZ zirkelite
SA alumina

Oxisols
Includes use on level 3 under soils(1). See list M.
BA soils

Oxnard
City in Ventura County in S.
BT California
   United States

Oxus
use Amu Darya

oxygen
Includes use on level 1 and 2 as a chemical element (list D). Also search O.
UF O
SA elements
   isotopes
   O-16
   O-18
   O-18/O-16
   oxidation
   ozone
   stable isotopes

oxymagnite
use maghemite

Ozark Highlands
use Ozark Mountains

Ozark Mountains
Eroded tableland. Also search Ozarks; Ozark uplift. Index states as applicable.
UF Ozark Highlands
   Ozark Uplift
   Ozarks
BT United States
SA Arkansas
   Missouri
   Oklahoma

Ozark Uplift
use Ozark Mountains

Ozarkodina
Genus.
BA conodonts

Ozarks
use Ozark Mountains

Ozernoye
Village in SE Kamchatka on Okhotsk Sea just N of Ozernovskiy in Khabarovsk Kray.
BT Russian Republic
   USSR

ozocerite
BA organic compounds
BT minerals

ozone
SA atmosphere
   oxygen

Ozun
use Uzon

P

P
use phosphorus

P-T conditions
Includes use on level 2 under metamorphism(1).
UF conditions, P-T
SA geologic barometry
   geologic thermometry
   grade
   inclusions
   metamorphism
   metasomatism
   phase equilibria
   pressure
   temperature

P-waves
Includes use on level 3 under seismology(1). Also search compressional wave; compressional waves; longitudinal wave.
UF compressional wave
   dilatational wave
   irrotational wave
   longitudinal wave
   pressure wave
   primary wave
   push wave
   push-pull wave
BT body waves
   elastic waves
SA S-waves
   seismology
   waves

Pa
use protactinium

Pa-231
Includes use on level 3 under isotopes(1).
SA isotopes
   protactinium

Pacaya
Extinct volcano in S central

Guatemala. Also search Pacaya Volcano.
- UF Pacaya Volcano
- BT Guatemala

Pacaya Volcano
use Pacaya

Pachelma
Town in Penza Oblast W of Kuibyshev in S central European USSR.
- BT Russian Republic
  - USSR

Pachuca
City NE of Mexico City.
- UF Pachuca de Soto
- BT Hidalgo
  - Mexico

Pachuca de Soto
use Pachuca

Pachypteris
- BA Caytoniales
    - gymnosperms
- BT Plantae
- SA Cycadales

Pacific Basin
Generally the entire ocean floor of both the North and South Pacific up to the continental shelf.
- BT Pacific Ocean
- SA North Pacific
  - Pacific Plate
  - South Pacific

Pacific Coast
Region comprising those states fronting on the Pacific Ocean. Index states as applicable. Also search West Coast. Introduced as level 1 area term in 1978.
- UF West Coast
- BA United States
- SA California
  - Oregon
  - Pacific region
  - Washington

Pacific mobile belt
Crustal regions of tectonic activity along the coastal fringes of continents and some entire islands of the Pacific. Index continents and islands as applicable.
- BT Pacific region
- SA Asia
  - New Guinea
  - New Zealand
  - North America
  - South America

Pacific Ocean
Includes use on level 1 or 2 as an area term (list O). If 1, see term set options under list B.
- NT Aleutian Ridge
    Aleutian Trench
    Arafura Sea
    Astoria Canyon
    Auckland Islands
    Austral Islands
    Bauer Deep
    Bay of Plenty
- NA Bering Sea
- NT Bismarck Sea
    Blanco fracture zone
    Bonin Islands
    Bowie Seamount
    Carnegie Ridge
    Caroline Islands
    Cascadia Basin
    Cascadia Channel
- NA Celebes Sea
- NT Chatham Rise
    Chile Ridge
    China Sea
    Circum-Pacific region
    Cobb Seamount
    Cocos Plate
    Cocos Ridge
    Cook Islands
- NA Coral Sea
    East China Sea
    East Pacific
- NT East Pacific Rise
    Easter Island
    Emperor Seamounts
- NA Equatorial Pacific
- NT Erimo Seamount
    Fanning Island
    Fiji Plateau
    Formosa Strait
- NA Galapagos Islands
- NT Gilbert Islands
    Gorda Rise
    Guadalupe Island
    Gulf of Alaska
- NA Gulf of California
- NT Gulf of Panama
    Hachijo-jima
- NZ Hawaii
- NT Hawaiian Ridge
    Iwo-Jima
- NA Japan Sea
- NT Japan Trench
    Java Sea
    Juan de Fuca Ridge
    Kermadec Islands
    Kuril Trench
    Lau Basin
    Line Islands
    Lord Howe Island
    Lord Howe Rise
    Loyalty Islands
    Macquarie Island
    Macquarie Ridge
    Malaita
    Malekula
    Manihiki Plateau
    Marion Island
- NA Melanesia
- NT Mendocino fracture zone
- NA Micronesia
- NT Middle America Trench
    Midway
    Monterey Canyon
    Monterey Fan
    Murray fracture zone
    Nazca Plate
- NA New Caledonia
- NT New Caledonia Basin
    Norfolk Island
    Norfolk Ridge
- NA North Pacific
    Northeast Pacific
    Northwest Pacific
    Oceania
    Okhotsk Sea
- NT Ontong Java Plateau
    Pacific Basin
    Panama Basin
    Peru-Chile Trench
- NA Philippine Sea
    Polynesia
- NT Raoul Island
    Redondo Canyon
    San Diego Trough
    Santa Barbara Basin
    Shatsky Rise
    Shikoku Basin
    Solomon Sea
- NA South China Sea
    South Pacific
    Southeast Pacific
    Southwest Pacific
- NT Sulu Sea
    Tasman Basin
- NA Tasman Sea
- NT Tonga Trench
    Torres Strait
    Wake
- NA West Pacific
    Yellow Sea
- SA Antarctic Ocean
    Mid-Pacific Rise
    New Hebrides
    Pacific region
    Ross Sea

Pacific Plate
Includes most of the Pacific Basin with the Eurasian Plate on the W and N; the American, Cocos, and Nazca Plates on the E; the Indian Plate on the SW; and the Antarctic Plate on the S.
- SA Pacific Basin
    plate tectonics

Pacific region
Includes use on level 1 or 2 as an area term (list O). If 1, see term set options under list B.
- CO S700000N630000
    W0700000E1300000
- NT Pacific mobile belt
- SA Circum-Pacific region
    Far East
    Pacific Coast
    Pacific Ocean

Pacific-Antarctic Ridge
Between the Antarctic continent and New Zealand.
- SA Antarctica

packing
Includes use on level 3 under sediments(1) and sedimentation(1). Before 1978, also used for atomic packing under crystal structure.
- SA atomic packing
    sedimentary rocks
    sedimentation
    textures

packstone
Includes use on level 3 under sedimentary rocks(1) carbonate rocks(2). See list I.
- BA carbonate rocks
- BT sedimentary rocks

Pacoima Dam
On Pacoima River in Los Angeles County NW of Los Angeles. Also search Pacoima.
- BT California
    United States

Paderborn
City in E.
- BT North Rhine-Westphalia
    West Germany

Padre Island
Narrow barrier island along Gulf of Mexico from SE of Corpus Christi to near the Mexican border.
- BT Texas
    United States

Padstow
Town in SW near N coast of Cornwall.
- BT England
    Great Britain
    United Kingdom

Padurea Craiului Mountains
Range in SW. Also search Padurea Craiului.
- BT Transylvania
    Romania

Paganzo
Village in central.
- BT La Rioja
    Argentina

Paganzo Group
- BT Paleozoic
- SA Argentina
    Carboniferous
    Permian

Pageland
Town in Chesterfield County in N.
- BT South Carolina
    United States

Pagoda Formation
Underlies Pentagon Shale; overlies Dearborn Limestone. NW Montana.
- UF Pagoda Limestone
- BT Middle Cambrian
    Cambrian
- SA Montana

Pagoda Limestone
use Pagoda Formation

Pahang
State on the Malay Peninsula in West Malaysia.
- BT Malaysia
- SA Malay Peninsula
    West Malaysia

pahoehoe
Includes use on level 3, e.g. under igneous rocks(1) volcanic(2).
- BT lava
- SA aa lava
    volcanism

Pahute Mesa
Tableland in Nye County in S.
- UF Paiute Mesa
- BT Nevada
    United States

Pai Khoi
use Pai-Khoi

Pai-Khoi
Mountain range on Yugor Peninsula between Barents and Kara seas. N extension of Northern Urals.
- CO N683000N691500
    E0650000E0604500
- UF Pai Khoi
- BT Russian Republic
    USSR

Paiute Mesa
use Pahute Mesa

Pajaro Valley
River valley S of San Jose in W California. Also search Pajaro River.
- BT California
    United States

Pakistan
Formerly consisted of East Pakistan and West Pakistan which were separated by about 1,000 miles of Indian territory. After East Pakistan became the independent state of Bangladesh in 1971, West Pakistan and Pakistan became coextensive. Also search West Pakistan. Includes use on level 1 as an area term (list O). For term set options see list B.
- CO N233500N373000
    E0751500E0601500
- BA Asia
- NT Baluchistan
    North-West Frontier Province
    Sind
    Sulaiman Range
- SA Hindu Kush
    Hyderabad
    Indian Plate
    Indus River
    Jammu and Kashmir
    Karakoram
    Ladakh
    Nari Series
    Punjab
    Ramnagar
    Siwalik Range
    Siwalik System

Paktia
Province in E.
- BT Afghanistan

Palaeofavosites
Genus. Includes use on level 3 under Coelenterata(1) Tabulata(2).
- BA Favositidae
    Tabulata
    Anthozoa
    Coelenterata
- BT Invertebrata

SA Favosites

Palaeoloxodon naumanni
Includes use on level 3 under Mammalia(1) Proboscidea(2).
BA Elephantidae
Proboscidea
Mammalia
BT Tetrapoda
Vertebrata

palagonite
Includes use on level 3 under igneous rocks(1) pyroclastics and glasses(2). See list H.
BA pyroclastics and glasses
BT igneous rocks

Palamau
District in coal mining area in W.
BT Bihar
India

Palatinate
Historical region in two parts. Lower or Rhine Palatinate on both sides of Rhine River S of the Main River; and the Upper Palatinate in E Bavaria. Index states as applicable.
BT Germany
SA Baden-Wurttemberg
Bavaria
Hesse
Rhineland-Palatinate
Upper Palatinate

Palawan
Long, narrow island between N Borneo and W Philippines.
UF Paragua
BT Philippine Islands

Palencia
Province in N Spain. Also a city.
BT Spain

paleo-oceanography
SA bathymetry
ocean basins
ocean floors
oceanography
paleocirculation
sea-floor spreading

paleobathymetry
SA bathymetry
ocean floors
paleocirculation
sea-floor spreading

paleobiology
SA biochemistry
biology
fossils
paleontology

paleobotany
For general discussion of fossil plants. See names of major floral groups (list F). Includes use on 1 (list A); on level 2 under area terms (list B). If level 1, term set options are: topic [applications, bibliography, catalogs, classification, concepts, education, evolution, history, instruments, methods, nomenclature, objectives, practice, principles, symposia, textbooks]
subtopic
UF botany, paleo-
SA bacteria
bibliography
biogeography
bryophytes
catalogs
flora
floral studies
fossilization
fossils
fungi
leaves
micropaleontology
organisms

paleoecology
paleontology
palynology
photosynthesis
Plantae
Protista
seeds
species
sporangia
staining
vegetation
wood

Paleocene
World. Lower Tertiary, above Gulfian (Cretaceous), below Eocene. Includes use on level 1 as an age term (list E); on level 2 under paleoterms, e.g. paleoecology, paleogeography; on level 3 under age terms(1).
BA Paleogene
Tertiary
Cenozoic
NT Beaufort Formation
Clayton Formation
NA lower Paleocene
middle Paleocene
Montian
NT Pinyon Conglomerate
Ravenscrag Formation
Sentinel Butte Formation
Tongue River Formation
NA upper Paleocene
SA Beaverhead Formation
Briançonnais Zone
Difunta Group
Fort Union Formation
Mardin Formation
Moreno Formation
Paskapoo Formation
Scaglia Formation
Sinjar Formation
Sparnacian
Wasatch Formation

paleochannels
Not a valid index term for GeoRef. See channels under sedimentary structures(1) planar bedding structures(2).

paleocirculation
SA circulation
ocean circulation
paleo-oceanography
paleobathymetry

paleoclimatology
Includes use on level 1 (list A). Used for treatments of the climate of a given period of time in the geologic past. If 1, term set options are:
age (single term from list E)
(area)
topic [applications, changes, concepts, cycles, evolution, indicators, interpretation, methods, patterns, temperature]
subtopic
UF climatology, paleo-
SA arctic environment
atmosphere
Boreal
boreal environment
C-13/C-12
climate
cycles
fluctuations
fossils
glacial geology
glaciation
indicators
interglacial environment
isotopes
O-18/O-16
paleoecology
paleogeography
S-34/S-32

Paleocopida
Includes use on level 2 under Ostracoda(1). See list F.
BA Ostracoda
BT Crustacea
Arthropoda
Invertebrata

paleocurrents
Includes use on level 3 under sedimentation(1) provenance(2) or paleogeography(1).
SA currents
paleogeography
provenance
sedimentation

paleoecology
Includes use on level 1 (list A); on level 2 under fossil terms (list F). Used for the study of the relationships between organisms and their environments, the death of organisms, and their burial and postburial history in the geologic past based on fossil fauna and flora and their stratigraphic position. 1, term set options are:
age [List E]
area
fossil group [List F]
age [List E]
topic [analysis, changes, indicators, interpretation, sedimentation]
subtopic [e.g. name of environment: beaches, deltas, estuaries, lagoons, lakes, marine, marshes, reefs, streams, terrestrial environment]
SA adaptation
archaeology
arctic environment
beaches
benthonic taxa
biogeography
biostratigraphy
biotypes
boreal environment
changes of level
communities
deltas
ecology
estuaries
fresh water
fresh-water environment
indicators
lagoons
lakes
marine environment
paleobotany
paleoclimatology
paleoenvironment
paleogeography
paleontology
peat bogs
ponds
reefs
salinity
salt marshes
sedimentation
taphonomy
terrestrial environment

paleoenvironment
Distinguish from paleoecology.
SA environment
environmental analysis
historical geology
paleoecology

Paleogene
As of 1978, includes use on level 1 and 2 (list E). An interval of geologic time incorporating the Oligocene, Eocene, and Paleocene of the Tertiary; the lower Tertiary.
BA Tertiary
Cenozoic
NT Duchesne River Formation
NA Eocene

Ilerdian
NT Kenai Group
NA Oligocene
Paleocene
NT Sinjar Formation
Wasatch Formation
SA Arikareean
lower Tertiary

paleogeographic controls
Includes use on level 3 under mineral deposits, genesis(1) controls(2).
SA controls
mineral deposits
paleogeography

paleogeographic maps
Term introduced in 1978. Before 1978, also search maps AND paleogeographic.
BA maps

paleogeography
Includes use on level 1 (list A). Used for the geography of ancient times, specifically the study and description of the physical geography of the geologic past. Term set options are:
age [List E; a single term]
(area)
topic [applications, changes, concepts, interpretation, maps, methods, patterns, principles]
subtopic (no area term)
SA alluvial plains
basins
biogeography
buried channels
buried valleys
changes of level
continental drift
geography
geomorphology
geosynclines
Mesogaea
ocean basins
paleoclimatology
paleocurrents
paleoecology
paleogeographic controls
reconstruction
sebkha environment
sedimentary basins
sedimentary structures
sedimentation
shorelines
syntectonic processes
transgression

paleolatitude
SA paleomagnetism
polar wandering

paleolimnology
SA glacial lakes
limnology

Paleolithic
Archaeologic classification.
UF Old Stone Age
BA Cenozoic
NA Acheulian
Magdalenian
Mousterian
SA Pleistocene
Quaternary
Tertiary

paleomagnetism
Used for the study of natural remanent magnetization. Includes use on level 1 (list A); on level 2 under crust(1) and geochronology(1); on level 3 under plate tectonics(1). If 1, term set options are:
age [List E]
area [applications, causes, changes, concepts, experimental studies, geochemistry, interpretation, methods, patterns, polar wandering, pole posi-

tions, reversals, stability]
topic
    subtopic
UF magnetism, paleo-
SA anhysteretic remanent magnetization
    chemical remanent magnetization
    coercivity
    continental drift
    crust
    Curie point
    declination
    demagnetization
    depositional remanent magnetization
    dipole moment
    domain structure
    Earth
    geochemistry
    geochronology
    geophysical methods
    geophysics
    induction
    intensity
    isothermal remanent magnetization
    lava
    magnetic anomalies
    magnetic domains
    magnetic field
    magnetic susceptibility
    magnetization
    natural remanent magnetization
    paleolatitude
    plate tectonics
    polar wandering
    pole positions
    remagnetization
    remanent magnetization
    reversals
    sea-floor spreading
    stability
    tectonophysics
    thermoremanent magnetization
    viscous remanent magnetization

paleontology
Used for discipline as a whole. Includes use on level 1 (list A); on level 2 under area terms (list B), age terms (list E), continental drift(1), symposia(1), bibliography(1), and automatic data processing(1). See also names of fossil groups (list F). If 1, term set options are:
topic [applications, bibliography, catalogs, classification, concepts, education, evolution (as a concept), fossilization, history, instruments, life, origin, nomenclature, practice, principles, symposia, taxonomy, (principles of), textbooks]
    subtopics
SA adaptation
    bibliography
    biocenoses
    biogeography
    biometry
    biostratigraphy
    bones
    catalogs
    collections
    continental drift
    fauna
    faunal provinces
    fossilization
    fossils
    geochronology
    glossaries
    historical geology
    isomorphism
    jaws
    lexicons
    life origin
    microfauna
    micropaleontology
    nutrition
    ontogeny
    ornamentation
    osteology
    paleobiology
    paleobotany
    paleoecology
    palynology
    phylogeny
    problematic fossils
    Protista
    Pterobranchia
    radiation
    reproduction
    sample preparation
    shells
    skeletons
    skulls
    species
    species diversity
    staining
    stratigraphy
    synonymy
    taxonomy
    thanatocenoses
    ultrastructure
    wood

paleorelief
SA landform evolution
    relief

paleosalinity
As of 1978, term is used on level 2 under fluid inclusions(1).
SA brackish water
    fluid inclusions
    geochemistry
    salinity
    salt
    sea water

Paleosols
UF fossil soils
BA soils

paleotectonics
A valid term through 1975. After 1975, see tectonics(1) and paleogeography(1). See also neotectonics.

paleotemperature
A valid term through 1973. After 1973, use temperature under paleoclimatology(1) and see geologic thermometry.

Paleozoic
Includes use on level 1 as an age term (list E); on level 2 under paleo- terms, e.g. paleoecology, paleogeography, paleomagnetism.
BA Phanerozoic
NT Antler Orogeny
    Arbuckle Group
    Baralaba Coal Measures
    Bedford Shale
    Berea Sandstone
    Bird Spring Formation
    Broken River Formation
    Caledonian Orogeny
NA Cambrian
    Carboniferous
NT Casper Formation
    Catskill Formation
    Chattanooga Shale
    Deadwood Formation
NA Devonian
NT Dunkard Group
    Geirud Formation
    Hercynian Orogeny
    Hidden Valley Dolomite
    Honaker Trail Formation
    Hunton Group
    Knox Group
    Lisburne Group
NA lower Paleozoic
    middle Paleozoic
NT Mikabu System
    Minnelusa Formation
NA Mississippian
    Ordovician
NT Paganzo Group
    Peel Sound Formation
NA Pennsylvanian
    Permian
NT Petersburg Granite
    Read Bay Formation
    Ringerike Sandstone
    Rondout Formation
    Sambagawa Belt
NA Silurian
NT Silvretta Group
    Supai Formation
    Taconic Orogeny
    Talchir Formation
    Talladega Group
    Tensleep Sandstone
    Tulghes Series
NA upper Paleozoic
NT Weber Sandstone
    Wissahickon Formation
    Woodford Shale
SA Beacon Supergroup
    Botucatu Formation
    Ceneri Zone
    Ferrar Group
    New York City Group
    Shuswap Complex
    Sparagmite Group

Palermo
City on Bay of Palermo in NW.
BT Sicily
    Italy

Palestine
Approximately coextensive with Israel and the W bank of the Jordan River. Index countries as applicable.
SA Israel
    Jordan

palingenesis
As of 1978, term is restricted to meaning in petrology. Term is not used in reference to paleontology.
SA anatexis
    magmas
    metasomatism
    petrology

palladium
Includes use on level 1 and 2 as a chemical element (list D).
UF Pd
SA elements

pallasites
Used as igneous rock until 1977. Now used as stony meteorites.
BA meteorites
SA stony irons

Palliser Formation
BT Upper Devonian
    Devonian
SA Alberta
    Canada

Palm Beach County
SE Florida.
BT Florida
    United States

Palm Springs
City in Riverside County in S.
BT California
    United States

Palma
City on Majorca.
UF Palma de Majorca
BT Balearic Islands
    Mediterranean Sea

Palma de Majorca
use Palma

Palmae
Family coextensive with order Palmales. Also search Palmales.
BA Monocotyledoneae
    angiosperms
BT Plantae

Palmatolepis
Genus.
BA conodonts

Palmer Peninsula
use Antarctic Peninsula

Palo Alto
City S of San Francisco in Santa Clara County.
BT California
    United States

Palos Verdes Hills
Occupy peninsula between Santa Monica Bay and San Pedro Bay S of Los Angeles.
UF San Pedro Hills
BT California
    United States

Paluxy Formation
In Trinity Group. SW Arkansas, SE Oklahoma, and E Texas.
BT Comanchean
    Cretaceous
SA Arkansas
    Lower Cretaceous
    Oklahoma
    Texas
    Trinity Group

palygorskite
UF attapulgite
BA sheet silicates
    silicates
BT minerals
SA clay minerals

palynology
Includes use on level 1 (list A). See list F. Used for the study of pollen of seed plants and spores of other embryophytic plants whether living or fossil, including their dispersal and applications. Term set options are:
topic [applications, bibliography, catalogs, classification, concepts, education, fossilization, history, instruments, methods, nomenclature, practice, principles, symposia, techniques, textbooks]
    subtopic
SA biogeography
    geochronology
    micropaleontology
    paleobotany
    paleontology
    palynomorphs
    pollen
    pollen analysis
    pollen diagrams
    sample preparation
    statistical methods
    stratigraphy

palynomorphs
Includes use on level 1 and 2 as a fossil term (list F).
NA acritarchs
    Chitinozoa
    Dinoflagellata
    megaspores
    miospores
    problematic palynomorphs
SA Artemisia
    exine
    floral studies
    monolete taxa
    palynology
    plankton
    Plantae
    pollen
    Pyrrhophyta
    spores
    Tasmanites

Pambak
NE section of city of Kirovakan in N

central Armenia.
UF Bambak
BT Armenia
USSR

Pamir (Range)
use Pamirs

Pamir Range
use Pamirs

Pamirs
A high altitude region of central Asia from which great mountain ranges extend. Also search Pamir; Pamir Range. Index Afghanistan and administrative units in China, India, and USSR as applicable.
CO N360000N400000
E0750000E0700000
UF Pamir (Range)
Pamir Range
BT Asia
SA Afghanistan
Jammu and Kashmir
Sinkiang Uighur
Tadzhikistan

Pamlico River
Bisects Beaufort County and flows into Pamlico Sound. Actually it is the estuary of the Tar River.
BT North Carolina
United States

Pamlico Sound
Between E North Carolina mainland and narrow islands off the coast.
BT North Carolina
United States

Pampa
use Pampas

Pampas
Vast, treeless, fertile plain extending N and S approximately 1000 miles from lower Parana River to Patagonia, and from the Atlantic to the Andean piedmont. The humid Pampas are in the E and the dry Pampas are in the W. Also search Pampa.
UF Pampa
BT Argentina
South America

Pampean Mountains
In the province of Cordoba.
BT Argentina

Pamplona
City in N part of country.
BT Navarre
Spain

Pan-African Orogeny
Term introduced in 1978. Includes use on level 3 under orogeny(1). Before 1978, also search orogeny AND Pan-African or Pan African.
SA orogeny

Panagyurishte
Plovdiv district, W Central Bulgaria in central Sredna Gora on Luda Yana River.
BT Bulgaria
SA Sredna Gora

Panama
Includes use on level 1 as an area term (list O). For term set options see list B.
BA Central America
NT Darien
Sansan

Panama Basin
SW of Panama and W of Colombia.
BT Pacific Ocean

Panama Canal Zone
10 mile strip of territory across Panama in which canal is located.
BT Central America

Panamint Range
W of Death Valley in Inyo County in E.
BT California
United States

Panasqueira
Village in S part of country.
BT Portugal

Panay Island
One of the Visayan Islands midway between Luzon and Mindanao. Also search Panay.
BT Philippine Islands
Malay Archipelago

Panch Mahals
District in W part of country. Also search Panchmahal; Panchmahals.
UF Panchmahal
Panchmahals
BT Gujarat
India

Panchet Series
Series of rocks overlying the Damuda Series in the Raniganj coal field.
BT Lower Triassic
Triassic
SA Bihar
India

Panchmahal
use Panch Mahals

Panchmahals
use Panch Mahals

Panderodus
BA conodonts

Pangaea
Name proposed for the supercontinent comprising all the landmasses or earth which existed about 300 million years ago prior to continental drift. Includes use on level 2 under continental drift(1).
UF Pangea
SA continental drift
Gondwana
Laurasia
Northern Hemisphere

Pangea
use Pangaea

Panhandle
An area or projection of land like the handle of a pan which occurs in a number of states. Index states as applicable.
BT United States
SA Florida
Oklahoma
Texas
West Virginia

Pannonia
Ancient Roman province S and W of the Danube. Index countries as applicable.
BT Europe
SA Austria
Hungary
Yugoslavia

Pannonian
Europe. Includes use on level 3 under age terms(1). See list E.
BA Neogene
Tertiary
BT Cenozoic
SA Miocene
Pliocene

Pannonian Basin
SE Austria, Hungary W of the Danube, and N central Yugoslavia. Index countries as applicable.
BT Europe
SA Austria
Hungary
Yugoslavia

Pantar
Island of the Alor group 60 miles NW of Timor.
BT Lesser Sunda Islands
Indonesia

Pantellaria
use Pantelleria

Pantelleria
Volcanic Italian island in the Mediterranean Sea between Sicily and Tunisia.
UF Pantellaria
BT Sicily
Italy

pantellerite
Includes use on level 3 under igneous rocks(1) andesite-rhyolite family(2). See list H.
BA andesite-rhyolite family
BT igneous rocks
SA comendite

Pantotheria
Includes use on level 2 under Mammalia(1). See list F.
BA Theria
Mammalia
BT Tetrapoda
Vertebrata

Papua
Former Australian territory. Now the southern half of new state of Papua New Guinea on the island of New Guinea. Includes use on level 3. Included use as a first order area term until 1976. Now use Papua New Guinea on level 1.
SA Papua New Guinea

Papua New Guinea
Eastern half of the island of New Guinea. Comprises Papua and the former Australian U.N. trusteeship of The Territory of New Guinea plus the Bismarck Archipelago, and Bougainville, Buka and Green islands of the W Solomon Islands. Became an independent state on September 16, 1976. Includes use on level 1 as an area term (list O). Also search Papua-New Guinea.
UF Papua-New Guinea
BA Australasia
NT Bismarck Archipelago
Bougainville
Huon Peninsula
Louisiade Archipelago
New Britain
New Ireland
Port Moresby
Talasea
SA Malay Archipelago
New Guinea
Papua
Solomon Islands

Papua-New Guinea
use Papua New Guinea

Para
State in N.
BT Brazil
NT Marajo
SA Barreiras Formation
Brazilian Shield
Tocantins River region

para-amphibolite
See list J.
BA amphibolites
BT metamorphic rocks
SA amphibolite

paracelsian
BA feldspar group
framework silicates
silicates
BT minerals
SA celsian

Paradox Basin
In the "Four Corners" area of the SW where the boundaries of 4 states intersect. Index states as applicable.
BT United States
SA Arizona
Colorado
New Mexico
Utah

Paradox Member
Of Hermosa Formation. W Colorado and SE Utah.
BT Middle Pennsylvanian
Pennsylvanian
SA Colorado
Utah

paragenesis
For detailed treatment of mineral sequences in metamorphosed or altered rocks and mineral deposits. Includes use on level 1 (list A). Term set options are:
kind of mineral deposit (first level term only)
area
kind of rock (first level term only)
area
topic [changes, evolution, interpretation, observations, patterns, processes, rates]
subtopic
UF mineral sequence
paragenetic sequence
SA clay mineralogy
crystallography
economic geology
igneous rocks
metamorphic rocks
metamorphism
metasomatism
mineral deposits
minerals
phase equilibria
rocks

paragenetic sequence
use paragenesis

paragneiss
BA metasedimentary rocks
BT metamorphic rocks
SA gneiss

paragonite
BA sheet silicates
silicates
BT minerals
SA mica group

Paragua
use Palawan

Paraguana Peninsula
Between the Caribbean Sea and the Gulf of Venezuela in NW part of country.
BT Falcon
Venezuela

Paraguay
Includes use on level 1 as an area term (list O). For term set options see list B.
CO S273000S191000
W0543000W0624500
BA South America
SA Botucatu Formation
Chaco
Parana Basin
Parana River
Serra Gerral Formation

Parahiba
use Paraiba

Parahyba
use Paraiba

Paraiba
State in NE Brazil.
UF Parahiba

Parahyba
BT Brazil
SA Borborema

parallel faults
Term introduced in 1978. Includes use on level 3 under faults(1) patterns(2). Before 1978, also search faults AND parallel.
BA faults

parallel folds
use concentric folds

parameters
A valid general term through 1978.

Paramushir
Large island in the N Kuril Islands near S tip of Kamchatka Peninsula.
BT Russian Republic
   USSR
SA Kuril Islands

Parana
State in S.
BT Brazil
SA Acungui Group
   Brazilian Shield
   Estrada Nova Formation
   Irati Formation
   Serra do Mar

Parana Basin
River basin. Index countries as applicable.
UF Parana River basin
BT South America
SA Argentina
   Brazil
   Paraguay

Parana River
Formed in S central Brazil by the confluence of the Rio Grande and the Paranaiba rivers. It flows SW and S into the Rio de la Plata. Index countries as applicable. Also search Parana.
BT South America
SA Argentina
   Brazil
   Paraguay

Parana River basin
use Parana Basin

Paranthropus
Genus. Includes use on level 3 under Mammalia(1) Primates(2).
BA Hominidae
   Primates
   Mammalia
BT Tetrapoda
   Vertebrata

pararammelsbergite
BA arsenides
   sulfides
BT minerals
SA rammelsbergite

parasites
SA bacteria
   fungi

paratacamite
BA halides
BT minerals
SA atacamite
   chlorides

Paratethys
Includes use on level 3 under continental drift(1).
SA continental drift
   Tethys

Paratunka
Village just W of Petropavlosk Kamchatskiy in S Kamchatka Oblast on Kamchatka Peninsula.
BT Russian Republic
   USSR

parent materials
Includes use on level 3 under soils(1) genesis(2) as type of factor; on level 3 under weathering(1).
SA factors
   horizons
   materials
   soils
   weathering

parent rocks
use protoliths

pargasite
BA amphibole group
   chain silicates
   silicates
BT minerals

Paria Peninsula
NE part of country. Along with island of Trinidad nearly encloses Gulf of Paria.
BT Venezuela
SA Sucre

Paricutin
Volcano on site of former village of Paricutin in SW part of country.
BT Michoacan
   Mexico

Paring Mountains
BT Romania

Paris
City on the Seine river. Also search Paris region.
UF Paris region
BT France

Paris Basin
Chief depression of N and N central France. Bounded by English Channel on NW, Amorican Massif on W, Massif Central on S, and plateaus of Langres and Lorraine on E. Also search Paris AND basin.
CO N480000N500000
   E0031500E0013000
BT France

Paris region
use Paris

Park City
City in Summir County in NE.
BT Utah
   United States

Park County
Index states as applicable.
BX Colorado
   Montana
   Wyoming
BT United States

Parker County
W of Fort Worth in N.
BT Texas
   United States

Parkfield
Village in Monterey County near Fremont County line in W. Location of earthquake.
BT California
   United States

Parkman Sandstone
In Montana Group. N Wyoming and S Montana:
BT Upper Cretaceous
   Cretaceous
SA Montana
   Montana Group
   Wyoming

Parma
City in N central part of country.
BT Emilia-Romagna
   Italy

Parnaiba Basin
River basin. Index states as applicable.

BT Brazil
SA Maranhao
   Piaui

Parnalee
Located 16 miles S of Madurai where meteorite impacted.
BT Tamil Nadu
   India

Parnassos
use Parnassus

Parnassus
One of highest massifs in central Greece. N of Gulf of Corinth.
UF Parnassos
BT Greece

Parras Basin
S Coahuila in NE central part of country. Also search Parras.
UF Parras de la Fuente
BT Coahuila
   Mexico

Parras de la Fuente
use Parras Basin

Parry Sound
Town in SE on E shore of Georgian Bay.
BT Ontario
   Canada

partial melting
Includes use on level 3 under magmas(1).
SA magmas
   mantle
   melting
   melts
   phase equilibria
   plate tectonics

partial pressure
BT pressure
SA fugacity

particle precipitation
Includes use on level 2 under ionosphere(1).
SA intensity
   ionosphere
   ions
   particle radiation
   particles
   precipitation

particle radiation
Includes use on level 2 under astrophysics and solar physics(1).
SA astrophysics and solar physics
   ionization
   ionosphere
   particle precipitation
   particles
   radiation
   solar wind

particle size
use size

particle track
Cosmic-ray track. Includes use on level 2 under geochronology(1).
UF track, particle
SA cosmic rays
   fission tracks
   geochronology
   radiation damage

particles
Includes use on level 2 and 3 under meteorology(1); on level 3 under interplanetary space(1) cosmic rays(2).
SA activation energy
   aerosols
   agglutinates
   consistency
   cosmic dust
   cosmic rays
   dust
   electromagnetic radiation

   electrons
   fines
   fission
   fragments
   grains
   granulometry
   intensity
   interplanetary dust
   interplanetary space
   ions
   meteorology
   Moon
   neutrons
   particle precipitation
   particle radiation
   precipitation
   rounding
   roundness
   scintillations
   shape
   size
   soils
   solar wind
   sorting
   sphericity
   trapped particles

parting lineation
Before 1978, also search current lineations.
UF current lineations
   current partings
BA bedding plane irregularities
   sedimentary structures
SA lineation

partition coefficients
UF coefficients, partition
SA crystal chemistry
   partitioning

partitioning
As of 1978, term is used on level 2 under crystal chemistry(1).
SA crystal chemistry
   geochemistry
   partition coefficients
   phase equilibria

partridgeite
use bixbyite

Pas-de-Calais
Department in extreme N.
CO N500000N511500
   E0031000E0014500
BT France
NT Boulogne
   Boulonnais
SA Nord-Pas-de-Calais Basin

Pasadena
City in Los Angeles County.
BT California
   United States

Pasamonte
Point of impact near village of Pasamonte in Union County, New Mexico. Also search Pasamonte Meteorite.
BT New Mexico
   United States
SA meteorites

Pasayten Group
Stratigraphically above Dewdney Creek Formation. S British Columbia and Central N Washington.
BT Lower Cretaceous
   Cretaceous
SA British Columbia
   Washington

Pascagoula River basin
SE Mississippi.
BT Mississippi
   United States

Pasco Basin
S central Washington.
BT Washington
   United States

Paskapoo Formation
SA Alberta
Canada
Cretaceous
Eocene
Paleocene
Saskatchewan
Upper Cretaceous

Passa Dois Group
BT Permian
SA Brazil
Irati Formation

Passau
City on the Danube River at the Austrian border.
BT Bavaria
West Germany

Patagonia
Region. A barren tableland between the Atlantic Ocean and the Andes extending from the Rio Negro River in the N to the Strait of Magellan in the S. Index provinces as applicable.
CO S520000S400000
W0650000W0670000
BT Argentina
SA Chile
Chubut
Rio Negro
Santa Cruz

Patagonia Cordillera
use Patagonian Andes

Patagonian Andes
Mountain range extending the length of Patagonia. Index countries as applicable.
UF Patagonia Cordillera
BT South America
SA Argentina
Chile

patch reefs
Includes use in level 3 under reefs(1).
BT reefs

paternoite
use kaliborite

Patom Plateau
Between Vitim and Lena rivers in NE Irkutsk Oblast N of Lake Baikal. Also search Patom.
CO N540000N590000
E1180000E1060000
BT Russian Republic
USSR

patterned ground
Includes use on level 3 under glacial geology(1) periglacial features(2) and under geomorphology(1).
UF ground, patterned
SA frost action
geomorphology
glacial geology
periglacial features
polygons

patterns
Includes use as a level 2 or 3 term appropriate to a large number of topics, e.g. on level 2 under faults(1), fractures(1), and under soils(1). See list G.
SA dispersion patterns
drainage patterns
Ekman spiral
regional patterns

Patuxent River
Rises in central Maryland and flows S and SE into Chesapeake Bay.
BT Maryland
United States

Pau
City in extreme SW part of country.
BT Pyrenees-Atlantiques
France

Pauni
Town in central.
BT Madhya Pradesh
India

Pavia
City and province S of Milan.
BT Lombardy
Italy

Pavlodar
Town and oblast in NE.
BT Kazakhstan
USSR

Pavlograd
City in Dnepropetrovsk Oblast in E central.
BT Ukraine
USSR

Pawhuska Rock Plain
Osage county in N.
BT Oklahoma
United States

Pawnee County
Index states as applicable.
BX Kansas
Nebraska
Oklahoma
BT United States

Payne County
N central Oklahoma.
BT Oklahoma
United States

Payson
City in Utah County in N central.
BT Utah
United States

Payun Matru
Volcano on Payun Plateau in SW Mendoza.
UF Payun Matru Volcano
Payun-Matru
Payun-Matru Volcano
BT Mendoza
Argentina

Payun Matru Volcano
use Payun Matru

Payun-Matru
use Payun Matru

Payun-Matru Volcano
use Payun Matru

Pb
use lead

Pb-204
Includes use on level 3 under isotopes(1).
SA isotopes
lead
Pb-206/Pb-204
Pb-207/Pb-204
Pb-208/Pb-204

Pb-204/Pb-206
use Pb-206/Pb-204

Pb-204/Pb-208
use Pb-208/Pb-204

Pb-206
Includes use on level 3 under isotopes(1).
SA isotopes
lead
Pb-206/Pb-204
Pb-207/Pb-206

Pb-206 Pb-207
use Pb-207/Pb-206

Pb-206/Pb-204
Includes use on level 3 under isotopes(1). Also search Pb-204/Pb-206; Pb-204 AND Pb-206.
UF Pb-204/Pb-206
SA isotopes
lead
Pb-204
Pb-206

Pb-206/Pb-207
use Pb-207/Pb-206

Pb-207
Includes use on level 3 under isotopes(1).
SA isotopes
lead
Pb-207/Pb-204
Pb-207/Pb-206

Pb-207/Pb-204
Includes use on level 3 under isotopes(1). Also Search Pb-207 AND Pb-204.
SA isotopes
lead
Pb-204
Pb-207

Pb-207/Pb-206
Includes use on level 3 under isotopes(1). Also search Pb-206/P-207; Pb-207 AND Pb-206.
UF Pb-206 Pb-207
Pb-206/Pb-207
SA isotopes
lead
Pb-206
Pb-207

Pb-208
Includes use on level 3 under isotopes(1).
SA isotopes
lead
Pb-208/Pb-204

Pb-208/Pb-204
Includes use on level 3 under isotopes(1). Also search Pb-204/Pb-208; Pb-204 AND Pb-208.
UF Pb-204/Pb-208
SA isotopes
lead
Pb-204
Pb-208

Pb-210
Includes use on level 3 under absolute age(1) methods(2).
SA absolute age
isotopes
lead
Pb/Pb
Pb/Th
U/Pb

Pb/Pb
Includes use on level 3 under absolute age(1) methods(2). Also search lead-lead.
SA absolute age
lead
Pb-210
Pb/Th
U/Pb

Pb/Th
Includes use on level 3 under absolute age(1) methods(2). Also search Pb AND Th.
SA absolute age
lead
Pb-210
Pb/Pb
thorium
U/Pb

Pd
use palladium

Peace River
Flows into Snake River just N of Lake Athabasca. Index provinces as applicable.
BT Canada
SA Alberta
British Columbia

Peak District
Plateau region in N central.
BT England

Great Britain
United Kingdom

pearceite
BA sulfarsenites
sulfosalts
BT minerals

Pearl Harbor
Inlet on S coast of Oahu 6 miles W of Honolulu.
BT Hawaii
United States

Pearlette Ash
use Pearlette Volcanic Ash

Pearlette Volcanic Ash
SW Kansas.
UF Pearlette Ash
BT Pleistocene
Quaternary

pearlite
use perlite

pearls
SA aragonite
gems
Mollusca

Peary Land
Region in extreme N on Arctic Ocean.
BT Greenland
Arctic region

peat
Includes use on level 1 as a commodity term (list C); on level 3 under sediments(1) organic sediments(2). See list N.
BA organic sediments
BT sediments
SA biogenic structures
Bog soils
brown coal
coal
humus
lignite
organic materials
peat bogs

peat bed
use peat bogs

peat bogs
Also search peat bed.
UF peat bed
peat moor
BT bogs
SA ecology
paleoecology
peat
swamps

peat moor
use peat bogs

pebbles
Includes use on level 3 under sediments(1) clastic sediments(2); on level 3 under geomorphology(1) and glacial geology(1). See list N.
BA clastic sediments
BT sediments
SA cobbles
gravel
rounding
shingle
terrigenous materials

Pechenga
Territory, formerly in N Finland, which was called Petsamo. Now it is part of Murmansk Oblast.
BT Russian Republic
USSR

Pechora
City in Komi Autonomous Soviet Socialist Republic on left bank of Pechora River in NE European USSR.
BT Russian Republic
USSR

Pechora Basin
Coal basin mainly in basin of Usa River in NE Komi Autonomous S.S.R. in NE European USSR. Also search Pechora.
　CO　N650000N680000
　　　E0600000E0500000
　BT　Russian Republic
　　　USSR

Pechora River
Rises in the Central Urals of N Perm Oblast and flows N through the Komi A.S.S.R., and the Nenets National Okrug into Pechora Bay. Also search Pechora.
　CO　N620000N670000
　　　E0593000E0520000
　BT　USSR

Pecopteris
Genus. Includes use on level 3 under pteridophytes(1) Filicopsida(2).
　BA　Filicopsida
　　　pteridophytes
　BT　Plantae
　SA　Glossopteris
　　　gymnosperms
　　　Pteridospermae

Pecos County
W Texas.
　BT　Texas
　　　United States

Pecos River valley
Index states as applicable. Also search Pecos River.
　BT　United States
　SA　New Mexico
　　　Texas

Pectinacea
Superfamily. Includes use on level 3 under Mollusca(1) Bivalvia(2).
　BA　Bivalvia
　　　Mollusca
　BT　Invertebrata

Pectinidae
Family. Includes use on level 3 under Mollusca(1) Bivalvia(2).
　BA　Bivalvia
　　　Mollusca
　BT　Invertebrata

pectolite
　BA　chain silicates
　　　silicates
　BT　minerals

pediments
Includes use on level 3 under geomorphology(1) erosion features(2).
　SA　erosion features
　　　erosion surfaces
　　　geomorphology

pediplains
　UF　desert pediplains
　　　desert plains
　　　pediplanes
　SA　erosion surfaces
　　　peneplains

pediplanes
　use pediplains

Pedregosa Basin
Index Mexican and U.S. states as applicable.
　SA　Arizona
　　　Chihuahua
　　　New Mexico
　　　Sonora

Peedee Formation
Variable gray to green argillaceous sands and impure limestones. E North Carolina and E South Carolina.
　BT　Upper Cretaceous
　　　Cretaceous
　SA　North Carolina
　　　South Carolina

Peekskill
City in Westchester County just N of New York City.
　BT　New York
　　　United States

Peel River
Flows into Mackenzie River. Index territories as applicable.
　BT　Canada
　SA　Northwest Territories
　　　Yukon Territory

Peel Sound Formation
Franklin District.
　BT　Paleozoic
　SA　Canada
　　　Devonian
　　　Northwest Territories
　　　Silurian

pegmatite
Includes use on level 1 or 2 as a commodity term (list C); on level 3 under igneous rocks(1) granite-granodiorite family(2). See list H. Also search pegmatites.
　BA　granite-granodiorite family
　BT　igneous rocks
　SA　graphic granite
　　　micropegmatite

Peichang
　use Peikang

Peikang
Town in W central.
　UF　Peichang
　BT　Taiwan

pelagic environment
Term introduced in 1978. Before 1978, search pelagic.
　SA　environment
　　　oceanography

Pelecypoda
A valid index term through 1975. After 1975, use Bivalvia.

pelite
　use shale

pelitic gneiss
A valid metamorphic rock term through 1977. After 1977, use gneiss and pelitic texture.

pelitic rocks
　use pelitic texture

pelitic schist
See list J.
　BA　schists
　BT　metamorphic rocks
　SA　schist

pelitic texture
Term introduced in 1978. Before 1978, search pelitic AND texture; pelitic rocks.
　UF　pelitic rocks
　SA　metamorphic rocks
　　　sedimentary rocks
　　　textures

pellets
　SA　oolite
　　　pisoliths

Pelmatozoa
Subphylum or other division of Echinodermata.
　BA　Echinodermata
　BT　Invertebrata

Pelomedusidae
Family. Includes use on level 3 under Reptilia(1) Anapsida(2).
　BA　Anapsida
　　　Reptilia
　BT　Tetrapoda
　　　Vertebrata

Pelona Schist
Precambrian(?). S California.
　BT　Precambrian
　SA　California

Peloponnese
　use Peloponnesus

Peloponnesos
　use Peloponnesus

Peloponnesus
Administrative region and peninsula forming S part of the mainland.
　CO　N361500N382000
　　　E0233000E0210000
　UF　Peloponnese
　　　Peloponnesos
　BT　Greece
　NT　Argolis
　　　Corinth
　　　Gavrovo Zone
　　　Megalopolis Basin
　　　Tripolis
　SA　Hellenides

Peloritani Mountains
NE Sicily.
　BT　Sicily
　　　Italy

Pelotas Basin
River basin. Index states as applicable.
　BT　Brazil
　SA　Rio Grande do Sul
　　　Santa Catarina

Pelvoux Massif
Mountain group which contains Barre das Ecrins, the highest peak in the Dauphine Alps. Index departments as applicable. Also search Pelvoux, and Ecrins-Pelvoux Massif.
　UF　Ecrins-Pelvoux Massif
　BT　France
　SA　Dauphine Alps
　　　Hautes-Alpes
　　　Isere

Pembroke
Town in Pembrokeshire in SW.
　BT　Wales
　　　Great Britain
　　　United Kingdom

Pembroke County
　use Pembrokeshire

Pembrokeshire
County in SW.
　CO　N513500N520500
　　　W0043000W0051500
　UF　Pembroke County
　BT　Wales
　　　Great Britain
　　　United Kingdom

Pench Valley
River valley in central Madhya Pradesh. Also search Pench.
　BT　Madhya Pradesh
　　　India

Pend Oreille County
Extreme NE.
　BT　Washington
　　　United States

Pendleton County
Index states as applicable.
　BX　Kentucky
　　　West Virginia
　BT　United States

peneplains
Includes use on level 3 under geomorphology(1) erosion features(2) and landform evolution(2).
　UF　base-level peneplain
　SA　erosion features
　　　erosion surfaces
　　　geomorphology
　　　pediplains

penetration
Not to be used for crystal growth.
　SA　engineering geology
　　　rock mechanics
　　　soil mechanics

Penghu Islands
Group of about 48 islands in Formosa Strait between Taiwan and the mainland of China.
　UF　Hoko Gunto
　　　Hoko Shoto
　　　Pescadores
　BT　Taiwan

Peninsular Ranges
Part of the Coast Ranges S of Los Angeles.
　BT　California
　　　United States
　SA　Coast Ranges

Peninsular Shield
　use Indian Shield

Pennine Alps
SW division of Central Alps extending from Great Saint Bernard Pass to Simplon Pass. Index countries as applicable. Also search Pennine.
　CO　N455000N461000
　　　E0081500E0070000
　BT　Alps
　　　Europe
　SA　Central Alps
　　　Italy
　　　Monte Rosa
　　　Piedmont Alps
　　　Switzerland

Pennine Chain
　use Pennines

Pennine Hills
　use Pennines

Pennine Range
　use Pennines

Pennines
Long hill range extending from the Cheviot Hills in the N to the S Midlands.
　UF　Pennine Chain
　　　Pennine Hills
　　　Pennine Range
　BT　England
　　　Great Britain
　　　United Kingdom

Pennington County
Index states as applicable.
　BX　Minnesota
　　　South Dakota
　BT　United States

Pennington Formation
Comprises Stony Gap Sandstone, Avis Limestone, Falls Mills Sandstone members. N Alabama, NW Georgia, E Kentucky, E Tennessee, and SW Virginia.
　BT　Upper Mississippian
　　　Mississippian
　SA　Alabama
　　　Georgia
　　　Kentucky
　　　Tennessee
　　　Virginia

penninite
　BA　sheet silicates
　　　silicates
　BT　minerals

Pennsylvania
Includes use on level 1 as an area term (list O). For term set options see list B.
　CO　N394500N423500
　　　W0744500W0803500
　BA　United States
　NX　Adams County
　NT　Allegheny County
　　　Altoona
　NX　Beaver County
　　　Bedford County
　NT　Berks County
　NX　Butler County

Carbon County
NT Centre County
NX Chester County
Columbia County
NT Conemaugh
NX Cumberland County
Delaware County
NT Dunkard Basin
NX Erie County
Fayette County
Franklin County
Fulton County
Greene County
NT Harrisburg
NX Jefferson County
Lawrence County
NT Lebanon County
Luzerne County
NX Mercer County
Monroe County
Montgomery County
NT Nittany Valley
Pennsylvania State University
NX Perry County
NT Philadelphia
NX Pike County
NT Pittsburgh
Pittsburgh coal basin
NX Presque Isle
NT Reading
Schuylkill County
NX Somerset County
Sullivan County
Union County
NT University of Pennsylvania
NX Warren County
Washington County
Wayne County
Waynesboro
Westmoreland County
Woodbury
York County
SA Allegheny Front
Allegheny Group
Allegheny Mountains
Allegheny Plateau
Ames Limestone
Antietam Formation
Appalachian Basin
Appalachian Plateau
Baltimore Gneiss
Bedford Shale
Berea Sandstone
Black River Group
Bloomsburg Formation
Blue Ridge Province
Catskill Delta
Catskill Formation
Clinton Group
Coeymans Formation
Conemaugh Group
Delaware River
Delaware River basin
Dunkard Group
Eastern U.S.
Genesee Group
Glenarm Series
Great Lakes region
Hamilton Group
Helderberg Group
Juniata Formation
Kinzers Formation
Lake Erie
Mahantango Formation
Martinsburg Formation
McKenzie Formation
Monongahela Group
Newark Basin
Newark-Gettysburg Basin
Ohio River
Ohio River valley
Onondaga Limestone
Oriskany Sandstone
Perry Formation
Piedmont
Pocono Formation
Pottsville Group

Rochester Formation
Rondout Formation
Sharon Conglomerate
Somerset
South Mountain
Susquehanna River
Susquehanna River basin
Tioga Bentonite
Tonoloway Limestone
Trenton Group
Tully Limestone
Tuscarora Formation
Valley and Ridge Province
Wissahickon Formation
York

Pennsylvania State University
University Park, Centre County.
BT Pennsylvania
United States

Pennsylvanian
Includes use on level 1 as an age term (list E); on level 2 under paleoterms, e.g. paleoecology, paleogeography, paleomagnetism.
BA Paleozoic
NT Bartlesville Sand
Bevier Coal
Conemaugh Group
Excello Shale
Francis Creek Shale
La Salle Limestone
NA Lower Pennsylvanian
NT Mansfield Formation
Marble Falls Group
Mattoon Formation
NA Middle Pennsylvanian
NT Minturn Formation
Monongahela Group
Morrow Formation
Pottsville Group
Red Fork Sandstone
Saginaw Formation
Sharon Conglomerate
Springfield Coal Member
Tennessee Sandstone
NA Upper Pennsylvanian
NT Wabash Formation
Wapanucka Limestone
Westerly Granite
SA Allegheny Orogeny
Baralaba Coal Measures
Bird Spring Formation
Carboniferous
Casper Formation
Dunkard Group
Honaker Trail Formation
Johns Valley Formation
Lisburne Group
Minnelusa Formation
Mississippian
Springer Formation
Stephanian
Supai Formation
Talladega Group
Tensleep Sandstone
Upper Carboniferous
Weber Sandstone

Penobscot Bay
Large inlet of Atlantic Ocean in S.
BT Maine
United States

Penokean Orogeny
A time of deformation and granite emplacement during the Precambrian. Index states as applicable. Before 1978, also search Penokean AND orogeny.
BT Precambrian
SA Michigan
Minnesota
orogeny

Penrhyn Slate
BT Precambrian
SA United Kingdom
Wales

Pensacola
City on Pensacola Bay in Escambia County in extreme W of the Florida panhandle.
BT Florida
United States

Pensacola Mountains
In the Horlick Mountains area SE of Ross Ice Shelf on the Pacific Ocean side.
BT Antarctica

Pentameracea
Superfamily. Includes use on level 3 under Brachiopoda(1) Pentamerida(2).
BA Pentamerida
Articulata
Brachiopoda
BT Invertebrata

Pentamerida
Includes use on level 2 under Brachiopoda(1). See list F.
BA Articulata
Brachiopoda
BT Invertebrata
NA Pentameracea

pentlandite
UF nicopyrite
BA sulfides
BT minerals
SA nickel

Penzhina Bay
NE extension of the Shelikhov Gulf of the Okhotsk Sea in NE Siberia. Also search Penzhina.
UF Penzhinskaya Bay
BT Russian Republic
USSR

Penzhinskaya Bay
use Penzhina Bay

Peoples Democratic Republic of Yemen
use Southern Yemen

People's Republic of the Congo
use Congo

Peoria Loess
Underlies Bignell Formation.
UF Peorian Loess
BT Pleistocene
SA Illinois
Iowa
Kansas
Nebraska

Peorian Loess
use Peoria Loess

Pequop Mountains
In Elko County in NE.
BT Nevada
United States

Perak
State on W coast of Malay Peninsula in West Malaysia.
BT Malaysia
NT Kinta Valley
SA West Malaysia

percolation
SA ground water
hydrology
infiltration
leaching
seepage
water

Peredovoy Range
In Kabardino-Balkarian A.S.S.R. in Northern Caucasus. Also search Peredovoy.
BT Russian Republic
USSR

Peremyshl
use Przemysl

Peribaltic Syneclise
A depressed structure of the continental platform in the Baltic Sea area produced by slow crustal downwarp. Also search Baltic Syneclise.
UF Baltic Syneclise
BT Europe
SA Baltic region

periclase
BA oxides
BT minerals

peridineans
use Dinoflagellata

Peridinium
BA Dinoflagellata
palynomorphs

peridot
Also search peridote.
UF peridote
peridote (olivine)
BA olivine group
orthosilicates
silicates
BT minerals
SA gems
tourmaline

peridote
use peridot

peridote (olivine)
use peridot

peridotite
Includes use on level 3 under igneous rocks(1) ultramafic family(2). See list H.
BA ultramafic family
BT igneous rocks
SA dunite
garnet peridotite
harzburgite
lherzolite
metaperidotite
plutonic rocks
wehrlite

periglacial environment
Term introduced in 1978. Before 1978, search periglacial.
SA environment
frost action
glacial geology

periglacial features
Includes use on level 2 under glacial geology(1). Also search periglacial.
UF features, periglacial
SA congelifraction
cryopedology
cryoturbation
extinct lakes
fans
firn
fossil ice wedges
Frozen ground
glacial features
glacial geology
glaciers
hummocks
ice
ice wedges
lakes
non-glacial ice
patterned ground
permafrost
pingos
polygons
solifluction

periglacial phenomena
A valid term through 1976. Use periglacial features.

periodicity
Includes use on level 2 under orogeny(1); on level 3 under earthquakes(1) and under volcanology(1).
SA earthquakes

        frequency
        orogeny
        sedimentation
        seismology
        volcanology

peripheral faults
Term introduced in 1978. Includes use on level 3 under faults(1) patterns(2). Before 1978, also search faults AND peripheral.
  BA faults

Perisphinctidae
Family. Includes use on level 3 under Mollusca(1) Cephalopoda(2).
  BA Ammonoidea
        Cephalopoda
        Mollusca
  BT Invertebrata

Perissodactyla
Includes use on level 2 under Mammalia(1). See list F.
  BA Mammalia
  BT Tetrapoda
        Vertebrata
  NA Coelodonta antiquitatis
        Equidae
        Rhinocerotidae
  SA Ungulata

peristerite
  BA feldspar group
        framework silicates
        silicates
  BT minerals

perlite
Includes use on level 3 under igneous rocks(1) pyroclastics and glasses(2); on level 3 as a commodity term under construction materials(1). See list C (commodities) and list H (igneous rocks). Also search pearlite.
  UF pearlite
  BA pyroclastics and glasses
  BT igneous rocks
  SA construction materials
        lava
        volcanic glass

Perm
City and oblast W of the Urals.
  BT Russian Republic
        USSR
  SA Molotov

permafrost
A level 1 term as of 1978. Used for geotechnical studies of permafrost; geomorphological studies should be under geomorphology. Term set options are:
  active layer
    subtopic
  classification
    subtopic
  creep
    subtopic
  engineering properties
    subtopic
  experimental studies
    subtopic
  frost action
    subtopi
  frost heaving
    subtopic
  site exploration
    subtopic
  solifluction
    subtopic
  theoretical studies
    subtopic
  SA active layer
        creep
        cryopedology
        engineering geology
        engineering properties
        frost action
        frost heaving
        Frozen ground
        geomorphology
        glacial geology
        ground ice
        hummocks
        ice
        ice wedges
        periglacial features
        pingos
        site exploration
        soils
        solifluction
        thawing
        thermal properties
        thermokarst
        tundra
        Tundra soils
        underground installations

permeability
Includes use on level 3 under engineering geology(1) and under area terms(1) hydrogeology(2). See list B.
  SA circulation
        Darcy's law
        diffusion
        engineering geology
        ground water
        hydraulic conductivity
        porosity
        rock mechanics
        roughness
        sediments
        thermal waters
        transmissivity

permeability coefficient
use hydraulic conductivity

Permian
Includes use on level 1 as an age term (list E); on level 2 under paleoterms, e.g. paleoecology, paleogeography, paleomagnetism. Above Carboniferous, below Triassic (Mesozoic).
  BA Paleozoic
  NT Akiyoshi Limestone
        Americus Limestone Member
        Appalachian Phase
  NA Asselian
  NT Blaine Formation
        Bulli Seam
        Castile Formation
        Chase Group
        Coconino Sandstone
        Cottonwood Limestone
        Cutler Formation
        Damuda Series
        Ecca Series
        Eskridge Shale
        Fort Riley Limestone
  NA Guadalupian
  NT Hennessey Formation
        Hughes Creek Shale
        Hutchinson Salt Member
        Illawarra Coal Measures
        Irati Formation
        Kaibab Formation
  NA Kungurian
        Lower Permian
        Middle Permian
  NT Passa Dois Group
        Phosphoria Formation
        Red Eagle Limestone
        Rio Bonito Formation
  NA Rotliegendes
  NT Saalian Phase
        San Andres Formation
        Stearns Shale
        Stone Corral Formation
        Toroweap Formation
  NA Upper Permian
  NT Val Gardena Sandstone
        Wellington Formation
        Wreford Limestone
        Yates Formation
  SA Allegheny Orogeny
        Baralaba Coal Measures
        Baskunchak Series
        Beacon Supergroup
        Bird Spring Formation
        Botucatu Formation
        Brianconnais Zone
        Casper Formation
        Dunkard Group
        Dwyka Series
        Estrada Nova Formation
        Geirud Formation
        Gondwana System
        Hercynian Orogeny
        Honaker Trail Formation
        Karroo System
        Korenevskaya Formation
        Lisburne Group
        Minnelusa Formation
        New Red Sandstone
        Paganzo Group
        Sadlerochit Formation
        Singleton Coal Measures
        Slovakian Karst
        Spearfish Formation
        Supai Formation
        Talchir Formation
        Tensleep Sandstone
        Weber Sandstone

Pernambuco
State in NE Brazil. Also old name for city of Recife.
  BT Brazil
  SA Barreiras Formation
        Sao Francisco Basin

Pernik coal basin
15 miles SW of Sofia. Also search Pernik.
  UF Dimitrovo coal basin
  BT Bulgaria

perofskite
use perovskite

perovskite
  UF perofskite
  BA oxides
  BT minerals
  SA perovskite structure

perovskite structure
  SA crystal structure
        perovskite

perrierite
  BA orthosilicates
        silicates
  BT minerals

Perris
City in Riverside County in S.
  BT California
        United States

Perry County
Index states as applicable.
  BX Alabama
        Arkansas
        Illinois
        Indiana
        Kentucky
        Mississippi
        Missouri
        Ohio
        Pennsylvania
        Tennessee
  BT United States

Perry Formation
Consists of thick sequence of coarse clastic sediments containing interbedded basalt flows. Central Pennsylvania.
  BT Upper Devonian
        Devonian
  SA Pennsylvania

Persani Mountains
SE Transylvania.
  BT Transylvania
        Romania

Pershing County
NW Nevada
  BT Nevada
        United States

Persia
use Iran

Persian Gulf
Between Arabian Peninsula on the W and S, and Iran on E. Includes use on level 1 as an area term (list O). For term set options see list B. Also search Arabian Gulf.
  CO N240000N310000
        E0570000E0480000
  UF Arabian Gulf
  BA Indian Ocean
  NT Khor al Bazam

Perth abyssal plain
Off SW Australia.
  BT Indian Ocean

Perth Basin
Along the Indian Ocean from the Perth area on the S to the Carnarvon Basin in the N.
  BT Western Australia
        Australia

Perth Stade
Europe. Term introduced in 1978. Before 1978, also search Perth.
  BA Weichselian
        upper Pleistocene
        Pleistocene
        Quaternary
  BT Cenozoic

perthite
  BA feldspar group
        framework silicates
        silicates
  BT minerals
  SA antiperthite
        cryptoperthite
        microperthite

Perthshire
County in central.
  BT Scotland
        Great Britain
        United Kingdom

Peru
Includes use on level 1 as an area term (list O). For term set options see list B.
  CO S181500N000000
        W070000W0811000
  BA South America
  NT Arequipa
        Ayacucho
        Cerro de Pasco
        Junin
        Lima
  SA Amazon Basin
        Amazon River
        Amazonas
        Andes
        Eastern Cordillera
        Lake Titicaca
        Magdalena
        Pucara Group

Peru-Chile Trench
Just off Peruvian and Chilean coasts.
  BT Pacific Ocean

Perugia
City and province in central part of country.
  BT Umbria
        Italy

Pervomaisk
City in Nikolayev Oblast in S.
  BT Ukraine
        USSR

Pesaro
City on Adriatic Sea.
  BT Marches
        Italy

Pescadores
  use Penghu Islands

Peschany Island
  use Peschanyy

Peschanyy
  Island in Caspian Sea off S shore of Apsheron Peninsula.
  UF Peschany Island
  BT Azerbaidzhan
      USSR

Peshawar
  City, district and division near the Khyber Pass.
  BT North-West Frontier Province
      Pakistan

pesticides
  Includes use on level 3 under environmental geology(1) waste disposal(2).
  SA environmental geology
      waste disposal

petalite
  BA nepheline group
      framework silicates
      silicates
  BT minerals
  SA aluminosilicates

Petersburg
  City S of Richmond. Within Dinwiddie County but independent of it.
  BT Virginia
      United States

Petersburg Granite
  Late Paleozoic. Underlies Triassic sediments. E Virginia.
  BT Paleozoic
  SA Virginia

Petrified Forest National Park
  In Painted Desert area in E.
  BT Arizona
      United States

petrified moss
  use tufa

petrofabric analysis
  use structural analysis

petrofabrics
  Includes use on level 3 under structural analysis(1). See fabric under igneous rocks(1), metamorphic rocks(1), and sedimentary rocks(1).
  SA deformation
      fabric
      orientation
      preferred orientation
      structural analysis

petrogenesis
  A valid term through 1973. Use igneous rocks and genesis; metamorphic rocks and genesis.

petrogeometry
  use structural analysis

petrography
  Includes use on level 2 under sedimentary rocks(1) for microscopic studies only; on level 2 under metamorphic rocks(1) and under igneous rocks(1).
  SA igneous rocks
      lithology
      macerals
      metamorphic rocks
      microfacies
      microscope methods
      petrology
      sedimentary rocks
      sediments
      thin sections
      ultrastructure

petroleum
  Includes use on level 1 and 2 as a commodity term (list (list C). Also search petroleum-gas; petroleum-natural gas.
  UF oil (petroleum)
  SA bitumens
      condensates
      Darcy's law
      depletion
      drilling
      energy sources
      geopressure
      giant fields
      heavy oil
      hydrocarbons
      kerogen
      migration
      natural gas
      oil and gas fields
      oil sands
      oil seeps
      oil shale
      oil spills
      oil-gas interface
      oil-water interface
      organic materials
      pipelines
      reservoir properties
      reservoir rocks
      secondary recovery
      seepage
      source rocks
      stratigraphic traps
      structural traps
      tertiary recovery
      traps

petroleum engineering
  Term introduced in 1978 on level 2 under engineering geology(1).
  UF engineering, petroleum
  SA engineering geology
      oil and gas fields
      secondary recovery
      subsurface reservoirs

petroleum seepage
  use oil seeps

petrologen
  use kerogen

petrology
  Treated as a whole. For studies on igneous or metamorphic rocks. Includes use on level 1 (list A); on level 2 under area terms(1), bibliography(1), continental shelf(1), education(1), intrusions(1), lava(1), and sediments(1). See list B. If level 1, term set options are:
  topic [applications, bibliography, catalogs, classification, concepts, education, experimental studies, general, history, instruments, methods, nomenclature, objectives, philosophy, practice, principles, symposia, textbooks, theoretical studies]
  SA subtopic
      banded structures
      bibliography
      catalogs
      crystallography
      education
      fluid inclusions
      homogenization
      igneous rocks
      inclusions
      intrusions
      lava
      lexicons
      magmas
      melts
      metamorphic rocks
      metamorphism
      metasomatism
      meteorites
      microscope methods
      mineralogy
      modal analysis
      palingenesis
      petrography
      sedimentary petrology
      sediments
      silicate rocks
      staining
      standard materials
      tektites
      volcanology

petromorphology
  use structural analysis

petrophysics
  SA physical properties
      reservoir rocks

Petrosani Basin
  Coal basin in Hunedoara County in the Transylvanian Alps. Also search Petrosani.
  UF Petroseni Basin
  BT Transylvania
      Romania

Petroseni Basin
  use Petrosani Basin

petrostratigraphy
  use lithostratigraphy

Petrovsk
  use Makhachkala

pH
  Includes use on level 3 under sea water(1). Also search acidity.
  UF acidity
  SA acidic composition
      acids
      alkalinity
      geochemistry
      physical properties
      sea water

Phacopida
  Includes use on level 2 under Trilobita(1). See list F.
  BA Trilobita
  BT Trilobitomorpha
      Arthropoda
      Invertebrata
  NA Phacopina
      Phacops

Phacopina
  Suborder or genus. Includes use on level 3 under Trilobita(1) Phacopida(2).
  BA Phacopida
      Trilobita
  BT Trilobitomorpha
      Arthropoda
      Invertebrata
  SA Phacops

Phacops
  Genus. Includes use on level 3 under Trilobita (1) Phacopida(2).
  BA Phacopida
      Trilobita
  BT Trilobitomorpha
      Arthropoda
      Invertebrata
  SA Phacopina

Phaeophyta
  Including brown algae and seaweed. Includes use on level 2 under algae(1). See list F.
  UF brown algae
      seaweed
  BA algae
  BT Plantae

Phanerozoic
  Includes use on level 1 and 2 as an age term (list E).
  NA Cenozoic
      Mesozoic
      Paleozoic

pharmacolite
  BA arsenates
  BT minerals

pharmacosiderite
  BA arsenates
  BT minerals

phase diagrams
  SA diagrams
      phase equilibria

phase equilibria
  Includes use on level 1 (list A). Includes use on level 2 under crystal chemistry(1) and crystal growth(1). Term set options are:
  material [e.g. igneous rocks, silicates; only 1st and 2nd level terms]
      *(system in chemical symbols) or *(system by names of end-member minerals) or topic
  topic [anomalies, concepts, experimental studies, interpretation, melting, theoretical studies]
      *(system) or subtopic
  SA carbon dioxide
      crystal chemistry
      crystal growth
      crystallography
      entropy
      equations of state
      equilibrium
      free energy
      gases
      geochemistry
      igneous rocks
      immiscibility
      inclusions
      low pressure
      low temperature
      magmas
      melting
      melts
      metamorphic rocks
      metamorphism
      metasomatism
      mineral deposits
      mineralogy
      miscibility gap
      P-T conditions
      paragenesis
      partial melting
      partitioning
      phase diagrams
      phase rule
      sedimentation
      solid solution
      transformations

phase, gaseous
  use gaseous phase

phase rule
  Includes use on level 3 under phase equilibria(1).
  UF Gibbs phase rule
  SA phase equilibria

phase, solid
  use solid phase

phases
  A valid general term through 1978. After 1978, see specific terms, e.g. liquid phase, solid phase, gaseous phase, etc.

phenacite
  use phenakite

phenakite
  Also search phenacite.
  UF phenacite
  BA orthosilicates
      silicates
  BT minerals

phengite
  BA sheet silicates
      silicates
  BT minerals
  SA mica group
      muscovite

phenocrysts
  SA crystals

igneous rocks
minerals
porphyritic texture

phenols
Includes use on level 2 under organic materials(1).
BA organic materials

phenomena, electrical
use electrical phenomena

phenomena, premonitory
use premonitory phenomena

phenomena, surface
use surface phenomena

phi grade scale
use phi scale

phi scale
Includes use on level 3 under sedimentary rocks(1), sediments(1), and sedimentary petrology(1).
UF phi grade scale
SA sedimentary rocks
     sediments
     size
     statistical methods

Philadelphia
City on the Delaware River in SE Pennsylvania. Coextensive with Philadelphia County.
BT Pennsylvania
     United States

Philip Island
In Indian Ocean E of Mornington Peninsula S of Melbourne.
BT Victoria
     Australia

Philippine Islands
Includes use on level 1 as an area term (list O). For term set options see list B.
CO N050000N190000
     E1263000E1170000
UF Philippines
BA Asia
NT Cebu
     Iloilo Basin
     Luzon
     Marinduque
     Mindanao
     Mindoro
     Palawan
     Panay Island
     Samar
SA Mansalay Formation

Philippine Sea
Includes use on level 1 as an area term (list O). That part of the W Pacific Ocean with the Philippines Islands on the W, Taiwan and the Ryukyus on the NW, and the U.S. Trust Territory of the Pacific Islands on the E and SE.
CO N000000N350000
     E1500000E1200000
BA Pacific Ocean
SA Philippine Sea Plate

Philippine Sea Plate
A tectonic plate roughly covering the same area as the Philippine Sea and virtually surrounded by the SE edge of the Eurasian Plate and the Pacific Plate.
SA Philippine Sea
     plate tectonics

Philippines
use Philippine Islands

philippinite
BA tektites

Phillips County
Index states as applicable.
BX Arkansas
     Colorado
     Kansas
     Montana
BT United States

phillipsite
BA zeolite group
     framework silicates
     silicates
BT minerals

philosophy
Use term under discipline. Includes use as a level 2 or 3 term appropriate to a large number of topics, e.g. on level 2 under geology(1). See list G.

Phlegraean Fields
Volcanic region W of Naples.
BT Campania
     Italy

phlogopite
UF amber mica
     brown mica
BA sheet silicates
     silicates
BT minerals
SA fluor-phlogopite
     mica group

Phobos
SA Deimos
     Mars
     satellites

Phoenix
City in Maricopa County in S central.
BT Arizona
     United States

phonolite
Includes use on level 3 under igneous rocks(1) trachyte-phonolite family(2). See list H.
BA trachyte-phonolite family
BT igneous rocks

phosgenite
BZ carbonates
     halides
BT minerals
SA chlorides

phosphate
Includes use on level 1 as a commodity term (list C). Use only when of economic value, otherwise use phosphates. Also search apatite ores.
UF apatite ores
SA phosphates
     soils

phosphate composition
Term introduced in 1978 on level 2 under nodules(1). Before 1978, also search nodules AND phosphate.
SA composition
     nodules

phosphate rocks
Refers to any rock containing phosphate; more general term than phosphorite. Includes use on level 3 under sedimentary rocks(1) chemically precipitated rocks(2). See list I.
BA chemically precipitated rocks
BT sedimentary rocks
SA phosphorite

phosphates
Includes use on level 2 under minerals(1). See list L. When of economic value, use phosphate (list C).
BT minerals
NA amblygonite
     apatite
     autunite
     barbosalite
     beraunite
NZ britholite
NA brushite
     cacoxenite
     carbonate apatite
     chlorapatite
     crandallite
NZ dahllite
NA eosphorite
     florencite
     fluorapatite
     francolite
     graftonite
     herderite
NZ hinsdalite
NA hureaulite
     hydroxylapatite
     lithiophilite
NZ minyulite
NA monazite
     monetite
     montebrasite
     newberyite
     plumbogummite
NZ pyromorphite
NA rhabdophane
     rockbridgeite
     scholzite
     strengite
     struvite
     triphylite
NZ triplite
NA turquoise
     variscite
     vivianite
     wavellite
     whitlockite
     xenotime
SA phosphate

phosphatization
SA diagenesis

phosphides
BA native elements and alloys
BT minerals
NZ sarcopside

Phosphoria Formation
Includes Sybille tongue (new), Forelle tongue, and Ervay tongue (new).
BT Permian
     Paleozoic
SA Idaho
     Montana
     Utah
     Wyoming

phosphorite
Includes use on level 3 under sedimentary rocks(1) chemically precipitated rocks(2). See list I.
BA chemically precipitated rocks
BT sedimentary rocks
SA phosphate rocks

phosphorus
Includes use on level 1 and 2 as a chemical element (list D); on level 3 under soils(1) chemistry(2). Also search P.
UF P
SA elements
     fertilizers
     KREEP
     soils

photogeologic maps
Term introduced in 1978. Before 1978, also search maps AND photogeologic.
BA maps

photogeologic methods
Includes use on level 2 under mineral exploration(1).
SA methods
     mineral exploration
     photogeology
     photography
     remote sensing

photogeology
Includes use on level 3 under engineering geology(1) and maps(1). See also mineral exploration(1) photogeologic methods(2) and photogrammetric studies(3) under engineering geology(1).
SA aerial photography
     cartography
     engineering geology
     exploration
     maps
     photogeologic methods
     photogrammetry
     photography

photogrammetry
Includes use on level 3 under engineering geology(1) methods(2). See also photogrammetric studies.
SA aerial photography
     cartography
     definition
     engineering geology
     photogeology
     photography
     remote sensing

photographs
A valid term through 1978. After 1978, use photography.

photography
Includes use on level 2 under micropaleontology(1). Before 1978, also search photographs.
SA aerial photography
     photogeologic methods
     photogeology
     photogrammetry

photometric
A valid term through 1977. After 1977, see photometry.

photometry
Includes use on level 3 under chemical analysis(1).
SA chemical analysis
     flame photometry
     spectroscopy

photosynthesis
Includes use on level 3 under geochemistry(1) processes(2).
SA geochemistry
     paleobotany

Phuket Group
May be the equivalent of the Megui Series, at least in part, of Burma. W side of peninsular Thailand, including island of Phuket, and lower Burma.
UF Phuket Series
BT Cambrian
SA Asia
     Burma
     Thailand

Phuket Series
use Phuket Group

phyllite
Includes use on level 3 under metamorphic rocks(1) phyllites(2). See list J.
BA phyllites
BT metamorphic rocks

phyllites
Includes use on level 2 under metamorphic rocks(1). See list J.
BT metamorphic rocks
NA phyllite

phyllonite
See list J.
BA phyllonites
BT metamorphic rocks
SA phyllonitization

phyllonites
Includes use on level 2 under metamorphic rocks(1).
BT metamorphic rocks
NA phyllonite
SA phyllonitization

phyllonitization
Term introduced in 1978.
BA metamorphism
SA mylonitization
    phyllonite
    phyllonites
    processes
    recrystallization

phyllosilicates
use sheet silicates

phylogeny
Includes use on level 3 under fossil group(1) and under paleontology(1). See list F.
SA ontogeny
    paleontology

physical
A valid term through 1977. After 1977, use physical properties.

physical geography
use geography

physical geology
Includes use on level 3.
SA geology
    historical geology

physical methods
Includes use on level 3 under soils(1) analysis(2).
SA methods
    soils

physical models
BA models

physical properties
Includes use on level 3 under igneous rocks(1) properties(2) and under seismology(1). See also appropriate entries under material name. See under Earth(1), crust(1), core(1), Mohorovicic discontinuity(1), and mantle(1).
SA acoustical properties
    bearing capacity
    capacity
    compressibility
    dilatancy
    friction
    hardness
    mechanical properties
    microhardness
    petrophysics
    pH
    porosity
    properties
    soils
    specific heat
    thermal properties

physical weathering
Term introduced in 1978. Before 1978, search weathering AND physical.
BT weathering
SA mechanical weathering

physiographic provinces
UF provinces, physiographic
SA geomorphology
    topography

physiography
A valid term through 1975. After 1975, see geomorphology.

physiology
Includes use under appropriate fossil term (list F).
SA morphology

phytogeography
use biogeography

phytoplankton
Includes use on level 3 under fossil groups (list F).
SA algae
    diatoms
    plankton

Piacenza
City and province in NW.
BT Emilia-Romagna
    Italy

Piatra Neamt
use Piatra-Neamt

Piatra-Neamt
City in Neamt County in N.
UF Piatra Neamt
BT Moldavia
    Romania

Piaui
State in NE.
BT Brazil
SA Parnaiba Basin

Piave Valley
River valley in N and NE Veneto. Also search Piave River.
BT Veneto
    Italy

Picardy
Historical region in extreme N France. Abuts on English Channel between Calais in N and Treport on S, and extends E along Somme river and upper Oise River to Belgian border.
BT France

Piceance Basin
use Piceance Creek basin

Piceance Creek basin
Rio Blanco County in NW Colorado. Also search Piceance Basin.
CO N393000N401500
    W1075500W1083000
UF Piceance Basin
BT Colorado
    United States

Picher
City in Ottawa County in extreme NE.
BT Oklahoma
    United States

pickeringite
BA sulfates
BT minerals

Pico Formation
Consists of three members: lower Pico marine, middle Pliocene; upper Pico marine, upper Pliocene; and continental upper Pliocene Sunshine Ranch member.
BT Pliocene
    Neogene
    Tertiary
SA California

picotite
use chrome spinel

picrite
Includes use on level 3 under igneous rocks(1) basalt family(2) and ultramafic family(2). See list H.
BA basalt family
BT igneous rocks
SA ultramafic family

picrite porphyry
See list H.
BA basalt family
BT igneous rocks
SA porphyry

Picuris Mountains
use Picuris Range

Picuris Range
N New Mexico.
UF Picuris Mountains
BT New Mexico
    United States

Piedmont
An upland belt in the U.S. lying E of the Blue Ridge and Appalachian mountains on the W to the coastal plain at the fall line on the E. Also an autonomous region in NW Italy. Index Italy or states of the United States as applicable.
BX Italy
    United States
SA Acceglio
    Acqui
    Alabama
    Alessandria
    Argentera
    Argentera Massif
    Asti
    Biella
    Canavese Zone
    Georgia
    Italy
    Ivrea
    Ivrea Zone
    Ivrea-Verbano Zone
    Langhe
    Lanzo Massif
    Maryland
    New Jersey
    New York
    North Carolina
    Novara
    Pennsylvania
    Po River
    Po Valley
    Sesia Valley
    Sesia Zone
    Sesia-Lanzo Zone
    Simplon region
    South Carolina
    Strona Valley
    Turin
    Vercelli
    Villafranca d'Asti
    Virginia

Piedmont Alps
Those alpine ranges such as the Maritime Alps, Cottian Alps, Graian Alps, Pennine Alps, and Lepontine Alps, which lie within or on the borders of Piedmont.
BT Italy
SA Alps
    Cottian Alps
    Lepontine Alps
    Maritime Alps
    Pennine Alps

piedmontite
use piemontite

piemontite
Also search piedmontite.
UF piedmontite
BA epidote group
    orthosilicates
    silicates
BT minerals

Pieniny Klippen Belt
In the Pieniny Mountains. Also search Pieniny.
BT Krakow
    Poland
SA Pieniny Mountains

Pieniny Mountains
Range of the Beskids in S part of the country. Also search Pieniny.
UF Pieniny Range
BT Krakow
    Poland
SA Pieniny Klippen Belt

Pieniny Range
use Pieniny Mountains

piercement
use diapirs

piercing fold
use diapirs

Pierre
City in Hughes County in central.
BT South Dakota
    United States

Pierre Formation
use Pierre Shale

Pierre Shale
In Montana Group. E Colorado, W Minnesota, E Montana, Nebraska, North Dakota, South Dakota, and E Wyoming. Also search Pierre Formation.
UF Pierre Formation
BT Upper Cretaceous
    Cretaceous
SA Colorado
    Minnesota
    Montana
    Montana Group
    Nebraska
    North Dakota
    South Dakota
    Wyoming

piers
BT marine installations

pigeonite
BA pyroxene group
    chain silicates
    silicates
BT minerals
SA clinopyroxene

pigments
BA organic materials
SA carotenoids

Pike County
Index states as applicable.
BX Alabama
    Arkansas
    Georgia
    Illinois
    Indiana
    Kentucky
    Mississippi
    Missouri
    Ohio
    Pennsylvania
BT United States

Pikes Peak
Mountain W of Colorado Springs in El Paso County in central Colorado. Before 1978, also included use for Pikes Peak Batholith.
BT Colorado
    United States
SA Pikes Peak Batholith

Pikes Peak Batholith
Term introduced in 1978. Before 1978, also search Pikes Peak for the batholith.
BT Colorado
    United States
SA Pikes Peak

Pilbara gold field
NW Western Australia. Also search Pilbara.
UF Pilbara Goldfield
BT Western Australia
    Australia

Pilbara Goldfield
use Pilbara gold field

piles
As of 1978, term is used on level 2 under foundations(1).
SA foundations

Pilgrim Formation
Threefold subdivision. Underlies Maywood or Red Lion Formation.
UF Pilgrim Limestone
BT Upper Cambrian
    Cambrian
SA Montana

Pilgrim Limestone
use Pilgrim Formation

Pilica
Village in N.
BT Krakow

Poland
pillars
Includes use as a general term.
pillow lava
Includes use on level 3 under lava(1).
SA basalt
lava
pillow structure
spilite
pillow structure
Includes use on level 3 under lava(1). Included use under sedimentary structures(1) soft sediment deformation(2) until 1976; now use ball-and-pillow.
SA ball-and-pillow
lava
pillow lava
structure
Pilsen Basin
Valley in W in which city of Pilsen is located. Also search Pilsen.
UF Plzen Basin
BT Bohemia
Czechoslovakia
Piltdown
Locality in East Sussex in SE.
BT England
Great Britain
United Kingdom
Pima County
On the Mexican border in S.
BT Arizona
United States
Pinaceae
Family.
BA Coniferales
gymnosperms
BT Plantae
NA Pinus
SA Taxodiaceae
Pinal County
S central Arizona.
BT Arizona
United States
Pinar del Rio
Easternmost province. Also a city.
BT Cuba
West Indies
NT Sierra de los Organos
Pinchi Lake
Central British Columbia.
BT British Columbia
Canada
Pindus Mountains
W central and NW Greece. Index administrative regions as applicable. Also search Pindus.
BT Greece
SA Epirus
Hellenides
Thessaly
Pine Barrens
Region of coastal plain of S and SE.
BT New Jersey
United States
Pine Valley Mountains
Washington County in SW.
BT Utah
United States
Pinedale
Town in Sublette County in W.
BT Wyoming
United States
Pinellas County
On Gulf of Mexico W of Tampa.
BT Florida
United States
pingos
Includes use on level 3 under glacial geology(1) periglacial features(2).
SA glacial geology
periglacial features
permafrost
pinnacle reefs
Includes use on level 3 under reefs(1).
UF coral pinnacle
reef pinnacle
BT reefs
Pinnacles National Monument
San Benito County in W.
BT California
United States
Pinon Range
Also search Pinon.
BT Nevada
United States
pinquite
use nontronite
Pinus
Genus. Includes use on level 3 under gymnosperms(1) Coniferales(2).
BA Pinaceae
Coniferales
gymnosperms
BT Plantae
Pinyon Conglomerate
Unconformably overlies Harebell Formation and unconformably underlies Colter Formation. Yellowstone National Park in NW Wyoming.
BT Paleocene
SA Wyoming
Pioneer Mountains
Custer and Blaine Counties in S central.
BT Idaho
United States
Pioneer 10
SA Jupiter
satellite methods
pipelines
As of 1978, term is used on level 2 under marine installations(1).
SA marine installations
natural gas
petroleum
transportation
pipes
Includes use on level 2 under intrusions(1).
UF tuffsite
BA intrusions
NT breccia pipes
SA diatremes
igneous rocks
plugs
volcanoes
Pirabas Formation
BT Miocene
Neogene
Tertiary
SA Brazil
piracy
use stream capture
Pisa
City and province in NW.
BT Tuscany
Italy
Pisces
Includes use on level 1 and 2 as a fossil term (list F).
BT Vertebrata
NA Agnatha
Chondrichthyes
Osteichthyes
Placodermi
SA fish
gastroliths
otoliths
Pisgah Crater
BT California
United States
Pishpek
use Frunze
pisolites
SA limestone
pisoliths
pisolitic texture
sedimentary rocks
pisoliths
SA pellets
pisolites
pisolitic texture
sedimentary rocks
sediments
pisolitic texture
Term introduced in 1978. Includes use on level 3 under sediments(1) and sedimentary rocks(1). Before 1978, also search pisolitic AND specific sediment type or sedimentary rocks.
SA pisolites
pisoliths
sedimentary rocks
sediments
textures
pistacite
use epidote
pitchblende
UF nasturan
BA oxides
BT minerals
SA uraninite
pitching folds
use plunging folds
pitchstone
Includes use on level 3 under igneous rocks(1) pyroclastics and glasses(2). See list H.
BA pyroclastics and glasses
BT igneous rocks
SA volcanic glass
Pitesti
City in Arges County in central.
BT Walachia
Romania
Pithecanthropus
Genus. Includes use on level 3 under Mammalia(1) Primates(2).
BA Hominidae
Primates
Mammalia
BT Tetrapoda
Vertebrata
Pitkin County
W central Colorado.
BT Colorado
United States
Pitkin Limestone
In Arkansas, underlies Cane Hill Member of Hale Formation. N Arkansas, and E Oklahoma.
BT Upper Mississippian
Mississippian
SA Arkansas
Oklahoma
Piton de la Fournaise
Volcanic peak on Reunion Island.
BT Mascarene Islands
Indian Ocean
Piton des Neiges
Highest peak on Reunion Island.
BT Mascarene Islands
Indian Ocean
Pittsburgh
City in Allegheny County in SW.
BT Pennsylvania
United States
Pittsburgh coal basin
Bituminous coal and steel producing region in the Pittsburgh area. Also search Pittsburgh Coal.
BT Pennsylvania
United States
Piute County
SW central.
BT Utah
United States
Placer County
W of Lake Tahoe in E.
BT California
United States
placers
Includes use on level 1 (list A). Used for a surficial mineral deposit formed by mechanical concentration of mineral particles from weathered debris. See also list C for type of commodity. Term set options are:
kind of placer [e.g. diamonds, gold, heavy minerals, platinum, tin; first level terms only]
area
topic [detection, distribution, exploration, genesis, identification, patterns, sampling]
subtopic
UF ore of sedimentation
SA diamonds
discoveries
gold
heavy minerals
mineral deposits
mineral exploration
monazite
ore deposits
platinum
sampling
tin
weathering
Placodermi
Includes use on level 2 under Pisces(1). See list F.
BA Pisces
BT Vertebrata
NA Acanthodii
Arthrodira
SA fish
plagioclase
BA feldspar group
framework silicates
silicates
BT minerals
SA albite
andesine
anorthite
bytownite
labradorite
oligoclase
plagiogranite
BA plutonic rocks
BT igneous rocks
SA quartz diorite
trondhjemite
plagionite
BA sulfantimonites
sulfosalts
BT minerals
plains
SA abyssal plains
alluvial plains
coastal plains
floodplains
geomorphology
landforms
outwash plains
plateaus
playas
steppes
topography
tundra

Plaisancian
Europe. Above Pontian (Miocene), below Astian. Includes use on level 3 under age terms(1). See list E.
UF Plaisanzian
BA lower Pliocene
    Pliocene
    Neogene
    Tertiary
BT Cenozoic

Plaisanzian
use Plaisancian

Plan de la Tour
Village in SE part of country.
BT Var
    France

Plana
Town in W.
BT Bohemia
    Czechoslovakia

planar bedding structures
Includes use on level 2 under sedimentary structures(1). See list K.
UF bedding structures, planar
BA sedimentary structures
NA bedding
    cross-bedding
    cross-laminations
    cross-stratification
    cut and fill
    cyclothems
    flaser bedding
    imbrication
    laminations
    massive bedding
    rhythmic bedding
    ripple drift-cross laminations
    sand bodies
    stratification
    varves
SA bars
    buried channels
    channels
    graded bedding

planation
SA abrasion
   erosion
   erosion surfaces

planation surfaces
use erosion surfaces

plancheite
BA chain silicates
   silicates
BT minerals

planes, fault
use fault planes

planetology
Use only in discussing the status of the planetary body within the solar system. For studies of more than one planet, or relationship between planets, or cosmochemistry. See also names of planets. Includes use on level 1 (list A). Term set options are:
topic [atmosphere, bibliography, concepts, cosmic dust, methods, nomenclature, principles, techniques, theoretical studies]
    subtopic
SA accretion
   asteroids
   astronomy
   astrophysics and solar physics
   atmosphere
   condensation
   cosmic dust
   cosmochemistry
   Earth
   exobiology
   extraterrestrial geology
   impacts
   interplanetary space
   Jupiter
   Mars

Mercury Planet
meteorites
meteors
Moon
Neptune
outer planets
planets
Pluto
Saturn
solar system
Sun
surface features
tektites
terrestrial planets
Uranus
Venus

planets
Includes use on level 3 under planetology(1).
SA asteroids
   condensation
   Earth
   extraterrestrial geology
   Jupiter
   Mars
   Mercury Planet
   meteoroids
   Moon
   Neptune
   outer planets
   planetology
   Pluto
   Saturn
   solar system
   Sun
   surface features
   terrestrial planets
   Uranus
   Venus

plankton
Includes use on level 3 under various fossil groups(1). See list F. Also search microplankton.
UF microplankton
NT zooplankton
SA algae
   diatoms
   Dinoflagellata
   foraminifera
   Ostracoda
   palynomorphs
   phytoplankton
   planktonic taxa
   Protista
   Radiolaria

planktonic
A valid term through 1977. After 1977, use planktonic taxa.

planktonic taxa
Term introduced in 1978. Includes use on level 3 under fossil group(1). Before 1978, also search planktonic.
UF taxa, planktonic
SA plankton

planning
As of 1978 term is used on level 2 under highways(1) and land use(1).
SA associations
   highways
   land use
   management
   programs
   regional planning
   urban planning

Planolites
Genus.
BA ichnofossils

Planorbis
Genus. Includes use on level 3 under Mollusca(1) Gastropoda(2).
BA Gastropoda
   Mollusca
BT Invertebrata

Planosols
Includes use on level 3 under soils(1). See list M.
BA soils
SA Intrazonal soils
   soil group

Plantae
Also use vegetation on level 3. Includes use on level 1 and 2 as a fossil term (list F). In indexing, used in a general sense when specific plants are unknown.
NT algae
   angiosperms
   bryophytes
   fungi
   gymnosperms
   lichens
   pteridophytes
   Pteropsida
   thallophytes
SA bacteria
   flora
   floral studies
   fossil wood
   leaves
   paleobotany
   palynomorphs
   vegetation

plasma
Not to be used for soil plasma.
SA interplanetary space

plasma instabilities
Includes use on level 2 under magnetosphere(1).
UF instabilities, plasma
SA corona
   magnetic field
   magnetosphere
   plasma motion
   plasmapause
   solar wind

plasma motion
Includes use on level 2 under magnetosphere(1).
UF motion, plasma
SA magnetosphere
   plasma instabilities
   solar wind

plasmapause
Includes use on level 2 under magnetosphere(1).
SA magnetosphere
   plasma instabilities

plasmasphere
Includes use on level 3 under magnetosphere(1) magnetic storms(2).
SA magnetic storms
   magnetosphere

plaster of paris
use gypsum

plaster stone
use gypsum

plastic flow
Includes use on level 3 under deformation(1) field studies(2). Also search plastic deformation; plastic.
UF plastic strain
SA deformation
   flow
   flow lines
   plasticity

plastic materials
Term introduced in 1978. Before 1978, also search plastic AND materials.
SA elastic materials
   engineering geology
   materials
   plasticity
   rock mechanics
   soil mechanics

plastic strain
use plastic flow

plasticity
As of 1978, term is used on level 2 under rock mechanics(1).
SA Atterberg limits
   compressibility
   deformation
   elasticity
   mechanical properties
   plastic flow
   plastic materials
   rock mechanics
   viscosity

plate tectonics
Includes use on level 1, as of 1976. Used for global tectonics based on an Earth model characterized by a small number of large, broad, thick plates. Before 1976, articles discussing plate tectonics were included under tectonophysics(1). See list B. Term set options are:
age
   age term (List E)
topic [concepts, effects, evolution, geometry, indicators, island arcs, mechanism, movement, obduction, processes, rifting, structure, subduction]
   subtopic
SA African Plate
   Antarctic Plate
   Arabian Plate
   asthenosphere
   Benioff zone
   Caribbean Plate
   Cocos Plate
   collision
   continental drift
   continental margin
   continental type
   convection
   convection cells
   convection currents
   convergence
   crust
   crustal shortening
   deformation
   Earth
   earthquakes
   Eurasian Plate
   fault zones
   faults
   fracture zones
   fracturing
   geodynamics
   geometry
   geosynclines
   heat flow
   hot spots
   Indian Plate
   indicators
   intraplate
   island arcs
   lithosphere
   mantle
   marginal basins
   marginal seas
   margins
   mechanics
   Median Tectonic Line
   melange
   metamorphic belts
   microplates
   Mohorovicic discontinuity
   movement
   Nazca Plate
   North American Plate
   obduction
   ocean basins
   oceanic type
   ophiolite
   orogeny
   Pacific Plate
   paleomagnetism

partial melting
Philippine Sea Plate
plates
plumes
polar wandering
rift valleys
rift zones
rifting
rotation
sea-floor spreading
seismology
seismotectonics
slabs
spherical harmonic analysis
spreading centers
subduction
subduction zones
taphrogeny
tectonics
tectonophysics
tectonosphere
transition zones
trenches
triple junctions
volcanic belts
volcanology
West Siberian Plate

plateaus
  SA geomorphology
      highlands
      landforms
      plains
      topography

plates
Includes use on level 3 under plate tectonics(1).
  SA Benioff zone
      crust
      intraplate
      mantle
      margins
      microplates
      plate tectonics
      subduction zones

platforms
Includes use on level 3 under tectonics(1) structure(2).
  SA gravity platforms
      marine platforms
      tectonics

platinum
Includes use on level 1 and 2 as a commodity term (list C) and as a chemical element (list D); on level 2 under placers(1).
  UF Pt
  SA elements
      heavy metals
      placers

Platte River
Formed by confluence of the North Platte and South Platte rivers in SW central Nebraska and flows E into the Missouri River below Omaha.
  BT Nebraska
      United States

Platteville Formation
Includes Mifflin Limestone Member, Quimbys Mill Member, Pecatonica Member, McGregor Member, and Carimona Member, Hidden Falls Member, Magnolia Member, Glenwood Shale Member. E Iowa, NW Illinois, S Minnesota, and SW Wisconsin.
  BT Middle Ordovician
      Ordovician
  SA Illinois
      Iowa
      Minnesota
      Wisconsin

Plattsburg Limestone
In Lansing Group. Includes Merriam Limestone, Hickory Creek Shale and Spring Hill Limestone members. SW Iowa, E Kansas, NW Missouri, and SE Nebraska.
  BT Missourian
      Upper Pennsylvanian
      Pennsylvanian
  SA Iowa
      Kansas
      Lansing Group
      Missouri
      Nebraska

Plattsmouth Limestone
use Plattsmouth Limestone Member

Plattsmouth Limestone Member
Of Oread Limestone. SW Iowa, E Kansas, NW Missouri, and SE Nebraska.
  UF Plattsmouth Limestone
  BT Virgilian
      Upper Pennsylvanian
      Pennsylvanian
  SA Iowa
      Kansas
      Missouri
      Nebraska
      Oread Limestone

playas
  SA basins
      coastal environment
      deserts
      geomorphology
      plains
      sebkha environment

Pleasanton Group
Comprises Warrensburg Channel Sandstone, Sni Mills Limestone, Dawson Coal Horizon, Wayside Sandstone, Exline Limestone, Knobtown Sandstone, Hepler Sandstone, Checkerboard Limestone, and Ovid Coal. SW Iowa, E Kansas, NW Missouri, and NE Oklahoma.
  BT Missourian
      Upper Pennsylvanian
      Pennsylvanian
  SA Iowa
      Kansas
      Missouri
      Oklahoma

Plecoptera
Includes use on level 2 under Insecta(1). See list F.
  BA Insecta
  BT Arthropoda
      Invertebrata

Plectoptera
Includes use on level 2 under Insecta(1). See list F.
  BA Insecta
  BT Arthropoda
      Invertebrata

Pleistocene
Glacial epoch. World. Above Pliocene (Tertiary), below Holocene. Includes use on level 1 as an age term (list E); on level 2 under paleo- terms, e.g. paleoecology, paleogeography; on level 3 under age terms(1).
  BA Quaternary
      Cenozoic
  NT Baltic Glaciation
      Bishop Tuff
      Crooked Creek Formation
  NA Gunz
  NT Key Largo Limestone
      Kingsdown Formation
  NA lower Pleistocene
  NT McPherson Formation
      Miami Limestone
  NA middle Pleistocene
      Mindel
      Mindel/Riss Interglacial
  NT Pearlette Volcanic Ash
      Peoria Loess
  NA Riss
      Riss/Wurm Interglacial
  NT Semata Formation
      Shimosa Group
  NA upper Pleistocene
  NT Wedron Formation
      Wicomico Formation
  NA Woodfordian
  SA Blancan
      Loup Fork Group
      Magdalenian
      Mousterian
      Osaka Group
      Paleolithic
      San Juan Formation
      Santa Maria Formation
      Siwalik System
      Tulare Formation
      Uonuma Group
      Verde Formation
      Villafranchian

plenargyrite
use matildite

pleochroism
Includes use on level 3 under minerals(1) optical properties(2).
  UF polychroism
  SA crystallography
      minerals
      optical properties

pleosponge
use Archaeocyatha

Plesiosauria
Suborder. Includes use on level 3 under Reptilia(1) Euryapsida(2).
  BA Euryapsida
      Reptilia
  BT Tetrapoda
      Vertebrata

plessite
  BA native elements and alloys
  BT minerals
  SA kamacite
      meteorites
      taenite

Plettenberg
City NE of Cologne in W.
  BT North Rhine-Westphalia
      West Germany

Pleven
Province in N Bulgaria. Also a city.
  BT Bulgaria
  NT Teteven

Pliensbachian
Europe. Above Sinemurian, below Toarcian. Includes use on level 3 under age terms(1). See list E.
  BA Lower Jurassic
      Jurassic
  BT Mesozoic
  SA Liassic

Pliocene
World. Above Miocene, below Pleistocene (Quaternary). Includes use on level 1 as an age term (list E); on level 2 under paleo- terms, e.g. paleoecology, paleogeography; on level 3 under age terms(1).
  BA Neogene
      Tertiary
      Cenozoic
  NT Ash Hollow Formation
      Bakhtiari Formation
      Bidahochi Formation
  NA Cimmerian
  NT Citronelle Formation
  NA Dacian
      Gauss epoch
      lower Pliocene
      middle Pliocene
  NT Narita Formation
      Ogallala Formation
      Pico Formation
      Saint George Formation
      San Diego Formation
  NA upper Pliocene
  NT Uvalde Gravel
      Valentine Formation
  SA Blancan
      Columbia River Basalt
      Cuddalore Series
      Hemphillian
      Loup Fork Group
      Meotian
      Neyveli Lignite
      Osaka Group
      Pannonian
      Santa Maria Formation
      Siwalik System
      Snoqualmie Batholith
      Tulare Formation
      Uonuma Group
      Verde Formation
      Warkalli Formation

pliomagmatic zone
use eugeosynclines

Pliomys
Includes use on level 3 under Mammalia(1) Rodentia(2).
  BA Cricetidae
      Rodentia
      Mammalia
  BT Tetrapoda
      Vertebrata

Plock
City in central part of country.
  UF Plokz
  BT Warsaw
      Poland

Ploesti
City N of Bucharest in Prahova County.
  UF Ploiesti
  BT Walachia
      Romania

Ploiesti
use Ploesti

Plokz
use Plock

Plovdiv
Province in S central part of country. Also a city.
  BT Bulgaria
  NT Madan

plugs
Includes use on level 2 under intrusions(1).
  BA intrusions
  SA breccia pipes
      differentiation
      lava
      pipes
      volcanoes

Plumas County
NE California.
  BT California
      United States

plumbogummite
  BA phosphates
  BT minerals

plumes
Includes use on level 3 under plate tectonics(1).
  SA hot spots
      plate tectonics

plunging folds
Term introduced in 1978. Folds defined on the basis of orientation in relation to the geographic horizontal plane. Includes use on level 3 under folds(1) orientation(2). Before 1978, also search folds AND plunging.
  UF pitching folds
  BA folds

SA orientation

Pluto
Includes use on level 1 and 2. See entry under Moon for term set options.
SA outer planets
planetology
planets
solar system

plutonic
A valid term through 1977. After 1977, use plutonic rocks on level 2 under igneous rocks(1).

plutonic rocks
Term introduced in 1978. Includes use on level 2 and under igneous rocks(1). Before 1978, also search igneous rocks AND plutonic.
BT igneous rocks
NA appinite
plagiogranite
SA granodiorite
hypabyssal rocks
igneous processes
ijolite
mangerite
metaplutonic rocks
monzodiorite
peridotite
quartz syenite
rocks

plutonium
Includes use on level 1 and 2 as a chemical element (list D).
UF Pu
SA elements
Pu-239
Pu-244

plutons
Includes use on level 2 under intrusions(1).
BA intrusions
SA batholiths
emplacement
igneous rocks
metasomatism

Plzen Basin
use Pilsen Basin

Pm
use promethium

Po
use polonium

Po Delta
On Adriatic Sea S of Venice.
UF Po River delta
BT Veneto
Italy

Po River
Rises in the Cottian Alps of W Piedmont and flows E into the Adriatic Sea. Index autonomous regions as applicable.
BT Italy
SA Emilia-Romagna
Lombardy
Piedmont
Veneto

Po River delta
use Po Delta

Po Valley
River valley. Index autonomous regions as applicable.
BT Italy
SA Emilia-Romagna
Lombardy
Piedmont
Veneto

Po-210
Includes use on level 3 under isotopes(1).
SA isotopes
polonium

Pobla de Segur
Town in NE part of country.
BT Lerida
Spain

Pocahontas County
Index states as applicable.
BX Iowa
West Virginia
BT United States

Pocahontas Formation
In Pottsville Group. SW Virginia and S West Virginia.
BT Lower Pennsylvanian
Pennsylvanian
SA Pottsville Group
Virginia
West Virginia

Pocatello Formation
Divided into two series: lower tillite series and upper varved slate series. SE Idaho.
BT Precambrian
SA Idaho

Pocono Formation
Includes Benezette Limestone, Patton Shale Member, Peters Mountain Sandstone, Second Mountain Member, Cove Mountain Member, Burgoon Sandstone Member. W Maryland, E Ohio, Pennsylvania, W Virginia, N West Virginia.
BT Lower Mississippian
Mississippian
SA Maryland
Pennsylvania
Virginia
West Virginia

Pocos de Caldas
Resort city in SW.
BT Minas Gerais
Brazil

Podhale
Highland basin in the Carpathians.
BT Krakow
Poland

Podocopida
Includes use on level 2 under Ostracoda(1). See list F.
BA Ostracoda
BT Crustacea
Arthropoda
Invertebrata
NA Bairdiacea
Cavellinidae
Cyprididae
Cytheracea
Cytherellidae
Darwinula
Healdiidae
Loxoconcha

Podolia
Region between the Dniester and the Southern Bug rivers nearly coextensive with Khmelnitskiy Oblast in W.
BT Ukraine
USSR

Podsols
use Podzols

Podzols
Includes use on level 3 under soils(1). See list M.
UF Podsols
BA soils
SA Zonal soils

Pogonip Group
Comprises House Limestone, Fillmore Limestone, Wahwah Limestone, Juab Limestone, Kanosh Shale, Lehman Formation, Mazourka Formation, Goodwin Limestone, Ninemile Formation, Antelope Valley Limestone. SE California, E and S Nevada, and W Utah.
BT Ordovician
SA Antelope Valley Limestone
California
Lower Ordovician
Middle Ordovician
Nevada
Utah

Pohang
City NNE of Pusan in SE South Korea. North Kyongsang.
BT South Korea

Poiana Rusca (Mountains)
use Poiana-Rusca Mountains

Poiana Rusca Massif
use Poiana-Rusca Mountains

Poiana-Rusca Massif
use Poiana-Rusca Mountains

Poiana-Rusca Mountains
Hunedoara County in SW Transylvania. Also search Poiana Rusca, Poiana Rusca Massif, and Poiana-Rusca Massif.
UF Poiana Rusca (Mountains)
Poiana Rusca Massif
Poiana-Rusca Massif
BT Transylvania
Romania

Point Arena
A promontory on the Pacific Ocean in Mendocino County in NW. Also a town.
BT California
United States

Point Barrow
Eskimo village, N Alaska, on Point Barrow headland.
UF Nuwuk
BT Alaska
United States

Point Loma
Peninsula sheltering San Diego Bay.
BT California
United States

Point Mugu
On the Pacific Ocean in Ventura County W of Los Angeles. Also site of Point Mugu Naval Missile center.
BT California
United States

Point Reyes
On the Pacific Ocean N of San Francisco in Marin County. Part of Point Reyes National Seashore.
BT California
United States

Poisson's ratio
BA elastic constants
SA elasticity
engineering geology
mechanical properties
strain
stress
tunnels

Poitiers
City in W central.
BT Vienne
France

Poitou
Historical region of W central. Index departments as applicable.
BT France
SA Deux-Sevres
Vendee
Vienne

Pokutye
Upland region between E Beskids on the S and the Dniester River on the N in SW.
BT Ukraine
USSR

Poland
Includes use on level 1 as an area term (list O). For term set options see list B.
CO N490000N544500
E0241500E0141500
BA Europe
NT Bialowieza
Bialystok
Borzeta
Boza Wola
Bydgoszcz
Central Polish Glaciation
Galezice
Gdansk
Grzybow
Kampinos Forest
Katowice
Kielce
Koszalin
Krakow
Krakow-Czestochowa Jura
Kujawy
Lodz
Lower Lusatia
Lower Silesia
Lublin
Olsztyn
Opole
Polish Lowland
Poznan
Rzeszow
Silesian coal basin
Sowie Mountains
Szczecin
Upper Silesia
Vistula River
Vistula River valley
Warsaw
Warta
Wroclaw
Zielona Gora
SA Beschady Mountains
Brest Basin
Bug region
Bug River
Carpathian Foredeep
Carpathian Foreland
Carpathians
European Platform
Galicia
Isergebirge
Izera Mountains
Karkonosze Mountains
Krosno Beds
Lusatia
North Sudetic Basin
Oder Valley
Pomerania
Poznan Clays
Roztocze
Silesia
Sniezknik
South Polish Glaciation
Spis
Subcarpathians
Sudeten Mountains
Sudetic Basin
Tatra Mountains
Western Carpathians

Polar Cap
An ice sheet centered at the South Pole. (A term incorrectly applied to the sea ice of the Arctic Ocean). Also search Polar AND cap; Polar Caps.
BT Antarctica
SA South Pole

Polar Continental Shelf
The continental shelf surrounding the Antarctic Continent. (Might apply also to continental shelves of Europe, Asia, and North America facing the Arctic Ocean and lying within the Arctic Circle).
BT Antarctica
SA continental shelf

polar migration
use polar wandering

polar regions
Term introduced in 1978. Includes use on level 3 under ionosphere(1) electrojet(2). Before 1978, also search polar.
SA   electrojet
     ionosphere
     magnetic storms

Polar regions
Region surrounding both geographic poles. Also search Polar and Polar region. Index poles as applicable.
SA   North Pole
     South Pole

Polar Urals
That section of the Northern Urals extending from the Arctic Circle to the Arctic Ocean.
CO   N650000N700000
     E0670000E0600000
BT   Russian Republic
     USSR
SA   Northern Urals
     Urals

polar wandering
As of 1978, term is used on level 2 under paleomagnetism(1).
UF   polar migration
     wandering, polar
SA   continental drift
     paleolatitude
     paleomagnetism
     plate tectonics
     pole positions
     rotation

polarization
A general term used mostly for optical properties. Not used for magnetization.
SA   birefringence
     induced polarization
     optical properties

polarographic analysis
use polarography

polarography
Includes use on level 3 under chemical element(1) analysis(2). See list D. For methods, see under chemical analysis(1). For data, see appropriate material.
UF   polarographic analysis
SA   analysis
     chemical analysis
     diffusion
     electrolysis
     quantitative analysis

pole positions
Includes use on level 2 under paleomagnetism(1).
UF   positions, pole
SA   magnetic field
     paleomagnetism
     polar wandering
     reversals

Polesie
use Polesye

Polesye
Lowland area, formerly in Poland, comprising the Pripet Marshes of the Pripet Basin. Also search Polesie. Index Soviet republics as applicable.
CO   N513000N530000
     E0320000E0250000
UF   Polesie
BT   USSR
SA   Byelorussia
     Pripet Basin
     Ukraine

polianite
use pyrolusite

policy
Used as a general term.
SA   international cooperation
     legislation
     programs

Polish Lowland
Eastern extension of the North German Plain in NE and N part of country. Also search Polish Lowlands.
CO   N513000N544500
     E0200000E0141500
BT   Poland
     Europe

polished sections
SA   microscope methods
     sections
     thin sections

polished surface
use slickensides

Polk County
Index states as applicable.
BX   Arkansas
     Florida
     Georgia
     Iowa
     Minnesota
     Missouri
     Nebraska
     North Carolina
     Oregon
     Tennessee
     Texas
     Wisconsin
BT   United States

Polkowice
Town in Lower Silesia in SW part of country.
BT   Zielona Gora
     Poland

pollen
SA   exine
     microfossils
     microorganisms
     palynology
     palynomorphs
     pollen analysis
     pollen diagrams
     spores
     sporopollenin

pollen analysis
Includes use on level 3 under palynology(1).
SA   analysis
     palynology
     pollen
     pollen diagrams

pollen diagrams
Includes use on level 3 under palynology(1).
SA   diagrams
     palynology
     pollen
     pollen analysis

pollucite
BA   framework silicates
     silicates
BT   minerals

pollutants
As of 1978, term is used on level 2 under pollution(1).
SA   pollution
     waste disposal

pollution
A level 1 term as of 1978. Used for geological studies on pollution of the environment. Includes use on level 2 under impact statements(1) and nuclear facilities(1). If 1, term set options are:
air
    subtopic
case studies
    subtopic
causes
    subtopic
concepts
    subtopic
controls
    subtopic
detection
    subtopic
effects
    subtopic
experimental studies
    subtopic
field studies
    subtopic
human ecology
    subtopic
impact statements
    subtopic
metals
    subtopic
oil spills
    subtopic
pollutants
    subtopic
waste disposal
    subtopic
water
    subtopic
SA   acid mine drainage
     air
     atmosphere
     case studies
     conservation
     contamination
     environmental geology
     fallout
     geochemistry
     geologic hazards
     human ecology
     hydrology
     impact statements
     industrial waste
     land use
     medical geology
     metals
     oil spills
     pollutants
     purification
     radioactive waste
     radioactivity
     reclamation
     sewage
     sulfur dioxide
     sulfuric acid
     thermal pollution
     toxic materials
     trace metals
     waste disposal
     water
     water supply

polonium
Includes use on level 1 and 2 as a chemical element (list D).
UF   Po
SA   elements
     Po-210

Polousnyy
A ridge in N Yakutia W of the Indigirka River.
BT   Russian Republic
     USSR

Poltava
City and Oblast in NE central.
UF   Pultova
     Pultowa
BT   Ukraine
     USSR

Poltava Series
In Donets Ridge, and W and NW Ukraine.
BT   Tertiary
SA   Miocene
     Oligocene
     Russian Republic
     Ukraine
     USSR

polybasite
BA   sulfantimonites
     sulfosalts
BT   minerals

Polychaetia
Includes use on level 2 under Annelida(1). See list F.
BA   Annelida
BT   Invertebrata
NA   Serpulidae
SA   Chaetopoda
     worms

polychroism
use pleochroism

polydymite
BA   sulfides
BT   minerals
SA   linnaeite

Polygnathus
Genus.
BA   conodonts

polygonal fractures
Term introduced in 1977. Includes use on level 3 under fractures(1) style(2). Before 1978, also search fractures AND polygonal.
BA   fractures

polygons
Includes use on level 3 under glacial geology(1) periglacial features(2).
SA   frost action
     patterned ground
     periglacial features

polyhalite
BA   sulfates
BT   minerals

polyhedra
NT   tetrahedra
SA   crystal structure

polylithionite
BA   sheet silicates
     silicates
BT   minerals
SA   mica group

polymerization
SA   geochemistry
     organic materials

polymetallic ores
Includes use on level 1 and 2 as a commodity term (list C).
UF   ores, polymetallic

polymetamorphism
Includes use on level 2 under metamorphism(1).
UF   superimposed metamorphism
BA   metamorphism
SA   prograde metamorphism
     retrograde metamorphism

polymorphism
Includes use on level 3 under fossil sets (list F) or under crystal structure(1) or minerals(1). Also search polymorphs.
UF   polymorphs
SA   crystal structure
     dimorphism
     minerals
     polytypism

polymorphs
use polymorphism

Polynesia
Collective name for islands of the central and SE. Index island and island groups as applicable. Includes use on level 1 or 2 as an area term (list O). If 1, see term set options under list B.
BA   Pacific Ocean
NT   Mururoa Atoll
NA   Samoa
     Society Islands

   Tahiti
   Tonga
 NT Tuamotu Islands
 SA Cook Islands
   Easter Island
   Hawaii
   Line Islands
   Oceania

polyphase processes
Also search polyphase.
 SA processes

Polyplacophora
Includes use on level 2 under Mollusca(1). See list F.
 BA Mollusca
 BT Invertebrata

polytypes
use polytypism

polytypism
Includes use on level 3 under crystal structure(1) or minerals(1). Also search polytypes.
 UF polytypes
 SA crystal structure
   minerals
   polymorphism

Polyzoa
use Bryozoa

Pomerania
Historical region on Baltic Sea extending from Stralsund, W of the Oder River, to the Vistula River on the E. Now primarily in Poland. Index countries as applicable.
 BT Europe
 SA East Germany
   Poland

Ponce
City in S on Caribbean Sea.
 BT Puerto Rico
   West Indies

Pondicherry
A centrally administered territory known as an Union Territory composed of the four scattered former French settlements of Karikal, Pondicherry, and Yanaon on the Bay of Bengal and Mahe on the Indian Ocean. The city of Pondicherry is the capital.
 BT India

ponds
Includes use on level 3 under ecology(1) and paleoecology(1).
 SA ecology
   irrigation
   lakes
   limnology
   paleoecology
   reservoirs
   surface water
   water resources
   water supply

Pongidae
Family. Includes use on level 3 under Mammalia(1) Primates(2).
 BA Primates
   Mammalia
 BT Tetrapoda
   Vertebrata

Pontevedra
Province and city in NW.
 BT Spain

Pontian
Europe. Above Sarmatian, below Plaisancian (Pliocene). Includes use on level 3 under age terms(1). See list E.
 BA upper Miocene
   Miocene
   Neogene
   Tertiary
 BT Cenozoic

Pontic Mountains
NW Anatolia.
 BT Turkey
   Middle East

Pontotoc County
Index states as applicable.
 BX Mississippi
   Oklahoma
 BT United States

Poona
City, district, and division in W Maharashtra.
 UF Pune
 BT Maharashtra
   India

Pope County
Index states as applicable.
 BX Arkansas
   Illinois
   Minnesota
 BT United States

Popigay
Village on Popigay River of E Taymyr National Orug in NW Siberia.
 BT Russian Republic
   USSR

popular and elementary geology
A valid term through 1978. After 1978, use terms separately, i.e. elementary geology.

popular geology
Term introduced in 1978. Before 1978, search popular and elementary geology.
 SA education
   elementary geology
   geology

populations
Used as a general term.
 SA statistical analysis
   statistical methods

porcelanite
use porcellanite

porcellanite
Includes use on level 3 under sedimentary rocks(1) clastic rocks(2). See list I.
 UF porcelanite
 BA clastic rocks
 BT sedimentary rocks

Porcupine Mountains
Range in Gogebic and Ontonagon Counties in NW extremity of Upper Peninsula.
 BT Michigan
   United States

pore pressure
Includes use on level 3 under soil mechanics(1) or engineering geology(1).
 UF neutral stress
 BT pressure
 SA engineering geology
   high pressure
   porous media
   soil mechanics
   stress

pore water
Includes use on level 2 under sediments(1). Also search interstitial water.
 UF interstitial water
 SA connate waters
   desiccation
   porous media
   sediments
   water

Porifera
Includes use on level 1 and 2 as a fossil term (list F).
 UF Spongiae
 BT Invertebrata

 NA Calcispongea
   Demospongea
   Hyalospongea
   Silicispongiae
 SA Receptaculitaceae
   spicules

porosity
Includes use on level 3 under rock name, material, e.g. under sedimentary rocks(1), ground water(1), soils(1) and sediments(1).
 SA compaction
   Darcy's law
   ground water
   permeability
   physical properties
   porous media
   sedimentary rocks
   sediments

porosity traps
use stratigraphic traps

porous materials
use porous media

porous media
Used as a general term. Also search porous materials.
 UF porous materials
 SA media
   pore pressure
   pore water
   porosity

porphyrins
Includes use on level 3 under organic materials(1).
 BA organic materials
 SA organic compounds

porphyrite
use porphyry

porphyritic texture
Term introduced in 1978. Before 1978, search porphyritic AND texture.
 SA igneous rocks
   phenocrysts
   porphyry
   textures
   vitrophyre

porphyroblastic texture
Term introduced in 1978. Includes use on level 3 under metamorphic rocks(1). Before 1978, search porphyroblasts; porphyroblastic AND texture.
 UF porphyroblasts
 SA metamorphic rocks
   textures

porphyroblasts
use porphyroblastic texture

porphyry
Also search porphyrite.
 UF porphyrite
 BA igneous rocks
 SA andesite porphyry
   dacite porphyry
   diabase porphyry
   diorite porphyry
   granite porphyry
   granodiorite porphyry
   liparite porphyry
   melaphyre
   picrite porphyry
   porphyritic texture
   quartz porphyry
   rhyolite porphyry
   syenite porphyry

porphyry copper
Includes use on level 3 under mineral deposits, genesis(1).
 SA copper
   mineral deposits
   ore deposits

Porsang Fjord
Inlet of Arctic Ocean on N coast.
 UF Porsanger Fjord
   Porsangerfjord
 BT Finnmark
   Norway

Porsangen Fjord
use Porsang Fjord

Porsanger Dolomite Formation
Uppermost Precambrian or lowermost Cambrian. Porsanger Dolomite found in upper part of Porsanger Series.
 SA Cambrian
   Finnmark
   Norway
   Precambrian

Porsangerfjord
use Porsang Fjord

Port au Port Peninsula
On the Gulf of Saint Lawrence between Port au Port Bay and Saint George Bay.
 BT Newfoundland
   Canada

Port Campbell
Town on Indian Ocean in SW.
 BT Victoria
   Australia

Port Elizabeth
City on Indian Ocean in SE.
 BT Cape Province
   South Africa

Port Jervis
City and summer resort at meeting of borders with New Jersey and Pennsylvania in S.
 BT New York
   United States

Port Moresby
City on Coral Sea in SE.
 BT Papua New Guinea
   Australasia

Port Phillip Bay
Harbor of Melbourne.
 BT Victoria
   Australia

Port Royal Sound
Inlet of Atlantic Ocean between islands of St. Helena and Hilton Head in Beaufort County.
 BT South Carolina
   United States

Port Valdez
use Valdez

Portage County
Index states as applicable.
 BX Ohio
   Wisconsin
 BT United States

Portage Lake Lava Series
Tops of conglomerate beds used for stratigraphic reference include: St. Louis Conglomerate, The Old Colony and Wolverine sandstones, and the Kingston, Calumet and Hecla, Houghton, Allouez, Pewabic West, and Hancock conglomerates. Upper Peninsula of Michigan. Also search Portage Lake.
 BT Keweenawan
   Precambrian
 SA Michigan

Portel
Village in S part of country.
 BT Aude
   France

Portland
City in Rockingham County in SE New Hampshire, and city on the Willamette River in NW Oregon. Index

states as applicable.
CO N453200N453200
   W1224000W1224000
BX New Hampshire
   Oregon
BT United States

Portlandian
Europe. Above Kimmeridgian, below Berriasian (Cretaceous). Includes use on level 3 under age terms(1). See list E.
BA Upper Jurassic
   Jurassic
BT Mesozoic
SA Malm

Porto Santo Island
One of two inhabited islands of Portuguese island group. Also search Porto Santo.
BT Madeira
   Atlantic Ocean

Portola Valley
Town in San Mateo County S of San Francisco.
BT California
   United States

Portugal
Includes use on level 1 as an area term (list O). For term set options see list B.
CO N370000N421000
   W0061000W0093000
BA Europe
NT Alentejo
   Algarve
   Alto Alentejo
   Baixo Alentejo
   Beira Baixa
   Braganca
   Coimbra
   Elvas
   Evora
   Leiria
   Panasqueira
   Setubal
   Tomar
   Tras-os-Montes
   Vila Real
   Viseu
SA Iberian Peninsula
   Tagus Basin
   Tagus River
   Timor

Portuguese Guinea
Not a valid index term for GeoRef since 1974. Use Guinea-Bissau. Before 1975, also search Portuguese Guinea.

Portuguese Timor
A valid index term through mid-1978. Included use on level 1 as an area term until 1977. After mid-1978, use Timor.

Portuguese West Africa
use Angola

Porulosida
Order. Includes use on level 3 under Radiolaria(1).
BA Radiolaria
BT Invertebrata
NA Acantharina

Posen
use Poznan

Posidonia Shale
BT Jurassic
SA Germany
   West Germany

positions, pole
use pole positions

posnjakite
BA sulfates
BT minerals

Possagno
Village in central.
BT Veneto
   Italy

possibilities
Includes use as a level 2 or 3 term appropriate to a large number of topics, e.g. on level 2 under energy sources(1). See list C (commodities) and list G (general terms).

Postglacial
Not a valid term for GeoRef. Use Holocene.

postglacial environment
Term introduced in 1978. Before 1978, search postglacial.
SA environment
   glaciers

Postmasburg
Village in N central.
BT Cape Province
   South Africa

potash
Includes use on level 1 as a commodity term (list C).

potash alum
use alum

potash feldspar
use K-feldspar

potash mica
use muscovite

potassic composition
Term introduced in 1978. Before 1978, search potassic.
SA composition
   potassium

potassium
Includes use on level 1 and 2 as a chemical element (list D). Also search K.
UF K
SA alkali metals
   elements
   isotopes
   K-40
   K/Ar
   KREEP
   potassic composition

potassium alum
use alum

potassium feldspar
use K-feldspar

Potenza
City and province in S part of country.
BT Basilicata
   Italy

potholes
Used as a general term under glacial geology(1) or geomorphology(1).
BT erosion features
SA erosion
   sinkholes

Potomac Group
Consists of Patuxent, Arundel, and Patapsco formations.
BT Cretaceous
SA Delaware
   Lower Cretaceous
   Maryland
   Upper Cretaceous
   Virginia

Potomac River
Formed by the confluence of the North and South branches of the Potomac River SE of Cumberland, Maryland. It flows SE into Chesapeake Bay. Index District of Columbia and states as applicable.
BT United States
SA District of Columbia
   Maryland
   Virginia
   West Virginia

Potomac River basin
Index states and District of Columbia as applicable.
BT United States
SA District of Columbia
   Maryland
   Virginia
   West Virginia

Potsdam
District N, W, and S of West Berlin. Also a city just outside West Berlin.
BT East Germany

Potsdam Sandstone
Underlies Ticonderoga Formation. Central and E New York, and Vermont.
BT Upper Cambrian
   Cambrian
SA New York
   Vermont

Pottawatomie County
Index states as applicable.
BX Kansas
   Oklahoma
BT United States

Pottsville Group
Includes Lee Formation, Norton Formation, Gladeville Sandstone, Wise Formation, Harlan Sandstone, Lookout Sandstone, Walden Sandstone, Tumbling Run Formation, Schuylkill Formation, Sharp Mountain Formation. Alabama, S Ohio, S Illinois, S Indiana, Kentucky, Maryland, NE Mississippi, Pennsylvania, Tennessee, Virginia, West Virginia.
BT Pennsylvanian
SA Alabama
   Breathitt Formation
   Illinois
   Kentucky
   Lower Pennsylvanian
   Maryland
   Middle Pennsylvanian
   Mississippi
   Ohio
   Pennsylvania
   Pocahontas Formation
   Sharon Conglomerate
   Tennessee
   Virginia
   West Virginia

Potwar Plateau
Lies between Indus and Jhelum rivers in N. Also search Potwar.
BT Punjab
   Pakistan

Pouilly-en-Auxois
Village in E central part of country.
BT Cote-d'Or
   France

Poway Conglomerate
Nearly horizontal, and in most places lies unconformably on Cretaceous Peninsular Range Batholith and accompanying metamorphic rocks. S California.
UF Poway Conglomerates
BT upper Eocene
   Eocene
SA California

Poway Conglomerates
use Poway Conglomerate

Powder River basin
Index statas as applicable.
BT United States
SA Montana
   Wyoming

Powell
Town in Park County in NW.
BT Wyoming
   United States

Powell County
Index states as applicable.
BX Kentucky
   Montana
BT United States

powellite
BA molybdates
BT minerals

power plants
SA energy sources
   nuclear facilities

Poza Rica
City in N Veracruz.
UF Poza Rica de Hidalgo
BT Veracruz
   Mexico

Poza Rica de Hidalgo
use Poza Rica

Poznan
Province in W central Poland. Also a city province and a city.
UF Posen
BT Poland
NT Adamow Mine
   Klodawa
   Konin
   Ksiaz

Poznan Clays
East Germany and Poland.
BT upper Tertiary
   Tertiary
SA Germany
   Poland

Pozzuoli
City on inlet of Gulf of Naples.
BT Campania
   Italy

Pr
use praesodymium

practice
To be used for geology as a profession after 1975. Includes use as a level 2 or 3 term appropriate to a large number of topics, e.g. on level 2 under geochemistry(1), geology(1), and economic geology(1). See list G.

praesodymium
Includes use on level 1 and 2 as a chemical element (list D.)
UF Pr
SA elements
   rare earths

Prague
City in central Bohemia.
UF Praha
BT Bohemia
   Czechoslovakia

Praha
use Prague

Prahova
County in NE.
BT Walachia
   Romania

Prairie Evaporite
Salt and anhydrite beds that form upper unit of Elk Point Group. Subsurface. Also search Prairie Evaporite Formation.
UF Prairie Evaporite Formation
BT Middle Devonian
   Devonian
SA Elk Point Group
   Manitoba
   North Dakota
   Saskatchewan

Prairie Evaporite Formation
use Prairie Evaporite

Pranhita-Godavari Valley
Combined valleys of the Pranhita and Godavari river systems. Index states as applicable.
BT India
SA Andhra Pradesh
Godavari Valley
Maharashtra

prasinite
use greenschist

Pratt County
S Kansas.
BT Kansas
United States

Prbram
use Pribram

Prealps
Mostly in Italy. Also search Alpine Foreland.
UF Alpine Foreland
SA Alps
Italy

Prebetic Zone
A geographic term with stratigraphic-tectonic connotations.
BT Spain
SA Betic Cordillera
Subbetic Zone

Preboreal
Europe. An interval of post-glacial time following the Younger Dryas of the late-glacial Arctic interval and preceding the Boreal. Also search Subarctic.
UF Subarctic
BA Holocene
Quaternary
BT Cenozoic
SA Boreal

Precambrian
Includes use on level 1 as an age term (list E); on level 2 under paleo- terms, e.g. paleoecology, paleogeography, paleomagnetism.
NT Acungui Group
NA Adelaidean
NT Adirondack Anorthosite
NA Algonkian
NT Altyn Limestone
Animikie Group
Assyntic Orogeny
Avalonian Orogeny
Baltimore Gneiss
Belt Supergroup
Bijawar System
Biwabik Iron Formation
Bonner Formation
Brockman Iron Formation
Catoctin Formation
Chuar Group
Copper Harbor Conglomerate
Cuddapah System
Delhi System
NA Eocambrian
NT Espanola Formation
Fig Tree Series
Flinton Group
Grenvillian Orogeny
Gunflint Iron Formation
Hamersley Group
Harbour Main Group
Hecla Hoek Formation
Hector Formation
NA Helikian
NT Highland Series
Hudsonian Orogeny
Johnnie Formation
Kaladgi System
Karelian Orogeny
Katangan Orogeny
Kenoran Orogeny
NA Keweenawan
NT Knife Lake Group
NA Lewisian
lower Precambrian
NT Manhattan Formation
NA middle Precambrian
NT Minas Series
Missi Group
Missoula Group
Moldanubian
Nipissing Diabase
Nonesuch Shale
Noonday Dolomite
North Shore Volcanics
NA Ocoee Series
NT Onverwacht Group
Osler Series
Pelona Schist
Penokean Orogeny
Penrhyn Slate
Pocatello Formation
Prichard Formation
Prince Albert Group
NA Proterozoic
NT Purcell System
Ravalli Group
Roan Supergroup
Sioux Quartzite
Sokoman Formation
Stillwater Complex
Stirling Quartzite
Swaziland Sequence
Swaziland System
Thomson Slate
Transvaal Supergroup
Twilight Gneiss
Unkar Group
NA upper Precambrian
Vendian
NT Vindhyan
Wallace Formation
Waterberg System
Wilcox Formation
Witwatersrand System
Wyman Formation
SA Baikalian Phase
Bhander Group
Cadomian Orogeny
Dalradian
Dore Lake Complex
Duluth Complex
Jacobsville Sandstone
Kaimur Sandstone
Kurnool System
Lardeau Group
New York City Group
Porsanger Dolomite Formation
Shuswap Complex
Sparagmite Group
Talladega Group
Wood Canyon Formation

Precambrian Shield
use Canadian Shield

precession
Used as a general term.
SA processes

precipitation
As of 1978, term refers to geochemistry. For meteorology, use atmospheric precipitation. Includes use on level 2 under sedimentation(1).
SA accumulation
atmospheric precipitation
climate
crystallization
drainage basins
geochemistry
hydrology
particle precipitation
particles
rivers and streams
runoff
sedimentation
storms
water

Precordillera
BT Argentina

precursors
Includes use on level 3 under seismology(1).
SA seismology

Predazzo
City and resort in Dolomites in S.
BT Trentino-Alto Adige
Italy

prediction
Includes use on level 2 under earthquakes(1).
SA earthquakes

preferred orientation
Includes use on level 2 under structural analysis(1).
SA fabric
lineation
orientation
petrofabrics
structural analysis
tectonite

prehnite
BA sheet silicates
silicates
BT minerals
SA aluminosilicates

prehnite-pumpellyite facies
BT facies
SA metamorphic rocks

premonitory phenomena
Term introduced in 1978.
UF phenomena, premonitory
SA premonitory phenomena
seismology

preobrazhenskite
BA borates
BT minerals

preparation
Includes use as a level 2 or 3 term appropriate to a large number of topics, e.g. on level 2 under coal(1). See list G. Also search preparations.
SA sample preparation

preservation
As of 1978, term is used on level 2 under land use(1).
SA land use

Presidente Vargas
use Itabira

Presidio County
W Texas.
BT Texas
United States

Presque Isle
City in Aroostook County in N Maine; small peninsula in Lake Erie in NW Pennsylvania; county in NE Michigan.
BX Maine
Michigan
Pennsylvania
BT United States

pressure
Includes use on level 3 under deformation(1); on level 3 under experimental studies(2) under appropriate headings in which pressure is a factor, e.g. faults(1), folds(1), fractures(1), foliation(1), metasomatism(1), magmas(1), and metamorphism(1).
NT atmospheric pressure
capillary pressure
high pressure
NA hydraulic pressure
NT hydrostatic pressure
low pressure
partial pressure
pore pressure
water pressure
SA compression
deformation
earth pressure
fluid pressure
geopressure
load pressure
P-T conditions
stress
tension

pressure solution
SA diagenesis
solution

pressure wave
use P-waves

Preston County
BT West Virginia
United States

Preston Peak
In Siskiyou Mountains in Siskiyou County in extreme N.
BT California
United States

Pretoria
City in S.
BT Transvaal
South Africa

prevention
A valid general term through mid-1978.

Priabonian
Europe. Includes use on level 3 under age terms(1). See list E.
BA upper Eocene
Eocene
Paleogene
Tertiary
BT Cenozoic
SA Auversian
Bartonian
Ludian

Pribram
City SW of Prague in central Bohemia. Also search Prbram.
UF Prbram
BT Bohemia
Czechoslovakia

Price Formation
Includes Cloyd Conglomerate Member. SW Virginia.
BT Mississippian
SA Virginia

Prichard Formation
Belt Series. NE Idaho and NW Montana.
BT Precambrian
SA Idaho
Montana

Pridolian
Europe.
BA Upper Silurian
Silurian
BT Paleozoic

Prikumsk
City in Stavropol Kray in the Northern Caucasus.
UF Budennovsk
BT Russian Republic
USSR

primary dispersion
Term introduced in 1978. Includes use on level 3 under mineral exploration(1). Before 1978, search primary AND dispersion.
SA dispersion
mineral exploration
secondary dispersion

primary structures
Includes use on level 2 under sedimentary structures(1). See list K.
BA sedimentary structures
SA ball-and-pillow
secondary structures
structures

primary wave
  use P-waves

Primates
  Includes use on level 2 under Mammalia(1). See list F.
  BA Mammalia
  BT Tetrapoda
      Vertebrata
  NA Australopithecinae
      Hominidae
      Pongidae
  SA fossil man
      man

Primorski Krai
  use Primorye

Primorye
  Primorye Kray (territory). New name for Primorski Krai which is the SE section of the Soviet Far East between Manchuria and the Japan Sea. Also search Maritime Territory.
  CO N420000N500000
      E1400000E1300000
  UF Maritime Kray
      Maritime Territory
      Primorski Krai
  BT Russian Republic
      USSR

Prince Albert Group
  BT Precambrian
  SA Canada
      Saskatchewan

Prince Charles Mountains
  In Mac-Robertson Land near the Amery Ice Shelf on Indian Ocean side.
  BT Antarctica

Prince Edward Island
  Island in the Gulf of Saint Lawrence constituting a province. Includes use on level 1 as an area term (list O). For term set options see list B.
  CO N455500N470500
      W0620000W0643000
  BA Canada
  SA Maritime Provinces

Prince Georges County
  NE, E, and SE of District of Columbia.
  BT Maryland
      United States

Prince of Wales Island
  Largest island of Alexander Archipelago in SE Alaska, W of Ketchikan; and an island between Victoria and Somerset Islands in District of Franklin, Northwest Territories. Index Alaska and/or Northwest Territories as applicable.
  SA Alaska
      Northwest Territories

Prince William Sound
  Inlet of Gulf of Alaska E of Kenai Pensula.
  BT Alaska
      United States

principles
  Includes use as a level 2 or 3 term appropriate to a large number of topics, e.g. on level 2 under geology(1). See list G.

Pripet Basin
  River basin most of which comprises the Pripet or Pinsk Marshes which is the largest tract of swamp in Europe. Largely coextensive with Polesye Lowland. Also search Pripet; Pripet Depression; Pripyat Basin. Index Soviet republics as applicable.
  CO N510000N540000
      E0330000E0240000
  UF Pripet Depression
      Pripyat Basin
  BT USSR
  SA Byelorussia
      Polesye
      Ukraine

Pripet Depression
  use Pripet Basin

Pripyat Basin
  use Pripet Basin

prismatine
  use kornerupine

Privas
  Town in SE central part of country.
  BT Ardeche
      France

probability
  Includes use on level 3 under mathematical geology(1).
  SA mathematical methods
      reliability
      sampling
      statistical analysis
      stochastic processes

probe, electron
  use electron probe

probe, ion
  use ion probe

problematic
  A valid term through 1977 used on level 1 in combination with fossils (i.e. fossils, problematic). After 1977, use problematic fossils on level 1.

problematic fossils
  Term introduced in 1978. Includes use on level 1 and 2. Before 1978, also search fossils AND problematic.
  NA Calcispherulidae
      Dacryoconarida
      Hyolithidae
      Receptaculitaceae
      Tentaculitida
  SA chemical fossils
      fossilization
      fossils
      nannofossils
      paleontology
      Stromatoporoidea

problematic palynomorphs
  Includes use on level 2 under palynomorphs(1). See list F.
  BA palynomorphs

problems
  A valid general term through mid-1978.

Proboscidea
  Includes use on level 2 under Mammalia(1). See list F.
  BA Mammalia
  BT Tetrapoda
      Vertebrata
  NA Elephantidae
      Mastodontidae

processes
  Includes use as a level 2 or 3 term appropriate to a large number of topics, e.g. on level 2 under geochemistry(1), sedimentation(1), and under mineral deposits, genesis(1). See list G. Also search process.
  SA adsorption
      aggradation
      albitization
      alteration
      autometamorphism
      autometasomatism
      bauxitization
      biological processes
      brecciation
      calcification
      calcitization
      chloritization
      coalification
      combustion
      compaction
      complexing
      contraction
      cyclic processes
      dedolomitization
      diagenesis
      diffusion
      dissociation
      dolomitization
      electrolysis
      endogene processes
      epigene processes
      epithermal processes
      erosion
      exhalative processes
      exogene processes
      filtration
      fractionation
      fragmentation
      geothermal processes
      granitization
      hydration
      hydrolysis
      hydrothermal processes
      hypogene processes
      igneous processes
      kaolinization
      mixing
      oxidation
      phyllonitization
      polyphase processes
      precession
      propylitization
      reduction
      sedimentary processes
      sedimentation
      separation
      serpentinization
      silicification
      stochastic processes
      supergene processes
      syntectonic processes

processing, data
  use data processing

production
  Includes use on level 2 and 3 under commodity terms, e.g. on level 2 under energy sources(1). See list C. When referring to mining geology before May 1978, also search extraction.
  SA production control

production control
  Includes use on level 2 under mining geology(1).
  UF control, production
  SA mining geology
      production
      technology

productivity
  Used as a general term.

products
  A valid general term through mid-1978.

Proetidae
  Family. Includes use on level 3 under Trilobita(1).
  BA Trilobita
  BT Arthropoda
      Invertebrata

profiles
  Includes use on level 3 under soils(1) morphology(2) and under geophysical surveys(1). Before 1976, also search analytic profiles.
  SA geophysical surveys
      horizon differentiation
      horizons
      micromorphology
      morphology
      soils

progradation
  SA littoral drift
      sedimentation
      shore features

prograde metamorphism
  Term introduced in 1978. Includes use on level 2 under metamorphism(1). Before 1978, also search metamorphism AND prograde.
  BA metamorphism
  SA polymetamorphism
      retrograde metamorphism

programs
  Includes use on level 2 under mineral exploration(1); on level 3 under automatic data processing(1) and education(1). Also search projects.
  UF projects
  SA associations
      automatic data processing
      computer programs
      computers
      education
      mineral exploration
      planning
      policy
      reclamation
      research

progress report
  Includes use on level 3 under associations(1), surveys(1), symposia(1).
  BT report
  SA annual report
      associations
      current research
      research
      surveys
      symposia

projects
  use programs

Prokopyevsk
  City at S end of Kuznetsk Basin in Kemerovo Oblast in S Siberia.
  BT Russian Republic
      USSR

proluvium
  BA clastic sediments
  BT sediments

prometheum
  use promethium

promethium
  Includes use on level 1 and 2 as a chemical element (list D).
  UF Pm
      prometheum
  SA elements
      rare earths

propagation
  Includes use on level 3 under seismology(1) elastic waves(2).
  SA elastic waves
      seismology
      wave propagation

properties
  Includes use in combination with materials (i.e. materials, properties) on level 2 under engineering geology(1); as a level 2 or 3 term appropriate to a large number of topics; for physical or chemical properties on level 2 or 3 under commodity terms(1). See list C (commodities) and list G (general terms).
  SA acoustical properties
      chemical properties
      dielectric properties
      elastic properties
      electrical properties
      electrochemical properties
      engineering properties
      magnetic properties
      materials
      mechanical properties
      optical properties
      physical properties
      radiometric properties
      reservoir properties

solubility
surface properties
thermal properties
thermochemical properties
thermodynamic properties
tortuosity

Propria Geosyncline
In Propria region of the lower Sao Francisco River near the Atlantic Ocean.
BT Sergipe
Brazil
SA geosynclines

propylite
As of 1978, term is used on level 2 under metasomatic rocks(1). Before 1978, included use on level 3 under igneous rocks(1) andesite-rhyolite family(2).
BA metasomatic rocks
SA andesite-rhyolite family
propylitization

propylitization
Includes use on level 3 under metasomatism(1) processes(2).
BA metasomatism
SA hydrothermal alteration
processes
propylite

Prosobranchia
Subclass. Includes use on level 3 under Mollusca(1) Gastropoda(2).
BA Gastropoda
Mollusca
BT Invertebrata

prosopite
BA halides
BT minerals
SA fluorides

Prospect Mountain Quartzite
use Stirling Quartzite

prospecting
Used as a general term.
SA exploration
geochemical prospecting
homogeneity
inhomogeneity
networks

protactinium
Includes use on level 1 and 2 as a chemical element (list D).
UF Pa
SA elements
Pa-231

proteins
Includes use on level 3 under organic materials(1).
BA organic materials
SA amino acids
collagen
conchiolin

Proterozoic
As of 1978, includes use on level 1 and 2 (list E).
UF Agnotozoic
BA Precambrian
NT Badami Series
Bambui Group
Bitter Springs Formation
Damara System
NA Huronian
NT Kursk Series
Lorrain Formation
NA lower Proterozoic
middle Proterozoic
NT Miette Group
Ovruch Series
Udokan Series
NA upper Proterozoic
NT Windermere System
Wolkberg Group
SA Algonkian

Helikian
Shuswap Complex
upper Precambrian

Protista
Includes use on level 1 and 2 as a fossil term (list F). Before 1976, also search Protozoa.
NA ebridians
Silicoflagellata
Thecamoeba
Tintinnidae
SA algae
bacteria
Buliminacea
foraminifera
Metazoa
microorganisms
paleobotany
paleontology
plankton
Radiolaria
zooxanthellae

protodolomite
Artificial mineral.
BA carbonates
BT minerals

protoenstatite
Artificial mineral.
BA pyroxene group
chain silicates
silicates
BT minerals

protoliths
Term introduced in 1978. Also search parent rocks.
UF parent rocks
SA igneous rocks
metamorphic rocks
metamorphism

protons
Used as a general term.
SA cosmic rays
electrons

Protopivskaya Formation
Central Ukraine and NW Donets Basin.
BT Upper Triassic
Triassic
SA Ukraine
USSR

Protorthoptera
Includes use on level 2 under Insecta(1). See list F.
BA Insecta
BT Arthropoda
Invertebrata

Protozoa
A valid index term through 1976. After 1976, use Protista.

proustite
BA sulfarsenites
sulfosalts
BT minerals

provenance
Includes use on level 2 under sedimentation(1) and under sediments(1).
UF source area
sourceland
SA paleocurrents
sedimentary rocks
sedimentation
sediments

Provence
Historical region in SE. Index departments as applicable.
CO N430000N444500
E0074500E0043000
BT France
SA Alpes-de-Haute Provence
Alpes-Maritimes
Bouches-du-Rhone
Var

Vaucluse

Provence Alps
In SE Provence. Index departments as applicable.
BT France
SA Alpes-de-Haute Provence
Alpes-Maritimes
Alps
Var

provinces
A valid general term through 1976. See specific term, e.g. metallogenic provinces, physiographic provinces, etc.

provinces, faunal
use faunal provinces

provinces, floral
use floral provinces

provinces, metallogenic
use metallogenic provinces

provinces, physiographic
use physiographic provinces

provincialism
use provinciality

provinciality
Used mostly for fossils. See list F.
UF provincialism
SA fossils

Provo
City S of Salt Lake City in Utah County.
BT Utah
United States

Prudhoe Bay
Inlet of the Arctic Ocean on the North Slope.
BT Alaska
United States
SA North Slope

Prut River
Rises in W central Carpathians and flows into Danube River below Galati. Index Romania and Soviet republics as applicable. Also search Prut.
UF Pruth River
BT Europe
SA Moldavia
Romania
Ukraine

Pruth River
use Prut River

Przemysl
City near Ukrainian border in SE part of country.
UF Peremyshl
BT Rzeszow
Poland

psammite
use sandstone

psammitic texture
use arenaceous texture

pseudobrookite
BA oxides
BT minerals
SA armalcolite
ferropseudobrookite

pseudogalena
use sphalerite

Pseudogleys
See list M.
BA soils
SA Gleys

pseudoleucite
BA framework silicates
silicates
BT minerals

pseudomorphism
Includes use on level 3 under mineralogy(1). Before 1976, also search replacement.

SA minerals

pseudotachylite
Includes use on level 3 under metamorphic rocks(1) mylonites(2). See list J.
BA mylonites
BT metamorphic rocks

psilomelane
BA oxides
BT minerals
SA manganese oxide

Psilophytales
use Psilopsida

Psilopsida
Including Psilophytales. Includes use on level 2 under pteridophytes(1). See list F.
UF Psilophytales
BA pteridophytes
BT Plantae

Pskem Range
Outlier of Kirghiz Range. Index Soviet republics as applicable.
BT USSR
SA Kazakhstan
Kirghizia

Pskov
City 155 miles SW of Leningrad. Also an oblast.
BT Russian Republic
USSR

Pt
use platinum

Pteranodon
Genus. Includes use on level 3 under Reptilia(1) Archosauria(2).
BA Archosauria
Reptilia
BT Tetrapoda
Vertebrata

Pteridophyllen
Includes use on level 2 under pteridophytes(1). See list F.
BA pteridophytes
BT Plantae

pteridophytes
Includes use on level 1 and 2 as a fossil term (list F).
BT Plantae
NA Filicopsida
Lycopsida
Psilopsida
Pteridophyllen
Sphenopsida
SA ferns
Juglans
Neuropteris
Pteropsida
sporangia
spores

Pteridospermae
Includes use on level 2 under gymnosperms(1). See list F.
BA gymnosperms
BT Plantae
NA Neuropteris
SA Callipteris
Dicroidium
Pecopteris
Ptilophyllum
Sphenopteris
Taeniopteris

Pterobranchia
Includes use on level 1. See list F.
BT Invertebrata
SA paleontology

Pteropoda
Includes use on level 3 under Mollusca(1) Gastropoda(2). See list F.
BA Gastropoda
Mollusca
BT Invertebrata

Pteropsida
Used as a general term.
BT Plantae
SA angiosperms
gymnosperms
pteridophytes
thallophytes

Pterosauria
Order. Includes use on level 3 under Reptilia(1) Archosauria(2).
BA Archosauria
Reptilia
BT Tetrapoda
Vertebrata

ptilolite
use mordenite

Ptilophyllum
Includes use on level 3 under gymnosperms(1) Cycadales(2).
BA Cycadales
gymnosperms
BT Plantae
SA Bennettitales
Pteridospermae

Ptychopariida
Includes use on level 2 under Trilobita(1). See list F.
BA Trilobita
BT Trilobitomorpha
Arthropoda
Invertebrata
NA Asaphidae
Cryptolithus
Olenidae
Scutelluidae

ptygma
use ptygmatic folds

ptygmatic folds
Term introduced in 1978. Includes use on level 3 under folds(1) style(2). Before 1978, also search folds AND ptygmatic.
UF ptygma
BA folds
SA migmatites

Pu
use plutonium

Pu-239
Includes use on level 3 under isotopes(1).
SA isotopes
plutonium

Pu-244
Includes use on level 3 under isotopes(1).
SA isotopes
plutonium

publications
Used as a general term.
SA associations
surveys

Pucara Group
Includes Paria Formation. W part of Country. Upper Triassic and Jurassic.
BT Mesozoic
SA Junin
Jurassic
Lima
Peru
Triassic
Upper Triassic

Puck Bay
Inlet of the Gulf of Danzig on the Baltic Sea. Also search Puck.
BT Gdansk
Poland

Puebla
State E of Mexico City. Also a city.
BT Mexico
SA Sierra Madre Oriental

Puebla de Guzman
Town in SW part of country.
BT Huelva
Spain

Pueblo County
SE central.
CO N374500N383000
W1040500W1050500
BT Colorado
United States

Puente Formation
Includes Papel Blanco Shale, Blanco Sandstone, Cubierto Shale, Hunter Sandstone and Conglomerate, Peculiar Shale, Mahala Sandstone and Conglomerate, Sycamore Canyon Member, La Vida Member, Soquel Member, Yorba Member. S California.
BT upper Miocene
Miocene
SA California

Puente Hills
S part of Los Angeles County.
BT California
United States

Puerto Cabello
City W of Caracas on the Caribbean Sea.
BT Carabobo
Venezuela

Puerto Deseado
Town and Bay in S part of country.
BT Argentina
SA Santa Cruz

Puerto Rico
Includes use on level 1 as an area term (list O). For term set options see list B.
CO N175000N183000
W0654000W0671500
BA West Indies
NT Anasco Bay
Arecibo
Cabo Rojo
Mayaguez Bay
Ponce
SA Greater Antilles
Robles Formation
San Juan
San Juan Formation

Puerto Rico Trench
NNE of Puerto Rico.
UF Puerto Rico Trough
BT Atlantic Ocean

Puerto Rico Trough
use Puerto Rico Trench

Puget Lowland
Area surrounding Puget Sound and the broad trough extending S.
BT Washington
United States

Puget Sound
Arm of Pacific Ocean extending S from E end of Juan de Fuca Strait.
CO N471000N483000
W1221000W1231500
BT Washington
United States

Puglia
use Apulia

Pulaski County
Index states as applicable.
BX Arkansas
Georgia
Illinois
Indiana
Kentucky
Missouri
Virginia
BT United States

Pulaski Fault
SW Virginia.
BT Virginia

United States

pulaskite
Includes use on level 3 under igneous rocks(1) syenite family(2). See list H.
BA syenite family
BT igneous rocks

pull apart structures
use boudinage

pulsating spring
use geysers

pulsations
Includes use on level 3 under Earth(1).
SA Earth
magnetic field
magnetosphere
micropulsations
vibration

Pultova
use Poltava

Pultowa
use Poltava

Pultusk
City in E central part of country.
BT Warsaw
Poland

pumice
Includes use on level 1 as a commodity term (list C); on level 3 under igneous rocks(1) pyroclastics and glasses(2) and volcanic(2). See list H.
BA pyroclastics and glasses
BT igneous rocks
SA abrasives
aggregate
scoria

pump tests
Term introduced in 1978. Before 1978, also search pumping; pumping tests.
UF pumping tests
SA ground water

pumpellyite
BA orthosilicates
silicates
BT minerals
SA aluminosilicates

pumping tests
use pump tests

punch cards
Includes use on level 3 under automatic data processing(1). Also search punched cards.
UF punched cards
SA automatic data processing

Punchbowl Formation
Described in Valyermo Quadrangle as consisting of two facies. S California.
BT upper Miocene
Miocene
SA California

punched cards
use punch cards

Pune
use Poona

Pungo River
Tidal estuary near the Pamlico River off Pamlico Sound.
BT North Carolina
United States

Punjab
Former province of British India. Since partition in August 1947, a state in India and a province of Pakistan. Index countries as applicable.
BT Asia
NT Attock
Chandigarh
Khewra
Mangla Dam
Mansehra
Mianwali
Potwar Plateau
Salt Range
Spiti
SA Beas River
India
Jumna River
Kasauli Series
Muth Quartzite
Pakistan
Sulaiman Range

Punta Mosquito Formation
BT Eocene
Paleogene
Tertiary
SA Venezuela

Pupillidae
Family. Includes use on level 3 under Mollusca(1) Gastropoda(2).
BA Gastropoda
Mollusca
BT Invertebrata

Purbeckian
Europe. Includes use on level 3 under age terms(1). See list E.
BA Lower Cretaceous
Cretaceous
BT Mesozoic

Purcell System
Also search Purcell Series.
BT Precambrian
SA British Columbia
Idaho
Montana

pure coal
use vitrain

purification
Includes use on level 3 under ground water(1) or hydrology(1).
SA beneficiation
contamination
ground water
hydrology
impurities
pollution
water

purines
BA organic materials

purple copper ore
use bornite

Purulia
City in E part of country.
BT West Bengal
India

push wave
use P-waves

push-pull wave
use P-waves

Putnam County
Index states as applicable.
BX Florida
Georgia
Illinois
Indiana
Missouri
New York
Ohio
Tennessee
West Virginia
BT United States

Putrid Sea
use Sivash

Puy-de-Dome
Department in S central.
CO N451500N461500
E0040000E0023000
BT France
NT Clermont-Ferrand
Mont-Dore
Monts Dome

SA Limagne
Pyramid Lake
Washoe County in NW.
BT Nevada
United States
pyrargyrite
BA sulfantimonites
sulfosalts
BT minerals
Pyrenean Orogenic Phase
use Pyrenean Orogeny
Pyrenean Orogeny
One of the 30 or more short-lived orogenies during Phanerozoic time identified by Stille, in this case during the late Eocene, between the Bartonian and Ludian stages. Before 1978, search Pyrenean AND orogeny.
UF Pyrenean Orogenic Phase
BT Eocene
SA orogeny
Pyrenees
Mountain range extending from the Bay of Biscay to the SW coast of the Gulf of Lion. Index countries as applicable. Includes use on level 1 as an area term (list O). For term set options see list B.
CO N420000N430000
E0031500W0020000
BA Europe
SA Andorra
France
Spain
Pyrenees-Atlantiques
Department in SW France. Also search Basses-Pyrenees.
UF Basses-Pyrenees
BT France
NT Biarritz
Pau
Pyrenees-Orientales
Department in E Pyrenees.
CO N421500N430000
E0031500E0013000
BT France
NT Agly Massif
pyrite
Includes use on level 1 and 2 as a commodity term (list C).
BA sulfides
BT minerals
SA bravoite
framboidal texture
pyritization
BA metasomatism
SA hydrothermal alteration
pyroaurite
BA carbonates
BT minerals
pyrochlore
BZ halides
oxides
BT minerals
SA betafite
microlite
niobates
tantalates
pyroclastics
Used both as sedimentary rock and as igneous rock. Includes use on level 3 under sedimentary rocks(1) clastic rocks(2). See lists I. Also search tephra.
UF tephra
BA clastic rocks
BT sedimentary rocks
SA agglomerate
ash falls
ash flows
bentonite
ejecta

eruptions
glasses
igneous rocks
lapilli
pyroclastics and glasses
scoria
terrigenous materials
pyroclastics and glasses
Includes use on level 2 under igneous rocks(1). See list H.
BT igneous rocks
NA andesite tuff
ash flows
ash-flow tuff
green tuff
ignimbrite
obsidian
palagonite
perlite
pitchstone
pumice
rhyolite tuff
tuff
tuffite
volcanic ash
volcanic glass
welded tuff
NZ zirkelite
SA glasses
lapilli
pyroclastics
pyrolusite
Also search polianite.
UF polianite
BA oxides
BT minerals
pyrolysis
Includes use on level 3 under geochemistry(1) processes(2).
SA geochemistry
pyrometasomatism
BA metasomatism
SA hydrothermal alteration
pyromorphite
BZ halides
phosphates
BT minerals
SA apatite
chlorides
pyrope
BA garnet group
orthosilicates
silicates
BT minerals
pyrophyllite
BA sheet silicates
silicates
BT minerals
pyroxene
BA pyroxene group
chain silicates
silicates
BT minerals
pyroxene andesite
See list H.
BA andesite-rhyolite family
BT igneous rocks
SA andesite
pyroxene granulite
See list J.
BA granulites
BT metamorphic rocks
SA granulite
pyroxene group
Includes use in combination with chain silicates (i.e. chain silicates,pyroxene group) on level 2 under minerals(1). See list L. Also search pyroxenes.
UF pyroxenes
BA chain silicates
silicates
BT minerals

NA acmite
aegirine
augite
bronzite
chrome diopside
clinoenstatite
clinohypersthene
clinopyroxene
diopside
enstatite
eulite
fassaite
ferrosilite
hedenbergite
hypersthene
jade
jadeite
johannsenite
kunzite
omphacite
orthoenstatite
orthopyroxene
pigeonite
protoenstatite
pyroxene
pyroxferroite
spodumene
titanaugite
pyroxenes
use pyroxene group
pyroxenite
Includes use on level 3 under igneous rocks(1) ultramafic family(2). See list H.
BA ultramafic family
BT igneous rocks
SA bronzitite
clinopyroxenite
garnet pyroxenite
metapyroxenite
websterite
pyroxferroite
BA pyroxene group
chain silicates
silicates
BT minerals
pyroxmangite
BA chain silicates
silicates
BT minerals
Pyrrhophyta
SA Dinoflagellata
palynomorphs
pyrrhotine
use pyrrhotite
pyrrhotite
Also search pyrrhotine.
UF dipyrite
pyrrhotine
BA sulfides
BT minerals
SA troilite

Q

Q
Ratio of the peak seismic energy in a cycle to the energy dissipated.
SA earthquakes
elastic waves
Love waves
seismology
Q waves
use Love waves
Qaraqum
use Karakum
Qatar
Includes use on level 1 as an area

term (list O). For term set options see list B.
BA Arabian Peninsula
NT Nakhla
SA Asia
Middle East
Near East
Qizil Qum
use Kyzylkum
qualitative analysis
SA analysis
chemical analysis
quantitative analysis
quality
A valid general term through 1978. After 1978 use specific term, e.g. water quality.
quantitative analysis
Includes use on level 3 under chemical analysis(1) methods(2).
SA analysis
chemical analysis
flame photometry
methods
neutron activation analysis
polarography
qualitative analysis
quantitative geomorphology
Includes use on level 3 under geomorphology(1).
SA geometry
geomorphology
quantitative methods
Term introduced in 1978. Before 1978, search quantitative, or methods AND quantitative.
SA methods
quarries
Includes use on level 3 under commodities (list C).
SA mines
mining geology
strip mining
quartz
BA silica minerals
framework silicates
silicates
BT minerals
SA agate
amethyst
chalcedony
cristobalite
light minerals
quartz crystal
quartz crystal
Includes use on level 1 as a commodity term (list C).
UF berg crystal
crystal, quartz
mountain crystal
rock crystal
SA quartz
quartz diabase
See list H.
BA basalt family
BT igneous rocks
SA diabase
quartz diorite
Includes use on level 3 under igneous rocks(1) diorite family(2). See list H.
BA diorite family
BT igneous rocks
SA diorite
plagiogranite
tonalite
quartz dolerite
See list H.
BA basalt family
BT igneous rocks
SA dolerite

quartz keratophyre
See list H.
BA trachyte-phonolite family
BT igneous rocks
SA keratophyre

quartz latite
Includes use on level 3 under igneous rocks(1) andesite-rhyolite family(2). See list H.
BA andesite-rhyolite family
BT igneous rocks
SA latite
rhyodacite

quartz monzonite
Includes use on level 3 under igneous rocks(1) granite-granodiorite family(2). See list H.
BA granite-granodiorite family
BT igneous rocks

quartz porphyry
Includes use on level 3 under igneous rocks(1) andesite-rhyolite family(2). See list H.
BA andesite-rhyolite family
BT igneous rocks
SA beresite
porphyry

quartz sand
See list N.
BA clastic sediments
BT sediments
SA sand

quartz syenite
Includes use on level 3 under igneous rocks(1) syenite family(2). See list H.
BA syenite family
BT igneous rocks
SA plutonic rocks
syenite

quartzite
Compositional term.
BA metamorphic rocks
SA ferruginous quartzite

Quaternary
Includes use on level 1 as an age term (list E); on level 2 under paleoterms, e.g. paleoecology, paleogeography, paleomagnetism. Consists of Pleistocene and Holocene.
BA Cenozoic
NA Holocene
Jaramillo Event
lower Quaternary
middle Quaternary
Pleistocene
upper Quaternary
SA Blancan
Magdalenian
Matuyama Epoch
Mousterian
Paleolithic
Tertiary
Tulare Formation
Uonuma Group
upper Cenozoic
Verde Formation
Villafranchian

Quebec
Includes use on level 1 as an area term (list O). For term set options see list B.
CO N450000N630000
W0570000W0790000
BA Canada
NT Abitibi
Anticosti Island
Chibougamau
Dore Lake Complex
Gaspe
Gaspe Peninsula
Gatineau Valley
Horne Mine
Hull
Kipawa Lake
Manicouagan
Matagami
Montreal
Morin Complex
Noranda
Quebec City
Rouyn
Saguenay Valley
Saint Lawrence Estuary
Schefferville
Sept-Iles
Thetford Mines
Val d'Or
SA Canadian Shield
Chaleur Bay
Champlain Valley
Charlevoix
Churchill Province
Cloridorme Formation
Frontenac County
Green Mountains
Grenville Front
Grenville Province
Hudson Bay Lowlands
Lake Champlain
Lake Timiskaming
Levis Shale
Lorrain Formation
Nipissing Diabase
Ottawa Valley
Restigouche Estuary
Saint Lawrence Lowlands
Saint Lawrence River
Saint Lawrence Valley
Sokoman Formation
Timiskaming
Ungava

Quebec City
City on the Saint Lawrence River near where the river widens into its estuary.
BT Quebec
Canada

Queen Alexandra Range
W of the S Ross Ice Shelf and NW of Queen Maud Range on the Pacific Ocean side.
BT Antarctica

Queen Charlotte Islands
Group of islands in Pacific Ocean separated from mainland by Hecate Strait.
BT British Columbia
Canada

Queen City Formation
In Claiborne Group. Includes Arp Member, Omen Glauconitic Sandstone Member, and unnamed upper sand member. NW Louisiana and E Texas.
BT middle Eocene
Eocene
SA Claiborne Group
Louisiana
Texas

Queen Elizabeth Islands
Large group in Arctic Ocean N of Parry Channel in District of Franklin.
BT Northwest Territories
Canada
SA Amund Ringnes Island

Queen Maud Mountains
use Queen Maude Range

Queen Maude Land
Large region between Coats Land and Enderby Land in Norwegian Sector on the Atlantic Ocean side. Also search Dronning Maud Land.
UF Dronning Maud Land
BT Antarctica

Queen Maude Range
SSW of S tip of Ross Ice Shelf on the Pacific Ocean side.
UF Queen Maud Mountains
BT Antarctica
SA Beardmore Glacier

Queensland
Includes use on level 1 as an area term (list O). For term set options see list B.
CO S290000S100000
E1530000E1380000
BA Australia
NT Adavale Basin
Bowen
Bowen Basin
Brisbane
Burdekin Delta
Burdekin River
Cape York Peninsula
Charters Towers
Darling Downs
Eromanga Basin
Heron Island
Ipswich
Maryborough Basin
Moreton
Moreton Bay
Mount Isa
Rockhampton
Selwyn Range
Springsure
Surat Basin
Townsville
Vale of Glamorgan
Weipa
Yarrol Basin
SA Baralaba Coal Measures
Broken River Formation
Carpentaria Basin
Cooper Basin
Georgina Basin
Great Artesian Basin
Ipswich Coal Measures
New England Batholith
Simpson Desert
Tasman Geosyncline

queenstownite
use Darwin glass

Quercus
Genus.
BA Dicotyledoneae
angiosperms
BT Plantae

Quercy
Region in SW France. Index departments as applicable.
BT France
SA Lot
Tarn-et-Garonne

Querigut Massif
In the eastern Pyrenees. Also search Querigut.
UF Querigut Mountains
BT Ariege
France

Querigut Mountains
use Querigut Massif

Quesnel Lake
SE British Columbia.
BT British Columbia
Canada

Questa Mine
Taos County in N New Mexico. Also search Questa.
BT New Mexico
United States

quick clay
Also search quickclays; quick clays.
UF quickclays
SA clays
engineering geology

quickclays
use quick clay

quicksilver
use mercury

Quillan
Town in S part of country.
BT Aude
France

Quinqueloculina
Genus. Includes use on level 3 under foraminifera(1) Miliolacea(2).
BA Miliolacea
Miliolina
foraminifera
BT Invertebrata

Quintana Roo
Territory on the Caribbean Sea on E Yucatan Peninsula.
BT Mexico
SA Gulf Coastal Plain
Yucatan Peninsula

Quirke Lake
N of Lake Huron.
BT Ontario
Canada

Quseir
use Kosseir

R

R waves
use Rayleigh waves

Ra
use radium

Ra-226
Includes use on level 3 under isotopes(1).
SA isotopes
radium

Raasay
Island of the Inner Hebrides NE of Skye Island off NW.
BT Scotland
Great Britain
United Kingdom
SA Inner Hebrides

Rabat
City on the Atlantic Ocean in NW.
BT Morocco

Rabaul Caldera
On New Britain Island.
BT Bismarck Archipelago
Papua New Guinea

Raca
use Racha

racemization
Includes use on level 2 under geochronology(1).
SA amino acids
geochronology

Racha
Village in central Serbia.
UF Raca
BT Serbia
Yugoslavia

Raciborz
City on Oder River in SW part of country.
UF Ratibor
BT Opole
Poland

radar
A valid term through 1977. After 1977, use radar methods.

radar methods
Term introduced in 1978. Includes use on level 3 under geophysical methods(1) or geophysical surveys(1): Before 1978 search radar.
- BA geophysical methods
  - geophysical surveys
- SA methods
  - remote sensing
  - sonar methods

radial faults
Term introduced in 1978. Includes use on level 3 under faults(1) patterns(2). Before 1978, also search faults AND radial.
- BA faults

radiation
As of 1978, term is restricted for use on level 3 under paleontology(1) for evolution. Before 1978, term could also have been used for radioactivity.
- SA electromagnetic radiation
  - paleontology
  - particle radiation
  - radioactivity

radiation damage
Includes use on level 2 under geochronology(1).
- UF damage, radiation
- SA fission tracks
  - fission-track dating
  - geochronology
  - metamict
  - metamictization
  - particle track
  - radioactivity

radio-frequency spectroscopy
Term introduced in 1978. Includes use on level 3 under spectroscopy(1) methods(2). Before 1978, also search spectroscopy AND radio frequency. See list D.
- BT spectroscopy
- SA analysis
  - methods

radio-wave methods
Term introduced in 1978. Before 1978, search radio waves.
- SA electromagnetic radiation
  - electromagnetic waves
  - methods

radioactivation analysis
use activation analysis

radioactive decay
Term introduced in 1978. Includes use on level 3 under isotopes(1) or absolute age(1). Before 1978, search decay AND radioactive.
- UF decay, radioactive
- SA absolute age
  - isotopes

radioactive elements
use radioactive isotopes

radioactive isotopes
Includes use on level 3 under isotopes(1). Also search radioactive elements; radionuclides.
- UF radioactive elements
- BA isotopes
- SA radioactivity

radioactive tracers
Includes use on level 3 under meteorology(1) aerosols(2); and under isotopes(1).
- SA aerosols
  - isotopes
  - radioactivity
  - tracer experiments
  - tracers

radioactive waste
As of 1978, term is used on level 2 under waste disposal(1).
- UF waste, radioactive
- SA engineering geology
  - environmental geology
  - industrial waste
  - liquid waste
  - pollution
  - radioactivity
  - solid waste
  - storage
  - waste disposal

radioactivity
For properties, term is used on level 2 and 3. Includes use on level 2 under well-logging(1); on level 3 under engineering geology(1) waste disposal(2), and under soils(1). Natural or artificial is indicated. Before 1978, also search radiation.
- SA alpha-ray spectroscopy
  - autoradiography
  - electromagnetic radiation
  - engineering geology
  - fallout
  - fission tracks
  - flux
  - gamma rays
  - gamma-gamma methods
  - gamma-ray methods
  - gamma-ray spectroscopy
  - isotopes
  - pollution
  - radiation
  - radiation damage
  - radioactive isotopes
  - radioactive tracers
  - radioactive waste
  - radioactivity methods
  - radioactivity surveys
  - radiometric properties
  - scintillations
  - tracer experiments
  - tracers
  - waste disposal
  - well-logging

radioactivity methods
Includes use on level 2 under geophysical methods(1).
- BA geophysical methods
- SA methods
  - radioactivity
  - radioactivity surveys

radioactivity surveys
Includes use on level 2 under geophysical surveys(1).
- BA geophysical surveys
- BT surveys
- SA radioactivity
  - radioactivity methods

radiography
use autoradiography

radiography, X-ray
use X-ray radiography

Radiolaria
- BT Invertebrata
- NA Osculosida
  - Porulosida
  - Spumellina
- SA plankton
  - Protista

radiolarite
Includes use on level 3 under sedimentary rocks(1) clastic rocks(2). See list I.
- BA clastic rocks
- BT sedimentary rocks
- SA nonterrigenous materials

radiometric properties
Term introduced in 1978. Includes use on level 3 under Moon(1) surface properties(2). Before 1978, also search radiometric; radiometry.
- UF radiometry
- SA absolute age
  - geochronology
  - isotopes
  - properties
  - radioactivity
  - surface properties

radiometry
use radiometric properties

radionuclides
A valid term through 1978. After 1978 use radioactive isotopes.

radium
Includes use on level 1 and 2 as a chemical element (list D).
- UF Ra
- SA elements
  - isotopes
  - Ra-226
  - uranium

Radnorshire
County in E.
- BT Wales
  - Great Britain
  - United Kingdom

radon
Includes use on level 1 and 2 as a chemical element (list D).
- UF Rn
- BT noble gases
- SA argon
  - elements
  - ionium
  - isotopes
  - krypton
  - neon
  - Rn-222

radon-222
use Rn-222

Radstock Bay
SW coast of Devon Island on Lancaster Sound in District of Franklin.
- BT Northwest Territories
  - Canada

Raft River
In N Utah and S Idaho. Index states as applicable.
- BT United States
- SA Idaho
  - Raft River basin
  - Utah

Raft River basin
Index states as applicable. Also search Raft River.
- BT United States
- SA Idaho
  - Raft River
  - Utah

rafting, ice
use ice rafting

Ragavapuram (Shales)
use Raghavapuram Shales

Raghavapuram Shales
Also search Ragavapuram.
- UF Ragavapuram (Shales)
- BT Upper Jurassic
  - Jurassic
- SA Andhra Pradesh
  - India

rain
use rainfall

Rainbow Mountain
Peak in Salmon River Mountains in Valley County in W central.
- BT Idaho
  - United States

raindrops
Includes use on level 3 under meteorology(1) electrical phenomena(2) and water(2).
- SA atmospheric precipitation
  - clouds
  - droplets
  - electrical phenomena
  - humidity
  - meteorology
  - rainfall
  - water

rainfall
Includes use on level 3 under hydrology(1). Also search rain.
- UF rain
- SA atmospheric precipitation
  - hydrology
  - raindrops
  - watersheds

Rainy Lake
Index Minnesota and/or Ontario.
- SA Minnesota
  - Ontario

Rainy River
Flows from Rainy Lake to Lake of the Woods. Index Minnesota and/or Ontario as applicable.
- SA Minnesota
  - Ontario

Raipur
City in SE.
- BT Madhya Pradesh
  - India

Rajahmundry
City in NE.
- BT Andhra Pradesh
  - India

Rajasthan
State in NW.
- BT India
- NT Ajmer
  - Alwar
  - Aravalli Range
  - Barmer
  - Bhilwara
  - Bikaner
  - Jaipur
  - Jaisalmer
  - Jhunjhunu
  - Jodhpur
  - Khetri
  - Khetri copper belt
  - Sirohi
  - Udaipur
- SA Aravalli System
  - Kaimur Sandstone
  - Lathi Formation
  - Trans-Aravalli Vindhyan Basin

Rajgir
Village in N central.
- BT Bihar
  - India

Rajmahal
Town in E.
- BT Bihar
  - India

Rajmahal Hills
Low range of hills S and W of the Ganges River in E.
- BT Bihar
  - India

Rakhov
Town in Transcarpathian Oblast in W.
- BT Ukraine
  - USSR

Rakhov Massif
In central Carpathian Mountains of Transcarpathian Oblast. Also search Rakhov.
- BT Ukraine
  - USSR

Raman spectroscopy
Term introduced in 1978. Includes use on level 3 under spectroscopy(1) methods(2) and under chemical element(1). Before 1978, also search spectroscopy AND Raman.
- BT spectroscopy
- SA analysis
  - spectral analysis

Ramapithecus
Genus. Includes use on level 3 under Mammalia(1) Primates(2).
BA Hominidae
   Primates
   Mammalia
BT Tetrapoda
   Vertebrata

Ramgarh coal field
S Bihar, India. Also search Ramgarh.
UF Ramgarh Coalfield
BT Bihar
   India

Ramgarh Coalfield
use Ramgarh coal field

Rammelsberg
Mountain in the Harz Mountains.
BT Lower Saxony
   West Germany

rammelsbergite
BA arsenides
   sulfides
BT minerals
SA pararammelsbergite

Ramnagar
Town in SE Uttar Pradesh, India, and village NW of Lahore in Punjab, Pakistan. Index countries as applicable.
BT Asia
SA India
   Pakistan

Rampart Group
Probably Lower Mississippian. Yukon-Tanana region in E central and E Alaska.
BT Mississippian
SA Alaska
   Lower Mississippian

Ramparts Formation
Mackenzie District. Also search Ramparts.
BT Devonian
SA Canada
   Northwest Territories

Rampur coal field
NW Orissa. Also search Rampur.
BT Orissa
   India

ramsdellite
BA oxides
BT minerals

Ran
use Rana Fjord

Rana Fjord
On the Norwegian Sea in central part of country. Also search Rana.
UF Ran
   Ranen Fjord
BT Nordland
   Norway

Ranchi
City and District in S.
BT Bihar
   India

rancieite
BA oxides
BT minerals

Rand
use Witwatersrand

Rand Mountains
Range in the Mohave Desert. Also search Rand.
BT California
   United States

Randolph County
Index states as applicable.
BX Alabama
   Arkansas
   Georgia
   Illinois
   Indiana
   Missouri
   North Carolina
   West Virginia
BT United States

random processes
use stochastic processes

Ranen Fjord
use Rana Fjord

range
Usage is restricted to stratigraphy. Also search stratigraphic range.
UF stratigraphic range
SA biogeography
   biostratigraphy
   fossils
   species
   stratigraphy
   taxonomy

range, basin
use basin range

Rangeley Lakes
Index states as applicable. Also search Rangeley.
BT United States
SA Maine
   New Hampshire

Rangely
Town in Rio Blanco County in NW.
BT Colorado
   United States

Rangely Anticline
In NW near Utah border. Also search Rangely.
BT Colorado
   United States

Rangely Field
Term introduced in 1978. In NW near Utah border. Before 1978, also search Rangely; Rangely oil field.
UF Rangely oil field
BT Colorado
   United States

Rangely oil field
use Rangely Field

Ranger
Town in Eastland County in central.
BT Texas
   United States

Rangitata River
E central South Island. Also search Rangitata.
BT South Island
   New Zealand

Raniganj
City on N bank of Damodar River.
BT West Bengal
   India

Raniganj (Stage)
use Raniganj Stage

Raniganj coal field
Also search Raniganj and Raniganj Coalfield.
CO N233500N233900
   E0870300E0870900
UF Raniganj Coalfield
BT West Bengal
   India

Raniganj Coalfield
use Raniganj coal field

Raniganj Stage
The highest division of The Damuda Series. Also search Raniganj.
UF Raniganj (Stage)
BT Upper Permian
   Permian
SA Damuda Series
   India
   West Bengal

rank
To be used only for classification of coal.
SA coal

Rankin County
S central.
BT Mississippi
   United States

Raoul Island
Largest of the volcanic Kermadec Islands which are a dependency of New Zealand 500 miles NE of Auckland.
BT Pacific Ocean
SA Kermadec Islands

rapakivi
Includes use on level 3 under igneous rocks(1) granite-granodiorite family(2). See list H.
UF wiborgite
BA granite-granodiorite family
BT igneous rocks

Raphidiodea
Includes use on level 2 under Insecta(1). See list F.
BA Insecta
BT Arthropoda
   Invertebrata
SA Diptera
   Hymenoptera
   Lepidoptera

Rapid City
City in Pennington County in SW.
BT South Dakota
   United States

Rapides Parish
Central.
BT Louisiana
   United States

Rappahannock River
Rises in the Blue Ridge Mountains and flows into Chesapeake Bay.
BT Virginia
   United States

Rarau Massif
Suceava County in N Moldavia. Also search Rarau.
UF Rarau Mountains
BT Moldavia
   Romania

Rarau Mountains
use Rarau Massif

rare earths
Includes use on level 1 as a commodity term (list C). Also search lanthanides.
UF inner transition elements
   lanthanide series
   lanthanides
   lanthanoans
SA cerium
   dysprosium
   elements
   erbium
   europium
   gadolinium
   holmium
   KREEP
   lanthanum
   lutetium
   neodymium
   praesodymium
   promethium
   samarium
   scandium
   thulium
   ytterbium
   yttrium

rare gases
use noble gases

rare metals
BT metals

Raritan Formation
Comprises Amboy Stonewane Clay, Old Bridge Sand Member, South Amboy Fire Clay, Sayreville Sand Member, Woodbridge Clay, Farrington Sand Member, and Raritan Fire Clay.
BT Upper Cretaceous
   Cretaceous
SA Delaware
   Maryland
   New Jersey

Ras al-Khaimah
Includes use on level 3 as an area term (list O).
BT United Arab Emirates

rasorite
use kernite

Rat Island
Small island in center of Rat Islands in W Aleutian Islands.
BT Alaska
   United States
SA Aleutian Islands

Rat Islands
Group in W Aleutian Islands.
BT Alaska
   United States
SA Aleutian Islands

rates
Includes use as a level 2 or 3 term appropriate to a large number of topics, e.g. on level 2 under sedimentation(1). See list G.

Ratibor
use Raciborz

ratios
Includes use on level 2 under isotopes(1); on level 3 under name of element(1) isotopes(2) and under absolute age(1). See list D. Before 1973, also search isotope ratios.
SA isotopes

Raton Basin
SE Colorado and NE New Mexico. Index states as applicable. Also search Raton.
BT United States
SA Colorado
   New Mexico

Rattlesnake Mountain
BT Montana
   United States

Raukumara Peninsula
NE North Island.
BT North Island
   New Zealand

Rauracian
Europe. Above Argovian, below Sequanian. Includes use on level 3 under age terms(1). See list E.
BA Upper Jurassic
   Jurassic
BT Mesozoic
SA Lusitanian

Ravalli Group
Belt Series. Includes Altyn Formation, Appekunny Formation, Grinnell Formation. NE Idaho, and NW Montana.
BT Precambrian
SA Altyn Limestone
   Belt Supergroup
   Idaho
   Montana

Ravenscrag Formation
BT Paleocene
SA Canada
   Saskatchewan

raw humus
use Mor

Rawlins County
NW Kansas.
BT Kansas

United States

Rayleigh waves
Includes use on level 3 under seismology(1).
UF R waves
BT surface waves
SA elastic waves
    seismology

rays, alpha
use alpha rays

rays, cosmic
use cosmic rays

rays, gamma
use gamma rays

Razdan
River which serves as an outlet of Lake Sevan.
UF Zanga
BT Armenia
    USSR

razor stone
use novaculite

Rb
use rubidium

Re
use rhenium

reactions
Used only for chemical reactions. Also search chemical reactions.
UF chemical reactions
SA endothermic reactions
    geochemistry

reactivation
Includes use on level 3 under faults(1) displacements(2).
SA displacements
    faults

Read Bay Formation
Middle Silurian to lower Devonian. Cornwallis Island in Franklin District. Also search Read; Read Bay.
BT Paleozoic
SA Canada
    Devonian
    Lower Devonian
    Middle Silurian
    Silurian
    Upper Silurian

Reading
City in Berks County in SE.
BT Pennsylvania
    United States

Reagan Sandstone
In Timbered Hills Group. Central S Oklahoma.
BT Upper Cambrian
    Cambrian
SA Oklahoma

reagents
SA geochemistry
    solutions

realgar
UF red arsenic
    red orpiment
    sandarac
BA sulfides
BT minerals

Reasi
use Riasi

Recent
Not a valid index term for GeoRef. See Holocene.

Receptaculitaceae
BA problematic fossils
SA algae
    Coelenterata
    Echinodermata
    foraminifera
    Porifera

recharge
Includes use on level 2 under ground water(1).
UF ground-water increment
    ground-water recharge
    ground-water replenishment
    increment
SA aquifers
    artificial recharge
    discharge
    ground water
    infiltration

Rechitsa
City in Gomel Oblast in S.
BT Byelorussia
    USSR

Recife-Joao Pessoa
Two city region in neighboring states in NE. Recife is in Pernambuco and Joao Pessoa is in Paraiba.
BT Brazil

Recita
use Resita

reclamation
A level 1 term as of 1978. Used for geological studies on the reclamation of the natural environment. Includes use on level 2 under impact statements(1) and land use(1). Also search land reclamation. If 1, term set options are:
  environment
  beaches
  drainage basins
  mines
  open-pit mining
  strip mining
  natural resources
  type (floods, land, soils, waste water, etc.)
  topic (experimental studies, methods, practice, programs, surveys)
  subtopic
SA conservation
    deforestation
    depletion
    environment
    environmental geology
    impact statements
    land leases
    land use
    mining geology
    natural resources
    pollution
    programs
    recovery
    soils
    strip mining
    tailings
    treatment
    utilization
    waste disposal

reclined
A valid term through 1976. After 1976, use recumbent folds.

reclined folds
use recumbent folds

Reconcavo Basin
Fertile coastal lowland surrounding Todos os Santos Bay W of Salvador. Also search Reconcavo.
BT Bahia
    Brazil

reconstruction
Includes use on level 3 under paleogeography(1) or continental drift(1).
SA continental drift
    paleogeography

recording
A valid general term through 1976.

recovery
A general term used mostly for mining geology(1) or commodity sets (list C). Use secondary or tertiary recovery for petroleum; use water recovery under ground water(1).
SA mining
    mining geology
    reclamation
    secondary recovery
    tertiary recovery
    water recovery

recreation
As of 1978, term is used on level 2 under land use(1).
SA land use
    soils

recrystallization
Includes use on level 3 under deformation(1) field studies(2).
SA authigenesis
    crystallization
    deformation
    grains
    lithification
    metamorphism
    phyllonitization

rectorite
Also search allevardite.
UF allevardite
BA sheet silicates
    silicates
BT minerals
SA clay minerals

recumbent
A valid term through 1977. After 1977, use recumbent folds.

recumbent folds
Term introduced in 1978. Includes use on level 3 under folds(1) orientation(2). Before 1978, search folds AND recumbent; folds AND reclined.
UF reclined folds
BA folds
SA orientation

red algae
use Rhodophyta

red arsenic
use realgar

red beds
Includes use on level 3 under sedimentary rocks(1) clastic rocks(2). See list I. Also search redbeds.
UF red rock
    redbeds
BA clastic rocks
BT sedimentary rocks
SA ferruginous composition
    sandstone
    terrigenous materials

Red Deer River
Rises in Banff National Park and flows into South Saskatchewan River near the Saskatchewan border.
BT Alberta
    Canada

Red Eagle Limestone
In Council Grove Group. Includes Glenrock Limestone Member, Bennett Shale Member, Howe Limestone Member.
BT Permian
SA Kansas
    Missouri
    Nebraska
    Oklahoma

Red Fork Sandstone
Said to lie higher than Glenn Sand, lower than Skinner Sand. NE Oklahoma.
BT Pennsylvanian
SA Oklahoma

Red Hill
Mountain in Haleakala National Park on Maui Island.
BT Hawaii
    United States

Red Lake
Beltrami County in N. Includes both Upper Red Lake and Lower Red Lake.
BT Minnesota
    United States

Red Mountain
Ridge primarily in Jefferson County in N central.
BT Alabama
    United States

Red Mountain Formation
Overlies Chickamauga Limestone; underlies Frog Mountain Sandstone. N Alabama, and NW Georgia.
BT Silurian
SA Alabama
    Georgia

red orpiment
use realgar

Red Peak Formation
In Chugwater Group. NW Wyoming.
BT Triassic
SA Wyoming

Red River
Rises in high plains of E New Mexico and flows E and SE into the Mississippi River. Index states as applicable.
BT United States
SA Arkansas
    Louisiana
    New Mexico
    Oklahoma
    Texas

Red River Formation
In Bighorn Group.
BT Upper Ordovician
    Ordovician
SA Manitoba
    Montana
    North Dakota
    South Dakota
    Wyoming

Red River valley
Index states as applicable.
BT United States
SA Arkansas
    Louisiana
    New Mexico
    Oklahoma
    Texas

red rock
use red beds

Red Sea
Between NE Africa and the Arabian Peninsula connecting the Mediterranean Sea with the Indian Ocean. Includes use on level 1 or 2 as an area term (list O). If 1, see term set options under list B.
CO N100000N300000
    E0430000E0340000
NT Atlantis II Deep
    Chain Deep
    Discovery Deep
    Gulf of Aqaba
    Gulf of Suez
SA Great Rift Valley
    Red Sea Basin
    Red Sea region

Red Sea Basin
Index Red Sea, Djibouti and countries as applicable.
BT Middle East
SA Djibouti
    Egypt
    Ethiopia
    Israel
    Jordan
    Red Sea

Red Sea region
Saudi Arabia
Southern Yemen
Sudan
Yemen

Red Sea region
Introduced as a level 1 and 2 area term (list O) in 1976. This is an artificial term used to indicate the coastal region immediately adjacent to the Red Sea and the immediate littoral zone. For term set options see list B.
SA Red Sea
Red Sea Basin

Red soils
BA soils
SA Zonal soils

red zinc ore
use zincite

redbeds
use red beds

Redlichiida
Includes use on level 2 under Trilobita(1). See list F.
BA Trilobita
BT Trilobitomorpha
Arthropoda
Invertebrata

Redon
Town in E Brittany.
BT France

Redonda Formation
Unconformably overlies Chinle Formation; unconformably underlies Entrada Sandstone. NE New Mexico.
BT Upper Triassic
Triassic
SA New Mexico

Redondo Canyon
Off S California.
UF Redondo submerged valley
BT Pacific Ocean

Redondo submerged valley
use Redondo Canyon

Redonian
North America.
BA upper Pliocene
Pliocene
Neogene
Tertiary
BT Cenozoic

redox
use Eh

redox potential
use Eh

reduction
Includes use on level 3 under geochemistry(1) processes(2).
SA Eh
geochemistry
oxidation
processes

Redwall Limestone
Comprises four unnamed members. N Arizona.
BT Mississippian
SA Arizona

reedmergnerite
BA framework silicates
silicates
BT minerals

reef pinnacle
use pinnacle reefs

reefs
Includes use on level 1 (list A); on level 3 for type of environment under sedimentation(1). Before 1974, also search coral reefs. If used on level 1, term set options are:
topic [age, distribution, ecology, evolution, lithofacies, paleoecology, structure]
subtopic (no area term)
NT atolls
barrier reefs
fringing reefs
patch reefs
pinnacle reefs
SA biogenic structures
bioherms
biostromes
changes of level
continental shelf
corals
ecology
geomorphology
hermatypic taxa
lagoons
lithofacies
marine geology
paleoecology
sedimentary rocks
sedimentary structures
sedimentation
shoals
spits

Reefton
Village in NW.
BT South Island
New Zealand

Reeves County
W Texas.
BT Texas
United States

refinement
As of 1978, term is used on level 2 under crystal structure(1).
SA crystal structure

reflectance
Used when discussing study of cores and organic materials (refers to a specific sample). For ore deposits, use reflectivity. When discussing methods, use microscope methods. Includes use on level 3 under methods(2) or properties(2) under appropriate material or discipline, e.g. under minerals(1), mineralogy(1), and crystallography(1).
SA microscope methods
mineralogy
minerals
organic materials
reflectivity

reflection
Includes use on level 3 under seismology(1) and under geophysical surveys(1). Used on level 2 under ocean waves(1) through 1977. After 1977, use reflection waves(2) under ocean waves(1).
SA diffraction
geophysical surveys
reflection waves
reflectograms
refraction
seismic surveys
seismology

reflection methods
Term introduced in 1978. Before 1978, search reflection AND seismic methods or geophysical methods.
BA seismic methods
BT geophysical methods
SA reflectograms

reflection waves
Term introduced in 1978. Includes use on level 2 under ocean waves(1). Before 1978, also search ocean waves AND reflection.
SA ocean waves
reflection
waves

reflectivity
For ore deposits. Does not include reflectivity of a bed of rocks to seismic waves. Includes use on level 3 under methods(2) or properties(2) under appropriate material or discipline, e.g. under minerals(1), and crystallography(1). When discussing methods, use microscope methods. Use reflectance when referring to a specific sample.
SA albedo
crystallography
microscope methods
mineralogy
minerals
ore deposits
reflectance

reflectograms
Term introduced in 1978.
SA reflection
reflection methods

refraction
Includes use on level 3 under seismology(1) and under geophysical surveys(1). Used on level 2 under ocean waves(1) through 1977. After 1977, use refraction waves(2) under ocean waves(1).
SA birefringence
diffraction
geophysical surveys
reflection
refraction waves
refractive index
seismic surveys
seismology

refraction methods
Term introduced in 1978. Before 1978, search refraction AND seismic methods or geophysical methods.
BA seismic methods
BT geophysical methods

refraction waves
Term introduced in 1978. Includes use on level 2 under ocean waves(1). Before 1978, also search ocean waves AND refraction.
SA ocean waves
refraction
waves

refractive index
Includes use on level 3 under minerals(1). Also search index of refraction; refractivity.
UF index of refraction
refractivity
SA birefringence
minerals
optical properties
refraction

refractivity
use refractive index

refractory clay
use fireclay

refractory materials
Includes use as level 3 commodity term under ceramic materials(1). See list C.
SA ceramic materials
clays
construction materials
fireclay
flint clay
graphite
industrial minerals
kaolin
magnesite
materials

Refugian
North America. Above Fresnian, below Zemorrian. Includes use on level 3 under age terms(1). See list E.

BA Eocene
Paleogene
Tertiary
BT Cenozoic
SA Oligocene

Regensburg
City on Danube River in E central.
BT Bavaria
West Germany

Reggio
use Reggio di Calabria

Reggio Calabria
use Reggio di Calabria

Reggio di Calabria
City and province on Strait of Messina.
UF Reggio
Reggio Calabria
BT Calabria
Italy

regimes, water
use water regimes

Regina
City in S.
BT Saskatchewan
Canada

regional
After 1977, used only as a general term for geography. A valid level 2 term through 1977 used in combination with metamorphism(1); after 1977, regional metamorphism is used on level 2.

regional metamorphism
Term introduced in 1978. Includes use on level 2 under metamorphism(1). Before 1978, also search metamorphism AND regional.
BA metamorphism
SA burial metamorphism
dynamic metamorphism

regional patterns
Includes use on level 2 under heat flow(1).
SA geothermal gradient
heat flow
heat sources
patterns

regional planning
As of 1978, term is used on level 2 under land use(1).
SA environmental geology
land use
planning
urban planning

regolith
Includes use on level 3 under planet sets(1).
SA debris
fragments
overburden
soils

regression
Term is used in stratigraphy for Pre-Quaternary sedimentation. For Quaternary, use changes of level. For statistical meaning, use regression analysis.
SA changes of level
regression analysis
sedimentation
stratigraphy
transgression

regression analysis
Includes use on level 3 under mathematical geology(1).
SA analysis
least-squares analysis
mathematical geology
regression
statistical analysis

Reguibat Ridge
NW Mauritania and NE Spanish Sahara. Index countries as applicable.
BT Africa
SA Mauritania

Rehamma
use Rehamna

Rehamna
Tribal area N of Marrakech in W central Morocco. Also search Rehamma.
UF Rehamma
BT Morocco

Reichenstein
use Zloty Stok

relation
A valid general term through 1977.

relative age
Includes use on level 3 under geochronology(1).
SA absolute age
    age
    geochronology

release fractures
Term introduced in 1978. Includes use on level 3 under fractures(1) genesis(2). Before 1978, also search fractures AND release.
BA fractures
SA joints
    stress

reliability
Used as a general term.
SA accuracy
    errors
    probability
    sampling
    statistical analysis

relief
Includes use on level 3 under geomorphology(1) or ocean floors(1).
SA geomorphology
    landform description
    landform evolution
    ocean floors
    paleorelief
    terrains

relief features
use landforms

remagnetization
SA natural remanent magnetization
    paleomagnetism

remanent magnetism
use remanent magnetization

remanent magnetization
Includes use on level 3 under paleomagnetism(1).
UF remanent magnetism
NA anhysteretic remanent magnetization
    chemical remanent magnetization
    depositional remanent magnetization
    isothermal remanent magnetization
    natural remanent magnetization
    thermoremanent magnetization
    viscous remanent magnetization
SA coercivity
    demagnetization
    magnetic field
    magnetic properties
    magnetization
    paleomagnetism

remobilization
A general term associated mostly with petrological terms such as metasomatism.

remote sensing
A level 1 term as of 1978. Used for both methods and applications. Includes use on level 2 under mineral exploration(1) and land use(1). If 1, term set options are:
aerial photography
    subtopic
applications
    subtopic
automatic data processing
    subtopic
imagery
    subtopic
instruments
    subtopic
interpretation
    subtopic
methods
    subtopic
photogeologic methods
    subtopic
UF sensing, remote
SA aerial photography
    airborne methods
    albedo
    Apollo
    Apollo 9
    Apollo 11
    Apollo 12
    Apollo 13
    Apollo 14
    Apollo 15
    Apollo 16
    Apollo 17
    automatic data processing
    engineering geology
    environmental geology
    geophysical methods
    geophysical surveys
    ground truth
    imagery
    infrared methods
    land use
    Landsat
    lineaments
    mineral exploration
    multispectral analysis
    photogeologic methods
    photogrammetry
    radar methods
    satellite methods
    Skylab
    sonar methods
    thermal emission

Rendzinas
Includes use on level 3 under soils(1). See list M.
BA soils

renierite
BA sulfogermanates
    sulfosalts
BT minerals

Renmark
Town on Murray River in SE.
BT South Australia
    Australia

Rennes
City in E Brittany.
BT France

Reno
City near the California border in Washoe County in W.
BT Nevada
    United States

Reno County
S central.
BT Kansas
    United States

Rensselaer County
E New York.
BT New York
    United States

replacement
A valid term through 1975. After 1975, use metasomatism or pseudomorphism under mineralogy(1).

report
Includes use on level 3 under symposia(1). Also search brief report.
NT progress report
SA annual report
    associations
    atlas
    current research
    symposia

reproduction
A general term used on level 3 under paleontology(1) or fossil sets (list F).
SA paleontology
    sexual dimorphism

Reptilia
Includes use on level 1 and 2 as a fossil term (list F).
BT Tetrapoda
    Vertebrata
NA Anapsida
    Archosauria
    dinosaurs
    Euryapsida
    Ichthyopterygia
    Lepidosauria
    Synapsida
SA eggs
    gastroliths

Republic County
On the Nebraska border in N.
BT Kansas
    United States

research
Includes use on level 2 under geology(1).
SA associations
    bibliography
    current research
    geology
    programs
    progress report

reserves
Includes use on level 2 and 3 under commodity terms(1), e.g. on level 2 under coal(1). See list C.
SA ore deposits
    resources

reservoir properties
Used only for petroleum or gas, not for water reservoirs.
SA Darcy's law
    natural gas
    oil and gas fields
    petroleum
    properties
    reservoirs

reservoir rocks
Includes use on level 3.
SA geopressure
    natural gas
    oil and gas fields
    petroleum
    petrophysics
    rocks
    roughness
    sedimentation

reservoirs
A level 1 term as of 1978. Before 1978, included use on level 2 under engineering geology(1). Used for geological studies on surface reservoirs only. For subsurface reservoirs, use petroleum engineering under engineering geology.
construction
    subtopic
dams
    subtopic
design
    subtopic
earthquakes
    subtopic
experimental studies
    subtopic
feasibility studies
    subtopic
field studies
    subtopic
maintenance
    subtopic
seepage
    subtopic
site exploration
    subtopic
storage
    subtopic
SA construction
    dams
    deposition
    design
    drawdown
    earthquakes
    embankments
    engineering geology
    feasibility studies
    floods
    gas storage
    geologic hazards
    hydrogeology
    hydrology
    levels
    maintenance
    Oroville Dam
    ponds
    reservoir properties
    rock mechanics
    seepage
    site exploration
    slope stability
    soil mechanics
    storage
    subsurface reservoirs
    surface reservoirs
    waste disposal
    water balance
    water storage
    water supply
    waterways

residues, insoluble
use insoluble residues

residues, organic
use organic residues

residuum
Includes use on level 3 under sediments(1) clastic sediments(2). See list N.
BA clastic sediments
BT sediments
SA terrigenous materials
    weathering

resinite
BA macerals
SA coal
    organic residues
    sedimentary rocks

resins
Also search fossil resins; mineral resins.
BA organic materials
SA amber

resistivity
Includes use on level 3 under geophysical methods(1) electrical methods(2) and under geophysical surveys(1) electrical surveys(2). Also search electrical resistivity.
SA conductivity
    electrical conductivity

Resita
electrical logging
electrical methods
electrical surveys
geophysical methods
geophysical surveys
well-logging

Resita
City in Caras-Severin County in W part of country.
UF Recita
BT Banat
     Romania

resonance
SA electron paramagnetic resonance
     nuclear magnetic resonance
     oscillations
     vibration

resources
Includes use on level 2 and 3 under commodity terms(1), e.g. on level 2 under energy sources(1). See list C.
SA conservation
     energy sources
     mineral resources
     natural resources
     reserves
     water resources

response
Used as a general term mostly under geophysics(1) or geophysical methods(1) or geophysical surveys(1).

Restigouche Estuary
W extremity of Chaleur Bay. Index provinces as applicable.
BT Canada
SA New Brunswick
     Quebec

retention
The amount of water from precipitation that has not escaped as runoff or through evapotranspiration.
SA atmospheric precipitation
     runoff
     water

Retezat Mountains
Group in W Transylvanian Alps in SW.
BT Transylvania
     Romania
SA Transylvanian Alps

retrieval, data
use data retrieval

retrograde
A valid term through 1977. After 1977, use retrograde metamorphism on level 2 under metamorphism(1).

retrograde metamorphism
Term introduced in 1978. Includes use on level 2 under metamorphism(1). Before 1978, also search metamorphism AND retrograde; diaphthoresis.
UF diaphthoresis
     retrogressive metamorphism
BA metamorphism
SA dynamic metamorphism
     polymetamorphism
     prograde metamorphism

retrogressive metamorphism
use retrograde metamorphism

Reunion
One of the Mascarene Islands and a French overseas territory 425 miles E of the Malagasy Republic. Includes use on level 1 as an area term (list O). For term set options see list B.
CO S205000S212500
     E0560000E0550000
BA Indian Ocean
SA Mascarene Islands
     Saint-Louis

Revelstoke
City in SF.
BT British Columbia
     Canada

reversals
Includes use on level 2 under paleomagnetism(1).
SA magnetic field
     paleomagnetism
     pole positions

reverse
A valid term through 1977. After 1977, use reverse faults.

reverse faults
Term introduced in 1978. Includes use on level 3 under faults(1) displacements(2). Before 1978, also search faults AND reverse.
BA faults
SA displacements
     thrust faults

reverse slip faults
use thrust faults

review
Used as a general term.
SA bibliography
     book reviews

revision
Used for revision of classification and taxonomy, especially in fossils (list F).
SA fossils
     new taxa
     species
     taxonomy

Rewa
City and former state in N.
BT Madhya Pradesh
     India

reworking
Used in relation to fossils.
SA deposition
     fossils
     sediments

Rexroad Formation
Underlies Angell Member of Ballard Formation.
BT upper Pliocene
     Pliocene
SA Kansas
     Oklahoma

Reykjanes Peninsula
WSW of Reykjavik. Also search Reykjanes.
CO N633000N643000
     W0220000W0240000
BT Iceland

Reykjanes Ridge
SW of Iceland.
BT Atlantic Ocean

Reynolds Creek
Flows into Snake River in extreme SW.
BT Idaho
     United States

Rh
use rhodium

rhabdite
use schreibersite

Rhabdomesidae
Family. Includes use on level 3 under Bryozoa(1) Cryptostomata(2).
BA Cryptostomata
     Bryozoa
BT Invertebrata

rhabdophane
BA phosphates
BT minerals

Rhaetian
Europe. Above Norian, below Hettangian (Jurassic). Includes use on level 3 under age terms(1). See list E.
BA Upper Triassic
     Triassic
BT Mesozoic

Rhaetian Alps
Division of Central Alps principally in Graubunden Canton, Switzerland. Index countries as applicable.
BT Alps
SA Austria
     Central Alps
     Italy
     Switzerland

Rhein River
use Rhine River

Rhein Valley
use Rhine Valley

Rheingau
Region on S slope of Rheingau Mountains on right bank of Rhine E of Rudesheim in extreme W.
BT Hesse
     West Germany

Rheinland
use Rhineland

Rhenish Hesse
Administrative division along the left bank of the Rhine River with Worms and Mainz the major cities.
BT Rhineland-Palatinate
     West Germany

Rhenish Schiefergebirge
Rhenish Slate Mountains. Extensive plateau with Belgium and Luxembourg on the W, the Lahn River on the E, the Bonn area in the N, and including the Hunsruck mountain region on the S. Also search Rhenish Slate Mountains and Schiefergebirge. Index states as applicable.
CO N503000N510000
     E0080000E0063000
UF Rhenish Slate Mountains
     Schiefergebirge
BT West Germany
SA Hesse
     North Rhine-Westphalia
     Rhineland-Palatinate
     Westerwald

Rhenish Slate Mountains
use Rhenish Schiefergebirge

rhenium
Includes use on level 1 and 2 as a chemical element (list D).
UF Re
SA elements

rheology
Includes use on level 3 under deformation(1).
SA deformation
     flow
     flow lines
     flow mechanism
     viscosity

rheomorphism
Includes use on level 2 under metamorphism(1).
SA metamorphism

Rhin River
use Rhine River

Rhin Valley
use Rhine Valley

Rhine Basin
Index Lake Constance and countries as applicable. Also search Rhine, and Rhine region.
UF Rhine region
BT Europe
SA Austria
     France
     Lake Constance
     Liechtenstein
     Lower Rhine Basin
     Netherlands
     Switzerland
     West Germany

Rhine Graben
Rift valley extending N from Basel to Mainz. Also search Rhine Trough.
UF Rhine Trough
BT Europe
NT Lower Rhine Graben
     Upper Rhine Graben
SA Germany
     Switzerland

Rhine region
use Rhine Basin

Rhine River
Formed by confluence of the Hinterrhein and the Vorderrhein rivers in SE Switzerland; it flows through Lake Constance and then W, N, and NW to the North Sea. Index Lake Constance and countries as applicable. Also search Rhine.
UF Rhein River
     Rhin River
     Rijn River
BT Europe
SA Austria
     France
     Lake Constance
     Liechtenstein
     Netherlands
     Switzerland
     West Germany

Rhine Trough
use Rhine Graben

Rhine Valley
Commonly recognized as that part of the river valley extending from Basel, Switzerland, to the North Sea. Index countries as applicable. Also search Rhine.
UF Rhein Valley
     Rhin Valley
     Rijn Valley
BT Europe
SA France
     Lower Rhine Basin
     Netherlands
     Switzerland
     Upper Rhine Valley
     West Germany

Rhineland
Region. The part of West Germany W of the Rhine River (left bank). Index states as applicable.
UF Rheinland
BT West Germany
SA North Rhine-Westphalia
     Rhineland-Palatinate

Rhineland-Palatinate
State in W.
CO N490000N510000
     E0083000E0061000
BT West Germany
NT Eifel
     Hunsruck
     Mainz
     Rhenish Hesse
SA Hunsruck Shale
     Lahn River
     Lahn River valley
     Mainz Basin
     Nahe
     Palatinate
     Rhenish Schiefergebirge
     Rhineland
     Saar Basin
     Saar-Nahe Basin
     Westerwald

rhinoceros
use Rhinocerotidae

Rhinocerotidae
Family. Includes use on level 3 under Mammalia(1) Perissodactyla(2). Also search rhinoceros; Rhinocerotoidea.
UF rhinoceros
BA Perissodactyla
Mammalia
BT Tetrapoda
Vertebrata

Rhipidistia
Suborder. Includes use on level 3 under Pisces(1) Osteichthyes(2).
BA Crossopterygii
Osteichthyes
Pisces
BT Vertebrata

Rhode Island
Includes use on level 1 as an area term (list O). For term set options see list B.
CO N411000N420200
W0710700W0715500
BA United States
NT Narragansett Bay
NX Newport
Washington County
SA Atlantic Coastal Plain
Cumberland
Eastern U.S.
New England
Westerly Granite

Rhodes
Largest island of the Dodecanese off SW Turkey.
BT Aegean Islands
Greece
SA Dodecanese

Rhodesia
Formerly Southern Rhodesia. Unilateral declaration of independence from Britain announced on November 11, 1965. Includes use on level 1 as an area term (list O). For term set options see list B. Before 1973, also search Southern Rhodesia.
CO S223000S153000
E0330000E0253000
BA Africa
NT Great Dyke
Kariba
Rhodesian Plateau
Salisbury
Selukwe
SA Bulawayan Group
Lake Kariba
Limpopo Basin

Rhodesian Plateau
Part of great South African plateau.
BT Rhodesia

rhodium
Includes use on level 1 and 2 as a chemical element (list D).
UF Rh
SA elements

rhodochrosite
Also search dialogite.
UF dialogite
BA carbonates
BT minerals

rhodonite
BA chain silicates
silicates
BT minerals

Rhodope
Department in central.
BT Thrace
Greece

Rhodope Massif
use Rhodope Mountains

Rhodope Mountains
Major mountain system primarily in S Bulgaria. Index countries as applicable. Also search Rhodope and Rhodope Massif.
CO N411000N421000
E0253000E0233000
UF Rhodope Massif
BT Europe
SA Bulgaria
Greece

Rhodophyta
Including red algae. Includes use on level 2 under algae(1). See list F.
UF red algae
BA algae
BT Plantae
NA Corallinaceae

Rhoen Mountains
use Rhon Mountains

Rhon Mountains
Index Suhl District and West German states as applicable. Also search Rhon; Rhoen Mountains.
UF Rhoen Mountains
BT West Germany
SA Bavaria
Hesse

Rhone
Department in E central.
BT France
NT Lyons
SA Monts du Lyonnais

Rhone Basin
use Rhone Valley

Rhone Delta
On Gulf of Lion W of Marseilles.
BT Bouches-du-Rhone
France

Rhone River
Rises in Rhone Glacier in S central Switzerland; it flows through Lake Geneva and then W and S into the Gulf of Lion. Index Lake Geneva and countries as applicable.
BT Europe
SA France
Lake Geneva
Switzerland

Rhone River valley
use Rhone Valley

Rhone Valley
Index Lake Geneva and countries as applicable. Also search Rhone Basin; Rhone River valley.
UF Rhone Basin
Rhone River valley
BT Europe
SA France
Lake Geneva
Switzerland

rhonite
BA chain silicates
silicates
BT minerals
SA aenigmatite

Rhum
Isle of Rhum. One of the Inner Hebrides S and SE of the Isle of Skye. Also search Rhum Island; Rum.
UF Rhum Island
Rum
BT Scotland
Great Britain
United Kingdom
SA Inner Hebrides

Rhum Island
use Rhum

rhyacolite
use sanidine

Rhynchonellida
Includes use on level 2 under Brachiopoda(1). See list F.
BA Articulata
Brachiopoda
BT Invertebrata
NA Rhynchonellidae

Rhynchonellidae
Family. Includes use on level 3 under Brachiopoda(1) Rhynchonellida(2).
BA Rhynchonellida
Articulata
Brachiopoda
BT Invertebrata

rhyodacite
Includes use on level 3 under igneous rocks(1) andesite-rhyolite family(2). See list H.
BA andesite-rhyolite family
BT igneous rocks
SA dellenite
quartz latite

rhyolite
Includes use on level 3 under igneous rocks(1) andesite-rhyolite family(2). See list H.
BA andesite-rhyolite family
BT igneous rocks
SA metarhyolite
rhyolitic composition
volcanic rocks

rhyolite porphyry
See list H.
BA andesite-rhyolite family
BT igneous rocks
SA porphyry

rhyolite tuff
See list H.
BA pyroclastics and glasses
BT igneous rocks
SA tuff

rhyolitic composition
Term introduced in 1978. Before 1978, search rhyolitic; rhyolitic AND composition.
SA composition
rhyolite

rhythmic bedding
Includes use on level 3 under sedimentary structures(1) planar bedding structures(2). See list K.
BA planar bedding structures
sedimentary structures
SA bedding
rhythmite

rhythmite
SA cyclothems
rhythmic bedding
sedimentation

Rians
Village in SE part of country.
BT Var
France

Riasi
Town and district in SW Jammu and Kashmir.
UF Reasi
BT Jammu and Kashmir
India

Ribeira de Iguape River
Flows into the Atlantic Ocean near Iguape in S.
BT Sao Paulo
Brazil

Rice County
Index states as applicable.
BX Kansas
Minnesota
BT United States

Rice Lake
Lake in SE Ontario, and lake in Manitoba. Index provinces as applicable.
BT Canada
SA Manitoba
Ontario

Rice Lake Group
BT Archean
lower Precambrian
Precambrian
SA Canada
Manitoba

Richards Island
Large island in Beaufort Sea at mouth of Mackenzie River, District of Mackenzie.
BT Northwest Territories
Canada

Richardson Mountains
Index territories as applicable.
BT Canada
SA Northwest Territories
Yukon Territory

Richat Mountain
E of S border of Spanish Sahara. Also search Richat.
BT Mauritania

Richland
City on the Columbia River in Benton County in S.
BT Washington
United States

Richmond
City on the James River in Henrico County in E central.
BT Virginia
United States

Richmond Group
Includes Queenston Shale, Arnheim Formation, Waynesville Formation, Liberty Formation, Whitewater Formation, Sequatchie Formation, Fernvale Limestone, Marnie Shale, Oswego Sandstone, Juniata Formation, Saluda Formation, Elkhorn Formation.
BT Upper Ordovician
Ordovician
SA Fernvale Formation
Illinois
Indiana
Juniata Formation
Kentucky
Missouri
Ohio

richterite
BA amphibole group
chain silicates
silicates
BT minerals

Ridder
use Leninogorsk

Riddle
Town in Douglas County in SW.
BT Oregon
United States

ridges
Used as a general term in geomorphology.
SA aseismic ridges
beach ridges
hills
mid-ocean ridges
mountains

Riding Mountain National Park
SW Manitoba. Also search Riding Mountain.
BT Manitoba
Canada

riebeckite
BA amphibole group
chain silicates
silicates
BT minerals
SA crocidolite

Ries Basin
use Ries Crater

Ries Crater
Term introduced in 1978 for the meteor crater. Before 1978, the terms Ries Basin and Nordlingen were used. Also search Ries Basin; Nordlingen; Nordlinger Ries; Nordlinger Ries Crater; Ries.
  UF Nordlingen
        Nordlinger Ries
        Nordlinger Ries Crater
        Ries Basin
  BT Bavaria
        West Germany

Rif
Hilly region constituting part of former Spanish Morocco extending along Mediterranean Sea from E of Melilla on E to Ceuta on W.
  UF Er Rif
        Er Riff
        Riff
  BT Morocco

Riff
  use Rif

Rifle
Town on Colorado River in Garfield County in W.
  BT Colorado
        United States

Rift Valley
  use Great Rift Valley

rift valleys
Includes use on level 2 under ocean floors(1); on level 3 under faults(1) systems(2) and under tectonics(1) structure(2), and under plate tectonics(1).
  SA faults
        grabens
        mid-ocean ridges
        ocean floors
        plate tectonics
        rifting
        sea-floor spreading
        systems
        valleys

rift zones
Includes use on level 3 under plate tectonics(1).
  UF zones, rift
  SA fault zones
        median valley
        plate tectonics
        tectonics
        tectonophysics

rifting
Includes use on level 2 under plate tectonics(1).
  SA median valley
        plate tectonics
        rift valleys
        sea-floor spreading
        taphrogeny

Riga
City on Gulf of Riga.
  BT Latvia
        USSR

right-lateral faults
Term introduced in 1978. Before 1978, search faults AND right-lateral; faults AND right-slip.
  UF dextral faults
        right-lateral slip faults
        right-slip faults
  BA faults

right-lateral slip faults
  use right-lateral faults

right-slip faults
  use right-lateral faults

rigidities, cutoff
  use cutoff rigidities

rigidity
  SA elasticity
        mechanical properties
        rock mechanics
        stress

rigidity modulus
  use shear modulus

Rijn River
  use Rhine River

Rijn Valley
  use Rhine Valley

Rila
Village in W part of country.
  BT Sofia
        Bulgaria

Rila Mountains
Range in SW at W end of the Rhodope Mountains.
  BT Bulgaria

Riley County
NE central.
  BT Kansas
        United States

rilles
Includes use on level 3 under Moon(1).
  SA Moon
        valleys

Rincon Formation
Bonaire.
  BT Upper Cretaceous
        Cretaceous
  SA West Indies

ring
A valid term through 1976. After 1976, use ring dikes.

ring complexes
As of 1978, term is used on level 2 under intrusions(1).
  BA intrusions
  SA complexes
        dikes
        ring dikes
        ring structures

ring dikes
Includes use on level 3 under intrusions(1) dikes(2). Before 1976, also search ring.
  UF ring-fracture intrusion
  BT dikes
        intrusions
  SA dike swarms
        ring complexes
        ring structures
        swarms

ring silicates
Includes use on level 2 under minerals(1). See list L. Also search cyclosilicates.
  UF cyclosilicates
  BA silicates
  BT minerals
  NA axinite
        barylite
        bertrandite
        beryl
        buergerite
        cordierite
        dravite
        ekanite
        elbaite
        emerald
  NZ eudialyte
  NA malayaite
        milarite
        osumilite
        roedderite
        tourmaline

ring structures
Includes use on level 3 under intrusions(1). Also search ring structure.
  SA dikes
        intrusions
        ring complexes
        ring dikes
        structures

ring-fracture intrusion
  use ring dikes

Ringerike
Region NW of Oslo.
  BT Norway

Ringerike Sandstone
Included in Ringerike Series. S Norway.
  BT Paleozoic
  SA Devonian
        Norway
        Silurian

rings, tree
  use tree rings

rinneite
  BA halides
  BT minerals
  SA chlorides

Rio Arriba County
On the Colorado border in NW.
  BT New Mexico
        United States

Rio Blanco Basin
River basin in NW part of country. Also search Rio Blanco.
  BT Salta
        Argentina

Rio Blanco County
On the Utah border in NW.
  CO N394000N402000
        W1070000W1090500
  BT Colorado
        United States

Rio Bonito Formation
  BT Permian
        Paleozoic
  SA Brazil
        upper Paleozoic

Rio Bravo
  use Rio Grande

Rio Bravo del Norte
  use Rio Grande

Rio Claro
City in E central.
  BT Sao Paulo
        Brazil

Rio de Janeiro
State in SE Brazil. Also a city on the Atlantic Ocean in Guanabara State.
  CO S230000S210000
        W0410000W0450000
  BT Brazil
  NT Angra dos Reis
        Itaborai
        Macacu River

Rio de la Plata
Estuary of the combined Parana and Uruguay rivers. Index countries as applicable.
  UF Rio de la Plata Estuary
        River Plate
  BT South America
  SA Argentina
        Uruguay

Rio de la Plata Estuary
  use Rio de la Plata

Rio Grande
River which rises in SW Colorado and flows S and then SE into the Gulf of Mexico. Index Mexican and U.S. states as applicable.
  UF Rio Bravo
        Rio Bravo del Norte
        Rio Grande River
  SA Chihuahua
        Coahuila
        Colorado
        New Mexico
        Nuevo Leon
        Tamaulipas
        Texas

Rio Grande Depression
Lower Rio Grande Valley. Index Mexican and U.S. states as applicable.
  SA Tamaulipas
        Texas

Rio Grande do Norte
State on the Atlantic Ocean in extreme NE.
  BT Brazil
  SA Borborema

Rio Grande do Sul
State in extreme S.
  CO S330000S270000
        W0500000W0580000
  BT Brazil
  NT Encruzilhada do Sul
        Sao Gabriel
  SA Pelotas Basin

Rio Grande Rift
Index Mexican and U.S. states as applicable. Also search Rio Grande Rift Zone.
  CO N313000N370000
        W1053000W1073000
  UF Rio Grande Rift Zone
  BT North America
  SA Chihuahua
        Chihuahua tectonic belt
        Coahuila
        New Mexico
        Texas

Rio Grande Rift Zone
  use Rio Grande Rift

Rio Grande River
  use Rio Grande

Rio Grande Valley
Index Mexican and U.S. states as applicable.
  SA Chihuahua
        Coahuila
        Colorado
        New Mexico
        Nuevo Leon
        Tamaulipas
        Texas

Rio Negro
Province in central.
  CO S420000S370000
        W0630000W0720000
  BT Argentina
  SA Neuquen Basin
        Patagonia

Rio Tinto
Town in copper mining area of W Sierra Moreno in SW part of country.
  UF Riotinto
  BT Huelva
        Spain

Rioni Basin
River basin in W Georgian Republic. Also search Rioni; Rioni River.
  BT Georgian Republic
        USSR

Riotinto
  use Rio Tinto

Riphean
Europe. See list E.
  BA upper Precambrian
        Precambrian
  SA Eocambrian
        Sparagmite Group

Ripley Formation
In Selma Group. Includes Coon Creek Tongue, Chiwapa Sandstone Member, McNairy Sand Member, Keownville Limestone Member. Alabama, Georgia, S Illinois, W Ken-

tucky, SE Missouri, N and central Mississippi, and W Tennessee.
BT Upper Cretaceous
  Cretaceous
SA Alabama
  Georgia
  Illinois
  Kentucky
  Mississippi
  Missouri
  Tennessee

ripple drift-cross laminations
Includes use on level 3 under sedimentary structures(1) planar bedding structures(2). See list K.
UF ripple-cross-laminations
BA planar bedding structures
  sedimentary structures
SA cross-laminations
  laminations

ripple marks
Asymmetrical, interference, etc. Includes use on level 3 under sedimentary structures(1) bedding plane irregularities(2). See list K.
BA bedding plane irregularities
  sedimentary structures
SA sand waves

ripple-cross-laminations
use ripple drift-cross laminations

rise, continental
use continental rise

risk, seismic
use seismic risk

Riss
Europe. Above Mindel, below Wurm. Includes use on level 3 under age terms (1). See list E.
BA Pleistocene
  Quaternary
BT Cenozoic

Riss-Wurm
use Riss/Wurm Interglacial

Riss-Wurm Interglacial
use Riss/Wurm Interglacial

Riss/Wurm Interglacial
Europe. A term applied in the Alps to the third interglacial stage of the Pleistocene Epoch, after the Riss glacial stage and before the Wurm. Also search Riss-Wurm Interglacial, and Riss-Wurm.
UF Riss-Wurm
  Riss-Wurm Interglacial
BA Pleistocene
  Quaternary
BT Cenozoic
SA Sangamonian

Rita Blanca Lake
Formed by Rita Blanca Dam on Rita Blanca Creek in Texas panhandle.
BT Texas
  United States

Riukiu Islands
use Ryukyu Islands

river capture
use stream capture

river piracy
use stream capture

river plain
use alluvial plains

River Plate
use Rio de la Plata

rivers
Includes use on level 3 under geomorphology(1) fluvial features(2).
SA channel geometry
  channels
  drainage basins
  drainage patterns
  estuaries
  floodplains
  fluvial features
  geomorphology
  hydrogeology
  hydrology
  hydrosphere
  meanders
  rivers and streams
  runoff
  stream capture
  stream transport
  streams
  surface water
  water supply
  watersheds
  waterways

rivers and streams
Includes use on level 2 under hydrology(1) and under waterways.
SA channel geometry
  drainage basins
  estuaries
  floods
  fluvial features
  hydrology
  hydrosphere
  meanders
  precipitation
  rivers
  runoff
  stream capture
  stream gradient
  stream transport
  streams
  valleys
  waterways

Riverside
City SE of Los Angeles in Riverside County.
BT California
  United States

Riverside County
Southern California.
BT California
  United States

Rn
use radon

Rn-222
Includes use on level 3 under isotopes(1). Also search thoron.
UF radon-222
  thoron
SA isotopes
  radon

road log
UF log, road
SA guidebook
  maps

Roan Supergroup
Interlayered with Carolina Gneiss in Spruce Pine District. N Georgia, W North Carolina, NW South Carolina, and E Tennessee.
BT Precambrian
SA Georgia
  North Carolina
  South Carolina
  Tennessee

Roane County
Index states as applicable.
BX Tennessee
  West Virginia
BT United States

Roanoke
City in Roanoke County in SW.
BT Virginia
  United States

Roaring Brook Valley
Valley in Hartford County and another in Tolland County.
BT Connecticut
  United States

robbery
use stream capture

Robertinacea
Superfamily. Includes use on level 2 under foraminifera(1).
BA Rotaliina
  foraminifera
BT Invertebrata

Roberts Mountains
Eureka County in central.
BT Nevada
  United States

Roberts Mountains Formation
Underlies Rabbit Hill Formation. E Nevada, and W Utah.
BT Middle Silurian
  Silurian
SA Nevada
  Utah

Robles Formation
BT Cretaceous
SA Puerto Rico
  West Indies

Roca Formation
Underlies Sallyards Limestone Member of Grenola Limestone; overlies Howe Limestone Member of Red Eagle Limestone. NE Kansas, SE Nebraska, and N Oklahoma.
BT Wolfcampian
  Lower Permian
  Permian
SA Kansas
  Nebraska
  Oklahoma

Roccamonfina
Village in extinct volcanic crater in NW.
BT Campania
  Italy

Rochechouart
Town in W central part of country.
BT Haute-Vienne
  France

Rochester
City in Monroe County on S Lake Ontario.
BT New York
  United States

Rochester Formation
Includes Keefer Sandstone Member, Decew Waterlime Bed.
BT Middle Silurian
  Silurian
SA Maryland
  New York
  Ontario
  Pennsylvania
  West Virginia

rock bursts
As of 1978, term is used on level 2 under geologic hazards(1).
UF bursts, rock
SA geologic hazards

Rock Creek Park
Also search Rock Creek.
BT District of Columbia
  United States

rock crystal
use quartz crystal

rock failure
use failure

rock glaciers
SA boulders
  glaciers

Rock Island
City in Rock Island County in NE.
BT Illinois
  United States

Rock Lake Shale
use Rock Lake Shale Member

Rock Lake Shale Member
Of Stanton Limestone. E Kansas, NW Missouri, and SE Nebraska.
UF Rock Lake Shale
BT Missourian
  Upper Pennsylvanian
  Pennsylvanian
SA Kansas
  Missouri
  Nebraska
  Stanton Formation

rock mechanics
A level 1 term as of 1978. Includes use on level 2 under nuclear facilities(1), tunnels(1), and underground installations(1). Used for geotechnical studies. If 1, term set options are:
applications
  subtopic
case studies
  subtopic
concepts
  subtopic
deformation
  subtopic
elasticity
  subtopic
excavations
  subtopic
experimental studies
  subtopic
field studies
  subtopic
frost action
  subtopic
failure
  subtopic
materials, properties
  subtopic
methods
  subtopic
plasticity
  subtopic
site exploration
  subtopic
techniques
  subtopic
theoretical studies
  subtopic
SA Atterberg limits
  case studies
  cohesive materials
  competent materials
  compressive strength
  congelifraction
  cracks
  cryoturbation
  dams
  deformation
  ductility
  earthquakes
  elastic materials
  elasticity
  engineering geology
  excavations
  explosions
  extensometers
  failure
  finite strain
  foundations
  frost action
  geologic hazards
  highways
  hydraulic fracturing
  land subsidence
  marine installations
  mechanics
  nuclear facilities
  penetration
  permeability
  plastic materials
  plasticity
  reservoirs
  rigidity
  shear strength
  site exploration
  slope stability

       soil mechanics
       strain
       strength
       stress
       structural analysis
       triaxial tests
       tunnels
       underground installations
       vibration

rock salt
    use halite

rock slides
    use rockslides

rock slip
    use rockslides

Rock Springs
    Town in Sweetwater County in SW.
    BT Wyoming
           United States

rock-stratigraphy
    use lithostratigraphy

Rockall Bank
    About 225 miles W of the Hebrides.
    BT Atlantic Ocean

Rockall Plateau
    Rise or submarine plateau just N of Rockall Bank W of the Hebrides.
    UF Rockall Rise
    BT Atlantic Ocean

Rockall Rise
    use Rockall Plateau

Rockall Trench
    use Rockall Trough

Rockall Trough
    SE of Rockall Bank W of the Hebrides.
    UF Rockall Trench
    BT Atlantic Ocean

rockbridgeite
    BA phosphates
    BT minerals

rockfalls
    As of 1978, term is used on level 2 under slope stability(1).
    SA slope stability

Rockhampton
    City in E.
    BT Queensland
           Australia

Rockingham County
    Index states as applicable.
    BX New Hampshire
           North Carolina
           Virginia
    BT United States

rocks
    Includes use on level 3 under engineering geology(1) materials, properties(2); on level 3 under inclusions(1).
    SA aggregate
           bedrock
           cap rocks
           carbonate rocks
           chemically precipitated rocks
           country rocks
           crystalline rocks
           extrusive rocks
           fluid inclusions
           host rocks
           hypabyssal rocks
           inclusions
           intrusive rocks
           materials
           metaigneous rocks
           metamorphic rocks
           metaplutonic rocks
           metasedimentary rocks
           metasomatic rocks
           metavolcanic rocks
           paragenesis
           plutonic rocks
           reservoir rocks
           sedimentary rocks
           silicate rocks
           source rocks
           standard rocks
           volcanic rocks

rockslides
    Also search rock slides.
    UF rock slides
           rock slip
    SA landslides
           mass movements

Rockwood Formation
    Underlies Hancock Limestone; overlies Sequatchie Formation. E Tennessee and NW Georgia.
    BT Silurian
    SA Georgia
           Lower Silurian
           Middle Silurian
           Tennessee

Rocky Mountain Trench
    A trough in the Northern Rockies, E of the interior plateau of British Columbia, between a series of parallel ranges extending WNW from Montana to the headwaters of the Yukon River. Index British Columbia, Washington, and Yukon Territory as applicable.
    BT North America
    SA British Columbia
           Montana
           Rocky Mountains
           Yukon Territory

Rocky Mountains
    Mountain system in W North America extending from N Alaska to the Mexican frontier. Index Alaska, and countries as applicable. Includes use on level 1 as an area term (list O). For term set options see list B.
    BA North America
    NA Central Rocky Mountains
           Northern Rocky Mountains
           Southern Rocky Mountains
    SA Absaroka Range
           Alaska
           Big Belt Mountains
           Bighorn Mountains
           Bitterroot Range
           Brooks Range
           Canada
           Canadian Cordillera
           Cariboo Mountains
           Elk Mountains
           Flint Creek Range
           Franklin Mountains
           Front Range
           Laramie Mountains
           Little Belt Mountains
           Mackenzie Mountains
           Medicine Bow Mountains
           Rocky Mountain Trench
           San Juan Mountains
           Sangre de Cristo Mountains
           Sawatch Range
           Selkirk Mountains
           Tobacco Root Mountains
           Uinta Mountains
           United States
           Wasatch Range
           Wet Mountains
           Wind River Range

Rocroi
    Town in N part of country.
    UF Rocroy
    BT Ardennes
           France

Rocroy
    use Rocroi

Rodentia
    Includes use on level 2 under Mammalia(1). See list F. Also search rodents.
    UF rodents
    BA Mammalia
    BT Tetrapoda
           Vertebrata
    NA Arvicolidae
           Castoridae
           Cricetidae
           Gliridae
           Muridae

rodents
    use Rodentia

Rodez Pass
    use Rodez Trough

Rodez Trough
    Between Segala Plateau in S and Causse du Comtal in N. Index departments as applicable. Also search Rodez; Rodez Pass.
    UF Rodez Pass
    BT France
    SA Aveyron
           Haute-Garonne
           Tarn

rodingite
    Includes use on level 3 under igneous rocks(1) gabbro family(2). See list H.
    BA gabbro family
    BT igneous rocks

Rodna
    Village in NE Cluj County in NE.
    BT Transylvania
           Romania

Rodna Mountains
    Range of the Carpathians along border between Maramures and Cluj Counties in N.
    UF Rodnei Mountains
    BT Transylvania
           Romania
    SA Carpathians

Rodnei Mountains
    use Rodna Mountains

Rodrigues Island
    use Rodriguez Island

Rodriguez Island
    A dependency of Mauritius in the Mascarene Islands about 500 miles E of the Malagasy Republic.
    UF Rodrigues Island
    BT Indian Ocean
    SA Mascarene Islands

roedderite
    Meteorite mineral.
    BA ring silicates
           silicates
    BT minerals

roemerite
    BA sulfates
    BT minerals

roestone
    use oolite

Rogaland
    County in SW.
    BT Norway
    NT Sandnes
           Sogndal
           Stavanger

roggianite
    BA orthosilicates
           silicates
    BT minerals

Rogue River
    SW Oregon. Also search Rogue.
    BT Oregon
           United States

Rokko Mountains
    NE of Kobe in S central.
    BT Honshu
           Japan

roll-type deposits
    Term introduced in 1978. Before 1978, search roll-type.
    SA deposits
           ore deposits
           uranium

Romagna
    Historical region in N central part of country on the Adriatic Sea.
    BT Emilia-Romagna
           Italy

Romanche Deep
    use Romanche Trench

Romanche fracture zone
    Cuts across the Mid-Atlantic Ridge between the bulge of South America and Liberia.
    BT Atlantic Ocean

Romanche Gap
    use Romanche Trench

Romanche Trench
    Extends in a NW-SE direction to the SW of the Romanche fracture zone between bulge of South America and Liberia.
    UF Romanche Deep
           Romanche Gap
    BT Atlantic Ocean

Romania
    Includes use on level 1 as an area term (list O). For term set options see list B.
    CO N434000N481000
           E0294500E0201500
    UF Roumania
           Rumania
    BA Europe
    NT Apuseni Mountains
           Arges River
           Bistrita Mountains
           Bucegi Mountains
           Dacian Basin
           Dobruja
           Fagaras Mountains
           Getic Nappe
           Haghimas Syncline
           Jiu River valley
           Mehedinti Plateau
           Olt River
           Olt River valley
           Paring Mountains
           Romanian Plain
           Transylvania
           Transylvanian Alps
           Vrancea
           Vulcan Mountains
           Walachia
    SA Alfold
           Balkan Peninsula
           Banat
           Black Sea region
           Bukovina
           Carpathian Foredeep
           Carpathian Foreland
           Carpathians
           Danube Delta
           Danube Plain
           Danube River
           Danube Valley
           Dobruja Basin
           Eastern Carpathians
           Moesia
           Moesian Platform
           Moldavia
           Prut River
           Somes Basin
           Subcarpathians
           Tisza River
           Tulghes Series

Romania Plain
    use Romanian Plain

Romanian
    Provincial Series, Europe.

BA upper Pliocene
   Pliocene
   Neogene
   Tertiary
BT Cenozoic

Romanian Plain
Area between the Transylvanian Alps on the N, and the Danube River on the S and E. Index regions as applicable. Also search Romanian.
UF Romania Plain
BT Romania
SA Moldavia
   Walachia

Romashkino Field
Term introduced in 1978. Orenburg Oblast in Southern Urals. Before 1978, also search Romashkino; Romashkino oil field.
UF Romashkino oil field
BT Russian Republic
   USSR

Romashkino oil field
use Romashkino Field

Rome
City on the Tiber River. Also search Roma.
BT Latium
   Italy

Rome Formation
Named for exposures south of Rome, Georgia.
BT Lower Cambrian
   Cambrian
SA Alabama
   Georgia
   North Carolina
   Tennessee
   Virginia

Ronda Massif
use Serrania de Ronda

Ronda, Serrania de
use Serrania de Ronda

Ronda Sierra
use Serrania de Ronda

Rondonia
Territory. In W part of country on Bolivian border.
UF Guapore
BT Brazil

Rondout Formation
Includes Fuyk Sandstone Member.
BT Paleozoic
SA Devonian
   Lower Devonian
   New York
   Pennsylvania
   Silurian
   Upper Silurian

Rooks County
NW central.
BT Kansas
   United States

Roorkee
Town in N Uttar Pradesh.
UF Rurki
BT Uttar Pradesh
   India

Roosevelt County
Index states as applicable.
BX Montana
   New Mexico
BT United States

root clay
use underclay

roquesite
BA sulfides
BT minerals

Roraima Formation
Tentatively dated from lower Proterozoic to as young as Miocene. In the Pacaraima Mountains.
SA Brazil
   Guyana
   South America
   Venezuela

Ros River valley
W central Ukraine. Also search Ros; Ros River.
BT Ukraine
   USSR

Rosario Formation
BT Upper Cretaceous
   Cretaceous
SA Baja California
   Campanian
   Maestrichtian
   Mexico

Roscommon County
N central Lower Peninsula.
BT Michigan
   United States

Rose Canyon Formation
Overlies Torrey Sand; underlies Poway Conglomerate. San Diego County.
UF Rose Canyon Shale
BT Eocene
SA California
   lower Eocene
   middle Eocene

Rose Canyon Shale
use Rose Canyon Formation

Rosebery
Village in W.
BT Tasmania
   Australia

Rosetown
Town in SW.
BT Saskatchewan
   Canada

Rosia
Village 25 miles SE of Oradea in Bihor County in W.
BT Transylvania
   Romania

Ross Barrier
use Ross Ice Shelf

Ross Dependency
Section lying S of 60° S and between 160° E and 150° W extending to the South Pole. Placed under jurisdiction of New Zealand in 1923 by act of British Parliament. Also search Ross.
BT Antarctica

Ross Ice Shelf
Covers S part of Ross Sea between Marie Byrd Land and Victoria Land with its S end at the foot of Queen Maude Range on the edge of the Antarctic continent. Also search Ross.
UF Ross Barrier
BT Antarctica
SA Beardmore Glacier
   McMurdo Ice Shelf
   Shackleton Glacier

Ross Island
In Ross Sea at W end of Ross Ice Shelf separated from Victoria Land by McMurdo Sound. Also search Ross.
BT Antarctica

Ross Sea
Arm of S Pacific Ocean just N of Ross Ice Shelf between Victoria Land and Edward VII Peninsula. Also search Ross.
BT Antarctic Ocean
SA Pacific Ocean

Ross-shire
A separate county until the 17th century but now part of Ross and Cromarty County in N.
CO N573000N580000
   W0043000W0053000
BT Scotland
   Great Britain
   United Kingdom

Rosses Granite
BT Donegal
   Ireland

Rossland
City in SE.
BT British Columbia
   Canada

Rostov
City near mouth of Don River. Also an oblast.
UF Rostov-na-Donu
   Rostov-on-Don
BT Russian Republic
   USSR

Rostov-na-Donu
use Rostov

Rostov-on-Don
use Rostov

Rotalia
Genus. Includes use on level 3 under foraminifera(1) Rotaliacea(2).
BA Rotaliacea
   Rotaliina
   foraminifera
BT Invertebrata

Rotaliacea
Includes use on level 2 under foraminifera(1). See list F.
BA Rotaliina
   foraminifera
BT Invertebrata
NA Ammonia
   Elphidium
   Heterostegina
   Miogypsinidae
   Nummulitidae
   Rotalia

Rotaliina
Includes use on level 2 under foraminifera(1). See list F.
BA foraminifera
BT Invertebrata
NA Buliminacea
   Cassidulinacea
   Discorbacea
   Globigerinacea
   Lagenidae
   Nodosariacea
   Orbitoidacea
   Robertinacea
   Rotaliacea
   Spirillinacea

Rotalipora
Genus. Includes use on level 3 under foraminifera(1) Globigerinacea(2).
BA Globigerinacea
   Rotaliina
   foraminifera
BT Invertebrata

rotation
Includes use on level 3 under plate tectonics(1).
SA plate tectonics
   polar wandering

rotational wave
use S-waves

Rotliegendes
Europe: lower and middle Permian. Below Zechstein. Includes use on level 3 under age terms(1). See list E.
BA Permian
BT Paleozoic
SA Autunian
   Lower Permian
   Middle Permian
   Saxonian

Rottleberode
Village in SW central part of country.
BT Halle
   East Germany

Rouen
City on the Seine River in N part of country.
BT Seine-Maritime
   France

Rouergue
Ancient province. Primarily in Aveyron Department in S central. Index departments as applicable.
BT France
SA Aveyron
   Tarn-et-Garonne

Rough Creek fault zone
W Kentucky.
BT Kentucky
   United States

roughness
SA flow regime
   hydraulics
   permeability
   reservoir rocks

Roumania
use Romania

Round Mountain
On E spur of Great Dividing Range in NE.
BT New South Wales
   Australia

rounding
Includes use on level 3 under sedimentary rocks(1).
SA particles
   pebbles
   roundness
   sand
   sedimentary rocks
   sorting

roundness
SA particles
   rounding
   sediments
   shape analysis
   sphericity

Roussillon
Region and former province roughly coextensive with Pyrenees-Orientales Department on the Spanish border.
BT France

Rouyn
Mining city in SW near N Ontario border.
BT Quebec
   Canada

Rowan County
Index states as applicable.
BX Kentucky
   North Carolina
BT United States

Rowne
use Zloty Stok

Royal Creek
BT Yukon Territory
   Canada

Rozdol
Town in Lvov Oblast in E.
BT Ukraine
   USSR

rozenite
BA sulfates
BT minerals

Roztocze
Mountain range. Index Poland and/or Ukraine as applicable.
UF Tomaszow-Lvov Ridge

  BT Europe
  SA Poland
    Ukraine

Ru
 use ruthenium

Ruanda
 use Rwanda

Ruapehu
 Volcano in Tongariro National Park in S central North Island.
  UF Ruapehu Volcano
  BT North Island
    New Zealand

Ruapehu Volcano
 use Ruapehu

rubblerock
 use breccia

rubidium
 Includes use on level 1 and 2 as a chemical element (list D). Also search Rb.
  UF Rb
  SA alkali metals
    elements
    isotopes
    Sr/Rb

Ruby Mountains
 Range in Elko and White Pine counties in NE Nevada. Also search Ruby Range.
  BT Nevada
    United States
  SA Ruby Range

Ruby Range
 N extension of Snowcrest Mountains in SW Montana, lies just W of Ruby River.
  BT Montana
    United States
  SA Ruby Mountains

Ruda
 Town in coal mining region in S part of country.
  BT Katowice
    Poland

Rudistae
 Includes use on level 3 under Mollusca(1) Bivalvia(2).
  BA Bivalvia
    Mollusca
  BT Invertebrata

rudists
 Included on level 3 under Hippuritoidea.
  BA Bivalvia
    Mollusca
  BT Invertebrata

Rudki
 Town in Lvov Oblast in E.
  BT Ukraine
    USSR

Rudnichny
 Town in W Chelyabinsk Oblast in Southern Urals of W Siberia.
  UF Rudnichnyy
  BT Russian Republic
    USSR

Rudnichnyy
 use Rudnichny

Rudny Altai
 Region in the Altai Mountains in Vostochno Kazakhstan Oblast and in Altai Kray. Index Soviet republics as applicable.
  CO N480000N520000
    E0873000E0823000
  UF Rudnyy Altai
    Rudnyy Altay
  BT USSR
  SA Kazakhstan
    Russian Republic

Rudnyy Altai
 use Rudny Altai

Rudnyy Altay
 use Rudny Altai

Rudolf (Lake)
 use Lake Turkana

Ruegen (Island)
 use Rugen Island

Rugen Island
 In the Baltic Sea just off Rostock District. Largest island of East Germany. Also search Ruegen; Rugen.
  UF Ruegen (Island)
  BT East Germany

Rugosa
 Includes use on level 2 under Coelenterata(1). See list F.
  BA Anthozoa
    Coelenterata
  BT Invertebrata
  SA corals
    Scleractinia
    Tabulata
    Tetracorallia

Ruhla
 Town in Thuringian Forest in SW part of country.
  BT East Germany
  SA Erfurt

Ruhr
 Major coal-mining and industrial region. Includes the Ruhr River valley and the Dusselforf area to the S along the Rhine River.
  UF Ruhr Basin
    Ruhr Valley
  BT North Rhine-Westphalia
    West Germany

Ruhr Basin
 use Ruhr

Ruhr Valley
 use Ruhr

Rum
 use Rhum

Rumania
 use Romania

Rumilly
 Town in E part of country.
  UF Rumilly-Albanais
  BT Haute-Savoie
    France

Rumilly-Albanais
 use Rumilly

Ruminantia
 Includes use on level 3 under Mammalia(1) Artiodactyla(2).
  BA Artiodactyla
    Mammalia
  BT Tetrapoda
    Vertebrata
  NA Bovidae
    Camelidae
    Cervidae

Runnels County
 W central.
  BT Texas
    United States

runoff
 Includes use on level 3 under hydrology(1) rivers and streams(2).
  SA discharge
    hydrology
    precipitation
    retention
    rivers
    rivers and streams
    sediment yield
    streams
    surface water
    water balance
    water yield
    watersheds

runout
 use water yield

Rupelian
 Europe. Above Tongrian, below Chattian. Includes use on level 3 under age terms(1). See list E.
  BA Oligocene
    Paleogene
    Tertiary
  BT Cenozoic
  SA Krosno Beds
    Stampian

Rurki
 use Roorkee

Rusca Montana
 Village in the Poiana-Rusca Mountains in Caras-Severin County in E part of country.
  UF Rusca-Montana
  BT Banat
    Romania

Rusca-Montana
 use Rusca Montana

Ruse
 Province in N.
  BT Bulgaria

Rush County
 Index states as applicable.
  BX Indiana
    Kansas
  BT United States

Rusophycus
  BA ichnofossils
  SA Trilobita

Russell County
 Index states as applicable.
  BX Alabama
    Kansas
    Kentucky
    Virginia
  BT United States

Russia
 use USSR

Russian
 Used to indicate language of a document.

Russian Basin
 use Russian Plain

Russian Plain
 Comprises most of European USSR W of the Urals and is bordered on the S by the Carpathians, the Crimean Mountains and the Caucasus. Index Soviet republics as applicable. Also search Russian AND plain; Russian Basin.
  UF East European Plain
    Russian Basin
  BT USSR
  SA Byelorussia
    Estonia
    Latvia
    Lithuania
    Moldavia
    Russian Platform
    Russian Republic
    Ukraine

Russian Platform
 Ancient platform of Precambrian crystalline rocks overlain by sedimentary deposits of the Russian or East European Plain. Also search East European Platform; Russian AND platform. Index Soviet republics as applicable.
  CO N523000N573000
    E0550000E0273000
  BT USSR
  SA Byelorussia
    Estonia
    Latvia
    Lithuania
    Moldavia
    Russian Plain
    Russian Republic
    Ukraine

Russian Republic
 Russian Soviet Federated Socialist Republic (RSFSR) in both European and Asiatic USSR.
  CO N370000N830000
    W1700000E0280000
  BT USSR
  NT Abakan
    Aldan
    Aldan Plateau
    Aldan River
    Aldan Shield
    Allarechenskiy
    Altai
    Anabar Bay
    Anabar River
    Anabar Shield
    Anadyr Basin
    Anadyr Range
    Anapa
    Angara River
    Angara-Lena Basin
    Arbarastakh
    Argun
    Arkhangelsk
    Arlan
    Astrakhan
    Avacha
    Azov
    Baikal Mountains
    Baikal region
    Bakal
    Balei
    Bashkiria
    Batenev Ridge
    Belaya Gora
    Belgorod
    Belomorsk
    Belozero
    Biryusa
    Blyava
    Bodaibo
    Bolshezemelskaya Tundra
    Bryansk
    Bureya
    Buryat
    Caucasus Foreland
    Chadobets Uplift
    Chaya Massif
    Chechen-Ingush
    Chekalin
    Chelyabinsk
    Chita
    Chukchi Peninsula
    Chuya
    Chuya Alps
    Chuya Basin
    Dagestan
    Darasun
    Don Basin
    Don River
    Dzhagdy Range
    Dzhida River
    Elbrus
    Franz Josef Land
    Gay
    Golovanevsk
    Gorki
    Gornaya Shoriya
    Gorny Altai
    Goryachiy Plyazh
    Grozny
    Igarka
    Ilmen Mountains
    Imandra
    Indigirka River
    Irkutsk
    Irkutsk Basin
    Itaka
    Iturup Island
    Iya River
    Kabardia

Kabardin-Balkar
Kalinin
Kaliningrad
Kalmyk
Kama River
Kamchatka
Kamchatka Peninsula
Kamenka River
Kamyshin
Kanin Peninsula
Kansk
Kansk-Achinsk Basin
Karaginskiy Island
Karelia
Kavalerovo
Kazan
Kemerovo
Khabarovsk
Khamar-Daban Range
Khapcheranga
Khatanga
Khatanga Basin
Khatanga River
Kheta River
Khibiny Mountains
Khovu-Aksy
Kirov
Kiya River
Kizel coal basin
Klichka
Klyuchevskaya Sopka
Kochkar
Kodar Range
Kola
Kola Peninsula
Kolyma River
Kolyma River basin
Kolyma Uplift
Komi
Kommunar
Komsomolsk
Koryak Range
Kostroma
Kotui
Kozhim
Krasnodar
Krasnoyarsk
Kuban
Kuibyshev
Kular Range
Kuma Basin
Kunashir Island
Kureyka River region
Kurgan
Kuril Islands
Kursk
Kursk magnetic anomaly
Kushva
Kuznetsk Alatau
Kuznetsk Basin
Laba Basin
Lake Baikal
Lake Ladoga
Lake Onega
Lena Basin
Lena River
Leningrad
Leningrad Mining Institute
Lovozero
Lovozero Massif
Lower Tunguska River
Magadan
Magnitogorsk
Maikop
Makhachkala
Malgobek
Mama
Mama River
Mariinsk
Maya River basin
Maymecha
Maymecha-Kotuy
Megion Field
Mendeleyev Volcano
Mezen
Mezen River basin
Mikhailovka
Minusinsk
Minusinsk Basin
Molotov
Monchegorsk
Moneron Island
Moscow
Moscow Basin
Moscow Syneclise
Moscow University
Moskva River
Mukhor-Tala
Murmansk
Narym
Nayba River basin
Nenets
Norilsk
Norilsk region
North Ossetia
Northern Caucasus
Northern Dvina River
Novaya Zemlya
Novgorod
Novosibirsk
Ob River
Oka
Okhotsk
Okhotsk Massif
Okhotsk-Chukchi
Okhotsk-Chukchi volcanic belt
Olekma
Olekma-Vitim Highlands
Olenek River
Omolon
Omolon Block
Omsk
Omulevka
Onega
Orenburg
Orlov
Orsk
Orulgan Mountains
Ozernoye
Pachelma
Pai-Khoi
Paramushir
Paratunka
Patom Plateau
Pechenga
Pechora
Pechora Basin
Penzhina Bay
Peredovoy Range
Perm
Polar Urals
Polousnyy
Popigay
Prikumsk
Primorye
Prokopyevsk
Pskov
Romashkino Field
Rostov
Rudnichny
Ryazan
Sadon
Sakhalin
Sakmara
Salair
Salair Ridge
Salym
Samara Bend
Sangilen Mountains
Saratov
Sarbay
Segozero
Selennyakh
Serov
Sette-Daban Range
Sheveluch
Shilka Valley
Shoriya Mountains
Sibay
Siberian Lowland
Siberian Platform
Sikhote-Alin Range
Slyudyanka
Smolensk
Sochi
Solikamsk
Soviet Arctic
Soviet Far East
Sredniy Vasyugan
Stanovoy Range
Stavropol region
Stony Tunguska River
Suchan Basin
Sunzha
Surgut
Sutam River
Sverdlovsk
Sysert
Tagil Basin
Taiga
Taman
Taman Peninsula
Tambov
Tareya
Tas-Khayakhtakh Range
Tatar
Tatar Arch
Tatar Strait
Tataria
Taygonos Peninsula
Taymyr
Taymyr Peninsula
Taz Basin
Tetyukhe
Timan Ridge
Timan-Pechora region
Tomsk
Transbaikalia
Tuapse
Tula
Tunguska
Tunguska Basin
Tunguska River
Tunguska Syneclise
Turukhan
Turukhansk
Tuva
Tuymazy
Tyrny-Auz
Tyumen
Uchaly
Uchur River basin
Uda River
Udachnaya
Udmurtia
Udokan Mountains
Ufa
Ufaley
Ulkan
Ulyanovsk
Unda
Ural region
Ural-Tau
Uralian Foreland
Urals
Urup
USSR Academy of Sciences
Ust-Kut
Ust-Yenisei Basin
Valdai
Vaygach Island
Verkhoyansk
Verkhoyansk Range
Vetrenyy Ridge
Vilyuy River
Vilyuy River basin
Vilyuy Syneclise
Vitim
Vitim Plateau
Vladimir
Vladivostok
Volga region
Volga River
Volga-Don region
Volga-Urals
Volgograd
Vologda
Vorkuta
Voronezh
Voronezh Anteclise
Vuoriyarvi
Vyatka River
Vyatka-Kama Interfluve
Vychegda River
White Sea
Wrangel Island
Yakutia
Yakutsk
Yamal
Yamal-Nenets
Yana
Yana-Indigirka Lowland
Yaroslavl
Yenisei Ridge
Yenisei River
Yessey
Yudoma
Yuryuzan
Zeya
Zeya-Bureya Basin
Zmeinogorsk
Zyryanka
SA Amur Basin
     Amur River
     Argun River
     Asiatic USSR
     Azov region
     Baikalian Phase
     Baltic region
     Baltic Shield
     Barabash Suite
     Baskunchak Series
     Caspian Basin
     Caspian Depression
     Caucasus
     Courland Spit
     Dnieper Basin
     Dnieper River
     Dnieper-Donets Basin
     Donets Basin
     Dvina River
     European USSR
     Greater Caucasus
     Irtysh River
     Ishim
     Khanka Lake
     Kuban River
     Kuban Valley
     Kulunda Steppe
     Kursk Series
     Maikop Series
     Ob-Irtysh Interfluve
     Ossetia
     Poltava Series
     Rudny Altai
     Russian Plain
     Russian Platform
     Scythian Platform
     Serebryanka Formation
     Siberia
     Steppes
     Tannu-Ola Range
     Terek River
     Tunguska Series
     Tura Formation
     Udokan Series
     Ural River
     Usa Series
     Uzen
     Vetluga Series
     Vodino
     Volga-Ural region
     Vyshkovo
     Yenisei Basin
     Yenisei-Khatanga basin
     Yudoma Series

ruthenium
Includes use on level 1 and 2 as a chemical element (list D).
UF Ru
SA elements

rutile
BA oxides
BT minerals
SA anatase
     brookite
     rutile structure

rutile structure
Term introduced in 1978.
SA   rutile

Rutland
City in Rutland County in W.
CO   N523500N524500
        W0003000W0004500
BT   Vermont
        United States

Rwanda
Formerly Ruanda which was part of the Belgium trust territory of Ruanda-Urundi. Includes use on level 1 as an area term (list O). For term set options see list B.
UF   Ruanda
BA   Africa
SA   East African Rift
        Lake Kivu

Ryazan
City and oblast. City is located 120 miles SE of Moscow.
BT   Russian Republic
        USSR

Rybnik
City in S part of country.
BT   Katowice
        Poland

Ryoke
Region midway between Yokohama and Nagoya.
BT   Honshu
        Japan

Ryoke Belt
UF   Ryoke Metamorphic Belt
BT   Honshu
        Japan

Ryoke Metamorphic Belt
use   Ryoke Belt

Ryujima
Island in Japan Sea off the Noto Peninsula.
BT   Honshu
        Japan

Ryukyu Islands
600 mile chain of islands in W Pacific Ocean between Taiwan and Kyushu. Returned to Japan from U.S. control in 1972.
CO   N240000N360000
        E1380000E1250000
UF   Riukiu Islands
BT   Japan
NT   Amami-O-shima
        Okinawa
        Oshima

Rzeszow
Province in NW part of country.
BT   Poland
NT   Dukla
        Przemsyl
        Tarnobrzeg

S

S
use   sulfur

S-waves
Includes use on level 3 under seismology(1).
UF   rotational wave
        secondary wave
        shake wave
        shear wave
        tangential wave
BT   body waves
        elastic waves
SA   P-waves
        seismology
        SH-waves

S-32
Includes use on level 3 under isotopes(1).
SA   isotopes
        S-34/S-32
        sulfur

S-32/S-34
use   S-34/S-32

S-34
Includes use on level 3 under isotopes(1).
SA   isotopes
        S-34/S-32
        sulfur

S-34/S-32
Includes use on level 3 under isotopes(1). Also search S-32/S-34; S-32 AND S-34.
UF   S-32/S-34
SA   geologic thermometry
        isotopes
        paleoclimatology
        S-32
        S-34
        stable isotopes
        sulfur

Saale glacial stage
use   Saalian

Saale Glaciation
use   Saalian

Saale River
Rises in the Fichtelgebirge Range in NE Bavaria and flows N into the Elbe River SE of Magdeburg. Index countries as applicable. Also search Saale.
BT   Europe
SA   East Germany
        West Germany

Saalian
Term applied in northern Europe to the middle glacial stage of the Pleistocene Epoch, after the Elster glacial stage and before the Warthe; equivalent to the Riss and Illinoian glacial stages.
UF   Saale glacial stage
        Saale Glaciation
BA   upper Pleistocene
        Pleistocene
        Quaternary
BT   Cenozoic
SA   Central Polish Glaciation

Saalian Orogenic Phase
use   Saalian Phase

Saalian Phase
Orogenic phase. Before 1977 also search Saalian AND orogeny.
UF   Saalian Orogenic Phase
BT   Permian
SA   orogeny

Saanich Inlet
Arm of Strait of Georgia off SE Victoria Island.
BT   British Columbia
        Canada

Saar
use   Saarland

Saar Basin
River basin. Index French departments and West German states as applicable. Also search Saar.
BT   Europe
SA   Bas-Rhin
        Moselle
        Rhineland-Palatinate
        Saarland

Saar Nahe (Basin)
use   Saar-Nahe Basin

Saar-Nahe Basin
Extension of Saar Basin to include basin of Nahe River eastward to the Rhine. Index French departments and West German states as applicable. Also search Saar Nahe; Saar-Nahe.
UF   Saar Nahe (Basin)
BT   Europe
SA   Bas-Rhin
        Moselle
        Nahe
        Rhineland-Palatinate
        Saarland

Saarbrucken Anticline
BT   Saarland
        West Germany

Saarland
State constituting an industrial region, formerly known as the Saar. Achieved statehood within West Germany in 1957. Also search Saar.
UF   Saar
BT   West Germany
NT   Saarbrucken Anticline
SA   Nahe
        Saar Basin
        Saar-Nahe Basin

Sabah
State of East Malaysia on NE Borneo. Formerly British North Borneo. Also called North Borneo.
UF   British North Borneo
        North Borneo
BT   Malaysia
SA   Borneo
        East Malaysia

Sabana de Bogota
Plateau 55 miles long and 25 miles wide with Bogota at its center.
BT   Colombia

Sabatini Mountains
N of Lake Bracciano NW of Rome. Also search Sabatini.
BT   Latium
        Italy

Sabine Lake
Formed by expansion of Sabine River which flows through the lake and Sabine Pass to Gulf of Mexico. Index states as applicable.
BT   United States
SA   Louisiana
        Texas

Sabkah
use   Sabkha

Sabkha
Village on right bank of Euphrates River in N central Syria.
UF   Sabkah
BT   Syria
        Middle East

sabkha environment
use   sebkha environment

Sable Island
Low, sandy island in North Atlantic 115 miles off mainland.
BT   Nova Scotia
        Canada

Sable Island Bank
Marine bank in vicinity of Sable Island approximately 115 miles off mainland of Nova Scotia.
BT   Atlantic Ocean

sabulous texture
use   arenaceous texture

Saco River
Flows into Atlantic Ocean from S Maine. Index states as applicable.
BT   United States
SA   Maine
        New Hampshire

Sacramento
City in Sacramento County in N central.
BT   California
        United States

Sacramento County
N central.
BT   California
        United States

Sacramento Mountains
Range in Otero County NE of El Paso in S.
BT   New Mexico
        United States

Sacramento Valley
River valley. Northern half of Central Valley. Extends from Lake Shasta to San Joaquin Valley. Also search Sacramento River.
BT   California
        United States
SA   Central Valley

Sadlerochit Formation
Permian and lower Triassic: N Alaska.
SA   Alaska
        Lower Triassic
        Permian
        Triassic

Sado Island
Mountainous island in E Japan Sea off NW coast of Honshu.
BT   Honshu
        Japan

Sadon
Town in lead-zinc-silver mining area in North Ossetian A.S.S.R. in Northern Caucasus.
BT   Russian Republic
        USSR

Safaga
City on N Red Sea.
BT   Egypt

safflorite
BA   arsenides
        sulfides
BT   minerals

Saga
Prefecture in NW.
BT   Kyushu
        Japan

Sagami Bay
Inlet of the Pacific Ocean SW of Yokohama.
UF   Sagami Sea
        Sagami-nada
BT   Honshu
        Japan

Sagami Sea
use   Sagami Bay

Sagami-nada
use   Sagami Bay

Saganaga Lake
Chain of lakes near Lake Superior. Index Minnesota and/or Ontario as applicable.
SA   Minnesota
        Ontario

Sagar
Town in NW.
BT   Mysore
        India

Saghalin
use   Sakhalin

Saginaw Formation
Includes Verne Limestone Member, Eaton, Ionia, and Woodville Sandstone members. Lower Peninsula.
BT   Pennsylvanian
SA   Michigan

Saguache County
S Colorado.
CO N374000N383000
W1053000W1070000
BT Colorado
United States

Saguenay Valley
River valley between Lake Saint Jean and Saint Lawrence River NE of Quebec City. Also search Saguenay; Saguenay River.
BT Quebec
Canada

Sahara
Vast arid region extending across North Africa from the Atlantic Ocean to the Red Sea. Index countries as applicable. Includes use on level 1 as an area term (list O). For term set options see list B. As of 1977, documents on Spanish Sahara are indexed under Sahara as the level 1 term.
UF Sahara Desert
BA Africa
SA Ahaggar
Algeria
Chad
Egypt
Libya
Mali
Mauritania
Morocco
Niger
Sudan
Tanezrouft
Tunisia

Sahara Desert
use Sahara

Sahel
Region. A transitional steppe belt just S of the Sahara. Index countries as applicable.
BT Africa
SA Chad
Mali
Mauritania
Niger
Senegal
Upper Volta

Sahul Shelf
SW of W Timor Island.
BT Timor Sea
Indian Ocean

Saihun River
use Syr Darya

Saint Catharines
City NW of Niagara Falls on Welland Canal in S Ontario. Also search St. Catharines.
UF St. Catherines
BT Ontario
Canada

Saint Clair County
Index states as applicable. Also search St. Clair County.
UF St. Clair County
BX Alabama
Illinois
Michigan
Missouri
BT United States

Saint Clair River
Connects Lake Huron with Lake Saint Claire. Index Michigan and/or Ontario as applicable. Also search St. Claire River.
UF St. Clair River
BT North America
SA Michigan
Ontario

Saint Clair River delta
Formed at river mouth in Lake Saint Clair. Index Michigan and/or Ontario as applicable. Also search St. Clair River delta.
UF St. Clair River delta
BT North America
SA Michigan
Ontario

Saint Croix
Term introduced in 1978. Largest and most populous of the U.S. Virgin Islands which are E of Puerto Rico. Before 1978, also search Saint Croix Island; St. Croix.
CO N173500N175000
W0643500W0645500
UF Saint Croix Island
St. Croix
St. Croix Island
BT West Indies
SA Virgin Islands

Saint Croix Island
use Saint Croix

Saint Elias Mountains
Range near the Pacific Ocean. Index Alaska and/or Yukon Territory as applicable. Also search St. Elias Mountains.
UF St. Elias Mountains
BT North America
SA Alaska
Yukon Territory

Saint Francois Mountains
WNW of Cape Girardeau in SE Missouri. Also search St. Francois Mountains.
CO N372000N380500
W0901000W0904000
UF St. Francois Mountains
BT Missouri
United States

Saint Gall
Canton in NE Switzerland.
UF Saint Gallen
St. Gallen
BT Switzerland

Saint Gallen
use Saint Gall

Saint George Formation
Underlies Battery Formation; overlies Jurassic rocks. NW California. Also search St. George Formation.
UF St. George Formation
BT Pliocene
SA California

Saint Gotthard
use Gotthard Massif

Saint Helena
British island in S Atlantic Ocean about 1200 miles from W coast of Africa. Also search St. Helena.
UF Saint Helena Island
St. Helena
St. Helena Island
BT Atlantic Ocean

Saint Helena Island
use Saint Helena

Saint John
City on Bay of Fundy. Also search St. John.
UF St. John
BT New Brunswick
Canada

Saint John's
City on the Atlantic Ocean in SE Newfoundland. Also search St. John's.
UF St. John's
BT Newfoundland
Canada

Saint Johns River basin
NE Florida. Also search St. Johns River; St. Johns River basin.
UF St. Johns River basin

BT Florida
United States

Saint Laurent Limestone
Regarded as lower Hamilton (Cazenovia Stage) and below Lingle Limestone. S Illinois and E Missouri. Also search St. Laurent Limestone.
UF St. Laurent Limestone
BT Middle Devonian
Devonian
SA Illinois
Missouri

Saint Lawrence County
On the Saint Lawrence River. Also search St. Lawrence County.
UF St. Lawrence County
BT New York
United States

Saint Lawrence Estuary
The lower, wide part of the river from below Quebec City to its mouth on the Gulf of Saint Lawrence. Also search St. Lawrence Estuary.
UF St. Lawrence Estuary
BT Quebec
Canada
SA Saint Lawrence Valley

Saint Lawrence Lowlands
Along both sides of river from Lake Ontario to Quebec. Index New York and Canadian provinces as applicable. Also search St. Lawrence Lowlands.
UF St. Lawrence Lowland
St. Lawrence Lowlands
BT North America
SA Laurentide Ice Sheet
New York
Ontario
Quebec
Saint Lawrence Valley

Saint Lawrence River
Flows NE out of Lake Ontario into the Gulf of Saint Lawrence. Index New York and Canadian provinces as applicable. Also search Saint Lawrence AND river; St. Lawrence River.
UF St. Lawrence River
BT North America
SA Canada
New York
Ontario
Quebec

Saint Lawrence Valley
River valley extending to the Gulf of Saint Lawrence. Index New York and Canadian provinces as applicable. Also search Saint Lawrence AND valley; St. Lawrence Valley; St. Lawrence River valley.
UF St. Lawrence River valley
St. Lawrence Valley
SA Laurentide Ice Sheet
New York
Ontario
Quebec
Saint Lawrence Estuary
Saint Lawrence Lowlands

Saint Louis
City on the Mississippi River in E Missouri. Also search St. Louis.
UF St. Louis
BT Missouri
United States

Saint Louis County
Index states as applicable. Also search St. Louis County.
UF St. Louis County
BX Minnesota
Missouri
BT United States

Saint Louis Limestone
Comprises Croton and Verdi members. Also search St. Louis Limestone.
UF St. Louis Limestone
BT Meramecian
Mississippian
SA Alabama
Georgia
Illinois
Indiana
Iowa
Kentucky
Missouri
Tennessee
Upper Mississippian
Virginia

Saint Lucia
Island. A self governing state in association with United Kingdom in Windward Islands of the Lesser Antilles. Also search St. Lucia.
UF St. Lucia
BT West Indies
NX Soufriere
SA Windward Islands

Saint Mary Parish
On Gulf of Mexico. Also search St. Mary Parish.
UF St. Mary Parish
BT Louisiana
United States

Saint Marys Formation
In Chesapeake Group. Delaware, E Maryland and E Virginia. Also search St. Marys Formation.
UF St. Marys Formation
BT Miocene
SA Delaware
Maryland
middle Miocene
upper Miocene
Virginia

Saint Paul
City on the Mississippi River E of Minneapolis in Ramsey County in E. One of the "Twin Cities". Also search St. Paul.
UF St. Paul
BT Minnesota
United States

Saint Paul Island
Most northerly of Pribilof Islands.
BT Bering Sea
Pacific Ocean

Saint Paul Rocks
Group of uninhabited volcanic, rocky islets belonging to Brazil about 600 miles NE of Natal.
UF Saint Paul's Rocks
St. Paul's Rock
St. Paul's Rocks
BT Atlantic Ocean

Saint Paul's Rocks
use Saint Paul Rocks

Saint Peter Sandstone
Overlies Prairie du Chien Group and underlies Glenwood Shale Member of Platteville Formation. Also search St. Peter Sandstone.
UF St. Peter Sandstone
BT Middle Ordovician
Ordovician
SA Arkansas
Illinois
Indiana
Iowa
Kansas
Kentucky
Michigan
Minnesota
Missouri
Ohio
Oklahoma
Wisconsin

Saint Severin
use Saint-Severin

Saint Thomas
Island in the U.S. Virgin Islands E of Puerto Rico. Also search St. Thomas.
UF St. Thomas
BT West Indies
SA Virgin Islands

Saint Vincent
Self governing British state in the Windward Islands comprising Saint Vincent Island and the northern Grenadines. Also search St. Vincent.
UF St. Vincent
BT West Indies
NX Soufriere
SA Windward Islands

Saint Vincent Bay
use Saint Vincent Gulf

Saint Vincent Gulf
Inlet of Indian Ocean between Yorke Peninsula and mainland. Also search Saint Vincent; St. Vincent.
UF Saint Vincent Bay
BT South Australia
Australia

Saint-Chinian
Village in S part of country.
BT Herault
France

Saint-Etienne Basin
use Saint-Etienne coal basin

Saint-Etienne coal basin
In E central part of country. Also search Saint-Etienne.
UF Saint-Etienne Basin
BT Loire
France

Saint-Girons
Town at foot of central Pyrenees.
BT Ariege
France

Saint-Louis
Town on Reunion Island E of the Malagasy Republic.
BT Mascarene Islands
Pacific Ocean
SA Reunion

Saint-Severin
Village in E part of country. Also search St. Severin.
UF Saint Severin
St. Severin
BT Charente
France

Saint-Sylvestre Massif
In the Monts d'Ambazac in E central part of country.
BT Haute-Vienne
France

Saint-Vallier
Village in Provence Alps in extreme SE part of country.
UF Saint-Vallier-de-Thiey
BT Alpes-Maritimes
France

Saint-Vallier-de-Thiey
use Saint-Vallier

Sainte Genevieve Limestone
Includes Fredonia Limestone, Rosiclare Sandstone, and Levias Limestone members. N Alabama, Georgia, S Illinois, Indiana, Iowa, Kentucky, E Missouri, Tennessee. Also search Ste. Genevieve Formation; Ste. Genevieve Limestone.
UF Ste. Genevieve Formation
Ste. Genevieve Limestone
BT Meramecian
Upper Mississippian
Mississippian
SA Alabama
Georgia
Illinois
Indiana
Iowa
Kentucky
Missouri
Tennessee

Sainte-Baume Massif
In lower Provence Alps in SE part of country. Also search Sainte-Baume.
BT Var
France

Sainte-Marie-aux-Mines
Town near crest of the Vosges Mountains in NE part of country.
BT Haut-Rhin
France

Sainte-Victoire Massif
use Sainte-Victoire Mountain

Sainte-Victoire Mountain
In S part of country.
UF Sainte-Victoire Massif
BT Bouches-du-Rhone
France

Saipan
United States island N of Guam.
BT Mariana Islands
Micronesia

Saitama
Prefecture N of Tokyo.
BT Honshu
Japan

Sakar Mountains
Between Maritsa and Tundzha rivers in SE Bulgaria. Also search Sakar; Sakar Mountain.
BT Bulgaria

sakhaite
BZ borates
carbonates
BT minerals

Sakhalin
Island N of Hokkaido in W Okhotsk Sea. Also search Sakhalin Island.
CO N450000N550000
E1450000E1420000
UF Karafuto
Saghalin
Sakhalin Island
BT Russian Republic
USSR
SA Soviet Far East

Sakhalin Island
use Sakhalin

Sakmara
Village in central Orenburg Oblast in Southern Urals in European USSR.
CO N520000N523000
E0552000E0522000
BT Russian Republic
USSR

Sakmarian
Europe. Above Stephanian (Carboniferous), below Artinskian. Includes use on level 3 under age terms(1). See list E.
BA Lower Permian
Permian
BT Paleozoic

Sakoa Basin
Main source of coal in SW Malagasy Republic. Also search Sakoa.
BT Malagasy Republic

Saksagan River
In Dnepropetrovsk Oblast in E central Ukraine. Also search Saksagan.
BT Ukraine
USSR

Sakura-jima
Peninsula. NW projection of Osumi Peninsula in S Kyushu. An island until 1914. Also search Sakurajima.
UF Sakurajima
BT Japan

Sakurajima
use Sakura-jima

Salado Basin
600 mile-long basin of the combined Desaguadero and Salado river systems. Index provinces as applicable. Also search Salado.
BT Argentina
SA La Pampa
La Rioja
Mendoza
San Luis

Salair
City on Salair Ridge in W Kemerovo Oblast in S Siberia.
BT Russian Republic
USSR

Salair Ridge
Along borders of Altai Kray and Kemerovo Oblast in S Siberia. Also search Salair.
BT Russian Republic
USSR

Salaj
County in NW.
BT Transylvania
Romania

Salamanca
Province on the Portuguese border in W. Also a city.
BT Spain

Salat Valley
River valley in S part of country. Index departments as applicable. Also search Salat River.
BT France
SA Ariege
Haute-Garonne

Salem Limestone
Includes Kidd, Fults, Chalfin, and Rocher members. S Illinois, S Indiana, SE Iowa, W and central Kentucky, and E Missouri. Also search Salem.
BT Meramecian
Upper Mississippian
Mississippian
SA Illinois
Indiana
Iowa
Kentucky
Missouri

Salentina Peninsula
SE Apulia. The heel of the Italian boot. Also search Salentina and Salentine Peninsula.
BT Apulia
Italy

Salerno
City on Gulf of Salerno SE of Naples.
BT Campania
Italy

Salida
City in Chaffee County in central.
BT Colorado
United States

Salina Group
Includes Vernon Shale, Camillus Shale, Bertie Formation, Akron Dolomite, Syracuse Formation. Michigan and N Ohio.
BT Upper Silurian
Silurian
SA Michigan
Ohio

Salinas
City in Monterey County in W.
BT California
United States

Salinas River valley
use Salinas Valley

Salinas Valley
River valley in San Luis Obispo and Monterey counties in W California. Also search Salinas River; Salinas River valley.
UF Salinas River valley
BT California
United States

saline composition
Term introduced in 1978. Before 1978, search saline.
SA composition
ground water
salinity

saline water
use salt water

salinity
Includes use on level 2 under soils(1) and under paleoecology(1).
SA brackish water
brackish-water environment
desalinization
paleoecology
paleosalinity
saline composition
salt
salt marshes
sea water
sedimentation
soils
Solonchak soils
springs
treatment

Salisbury
City in NE.
BT Rhodesia

Salmo
Village in S.
BT British Columbia
Canada

Salmon
City in Lemhi County in E.
BT Idaho
United States

Salmon River
Rises in central part of state and flows N then W, and again N to empty into Snake River in W Idaho. Also search Salmon.
BT Idaho
United States
SA Salmon River breaks

Salmon River breaks
Large gorge or canyon in lower course of Salmon River.
BT Idaho
United States
SA Salmon River

Salonika
City on Gulf of Salonika in N part of country.
UF Saloniki
Thessaloniki
BT Greece
SA Macedonia

Saloniki
use Salonika

Salsigne Mine
N of Carcassonne in S part of country. Also search Salsigne.
BT Aude
France

salt
Includes use on level 1 and 2 as a commodity term (list C); on level 2 under mineral deposits, genesis(1).
SA brines
bromine
desalinization
diapirs
evaporites
fractional crystallization
halite

iodine
paleosalinity
salinity
salt domes
salt tectonics
sea water
sodium chloride

Salt Creek
Central.
BT Wyoming
    United States

salt domes
Includes use on level 3 under salt tectonics(1), salt(1) and under folds(1) style(2).
SA cap rocks
   diapirism
   diapirs
   domes
   folds
   salt
   salt tectonics

Salt Lake City
City in Salt Lake County in N.
BT Utah
    United States

Salt Lake County
N Utah.
BT Utah
    United States

Salt Lakes
Region in SW central Western Australia.
BT Western Australia
    Australia

salt marshes
Includes use on level 3 under sedimentation(1) or paleoecology(1) or ecology(1).
SA brackish-water environment
   marshes
   paleoecology
   salinity
   sebkha environment
   sedimentation

Salt Range
Between the Indus and the Jhelum rivers.
CO N330000N340000
   E0730000E0720000
BT Punjab
    Pakistan

Salt River
Rises in E part of state and flows W into Gila River W of Phoenix.
BT Arizona
    United States

salt tectonics
Includes use on level 1 (list A). Used for the study of the structure and mechanism of emplacement of salt domes. Term set options are: topic [causes, evolution, interpretation, mechanism, processes] subtopic (no area term)
UF halokinesis
SA cap rocks
   deformation
   diapirism
   diapirs
   emplacement
   faults
   folds
   salt
   salt domes
   tectonics
   tectonophysics

salt water
Includes use on level 3 under ground water(1) when comparing with fresh water. Also search saline water.
UF saline water

SA brackish water
   brines
   fresh water
   ground water
   salt-water intrusion
   sea water
   water

salt-water contamination
Not a valid index term for GeoRef. See contamination or salt-water intrusion.

salt-water intrusion
Includes use on level 2 under ground water(1).
UF encroachment (ground water)
   intrusion (ground water)
   sea-water encroachment
   sea-water intrusion
SA brines
   contamination
   ground water
   salt water

Salta
Province in NNW.
BT Argentina
NT Rio Blanco Basin
SA Yacoraite Formation

saltation
SA sedimentation
   transport

Salton Sea
Shallow saline lake just N of Imperial Valley in S.
CO N331000N333000
   W1153500W1160500
BT California
    United States

Salton Trough
Depression including from NW to SE, the Coachella Valley, the Salton Sea, and the Imperial Valley in S.
CO N310000N340000
   W1140000W1170000
BT California
    United States

Saltville Fault
SW Virginia. Also search Saltville.
BT Virginia
    United States

Salvador
City on the Atlantic Ocean in E part of country. Formerly Sao Salvador or Bahia.
UF Sao Salvador
BT Bahia
    Brazil

Salym
Stream in NW Novosibirsk Oblast in S Siberia.
BT Russian Republic
    USSR

Salzach River
Primarily in Salzburg. Index Austrian and German states as applicable. Also search Salzach.
BT Europe
SA Bavaria
   Salzburg
   Upper Austria

Salzburg
State in W central Austria. Also a city.
BT Austria
NT Bad Gastein
   Mitterberg
SA Salzach River
   Salzkammergut
   Venediger Group

Salzkammergut
Lake and mountain region of Eastern Alps. Index states as applicable.
BT Austria
SA Eastern Alps

Salzburg
Styria
Upper Austria

Samar
Island on E side of central part of archipelago.
BT Philippine Islands

Samara Bend
Region in Kuibyshev Oblast within oxbow of the middle Volga River where it reaches its easternmost point. Also search Samara.
BT Russian Republic
    USSR

Samaria
Region extending from the Mediterranean Sea to the Jordan River S of Galilee and N of Judaea. Includes most of the Israeli occupied West bank of the Jordan River. Index countries as applicable.
BT Middle East
SA Israel
   Jordan

samarium
Includes use on level 1 and 2 as a chemical element (list D).
UF Sm
SA elements
   rare earths

Samarkand
City and oblast in central Uzbekistan.
UF Samarqand
BT Uzbekistan
    USSR

Samarqand
use Samarkand

samarskite
BA oxides
BT minerals
SA niobates

Sambagawa Belt
Innermost and oldest of three belts of different metamorphic grades of schist in a Paleozoic Group. SW Honshu, Kyushu, and Shikoku. Also search Sambagawa; Sambagawa Schist; Sanbagawa Belt.
UF Sanbagawa Belt
BT Paleozoic
SA Honshu
   Kyushu
   Shikoku

Sambalpur
Town in N.
BT Orissa
    India

Samoa
Group of volcanic islands in SW central Pacific Ocean. American Samoa is in E part of group, and independent Western Samoa comprises the W part. Includes use on level 1 as an area term (list O). For term set options see list B.
BA Polynesia

Samos
Island in the Aegean Sea off W coast of Turkey.
BT Aegean Islands
    Greece

sample preparation
Includes use on level 3 under chemical analysis(1) techniques(2) and under micropaleontology(1); on level 3 under soils(1) methods, analytical(2), under palynology(1), and under paleontology(1).
SA chemical analysis
   chemical methods
   decrepitation
   differential thermal analysis

micropaleontology
paleontology
palynology
preparation
samples
sampling
soil sampling
soils
staining
techniques
thermal analysis

samples
Used for physical samples, not for statistical meaning.
SA lunar samples
   sample preparation
   sampling
   terrestrial materials
   thin sections

sampling
Includes use on level 2 under placers(1); on level 3 under soils(1) methods(2).
SA analysis
   field studies
   placers
   probability
   reliability
   sample preparation
   samples
   sediment sampling
   soil sampling
   techniques

San Andreas Fault
A fault system or zone extending for more than 600 miles from the Pacific Ocean at Point Arena through the San Francisco Peninsula and on into S California. Also search San Andreas; San Andreas fault system; San Andreas fault zone; San Andreas System.
UF San Andreas fault system
   San Andreas fault zone
   San Andreas System
BT California
    United States

San Andreas fault system
use San Andreas Fault

San Andreas fault zone
use San Andreas Fault

San Andreas System
use San Andreas Fault

San Andres Formation
In Manzano Group. Central and SE New Mexico. Also search San Andres.
BT Permian
SA Guadalupian
   Leonardian
   Lower Permian
   New Mexico
   Upper Permian

San Antonio
City in Bexar County in S central.
BT Texas
    United States

San Benito County
W California.
BT California
    United States

San Bernardino
City in extreme SW San Bernardino County in S.
BT California
    United States

San Bernardino County
S California.
CO N340000N354500
   W1141500W1174000
BT California
    United States

San Bernardino Mountains
In San Bernardino and Riverside counties in S.
BT California
United States
SA San Giorgio Mountain
Transverse Ranges

San Buenaventura
use Ventura

San Carlos Indian Reservation
SE central Arizona. Also search San Carlos.
BT Arizona
United States

San Clemente Island
One of the Santa Barbara Islands in the Pacific Ocean W of San Diego. Also search San Clemente.
BT California
United States

San Diego
City of the Pacific Ocean in San Diego County in S.
CO N324500N324500
W1171000W1171000
BT California
United States

San Diego County
S California.
CO N323000N333000
W1161000W1174000
BT California
United States

San Diego Formation
Rests with angular unconformity upon Rose Canyon shale member of the La Jolla, or on Poway Formation, or overlaps them and rests with marked unconformity upon Black Mountain volcanics; unconformably underlies Sweitzer Formation. S California.
BT Pliocene
SA California

San Diego Trough
Just off San Diego.
BT Pacific Ocean

San Felipe Formation
Also search San Felipe.
BT Cretaceous
SA Mexico

San Fernando
City in Los Angeles County. An enclave of Los Angeles.
BT California
United States

San Fernando Valley
Fertile basin about 20 miles NW of downtown Los Angeles.
BT California
United States

San Francisco
City on the Pacific Ocean in W. Coextensive with San Francisco County.
BT California
United States

San Francisco Bay
Inlet connected to the Pacific Ocean via the Golden Gate in W.
BT California
United States
SA San Francisco Bay region

San Francisco Bay region
Term Introduced in 1978.
SA San Francisco Bay

San Francisco County
Coextensive with city of San Francisco.
BT California
United States

San Francisco de la Selva
use Copiapo

San Francisco Mountain
One of the three peaks N of Flagstaff in Coconino County known as San Francisco Peaks.
UF Humphreys Peak
BT Arizona
United States
SA San Francisco Peaks

San Francisco Mountains
Range primarily in Catron County, New Mexico. Index states as applicable.
BT United States
SA Arizona
New Mexico

San Francisco Peaks
An eroded volcano 10 miles N of Flagstaff in Coconino County with Agassiz, Fremont and San Francisco (Humphreys) peaks on its rim. Also search San Francisco Volcanic Field.
UF San Francisco Volcanic Field
BT Arizona
United States
SA San Francisco Mountain

San Francisco Peninsula
Extends S from San Francisco the length of San Francisco Bay.
BT California
United States

San Francisco Volcanic Field
use San Francisco Peaks

San Gabriel Mountains
Range SW of the Mojave Desert primarily in Los Angeles County in S. Also search San Gabriel.
BT California
United States
SA Transverse Ranges

San Giorgio Mountain
Highest peak of San Bernardino Mountains in San Bernardino County in S.
BT California
United States
SA San Bernardino Mountains

San Jacinto Fault
Riverside County in S California. Also search San Jacinto.
BT California
United States

San Joaquin County
In N San Joaquin Valley in central. Also search San Joaquin.
BT California
United States

San Joaquin Valley
Southern half of Central Valley. Extends from Buena Vista Lake in S to Sacramento Valley in N. Also search San Joaquin.
BT California
United States
SA Central Valley

San Jose
City of San Francisco Bay in Santa Clara County in W.
BT California
United States

San Jose Formation
Lithology of formation is highly variable, both vertically and horizontally. N New Mexico, and S Colorado.
BT lower Eocene
Eocene
SA Colorado
New Mexico

San Juan
Province in W Argentina, and city on the Atlantic Ocean in N Puerto Rico. Index Argentina and/or Puerto Rico as applicable.
SA Argentina
Barreal
Puerto Rico
Talacasto

San Juan Basin
River basin. Index states as applicable.
BT United States
SA Arizona
Colorado
New Mexico
Utah

San Juan Bautista
Town in San Benito County in W.
BT California
United States

San Juan County
Index states as applicable.
BX Colorado
New Mexico
Utah
Washington
BT United States

San Juan Formation
Middle and late Tertiary in SW Colorado; Pleistocene in Puerto Rico.
BT Cenozoic
SA Colorado
middle Tertiary
Pleistocene
Puerto Rico
upper Tertiary

San Juan Islands
Group of islands off NW Washington and E of Vancouver Island.
BT Washington
United States

San Juan Mountains
Range of the Rocky Mountains in SW.
CO N370000N374500
W1063000W1073000
BT Colorado
United States
SA Rocky Mountains

San Juan River
Rises in S Colorado and flows SW, bends W, then NW emptying into the Colorado River in SE Utah. Index states as applicable.
BT United States
SA Colorado
New Mexico
Utah

San Juan volcanic field
In San Juan Mountains of SW Colorado. Also search San Juan Volcanics.
BT Colorado
United States

San Lorenzo Formation
Includes Twobar Shale and Rices Mudstone. In Santa Cruz Mountain region of S California. Also search San Lorenzo.
BT Tertiary
SA California
Eocene
Oligocene

San Luis
Province in W central.
BT Argentina
SA Salado Basin

San Luis Obispo
City in San Luis Obispo County in SW.
BT California
United States

San Luis Obispo County
SW California.
BT California
United States

San Luis Potosi
State in E central Mexico. Also city.
BT Mexico
NT Valles
SA Sierra Madre Oriental

San Luis Valley
Once bottom of extensive lake. In Saguache, Alamosa, and Conejos counties in S Colorado and in Taos County in New Mexico. Index states as applicable.
BT United States
SA Colorado
New Mexico

San Manuel
Village in Pinal County in S central.
BT Arizona
United States

San Marcos Arch
S central Texas. Also search San Marcos.
BT Texas
United States

San Mateo County
Just S of San Francisco.
BT California
United States

San Mateo Mountains
SW Socorro County in SW central.
BT New Mexico
United States

San Miguel Island
Northernmost of Santa Barbara Islands in Pacific Ocean SW of Santa Barbara. Also search San Miguel AND appropriate area.
BT California
United States

San Nicolas Island
In central Santa Barbara Islands in the Pacific Ocean off SW California. Also search San Nicolas.
BT California
United States

San Onofre Breccia
Underlies Monterey Formation; overlies Cozy Dell. San Diego County in S California.
BT middle Miocene
Miocene
SA California

San Pedro
Harbor in Los Angeles County. Former city which was annexed to Los Angeles in 1909.
BT California
United States

San Pedro Hills
use Palos Verdes Hills

San Pedro Valley
In San Mateo County in W California. Also search San Pedro AND California.
BT California
United States

San Rafael Swell
E central Utah. Also search San Rafael.
BT Utah
United States

San Saba County
Central.
BT Texas
United States

San Salvador
Island in E central Bahamas, and a city in El Salvador. Index Bahamas

and/or El Salvador as applicable.
SA Bahamas
El Salvador

San San
use Sansan

San Sebastian
City on Bay of Biscay in Basque country.
BT Guipuzcoa
Spain

Sanbagawa Belt
use Sambagawa Belt

sand
For sand used as construction material, see gravel. Includes use on level 1 as a commodity term (list C) for glass, ceramic, chemical use, etc.; on level 3 under sediments(1) carbonate sediments(2) and clastic sediments(2). See list N.
BA clastic sediments
BT sediments
SA aggregate
 alluvium
 arenaceous texture
 carbonate sediments
 construction materials
 grains
 gravel
 greensand
 loam
 quartz sand
 rounding
 sandstone
 silica
 silt
 soils
 terrigenous materials

sand bodies
Includes use on level 3 under sedimentary structures(1) planar bedding structures(2). See list K.
BA planar bedding structures
 sedimentary structures
SA bedding plane irregularities

Sand Hills
Belt, 20 to 40 miles wide, of low, sandy hills extending along inner border of coastal plain from central North Carolina to central Georgia. Also large area of stable dunes covered with vegetation in NW central Nebraska. Index states as applicable.
BT United States
SA Georgia
 Nebraska
 North Carolina
 South Carolina

sand waves
Includes use on level 3 under sedimentary structures(1) bedding plane irregularities(2). See list K.
BA bedding plane irregularities
 sedimentary structures
SA antidunes
 dunes
 megaripples
 ripple marks

Sandakan
City on Sandakan Harbor on Sulu Sea in Sabah in East Malaysia.
BT Malaysia

sandarac
use realgar

Sandelzhausen
Village N of Munich.
BT Bavaria
 West Germany

Sandia Mountains
NE of Albuquerque in N central.
BT New Mexico
 United States

Sandnes
Town S of Stavanger in SW part of country.
BT Rogaland
 Norway

Sandomierz
Town on Vistula River in SE central part of country.
UF Sandomir
BT Kielce
 Poland

Sandomir
use Sandomierz

Sandoval County
NW central.
BT New Mexico
 United States

sandstone
Includes use on level 1 as a commodity term (list C); on level 3 under sedimentary rocks (1) clastic rocks(2). See list I. Also search psammite.
UF psammite
BA clastic rocks
BT sedimentary rocks
SA arenaceous texture
 arkose
 building stone
 construction materials
 dimension stone
 gaize
 graywacke
 greensand
 metasandstone
 microbreccia
 orthoquartzite
 red beds
 sand
 sandstone dikes
 subgraywacke
 terrigenous materials

sandstone dikes
Includes use on level 3 under sedimentary structures(1) soft sediment deformation(2). See list K.
BA soft sediment deformation
 sedimentary structures
SA clastic dikes
 sandstone

Sandur
Town in W central.
BT Mysore
 India

sandurs
use outwash plains

Sandy Hook
Peninsula in NE Monmouth County constituting S side of entrance to lower New York Bay.
CO N402300N402900
 W0735500W0770500
BT New Jersey
 United States

sandy texture
use arenaceous texture

Sangamon
River which flows into the Illinois River in central.
BT Illinois
 United States

Sangamonian
Pertaining to third interglacial stage of Pleistocene Epoch in North America.
BA upper Pleistocene
 Pleistocene
 Quaternary
BT Cenozoic
SA Riss/Wurm Interglacial

Sangilen Mountains
S Tuva A.S.S.R. just N of Mongolian border. Also search Sangilen.

BT Russian Republic
 USSR

Sangre de Cristo Mountains
Range of the Rocky Mountains. Index states as applicable. Also search Sangre de Cristo Range.
UF Sangre de Cristo Range
 Sangre de Cristo Uplift
BT United States
SA Colorado
 New Mexico
 Rocky Mountains

Sangre de Cristo Range
use Sangre de Cristo Mountains

Sangre de Cristo Uplift
use Sangre de Cristo Mountains

Sangun
Village near the Afghanistan border in NE part of country.
BT Iran

sanidine
UF glassy feldspar
 rhyacolite
BA feldspar group
 framework silicates
 silicates
BT minerals
SA K-feldspar
 orthoclase

sanitary landfills
BT landfills
SA waste disposal

Sannoisian
Europe. Lower Oligocene in France overlain by Stampian; same as Lattorfian. Includes use on level 3 under age terms(1). See list E.
BA lower Oligocene
 Oligocene
 Paleogene
 Tertiary
BT Cenozoic
SA Tongrian

Sanpete County
Central Utah. Also search Sanpete.
BT Utah
 United States

Sansan
Village in NW Panama.
UF San San
BT Panama

Santa Ana
City in Orange County at base of Santa Ana Mountains in S.
BT California
 United States

Santa Ana Mountains
Range along border between Orange and Riverside counties in S California. Also search Santa Ana AND California.
BT California
 United States

Santa Barbara
City on Santa Barbara Channel in Santa Barbara County in SW.
BT California
 United States

Santa Barbara Basin
Undersea feature between Santa Barbara County and the N Santa Barbara Islands.
BT Pacific Ocean

Santa Barbara Channel
Between Santa Barbara County and N Santa Barbara Islands.
BT California
 United States

Santa Barbara County
SW California.
CO N342500N351000
 W1192500W1204500
BT California
 United States

Santa Catalina Island
Island in central Santa Barbara Islands in Pacific Ocean off Long Beach. Also search Santa Catalina AND California.
BT California
 United States

Santa Catalina Mountains
Small range in NE Pima County in S Arizona. Also search Santa Catalina AND Arizona.
BT Arizona
 United States

Santa Catarina
State in S.
BT Brazil
NT Santa Catarina Island
SA Brazilian Shield
 Estrada Nova Formation
 Pelotas Basin
 Serra do Mar

Santa Catarina Island
Island in Atlantic Ocean just off central Santa Catarina. Also search Santa Catarina.
BT Santa Catarina
 Brazil

Santa Clara
City in Santa Clara County in W.
BT California
 United States

Santa Clara County
S of San Francisco Bay.
BT California
 United States

Santa Clara Valley
River valley in Los Angeles and Ventura counties. Also the S extension of the San Francisco Bay depression in Santa Clara and San Benito Counties is called the Santa Clara Valley.
BT California
 United States

Santa Cruz
Province in S Argentina and department in E Bolivia. Index countries as applicable.
BT South America
SA Argentina
 Baquero Formation
 Bolivia
 Patagonia
 Puerto Deseado

Santa Cruz County
Index states as applicable. Also search Santa Cruz AND appropriate area.
BX Arizona
 California
BT United States

Santa Cruz Island
Easternmost of northern islands in Santa Barbara Islands off Ventura County. Also search Santa Cruz AND California.
CO N335000N341000
 W1193000W1195500
BT California
 United States

Santa Cruz Mountains
One of the Coast Ranges extending along the W side of Santa Clara Valley just S of San Francisco Bay. Also search Santa Cruz AND California.
BT California
 United States
SA Coast Ranges

Santa Eulalia
Town on Santa Eulalia River with water reservoir and hydroelectric plant

Santa Eulalia • Sargent

servicing Lima.
BT Lima
    Peru

Santa Fe
Province in NE central Argentina, and a city in N central New Mexico. Index Argentina and/or New Mexico as applicable.
SA Argentina
    New Mexico

Santa Fe County
N central.
BT New Mexico
    United States

Santa Isabel
Volcanic island of the British Solomon Islands in E central.
BT Solomon Islands
    Australasia
SA Melanesia

Santa Lucia Range
One of the Coast Ranges in Monterey and San Luis Obispo counties in W.
BT California
    United States
SA Coast Ranges

Santa Margarita Formation
Includes Quatal Red Clay Member. S California. Also search Santa Margarita.
BT upper Miocene
    Miocene
    United States

Santa Maria
City in Santa Barbara County in SW.
BT California
    United States

Santa Maria Formation
S California.
BT Cenozoic
SA California
    Pleistocene
    Pliocene

Santa Marta
City on Caribbean Sea.
BT Colombia

Santa Monica Mountains
E-W range paralleling N shore of Santa Monica Bay in S. Also search Santa Monica AND California.
BT California
    United States

Santa Rita
Village in Grant County in SW.
BT New Mexico
    United States

Santa Rita Mountains
Santa Cruz County in SE. Also search Santa Rita.
BT Arizona
    United States

Santa Rosa
City in Sonoma County N of San Francisco.
BT California
    United States

Santa Rosa Mountains
SE California. Range along W side of Coachella Valley.
BT California
    United States

Santa Rosa Range
In NE Humboldt County in N.
BT Nevada
    United States

Santa Ynez Mountains
E-W coastal range bordering Santa Barbara Channel in SW. Also search Santa Ynez.
BT California
    United States

Santana Formation
BT Cretaceous
SA Brazil

Santander
Province in N Spain, and department in N central Colombia. Index countries as applicable.
NT Santander Massif
SA Cantabrian Basin
    Colombia
    Spain

Santander Massif
In the E central Cantabrian Mountains in N part of country.
BT Santander
    Spain

Santee River
Flows into Atlantic Ocean. SE central South Carolina. Also search Santee.
BT South Carolina
    United States

Santiago
Province in central Chile. Also a city.
BT Chile

Santiago de Compostela
City in NW part of country.
BT La Coruna
    Spain

Santiago del Estero
Province in N.
BT Argentina

Santiaguito
Village just SW of Mexico City.
BT Federal District
    Mexico

Santo Domingo
City on the Caribbean Sea.
UF Ciudad Trujillo
BT Dominican Republic

Santonian
Europe. Above Coniacian, below Campanian. Includes use on level 3 under age terms(1). See list E.
BA Upper Cretaceous
    Cretaceous
BT Mesozoic
SA Senonian

Santorin
Volcano. A portion of the crater forms the island of Thera, southernmost of the Cyclades, in the Aegean Sea. Also search Santorin Volcano.
UF Santorin Volcano
    Santorini Volcano
BT Aegean Islands
    Greece
SA Thera

Santorin Volcano
use Santorin

Santorini
use Thera

Santorini Volcano
use Santorin

Sao Francisco Basin
Drainage basin of Sao Francisco River. Index states as applicable. Also search Sao Francisco; Sao Francisco River.
BT Brazil
SA Alagoas
    Bahia
    Minas Gerais
    Pernambuco
    Sergipe

Sao Gabriel
City in extreme S part of country.
BT Rio Grande do Sul
    Brazil

Sao Miguel Island
Largest island in group. In Portugal's Ponta Delgado District. Also search Sao Miguel.
BT Azores
    Atlantic Ocean

Sao Paulo
State in S Brazil. Also a city 220 miles WSW of Rio de Janeiro.
CO S253000S195000
    W0443000W0533000
BT Brazil
NT Amparo
    Marilia
    Ribeira de Iguape River
    Rio Claro
SA Bauru Formation
    Botucatu Formation
    Brazilian Shield
    Estrada Nova Formation
    Irati Formation
    Serra do Mar
    Serra Gerral Formation
    Tubarao Group

Sao Salvador
use Salvador

Saone Valley
River valley in E central part of country extending from the Vosges Mountains to the Rhone river at Lyon. Also search Saone.
BT France

Saone-et-Loire
Department in E central.
BT France
NT Autun
SA Morvan

Saoura
Department on the Moroccan border in W.
BT Algeria
NT Bechar

Sapelo Island
In Atlantic Ocean off McIntosh County.
BT Georgia
    United States

saponite
BA sheet silicates
    silicates
BT minerals
SA clay minerals
    smectite

Sappada
Village on Piave River in NE part of country.
BT Veneto
    Italy

sapphire
BA oxides
BT minerals
SA corundum
    gems

sapphirine
BA orthosilicates
    silicates
BT minerals
SA aluminosilicates

saprolite
Includes use on level 3 under sedimentary rocks(1) clastic rocks(2). See list I.
BA clastic rocks
BT sedimentary rocks

sapropel
Includes use on level 3 under organic materials(1).
BA organic materials
SA anaerobic environment
    gyttja
    hydrocarbons
    ooze

sapropelic coal
use sapropelite

sapropelite
UF sapropelic coal
BA organic residues
BT sedimentary rocks
SA coal

Saragossa
Province in NE central Spain. Also a city. When referring to Mexico, use Zaragoza.
BT Spain
SA Calatayud-Teruel Basin
    Zaragoza

Sarajevo
City in E Bosnia and Herzegovina.
UF Serajevo
BT Yugoslavia

Saratoga Chalk
Includes a lower chalk member and an upper argillaceous-arenaceous unit. SW Arkansas.
BT Upper Cretaceous
    Cretaceous
SA Arkansas

Saratov
City on W bank of the Volga River. Also an oblast.
BT Russian Republic
    USSR

Sarawak
State of East Malaysia in NW Borneo. Former British protectorate governed by the Brooke family. Joined Malaysia in 1963. Included use as a first order area term until 1976. Now includes use on level 3. See under Malaysia(1).
BT Malaysia
SA Borneo
    East Malaysia
    Malay Archipelago

Sarbay
River in Kuybyshev Oblast in the Middle Volga region.
BT Russian Republic
    USSR

sarcopside
BZ halides
    phosphides
BT minerals
SA fluorides

Sardinia
Island and autonomous region in the Mediterranean Sea. Includes use on level 1 as an area term (list O). For term set options see list B.
CO N385000N413000
    E0094500E0081500
BA Italy
    Europe
NT Bosano
    Gerrei
    Iglesiente
    Logudoro
    Orosei
    Sarrabus
    Sulcis
SA Mediterranean region
    Mediterranean Sea

Sargasso Sea
Large tract of relatively still water between the West Indies and Bermuda. Named after Sargasso weed which floats there.
BT Atlantic Ocean

Sargent
Village in S Santa Clara County in W.
BT California
    United States

Sarmatian
Europe. Above Tortonian, below Pontian. Includes use on level 3 under age terms(1). See list E.
BA Miocene
    Neogene
    Tertiary
BT Cenozoic

Sarrabus
Region in SE.
BT Sardinia
    Italy

Sarthe
Department in NW.
BT France
NT Le Mans

Sary-Su
use Sarysu

Sary-Su River
use Sarysu

Sarysu
River in central Kazakhstan. Flows S into desert but becomes dry before reaching Syr Darya. Also search Sary-Su; Sary-Su River.
UF Sary-Su
    Sary-Su River
BT Kazakhstan
    USSR

Sasca-Montana
Village in SW.
BT Banat
    Romania

Sasebo
City on inlet of east China Sea in NW.
BT Kyushu
    Japan

Saskatchewan
Includes use on level 1 as an area term (list O). For term set options see list B.
CO N490000N600000
   W1012000W1100000
BA Canada
NT Amisk Lake
   Beaverlodge
   Coronation Mine
   Delmas
   Eldorado
   Esterhazy
   Estevan
   Hanson Lake
   La Ronge
   Lac La Ronge
   Leoville
   Regina
   Rosetown
   Saskatoon
   Tazin Lake
   Wapawekka Lake
   Wollaston Lake Belt
SA Assiniboine River
   Belly River Formation
   Canadian Shield
   Carswell Structure
   Churchill Province
   Cypress Hills Formation
   Edmonton Formation
   Elk Point Basin
   Elk Point Group
   Great Plains
   Missi Group
   Missouri River basin
   North Saskatchewan River
   Paskapoo Formation
   Prairie Evaporite
   Prince Albert Group
   Ravenscrag Formation
   Saskatchewan River
   Shaunavon Formation
   Slave Province
   Souris River basin
   South Saskatchewan River
   Whitemud Formation
   Williston Basin
   Winnipeg Formation
   Wood Mountain Formation
   Wynyard

Saskatchewan River
Formed by confluence of the North and South Saskatchewan rivers in central Saskatchewan. Empties into N Lake Winnipeg. Index provinces as applicable.
BT Canada
SA Manitoba
   North Saskatchewan River
   Saskatchewan
   South Saskatchewan River

Saskatoon
City in S central.
BT Saskatchewan
   Canada

satellite
A valid term through 1977. After 1977, use satellite methods.

satellite measurements
Includes use on level 2 under geodesy(1).
SA geodesy
   geodetic coordinates
   measurement

satellite methods
Term introduced in 1978. Used for artificial satellites. For natural satellites, see satellites. Includes use on level 3 under geophysical methods(1). Before 1978, search satellite AND methods.
BA geophysical methods
SA Apollo
   Apollo 9
   Apollo 11
   Apollo 12
   Apollo 13
   Apollo 14
   Apollo 15
   Apollo 16
   Apollo 17
   Explorer 35
   geophysical surveys
   Landsat
   Luna 16
   Luna 20
   Mariner 6
   Mariner 7
   Mariner 9
   Mariner 10
   methods
   Pioneer 10
   remote sensing
   satellites
   Skylab
   Surveyor 3
   Viking

satellites
Used for natural satellites. Includes use on level 3 under Moon(1) or specific planets, e.g. Jupiter(1). For artificial satellites, use satellite methods.
SA Deimos
   Galilean satellites
   Jupiter
   Mars
   Phobos
   satellite methods
   Saturn
   Tethys Satellite

Satna
Town in NE Madhya Pradesh.
UF Sutna
BT Madhya Pradesh
   India

Satpura Range
Line of hills forming N limit of Deccan Plateau. Primarily in Madhya Pradesh. Index states as applicable. Also search Satpura.
BT India
SA Madhya Pradesh
   Maharashtra
   Uttar Pradesh

Satsuma Peninsula
Between East China Sea and Kagoshima in SW Kyushu. Also search Satsuma.
BT Kyushu
   Japan

Satu Mare
County in NW.
BT Transylvania
   Romania

saturated materials
Term introduced in 1978. Before 1978, search saturated.
SA materials

saturation
Used as a general term, e.g. under sea water(1) or springs(1).
SA concentration
   geochemistry
   sea water
   soils
   solubility
   springs
   water

Saturn
Includes use on level 1 and 2. See entry under Moon(1) for term set options.
SA Iapetus Satellite
   outer planets
   planetology
   planets
   satellites
   solar system
   Tethys Satellite

Sau Alpe
use Sau Alps

Sau Alps
Mountains in NW Carinthia. Also search Saualpe.
UF Sau Alpe
   Saualpe
BT Carinthia
   Austria
SA Alps

Saualpe
use Sau Alps

Sauce Grande River valley
S Buenos Aires Province. Also search Sauce Grande; Sauce Grande River.
BT Buenos Aires Province
   Argentina

Saucesian
North America. Above Zemorrian, below Relizian. Includes use on level 3 under age terms(1). See list E.
BA lower Miocene
   Miocene
   Neogene
   Tertiary
BT Cenozoic
SA Oligocene

Saudi Arabia
Includes use on level 1 as an area term (list O). For term set options see list B.
CO N170000N323000
   E0570000E0344500
BA Arabian Peninsula
NT Hail
SA Arabian Plate
   Arabian Shield
   Asia
   Red Sea Basin

Sauerland
Region S and E of the Ruhr River.
BT North Rhine-Westphalia
   West Germany

Sault Sainte Marie
Twin cities on Saint Mary's River between Lake Huron and Lake Superior. Near Sault Saint Marie Canals. Index Michigan and/or Ontario. Also search Sault Ste. Marie.
UF Sault Ste. Marie
SA Michigan
   Ontario

Sault Ste. Marie
use Sault Sainte Marie

Saurashtra
Region and former state comprising greater part of Kathiawar Peninsula in W part of country.
BT Gujarat
   India

Saurischia
Order. Includes use on level 3 under Reptilia(1) Archosauria(2).
BA Archosauria
   Reptilia
BT Tetrapoda
   Vertebrata
NA Sauropoda
   Theropoda

Sauropoda
Infraorder. Includes use on level 3 under Reptilia(1) Archosauria(2).
BA Saurischia
   Archosauria
   Reptilia
BT Tetrapoda
   Vertebrata

sausage structure
use boudinage

Sausar Group
use Sausar Series

Sausar Series
Indian Shield region of central S and S India. Also search Sausar Group.
UF Sausar Group
BT Archean
SA India

Savage River
In Garrett County in extreme W.
BT Maryland
   United States

Savanna Formation
In Krebs Group. Includes Spaniard Limestone Member, Spiro Sandstone, Sam Creek Limestone, Doneley Limestone. W Arkansas and E and S Oklahoma. Also search Savanna.
BT Desmoinesian
   Middle Pennsylvanian
   Pennsylvanian
SA Arkansas
   Oklahoma

Savannah River
Serves as boundary between Georgia and South Carolina, and flows into Atlantic Ocean. Index states as applicable.
BT United States
SA Georgia
   South Carolina

Savannah River Plant
U.S. Atomic Energy Commission plant in Aiken and Barnwell counties on Savannah River.
BT South Carolina
   United States

Savoia
use Savoy

Savoie
Department in SE.

Savoie
CO N450000N455000
 E0070000E0053000
BT France
NT Ambin Massif
 Tarentaise
 Vanoise

Savona
City and province near Genoa in NW part of country.
BT Liguria
 Italy

Savoy
Historical region of SE France and NW Italy, now chiefly in French departments of Haute-Savoie and Savoie. Index countries as applicable.
UF Savoia
BT Europe
SA France
 Italy

Savoy Alps
NW offshoots of Graian Alps S of Lake Geneva. Also search Savoy.
BT Haute-Savoie
 France
SA Alps
 Mont Blanc
 Vanoise

Sawatch Mountains
use Sawatch Range

Sawatch Range
Range of the Rocky Mountains in central.
UF Sawatch Mountains
BT Colorado
 United States
SA Rocky Mountains

Sawtooth Range
Large group of mountain ranges in S central Idaho.
BT Idaho
 United States

Saxonian
Europe. Above Autunian, below Thuringian. Includes use on level 3 under age terms(1). See list E.
BA Middle Permian
 Permian
BT Paleozoic
SA Rotliegendes

Saxony
Former German state now part of East Germany. Index districts as applicable.
BT East Germany
SA Cottbus
 Dresden
 Karl-Marx-Stadt
 Leipzig
 Saxony-Thuringia

Saxony-Anhalt
Former East German state. Index districts as applicable.
BT East Germany
SA Halle
 Magdeburg

Saxony-Thuringia
Region in S and SE East Germany including the former German states of Saxony and Thuringia.
BT East Germany
SA Saxony
 Thuringia

Sayak
Village N of W Lake Balkash in W central.
BT Kazakhstan
 USSR

Sayan
E-W range just N of Mongolia extending from S Krasnoyarsk Kray on W across Tuva A.S.S.R. into W Irkutsk Oblast on the E. Sayan Mountains used from 1976-1978. Also search Sayan Mountains; Sayan Range.
UF Sayan Mountains
 Sayan Range
BT USSR
NA Eastern Sayan
NT Western Sayan

Sayan Mountains
use Sayan

Sayan Range
use Sayan

Sb
use antimony

Sc
use scandium

Scaglia Formation
Cretaceous to Eocene. In the Apennines and NE Italy.
SA Cretaceous
 Eocene
 Italy
 Paleocene

scales, time
use time scales

Scandinavia
Region comprising Denmark, Norway, and Sweden. Finland and Iceland sometimes included. Index countries as applicable. Includes use on level 1 as an area term (list O). For term set options see list B.
CO N543000N710000
 E0320000E0050000
BA Europe
SA Caledonides
 Denmark
 Norway
 Sweden

scandium
Includes use on level 1 and 2 as a chemical element (list D).
UF Sc
SA elements
 rare earths

Scania
use Skane

scanning
A valid term through 1978 on level 3 under electron microscopy(1). Use scanning method.

scanning electromicroscopy data
use SEM data

scanning method
Term introduced in 1978. Includes use on level 3 under electron microscopy(1). Before 1978, also search scanning.
SA electron microscopy
 methods
 SEM data
 TEM data

Scaphites
Genus. Includes use on level 3 under Mollusca(1) Cephalopoda(2).
BA Ammonoidea
 Cephalopoda
 Mollusca
BT Invertebrata

Scaphopoda
Includes use on level 2 under Mollusca(1). See list F.
BA Mollusca
BT Invertebrata
NA Dentalium

scapolite
BA scapolite group
 framework silicates
 silicates
BT minerals

scapolite group
Includes use in combination with framework silicates (i.e. framework silicates, scapolite group) on level 2 under minerals(1). See list L.
BA framework silicates
 silicates
BT minerals
NA kenyaite
 magadiite
 scapolite
SA meionite

scarps
Also search escarpments.
UF escarpments
SA cliffs
 fault scarps
 faults
 geomorphology
 slopes

scattering
A general term used for properties of waves.
SA absorption and scattering
 dissipation
 elastic waves
 electron microscopy
 optical properties
 spectroscopy

scawtite
BA chain silicates
 silicates
BT minerals

Schaffhausen
Northernmost canton. Also a city.
BT Switzerland

schapbacite
use matildite

scheelite
BA tungstates
BT minerals

Schefferville
Town in NE near Labrador border.
BT Quebec
 Canada

Schela
Village in SW.
BT Walachia
 Romania

Schelde River
use Scheldt River

Scheldt River
Rises in N France and empties into North Sea through two estuaries, the East and West Scheldt. Index countries as applicable.
UF Schelde River
BT Europe
SA Belgium
 France
 Netherlands

Schenectady Formation
Underlies Indian Ladder Beds; overlies Snake Hill Formation. E central New York. Also search Schenectady.
BT Middle Ordovician
 Ordovician
SA New York

Schiefergebirge
use Rhenish Schiefergebirge

schist
Includes use on level 3 under metamorphic rocks(1) schists(2). See list J. Also search crystalline schist; search hornblende schist.
UF bookstone
 crystalline schist
BA schists
BT metamorphic rocks
SA biotite schist
 calc-schist
 chlorite schist
 glaucophane schist
 greenschist
 mica schist
 pelitic schist

schistosity
Includes use on level 3 under foliation(1) style(2) and under structural analysis(1).
BA foliation
SA axial-plane structures
 cleavage
 flow cleavage
 metamorphic rocks
 orientation
 schists
 slaty cleavage
 slip cleavage
 structural analysis

schists
Includes use on level 2 under metamorphic rocks(1). See list J.
BT metamorphic rocks
NA biotite schist
 blueschist
 calc-schist
 chlorite schist
 epidiorite
 glaucophane schist
 greenschist
 greenstone
 listwanite
 mica schist
 pelitic schist
 schist
SA schistosity

schizomycetes
use bacteria

Schizophoria
Genus. Includes use on level 3 under Brachiopoda(1) Orthida(2).
BA Enteletacea
 Orthida
 Articulata
 Brachiopoda
BT Invertebrata

Schizophyta
use Cyanophyta

Schleswig-Holstein
State in extreme N.
CO N531500N550000
 E0112500E0081000
BT West Germany
NT Helgoland
 Holstein
 Kiel
 Sylt
SA Jutland

schlieren
SA igneous rocks

Schlotheim
Town in E.
BT Halle
 East Germany

Schneeberg
Town in Erzgebirge.
BT Karl-Marx-Stadt
 East Germany

Schoharie County
E central.
BT New York
 United States

scholzite
BA phosphates
BT minerals

schorlomite
BA garnet group
 orthosilicates
 silicates
BT minerals

schreibersite
Meteorite mineral. Also search rhabdite.

UF rhabdite
BZ native elements and alloys
      sulfides
BT minerals

Schuler Formation
In Cotton Valley Group. Includes Morgan Sands Zone, Jones Sand, Dorcheat and Shongaloo members. Subsurface.
BT Upper Jurassic
      Jurassic
SA Arkansas
      Cotton Valley Group
      Louisiana
      Texas

Schuylkill County
E Central.
BT Pennsylvania
      United States

Schwagerina
Genus. Includes use on level 3 under foraminifera(1) Fusulinidae(2).
BA Fusulinidae
      Fusulinina
      foraminifera
BT Invertebrata

Schwartzburg Anticline
use Schwarzburg Anticlinorium

Schwarzburg Anticlinorium
In Thuringia.
UF Schwartzburg Anticline
BT Gera
      East Germany

Schwarzenberg
City in W.
BT Karl-Marx-Stadt
      East Germany

Schwechat Valley
River valley SE of Vienna in E.
BT Lower Austria
      Austria

Schwerin
District in NW. Also a city.
BT East Germany

scintillations
Includes use on level 2 under ionosphere(1).
SA gamma rays
      ionosphere
      particles
      radioactivity

Scioto drainage basin
use Scioto River basin

Scioto River basin
Central and S central.
UF Scioto drainage basin
BT Ohio
      United States

Scleractinia
Includes use on level 2 under Coelenterata(1). See list F. Also search Madreporaria; Madrepores.
UF Madreporaria
BA Anthozoa
      Coelenterata
BT Invertebrata
NA Montastrea
SA corals
      Rugosa
      Tabulata

sclerites
SA Echinodermata
      fossils
      Holothuroidea

scolecite
BA zeolite group
      framework silicates
      silicates
BT minerals

scolecodonts
Includes use on level 2 under worms(1). See list F.
BA worms
SA Annelida

Scoresby Land
Region in E on Greenland Sea between King Oscar Fjord on N and Scoresby Sound on the S.
BT Greenland
      Arctic region

Scoresby Sound
Deep inlet and fjord system of Greenland Sea on central E coast. Also search Scoresby Sund.
UF Scoresby Sund
BT Greenland
      Arctic region

Scoresby Sund
use Scoresby Sound

scoria
BT volcanic rocks
SA pumice
      pyroclastics

scorodite
BA arsenates
BT minerals

Scotia Arc
use Scotia Ridge

Scotia Ridge
Extends in an arc in the South Atlantic Ocean from S of the Falkland Islands eastward to the South Sandwich Islands and then westward to S of the South Orkney Islands. Also search Scotia Arc.
UF Scotia Arc
BT Atlantic Ocean

Scotia Sea
Part of the South Atlantic Ocean within the arc of the Scotia Ridge and E of Drake Passage.
BT Antarctic Ocean
SA Atlantic Ocean

Scotian Shelf
Off Nova Scotia. Also search Nova Scotia Shelf.
UF Nova Scotia Shelf
      Nova Scotian Shelf
BT Atlantic Ocean

Scotland
Includes use on level 1 as an area term (list O). For term set options see list B.
CO N544000N610000
      W0004500W0083000
BT Great Britain
      United Kingdom
BA Europe
NT Aberdeen
      Aberdeenshire
      Angus
      Ardnamurchan
      Argyllshire
      Arran
      Ayrshire
      Banffshire
      Berwickshire
      Cairngorm Mountains
      Caithness
      Dundee
      East Lothian
      Edinburgh
      Fair Isle
      Fife
      Firth of Clyde
      Forth Valley
      Gairloch
      Galloway
      Girvan
      Glasgow
      Glen Coe
      Great Glen Fault
      Harris
      Hebrides
      Highlands
      Huntly
      Inner Hebrides
      Inverness
      Inverness-shire
      Islay
      Isle of Skye
      Kirkcudbrightshire
      Loch Lomond
      Lowlands
      Midlothian
      Orkney Islands
      Outer Hebrides
      Perthshire
      Raasay
      Rhum
      Ross-shire
      Scourie
      Southern Uplands
      Sutherland
      Tay Estuary
      Unst
SA Cheviot Hills
      Esk Trough
      North Sea Coast
      Shetland Islands

Scott County
Index states as applicable.
BX Arkansas
      Illinois
      Indiana
      Iowa
      Kansas
      Kentucky
      Minnesota
      Mississippi
      Missouri
      Tennessee
      Virginia
BT United States

scour casts
Includes use on level 3 under sedimentary structures(1) bedding plane irregularities(2). See list K.
BA bedding plane irregularities
      sedimentary structures
SA flute casts

scour marks
Includes use on level 3 under sedimentary structures(1) bedding plane irregularities(2). See list K.
BA bedding plane irregularities
      sedimentary structures
SA current markings
      flute casts

Scourie
Village on North Minch in NW.
BT Scotland
      Great Britain
      United Kingdom

Scripps Institution of Oceanography
In La Jolla, a NW section of San Diego, in San Diego County.
BT California
      United States

Sculptured Hills
BT Moon

Scutelluidae
Includes use on level 3 under Trilobita(1) Ptychopariida(2).
BA Ptychopariida
      Trilobita
BT Trilobitomorpha
      Arthropoda
      Invertebrata

Scyphozoa
Includes use on level 2 under Coelenterata(1). See list F.
BA Coelenterata
BT Invertebrata

Scythian
Europe. Above Tatarian (Permian), below Anisian. Includes use on level 3 under age terms(1). See list E.
UF Skythian
BA Lower Triassic
      Triassic
BT Mesozoic
SA Werfenian

Scythian Platform
Broad, sedimentary plain N and NE of the Black Sea. Index Soviet republics as applicable.
BT USSR
SA Russian Republic
      Ukraine

Se
use selenium

sea fan
use submarine fans

sea floor spreading
use sea-floor spreading

sea ice
Includes use on level 2 under oceanography(1).
SA ice
      oceanography
      sea water

sea level
Not a valid term for GeoRef. Use changes of level for recent sea level. Use regression or transgression when referring to older sea levels.

sea mounts
use seamounts

Sea of Azof
use Azov Sea

Sea of Azov
use Azov Sea

Sea of Crises
Also search Crisium.
UF Mare Crisium
BT Moon

Sea of Fertility
BT Moon

Sea of Galilee
Freshwater lake lying in N part of the Great Rift Valley. Also search Lake Tiberias; Lake Kinneret.
UF Lake Kinneret
      Lake Tiberias
BT Israel
SA Great Rift Valley

Sea of Japan
Not a valid index term for GeoRef since 1973. Use Japan Sea. Before 1974, also search Sea of Japan.

Sea of Nectar
Also search Mare Nectaris.
UF Mare Nectaris
BT Moon

Sea of Okhotsk
Not a valid index term for GeoRef since 1973. Use Okhotsk Sea. Before 1974, also search Sea of Okhotsk.

Sea of Rains
Also search Mare Imbrium.
UF Mare Imbrium
BT Moon

Sea of Serenity
Also search Mare Serenitatis.
UF Mare Serenitatis
BT Moon

Sea of Tranquillity
Also search Mare Tranquillitatis.
UF Mare Tranquillitatis
BT Moon

Sea of Waves
UF Mare Undarum
BT Moon

sea water
Includes use on level 1 (list A); on level 2 under isotopes(1). Used for

sea water ● sedimentary rocks

studies on the composition and properties of the water of the oceans. For ancient water, see sets such as sedimentation or geochemistry. Term set options are:
composition
   subtopic
evolution
   subtopic
experimental studies
   subtopic
genesis
   subtopic
geochemistry
   subtopic
theoretical studies
   subtopic
SA atmosphere
   brackish water
   brines
   density
   desalinization
   geochemistry
   halmyrolysis
   ion exchange
   isotopes
   marine geology
   nepheloid layer
   oceanography
   paleosalinity
   pH
   salinity
   salt
   salt water
   saturation
   sea ice
   solubility
   suspended materials
   upwelling

sea-floor spreading
Includes use on level 1 (list A); on level 2 under tectonophysics(1). Used for topics related to the hypothesis that the oceanic crust is increasing by convective upwelling of magma along the mid-ocean ridges or world rift system. If 1, term set options are:
topic [causes, concepts, evolution, mechanism, rates]
   subtopic (no area term)
UF ocean-floor spreading
   sea floor spreading
   spreading concept
   spreading-floor hypothesis
SA Benioff zone
   continental drift
   crust
   fracture zones
   geosynclines
   heat flow
   magnetic anomalies
   mantle
   marine geology
   mid-ocean ridges
   Mohorovicic discontinuity
   ocean basins
   ocean floors
   paleo-oceanography
   paleobathymetry
   paleomagnetism
   plate tectonics
   rift valleys
   rifting
   spreading centers
   tectonophysics

sea-floor trench
use trenches

sea-water encroachment
use salt-water intrusion

sea-water intrusion
use salt-water intrusion

seamounts
Includes use on level 2 under ocean floors(1). Also search guyots.
UF guyots
   sea mounts
SA ocean floors

seams, coal
use coal seams

Searles Lake
Dry lake in NW San Bernardino County in S.
BT California
   United States

seas, marginal
use marginal seas

seasonal
A valid term through 1977. After 1977, use seasonal variations.

seasonal variations
Term introduced in 1978. Before 1978, search seasonal AND variations.
SA variations

seat clay
use underclay

seat earth
use underclay

Seattle
City on Puget Sound in King County in NW.
BT Washington
   United States

seawalls
Term introduced in 1978 on level 2 under shorelines(1).
SA shorelines

seaweed
use Phaeophyta

Sebes
Town in Alba County in SW.
BT Transylvania
   Romania

Sebes Mountains
SW Transylvania.
BT Transylvania
   Romania

Sebkha el Melah
BT Tunisia

sebkha environment
Term introduced in 1978. Before 1978, search sebkha.
UF sabkha environment
SA environment
   paleogeography
   playas
   salt marshes
   sedimentation
   supratidal environment

Sebkra de Tindouf
use Tindouf Basin

secondary dispersion
Term introduced in 1978 on level 3 under mineral exploration(1).
SA dispersion
   mineral exploration
   primary dispersion

secondary minerals
BT minerals

secondary recovery
SA petroleum
   petroleum engineering
   recovery
   tertiary recovery

secondary structures
Includes use on level 2 under sedimentary structures(1). See list K.
BA sedimentary structures
NA armored mud balls
   concretions
   cone-in-cone
   geodes
   microstylolites
   nodules
   septaria
   stylolites
SA primary structures
   structures

secondary wave
use S-waves

sections
Not to be used for thin sections. Used for cross-sections in stratigraphy.
SA polished sections
   stratigraphy
   thin sections
   type sections

secular variations
Includes use on level 3 under Earth(1).
UF geomagnetic secular variation
SA Earth
   magnetic field
   variations

Sedan
City near Belgian border.
BT Ardennes
   France

Sedgwick Basin
S central.
BT Kansas
   United States

Sedgwick County
Index states as applicable.
BX Colorado
   Kansas
BT United States

sediment sampling
Term introduced in 1978. Includes use on level 3 under mineral exploration(1).
SA mineral exploration
   sampling
   sediments
   soil sampling
   stream sediments

sediment transport
A valid term through 1976. Use transport under sedimentation(1).

sediment yield
SA erosion
   runoff
   sediments
   streams
   yields

sediment-water interface
Term introduced in 1978. Before 1978, search interface AND sediment-water.
UF interface, sediment-water
SA sedimentation
   sediments
   water

sedimentary basins
Includes use on level 3 under paleogeography(1). Term should be used in paleogeographic papers, not for recent basins.
SA basins
   paleogeography

sedimentary cover
UF cover, sedimentary
SA basement
   tectonics

sedimentary fault
use growth faults

sedimentary petrology
Treated as a whole. Includes use on level 1 (list A); on level 2 under area terms(1), bibliography(1), education(1), and symposia(1). If level 1, term set options are:
topic [applications, bibliography, catalogs, classification, concepts, education, history, instruments, methods, nomenclature, philosophy, practice, principles, symposia, techniques, textbooks]
   subtopic
SA agglomerate
   bibliography
   catalogs
   clay mineralogy
   diagenesis
   domains
   glossaries
   lexicons
   marine geology
   oceanography
   petrology
   sedimentary rocks
   sedimentary structures
   sedimentation
   sedimentology
   sediments
   shape analysis

sedimentary processes
Includes use on level 3 under mineral deposits, genesis(1) processes(2).
SA deposition
   mineral deposits
   processes

sedimentary quartzite
use orthoquartzite

sedimentary rocks
See list I for rock groups and names. Includes use on level 1 (list A); on level 2 under paragenesis(1) and weathering (1). If level 1, term set options are:
rock group [carbonate rocks, chemically precipitated rocks, clastic rocks, organic residues]
   rock name (List I)
topic [classification, composition, diagenesis, environmental analysis, geochemistry, lithofacies, lithostratigraphy, petrography, petrology, pore water, properties, provenance, textures]
   subtopic [e.g. fabric, grains, grain size, rounding, shape, size, sorting, surface textures, etc.]
NT carbonate rocks
   chemically precipitated rocks
   clastic rocks
NA grapestone
NT organic residues
SA anhydrite
   arenaceous texture
   argillaceous texture
   arkosic composition
   authigenesis
   bauxite
   boulders
   calcareous composition
   catagenesis
   cement
   clasts
   clay mineralogy
   coal measures
   diagenesis
   dolomitic composition
   dolomitization
   environmental analysis
   exinite
   fabric
   ferruginous composition
   fossiliferous materials
   fusinite
   glauconitic composition
   grains
   greensand
   gypsum
   inertinite
   insoluble residues
   iron-rich composition
   kerogen
   lenses
   lithic texture

lithofacies
lithology
lithostratigraphy
macerals
metasedimentary rocks
micrinite
nodules
nonterrigenous materials
oolite
oolitic texture
organic materials
packing
pelitic texture
petrography
phi scale
pisolites
pisoliths
pisolitic texture
porosity
provenance
reefs
resinite
rocks
rounding
sedimentary petrology
sedimentary structures
sedimentation
sedimentology
sediments
shape
shape analysis
shells
siliceous composition
size
size distribution
sorting
sporinite
terrigenous materials
vitrinite
wildflysch

sedimentary structures
See list K for types and names of structures. Includes use on level 1 (list A). Used for a structure in a sedimentary rock, formed either contemporaneously with deposition (primary) or subsequently to deposition (secondary). Term set options are:
topic [classification, environmental analysis, genesis, interpretation, nomenclature, patterns]
    subtopic
type of structure [bedding plane irregularities, biogenic structures, cylindrical structures, planar bedding structures, primary structures, secondary structures, soft sediment deformation, turbidity current structures]
    specific type (List K)
NA bedding plane irregularities
    biogenic structures
    cylindrical structures
    planar bedding structures
    primary structures
    secondary structures
    soft sediment deformation
    turbidity current structures
SA bars
    channels
    deposition
    diagenesis
    dunes
    environmental analysis
    fluvial features
    glacial geology
    ice wedges
    paleogeography
    reefs
    sedimentary petrology
    sedimentary rocks
    sedimentation
    sediments
    solution cavities
    spherules
    stromatolites
    structures

sedimentation
Includes use on level 1 (list A); on level 2 under paleoecology(1). Used for the act or process of forming or accumulating sediment in layers, including all processes from transport through diagenesis. Before 1976, also search aggregation. If level 1, term set options are:
environment
    type of environment [basins, beaches, coastal environment, deltas, dunes, estuaries, lagoons, lakes, marine environment, reefs, swamps, terrestrial environment, tidal environment]
topic [controls, cyclic processes, deposition, diagenesis, flow regime, precipitation, processes, provenance, rates]
    subtopic
transport
    type of transport [glacial transport, ice-rafting, marine transport, stream transport, turbidity currents, wind transport]
SA ablation
    accretion
    barriers
    base surges
    basins
    beaches
    biofacies
    bioturbation
    braided streams
    cementation
    changes of level
    channels
    coastal environment
    continental shelf
    continental slope
    controls
    current directions
    cyclic processes
    cyclothems
    deltas
    deposition
    diagenesis
    dolomitization
    engineering geology
    environment
    environmental analysis
    estuaries
    flow regime
    flow structures
    fresh water
    geochemical controls
    geopressure
    glacial geology
    glacial transport
    heavy minerals
    high-energy environment
    hydrology
    ice rafting
    lagoons
    lakes
    lithification
    littoral environment
    localization
    low-energy environment
    marine environment
    marine geology
    marine transport
    nodules
    ocean circulation
    ocean floors
    oceanography
    organic materials
    packing
    paleocurrents
    paleoecology
    paleogeography
    periodicity
    phase equilibria
    precipitation
    processes
    progradation
    provenance
    reefs
    regression
    reservoir rocks
    rhythmite
    salinity
    salt marshes
    saltation
    sebkha environment
    sediment-water interface
    sedimentary petrology
    sedimentary rocks
    sedimentary structures
    sedimentology
    sediments
    shallow-water environment
    shelf environment
    silicification
    slope environment
    slumping
    stream transport
    structural controls
    swamps
    syntectonic processes
    taiga environment
    terrestrial environment
    tidal environment
    tidal flats
    transport
    turbidity currents
    uniformitarianism
    wind transport

sedimentology
For the discipline.
SA geomorphology
    marine geology
    sedimentary petrology
    sedimentary rocks
    sedimentation
    sediments

sediments
See list N for sediment groups and names. Includes use on level 1 (list A); on level 2 under weathering(1). Used for unconsolidated solid fragmental material, or a mass of such material, that originates from weathering of rocks. If 1, term set options are:
sediment group [carbonate sediments, chemically precipitated sediments, clastic sediments, organic sediments]
    sediment type (List N)
topic [classification, composition, diagenesis, distribution, environmental analysis, genesis, geochemistry, lithofacies, lithostratigraphy, petrography, petrology, pore water, properties, provenance, textures]
    subtopic [e.g. grain size, shape, ...]
NA carbonate sediments
    chemically precipitated sediments
    clastic sediments
    organic sediments
SA anhydrite
    arenaceous texture
    argillaceous texture
    arkosic composition
    biofacies
    breccia
    calcareous composition
    cement
    clasts
    clay mineralogy
    coccoliths
    dehydration
    deposition
    diagenesis
    dolomitic composition
    dolomitization
    engineering properties
    entropy
    environmental analysis
    evaporites
    fabric
    ferruginous composition
    filtration
    fines
    fossiliferous materials
    glacial geology
    glauconitic composition
    grain size
    grains
    gypsum
    halite
    heavy minerals
    hydrology
    iron-rich composition
    liquefaction
    lithic texture
    lithification
    lithofacies
    lithostratigraphy
    littoral drift
    marine geology
    nonterrigenous materials
    ocean floors
    oolitic texture
    organic materials
    organic residues
    permeability
    petrography
    petrology
    phi scale
    pisoliths
    pisolitic texture
    pore water
    porosity
    provenance
    reworking
    roundness
    sediment sampling
    sediment yield
    sediment-water interface
    sedimentary petrology
    sedimentary rocks
    sedimentary structures
    sedimentation
    sedimentology
    shape
    shape analysis
    shells
    siliceous composition
    size
    size distribution
    soils
    sorting
    sphericity
    spherules
    stream sediments
    suspension
    terrigenous materials
    unconsolidated materials
    volcanic ash
    weathering
    wildflysch

seeds
SA paleobotany

Seeland
use Zealand

seepage
Includes use on level 2 under engineering geology(1).
SA drainage
    engineering geology
    foundations
    ground water
    nuclear facilities
    oil seeps
    percolation
    petroleum
    reservoirs
    soil mechanics
    soils
    tunnels
    underground installations
    waste disposal

water
waterways

seeps, oil
use oil seeps

Segovia
Province in central.
BT Spain

Segozero
Lake in S central Karelian A.S.S.R. in NW European USSR.
BT Russian Republic
USSR

Segre Valley
River valley in Catalonia. Also search Segre; Segre River.
BT Lerida
Spain

seiches
SA limnology
waves

Seiland
Island in Arctic Ocean off NW.
BT Finnmark
Norway

Seine
Former department which constituted Paris proper and a ring of industrial and residential suburbs. Department officially abolished January 1, 1960.
BT France

Seine Estuary
The tidal mouth of the Seine River emptying into the Bay of the Seine at Le Havre. Index departments as applicable.
BT France
SA Calvados
Eure
Seine-Maritime

Seine River
Rises in Cote-d'Or Department in E and flows NW through Paris and into the English Channel near Le Havre.
BT France

Seine Valley
River valley in N central and N.
BT France

Seine-et-Marne
Department in N central.
CO N481000N491000
E0033000E0021500
BT France
NT Fontainebleau
SA Yonne Valley

Seine-et-Oise
Former department outside Paris suburbs which completely surrounded former Seine Department. Officially abolished January 1, 1968.
BT France

Seine-Maritime
Department in NW.
CO N491500N500500
E0015000E0000000
BT France
NT Caux
Dieppe
Le Havre
Rouen
SA Normandy
Seine Estuary

Seislog
Term introduced in 1978.
SA instruments
seismic methods
seismology

seismic data
A valid term through 1977. Use data and earthquakes.

seismic effects
A valid term through 1977. Use effects and earthquakes.

seismic focus
use focus

seismic methods
Includes use on level 2 under geophysical methods(1).
BA geophysical methods
NA reflection methods
refraction methods
SA common-depth-point method
elastic waves
methods
ocean bottom seismographs
Seislog
seismic surveys
Vibroseis

seismic noise
A valid term through 1978. After 1978, use noise.

seismic risk
Includes use on level 3 under nuclear facilities(1) or geologic hazards(1).
UF risk, seismic
SA geologic hazards
nuclear facilities

seismic sea waves
use tsunamis

seismic sounding
use deep seismic sounding

seismic sources
Includes use on level 2 under seismology(1).
UF sources, seismic
SA earthquakes
elastic waves
explosions
seismology

seismic studies
A valid term through 1976. Use seismic surveys and theoretical studies or experimental studies.

seismic surge
use tsunamis

seismic surveys
Includes use on level 2 under geophysical surveys(1), crust(1), and under Mohorovicic discontinuity(1).
BA geophysical surveys
BT surveys
SA crust
explosions
reflection
refraction
seismic methods
seismology
Vibroseis

seismic waves
A valid term through 1975. After 1975, use elastic waves.

seismicity
Includes use on level 2 under seismology(1).
SA seismology
seismotectonics

seismograms
Includes use on level 3 under earthquakes(1) or seismology(1).
UF earthquake record
SA earthquakes
seismographs
seismology

seismographs
Includes use on level 3 under seismology(1).
SA earthquakes
instruments
ocean bottom seismographs
seismograms
seismology

seismometers

seismology
For general treatments of the subject. See earthquakes for specific treatments. Includes use on level 1 (list A); on level 2 under area terms(1), bibliography(1) and symposia(1). See list B. If level 1, term set options are:
topic [core, crust, earthquakes, elastic waves, experimental studies, explosions, interior, mantle, methods, microearthquakes, microseisms, observatories, properties, seismicity, seismic sources, theoretical studies, tsunamis, volcanology]
subtopic [e.g. elastic properties, physical properties, P-waves, structure, S-waves]
SA accelerograms
aftershocks
amplitude
anelastic media
anelasticity
anisotropy
arrays
arrival time
attenuation
Benioff zone
body waves
coda waves
core
crust
damping
deconvolution
deep seismic sounding
deformation
diffraction
dilatancy
discontinuities
dislocations
Earth
earthquakes
elastic properties
elastic waves
epicenters
explosions
faults
filters
focus
geodesy
geophysical methods
geophysics
ground motion
half-space
hydraulic fracturing
interior
inverse problem
isochrons
isoseismic maps
long-period waves
Love waves
mantle
marker beds
media
microearthquakes
microseisms
modified Mercalli scale
Mohorovicic discontinuity
neotectonics
networks
noise
nuclear explosions
observatories
ocean bottom seismographs
P-waves
periodicity
plate tectonics
precursors
premonitory phenomena
propagation
Q
Rayleigh waves
reflection
refraction
S-waves
Seislog

seismic sources
seismic surveys
seismicity
seismograms
seismographs
seismometers
seismotectonics
SH-waves
shock waves
signals
sounding
spectral analysis
strain
stress
strong motion
surface waves
swarms
tectonophysics
teleseismic signals
three-dimensional models
tilt
tiltmeters
traveltime
traveltime curves
tsunamis
two-dimensional models
vibration
Vibroseis
volcanology

seismometers
SA accelerograms
earthquakes
instruments
motions
seismographs
seismology

seismotectonics
SA plate tectonics
seismicity
seismology
tectonics

selachians
use Selachii

Selachii
Order. Includes use on level 3 under Pisces(1) Chondrichthyes(2). Also search selachians.
UF selachians
BA Elasmobranchii
Chondrichthyes
Pisces
BT Vertebrata

Selangor
State on W coast of Malay Peninsula in West Malaysia.
BT Malaysia
NT Kuala Lumpur
SA West Malaysia

selection, natural
use natural selection

selenates and selenites
Includes use on level 2 under minerals(1). See list L.
UF selenites, selenates and
BT minerals

Selenga
Province on the Buryat A.S.S.R. border in N Mongolia.
UF Selenge
BT Mongolia

Selenga River valley
Extends to SE Lake Baikal in Russian Republic. Also search Selenga. Index countries as applicable.
UF Selenge River valley
BT Asia
SA Mongolia
USSR

Selenge
use Selenga

Selenge River valley
use Selenga River valley

selenides
Includes use on level 3 under minerals(1) sulfides(2). See list L.
BA sulfides
BT minerals
NA eskebornite
    umangite

selenite
BA sulfates
BT minerals
SA gypsum

selenites, selenates and
use selenates and selenites

selenium
Includes use on level 1 and 2 as chemical element (list D).
UF Se
SA elements
    onofrite

Selennyakh
River which rises in the N Cherskiy Range and flows into the Indigirka River in NE Yakutia.
UF Selenyak
BT Russian Republic
    USSR

selenodesy
SA geodesy
    Moon

Selenyak
use Selennyakh

self-potential methods
Term introduced in 1978. Includes use on level 3 under geophysical surveys(1) electrical surveys(2). Before 1978, also search self-potential.
SA electrical surveys
    geophysical surveys
    methods

seligmannite
BA sulfarsenites
    sulfosalts
BT minerals

Selkirk Mountains
Range in the Rocky Mountains in SE.
UF Selkirks Mountains
BT British Columbia
    Canada
SA Esplanade Range
    Rocky Mountains

Selkirks Mountains
use Selkirk Mountains

Selukwe
Town in central.
BT Rhodesia

Selva
use Copiapo

Selwyn Range
W central.
BT Queensland
    Australia

SEM data
Term introduced in 1978. For the method, see scanning method.
UF scanning electromicroscopy data
SA data
    electron microscopy
    scanning method
    TEM data
    transmission method

Semata Formation
BT Pleistocene
    Quaternary
SA Japan

Semenic Mountains
Central Banat. Also search Semenic.
BT Banat
    Romania

semi-arid environment
Term introduced in 1978. Before 1978, also search semi-arid; semiarid; semiarid regions.
UF semiarid regions
SA arid environment
    ecology
    environment
    geomorphology

semiarid regions
use semi-arid environment

Seminole Formation
In Skiatook Group. Includes Lenapah Limestone, Nowata Shale, Memorial Shale. NE, central, and central S Oklahoma.
BT Missourian
    Upper Pennsylvanian
    Pennsylvanian
SA Oklahoma

Semipalatinsk
City and oblast in NE.
BT Kazakhstan
    USSR

semitropical environment
use subtropical environment

Semmering
Resort area in Eastern Alps.
BT Austria
SA Lower Austria
    Styria

Semri Series
BT Cambrian
SA Bihar
    India
    Madhya Pradesh

semseyite
BA sulfantimonites
    sulfosalts
BT minerals

senarmontite
BA oxides
BT minerals

Sendai
City near E coast in N.
BT Honshu
    Japan

Seneca Lake
One of the Finger Lakes in Yates and Seneca counties. Also search Seneca.
CO N422000N424500 W0764500W0770000
BT New York
    United States

Senegal
Formerly a republic in French Community. Achieved independence in 1960. Includes use on level 1 as an area term (list O). For term set options see list B.
CO N123000N163000 W0120000W0170000
BA Africa
NT Dakar
SA Sahel
    Senegal Basin
    Senegal River
    West Africa

Senegal Basin
River basin. Index countries as applicable. Also search Senegal.
UF Senegal River basin
BT Africa
SA Guinea
    Mali
    Mauritania
    Senegal

Senegal River
Rises in the highlands of Guinea and flows N and then NW into the Atlantic Ocean at Saint-Louis in Senegal. Index countries as applicable. Also search Senegal.
BT Africa
SA Guinea
    Mali
    Mauritania
    Senegal

Senegal River basin
use Senegal Basin

Senonian
Europe. Above Turonian, below Danian. Includes use on level 3 under age terms(1). See list E.
BA Upper Cretaceous
    Cretaceous
BT Mesozoic
NT Futaba Group
SA Campanian
    Chalk
    Coniacian
    Izumi Group
    Maestrichtian
    Santonian

sensing, remote
use remote sensing

Sentinel Butte Formation
Overlies Tongue River Member. NE Montana, and SW North Dakota.
UF Sentinel Butte Shale Member
BT Paleocene
SA Montana
    North Dakota

Sentinel Butte Shale Member
use Sentinel Butte Formation

separation
Used as a general term, e.g. on level 3 under micropaleontology(1).
SA dewatering
    mixing
    processes

sepiolite
Also search meerschaum.
UF meerschaum
BA sheet silicates
    silicates
BT minerals
SA clay minerals

Sept-Iles
City on the Saint Lawrence River in SE.
BT Quebec
    Canada

septa
UF septae
SA fossils
    morphology

septae
use septa

septaria
Includes use on level 3 under sedimentary structures(1) secondary structures(2). See list K.
BA secondary structures
    sedimentary structures
SA concretions
    cone-in-cone

Sequanian
Europe.
BA Upper Jurassic
    Jurassic
BT Mesozoic

Sequatchie River valley
use Sequatchie Valley

Sequatchie Valley
River valley in SE central Tennessee. Also search Sequatchie River valley.
UF Sequatchie River valley
BT Tennessee
    United States

Sequoia
Genus. Includes use on level 3 under gymnosperms(1) Coniferales(2).
BA Taxodiaceae
    Coniferales
    gymnosperms
BT Plantae

Serajevo
use Sarajevo

Serbia
Constituent republic in E and NE Yugoslavia. Formerly Servia.
UF Servia
BT Yugoslavia
NT Aleksinac
    Belgrade
    Bor
    Kopaonik
    Kostolac
    Majdanpek
    Metohija
    Racha
    Soko Banja
    Timok Basin
    Trepca Mine
    Vojvodina
    Zapadna Morava
    Zlatibor Mountains
SA Moesia
    Morava River valley
    Serbo-Macedonian Massif

Serbo-Macedonian Massif
Macedonia and S Serbia. Index republics as applicable.
BT Yugoslavia
SA Macedonia
    Serbia

Serebryanka Formation
Lower and middle Triassic (?). SE and E Ukraine and adjoining area in Russian Republic.
BT Lower Triassic
    Triassic
SA Middle Triassic
    Russian Republic
    Ukraine

Sereth River
use Siret River

Sergipe
State in NE.
BT Brazil
NT Aracaju
    Propria Geosyncline
SA Sao Francisco Basin
    Sergipe-Alagoas Basin

Sergipe Basin
use Sergipe-Alagoas Basin

Sergipe-Alagoas Basin
In NE part of country. Index states as applicable. Also search Sergipe Basin.
UF Sergipe Basin
BT Brazil
SA Alagoas
    Sergipe

sericite
BA sheet silicates
    silicates
BT minerals
SA mica group
    muscovite

sericitization
BA metasomatism
SA hydrothermal alteration

Serov
City in Sverdlovsk Oblast E of Urals.
BT Russian Republic
    USSR

Serozem
use Sierozems

serpentine
BA sheet silicates
    silicates
BT minerals

SA serpentine group
serpentine group
Includes use in combination with sheet silicates (i.e. sheet silicates, serpentine group) on level 2 under minerals(1). See list L.
BA sheet silicates
 silicates
BT minerals
SA antigorite
 cerolite
 chrysotile
 garnierite
 lizardite
 serpentine

serpentine rock
use serpentinite

serpentinite
Compositional term.
UF serpentine rock
BZ metaigneous rocks
 metasomatic rocks
SA serpentinization

serpentinization
Includes use on level 3 under metasomatism(1) processes(2).
BA metasomatism
SA alteration
 hydrothermal alteration
 processes
 serpentinite

Serpulidae
Family. Includes use on level 3 under Annelida(1) Polychaetia(2). Also search serpulids.
UF serpulids
BA Polychaetia
 Annelida
BT Invertebrata

serpulids
use Serpulidae

Serra do Mar
Coastal mountain range in S and SE. Index states as applicable.
BT Brazil
SA Parana
 Santa Catarina
 Sao Paulo

Serra do Navio
Town in central.
BT Amapa
 Brazil

Serra Gerral Formation
BT Mesozoic
SA Brazil
 Cretaceous
 Jurassic
 Paraguay
 Sao Paulo

Serrania de Cuenca
Mountain range in E central. Index provinces as applicable.
BT Spain
SA Cuenca
 Guadalajara

Serrania de Ronda
Spur of the Cordillera Penibetica in Andalusia. Index provinces as applicable.
UF Ronda Massif
 Ronda, Serrania de
 Ronda Sierra
 Sierra de Ronda
BT Spain
SA Cadiz
 Malaga

Serravallian
Europe. See list E.
BA middle Miocene
 Miocene
 Neogene
 Tertiary

BT Cenozoic

Servia
use Serbia

Sesia Valley
River valley E of Lake Maggiore in N.
BT Italy
SA Piedmont

Sesia Zone
N Piedmont.
BT Italy
SA Piedmont

Sesia-Lanzo Zone
N Piedmont and N Lombardy. Index autonomous regions as applicable.
BT Italy
SA Alps
 Lombardy
 Piedmont

Sespe Formation
Overlying Coldwater Sandstone Member of Tejon Formation and underlying Vaqueros Formation. In S California.
BT Tertiary
SA California
 Eocene
 Oligocene
 upper Eocene

Sete
City on Gulf of Lion in S Herault.
UF Cette
BT Herault
 France

Seto Inland Sea
use Inland Sea

Seto Sea
use Inland Sea

Seto-chi-umi
use Inland Sea

Seto-Naikai
use Inland Sea

Setouchi
use Inland Sea

Sette Daban (Range)
use Sette-Daban Range

Sette-Daban Range
NW Khabarovsk Kray and SE Yakutia. Also search Sette Daban; Sette-Daban.
UF Sette Daban (Range)
BT Russian Republic
 USSR

Setting Lake
Central.
BT Manitoba
 Canada

settlement
As of 1978, term is used on level 2 under foundations(1), land subsidence(1), and soil mechanics(1).
SA engineering geology
 foundations
 land subsidence
 soil mechanics

Setubal
District in SW.
BT Portugal

Sevan
Town on shore of Lake Sevan.
BT Armenia
 USSR
SA Lake Sevan

Seven Devils Mountains
Range in Adams and Idaho counties in W.
BT Idaho
 United States

Severin
Former province whose capital was Caransebes. Much of the province is now Caras-Severin County.
BT Banat
 Romania

Severn Valley
River valley. Index political divisions as applicable. Also search Severn River.
BT United Kingdom
SA England
 Wales

Sevier County
Index states as applicable. Also search Sevier AND appropriate state.
BX Arkansas
 Tennessee
 Utah
BT United States

Sevier orogenic belt
The Sevier Plateau area in Garfield, Piute, and Sevier counties in S central Utah. Also search Sevier AND Utah.
UF Sevier Plateau
BT Utah
 United States

Sevier Plateau
use Sevier orogenic belt

Sevilla
use Seville

Seville
Province in SW Spain. Also a city.
UF Sevilla
BT Spain
NT Carmona
SA Andalusia

sewage
Includes use on level 3 under waste disposal(1) or pollution(1).
SA pollution
 waste disposal
 water quality

Seward County
Index states as applicable.
BX Kansas
 Nebraska
BT United States

Seward Peninsula
On Bering Strait between Kotzebue and Norton sounds in W Alaska. Also search Seward AND Alaska.
BT Alaska
 United States

sexual dimorphism
Includes use on level 3 under fossil groups (list F).
SA dimorphism
 reproduction

Seychelles
Level 1 area term as of 1978. Island group about 700 miles NE of Malagasy Republic.
BA Indian Ocean
NT Aldabra Island
SA Mahe Island

SH waves
use SH-waves

SH-waves
Also search SH waves.
UF SH waves
SA elastic waves
 S-waves
 seismology
 waves

Shaba
Province in S Zaire.
CO S130000S050000
 E0310000E0220000
UF Katanga
BT Zaire
NT Kamoto

Shackleton Glacier
One of glaciers feeding Ross Ice Shelf from Queen Maud Range. Also Shackleton Ice Shelf off Queen Mary Coast extending into Indian Ocean is sometimes called Shackleton Glacier.
BT Antarctica
SA Ross Ice Shelf

Shackleton Range
SE of the Filchner Ice Shelf which is on the Weddell Sea on the Atlantic Ocean side.
BT Antarctica

Shadow Mountains
Index states as applicable.
BT United States
SA California
 Nevada

Shady Dolomite
Equivalent to Tomstown Dolomite. N Alabama, NW Georgia, W North Carolina, E Tennessee, and SW Virginia.
BT Lower Cambrian
 Cambrian
SA Alabama
 Chilhowee Group
 Georgia
 North Carolina
 Tennessee
 Virginia

shake wave
use S-waves

Shakh-Dag Range
use Shakhdag Range

Shakhdag Range
In the Lesser Caucasus on N shore of Lake Sevan. Index Soviet republics as applicable.
UF Shakh-Dag Range
BT Armenia
 Azerbaidzhan
SA Lesser Caucasus

Shakopee Dolomite Member
use Shakopee Formation

Shakopee Formation
In Prairie du Chien Group. Overlies and interbedded with Root Valley Sandstone. N Illinois, Iowa, S Minnesota, and S Wisconsin.
UF Shakopee Dolomite Member
BT Lower Ordovician
 Ordovician
SA Illinois
 Iowa
 Minnesota
 Wisconsin

shale
Does not include oil shale. Includes use on level 3 as a commodity term under clays(1), for shale as brick clay; on level 3 under construction materials(1) for bloating shale; on level 3 under sedimentary rocks(1) clastic rocks(2). See list C (commodities) and list I (sed. rocks). Before 1976, also search pelite.
UF bloating shale
 pelite
BA clastic rocks
BT sedimentary rocks
SA argillaceous texture
 bituminous shale
 black shale
 clays
 construction materials
 oil shale
 slates
 terrigenous materials
 torbanite

shale oil
use oil shale

shallow
A valid term through 1977. After 1977, use shallow-water environment.

shallow earthquake
use shallow-focus earthquakes

shallow-focus earthquakes
Term introduced in 1978. Includes use on level 3 under earthquakes(1). Before 1978, search earthquakes AND shallow focus.
- UF normal earthquake
  shallow earthquake
- SA deep-focus earthquakes
  earthquakes
  focus
  intermediate-focus earthquakes

shallow-water environment
Term introduced in 1978. Before 1978, search shallow-water or shallow.
- SA ecology
  environment
  sedimentation

Shamlug
Town in NE.
- BT Armenia
  USSR

Shan State
E Burma.
- UF Federated Shan States
  Shan States
- BT Burma

Shan States
use Shan State

Shannon County
Index states as applicable.
- BX Missouri
  South Dakota
- BT United States

Shansi
Province in N central.
- BT China

Shap Granite
The Shap Granite and associated igneous and metamorphic rocks have been called Shap Andesites and Shap Rhyolites. English Lake District in NW England.
- UF Shap Rhyolites
- BT Ordovician
- SA England
  United Kingdom

Shap Rhyolites
use Shap Granite

shape
Includes use on level 2 under core(1) and Earth(1); on level 3 under sedimentary rocks(1) and under sediments(1).
- SA dimensions
  Earth
  geodesy
  particles
  sedimentary rocks
  sediments
  size

shape analysis
- SA analysis
  grains
  granulometry
  roundness
  sedimentary petrology
  sedimentary rocks
  sediments
  sphericity
  textures

Sharjah
Emirate. One of federation of 7 states at S end of Persian Gulf. Includes use on level 3 as an area term (list O).
- BT United Arab Emirates
  Arabian Peninsula

Shark Bay
Large inlet of Indian Ocean. Also search Sharks Bay.
- UF Sharks Bay
- BT Western Australia
  Australia

Sharks Bay
use Shark Bay

Sharon Conglomerate
In Pottsville Group. Maryland, E Ohio, W Pennsylvania, and N West Virginia.
- BT Pennsylvanian
- SA Maryland
  Ohio
  Pennsylvania
  Pottsville Group
  West Virginia

Sharps
Point of impact near Sharps in Richmond County in E Virginia. Also search Sharps Meteorite.
- BT Virginia
  United States
- SA meteorites

Shasta County
N central.
- BT California
  United States

Shatskiy Rise
use Shatsky Rise

Shatsky Rise
In mid Northwest Pacific Basin E of Tokyo.
- UF Shatskiy Rise
- BT Pacific Ocean

shatter cones
Includes use on level 3 under engineering geology(1) explosions(2); on level 3 under metamorphism(1) shock(2) and deformation(1) shock(2); on level 3 under geomorphology(1) cryptoexplosion features(2) and under seismology(1).
- UF cones, shatter
- SA cryptoexplosion features
  deformation
  explosions
  metamorphism
  meteor craters
  meteorites
  nuclear explosions
  shock waves

shattuckite
- BA chain silicates
  silicates
- BT minerals

Shaunavon Formation
Subsurface.
- BT Jurassic
- SA Canada
  Saskatchewan

Shawangunk Mountains
Part of the Kittatinny Mountains in Ulster County in S.
- BT New York
  United States

Shawnee County
NE Kansas.
- BT Kansas
  United States

Shawnee Group
Comprises Oread Limestone, Kanwaka Shale, Lecompton Limestone, Tecumseh Shale, Deer Creek Limestone, Calhoun Shale, Topeka Limestone, Severy Shale, Howard Limestone, White Cloud Shale, Happy Hollow Limestone, Cedar Vale Shale, Rulo Limestone, Silver Lake-Auburn Shale Interval, Reading Limestone, Harveyville Shale, Elmont or Preston limestone, and Williard Shale.
- BT Virgilian
  Upper Pennsylvanian
  Pennsylvanian
- SA Kansas
  Lecompton Limestone
  Nebraska
  Oread Limestone

shear
As of 1978, used only to indicate the mechanics of deformation. Before 1978, used to indicate style of structure. Now use similar folds for style of structure.
- UF shear strain
- SA deformation
  mechanics
  shear zones
  similar folds
  strain
  stress

shear cleavage
use slip cleavage

shear folds
use similar folds

shear modulus
- UF Coulombs' modulus
  modulus of rigidity
  modulus, shear
  rigidity modulus
  torsion modulus
- BA elastic constants
- SA bulk modulus
  elasticity

shear strain
use shear

shear strength
Includes use on level 3 under soil mechanics(1), rock mechanics(1), or engineering geology(1). Also search cohesion.
- UF cohesion
- SA mechanical properties
  rock mechanics
  soil mechanics
  strain
  strength
  stress
  tensile strength

shear stress
Includes use on level 3 under engineering geology(1) or deformation(1).
- UF tangential stress
- SA deformation
  engineering geology
  stress

shear wave
use S-waves

shear zones
Includes use on level 3 under faults(1) effects(2) and under tectonics(1) structure(2).
- UF zones, shear
- SA breccia
  faults
  fractures
  shear
  strain
  structural analysis
  tectonics

shear-cleavage folds
use cleavage folds

shearing
A valid term through 1975. After 1975, see shear. See also fractures, foliation, and structural analysis.

sheet silicates
Includes use on level 2 under minerals(1); in combination with chlorite group, clay minerals, mica group, and serpentine group (i.e. sheet silicates, chlorite group) to form terms on level 2 under minerals(1). See list L. Also search phyllosilicates.
- UF phyllosilicates
- BA silicates
- BT minerals
- NA allophane
  antigorite
  apophyllite
  astrophyllite
  bavenite
  beidellite
  berthierine
  biotite
  celadonite
  cerolite
  chamosite
  chlorite
  chlorite group
  chrysocolla
  chrysotile
  clay minerals
  clinochlore
  clintonite
  cookeite
  corrensite
  cymrite
  dickite
  fluor-phlogopite
  garnierite
  gillespite
  glauconite
  gyrolite
  halloysite
  hectorite
  hisingerite
  hydrobiotite
  hydromica
  hydromuscovite
  illite
  imogolite
  kammererite
  kaolinite
- NZ lepidolite
- NA lepidomelane
  leucosphenite
  listvenite
  lizardite
  margarite
  mica
  mica group
  montmorillonite
  muscovite
  nacrite
  nontronite
  palygorskite
  paragonite
  penninite
  phengite
  phlogopite
  polylithionite
  prehnite
  pyrophyllite
  rectorite
  saponite
  sepiolite
  sericite
  serpentine
  serpentine group
  smectite
  stevensite
  stilpnomelane
  sudoite
  talc
  tosudite
  vermiculite
- NZ zinnwaldite
- NA zussmanite
- SA asbestos
  bentonite
  clay mineralogy

sheets
As of 1978, includes use on level 2 under intrusions(1).
SA ice sheets
 intrusions

Sheffield
City in Yorkshire in N.
BT England
 Great Britain
 United Kingdom

shelf
A valid term through 1978 used to distinguish older sedimentation from more recent. After 1978, use shelf environment.

shelf, continental
use continental shelf

shelf environment
Term introduced in 1978. Use this term or slope environment for older sedimentation. For recent sedimentation, use continental shelf or continental slope. Before 1978, search shelf.
SA continental shelf
 environment
 sedimentation
 slope environment

shells
Includes use on level 3 under fossil group(1) for type of material. See list F. Before 1975, also search shell.
SA fossils
 Invertebrata
 ornamentation
 paleontology
 sedimentary rocks
 sediments
 tests

shelves, ice
use ice shelves

Shemakha
City on E Caucasus.
BT Azerbaidzhan
 USSR

Shemya Island
One of the Semichi Islands at W end of Aleutian Islands. Also search Shemya.
BT Alaska
 United States

Shenandoah Valley
River valley between the Allegheny and Blue Ridge mountains. Primarily in Virginia. Index states as applicable.
BT United States
SA Virginia
 West Virginia

Shensi
Province in N central.
BT China
SA Han River basin

Sheridan County
Index states as applicable.
BX Kansas
 Montana
 Nebraska
 North Dakota
 Wyoming
BT United States

Sherman County
Index states as applicable.
BX Kansas
 Nebraska
 Oregon
 Texas
BT United States

Shetland Islands
Archipelago off N Scotland 50 miles NE of Orkney Islands. Also search Shetland. Includes use on level 1 as an area term (list O). For term set options see list B.
CO N595000N610000
 W0004500W0014500
UF Shetlands
 Zetland
BA Atlantic Ocean
SA Fair Isle
 Scotland
 Unst

Shetlands
use Shetland Islands

Sheveluch
Volcano on E central Kamchatka Peninsula. Also search Sheveluch Volcano.
UF Sheveluch Volcano
 Shiveluch Volcano
BT Russian Republic
 USSR

Sheveluch Volcano
use Sheveluch

shield volcanoes
Also search lava domes.
UF basaltic domes
 lava domes
BA volcanoes
SA stratovolcanoes
 volcanism

shields
Usage restricted to tectonics.
SA basement
 cratons
 crust
 massifs
 structural complexes
 tectonics

Shiga
Prefecture in S.
BT Honshu
 Japan

Shikoku
Smallest of the 4 major Japanese islands in S.
BT Japan
NT Ehime
 Kagawa
 Kochi
 Matsuyama
 Tokushima
 Tosa Bay
 Yoshino River
SA Izumi Group
 Nankaido
 Sambagawa Belt
 Shimanto Group

Shikoku Basin
Undersea feature between Shikoku on the NW and the Bonin and Volcano islands to the SE.
BT Pacific Ocean

Shikotsu
Lake in SW.
BT Hokkaido
 Japan

Shilka Valley
River valley in SW central Chita Oblast in Transbaikalia. Also search Shilka; Shilka River.
BT Russian Republic
 USSR
SA Argun River

Shillong
City in W central.
BT Assam
 India

Shillong Plateau
Undulating tableland in W.
BT Assam
 India

Shimane
Prefecture in SW.
BT Honshu
 Japan

Shimanto Group
Includes Terazoma Series, Nishigawa Series, and Higashigawa Series. Trias-Cretaceous, presumably. Kyushu and S Shikoku. Also search Shimanto.
BT Mesozoic
SA Cretaceous
 Japan
 Kyushu
 Shikoku
 Triassic

Shimokita Peninsula
In Mutsu Bay area in extreme N.
BT Honshu
 Japan

Shimosa Group
BT Pleistocene
 Quaternary
SA Japan

Shinano River
W central Honshu. Flows N into Japan Sea. Longest river in Japan.
BT Honshu
 Japan

shingle
Includes use on level 3 under sediments(1) clastic sediments(2). See list N.
BA clastic sediments
BT sediments
SA cobbles
 gravel
 pebbles

Shinji Lake
SW Honshu.
BT Honshu
 Japan

Shiobara
Town in central.
BT Honshu
 Japan

Shiraki Steppe
Semidesert plain in E Georgian Republic. Also search Shiraki.
BT Georgian Republic
 USSR

Shiranish Formation
BT Cretaceous
SA Iraq

Shirasu
Stream NW of Tokyo in central.
BT Honshu
 Japan

Shirley Basin
NW Carbon County in S central.
BT Wyoming
 United States

Shiveluch Volcano
use Sheveluch

Shizukuishi
Town in N.
BT Honshu
 Japan

Shizuoka
Prefecture. Also a city 55 miles SW of Tokyo.
BT Honshu
 Japan

shoaling
Includes use on level 2 under ocean waves(1).
SA ocean waves
 waves

shoals
SA fluvial features
 geomorphology
 lacustrine features
 oceanography
 reefs
 shore features
 spits

shock
A valid term through 1977. After 1977, use shock metamorphism on level 2 under metamorphism(1).

shock metamorphism
Term introduced in 1978. Includes use on level 2 under metamorphism(1). Before 1978, also search metamorphism AND shock.
BA metamorphism
SA meteor craters

shock waves
Includes use on level 2 under interplanetary space(1); on level 3 under deformation(1) experimental studies(2), and under seismology(1).
SA bow shock waves
 deformation
 foreshocks
 interplanetary space
 metamorphism
 seismology
 shatter cones
 waves

Shonai River
In the Nagoya area of S central.
BT Honshu
 Japan

Shonkin Sag Laccolith
In Highwood Mountains area of N central.
BT Montana
 United States

shonkinite
Includes use on level 3 under igneous rocks(1) syenite family(2) and alkali gabbro family(2). See list H.
BA syenite family
BT igneous rocks
SA alkali gabbro family

shooting star
use meteors

shore drift
use littoral drift

shore features
Includes use on level 2 under geomorphology(1).
UF features, shore
NT mud flats
SA bars
 bays
 beach ridges
 beaches
 benches
 cliffs
 deltas
 estuaries
 fjords
 geomorphology
 inlets
 mud banks
 progradation
 shoals
 shorelines
 spits
 terraces

shore reef
use fringing reefs

shorelines
A level 1 term as of 1978. Used for geological studies on the engineering aspects of shorelines. Before 1978, included use on level 2 under engineering geology(1). Term set options are:
 barrier islands
  subtopic
 beaches
  subtopic
 changes

               subtopic
           construction
               subtopic
           design
               subtopic
           dynamics
               subtopic
           erosion
               subtopic
           hydraulics
               subtopic
           management
               subtopic
           seawalls
               subtopic
           stabilization
               subtopic
       SA barrier islands
           bays
           beach ridges
           beaches
           breakwaters
           changes of level
           coastal environment
           construction
           design
           dynamics
           engineering geology
           erosion
           geomorphology
           hydraulics
           inlets
           management
           marine installations
           ocean circulation
           ocean waves
           offshore
           onshore
           paleogeography
           seawalls
           shore features
           slope stability
           stabilization
           tidal inlets
           waterways

Shoriya Mountains
At SW end of Kuznetsk Alatau just S of Kuznetsk Basin. Also search Shoriya.
  UF Shoriya Range
  BT Russian Republic
      USSR

Shoriya Range
  use Shoriya Mountains

shortening, crustal
  use crustal shortening

shortite
  BA carbonates
  BT minerals

Shorty Crater
  BT Moon

Shoshone County
In N Idaho.
  BT Idaho
      United States

Shoshone Mountains
W of Toiyabe Range and Reese River in central Nevada. Also search Shoshone Range.
  UF Shoshone Range
  BT Nevada
      United States

Shoshone Range
  use Shoshone Mountains

shoshonite
Includes use on level 3 under igneous rocks(1) basalt family(2). See list H.
  BA basalt family
  BT igneous rocks

shrinkage cracks
Includes use on level 3 under sedimentary structures(1) bedding plane irregularities(2). See list K.
  BA bedding plane irregularities
      sedimentary structures
  SA cracks
      mudcracks

Shropshire
County on Welsh border.
  CO N522000N530000
      W0021500W0031500
  BT England
      Great Britain
      United Kingdom

Shugnan
Village in Gorno Badakhshan Autonomous Oblast in Pamirs in E.
  BT Tadzhikistan
      USSR

Shuswap Complex
Underlies the southern culmination of the Omineca Geanticline. Comprises rocks ranging in age from Proterozoic to early Mesozoic. S British Columbia. Also search Shuswap metamorphic complex.
  UF Shuswap metamorphic complex
  BT British Columbia
      Canada
  SA Mesozoic
      Paleozoic
      Precambrian
      Proterozoic

Shuswap metamorphic complex
  use Shuswap Complex

Shventoyi River
  use Sventoji River

Si
  use silicon

Si-32
Includes use on level 3 under isotopes(1).
  SA isotopes
      silicon

sial
  use granitic layer

Siam
  use Thailand

Sibay
Town in the Southern Urals in Bashkir A.S.S.R. in SE European USSR.
  BT Russian Republic
      USSR

Siberia
Region in Asia extending from the Urals to the Pacific Ocean, and from the Arctic Ocean to the Chinese and Mongolian borders and including N Kazakhstan. Index Soviet republics as applicable.
  CO N500000N800000
      W1700000E0600000
  BT USSR
  SA Kazakhstan
      Russian Republic

Siberia-Soviet Far East
  use Soviet Far East

Siberian Lowland
Region comprising the West Siberian Plain which extends from the Urals in the W to the Yenisei River in the E, and from the Kara Sea in the N to the Kazakh Hills in the S. Also search Siberian, West Siberian Basin, West Siberian Lowland and West Siberian Plain.
  UF West Siberian Basin
      West Siberian Lowland
      West Siberian Plain
  BT Russian Republic
      USSR

Siberian Platform
Region comprising the Central Siberian Plateau which lies between the Yenisei River on the W, and the Lena River on the E. It is bounded on the N by the Arctic Ocean and on the S by the Upper Tunguska and the upper Lena Rivers. Also search Siberian, and West Siberian Platform.
  CO N550000N720000
      E1260000E0900000
  UF West Siberian Platform
  BT Russian Republic
      USSR
  SA Anabar Shield
      West Siberian Plate

Sibiu
City and county in S.
  BT Transylvania
      Romania

Sicilia
  use Sicily

Sicilian
Europe. See list E.
  BA upper Pleistocene
      Pleistocene
      Quaternary
  BT Cenozoic

Sicily
Island in Mediterranean Sea and autonomous region.
  CO N364500N381500
      E0150000E0120000
  UF Sicilia
  BT Italy
  NT Caltanissetta
      Enna
      Lipari Island
      Lipari Islands
      Madonie Mountains
      Messina
      Mount Etna
      Nebrodi Mountains
      Palermo
      Pantelleria
      Peloritani Mountains
      Stromboli
      Trapani
      Ustica Island
      Vulcano
  SA Mediterranean region
      Syracuse

sideraerolite
  use stony irons

siderite
Includes use on level 3 under minerals (1). Also search chalybite.
  BA carbonates
  BT minerals
  SA chalybite

siderite (meteorite)
  use iron meteorites

siderolite
  use stony irons

siderophile elements
  SA elements

Sidhi
Village and district in E.
  BT Madhya Pradesh
      India

Sidobre Massif
In S Central Massif NE of Castres. Also search Sidobre.
  BT Tarn
      France
  SA Central Massif

Siebengebirge
Hills in the Westerwald on right bank of Rhine River SSE of Bonn.
  BT North Rhine-Westphalia
      West Germany

Siedlce
City in E part of country.
  UF Syedlets
  BT Warsaw
      Poland

Siegenian
Europe. Above Gedinnian, below Emsian. Includes use on level 3 under age terms(1). See list E.
  BA Lower Devonian
      Devonian
  BT Paleozoic

Siena
City and province in S central.
  BT Tuscany
      Italy

Sierozems
Includes use on level 3 under soils(1). See list M.
  UF Cerozem
      Gray desert soil
      Gray earth
      Serozem
  BA soils
  SA Zonal soils

Sierra Ancha
Ridge in central Gila County in E central.
  BT Arizona
      United States

Sierra Blanca
Range in Otero and Lincoln counties in S central.
  BT New Mexico
      United States

Sierra County
Index states as applicable.
  BX California
      New Mexico
  BT United States

Sierra de Gador
Range in SE part of country.
  BT Almeria
      Spain

Sierra de Gredos
Range W of Madrid in W central.
  BT Spain

Sierra de Guadarrama
Range N of Madrid separating Old Castile from New Castile.
  BT Spain

Sierra de la Demanda
Range of the Cordillera Iberica in Old Castile in N.
  BT Spain

Sierra de los Filabres
Range in W central.
  BT Almeria
      Spain

Sierra de los Organos
Range in W.
  BT Pinar del Rio
      Cuba

Sierra de Ronda
  use Serrania de Ronda

Sierra Gorda
E range of S Sierra Madre Occidental in central part of country.
  BT Guanajuato
      Mexico

Sierra Leone
Former British colony and protectorate. Became independent in 1961. Includes use on level 1 as an area term (list O). For term set options see list B.
  CO N070000N100000
      W0110000W0133000
  BA Africa
  SA West Africa

Sierra Madre
Chief mountain system extending 1500 miles SE from U.S. border in-

cluding 3 separate ranges which enclose the great central plateau. Index states as applicable. Before 1978, this term also included Sierra Madre Range.
  BT Mexico
  SA Sierra Madre del Sur
        Sierra Madre Occidental
        Sierra Madre Oriental
        Sierra Madre Range

Sierra Madre del Sur
Coastal range in SW Mexico. Index states as applicable.
  BT Mexico
  SA Guerrero
        Oaxaca
        Sierra Madre

Sierra Madre Occidental
Range running for 700 miles parallel to the Pacific Ocean. Index states as applicable.
  BT Mexico
  SA Chihuahua
        Durango
        Sierra Madre
        Sonora
        Zacatecas

Sierra Madre Oriental
Range running parallel to the Gulf of Mexico. Index states as applicable.
  BT Mexico
  SA Coahuila
        Hidalgo
        Nuevo Leon
        Puebla
        San Luis Potosi
        Sierra Madre
        Tamaulipas
        Veracruz

Sierra Madre Range
Term introduced in 1978 for range in Continental Divide, S Wyoming, just W of North Platte River. Before 1978, also search Sierra Madre AND Wyoming.
  BT Wyoming
        United States
  SA Sierra Madre

Sierra Morena
A valid term through mid-1978. See Betic Cordillera.

Sierra Nevada
Mountain Range. Index California and/or Spain as applicable.
  CO N353000N400000
        W1174500W1210000
  SA California
        Cascade Range
        Spain

Sierra Nevada de Santa Marta
Mountain range in N on Caribbean coast.
  BT Colombia

Sierrita Mountains
  BT Arizona
        United States

Sifton Basin
N British Columbia. Also search Sifton.
  BT British Columbia
        Canada

Sigillaria
Genus.
  BA Lycopsida
        pteridophytes
  BT Plantae

Signal Hill Formation
Unconformably underlies Random Formation and overlies Avalonian Formation.
  BT Huronian
        Proterozoic
        Precambrian
  SA Canada
        Newfoundland

signals
Includes use on level 3 under seismology(1).
  SA geophysical methods
        noise
        seismology
        surveys
        teleseismic signals

Signy Island
One of the South Orkney Islands SE of the Falkland Islands.
  BT Atlantic Ocean
  SA South Orkney Islands

Sigsbee Deep
In SW central part of Gulf of Mexico.
  UF Mexico Basin
  BT Gulf of Mexico

Sikhote Alin
A valid term through 1976. After 1976, use Sikhote-Alin Range.

Sikhote-Alin Range
Mountain range in Primorye and Khabarovsk krays along the Japan Sea and the Tatar Strait. Also search Sikhote-Alin.
  BT Russian Republic
        USSR
  SA Soviet Far East

Sikkim
Former British and Indian protectorate between Bhutan and Nepal. Became an associated Indian state on Sept. 7, 1974. Included use on level 1 as an area term until 1977. Now use term on level 3; use India on level 1.
  BT India
        Asia
  SA Himalayas
        Lesser Himalayas
        Siwalik Range
        Siwalik System

Sila Massif
In the Southern Apennines forming central part of toe of the Italian boot.
  BT Calabria
        Italy

silcrete
  BA chemically precipitated sediments
  BT sediments
  SA clastic sediments
        terrigenous materials

Silesia
Region in E central Europe lying mostly in SW Poland. It comprises both Upper and Lower Silesia plus small areas in SE East Germany and N Moravia. Index countries as applicable.
  BT Europe
  SA Czechoslovakia
        East Germany
        Lower Silesia
        Poland
        Upper Silesia

Silesian
  BA Carboniferous
  BT Paleozoic

Silesian Basin
use Silesian coal basin

Silesian coal basin
In Lower Silesia in Wroclaw Province, and in Upper Silesia in Katowice Province. Index provinces as applicable.
  UF Silesian Basin
  BT Poland
  SA Katowice
        Lower Silesia
        Upper Silesia
        Upper Silesian coal basin
        Wroclaw

silexite
Restricted to usage as igneous rock. Includes use on level 3 under igneous rocks(1) granite-granodiorite family(2). See list H.
  BA granite-granodiorite family
  BT igneous rocks

silica
Use only when of economic value. Includes use as level 3 commodity term under sand(1). See list C.
  SA abrasives
        sand
        silicates
        siliceous composition
        silicon

silica group
use silica minerals

silica minerals
Includes use in combination with framework silicates (i.e. framework silicates, silica minerals) on level 2 under minerals(1). See list L. Also search silica group.
  UF silica group
  BA framework silicates
        silicates
  BT minerals
  NA agate
        amethyst
        chalcedony
        coesite
        cristobalite
        hyalite
        jasper
        keatite
        lechatelierite
        melanophlogite
        opal
        quartz
        stishovite
        tridymite

silicate minerals
use silicates

silicate rocks
  SA petrology
        rocks
        siliceous composition
        standard materials

silicates
In indexing, use this term only for broad treatments of the entire class of minerals; otherwise, use a narrower term, e.g. orthosilicates, ring silicates, chain silicates, etc. Includes use on level 2 under minerals(1). See list L. As of 1978, the term silicates is autoposted to all individual silicate minerals. Before 1978 also search individual silicate groups. Also search the term silicate minerals.
  UF silicate minerals
  BT minerals
  NA aluminosilicates
        asbestos
        borosilicates
  NZ britholite
  NA chain silicates
        chlorophaeite
        framework silicates
  NZ helvite
  NA orthosilicates
        ring silicates
        sheet silicates
  SA clay mineralogy
        silica
        silicon

siliceous composition
Term introduced in 1978. Includes use on level 3 under sediments(1). Before 1978 also search siliceous in combination with specific sediment type. Also search siliceous rocks.
  SA chemically precipitated sediments
        clastic rocks
        composition
        insoluble residues
        jasperoid
        sedimentary rocks
        sediments
        silica
        silicate rocks
        siliceous sinter

siliceous rocks
A valid term through 1977. After 1977, use siliceous composition.

siliceous sinter
Includes use on level 3 under sediments (1) chemically precipitated sediments (2). See list N and list I. Also search sinter AND siliceous.
  BA chemically precipitated sediments
  BT sediments
  SA chemically precipitated rocks
        siliceous composition

silicides
  BA native elements and alloys
  BT minerals
  SA ferrosilicon

silicification
Includes use on level 3 under diagenesis(1), under metasomatism(1) processes(2), or under sedimentation(1).
  UF silification
  SA chertification
        diagenesis
        fossilization
        metasomatism
        processes
        sedimentation

Silicispongiae
  UF Noncalcarea
  BA Porifera
  BT Invertebrata

Silicoflagellata
Includes use on level 2 under Protista(1). See list F.
  BA Protista
  NA Dictyocha
        Distephanus

silicon
Includes use on level 1 and 2 as a chemical element (list D). Also search Si.
  UF Si
  SA elements
        Si-32
        silica
        silicates

silification
use silicification

Siljan
Lake in central part of country.
  BT Sweden

sillimanite
Includes use as level 3 commodity term under ceramic materials(1). See list C. Also search fibrolite.
  UF fibrolite
  BA orthosilicates
        silicates
  BT minerals
  SA andalusite
        ceramic materials
        kyanite

sillimanite gneiss
See list J.
  BA gneisses
  BT metamorphic rocks
  SA gneiss

sills
Includes use on level 2 under intrusions(1).
BA intrusions
SA differentiation
dikes

silt
Includes use on level 3 under sediments(1) clastic sediments(2). See list N.
BA clastic sediments
BT sediments
SA alluvium
clay
loam
loess
mud
sand
siltstone
terrigenous materials

siltite
use siltstone

siltstone
Includes use on level 3 under sedimentary rocks(1) clastic rocks(2). See list I. Also search siltite.
UF siltite
BA clastic rocks
BT sedimentary rocks
SA metasiltstone
mudstone
silt
terrigenous materials

Silurian
Includes use on level 1 as an age term (list E); on level 2 under paleoterms, e.g. paleoecology; paleogeography, paleomagnetism. Above Ordovician, below Devonian. Also search Gotlandian.
UF Gothlandian
Gotlandian
BA Paleozoic
NT Aberystwyth Grits
NA Lower Silurian
Middle Silurian
NT Red Mountain Formation
Rockwood Formation
NA Upper Silurian
NT Waterville Formation
SA Broken River Formation
Hidden Valley Dolomite
Hunton Group
Peel Sound Formation
Read Bay Formation
Ringerike Sandstone
Rondout Formation
Taconic Orogeny
Talladega Group

silvanite
use sylvanite

silver
Includes use on level 1 and 2 as a commodity term (list C) and as a chemical element (list D); on level 2 under mineral deposits, genesis(1) and paragenesis(1).
UF Ag
SA argentite
elements
heavy metals
native elements and alloys

Silver Bow County
SW Montana.
BT Montana
United States

Silver City
Town in Grant County in SW.
BT New Mexico
United States

Silver Peak Mountains
Small range in W Esmeralda County in SW Nevada. Also search Silver Peak.
UF Silver Peak Range
BT Nevada
United States

Silver Peak Range
use Silver Peak Mountains

Silverton Caldera
Near town of Silverton in San Juan County in San Juan Mountains of SW. Also search Silverton.
BT Colorado
United States

Silvretta Group
Mountain group. Also search Silvretta.
UF Silvretta Massif
BT Paleozoic
SA Switzerland
Tyrol

Silvretta Massif
use Silvretta Group

sima
use basaltic layer

Simbirsk
use Ulyanovsk

Simferopol
City in the S central Crimea.
BT Ukraine
USSR

Simi Hills
In Ventura County NW of Los Angeles.
BT California
United States

similar folds
Term introduced in 1978. Includes use on level 3 under folds(1) style(2). Before 1978, search similar AND folds.
UF shear folds
BA folds
SA concentric folds
shear

Simla
Town and hill resort in S.
BT Himachal Pradesh
India

Simla Hills
Hill and mountain area of the outer Kumaun Himalayas around Simla. Also search Simla.
BT Himachal Pradesh
India

Simleu Basin
In Bihor Mountains near town of Simleu Silvaniei in NW. Also search Simleu.
BT Transylvania
Romania

Simplon region
Includes Simplon Pass, Simplon Road, and Simplon Tunnel area. Index Italy and/or Switzerland. Also search Simplon.
BT Europe
SA Italy
Piedmont
Switzerland

Simpson Desert
Primarily in SE Northern Territory. Index Northern Territory and states as applicable.
BT Australia
SA Northern Territory
Queensland
South Australia

Simpson Group
Comprises Joins, Oil Creek, McLish, Tulip Creek, Bromide, and Corbin Ranch formations. Central and S Oklahoma.
BT Middle Ordovician
Ordovician
SA Oklahoma

simulation
Includes use on level 3 under automatic data processing(1) or mathematical geology(1).
SA analog simulation
automatic data processing
digital simulation
mathematical geology
models

Sinai
Peninsula extending from the Mediterranean Sea to the Red Sea. It is bounded on the W by the Suez Canal and the Gulf of Suez, and on the E by Israel and the Gulf of Aqaba. Also search Sinai Peninsula.
UF Sinai Peninsula
BT Egypt
SA Israel

Sinai Peninsula
use Sinai

Sinaia
Town and health resort at SE foot of Bucegi Mountains of the Transylvanian Alps in Prahova County.
BT Walachia
Romania

Sinaloa
State in W.
BT Mexico
NT Mazatlan

Sind
Province in SE.
BT Pakistan
NT Chor
SA Nari Series

Sinemurian
Europe. Above Hettangian, below Pliensbachian. Includes use on level 3 under age terms(1). See list E.
BA Lower Jurassic
Jurassic
BT Mesozoic
SA Liassic

Singapore
Island republic and city off the southern tip of the Malay Peninsula. Includes use on level 1 as an area term (list O). For term set options see list B.
BA Asia
SA Malay Archipelago

Singhbhum
District, Chota Nagpur division, S Bihar.
BT Bihar
India
SA Singhbhum Granite

Singhbhum Granite
Also search Singhbhum.
UF Singhbhum Shear Zone
BT Archean
lower Precambrian
Precambrian
SA Bihar
India
Orissa
Singhbhum

Singhbhum shear zone
Index states as applicable. Also search Singhbhum.
BT India
SA Bihar
Orissa

Singhbhum Shear Zone
use Singhbhum Granite

single-crystal method
Term introduced in 1978. Before 1978, search single crystal.
SA crystal growth
crystal structure
methods
minerals

Singleton Coal Measures
E New South Wales.
SA Australia
Carboniferous
New South Wales
Permian
Triassic

Sini
Village in Singhbhum District. Site of limestone quarries for Jamshedpur iron and steel works.
BT Bihar
India

Sinjar Formation
Paleocene-lower Eocene: In the Jabel Sinjar Mountains.
UF Sinjar Limestone Formation
BT Paleogene
Tertiary
Cenozoic
SA Eocene
Iraq
lower Eocene
Middle East
Paleocene

Sinjar Limestone Formation
use Sinjar Formation

sinkholes
Includes use on level 3 under geomorphology(1) solution features(1).
SA dolines
geomorphology
karst
potholes
solution features

Sinkiang
use Sinkiang Uighur

Sinkiang Province
use Sinkiang Uighur

Sinkiang Uighur
Autonomous region in NW China. Also search Sinkiang.
UF Sinkiang
Sinkiang Province
BT China
NT Dzhungaria
SA Dzhungarian Alatau
Ili Basin
Irtysh River
Pamirs
Tarbagatay Range
Tien Shan
Turkestan

sinking
use subsidence

sinks
SA geochemistry

sinnerite
BA sulfides
BT minerals

sinter
A valid term through 1977. After 1977, use siliceous sinter.

sinuosity
SA channels
fluvial features
meanders
streams

Sioux County
Index states as applicable.
BX Iowa
Nebraska
North Dakota
BT United States

Sioux Quartzite
Has been correlated with Baraboo Quartzite in Wisconsin. NW Iowa, SW Minnesota, NE Nebraska, and

SE South Dakota.
BT Precambrian
SA Iowa
   Minnesota
   Nebraska
   South Dakota

Sir Darya
use Syr Darya

Sirenia
Includes use on level 2 under Mammalia(1). See list F.
BA Mammalia
BT Tetrapoda
   Vertebrata

Siret River
Rises on the E slopes of the Carpathians in the Ukraine and flows SE into the Danube River just N of Galati. Index Moldavia and/or Ukraine. Also search Siret.
UF Sereth River
   Siretul River
BT Europe
SA Moldavia
   Ukraine

Siretul River
use Siret River

Sirohi
Former Indian state now part of Rajasthan. Also a town in SW.
BT Rajasthan
   India

Sirte Basin
On the Gulf of Sidra in N Libya. Also search Sirte.
UF Syrte
BT Libya

Sisian
Village in S.
BT Armenia
   USSR

Siskiyou County
On Oregon border.
CO N410000N420000
   W1213000W1214500
BT California
   United States

site exploration
Includes use on level 2 under engineering geology(1). Before 1978, also search site selection.
UF site selection
SA dams
   engineering geology
   exploration
   explosions
   foundations
   geologic hazards
   highways
   land subsidence
   marine installations
   nuclear facilities
   permafrost
   reservoirs
   rock mechanics
   slope stability
   soil mechanics
   tunnels
   underground installations
   waste disposal

site selection
use site exploration

sites, archaeological
use archaeological sites

sites, landing
use landing sites

Sitka Sound
On W side of Baranof Island in SE Alaska. Entrance to Sitka from Gulf of Alaska. Also search Sitka.
BT Alaska
   United States

Sivamalai
A hill in the Coimbatore District in E.
BT Tamil Nadu
   India

Sivash
Salt lagoons and marshes in N and NE Crimea.
UF Putrid Sea
BT Ukraine
   USSR

Siwalik Formation
use Siwalik System

Siwalik Group
use Siwalik System

Siwalik Hills
use Siwalik Range

Siwalik Range
Range of foothills parallel with the main Himalayan system and extending 1000 miles SE from N Punjab in Pakistan to Sikkim. Also search Siwalik and Siwaliks. Index Sikkim and countries as applicable.
UF Siwalik Hills
   Siwaliks
BT Asia
SA India
   Nepal
   Pakistan
   Sikkim

Siwalik Sandstone
use Siwalik System

Siwalik Series
use Siwalik System

Siwalik System
Middle Miocene-Pleistocene: Extends beyond limits of Siwalik Range to include parts in Baluchistan and Northwest Frontier Province in Pakistan, Tipam Series in Assam, and Irrawaddy Series in Burma. Includes Lower (Kamlial and Chinji), Middle (Nagri and Dhokpathan) and Upper (Tatrot, Pinjor and Boulder conglomerates) Siwalik Sub-groups. Also search Siwalik, Siwalik Series, Siwalik Group, Siwalik Formation, and Siwalik Sandstone.
UF Siwalik Formation
   Siwalik Group
   Siwalik Sandstone
   Siwalik Series
BT Cenozoic
SA Burma
   India
   middle Miocene
   Miocene
   Neogene
   Pakistan
   Pleistocene
   Pliocene
   Sikkim
   Tertiary
   upper Miocene
   upper Tertiary

Siwaliks
use Siwalik Range

Sixes River
N Curry County in SW Oregon. Flows into the Pacific Ocean.
BT Oregon
   United States

size
Includes use on level 3 under sedimentary rocks(1), sediments(1), and sedimentary petrology(1). Use grain size under sediments(1).
UF particle size
SA dimensions
   grain size
   granulometry
   particles
   phi scale
   sedimentary rocks
   sediments
   shape
   size distribution
   sorting

size analysis
A valid term through mid-1978. Use granulometry.

size distribution
SA distribution
   granulometry
   sedimentary rocks
   sediments
   size
   textures

Sjaeland
use Zealand

sjogrenite
BA carbonates
BT minerals

Skaergaard Intrusion
BT Greenland

Skagerrak
Arm of the North Sea between Norway and Denmark.
BT North Sea
   Atlantic Ocean

Skagit Valley
River valley. Primarily in NW Washington. Index British Columbia and/or Washington as applicable. Also search Skagit; Skagit River.
SA British Columbia
   Washington

Skamania County
On the Oregon border in SW.
BT Washington
   United States

Skane
Region in S Sweden. Index counties as applicable. Also search Scania.
UF Scania
BT Sweden
SA Kristianstad
   Malmohus

skarn
Compositional term. As of 1978, includes use on level 2 under metasomatic rocks(1). Before 1978, included use on level 3 under metamorphic rocks(1).
BA metasomatic rocks
SA metamorphic rocks

Skeena Mountains
NW central British Columbia. E of Coast Mountains.
BT British Columbia
   Canada

Skeidararjokull
Glacier.
BT Iceland

skeletons
Includes use on level 3 under fossil group(1) for type of material; on level 3 under paleontology(1). See list F.
SA bones
   fossils
   paleontology
   skulls
   teeth
   Vertebrata

Skellefte
River in N Sweden. Flows SE into Gulf of Bothnia. Index counties as applicable.
BT Sweden
SA Norrbotten
   Vasterbotten

skewness
SA kurtosis
   statistical analysis
   statistical methods

Skiddaw Group
use Skiddaw Slates

Skiddaw Slates
Arenig and Llanvirn Series. Cumberland County in NW England. Also search Skiddaw Group.
UF Skiddaw Group
BT Ordovician
SA England
   United Kingdom

Skolithos
BA ichnofossils

Skomer Island
In Saint Georges Channel off Pembrokeshire.
BT Wales
   Great Britain
   United Kingdom

Skopje
City on Vardar River in N Macedonia.
UF Skoplje
BT Macedonia
   Yugoslavia

Skoplje
use Skopje

skulls
Includes use on level 3 under fossil group(1) for type of material. See list F.
SA bones
   fossils
   jaws
   paleontology
   skeletons
   teeth
   Vertebrata

skutterudite
BA arsenides
   sulfides
BT minerals

Skye
use Isle of Skye

Skylab
SA remote sensing
   satellite methods

skystones
use meteorites

Skythian
use Scythian

slabs
Includes use on level 3 under plate tectonics(1).
SA plate tectonics

Slak Dolny
use Wroclaw

Slanic
Town in Prahova County in N central.
BT Walachia
   Romania

Slany
Town in W central.
BT Bohemia
   Czechoslovakia

slate
Includes use on level 1 as a commodity term (list C); on level 3 under metamorphic rocks (1) slates(2). See list J.
BA slates
BT metamorphic rocks
SA slaty cleavage

slates
Includes use on level 2 under metamorphic rocks(1). See list J.
BT metamorphic rocks
NA slate

SA shale

Slatina
City in Olt County in SW.
BT Walachia
Romania

slaty cleavage
Includes use on level 3 under foliation(1) style(2).
BA foliation
SA axial-plane structures
cleavage
flow cleavage
metamorphism
schistosity
slate
slip cleavage

Slave Province
Region comprising the Lake Athabasca, Slave River, and Great Slave Lake area. Index Northwest Territories and provinces as applicable.
CO N600000N650000
W1050000W1200000
BT Canada
SA Alberta
Northwest Territories
Saskatchewan

slickensides
Includes use on level 3 under faults(1) effects(2) and under lineation(1) style(2).
UF polished surface
SA breccia
faults
lineation

sliding, gravity
use gravity sliding

Sligo
County in NW. Also a town.
BT Ireland

slip cleavage
Term introduced in 1978. Before 1978, search slip AND cleavage; shear cleavage; crenulation cleavage.
UF crenulation cleavage
shear cleavage
strain-slip cleavage
BA foliation
SA cleavage
schistosity
slaty cleavage

Slocan mining camp
SE British Columbia. Also search Slocan.
BT British Columbia
Canada

slope, continental
use continental slope

slope environment
Term introduced in 1978. Use this term or shelf environment for older sedimentation. For recent sedimentation, use continental slope or continental shelf. Before 1978, search slope.
SA continental shelf
continental slope
environment
sedimentation
shelf environment

slope, outer
use outer slope

slope stability
A level 1 term as of 1978. Used for geological studies on the engineering aspects of mass movements. For other aspects, see geomorphology. Before 1978, included use on level 2 under engineering geology(1). Term set options are:
creep

subtopic
debris flows
subtopic
earthflows
subtopic
embankments
subtopic
erosion
subtopic
excavations
subtopic
experimental studies
subtopic
failure
subtopic
field studies
subtopic
landslides
subtopic
mass movements
subtopic
mudflows
subtopic
rockfalls
subtopic
site exploration
subtopic
stabilization
subtopic
talus
subtopic
theoretical studies
subtopic
SA avalanches
controls
creep
dams
debris flows
earthflows
earthquakes
embankments
engineering geology
erosion
excavations
explosions
failure
foundations
geologic hazards
geomorphology
highways
land subsidence
landslides
mass movements
mudflows
reservoirs
rock mechanics
rockfalls
shorelines
site exploration
slopes
slumping
soil mechanics
solifluction
stability
stabilization
talus slopes
tunnels
underground installations

slopes
Includes use on level 3 under geomorphology(1).
SA cliffs
continental slope
embankments
fault scarps
geomorphology
glacis
landslides
scarps
slope stability
talus slopes

Slovakia
Region in E now comprising the Slovak Socialist Republic.
UF Slovensko
BT Czechoslovakia
NT Banska Stiavnica
Bojnice
Bratislava
Brezno
Gemer
Hodrusa
Korytnica
Kremnica
Kremnica Mountains
Low Tatra Mountains
Novaky
Orava Valley
Slovakian Karst
Spis-Gemer
Sturovo
Vah Valley
SA Morava River valley
Spis
Tokaj-Eperjes Mountains

Slovakian Karst
Near the Hungarian border in SE.
BT Slovakia
Czechoslovakia
SA Permian
Triassic

Slovenia
Constituent republic in NE Yugoslavia.
CO N451500N465000
E0170000E0133000
UF Slovenija
BT Yugoslavia
NT Celje
Ljubljana
SA Istria

Slovenija
use Slovenia

Slovenske Rudohorie
use Spis-Gemer

Slovensko
use Slovakia

sludging
use solifluction

slump faults
Not a valid term for GeoRef. See normal faults or growth faults.

slump structures
Includes use on level 3 under sedimentary structures(1) soft sediment deformation(2). See list K. Also search slump structure.
BA soft sediment deformation
sedimentary structures
SA collapse structures
convoluted beds
slumping
structures

slumping
Includes use on level 3 under sedimentation(1). Also search slump; slumps.
UF slumps
SA continental shelf
continental slope
creep
erosion
geomorphology
landslides
liquefaction
mass movements
sedimentation
slope stability
slump structures

slumps
use slumping

Slyudyanka
City at SW end of Lake Baikal in S Irkutsk Oblast.
BT Russian Republic
USSR

Sm
use samarium

Smackover Formation
Overlies Eagle Mills Formation; underlies Buckner Formation. Subsurface.
BT Upper Jurassic
Jurassic
SA Alabama
Arkansas
Louisiana
Mississippi
Texas

Smaland
Plateau region S of Lake Vattern in S Sweden. Index counties as applicable.
BT Sweden
SA Kalmar

smaragd
use emerald

smectite
BA sheet silicates
silicates
BT minerals
SA clay minerals
fuller's earth
montmorillonite
saponite

Smith County
Index states as applicable.
BX Kansas
Mississippi
Tennessee
Texas
BT United States

Smithers
Village in W central.
BT British Columbia
Canada

Smithsonian Institution
BT District of Columbia
United States

smithsonite
UF zinc spar
BA carbonates
BT minerals
SA azurite

Smithville Formation
Above Powell Formation and below Black Rock Limestone in N Arkansas, and below Everton Formation in SE Missouri. N Arkansas and SE Missouri.
BT Lower Ordovician
Ordovician
SA Arkansas
Missouri

Smokies
use Great Smoky Mountains

Smoky Hill Chalk Member
Of Niobrara Formation. E Colorado, W Kansas, NE New Mexico, and SE South Dakota.
BT Upper Cretaceous
Cretaceous
SA Colorado
Kansas
New Mexico
Niobrara Formation
South Dakota

Smoky Hill River basin
W central and central Kansas. Also search Smoky Hill River.
BT Kansas
United States

Smoky River
Rises in Jasper National Park and flows NNE into Peace River.
BT Alberta
Canada

Smolensk
City on left bank of upper Dnieper River SW of Moscow. Also an oblast.

Smolensk ● soil mechanics

    BT   Russian Republic
           USSR

smythite
    BA   sulfides
    BT   minerals

Sn
    use tin

Snaefellsnes Peninsula
    Juts into Denmark Strait between Breidi Fjord and Faxa Bay in W Iceland. Also search Snaefellsnes.
    BT   Iceland

Snake Range
    Mountain range in E White Pine County in E.
    BT   Nevada
           United States

Snake River
    Rises in Yellowstone National Park and flows S, then W, N, and again W into the Columbia River in SE Washington. Index states as applicable.
    BT   United States
    SA   Idaho
           Oregon
           Washington
           Wyoming

Snake River basin
    Index states as applicable.
    BT   United States
    SA   Idaho
           Nevada
           Oregon
           Utah
           Washington
           Wyoming

Snake River canyon
    Grand Canyon of the Snake River. Extends N-S between Wallowa Mountains, Oregon, and Seven Devils Mountains, Idaho. Index states as applicable.
    UF   Hell's Canyon
    BT   United States
    SA   Idaho
           Oregon

Snake River plain
    Crescent-shaped lava tableland across S central.
    CO   N421500N441500
           W1113000W1160000
    BT   Idaho
           United States

Snieznik
    Highest peak in Kralicky Sneznik Mountains in NE Bohemia and Lower Silesia. Index countries as applicable.
    BT   Europe
    SA   Czechoslovakia
           Poland

Snohomish County
    NNE of Seattle.
    BT   Washington
           United States

Snoqualmie Batholith
    Probably intermediate in age between lower part of Keechelus Andesitic Series and Fifes Peak Andesite. W central Washington.
    BT   Tertiary
    SA   Miocene
           Pliocene
           Washington

snow
    Includes use on level 2 under hydrology(1); as level 3 under glacial geology(1) glaciers(2).
    SA   accumulation
           atmospheric precipitation
           firn
           glacial geology
           glaciers
           hydrology
           hydrosphere
           ice
           meltwater
           thawing
           water

Snow Lake
    Village in gold-mining region 70 miles E of Flin Flon in W.
    BT   Manitoba
           Canada

snow-patch erosion
    use nivation

Snowdon
    use Snowdonia

Snowdonia
    Highest mountain in Wales, in Caenarvonshire 10 miles SE of Caernarvon; consists of 5 peaks, separated by passes. Also search Snowdon.
    UF   Snowdon
    BT   Wales
           Great Britain
           United Kingdom

Snowy Mountains
    Range of the Australian Alps. Index states as applicable.
    BT   Australia
    SA   New South Wales
           Victoria

soap clay
    use bentonite

soapstone
    Includes use as level 3 commodity term under talc(1). See list C.
    SA   talc

Sob River basin
    Near border of Moldavia, USSR, in SW Ukraine. Also search Sob.
    BT   Ukraine
           USSR

Sobotka
    Town in N.
    BT   Bohemia
           Czechoslovakia

Sochi
    City in S Krasnodar Kray on Black Sea near Georgian border.
    BT   Russian Republic
           USSR

Society Islands
    Includes use on level 1 as an area term (list O). For term set options see list B.
    BA   Polynesia
    SA   Tahiti

Socorro
    City in Socorro County in W central.
    BT   New Mexico
           United States

Socorro County
    W central.
    BT   New Mexico
           United States

sodalite
    BA   sodalite group
           framework silicates
           silicates
    BT   minerals

sodalite group
    Includes use in combination with framework silicates (i.e. framework silicates, sodalite group) on level 2 under minerals(1). See list L.
    BA   framework silicates
           silicates
    BT   minerals
    NA   hydrosodalite
           nosean
           sodalite
           tugtupite
    SA   genthelvite
           hauyne
           lazurite

sodium
    Includes use on level 1 and 2 as a chemical element (list D). Also search Na.
    UF   Na
    SA   alkali metals
           elements
           Na-22
           Na-24

sodium carbonate
    Includes use on level 1 as a commodity term (list C).
    SA   carbonates
           trona

sodium chloride
    Includes use on level 3 under geochemistry(1).
    SA   chlorides
           evaporites
           geochemistry
           halides
           salt

sodium sulfate
    Includes use on level 1 as a commodity term (list C).
    SA   sulfates

Sofia
    Province in W Bulgaria. Also a city.
    UF   Sofiya
           Sophia
    BT   Bulgaria
    NT   Rila
           Yetropole

Sofiya
    use Sofia

soft coal
    use bituminous coal

soft sediment deformation
    Includes use on level 2 under sedimentary structures(1). See list K.
    BA   sedimentary structures
    NA   ball-and-pillow
           boudinage
           clastic dikes
           convoluted beds
           flame structures
           flow structures
           olistoliths
           olistostromes
           sandstone dikes
           slump structures
    SA   flute casts
           load casts
           sole marks
           tool marks

Sogn
    Mountain region around Sogne Fjord in SW part of country.
    BT   Norway

Sogndal
    Village on North Sea.
    BT   Rogaland
           Norway

soil flow
    use solifluction

soil fluction
    use solifluction

soil group
    Includes use on level 2 under soils(1). See list M.
    SA   Bog soils
           Brown forest soils
           Chernozems
           Chestnut soils
           Desert soils
           Planosols
           soils
           Solonchak soils
           Solonetz soils

soil horizon
    use horizons

soil management
    Includes use on level 3 under soils(1) utilization(2).
    SA   fertilizers
           management
           soils
           tillage
           treatment
           utilization

soil mechanics
    A level 1 term as of 1978. Used for geotechnical studies. Includes use on level 2 under tunnels(1), underground installations(1), and nuclear facilities(1). If 1, term set options are:
    applications
        subtopic
    case studies
        subtopic
    concepts
        subtopic
    deformation
        subtopic
    earth pressure
        subtopic
    elasticity
        subtopic
    experimental studies
        subtopic
    frost action
        subtopic
    materials, properties
        subtopic
    methods
        subtopic
    settlement
        subtopic
    site exploration
        subtopic
    techniques
        subtopic
    theoretical studies
        subtopic
    SA   alluvium
           Atterberg limits
           bearing capacity
           case studies
           clays
           cohesive materials
           compaction
           competent materials
           compressibility
           compressive strength
           congelifraction
           consolidation
           cryoturbation
           dams
           deformation
           earth pressure
           earthquakes
           elasticity
           engineering geology
           engineering properties
           explosions
           finite strain
           foundations
           frost action
           geologic hazards
           granular materials
           highways
           hydraulics
           land subsidence
           liquefaction
           loading
           loess
           marine installations
           mechanical properties
           mechanics
           nuclear facilities
           overconsolidated materials
           penetration
           plastic materials
           pore pressure

reservoirs
rock mechanics
seepage
settlement
shear strength
site exploration
slope stability
soils
triaxial tests
tunnels
unconsolidated materials
underground installations
water pressure

soil sampling
Term introduced in 1978. Includes use on level 3 under mineral exploration.
SA sample preparation
    sampling
    sediment sampling
    soils

soil ulmin
use humus

soil zone
use horizons

soils
Includes use on level 1 (list A); on level 2 under area terms(1), education(1), bibliography(1), and symposia(1). Used for general pedology as well as specific topics. If level 1, term set options are:
analysis
  biological methods
  chemical methods
  physical methods
  sample preparation
biota
  bacteria
  fungi
  Invertebrata
  microorganisms (for several types)
  Protozoa
  Vertebrata
chemistry
  ion exchange
  nitrogen
  phosphorus
  (This heading of second rank picks up papers on cation exchange relations, the nitrogen status of soils, transformations of nitrogen, phosphorus fixation, potassium fixation and the like. Listing of nitrogen and phosphorus is illustrative; names of all other elements are allowed.
classification
  soil groups [List M; names of groups in different systems may be used; paleosols may also be used]
composition
  chemical composition
  mineral composition
  organic materials
conservation
  erosion control
  fertility maintenance
  physical properties
  (This covers maintenance of structure, pore space, good tilth and the like so that root penetration would be facilitated. Included should be papers on deterioration in the physical properties of soils and its prevention.)
correlation
  subtopic
erosion
  landslides
  water erosion
  wind erosion
fertilizers
  farms
field studies
  methods or type of methods (e.g. remote sensing, sampling, applications, programs, techniques]
genesis
  factors or type of factor (e.g. climate, geomorphology, organisms, parent materials, time)
  horizon differentiation
  leaching
  migration
  organic materials (gains, losses, etc.)
  transformations
morphology
  color
  consistency
  horizons
  micromorphology
  profiles
  structure
  textures
nutrients
  major elements
  toxic substances
  trace elements
organic materials
  abundance
  composition
  distribution
  functions
patterns
  subtopic
properties
  subtopic
salinity
  occurrence
  treatment
soil group
  name of group
surveys (only for detailed study of a particular area)
  area
utilization
  agriculture
  fertilization
  field crops
  forestry
  fruit crops
  housing
  reclamation
  recreation
  soil management
  tillage
  vegetable crops
waste disposal
  animal waste
  human waste
water regimes
  characterization
  drainage
  irrigation
  movement (in soil)
  storage (storage in soil)
yields
  field crops
  forests
  fruits
  vegetables
NA Alfisols
    Alluvial soils
    Andosols
    Aridisols
    Bog soils
    Brown forest soils
    Brown soils
    Calcareous soils
    Chernozems
    Chestnut soils
    Clay soils
    Desert soils
    Entisols
    Frozen ground
    Gleys
    Gray forest soils
    Histosols
    Hydromorphic soils
    Inceptisols
    Intrazonal soils
    laterites
    Latosols
    loam
    Luvisols
    Meadow soils
    Mollisols
    Mor
    Mull
    Oxisols
    Paleosols
    Planosols
    Podzols
    Pseudogleys
    Red soils
    Rendzinas
    Sierozems
    Solonchak soils
    Solonetz soils
    Spodosols
    Terra rossa
    Tundra soils
    Ultisols
    Vertisols
    Zonal soils
SA active layer
    agriculture
    alluvium
    bacteria
    bedrock
    biota
    calcification
    calcrete
    caliche
    capillarity
    catenas
    characterization
    chemical composition
    chemical methods
    chemistry
    clay mineralogy
    climate
    colluvium
    color
    conservation
    consistency
    creep
    degradation
    desalinization
    desiccation
    drainage
    duricrust
    ecology
    eluvium
    engineering geology
    engineering properties
    environmental geology
    erosion
    erosion control
    expansive materials
    factors
    fertilization
    fertilizers
    field crops
    field studies
    forestry
    forests
    foundations
    fruits
    fuller's earth
    geomorphology
    glacial geology
    gyttja
    highways
    horizon differentiation
    horizons
    humus
    ion exchange
    irrigation
    land use
    landslides
    laterization
    leaching
    micromorphology
    microorganisms
    migration
    moisture
    morphology
    movement
    nitrogen
    nutrients
    organic materials
    organisms
    parent materials
    particles
    permafrost
    phosphate
    phosphorus
    physical methods
    physical properties
    profiles
    reclamation
    recreation
    regolith
    salinity
    sample preparation
    sand
    saturation
    sediments
    seepage
    soil group
    soil management
    soil mechanics
    soil sampling
    solifluction
    storage
    surficial geology
    surveys
    tillage
    time
    topography
    transformations
    treatment
    tundra
    unconsolidated materials
    utilization
    vegetation
    waste disposal
    water erosion
    water regimes
    weathering
    weathering crust
    wind erosion
    yields

Sokh
Village S of Kokand near Kirghiz border in E.
BT Uzbekistan
    USSR

Soko Banja
Village near Nis in E Serbia.
UF Soko Banya
BT Serbia
    Yugoslavia

Soko Banya
use Soko Banja

Sokolov
Town on Ohre River in W Bohemia.
UF Falkenau
    Falknov
BT Bohemia
    Czechoslovakia

Sokolov Basin
W Bohemia. Also search Sokolov.
BT Bohemia
    Czechoslovakia

Sokoman Formation
In Knob Lake Group.
UF Sokoman Iron Formation
BT Precambrian
SA Canada
    Labrador
    Quebec

Sokoman Iron Formation
use Sokoman Formation

Sokoto Basin
River basin in N Nigeria. Also search

Sokoto.
  BT Nigeria
solar activity
  Term includes use under a variety of topics as a cause of whatever is being discussed in the document.
  SA disturbances
    solar flares
    Sun
solar cycles
  Includes use on level 3 under interplanetary space(1) cosmic rays(2).
  SA astrophysics and solar physics
    cosmic rays
    cycles
    disturbances
    intensity
    interplanetary space
    ionosphere
    magnetic storms
    solar wind
    Sun
solar energy
  Includes use on level 3 under energy sources(1).
  SA energy
    energy sources
    Sun
solar flares
  Term introduced on level 2 under astrophysics and solar physics(1) in 1978. Before 1978, search flares.
  SA astrophysics and solar physics
    cosmic rays
    disturbances
    emissions
    interplanetary space
    magnetic storms
    Moon
    solar activity
    solar wind
    Sun
solar nebula
  UF nebula, solar
  SA Sun
solar physics, astrophysics and
  use astrophysics and solar physics
solar system
  Includes use on level 3 under planetology(1). See also names of planets.
  SA asteroids
    comets
    Earth
    extraterrestrial geology
    Jupiter
    Mars
    Mercury Planet
    meteorites
    Moon
    Neptune
    planetology
    planets
    Pluto
    Saturn
    Sun
    Uranus
    Venus
solar wind
  Includes use on level 2 under interplanetary space(1) and under magnetosphere(1).
  UF wind, solar
  SA aurora
    comets
    corona
    electromagnetic radiation
    interplanetary space
    ionization
    magnetic field
    magnetic storms
    magnetic tail
    magnetopause
    magnetosphere

    Moon
    particle radiation
    particles
    plasma instabilities
    plasma motion
    solar cycles
    solar flares
    Sun
sole markings
  A valid term through 1971. After 1971, use sole marks.
sole marks
  Includes use on level 3 under sedimentary structures(1) turbidity current structures(2) and bedding plane irregularities(2). See list K. Before 1972, also search sole markings.
  BA turbidity current structures
    sedimentary structures
  SA bedding plane irregularities
    flute casts
    load casts
    soft sediment deformation
Solenhofen
  use Solnhofen
Solenhofen Limestone
  use Solnhofen Limestone
solfataras
  BT fumaroles
  SA gases
    sublimates
    volcanism
solid phase
  Term introduced in 1978. Icludes use as general term under geochemistry(1). Also search solids.
  UF phase, solid
    solids
  SA fluid phase
    gaseous phase
    geochemistry
    solid solution
solid solution
  Includes use on level 3 under phase equilibria(1). Also search mixed crystals.
  UF mixed crystals
  SA phase equilibria
    solid phase
    solution
    solutions
solid waste
  As of 1978, term is used on level 2 under waste disposal(1).
  UF waste, solid
  SA engineering geology
    environmental geology
    industrial waste
    liquid waste
    radioactive waste
    waste disposal
solidification
  use consolidation
solids
  use solid phase
solifluction
  As of 1978, term is used on level 2 under permafrost(1).
  UF sludging
    soil flow
    soil fluction
    solifluxion
  SA creep
    engineering geology
    glacial geology
    mass wasting
    periglacial features
    permafrost
    slope stability
    soils
solifluxion
  use solifluction

Solikamsk
  City in Perm Oblast W of the Urals.
  BT Russian Republic
    USSR
solitary waves
  Term introduced in 1978. Includes use on level 2 under ocean waves(1). Before 1978, search ocean waves AND solitary.
  SA ocean waves
    waves
Solnhofen
  Village in W central Bavaria. Also search Solenhofen.
  UF Solenhofen
  BT Bavaria
    West Germany
Solnhofen Limestone
  In Bavaria, West Germany. Also search Solenhofen, Solnhofen Limestone, and Solnhofen.
  UF Solenhofen Limestone
  BT Jurassic
  SA Bavaria
    West Germany
Solomon Islands
  Group of islands E of New Guinea in the SW Pacific Ocean. Bougainville, Buka, and Green Islands in the W are part of Papua New Guinea while the remaining 10 large islands and 4 groups of small islands remain a protectorate of the United Kingdom. Includes use on level 1 as an area term (list O). For term set options see list B.
  CO S120000S070000
    E1680000E1570000
  BA Melanesia
  NT Santa Isabel
  SA Australasia
    Bougainville
    Malaita
    Melanesia
    Papua New Guinea
Solomon Sea
  Enclosed on the W by New Guinea, on the N by the Bismarck Archipelago, on the E by the Solomon Islands and on the S by the Coral Sea.
  BT Pacific Ocean
Solonchak soils
  Saline. Includes use on level 3 under soils(1). See list M. Also search saline soils.
  BA soils
  SA Intrazonal soils
    salinity
    soil group
    Solonetz soils
Solonetz soils
  Includes use on level 3 under soils(1). See list M.
  BA soils
  SA Intrazonal soils
    soil group
    Solonchak soils
solongoite
  BA borates
  BT minerals
solubility
  Includes use on level 3 under geochemistry(1) properties(2).
  SA chemical analysis
    concentration
    geochemistry
    immiscibility
    insoluble residues
    properties
    saturation
    sea water
    solution

Solund Islands
  Island group in the North Sea at mouth of Sogne Fjord. Also search Solund.
  BT Norway
solutes
  SA geochemistry
    solution
solution
  Includes use on level 3 under geochemistry(1) properties(2). Used as process, property, and mixture. Before 1978, also search dissolution.
  SA adsorption
    aqueous solutions
    caves
    concentration
    electrolysis
    exsolution
    geochemistry
    leaching
    lysoclines
    pressure solution
    solid solution
    solubility
    solutes
    solution features
    solutions
solution cavities
  Includes use on level 3 under geomorphology(1) solution features(2).
  UF cavities, solution
  SA caves
    geomorphology
    karst
    sedimentary structures
    solution features
solution features
  Includes use on level 2 under geomorphology(1) and land subsidence(1).
  UF features, solution
  NT karren
    karst
  SA caverns
    caves
    geomorphology
    land subsidence
    sinkholes
    solution
    solution cavities
    speleology
solution mining
  Includes use on level 3 under mining geology(1).
  SA leaching
    mines
    mining
    mining geology
solutions
  Used for material or mixture. Singular form used to indicate process.
  SA aqueous solutions
    geochemistry
    hydrothermal solutions
    reagents
    solid solution
    solution
Somali Basin
  E of the Somali Republic and N of the Seychelles.
  BT Indian Ocean
Somali Republic
  Comprises former British Somaliland and Trust Territory of Somalia (formerly Italian Somaliland). Also search Somalia. Includes use on level 1 as an area term (list O). For term set options see list B.
  UF Somalia
  BA Africa
Somalia
  use Somali Republic

Somasteroidea
Includes use on level 3 under Echinodermata(1) Stelleroidea(2). See list F.
BA Stelleroidea
Asterozoa
Echinodermata
BT Invertebrata

Somerset
A county in SE England. Also a city in Kentucky and Pennsylvania. Index England or states as applicable.
CO N505000N513000
W0021500W0034500
SA England
Kentucky
Pennsylvania

Somerset County
Index states as applicable.
BX Maine
Maryland
New Jersey
Pennsylvania
BT United States

Somerset Island
In central District of Franklin N of Boothia Peninsula.
CO N720000N740000
W0900000W0960000
BT Northwest Territories
Canada

Somes Basin
River basin in NE Hungary and NW Romania. Index countries as applicable. Also search Somes; Somes River.
UF Somesul Basin
BT Europe
SA Hungary
Romania

Somesul Basin
use Somes Basin

Somme
Department in N.
CO N493000N503000
E0031000E0011500
BT France
SA Somme River valley

Somme River valley
Index departments as applicable. Also search Somme Valley.
UF Somme Valley
BT France
SA Aisne
Somme

Somme Valley
use Somme River valley

Son Valley
River valley. Index states as applicable. Also search Son River.
BT India
SA Bihar
Madhya Pradesh

sonar methods
Term introduced in 1978. Includes use on level 3 under geophysical methods(1) or geophysical surveys(1). Before 1978, search sonar.
BA geophysical methods
SA acoustical methods
acoustical surveys
echo sounding
geophysical surveys
methods
radar methods
remote sensing
sounding

Sondre Strom Fjord
Inlet of Davis Strait on Arctic Circle in SW.
BT Greenland
Arctic region

Sondrio
City and province in N.
BT Lombardy
Italy

Songhor
Village in Kermanshahan, W Iran.
UF Sonqor
Sunqur
BT Iran

sonic logging
Not a valid term for GeoRef. See acoustical logging.

sonic waves
use acoustical waves

sonobuoys
SA geophysical methods
geophysical surveys
marine methods

Sonoma County
W California. Also search Sonoma.
BT California
United States

Sonora
State in NW.
CO N263000N323000
W1083000W1150500
BT Mexico
NT Cananea
Moctezuma
Nacozari de Garcia
SA Colorado River
Colorado River delta
Pedregosa Basin
Sierra Madre Occidental
Sonoran Desert

Sonoran Desert
Index Sonora and U.S. states as applicable.
SA Arizona
California
Sonora

Sonqor
use Songhor

Sonyea Formation
use Sonyea Group

Sonyea Group
Includes Middlesex Shale, Pultenay Shale, Rock Stream Siltstone, and Cashaqua Shale members. W New York. Also search Sonyea Formation.
UF Sonyea Formation
BT Upper Devonian
Devonian

Sophia
use Sofia

Sor Rondane Mountains
use Sor-Rondane Mountains

Sor-Rondane Mountains
In Queen Maud Land near Princess Ragnhild Coast in Norwegian Sector on Atlantic Ocean side. Also search Sor Rondane Mountains; Sor-Rondane.
UF Sor Rondane Mountains
BT Antarctica

Sor-Trondelag
County S of Trondheim Fjord in central.
BT Norway
NT Trondheim
SA Trondelag

Sorachi
River in W central.
BT Hokkaido
Japan

Soria
Province in N central Spain. Also a city.
BT Spain

Soroka
use Belomorsk

sorosilicates
use orthosilicates

Soroy
Island in Norwegian Sea off NW. Finnmark County.
BT Norway

sorption
Includes use on level 3 under geochemistry(1) processes(2).
SA absorption
adsorption
geochemistry

Sorrento Peninsula
On S side of Bay of Naples. Also search Sorrento.
BT Campania
Italy

sorting
Includes use on level 3 under sedimentary rocks(1). See also textures under appropriate rock types.
SA grains
particles
rounding
sedimentary rocks
sediments
size

Sosnowice
use Sosnowiec

Sosnowiec
City in S part of country.
UF Sosnowice
BT Katowice
Poland

Soufriere
Volcano. Index islands as applicable. Also search Soufriere Volcano.
UF Soufriere Volcano
BX Guadeloupe
Montserrat
Saint Lucia
Saint Vincent
BT West Indies

Soufriere Volcano
use Soufriere

Souk-el-Arba salt works
In Souk-el-Arba area in NW Tunisia.
UF Souk-el-Arba Works
BT Tunisia

Souk-el-Arba Works
use Souk-el-Arba salt works

sound waves
use acoustical waves

sounding
Includes use on level 3 under geophysical methods(1) or geophysical surveys(1).
SA deep seismic sounding
deep sounding
echo sounding
geophysical methods
geophysical surveys
seismology
sonar methods

Sounds National Park
use Fiordland National Park

source area
use provenance

source rocks
SA natural gas
petroleum
rocks

sourceland
use provenance

sources
A valid general term through 1978.

sources, energy
use energy sources

sources, heat
use heat sources

sources, ore
use ore sources

sources, seismic
use seismic sources

Souris River basin
Index North Dakota and Canadian provinces as applicable. Also search Souris; Souris River.
SA Alberta
North Dakota
Saskatchewan

south
General term used after any main geographic term (list O).
UF southern
SA south-central
southeast
southwest

South Africa
Includes use on level 1 as an area term (list O). For term set options see list B.
CO S350000S220000
E0330000E0160000
BA Africa
NT Cape Province
Copperbelt
Orange Free State
Transvaal
Vaal River
SA Dwyka Series
Ecca Series
Fig Tree Series
Kalahari Desert
Karroo System
Nama System
Natal
Onverwacht Group
South-West Africa
Stormberg Series
Swaziland Sequence
Swaziland System
Transvaal Supergroup
Waterberg System
Witwatersrand System
Wolkberg Group

South America
Includes use on level 1 or 2 as an area term (list O). If 1, see term set options under list B. To retrieve all documents, individual countries and physiographic regions should also be searched (see list O).
CO S550000N130000
W0350000W0820000
NT Amazon Basin
Amazon River
Amazonas
NA Andes
Argentina
Bolivia
Brazil
NT Chaco
NA Chile
Colombia
NT Eastern Cordillera
NA Ecuador
French Guiana
NT Guiana Basin
Guianas
NA Guyana
NT Guyana Shield
Lake Titicaca
Llanos
Orinoco River
NA Paraguay
NT Parana Basin
Parana River
Patagonian Andes
NA Peru
NT Rio de la Plata
Santa Cruz
NA Surinam

NT Tierra del Fuego
NA Uruguay
     Venezuela
SA Botucatu Formation
     Galapagos Islands
     Gondwana
     Pacific mobile belt
     Roraima Formation
	Lee County
	Marion County
	Oconee County
		Colorado Group
		Dakota Formation
			BT Korea
			NT Kyongsang Basin

South America • South Sandwich Islands
288

- NT Tierra del Fuego
- NA Uruguay
-     Venezuela
- SA Botucatu Formation
-     Galapagos Islands
-     Gondwana
-     Pacific mobile belt
-     Roraima Formation

South Arcot
Region on Coromandel Coast in E.
- BT Tamil Nadu
-     India

South Atlantic
Term introduced in 1978. Before 1978, search Atlantic Ocean AND south.
- BA Atlantic Ocean
- SA East Atlantic
-     North Atlantic
-     Northeast Atlantic
-     Northwest Atlantic
-     Southeast Atlantic
-     Southwest Atlantic
-     West Atlantic

South Australia
Includes use on level 1 as an area term (list O). For term set options see list B.
- CO S380000S260000
-     E1410000E1290000
- BA Australia
- NT Adelaide
-     Adelaide Geosyncline
-     Balcanoona
-     Beltana
-     Coorong Lagoon
-     Encounter Bay
-     Eyre Peninsula
-     Fleurieu Peninsula
-     Flinders Ranges
-     Kangaroo Island
-     Lake Bonney
-     Lake Eyre
-     Lake Torrens
-     Mount Lofty Ranges
-     Renmark
-     Saint Vincent Gulf
-     Spilsby Island
-     Weekeroo Station
-     Willunga
-     Yorke Peninsula
- SA Cooper Basin
-     Eucla Basin
-     Gambier Embayment
-     Giles Complex
-     Great Artesian Basin
-     Murray Basin
-     Murray River
-     Musgrave Ranges
-     Nullarbor Plain
-     Officer Basin
-     Otway Basin
-     Simpson Desert

South Canadian River
use Canadian River

South Carolina
Includes use on level 1 as an area term (list O). For term set options see list B.
- CO N320400N351200
-     W0783200W0831500
- BA United States
- NT Aiken
-     Aiken County
- NX Anderson County
-     Calhoun County
-     Charleston
- NT Charleston County
- NX Cherokee County
-     Chester County
-     Columbia
- NT Georgetown County
- NX Greenwood County
- NT Horry County
- NX Jasper County

-     Lee County
-     Marion County
-     Oconee County
- NT Pageland
-     Port Royal Sound
-     Santee River
-     Savannah River Plant
- NX Union County
-     York County
- SA Atlantic Coastal Plain
-     Blue Ridge Province
-     Carolina slate belt
-     Castle Hayne Limestone
-     Duplin Formation
-     Eastern U.S.
-     Georgetown
-     Hawthorn Formation
-     Kings Mountain
-     Peedee Formation
-     Piedmont
-     Roan Supergroup
-     Sand Hills
-     Savannah River
-     Tuscaloosa Formation
-     Waccamaw Formation
-     Wicomico Formation

South Carpathians
use Transylvanian Alps

South Cascade Glacier
- BT Washington
-     United States

south central
use south-central

South China Sea
Bounded on N by China and Taiwan, on the E by the Philippine Islands, on the S by Malaysia and on the W by Vietnam. As of 1977, includes use as level 1 area term (list O). For term set options see list B.
- BA Pacific Ocean
- NT Gulf of Siam
-     Manila Trench
- SA China Sea

South Dakota
Includes use on level 1 as an area term (list O). For term set options see list B.
- CO N423000N455500
-     W0962700W1040500
- BA United States
- NT Badlands National Monument
- NX Butte County
-     Campbell County
-     Clark County
-     Clay County
-     Custer County
-     Dewey County
-     Douglas County
- NT Fall River County
- NX Grant County
-     Harding County
- NT Homestake Mine
- NX Jackson County
-     Lake County
-     Lawrence County
-     Lincoln County
-     Marshall County
-     McPherson County
-     Meade County
-     Pennington County
- NT Pierre
-     Rapid City
- NX Shannon County
-     Union County
- SA Arikaree Group
-     Ash Hollow Formation
-     Badlands
-     Bald Mountain
-     Benton Formation
-     Black Hills
-     Brule Formation
-     Carlile Shale
-     Chadron Arch
-     Chadron Formation

-     Colorado Group
-     Dakota Formation
-     Deadwood Formation
-     Elk Point Group
-     Fall River Formation
-     Fort Hays Limestone Member
-     Fort Union Formation
-     Fox Hills Formation
-     Grand River
-     Graneros Shale
-     Great Plains
-     Greenhorn Limestone
-     Hell Creek Formation
-     Lance Formation
-     Little Missouri River basin
-     Loup Fork Group
-     Midcontinent
-     Midwest
-     Minnelusa Formation
-     Missouri River
-     Missouri River basin
-     Missouri River valley
-     Montana Group
-     Morrison Formation
-     Mowry Shale
-     Niobrara Formation
-     Pierre Shale
-     Red River Formation
-     Sioux Quartzite
-     Smoky Hill Chalk Member
-     Spearfish Formation
-     Sundance Formation
-     Tongue River Formation
-     Valentine Formation
-     White River Group
-     Williston Basin
-     Winnipeg Formation
-     Winnipegosis Formation

South Georgia
Island 800 miles E of the Falkland Islands of which it is a dependency.
- UF South Georgia Island
- BT Atlantic Ocean
- SA Falkland Islands

South Georgia Island
use South Georgia

South Greenland
Term introduced in 1978. Before 1978, search Greenland AND south or southern.
- BA Greenland
- SA East Greenland
-     West Greenland

South Island
Largest Island of New Zealand S and SW of North Island.
- CO S473000S403000
-     E1743000E1663000
- BT New Zealand
- NT Canterbury
-     Christchurch
-     Dunedin
-     Fiordland National Park
-     Fox Glacier
-     Haast River
-     Inangahua
-     Kaikoura
-     Kakanui
-     Oamaru
-     Otago
-     Otago Peninsula
-     Rangitata River
-     Reefton
-     Southern Alps
-     Southland
-     Westland
- SA Tasman orogenic zone

South Korea
Officially Republic of Korea. Bounded on N by North Korea, on E by Japan Sea, on S by the Korea Strait, and on W by the Yellow Sea.
- CO N330000N383000
-     E1293000E1260000

- BT Korea
- NT Kyongsang Basin
-     Pohang
-     Ulsan
-     Yongyang
- SA Kangwon
-     North Korea

South Massif
use Causses

South Mountain
Ridge in W Maryland and S Pennsylvania. Index states as applicable.
- BT United States
- SA Maryland
-     Pennsylvania

South Nahanni River
SW District of Mackenzie.
- BT Northwest Territories
-     Canada

South Orkney Islands
British islands in S Atlantic Ocean S of Scotia Sea and 850 miles NE of Antarctic Peninsula.
- UF South Orkneys
- BT Antarctica
- SA Falkland Islands
-     Signy Island

South Orkneys
use South Orkney Islands

South Pacific
Term introduced in 1978. Before 1978, search Pacific Ocean AND south.
- BA Pacific Ocean
- SA East Pacific
-     Equatorial Pacific
-     North Pacific
-     Northeast Pacific
-     Northwest Pacific
-     Pacific Basin
-     Southeast Pacific
-     Southwest Pacific
-     West Pacific

South Pass
One of the channels at the mouth of the Mississippi River.
- BT Louisiana
-     United States

South Platte River valley
Index states as applicable. Also search South Platte; South Platte River.
- BT United States
- SA Colorado
-     Nebraska

South Pole
The S extremity of the Earth's axis at 90° S latitude, and the point from which all directions are N.
- BT Antarctica
- SA Polar Cap
-     Polar regions

South Polish Glaciation
Refers to glaciation in S Poland which was roughly the southern limit of Pleistocene glaciers in central Europe.
- BT Cromerian
-     upper Pleistocene
-     Pleistocene
- SA Poland

South Ray Crater
- BT Moon

South Sandwich Islands
Group of small volcanic islands at E end of Scotia Sea about 1350 miles E of Cape Horn. Part of Falkland Island Dependencies.
- CO S540000S580000
-     W0250000W0300000
- BT Antarctic Ocean
- SA Atlantic Ocean
-     Falkland Islands

South Saskatchewan River
Rises in the Rocky Mountains of W Alberta. Index provinces as applicable.
- BT Canada
- SA Alberta
  - North Saskatchewan River
  - Saskatchewan
  - Saskatchewan River

South Shetland Islands
N of the Antarctic Peninsula and S of Drake Passage. Part of British Antarctic Territory.
- UF South Shetlands
- BT Antarctica
- SA Deception Island
  - King George Island
  - Livingston Island

South Shetlands
use South Shetland Islands

South Victoria Land
use Victoria Land

South Wales coal field
Concentrated industrial area N of Bristol Channel. Also search South Wales.
- UF South Wales Coalfield
- BT Wales
  - Great Britain
  - United Kingdom

South Wales Coalfield
use South Wales coal field

south-central
General term used after any of the main geographic terms (list O).
- UF south central
- SA central
  - south

south-east
use southeast

south-west
use southwest

South-West Africa
In 1968, the U.N. gave area the name Namibia which is not recognized by South Africa. It was formerly German Southwest Africa. Includes use on level 1 as an area term (list O). For term set options see list B.
- CO S280000S170000
  - E0251500E0120000
- BA Africa
- NT Gibeon
  - Karibib
  - Luderitz
  - Namib Desert
  - Otavi
  - Tsumeb
  - Walvis Bay
  - Windhoek
- SA Damara System
  - Dwyka Series
  - Kalahari Desert
  - Karroo System
  - Nama System
  - Namaqualand
  - Orange River
  - South Africa
  - Zambezi Valley

southeast
General term used after any of the main geographic terms (list O).
- UF south-east
  - southeastern
- SA east
  - south

Southeast Asia
Not a valid term for GeoRef. Use Asia AND southeast.

Southeast Atlantic
Term introduced in 1978. Before 1978, search Atlantic Ocean AND southeast.
- BA Atlantic Ocean
- SA East Atlantic
  - North Atlantic
  - Northeast Atlantic
  - Northwest Atlantic
  - South Atlantic
  - Southwest Atlantic
  - West Atlantic

Southeast Pacific
Term introduced in 1978. Before 1978, search Pacific Ocean AND southeast.
- BA Pacific Ocean
- SA East Pacific
  - Equatorial Pacific
  - North Pacific
  - Northeast Pacific
  - Northwest Pacific
  - South Pacific
  - Southwest Pacific
  - West Pacific

southeastern
use southeast

southern
use south

Southern Africa
Term introduced in 1978. Before 1978, search Africa AND southern.
- CO S350000S050000
  - E0510000E0120000
- BA Africa
- SA Central Africa
  - North Africa
  - West Africa

Southern Alps
W central.
- BT South Island
  - New Zealand
- SA Alps

Southern Andes
Term introduced in 1978. Before 1978, also search Andes AND south or southern.
- BA Andes
- BT South America
- SA Central Andes
  - Northern Andes

Southern Appalachians
Term introduced in 1978. Before 1978, also search Appalachians AND south or southern.
- BA Appalachians
- BT North America
- SA Central Appalachians
  - Northern Appalachians

Southern California
Term introduced in 1978. Before 1978, search California AND south.
- BT California
  - United States

Southern California Batholith
- BT California
  - United States

Southern Carpathians
use Transylvanian Alps

Southern Cook Islands
use Cook Islands

Southern Great Plains
Term introduced in 1978. Before 1978, search Great Plains AND south or southern; also search Southern High Plains.
- UF Southern High Plains
- BA Great Plains
- BT North America
- SA Northern Great Plains

Southern Hemisphere
Used when discussing many large areas too numerous to mention. Includes use on level 1 and 2 as an area term (list O). If 1, see list B for term set options.
- SA Eastern Hemisphere
  - Gondwana
  - Northern Hemisphere
  - Western Hemisphere

Southern High Plains
use Southern Great Plains

Southern Highlands
Highland region. Former province of Tanganyika bounded on SW by Zambia, S by Malawi and Lake Nyasa.
- BT Tanzania

Southern Peninsula
use Lower Peninsula

Southern Province
In S Sri Lanka. Known for precious stones, anthracite and graphite mining.
- CO N460000N500000
  - W0790000W0900000
- BT Sri Lanka

Southern Rhodesia
A valid index term through 1972. After 1972, use Rhodesia.

Southern Rocky Mountains
Term introduced in 1978. Before 1978, also search Rocky Mountains AND south or southern.
- BA Rocky Mountains
- BT North America
- SA Central Rocky Mountains
  - Northern Rocky Mountains

Southern Uplands
Between the Scottish lowlands to the N and the Cheviot Hills on the English border to the S.
- BT Scotland
  - Great Britain
  - United Kingdom

Southern Urals
Southern third of the Urals extending S from roughly the Zlatoust area to the Kazakh border.
- CO N500000N570000
  - E0600000E0570000
- BA Urals
- BT Russian Republic
  - USSR
- SA Ural-Tau

Southern Yemen
Peoples Democratic Republic of Yemen. Includes use on level 1 as an area term (list O). For term set options see list B.
- UF Peoples Democratic Republic of Yemen
- BA Arabian Peninsula
- NT Aden
- SA Arabian Shield
  - Asia
  - Middle East
  - Near East
  - Red Sea Basin
  - Yemen

Southland
A land district in SW.
- BT South Island
  - New Zealand

southwest
General term used after any of the main geographic terms (list O).
- UF south-west
  - southwestern
- SA south
  - west

Southwest Atlantic
Term introduced in 1978. Before 1978, search Atlantic Ocean AND southwest.
- BA Atlantic Ocean
- SA East Atlantic
  - North Atlantic
  - Northeast Atlantic
  - Northwest Atlantic
  - South Atlantic
  - Southeast Atlantic
  - West Atlantic

Southwest Pacific
Term introduced in 1978. Before 1978, search Pacific Ocean AND southwest.
- BA Pacific Ocean
- SA East Pacific
  - Equatorial Pacific
  - North Pacific
  - Northeast Pacific
  - Northwest Pacific
  - South Pacific
  - Southeast Pacific
  - West Pacific

southwestern
use southwest

Southwestern U.S.
Term introduced in 1978. Includes use on level 1 as an area term. Before 1978, also search United States AND southwest or southwestern.
- BA United States
- SA Arizona
  - New Mexico
  - Oklahoma
  - Texas

Southwestern USSR
Term introduced in 1978. Before 1978, search USSR AND southwest.
- BT USSR

Soviet Arctic
That area N of the Arctic Circle in European USSR and Siberia, including Soviet islands in the Arctic Ocean. In Siberia arctic climatic conditions prevail far S of the Arctic Circle because of its great continental expanse. Also search USSR AND Arctic region.
- CO N700000N900000
  - W1700000E0300000
- BT Russian Republic
  - USSR
- SA Arctic region

Soviet Far East
Region occupying easternmost Siberia including Amur Oblast, Khabarovsk Kray, Primorye Kray, and Sakhalin Oblast. Also search USSR AND Far East.
- CO N420000N750000
  - W1700000E1300000
- UF Siberia-Soviet Far East
- BT Russian Republic
  - USSR
- SA Amur Basin
  - Far East
  - Kamchatka Peninsula
  - Kuril Islands
  - Okhotsk-Chukchi
  - Sakhalin
  - Sikhote-Alin Range

sovite
Includes use on level 3 under igneous rocks(1) lamprophyre and carbonatite family(2). See list H.
- BA lamprophyre and carbonatite family
- BT igneous rocks

Sowerbyella
- BA Strophomenida
  - Articulata
  - Brachiopoda
- BT Invertebrata

Sowie Mountains
Range of the Sudeten Mountains in Lower Silesia in SW Poland.

UF Eulengebirge
BT Poland
SA Sudeten Mountains

SO2
 use sulfur dioxide

space
 use interplanetary space

space groups
 Includes use on level 3 under crystal structure(1).
 SA crystal structure

space lattice
 use lattice

space, underground
 use underground space

Spain
 Includes use on level 1 as an area term (list O). For term set options see list B.
 CO N360000N434500
    E0043000W0093000
 BA Europe
 NT Alava
    Albacete
    Alicante
    Almeria
    Alpujarras
    Andalusia
    Aragon
    Asturias
    Avila
    Badajoz
    Barcelona
    Basque Provinces
    Betic Cordillera
    Betic Zone
    Burgos
    Caceres
    Cadiz
    Calatayud-Teruel Basin
    Cantabrian Basin
    Cantabrian Mountains
    Castellon de la Plana
    Castile
    Catalonia
    Ciudad Real
    Cuenca
    Ebro Basin
    Ebro River
    Estremadura
    Gerona
    Granada
    Guadalajara
    Guadalquivir
    Guadalquivir Basin
    Guipuzcoa
    Huelva
    Huesca
    Iberian Mountains
    Jaen
    La Coruna
    Lerida
    Logrono
    Madrid
    Maestrazgo
    Malaga
    Meseta
    Montes de Toledo
    Murcia
    Orense
    Oviedo
    Palencia
    Pontevedra
    Prebetic Zone
    Salamanca
    Saragossa
    Segovia
    Serrania de Cuenca
    Serrania de Ronda
    Seville
    Sierra de Gredos
    Sierra de Guadarrama
    Sierra de la Demanda
    Soria
    Subbetic Zone
    Tarragona
    Ter River basin
    Teruel
    Valencia
    Vizcaya
    Zamora
 SA Ager Formation
    Andalusian
    Canary Islands
    Cordoba
    Galicia
    Iberian Peninsula
    Lancara Formation
    Leon
    Mediterranean region
    Navarre
    Pyrenees
    Santander
    Sierra Nevada
    Tagus Basin
    Tagus River
    Toledo
    Vitoria

spallation
 SA cosmic rays
    isotopes

Spanish Guinea
 A valid term through 1975. After 1975, use Equatorial Guinea.

Spanish Peaks
 Two mountains in Huerfano and Las Animas counties in S.
 BT Colorado
    United States

Spanish Sahara
 A valid level 1 index term through 1976. Documents on Spanish Sahara are now indexed using Sahara and west under Sahara as the level 1 term. See entry under Sahara. Presently referred to as Western Sahara.

sparagmite
 Includes use on level 3 under sedimentary rocks(1) clastic rocks(2). See list I.
 BA clastic rocks
 BT sedimentary rocks
 SA terrigenous materials

Sparagmite Division
 use Sparagmite Group

Sparagmite Group
 Late Precambrian-early Paleozoic.
 UF Sparagmite Division
    Sparagmite Series
 SA Norway
    Paleozoic
    Precambrian
    Riphean
    Sweden
    upper Precambrian

Sparagmite Series
 use Sparagmite Group

Sparnacian
 Europe. Above Thanetian, below Ypresian of France. Includes use on level 3 under age terms(1). See list E.
 BA lower Eocene
    Eocene
    Paleogene
    Tertiary
 BT Cenozoic
 SA Paleocene

Spathognathodus
 Genus.
 BA conodonts

spatial variations
 Used as a general term.
 SA variations

Spearfish Formation
 Underlies Gypsum Spring Formation, geographically extended into Black Hills area. NW Nebraska, W South Dakota, and E Wyoming.
 SA Nebraska
    Permian
    South Dakota
    Triassic
    Wyoming

species
 Includes use on level 3 under paleontology(1) or under fossil groups (list F). Not to be used for mineral species.
 SA diversity
    paleobotany
    paleontology
    range
    revision
    taxonomy

species diversity
 SA diversity
    evolution
    fossils
    paleontology

specific gravity
 Includes use on level 3 under appropriate material name, e.g. under soils(1) and sediments(1).
 SA density

specific heat
 UF heat, specific
 SA geochemistry
    heat capacity
    physical properties
    temperature
    thermal conductivity
    thermal properties
    thermodynamic properties

specimens, type
 use type specimens

spectra
 As of 1978, term is used on level 2 under minerals(1). Also search spectrum.
 UF spectrum
 SA EPR spectra
    geochemistry
    infrared spectra
    minerals
    Mossbauer spectra
    optical spectra
    spectral analysis
    spectroscopy
    ultraviolet spectra
    X-ray fluorescence spectra

spectral analysis
 Includes use on level 3 under seismology(1).
 SA analysis
    Raman spectroscopy
    seismology
    spectra
    X-ray analysis
    X-ray spectroscopy

spectrography
 use spectroscopy

spectrometry
 A valid term through 1974. After 1974, use spectroscopy.

spectrophotometry
 use spectroscopy

spectroscopy
 Used for methodology. For data, see under appropriate material. Includes use on level I; on level 3 in combination with terms in list D under chemical element(1) analysis(2). Before 1973, also search mass spectroscopy; before 1975, also search spectrometry. If used on level I, term set options are:
 methods
    name of method [alpha-ray spectroscopy, atomic absorption, electron probe, flame photometry, gamma-ray spectroscopy, infrared spectroscopy, ion probe, laser probe, mass spectroscopy, microwave spectroscopy, Mossbauer spectroscopy, neutron spectroscopy, nuclear magnetic resonance, optical spectroscopy, radio-frequency spectroscopy, Raman spectroscopy, ultraviolet spectroscopy, X-ray spectroscopy, X-ray fluorescence]
 techniques
    topic [e.g. sample preparation]
 UF spectrography
    spectrophotometry
 NT alpha-ray spectroscopy
    emission spectroscopy
    gamma-ray spectroscopy
    infrared spectroscopy
    mass spectroscopy
    microwave spectroscopy
    Mossbauer spectroscopy
    optical spectroscopy
    radio-frequency spectroscopy
    Raman spectroscopy
    ultraviolet spectroscopy
 SA X-ray spectroscopy
    absorption
    alpha rays
    analysis
    atomic absorption
    chemical analysis
    chromatography
    clay mineralogy
    colorimetry
    crystallography
    differential thermal analysis
    electron microscopy
    electron paramagnetic resonance
    electron probe
    flame photometry
    geochemistry
    infrared spectra
    ion probe
    ion probe data
    major-element analyses
    microwave methods
    nuclear magnetic resonance
    optical spectra
    photometry
    scattering
    spectra
    standard materials
    surface properties
    thermal analysis
    trace-element analyses
    X-ray analysis
    X-ray diffraction analysis
    X-ray fluorescence

spectrum
 use spectra

specularite
 UF gray hematite
 BA oxides
 BT minerals
 SA hematite

Speeton Clay
 Cretaceous: Yorkshire.
 BT Lower Cretaceous
    Cretaceous
 SA Barremian
    England
    United Kingdom

speleology
 Includes use on level 3 under geomorphology(1) solution features(2).
 SA caves
    exploration
    geomorphology
    solution features
    speleothems

speleothems
SA caves
speleology

sperrylite
BA arsenides
sulfides
BT minerals

Spessart
Low mountain range between the Odenwald and the Rhon Mountains in NW Bavaria. Also search The Spessart.
UF The Spessart
BT Bavaria
West Germany

spessartine
BA garnet group
orthosilicates
silicates
BT minerals

spessartite
Includes use on level 3 under igneous rocks(1) lamprophyre and carbonatite family(2). See list H.
BA lamprophyre and carbonatite family
BT igneous rocks
SA garnet

Sphaeroidinella dehiscens
Includes use on level 3 under foraminifera(1) Globigerinacea(2).
BA Globigerinidae
Globigerinacea
Rotaliina
foraminifera
BT Invertebrata

sphaerolites
use spherulites

sphalerite
UF pseudogalena
zinc blende
BA sulfides
BT minerals

Sphenolithus
BA nannofossils
algae
BT Plantae
SA discoasters

Sphenophyllum
Genus of Paleozoic fossil plants.
BA Sphenopsida
pteridophytes
BT Plantae

Sphenopsida
Including Articulatae. Includes use on level 2 under pteridophytes(1). See list F.
UF horsetail
BA pteridophytes
BT Plantae
NA Articulatae
Equisetales
Sphenophyllum
NZ Sphenopteris

Sphenopteris
Genus. Includes use on level 3 under pteridophytes(1) Filicopsida(2) or Sphenopsida(2).
BZ Filicopsida
Sphenopsida
BT pteridophytes
Plantae
SA Articulatae
Pteridospermae

spherical harmonic analysis
Includes use on level 3 under plate tectonics(1).
SA analysis
Earth
magnetic field
plate tectonics

sphericity
SA particles
roundness
sediments
shape analysis

spherules
Includes use on level 3 under tektites(1).
SA sedimentary structures
sediments
tektites

spherulites
UF sphaerolites
SA orbicular texture

Sphinctozoa
Includes use on level 3 under Porifera(1) Calcispongea(2).
BA Calcispongea
Porifera
BT Invertebrata

Spice Islands
use Moluccas

spicules
SA Porifera

spilite
Includes use on level 3 under igneous rocks(1) alkali basalt family(2). See list H.
BA alkali basalt family
BT igneous rocks
SA pillow lava
spilitization

spilitization
Includes use on level 3 under metasomatism(1).
BA metasomatism
SA albitization
spilite

spills, oil
use oil spills

Spilsby Island
Largest of Sir Joseph Banks Islands in Spencer Gulf about 5 miles off SE coast of Eyre Peninsula. Also search Spilsby.
BT South Australia
Australia

spinel
BA oxides
BT minerals
SA chrome spinel
spinel group

spinel group
BA oxides
BT minerals
SA gahnite
hercynite
jacobsite
maghemite
magnesioferrite
magnetite
spinel
ulvospinel

spinel lherzolite
BA ultramafic family
BT igneous rocks
SA lherzolite

spinifex
SA igneous rocks
textures

spirals, growth
use growth spirals

Spiriferida
Includes use on level 2 under Brachiopoda(1). See list F.
BA Articulata
Brachiopoda
BT Invertebrata
NA Spiriferidina

Spiriferidae
Family. Includes use on level 3 under Brachiopoda(1) Spiriferida(2).
BA Spiriferidina
Spiriferida
Articulata
Brachiopoda
BT Invertebrata

Spiriferidina
Suborder. Includes use on level 3 under Brachiopoda(1) Spiriferida(2).
BA Spiriferida
Articulata
Brachiopoda
BT Invertebrata
NA Spiriferidae
Spiriferina

Spiriferina
Genus. Includes use on level 3 under Brachiopoda(1) Spiriferida(2).
BA Spiriferidina
Spiriferida
Articulata
Brachiopoda
BT Invertebrata

Spirillinacea
Includes use on level 2 under foraminifera(1). See list F.
BA Rotaliina
foraminifera
BT Invertebrata

Spis
Region. An historic area in dispute before WWI between Hungary, Austria, and Russia. Index Poland and/or Slovakia as applicable.
BT Europe
SA Poland
Slovakia

Spis-Gemer
The Slovak Ore Mountains. A range of the Carpathians in S Slovakia. Also search Spis-Gemer Mountains.
UF Slovenske Rudohorie
Spis-Gemer Mountains
BT Slovakia
Czechoslovakia
SA Carpathians

Spis-Gemer Mountains
use Spis-Gemer

Spiti
Region in NE.
BT Punjab
India

spits
SA beaches
coastal environment
geomorphology
marine environment
reefs
shoals
shore features

Spitsbergen
Norwegian archipelago, 360 miles N of Norway, including the main island of Spitsbergen plus North East Land, Edge Island, and Barents Island. Part of the Svalbard Island group. Includes use on level 1 as an area term (list O). For term set options see list B.
BA Arctic region
NT Brogger Peninsula
Hornsund
Kings Bay
Nordauslandet
Ny Friesland
Vestspitsbergen
SA Hecla Hoek Formation
Svalbard

Spitz
Village on left bank of Danube River.
BT Lower Austria
Austria

Split
City on the central Dalmatian Coast on the Adriatic Sea.
BT Croatia
Yugoslavia

Spodosols
Includes use on level 3 under soils(1). See list M.
BA soils

spodumene
BA pyroxene group
chain silicates
silicates
BT minerals
SA aluminosilicates
kunzite

Spokane
City in Spokane County in E.
BT Washington
United States

Spokane County
BT Washington
United States

Spoleto
City in Perugia Province in central part of country.
BT Umbria
Italy

Spongiae
use Porifera

spongolite
Includes use on level 3 under sedimentary rocks(1) clastic rocks(2). See list I.
BA clastic rocks
BT sedimentary rocks
SA nonterrigenous materials

spontaneous fission-track dating
use fission-track dating

Spor Mountain
BT Utah
United States

sporangia
Also search sporangium.
UF sporangium
SA angiosperms
gymnosperms
paleobotany
pteridophytes
spores

sporangium
use sporangia

spores
Includes use on level 3 under palynomorphs(1) miospores(2) or megaspores(2).
SA exine
megaspores
miospores
palynomorphs
pollen
pteridophytes
sporangia
sporopollenin

sporinite
BA macerals
SA coal
organic residues
sedimentary rocks

sporopollenin
UF sporopollenine
BA organic materials
SA exine
miospores
pollen
spores

sporopollenine
use sporopollenin

spreading centers
Includes use on level 3 under plate tectonics(1).

spreading centers ● standard deviation

UF centers, spreading
SA plate tectonics
    sea-floor spreading

spreading concept
use sea-floor spreading

spreading-floor hypothesis
use sea-floor spreading

Spring Mountains
In W Clark County near California line.
BT Nevada
    United States

Springdale
Village in Washington County in SW Utah. Gateway to Zion National Park.
BT Utah
    United States

Springer Formation
Chester and Morrow Series. Conformably overlies Goddard Shale. Central S Oklahoma.
BT Carboniferous
SA Chesterian
    Mississippian
    Morrowan
    Oklahoma
    Pennsylvanian

Springfield
City. Index states as applicable.
BT United States
SA Illinois
    Massachusetts
    Missouri
    Ohio

Springfield Coal Member
Of Carbondale Formation. W and N Illinois.
BT Pennsylvanian
SA Carbondale Formation
    Illinois

springs
Includes use on level 1 (list A). Used for papers stressing spring hydrology. Term set options are:
topic [composition, genesis, geochemistry, hot springs, mineral waters, temperature]
    subtopic (no area term)
NT hot springs
SA circulation
    fumaroles
    geysers
    ground water
    hydrogeology
    hydrology
    mineral waters
    salinity
    saturation
    thermal waters
    water
    water resources

Springsure
Village 165 miles WSW of Rockhampton in E central.
BT Queensland
    Australia

Springwater
Village in Livingston County in W central.
BT New York
    United States

Spumellaria
use Spumellina

Spumellina
Includes use on level 2 under Radiolaria(1). See list F. Also search Spumellaria.
UF Spumellaria
BA Radiolaria
BT Invertebrata

spurrite
BZ orthosilicates
    carbonates
BT minerals

Squamata
Order. Includes use on level 3 under Reptilia(1) Lepidosauria(2).
BA Lepidosauria
    Reptilia
BT Tetrapoda
    Vertebrata

Sr
use strontium

Sr-86
Includes use on level 3 under isotopes(1).
SA isotopes
    Sr-87/Sr-86
    strontium

Sr-86/Sr-87
use Sr-87/Sr-86

Sr-87
Includes use on level 3 under isotopes(1).
SA isotopes
    Sr-87/Sr-86
    strontium

Sr-87/Sr-86
Includes use on level 3 under isotopes(1). Also search Sr-86/Sr-87; Sr-86 AND Sr-87.
UF Sr-86/Sr-87
SA isotopes
    Sr-86
    Sr-87
    strontium

Sr-90
Includes use on level 3 under isotopes(1).
SA isotopes
    strontium

Sr/Rb
Includes use on level 3 under absolute age(1) methods(2).
SA absolute age
    rubidium
    strontium

Sredna Gora
Mountain range between the Balkan and Rhodope mountains in central Bulgaria.
CO N421500N424500
    E0250000E0233000
UF Sredna Gora Mountains
BT Bulgaria
SA Panagyurishte

Sredna Gora Mountains
use Sredna Gora

Sredniy Vasyugan
Village on the Vasyugan River in NW Tomsk Oblast in S central Siberia.
BT Russian Republic
    USSR

Sri Lanka
Formerly Ceylon. Includes use on level 1 as an area term (list O). For term set options see list B. Before 1974, also search Ceylon.
CO N060000N100000
    E0823000E0790000
BA Asia
NT Southern Province
SA Highland Series
    Indian Shield

Srikakulam
City on Bay of Bengal in NE.
BT Andhra Pradesh
    India

Srinagar
City and district in the Vale of Kashmir.

BT Jammu and Kashmir
    India

St. Catherines
use Saint Catharines

St. Clair County
use Saint Clair County

St. Clair River
use Saint Clair River

St. Clair River delta
use Saint Clair River delta

St. Croix
use Saint Croix

St. Croix Island
use Saint Croix

St. Elias Mountains
use Saint Elias Mountains

St. Francois Mountains
use Saint Francois Mountains

St. Gallen
use Saint Gall

St. George Formation
use Saint George Formation

St. Gotthard
use Gotthard Massif

St. Helena
use Saint Helena

St. Helena Island
use Saint Helena

St. John
use Saint John

St. John's
use Saint John's

St. Johns River basin
use Saint Johns River basin

St. Laurent Limestone
use Saint Laurent Limestone

St. Lawrence County
use Saint Lawrence County

St. Lawrence Estuary
use Saint Lawrence Estuary

St. Lawrence Lowland
use Saint Lawrence Lowlands

St. Lawrence Lowlands
use Saint Lawrence Lowlands

St. Lawrence River
use Saint Lawrence River

St. Lawrence River valley
use Saint Lawrence Valley

St. Lawrence Valley
use Saint Lawrence Valley

St. Louis
use Saint Louis

St. Louis County
use Saint Louis County

St. Louis Limestone
use Saint Louis Limestone

St. Lucia
use Saint Lucia

St. Mary Parish
use Saint Mary Parish

St. Marys Formation
use Saint Marys Formation

St. Paul
use Saint Paul

St. Paul's Rock
use Saint Paul Rocks

St. Paul's Rocks
use Saint Paul Rocks

St. Peter Sandstone
use Saint Peter Sandstone

St. Severin
use Saint-Severin

St. Thomas
use Saint Thomas

St. Vincent
use Saint Vincent

stability
Includes use on level 2 under paleomagnetism(1), foundations(1), land subsidence(1), and tunnels(1).
SA engineering geology
    foundations
    geochemistry
    land subsidence
    paleomagnetism
    slope stability
    stabilization
    thermodynamics
    tunnels
    underground installations

stabilization
As of 1978, term is used on level 2 under shorelines(1) and slope stability(1).
SA shorelines
    slope stability
    stability

stable isotopes
BA isotopes
SA C-13/C-12
    carbon
    D/H
    deuterium
    hydrogen
    O-18/O-16
    oxygen
    S-34/S-32
    sulfur

Stablo
use Stavelot

Stafford County
Index states as applicable.
BX Kansas
    Virginia
BT United States

Staffordshire
County in W central.
CO N523000N531500
    W0013000W0023000
BT England
    Great Britain
    United Kingdom

stage, universal
use universal stage

stages
A valid general term through 1977. Use specific term, e.g. stratigraphic units.

stainierite
use heterogenite

staining
SA paleobotany
    paleontology
    petrology
    sample preparation

Staked Plain
use Llano Estacado

Stalin
use Brasov

Stalinabad
use Dushanbe

Stalingrad
use Volgograd

Stampian
France. Includes use on level 3 under age terms(1). See list E.
BA lower Oligocene
    Oligocene
    Tertiary
BT Cenozoic
SA Rupelian

standard deviation
UF deviation, standard

SA statistical analysis
standard materials
Level 1 term introduced in 1976. Used for rocks or minerals or other materials that have been designated as standard by geological laboratories. Before 1976, search standard rocks. Includes use on level 1 (list A). Term set options are:
material name (1st level terms only)
 *type
topic [age, analysis, alteration, catalogs, classification, experimental studies, identification, nomenclature, observations, preparation, properties], laboratory
 subtopic
SA chemical analysis
 crystallography
 geochemistry
 igneous rocks
 isotopes
 materials
 minerals
 petrology
 silicate rocks
 spectroscopy
 standard rocks
 thermal analysis
standard rocks
Includes use on level 3 under standard materials (1). A first order term until 1976; now use standard materials on level 1.
SA rocks
 standard materials
standardization
Used as a general term.
Stanislaus County
W central.
BT California
 United States
Stanislav
use Ivano-Frankovsk
Stanislawow
use Ivano-Frankovsk
Stanley Group
Includes Hatton Tuff Lentil, Ten Mile Creek Formation, Moyers Formation, Chickasaw Creek Formation. W Arkansas, and central S and SE Oklahoma.
BT Mississippian
SA Arkansas
 Oklahoma
stannite
UF tin pyrites
BA sulfostannates
 sulfosalts
BT minerals
stannoidite
BA sulfides
BT minerals
Stanovoi Range
use Stanovoy Range
Stanovoy Range
Mountain range between Yakutia and Amur Oblast with Khabarovsk Kray on the E. Also search Stanovoi Range; Stanovoy.
CO N550000N573000
 E1330000E1230000
UF Stanovoi Range
 Stanovoy Upland
BT Russian Republic
 USSR
Stanovoy Upland
use Stanovoy Range
Stansbury Mountains
In Wasatch National Forest in NW.
BT Utah
 United States

Stanton County
Index states as applicable.
BX Kansas
 Nebraska
BT United States
Stanton Formation
In Lansing Group. Includes Captain Creek Limestone, Eudora Shale, Stoner Limestone, Rock Lake Shale, South Bend Limestone members. SW Iowa, E Kansas, NW Missouri, SE Nebraska, and NE Oklahoma.
UF Stanton Limestone Member
BT Missourian
 Pennsylvanian
SA Iowa
 Kansas
 Lansing Group
 Missouri
 Nebraska
 Oklahoma
 Rock Lake Shale Member
Stanton Limestone Member
use Stanton Formation
Stark County
Index states as applicable.
BX Illinois
 North Dakota
 Ohio
BT United States
Starobin
Town in S central.
BT Byelorussia
 USSR
Stassfurt
Salt mining city in W part of country.
BT Magdeburg
 East Germany
Staten Island
S of Manhattan and across The Narrows from Long Island on the E. Constitutes Borough of Richmond in New York City.
BT New York
 United States
statistical analysis
Used as a general term, e.g. on level 3 under automatic data processing(1) or mathematical geology(1).
SA analysis
 automatic data processing
 cluster analysis
 correlation coefficient
 correspondence analysis
 dendrograms
 histograms
 kurtosis
 mathematical geology
 multivariate analysis
 numerical analysis
 populations
 probability
 regression analysis
 reliability
 skewness
 standard deviation
 statistical methods
 variance analysis
statistical methods
Includes use on level 2 under mineral exploration(1); on level 3 under fossil groups (list F); on level 3 under appropriate material and discipline, i.e. seismology(1) methods(2).
NA autocorrelation
 canonical analysis
 cluster analysis
 correspondence analysis
 discriminant analysis
 factor analysis
 finite element analysis
 least-squares analysis
 trend-surface analysis
SA analysis

 biometry
 correlation coefficient
 equations
 geometry
 mathematical geology
 mathematical methods
 methods
 mineral exploration
 palynology
 phi scale
 populations
 skewness
 statistical analysis
Staunton Formation
Overlies Brazil Formation. SW Indiana.
BT Middle Pennsylvanian
 Pennsylvanian
SA Indiana
staurolite
BA orthosilicates
 silicates
BT minerals
Stavanger
City on Stavanger Fjord on North Sea S of Bergen.
BT Rogaland
 Norway
Stavelot
Town in the N Ardennes.
UF Stablo
BT Liege
 Belgium
Stavelot-Venn Massif
Primarily in the N Ardennes of Belgium. Index countries as applicable.
BT Europe
SA Belgium
 West Germany
Stavers Island
use Vostok
Stavropol region
In Stavropol Kray (formerly Ordzhonikidze Kray) in the Northern Caucasus. Also search Stavropol.
UF Voroshilovsk (region)
BT Russian Republic
 USSR
Ste. Genevieve Formation
use Sainte Genevieve Limestone
Ste. Genevieve Limestone
use Sainte Genevieve Limestone
Stearns Shale
In Council Grove Group. Kansas, and SE Nebraska.
BT Permian
SA Kansas
 Nebraska
Stebnik
Town in Lvov Oblast in SW.
BT Ukraine
 USSR
Steele Glacier
Emanates from Steele Mountain in Saint Elias Mountains near Alaska border in SW.
BT Yukon Territory
 Canada
Steens Mountain
Mountain mass in SW Harney County in SE.
CO N420000N431000
 W1181000W1185000
BT Oregon
 United States
Stegodon
Genus. Includes use on level 3 under Mammalia(1) Proboscidea(2).
BA Elephantidae
 Proboscidea
 Mammalia
BT Tetrapoda

 Vertebrata
Steinach
Town in Thuringian Forest in SE Suhl.
BT East Germany
Steinbach
Town 39 miles SE of Winnipeg.
BT Manitoba
 Canada
Steinheim
Town in two different states. Index states as applicable.
BT West Germany
SA Hesse
 North Rhine-Westphalia
Steinheim Basin
BT West Germany
stellerite
BA zeolite group
 framework silicates
 silicates
BT minerals
Stelleroidea
Class. Includes use on level 2 under Echinodermata(1). See list F.
BA Asterozoa
 Echinodermata
BT Invertebrata
NA Asteroidea
 Ophiuroidea
 Somasteroidea
step faults
Term introduced in 1978. Includes use on level 3 under faults(1) systems(2). Before 1978, also search faults AND step.
BA faults
SA thrust faults
Stephanian
Europe. Above Westphalian, below Sakmarian of Permian. Includes use on level 3 under age terms(1). See list E.
BA Upper Carboniferous
 Carboniferous
BT Paleozoic
SA Pennsylvanian
stephanite
BA sulfantimonites
 sulfosalts
BT minerals
Stepnyak
City in Kokchetav Oblast in N.
BT Kazakhstan
 USSR
Steppe
use Steppes
steppes
SA geomorphology
 plains
 taiga environment
Steppes
An extensive, treeless, semi-arid, grassland area of the mid-latitudes extending from the western border of the Soviet Union to the Altai Mountains in the E. Index Soviet republics as applicable. Also search Steppe.
UF Steppe
BT USSR
SA Kazakhstan
 Russian Republic
 Ukraine
stereochemistry
use crystal chemistry
Sterling County
W central Texas. Also search Sterling.
BT Texas
 United States

sternbergite
   BA sulfides
   BT minerals

Stettin
   use Szczecin

Stevens County
   Index states as applicable.
   BX Kansas
         Minnesota
         Washington
   BT United States

stevensite
   BA sheet silicates
         silicates
   BT minerals
   SA clay minerals

Stewart Valley
   River valley in W central Yukon Territory. Also search Stewart River.
   BT Yukon Territory
         Canada

stibiotantalite
   BA oxides
   BT minerals
   SA niobates

stibnite
   Also search antimonite.
   UF antimonite
   BA sulfides
   BT minerals
   SA metastibnite

stick-slip
   Includes use on level 3 under faults(1) mechanics(2); on level 3 under earthquakes(1) mechanism(2).
   SA earthquakes
         faults
         mechanics

Stigmaria
   Genus.
   BA Lycopsida
         pteridophytes
   BT Plantae

stilbite
   Also search desmine.
   UF desmine
   BA zeolite group
         framework silicates
         silicates
   BT minerals

Stillwater
   City in Payne County in N central.
   BT Oklahoma
         United States

Stillwater Complex
   Consists of four zones: basal, ultramafic zone, banded zone and upper zone.
   BT Precambrian
   SA Montana

stillwellite
   BA orthosilicates
         silicates
   BT minerals

stilpnomelane
   BA sheet silicates
         silicates
   BT minerals
   SA aluminosilicates

Stirling Quartzite
   Considered synonym for Prospect Mountain Quartzite. E California and SE Nevada. Also search Prospect Mountain Quartzite.
   UF Prospect Mountain Quartzite
   BT Precambrian
   SA California
         Nevada

stishovite
   BA silica minerals
         framework silicates
         silicates
   BT minerals

stochastic methods
   use stochastic processes

stochastic models
   use stochastic processes

stochastic processes
   Includes use on level 3 under mathematical geology(1). Also search random process; random processes; stochastic methods; stochastic models.
   UF random processes
         stochastic methods
         stochastic models
   SA probability
         processes

Stockholm
   County. Also a city on the Baltic Sea.
   BT Sweden

stocks
   Includes use on level 2 under intrusions(1).
   BA intrusions
   SA batholiths
         differentiation
         igneous rocks

stockwork deposits
   Term introduced in 1978. Includes use on level 3 under commodities (list C) or under mineral deposits, genesis(1). Before 1978, search ore deposits AND stockwork, or search deposits AND stockwork.
   UF network deposits
   SA deposits
         mineral deposits
         ore deposits

stoichiometry
   SA geochemistry

Stokes County
   On the Virginia border in NW North Carolina.
   BT North Carolina
         United States

stolzite
   BA tungstates
   BT minerals

Stomachorda
   use Hemichordata

Stomiosphaera
   BA Tintinnidae
         Protista

stone, building
   use building stone

Stone Corral Formation
   In Sumner Group. E Kansas.
   BT Permian
   SA Kansas

stone, dimension
   use dimension stone

Stone Mountain
   Huge gray granite monadnock in De Kalb County near Atlanta.
   BT Georgia
         United States

Stonehenge
   Prehistoric assemblage of stones on the Salisbury Plain 7 miles N of Salisbury in S.
   BT England
         Great Britain
         United Kingdom

stony irons
   Includes use on level 3 under meteorites(1).
   UF iron-stony meteorite
         lithosiderite
         sideraerolite
         siderolite
   BA meteorites
   SA achondrites
         chondrites
         howardites
         mesosiderites
         pallasites

Stony Tunguska River
   Rises in SE Evenk National Okrug and flows WNW into mid course of the Yenisei River.
   UF Middle Tunguska River
   BT Russian Republic
         USSR
   SA Tunguska
         Tunguska River

storage
   As of 1978, term is used on level 2 under reservoirs(1).
   SA data storage
         drainage basins
         gas storage
         ground water
         liquid waste
         radioactive waste
         reservoirs
         soils
         underground installations
         waste disposal
         water regimes
         water storage

Stormberg Series
   BT Upper Triassic
         Triassic
   SA South Africa

storms
   Includes use on level 2 under meteorology(1) and geologic hazards(1).
   SA atmosphere
         climate
         dust storms
         geologic hazards
         hurricanes
         magnetic storms
         meteorology
         monsoons
         precipitation
         substorms
         thunderstorms
         winds

strain
   Includes use on level 3 under deformation(1) and under engineering geology(1).
   SA deformation
         elastic limit
         elastic strain
         elasticity
         engineering geology
         finite strain
         Hooke's law
         mechanical properties
         Poisson's ratio
         rock mechanics
         seismology
         shear
         shear strength
         shear zones
         strainmeters
         stress
         structural analysis
         viscoelasticity

strain-slip cleavage
   use slip cleavage

strainmeters
   SA deformation
         extensometers
         geodesy
         strain

Strait of Georgia
   Channel between Vancouver Island on the W and the mainland of British Columbia and Washington on the E. Index British Columbia and/or Washington as applicable.
   SA British Columbia
         Washington

Strait of Gibraltar
   Passage connecting Mediterranean Sea and the Atlantic Ocean between Spain and Morocco.
   BT Atlantic Ocean
   SA Gibraltar

Strait of Juan de Fuca
   use Juan de Fuca Strait

Strait of Malacca
   Channel between the S Malay Peninsula and the island of Sumatra connecting the Indian Ocean with the South China Sea. Also search Malacca Strait; Malacca Straits.
   UF Malacca Strait
         Malacca Straits
         Straits of Malacca
   BT Asia

Strait of Sicily
   Between Sicily and Tunisia.
   BT Mediterranean Sea

Straits of Florida
   Wide channel between Florida Keys and Cuba connecting the Atlantic Ocean with the Gulf of Mexico. Also search Florida Strait; Florida Straits.
   UF Florida Strait
         Florida Straits
   BT Atlantic Ocean

Straits of Malacca
   use Strait of Malacca

Stramberk
   Town SSW of Ostrava in NE.
   BT Moravia
         Czechoslovakia

Strandzha Mountains
   use Istranca Mountains

Strasbourg
   City on the Rhine River in Alsace.
   UF Strassburg
   BT Bas-Rhin
         France

Strassburg
   use Strasbourg

strata
   use stratigraphic units

strata-bound deposits
   use stratabound deposits

stratabound deposits
   Term introduced in 1978. Includes use on level 3 under commodities (list C) or mineral deposits, genesis(1). Before 1978, search ore deposits AND stratabound; deposits AND stratabound; strata-bound.
   UF strata-bound deposits
   SA deposits
         mineral deposits
         ore deposits
         stratiform deposits

Strathcona Mine
   Near Edmonton in central.
   BT Alberta
         Canada

stratification
   Includes use on level 3 under sedimentary structures(1) planar bedding structures(2). See list K.
   BA planar bedding structures
         sedimentary structures
   SA bedding
         cross-stratification
         deposition
         laminations

stratified volcanic cone
   use stratovolcanoes

stratified volcano
   use stratovolcanoes

stratiform deposits
Term introduced in 1978. Includes use on level 3 under commodities (list C) or mineral deposits, genesis(1). Before 1978, search ore deposits AND stratiform; deposits AND stratiform.
UF stratiform ore deposits
SA deposits
  mineral deposits
  ore deposits
  stratabound deposits

stratiform ore deposits
use stratiform deposits

stratigraphic
A valid term through 1977. After 1977, use stratigraphic maps.

stratigraphic controls
Includes use on level 3 under mineral deposits, genesis(1) controls(2).
SA controls
  lithologic controls
  mineral deposits
  structural controls

stratigraphic geology
use stratigraphy

stratigraphic maps
Term introduced in 1978. Before 1978, search maps AND stratigraphic.
BA maps
SA structural maps

stratigraphic range
use range

stratigraphic traps
Includes use on level 3 under oil and gas fields(1).
UF porosity traps
BA traps
SA natural gas
  oil and gas fields
  petroleum
  structural traps

stratigraphic units
Term introduced in 1978. Includes use on level 3 under stratigraphy(1). Also search strata.
UF strata
  units, stratigraphic
SA stratigraphy

stratigraphy
Used for the discipline as a whole. See under age terms (list E). Includes use on level 1 (list A); on level 2 under area terms(1), bibliography(1), education(1), continental shelf(1), and symposia(1). If level 1, term set options are:
topic [applications, bibliography, catalogs, classification, concepts, education, history, methods, nomenclature, objectives, philosophy, practice, principles, textbooks]
  subtopic
UF stratigraphic geology
NT magnetostratigraphy
SA bibliography
  biogeography
  biostratigraphy
  boundary
  catalogs
  changes of level
  chronostratigraphy
  correlation
  diachronism
  education
  geochronology
  glacial geology
  glossaries
  historical geology
  index fossils
  isopach maps
  key beds
  lexicons
  lithostratigraphy
  micropaleontology
  paleontology
  palynology
  range
  regression
  sections
  stratigraphic units
  stratotypes
  transgression
  type sections
  unconformities
  uniformitarianism

stratosphere
Includes use on level 3 under aeronomy(1) composition(2).
SA aeronomy
  atmosphere
  ionosphere
  troposphere

stratotypes
SA boundary
  stratigraphy
  type sections

stratovolcanoes
UF bedded volcano
  composite cone
  composite volcano
  stratified volcanic cone
  stratified volcano
BA volcanoes
SA shield volcanoes

Strawn Series
Comprises Millsap Lake and Lone Camp groups. Central and N Texas.
BT Middle Pennsylvanian
  Pennsylvanian
SA Texas

stream action
Not a valid index term for GeoRef. See geomorphology(1) fluvial features(2).

stream capture
Includes use on level 3 under geomorphology(1). Also search stream piracy; capture.
UF capture
  piracy
  river capture
  river piracy
  robbery
  stream piracy
  stream robbery
SA rivers
  rivers and streams

stream gradient
Term introduced in 1978. Before 1978, search streams AND gradient.
SA rivers and streams
  streams

stream order
Includes use on level 3 under geomorphology(1) fluvial features(2).
UF channel order
SA drainage
  fluvial features
  geomorphology
  streams

stream piracy
use stream capture

stream robbery
use stream capture

stream sampling
use stream sediments

stream sediments
Includes use on level 3 under mineral exploration(1). Also search stream sampling.
UF stream sampling
SA mineral exploration
  sediment sampling
  sediments
  streams

stream transport
Includes use on level 3 under sedimentation(1). Also search fluvial transport.
UF fluvial transport
SA flow regime
  rivers
  rivers and streams
  sedimentation
  streams
  transport

streams
Includes use on level 3 under geomorphology(1) and sedimentation(1). Also search alluvial.
NA braided streams
  ephemeral streams
SA bedload
  canals
  cascades
  channels
  drainage
  efficiency
  fluvial features
  hydrosphere
  levees
  meanders
  rivers
  rivers and streams
  runoff
  sediment yield
  sinuosity
  stream gradient
  stream order
  stream sediments
  stream transport
  surface water
  watersheds

strengite
BA phosphates
BT minerals

strength
SA compressive strength
  deformation
  fracture strength
  mechanical properties
  rock mechanics
  shear strength
  stress
  tensile strength
  yield strength

stress
Includes use on level 3 under deformation(1).
SA bearing capacity
  creep
  deformation
  elasticity
  extensometers
  failure
  finite strain
  Hooke's law
  Poisson's ratio
  pore pressure
  pressure
  release fractures
  rigidity
  rock mechanics
  seismology
  shear
  shear strength
  shear stress
  strain
  strength
  structural analysis
  tension
  torsion
  viscoelasticity
  yield strength

stretch modulus
use Young's modulus

striations
Includes use on level 3 under sedimentary structures(1) bedding plane irregularities(2). See list K.
BA bedding plane irregularities
  sedimentary structures
SA glacial features
  grooves

strike
Includes use on level 3 under faults(1) and fractures(1). Term used to indicate strike faults through 1977. After 1977, use strike faults when discussing type of faults.
UF line of strike
SA bedding
  dip
  faults
  folds
  fractures
  joints
  orientation
  strike faults
  strike-slip faults

strike faults
Term introduced in 1978. Includes use on level 3 under faults(1) orientation(2). Before 1978, also search faults AND longitudinal; faults AND strike.
UF longitudinal faults
BA faults
SA longitudinal orientation
  strike

strike-shift faults
use strike-slip faults

strike-slip faults
Term introduced in 1978. Includes use on level 3 under faults(1) displacements(2). Before 1978, also search strike-slip.
UF strike-shift faults
BA faults
SA strike
  transcurrent faults
  transform faults

strip mining
Includes use on level 3 under mining geology(1) or land use(1).
SA environmental geology
  land use
  mining
  mining geology
  open-pit mining
  quarries
  reclamation

stromatolites
Includes use on level 2 under algae(1); on level 3 under sedimentary structures(1) biogenic structures(2). See list F (fossils) and list K (sedimentary structures).
BZ biogenic structures
  algae
NA Collenia
SA algal mats
  calcareous algae
  oncolites
  sedimentary structures

Stromatoporoidea
Includes use on level 2 under Coelenterata(1). See list F.
BA Coelenterata
BT Invertebrata
SA problematic fossils

Stromboli
Active volcano on Stromboli Island of Lipari group in Tyrrhenian Sea.
BT Sicily
  Italy

stromeyerite
BA sulfides
BT minerals

Strona Valley
River valley W of Lake Maggiore in N.
BT Italy
SA Piedmont

strong motion
Includes use on level 3 under seismology(1).
UF motion, strong
SA ground motion
   seismology

strontianite
BA carbonates
BT minerals

strontium
Includes use on level 1 and 2 as a chemical element (list D). Also search Sr.
UF Sr
SA elements
   isotopes
   Sr-86
   Sr-87
   Sr-87/Sr-86
   Sr-90
   Sr/Rb

Strophomena
Genus. Includes use on level 3 under Brachiopoda(1) Articulata(2).
BA Strophomenida
   Articulata
   Brachiopoda
BT Invertebrata

Strophomenida
Includes use on level 2 under Brachiopoda(1). See list F.
BA Articulata
   Brachiopoda
BT Invertebrata
NA Sowerbyella
   Strophomena

structural
A valid term through 1977. After 1977, use structural maps.

structural analysis
For small-scale analysis. Larger-scale treatments are assigned to faults, folds, fractures, or tectonics. Includes use on level 1 (list A). Before 1973, also search fabric analysis; before 1975, also search structural petrology; before 1976, also search microtectonics. Term set options are:
concepts
   subtopic
experimental studies
   subtopic
faults
   subtopic
folds
   subtopic
foliation
   subtopic
fractures
   subtopic
interpretation
   topic [e.g. axial-plane structures, boudinage, cleavage, elongate minerals, fold axes, folds, foliation, fractures, joints, laminations, layering, linear deformation, lineation, melange, mullions, petrofabrics, planar deformation, preferred orientation, schistosity, slickensides, etc...]
lineation
   subtopic
methods
   subtopic [e.g. electron microscopy, universal stage, X-ray analysis]
preferred orientation
   subtopic
principles
   subtopic
theoretical studies
   subtopic
UF petrofabric analysis
   petrogeometry
   petromorphology
SA analysis
   axial-plane structures
   boudinage
   boudins
   breccia
   cleavage
   deformation
   dip
   electron microscopy
   elongate minerals
   engineering geology
   fabric
   faults
   flow cleavage
   fold axes
   folds
   foliation
   fractures
   geometry
   geophysics
   joints
   lineation
   melange
   microfractures
   mullions
   orientation
   petrofabrics
   preferred orientation
   rock mechanics
   schistosity
   shear zones
   strain
   stress
   structural geology
   tectonics
   tectonophysics
   universal stage
   X-ray analysis
   X-rays

structural basins
use basins

structural complexes
Used for general treatments of the structures of an area and for combinations of structures. Includes use on level 3 under tectonics(1) area(2).
SA complexes
   faults
   folds
   shields
   tectonics

structural controls
Includes use on level 3 under mineral deposits, genesis(1) controls(2).
SA controls
   mineral deposits
   sedimentation
   stratigraphic controls

structural features
A valid term through mid-1978, used under geomorphology, volcanics, structural geology, and igneous rocks.

structural geology
Used for the discipline as a whole. Includes use on level 1 (list A); on level 2 under area terms(1). If 1, term set options are:
topic [applications, bibliography, catalogs, classification, concepts, education, experimental studies, history, instruments, methods, nomenclature, philosophy, practice, principles, textbooks, theoretical studies]
   subtopic
SA allochthons
   bibliography
   catalogs
   dislocations
   faults
   folds
   foliation
   fractures
   geology
   glossaries
   lexicons
   lineation
   maps
   orogeny
   structural analysis
   tectonics

structural maps
Term introduced in 1978. Before 1978, search maps AND structural. Also search structure-contour maps.
UF structure maps
   structure-contour maps
BA maps
SA stratigraphic maps
   structure contour maps

structural petrology
A valid term through 1974. After 1974, use structural analysis.

structural traps
BA traps
SA natural gas
   petroleum
   stratigraphic traps

structure
Includes use as a level 2 or 3 term appropriate to a large number of topics e.g. on level 2 under crust(1), tectonics(1), and tectonophysics(1). Includes use on level 3 under commodity terms (list C). See list G. For use under crust, also search crustal structure.
SA crystal structure
   domain structure
   molecular structure
   pillow structure
   velocity structure

structure contour maps
BA maps
SA contour maps
   structural maps
   tectonic maps

structure maps
use structural maps

structure-contour maps
use structural maps

structures
As of 1978, term is used on level 2 under foundations(1).
SA axial-plane structures
   banded structures
   biogenic structures
   buildings
   collapse structures
   cylindrical structures
   deep-seated structures
   flame structures
   flow structures
   foundations
   kink-band structures
   primary structures
   ring structures
   secondary structures
   sedimentary structures
   slump structures
   turbidity current structures

Struma River valley
Index countries as applicable. Also search Struma; Struma River.
UF Struma Valley
BT Europe
SA Bulgaria
   Greece

Struma Valley
use Struma River valley

struvite
BA phosphates
BT minerals

Strzegom
City in Lower Silesia in central.
BT Wroclaw
   Poland

Strzegom-Sobotka Granitoid Massif
use Strzegom-Sobotka Massif

Strzegom-Sobotka Massif
Granitoid massif in Lower Silesia. Also search Strzegom-Sobotka.
UF Strzegom-Sobotka Granitoid Massif
BT Wroclaw
   Poland

Strzelin
City of Lower Silesia in SE.
BT Wroclaw
   Poland

Stubai Alps
A NE group of Otztal Alps in central Tyrol. Also search Stubai.
UF Stubai Massif
   Stubai Mountains
BT Tyrol
   Austria
SA Alps
   Otztal Alps

Stubai Massif
use Stubai Alps

Stubai Mountains
use Stubai Alps

studies, areal
use areal studies

studies, case
use case studies

studies, experimental
use experimental studies

studies, faunal
use faunal studies

studies, feasibility
use feasibility studies

studies, field
use field studies

studies, floral
use floral studies

studies, laboratory
use laboratory studies

studies, theoretical
use theoretical studies

Stump Formation
Overlies Preuss Sandstone and underlies Bechler and Ephraim conglomerates undifferentiated. SE Idaho and W Wyoming. Also search Stump Sandstone.
UF Stump Sandstone
BT Upper Jurassic
   Jurassic
SA Idaho
   Wyoming

Stump Sandstone
use Stump Formation

Sturovo
Town on left bank of Danube River in SW.
BT Slovakia
   Czechoslovakia

Stutsman County
E central.
BT North Dakota
   United States

Stuttgart
City in Arkansas, and a city in Baden-Wurttemberg. Index Arkansas and/

or West Germany as applicable.
SA Arkansas
   West Germany

style
Includes use on level 2 under folds(1), foliation(1), fractures(1) and lineation(1).
SA folds
   foliation
   fractures
   lineation
   orientation

stylolites
Includes use on level 3 under sedimentary rocks(1) diagenesis(2); on level 3 under sedimentary structures(1) secondary structures(2); on level 3 under sediments(1) and sedimentation(1). See list K.
BA secondary structures
   sedimentary structures
SA microstylolites

Styria
State in SE.
CO N463500N474500
   E0161500E0133000
BT Austria
NT Graz
   Hartberg
SA Salzkammergut
   Semmering
   Totes Gebirge
   Wechsel

suanite
BA borates
BT minerals

subaerial environment
Term introduced in 1978. Before 1978 search subaerial.
SA environment

subantarctic regions
Term introduced in 1978. Pertaining or relating to the regions immediately outside of the Antarctic circle. Before 1978, also search subantarctic.
SA Antarctic Ocean
   Antarctica

Subarctic
use Preboreal

subarctic regions
Term introduced in 1978. Pertaining or relating to the regions immediately outside of the Arctic circle or to areas that have characteristics such as climate, vegetation, and animals similar to these regions. Before 1978, also search subarctic.
SA Arctic region

subarkose
Includes use on level 3 under sedimentary rocks(1) clastic rocks(2). See list I.
BA clastic rocks
BT sedimentary rocks
SA arkose

Subbetic Zone
A geographic term with stratigraphic-tectonic connotations.
BT Spain
SA Betic Cordillera
   Prebetic Zone

Subboreal
Europe. A term used primarily in Europe for an interval of postglacial time following the Atlantic and preceding the Subatlantic.
BA Holocene
   Quaternary
BT Cenozoic

Subcarpathians
Sub ranges of the Carpathians. Index Ukraine and countries as applicable.
BT Europe
SA Carpathian Foreland
   Carpathians
   Czechoslovakia
   Hungary
   Poland
   Romania
   Ukraine

subduction
Includes use on level 2 under plate tectonics(1); on level 3 under tectonophysics(1).
SA Benioff zone
   marginal basins
   marginal seas
   obduction
   plate tectonics
   subduction zones
   tectonophysics
   trenches

subduction zones
Includes use on level 3 under plate tectonics(1) and tectonophysics(1).
UF zones, subduction
SA Benioff zone
   plate tectonics
   plates
   subduction
   tectonophysics
   trenches

subgraywacke
Includes use on level 3 under sedimentary rocks(1) clastic rocks(2). See list I.
BA clastic rocks
BT sedimentary rocks
SA graywacke
   sandstone
   terrigenous materials

Sublette County
W Wyoming.
BT Wyoming
   United States

sublimates
SA fumaroles
   gases
   solfataras
   sublimation
   volcanism
   volcanoes

sublimation
SA evaporation
   geochemistry
   mineral deposits
   sublimates

sublittoral environment
Term introduced in 1978. Before 1978 search sublittoral.
SA environment
   littoral environment

submarine
A valid term through 1977. After 1977, use submarine environment.

submarine canyons
Includes use on level 2 under ocean floors(1). Before 1976, also search canyons for submarine canyons.
UF submarine valleys
SA bottom features
   canyons
   continental margin
   continental shelf
   continental slope
   marine geology
   ocean floors
   submarine fans
   turbidity currents

submarine cone
use submarine fans

submarine delta
use submarine fans

submarine environment
Term introduced in 1978. Before 1978, search submarine.
SA environment

submarine fans
Also search deep sea fans; sea fan.
UF abyssal cones
   abyssal fans
   deep-sea fans
   sea fan
   submarine cone
   submarine delta
SA fans
   ocean floors
   submarine canyons
   turbidity currents

submarine geology
Not a valid index term for GeoRef. See marine geology, oceanography, ocean floors, and ocean basins.

submarine installations
As of 1978, term is used on level 2 under marine installations(1). Term used only on level 3 from 1976 through 1977.
UF installations, submarine
BT marine installations
SA engineering geology
   tunnels

submarine valleys
use submarine canyons

submarine weathering
use halmyrolysis

submergence
SA changes of level

submersibles
SA FAMOUS
   oceanography

subsequent folds
use superposed folds

subsidence
As of 1978, term is used on level 2 under marine installations(1).
UF sinking
SA land subsidence
   marine installations
   neotectonics
   tectonics
   uplifts
   vertical tectonics

substitution
Includes use on level 3 under crystal chemistry(1).
SA crystal chemistry
   lattice
   order-disorder

substorms
Includes use on level 3 under magnetosphere(1) magnetic storms(2).
SA disturbances
   magnetic storms
   magnetosphere
   storms

substrates
Includes use on level 3 under appropriate fossil group (list F).

subsurface
A valid term through mid-1978 under engineering geology(1). Use subsurface reservoirs.

subsurface reservoirs
Term introduced in 1978. To index, use petroleum engineering under engineering geology(1). Before 1978, search reservoirs AND subsurface.
SA engineering geology
   petroleum engineering
   reservoirs

subtidal environment
Term introduced in 1978. Before 1978, search subtidal.
SA environment
   tidal environment

subtropical environment
Term introduced in 1978. Before 1978, search subtropical.
UF semitropical environment
SA environment
   tropical environment

subways
Term introduced in 1978 on level 2 under tunnels(1).
SA tunnels

Suceava
Town in Suceava County in N.
BT Moldavia
   Romania

Suchan Basin
Coal basin in Primorye Kray 60 miles ENE of Vladivostok. Also search Suchan.
BT Russian Republic
   USSR

Sucre
A department in N Colombia, a state in N Venezuela and a city in S central Bolivia. Index countries as applicable.
BX Bolivia
   Colombia
   Venezuela
BT South America
SA Araya Peninsula
   Cumana
   Paria Peninsula

Sudan
Includes use on level 1 as an area term (list O). For term set options see list B.
CO N030000N220000
   E0383000E0220000
BA Africa
NT Darfur
   Kapoeta
   Kordofan
   Malakal
SA Blue Nile
   Lake Turkana
   Libyan Desert
   Near East
   Nile River
   Nile Valley
   Nubia
   Nubian Sandstone
   Red Sea Basin
   Sahara

Sudbury
Mining city N of Georgian Bay in SE.
BT Ontario
   Canada

Sudbury Basin
Mining basin N of Georgian Bay in SE Ontario. Also search Sudbury.
BT Ontario
   Canada

Sudbury Irruptive
An intrusive region around Sudbury in SE Ontario.
UF Sudbury Nickel Irruptive
BT Ontario
   Canada

Sudbury Nickel Irruptive
use Sudbury Irruptive

sudden commencements
Includes use on level 3 under magnetosphere(1) magnetic storms(2).
UF commencements, sudden
SA disturbances
   intensity
   magnetic storms
   magnetosphere

Sudeten
Region. All the borderlands of Bohemia and Moravia formerly inhabited by German speaking people (Sudeten Germans). Index Bohemia and/or Moravia as applicable.
UF Sudetenland
BT Czechoslovakia
SA Bohemia
   Moravia

Sudeten Mountains
Ranges between NW Czechoslovakia and SW Poland. Also search Sudeten and Sudetes. Index Poland and Czechoslovak regions as applicable.
UF Sudetes
   Sudetes Mountains
   Sudetic Mountains
   Sudety Mountains
BT Europe
SA Bohemia
   Isergebirge
   Izera Mountains
   Karkonosze Mountains
   Moravia
   North Sudetic Basin
   Poland
   Sowie Mountains

Sudetenland
use Sudeten

Sudetes
use Sudeten Mountains

Sudetes Mountains
use Sudeten Mountains

Sudetic Basin
In the Sudeten Mountains. Index countries as applicable. Also search Sudetic.
UF Sudetic Depression
BT Europe
SA Czechoslovakia
   Moravia
   North Sudetic Basin
   Poland

Sudetic Depression
use Sudetic Basin

Sudetic Mountains
use Sudeten Mountains

Sudety Mountains
use Sudeten Mountains

sudoite
BA sheet silicates
   silicates
BT minerals
SA aluminosilicates
   chlorite group

suevite
SA breccia
   cryptoexplosion features
   impact features
   impactite
   metamorphic rocks
   meteorites

Suez (Canal)
use Suez Canal

Suez Canal
Sea level canal crossing the Isthmus of Suez between the E Mediterranean Sea and the Gulf of Suez. Also search Suez.
UF Suez (Canal)
BT Egypt
   Africa

Suffield
Town in Hartford County in N.
BT Connecticut
   United States

Suffolk
City in SE.
CO N515500N523500
   E0014500W0003000

BT England
   Great Britain
   United Kingdom

Suffolk County
Index states as applicable.
BX Massachusetts
   New York
BT United States

Sugarloaf Mountain
Granitic peak in Rio de Janeiro.
BT Guanabara
   Brazil

sugars
BA organic materials

Suidae
Family. Includes use on level 3 under Mammalia(1) Artiodactyla(2).
BA Artiodactyla
   Mammalia
BT Tetrapoda
   Vertebrata

Sulaiman Range
W of the Indus River. Index provinces as applicable.
BT Pakistan
SA Baluchistan
   North-West Frontier Province
   Punjab

Sulawesi
use Celebes

Sulcis
Region in extreme SW.
BT Sardinia
   Italy

sulfantimonates
BA sulfosalts
BT minerals
NA cylindrite
   famatinite

sulfantimonites
BA sulfosalts
BT minerals
NA boulangerite
   bournonite
   chalcostibite
   freibergite
   freieslebenite
NZ geocronite
NA heteromorphite
   jamesonite
   kobellite
   meneghinite
   miargyrite
   plagionite
   polybasite
   pyrargyrite
   semseyite
   stephanite
   tetrahedrite
   zinckenite

sulfarsenites
BA sulfosalts
BT minerals
NA dufrenoysite
NZ geocronite
NA gratonite
   jordanite
   pearceite
   proustite
   seligmannite
   tennantite

sulfates
Includes use on level 2 under minerals(1). See list L.
BT minerals
NA alabaster
   alum
   alunite
   anglesite
   anhydrite
   barite
   bassanite
NZ beudantite

NA brochantite
   celestite
   chalcanthite
   copiapite
   coquimbite
   epsomite
   ettringite
   glauberite
   gypsum
   halotrichite
NZ hauyne
NA hexahydrite
NZ hinsdalite
NA jarosite
   jouravskite
NZ kainite
NA kieserite
   langbeinite
NZ lazurite
NA melanterite
   mirabilite
   pickeringite
   polyhalite
   posnjakite
   roemerite
   rozenite
   selenite
   svanbergite
NZ thaumasite
NA thenardite
NZ voltaite
   wenkite
SA sodium sulfate

sulfides
Includes antimonides, arsenides, bismuthides, oxysulfides, selenides, and tellurides. Includes use on level 2 under minerals(1). See list L.
UF sulphides
BT minerals
NA acanthite
   aikinite
   alabandite
   alloclasite
   anilite
   antimonides
   argentite
   arsenides
   arsenosulfides
   berndtite
   betekhtinite
   bismuthides
   bismuthinite
   bohdanowiczite
   bornite
   bravoite
   briartite
   carrollite
   cattierite
   chalcocite
   chalcopyrite
   cinnabar
   copper sulfides
   covellite
   cubanite
   digenite
   djurleite
   galena
NZ genthelvite
NA gersdorffite
   greenockite
   greigite
   gudmundite
   heazlewoodite
   hedleyite
NZ helvite
NA herzenbergite
   idaite
   iron sulfides
   jordisite
   kesterite
NZ lazurite
NA linnaeite
   mackinawite
   marcasite
   mawsonite
   metacinnabar

   metastibnite
   michenerite
   millerite
   molybdenite
NZ niggliite
NA niningerite
   onofrite
   orpiment
   pentlandite
   polydymite
   pyrite
   pyrrhotite
   realgar
   roquesite
NZ schreibersite
NA selenides
   sinnerite
   smythite
   sphalerite
   stannoidite
   sternbergite
   stibnite
   stromeyerite
   talnakhite
   tellurides
   troilite
   tungstenite
   ullmannite
   ultrabasite
   vaesite
   valleriite
   violarite
NZ voltaite
NA willyamite
   wurtzite
SA hydrogen sulfide

sulfobismuthites
Includes use on level 3 under minerals(1) sulfosalts(2). See list L.
BA sulfosalts
BT minerals
NA cosalite
   emplectite
   galenobismutite
   hammarite
   lillianite
   wittichenite

sulfogermanates
BA sulfosalts
BT minerals
NZ canfieldite
NA germanite
   renierite

sulfosalts
Includes sulfantimonates, sulfantimonites, sulfarsenates, sulfarsenites, sulfobismuthites, sulfogermanates, sulfostannates, sulfovanadates. Includes use on level 2 under minerals(1). See list L.
BT minerals
NA berryite
   enargite
   gustavite
   luzonite
   matildite
   sulfantimonates
   sulfantimonites
   sulfarsenites
   sulfobismuthites
   sulfogermanates
   sulfostannates
   sulfovanadates
   vrbaite

sulfostannates
BA sulfosalts
BT minerals
NZ canfieldite
NA stannite

sulfovanadates
BA sulfosalts
BT minerals
NA sulvanite

sulfur
Includes use on level 1 as a com-

modity term (list C) and as a chemical element (list D); on level 2 under mineral deposits, genesis(1). Also search S.
UF S
    sulphur
SA elements
   isotopes
   native elements and alloys
   S-32
   S-34
   S-34/S-32
   stable isotopes

sulfur dioxide
UF SO2
SA atmosphere
   pollution

sulfuric acid
UF H2SO4
SA acid mine drainage
   acids
   pollution
   waste disposal

Sulina
Town on the Black Sea in Tulcea County in NE.
BT Dobruja
   Romania

Sulitjelma
Peak in Kjolen Mountains on Norwegian-Swedish border in N part of Scandinavian Peninsula. Index countries as applicable.
BT Europe
SA Norway
   Sweden

Sullivan County
Index states as applicable.
BX Indiana
   Missouri
   New Hampshire
   New York
   Pennsylvania
   Tennessee
BT United States

sulphides
use sulfides

sulphur
use sulfur

Sulu Sea
Large interisland sea between the Philippine Islands and NE Borneo with Palawan Island on the NW and the Sulu Archipelago on the SW. Also search Sulu.
BT Pacific Ocean

sulvanite
BA sulfovanadates
   sulfosalts
BT minerals

Sumatra
One of the islands of the Malay Archipelago SE and S of the Malay Peninsula. The second largest Indonesian island.
BT Indonesia
NT Toba Lake
SA Malay Archipelago

Summit County
Index states as applicable.
BX Colorado
   Ohio
   Utah
BT United States

Sumner County
Index states as applicable.
BX Kansas
   Tennessee
BT United States

Sun
Includes use on level 1. See entry under Moon(1) for term set options.
SA corona
   planetology
   planets
   solar activity
   solar cycles
   solar energy
   solar flares
   solar nebula
   solar system
   solar wind

Sunda Shelf
The continental shelf between Borneo and Java.
UF Borneo-Java Shelf
BT Java Sea
   Pacific Ocean

Sundance Formation
Includes Canyon Springs Sandstone, Stockade Beaver Shale, Hulett Sandstone, Lak, and Redwater Shale members. Central N Colorado, central S Montana, NW Nebraska, SW South Dakota, and SW Wyoming.
BT Upper Jurassic
   Jurassic
SA Colorado
   Montana
   Nebraska
   South Dakota
   Wyoming

Sunlight
Peak in SW Colorado and mountain in NW Wyoming. Index states as applicable.
BT United States
SA Colorado
   Wyoming

Sunqur
use Songhor

Sunrise Formation
Overlies Gabbs Formation; unconformably underlies Dunlap Formation. SW Nevada.
BT Lower Jurassic
   Jurassic
SA Nevada

sunspots
As of 1978, term is used on level 2 under geologic hazards(1).
SA geologic hazards

Sunzha
Mountain range. N outlier of the central Greater Caucasus in Northern Caucasus.
BT Russian Republic
   USSR
SA Greater Caucasus

Supai Formation
In Aubrey Group. Includes Esplanade Sandstone Member, Fort Apache Limestone, Kinishba Beds, Amos Wash Member, Big "A" Member, Apache Member, Corduroy Member, Packard Member, Oak Creek Member. N Arizona, E California, E Nevada, W New Mexico, and Sutah.
BT Paleozoic
SA Arizona
   California
   Nevada
   New Mexico
   Pennsylvanian
   Permian
   Utah

supergene processes
Includes use on level 3 under mineral deposits, genesis(1) processes(2).
SA enrichment
   hypogene processes
   mineral deposits
   processes

superimposed folds
use superposed folds

superimposed metamorphism
use polymetamorphism

Superior Province
CO N480000N550000
   W0850000W0950000
BT Ontario
   Canada

superposed folds
Term introduced in 1978. Includes use as orientation of folds relative to spatially associated macroscopic structures such as large folds, fold systems, and orogenic zones. Includes use on level 3 under folds(1) orientation(2). Before 1978, also search folds AND superposed; folds AND cross.
UF cross folds
   subsequent folds
   superimposed folds
   transverse folds
BA folds
SA orientation

Superstition Mountains
S central Arizona. Also search Superstition.
BT Arizona
   United States

superstructure
SA crystal structure

supply, water
use water supply

supralittoral environment
use supratidal environment

supratidal environment
Term introduced in 1978. Before 1978, search supratidal.
UF supralittoral environment
SA environment
   sebkha environment
   tidal environment
   tidal flats

Sur fault zone
In the Point Sur area in W California. Also search Sur-Nacimiento fault zone.
UF Sur-Nacimiento fault zone
BT California
   United States

Sur-Nacimiento fault zone
use Sur fault zone

Surakhany
Town on central Apsheron Peninsula in oil fields near Baku. Also search Surakhany oil field.
BT Azerbaidzhan
   USSR

Surat Basin
Sedimentary basin in SE.
CO S290000S270000
   E1510000E1470000
BT Queensland
   Australia

surface
A valid term through 1977. After 1977, use surface reservoirs for engineering geology, and use surficial geology for maps.

surface features
Also see under individual planets, e.g. Mars(1).
UF features, surface
SA geomorphology
   Moon
   planetology
   planets
   surface properties

surface phenomena
Includes use on level 2 under astrophysics and solar physics(1).
UF phenomena, surface
SA astrophysics and solar physics

surface properties
For photographic characteristics, electromagnetic responses, etc. Includes use on level 2 under Moon(1).
SA electrical properties
   Moon
   properties
   radiometric properties
   spectroscopy
   surface features

surface reservoirs
Term introduced in 1978. Includes use on level 3 under engineering geology(1) and under reservoirs(1). Before 1978, also search surface ANd reservoirs.
SA engineering geology
   Oroville Dam
   reservoirs

surface water
Includes use on level 3 under environmental geology(1).
SA Earth
   environmental geology
   floods
   ground water
   hydrology
   hydrosphere
   irrigation
   lakes
   ponds
   rivers
   runoff
   streams
   swamps
   water
   water supply
   watersheds

surface waves
Includes use on level 3 under seismology(1).
UF L waves
BT elastic waves
NT Love waves
   Rayleigh waves
SA earthquakes
   engineering geology
   ground motion
   seismology
   waves

surfaces, erosion
use erosion surfaces

surficial geology
Not valid term through 1977. After 1977, includes use on level 3 under geomorphology(1) and maps(1). Before 1978, also search maps AND surficial; search surface AND maps.
SA geology
   geomorphology
   maps
   soils

surges
As of 1978, term is restricted for use with ocean waves. Before 1978, term was also used for glacier surges.
SA base surges
   glacier surges
   ocean waves
   waves

Surgut
Town on right bank of Ob River in S Khanty-Mansino National Okrug in W Siberia.
BT Russian Republic
   USSR

Surinam
Includes use on level 1 as an area term (list O). For term set options see list B.
UF Dutch Guiana
    Netherlands Guiana
BA South America
NT Wilhelmina Mountains
SA Guiana Basin
    Guianas
    Guyana Shield

Surkhan Darya
River which rises in the Hissar Range of Tadzhikistan and flows SSW into the Amu Darya near Termez. Index Soviet republics as applicable.
BT USSR
SA Tadzhikistan
    Uzbekistan

Surkhan Darya basin
River basin, primarily in Surkhan Darya Oblast in S Uzbekistan. Index Soviet republics as applicable.
UF Surkhan-Darya basin
BT USSR
SA Tadzhikistan
    Uzbekistan

Surkhan-Darya basin
use Surkhan Darya basin

Surrey
County SW of London.
CO `N511000N513000
    E0001000W0004500
BT England
    Great Britain
    United Kingdom

Surry County
Index states as applicable. Also search Surry.
BX North Carolina
    Virginia
BT United States

Surtsey
Volcanic island off S.
BT Iceland

Suruga Bay
Inlet of Pacific Ocean SW of Yokohama on SE coast.
BT Honshu
    Japan

Surveyor 3
Includes use on level 3 under Moon(1). Also search Surveyor III.
SA Moon
    satellite methods

surveys
When level 1, it is used for the listing of the work of geological surveys, national or local. It is not to be confused with usage of the term surveys in other sets where it indicates actual surveying and its results. Includes use on level 1 (list A); as a level 2 or 3 term appropriate to a large number of topics, e.g. on level 2 under soils(1), engineering geology(1), and ground water(1). If 1, term set options are:
topic [annual report, current research, history, organization, research]
        name of survey [e.g. U. S. Geological Survey, etc.]
UF geologic surveys
NT geophysical surveys
SA annual report
    associations
    bibliography
    catalogs
    current research
    engineering geology
    environmental geology
    exploration
    geodesy

Geological Survey of Canada
geology
ground water
Illinois State Geological Survey
maps
networks
organization
progress report
publications
signals
soils
triangulation
U. S. Geological Survey

susceptibility
A valid term through 1977. Use magnetic susceptibility.

Susitna River basin
N of Anchorage in S Alaska. Also search Susitna; Susitna River.
BT Alaska
    United States

suspended materials
Before 1978, search materials AND suspended.
SA dissolved materials
    flocculation
    materials
    sea water

suspension
SA geochemistry
    hydrology
    sediments
    transport

suspension current
use turbidity currents

Susquehanna River
Rises in Otsego Lake in central New York and flows S emptying into N Chesapeake Bay. Index states as applicable.
BT United States
SA Maryland
    New York
    Pennsylvania

Susquehanna River basin
Primarily in Pennsylvania. Index states as applicable. Also search Susquehanna Valley.
UF Susquehanna Valley
BT United States
SA Maryland
    New York
    Pennsylvania

Susquehanna Valley
use Susquehanna River basin

Sussex
County on English Channel. Divided into two administrative counties: East and West Sussex.
CO N504500N511500
    E0004500W0010000
BT England
    Great Britain
    United Kingdom

Sussex County
Index states as applicable.
BX Delaware
    New Jersey
    Virginia
BT United States

Sustut Basin
River basin in N central.
BT British Columbia
    Canada

Sutam River
S Yakutia near N border of Amur Oblast in SE Siberia. Also search Sutam.
BT Russian Republic
    USSR

Sutherland
County in highlands of extreme N.
CO N575000N583000
    W0033500W0052500
BT Scotland
    Great Britain
    United Kingdom

Sutna
use Satna

Sutton County
In broken uplands of Edwards Plateau in W Texas. Also search Sutton.
BT Texas
    United States

sutures
SA fossils
    morphology

Suva
Town with one of best harbors in South Pacific on SE coast of Viti Levu Island.
BT Fiji
    Pacific Ocean

Suvalkai
use Suwalki

Suvalki
use Suwalki

Suwalki
Region E of the Masurian Lakes in extreme NE part of country. Also a city.
UF Suvalkai
    Suvalki
BT Bialystok
    Poland

Suwannee Limestone
Upper part of formation considered to be equivalent to lower Chickasawhay Marl of E Mississippi, and lower part possibly equivalent of Byram Formation and Marianna Limestone of Florida. S central Georgia, and E Florida.
BT upper Oligocene
    Oligocene
SA Florida
    Georgia

Suzak
Village in central.
BT Kirghizia
    USSR

Svalbard
Norwegian Island group including Spitsbergen group and Bear Island.
BT Arctic Ocean
SA Bear Island
    Spitsbergen
    Vestspitsbergen

svanbergite
BA sulfates
BT minerals

Svanetia
Region on S slopes of the Greater Caucasus in NW.
BT Georgian Republic
    USSR

Svecofennian
Europe.
BA middle Precambrian
    Precambrian

Sventoji River
Central Lithuania. Also search Sventoji.
UF Shventoyi River
BT Lithuania
    USSR

Sverdlovsk
City and oblast in foothills of the Central Urals in W Siberia.
UF Ekaterinburg
    Yekaterinburg
BT Russian Republic
    USSR

Sverdrup Basin
In the Sverdrup Channel area of the Sverdrup Islands in N District of Franklin.
BT Northwest Territories
    Canada

Svinita
Village in W part of country in S.
BT Banat
    Romania

Svratka
Village in W central.
BT Moravia
    Czechoslovakia

Swabia
Region. Duchy of medieval Germany. Nearly coextensive with Baden-Wurttemberg, Hesse, and W Bavaria. Index states as applicable.
BT West Germany
SA Baden-Wurttemberg
    Bavaria
    Hesse

Swabian Alb
Mountain range between the Neckar and the upper Danube rivers.
UF Swabian Jura
BT Baden-Wurttemberg
    West Germany

Swabian Jura
use Swabian Alb

swamps
SA bogs
    brackish-water environment
    ecology
    geomorphology
    mangrove swamps
    marshes
    peat bogs
    sedimentation
    surface water

Swan Hills
Hilly region S of Lesser Slave Lake in central.
BT Alberta
    Canada

Swan Peak Formation
Includes Watson Ranch Tongue. SE Idaho and NE Utah.
BT Middle Ordovician
    Ordovician
SA Idaho
    Utah

Swansea
City on Bristol Channel in S.
BT Wales
    Great Britain
    United Kingdom

swarms
Includes use on level 3 under earthquakes(1) and under intrusions(1).
SA dike swarms
    earthquakes
    intrusions
    ring dikes
    seismology

Swartkrans
Mountains near Lesotho border in SE.
BT Orange Free State
    South Africa

Swat
District in the valley of the Swat River NNE of Peshawar. Former princely state.
BT North-West Frontier Province
    Pakistan

Swauk Formation
Unconformably overlies metamorphic basement of unknown age. Central and central N Washington.
BT Eocene

SA Washington
Swaziland
Borders on Mozambique and South Africa. Administered by a British High Commissioner until independence in 1968. Includes use on level 1 as an area term (list O). For term set options see list B.
CO S273000S260000
    E0321500E0310000
BA Africa
SA Swaziland Sequence
    Swaziland System

Swaziland Sequence
In South Africa, and Swaziland.
BT Precambrian
SA South Africa
    Swaziland

Swaziland System
In South Africa and Swaziland.
BT Precambrian
SA South Africa
    Swaziland

Sweden
Includes use on level 1 as an area term (list O). For term set options see list B.
CO N551500N691500
    E0241500E0110000
BA Europe
NT Alno
    Blekinge
    Dalarna
    Gota Valley
    Goteborg
    Gotland
    Halland
    Jamtland
    Kalmar
    Kristianstad
    Malmohus
    Norrbotten
    Nykoping
    Orebro
    Siljan
    Skane
    Skellefte
    Smaland
    Stockholm
    Uppsula
    Varmland
    Vasterbotten
SA Arctic region
    Baltic region
    Baltic Shield
    Fennoscandia
    Lapland
    North Sea Coast
    Oresund
    Scandinavia
    Sparagmite Group
    Sulitjelma

Sweetgrass Arch
Index Alberta and/or Montana as applicable.
SA Alberta
    Montana

Sweetwater County
SW Wyoming.
CO N410000N422000
    W1073000W1101000
BT Wyoming
    United States

Swietokrzyskie Mountains
use Swiety Krzyz Mountains

Swiety Krzyz Mountains
Mountain group between the Vistula and Pilica rivers in SE central part of country. Also search Holy Cross Mountains; Swiety Krzyz.
UF Holy Cross Mountains
    Swietokrzyskie Mountains
BT Kielce
    Poland

SA Galezice
Swiss Alps
Those ranges of the Alps mountain system which lie within the country.
BT Switzerland
SA Alps

Switzerland
Includes use on level 1 as an area term (list O). For term set options see list B.
CO N454500N474500
    E0103000E0055000
BA Europe
NT Aar Massif
    Aar Valley
    Aargau
    Basel
    Bergell Massif
    Bern
    Bernese Alps
    Geneva
    Glarus
    Glarus Alps
    Gotthard Massif
    Graubunden
    Lake of Zurich
    Lucerne
    Neuchatel
    Saint Gall
    Schaffhausen
    Swiss Alps
    Thur Valley
    Ticino
    Uri
    Valais
    Vaud
    Zurich
SA Alps
    Arve Valley
    Ceneri Zone
    Central Alps
    Inn Valley
    Jura Mountains
    Lago Maggiore
    Lake Constance
    Lake Geneva
    Lepontine Alps
    Molasse Basin
    Monte Rosa
    Pennine Alps
    Rhaetian Alps
    Rhine Basin
    Rhine Graben
    Rhine River
    Rhine Valley
    Rhone River
    Rhone Valley
    Silvretta Group
    Simplon region
    Upper Rhine Valley
    Western Alps

sychnodymite
use carrollite

Sydney
City on the Pacific Ocean in E.
CO S335500N335500
    E1511000E1511000
BT New South Wales
    Australia

Sydney Basin
Including the Sydney area, it extends from NW of Newcastle on the N to W of Wollongong on the S.
BT New South Wales
    Australia

Syedlets
use Siedlce

syenite
Includes use on level 3 under igneous rocks(1) syenite family(2). See list H.
BA syenite family
BT igneous rocks
SA microsyenite

nepheline syenite
quartz syenite

syenite family
Includes use on level 2 under igneous rocks(1). See list H.
BT igneous rocks
NA agpaite
    albitophyre
    alkali syenite
    bostonite
    foyaite
NZ granosyenite
NA kakortokite
    lujavrite
    miaskite
    microsyenite
    monzonite
    naujaite
    nepheline syenite
    pulaskite
    quartz syenite
    shonkinite
    syenite
    syenite porphyry
SA appinite

syenite porphyry
See list H.
BA syenite family
BT igneous rocks
SA porphyry

syenodiorite
Includes use on level 3 under igneous rocks(1) diorite family(2). See list H.
BA diorite family
BT igneous rocks

Sylmar Fault
Named after town in San Fernando Valley in Los Angeles County. Also search Sylmar.
BT California
    United States

Sylt
Main island of the North Frisian group just off the mainland in the North Sea.
UF Sylt Island
BT Schleswig-Holstein
    West Germany

Sylt Island
use Sylt

Sylvania Formation
Underlies Detroit River Formation; overlies Bois Blanc Formation. SE Michigan, NW Ohio, and Ontario.
BT Middle Devonian
    Devonian
SA Michigan
    Ohio
    Ontario

sylvanite
UF silvanite
BA tellurides
    sulfides
BT minerals

sylvine
use sylvite

sylvite
Also search sylvine.
UF leopoldite
    sylvine
BA halides
BT minerals
SA chlorides

symbiosis
Includes use on level 3 under appropriate fossil group (list F).

symmetric folds
Term introduced in 1978. Includes use on level 3 under folds(1) style(2). Before 1978, search folds AND symmetric.
BA folds

SA asymmetric folds
    normal folds

Symmetrodonta
Order. Includes use on level 2 under Mammalia(1).
BA Theria
    Mammalia
BT Tetrapoda
    Vertebrata

symmetry
Used as a general term, e.g. on level 3 under crystal growth(1) or crystal structure(1) or under fossil groups (list F). Also search asymmetry.
UF asymmetry

symmicton
use diamicton

symposia
Includes use on level 1 (list A); as a level 2 or 3 term appropriate to a large number of topics, e.g. on level 2 under sedimentary petrology(1). Also search meetings and symposium. See list G. Also search symposium. If level 1, term set options are: topic [list B or extraterrestrial geology, general, geophysics]
    subtopic
UF colloquia
    symposium
SA annual report
    associations
    geology
    organization
    progress report
    report

symposium
use symposia

Synapsida
Includes use on level 2 under Reptilia(1). See list F.
BA Reptilia
BT Tetrapoda
    Vertebrata
NA Lystrosaurus
    Therapsida

synchisite
UF synchysite
BZ carbonates
    halides
BT minerals
SA fluorides

synchysite
use synchisite

synclinal
A valid term through 1977. After 1977, use synclines.

synclines
Term introduced in 1978. Includes use on level 3 under folds(1) style(2). Before 1978, also search folds AND synclinal.
BA folds
SA anticlines
    synform folds

synclinoria
Includes use on level 3 under folds(1) systems(2).
BA folds
SA anticlinoria
    geanticlines
    geosynclines
    systems

synform folds
Term introduced in 1978. Includes use on level 3 under folds(1) style(2). Before 1978, search folds AND synform.
BA folds
SA antiform folds
    synclines

syngenesis
Includes use on level 3 under mineral deposits, genesis(2) processes(2).
BA diagenesis
SA mineral deposits

synonymy
SA paleontology
    taxonomy

syntectonic processes
Term introduced in 1978. Before 1978, search syntectonic.
SA paleogeography
    processes
    sedimentation
    tectonics

synthesis
As of 1978, term is used on level 2 under crystal growth(1). Before 1971, also search mineral synthesis.
SA crystal growth
    minerals
    synthetic materials

synthetic
A valid term through 1977 used in combination with the name of the mineral on level 3 to index artificial minerals. After 1977, use synthetic materials.

synthetic materials
Term introduced in 1978 to index artificial minerals. Before 1978, also search synthetic AND name of mineral.
SA materials
    minerals
    natural materials
    synthesis

Syr Darya
River which rises in the Tien Shan Mountains and flows W and NW into the Aral Sea. Index Soviet republics as applicable.
UF Jaxartes River
    Saihun River
    Sir Darya
    Syr Darya River
    Syr-Darya
BT USSR
SA Kazakhstan
    Kirghizia
    Tadzhikistan
    Uzbekistan

Syr Darya River
use Syr Darya

Syr-Darya
use Syr Darya

Syracuse
City in Onondaga County in New York, and city in SE Sicily. Index New York and/or Sicily as applicable.
SA New York
    Sicily

Syria
Includes use on level 1 as an area term (list O). For term set options see list B.
CO N323000N371500
    E0423000E0353000
BA Middle East
NT Damascus
    Golan Heights
    Sabkha
SA Asia
    Euphrates River
    Great Rift Valley
    Jordan River
    Jordan Valley
    Levant
    Mediterranean region
    Mesopotamia
    Near East

Syrte
use Sirte Basin

Sysert
City 25 miles SSE of Sverdlovsk in Sverdlovsk Oblast in W Siberia.
BT Russian Republic
    USSR

systematics
A valid term through 1977. Use taxonomy.

systems
Includes use on level 2 under faults(1), folds(1), and fractures(1).
SA anticlinoria
    block structures
    closed systems
    coordinate systems
    en echelon faults
    faults
    folds
    fractures
    geothermal systems
    grabens
    horsts
    information systems
    open systems
    rift valleys
    synclinoria

systems analogs
Includes use on level 2 under ground water(1).
SA ground water

szaibelyite
Also search ascharite.
UF ascharite
BA borates
BT minerals

Szczecin
Province in NW Poland. Also a city.
UF Stettin
BT Poland
NT Wolin

Szechuan
use Szechwan

Szechwan
Province in S central China.
UF Szechuan
BT China

Szolnok
County in E central Hungary. Also a city.
BT Hungary

T

Ta
use tantalum

taaffeite
BA oxides
BT minerals

Taal
On Volcano Island in center of Lake Taal in Batangas Province in S Luzon. Also search Taal Volcano.
UF Taal Volcano
BT Luzon
    Philippine Islands

Taal Volcano
use Taal

Tabasco
State in SE.
BT Mexico
SA Gulf Coastal Plain

Tabianian
Europe.
BA lower Pliocene
    Pliocene
    Neogene
    Tertiary
BT Cenozoic

Tabulata
Includes use on level 2 under Coelenterata(1). See list F.
BA Anthozoa
    Coelenterata
BT Invertebrata
NA Chaetetidae
    Favositidae
    Heliolitidae
SA corals
    Rugosa
    Scleractinia

Tachira
State in W.
BT Venezuela

Taconian Orogeny
use Taconic Orogeny

Taconic Allochthon
In Northern Appalachians. Index countries as applicable. Also search Taconic.
SA Canada
    United States

Taconic Orogeny
Term introduced in 1978. An orogeny in the latter part of the Ordovician period, named for the Taconic Range of eastern New York State and well developed through most of the northern Appalachians in U.S. and Canada. Before 1978, search Taconic AND orogeny; Taconian; Taconian Orogeny.
UF Taconian Orogeny
BT Paleozoic
SA Canada
    Ordovician
    orogeny
    Silurian
    United States

taconite
Includes use on level 3 under sedimentary rocks(1) chemically precipitated rocks(2). See list I.
BA chemically precipitated rocks
BT sedimentary rocks
SA iron formations

tactite
Compositional term. Before 1978, included use on level 3 under metamorphic rocks(1).
BA metasomatic rocks

Tadzhik Basin
use Tadzhik Depression

Tadzhik Depression
North section of Afghan-Tadzhik Depression in S and SW Tadzhikistan. Also search Tadzhik; Tadzhik Basin; Tadzhikistan Depression.
UF Tadzhik Basin
    Tadzhikistan Depression
BT Tadzhikistan
    USSR
SA Afghan-Tadzhik Depression

Tadzhikhistan
use Tadzhikistan

Tadzhikistan
Tadzhik Soviet Socialist Republic in Soviet Central Asia. Also search Tadzhikhistan; Tadzhikstan.
CO N360000N410000
    E0750000E0670000
UF Tadzhikhistan
    Tadzhikstan
    Tajikistan
BT USSR
NT Altyn-Topkan
    Darvaz
    Darvaza Range
    Dushanbe
    Garm
    Hissar Range
    Kara-Mazar
    Karamazar
    Kategin Range
    Kulyab
    Leninabad
    Nurek
    Shugnan
    Tadzhik Depression
    Tary-Ekam
    Turkestan Range
    Vakhsh
    Vanchskiy Range
    Zeravshan Range
    Zeravshan-Hissar
SA Afghan-Tadzhik Depression
    Amu Darya
    Asiatic USSR
    Badakhshan
    Central Asia
    Fergana Basin
    Hindu Kush
    Kurama Range
    Pamirs
    Surkhan Darya
    Surkhan Darya basin
    Syr Darya
    Zeravshan River

Tadzhikistan Depression
use Tadzhik Depression

Tadzhikstan
use Tadzhikistan

taele
use Frozen ground

Taeniopteris
Genus. Includes use on level 3 under pteridophytes(1) Filicopsida(2).
BA Filicopsida
    pteridophytes
BT Plantae
SA Cycadales
    Glossopteris
    gymnosperms
    Pteridospermae

taenite
BA native elements and alloys
BT minerals
SA kamacite
    meteorites
    plessite

Tafilalet
use Tafilalt

Tafilalt
Oasis in SE Morocco.
UF Tafilalet
    Tafilelt
    Tafilet
BT Morocco

Tafilelt
use Tafilalt

Tafilet
use Tafilalt

Tagil Basin
River basin in Sverdlovsk Oblast E of the Central Urals. Also search Tagil; Tagil River.
BT Russian Republic
    USSR

Tagliamento Valley
River valley extending from Carnic Alps to the head of the Gulf of Venice. Index autonomous regions as applicable. Also search Tagliamento; Tagliamento River.
BT Italy
SA Friuli-Venezia Giulia
    Veneto

Tagus Basin
River basin in central Spain and S central Portugal. Index countries as applicable. Also search Tagus.

UF Tagus River basin
BT Europe
SA Portugal
    Spain

Tagus River
Longest river in the Iberian Peninsula. Rises in E central Spain and flows Wand SW entering the Atlantic Ocean at Lisbon. Index countries as applicable.
BT Europe
SA Portugal
    Spain

Tagus River basin
use Tagus Basin

Tahiti
Island of E group of Society Islands in French Polynesia. Includes use on level 1 as an area term (list O). For term set options see list B.
BA Polynesia
SA Society Islands

Taiga
City in NW Kemerovo Oblast.
UF Tayga
BT Russian Republic
    USSR

taiga environment
Term introduced in 1978. Before 1978, also search taiga.
SA ecology
    environment
    geomorphology
    sedimentation
    steppes
    tundra

Taigonos Peninsula
use Taygonos Peninsula

tailings
SA mining
    reclamation
    waste disposal

Taimyr
use Taymyr

Taimyr (Peninsula)
use Taymyr Peninsula

Taimyr Peninsula
use Taymyr Peninsula

Tainan
City on SW coast.
BT Taiwan

Taipei
City at N end of island.
BT Taiwan

Taishu Group
Chiefly composed of sandstone and shale in alternation, and intercalating arkose sandstone and red-coloured conglomerates. On Tsushima Island a part of Nagasaki Prefecture, Kyushu.
BT lower Tertiary
    Tertiary
SA Japan
    Kyushu

Taito
use Taitung

Taitung
City on SE coast.
UF Taito
BT Taiwan

Taiwan
Island off Fuklen Province of China. Seat of Chinese Nationalist government known as the Republic of China. Includes use on level 1 as an area term (list O). For term set options see list B. Before 1977, also search Formosa.
CO N220000N253000
    E1230000E1200000
BA Asia
NT Chiayi-Hsinying area
    Chinkuashih
    Chinkuashih Mine
    Hsinchu
    Miaoli
    Peikang
    Penghu Islands
    Tainan
    Taipei
    Taitung
    Taoyuan
    Tatun Shan
    Tiehchenshan
SA Far East

Taiwan Strait
use Formosa Strait

Tajikistan
use Tadzhikistan

Takanuki
Village NE of Utsunomiya in E central.
BT Honshu
    Japan

Tal Formation
According to the Indian Lexique, the Tal Beds may represent partly the Jurassic or even Cretaceous. Tal Series consist of sandstones, black shales, and arenaceous blue limestones. In NW Uttar Preadesh.
BT Mesozoic
SA Cretaceous
    India
    Jurassic
    Uttar Pradesh

Talacasto
Village in central.
BT Argentina
SA San Juan

Talas Ala-Tau
use Talas Range

Talas Range
Branch of the Tien Shan in NW Kirghizia. Also search Talas.
UF Talas Ala-Tau
BT Kirghizia
    USSR
SA Tien Shan

Talasea
Settlement on E side of Willaumez Peninsula on N coast of New Britain, Bismarck Archipelago.
BT Papua New Guinea
    Australasia

talc
Includes use on level 1 as a commodity term (list C).
BA sheet silicates
    silicates
BT minerals
SA ore deposits
    soapstone

talc rock
Compositional term. Before 1978, included use on level 3 under metamorphic rocks(1).
BA metasomatic rocks

Talcher
use Talchir

Talcher coal field
use Talchir coal field

Talcher Coalfield
use Talchir coal field

Talchir
Region. Former princely state in Orissa States. Also search Talcher.
UF Talcher
BT Orissa
    India

Talchir coal field
E central Orissa. Also search Talcher coal field, Talcher Coalfield, and Talchir.
UF Talcher coal field
    Talcher Coalfield
BT Orissa
    India

Talchir Formation
Upper Carboniferous to Permo-Carboniferous. Consists of boulder bed with boulders set in an ill-sorted matrix, often containing angular clastics, and is followed by greenish shales, silty shales, and sandstones. In peninsular India.
BT Paleozoic
SA Carboniferous
    India
    Permian
    Upper Carboniferous

Talchir Series
Upper Carboniferous. At the base is a glacial boulder bed (Talchir Boulder Bed). Overlain by Damuda Series. In Orissa.
BT Upper Carboniferous
    Carboniferous
SA India
    Orissa

Talin
use Tallin

Talladega Front
E Talladega County in E central.
BT Alabama
    United States

Talladega Group
According to the latest published Lexique, the age of Talladega Formation is Precambrian (?) to Carboniferous (?). In NW Georgia, the Talladega Series has been grouped into 10 formations (ascending): Pinelog Quartzite, Hiawassee Slate, Great Smoky Formation, Natahala Schist, Tusquitee Quartzite, Brasstown Schist, Valleytown Schist, Murphy Marble, Andrews Schist and Nottley Quartzite. In E Alabama, North Carolina, NW Georgia, and Tennessee.
BT Paleozoic
SA Alabama
    Cambrian
    Devonian
    Georgia
    Mississippi
    North Carolina
    Ordovician
    Pennsylvanian
    Precambrian
    Silurian
    Tennessee

Tallapoosa County
E central.
BT Alabama
    United States

Tallin
City on the Gulf of Finland. Also search Tallinn.
UF Talin
    Tallinn
BT Estonia
    USSR

Tallinn
use Tallin

talnakhite
BA sulfides
BT minerals

talus
A valid term through 1978. Use talus slopes.

talus fan
use alluvial fans

talus slopes
Term introduced in 1978 on level 2 under slope stability(1). Before 1978, search talus; debris slopes.
UF debris slopes
SA cliffs
    colluvium
    erosion
    erosion features
    geomorphology
    landslides
    slope stability
    slopes

Talysh Mountains
NW extremity of the Elburz in NW Iran. Also search Talysh.
BT Iran

Tamagawa
River flowing into Tokyo Bay between Tokyo and Yokohama.
BT Honshu
    Japan

Taman
Village on N shore of Taman Peninsula near Kerch Strait in W Krasnodar Kray in Northern Caucasus.
BT Russian Republic
    USSR

Taman Peninsula
Peninsula of W Krasnodar Kray jutting into Kerch Strait connecting Black Sea and Azov Sea. Also search Taman.
BT Russian Republic
    USSR

Tamar Valley
River valley in Cornwall and Devon in SW England. Also search Tamar River.
BT England
    Great Britain
    United Kingdom

Tamaulipas
State in E.
BT Mexico
SA Gulf Coastal Plain
    Rio Grande
    Rio Grande Depression
    Rio Grande Valley
    Sierra Madre Oriental

Tamba Plateau
In the Kyoto area. Also search Tamba.
BT Honshu
    Japan

Tambov
City 260 miles SE of Moscow. Also an oblast.
BT Russian Republic
    USSR

Tamdy-Tau
use Tamdytau

Tamdytau
Mountains in the Kyzylkum SE of the Aral Sea.
UF Tamdy-Tau
    Tamdytau Range
BT Uzbekistan
    USSR

Tamdytau Range
use Tamdytau

Tamil Nadu
State in extreme S and SE. Formerly Madras.
BT India
    Asia
NT Ariyalur
    Coimbatore
    Kondapalli
    Madras

Neyveli
North Arcot
Parnalee
Sivamalai
South Arcot
Tiruchirapalli
Tirupati
Vridhachalam
SA Cauvery Basin
Cuddalore Series
Neyveli Lignite
Niniyur Group

Tampa Bay
Inlet of Gulf of Mexico on W coast of Hillsborough County in W.
BT Florida
United States

Tampere
City in SW part of country.
BT Finland

Tamworth
Town in E central.
BT New South Wales
Australia

Tana Fjord
Inlet of Arctic Ocean which receives the Tana River in NE.
BT Finnmark
Norway

Tananarive
City on the Indian Ocean in E.
BT Malagasy Republic

Tanco Pegmatite
BT Manitoba
Canada

Tanega-shima
Largest of the Osumi Islands off S Kyushu.
UF Tanegashima Island
BT Kyushu
Japan

Tanegashima Island
use Tanega-shima

Tanezrouft
Section of the Sahara primarily in Algeria. Index countries as applicable.
BT Africa
SA Algeria
Mali
Sahara

Tanganyika
Not a valid index term. Former British U.N. Trust Territory which became independent in 1961. See Tanzania.

tangential stress
use shear stress

tangential wave
use S-waves

Tanger
use Tangier

Tangier
City at W end of the Strait of Gibraltar.
UF Tanger
Tangiers
BT Morocco

Tangiers
use Tangier

Tannu Tuva
use Tuva

Tannu-Ola Range
On the Mongolian-Tuva A.S.S.R. border. Index Mongolia and/or Russian Republic as applicable. Also search Tannu-Ola.
BT Asia
SA Mongolia
Russian Republic

Tanquary Fiord
W central Ellesmere Island in District of Franklin.
BT Northwest Territories
Canada

tantalates
Includes use on level 3 under minerals(1) oxides(2). See list L.
BA oxides
BT minerals
SA columbite
euxenite
fergusonite
microlite
nenadkevichite
pyrochlore
tapiolite
wodginite

tantalite
BA oxides
BT minerals
SA niobates

tantalum
Includes use on level 1 as a commodity term (list C) and as a chemical element (list D); on level 2 under mineral deposits, genesis(1).
UF Ta
SA elements
heavy metals
niobium

Tanzania
Former British U.N. Trust Territory of Tanganyika, which became independent in 1961. United with the island of Zanzibar to become Tanzania in 1964. Includes use on level 1 as an area term (list O). For term set options see list B.
BA Africa
NT Kilimanjaro
Oldoinyo Lengai
Olduvai Gorge
Southern Highlands
SA East African Rift
Gregory Rift
Lake Malawi
Lake Natron
Lake Tanganyika
Lake Victoria
Umba
Zanzibar

tanzanite
BA epidote group
orthosilicates
silicates
BT minerals
SA zoisite

Tanzawa Massif
use Tanzawa Mountains

Tanzawa Mountainland
use Tanzawa Mountains

Tanzawa Mountains
Between Fujiyama and Yokohama. Also search Tanzawa.
UF Tanzawa Massif
Tanzawa Mountainland
BT Honshu
Japan

Taos County
On the Colorado border in N New Mexico. Also search Taos.
BT New Mexico
United States

Taoudeni
use Taoudenni

Taoudeni Basin
use Taoudenni

Taoudenni
An oasis in the Sahara in NW Mali.
UF Taoudeni
Taoudeni Basin
Taoudenni Basin

BT Mali

Taoudenni Basin
use Taoudenni

Taourirt
Town in NE Morocco, and village in W central Algeria. Index countries as applicable.
BT Africa
SA Algeria
Morocco

Taoyuan
Town in N.
BT Taiwan

taphocenoses
use thanatocenoses

taphocoenosis
use thanatocenoses

taphonomy
Includes use on level 3 under fossil group(1). See list F.
SA fossilization
paleoecology
thanatocenoses

taphrogeny
Implies rifting. Includes use on level 3 under plate tectonics(1).
SA orogeny
plate tectonics
rifting
tectonics

tapiolite
BA oxides
BT minerals
SA niobates
tantalates

tar sands
use oil sands

Taranaki
Provincial district in W.
BT North Island
New Zealand

Taranto
City on N Gulf of Taranto on heel of Italian boot.
BT Apulia
Italy

Tarapaca
Northernmost province.
BT Chile

Tarawera Complex
use Tarawera volcanic complex

Tarawera volcanic complex
N central North Island.
UF Tarawera Complex
BT North Island
New Zealand

Tarbagatai Range
use Tarbagatay Range

Tarbagatay Mountains
use Tarbagatay Range

Tarbagatay Range
Northern outlier of the Tien Shan. Index Kazakhstan and/or Sinkiang Uighur as applicable. Also search Tarbagatai Range; Tarbagatay.
UF Tarbagatai Range
Tarbagatay Mountains
BT Asia
SA Kazakhstan
Sinkiang Uighur
Tien Shan

Tarentaise
Alpine valley of upper Isere River in Savoy Alps.
BT Savoie
France

Tareya
Village in NE Taymyr National Okrug in NW Siberia.
BT Russian Republic
USSR

Tarfaia
use Tarfaya

Tarfaya
Town in extreme SW Morocco. A former Spanish enclave. Also search Tarfaia.
UF Tarfaia
BT Morocco

Tarkhankut Peninsula
In extreme W Crimea jutting into Black Sea. Also search Tarkhankut.
BT Ukraine
USSR

Tarn
Department in S.
BT France
NT Sidobre Massif
SA Montagne Noire
Rodez Trough

Tarn-et-Garonne
Department in SW.
CO N434000N443000
E0020000E0004500
BT France
NT Orgueil
SA Quercy
Rouergue

Tarnobrzeg
Town on the Vistula River in SE part of country.
BT Rzeszow
Poland

Tarnow
City in E.
BT Krakow
Poland

Taro Valley
River valley extending in a SW-NE direction from the N Apennines to the Po River. Also search Taro.
BT Emilia-Romagna
Italy

Tarragona
Province in Catalonia in NE.
BT Spain
SA Catalonia

Tarrant County
County in which Fort Worth in located.
BT Texas
United States

Tartary
use Tataria

Tartu
City on Ema River W of Lake Peipus.
UF Dorpat
BT Estonia
USSR

Tarumae
Active volcano. S of Sapporo in S Hokkaido. Also search Tarumae Volcano.
UF Tarumae Volcano
BT Hokkaido
Japan

Tarumae Volcano
use Tarumae

Tarvisio
Town near Austrian and Yugoslav borders.
BT Friuli-Venezia Giulia
Italy

Tary Ekan
use Tary-Ekam

Tary-Ekam
Village in Leninabad Oblast in N Tadzhikistan. Also search Tary-Ekan.
UF Tary Ekan
Tary-Ekan
BT Tadzhikistan

USSR

Tary-Ekan
use Tary-Ekam

Tas-Khayakhtakh Range
Northern section of Cherski Range N of Okhotsk Sea in NE Yakutia. Also search Tas-Khayakhtakh.
BT Russian Republic
   USSR

Tashkent
City and oblast in NE.
BT Uzbekistan
   USSR

Tasman Basin
Extends S of Tasman Sea proper between South Tasmanian Rise and Macquarie Ridge SW of New Zealand. Also search Tasman.
UF Tasman-Becken
   Tasmania Basin
BT Pacific Ocean

Tasman Geosyncline
Index states as applicable. Also search Tasman AND geosynclines.
BT Australia
SA geosynclines
   New South Wales
   Queensland
   Tasmania
   Victoria

Tasman orogenic zone
Index islands as applicable. Also search Tasman.
BT New Zealand
SA North Island
   South Island

Tasman Sea
Between SE Australia and Tasmania on W and New Zealand on the E. Also search Tasman. As of 1977, includes use as a level 1 area term (list O). For set options see list B.
BA Pacific Ocean
NT Dampier Ridge

Tasman-Becken
use Tasman Basin

Tasmania
Island and state S of Victoria. Includes use on level 1 as an area term (list O). For term set options see list B.
CO S434000S393000
   E1483000E1435000
BA Australia
NT Burnie
   Coles Bay
   Flinders Island
   Great Lake
   Hobart
   Rosebery
   Zeehan
SA Heemskirk Granite
   Tasman Geosyncline
   Wynyard

Tasmania Basin
use Tasman Basin

Tasmanites
Includes use on level 2 under algae(1). See list F.
BA algae
BT Plantae
SA palynomorphs

Tassili des Adjer
use Tassili n'Ajjer

Tassili n'Ajjer
An arid plateau region which is a NE extension of the Ahaggar Mountains in SE Algeria. Also search Tassili.
UF Tassili des Adjer
BT Algeria

Tatar
Tatar Autonomous Soviet Socialist Republic. In area around great bend of Volga River where it heads S at Kazan.
CO N540000N560000
   E0540000E0480000
UF Tatarstan
BT Russian Republic
   USSR

Tatar Arch
E central European USSR centering on Tatar A.S.S.R. Also search Tatar.
BT Russian Republic
   USSR

Tatar Strait
Between E coast of Primorye Kray and W coast of Sakhalin Island at the N end of Japan Sea. Also search Tatar.
BT Russian Republic
   USSR

Tataria
Originally applied to area of Mongol Empire at its height. Now, since displacement of Crimean Tatars, it can be applied to the Tartar A.S.S.R and surrounding areas inhabited by Tatars.
UF Tartary
   Tatary
BT Russian Republic
   USSR

Tatarian
Europe. Above Kazanian, below Scythian (Triassic). Includes use on level 3 under age terms(1). See list E.
BA Upper Permian
   Permian
BT Paleozoic

Tatarstan
use Tatar

Tatary
use Tataria

Tatra Mountains
Chief mountain group of central Carpathians. Also search High Tatra; High Tatra Mountains; High Tatras; Tatras. Index countries as applicable.
UF High Tatra
   High Tatra Mountains
   High Tatras
   Tatras
BT Carpathians
   Europe
SA Czechoslovakia
   Low Tatra Mountains
   Poland

Tatras
use Tatra Mountains

Tatum Basin
SE New Mexico.
BT New Mexico
   United States

Tatun Shan
Mountain 9 miles N of Taipei. Also search Tatun.
BT Taiwan

Tauern Tunnel
Railroad tunnel through the Hohe Tauern in SW Austria. Also search Tauern.
BT Austria
SA Hohe Tauern

Taunus
Mountain range E of the Rhine River and N of the lower Main River.
UF Taunus Mountains
BT Hesse
   Germany

Taunus Mountains
use Taunus

Taupo
Town on Lake Taupo in central.
BT North Island
   New Zealand

Taupo volcanic zone
Mountain region surrounding Lake Taupo in central North Island. Also search Taupo.
UF Taupo Zone
BT North Island
   New Zealand

Taupo Zone
use Taupo volcanic zone

Taurus Littrow
use Taurus-Littrow

Taurus Mountains
Mountain chain running parallel to the Mediterranean Sea across much of S Anatolia. Also search Taurus AND appropriate area term.
CO N363000N373000
   E0343000E0303000
BT Turkey
   Middle East

Taurus-Littrow
Includes use on level 3 under Moon(1). Also search Taurus Littrow; Taurus-Littrow Valley; Tranquillity Base.
UF Taurus Littrow
   Taurus-Littrow Valley
BT Moon

Taurus-Littrow Valley
use Taurus-Littrow

tavistockite
use carbonate apatite

Tavua
Village on Viti Levu Island.
BT Fiji
   Pacific Ocean

Tawi River
use Tawi Valley

Tawi Valley
River valley in SW Jammu and Kashmir. Also search Tawi River.
UF Tawi River
BT Jammu and Kashmir
   India

taxa, ahermatypic
use ahermatypic taxa

taxa, anaerobic
use anaerobic taxa

taxa, benthonic
use benthonic taxa

taxa, colonial
use colonial taxa

taxa, endemic
use endemic taxa

taxa, extinct
use extinct taxa

taxa, hermatypic
use hermatypic taxa

taxa, heteromorphic
use heteromorphic taxa

taxa, new
use new taxa

taxa, planktonic
use planktonic taxa

Taxodiaceae
Family. Includes use on level 3 under gymnosperms(1) Coniferales(2).
BA Coniferales
   gymnosperms
BT Plantae
NA Sequoia
SA Pinaceae
   Taxodium

Taxodium
BA Coniferales
   gymnosperms
BT Plantae
SA Taxodiaceae

taxonomy
Includes use on level 2 under fossil group(1); on level 2 under paleontology(1). See list F.
SA homonymy
   new names
   new taxa
   paleontology
   range
   revision
   species
   synonymy

taxonomy review
A valid term through 1973. Use taxonomy.

Tay Estuary
N of the Firth of Forth on the North Sea. Also search Firth of Tay.
UF Firth of Tay
BT Scotland
   Great Britain
   United Kingdom

Tayga
use Taiga

Taygonos Peninsula
On N Okhotsk Sea between Gizhiga and Penzhina bays. Also search Taigonos; Taigonos Peninsula; Taygonos.
UF Taigonos Peninsula
BT Russian Republic

Taylor County
Index states as applicable.
BX Florida
   Georgia
   Iowa
   Kentucky
   Texas
   West Virginia
   Wisconsin
BT United States

Taylor Glacier
In the McMurdo Sound area in S Victoria Land just NW of Ross Ice Shelf. Also search Taylor.
BT Antarctica

Taylor Valley
At McMurdo Sound in S Victoria Land in the Taylor Glacier area NW of Ross Ice Shelf. Also search Taylor.
BT Antarctica

Taymyr
National Okrug. A national district approximately coextensive with the Taymur Peninsula in northernmost Siberia. Also search Taimyr.
CO N670000N773000
   E1140000E0780000
UF Taimyr
BT Russian Republic
   USSR

Taymyr Peninsula
Between the Yenisei and the Khatanga rivers in the Taymyr National Okrug in northernmost Siberia. Also search Taimyr, Taimyr Penisnula, and Taymyr.
UF Taimyr (Peninsula)
   Taimyr Peninsula
BT Russian Republic
   USSR

Taz Basin
River basin in E and NE Yamal Nenets National Okrug in NW Siberia. Also search Taz; Taz River.
UF Taz River
BT Russian Republic
   USSR

Taz River
use Taz Basin

Taza
Town E of Fez in N.
BT Morocco

Tazewell County
SW Virginia. Also search Tazewell.
BT Virginia
United States

Tazin Lake
Extreme NW.
BT Saskatchewan
Canada

Tb
use terbium

Tbilisi
City in SE Georgian Republic, USSR. Also search Tiflis.
UF Tiflis
BT Georgian Republic
USSR

Tc
use technetium

Tchornozem
use Chernozems

Te
use tellurium

Te Aroha
Town 70 miles SE of Auckland in N central.
BT North Island
New Zealand

Teapot Dome
Near Casper in central.
BT Wyoming
United States

technetium
Includes use on level 1 and 2 as chemical element (list D).
UF Tc
SA elements

techniques
This term deals with sampling. Includes use as a level 2 or 3 term appropriate to a large number of topics, e.g. on level 2 under meteorology(1), mineral exploration(1), and spectroscopy(1). See list G.
SA analog techniques
digital techniques
instruments
methods
sample preparation
sampling

technology
Includes use on level 2 under mining geology(1).
SA mining geology
production control

tectite
use tektites

tectonic
A valid term through 1977. After 1977, use tectonic maps.

tectonic controls
A valid term through mid-1978. Use structural controls, e.g. under sedimentation(1) or mineral deposits, genesis(1).

tectonic maps
Term introduced in 1978. Includes use on level 3 under maps(1). Before 1978, also search maps AND tectonic.
BA maps
SA structure contour maps
tectonic

tectonics
Used for large-scale treatments. For "microtectonics" see structural analysis. Includes use on level 1. Term set options are:
concepts
topic
evolution
(type of structure, e.g. Alpine Orogeny, Appalachian Orogeny, basement, basin range structure, Caledonian Orogeny, geanticlines, Hercynian Orogeny, lineaments, platforms, rift zones, shear zones, etc.]
gravity sliding
topic
structure
type of structure
vertical tectonics
topic (subsidence, uplifts)
UF geotectonics
SA allochthons
Alpine Orogeny
alpine-type
Andean Orogeny
Appalachian Phase
aulacogens
autochthons
basement
basin range
basins
breccia
changes of level
continental drift
continental margin
Cordilleran Orogeny
cratons
crust
crustal shortening
decollement
deep-seated structures
deformation
diastrophism
epeirogeny
eugeosynclines
faults
fold belts
folds
foliation
fractures
geanticlines
geodynamics
geosynclines
gravity sliding
imbrication
isostasy
klippen
lineaments
lineation
massifs
median masses
Median Tectonic Line
meteor craters
miogeosynclines
mobile belts
mobility
nappes
neotectonics
orogeny
plate tectonics
platforms
rift zones
salt tectonics
sedimentary cover
seismotectonics
shear zones
shields
structural analysis
structural complexes
structural geology
subsidence
syntectonic processes
taphrogeny
tectonic maps
tectonophysics
tectonosphere
undation
uplifts
vertical tectonics
volcanology

tectonite
SA deformation
fabric
preferred orientation

tectonophysics
Used for treatments of the application of physics in tectonics. Until 1976 this set included documents dealing with plate tectonics(1). Includes use on level 1; on level 2 under area terms (1), bibliography (1), or symposia (1). See list B. If level 1, term set options are:
topic [bibliography, concepts, convection, convection currents, experimental studies, methods, observations, practice, processes, structure, symposia, theoretical studies]
subtopic
SA asthenosphere
bibliography
changes of level
continental drift
continental margin
convection
convection currents
core
crust
deformation
Earth
Earth tides
faults
folds
foliation
fractures
geophysics
geosynclines
heat flow
island arcs
isostasy
lithosphere
mantle
Mohorovicic discontinuity
neotectonics
ocean basins
paleomagnetism
plate tectonics
rift zones
salt tectonics
sea-floor spreading
seismology
structural analysis
subduction
subduction zones
tectonics

tectonosphere
SA biosphere
hydrosphere
lithosphere
plate tectonics
tectonics

tectosilicates
use framework silicates

teeth
Includes use on level 3 under fossil group(1) for type of material. See list F.
SA bones
jaws
skeletons
skulls
Vertebrata

Tehachapi Mountains
Range running E-W between S end of Sierra Nevada and the Coast Ranges in Kern County in S California. Also search Tehachapi.
BT California
United States

Teheran
Province in NW central Iran. Also a city at S foot of the Elburz.
UF Tehran
BT Iran

Tehran
use Teheran

Tehri
use Tehri Garhwal

Tehri Garhwal
District in Himalayas in NW Uttar Pradesh. Also search Tehri.
UF Tehri
BT Uttar Pradesh
India

Teign Valley
River valley in Devonshire in SW England. Also search Teign River.
BT England
Great Britain
United Kingdom

Tejon Formation
Subdivided into 4 members (ascending): Uvas Conglomerate, Liveoak, Metralla Sandstone, and Reed Canyon Silt. Underlies Tecuya Formation; overlies basement complex. In W California.
BT upper Eocene
Eocene
SA California

Tekeli
Town in Taldy Kurgan Oblast near Sinkiang Uighur border in E.
BT Kazakhstan
USSR

tektites
Includes use on level 1 (list A). Used for the descriptions of small, rounded, pitted, black to green or yellow bodies of silicate glass of nonvolcanic origin. Term set options are:
topic [age, classification, composition, distribution, genesis, mineral composition, properties]
(type of tektite) or subtopic (no area term)
UF tectite
NA australite
Darwin glass
indochinite
microtektites
moldavite
muong nong type
philippinite
SA asteroids
comets
glasses
isotopes
meteorites
petrology
planetology
spherules

tele
use Frozen ground

Teleajen River
use Teleajen Valley

Teleajen Valley
River valley in Prahova County in NE Moldavia. Also search Teleajen; Teleajen River.
UF Teleajen River
BT Moldavia
Romania

Telemark
County in S.
BT Norway
NT Bamble
Langesund Fjord

Teleosauridae
Family. Includes use on level 3 under Reptilia(1) Archosauria(2).
BA Crocodilia
Archosauria
Reptilia

BT Tetrapoda
   Vertebrata

Teleostei
Includes use on level 3 under Pisces(1) Osteichthyes(2). See list F.
BA Osteichthyes
   Pisces
BT Vertebrata
SA Cyprinidae
   fish

teleseismic signals
SA earthquakes
   seismology
   signals

Tell
The Mediterranean coastal region of Algeria which is favored by a Mediterranean type climate.
BT Algeria

tellurates and tellurites
Includes use on level 2 under minerals(1). See list L.
UF tellurites, tellurates and
BT minerals
NA mroseite

tellurbismuth
use tellurbismuthite

telluric currents
Not a valid index term for GeoRef. See geophysical methods(1) Earth-current methods(2).

telluric methods
use Earth-current methods

telluric surveys
use Earth-current surveys

tellurides
Includes use on level 3 under minerals(1) sulfides(2).
BA sulfides
BT minerals
NA altaite
   calaverite
   coloradoite
   hessite
   joseite
   sylvanite
   tellurobismuthite
   tetradymite
SA niggliite
   wehrlite

tellurites, tellurates and
use tellurates and tellurites

tellurium
Includes use on level 1 and 2 as a chemical element (list D).
UF Te
SA elements
   heavy metals

tellurobismuthite
Also search tellurbismuth.
UF tellurbismuth
BA tellurides
   sulfides
BT minerals

TEM data
Term introduced in 1978. For the method, see transmission method.
UF transmission electromicroscopy data
SA data
   electron microscopy
   scanning method
   SEM data
   transmission method

Temagami Mine
Near village of Temegami on E end of Lake Temagami near Quebec border. Also search Temagami.
UF Timagami Mine
BT Ontario
   Canada

Temblor Range
Along SW San Joaquin Valley in W and SW Kern County.
BT California
   United States

Temiscamingue
use Timiskaming

Temiskaming
use Timiskaming

temperate environment
Term introduced in 1978. Before 1978, search temperate.
SA environment

temperature
For temperatures of mineral formations, see phase equilibria(1). Includes use on level 2 under core(1), Earth(1), heat flow(1), lava(1), mantle(1) and meteorology(1); on level 2 under paleoclimatology(1) for paleotemperatures. Before 1974, also search paleotemperatures. Also includes use as a level 2 or 3 term appropriate to a large number of topics. See list G. Also search temperatures.
NT high temperature
   low temperature
SA cooling
   densities and temperatures
   fluid inclusions
   geologic thermometry
   ion densities and temperatures
   Moon
   P-T conditions
   specific heat
   thermal properties

Tenerife
Largest island in the Spanish controlled Canary Islands off NW Africa.
UF Teneriffe
BT Canary Islands
   Atlantic Ocean

Teneriffe
use Tenerife

Tengiz
Salt lake NW of Karaganda in N central Kazakhstan. Also search Teniz.
UF Teniz
BT Kazakhstan
   USSR

Teniz
use Tengiz

Tennant Creek
Village 280 miles N of Alice Springs in central.
BT Northern Territory
   Australia

tennantite
BA sulfarsenites
   sulfosalts
BT minerals

Tennessee
Includes use on level 1 as an area term (list O). For term set options see list B.
CO N350000N364500
   W0814000W0901500
BA United States
NX Anderson County
   Bedford County
   Benton County
   Blount County
NT Buffalo River
NX Campbell County
NT Cannon County
NX Carroll County
   Carter County
NT Central Basin
NX Chester County
   Clay County
   Coffee County
   Columbia
   Cumberland County
   Davidson County
NT Ducktown
NX Fayette County
   Franklin County
   Greene County
   Grundy County
   Hamilton County
   Hancock County
   Hardeman County
   Hardin County
NT Henry County
NX Jackson County
   Jefferson County
   Johnson County
   Lake County
   Lawrence County
   Lincoln County
   Macon County
   Madison County
   Marion County
   Marshall County
   Monroe County
   Montgomery County
   Moore County
   Morgan County
NT Overton County
NX Perry County
   Polk County
   Putnam County
   Roane County
   Scott County
NT Sequatchie Valley
NX Sevier County
   Smith County
   Sullivan County
   Sumner County
NT Trousdale County
NX Union County
   Warren County
   Washington County
   Wayne County
   White County
   Williamson County
   Wilson County
   Woodbury
SA Appalachian Basin
   Appalachian Plateau
   Ashe Formation
   Bangor Limestone
   Bigby-Cannon Limestone
   Brassfield Formation
   Chattanooga Shale
   Chesterian
   Chickamauga Group
   Chilhowee Group
   Clayton Formation
   Cumberland Plateau
   Fernvale Formation
   Fort Payne Formation
   Great Smoky Mountains
   Juniata Formation
   Kingsport Formation
   Knox Group
   Lee Formation
   Lincolnshire Limestone
   Mascot Dolomite
   Midcontinent
   Mississippi Embayment
   Mississippi River
   Mississippi Valley
   Monteagle Limestone
   Murphy Marble
   Newman Limestone
   Pennington Formation
   Pottsville Group
   Ripley Formation
   Roan Supergroup
   Rockwood Formation
   Rome Formation
   Saint Louis Limestone
   Sainte Genevieve Limestone
   Shady Dolomite
   Talladega Group
   Tennessee River
   Tennessee Valley
   Tuscaloosa Formation
   Valley and Ridge Province
   Waldron Shale
   Warsaw Formation
   Wilcox Group

Tennessee River
Formed near Knoxville in E Tennessee. It flows SW, W, and then N entering the Ohio River at Paducah in W Kentucky. Index states as applicable.
BT United States
SA Alabama
   Kentucky
   Tennessee

Tennessee Sandstone
Name has either been abandoned by author or rejected for use by U.S. Geological Survey. Typically exposed in Tennessee Ridge, Sebastian County, in W Arkansas.
BT Pennsylvanian
SA Arkansas

Tennessee Valley
River valley. Index states as applicable.
BT United States
SA Alabama
   Kentucky
   Tennessee

tensile strength
SA deformation
   elastic properties
   engineering geology
   failure
   mechanical properties
   shear strength
   strength
   tension

tension
Includes use on level 3 under deformation(1) field studies(2) and under fractures(1) genesis(2).
SA compression
   deformation
   fractures
   pressure
   stress
   tensile strength
   torsion
   yield strength

Tensleep Formation
use Tensleep Sandstone

Tensleep Sandstone
In Montchauve Group. Pennsylvanian and lower Permian. Overlies Amsden Formation. Also search Tensleep Formation.
UF Tensleep Formation
BT Paleozoic
SA Lower Permian
   Montana
   Pennsylvanian
   Permian
   Wyoming

Tentaculites
Genus. Includes use on level 3 under problematic fossils(1).
BA Tentaculitidae
   Tentaculitida
   problematic fossils
SA Mollusca

Tentaculitida
Order.
BA problematic fossils
NA Tentaculitidae
SA Mollusca

Tentaculitidae
Family. Includes use on level 3 under problematic fossils(1).
BA Tentaculitida
   problematic fossils
NA Tentaculites
SA Mollusca

tephra
use pyroclastics

tephrite
Includes use on level 3 under igneous rocks(1) alkali basalt family(2). See list H.
BA alkali basalt family
BT igneous rocks

tephrochronology
Includes use on level 2 under geochronology(2). See list B.
SA geochronology

tephroite
BA olivine group
    orthosilicates
    silicates
BT minerals

Teplice
City in the Erzgebirge NNW of Prague.
UF Teplice-Sanov
BT Bohemia
    Czechoslovakia

Teplice-Sanov
use Teplice

Ter River basin
In Catalonia. Index provinces as applicable. Also search Ter River.
BT Spain
    Barcelona
SA Gerona

terbium
Includes use on level 1 and 2 as a chemical element (list D).
UF Tb
SA elements

Terceira Island
Central island of Portuguese controlled group. Site of U. S. Lajes Air Force Base. Also search Terceira.
BT Azores
    Atlantic Ocean

Tereblya
Village in SW near the Romanian border.
BT Ukraine
    USSR

Terebratulida
Includes use on level 2 under Brachiopoda(1). See list F.
BA Articulata
    Brachiopoda
BT Invertebrata
NA Terebratulidae
    Terebratulina

Terebratulidae
Family. Includes use on level 3 under Brachiopoda(1) Terebratulida(2).
BA Terebratulida
    Brachiopoda
BT Invertebrata

Terebratulina
Genus. Includes use on level 3 under Brachiopoda(1) Terebratulida(2).
BA Terebratulida
    Brachiopoda
BT Invertebrata

Terek River
Flows into the Caspian Sea N of Makhachkala in the Northern Caucasus. Index Soviet republics as applicable. Also search Terek.
BT USSR
SA Georgian Republic
    Russian Republic

Terlingua
Village in Brewster County in W.
BT Texas
    United States

terminology
A valid term through mid-1978. Use nomenclature.

Terni
City NE of Rome.
BT Umbria
    Italy

Terra rosa
use Terra rossa

Terra rossa
See list M. Also search Terra rosa.
UF Terra rosa
BA soils

terraces
Includes use on level 3 under appropriate headings in geomorphology(1) such as erosion features(2), fluvial features(2), landform description(2), landform evolution(2), and shore features(2).
SA benches
    changes of level
    erosion features
    floodplains
    fluvial features
    lacustrine features
    shore features

terrain classification
Term introduced in 1978. Before 1978, search classification AND terrains.
SA classification
    geomorphology
    terrains

terrains
Also search terranes.
UF terranes
SA geomorphology
    relief
    terrain classification
    topography

terranes
use terrains

terrestrial
A valid index term through 1977. After 1977, use terrestrial environment for type of environment, and use terrestrial samples or terrestrial materials for Moon studies to distinguish from lunar samples or extraterrestrial materials.

terrestrial environment
Term introduced in 1978. Includes use on level 3 under ecology(1), paleoecology(1), and sedimentation(1). Before 1978, also search terrestrial.
SA ecology
    environment
    paleoecology
    sedimentation

terrestrial materials
Term introduced in 1978. Use only for studies of the Moon, to distinguish from extraterrestrial materials. Before 1978, also search terrestrial AND materials; terrestrial samples.
UF terrestrial samples
SA materials
    Moon
    samples

terrestrial planets
SA Earth
    Mars
    Mercury Planet
    outer planets
    planetology
    planets
    Venus

terrestrial samples
use terrestrial materials

terrigenous
A valid level 2 term through 1977 used in combination with clastic rocks or clastic sediments. After 1977, use terrigenous materials only on level 3.

terrigenous materials
Term introduced in 1978. Includes use on level 3 under sediments(1) and sedimentary rocks(1). Before 1978, also search terrigenous AND clastic rocks or clastic sediments.
SA arenite
    argillite
    arkose
    bentonite
    black shale
    boulders
    breccia
    cinerite
    clastic rocks
    clastic sediments
    clay
    claystone
    cobbles
    colluvium
    conglomerate
    contourite
    coquina
    diamictite
    drift
    dust
    eluvium
    eolianite
    fanglomerate
    flint clay
    flysch
    gravel
    graywacke
    loess
    marl
    materials
    microbreccia
    molasse
    mud
    mudstone
    nonterrigenous materials
    orthoquartzite
    outwash
    pebbles
    pyroclastics
    red beds
    residuum
    sand
    sandstone
    sedimentary rocks
    sediments
    shale
    silcrete
    silt
    siltstone
    sparagmite
    subgraywacke
    till
    tillite
    tilloid
    tonstein
    turbidite
    volcanic ash

Tertiary
Includes use on level 1 as an age term (list E); on level 2 under paleo-terms, e.g. paleoecology, paleogeography, paleomagnetism.
BA Cenozoic
NA Arikareean
NT Barreiras Formation
    Columbia River Basalt
    Cuddalore Series
    John Day Formation
NA lower Tertiary
NT Maikop Series
NA middle Tertiary
    Neogene
NT Neyveli Lignite
NA Paleogene
NT Poltava Series
    San Lorenzo Formation
    Sespe Formation
    Snoqualmie Batholith
    Twin River Formation
    Twin Sisters Dunite
NA upper Tertiary
NT Warkalli Formation
    Yakima Basalt
SA Bauru Formation
    Blancan
    Brianconnais Zone
    Cretaceous
    Ewekoro Formation
    Ionian Zone
    Laramide Orogeny
    Matuyama Epoch
    Muro Group
    Paleolithic
    Quaternary
    Siwalik System
    Tulare Formation
    Uonuma Group
    Verde Formation
    Villafranchian

tertiary recovery
Term introduced in 1978 to be used in relation to petroleum.
SA petroleum
    recovery
    secondary recovery

Teruel
Province in NE central.
BT Spain
SA Calatayud-Teruel Basin
    Maestrazgo

teschenite
Includes use on level 3 under igneous rocks(1) alkali gabbro family(2). See list H.
BA alkali gabbro family
BT igneous rocks

tests
Includes use on level 3 under fossil group(1) for type of material. See list F. As of 1976, term is only to be used with fossils. For general procedures (i.e. under engineering geology) use experimental studies or procedures.
SA Invertebrata
    shells
    triaxial tests
    uniaxial tests

Tete
Town on the Zambezi River in NW.
BT Mozambique

Teterev River valley
NW Ukraine. Also search Teterev; Teterev River.
UF Teterev Valley
BT Ukraine
    USSR

Teterev Valley
use Teterev River valley

Teteven
Town in N part of country.
BT Pleven
    Bulgaria

Tethys
As of 1978, term is used on level 2 under continental drift(1). An elongated east-west sea, similar to the Mediterranean Sea, that separated Europe and Africa and extended across southern Asia in Pre-Tertiary time. Index continents as applicable.
UF Tethys Sea
SA Africa
    Asia
    continental drift
    Europe
    Mesogaea
    Paratethys

Tethys Satellite
Term introduced in 1978.
SA satellites
     Saturn

Tethys Sea
use Tethys

Teton County
Index states as applicable.
BX Idaho
     Montana
     Wyoming
BT United States

Teton National Forest
BT Wyoming
     United States

Tetracorallia
Includes use on level 3 under Coelenterata(1) Anthozoa(2). See list F. Also search tetracorals.
UF tetracorals
BA Anthozoa
     Coelenterata
BT Invertebrata
SA Rugosa
     Zoantharia

tetracorals
use Tetracorallia

tetradymite
BA tellurides
     sulfides
BT minerals

tetrahedra
BT polyhedra

tetrahedrite
BA sulfantimonites
     sulfosalts
BT minerals
SA freibergite

Tetrapoda
Also search tetrapods.
UF tetrapods
BT Vertebrata
NT Amphibia
     Aves
     Mammalia
     Reptilia

tetrapods
use Tetrapoda

Tetyukhe
Town in E Primorye Kray 220 miles NE of Vladivostok.
BT Russian Republic
     USSR

Teutoburg Forest
Range of hills S of Osnabruck in NW part of country. Index states as applicable.
UF Teutoburger Wald
BT West Germany
SA Lower Saxony
     North Rhine-Westphalia

Teutoburger Wald
use Teutoburg Forest

Texas
Includes use on level 1 as an area term (list O). For term set options see list B.
CO N254500N363000
     W0933000W1063000
BA United States
NX Anderson County
NT Ashmore
     Austin
     Azalea Field
     Baylor County
     Beaumont
     Big Bend National Park
     Brazos River
     Brenham
     Brewster County
     Burleson County
     Burnet County
NX Calhoun County
     Cass County
     Cherokee County
NT Christmas Mountains
NX Clay County
     Comanche County
NT Corpus Christi Bay
     Culberson County
     Dallas
NX Dallas County
NT Davis Mountains
     Del Rio
     Diablo Platform
     Duval County
     East Texas Field
     Ector County
NX Edwards County
NT Edwards Plateau
     El Paso
NX El Paso County
     Ellis County
     Fayette County
     Floyd County
NT Fort Worth Basin
NX Franklin County
NT Galveston
     Galveston Bay
     Galveston County
     Galveston Island
     Glass Mountains
     Guadalupe River
NX Hamilton County
     Hardeman County
     Hardin County
     Harris County
     Hidalgo County
NT Houston
     Hudspeth County
NX Jackson County
     Jasper County
     Jefferson County
     Johnson County
NT Karnes County
NX King County
     Lee County
     Limestone County
NT Llano
     Llano County
     Llano Uplift
     Lubbock
NX Madison County
NT Marathon Basin
NX Marion County
     Mason County
NT Matagorda Bay
     Maverick County
     Midland
     Midland Basin
NX Midland County
NT Milam County
NX Mitchell County
     Montgomery County
     Moore County
NT Mule Ear Diatreme
     Mustang Island
NX Newton County
NT Nueces River
NX Orange County
NT Padre Island
     Parker County
     Pecos County
NX Polk County
NT Presidio County
     Ranger
     Reeves County
     Rita Blanca Lake
     Runnels County
     San Antonio
     San Marcos Arch
     San Saba County
NX Sherman County
     Smith County
NT Sterling County
     Sutton County
     Tarrant County
NX Taylor County
NT Terlingua
     Travis County
NX Trinity County
NT Trinity River
     Upton County
     Val Verde Basin
     Val Verde County
     Van Horn
     Vinton Canyon
     Waco
NX Ward County
     Washington County
     Wheeler County
     Wichita County
     Williamson County
     Wilson County
NT Winkler County
NX Wise County
     Wood County
SA Anadarko Basin
     Antlers Sands
     Austin Group
     Basin and Range Province
     Bell Canyon Formation
     Blaine Formation
     Blancan
     Bossier Formation
     Buckner Formation
     Canadian River
     Canyon Group
     Capitan Formation
     Castile Formation
     Chihuahua tectonic belt
     Citronelle Formation
     Claiborne Group
     Cook Mountain Formation
     Cotton Valley Group
     Delaware Basin
     Dockum Group
     Edwards Formation
     Escondido Formation
     Fleming Formation
     Franklin Mountains
     Fredericksburg Group
     Frio Formation
     Glen Rose Formation
     Great Plains
     Guadalupe Mountains
     Gulf Coastal Plain
     Haymond Formation
     Jackson Group
     Leon
     Llano Estacado
     Loup Fork Group
     Marble Falls Group
     Midcontinent
     Norphlet Formation
     Ogallala Formation
     Paluxy Formation
     Panhandle
     Pecos River valley
     Queen City Formation
     Red River
     Red River valley
     Rio Grande
     Rio Grande Depression
     Rio Grande Rift
     Rio Grande Valley
     Sabine Lake
     Schuler Formation
     Smackover Formation
     Southwestern U.S.
     Strawn Series
     Trans-Pecos
     Trinity Group
     Uvalde Gravel
     Vicksburg Group
     Washita Group
     Washita River valley
     Wilberns Formation
     Wilcox Group
     Woodbine Formation
     Yates Formation

Texas County
Index states as applicable.
BX Missouri
     Oklahoma
BT United States

text, explanatory
use explanatory text

textbooks
Use under disciplines. Includes use as a level 2 or 3 term appropriate to a large number of topics, e.g. on level 2 under economic geology(1), geology(1), and geochemistry(1). See list G.
SA book reviews
     education
     glossaries
     manuals

Textulariina
Includes use on level 2 under foraminifera(1). See list F.
BA foraminifera
BT Invertebrata
NA Ammodiscacea
     Lituolacea

textures
Includes use as a level 2 or 3 term appropriate to a large number of topics, e.g. on level 2 under metamorphic rocks(1). See list G.
SA aphanitic texture
     arenaceous texture
     argillaceous texture
     framboidal texture
     gneissic texture
     graphic texture
     intraclasts
     lithic texture
     megacrysts
     oolitic texture
     orbicular texture
     packing
     pelitic texture
     pisolitic texture
     porphyritic texture
     porphyroblastic texture
     shape analysis
     size distribution
     spinifex

Th
use thorium

Th-228
Includes use on level 3 under isotopes(1).
SA isotopes
     thorium

Th-230
Includes use on level 3 under isotopes(1).
SA isotopes
     thorium

Th-230/Th-232
use Th-232/Th-230

Th-232
Includes use on level 3 under isotopes(1).
SA isotopes
     thorium

Th-232/Th-230
Includes use on level 3 under isotopes(1). Also search Th-230/Th-232; Th-230 AND Th-232.
UF Th-230/Th-232
SA isotopes
     thorium

Th-234
Includes use on level 3 under isotopes(1).
SA isotopes
     thorium

Th/Th
Includes use on level 3 under absolute age(1) methods(2).
SA absolute age
     Th/U
     thorium

Th/U
Includes use on level 3 under absolute age(1) methods(2).
- SA absolute age
  - Th/Th
  - thorium
  - U/Th/Pb
  - uranium

Thailand
Includes use on level 1 as an area term (list O). For term set options see list B.
- CO N054500N203000
  - E1060000E0963000
- UF Siam
- BA Asia
- NT Bangkok
  - Changwat Nakhon Phanom
  - Khorat Plateau
- SA Indochina
  - Khorat Group
  - Malay Peninsula
  - Phuket Group

Thalassinoides
- BA ichnofossils

thalenite
- BA orthosilicates
  - silicates
- BT minerals

thallium
Includes use on level 1 and 2 as a chemical element (list D).
- UF Tl
- SA elements

thallophytes
Includes use on level 1 and 2 as a fossil term (list F).
- BT Plantae
- SA bryophytes
  - lichens
  - Pteropsida

Thames Estuary
The broad tidal mouth of the Thames River in SE England. Also search Thames.
- BT England
  - Great Britain
  - United Kingdom

Thames River
Flows from the E slope of the Cotswold Hills E to the North Sea in SE England. Also search Thames.
- UF The Thames
- BT England
  - Great Britain
  - United Kingdom

thanatocenoses
Includes use on level 3 under fossil group(1) and under paleontology(1). See list F.
- UF death assemblages
  - taphocenoses
  - taphocoenosis
  - thanatocoenosis
- SA biocenoses
  - paleontology
  - taphonomy

thanatocoenosis
use thanatocenoses

Thanetian
Europe. Above Montian, below Ypresian (Eocene). Includes use on level 3 under age terms(1). See list E.
- BA upper Paleocene
  - Paleocene
  - Paleogene
  - Tertiary
- BT Cenozoic

Tharsis
As of 1978, term is restricted to morphologic features on Mars. Before 1978 term was also used for village in Spain.
- BT Mars

thaumasite
- BZ carbonates
  - orthosilicates
  - sulfates
- BT minerals

thawing
Includes use on level 3 under glacial geology(1) or permafrost(1).
- SA glacial geology
  - ice
  - permafrost
  - snow

Thaynes Formation
Consists of limestone, calcareous sandstone, sandstone, shale, and in the middle, red shale member. In SE Idaho, SW Montana, NE Utah, and SW Wyoming.
- BT Lower Triassic
  - Triassic
- SA Idaho
  - Montana
  - Utah
  - Wyoming

The Banks
use Outer Banks

The Coorong
use Coorong Lagoon

The Geysers
Eight miles NE of Geyserville in Sonoma County, W California.
- UF Geysers, The
- BT California
  - United States

The Great Oasis
use Kharga Oasis

The Himalaya
use Himalayas

The Marches
use Marches

The Rand
use Witwatersrand

The Spessart
use Spessart

The Thames
use Thames River

The Weald
Wooded region in Kent, Surrey, and Sussex counties of SE England. Also search Weald; Weald region.
- CO N504500N513000
  - E0013000W0010000
- UF Weald
  - Weald region
- BT England
  - Great Britain
  - United Kingdom

Thecamoeba
Includes use on level 2 under Protista(1). See list F.
- BA Protista

Thecideidae
Family. Includes use on level 3 under Brachiopoda(1) Articulata(2).
- BA Thecideidina
  - Articulata
  - Brachiopoda
- BT Invertebrata

Thecideidina
Order uncertain. Classed tentatively as a suborder equal to the Strophomenidina. Includes use on level 3 under Brachiopoda(1) Articulata(2).
- BA Articulata
  - Brachyopoda
- BT Invertebrata
- NA Thecideidae

Theiss River
use Tisza River

thenardite
- BA sulfates
- BT minerals

theoretical studies
Includes use as a level 2 or 3 term appropriate to a large number of topics, e.g. on level 2 under faults(1), folds(1), fractures(1), heat flow(1), and meteorology(1). See list G.
- UF studies, theoretical
- SA experimental studies
  - laboratory studies
  - undation

theory
A valid term through mid-1978. Use theoretical studies.

Thera
Volcanic island. The southernmost of the Cyclades in the Aegean Sea. Also search Santorin.
- UF Santorini
  - Thira
- BT Aegean Islands
  - Greece
- SA Cyclades
  - Santorin

Therapsida
Order. Includes use on level 3 under Reptilia(1) Synapsida(2).
- BA Synapsida
  - Reptilia
- BT Tetrapoda
  - Vertebrata
- NA Cynodontia

Theria
Subclass of Order Pantotheria.
- BA Mammalia
- BT Tetrapoda
  - Vertebrata
- NA Pantotheria
  - Symmetrodonta

thermal
A valid term through 1977. After 1977, use thermal metamorphism on level 2 under metamorphism(1), and use thermal circulation on level 2 under ocean circulation(1).

thermal alteration
Before 1978, also search thermal AND alteration.
- SA alteration
  - hydrothermal alteration

thermal analysis
Level 1 term introduced in 1978. Used for methodology and not for data. For data, use thermal analysis data. Before 1978, thermogravimetric analysis and differential thermal analysis were used as level 1 terms. Term set options are:
- differential thermal analysis
  - subtopic
- methods
  - subtopic
- techniques
  - subtopic
- thermogravimetric analysis
  - subtopic
- thermomagnetic analysis
  - subtopic
- SA analysis
  - chemical analysis
  - clay mineralogy
  - differential thermal analysis
  - DTA data
  - electron microscopy
  - endothermic reactions
  - sample preparation
  - spectroscopy
  - standard materials
  - thermogravimetric analysis
  - thermomagnetic analysis
  - X-ray analysis

thermal aureole
use aureoles

thermal circulation
As of 1978 term is used on level 2 under ocean circulation(1). Before 1978, also search oceans AND circulation AND thermal.
- SA circulation
  - ocean circulation

thermal conductivity
Includes use on level 2 under heat flow(1). Introduced as level 2 term in 1976. Before 1976, also search conductivity.
- SA conductivity
  - electrical conductivity
  - geothermal gradient
  - heat flow
  - measurement
  - specific heat
  - thermal diffusivity

thermal diffusivity
- UF diffusivity, thermal
- SA heat flow
  - thermal conductivity

thermal discharge
use thermal pollution

thermal emission
- UF emission, thermal
- SA infrared methods
  - remote sensing

thermal history
Includes use on level 3 under geosynclines(1) processes(2).
- SA geosynclines

thermal metamorphism
Not a valid index term through 1977. After 1977, includes use on level 2 under metamorphism(1). Before 1978, also search metamorphism AND thermal. Also search thermometamorphism.
- UF thermometamorphism
- BA metamorphism

thermal pollution
Term introduced in 1978 on level 2 under pollution(1). Before 1978, also search pollution AND thermal; thermal discharge.
- UF thermal discharge
- SA pollution
  - waste disposal

thermal properties
- SA engineering geology
  - heat capacity
  - permafrost
  - physical properties
  - properties
  - specific heat
  - temperature
  - thermodynamic properties

thermal waters
Includes use on level 1 (list A). Used for water of a spring or geyser, whose temperature is appreciably above the local mean annual air temperature. Term set options are:
area
  - locality, topic [fumaroles, geochemistry, genesis, geysers, temperature]
topic
  - subtopic (no area term)
- SA fractures
  - fumaroles
  - geophysical surveys
  - geothermal energy
  - geothermal gradient
  - geysers
  - ground water
  - heat flow

heat sources
hot springs
hydrogeology
hydrothermal alteration
migration of elements
mineral waters
permeability
springs
trace elements
wallrock alteration
water
water resources

thermochemical properties
SA chemical properties
geochemistry
properties

thermodynamic properties
Includes use on level 3 under geochemistry(1) properties(2) and under appropriate material name.
SA enthalpy
entropy
free energy
fugacity
geochemistry
minerals
properties
specific heat
thermal properties
thermodynamics

thermodynamics
Includes use on level 3.
SA enthalpy
entropy
fugacity
heat sources
stability
thermodynamic properties

thermography
use differential thermal analysis

thermogravimetric analysis
A valid level 1 term through 1977 used for methods. After 1977, use on level 2 under thermal analysis(1).
UF thermogravimetry
SA analysis
chemical analysis
differential thermal analysis
thermal analysis

thermogravimetry
use thermogravimetric analysis

thermohaline circulation
Term introduced in 1978 on level 2 under ocean circulation(1). Before 1978, also search oceans AND circulation AND thermohaline.
SA circulation
ocean circulation

thermokarst
Also search cryokarst.
UF cryokarst
SA ground ice
karst
permafrost

thermoluminescence
Includes use on level 2 under geochronology(1); on level 3 under material name or discipline.
SA fluorescence
geochronology
luminescence

thermomagnetic analysis
As of 1978, term is used on level 2 under thermal analysis(1).
SA analysis
thermal analysis

thermometamorphism
use thermal metamorphism

thermometry
A valid term through mid-1978. Use geologic thermometry.

thermonatrite
BA carbonates
BT minerals

thermoremanence
use thermoremanent magnetization

thermoremanent magnetization
Includes use on level 3 under paleomagnetism(1). Also search thermoremanence; TRM.
UF thermoremanence
TRM
BA remanent magnetization
SA Curie point
magnetic field
magnetization
paleomagnetism

Theropoda
Suborder. Includes use on level 3 under Reptilia(1) Archosauria(2).
BA Saurischia
Archosauria
Reptilia
BT Tetrapoda
Vertebrata

Thessaloniki
use Salonika

Thessaly
Administrative region in E central.
BT Greece
NT Olympus
SA Othrys
Pindus Mountains

Thetford Mines
City about 50 miles S of Quebec. Also search Thetford.
BT Quebec
Canada

thickness
Includes use on level 2 under crust(1).
SA crust

thin sections
Includes use on level 3 under electron microscopy(1).
SA electron microscopy
microfacies
microscope methods
petrography
polished sections
samples
sections

Thira
use Thera

tholeiite
Includes use on level 3 under igneous rocks(1) basalt family(2). See list H.
BA basalt family
BT igneous rocks
SA alkali basalt
olivine tholeiite
tholeiitic composition

tholeiitic basalt
See list H.
BA basalt family
BT igneous rocks
SA basalt

tholeiitic composition
Term introduced in 1978. Before 1978, search tholeiitic.
SA basalt
composition
tholeiite

tholeiitic dolerite
See list H.
BA basalt family
BT igneous rocks
SA dolerite

Thomar
use Tomar

thomsenolite
BA halides
BT minerals
SA fluorides

Thomson Slate
Has been referred to in literature as Thomson, St. Louis, Cloquet, and Carlton slates. Has usually been considered equivalent of Animikie-Virginia Slate of Mesabi Range. In NE Minnesota and NW Wisconsin. Also search Thomson Formation.
BT Precambrian
SA Minnesota
Wisconsin

thomsonite
BZ carbonates
framework silicates
BT minerals
SA zeolite group

thorianite
BA oxides
BT minerals

thorite
BA orthosilicates
silicates
BT minerals

thorium
Includes use on level 1 and 2 as a commodity (list C) and as a chemical element (list D). Also search Th.
UF Th
SA elements
Io/Th
isotopes
Pb/Th
Th-228
Th-230
Th-232
Th-232/Th-230
Th-234
Th/Th
Th/U
U/Th/Pb

Thorn
use Torun

thoron
use Rn-222

thortveitite
BA orthosilicates
silicates
BT minerals

Thrace
Administrative region in extreme NE Greece (Western Thrace), and all of Turkey in Europe (Eastern Thrace). Index countries as applicable.
BT Europe
NT Rhodope
SA Greece
Turkey

Three Forks
Town in Gallatin County in SW.
BT Montana
United States

three-dimensional models
Term introduced in 1978. Includes use on level 3 under seismology(1). Before 1978, search three-dimensional AND models.
BA models
SA one-dimensional models
seismology
two-dimensional models

thrust
A valid term through 1977. After 1977, use thrust faults.

Thrust Belt
use Overthrust Belt

thrust faults
Term introduced in 1978. Includes use on level 3 under faults(1) displacements(2). Before 1978, also search faults AND thrust.
UF reverse slip faults
thrust slip faults
BA faults
SA crustal shortening
overthrust faults
reverse faults
step faults

thrust slip faults
use thrust faults

Thule
Settlement on coast of Hayes Peninsula N of Cape York in NW. U.S. Air Force base nearby.
BT Greenland
Arctic region

thulium
Includes use on level 1 and 2 as a chemical element (list D).
UF Tm
SA elements
rare earths

Thunder Bay
Inlet of NW Lake Superior. Also a city on the bay. Search Thunder.
BT Ontario
Canada

thunderstorms
Includes use on level 3 under meteorology(1) electrical phenomena(2).
SA electrical phenomena
lightning
meteorology
storms

Thur Valley
River valley SSW of Lake Constance extending to the Rhine River. Also search Thur River.
BT Switzerland

Thuringer Wald
use Thuringian Forest

Thuringia
A former German state. Now a region around the Thuringian Forest in SE East Germany.
BT East Germany
SA Erfurt
Gera
Saxony-Thuringia

Thuringia Basin
use Thuringian Basin

Thuringian Basin
Thuringian Forest region in Thuringia. Index districts as applicable. Also search Thuringian.
UF Thuringia Basin
BT East Germany
SA Gera

Thuringian Forest
Forested mountain range extending NW-SE in Thuringia. Also search Thuringian.
UF Thuringer Wald
BT East Germany
SA Erfurt
Gera

Ti
use titanium

Tiber River
use Tiber Valley

Tiber Valley
River valley. Index autonomous regions as applicable. Also search Tiber; Tiber River.
UF Tiber River
BT Italy
SA Latium
Tuscany
Umbria

Tibesti
Habitation in S central.
BT Libya

Tibesti Massif
Highest mountain group of the Sahara in NW Chad. Also search Tibesti Mountains.
UF Tibesti Mountains
BT Chad

Tibesti Mountains
use Tibesti Massif

Tibet
Autonomous region in extreme SW.
BT China
SA Brahmaputra River
Himalayas
Indian Plate
Indus River
Kumaun Himalayas

Ticino
Canton in Lepontine Alps in S.
BT Switzerland

Ticonderoga
Village on N outlet of Lake George in Essex County in NE.
BT New York
United States

tidal channels
SA channels
tidal flats
tidal inlets

tidal environment
Term introduced in 1978. Includes use on level 3 under ecology(1) and sedimentation(1). Before 1978, also search tidal.
SA ecology
environment
sedimentation
subtidal environment
supratidal environment
tidal flats
tides

tidal flats
Includes use on level 3 under sedimentation(1).
UF tide flat
SA brackish-water environment
coastal environment
littoral environment
marine environment
mud flats
sedimentation
supratidal environment
tidal channels
tidal environment
tides

tidal inlets
Includes use on level 3 under shorelines(1).
UF tidal outlets
SA inlets
shorelines
tidal channels

tidal outlets
use tidal inlets

tidal wave
use tsunamis

tide flat
use tidal flats

tides
Includes use on level 2 under aeronomy(1) and under ocean circulation(1).
SA aeronomy
Earth tides
ocean circulation
tidal environment
tidal flats

Tieh-Chen Shan
use Tiehchenshan

Tiehchenshan
Hill region in NW Taiwan.
UF Tieh-Chen Shan
BT Taiwan

Tien Shan
Mountain system. Index Kirghizia and/or Sinkiang Uighur as applicable. Also search Tien-Shan.
CO N400000N420000
E0800000E0680000
UF Tien Shan Mountains
Tien Shan Range
Tien-Shan
Tien-Shan Range
BT Asia
SA Chatkal Range
Chu-Ili Mountains
Kirghizia
Kurama Range
Sinkiang Uighur
Talas Range
Tarbagatay Range
Zeravshan Range

Tien Shan Mountains
use Tien Shan

Tien Shan Range
use Tien Shan

Tien-Shan
use Tien Shan

Tien-Shan Range
use Tien Shan

Tierra Amarilla
Village in Rio Arriba County in N.
BT New Mexico
United States

Tierra del Fuego
Archipelago comprising all islands S of Strait of Magellan to Drake Passage. Also name of main island. Index countries as applicable.
BT South America
SA Argentina
Chile

Tiflis
use Tbilisi

Tigre
Province in N.
BT Ethiopia

till
Includes use on level 3 under sediments(1) clastic sediments(2); on level 3 under glacial geology(1) glacial features(2). See list N.
BA clastic sediments
BT sediments
SA diamicton
drift
drumlins
glacial features
glacial geology
glacial transport
moraines
terrigenous materials
tillite
tilloid

tillage
Includes use on level 3 under soils(1) utilization(2).
SA field crops
soil management
soils
utilization
yields

Tillamook County
NW Oregon. Also search Tillamook.
BT Oregon
United States

tillite
Includes use on level 3 under sedimentary rocks(1) clastic rocks(2). See list I.
BA clastic rocks
BT sedimentary rocks
SA glacial features
terrigenous materials
till

Tillodontia
Order. Includes use on level 2 under Mammalia(1).
BA Mammalia
BT Tetrapoda
Vertebrata

tilloid
Includes use on level 3 under sedimentary rocks(1) clastic rocks(2). See list I.
BA clastic rocks
BT sedimentary rocks
SA terrigenous materials
till

tilt
Includes use on level 3 under seismology(1).
SA Earth
geodesy
seismology
tiltmeters

tiltmeters
Includes use on level 3 under seismology(1).
SA earthquakes
instruments
seismology
tilt
volcanology

Timagami Mine
use Temagami Mine

Timan Mountains
use Timan Ridge

Timan Range
use Timan Ridge

Timan Ridge
An eroded range of ancient sedimentary rocks in N European USSR dividing the eastern Pechora Basin from the plains to the W. Also search Timan.
CO N620000N673000
E0550000E0473000
UF Timan Mountains
Timan Range
BT Russian Republic
USSR

Timan-Pechora Basin
use Timan-Pechora region

Timan-Pechora Province
use Timan-Pechora region

Timan-Pechora region
Includes the Timan Ridge area; and the Pechora Basin in Komi A.S.S.R., and Nenets National Okrug. In N and NE European USSR. Also search Timan-Pechora; Timan-Pechora Basin; Timan-Pechora Province.
UF Timan-Pechora Basin
Timan-Pechora Province
BT Russian Republic
USSR

time
Includes use on level 3 under absolute age(1) and geochronology(1). See age terms (list E). Also includes use on level 3 under soils(1) genesis(2) as type of factor.
SA absolute age
arrival time
chronology
factors
geochronology
soils
time scales
traveltime

time scales
Includes use on level 2 and 3 under geochronology(1). See list B and list E (age terms).
UF scales, time
SA geochronology
time

time variations
Includes use on level 3 under aurora(1) morphology(2).
SA aurora
variations

Timiskaming
Lake in SW Quebec and SE Ontario. Index provinces as applicable.
UF Temiscamingue
Temiskaming
BT Canada
SA Ontario
Quebec

Timmins
Mining town 135 miles N of Sudbury between Lake Huron and James Bay.
BT Ontario
Canada

Timna
Ancient site in Asaylan area in E.
BT Arabian Peninsula

Timok Basin
River and coal basin near the Bulgarian and Romanian borders in E Serbia. Also search Timok; Timok River; Timok River basin.
UF Timok River basin
BT Serbia
Yugoslavia

Timok River basin
use Timok Basin

Timor
Island in the Lesser Sundas separated from Australia by the Timor Sea. W half of island belongs to Indonesia. The E half belongs to Portugal although it was occupied by Indonesia in November 1975. Index countries as applicable. Before 1978, also search Portuguese Timor.
SA Indonesia
Lesser Sunda Islands
Portugal

Timor Sea
Between the island of Timor and NW Australia.
BT Indian Ocean
NT Bonaparte Gulf basin
Sahul Shelf
Timor Trough

Timor Trench
use Timor Trough

Timor Trough
Undersea feature off the S coast of Timor.
UF Timor Trench
BT Timor Sea
Indian Ocean

tin
Includes use on level 1 and 2 as a commodity (list C) and as a chemical element (list D); on level 2 under mineral deposits, genesis(1) and under placers(1).
UF Sn
SA elements
placers

tin pyrites
use stannite

Tinaquillo
Town in NW part of country.
BT Cojedes
Venezuela

Tindouf
Village and oasis in extreme W.

BT Algeria

Tindouf Basin
Salt flats in extreme W Algeria. Also search Tindouf.
UF Sebkra de Tindouf
BT Algeria

tinguaite
Includes use on level 3 under igneous rocks(1) trachyte-phonolite family(2). See list H.
BA trachyte-phonolite family
BT igneous rocks

Tintic District
Area around East Tintic Creek in Juab County in W central Utah. Also search Tintic AND Utah.
BT Utah
United States

Tintic Quartzite
Lower and middle Cambrian. Underlies Ophir Formation and unconformably overlies crystalline complex. Central and N Utah.
BT Cambrian
SA Lower Cambrian
Middle Cambrian
Utah

Tintina fault zone
Index Alaska and/or Yukon Territory as applicable.
SA Alaska
Yukon Territory

tintinids
use Tintinnidae

Tintinnidae
Family. Including Calpionellidae. Includes use on level 2 under Protista(1). See list F.
UF tintinids
BA Protista
NA Calpionella
Calpionellidae
Calpionellites
Stomiosphaera

Tioga Bentonite
According to latest published Lexique, Tioga Bentonite Bed is in Seneca Member of Onondaga Limestone. It is subsurface in Pennsylvania.
BT Middle Devonian
Devonian
SA New York
Ohio
Pennsylvania
Virginia
West Virginia

Tippecanoe County
W central.
BT Indiana
United States

Tipperary
County in S.
BT Ireland
NT Cashel

Tiraspol
City on the Dniester River in SE.
BT Moldavia
USSR

tirodite
BA amphibole group
chain silicates
silicates
BT minerals

Tirol
use Tyrol

Tiruchirapalli
City 200 miles SSW of Madras. Also search Trichinopoly.
UF Trichinopoly
BT Tamil Nadu
India

Tirupati
City in S part of country in central.
BT Tamil Nadu
India

Tisa River
use Tisza River

Tishomingo
Village and county in extreme NE.
BT Mississippi
United States

Tismana
Village in S foothills of Transylvanian Alps in NW.
BT Walachia
Romania

Tisza River
Rises in Carpathians in W Ukraine and flows W, SW and then S, into the Danube River N of Belgrade. Index countries as applicable. Also search Tisza.
UF Theiss River
Tisa River
BT Europe
SA Hungary
Romania
Ukraine

titanaugite
BA pyroxene group
chain silicates
silicates
BT minerals
SA augite

titanite
BA orthosilicates
silicates
BT minerals

titanium
Includes use on level 1 and 2 as a commodity term (list C) and as a chemical element (list D); on level 2 under (1). Also search Ti.
UF Ti
SA elements

titanoclinohumite
BA humite group
orthosilicates
silicates
BT minerals
SA clinohumite

titanomaghemite
BA oxides
BT minerals
SA maghemite

titanomagnetite
BA oxides
BT minerals
SA magnetic minerals
magnetite

Tithonian
Europe. Includes use on level 3 under age terms(1). See list E.
BA Upper Jurassic
Jurassic
BT Mesozoic

tjaele
use Frozen ground

Tkibuli
City in W.
UF Tkvibuli
BT Georgian Republic
USSR

Tkvibuli
use Tkibuli

Tl
use thallium

Tm
use thulium

Toarcian
Europe. Above Pliensbachian, below Bajocian. Includes use on level 3 under age terms(1). See list E.
BA Lower Jurassic
Jurassic
BT Mesozoic
SA Liassic

Toba Lake
In Barisan Mountains in N central Sumatra. Thought to occupy the crater of an extinct volcano. Also search Toba.
BT Sumatra
Indonesia

Tobacco Root Mountains
Range of Rocky Mountains in SW.
BT Montana
United States
SA Rocky Mountains

Tobago
Small island NE of Trinidad. A constituent part of Trinidad and Tobago.
BT Trinidad and Tobago
West Indies

tobermorite
BA chain silicates
silicates
BT minerals

Tocantins River region
N central and N part of country. Index states as applicable. Also search Tocantins.
BT Brazil
SA Goias
Maranhao
Para

Tochigi
Town near Utsonomiya in Tochigi Prefecture in E central.
BT Honshu
Japan

Todilto Formation
In San Rafael Group.
UF Todilto Limestone
BT Upper Jurassic
Jurassic
SA Arizona
Colorado
New Mexico

Todilto Limestone
use Todilto Formation

todorokite
BA oxides
BT minerals

Todos Santos Bay
Just S and SW of Ensenada. Also search Todos Santos.
BT Baja California
Mexico

Tofino Basin
SW Vancouver Island. Also search Tofino.
BT British Columbia
Canada

Tofua
Volcanic island. The largest of Haapai group in central.
BT Tonga
Pacific Ocean

Togo
Formerly French Togo. Includes use on level 1 as an area term (list O). For term set options see list B.
BA Africa
SA Volta Basin
West Africa

Tohoku
Region, comprising N Honshu, which represents a transitional zone between cold Hokkaido and temperate and subtropical areas farther S.
BT Honshu
Japan

Tokachi
Volcanic peak in central Hokkaido. Also a sub prefecture.
BT Hokkaido
Japan

Tokachi Plain
Central.
BT Hokkaido
Japan

Tokai
use Tokaj

Tokaj
Town in NE part of country. Also search Tokai.
UF Tokai
Tokay
BT Hungary

Tokaj Mountains
In NE Hungary. Hungarian section of Tokaj-Eperjes Mountains. Also search Tokaj.
BT Hungary
SA Tokaj-Eperjes Mountains

Tokaj-Eperjes Mountains
Outlier of the Carpathians. Index Hungary and/or Slovakia as applicable.
BT Europe
SA Carpathians
Hungary
Slovakia
Tokaj Mountains

Tokay
use Tokaj

Toki
City NE of Nagoya in central.
BT Honshu
Japan

Tokio
use Tokyo

Tokrau Synclinorium
N of Lake Balkhash. Also search Tokrau.
BT Kazakhstan
USSR

Tokushima
City on E coast.
BT Shikoku
Japan

Tokyo
City on Tokyo Bay in Tokyo Prefecture in SE Honshu.
UF Edo
Medo
Tokio
Yeddo
BT Honshu
Japan

Tokyo Imperial University
use University of Tokyo

Toledo
Province and city in central Spain; and a city in Lucas County, NW Ohio on the Maumee River. Index Ohio and/or Spain.
SA Montes de Toledo
Ohio
Spain

Tolfa Hills
ENE of Civitavechia in NW Latium. Also search Tolfa.
BT Latium
Italy

Toluca
Point of impact near Xiquipilco in Toluca area W of Mexico City. Also search Toluca Meteorite.
BT Mexico
SA meteorites

Tomales Bay
Inlet of Pacific Ocean on NW coast

of Marin County N of San Francisco.
BT   California
     United States

Tomar
City in W central part of country.
UF   Thomar
BT   Portugal

Tomaszow Lubelski
Town in SE part of country.
BT   Lublin
     Poland

Tomaszow Mazowiecki
Town SE of Lodz in central part of country.
BT   Lodz
     Poland

Tomaszow-Lvov Ridge
use Roztocze

Tombstone
City in Cochise County in SE.
BT   Arizona
     United States

Tomsk
City on right bank of Tom River near its junction with Ob River in S central Siberia. Also an oblast.
BT   Russian Republic
     USSR

tonalite
Includes use on level 3 under igneous rocks(1) diorite family(2). See list H.
BA   diorite family
BT   igneous rocks
SA   quartz diorite

tonalite gneiss
See list J.
BA   gneisses
BT   metamorphic rocks
SA   gneiss

Tonga
An archipelago of about 150 islands NE of New Zealand. Formerly a British protectorate which became independent in 1970. Also search Tonga Islands; Tonga island arc. Includes use on level 1 as an area term (list O). For term set options see list B.
UF   Friendly Islands
     Tonga Archipelago
     Tonga Islands
BA   Polynesia
NT   Ata Caldera
     Eua Island
     Tofua
     Tonga island arc
SA   Tonga Trench

Tonga Arc
use Tonga island arc

Tonga Archipelago
use Tonga

Tonga island arc
The islands of Tonga. Also search Tonga.
UF   Tonga Arc
BT   Tonga
     Pacific Ocean
SA   Tonga Trench

Tonga Islands
use Tonga

Tonga Trench
E of the Tongtapu group of the S Tonga Islands.
BT   Pacific Ocean
SA   Tonga
     Tonga island arc

Tonganoxie Sandstone
Of Stranger Formation. Underlies Westphalia Limestone Member. In E Kansas and NW Missouri. Also search Tonganoxie Sandstone Member.
UF   Tunganoxie Sandstone Member
BT   Virgilian
     Upper Pennsylvanian
     Pennsylvanian
SA   Kansas
     Missouri

Tongrian
Lattorfian. Europe. Above Ludian (Eocene), below Rupelian. Includes use on level 3 under age terms(1). See list E. Also search Lattorfian.
BA   upper Eocene
     Eocene
     Paleogene
     Tertiary
BT   Cenozoic
SA   Krosno Beds
     Lattorfian
     Sannoisian

Tongue of the Ocean
Strait between Andros Island on the W and New Providence Island and Exuma Cays on the E.
BT   Bahamas
     West Indies

Tongue River Formation
In Fort Union Group. In Montana and North Dakota, composed of light-yellow, tan, and gray sandstones and shales; thin lenses of limestone; and numerous beds of lignite. In E Montana, SW North Dakota, South Dakota and NW Wyoming.
BT   Paleocene
SA   Montana
     North Dakota
     South Dakota
     Wyoming

Tonnerre
Town in central part of country.
BT   Yonne
     France

Tono
Village in SW Washington, and village in N Honshu. Index Honshu and/or Washington as applicable.
SA   Honshu
     Washington

Tonoloway Limestone
In Cayuga Group. Underlies Keyser Limestone. Also search Tonoloway Formation.
BT   Upper Silurian
     Silurian
SA   Maryland
     Pennsylvania
     Virginia
     West Virginia

Tonopah
Village in Nye County in S central.
BT   Nevada
     United States

Tons Member
According to the India Lexique, the Tons Series is of Cambrian age. The Tons Series and Son Series are united to make up the Ken Subsystem, or lower main division of the Vindhyan System. In central India.
BT   Cambrian
SA   India

tonstein
Includes use on level 3 under sedimentary rocks(1) clastic rocks(2). See list I.
BA   clastic rocks
BT   sedimentary rocks
SA   terrigenous materials

Tonto Basin
Valley in N Gila County in central.
BT   Arizona
     United States

Tooele County
NW Utah.
BT   Utah
     United States

tool marks
Includes use on level 3 under sedimentary structures(1) bedding plane irregularities(2). See list K.
BA   bedding plane irregularities
     sedimentary structures
SA   current markings
     soft sediment deformation

topaz
BA   orthosilicates
     silicates
BT   minerals
SA   fluorides
     gems

Topeka
City in Shawnee County in NE.
BT   Kansas
     United States

tophus
use tufa

Topley Intrusions
Central.
BT   British Columbia
     Canada

topographic maps
Term introduced in 1978. Includes use on level 3 under maps(1). Before 1978, also search maps AND topographic.
BA   maps
SA   contour maps

topography
Includes use on level 3 under geodesy(1).
SA   cartography
     geodesy
     geomorphology
     hills
     ocean floors
     physiographic provinces
     plains
     plateaus
     soils
     terrains

Tor Formation
Overlies Antelope Valley Formation. In central Nevada. Also search Tor Limestone.
BT   Devonian
SA   Nevada

torbanite
See list I.
UF   bitumenite
BA   organic residues
BT   sedimentary rocks
SA   oil shale
     shale

Torino
use Turin

Torlesse Supergroup
BT   Upper Triassic
     Triassic
SA   New Zealand

Toronto
City at NW end of Lake Ontario.
CO   N434000N434000
     W0792800W0792800
BT   Ontario
     Canada

Toroweap Formation
In Aubrey Group. In Arizona can be subdivided into 3 main units (ascending): red to buff sandstone; calcareous sandstone and arenaceous limestone; and alternating red and buff sandstone, siltstone, and some shale. In NW Arizona, SE Nevada, and SW Utah.
BT   Permian
SA   Arizona
     Nevada
     Utah

Torquay
Resort city on English Channel in SW.
BT   England
     Great Britain
     United Kingdom

Torres Strait
Between the island of New Guinea and the N tip of Cape York Peninsula, Queensland. Connects the Arafura Sea with the Coral Sea. Also search Torres.
BT   Pacific Ocean

Torridonian
Europe.
BA   upper Precambrian
     Precambrian

tors
SA   geomorphology
     hills

torsion
Includes use on level 3 under deformation(1) field studies(2).
SA   deformation
     stress
     tension

torsion faults
use wrench faults

torsion modulus
use shear modulus

Tortonian
Europe. Above Helvetian, below Sarmatian. Includes use on level 3 under age terms(1). See list E.
BA   upper Miocene
     Miocene
     Neogene
     Tertiary
BT   Cenozoic
SA   Helvetian

tortuosity
Property of materials.
SA   geometry
     properties

Torun
City in N central part of country.
UF   Thorn
BT   Bydgoszcz
     Poland

Tosa Bay
Inlet of the Pacific Ocean on S coast. Also search Tosa.
BT   Shikoku
     Japan

tosudite
BA   sheet silicates
     silicates
BT   minerals
SA   clay minerals

total rock
use whole rock

Totes Gebirge
Mountain range in W central part of country. Index states as applicable.
BT   Austria
SA   Styria
     Upper Austria

Tottori
City on Japan Sea NW of Kyoto.
BT   Honshu
     Japan

Toulon
City on the Mediterranean Sea.

BT   Var
     France
Toulouse
City on the Garonne River in S part of country.
BT   Haute-Garonne
     France
Toumodi
Village NW of Abidjan in S central.
BT   Ivory Coast
Touraine
Historical region of NW central France. The area now comprises Indre-et-Loire Department and part of Indre Department. Index departments as applicable.
BT   France
SA   Indre
     Indre-et-Loire
tourmaline
BA   ring silicates
     silicates
BT   minerals
SA   dravite
     elbaite
     peridot
Tournai
City on Scheldt River near the French border.
UF   Doornik
     Tournay
BT   Hainaut
     Belgium
Tournaisian
Europe. Above Famennian of Devonian, below Visean. Includes use on level 3 under age terms(1). See list E.
BA   Lower Carboniferous
     Carboniferous
BT   Paleozoic
SA   Dinantian
Tournay
use Tournai
Tours
City on the Loire River in NW central part of country.
BT   Indre-et-Loire
     France
Towada
In Towada-Hachimantai National Park in N Honshu. Also search Towada Volcano.
UF   Towada Volcano
BT   Honshu
     Japan
Towada Volcano
use Towada
Towns County
On North Carolina border in N.
BT   Georgia
     United States
Townsville
City on Halifax Bay on the Coral Sea.
BT   Queensland
     Australia
toxic materials
Term introduced in 1978. Includes use on level 3 under environmental geology(1). Before 1978, also search toxic AND materials.
SA   environmental geology
     materials
     pollution
     toxicity
     waste disposal
toxicity
Includes use on level 3 under environmental geology(1) waste disposal(2).
SA   environmental geology
     toxic materials

waste disposal
Toyama
City on S shore of Toyama Bay on Japan Sea.
BT   Honshu
     Japan
Toyoma
Town SE of Taira on the Pacific Ocean in central.
BT   Honshu
     Japan
trace elements
Includes use on level 2 under organic materials(1); on level 3 under soils(1) nutrients(2). Also search trace-elements.
UF   guest element
     microelement
     trace-elements
SA   elements
     major elements
     minor elements
     nutrients
     thermal waters
     trace metals
     trace-element analyses
trace fossils
A valid term through 1972. After 1972, use ichnofossils.
trace metals
Includes use on level 3 under commodities (list C).
SA   environmental geology
     geochemistry
     metals
     pollution
     trace elements
trace-element analyses
For data, see appropriate material. For methods, includes use on level 3 under chemical analysis(1) methods(2) and under spectroscopy(1).
SA   analysis
     chemical analysis
     elements
     major-element analyses
     minor-element analyses
     spectroscopy
     trace elements
trace-elements
use trace elements
tracer experiments
Includes use on level 2 and 3 under isotopes(1).
UF   experiments, tracer
SA   D/H
     ground water
     isotopes
     radioactive tracers
     radioactivity
     tracers
tracers
Includes use on level 2 under isotopes(1).
SA   D/H
     isotopes
     radioactive tracers
     radioactivity
     tracer experiments
trachyandesite
Includes use on level 3 under igneous rocks(1) andesite-rhyolite family(2). See list H.
BA   andesite-rhyolite family
BT   igneous rocks
trachybasalt
Includes use on level 3 under igneous rocks(1) alkali basalt family(2). See list H.
BA   alkali basalt family
BT   igneous rocks
SA   latite
     trachydolerite

trachydolerite
Includes use on level 3 under igneous rocks(1) alkali basalt family(2). See list H.
BA   alkali basalt family
BT   igneous rocks
SA   trachybasalt
Trachyleberididae
Family. Includes use on level 3 under Ostracoda(1) Podocopida(2). Also search Trachyleberidinae.
UF   Trachyleberidinae
BA   Cytheracea
     Podocopida
     Ostracoda
BT   Crustacea
     Arthropoda
     Invertebrata
Trachyleberidinae
use Trachyleberididae
trachyte
Includes use on level 3 under igneous rocks(1) trachyte-phonolite family(2). See list H.
BA   trachyte-phonolite family
BT   igneous rocks
SA   volcanic rocks
trachyte-phonolite family
Includes use on level 2 under igneous rocks(1). See list H.
BT   igneous rocks
NA   albitite
     keratophyre
     lamproite
     phonolite
     quartz keratophyre
     tinguaite
     trachyte
track, particle
use particle track
tracks
Includes use on level 3 under sedimentary structures(1) biogenic structures(2). See list K. Before 1977, also search tracks and trails.
BA   biogenic structures
     sedimentary structures
SA   fission tracks
     ichnofossils
     lebensspuren
     trails
tracks and trails
Not a valid index term for GeoRef. Use tracks or trails as separate terms. Before 1977, also search tracks and trails.
trails
Includes use on level 3 under sedimentary structures(1) biogenic structures(2). See list K. Before 1977, also search tracks and trails.
BA   biogenic structures
     sedimentary structures
SA   ichnofossils
     tracks
Tranquillity Base
A valid term through 1977. Use Taurus-Littrow.
Trans-Alaska Pipeline
Pipeline constructed for the movement of crude oil from the North Slope at Prudhoe Bay on the Arctic Ocean in the NE to the port of Valdez off Prince William Sound in the SE.
BT   Alaska
     United States
Trans-Aravalli Vindhyan Basin
Index states as applicable.
BT   India
SA   Madhya Pradesh
     Rajasthan

Trans-Pecos
Region W of the Pecos and E of the Rio Grande rivers in W Texas and S central New Mexico. Index states as applicable.
CO   N290000N320000
     W1010000W1063000
BT   United States
SA   New Mexico
     Texas
Transantarctic Mountains
A series of mountain ranges extending across the continent from the Filchner Ice Shelf on the Atlantic Ocean side through Victoria Land on the Pacific Ocean side separating East Antarctica from West Antarctica.
BT   Antarctica
Transbaikal
use Transbaikalia
Transbaikalia
Region of SE Siberia E of Lake Baikal including Buryat A.S.S.R.; and Chita and Amur oblasts. Also search Transbaikal.
CO   N490000N573000
     E1250000E1023000
UF   Transbaikal
BT   Russian Republic
     USSR
Transcarpathia
Region in extreme W which was, prior to World War II, that part of Czechoslovakia known as Ruthenia or the Carpatho-Ukraine. Now coextensive with the Transcarpathian Oblast.
CO   N460000N490000
     E0294500E0223000
BT   Ukraine
     USSR
Transcaucasia
Region S of the Caucasus. Index Soviet republics as applicable.
BT   USSR
SA   Armenia
     Azerbaidzhan
     Georgian Republic
     Northern Caucasus
transcurrent faults
Term introduced in 1978. Includes use on level 3 under faults(1) displacements(2). Before 1978, also search faults AND transcurrent.
BA   faults
SA   displacements
     strike-slip faults
     transverse faults
Transdanubia
Fertile, hilly region between the Danube River and the Austrian border. Also search Dunantul.
UF   Dunantul
BT   Hungary
Transdanubian Central Mountains
use Bakony Mountains
transfer, heat
use heat transfer
transfer, mass
use mass transfer
transform
A valid term through 1977. After 1977, use transform faults.
transform faults
Term introduced in 1978. Includes use on level 3 under faults(1) displacements(2). Before 1978, also search faults AND transform.
BA   faults
SA   displacements
     strike-slip faults

transformations
As of 1978, term is used on level 2 under ocean waves(1). Includes use on level 3 under phase equilibria(1), minerals(1), crystal chemistry(1), or soils(1). Before 1978, transformation was used under ocean waves. Also search transformation.
- SA changes
    - clay mineralogy
    - crystal chemistry
    - minerals
    - ocean waves
    - phase equilibria
    - soils

transformations, Laplace
use Laplace transformations

transgression
Term is used in stratigraphy for Pre-Quaternary sedimentation. For Quaternary, use changes of level.
- SA changes of level
    - paleogeography
    - regression
    - stratigraphy

transition
A valid general term through 1977.

transition zones
Includes use on level 3 under plate tectonics(1). Also search transition zone.
- UF zones, transition
- SA core
    - mantle
    - plate tectonics

translation lattice
use lattice

transmissibility coefficient
Term introduced in 1978. Also search coefficient of transmissibility.
- UF coefficient of transmissibility
- SA aquifers
    - transmissivity
    - water

transmission
A valid term through 1978. After 1978, use transmission method, e.g. under electron microscopy(1).

transmission electromicroscopy data
use TEM data

transmission method
Term introduced in 1978. Includes use on level 3 under electron microscopy(1). Before 1978 search transmission.
- UF method, transmission
- SA electron microscopy
    - SEM data
    - TEM data

transmissivity
Restricted to usage in relation to hydrology. Not to be used for remote sensing.
- SA aquifers
    - ground water
    - hydrogeology
    - hydrology
    - permeability
    - transmissibility coefficient

transport
Includes use on level 2 under sedimentation(1).
- SA competence
    - degradation
    - flow regime
    - glacial transport
    - ice rafting
    - marine transport
    - ore transport
    - saltation
    - sedimentation
    - stream transport
    - suspension
    - transportation
    - wind transport
    - winds

transportation
As of 1978, term is restricted to non-geological meaning. Includes use on level 3 under pipelines(1).
- SA pipelines
    - transport

Transvaal
Province in NE.
- CO S273000S223000 E0320000E0250000
- BT South Africa
- NT Barberton
    - Barberton Mountain Land
    - Bushveld Complex
    - Johannesburg
    - Pretoria
    - Ventersdorp
    - Witwatersrand
- SA Fig Tree Series
    - Lebombo Mountains
    - Limpopo Basin
    - Vaal River
    - Waterberg System
    - Witwatersrand System
    - Wolkberg Group

Transvaal Supergroup
- BT Precambrian
- SA South Africa

transverse faults
Term introduced in 1978. Includes use on level 3 under faults(1) orientation(2). Before 1978, also search faults AND transverse.
- BA faults
- SA orientation
    - transcurrent faults

transverse folds
use superposed folds

Transverse Ranges
In S California.
- CO N333000N350000 W1153000W1193000
- BT California
- SA San Bernardino Mountains
    - San Gabriel Mountains

Transylvania
Region bounded by the Carpathians on the E and the Transylvanian Alps on the S in NW and central part of country.
- BT Romania
- NT Alba-Iulia
    - Almas Valley basin
    - Aries Valley
    - Baia Mare
    - Baita
    - Baraolt
    - Baraolt Basin
    - Beius Basin
    - Bihor
    - Bihor Mountains
    - Bistrita
    - Brad
    - Brasov
    - Calimani Mountains
    - Cavnic
    - Chiuzbaia
    - Cluj
    - Covasna
    - Crai
    - Crisana-Maramures
    - Ditrau
    - Gilau
    - Gurghiu
    - Gurghiu Mountains
    - Harghita
    - Harghita Mountains
    - Hateg
    - Herja
    - Hoghiz
    - Maramures
    - Muntii Metalici
    - Mures
    - Padurea Craiului Mountains
    - Persani Mountains
    - Petrosani Basin
    - Poiana-Rusca Mountains
    - Retezat Mountains
    - Rodna
    - Rodna Mountains
    - Rosia
    - Salaj
    - Satu Mare
    - Sebes
    - Sebes Mountains
    - Sibiu
    - Simleu Basin
    - Transylvanian Basin
    - Uzon
    - Vladeasa Mountain
- SA Fagaras Mountains
    - Getic Nappe
    - Jiu River valley
    - Olt River
    - Olt River valley
    - Transylvanian Alps

Transylvania Basin
use Transylvanian Basin

Transylvanian Alps
Continuation of the Carpathians extending E-W in central part of country. Also search South Carpathians an Southern Carpathians. Index region as applicable.
- UF South Carpathians
    - Southern Carpathians
- BT Romania
- SA Bucegi Mountains
    - Carpathians
    - Fagaras Mountains
    - Retezat Mountains
    - Transylvania
    - Vulcan Mountains
    - Walachia

Transylvanian Basin
Primarily a plateau. Also search Transylvania Basin; Transylvanian Depression; Transylvanian Plain.
- UF Transylvania Basin
    - Transylvanian Depression
    - Transylvanian Plain
- BT Transylvania
    - Romania

Transylvanian Depression
use Transylvanian Basin

Transylvanian Plain
use Transylvanian Basin

trap rock
Includes use as level 3 commodity term under construction materials(1). See list C.
- BA construction materials
- SA basalt
    - diabase
    - igneous rocks
    - trap rocks

trap rocks
Includes use on level 3 under igneous rocks(1). See list H.
- BA basalt family
- BT igneous rocks
- SA basalt
    - construction materials
    - diabase
    - trap rock
    - traps
    - volcanic rocks

Trapani
Province in W.
- BT Sicily
    - Italy

trapped particles
Includes use on level 2 under magnetosphere(1); on level 3 under magnetosphere(1) magnetic storms(2).
- SA ions
    - magnetic field
    - magnetic storms
    - magnetosphere
    - particles

traps
As of 1978, term is restricted to use in relation to petroleum and natural gas, e.g. on level 3 under oil and gas fields(1).
- NA stratigraphic traps
    - structural traps
- SA natural gas
    - oil and gas fields
    - petroleum
    - trap rocks

Tras-os-Montes
Former province in NE Portugal. Now it comprises Braganca and Vila Real districts. Index districts as applicable.
- BT Portugal
- SA Braganca
    - Vila Real

travel time
use traveltime

traveltime
Includes use on level 3 under seismology(1). Also search travel time.
- UF travel time
- SA elastic waves
    - seismology
    - time
    - traveltime curves

traveltime curves
Also search hodograph(s).
- UF curves, traveltime
    - hodographs
- SA elastic waves
    - seismology
    - traveltime

Traverse Group
Middle and upper Devonian. Comprises (ascending) Belle Shale, Rockport Quarry Limestone, Ferron Point Shale, Genshaw Formation, Koehler Limestone, Gravel Point Formation, and Beebe School Formation. In S Michigan.
- BT Devonian
- SA Michigan
    - Middle Devonian
    - Upper Devonian

travertine
Includes use on level 3 under sedimentary rocks(1) chemically precipitated rocks(2) and carbonate rocks(2); on level 3 under sediments(1) chemically precipitated sediments(2) or carbonate sediments(2). See list I (sed. rocks) and list N (sediments).
- UF calc-sinter
    - calcareous sinter
- BA carbonate rocks
- BT sedimentary rocks
- SA carbonate sediments
    - chemically precipitated rocks
    - chemically precipitated sediments
    - limestone
    - tufa

Travis County
Central.
- BT Texas
    - United States

treatment
Includes use on level 3 under soils(1) salinity(2).
- SA desalinization
    - fertilization

irrigation
reclamation
salinity
soil management
soils
utilization

tree rings
Includes use on level 2 under geochronology(1).
UF annual growth rings
rings, tree
SA geochronology

Trego County
NW central.
BT Kansas
United States

Tremadoc Bay
Inlet of Saint Georges Channel between Caernarvon and Merioneth counties in NW Wales. Also search Tremadoc.
BT Wales
Great Britain
United Kingdom

Tremadocian
Europe. Above Dolgellian (Cambrian), below Arenigian. Includes use on level 3 under age terms(1). See list E.
BA Lower Ordovician
Ordovician
BT Paleozoic

tremolite
BA amphibole group
chain silicates
silicates
BT minerals

Tremp
Town in Catalonia in NE part of country.
BT Lerida
Spain

Trempealeauan
North America. Upper Cambrian, above Franconian, below Lower Ordovician.
BA Upper Cambrian
Cambrian
BT Paleozoic

trenches
Includes use on level 2 under ocean floors(1); on level 3 under plate tectonics(1).
UF marginal trench
oceanic trench
sea-floor trench
SA Benioff zone
bottom features
island arcs
ocean floors
plate tectonics
subduction
subduction zones

trend surface analysis
use trend-surface analysis

trend-surface analysis
Includes use on level 3 under mathematical geology(1). Also search trend surface analysis.
UF trend surface analysis
BA statistical methods
SA analysis
automatic data processing
mathematical geology

Trent
use Trento

Trent Valley
River valley in central England, SE North Carolina, and SE Ontario. Index England, North Carolina, or Ontario as applicable. Also search Trent.
SA England

North Carolina
Ontario

Trentino
S part of Trentino-Alto Adige autonomous region.
BT Trentino-Alto Adige
Italy

Trentino-Alto Adige
Autonomous region in NE Italy. Formerly Venezia-Tridentina.
UF Venezia-Tridentina
BT Italy
NT Bolzano
Cima d'Asta
Latemar Massif
Merano
Predazzo
Trentino
Trento
Valsugana
SA Dolomites
Marmolada Glacier
Venetia

Trento
City in S Trentino-Alto Adige. Also search Trent.
UF Trent
BT Trentino-Alto Adige
Italy

Trenton
City on the Delaware River in Mercer County in W.
BT New Jersey
United States

Trenton Group
Standard section of group comprises (ascending) Rockland Limestone, Hull Formation, Sherman Fall Formation, Cobourg Formation, Collingwood Shale, and Gloucester Shale. In Georgia, Michigan, N Ohio, New York, Vermont, and W Virginia. Also search Trenton Limestone.
BT Middle Ordovician
Ordovician
SA Georgia
Michigan
New York
Ohio
Pennsylvania
Trentonian
Vermont
Virginia

Trentonian
North America. Upper Mohawkian, above Blackriverian. Includes use on level 3 under age terms(1). See list E.
BA Middle Ordovician
Ordovician
BT Paleozoic
SA Champlainian
Trenton Group

Trepca Mine
In the Kosovo region of S Serbia. Also search Trepca.
UF Trepcha Mine
BT Serbia
Yugoslavia

Trepcha Mine
use Trepca Mine

Trepostomata
Includes use on level 2 under Bryozoa(1). See list F.
BA Bryozoa
BT Invertebrata

trevorite
BA oxides
BT minerals

Trialet Mountains
use Trialet Range

Trialet Range
N range of the Lesser Caucasus in central Georgian Republic, USSR. Also search Trialet.
UF Trialet Mountains
BT Georgian Republic
USSR
SA Lesser Caucasus

triangulation
Includes use on level 3 under geodesy(1) or neotectonics(1).
SA cartography
geodesy
neotectonics
networks
surveys

Triassic
Includes use on level 1 as an age term (list E); on level 2 under paleo- terms, e.g. paleoecology, paleogeography, paleomagnetism. Above Permian (of Paleozoic), below Triassic.
BA Mesozoic
NA Andalusian
NT Chanares Formation
Fremouw Formation
Hallstatt Limestone
Hawkesbury Sandstone
Ipswich Coal Measures
NA Lower Triassic
Middle Triassic
NT Moenkopi Formation
Narrabeen Group
Nicola Group
Red Peak Formation
NA Upper Triassic
NT Wetterstein Limestone
SA Baskunchak Series
Beacon Supergroup
Botucatu Formation
Estrada Nova Formation
Gondwana System
Hercynian Orogeny
Karroo System
Khorat Group
Korenevskaya Formation
New Red Sandstone
Novorayskoe Formation
Pucara Group
Sadlerochit Formation
Shimanto Group
Singleton Coal Measures
Slovakian Karst
Spearfish Formation

triaxial tests
SA deformation
rock mechanics
soil mechanics
tests
uniaxial tests

Trichinopoly
use Tiruchirapalli

Tridacna
Genus. Includes use on level 3 under Mollusca(1) Bivalvia(2).
BA Bivalvia
Mollusca
BT Invertebrata

tridymite
BA silica minerals
framework silicates
silicates
BT minerals
SA cristobalite

Trieste
City at head of the Adriatic Sea on Gulf of Trieste.
BT Friuli-Venezia Giulia
Italy

Trigonia
Genus. Includes use on level 3 under Mollusca(1) Bivalvia(2).
BA Trigoniidae

Bivalvia
Mollusca
BT Invertebrata

Trigoniidae
Family. Includes use on level 3 under Mollusca(1) Bivalvia(2).
BA Bivalvia
Mollusca
BT Invertebrata
NA Trigonia

Trilobita
Class. Includes use on level 1 and 2 as a fossil term (list F).
BT Trilobitomorpha
Arthropoda
Invertebrata
NA Agnostida
Corynexochida
Cryptolithus
Dechenellidae
Lichida
Odontopleurida
Phacopida
Proetidae
Ptychopariida
Redlichiida
SA Crustacea
Cruziana
Rusophycus

Trilobitoidea
Class. Includes use on level 3 under Arthropoda(1) Trilobitomorpha(2). See list F.
BA Trilobitomorpha
Arthropoda
BT Invertebrata

Trilobitomorpha
Superclass. Exclude Trilobita but include Trilobitoidea on level 3. Includes use on level 2 under Arthropoda(1). See list F.
BA Arthropoda
BT Invertebrata
NT Trilobita
NA Trilobitoidea

Trinidad
Island off NE coast of Venezuela in the Lesser Antilles. Major part of independent state of Trinidad and Tobago. Included use on level 1 as an area term until 1977. Now index under Trinidad and Tobago(1).
BT Trinidad and Tobago
West Indies
SA Lesser Antilles

Trinidad and Tobago
Comprises the islands of Trinidad and Tobago, Atlantic Ocean, off NE coast of Venezuela. As of 1977, includes use as a level 1 area term (list O). For term set options see list B.
BA West Indies
NT Northern Range
Tobago
Trinidad

Trinity County
Index states as applicable.
BX California
Texas
BT United States

Trinity Group
Lower Cretaceous. Subdivided into (ascending) Pearsall, Rodessa, Ferry Lake, and Rusk time-stratigraphic units. In SW Arkansas, NW Louisiana, central S and SE Oklahoma, and Texas.
BT Comanchean
Cretaceous
SA Antlers Sands
Arkansas
Glen Rose Formation
Louisiana

Trinity Group • Tucson Mountains

Oklahoma
Paluxy Formation
Texas

Trinity River
Flows into Trinity Bay off Galveston Bay on the Gulf of Mexico.
BT Texas
United States

triphylite
BA phosphates
BT minerals

triple junctions
Includes use on level 3 under plate tectonics(1) geometry(2) or structure(2).
SA plate tectonics

triplite
BZ phosphates
halides
BT minerals
SA fluorides

Tripoli
City on the Mediterranean Sea in NW.
BT Libya

Tripolis
City in central.
BT Peloponnesus
Greece

Tripolitania
NW part of country which was a province until 1963. Under the Italians, it was a major province covering the entire W part of colony.
BT Libya
NT Garian

tripolite
use diatomaceous earth

Tripura
State E of Bangladesh in NE.
BT India
SA Bengal

Tristan da Cunha
A group of 5 British volcanic islands in the central South Atlantic Ocean. Also the name of the main island of the group.
UF Tristan da Cunha Islands
BT Atlantic Ocean

Tristan da Cunha Islands
use Tristan da Cunha

Triticites
Genus. Includes use on level 3 under foraminifera(1) Fusulinidae(2).
BA Fusulinidae
Fusulinina
foraminifera
BT Invertebrata
NA Triticites ventricosus

Triticites ventricosus
Includes use on level 3 under foraminifera(1) Fusulinidae(2).
BA Triticites
Fusulinidae
Fusulinina
foraminifera
BT Invertebrata

tritium
Includes use on level 1 and 2. See list D (chemical elements). Also search H-3.
UF H-3
SA elements
hydrogen
isotopes

TRM
use thermoremanent magnetization

troctolite
Includes use on level 3 under igneous rocks(1) gabbro family(2). See list H.
BA gabbro family
BT igneous rocks

Troia
use Troy

troilite
BA sulfides
BT minerals
SA meteorites
pyrrhotite

Troja
use Troy

Trompia Valley
Valley of upper Mella River in E.
BT Lombardy
Italy

Troms
County in far N.
BT Norway
NT Lyngen Peninsula
SA Vesteralen

trona
Includes use as level 3 commodity term under sodium carbonate(1). See list C.
BA carbonates
BT minerals
SA sodium carbonate

Trona Village
Village in San Bernardino County in Mojave Desert in S. Before 1977, Trona was used to indicate the village.
BT California
United States

Trondelag
Region in central part of country between Norwegian Sea and Swedish border around Trondheim Fjord. Index counties as applicable.
BT Norway
SA Nord-Trondelag
Sor-Trondelag

Trondheim
City on Norwegian Sea.
BT Sor-Trondelag
Norway

trondhjemite
Includes use on level 3 under igneous rocks(1) diorite family(2). See list H.
BA diorite family
BT igneous rocks
SA plagiogranite

Troodos Massif
Mountainous mass SW of Nicosia which includes Mt. Troodos, the highest peak on the island.
UF Troodos Mountain
Troodos Mountains
BT Cyprus
Middle East

Troodos Mountain
use Troodos Massif

Troodos Mountains
use Troodos Massif

tropical
A valid term through 1977. After 1977, use tropical environment.

tropical environment
Term introduced in 1978. Before 1978, search tropical.
SA environment
subtropical environment

troposphere
Includes use on level 3 under aeronomy(1) composition(2).
SA aeronomy
atmosphere
stratosphere

Trotus Valley
River valley SSW of Bacau in W Moldavia. Also search Trotus; Trotus River.
BT Moldavia
Romania

Trou Sans Fond
Submarine canyon off SW Nigeria in the Bight of Benin.
UF Le Trou Sans Fond
Trou-Sans-Fond
BT Atlantic Ocean
SA Ivory Coast

Trou-Sans-Fond
use Trou Sans Fond

troughs
Includes use on level 2 under ocean floors(1).
SA aulacogens
bottom features
channels
ocean floors

Trousdale County
N Tennessee.
BT Tennessee
United States

Troy
Ancient ruined city. An archeological site, named Hissarlik, on the Menderes River S of the Dardanelles in NW Anatolia.
UF Ilion
Ilium
Troia
Troja
BT Turkey

Trucial Coast
A 350 mile section of S Persian Gulf littoral extending from Qatar to Ruus al Jubal. Often formerly used interchangeably with Trucial Oman and Trucial States. A valid term through 1974. After 1974, use United Arab Emirates.

Trucial Oman
Former name of 7 British protected states on the Trucial Coast now known as the United Arab Emirates. A valid term through 1972. After 1972, use United Arab Emirates.

Truckee River
Rises in Lake Tahoe in E California and flows NE and E into Pyramid Lake in NW Nevada. Index states as applicable. Also search Truckee.
BT United States
SA California
Nevada

Trujillo
State in NW.
BT Venezuela
SA Maracaibo Basin

Tsaritsyn
use Volgograd

tschermakite
BA amphibole group
chain silicates
silicates
BT minerals

Tschernosem
use Chernozems

Tschernosiom
use Chernozems

Tsugaru Strait
Channel between islands of Honshu and Hokkaido connecting the Pacific Ocean with the Japan Sea. Index islands as applicable. Also search Tsugaru.
BT Japan
SA Hokkaido
Honshu

Tsumeb
Town in N central.
CO S190000S180000
E0180000E0170000
BT South-West Africa

tsunamis
Includes use on level 2 under seismology(1).
UF earthquake sea wave
seismic sea waves
seismic surge
tidal wave
SA catastrophic waves
earthquakes
geologic hazards
ocean waves
seismology
waves

Tsushima
Channel between Tsushima Island and NW Kyushu connecting the Japan Sea with the East China Sea.
BT Kyushu
Japan

Tuamotu Islands
Group of about 80 small islands in French Polynesia E of Society Islands and S of Marquesas Islands. Also search Tuamotu.
UF Dangerous Islands
Low Archipelago
BT Polynesia
Pacific Ocean
SA Mururoa Atoll

Tuapse
City on the Black Sea in S Krasnodar Kray in the Northern Caucasus.
BT Russian Republic
USSR

Tuar-Kyr
Village near Kara-Bogaz-Gol Gulf in W Turkmenia.
UF Tuarkyr
BT Turkmenia
USSR

Tuarkyr
use Tuar-Kyr

Tubarao Group
BT Carboniferous
Paleozoic
SA Brazil
Sao Paulo

tubes, lava
use lava tubes

Tuboidea
Includes use on level 2 under Graptolithina(1). See list F.
BA Graptolithina
BT Invertebrata
SA Hemichordata

Tubuai Islands
use Austral Islands

Tucano Basin
NE Bahia. Also search Tucano.
BT Bahia
Brazil

Tucson
City in Pima County in S.
CO N322000N322000
W1110000W1110000
BT Arizona
United States

Tucson Basin
S Arizona.
BT Arizona
United States

Tucson Mountains
W of Tucson in Pima County.
BT Arizona
United States

Tucuman
Province in NW.
BT Argentina

Tucumcari
City in Quay County in NE.
BT New Mexico
    United States

tufa
Includes use on level 3 under sedimentary rocks(1) chemically precipitated rocks(2) and carbonate rocks(2); on level 3 under sediments(1) chemically precipitated sediments(2) and carbonate sediments(2). See list I (sed. rocks) and list N (sediments). Also search calcareous tufa.
UF calc-tufa
    calcareous tufa
    petrified moss
    tophus
    tuft
BA chemically precipitated rocks
BT sedimentary rocks
SA calcite
    carbonate rocks
    carbonate sediments
    chemically precipitated sediments
    travertine

tuff
Includes use on level 3 under igneous rocks(1) pyroclastics and glasses(2). See list H.
BA pyroclastics and glasses
BT igneous rocks
SA andesite tuff
    ash
    ash flows
    ash-flow tuff
    bentonite
    green tuff
    ignimbrite
    metatuff
    rhyolite tuff
    tuffite
    volcanic ash
    welded tuff

tuff lava
use welded tuff

tuffite
BA pyroclastics and glasses
BT igneous rocks
SA tuff

tuffsite
use pipes

tuft
use tufa

Tugela Basin
River basin in central Natal. Also search Tugela and Tugela River.
BT Natal
    South Africa

tugtupite
BA sodalite group
    framework silicates
    silicates
BT minerals

Tuimazy
use Tuymazy

Tuktoyaktuk Peninsula
Extends into the Beaufort Sea E of the Mackenzie River delta in District in Mackenzie. Also search Tuktoyaktuk.
CO N683000N703000
    W1293000W1350000
BT Northwest Territories
    Canada

Tula
City S of Moscow. Also an oblast.
BT Russian Republic
    USSR

Tulameen coal area
S British Columbia.
BT British Columbia
    Canada

Tulare County
S central.
BT California
    United States

Tulare Formation
Pliocene and Pleistocene (?). Consists primarily of sandstone and conglomerate. In southern California.
BT Cenozoic
SA California
    Pleistocene
    Pliocene
    Quaternary
    Tertiary

Tulcea
City in the Danube Delta area in N.
BT Dobruja
    Romania

Tulear Basin
Before 1978, also search Tulear.
BT Malagasy Republic

Tulghes Series
BT Paleozoic
SA Eastern Carpathians
    Romania

Tully Limestone
In Susquehanna Group. Divided into several members in different areas: Apulia, Laurens, New Lisbon, Tinkers Falls, and West Brook.
BT Middle Devonian
    Devonian
SA New York
    Pennsylvania

Tulsa
City in Tulsa County in NE central.
BT Oklahoma
    United States

Tulsa County
NE central.
CO N355000N362500
    W0954500W0961500
BT Oklahoma
    United States

tundra
SA ecology
    geomorphology
    permafrost
    plains
    soils
    taiga environment

Tundra soils
Includes use on level 3 under soils(1). See list M.
BA soils
SA permafrost
    Zonal soils

Tunganoxie Sandstone Member
use Tonganoxie Sandstone

tungstates
Includes use on level 2 under minerals(1). See list L.
BT minerals
NA ferberite
    scheelite
    stolzite
    wolframite
SA molybdates

tungsten
Includes use on level 1 as a commodity term (list C) and as a chemical element (list D); on level 2 under mineral deposits, genesis(1) and paragenesis(1).
UF W

tungstenite
BA sulfides
BT minerals

Tunguska
In GeoRef indexing, this term is used to indicate a general area. Name of three rivers flowing into the Yenisei River in central Siberia: the Lower Tunguska, Stony Tunguska, and the lower course of the Angara river which was called the Upper Tunguska. Index rivers as applicable.
BT Russian Republic
    USSR
SA Angara River
    Lower Tunguska River
    Stony Tunguska River
    Tunguska River

Tunguska Basin
Large coal basin in central Siberia between the Yenisei and Lena rivers in an area drained by the Lower Angara (Upper Tunguska), Stony Tunguska, and Lower Tunguska rivers. Also search Tunguska.
BT Russian Republic
    USSR

Tunguska River
BT Russian Republic
    USSR
SA Angara River
    Lower Tunguska River
    Stony Tunguska River
    Tunguska

Tunguska Series
Composed of 2 sections (ascending): productive section with coal beds and a section of tufa containing intercalary beds of normal sedimentary rock. In Tunguska Basin of central Siberia.
BT upper Paleozoic
    Paleozoic
SA Russian Republic
    USSR

Tunguska Syneclise
The Tunguska Basin area in central Siberia. Also search Tunguska.
BT Russian Republic
    USSR

Tunis
City on the Gulf of Tunis in N.
BT Tunisia

Tunisia
Includes use on level 1 as an area term (list O). For term set options see list B.
CO N303000N373000
    E0120000E0073000
BA Africa
NT Sebkha el Melah
    Souk-el-Arba salt works
    Tunis
    Zarzis
SA Atlas Mountains
    Mediterranean region
    North Africa
    Sahara

tunnels
A valid level 1 term as of 1978. Used for geological studies on manmade tunnels. Before 1978, included use on level 2 under engineering geology(1). Term set options are:
construction
    subtopic
design
    subtopic
excavations
    subtopic
experimental studies
    subtopic
feasibility studies
    subtopic
instruments
    subtopic
rock mechanics
    subtopic
seepage
    subtopic
site exploration
    subtopic
soil mechanics
    subtopic
stability
    subtopic
subways
    subtopic
theoretical studies
    subtopic
SA construction
    design
    engineering geology
    excavations
    feasibility studies
    foundations
    geologic hazards
    highways
    land subsidence
    lava tunnels
    Poisson's ratio
    rock mechanics
    seepage
    site exploration
    slope stability
    soil mechanics
    stability
    submarine installations
    subways
    underground installations

Tunsbergdalsbreen
Glacier emanating from larger Jostedalsbreen Glacier in W part of country.
BT Norway

Tuolumne County
E central.
BT California
    United States

Tura Formation
According to USSR Lexique, the age of Tura Series is upper Silurian. Central Urals and West Siberian Plain. Also search Tura.
BT Upper Silurian
    Silurian
SA Russian Republic

Turan
Desert lowland N, S, and SE of the Aral Sea. Index Soviet republics as applicable.
BT USSR
SA Kazakhstan
    Turkmenia
    Uzbekistan

Turanian Platform
Extends from the Caspian Sea on the W to the Tien Shan in the E. Index Soviet republics as applicable.
BT USSR
SA Kazakhstan
    Turkmenia
    Uzbekistan

turbidite
Includes use on level 3 under sedimentary rocks(1) clastic rocks(2). See list I. Also search turbidites.
BA clastic rocks
BT sedimentary rocks
SA nonterrigenous materials
    terrigenous materials
    turbidity currents

turbidity
Includes use as level 3 under meteorology(1) aerosols(2).
SA aerosols
    meteorology

turbidity current structures
Includes use on level 2 under sedimentary structures(1). See list K.
BA sedimentary structures

NA graded bedding
         load casts
         sole marks
   SA convoluted beds
         flow structures
         flute casts
         olistoliths
         olistostromes
         structures
         turbidity currents

turbidity currents
Includes use on level 3 under sedimentation(1) transport(2). For structures, see under sedimentary structures (list K).
   UF suspension current
   BT currents
   SA density currents
         sedimentation
         submarine canyons
         submarine fans
         turbidite
         turbidity current structures

turbulence
Includes use on level 2 under aeronomy(1), meteorology(1) and under ocean circulation(1).
   SA aeronomy
         boundary layer
         meteorology
         ocean circulation
         oceanography
         waterways

Turgai
   use Turgay

Turgai Basin
   use Turgay Basin

Turgai Depression
   use Turgay Basin

Turgai Downwarp
   use Turgay Basin

Turgai Gates
   use Turgay Basin

Turgay
Oblast (region) NE of Aral Sea. Also a village. Search Turgai.
   UF Turgai
   BT Kazakhstan
         USSR

Turgay Basin
An elongated depression joining the Turan Lowland and West Siberian Plain. Also search Turgai, Turgai Basin, Turgai Downwarp, and Turgay.
   UF Turgai Basin
         Turgai Depression
         Turgai Downwarp
         Turgai Gates
         Turgay Depression
   BT Kazakhstan
         USSR

Turgay Depression
   use Turgay Basin

Turin
City in W central Piedmont. Also search Torino.
   UF Torino
   BT Italy
   SA Piedmont

Turkana District
Administrative district W of Lake Turkana (Rudolf) in NW.
   BT Kenya

Turkestan
Region including S Kazakhstan and the rest of Soviet Central Asia, plus a small section of NE Afghanistan and W and SW Sinkiang Uighur. Index Afghanistan, Sinkiang Uighur, and Soviet Central Asia as applicable.
   UF Turkistan

   BT Asia
   SA Afghanistan
         Sinkiang Uighur

Turkestan Range
W Tadzhikistan.
   BT Tadzhikistan
         USSR

Turkey
Includes both Anatolia in Asia and Turkey in Europe. Includes use on level 1 as an area term (list O). For term set options see list B.
   CO N355000N420000
         E0444500E0260000
   BA Middle East
   NT Adana
         Alanya
         Amanos Mountains
         Amasra Basin
         Anatolia
         Ankara
         Antalya
         Ararat
         Bosporus
         Elbistan
         Eskisehir
         Hatay
         Istanbul
         Keban Mine
         Kocaeli
         Konya
         Lycian Taurus
         Malatya
         Mudurnu
         Pontic Mountains
         Taurus Mountains
         Troy
         Tuz Golu
         Zonguldak
   SA Balkan Peninsula
         Black Sea region
         Euphrates River
         Istranca Mountains
         Kura River
         Levant
         Mardin Formation
         Maritsa River
         Mediterranean region
         Mediterranean Sea
         Near East
         Thrace

Turkistan
   use Turkestan

Turkmenia
Turkmen Soviet Socialist Republic. In Soviet Central Asia.
   CO N350000N430000
         E0660000E0520000
   UF Turkmenistan
   BT USSR
   NT Ashkhabad
         Badkhyz
         Balkhan
         Chardzhov
         Cheleken
         Cheleken Peninsula
         Darvaza
         Gaurdak
         Kara-Bogaz Gulf
         Kara-Bogaz-Gol
         Karakum
         Kizyl-Dere
         Kugitang-Tau
         Kurt
         Tuar-Kyr
   SA Amu Darya
         Aral region
         Asiatic USSR
         Caspian Basin
         Central Asia
         Kopet-Dag Range
         Murgab Basin
         Turan
         Turanian Platform

Turkmenistan
   use Turkmenia

Turku
Port city in SW part of country.
   UF Abo
   BT Finland

Turnyauz
   use Tyrny-Auz

Turonian
Europe. Above Cenomanian, below Coniacian. Includes use on level 3 under age terms(1). See list E.
   BA Upper Cretaceous
         Cretaceous
   BT Mesozoic
   SA Chalk
         Middle Cretaceous

turquoise
   BA phosphates
   BT minerals

Turritella
Genus. Includes use on level 3 under Mollusca(1) Gastropoda(2).
   BA Turritellidae
         Gastropoda
         Mollusca
   BT Invertebrata

Turritellidae
Family. Includes use on level 3 under Mollusca(1) Gastropoda(2).
   BA Gastropoda
         Mollusca
   BT Invertebrata
   NA Turritella

Turukhan
River which flows into Yenisei River just N of Turukhansk in N Krasnoyarsk Kray in central Siberia.
   BT Russian Republic
         USSR

Turukhansk
Town on Yenisei River at the mouth of the Lower Tunguska River in N Krasnoyarsk Kray in central Siberia.
   BT Russian Republic
         USSR

Tusas Mountains
   BT New Mexico
         United States

Tuscaloosa County
W central.
   BT Alabama
         United States

Tuscaloosa Formation
Irregularly or obscurely bedded quartzitic and miracaceous sands interbedded with heterogeneous clays. Also lenticular pebble beds. Coastal plain from W Tennessee, NE Mississippi, and NW Alabama across Alabama, Georgia, South Carolina, and North Carolina.
   BT Upper Cretaceous
         Cretaceous
   SA Alabama
         Georgia
         Mississippi
         North Carolina
         South Carolina
         Tennessee

Tuscany
Autonomous region in W.
   CO N422000N442000
         E0122500E0094000
   BT Italy
   NT Apuane Alps
         Arno River Basin
         Elba
         Florence
         Grosseto
         Larderello
         Livorno
         Lucca
         Monte Amiata
         Monte Capanne
         Monte Pisano
         Pisa
         Siena
   SA Tiber Valley

Tuscarawas County
E central.
   BT Ohio
         United States

Tuscarora Formation
In central Pennsylvania, it is gradational with overlying Rose Hill Formation and transitional with underlying Juniata Formation. In W Maryland, central S and E Pennsylvania, W Virginia, and E West Virginia. Also search Tuscarora Quartzite.
   UF Tuscarora Quartzite
   BT Lower Silurian
         Silurian
   SA Maryland
         Pennsylvania
         Virginia
         West Virginia

Tuscarora Quartzite
   use Tuscarora Formation

Tuttle Creek Dam
On Tuttle Creek forming reservoir in NE Kansas. Also search Tuttle Creek.
   BT Kansas
         United States

Tuva
Tuva Autonomous Soviet Socialist Republic. On the Mongolian border between the Altai Mountains on the W and the Sayan Mountains on the NE.
   CO N493000N530000
         E0990000E0880000
   UF Tannu Tuva
   BT Russian Republic
         USSR

Tuymazy
Town W of Ufa in W Bashkiria in SE European USSR. Also search Tuimazy.
   UF Tuimazy
   BT Russian Republic
         USSR

Tuz Golu
Salt lake in W central Anatolia.
   UF Tuz Lake
   BT Turkey
         Middle East

Tuz Lake
   use Tuz Golu

Twente
Region in SE Overijssel. Also search De Twente.
   UF De Twente
   BT Overijssel
         Netherlands

Twiggs Clay
Consists of pale green hackly fuller's earth clay, green hackly clay, gray marl, and calcareous sand at base of member. In E Georgia.
   UF Twiggs Clay Member
   BT upper Eocene
         Eocene
   SA Georgia

Twiggs Clay Member
   use Twiggs Clay

Twiggs County
Central.
   BT Georgia
         United States

Twilight Gneiss
According to latest published Lexique, the age of Twilight Granite is

Precambrian. IN SE Colorado.
BT Precambrian
SA Colorado

Twin Lakes
N California and central Colorado. Index states as applicable.
BT United States
SA California
  Colorado

Twin River Formation
Upper Eocene to upper Oligocene. Three mappable sequences recognized: lower member consisting of thin-bedded sandstone and siltstone; middle member of massive siltstone that grades westward into bedded siltstone and sandstone; and upper member composed chiefly of massive mudstone. In NW Washington.
BT Tertiary
SA Eocene
  Oligocene
  upper Eocene
  upper Oligocene
  Washington

Twin Sisters Dunite
An ultrabasic intrusive. In NW Washington.
BT Tertiary
SA Washington

twinning
As of 1978, term is used on level 2 under crystal growth(1).
SA crystal growth
  epitaxy
  lattice

Two Creeks
Village in SW.
BT Manitoba
  Canada

Two Harbors
City on Lake Superior in Lake County in NE.
BT Minnesota
  United States

two-dimensional models
Term introduced in 1978. Includes use on level 3 under seismology(1). Before 1978, search two-dimensional AND models.
BA models
SA one-dimensional models
  seismology
  three-dimensional models

Tyee Formation
Overlies Siletz River Volcanic Series. It consists of thick series of rhythmically bedded sandstone and intercalated siltstone. In W Oregon.
BT middle Eocene
  Eocene
SA Oregon

Tynagh
Village in S County Galway in W.
BT Ireland

Tyne River
Flows by Newcastle and enters the North Sea at Tynemouth in NE.
BT England
  Great Britain
  United Kingdom

type sections
SA sections
  stratigraphy
  stratotypes

type specimens
Used as a general term.
UF specimens, type
SA neotypes

types
A valid general term through 1977.

typomorphism
SA fossils
  minerals

Tyrny-Auz
Town NW of Ordzhonikidze in the Northern Caucasus. Also search Turnyauz.
UF Turnyauz
BT Russian Republic
  USSR

Tyrol
State between Germany and Italy in W.
UF Tirol
BT Austria
NT Baumkirchen
  Innsbruck
  Lanersbach
  Lechtal Alps
  Stubai Alps
  Vernagt Glacier
SA Allgau Alps
  Bavarian Alps
  Isar Valley
  Silvretta Group
  Venediger Group
  Wetterstein Limestone

Tyrone
County in W.
BT Northern Ireland
  United Kingdom

Tyrrhenian
Europe. Upper Pleistocene, Quaternary.
BA upper Pleistocene
  Pleistocene
  Quaternary
BT Cenozoic

Tyrrhenian Sea
Between Corsica and Sardinia on the W, the mainland of Italy on the E, and Sicily on the S. Also search Tyrrhenian. Includes use on level 1 as an area term (list O). For term set options see list B.
CO N380000N430000
  E0160000E0093000
BA Mediterranean Sea

Tyumen
City E of Sverdlovsk in W Siberia. Also an oblast.
BT Russian Republic
  USSR

U

U
use uranium

U. S. Geological Survey
Includes use on level 3 under surveys(1). Also search United States Geological Survey; USGS; U.S.G.S.; U. S. Geological Survey.
UF United States Geological Survey
  U.S. Geological Survey
  U.S.G.S.
  USGS
SA surveys

U-stage
use universal stage

U-234
Includes use on level 3 under isotopes(1).
SA isotopes
  U-238/U-234
  uranium

U-234/U-238
use U-238/U-234

U-235
Includes use on level 3 under isotopes(1).
SA isotopes
  uranium

U-238
Includes use on level 3 under isotopes(1).
SA isotopes
  U-238/U-234
  uranium

U-238/Pb-206
Includes use on level 3 under absolute age(1) methods(2).
SA absolute age
  lead
  methods
  U/He
  U/Pb
  U/Th/Pb
  uranium

U-238/U-234
Includes use on level 3 under isotopes(1). Also search U-234/U-238; U-238 AND U-234.
UF U-234/U-238
SA isotopes
  U-234
  U-238
  uranium

U/He
Includes use on level 3 under absolute age(1) methods(2).
SA absolute age
  helium
  methods
  U-238/Pb-206
  uranium

U/Pb
Includes use on level 3 under absolute age(1) methods(2).
SA absolute age
  lead
  methods
  Pb-210
  Pb/Pb
  Pb/Th
  U-238/Pb-206
  uranium

U/Th/Pb
Includes use on level 3 under absolute age(1) methods(2).
SA absolute age
  lead
  methods
  Th/U
  thorium
  U-238/Pb-206
  uranium

Uaua
Village in NE.
BT Bahia
  Brazil

Ube coal field
E of Shimonoseki in SW Honshu.
UF Ube Coalfield
BT Honshu
  Japan

Ube Coalfield
use Ube coal field

Uchaly
Village in E Bashkiria in the Southern Urals in SE European USSR.
BT Russian Republic
  USSR

Uchi Lake
Village on Uchi Lake in NW Ontario. Also search Uchi.
BT Ontario
  Canada

Uchur River basin
Central W Khabarovsk Kray and SE Yakutia in E Siberia. Also search Uchur; Uchur River.
BT Russian Republic
  USSR

Uda River
Rises in Irkutsk Oblast and flows into Angara River just above its junction with the Yenisei River in S central Siberia. Also search Uda.
UF Chuma River
BT Russian Republic
  USSR

Udachnaya
Village just S of the Arctic Circle in W central Yakutia.
BT Russian Republic
  USSR

Udaipur
City in S.
BT Rajasthan
  India

Udmurt
use Udmurtia

Udmurtia
Udmurt Autonomous Soviet Sovialist Republic. W of the Urals and NE of Kazan. Also search Udmurt.
UF Udmurt
BT Russian Republic
  USSR

Udokan Mountains
In area where Amur Oblast, Chita Oblast and Yakutia meet NE of Lake Baikal. Also search Udokan.
BT Russian Republic
  USSR

Udokan Series
Lower Proterozoic. An aggregate of metamorphised sedimentary rocks including schists, phyllites, sandstones, and crystalline limestone. In NW Amur Oblast, N Chite Oblast, and S Yakutia NE of Lake Baikal.
BT Proterozoic
  Precambrian
SA Russian Republic
  USSR

Ufa
City in Bashkiria W of the Southern Urals.
BT Russian Republic
  USSR

Ufaley
Town in NW Chelyabinsk Oblast in the Central Urals of W Siberia.
UF Nizhniy Ufaley
BT Russian Republic
  USSR

Uganda
Former British protectorate which became independent in 1962. Includes use on level 1 as an area term (list O). For term set options see list B.
CO S020000N040000
  E0350000E0290000
BA Africa
NT Bukusu
  Kakuto
  Moroto Mountain
SA East African Rift
  Lake Albert
  Lake Edward
  Lake Victoria
  Nile River

Uinta Basin
NW Colorado and NE Utah. Index states as applicable.
UF Uintah Basin
BT United States
SA Colorado

Utah
Uinta County
 Extreme SW.
 BT Wyoming
  United States
Uinta Mountains
 Range chiefly in NE Utah. Index states as applicable.
 BT United States
 SA Rocky Mountains
  Utah
  Wyoming
Uintah Basin
 use Uinta Basin
Uintah County
 NE Utah.
 BT Utah
  United States
Ukraine
 Ukrainian Soviet Socialist Republic.
 CO N443000N523000
  E0400000E0220000
 BT USSR
 NT Alma River
  Artemovsk
  Bakhchisarai
  Bar
  Belozerka
  Beregovo
  Bitkov
  Boltyshka Depression
  Borislav
  Borkut Deposit
  Borshchev
  Cherkassy
  Chernigov
  Crimea
  Crimean Mountains
  Crimean Plain
  Dnepropetrovsk
  Dolina
  Glinsk
  Gorlovka Basin
  Il'Intsa
  Ingul River
  Ingulets River
  Ivano-Frankovsk
  Kanev
  Kerch
  Kerch Peninsula
  Kharkov
  Kiev
  Kirovograd
  Konstantinovka
  Korosten
  Korsun
  Kosov
  Kremenchug
  Krivoy Rog
  Krivoy Rog Basin
  Lebedin
  Lvov
  Lvov Basin
  Lvov-Volyn Basin
  Nikitovka
  Nikopol
  Odessa
  Pavlograd
  Pervomaisk
  Podolia
  Pokutye
  Poltava
  Rakhov
  Rakhov Massif
  Ros River valley
  Rozdol
  Rudki
  Saksagan River
  Simferopol
  Sivash
  Sob River basin
  Stebnik
  Tarkhankut Peninsula
  Tereblya
  Teterev River valley
  Transcarpathia
  Ukrainian Carpathians
  Ukrainian Shield
  Verkhovtsevo
  Volhynia
  Volyn
  Volyn-Podolia
  Yalta
  Zaporozhe
  Zhdanov
  Zhitomir
 SA Azov region
  Beshchady Mountains
  Brest Basin
  Bug region
  Bug River
  Bukovina
  Carpathian Foredeep
  Carpathian Foreland
  Carpathians
  Danube Delta
  Danube River
  Danube Valley
  Dnieper Basin
  Dnieper River
  Dnieper-Donets Basin
  Dniester River
  Dniester-Prut Interfluve
  Donets Basin
  Eastern Carpathians
  European USSR
  Galicia
  Kiev Member
  Korenevskaya Formation
  Krivoy Rog Series
  Maikop Series
  Novorayskoe Formation
  Ovruch Series
  Polesye
  Poltava Series
  Pripet Basin
  Protopivskaya Formation
  Prut River
  Roztocze
  Russian Plain
  Russian Platform
  Scythian Platform
  Serebryanka Formation
  Siret River
  Steppes
  Subcarpathians
  Tisza River
  Vodino
Ukrainian Carpathians
 E extension of the Carpathians in SW.
 BT Ukraine
  USSR
 SA Carpathians
Ukrainian Shield
 Extends from the S Polesye in the NW to the Azov Sea.
 CO N450000N510000
  E0390000E0280000
 UF Volyno-Azov Massif
 BT Ukraine
  USSR
Ulan Bator
 City in N central Mongolia.
 UF Urga
 BT Mongolia
Ulatisian
 North America. Above Penutian, below Narizian. Includes use on level 3 under age terms(1). See list E.
 BA middle Eocene
  Eocene
  Paleogene
  Tertiary
 BT Cenozoic
ulexite
 UF boronatrocalcite
  natroborocalcite
 BA borates
 BT minerals
Ulianovsk
 use Ulyanovsk
Ulkan
 Village NW of N Lake Baikal in Irkutsk Oblast.
 BT Russian Republic
  USSR
ullmannite
 UF nickel-antimony glance
 BA sulfides
 BT minerals
 SA willyamite
Ulmaceae
 BA Dicotyledoneae
  angiosperms
 BT Plantae
Ulsan
 Town in S part of country.
 BT South Korea
Ulster
 use Northern Ireland
Ulster County
 S central.
 BT New York
  United States
Ultisols
 Includes use on level 3 under soils(1). See list M.
 BA soils
ultrabasic composition
 Term introduced in 1978. Before 1978, search ultrabasic.
 SA composition
  igneous rocks
  ultramafic composition
ultrabasic rocks
 Not a valid term for GeoRef. Use ultramafic family or ultramafic composition.
ultrabasite
 Also search diaphorite.
 UF diaphorite
 BA sulfides
 BT minerals
ultramafic
 A valid term through 1977. After 1977, use ultramafic composition.
ultramafic composition
 Term introduced in 1978. Before 1978, search ultramafic.
 SA composition
  igneous rocks
  mafic composition
  ultrabasic composition
ultramafic family
 Includes use on level 2 under igneous rocks(1). See list H.
 BT igneous rocks
 NA ariegite
  augitite
  bronzitite
  chromitite
  clinopyroxenite
  diallagite
  dunite
  eulysite
  garnet lherzolite
  garnet peridotite
  garnet pyroxenite
  griquaite
  harzburgite
  hornblendite
  kimberlite
  komatiite
  lherzolite
  limburgite
  meimechite
  olivinite
  ophiolite
  peridotite
  pyroxenite
  spinel lherzolite
  websterite
  wehrlite
 SA ankaramite
  leucitite
  melilitite
  nephelinite
  picrite
ultramafic rocks
 A valid term through 1976. Use ultramafic family.
ultrametamorphism
 Includes use on level 2 under metamorphism(1).
 BA metamorphism
ultrastructure
 Includes use on level 3 under fossil groups (list F). Before 1978, also search microstructure.
 SA paleontology
  petrography
ultraviolet spectra
 Term introduced in 1978. Before 1978, also search ultraviolet.
 UF ultraviolet spectrum
 SA spectra
  ultraviolet spectroscopy
ultraviolet spectroscopy
 Not a valid term from 1975 through 1977. After 1977, includes use on level 3 under spectroscopy(1) and under chemical element(1). Before 1978, also search spectroscopy AND ultraviolet.
 BT spectroscopy
 SA analysis
  ultraviolet spectra
ultraviolet spectrum
 use ultraviolet spectra
Ulu-Tau
 Range W of the Kazakh Hills in Dzhezkazgan Oblast in central Kazakhstan. Also search Ulutau.
 UF Ulutau
 BT Kazakhstan
  USSR
Ulutau
 use Ulu-Tau
ulvospinel
 BA oxides
 BT minerals
 SA spinel group
Ulyanovsk
 City on the right bank of the Volga River S of Kazan in Ulyanovsk Oblast.
 UF Simbirsk
  Ulianovsk
 BT Russian Republic
  USSR
Umanak
 Settlement on S shore of Umanak Fjord midway up W coast.
 BT Greenland
  Arctic region
umangite
 BA selenides
  sulfides
 BT minerals
Umaria
 Town in NE.
 BT Madhya Pradesh
  India
Umba
 River in Kenya which flows into Indian Ocean just N of Tanzanian boundary. Index countries as applicable.
 BT Africa
 SA Kenya
  Tanzania
Umbria
 Autonomous region in central.

CO N422000N434000
E0131500E0115000
BT Italy
NT Perugia
Spoleto
Terni
SA Tiber Valley

Umm al Qaiwain
use Umm al-Qaiwain

Umm al-Qaiwain
Emirate. One of federation of 7 states at S end of Persian Gulf.
UF Umm al Qaiwain
BT United Arab Emirates
Arabian Peninsula

umohoite
BA molybdates
BT minerals

Umpqua Formation
Lower to middle Eocene. Predominantly medium-grained sandstone with some shaly and conglomerate layers. In N California and SW Oregon.
BT Eocene
SA California
lower Eocene
middle Eocene
Oregon

Umrer
Town in central.
BT Madhya Pradesh
India

unconfined aquifers
BT aquifers
SA ground water

unconformities
Includes use on level 3 under age terms(1). See list E. Also search unconformity.
SA stratigraphy

unconsolidated materials
Term introduced in 1978. Before 1978, search unconsolidated.
SA materials
sediments
soil mechanics
soils

Unda
Village in S Chita Oblast in Transbaikalia.
BT Russian Republic
USSR

undation
SA crust
tectonics
theoretical studies

underclay
Also search seat earth.
UF coal clay
root clay
seat clay
seat earth
underearth
SA clay
coal seams
fireclay

underearth
use underclay

underground
A valid term through 1977. After 1977, use underground space.

underground installations
A valid level 1 term as of 1978. Used for geological studies on underground cavities (natural or otherwise), excluding tunnels. Before 1978, included use on level 2 under engineering geology(1). Term set options are:
construction
subtopic

design
subtopic
excavations
subtopic
experimental studies
subtopic
feasibility studies
subtopic
instruments
subtopic
mines
subtopic
rock mechanics
subtopic
seepage
subtopic
site exploration
subtopic
soil mechanics
subtopic
stability
subtopic
theoretical studies
subtopic
underground space
subtopic
waste disposal
subtopic
UF installations, underground
SA construction
design
Earth tides
engineering geology
excavations
feasibility studies
foundations
gas storage
geologic hazards
land subsidence
mines
permafrost
rock mechanics
seepage
site exploration
slope stability
soil mechanics
stability
storage
tunnels
underground space
utilization
waste disposal

underground space
Term introduced in 1978 on level 2 under land use(1) and underground installations(1).
UF space, underground
SA land use
underground installations

underthrust faults
Term introduced in 1978. Before 1978, search underthrust AND faults.
BA faults

Ungava
Region including Quebec N of Eastmain River and SW Labrador. Also search New Quebec. Index Labrador and/or Quebec as applicable.
CO N580000N623000
W0690000W0780000
UF New Quebec
BT Canada
SA Labrador
Quebec

Ungulata
Includes use on level 2 under Mammalia(1). Also search ungulates.
UF ungulates
BA Mammalia
BT Tetrapoda
Vertebrata
SA Artiodactyla
Bison latifrons
Bison occidentalis
Condylarthra

Equidae
Perissodactyla

ungulates
use Ungulata

uniaxial tests
Term introduced in 1978. Before 1978, search uniaxial.
SA engineering geology
tests
triaxial tests

uniformitarianism
Includes use on level 3 under geology(1).
SA catastrophism
deposition
geology
history
sedimentation
stratigraphy

Union County
Index states as applicable.
BX Arkansas
Florida
Georgia
Illinois
Indiana
Iowa
Kentucky
Mississippi
New Jersey
New Mexico
North Carolina
Ohio
Oregon
Pennsylvania
South Carolina
South Dakota
Tennessee
BT United States

Unionidae
Family. Includes use on level 3 under Mollusca(1) Bivalvia(2).
BA Bivalvia
Mollusca
BT Invertebrata

unit cell
UF cell, unit
SA crystal structure
lattice
lattice parameters
minerals

United Arab Emirates
Includes use on level 1 as an area term (list O). Federation of 7 states which was achieved in 1972. Formerly known as Trucial States, Trucial Oman, or Trucial Coast. Not strictly equivalent to Trucial Coast, but replaces Trucial Coast as first order term. For term set options see list B. Before 1973, also search Trucial Oman. Before 1975, also search Trucial Coast.
BA Arabian Peninsula
NT Abu Dhabi
Ajman
Dubai
Fujairah
Ras al-Khaimah
Sharjah
Umm al-Qaiwain
SA Asia
Middle East
Near East

United Kingdom
Index political divisions as applicable. Comprising Great Britain and Northern Ireland. Includes use on level 1 as an area term (list O). For term set options see list B.
CO N500000N610000
E0013000W0110000
BA Europe
NT Bristol Channel

Cheviot Hills
Esk Trough
Great Britain
Isle of Man
Northern Ireland
Severn Valley
Welsh Borderland
Wye Valley
SA Aberystwyth Grits
Barton Beds
Bembridge Marls
British Virgin Islands
London Clay
Lower Greensand
Manchester
Oxford Clay
Penrhyn Slate
Shap Granite
Skiddaw Slates
Speeton Clay
Upper Old Red Sandstone
Weald Clay
Wenlock Limestone

United Nations
SA associations

United States
Includes use on level 1 or 2 as an area term (list O). If 1, see term set options under list B.
BT North America
NT Absaroka Range
NA Alabama
Alaska
NT Allegheny Front
Allegheny Mountains
Allegheny Plateau
Amargosa Desert
Amite River
Anacostia River basin
Anadarko Basin
Appalachian Basin
Appalachian Plateau
NA Arizona
Arkansas
NT Arkansas River
Arkansas River valley
Arkoma Basin
Badlands
Bald Mountain
NA Basin and Range Province
NT Bear Lake
Bear River basin
Bear River Range
Beartooth Mountains
Beaverhead Mountains
Bighorn Basin
Bighorn Mountains
Bitterroot Range
Black Hills
Black Warrior Basin
Blue Ridge Mountains
Cache Valley
NA California
NT Cambridge Arch
Canadian River
Cascade Range
Catskill Delta
Chadron Arch
Chesapeake Bay
Choptank River
Cincinnati Arch
Coast Ranges
NA Colorado
Colorado Plateau
Columbia Plateau
NT Columbia River estuary
NA Connecticut
NT Connecticut River
Connecticut Valley
Cumberland Plateau
Dan River basin
NA Delaware
NT Delaware Basin
Delaware Bay
Delaware River
Delaware River basin

Delmarva Peninsula
Denver Basin
NA District of Columbia
NA Eastern U.S.
NT Flint Hills
NA Florida
NT Forest City Basin
Four Corners
Gallatin Range
NA Georgia
NT Gila River
NA Great Basin
NT Great Smoky Mountains
Green Bay
Green River
Green River basin
Guadalupe Mountains
NZ Hawaii
NT Hudson River
Hudson Valley
Hugoton Embayment
Hurricane Ridge Syncline
NA Idaho
NT Idaho Batholith
NA Illinois
NT Illinois Basin
NA Indiana
Iowa
Kansas
Kentucky
NT Kings Mountain
Klamath Mountains
Lake Bonneville
Lake Chatuge
Lake Mead
Lake Powell
Lake Tahoe
Laramie Mountains
Little Missouri River basin
Llano Estacado
NA Louisiana
Maine
Maryland
Massachusetts
NT Maumee River valley
Medicine Bow Mountains
Merrimack River valley
Merrimack Synclinorium
NA Michigan
NT Midcontinent
NA Midwest
Minnesota
Mississippi
NT Mississippi Embayment
Mississippi River
NA Mississippi Valley
Missouri
NT Missouri River
Missouri River valley
NA Montana
NT Navajo Indian Reservation
NA Nebraska
NT Nemaha Ridge
NA Nevada
New England
New Hampshire
New Jersey
New Mexico
NT New River
NA New York
NT Newark Basin
Newark-Gettysburg Basin
NA North Carolina
North Dakota
Ohio
NT Ohio River
Ohio River valley
Okefenokee Swamp
NA Oklahoma
Oregon
NT Ouachita Mountains
Overthrust Belt
Ozark Mountains
NA Pacific Coast
NT Panhandle
Paradox Basin
Pecos River valley
NA Pennsylvania
NX Piedmont
NT Potomac River
Potomac River basin
Powder River basin
Raft River
Raft River basin
Rangeley Lakes
Raton Basin
Red River
Red River valley
NA Rhode Island
NT Sabine Lake
Saco River
San Francisco Mountains
San Juan Basin
San Juan River
San Luis Valley
Sand Hills
Sangre de Cristo Mountains
Savannah River
Shadow Mountains
Shenandoah Valley
Snake River
Snake River basin
Snake River canyon
NA South Carolina
South Dakota
NT South Mountain
South Platte River valley
NA Southwestern U.S.
NT Springfield
Sunlight
Susquehanna River
Susquehanna River basin
NA Tennessee
NT Tennessee River
Tennessee Valley
NA Texas
NT Trans-Pecos
Truckee River
Twin Lakes
Uinta Basin
Uinta Mountains
NA Utah
NT Valley and Ridge Province
Verdigris River valley
NA Vermont
NT Virgin River valley
NA Virginia
NT Wasatch Front
Wasatch Range
Washakie Basin
NA Washington
NT Washita River valley
NA West Virginia
NA Western U.S.
NT White Mountains
Whitewater River valley
Wilmington
NA Wisconsin
Wyoming
NT Yellowstone National Park
SA Appalachians
Atlantic Coastal Plain
Canyon Group
conterminous regions
Great Lakes
Great Lakes region
Great Plains
Gulf Coastal Plain
Hanover
Huronian
Lake Michigan
Manchester
Mississippi River basin
Rocky Mountains
Taconic Allochthon
Taconic Orogeny
Virgin Islands
Western Interior

United States Geological Survey
use U. S. Geological Survey

units, stratigraphic
use stratigraphic units

universal stage
Includes use on level 3 under structural analysis(1) methods(2).
UF Fedorov stage
stage, universal
U-stage
SA microscope methods
ore microscopy
structural analysis

University of Arizona
In Tucson.
BT Arizona
United States

University of Bonn
In Bonn.
BT North Rhine-Westphalia
West Germany

University of California
In Berkeley.
BT California
United States
SA Los Alamos Scientific Laboratory

University of Cambridge
use Cambridge University

University of Louvain
In Louvain E of Brussels.
BT Brabant
Belgium

University of Lund
In Lund in extreme S part of country.
BT Malmohus
Sweden

University of Michigan
In Ann Arbor W of Detroit.
BT Michigan
United States

University of Pennsylvania
In Philadelphia.
BT Pennsylvania
United States

University of Rome
In Rome.
BT Latium
Italy

University of Tokyo
In Tokyo.
UF Tokyo Imperial University
BT Honshu
Japan

University of Wisconsin
In Madison.
BT Wisconsin
United States

Unkar Group
Grand Canyon Series. Divided into (descending) Dox Sandstone, Shinumo Quartzite, Hakatai Shale, Bass Limestone, and Hotuata Conglomerate. In N Arizona.
BT Precambrian
SA Arizona

Unst
Northernmost large island of the Shetland Islands off N.
BT Scotland
Great Britain
United Kingdom
SA Shetland Islands

Unstrut River
Flows into the Saale River in SW East Germany. Also search Unstrut. Index districts as applicable.
BT East Germany
SA Erfurt
Halle

Unzen
Active volcano. In Unzen National Park on central Shimabar Peninsula E of Nagasaki. Also search Unzen Volcano.
UF Unzen Volcano
BT Kyushu
Japan

Unzen Volcano
use Unzen

Uonuma Group
Upper Pliocene to lower Pleistocene. Composed of Oguni and Tsukayama formations. In NW Honshu.
BT Cenozoic
SA Honshu
Japan
lower Pleistocene
Pleistocene
Pliocene
Quaternary
Tertiary
upper Pliocene

uplifts
As of 1978, term is used on level 2 under marine installations(1).
SA domes
epeirogeny
marine installations
neotectonics
subsidence
tectonics
vertical tectonics

upper atmosphere
BA atmosphere

Upper Austria
State in NW.
BT Austria
NT Gosau
Linz
Weyer
SA Enns Valley
Salzach River
Salzkammergut
Totes Gebirge

Upper Bavaria
An administrative division in the extreme S in which Munich, the Bavarian Alps and Salzburg Alps are located.
BT Bavaria
West Germany

upper boundary
A valid general term through 1976. Now use boundary for stratigraphic meaning.

Upper Cambrian
BA Cambrian
BT Paleozoic
NT Bonneterre Formation
Franconia Formation
Little Falls Formation
Pilgrim Formation
Potsdam Sandstone
Reagan Sandstone
NA Trempealeauan
NT Wilberns Formation
SA Arbuckle Group
Deadwood Formation
Knox Group
Lower Cambrian
Middle Cambrian

Upper Carboniferous
BA Carboniferous
BT Paleozoic
NA Bashkirian
Coal Measures
NT Karharbari Stage
NA Millstone Grit
Moscovian
Stephanian
NT Talchir Series
NA Uralian
SA Gondwana System
Lower Carboniferous
Namurian
Pennsylvanian
Talchir Formation

upper Cenozoic
- BA Cenozoic
- SA Quaternary

Upper Cretaceous
- BA Cretaceous
- BT Mesozoic
- NT Almond Formation
  - Ariyalur Stage
  - Bagh Beds
  - Bearpaw Formation
  - Belly River Formation
  - Blackhawk Formation
- NA Campanian
- NT Carlile Shale
  - Chalk
  - Cody Shale
- NA Coniacian
- NT Djadokhta Formation
  - Edmonton Formation
  - Ferron Sandstone Member
  - Fort Hays Limestone Member
  - Fox Hills Formation
  - Frontier Formation
  - Gallup Sandstone
  - Greenhorn Limestone
- NA Gulfian
- NT Harebell Formation
  - Hell Creek Formation
  - Izumi Group
  - Judith River Formation
  - La Ventana Sandstone
  - Lance Formation
  - Laramie Formation
  - Lewis Shale
- NA Maestrichtian
- NT Magothy Formation
  - Marca Shale Member
  - Marshalltown Formation
  - Menefee Formation
  - Mesaverde Group
  - Milk River Formation
  - Monmouth Group
  - Montana Group
  - Moreno Formation
  - Mowry Shale
  - Navesink Formation
  - Niobrara Formation
  - Ojo Alamo Sandstone
  - Oldman Formation
  - Parkman Sandstone
  - Peedee Formation
  - Pierre Shale
  - Raritan Formation
  - Rincon Formation
  - Ripley Formation
  - Rosario Formation
- NA Santonian
- NT Saratoga Chalk
- NA Senonian
- NT Smoky Hill Chalk Member
- NA Turonian
- NT Tuscaloosa Formation
  - Woodbury Clay
- SA Bauru Formation
  - Beaverhead Formation
  - Benton Formation
  - Blairmore Group
  - Cenomanian
  - Colorado Group
  - Comanchean
  - Dakota Formation
  - Deccan Traps
  - Difunta Group
  - Ewekoro Formation
  - Fort Union Formation
  - Graneros Shale
  - Great Valley Sequence
  - Intertrappean Beds
  - Lower Cretaceous
  - Mancos Shale
  - Middle Cretaceous
  - Niniyur Formation
  - Paskapoo Formation
  - Potomac Group
  - Vraconian
  - Washita Group
  - Woodbine Formation

upper crust
- BA crust
- SA lower crust

Upper Devonian
- BA Devonian
- BT Paleozoic
- NA Famennian
  - Frasnian
- NT Greenland Gap Group
  - Lime Creek Formation
  - New Albany Shale
  - Ohio Shale
  - Olentangy Shale
  - Palliser Formation
  - Perry Formation
  - Sonyea Group
- SA Catskill Formation
  - Chattanooga Shale
  - Genesee Group
  - Lost Burro Formation
  - Lower Devonian
  - Martin Formation
  - Middle Devonian
  - Muth Quartzite
  - Traverse Group

upper Eocene
- BA Eocene
  - Paleogene
  - Tertiary
- BT Cenozoic
- NA Auversian
- NT Jackson Group
- NA Lattorfian
  - Ludian
- NT Ocala Group
  - Poway Conglomerate
- NA Priabonian
- NT Tejon Formation
- NA Tongrian
- NT Twiggs Clay
  - Yazoo Clay
- SA Biarritzian
  - Castle Hayne Limestone
  - lower Eocene
  - middle Eocene
  - Narizian
  - Sespe Formation
  - Twin River Formation

Upper Franconia
Administration division in NE Bavaria. Part of old historical region of Franconia.
- BT Bavaria
  - West Germany
- SA Franconia

upper Holocene
- BA Holocene
  - Quaternary
- BT Cenozoic

Upper Jurassic
- BA Jurassic
- BT Mesozoic
- NA Argovian
- NT Bossier Formation
  - Buckner Formation
- NA Callovian
- NT Cotton Valley Group
  - Jabalpur Series
- NA Kimmeridgian
  - Lusitanian
  - Malm
- NT Morrison Formation
- NA Oxfordian
  - Portlandian
- NT Raghavapuram Shales
- NA Rauracian
- NT Schuler Formation
- NA Sequanian
- NT Smackover Formation
  - Stump Formation
  - Sundance Formation
- NA Tithonian
- NT Todilto Formation
- NA Volgian
- NT Zuloaga Limestone

- SA Bowser Formation
  - Lower Jurassic
  - Middle Jurassic

Upper Lusatia
That part of Lusatia in SE East Germany. Index districts as applicable.
- BT East Germany
- SA Cottbus
  - Dresden
  - Lower Lusatia
  - Lusatia

upper mantle
Includes use on level 3 under mantle(1).
- UF outer mantle
- BA mantle
- SA asthenosphere
  - lower crust
  - lower mantle

upper Mesozoic
- BA Mesozoic
- SA Caracas Group
  - middle Mesozoic

upper Miocene
- BA Miocene
  - Neogene
  - Tertiary
- BT Cenozoic
- NT Duplin Formation
- NA Messinian
- NT Montesano Formation
- NA Pontian
- NT Puente Formation
  - Punchbowl Formation
  - Santa Margarita Formation
- NA Tortonian
- NT Yorktown Formation
- SA Capistrano Formation
  - Cuddalore Series
  - lower Miocene
  - Meotian
  - middle Miocene
  - Monterey Formation
  - Neyveli Lignite
  - Saint Marys Formation
  - Siwalik System
  - Warkalli Formation

Upper Mississippi Valley
That part of the Mississippi Valley N of Cairo, Illinois. Index states as applicable.
- BA Mississippi Valley
- BT United States
- SA Illinois
  - Iowa
  - Minnesota
  - Missouri
  - Wisconsin

Upper Mississippian
- BA Mississippian
- BT Paleozoic
- NT Bangor Limestone
- NA Chesterian
- NT Fayetteville Formation
- NA Meramecian
- NT Monteagle Limestone
  - Pennington Formation
  - Pitkin Limestone
- SA Bird Spring Formation
  - Borden Group
  - Leadville Formation
  - Lisburne Group
  - Lower Mississippian
  - Madison Group
  - Monte Cristo Limestone
  - Saint Louis Limestone
  - Valmeyeran
  - Warsaw Formation

Upper Old Red Sandstone
Represented by Nairn, Boghole, Alves-Scaat Craig, Rosebrae, Plateau, and Portishead beds; by Quartz Conglomerate; and by Farlow Sandstone. In Great Britain.

- BT Devonian
- SA Great Britain
  - United Kingdom

upper Oligocene
- BA Oligocene
  - Paleogene
  - Tertiary
- BT Cenozoic
- NA Chattian
- NT Chickasawhay Formation
- NA Egerian
- NT Suwannee Limestone
- SA Arikareean
  - Brule Formation
  - John Day Formation
  - lower Oligocene
  - middle Oligocene
  - Twin River Formation

Upper Ordovician
- BA Ordovician
- BT Paleozoic
- NA Ashgillian
  - Caradocian
  - Cincinnatian
  - Edenian
- NT Fairview Formation
  - Fernvale Formation
  - Juniata Formation
  - Maquoketa Formation
  - Red River Formation
  - Richmond Group
  - Utica Shale
- SA Bala
  - Chickamauga Group
  - Eden Shale
  - Lower Ordovician
  - Martinsburg Formation
  - Middle Ordovician
  - Viola Limestone

Upper Palatinate
That part of the historical region of the Palatinate located in E.
- BT Bavaria
  - West Germany
- SA Palatinate

upper Paleocene
- BA Paleocene
  - Paleogene
  - Tertiary
- BT Cenozoic
- NA Landenian
  - Thanetian
- SA lower Paleocene

upper Paleozoic
- BA Paleozoic
- NT Dwyka Series
  - Tunguska Series
- SA Rio Bonito Formation

Upper Peninsula
N part of state between Lake Michigan and Lake Superior.
- BT Michigan
  - United States
- SA Lower Peninsula

Upper Pennsylvanian
- BA Pennsylvanian
- BT Paleozoic
- NT Ames Limestone
  - Canyon Group
- NA Missourian
  - Virgilian
- SA La Salle Limestone
  - Lower Pennsylvanian
  - Middle Pennsylvanian

Upper Permian
- BA Permian
- BT Paleozoic
- NT Barabash Suite
- NA Kazanian
- NT Raniganj Stage
- NA Tatarian
- NT Usuginu Conglomerate
  - Vetluga Series
- NA Zechstein

SA Estrada Nova Formation
     Guadalupian
     Lower Permian
     Middle Permian
     San Andres Formation
upper Pleistocene
  BA Pleistocene
     Quaternary
  BT Cenozoic
  NA Brunhes Epoch
     Cromerian
     Devensian
     Eemian
     Holsteinian
     Hoxnian
     Illinoian
     Ipswichian
     Saalian
     Sangamonian
     Sicilian
     Tyrrhenian
     Weichselian
     Wisconsinan
     Wurm
  SA lower Pleistocene
upper Pliocene
  BA Pliocene
     Neogene
     Tertiary
  BT Cenozoic
  NA Akchagylian
     Astian
     Levantinian
     Redonian
  NT Rexroad Formation
  NA Romanian
  SA lower Pliocene
     Matuyama Epoch
     middle Pliocene
     Uonuma Group
     Yakima Basalt
upper Precambrian
  BA Precambrian
  NA Hadrynian
     Infracambrian
     Jotnian
     Laxfordian
     Moinian
     Riphean
     Torridonian
  SA Adelaidean
     Algonkian
     lower Precambrian
     Proterozoic
     Sparagmite Group
     Vendian
upper Proterozoic
  BA Proterozoic
     Precambrian
  NA Brioverian
  SA Adelaidean
upper Quaternary
  BA Quaternary
  BT Cenozoic
Upper Rhine
  use Upper Rhine Valley
Upper Rhine Graben
  BT Rhine Graben
     Europe
Upper Rhine Valley
  That part of the Rhine Valley between Basel and Mainz. Index countries as applicable. Also search Upper Rhine.
  UF Upper Rhine
  SA France
     Germany
     Rhine Valley
     Switzerland
Upper Silesia
  The eastern part of former German Silesia plus Polish Silesia lying SE of Lower Silesia centering on Katowice and Krakow.
  UF Upper Silesian area
  BT Poland
  SA Lower Silesia
     Silesia
     Silesian coal basin
Upper Silesia coal basin
  use Upper Silesian coal basin
Upper Silesian area
  use Upper Silesia
Upper Silesian Basin
  use Upper Silesian coal basin
Upper Silesian coal basin
  N foot of the W Beskid Mountains in Upper Silesia.
  UF Upper Silesia coal basin
     Upper Silesian Basin
     Upper Silesian coal field
  BT Katowice
     Poland
  SA Silesian coal basin
Upper Silesian coal field
  use Upper Silesian coal basin
Upper Silurian
  BA Silurian
  BT Paleozoic
  NT Bloomsburg Formation
  NA Cayugan
     Ludlovian
     Pridolian
  NT Salina Group
     Tonoloway Limestone
     Tura Formation
  SA Downtonian
     Lower Silurian
     Middle Silurian
     Read Bay Formation
     Rondout Formation
     Williamsport Sandstone
upper Tertiary
  BA Tertiary
  BT Cenozoic
  NT Poznan Clays
  SA Akchagylian
     lower Tertiary
     middle Tertiary
     Neogene
     San Juan Formation
     Siwalik System
Upper Triassic
  BA Triassic
  BT Mesozoic
  NA Carnian
  NT Chinle Formation
     Dockum Group
     Kayenta Formation
  NA Keuper
  NT Newark Group
  NA Norian
  NT Protopivskaya Formation
     Redonda Formation
  NA Rhaetian
  NT Stormberg Series
     Torlesse Supergroup
  SA Hallstatt Limestone
     Lower Triassic
     Middle Triassic
     Navajo Sandstone
     Novorayskoe Formation
     Pucara Group
Upper Tunguska
  use Angara River
Upper Volta
  Former French protectorate which achieved independence in 1960. Includes use on level 1 as an area term (list O). For term set options see list B.
  UF Voltaic Republic
  BA Africa
  SA Sahel
     Volta Basin
Uppsula
  County in E Sweden. Also a city.
  UF Upsula
  BT Sweden
Upsula
  use Uppsula
Upton County
  W Texas.
  BT Texas
     United States
upwelling
  Includes use on level 3 under ocean circulation(1).
  SA currents
     ocean circulation
     oceanography
     sea water
Ural Mountains
  use Urals
Ural Range
  use Urals
Ural region
  A mining and industrial region on both sides of the Central Urals comprising the Chelyabinsk, Sverdlovsk, Kurgan, Orenburg, and Perm oblasts; and Bashkiria and Udmurtia. Also search Ural.
  UF Urals Region
  BT Russian Republic
     USSR
Ural River
  Rises at S end of Urals and flows into the Caspian Sea. Index Soviet republics as applicable. Also search Ural.
  BT USSR
  SA Kazakhstan
     Russian Republic
Ural-Tau
  Low, water-divide mountain range in E Southern Urals.
  BT Russian Republic
     USSR
  SA Southern Urals
Ural-Volga Interfluve
  use Volga-Ural region
Ural-Volga region
  use Volga-Ural region
Uralian
  Russia. Above Gzhelian, below Sakmarian of Permian. Includes use on level 3 under age terms(1). See list E.
  BA Upper Carboniferous
     Carboniferous
  BT Paleozoic
Uralian Foreland
  The western slope and foothills of the Urals.
  BT Russian Republic
     USSR
uralitization
  BA metasomatism
  SA hydrothermal alteration
     metamorphism
Urals
  Mountain range extending from the Kara Sea in the N to W Kazakhstan border in the S separating Europe from Asia. Comprises the Northern, Central, and Southern Urals. Index ranges as applicable.
  CO N500000N680000
     E0680000E0570000
  UF Ural Mountains
     Ural Range
  BT Russian Republic
     USSR
  NA Northern Urals
     Southern Urals
  SA Polar Urals
Urals Region
  use Ural region
uraninite
  BA oxides
  BT minerals
  SA pitchblende
uranium
  Includes use on level 1 and 2 as a commodity term (list C) and as a chemical element (list D); on level 2 under mineral deposits, genesis(1) and paragenesis(1). Also search U.
  UF U
  SA elements
     Io/U
     isotopes
     radium
     roll-type deposits
     Th/U
     U-234
     U-235
     U-238
     U-238/Pb-206
     U-238/U-234
     U/He
     U/Pb
     U/Th/Pb
     uranium disequilibrium
     uranium minerals
uranium disequilibrium
  SA absolute age
     isotopes
     uranium
uranium minerals
  Includes use under commodities (list C).
  BT minerals
  SA miscellaneous minerals
     uranium
uranophane
  BA orthosilicates
     silicates
  BT minerals
Uranus
  Includes use on level 1 and 2. See entry under Moon(1) for term set options.
  SA outer planets
     planetology
     planets
     solar system
urban planning
  As of 1978, term is used on level 2 under land use(1).
  SA environmental geology
     land use
     planning
     regional planning
     urbanization
Urbana
  City in E Illinois and in W central Ohio. Index states as applicable.
  BX Illinois
     Ohio
  BT United States
urbanization
  Includes use on level 3 under environmental geology(1).
  SA conservation
     environmental geology
     land use
     urban planning
ureilites
  BA meteorites
Urga
  use Ulan Bator
Urgonian
  Europe. See list E.
  BA Lower Cretaceous
     Cretaceous
  BT Mesozoic
Uri
  Canton in central.
  BT Switzerland

Ursidae
Family. Includes use on level 3 under Mammalia(1) Carnivora(2).
BA Carnivora
 Mammalia
BT Tetrapoda
 Vertebrata
NA Ursus

Ursus
Genus. Includes use on level 3 under Mammalia(1) Carnivora(2).
BA Ursidae
 Carnivora
 Mammalia
BT Tetrapoda
 Invertebrata
NA Ursus spelaeus

Ursus spelaeus
Includes use on level 3 under Mammalia(1) Carnivora(2).
BA Ursus
 Ursidae
 Carnivora
 Mammalia
BT Tetrapoda
 Invertebrata

urtite
Includes use on level 3 under igneous rocks(1) alkali gabbro family(2). See list H.
BA alkali gabbro family
BT igneous rocks
SA ijolite
 melteigite

Uruguay
Includes use on level 1 as an area term (list O). For term set options see list B.
CO S350000S300000
 W0530000W0583000
BA South America
NT Montevideo
SA Rio de la Plata

Urup
Volcanic island in S central Kuril Islands NE of Japan.
UF Urup Island
BT Russian Republic
 USSR
SA Kuril Islands

Urup Island
use Urup

U.S. Geological Survey
use U. S. Geological Survey

Usa Series
Includes marble plus dolomites interspersed with beds of siliceous schist. In the Kuznetsk Alatau mountain region of Kemerovo Oblast and Khakass Autonomous Oblast E of Novosibirsk.
BT Lower Cambrian
 Cambrian
SA Russian Republic
 USSR

uses
A valid term through 1976. Use applications or utilization.

U.S.G.S.
use U. S. Geological Survey

USGS
use U. S. Geological Survey

Uspenski
use Uspenskiy

Uspenskiy
Town in Karaganda Oblast in E central Kazakhstan. Also search Uspenski.
UF Uspenski
BT Kazakhstan
 USSR

USSR
Includes use on level 1 or 2 as an area term (list O). If 1, see term set options under list B.
CO N360000N820000
 W1700000E0200000
UF Russia
NT Aral region
 Aral Sea
 Armenia
 Asiatic USSR
 Azerbaidzhan
 Azov region
 Azov Sea
 Byelorussia
 Caspian Depression
 Caucasus
 Central Asia
 Chirchik River
 Courland Spit
 Dnieper Basin
 Dnieper River
 Dnieper-Donets Basin
 Dniester River
 Dniester-Prut Interfluve
 Donets Basin
 Dvina River
 Estonia
 European USSR
 Fergana Basin
 Georgian Republic
 Greater Caucasus
 Ishim
 Kazakhstan
 Kirghizia
 Kuban River
 Kuban Valley
 Kulunda Steppe
 Kurama Range
 Kyzylkum
 Latvia
 Lesser Caucasus
 Lithuania
 Neman River basin
NA Northeastern USSR
 Northwestern USSR
NT Ob-Irtysh Interfluve
 Ossetia
 Pechora River
 Polesye
 Pripet Basin
 Pskem Range
 Rudny Altai
 Russian Plain
 Russian Platform
 Russian Republic
 Sayan
 Scythian Platform
 Siberia
 Southwestern USSR
 Steppes
 Surkhan Darya
 Surkhan Darya basin
 Syr Darya
 Tadzhikistan
 Terek River
 Transcaucasia
 Turan
 Turanian Platform
 Turkmenia
 Ukraine
 Ural River
 Ustyurt
 Uzbekistan
 Uzen
 Vodino
 Volga-Ural region
 Vyshkovo
 Zangezur
 Zeravshan River
SA Altai Mountains
 Altai-Sayan region
 Arctic region
 Baikalian Phase
 Barabash Suite
 Baskunchak Series
 Black Sea region
 Caspian Sea
 Eurasia
 European Platform
 Far East
 Ilimaussaq
 Kiev Member
 Korenevskaya Formation
 Krivoy Rog Series
 Kursk Series
 Lapland
 Maikop Series
 Moldavia
 Novorayskoe Formation
 Oktemberyan Series
 Ovruch Series
 Poltava Series
 Protopivskaya Formation
 Selenga River valley
 Tunguska Series
 Udokan Series
 Usa Series
 Vetluga Series
 Yudoma Series

USSR Academy of Sciences
In Moscow.
BT Russian Republic
 USSR

Ust-Kut
Town in N central Irkutsk Oblast NW of Lake Baikal.
BT Russian Republic
 USSR

Ust-Urt
use Ustyurt

Ust-Urt Plateau
use Ustyurt

Ust-Yenisei Basin
That part of the lower Yenisei Basin from Ust-Port to Yenisei Bay which constitutes a drowned river mouth or estuary. Located in NW Taymyr-Nenets National Okrug.
BT Russian Republic
 USSR

Ustica Island
Small island in the Tyrrhenian Sea NW of Sicily. Also search Ustica.
BT Sicily
 Italy

Ustron
Village in central.
BT Byelorussia
 USSR

Ustyurt
Desert plateau extending from the Caspian Sea to the Aral Sea. Also search Ust-Urt; Ust-Urt Plateau. Index Soviet republics as applicable.
CO N410000N450000
 E0590000E0550000
UF Ust-Urt
 Ust-Urt Plateau
BT USSR
SA Kazakhstan
 Uzbekistan

Usu
On NE Uchiura Bay in SW Hokkaido. Also search Usu Volcano.
UF Usu Volcano
BT Hokkaido
 Japan

Usu Volcano
use Usu

Usuginu Conglomerate
Includes large boulders of limestone, compact slate, sandstone, granite, and porphyrite. In N Honshu.
BT Upper Permian
 Permian
SA Honshu
 Japan

Usuki
Town on Bungo Strait in NE.
BT Kyushu
 Japan

Utah
Includes use on level 1 as an area term (list O). For term set options see list B.
CO N370000N420000
 W1090500W1140500
BA United States
NT Arches
NX Beaver County
NT Bingham
 Bonneville Salt Flats
 Box Elder County
NX Carbon County
NT Duchesne County
 Emery
 Emery County
NX Eureka
NT Fairfield
NX Garfield County
 Grand County
NT Great Salt Lake
 Henry Mountains
 House Range
NX Iron County
NT Juab County
 Kaiparowits Plateau
NX Kane County
NT Marysvale
 Millard County
 Moab
 Monticello
NX Morgan County
NT Oquirrh Mountains
 Park City
 Payson
 Pine Valley Mountains
 Piute County
 Provo
 Salt Lake City
 Salt Lake County
NX San Juan County
NT San Rafael Swell
 Sanpete County
NX Sevier County
NT Sevier orogenic belt
 Spor Mountain
 Springdale
 Stansbury Mountains
NX Summit County
NT Tintic District
 Tooele County
 Uintah County
 Utah County
 Wasatch County
 Wasatch Fault
 Wasatch Plateau
NX Washington County
 Wayne County
NT Weber County
SA Bald Mountain
 Basin and Range Province
 Bear Lake
 Bear River basin
 Bear River Range
 Bird Spring Formation
 Blackhawk Formation
 Cache Valley
 Cedar Mountain Formation
 Chinle Formation
 Coconino Sandstone
 Colorado Plateau
 Colorado River
 Columbia River basin
 Cutler Formation
 Duchesne River Formation
 Ferron Sandstone Member
 Fillmore Formation
 Four Corners
 Frontier Formation
 Great Basin
 Green River
 Green River basin
 Green River Formation

Honaker Trail Formation
Kaibab Formation
Kayenta Formation
Lake Bonneville
Lake Powell
Lodgepole Formation
Madison Group
Mancos Shale
Mesaverde Group
Moenkopi Formation
Montana Group
Morrison Formation
Navajo Indian Reservation
Navajo Sandstone
Overthrust Belt
Paradox Basin
Paradox Member
Phosphoria Formation
Pogonip Group
Raft River
Raft River basin
Roberts Mountains Formation
San Juan Basin
San Juan River
Snake River basin
Supai Formation
Swan Peak Formation
Thaynes Formation
Tintic Quartzite
Toroweap Formation
Uinta Basin
Uinta Mountains
Virgin River valley
Wasatch Formation
Wasatch Front
Wasatch Range
Weber Sandstone
Western Interior

Utah County
N central.
BT   Utah
     United States

utahite
use   jarosite

Ute Creek
Flows into the Canadian River in NE.
BT   New Mexico
     United States

Utica
City in SE Michigan; city in Oneida Co., New York.
BX   Michigan
     New York
BT   United States

Utica Shale
Consists of Nowadaga, Loyal Creek, Hopkinson, Holland, and Patent members.
BT   Upper Ordovician
     Ordovician
SA   Michigan
     New York
     Ontario

utilization
Includes use on level 2 under soils(1), and on level 3 under commodity terms (list C).
SA   agriculture
     fertilization
     field crops
     forestry
     irrigation
     reclamation
     soil management
     soils
     tillage
     treatment
     underground installations

Utrecht
Province. Also a city S of Amsterdam.
BT   Netherlands

Uttar Pradesh
State in N central.
CO   N240000N320000
     E0840000E0770000
BT   India
NT   Almora
     Dehra Dun
     Garhwal
     Jhansi
     Meerut
     Mirzapur
     Mussoorie
     Naini Tal
     Roorkee
     Tehri Garhwal
SA   Bundelkhand
     Jumna River
     Kaimur Sandstone
     Kumaun Himalayas
     Muth Quartzite
     Satpura Range
     Tal Formation

Uvalde Gravel
Consists almost wholly of rounded flat cobbles, boulders, and occasional limestone pebbles. In Texas.
BT   Pliocene
SA   Texas

uvarovite
UF   onvarovite
BA   garnet group
     orthosilicates
     silicates
BT   minerals

Uvigerina
Genus. Includes use on level 3 under foraminifera(1) Buliminacea(2).
BA   Buliminacea
     Rotaliina
     foraminifera
BT   Invertebrata

Uvigerinidae
Family. Includes use on level 3 under foraminifera(1) Buliminacea(2).
BA   Buliminacea
     Rotaliina
     foraminifera
BT   Invertebrata

Uzbekistan
Uzbek Soviet Socialist Republic. In Soviet Central Asia.
CO   N370000N450000
     E0720000E0550000
BT   USSR
NT   Almalyk
     Angren
     Bukantau
     Bukhara
     Bukhara-Khiva
     Chadak
     Chirchik
     Dengizkul
     Karshi
     Karshi Steppe
     Kermine
     Khiva
     Kul'dzhuktau
     Lyangar
     Mubarek
     Nura-Tau
     Oktyabr
     Samarkand
     Sokh
     Tamdytau
     Tashkent
     Ziaetdin
     Zirabulak
SA   Amu Darya
     Aral region
     Asiatic USSR
     Central Asia
     Chirchik River
     Fergana Basin
     Kurama Range
     Kyzylkum
     Surkhan Darya
     Surkhan Darya basin
     Syr Darya
     Turan
     Turanian Platform
     Ustyurt
     Zeravshan River

Uzen
Two steppe rivers, the Greater Uzen and Lesser Uzen, which flow parallel to one another in Saratov Oblast and NW Kazakhstan. Index Soviet republics as applicable.
BT   USSR
SA   Kazakhstan
     Russian Republic

Uzon
Village NE of Brasov in SE Transylvania. Also search Ozun.
UF   Ozun
BT   Transylvania
     Romania

v
use   vanadium

Vaal River
Forms boundary between Transvaal and Orange Free State, and empties into Orange River in N Cape Province. Index provinces as applicable.
BT   South Africa
SA   Cape Province
     Orange Free State
     Transvaal

vacuum fusion analysis
Term introduced in 1978. Includes use on level 2 under chemical analysis(1).
SA   analysis
     chemical analysis

Vadia
use   Wadia

vaesite
BA   sulfides
BT   minerals

Vah River valley
use   Vah Valley

Vah Valley
River valley in W Slovakia. Also search Vah; Vah River.
UF   Vah River valley
BT   Slovakia
     Czechoslovakia

Vaigach Island
use   Vaygach Island

Vail Pass
In Summit and Eagle counties W of Denver.
BT   Colorado
     United States

Vakhsh
River. A tributary of the Amu Darya in SW Tadzhikistan.
UF   Vaksh
BT   Tadzhikistan
     USSR

Vaksh
use   Vakhsh

Val d'Aosta
use   Valle d'Aosta

Val d'Or
Town in SW.
BT   Quebec
     Canada

Val Gardena Sandstone
NE Italy.
BT   Permian
SA   Italy

Val Verde Basin
SW Texas.
BT   Texas
     United States

Val Verde County
SW Texas.
CO   N291000N301500
     W1005000W1014500
BT   Texas
     United States

Val-d'Oise
Department just NNW of Paris.
BT   France
NT   Cormeilles-en-Parisis
SA   Oise River valley

Valais
Canton in SW.
BT   Switzerland
NT   Binnental
     Zermatt

Valanginian
Europe. Above Berriasian, below Hauterivian. Includes use on level 3 under age terms(1). See list E.
BA   Lower Cretaceous
     Cretaceous
BT   Mesozoic
SA   Neocomian

Valdai
Moraine region between Moscow and Leningrad. Also search Valdai Hills.
UF   Valdai Hills
     Valdai Uplands
BT   Russian Republic
     USSR

Valdai Hills
use   Valdai

Valdai Uplands
use   Valdai

Valdez
Town and port at head of an inlet off Prince William Sound in S Alaska. Terminal point of Trans-Alaska pipeline. Also search Port Valdez.
UF   Port Valdez
BT   Alaska
     United States

Valdres
Mountainous region W and SW of Fagernes in S central part of country.
BT   Norway

Vale of Glamorgan
In Brisbane area.
BT   Queensland
     Australia

Vale of Kashmir
use   Kashmir Valley

Valence
City on left bank of Rhone River in SE part of country.
BT   Drome
     France

Valencia
Province in E Spain. Also a city.
BT   Spain
NT   Bunol

Valencia County
W New Mexico.
BT   New Mexico
     United States

Valenciennes
City near the Belgian frontier.
BT   Nord
     France

valency
Term introduced in 1978.
SA   chemical properties
     geochemistry

Valentine Formation
In Ogallala Group. Composed largely of unconsolidated layers of greenish silty marl sand and very few local lenses of diatomaceous marl and light bluish-gray volcanic ash. In W Kansas, NW Nebraska, and SW South Dakota.
BT Pliocene
SA Kansas
 Nebraska
 Ogallala Formation
 South Dakota

valentinite
UF white antimony
BA oxides
BT minerals

Valle d'Aosta
Autonomous region in extreme NW Italy. Also search Aosta Valley; Val d'Aosta.
UF Aosta Valley
 Val d'Aosta
BT Italy

Valle Grande Mountains
NW of Santa Fe in N New Mexico. Also search Jemez Mountains; Valle Grande.
UF Jemez Mountains
BT New Mexico
 United States

valleriite
BA sulfides
BT minerals

Valles
City in E central part of country.
UF Ciudad de Valles
BT San Luis Potosi
 Mexico

Valles Caldera
Also search Valles.
BT New Mexico
 United States

Vallesian
Europe. See list E.
BA Miocene
 Neogene
 Tertiary
BT Cenozoic

Valley and Ridge Province
Folded mountain ridges and parallel valleys between the Appalachian Plateau on the W and the Older Appalachians on the E. Extends from along the Hudson River in the N into Alabama in the S. Index states as applicable. Also search Valley and Ridge.
BT United States
SA Alabama
 Georgia
 Great Valley Sequence
 Maryland
 New Jersey
 New York
 North Carolina
 Pennsylvania
 Tennessee
 Virginia
 West Virginia

Valley County
Index states as applicable.
BX Idaho
 Montana
 Nebraska
BT United States

valley, median
use median valley

Valley of Mexico
Large oval basin 50 miles by 40 miles which is a subdivision of the plateau of Anahuac in central part of country. Index Federal District and/or Mexico state as applicable.
BT Mexico
SA Federal District
 Mexico state

Valley of Ten Thousand Smokes
Volcanic region in Katmai National Monument W of Mt. Katmai in SW.
BT Alaska
 United States

valleys
Includes use on level 3 under geomorphology(1) erosion features(2), fluvial features(2), landform descriptions(2), and landform evolution(2).
SA buried valleys
 drainage
 drainage basins
 erosion
 erosion features
 fluvial features
 geomorphology
 landforms
 rift valleys
 rilles
 rivers and streams

Valmeyeran
Provincial series, Illinois. Lower and Upper Mississippian. Equivalent to Osagian and Meramecian elsewhere. Includes use on level 3 under age terms(1). See list E.
BA Mississippian
BT Paleozoic
SA Lower Mississippian
 Meramecian
 Osagian
 Upper Mississippian

Valpacos
Town in N central part of country.
BT Vila Real
 Portugal

Valparaiso
Province in central part of country. Also a city on the Pacific Ocean.
BT Chile

Valsequillo
Village in NW.
BT Cordoba
 Spain

Valsugana
Valley of upper Brenta River in SE.
BT Trentino-Alto Adige
 Italy

Valtellina
Valley of upper Adda River in N.
BT Lombardy
 Italy

valves
Includes use on level 3 under appropriate fossil group (list F).
SA morphology

Van Horn
Village in Culberson County in W.
BT Texas
 United States

vanadates
Includes use on level 2 under minerals(1). See list L.
BT minerals
NA carnotite
 descloizite
NZ vanadinite
NA volborthite

vanadinite
BZ halides
 vanadates
BT minerals
SA chlorides

vanadium
Includes use on level 1 and 2 as a commodity term (list C) and as a chemical element (list D).
UF V
SA elements

vanadium garnet
BA garnet group
 orthosilicates
 silicates
BT minerals
SA grossular

Vanch Range
use Vanchskiy Range

Vanchskiy Mountains
use Vanchskiy Range

Vanchskiy Range
Near Afghan border in S central Tadzhikistan. Also search Vanch; Vanch Range.
UF Vanch Range
 Vanchskiy Mountains
BT Tadzhikistan
 USSR

Vancouver
City in SW British Columbia and a city on the Columbia River in SW Washington. Index state or province as applicable.
SA British Columbia
 Washington

Vancouver Island
Large island off SW coast of mainland British Columbia.
CO N483000N510000
 W1233000W1283000
BT British Columbia
 Canada

Vanoise
High mountain group of the Savoy Alps. Also search Vanoise Massif.
UF Vanoise Massif
BT Savoie
 France
SA Alps
 Savoy Alps

Vanoise Massif
use Vanoise

vapor, water
use water vapor

vaporization
use evaporation

Var
Department in SE.
CO N430000N434500
 E0070000E0053000
BT France
NT Maures Massif
 Plan de la Tour
 Rians
 Sainte-Baume Massif
 Toulon
SA Esterel
 Provence
 Provence Alps

Varanger Fjord
Inlet of the Barents Sea S of the Varanger Peninsula in extreme NE part of country.
UF Varangerfjord
BT Finnmark
 Norway

Varanger Halvoy
use Varanger Peninsula

Varanger Peninsula
Extreme NE part of country jutting into the Barents Sea.
UF Varanger Halvoy
 Varangerhalvoeya
 Varangerhalvoya
BT Finnmark
 Norway

Varangerfjord
use Varanger Fjord

Varangerhalvoeya
use Varanger Peninsula

Varangerhalvoya
use Varanger Peninsula

Vardar River
Rises in S Yugoslavia and flows into Gulf of Salonika. Index countries as applicable.
BT Europe
SA Greece
 Yugoslavia

Varese
City in NW.
BT Lombardy
 Italy

variance analysis
SA analysis
 statistical analysis

variations
Includes use on level 2 under magnetosphere(1); on level 3 under Earth(1). Also search variation.
SA disturbances
 diurnal variations
 Earth
 magnetic field
 magnetosphere
 seasonal variations
 secular variations
 spatial variations
 time variations

varieties
Includes use as level 2 or 3 term appropriate to a large number of topics. See list G.

Variscan
A valid index term through 1976. After 1976, use Hercynian Orogeny.

Variscan Orogeny
use Hercynian Orogeny

Variscides
Mountain system raised in the latter part of the Paleozoic era, particularly in central Europe; more or less equivalent to Hercynian.
BT Europe
SA Hercynian Orogeny

variscite
BA phosphates
BT minerals

Varmland
County in SW.
BT Sweden

Varna
Province along the Black Sea. Also a city.
BT Bulgaria

Varpalota
Town N of E Lake Balaton.
BT Veszprem
 Hungary

Varta
use Warta

varved clay
A valid term through 1975. After 1975, use varves.

varves
As of 1978, includes use on level 2 under geochronology(1). See list K. Before 1976, also search varved clay.
BA planar bedding structures
 sedimentary structures
SA geochronology
 glacial features
 glacial lakes
 graded bedding

Vasterbotten
County in N.
BT Sweden

    NT   Boliden
    SA   Skellefte

Vastervik
City on Baltic Sea.
    BT   Kalmar
            Sweden

vaterite
Artificial mineral.
    BA   carbonates
    BT   minerals

Vatnajokull
Glacier region and snow field in SE.
    BT   Iceland

Vatukoula
Town on N coast of Viti Levu island in S.
    BT   Fiji
            Pacific Ocean

Vaucluse
Department in SE.
    CO   N434000N442000
            E0054500E0044500
    BT   France
    NT   Apt
            Mormoiron
    SA   Provence

Vaud
Canton in W.
    BT   Switzerland

Vaygach Island
Between NE Nenets National Okrug and island of Novaya Zemlya off N European USSR. Also search Vaigach; Vaigach Island; Vaygach.
    UF   Vaigach Island
    BT   Russian Republic
            USSR

vegetation
Includes use on level 3 under various fossil groups(1) and under paleoecology(1). Also use Plantae on level 1. See list F.
    SA   animals
            ecology
            floral studies
            grasslands
            leaves
            paleobotany
            Plantae
            soils
            wetlands

veins
General term not restricted to ore deposits. Includes use on level 3 under material, e.g. under gold(1).
    SA   fractures
            host rocks
            intrusions
            mineral deposits
            ore deposits

Vejer de la Frontera
City near the Atlantic Ocean in SW.
    BT   Cadiz
            Spain

Velay
Region of the Central Massif in W and central Haute-Loire. Also search Velay Massif; Velay Plateau.
    BT   Haute-Loire
            France

Velebit Mountains
Range of Dinaric Alps along Adriatic Sea opposite Pag Island in W Croatia. Also search Velebit.
    BT   Croatia
            Yugoslavia
    SA   Dinaric Alps

Velez Rubio
Town in Andalusia in N.
    BT   Almeria
            Spain

velocity
Includes use as a general term, e.g. under seismology(1). Also search velocities.
    SA   Conrad discontinuity
            flows
            half-space
            kinematics
            kinetics
            velocity structure
            vibration

velocity structure
Term introduced in 1978.
    SA   structure
            velocity

Vema fracture zone
Crosses Mid Atlantic Ridge near 11° N.
    BT   Atlantic Ocean

Vendee
Department on Bay of Biscay in W.
    CO   N462500N471000
            W0004500W0020500
    BT   France
    SA   Poitou

Vendian
Europe. See list E.
    BA   Precambrian
    SA   Eocambrian
            Infracambrian
            upper Precambrian

Vendsyssel
Region of Jutland Peninsula N of Lim Fjord on the Skagerrak.
    BT   Denmark

Venediger Group
Mountain group in SE Tyrol and SW Salzburg just S of W Hohe Tauern. Index states as applicable.
    UF   Venediger Gruppe
            Venediger Massif
    SA   Salzburg
            Tyrol

Venediger Gruppe
use Venediger Group

Venediger Massif
use Venediger Group

Venetia
Region E of Lombardy. Also a former autonomous region now known as Veneto. Index autonomous regions as applicable.
    BT   Italy
    SA   Friuli-Venezia Giulia
            Trentino-Alto Adige
            Veneto

Veneto
Autonomous region in NE Italy. Formerly Venetia autonomous region.
    UF   Venezia Euganea
    BT   Italy
    NT   Alpago Basin
            Belluno
            Berici Hills
            Cortina D'Ampezzo
            Euganean Hills
            Monte Baldo
            Monti Berici
            Monti Lessini
            Piave Valley
            Po Delta
            Possagno
            Sappada
            Venice
            Verona
            Vicenza
    SA   Dolomites
            Marmolada Glacier
            Po River
            Po Valley
            Tagliamento Valley
            Venetia

Venezia
use Venice

Venezia Euganea
use Veneto

Venezia-Tridentina
use Trentino-Alto Adige

Venezuela
Includes use on level 1 as an area term (list O). For term set options see list B.
    CO   N004500N121000
            W0595500W0731500
    BA   South America
    NT   Anzoategui
            Aragua
            Araya Peninsula
            Barinas
            Bolivar
            Carabobo
            Cojedes
            Cordillera de la Costa
            Cumana
            Falcon
            Guarico
            Lara
            Maracaibo Basin
            Miranda
            Nueva Esparta
            Paria Peninsula
    NX   Sucre
    NT   Tachira
            Trujillo
            Yaracuy
            Zulia
    SA   Amazon Basin
            Amazonas
            Andes
            Caracas Group
            Caribbean region
            Llanos
            Merida
            Mirador Formation
            Orinoco River
            Punta Mosquito Formation
            Roraima Formation

Venezuela Basin
use Venezuelan Basin

Venezuelan Basin
Marine basin between island of Hispaniola on the N and Venezuela on S. Also search Venezuela Basin.
    UF   Venezuela Basin
    BT   Caribbean Sea

Venice
City on Gulf of Venice off N Adriatic Sea.
    UF   Venezia
    BT   Veneto
            Italy

Ventersdorp
Town in SW.
    BT   Transvaal
            South Africa

vents
    SA   volcanology

Ventura
City WNW of Los Angeles in Ventura County.
    UF   San Buenaventura
    BT   California
            United States

Ventura Basin
WNW of Los Angeles in Ventura and Santa Barbara counties.
    BT   California
            United States

Ventura County
WNW of Los Angeles.
    BT   California
            United States

Venus
Includes use on level 1 and 2. See entry under Moon(1) for term set options.
    SA   landing sites
            planetology
            planets
            solar system
            terrestrial planets

Veracruz
State on the Gulf of Mexico in E Mexico. Also a city.
    CO   N170000N221500
            W0930000W0990000
    BT   Mexico
    NT   Poza Rica
    SA   Gulf Coastal Plain
            Sierra Madre Oriental

Vercelli
City in E.
    BT   Italy
    SA   Piedmont

Vercors
Limestone massif of the Dauphine Alps SE part of country. Index departments as applicable.
    UF   Vercors Massif
    BT   France
    SA   Dauphine Alps
            Drome
            Isere

Vercors Massif
use Vercors

Verde Formation
Pliocene (?) or Pleistocene. Unconformably overlies Hickey Formation in Jerome area. Central Arizona.
    BT   Cenozoic
    SA   Arizona
            Pleistocene
            Pliocene
            Quaternary
            Tertiary

Verde Valley
River valley in central.
    BT   Arizona
            United States

Verdigris River valley
SE Kansas, and NE Oklahoma. Index states as applicable. Also search Verdigris River.
    BT   United States
    SA   Kansas
            Oklahoma

Verdon Valley
River valley in SE part of country. Also search Verdon.
    BT   Alpes-de-Haute Provence
            France

Verdun
City on the Meuse River in NE part of country.
    UF   Verdun-sur-Meuse
    BT   Meuse
            France

Verdun-sur-Meuse
use Verdun

Verkhovtsevo
Town in W central Dnepropetrovsk Oblast in S central.
    BT   Ukraine
            USSR

Verkhoyansk
Town N of Arctic Circle on the Yana River in N central Yakutia in N central Siberia.
    BT   Russian Republic
            USSR

Verkhoyansk Mountains
use Verkhoyansk Range

Verkhoyansk Range
Extends from Lena River delta along the right banks of the Lena and lower Aldan rivers in N central Yakutia in N central Siberia. Also search

Verkhoyansk.
CO N630000N700000
E1400000E1240000
UF Verkhoyansk Mountains
Verkhoyanski Mountains
BT Russian Republic
USSR

Verkhoyanski Mountains
use Verkhoyansk Range

vermiculite
Includes use on level 1 as a commodity term (list C).
BA sheet silicates
silicates
BT minerals
SA clay mineralogy
clay minerals
mica
mica group

Vermilion Parish
On the Gulf of Mexico. Also search Vermilion AND Louisiana.
BT Louisiana
United States

Vermilion Range
Iron mining area in Saint Louis and Lake counties in NE. Also search Vermilion AND Minnesota.
BT Minnesota
United States

Vermont
Includes use on level 1 as an area term (list O). For term set options see list B.
CO N424500N450000
W0713000W0732500
BA United States
NT Caledonia County
Castleton
Chittenden County
NX Essex County
Franklin County
Orange County
NT Rutland
NX Washington County
Windham County
SA Barre Granite
Champlain Valley
Connecticut River
Connecticut Valley
Eastern U.S.
Green Mountains
Lake Champlain
Littleton Formation
New England
Normanskill Formation
Potsdam Sandstone
Trenton Group
Wilcox Formation

Vernagt Glacier
In Otztal Alps S of the Wildspitze in S.
BT Tyrol
Austria

Vernyi
use Alma-Ata

Verona
City in Verona Province in W.
BT Veneto
Italy

Versailles
City in Yvelines Department of N France WSW of Paris; and a city in numerous U. S. states. Index France and states as applicable.
SA France
Illinois
Indiana
Kentucky
Missouri

Versilian
Europe. See list E.
BA Holocene

Quaternary
BT Cenozoic

Vertebrata
Includes use on level 1 and 2 as a fossil term (list F); on level 3 under soils(1) biota(2). Term is to be used only when more specific terms do not apply, or are too numerous to be recorded.
NT Pisces
Tetrapoda
SA biota
bones
Chordata
fossil man
jaws
man
skeletons
skulls
teeth

Vertes
Town in E part of country.
BT Hungary

Vertes Mountains
N Central Hungary. Also search Vertes.
BT Hungary

vertical
A valid index term through 1977. After 1977, use vertical orientation for folds.

vertical movements
Used under many topics, e.g. includes use on level 3 under neotectonics(1).
UF movements, vertical
SA neotectonics
vertical tectonics

vertical orientation
Term introduced in 1978. Reference to folds defined on the basis of orientation in relation to the geographic horizontal plane. Includes use on level 3 under folds(1) orientation(2). Before 1978, also search folds AND vertical.
SA folds
orientation

vertical tectonics
Includes use on level 2 under tectonics(1).
SA isostasy
subsidence
tectonics
uplifts
vertical movements

Vertisols
Includes use on level 3 under soils(1). See list M.
BA soils

Vesdre Valley
River valley in E part of country.
BT Liege
Belgium

Vest Agder
use Vest-Agder

Vest-Agder
County in extreme S Norway.
UF Vest Agder
BT Norway
Europe

Vesteraalen
use Vesteralen

Vesteralen
Island group N of the Lofoten Islands in the Norwegian Sea off NW mainland. Index counties as applicable.
UF Vesteraalen
BT Norway
SA Nordland
Troms

Vestfold
County in SE.
UF Jarlsberg
BT Norway
NT Larvik

Vestman Islands
use Vestmannaeyjar

Vestmann Islands
use Vestmannaeyjar

Vestmannaeyjar
Small island group of volcanic origin S of Iceland.
UF Vestman Islands
Vestmann Islands
Westman Islands
BT Iceland

Vestspitsbergen
Old name for Spitsbergen, the main island, of the Spitsbergen Archipelago of the Svalbard Island group.
UF West Spitsbergen
BT Spitsbergen
Arctic region
SA Svalbard

Vesuvian garnet
use leucite

vesuvianite
Also search idocrase.
UF idocrase
BA orthosilicates
silicates
BT minerals

Vesuvio
use Vesuvius

Vesuvius
Active volcano on E side of the Bay of Naples. Also search Vesuvius Volcano.
UF Vesuvio
Vesuvius Volcano
BT Campania
Italy

Vesuvius Volcano
use Vesuvius

Veszprem
County in W Hungary. Also a city N of Lake Balaton.
BT Hungary
NT Varpalota

Vetluga Series
According to the USSR Lexique, the age of the Vetluga Horizon is upper Permian. It is composed of red spotted clay, and maroon and gray sandstone in an interwinning stratification not characterized paleontologically. NE of Moscow.
BT Upper Permian
Permian
SA Russian Republic
USSR

Vetrenyy Ridge
S of Onega Bay in W Arkhangelsk Oblast in NW European USSR. Also search Vetrenyy.
BT Russian Republic
USSR

Viatka River
use Vyatka River

vibration
Includes use on level 3 under rock mechanics(1) or seismology(1).
SA oscillations
pulsations
resonance
rock mechanics
seismology
velocity

Vibroseis
SA elastic waves

seismic methods
seismic surveys
seismology

Vicenza
City W of Venice in Vicenza Province.
BT Veneto
Italy

Vich
City NE of Barcelona in NE part of country.
BT Barcelona
Spain

Vicksburg Group
Includes Mariana Limestone and Byram Formation. Gulf Coastal Plain. Vicksburg.
BT middle Oligocene
Oligocene
SA Alabama
Florida
Louisiana
Marianna Limestone
Mississippi
Texas

Victoria
Includes use on level 1 as an area term (list O). For term set options see list B.
CO S390000S340000
E1500000E1400000
BA Australia
NT Gippsland
Gippsland Basin
Green Gully
Melbourne
Mildura
Philip Island
Port Campbell
Port Phillip Bay
SA Gambier Embayment
Lilydale Limestone
Murray Basin
Murray River
Otway Basin
Snowy Mountains
Tasman Geosyncline

Victoria Land
W of Ross Sea and W of NW Ross Ice Shelf, primarily in Ross Dependency on the Pacific Ocean side S of New Zealand.
UF South Victoria Land
BT Antarctica

Victoria Valley
River valley in NW.
BT Northern Territory
Australia

Vidda
use Hardangervidda

Vidin
Province in extreme NW Bulgaria. Also a city on the Danube River near the Yugoslav border.
BT Bulgaria
NT Lom Depression

Vienerwald
use Wiener Wald

Vienna
City on the Danube River in NE part of country.
UF Wien
BT Lower Austria
Austria

Vienna Basin
Index countries as applicable.
BT Europe
SA Austria
Czechoslovakia

Vienne
Department in W central.
CO N460500N471000
E0011500W0000500

Vienne ● Virginia

BT France
NT Poitiers
SA Poitou

Vietnam
North Vietnam (Democratic Republic of Vietnam) and South Vietnam (Republic of Vietnam) were combined into an unified nation on June 24, 1976. Apparently, the new country will be known as the Democratic Republic of Vietnam. Includes use on level 1 as an area term (list O). Use north or south on level 3. For term set options see list B. Before 1977, also search Vietnam AND north; search Vietnam AND south.
CO N083000N231500
E1092000E1020000
BA Asia
NT Baoha
Dalat
Mekong Delta
Nong-Son
Yenbay
SA Indochina

Vigarano
Point of impact near village of Vigarano in Ferrara Province in NE Emilia-Romagna. Also search Vigarano Meteorite.
UF Vigarano Meteorite
BT Emilia-Romagna
Italy
SA meteorites

Vigarano Meteorite
use Vigarano

Viking
SA landing sites
Mars
satellite methods

Viking Formation
Consists of sandstone which grades eastward into siltstone and shale and into the marine shales of the Ashville Formation. In SW Alberta, the sandstones grade into dark marine shales and sandstones of the Bow Island Formation. Oil reservoirs in beds equivalent to the lower Colorado are found in sandstone of the Viking Formation. NE and SW Alberta.
BT Cretaceous
SA Alberta

Vila Real
District in N.
BT Portugal
NT Valpacos
Vilarandelo
SA Tras-os-Montes

Vilaine Bay
At the mouth of the Vilaine River on the Bay of Biscay in SE Brittany. Also search Vilaine.
BT Morbihan
France

Vilarandelo
Village in N part of country.
BT Vila Real
Portugal

Villach
City W of Klagenfurt on the Drava River.
UF Beljak
BT Carinthia
Austria

Villafranca d'Asti
Village E of Turin in central.
BT Italy
SA Piedmont

Villafranchian
Europe. Terrestrial equivalent (in France and Italy) of the marine Calabrian. Includes use on level 3 under age terms(1). See list E.
BA Cenozoic
SA Calabrian
Pleistocene
Quaternary
Tertiary

Villany Mountains
In S part of county. In Baranya.
BT Hungary

villiaumite
BA halides
BT minerals
SA fluorides

Vilna
City near the Byelorussian border in SE Lithuania.
UF Vilnius
Vilno
Vilnyus
Wilna
Wilno
BT Lithuania
USSR

Vilnius
use Vilna

Vilno
use Vilna

Vilnyus
use Vilna

Vilyui Basin
use Vilyuy River basin

Vilyui River
use Vilyuy River

Vilyui Syneclise
use Vilyuy Syneclise

Vilyuy Basin
use Vilyuy River basin

Vilyuy River
Rises on the Central Siberian Plateau of E Evenk National Okrug and flows through W and W central Yakutia into the Lena River. Also search Vilyui; Vilyuy.
UF Vilyui River
BT Russian Republic
USSR

Vilyuy River basin
In E Evenk National Okrug and W and W central Yakutia in central Siberia.
UF Vilyui Basin
Vilyuy Basin
BT Russian Republic
USSR

Vilyuy Syneclise
The lower Vilyuy River basin in the area of the river's flow into Lena River in central Yakutia. Also search Vilyui Syneclise.
UF Vilyui Syneclise
Vilyuy Syneclise
BT Russian Republic
USSR

vimsite
BA borates
BT minerals

Vindhya Basin
use Vindhyan Basin

Vindhyan
Precambrian-Cambrian. Includes series of sandstones, shales, and limestones extensively exposed in the highlands of central India. Also search Vindhyan System.
UF Vindhyan System
BT Precambrian
SA Cambrian
India

Vindhyan Basin
Index countries as applicable.
UF Vindhya Basin
BT Asia
SA Bangladesh
India

Vindhyan System
use Vindhyan

Vindobonian
Europe. Includes use on level 3 under age terms(1). See list E.
BA Miocene
Neogene
Tertiary
BT Cenozoic

Vinton Canyon
BT Texas
United States

Viola Limestone
Middle and Upper Ordovician. In Patterson Ranch Group. Overlies Bromide Formation and underlies Fernvale. Central S and SW Oklahoma.
BT Ordovician
SA Middle Ordovician
Oklahoma
Upper Ordovician

violarite
BA sulfides
BT minerals

Virgilian
Provincial series, North America. Above Missourian, below Wolfcampian (Permian). Includes use on level 3 under age terms(1). See list E. Also search Virgilian Series.
UF Virgilian Series
BA Upper Pennsylvanian
Pennsylvanian
BT Paleozoic
NT Beil Limestone Member
Douglas Group
Ervine Creek Limestone
Lawrence Formation
Lecompton Limestone
Oread Limestone
Plattsmouth Limestone Member
Shawnee Group
Tonganoxie Sandstone
Wabaunsee Group

Virgilian Series
use Virgilian

Virgin Islands
An island group about 60 miles E of Puerto Rico. Before 1978, term included British Virgin Islands and the Virgin Islands of the United States, an unincorporated territory. After 1978 use this term only for the Virgin Islands of the United States; see British Virgin Islands for the others.
BT West Indies
SA British Virgin Islands
Saint Croix
Saint Thomas
United States

Virgin River valley
Index states as applicable. Also search Virgin River.
BT United States
SA Arizona
Nevada
Utah

Virginia
Includes use on level 1 as an area term (list O). For term set options see list B.
CO N363500N392800
W0751500W0833500
BA United States
NT Accomack County
NX Alleghany County
NT Amelia
Augusta County
NX Bedford County
NT Blacksburg
NX Buchanan County
NT Buckingham County
NX Campbell County
Carroll County
Clarke County
Cumberland County
NT Danville
Dickenson County
NX Essex County
Floyd County
Franklin County
Greene County
NT Hanover County
NX Highland County
NT James River
NX Lee County
Madison County
Middlesex County
Montgomery County
Nelson County
NT Norfolk
NX Orange County
NT Petersburg
NX Pulaski County
NT Pulaski Fault
Rappahannock River
Richmond
Roanoke
NX Rockingham County
Russell County
NT Saltville Fault
NX Scott County
NT Sharps
NX Stafford County
Surry County
Sussex County
NT Tazewell County
Virginia Beach
Wachapreague Inlet
NX Warren County
Washington County
Waynesboro
Westmoreland County
Wise County
NT Wythe County
NX York County
NT York River
SA Allegheny Front
Allegheny Group
Allegheny Mountains
Allegheny Plateau
Antietam Formation
Appalachian Basin
Appalachian Plateau
Aquia Formation
Ashe Formation
Atlantic Coastal Plain
Baltimore Gneiss
Bloomsburg Formation
Blue Ridge Mountains
Blue Ridge Province
Calvert Formation
Catoctin Formation
Catskill Formation
Charlotte County
Chesapeake Bay
Chickamauga Group
Chilhowee Group
Choptank Formation
Clinton Group
Coeymans Formation
Conemaugh Group
Cumberland
Cumberland Plateau
Dan River basin
Delmarva Peninsula
Eastern U.S.
Genesee Group
Glenarm Series
Greenland Gap Group
Helderberg Group
Hurricane Ridge Syncline

Juniata Formation
Kingsport Formation
Kinzers Formation
Knox Group
Lee Formation
Lincolnshire Limestone
Martinsburg Formation
Mascot Dolomite
McKenzie Formation
Monongahela Group
Nanjemoy Formation
New River
Newark Group
Newman Limestone
Onondaga Limestone
Oriskany Sandstone
Pennington Formation
Petersburg Granite
Piedmont
Pocahontas Formation
Pocono Formation
Potomac Group
Potomac River
Potomac River basin
Pottsville Group
Price Formation
Rome Formation
Saint Louis Limestone
Saint Marys Formation
Shady Dolomite
Shenandoah Valley
Tioga Bentonite
Tonoloway Limestone
Trenton Group
Tuscarora Formation
Valley and Ridge Province
Wicomico Formation
Williamsport Sandstone
Wissahickon Formation
Yorktown Formation

Virginia Beach
Independent city on the Atlantic Ocean 18 miles E of Norfolk in SE.
BT Virginia
United States

viridine
BA orthosilicates
silicates
BT minerals

Viruan
Europe.
BA Ordovician
BT Paleozoic

Viruddhachalam
use Vridhachalam

Visagapatnam
use Visakhapatnam

Visakhapatnam
City on the Coromandel Coast of the Bay of Bengal in NE Andhra Pradesh.
UF Visagapatnam
Vishakhapatnam
Vizagapatam
Vizagapatnam
BT Andhra Pradesh
India

viscoelasticity
Includes use on level 3 under deformation(1) field studies(2).
SA deformation
elastic limit
elastic strain
elasticity
strain
stress

viscosity
Includes use on level 2 under magmas(1) and lava(1); on level 3 under deformation(1) field studies(2).
SA deformation
flow
flow mechanism
lava
magmas
plasticity
rheology
viscous materials

viscous materials
SA materials
viscosity

viscous remanent magnetization
BA remanent magnetization
SA paleomagnetism

Visean
Europe. Above Tournaisian, below lower Namurian. Includes use on level 3 under age terms(1). See list E.
BA Lower Carboniferous
Carboniferous
BT Paleozoic
NT Great Scar Limestone
SA Dinantian

Viseu
District in N central Portugal. Also a city NE of Coimbra.
UF Vizeu
BT Portugal

Vishakhapatnam
use Visakhapatnam

Visla River
use Vistula River

Vistula River
Rises on the N slope of the Carpathian Mountains and flows into the Baltic Sea at Gdansk. Also search Vistula.
UF Visla River
Weichsel River
Wisla River
BT Poland

Vistula River valley
S, E central, and N part of country. Also search Vistula and Vistula Valley.
UF Vistula Valley
BT Poland

Vistula Valley
use Vistula River valley

Viterbo
City and province.
BT Latium
Italy

Viti Levu
Largest island in Fiji group.
BT Fiji
Pacific Ocean

Vitim
River which rises in central Buryatia and flows across NE Irkutsk Oblast into the Lena River on the SW border of Yakutia.
UF Vitim River
BT Russian Republic
USSR

Vitim Plateau
Gold mining area between the Vitim and Barguzin rivers E of Lake Baikal in NE Buryatia. Also search Vitim.
BT Russian Republic
USSR

Vitim River
use Vitim

Vitoria
City on the Atlantic Ocean 250 miles NE of Rio de Janeiro in Espirito Santo and city in Alava in N Spain. Index Brazil and/or Spain as applicable.
SA Brazil
Spain

vitrain
Includes use on level 3 under sedimentary rocks (1) organic residues(2). See list I.
UF pure coal
BA organic residues
BT sedimentary rocks
SA coal

vitrinite
BA macerals
SA coal
organic residues
sedimentary rocks

vitrophyre
See list H.
BA andesite-rhyolite family
BT igneous rocks
SA porphyritic texture

Vivarais
Ancient region now mostly in Ardeche department in SE part of country.
BT Ardeche
France

vivianite
BA phosphates
BT minerals

Viviparus
Genus. Includes use on level 3 under Mollusca(1) Gastropoda(2).
BA Gastropoda
Mollusca
BT Invertebrata

Vizagapatam
use Visakhapatnam

Vizagapatnam
use Visakhapatnam

Vizcaya
One of the Basque Provinces in N.
BT Spain
NT Bilbao
SA Basque Provinces

Vizeu
use Viseu

Vladeasa Massif
use Vladeasa Mountain

Vladeasa Mountain
In the Bihor Mountains of W central Transylvania. Also search Vladeasa.
UF Vladeasa Massif
BT Transylvania
Romania
SA Bihor Mountains

Vladimir
City 110 miles of Moscow. Also an oblast, and a former principality of central Russia.
BT Russian Republic
USSR

Vladivostok
City on the southern tip of a peninsula extending into Peter the Great Bay on the Japan Sea in S Primorye Kray.
BT Russian Republic
USSR

VLF
Includes use on level 3 under ionosphere(1) wave propagation(2).
SA emissions
HF
ionosphere
magnetic field
wave propagation
waves
whistlers

Vocontian Trough
BT France

Vodino
Village in 3 locations in the Ukraine; Also a village in Kuybyshev and Volgograd oblasts of the Russian Republic in the European USSR, and a village in Novosibirsk Oblast of the Russian Republic in Siberia. Index Soviet republics as applicable.
BT USSR
SA Russian Republic
Ukraine

Vogels Berg
use Vogelsberg

Vogelsberg
Mountain range in central Hesse.
UF Vogels Berg
BT Hesse
West Germany

Vogtland
Region NW of the W edge of the Erzgebirge around Plauen in S part of country.
BT Karl-Marx-Stadt
East Germany

Voivodina
use Vojvodina

Vojvodina
Autonomous province on the Hungarian and Romanian borders in N Serbia.
UF Voivodina
Voyvodina
BT Serbia
Yugoslavia

volatile combustible
use volatiles

volatile elements
SA elements
volatiles

volatile matter
use volatiles

volatiles
Includes use as a general term on level 3, e.g. under geochemistry(1).
UF volatile combustible
volatile matter
SA coal
geochemistry
volatile elements

volborthite
BA vanadates
BT minerals

volcanic
A valid term through 1977. After 1977, use volcanic rocks on level 2 under igneous rocks(1).

volcanic arcs
use island arcs

volcanic ash
Includes use on level 3 under igneous rocks(1) and glasses(2); on level 3 under sediments(1) clastic sediments(2). See list H and list N. Before 1978, also search ash.
BA pyroclastics and glasses
BT igneous rocks
SA ash
ash falls
ash flows
bentonite
clastic sediments
dust
ejecta
sediments
terrigenous materials
tuff

volcanic belts
Includes use on level 3 under plate tectonics(1).
UF belts, volcanic
SA plate tectonics
volcanoes

volcanic clay
use bentonite

volcanic features
Includes use on level 2 under geo-

morphology(1).
UF features, volcanic
NT calderas
    cinder cones
    lava channels
    lava fields
    lava lakes
    lava tunnels
    maars
SA cauldrons
    caves
    diatremes
    geomorphology
    lava
    lava flows
    lava tubes
    volcanism
    volcanoes
    volcanology

volcanic glass
Includes use on level 3 under igneous rocks(1) pyroclastics and glasses(2). See list H. For basalt glass, use volcanic glass and basalt.
UF glass, volcanic
BA pyroclastics and glasses
BT igneous rocks
SA crystallites
    devitrification
    glasses
    ignimbrite
    obsidian
    perlite
    pitchstone

volcanic rocks
Not a valid term from 1975 through 1977. After 1977, includes use on level 2 under igneous rocks(1). Before 1978, also search igneous rocks AND volcanic.
BT igneous rocks
NA diorite porphyry
    granophyre
NT scoria
SA basalt
    breccia pipes
    lava
    metavolcanic rocks
    rhyolite
    rocks
    trachyte
    trap rocks
    volcanism
    volcanoes
    volcanology

volcanicity
use volcanism

volcanics
A valid term through 1976. Use volcanic rocks.

volcanism
Includes use on level 2 under volcanology(1); on level 3 under igneous rocks(1) volcanic rocks(2) and under lava(1); on level 3 under plate tectonics(1).
UF volcanicity
    vulcanism
SA aa lava
    ash falls
    ash flows
    breccia pipes
    calderas
    cauldrons
    cinder cones
    eruptions
    fumaroles
    heat sources
    lahars
    lava
    maars
    pahoehoe
    shield volcanoes
    solfataras
    sublimates
    volcanic features
    volcanic rocks
    volcanoes
    volcanology

Volcano Island
use Vulcano

volcanoes
Used when discussing specific volcanoes. Includes use on level 2 under volcanology(1).
NA shield volcanoes
    stratovolcanoes
SA ash falls
    breccia pipes
    calderas
    eruptions
    geologic hazards
    lava
    mud volcanoes
    pipes
    plugs
    sublimates
    volcanic belts
    volcanic features
    volcanic rocks
    volcanism
    volcanology

volcanology
Used for the discipline and specific treatments. Includes use on level 1 (list A); on level 2 under area terms(1), bibliography(1), education(1), and symposia(1). If level 1, term set options are:
topic [applications, bibliography, catalogs, classification, concepts, education, history, instruments, methods, nomenclature, practice, research, symposia, textbooks, theoretical studies]
    subtopic (no area term)
volcanism [for ancient volcanoes and processes]
    subtopic [causes, classification, concepts, eruptions, fumaroles, processes, volcanoes]
volcanoes [for recent activity only]
    area (list O)
UF vulcanology
SA ash falls
    atmosphere
    bibliography
    calderas
    catalogs
    cauldrons
    continental drift
    craters
    crust
    ejecta
    eruptions
    extrusive rocks
    fumaroles
    igneous activity
    igneous rocks
    intrusions
    lava
    lava flows
    magmas
    mud volcanoes
    orogeny
    periodicity
    petrology
    plate tectonics
    seismology
    tectonics
    tiltmeters
    vents
    volcanic features
    volcanic rocks
    volcanism
    volcanoes

Volga
A valid term through 1976. Use Volga River or Volga region.

Volga region
The middle and lower courses of the Volga River and its tributaries. It includes the Volga drainage basin with the exception of parts of the NE, NW, and SW. Also search Volga.
BT Russian Republic
    USSR

Volga River
Longest river in Europe. It rises in the Valdai Hills NE of Moscow and flows SE and then S into the N Caspian Sea. Also search Volga.
BT Russian Republic
    USSR

Volga-Don Interfluve
use Volga-Don region

Volga-Don region
Between the Don and Volga rivers in the Volga-Don Canal area.
UF Volga-Don Interfluve
BT Russian Republic
    USSR

Volga-Ural Basin
use Volga-Ural region

Volga-Ural Interfluve
use Volga-Ural region

Volga-Ural region
Between the parallel courses of the lower Volga and lower Ural rivers. Also search Ural-Volga; Ural-Volga Interfluve; Ural-Volga Region; Volga-Ural; Volga-Ural Basin. Index Soviet republics as applicable.
CO N560000N600000
    E0580000E0400000
UF Ural-Volga Interfluve
    Ural-Volga region
    Volga-Ural Basin
    Volga-Ural Interfluve
BT USSR
SA Kazakhstan
    Russian Republic

Volga-Urals
Region between the middle course of the Volga River and the Urals.
BT Russian Republic
    USSR

Volgian
Europe. See list E.
BA Upper Jurassic
    Jurassic
BT Mesozoic
SA Malm

Volgograd
City in S Volgograd Oblast on the right bank of the Volga River about 280 miles from its mouth.
UF Stalingrad
    Tsaritsyn
BT Russian Republic
    USSR

Volhynia
Forested region with marshland and many lakes S and SW of the Pripet Marshes in NW Ukraine.
UF Volynia
BT Ukraine
    USSR
SA Volyn

Vologda
City on the Vologda River in S Vologda Oblast 330 miles E of Leningrad.
BT Russian Republic
    USSR

Volta Basin
River basin including Lake Volta. Index countries as applicable. Also search Volta.
BT Africa
SA Ghana
    Ivory Coast
    Togo

Upper Volta

Voltaic Republic
use Upper Volta

voltaite
BZ borates
    sulfates
    sulfides
BT minerals

Voltzia Sandstone
BT Bunter
    Triassic
SA France
    Vosges

volume
Used as a general term.
SA capacity

volume elasticity
use bulk modulus

volume susceptibility (magnetic)
use magnetic susceptibility

Volyn
Oblast in NW Ukraine. A province of Poland prior to WW II.
UF Wolyn
BT Ukraine
    USSR
SA Volhynia

Volyn-Podolia
An upland region between the Dniester and Dnieper rivers in W Ukraine. Also search Volyn-Podolian Platform; Volyn-Podolian Upland.
UF Volyn-Podolian Platform
    Volyn-Podolian Upland
BT Ukraine
    USSR

Volyn-Podolian Platform
use Volyn-Podolia

Volyn-Podolian Upland
use Volyn-Podolia

Volynia
use Volhynia

Volyno-Azov Massif
use Ukrainian Shield

vonsenite
BA borates
BT minerals

Vorarlberg
State in extreme W.
BT Austria

Voring Plateau
Underwater feature S of the Lofoten Basin off the central Norwegian coast.
BT Norwegian Sea
    Atlantic Ocean

Vorkuta
Town on the Vorkuta River in extreme NE Komi A.S.S.R. at N end of the Urals.
BT Russian Republic
    USSR

Voronezh
City on the Voronezh River near its junction with the Don River in Voronezh Oblast N of E Ukraine.
CO N503000N530000
    E0400000E0384500
BT Russian Republic
    USSR

Voronezh Anteclise
Near the Ukrainian border in SW Voronezh Oblast. Also search Voronezh Massif.
UF Voronezh Anticline
    Voronezh Anticlise
    Voronezh Massif
BT Russian Republic
    USSR

Voronezh Anticline
use Voronezh Anteclise

Voronezh Anticlise
use Voronezh Anteclise

Voronezh Massif
use Voronezh Anteclise

Voroshilovsk (region)
use Stavropol region

Vosges
Department in NE.
CO N474500N483000
    E0071500E0053000
BT France
SA Voltzia Sandstone
    Vosges Mountains

Vosges Mountains
Range extending from Belfort Gap to West German border. Index departments as applicable. Also search Vosges.
BT France
SA Haut-Rhin
    Vosges

Vostok
Small British uninhabited coral island in S.
UF Stavers Island
BT Line Islands
    Pacific Ocean

Vostok Station
Two soviet IGY stations. Vostok 1 Station is S of the Queen Mary Coast and Vostok 2 Station is midway between the Queen Mary Coast and the South Pole near the South Geomagnetic Pole. Both stations are on the Indian Ocean side of the continent.
BT Antarctica

Vourinos
Mountain SW of Kozane in W Macedonia.
UF Burono
    Vourinos Massif
BT Greece
SA Macedonia

Vourinos Massif
use Vourinos

Voyvodina
use Vojvodina

Vraca
use Vratsa

Vraconian
Europe. Lower Cretaceous or Upper Cretaceous. Includes use on level 3 under age terms (1). See list E.
BA Cretaceous
BT Mesozoic
SA Lower Cretaceous
    Upper Cretaceous

Vrancea
Range of the Carpathians near their junction with the Transylvanian Alps.
UF Vrancei Mountains
BT Romania
SA Carpathians

Vrancei Mountains
use Vrancea

Vratsa
Province in NW Bulgaria. Also a city NNE of Sofia.
UF Vraca
    Vrattsa
BT Bulgaria

Vrattsa
use Vratsa

vrbaite
BA sulfosalts
BT minerals

Vredefort Dome
N Orange Free State. Also search Vredefort.
BT Orange Free State
    South Africa

Vriddhachalam
use Vridhachalam

Vridhachalam
Town SW of Cuddalore in NE Tamil Nadu.
UF Viruddhachalam
    Vriddhachalam
BT Tamil Nadu
    India

Vries Island
use O-shima

Vulcan Mountains
Section of the Transylvanian Alps in W central Romania. Also search Vulcan.
BT Romania
SA Transylvanian Alps

vulcanism
use volcanism

Vulcano
Southernmost of the Lipari Islands in the Tyrrhenian Sea off NE Sicily.
UF Volcano Island
    Vulcano Island
BT Sicily
    Italy
SA Lipari Islands

Vulcano Island
use Vulcano

vulcanology
use volcanology

Vulture Mountain
Peak in Glacier National Park in NW.
BT Montana
    United States

Vuori-Yarvi
use Vuoriyarvi

Vuoriyarvi
Village near the Finnish border in NW Karelian A.S.S.R.
UF Vuori-Yarvi
BT Russian Republic
    USSR

Vyatka
use Kirov

Vyatka River
Rises in W foothills of the Central Urals and flows into the Kama River in N Tatar A.S.S.R. Also search Vyatka.
UF Viatka River
BT Russian Republic
    USSR

Vyatka-Kama Interfluve
Region between the lower reaches of the Vyatka and Kama rivers in the Izhevsk area W of the Urals. Also search Vyatka-Kama.
CO N550000N600000
    E0570000E0480000
BT Russian Republic
    USSR

Vychegda River
Rises in the Komi A.S.S.R. and flows W into the Northern Dvina at Kotlas in SE Arkhangelsk Oblast in N European USSR. Also search Vychegda.
BT Russian Republic
    USSR

Vyernyi
use Alma-Ata

Vyshkovo
Village N of Minsk in Byelorussia, and village in Kalinin Oblast NW of Moscow. Index Soviet republics as applicable.
BT USSR
SA Byelorussia
    Russian Republic

W
use tungsten

Wabash Formation
Comprises (ascending) Parker Limestone, Buffkin Limestone, Graysville Limestone, and Livingston Limestone. In SW Indiana.
BT Pennsylvanian
SA Indiana

Wabaunsee County
NE central.
BT Kansas
    United States

Wabaunsee Group
Includes (ascending) Severy Shale, Howard Limestone, Scranton Shale, Bern Limestone, Auburn Shale, Emporia Limestone, Willard Shale, Zeandale Limestone, Pillsbury Shale, Stotler Limestone, Root Shale, and Wood Siding Formation. E Kansas, SW Iowa, NW Missouri, SE Nebraska, and N Oklahoma.
BT Virgilian
    Upper Pennsylvanian
    Pennsylvanian
SA Iowa
    Kansas
    Missouri
    Nebraska
    Oklahoma

Wabush
Lake near the Quebec border in SW.
BT Labrador
    Canada

Waccamaw Formation
Consists of soft limestones and loose gray to buff fine quartz sands in which occasional small quartz pebbles are present. S North Carolina and S and E South Carolina.
BT lower Pliocene
    Pliocene
SA North Carolina
    South Carolina

Wachapreague Inlet
Inlet connecting small bay off town of Wachapreague with Atlantic Ocean. In Accomack County on the eastern shore. Also search Wachapreague.
BT Virginia
    United States

Wachusett-Marlborough Tunnel
Rock tunnel between the Wachusett Reservoir and Marlborough in the E central part of the state. Part of the tunnel system which provides water to Boston.
BT Massachusetts
    United States

wackestone
BA carbonate rocks
BT sedimentary rocks

Waco
City on the Brazos River in McClennan County in central.
CO N313300N313300
    W0971000W0971000
BT Texas
    United States

Wadden Sea
use Wadden Zee

Wadden Zee
Outer section of the former Zuider Zee between the West Frisian Islands and the dike enclosing IJsselmeer.
UF Wadden Sea
    Waddenzee
BT Netherlands

Waddenzee
use Wadden Zee

Wadi Araba
Depression along a valley extending from the Dead Sea to the Gulf of Aqaba. Part of the Great Rift Valley. Index countries as applicable.
UF Wadi Arabah
BT Middle East
SA Great Rift Valley
    Jordan

Wadi Arabah
use Wadi Araba

Wadia
Former Western Kathiawar state of Western India States Agency; merged in 1948 with Saurashtra. Now part of Gujarat.
UF Vadia
BT Gujarat
    India

Wadowice
Town on Skawa River in S part of country.
BT Krakow
    Poland

Wager
Bay off Roes Welcome Sound of N Hudson Bay in NE District of Keewatin.
BT Northwest Territories
    Canada

Wagga Wagga
Town on the Murrumbidgee River in S.
BT New South Wales
    Australia

Waianae
Mountain range extending along SW side of Oahu Island. Also search Waianae Range.
UF Waianae Range
BT Hawaii
    United States

Waianae Range
use Waianae

Waikato Basin
River basin in central.
BT North Island
    New Zealand

Waimanalo
Village on SE coast of Oahu.
BT Hawaii
    United States

Wairakei
Geothermal field N of Lake Taupo in central North Island. Also a village. Also search Wairakei geothermal field.
UF Wairakei geothermal field
BT North Island
    New Zealand

Wairakei geothermal field
use Wairakei

wairakite
BA zeolite group
    framework silicates
    silicates
BT minerals

Wairarapa
Lake E of Wellington in S.

BT North Island
    New Zealand

Waitemata Group
Comprises Manukau Breccias, Parnell Grit, Orakei Bay Greensand, Pakaurangi Beds, Albany Conglomerate, and Oneroa Beds all of which are known to be interbedded with the Waitemata Sandstone. In N North Island, New Zealand.
BT lower Miocene
    Miocene
SA New Zealand
    North Island

Wakayama
Prefecture in SW Honshu. Also a city on Kii Channel SSW of Osaka.
BT Honshu
    Japan

Wake
Coral atoll and 3 islets (Wake, Peale, and Wilkes) between Hawaii and Guam in the N Pacific Ocean. Also search Wake Island.
UF Wake Island
BT Pacific Ocean

Wake Island
use Wake

Walachia
Region between the Danube River and the Transylvanian Alps.
UF Wallachia
BT Romania
NT Braila
    Bucharest
    Buzau
    Buzau River
    Muntenia
    Oltenia
    Pitesti
    Ploesti
    Prahova
    Schela
    Sinaia
    Slanic
    Slatina
    Tismana
SA Fagaras Mountains
    Jiu River valley
    Mehedinti Plateau
    Olt River
    Olt River valley
    Romanian Plain
    Transylvanian Alps

Walbrzych
City on the Bobr River in SW part of country.
UF Waldenburg
    Waldenburg in Schlesien
BT Wroclaw
    Poland

Waldeck
Former German state in central part of country. Also a town in N.
BT Hesse
    West Germany

Waldenburg
use Walbrzych

Waldenburg in Schlesien
use Walbrzych

Waldron Shale
A gray or greenish-gray calcareous shale with occasional thin beds of limestone or argillaceous shale. S Indiana, W central Kentucky, and central Tennessee.
BT Middle Silurian
    Silurian
SA Indiana
    Kentucky
    Tennessee

Wales
Includes use on level 1 as an area term (list O). For term set options see list B.
CO N513000N533000
    W0024000W0051500
BT Great Britain
    United Kingdom
BA Europe
NT Aberystwyth
    Anglesey
    Arenig
    Bala District
    Brecknockshire
    Caernarvonshire
    Cardiff
    Cardigan Bay
    Cardiganshire
    Carmarthenshire
    Denbighshire
    Glamorgan
    Gower Peninsula
    Harlech
    Llandovery
    Lleyn Peninsula
    Merionethshire
    Monmouthshire
    Pembroke
    Pembrokeshire
    Radnorshire
    Skomer Island
    Snowdonia
    South Wales coal field
    Swansea
    Tremadoc Bay
SA Aberystwyth Grits
    Bristol Channel
    Penrhyn Slate
    Severn Valley
    Welsh Borderland
    Wye Valley

Walfischbai
use Walvis Bay

Walfish Bay
use Walvis Bay

Walker River
Flows through Walker River Indian Reservation into Walker Lake in W central Nevada.
BT Nevada
    United States

wall-rock alteration
use wallrock alteration

Wallace County
On the Colorado border in W.
BT Kansas
    United States

Wallace Formation
In Piegan Group. Belt Series. Thin-bedded bluish and greenish more or less calcareous shales, underlain by rapidly alternating thin beds of argillite, calcareous sandstone, and impure limestone. NE Idaho, and W Montana.
BT Precambrian
SA Idaho
    Montana

Wallachia
use Walachia

Wallowa Mountains
Range in NE.
BT Oregon
    United States

wallrock alteration
Includes use on level 3 under metasomatism(1).
UF wall-rock alteration
BA metasomatism
SA alteration
    hydrothermal alteration
    mineral deposits
    ore deposits
    thermal waters

Walsh County
On the Minnesota border in NE.
BT North Dakota
    United States

Walton County
Index states as applicable.
BX Florida
    Georgia
BT United States

Walvis Bay
Inlet of the Atlantic Ocean on the central W coast. Also a town which, along with the bay and immediate vicinity, constitutes an exclave of Cape Province, South Africa.
UF Walfischbai
    Walfish Bay
BT South-West Africa

Walvis Ridge
Extends in a NE-SW direction off South-West Africa and South Africa.
UF Walvish Ridge
BT Atlantic Ocean

Walvish Ridge
use Walvis Ridge

wandering, polar
use polar wandering

Wanganui
City at mouth of the Wanganui River in SW.
BT North Island
    New Zealand

Wanganui Valley
River valley in SW central North Island. Also search Wanganui and Wanganui River.
BT North Island
    New Zealand

Wann Formation
In Ochelata Group. Includes both Clem Creek and Washington Irving sandstones. In NE Oklahoma.
BT Missourian
    Upper Pennsylvanian
    Pennsylvanian
SA Oklahoma

Wapanucka Limestone
Underlies Atoka Formation in McAlestar Basin and Ouachita Front. Overlies Bloyd Formation in McAlestar Basin and Caney Shale and Union Valley (sandstone) in Ouachita Front. In central S and SE Oklahoma.
BT Pennsylvanian
SA Oklahoma

Wapawekka Lake
At foot of the Wapawekka Hills in central Saskatchewan. Also search Wapawekka.
BT Saskatchewan
    Canada

Warburton
Point of impact in Warburton Range. Also search Warburton Meteorite.
UF Warburton Meteorite
BT Western Australia
    Australia
SA meteorites

Warburton Meteorite
use Warburton

Ward County
Index states as applicable.
BX North Dakota
    Texas
BT United States

Wardak
Province in E central.
BT Afghanistan

Wardha River valley
In central part of country in E Maharashtra. Also search Wardha.
BT Maharashtra
    India

Ware
River in central.
BT Massachusetts
    United States

Warkalli Formation
According to "Indian Stratigraphical Nomenclature", the age of Warkalli Beds ranges from upper Miocene to Pliocene. They consist of sandy clays and lignite seams. In Kerala in S India.
BT Tertiary
SA India
    Kerala
    Miocene
    Pliocene
    upper Miocene

Warren County
Index states as applicable.
BX Georgia
    Illinois
    Indiana
    Iowa
    Kentucky
    Mississippi
    Missouri
    New Jersey
    New York
    North Carolina
    Ohio
    Pennsylvania
    Tennessee
    Virginia
BT United States

Warsaw
Province. Also a City province, and a city on both banks of the Vistula River in E central part of country.
UF Warschau
    Warszawa
BT Poland
NT Plock
    Pultusk
    Siedlce

Warsaw Formation
In Osage Group. Lower division consists of massive fine-grained earthy geode-bearing limestone below, thin bed of locally brownish dolomite cherty limestone in middle and bluish-gray slightly calcareous geode-bearing shale above. N Alabama, Illinois, Indiana, Iowa, Kentucky, NE Mississippi, E Missouri, and Tennessee.
BT Meramecian
    Upper Mississippian
    Mississippian
SA Alabama
    Illinois
    Indiana
    Iowa
    Kentucky
    Mississippi
    Missouri
    Tennessee
    Upper Mississippian

Warschau
use Warsaw

Warszawa
use Warsaw

Warta
River which rises NW of Krakow and flows NW and W into the Oder River at Kostrzyn on the East German border.
UF Varta
    Warthe
BT Poland

Warthe
use Warta

Warwick County
use Warwickshire

Warwickshire
County in central England.
CO N515500N524000
W0011500W0020000
UF Warwick County
BT England
Great Britain
United Kingdom

Wasatch County
N central.
BT Utah
United States

Wasatch Fault
N Utah.
BT Utah
United States

Wasatch Formation
Paleocene and Eocene. In Wasatch Plateau, consists of three members: lower member of sandstone, varicolored shale, conglomerate and small amounts of fresh-water limestone; and an upper member of varicolored shale and sandstone. W Colorado, central S and SE Montana, NE New Mexico, SW North Dakota, Utah, and W Wyoming.
BT Paleogene
Tertiary
SA Colorado
Eocene
Montana
New Mexico
North Dakota
Paleocene
Utah
Wyoming

Wasatch Front
Outer slope of the Wasatch Range. Index states as applicable.
BT United States
SA Idaho
Utah
Wasatch Range

Wasatch Mountains
use Wasatch Range

Wasatch Plateau
High tableland at S end of Wasatch Range in central.
BT Utah
United States

Wasatch Range
Range of Rocky Mountains extending from SE Idaho to central Utah. Index states as applicable. Also search Wasatch Mountains.
UF Wasatch Mountains
BT United States
SA Idaho
Rocky Mountains
Utah
Wasatch Front

Washakie Basin
Primarily in S Wyoming but also just within NW Colorado. Index states as applicable.
BT United States
SA Colorado
Wyoming

Washington
Includes use on level 1 as an area term (list O). For term set options see list B.
CO N453000N490000
W1165500W1244500
BA United States
NX Adams County
Benton County
NT Blue Glacier
NX Clark County
Columbia County
NT Cowlitz County
NX Douglas County
Franklin County
Garfield County
Grant County
NT Grays Harbor
Hanford Reservation
NX Jefferson County
King County
NT Kitsap Peninsula
NX Lincoln County
Mason County
NT Metaline Falls
Methow River valley
Mount Rainier
Mount Rainier National Park
Mount Saint Helens
Nisqually Glacier
Okanogan County
Olympic Mountains
Olympic Peninsula
Pasco Basin
Pend Oreille County
Puget Lowland
Puget Sound
Richland
NX San Juan County
NT San Juan Islands
Seattle
Skamania County
Snohomish County
South Cascade Glacier
Spokane
Spokane County
NX Stevens County
NT Whatcom County
SA Belt Supergroup
Blue Mountains
Cascade Range
Chuckanut Formation
Coast Ranges
Columbia Plateau
Columbia River
Columbia River Basalt
Columbia River basin
Columbia River estuary
Cordillera
Juan de Fuca Strait
Lincoln Creek Formation
Montesano Formation
Okanagan Valley
Pacific Coast
Pasayten Group
Skagit Valley
Snake River
Snake River basin
Snoqualmie Batholith
Strait of Georgia
Swauk Formation
Tono
Twin River Formation
Twin Sisters Dunite
Vancouver
Western U.S.
Yakima Basalt

Washington County
Index states as applicable.
BX Alabama
Arkansas
Colorado
Florida
Georgia
Idaho
Illinois
Indiana
Iowa
Kansas
Kentucky
Maine
Maryland
Minnesota
Mississippi
Missouri
Nebraska
New York
North Carolina
Ohio
Oklahoma
Oregon
Pennsylvania
Rhode Island
Tennessee
Texas
Utah
Vermont
Virginia
Wisconsin
BT United States

Washita Group
Lower and Upper Cretaceous. In Oklahoma, includes (ascending) Duck Creek Formation, Fort Worth Limestone, Denton Clay, Weno Clay, Pawpaw Formation, Main Street Limestone, and Grayson Shale. SW Arkansas, NW Louisiana, S Oklahoma, and Texas.
BT Cretaceous
SA Arkansas
Louisiana
Lower Cretaceous
Oklahoma
Texas
Upper Cretaceous

Washita River valley
E Texas panhandle, and W and S central Oklahoma. Index states as applicable. Also search Washita River.
BT United States
SA Oklahoma
Texas

Washoe County
NW Nevada.
BT Nevada
United States

washover
use overwash

waste, agricultural
use agricultural waste

waste disposal
A level 1 term as of 1978. Includes use on level 2 under impact statements(1), pollution(1), and underground installation(1). Term set options are:
liquid waste
  topic (effects, experimental studies, impact statements, methods, pollution, storage)
radioactive waste
  subtopic
seepage
  subtopic
site exploration
  subtopic
solid waste
  subtopic
thermal pollution
  subtopic
topic
  subtopic
UF disposal, waste
SA agricultural waste
conservation
controls
detergents
engineering geology
environmental geology
fluid injection
geologic hazards
ground water
human waste
impact statements
industrial waste
injection
land use
landfills
liquid waste
marine installations
pesticides
pollutants
pollution
radioactive waste
radioactivity
reclamation
reservoirs
sanitary landfills
seepage
sewage
site exploration
soils
solid waste
storage
sulfuric acid
tailings
thermal pollution
toxic materials
toxicity
underground installations
waste water

waste, human
use human waste

waste, industrial
use industrial waste

waste, liquid
use liquid waste

waste, radioactive
use radioactive waste

waste, solid
use solid waste

waste water
Includes use on level 3 under environmental geology(1) waste disposal(2).
UF water, waste
SA detergents
environmental geology
industrial waste
liquid waste
waste disposal
water

wasting, mass
use mass wasting

water
Includes use on level 2 under meteorology(1), under isotopes(1), and under pollution(1).
SA artesian waters
atmospheric precipitation
bottom water
brackish water
clouds
competence
connate waters
drainage
droplets
environmental geology
fresh water
gauging
geochemistry
ground water
hardness
humidity
hydration
hydrologic cycle
hydrological methods
hydrology
hydrosphere
hydrothermal processes
ice
impurities
irrigation
meltwater
meteorology
mineral waters
moisture
oil-water interface
percolation
pollution
pore water
precipitation
purification
raindrops
retention
salt water

saturation
sediment-water interface
seepage
snow
springs
surface water
thermal waters
transmissibility coefficient
waste water
water pressure
water regimes
water storage
water supply
water vapor
water yield
waterways

water balance
Also search water budget.
- UF balance, water
  hydrologic budget
  water budget
- SA aquifers
  drainage basins
  hydrology
  lakes
  reservoirs
  runoff

water biscuits
use algal biscuits

water budget
use water balance

water crop
use water yield

water cycle
use hydrologic cycle

water erosion
Includes use on level 3 under soils(1) erosion(2).
- BT erosion
- SA denudation
  erosion control
  erosion features
  soils
  wind erosion

water opal
use hyalite

water pressure
- BT pressure
- SA soil mechanics
  water

water quality
Includes use on level 3 under ground water(1), hydrology(1) and water resources(1).
- SA fresh water
  ground water
  hydrology
  irrigation
  sewage
  water resources
  water supply

water recovery
Term introduced in 1978 for recovery in relation to ground water. Before 1978 search recovery AND ground water.
- SA discharge
  ground water
  recovery
  wells

water regimes
Includes use on level 2 under soils(1).
- UF regimes, water
- SA characterization
  drainage
  irrigation
  movement
  soils
  storage
  water

water resources
Includes use on level 1 as a commodity term (list C) for economically oriented papers.
- SA economic geology
  energy sources
  environmental geology
  fresh water
  ground water
  hydrogeology
  hydrology
  ponds
  resources
  springs
  thermal waters
  water quality
  water storage
  water supply

water storage
Includes use on level 3 under engineering geology(1) reservoirs(2).
- SA dams
  engineering geology
  reservoirs
  storage
  water
  water resources
  water supply

water supply
- UF supply, water
- SA aquifers
  hydrology
  irrigation
  lakes
  pollution
  ponds
  reservoirs
  rivers
  surface water
  water
  water quality
  water resources
  water storage

water vapor
- UF vapor, water
- SA evaporation
  geochemistry
  humidity
  moisture
  water

water, waste
use waste water

water yield
- UF runout
  water crop
- SA drainage basins
  runoff
  water
  yields

water-break
use breakwaters

Waterberg System
Composed primarily of sedimentary rocks. In Transvaal, South Africa.
- BT Precambrian
- SA South Africa
  Transvaal

waterfalls
- BT fluvial features

Waterloo
Town in central Belgium just S of Brussels, town in NE central Iowa, and town in SE Ontario. Index Belgium, Iowa and Ontario as applicable.
- SA Belgium
  Iowa
  Ontario

watersheds
Includes use on level 3 under hydrology(1).
- SA drainage basins
  floods
  hydrology
  lakes
  rainfall
  rivers
  runoff
  streams
  surface water
  waterways

Waterville Formation
According to the latest published Lexique, the age of Waterville Shale is Silurian. A series of shales, fine grained sandstones, and impure limestones; often pyritiferous. In S central Maine.
- BT Silurian
- SA Maine

waterways
A valid level 1 term as of 1978. Used for geological studies on man-made or man-modified water channels. Before 1978, included use on level 2 under engineering geology(1). Term set options are:
canals
    subtopic
channels
    subtopic
design
    subtopic
erosion
    subtopic
floods
    subtopic
harbors
    subtopic
hydraulics
    subtopic
irrigation
    subtopic
rivers and streams
    subtopic
seepage
    subtopic
- SA canals
  channels
  controls
  design
  engineering geology
  environmental geology
  erosion
  estuaries
  floods
  geometry
  geomorphology
  harbors
  hydraulics
  hydrology
  irrigation
  marine installations
  ocean circulation
  reservoirs
  rivers
  rivers and streams
  seepage
  shorelines
  turbulence
  water
  watersheds

Waterways Formation
- BT Devonian
  Paleozoic
- SA Alberta

wave propagation
Includes use on level 2 under ionosphere(1) and magnetosphere(1).
- SA ELF
  HF
  ionosphere
  magnetosphere
  propagation
  VLF
  waves

wavellite
- BA phosphates
- BT minerals

waves
Includes use on level 2 under aeronomy(1) and meteorology(1).
- SA acoustical waves
  aeronomy
  airy waves
  bow shock waves
  breaking waves
  catastrophic waves
  coda waves
  elastic waves
  electromagnetic waves
  ELF
  generation
  HF
  ideal waves
  internal waves
  long-period waves
  Love waves
  meteorology
  ocean waves
  P-waves
  reflection waves
  refraction waves
  seiches
  SH-waves
  shoaling
  shock waves
  solitary waves
  surface waves
  surges
  tsunamis
  VLF
  wave propagation

Wawa
Village NE of Manila in Rizal Province.
- BT Luzon
  Philippine Islands

Wayne County
Index states as applicable.
- BX Georgia
  Illinois
  Indiana
  Iowa
  Kentucky
  Michigan
  Mississippi
  Missouri
  Nebraska
  New York
  North Carolina
  Ohio
  Pennsylvania
  Tennessee
  Utah
  West Virginia
- BT United States

Waynesboro
Borough in Franklin County, S Pennsylvania, and city in Augusta County in N central Virginia. Index states as applicable.
- BX Pennsylvania
  Virginia
- BT United States

Weald
use The Weald

Weald Clay
Divided into 3 lithological groups: Group 1- buff grey, Group 2- variegated and Group 3- yellow. In Sussex in SE England.
- BT Cretaceous
- SA England
  United Kingdom

Weald region
use The Weald

Wealden
Europe. Above Purbeckian, below Gault. Includes use on level 3 under age terms(1). See list E.
- BA Lower Cretaceous
  Cretaceous

BT Mesozoic

weathering
For treatments emphasizing the process. Includes use on level 1 (list A). See mineral deposits, genesis(1) processes(2). Term set options are:
topic [analysis, classification, environment, experimental studies, rates, textbooks]
  subtopic
  type of material [igneous rocks, metamorphic rocks, minerals, sedimentary rocks, sediments]
    specific type (e.g. rock name)
NT chemical weathering
  differential weathering
  mechanical weathering
  physical weathering
SA alteration
  clay mineralogy
  degradation
  denudation
  detritus
  diagenesis
  duricrust
  ecology
  engineering geology
  enrichment
  erosion
  etching
  exfoliation
  geochemistry
  geomorphology
  hydrolysis
  insolation
  kaolinization
  leaching
  mechanical properties
  migration of elements
  mineral deposits
  parent materials
  placers
  residuum
  sediments
  soils
  weathering crust

weathering crust
Includes use on level 3 under sediments(1) chemically precipitated sediments(2); on level 3 under soils(1) and under weathering(1). See list N.
UF crust, weathering
BA chemically precipitated sediments
BT sediments
SA calcrete
  caliche
  duricrust
  soils
  weathering

Weber County
E of Great Salt Lake in N.
BT Utah
  United States

Weber Sandstone
Pennsylvanian and Permian. In Duchesne River area of Utah, consists mainly of fine grained gray and white sandstone that weathers buff. W Colorado, and NE Utah.
BT Paleozoic
SA Colorado
  Pennsylvanian
  Permian
  Utah

weberite
BA halides
BT minerals
SA fluorides

Webster County
Index states as applicable.
BX Georgia
  Iowa
  Kentucky
  Mississippi
  Missouri
  Nebraska
  West Virginia
BT United States

websterite
Includes use on level 3 under igneous rocks(1) ultramafic family(2). See list H.
BA ultramafic family
BT igneous rocks
SA pyroxenite

Wechsel
An outlier of the Eastern Alps in E part of country. Index states as applicable.
BT Austria
SA Eastern Alps
  Lower Austria
  Styria

Weddell Sea
Arm of the S Atlantic Ocean between the Antarctic Peninsula on the W and Coats Land on the SE.
BT Antarctic Ocean

weddellite
BA carbonates
BT minerals
SA whewellite

wedges, fossil ice
use fossil ice wedges

wedges, ice
use ice wedges

Wedron Formation
Glacial till. In association with Wadsworth Till Member, Shorewood Till Member, Manitowoc Till Member, and Two Rivers Till Member. Indiana, Illinois, and under Lake Michigan.
BT Pleistocene
SA Illinois
  Indiana
  Lake Michigan

Weekeroo Station
Point of impact at Weekeroo Station near Mannahill in Australia. Also search Weekeroo Station iron meteorite.
BT South Australia
  Australia
SA meteorites

Wegener hypothesis
use continental drift

wehrlite
Includes use on level 3 under igneous rocks(1) ultramafic family(2). See list H. Also a mineral.
BA ultramafic family
BT igneous rocks
SA peridotite
  tellurides

Weichsel
use Weichselian

Weichsel River
use Vistula River

Weichselian
Term applied in Northern Europe to fourth and last glacial stage of the Pleistocene Epoch.
UF Weichsel
BA upper Pleistocene
  Pleistocene
  Quaternary
BT Cenozoic
NA Allerod
  Brandenburg Stade
  Brandon Stade
  Frankfurt Stade
  Older Dryas
  Oldest Dryas
  Perth Stade
  Younger Dryas
SA Wisconsinan
  Wurm

Weimar
City in SW part of the country.
BT East Germany
SA Erfurt

Weinheim
City NE of Mannheim in NE.
BT Baden-Wurttemberg
  West Germany

Weipa
Habitation in an aboriginal reserve on the W coast of Cape York Peninsula in N.
BT Queensland
  Australia

Weisse Elster Basin
River basin in extreme W Bohemia and S central East Germany. Index Bohemia and/or East Germany as applicable. Also search Weisse Elster.
UF White Elster Basin
BT Europe
SA Bohemia
  East Germany

Weissemburg in Bayern
use Weissenburg

Weissenberg
Town just N of Lobau near the Polish and Czechoslovak borders in Upper Lusatia.
BT East Germany

Weissenburg
Town about 30 miles S of Nuremberg in W central Bavaria.
UF Weissemburg in Bayern
  Weissenburg-am-Sand
BT Bavaria
  West Germany

Weissenburg-am-Sand
use Weissenburg

Weld County
N Colorado.
CO N400000N410000
  W1033000W1051000
BT Colorado
  United States

welded tuff
Includes use on level 3 under igneous rocks(1) pyroclastics and glasses(2). See list H.
UF tuff lava
BA pyroclastics and glasses
BT igneous rocks
SA ignimbrite
  tuff

well logging
use well-logging

well-logging
For treatments that stress methodology. Includes use on level 1 (list A). Also search well logging. Term set options are:
topic [applications, automatic data processing, instruments, interpretation, techniques]
  subtopic
  type [acoustical logging, caliper logging, dipmeter logging, electrical logging, electromagnetic logging, general, radioactivity]
    elaboration of type
UF well logging
NA acoustical logging
  caliper logging
  dipmeter logging
  electrical logging
  electromagnetic logging
SA automatic data processing
  boreholes
  cores
  cuttings
  drilling
  gamma-gamma methods
  gamma-ray methods
  gamma-ray spectroscopy
  geophysical methods
  geophysical surveys
  induced polarization
  induction
  radioactivity
  resistivity
  wells

Welland Canal
Ship canal connecting Lake Erie with Lake Ontario in SE Ontario. Also search Welland.
BT Ontario
  Canada

Wellington
City on Port Nicholson an inlet of Cook Strait on S.
BT North Island
  New Zealand

Wellington Formation
In Sumner Group. In Kansas, includes Milan Limestone Member at top; Hutchinson Salt Member in middle part but not exposed; Carlton Limestone Member below Hutchinson; Hollenburg Limestone Member near base. Central and S Kansas, and N Oklahoma.
BT Permian
SA Hutchinson Salt Member
  Kansas
  Oklahoma

wells
Includes use on level 3 under well-logging (1) and under engineering geology (1).
UF wells and drill holes
SA boreholes
  cores
  cuttings
  drilling
  engineering geology
  water recovery
  well-logging

wells and drill holes
use wells

weloganite
BA carbonates
BT minerals

Welsh Borderland
The borderland between England and Wales. (The area of the Anglo-Saxon and Norman border marches during their attempted invasions of Wales in the late Middle Ages). Index political divisions as applicable.
BT United Kingdom
SA England
  Wales

wenkite
BZ framework silicates
  sulfates
BT minerals

Wenlock Edge
A limestone ridge in Shropshire in W.
BT England
  Great Britain
  United Kingdom

Wenlock Limestone
Upper, thin-bedded, lenticular and lower, more massive and crystalline limestones, underlying the "Upper Ludlow Rock" and above the "Lower Ludlow Rock". In Shropshire in W England.
BT Wenlockian
  Middle Silurian
  Silurian

SA England
    Great Britain
    United Kingdom

Wenlockian
Europe. Above Llandoverian, below Ludlovian. Includes use on level 3 under age terms(1). See list E.
BA Middle Silurian
    Silurian
BT Paleozoic
NT Wenlock Limestone

Werfenian
Triassic. Includes use on level 3 under age terms(1). See list E.
BA Lower Triassic
    Triassic
BT Mesozoic
SA Scythian

Werillup Formation
BT Eocene
    Paleogene
    Tertiary
SA Western Australia

Werra
River which rises in Suhl District, East Germany, and unites with the Fulda River in Hesse, West Germany, to form the Weser River. Index countries as applicable.
BT Europe
SA East Germany
    West Germany

Weser River
Formed by the confluence of the Fulda and Werra rivers. Flows primarily through Lower Saxony into the North Sea. Index states as applicable. Also search Weser.
BT West Germany
SA Hesse
    Lower Saxony
    North Rhine-Westphalia

Weser-Ems
Region between the Weser and Ems rivers in W and W central.
BT Lower Saxony
    West Germany
SA Ems River

west
General term used after any of the main geographic terms (list O).
UF western
SA northwest
    southwest

West Africa
The bulge of Africa including that part of Morocco facing the Atlantic Ocean and including W Algeria. Index countries as applicable.
CO N040000N370000
    E0120000W0180000
BA Africa
SA Algeria
    Benin
    Gambia
    Ghana
    Guinea
    Guinea-Bissau
    Ivory Coast
    Liberia
    Mali
    Mauritania
    Morocco
    Niger
    Nigeria
    North Africa
    Senegal
    Sierra Leone
    Southern Africa
    Togo

West African Craton
use West African Shield

West African Shield
Index countries as applicable. Also search West African Craton.
UF West African Craton
BT Africa
SA Mali
    Niger

West Atlantic
Term introduced in 1978. Before 1978, also search Atlantic Ocean AND west or western.
BA Atlantic Ocean
SA East Atlantic
    North Atlantic
    Northeast Atlantic
    Northwest Atlantic
    South Atlantic
    Southeast Atlantic
    Southwest Atlantic

West Bengal
State bordering Bangladesh in E.
BT India
NT Bankura
    Barakar
    Burdwan
    Calcutta
    Cooch Behar
    Darjeeling
    Digha
    Purulia
    Raniganj
    Raniganj coal field
SA Ajay River
    Bengal
    Damodar Valley
    Raniganj Stage

West Carpathians
use Western Carpathians

west central
use west-central

West Coast
use Pacific Coast

West Germany
Officially known as Federal Republic of Germany or Bundesrepublik Deutschland. In W central Europe, bounded on N by North Sea and Denmark, on E by East Germany and Czechoslovakia, on SE by Austria, on S by Austria and Switzerland, on SW by France, and on W by Luxembourg, Belgium and the Netherlands. Introduced as a level 1 area term in 1978.
CO N472000N550000
    E0134500E0055500
BA Germany
    Europe
NT Baden-Wurttemberg
    Bavaria
    Bergisches Land
    Bergstrasse
    Ems River
    Franconia
    Franconian Forest
    Hamburg
    Lahn River
    Lahn River valley
    Lower Saxony
    Main River
    Nahe
    North Rhine-Westphalia
    Rhenish Schiefergebirge
    Rhineland
    Rhineland-Palatinate
    Rhon Mountains
    Saarland
    Schleswig-Holstein
    Steinheim
    Steinheim Basin
    Swabia
    Teutoburg Forest
    Weser River
    Westerwald
    Westphalia
SA Berlin
    Brunswick
    Central Alps
    Elbe River
    Elbe Valley
    Essen Beds
    Frankfurt
    Harz Foreland
    Harz Mountains
    Harz region
    Hesse
    Lower Rhine Basin
    Moselle River
    North German Plain
    Posidonia Shale
    Rhine Basin
    Rhine River
    Rhine Valley
    Saale River
    Solnhofen Limestone
    Stavelot-Venn Massif
    Stuttgart
    Werra
    Zechstein

West Greenland
Term introduced in 1978. Before 1978, search Greenland AND west.
BA Greenland
SA South Greenland

West Indian Ocean
Term introduced in 1978. Before 1978, also search Indian Ocean AND west or western.
BA Indian Ocean

West Indies
Islands between SE North America and N South America enclosing the Caribbean Sea. Index island groups as applicable. Includes use on level 1 or 2 as an area term (list O). If 1, see term set options under list B.
CO N100000N253000
    W0590000W0850000
NT Anegada Island
    Antilles
NA Bahamas
    Barbados
NT British Virgin Islands
    Caribbean Mountain Range
    Carriacou
NA Cuba
    Greater Antilles
    Guadeloupe
NT Hispaniola
NA Jamaica
NT Leeward Islands
NA Lesser Antilles
NT Montserrat
NA Puerto Rico
NT Saint Croix
    Saint Lucia
    Saint Thomas
    Saint Vincent
NA Trinidad and Tobago
NT Virgin Islands
    Windward Islands
SA Curacao
    Grenada
    Grenville
    Netherlands Antilles
    Rincon Formation
    Robles Formation

West Irian
use Irian Jaya

West Malaysia
That part of the Federation of Malaysia which is comprised of eleven states on the Malay Peninsula; Johore, Kedah, Kelantan, Malacca, Negri Sembilan, Pahang, Penang, Perak, Perlis, Selangor, and Trengganu.
BA Malaysia
SA Johore
    Kelantan
    Malay Peninsula
    Pahang
    Perak
    Selangor

West Mediterranean
Term introduced in 1978. Before 1978, search Mediterranean Sea AND west.
BT Mediterranean Sea
SA East Mediterranean

West New Guinea
use Irian Jaya

West Pacific
Term introduced in 1978. Before 1978, search Pacific Ocean AND west.
BA Pacific Ocean
SA East Pacific
    Equatorial Pacific
    North Pacific
    Northeast Pacific
    Northwest Pacific
    South Pacific
    Southeast Pacific
    Southwest Pacific

West Pakistan
The west wing of Pakistan prior to the independence of East Pakistan which became Bangladesh. A valid term through 1972. After 1972, use Pakistan. Also search Pakistan AND west.

West Siberian Basin
use Siberian Lowland

West Siberian Lowland
use Siberian Lowland

West Siberian Plain
use Siberian Lowland

West Siberian Plate
SA plate tectonics
    Siberian Platform

West Siberian Platform
use Siberian Platform

West Spitsbergen
use Vestspitsbergen

West Virginia
Includes use on level 1 as an area term (list O). For term set options see list B.
CO N371500N404000
    W0774500W0823000
BA United States
NT Blackwater
NX Boone County
    Calhoun County
    Charleston
    Clay County
NT Eskdale
NX Fayette County
    Grant County
NT Greenbrier County
NX Hampshire County
    Hancock County
    Jackson County
    Jefferson County
NT Kanawha County
NX Lincoln County
    Marion County
    Marshall County
    Mason County
    Mercer County
    Mineral County
    Monroe County
    Morgan County
    Pendleton County
    Pocahontas County
NT Preston County
NX Putnam County
    Randolph County
    Roane County
    Taylor County
    Wayne County
    Webster County
NT Wheeling
NX Wood County

SA Allegheny Front
   Allegheny Group
   Allegheny Mountains
   Allegheny Plateau
   Ames Limestone
   Antietam Formation
   Appalachian Basin
   Appalachian Plateau
   Berea Sandstone
   Bloomsburg Formation
   Blue Ridge Mountains
   Blue Ridge Province
   Catoctin Formation
   Catskill Delta
   Clinton Group
   Coeymans Formation
   Conemaugh Group
   Cumberland Plateau
   Dunkard Group
   Genesee Group
   Greenland Gap Group
   Hamilton Group
   Helderberg Group
   Hurricane Ridge Syncline
   Juniata Formation
   Martinsburg Formation
   McKenzie Formation
   Monongahela Group
   New River
   Ohio River
   Ohio River valley
   Ohio Shale
   Onondaga Limestone
   Oriskany Sandstone
   Panhandle
   Pocahontas Formation
   Pocono Formation
   Potomac River
   Potomac River basin
   Pottsville Group
   Rochester Formation
   Sharon Conglomerate
   Shenandoah Valley
   Tioga Bentonite
   Tonoloway Limestone
   Tuscarora Formation
   Valley and Ridge Province
   Williamsport Sandstone

west-central
General term used after any of the main geographic terms (list O).
UF west central
SA central

Westchester County
Just N of New York City.
BT New York
   United States

Westerly Granite
Pennsylvanian or younger. Finely crystalline gray rock, which shows minor variations in color and texture but which is petrographically the same. SE Connecticut and SW Rhode Island.
BT Pennsylvanian
SA Connecticut
   Rhode Island

western
use west

Western Alps
Ranges of the Alps in SE France, NW Italy, and SW Switzerland. Index countries as applicable.
BT Alps
   Europe
SA Cottian Alps
   France
   Italy
   Maritime Alps
   Switzerland

Western Australia
Includes use on level 1 as an area term (list O). For term set options see list B.
CO S350000S140000
   E1290000E1130000
BA Australia
NT Barrow Island
   Canning Basin
   Carnarvon Basin
   Cue
   Dalgaranga
   Darling Range
   Eastern Goldfields
   Fraser Range
   Hamersley Basin
   Hamersley Range
   Kalgoorlie
   Kambalda
   Mundrabilla
   Norseman
   Perth Basin
   Pilbara gold field
   Salt Lakes
   Shark Bay
   Warburton
   Widgiemooltha
   Wiluna
   Wittenoom Gorge
   Wolf Creek
   Yilgarn
   Yilgarn Block
SA Amadeus Basin
   Brockman Iron Formation
   Eucla Basin
   Giles Complex
   Hamersley Group
   Joseph Bonaparte Gulf
   Kalgoorlie System
   Nullarbor Plain
   Officer Basin
   Werillup Formation

Western Carpathians
Ranges of the Carpathians in S and SE Poland, and in Slovakia. Index countries as applicable. Also search West Carpathians.
UF West Carpathians
BT Carpathians
   Europe
SA Czechoslovakia
   Poland

Western Desert
Part of the Libyan Desert in W and W central.
BT Egypt
SA Libyan Desert

Western Europe
BA Europe

Western Ghat Mountains
use Western Ghats

Western Ghats
Mountain range extending 800 miles along SW and W coast as far N as mouth of Tapti River on the Gulf of Cambay.
UF Western Ghat Mountains
BT India
SA Ghats

Western Hemisphere
Used when discussing many large areas too numerous to mention. Includes use on level 1 and 2 as an area term (list O). If 1, see list B for term set options.
SA Eastern Hemisphere
   Northern Hemisphere
   Southern Hemisphere

Western Interior
Level 1 area term as of 1978. Tremendous region in North America including the Great Plains, the Rocky Mountains, the Basin and Range Province of the U.S., and the interior plateaus of Canada. Index countries as applicable.
BA North America
SA Canada
   Idaho
   Montana
   United States
   Utah
   Wyoming

Western Islands
use Hebrides

Western Morava River
use Zapadna Morava

Western Sayan
Term introduced in 1978. Before 1978, search Sayan AND west or western.
BT Sayan
   USSR

Western U.S.
Term introduced in 1978. Includes use as level 1 area term. Before 1978, also search United States AND west or western.
CO N400000N510000
   W1203000W1230000
BA United States
SA Alaska
   California
   Hawaii
   Nevada
   Oregon
   Washington

Westerwald
Mountainous region extending NE from near Koblenz for about 70 miles between the Rhine, Sieg, and Lahn rivers. Geologically it is considered part of the Rhenish Schiefergebirge (Rhenish Slate Mountains). Index states as applicable.
BT West Germany
SA Hesse
   Rhenish Schiefergebirge
   Rhineland-Palatinate

Westland
Provincial district along the Tasman Sea in W.
CO S442000S420000
   E1723000E1680000
BT South Island
   New Zealand

Westman Islands
use Vestmannaeyjar

Westmoreland County
Index states as applicable. Also search Westmoreland AND appropriate area.
BX Pennsylvania
   Virginia
BT United States

Westmorland
County in NW.
CO N541500N544500
   W0021000W0031000
BT England
   Great Britain
   United Kingdom

Weston County
On the South Dakota border in NE Wyoming. Also search Weston AND Wyoming.
BT Wyoming
   United States

Westphalia
Region and former Prussian province now in parts of 3 West German states. Index states as applicable.
BT West Germany
SA Hesse
   Lower Saxony
   North Rhine-Westphalia

Westphalian
Europe. Above upper Namurian, below Stephanian. Includes use on level 3 under age terms(1). See list E.
BA Middle Carboniferous
   Carboniferous
BT Paleozoic
SA Essen Beds

wet
A valid term through 1977. After 1977, use wet methods.

wet methods
Term introduced in 1978. Includes use on level 3 under chemical analysis(1) methods(2). Before 1978, also search chemical analysis AND wet.
SA analysis
   chemical analysis
   methods

Wet Mountains
Range of Rocky Mountains in S central.
BT Colorado
   United States
SA Rocky Mountains

wetlands
SA ecology
   geomorphology
   grasslands
   vegetation

Wetterau
Region NNE of Frankfurt much of it along the Wetter River.
BT Hesse
   West Germany

Wetterstein Limestone
According to the latest published Austrian Lexique, the age of Wetterstein Dolomite is Triassic. In Tyrol in W Austria.
BT Triassic
SA Austria
   Tyrol

Wewoka Formation
In Marmaton Group. Conformably overlies Wetunka Formation and conformably underlies Holdenville Shale.
BT Desmoinesian
   Middle Pennsylvanian
   Pennsylvanian
SA Marmaton Group
   Oklahoma

Wexford
County in SE Ireland. Also a city on Wexford Harbour off Saint George's Channel.
CO N521000N524500
   W0061000W0070000
BT Ireland

Weyer
Village on the Enns River in SE.
BT Upper Austria
   Austria

Wharton Basin
SSW of the Java Trench in E.
BT Indian Ocean

Whatcom County
On the British Columbia border E of the Strait of Georgia.
BT Washington
   United States

wheel ore
use bournonite

Wheeler County
Index states as applicable.
BX Georgia
   Nebraska
   Oregon
   Texas
BT United States

Wheeling
City on the Ohio River in the N panhandle in Ohio County.
BT West Virginia
   United States

Whetstone Lake
BT Ontario
Canada

whewellite
BA carbonates
BT minerals
SA weddellite

whistlers
Includes use on level 2 under magnetosphere(1).
SA electromagnetic radiation
lightning
magnetic field
magnetosphere
VLF

white antimony
use valentinite

White Bay
Large inlet of the Atlantic Ocean in N Newfoundland.
BT Newfoundland
Canada

White County
Index states as applicable.
BX Arkansas
Georgia
Illinois
Indiana
Tennessee
BT United States

White Elster Basin
use Weisse Elster Basin

White Island
In Bay of Plenty off N central North Island. An active volcano is located on the island.
BT North Island
New Zealand

White Limestone
Name has either been abandoned by author or rejected for use by the U.S. Geological Survey. It was a color term applied in a titular sense to 160 feet of light-gray to dark-gray dolomitic limestone with shaly layers. Leadville District, Colorado.
BT Lower Ordovician
Ordovician
SA Colorado

white mica
use muscovite

White Mountain
Peak in the Sierra Nevada Mountains in E central California, a hill W of Prague, and a village on S Seward Peninsula in W Alaska. Index Bohemia and states as applicable.
SA Alaska
Bohemia
California

White Mountains
Range of the Appalachians in N central New Hampshire, range in E California and SW Nevada, and a range in Fort Apache Indian Reservation in E Arizona. Index states as applicable. Also search White-Inyo Mountains or search White-Inyo Range for California mountains.
BT United States
SA Arizona
California
Nevada
New Hampshire

White Pine
Village on the Iron River in Ontonogan County in W Upper Penninsula.
BT Michigan
United States

White Pine County
On the Utah border in E Nevada. Also search White Pine.
BT Nevada
United States

White Pine Mine
Copper mine in the Porcupine Mountains near Lake Superior in W Upper Peninsula. Also search White Pine.
BT Michigan
United States

white pyrites
use arsenopyrite

White River
River in Alaska and SW Yukon Territory; a river in Arkansas and S Missouri; and a river in SW Indiana. Index Yukon Territory and states as applicable.
BT North America
SA Alaska
Arkansas
Indiana
Missouri
Yukon Territory

White River Group
In Wyoming, includes Chadron, Brule, and Arikaree formations. NE Colorado, E Montana, Nebraska, North Dakota, South Dakota, Wyoming.
BT Oligocene
SA Brule Formation
Chadron Formation
Colorado
Montana
Nebraska
North Dakota
South Dakota
Wyoming

White River Plateau
NW central Colorado.
UF White River Uplift
BT Colorado
United States

White River Uplift
use White River Plateau

White Russia
use Byelorussia

White Sands
National Monument constituting a great expanse of white gypsum sand and dunes in Tularosa Basin in S New Mexico. Also a proving ground for testing rockets and guided missiles SW of White Sands National Monument.
BT New Mexico
United States

White Sea
Large inlet of the Barents Sea S of the Kola Peninsula in NW European USSR.
CO N640000N660000
E0440000E0350000
BT Russian Republic
USSR

Whitehorse
City on the Yukon River in S.
BT Yukon Territory
Canada

Whitemud Formation
Clay for china, ball, fire, and stoneware products are obtained from this formation. S and SW Saskatchewan.
BT Cretaceous
SA Saskatchewan

Whitewater River valley
Primarily E and SE Indiana, but also in extreme SW Ohio. Also search Whitewater River.
BT United States
SA Indiana
Ohio

whitlockite
BA phosphates
BT minerals

Whittier Fault
E of Los Angeles in S.
BT California
United States

whole rock
Includes use on level 3 under absolute age(1). Also search whole-rock.
UF total rock
whole-rock
SA absolute age

whole-rock
use whole rock

wiborgite
use rapakivi

Wichita
City on the Arkansas River in Sedgwick County in S central.
BT Kansas
United States

Wichita County
Index states as applicable.
BX Kansas
Texas
BT United States

Wichita Mountains
Range in SW.
CO N343000N350000
W0983000W0990000
BT Oklahoma
United States

Wicklow
County in E Ireland. Also a city on the Irish Sea, and a mountain range extending along E coast.
CO N524000N531500
W0060000W0064500
BT Ireland

Wicomico Formation
In Columbia Group. In Prince Georges County, Maryland, consists of coarse gravel bed at base and finer sand and silt above; color of silt ranges from yellow to drab to dirty white. Atlantic Coastal Plain from Delaware to Florida.
BT Pleistocene
SA Delaware
District of Columbia
Florida
Georgia
Maryland
North Carolina
South Carolina
Virginia

Widawka Basin
River basin in central part of country.
UF Widawka River basin
BT Lodz
Poland

Widawka River basin
use Widawka Basin

Widgiemooltha
Village in S central.
BT Western Australia
Australia

Wiehen Mountains
Low range of the Weser Mountains in NE.
BT North Rhine-Westphalia
West Germany

Wieliczka
Town SE of Krakow.
BT Krakow
Poland

Wielun
Town in central part of country.
BT Lodz
Poland

Wien
use Vienna

Wiener Wald
A spur of the Eastern Alps W and NW of Vienna and S of the Danube River.
CO N480000N482000
E0161500E0154000
UF Vienerwald
BT Lower Austria
Austria
SA Eastern Alps

Wiesbaden
City on the Rhine River 20 miles W of Frankfurt.
BT Hesse
West Germany

Wilberns Formation
Includes four members: Welge Sandstone, Morgan Creek Limestone, Point Peak Shale, and San Saba Limestone. In central Texas.
BT Upper Cambrian
Cambrian
SA Texas

Wilcox Formation
Predominantly green, white, and black schist enclosing thin dolomite beds near base of schist. Pegmatitic quartzose gneiss occurs near middle of the formation. Above the gneiss, dark schistose grits contrast with strictly argillaceous types below. In W central Vermont.
BT Precambrian
SA Vermont

Wilcox Group
In Louisiana, comprises (ascending) Converse, Lime Hill, "Hall Summit", Marthaville, Pendleton, Sabinetown, and Carrizo formations. Gulf Coastal Plain from Georgia to S Texas plus SW Illinois, W Kentucky, SE Missouri, and W Tennessee.
BT lower Eocene
Eocene
SA Florida
Georgia
Illinois
Kentucky
Louisiana
Mississippi
Missouri
Tennessee
Texas

Wildenfels
Town at the foot of the Erzgebirge in SW.
BT Karl-Marx-Stadt
East Germany

wildflysch
SA clastic rocks
clastic sediments
flysch
sedimentary rocks
sediments

Wilhelmina Mountains
Range in central.
BT Surinam

Wilkes Land
Coastal region from the Shackleton Ice Shelf to the George V Coast on the Indian Ocean side.
BT Antarctica

Wilkinson Basin
Off the coast of Massachusetts.
BT Atlantic Ocean

Wilkinson County
Index states as applicable.
BX Georgia
Mississippi
BT United States

Will County
S and SW of Chicago.
BT Illinois
United States

Willamette River valley
use Willamette Valley

Willamette Valley
River valley in NW Oregon.
UF Willamette River valley
BT Oregon
United States

willemite
BA orthosilicates
silicates
BT minerals

Williamson County
Index states as applicable.
BX Illinois
Tennessee
Texas
BT United States

Williamsport Sandstone
In Maryland, includes Cedar Hill Limestone Member (reallocated). Abruptly overlies calcareous shales of Mackenzie Formation. W Maryland, W Virginia, and N West Virginia. Also search Williamsport Formation.
BT Cayugan
Upper Silurian
Silurian
SA Maryland
Upper Silurian
Virginia
West Virginia

Williston Basin
Index states and provinces as applicable. Also search Williston.
SA Manitoba
Montana
North Dakota
Saskatchewan
South Dakota

Willow Creek
Rises in Blue Mountains in NE and flows NW into Columbia River.
BT Oregon
United States

Willunga
Village S of Adelaide in SE.
BT South Australia
Australia

Willwood Formation
Variegated shales and hornblende-bearing sandstones conformably overlying Polecat Bench Formation near center of Bighorn Basin. In N Wyoming.
BT lower Eocene
Eocene
SA Wyoming

Willyama Complex
Great igneous and metamorphic complex comprised of slaty and schistose rocks, gneisses, amphibolites, and granites. In extreme W New South Wales.
BT Archean
SA New South Wales

willyamite
BA sulfides
BT minerals
SA ullmannite

Wilmington
City in New Castle County in N Delaware, a city on Los Angeles Harbor in Los Angeles County, California, and a city on Cape Fear River in New Hanover County, North Carolina. Index states as applicable.
BT United States
SA California
Delaware
North Carolina

Wilmington Canyon
Off the coast of Delaware. Also search Wilmington.
UF Wilmington submarine valley
BT Atlantic Ocean

Wilmington oil field
In the Wilmington area around Los Angeles Harbor in Los Angeles County. Also search Wilmington.
BT California
United States

Wilmington submarine valley
use Wilmington Canyon

Wilna
use Vilna

Wilno
use Vilna

Wilson County
Index states as applicable. Also search Wilson AND appropriate state.
BX Kansas
North Carolina
Tennessee
Texas
BT United States

Wilson Lake
A reservoir in Jerome County in S.
BT Idaho
United States

Wilts County
use Wiltshire

Wiltshire
County in S England.
CO N510000N514000
W0013000W0022000
UF Wilts County
BT England
Great Britain
United Kingdom

Wiluna
Town in a gold mining area in W central.
BT Western Australia
Australia

wind erosion
Includes use on level 3 under soils(1) erosion(2).
BT erosion
SA ablation
eolian features
erosion control
soils
water erosion
wind transport
winds

Wind River
Rises in NW Fremont County and flows SE uniting with the Popo Agie River to form the Big Horn River in W central.
BT Wyoming
United States

Wind River basin
River basin along E slopes of Wind River Range in Fremont County in W central.
BT Wyoming
United States

Wind River Formation
Divides Lysite and Lost Cabin members, each of which consists of two facies. In W Wyoming.
BT lower Eocene
Eocene
SA Wyoming

Wind River Mountains
use Wind River Range

Wind River Range
Range of the Rocky Mountains in W central Wyoming. Also search Wind River Mountains.
CO N424500N433000
W1083000W1093000
UF Wind River Mountains
BT Wyoming
United States
SA Rocky Mountains

wind, solar
use solar wind

wind transport
Includes use on level 3 under sedimentation(1).
SA ablation
eolian features
sedimentation
transport
wind erosion
winds

Windermere System
Unconformably overlays the Purcell System in W and N Purcell Mountains of SE British Columbia.
BT Proterozoic
Precambrian
SA British Columbia

Windham County
Index states as applicable.
BX Connecticut
Vermont
BT United States

Windhoek
City in central.
BT South-West Africa

winds
Includes use on level 2 under aeronomy(1) and under meteorology(1).
SA aeronomy
circulation
competence
Ekman spiral
eolian features
hurricanes
meteorology
storms
transport
wind erosion
wind transport

Windsor
City on Detroit River across from Detroit.
BT Ontario
Canada

Windsor Group
Consists of thick members of massive to poorly bedded, red or red and greenish gray, mottled siltstone, shale, and sandstone intercalated with gypsum, salt, and thin tabular sequences of limestone and dolomite. New Brunswick and Nova Scotia.
BT Mississippian
SA Canada
New Brunswick
Nova Scotia

Windward Islands
The S chain of the Lesser Antilles extending from Martinique in the N to Grenada in the S.
BT West Indies
SA Carriacou
Grenada
Lesser Antilles
Martinique
Saint Lucia
Saint Vincent

Winkler County
At the SE corner of New Mexico in W.
BT Texas
United States

Winnebago County
Index states as applicable.
BX Illinois
Iowa
Wisconsin
BT United States

Winnemucca
City on the Humboldt River in Humboldt County in NW.
BT Nevada
United States

Winnfield salt dome
In Winnfield area in Winn County in N central Louisiana.
UF Winnfield salt mine
BT Louisiana
United States

Winnfield salt mine
use Winnfield salt dome

Winnipeg
City S of Lake Winnipeg in SE.
BT Manitoba
Canada

Winnipeg Formation
Defined as the shale and sandstone section which underlies the limestone of Red River Formation and which rests upon Precambrian basement complex in Manitoba. Surface and subsurface in Manitoba, and subsurface in Saskatchewan; and subsurface in Montana, North Dakota, and South Dakota.
BT Middle Ordovician
Ordovician
SA Manitoba
Montana
North Dakota
Saskatchewan
South Dakota

Winnipegosis Formation
In Elk Point Group. In outcrop, overlies Elm Point Formation and underlies Dawson Bay Formation. Redefined in subsurface to include Elm Point Limestone. Surface and subsurface in Manitoba, and subsurface in NE Montana, W North Dakota, and NW South Dakota.
BT Middle Devonian
Devonian
SA Elk Point Group
Manitoba
Montana
North Dakota
South Dakota

Wisconsin
Includes use on level 1 as an area term (list O). For term set options see list B.
CO N423000N470000
W0864500W0925000
BA United States
NX Adams County
NT Baraboo
NX Clark County
Columbia County
Douglas County
Grant County
NT Green Lake
NX Iowa County
Iron County
Jackson County
Jefferson County
Lafayette County
NT Lake Mendota
NX Lincoln County
NT Madison
Marathon County
Milwaukee
NX Monroe County
Oneida County
Polk County

Portage County
Taylor County
NT University of Wisconsin
NX Washington County
Winnebago County
NT Wolf River Batholith
NX Wood County
SA Animikie Group
Baraboo Quartzite
Canadian Shield
Franconia Formation
Great Lakes region
Green Bay
Keweenawan
Lake Michigan
Lake Superior
Maquoketa Formation
Midcontinent
Midwest
Mississippi River
Mississippi Valley
Nonesuch Shale
Platteville Formation
Saint Peter Sandstone
Shakopee Formation
Thomson Slate
Upper Mississippi Valley

Wisconsin Range
N of the Horlick Mountains and S of Marie Byrd Land.
BT Antarctica

Wisconsinan
Illinois and Wisconsin. Includes use on level 3 under age terms (1). See list E.
BA upper Pleistocene
Pleistocene
Quaternary
BT Cenozoic
SA Weichselian
Woodfordian

Wise County
Index states as applicable.
BX Texas
Virginia
BT United States

Wisla River
use Vistula River

Wissahickon Formation
Lower Paleozoic (?). Prevalent structure is schistose. Delaware, N Maryland, SE Pennsylvania, and Virginia.
BT Paleozoic
SA Delaware
Glenarm Series
lower Paleozoic
Maryland
Pennsylvania
Virginia

witherite
BA carbonates
BT minerals
SA aragonite

Wittenoom Gorge
Village N of the SE end of the Hamersley Range in NW.
BT Western Australia
Australia

wittichenite
BA sulfobismuthites
sulfosalts
BT minerals

Witwatersrand
Region on a ridge of auriferous rock about 62 miles long and 23 miles wide with Johannesburg nearly at its center in S Transvaal. Also search Witwatersrand Basin.
CO S270000S260000
  E0300000E0260000
UF Rand
The Rand
Witwatersrand Basin

BT Transvaal
South Africa

Witwatersrand Basin
use Witwatersrand

Witwatersrand System
Contains detrital gold, uranium, and pyrite concentration. In Transvaal, South Africa.
BT Precambrian
SA South Africa
Transvaal

wodginite
BA oxides
BT minerals
SA tantalates

Wolf Creek
Point of Impact at Wolf Creek S of Hallis Creek in NE Western Australia. Also search Wolf Creek Meteorite.
UF Wolf Creek Meteorite
BT Western Australia
Australia
SA meteorites

Wolf Creek Meteorite
use Wolf Creek

Wolf River Batholith
BT Wisconsin
United States

Wolfcampian
Provincial series, North America. Above Virgilian (Pennsylvanian), below Leonardian. Includes use on level 3 under age terms(1). See list E.
BA Lower Permian
Permian
BT Paleozoic
NT Roca Formation

wolfram
use wolframite

wolframite
UF wolfram
BA tungstates
BT minerals

Wolfsberg
Town in S part of country.
BT Carinthia
Austria

Wolin
Island in the Baltic Sea just off the extreme NW part of country. Also a town on the SE shore of the island.
UF Wolin Island
Wollin
BT Szczecin
Poland

Wolin Island
use Wolin

Wolkberg Group
Composed of sedimentary rocks. In Transvaal, South Africa.
BT Proterozoic
Precambrian
SA South Africa
Transvaal

Wollaston Lake Belt
Fold Belt in NE. Also search Wollaston Lake.
BT Saskatchewan
Canada

wollastonite
BA chain silicates
silicates
BT minerals

Wollin
use Wolin

Wolverhampton
County borough NW of Birmingham in W central.
BT England
Great Britain

United Kingdom

Wolyn
use Volyn

wood
SA fossil wood
paleobotany
paleontology

Wood Canyon Formation
Overlies Stirling Quartzite; underlies Cadiz Formation.
SA California
Cambrian
Lower Cambrian
Nevada
Precambrian

Wood County
Index states as applicable.
BX Ohio
Texas
West Virginia
Wisconsin
BT United States

Wood Mountain Formation
Fluvial deposits of gravels. In S Saskatchewan.
BT Miocene
SA Saskatchewan

Wood River
City NE of St. Louis in Madison County in SW.
BT Illinois
United States

Woodbine Formation
In Oklahoma, lower member is principally crossbedded dark tuffaceous sand, red clay, and gravel lentils. Upper member mostly gray to brown crossbedded quartz and sandy gravel. SW Arkansas, W Louisiana, central S and SE Oklahoma, and Texas.
BT Gulfian
Upper Cretaceous
Cretaceous
SA Arkansas
Louisiana
Oklahoma
Texas
Upper Cretaceous

Woodbury
Town in central Tennessee. Borough in Bedford Co., S Pennsylvania. Town in W Connecticut. City in SW New Jersey. Index states as applicable.
BX Connecticut
New Jersey
Pennsylvania
Tennessee
BT United States

Woodbury Clay
In Matawan Group. Thick black clay which weathers to dove or light chocolate color. In New Jersey.
BT Upper Cretaceous
Cretaceous
SA New Jersey

Woodford Shale
Devonian and Mississippian. Upper part consists of alternating beds of black papery shale that weather light gray, and black chert in beds 1 to 4 inches thick. Central S and SE Oklahoma.
BT Paleozoic
SA Devonian
Mississippian
Oklahoma

Woodfordian
Substage in Illinois which is part of Wisconsinan Stage. Includes Peoria loess and Richland loess.
BA Pleistocene
Quaternary

BT Cenozoic
SA Illinois
Wisconsinan

Woods County
On the Kansas border in N.
BT Oklahoma
United States

Woods Hole
Village in Falmouth town, Barnstable Co., SE Massachusetts. U. S. fish and wildlife station located here. Also site of important marine biological institute and of Oceanographic Institution.
BT Massachusetts
United States

Woods Hole Oceanographic Institution
Located in village of Woods Hole in town of Falmouth at SW tip of Cape Cod in Barnstable County.
BT Massachusetts
United States

Woodson County
SE Kansas.
BT Kansas
United States

Woodstock
Town on the Saint John River in W.
BT New Brunswick
Canada

Woodward
City SE of the panhandle in Woodward County.
BT Oklahoma
United States

Worcester
City in Worcester County in central.
BT Massachusetts
United States

Worcester County
Index states as applicable.
BX Maryland
Massachusetts
BT United States

Worcestershire
County in W central.
CO N515500N523000
  W0014500W0023000
BT England
Great Britain
United Kingdom

world
use global

world ocean
Used as a general term. Also see entries for specific oceans.
UF ocean, world
SA global

worm borings
use borings

worms
Includes use on level 1 and 2 as a fossil term (list F).
NA scolecodonts
SA Annelida
Chaetopoda
Polychaetia

Wrangel Island
About 100 miles off N coast of Chukchi National Okrug in the Arctic Ocean NW of Alaska.
BT Russian Republic
USSR

Wrangell Mountains
Range near the Canadian border at the N end of the Alaskan panhandle.
BT Alaska
United States

Wreford Limestone
In Chase Group. In Kansas, includes

(ascending) Threemile Limestone, Havensville Shale, and Schroyer Limestone members. E Kansas, SE Nebraska, and central Oklahoma. Also search Wreford Megacyclothem.
BT Permian
SA Chase Group
Kansas
Nebraska
Oklahoma

wrench faults
Term introduced in 1978. Includes use on level 3 under faults(1) displacements(2). Before 1978, also search faults AND wrench.
UF basculating faults
torsion faults
BA faults
SA displacements

Wright Valley
In the Wright Glacier area W of McMurdo Sound in Victoria Land in Ross Dependency on the Pacific Ocean side. Also search Wright AND appropriate area.
BT Antarctica

Wroclaw
Province in SW Poland. Also a city on the Oder River.
UF Breslau
Slak Dolny
BT Poland
NT Bardo Mountains
Boguszow
Boleslawiec
Bystrzyca
Lesna
Lubin
Lubin Legnicki
Mirsk
Nowa Ruda
Strzegom
Strzegom-Sobotka Massif
Strzelin
Walbrzych
Zloty Stok
SA Silesian coal basin

Wuerttemberg
use Wurttemberg

wulfenite
BA molybdates
BT minerals

Wurm
Europe. Above Riss, below Holocene. Includes use on level 3 under age terms(1). See list E.
BA upper Pleistocene
Pleistocene
Quaternary
BT Cenozoic
SA Weichselian

Wurttemberg
Former German state now part of Baden-Wurttemberg. Also search Wuerttemberg.
UF Wuerttemberg
BT Baden-Wurttemberg
West Germany

wurtzite
BA sulfides
BT minerals

wustite
Artificial mineral. Also search iozite.
UF iozite
BA oxides
BT minerals

Wutach River valley
use Wutach Valley

Wutach Valley
River valley in SW Baden-Wurttemberg. Also search Wutach; Wutach River.

UF Wutach River valley
BT Baden-Wurttemberg
West Germany

Wyandotte County
Just W of Kansas City, Missouri.
BT Kansas
United States

Wyandotte Limestone
In Kansas City Group. Includes (ascending) Frisbie Limestone, Quindaro Shale, Argentine Limestone, Island Creek Shale, and Farley Limestone members. SW Iowa, E Kansas, NW Missouri, and SE Nebraska. Also search Wyandotte Limestone.
BT Missourian
Upper Pennsylvanian
Pennsylvanian
SA Iowa
Kansas
Kansas City Group
Missouri
Nebraska

Wye Valley
River valley in E Wales, and W England. Index political divisions as applicable. Also search Wye River.
BT United Kingdom
SA England
Wales

Wyman Formation
Spotted schists and phyllites with a few interbedded dolomite. In S California.
BT Precambrian
SA California

Wynyard
Town in S central Saskatchewan, and a town on Bass Strait in NW Tasmania. Index province and/or state as applicable.
SA Saskatchewan
Tasmania

Wyoming
Includes use on level 1 as an area term (list O). For term set options see list B.
CO N410000N450000
W1040500W1110500
BA United States
NX Albany County
NT Beaver Creek
NX Big Horn County
Campbell County
Carbon County
NT Casper
Converse County
NX Crook County
Fremont County
NT Gas Hills
Goshen County
Granite Mountains
Hanna Basin
Heart Mountain Fault
Jackson Hole
NX Johnson County
NT Laramie Basin
Laramie County
NX Lincoln County
NT Natrona County
Old Faithful Geyser
NX Park County
NT Pinedale
Powell
Rock Springs
Salt Creek
NX Sheridan County
NT Shirley Basin
Sierra Madre Range
Sublette County
Sweetwater County
Teapot Dome
NX Teton County
NT Teton National Forest
Uinta County

Weston County
Wind River
Wind River basin
Wind River Range
SA Absaroka Range
Almond Formation
Arikaree Group
Ash Hollow Formation
Bald Mountain
Bearpaw Formation
Beartooth Mountains
Benton Formation
Bighorn Basin
Bighorn Mountains
Black Hills
Brule Formation
Carlile Shale
Casper Formation
Chadron Formation
Cody Shale
Colorado Group
Columbia River basin
Dakota Formation
Deadwood Formation
Denver Basin
Fall River Formation
Flathead Sandstone
Fort Union Formation
Fox Hills Formation
Frontier Formation
Gallatin Range
Graneros Shale
Great Plains
Green River
Green River basin
Green River Formation
Greenhorn Limestone
Harebell Formation
Lance Formation
Laney Shale Member
Laramie Mountains
Lewis Shale
Little Missouri River basin
Lodgepole Formation
Loup Fork Group
Madison Group
Mancos Shale
Medicine Bow Mountains
Mesaverde Group
Midcontinent
Minnelusa Formation
Missouri River basin
Montana Group
Morrison Formation
Mowry Shale
Muddy Sandstone
Niobrara Formation
Ogallala Formation
Overthrust Belt
Parkman Sandstone
Phosphoria Formation
Pierre Shale
Pinyon Conglomerate
Powder River basin
Red Peak Formation
Red River Formation
Snake River
Snake River basin
Spearfish Formation
Stump Formation
Sundance Formation
Sunlight
Tensleep Sandstone
Thaynes Formation
Tongue River Formation
Uinta Mountains
Wasatch Formation
Washakie Basin
Western Interior
White River Group
Willwood Formation
Wind River Formation
Yellowstone National Park

Wythe County
SW Virginia.
BT Virginia
United States

X-ray
A valid term through 1977. After 1977, use X-ray spectroscopy or X-ray analysis.

X-ray analysis
Level 1 term introduced in 1978. Used for methodology not data. For data, use X-ray data. X-ray diffraction analysis was used on level 1 through 1977. Before 1978, also search X-ray. Term set options are:
methods
topic or instruments or application
techniques
topic
X-ray diffraction analysis
subtopic
X-ray fluorescence
subtopic
X-ray radiography
subtopic
X-ray spectroscopy
subtopic
SA analysis
chemical analysis
clay mineralogy
crystal structure
crystallography
electron microscopy
spectral analysis
spectroscopy
structural analysis
thermal analysis
X-ray data
X-ray diffraction analysis
X-ray fluorescence
X-ray radiography
X-ray spectroscopy

X-ray data
For methodology, use X-ray analysis.
SA data
X-ray analysis

X-ray diffraction analysis
A valid level 1 term through 1977 used for methodology. After 1977, used on level 2 under X-ray analysis(1).
SA analysis
chemical analysis
differential thermal analysis
electron diffraction analysis
neutron diffraction analysis
spectroscopy
X-ray analysis
X-ray fluorescence

X-ray fluorescence
As of 1978, term is used on level 2 under X-ray analysis(1) for technique or applications.
SA analysis
chemical analysis
fluorescence
luminescence
spectroscopy
X-ray analysis
X-ray diffraction analysis
X-ray fluorescence spectra

X-ray fluorescence spectra
Term introduced in 1978. Refers to data; to be distinguished from X-ray fluorescence as a method.
SA spectra
X-ray fluorescence

X-ray radiography
As of 1978, term is used on level 2 under X-ray analysis(1).
UF radiography, X-ray

X-ray spectroscopy
  SA X-ray analysis
X-ray spectroscopy
  Term introduced in 1978. Includes use on level 2 under X-ray analysis(1) and on level 3 under chemical element(1) analysis(2). Before 1978, also search spectroscopy AND X-ray.
  BT spectroscopy
  SA analysis
      spectral analysis
      X-ray analysis
X-rays
  Use the plural when referring to the rays themselves. Includes use on level 3 under spectroscopy(1), structural analysis(1) and chemical element(1); on level 3 under aeronomy(1) ionization(2).
  SA electromagnetic radiation
      gamma rays
      structural analysis
xanthochroite
  use greenockite
xanthophyllite
  use clintonite
xanthosiderite
  use goethite
Xe
  use xenon
Xe-124
  Includes use on level 3 under isotopes(1).
  SA isotopes
      xenon
Xe-126
  Includes use on level 3 under isotopes(1).
  SA isotopes
      xenon
Xe-128
  Includes use on level 3 under isotopes(1).
  SA isotopes
      xenon
Xe-129
  Includes use on level 3 under isotopes(1).
  SA isotopes
      xenon
Xe-131
  Includes use on level 3 under isotopes(1).
  SA isotopes
      xenon
Xe-132
  Includes use on level 3 under isotopes(1).
  SA isotopes
      xenon
xenoliths
  Includes use on level 2 under inclusions(1).
  UF exogenous inclusions
  BT inclusions
  SA fluid inclusions
      host rocks
      mineral inclusions
xenon
  Includes use on level 1 and 2 as a chemical element (list D).
  UF Xe
  BT noble gases
  SA argon
      elements
      isotopes
      krypton
      neon
      Xe-124
      Xe-126
      Xe-128
      Xe-129
      Xe-131
      Xe-132
xenotime
  BA phosphates
  BT minerals
Xiphosura
  Subclass. Includes use on level 3 under Arthropoda(1) Merostomata(2).
  BA Merostomata
      Chelicerata
      Arthropoda
  BT Invertebrata
xonotlite
  BA chain silicates
      silicates
  BT minerals

Y
  use yttrium
Yacoraite Formation
  In Jujuy and Salta, Argentina.
  BT Cretaceous
  SA Argentina
      Jujuy
      Salta
Yahagi River
  Flows into Chita Bay on the Pacific Ocean S of Nagoya in central Honshu. Also search Yahagi.
  BT Honshu
      Japan
Yake-Dake
  Peak NE of Kure in W Honshu.
  UF Yakedake
      Yakeyama
  BT Honshu
      Japan
Yakedake
  use Yake-Dake
Yakeyama
  use Yake-Dake
Yakima Basalt
  Upper Miocene and lower Pliocene. Typical plateau basalt. In Yakima East Quadrangle, it is exposed in four southeast-trending strips that coincide with crestal portions of four anticlinal axes. In E Washington.
  BT Tertiary
  SA lower Pliocene
      upper Pliocene
      Washington
Yakut A.S.S.R.
  use Yakutia
Yakutia
  Yakut Autonomous Soviet Socialist Republic. Located in E central Siberia. Also search Yakut.
  CO N550000N730000
      E1620000E1080000
  UF Yakut A.S.S.R.
  BT Russian Republic
      USSR
Yakutsk
  City on the bend of the Lena River in central Yakutia in E central Siberia.
  BT Russian Republic
      USSR
Yalta
  City on the Black Sea in S Crimea.
  BT Ukraine
      USSR
Yamagata
  City on the Mogami river W of Sendai in Yamagata Prefecture in N.
  BT Honshu
      Japan
Yamaguchi
  City NE of Shimonoseki in Yamaguchi Prefecture in W.
  BT Honshu
      Japan
Yamal
  Peninsula between the Kara Sea on the W and the Gulf of Ob on the E in the Yamal-Nenets National Okrug in NW Siberia.
  UF Yamal Peninsula
  BT Russian Republic
      USSR
Yamal Peninsula
  use Yamal
Yamal-Nenets
  National Okrug in N Tyumen Oblast in NW Siberia.
  UF Yamalo-Nenets
  BT Russian Republic
      USSR
Yamalo-Nenets
  use Yamal-Nenets
Yamanashi
  Prefecture W of Tokyo in central.
  BT Honshu
      Japan
Yamato Mountains
  S of Nara in S central Honshu; also a range in Victoria Land, Antarctica. Also search Yamato.
  SA Antarctica
      Honshu
      Japan
Yamhill Formation
  Comprises Mill Creek Beds, of predominantly dark-gray shale and siltstone with occasional beds of lime-cemented sandstone, and an overly of a sequence of massive sandstone beds which grades upward into more agrillaceous rock. In W Oregon.
  BT Eocene
  SA Oregon
Yamuna River
  use Jumna River
Yana
  River which rises in the Verkhoyansk Range of N central Yakutia and flows N into the Laptev Sea. Also search Yana River; Yana River region; Yana River basin.
  UF Yana River
  BT Russian Republic
      USSR
Yana River
  use Yana
Yana-Indigirka Depression
  use Yana-Indigirka Lowland
Yana-Indigirka Lowland
  Coastal lowland comprising the area around the lower courses and the lower interfluve of the Yana and the Indigirka rivers in N central Yakutia. Also search Yana-Indigirka Depression.
  UF Yana-Indigirka Depression
  BT Russian Republic
      USSR
Yaquina Bay
  Inlet of the Pacific Ocean at the mouth of the Yaquina River in W Oregon.
  UF Yaquina Estuary
  BT Oregon
      United States
Yaquina Estuary
  use Yaquina Bay
Yaracuy
  State in NW.
  BT Venezuela
Yarmouth
  Town on the Atlantic Ocean in SW.
  BT Nova Scotia
      Canada
Yaroslavl
  City NE of Moscow on the Volga River in Yaroslavl Oblast.
  BT Russian Republic
      USSR
Yarrol Basin
  BT Queensland
      Australia
Yass
  River basin in SE New South Wales. Also search Yass Basin.
  UF Yass Basin
  BT New South Wales
      Australia
Yass Basin
  use Yass
Yates Formation
  In Artesia Group. A 50-foot sandstone that is subsurface in Texas, and subsurface and surface in New Mexico.
  BT Permian
  SA New Mexico
      Texas
Yatsudake Mountain
  use Yatsugatake
Yatsugatake
  Mountain peak NW of Tokyo in central Honshu.
  UF Yatsudake Mountain
  BT Honshu
      Japan
Yavapai County
  Central.
  BT Arizona
      United States
Yazoo Clay
  In Jackson Group. Divided into (ascending) North Creek Clay, Cocoa Sand, Pachuta Clay, and Shubuta Clay members. SW Alabama, central N Louisiana and Mississippi. Also search Yazoo Formation.
  BT upper Eocene
      Eocene
  SA Alabama
      Jackson Group
      Louisiana
      Mississippi
Yb
  use ytterbium
Yeddo
  use Tokyo
Yekaterinburg
  use Sverdlovsk
yellow arsenic
  use orpiment
Yellow Sea
  Between NE China and the Korean Peninsula. As of 1977, includes use as level 1 area term (list O). For term set options see list B.
  BA Pacific Ocean
Yellowknife
  Town on NW shore of Great Slave Lake at the mouth of the Yellowknife River in S District of Mackenzie.
  BT Northwest Territories
      Canada
Yellowknife Group
  Composed of quartzite, conglomer-

ate, graywacke, meta-basalt, meta-andesite, dacite, rhyolite, tuff, agglomerate; undifferentiated gabbro. In District of Mackenzie, Northwest Territories.
BT Archean
   lower Precambrian
   Precambrian
SA Canada
   Northwest Territories

Yellowstone National Park
Largest and oldest of U. S. national parks primarily in extreme NW Wyoming. Index states as applicable. Also search Yellowstone.
CO N441000N450000
   W1095000W1110500
UF Yellowstone Park
BT United States
SA Idaho
   Montana
   Wyoming

Yellowstone Park
use Yellowstone National Park

Yemen
Yemen Arab Republic. Includes use on level 1 as an area term (list O). For term set options see list B.
BA Arabian Peninsula
SA Arabian Shield
   Asia
   Middle East
   Near East
   Red Sea Basin
   Southern Yemen

Yen Bai
use Yenbay

Yen Bay
use Yenbay

Yenbay
Town on the Red River NW of Hanoi.
UF Yen Bai
   Yen Bay
BT Vietnam

Yenisei
Not a valid term for GeoRef. Use Yenisei Basin, Yenisei River, or Yenisei Ridge.

Yenisei Basin
River basin which extends into N Mongolia in the S and drains the mountain areas W of Lake Baikal and much of the Central Siberian Plateau. Also search Yenisei. Index Mongolia and/or Russian Republic as applicable.
BT Asia
SA Mongolia
   Russian Republic
   Yenisei-Khatanga basin

Yenisei Range
use Yenisei Ridge

Yenisei Ridge
Upland which extends N-S for 400 miles along the right bank of the Yenisei River between the Trans-Siberian R.R. and the Stony Tunguska River. A major gold-mining region. Also search Yenisei, Yenisei Range, and Yenisey Ridge.
CO N560000N610000
   E0950000E0900000
UF Enisei Ridge
   Yenisei Range
   Yenisey Ridge
BT Russian Republic
   USSR

Yenisei River
Rises in the E Sayan Mountains of E Tuva A.S.S.R.; it flows W for a short distance and then N across central Siberia into the Kara Sea via Yenisei Bay and Yenisei Gulf. Also search Yenisei.
UF Enisei River
   Yenisey River
BT Russian Republic
   USSR

Yenisei-Khatanga basin
Combined drainage basin of the Khatanga and the Yenisei rivers. The Khatanga basin drains much of the northern Central Siberian Plateau S of the Taymyr Peninsula. Index Mongolia and/or Russian Republic as applicable.
BT Asia
SA Khatanga Basin
   Mongolia
   Russian Republic
   Yenisei Basin

Yenisey Ridge
use Yenisei Ridge

Yenisey River
use Yenisei River

Yeoval
Village NW of Sydney in E central.
BT New South Wales
   Australia

Yerevan
City on the Razdan River in W Armenia.
UF Erevan
   Erivan
BT Armenia
   USSR

Yerington
City in Lyon County in W.
BT Nevada
   United States

Yessei
use Yessey

Yessey
Village N of the Arctic Circle in N Evenk National Okrug in N central Siberia.
UF Essei
   Yessei
BT Russian Republic
   USSR

Yetropole
Town ENE of Sofia in W central part of country.
BT Sofia
   Bulgaria

yield strength
Includes use on level 3 under deformation(1) field studies(2).
SA compression
   deformation
   elastic limit
   elasticity
   fracture strength
   mechanical properties
   strength
   stress
   tension

yields
Includes use on level 2 under soils(1). Before 1978, also search yield.
SA fertilizers
   field crops
   forests
   fruits
   sediment yield
   soils
   tillage
   water yield

Yilgarn
Gold Field ENE of Perth in SW central Western Australia. Also search Yilgarn Goldfield.
CO S300000S300000
   E1180000E1180000
UF Yilgarn goldfield
BT Western Australia
   Australia

Yilgarn Block
UF Yilgarn Shield
BT Western Australia
   Australia

Yilgarn goldfield
use Yilgarn

Yilgarn Shield
use Yilgarn Block

Ylojarvi
Village NW of Tampere in SW part of country. Copper and wolfram mines in area.
BT Finland

Yokohama
City on W shore of Tokyo Bay in Kanagawa Prefecture in SE.
BT Honshu
   Japan

Yonezawa
City in Yamagata Prefecture in N.
BT Honshu
   Japan

Yongyang
Village N of Taegu in E central part of country. North Kyongsang.
BT South Korea

Yonne
Department in NE central.
BT France
NT Auxerre
   Avallon
   Tonnerre
SA Morvan
   Yonne Valley

Yonne River valley
use Yonne Valley

Yonne Valley
River valley in central part of country. Index departments as applicable.
UF Yonne River valley
BT France
SA Nievre
   Seine-et-Marne
   Yonne

York
City S of Harrisburg in York County in S Pennsylvania, and a city in Yorkshire in N England. Index England and/or Pennsylvania as applicable.
SA England
   Pennsylvania

York County
Index states as applicable.
BX Maine
   Nebraska
   Pennsylvania
   South Carolina
   Virginia
BT United States

York River
An estuary receiving the Pamunkey and Mattaponi rivers at West Point and flowing SE to Chesapeake Bay.
BT Virginia
   United States

Yorke Peninsula
Between Spencer Gulf on W and Gulf of Saint Vincent on E in SE.
BT South Australia
   Australia

Yorkshire
County in N.
CO N532000N544000
   E0001000W0023000
BT England
   Great Britain
   United Kingdom

Yorktown Formation
In Chesapeake Group. Divisible into 2 major zones. Zone 1 (at base) corresponds to beds exposed at Raysor Bridge, South Carolina; Zone 2 (upper) includes: Uppermost Yorktown beds at Suffolk, Virginia, the beds at Yorktown, Virginia, and the Chama-bearing bed which is correlated with the aluminous clay of Florida. E Maryland, E Virginia, and North Carolina.
BT upper Miocene
   Miocene
SA Maryland
   North Carolina
   Virginia

Yosemite National Park
On W slope of the Sierra Madre in E central California. Also search Yosemite.
UF Yosemite Park
BT California
   United States

Yosemite Park
use Yosemite National Park

Yoshino River
Rises in mountains of NW and flows E to Kii Channel at Tokushima. Also search Yoshino.
BT Shikoku
   Japan

Younger Dryas
Term used primarily in Europe for an interval of late-glacial time following the Allerod and preceding the Preboreal. Before 1978, search Dryas.
BA Weichselian
   upper Pleistocene
   Pleistocene
   Quaternary
BT Cenozoic
SA Dryas

Younger Granites
In S Niger and N Nigeria.
BT Jurassic
SA Niger
   Nigeria

Young's modulus
UF stretch modulus
BA elastic constants
SA elasticity

Ypresian
Europe. Above Thanetian (Paleocene), below Cuisian. Includes use on level 3 under age terms(1). See list E.
BA lower Eocene
   Eocene
   Paleogene
   Tertiary
BT Cenozoic
NT London Clay

ytterbium
Includes use on level 1 and 2 as a chemical element (list D).
UF Yb
SA elements
   rare earths

yttrialite
BA orthosilicates
   silicates
BT minerals

yttrium
Includes use on level 1 and 2 as a chemical element (list D).
UF Y
SA elements
   rare earths

Yuba County
N central.
BT California

United States

Yucatan
State on N Yucatan Peninsula in SE.
BT Mexico
SA Gulf Coastal Plain
Yucatan Peninsula

Yucatan Basin
Undersea feature between the Yucatan Peninsula and Belize on the W and S Cuba on the NE and E. Also search Yucatan.
BT Caribbean Sea

Yucatan Channel
Strait between W tip of Cuba and NE Yucatan Peninsula connecting the Gulf of Mexico and the Caribbean Sea.
BT Atlantic Ocean

Yucatan Peninsula
Separates the Gulf of Mexico from the Caribbean Sea in SE Mexico. Index Quintana Roo and states as applicable. Also search Yucatan.
BT Mexico
SA Campeche
Caribbean region
Quintana Roo
Yucatan

Yucatan Shelf
N and W of the Yucatan Peninsula.
BT Gulf of Mexico
SA Campeche Bank

Yucca Flat
NNW of Frenchman Flat in the Nellis Air Force Range and AEC Nuclear Testing Site.
BT Nevada
United States

Yudoma
River which rises in extreme N Khabarovsk Kray in East Siberia and flows S and SW into the Maya River.
BT Russian Republic
USSR

Yudoma Series
Homogeneous bed of gray dolomite and of dolomite limestone. In N Khabarovsk Kray and S Yakutia in E Siberia.
BT Lower Cambrian
Cambrian
SA Russian Republic
USSR

yugawaralite
BA zeolite group
framework silicates
silicates
BT minerals

Yugoslavia
Includes use on level 1 as an area term (list O). For term set options see list B.
CO N410000N470000
E0230000E0133000
BA Europe
NT Banja Luka
Bosnia
Croatia
Dalmatia
Dinaric Alps
Herzegovina
Istria
Montenegro
Neretva Valley
Sarajevo
Serbia
Serbo-Macedonian Massif
Slovenia
SA Adriatic region
Alfold
Alps
Balkan Peninsula
Banat
Danube River
Danube Valley
Eastern Alps
Julian Alps
Karawanken
Karst region
Krajiste
Macedonia
Mediterranean region
Osogovo Mountains
Pannonia
Pannonian Basin
Vardar River

Yukon River
Formed by confluence of Lewes and Pelly rivers in SW Yukon Territory. It flows NW into Alaska and then SE into Bering Sea S of Norton Sound. Index Alaska and/or Yukon Territory as applicable. Also search Yukon.
SA Alaska
Yukon Territory

Yukon Territory
Includes use on level 1 as an area term (list O). For term set options see list B.
CO N600000N700000
W1250000W1410000
BA Canada
NT Beaver River
Blow River
Dawson
Donjek Glacier
Donjek River
Hess Mountains
Kaskawulsh Glacier
Keno Hill
Klondike
Kluane Lake
Mount Wood
Ogilvie Mountains
Old Crow
Royal Creek
Steele Glacier
Stewart Valley
Whitehorse
SA Chitistone Pass
Coast Mountains
Cordillera
Liard River
Mackenzie Mountains
Michelle Formation
Nahanni Formation
Peel River
Richardson Mountains
Rocky Mountain Trench
Saint Elias Mountains
Tintina fault zone
White River
Yukon River

Yukon-Tanana Upland
Mountainous area between the Yukon and Tanana rivers in SE central Alaska.
UF Yukon-Tanana Uplands
BT Alaska
United States

Yukon-Tanana Uplands
use Yukon-Tanana Upland

Yule Marble
Former trade name for marble quarried from Leadville Limestone in Gunnison County, Colorado. In 1928 the trade name became Yule Colorado Marble.
BT Mississippian
SA Colorado
Leadville Formation

Yuma
City on the Colorado River in Yuma County in extreme SW.
BT Arizona
United States

Yuma County
Index states as applicable.
BX Arizona
Colorado
BT United States

Yunnan
Province in SW.
BT China

Yuryuzan
City on the Yuryuzan River in W Chelyabinsk Oblast in the Southern Urals.
BT Russian Republic
USSR

Z

Zabrze
City in S part of country.
UF Hindenburg
Hindenburg in Oberschlesien
BT Katowice
Poland

Zacatecas
State in central.
BT Mexico
SA Sierra Madre Occidental
Zuloaga Limestone

Zagros
Mountain system primarily in W and SW Iran. Index Iran and/or Iraq as applicable. Also search Zagros Mountains.
UF Zagros Mountains
Zagros Range
BT Asia
SA Iran
Iraq

Zagros Mountains
use Zagros

Zagros Range
use Zagros

Zaire
Formerly Belgian Congo and now officially the Democratic Republic of the Congo. Includes use on level 1 as an area term (list O). For term set options see list B.
CO S130000N051500
E0320000E0120000
UF Belgian Congo
BA Africa
NT Bambu
Kinshasa
Kivu
Kundelungu Plateau
Lukuga River valley
Matadi
Shaba
SA Congo
Congo Basin
Congo River
Kama
Kasai River
Katangan Orogeny
Lake Albert
Lake Edward
Lake Kivu
Lake Tanganyika

Zaire River
use Congo River

Zaisan
Lake primarily in Vostochno Kazakhstan Oblast in extreme E Kazakhstan. Also a town SE of the lake.
UF Zaisan Lake
Zaysan
BT Kazakhstan
USSR

Zaisan Basin
Includes Zaisan Lake and surrounding area between Narym Range of the Altai Mountains on the N and Tarbagatai Range on the S in Vostochno Oblast and Semipalatinsk Oblast in extreme E Kazakhstan. Also search Zaisan.
UF Zaisan Depression
Zaysan Basin
Zaysan Depression
BT Kazakhstan
USSR

Zaisan Depression
use Zaisan Basin

Zaisan Lake
use Zaisan

Zakopane
Town at the N foot of the Tatra Mountains in S part of country. The chief summer resort and winter sports center in Poland.
BT Krakow
Poland

Zambesi Valley
use Zambezi Valley

Zambeze Valley
use Zambezi Valley

Zambezi Valley
River valley in S central and SE Africa. Index South-West Africa and countries as applicable. Also search Zambezi; Zambezi River.
UF Zambesi Valley
Zambeze Valley
BT Africa
SA Angola
Botswana
Mozambique
South-West Africa
Zambia

Zambezia
District on the Mozambique Channel in E.
BT Mozambique

Zambia
Formerly Northern Rhodesia. Includes use on level 1 as an area term (list O). For term set options see list B.
CO S170000S080000
E0333000E0220000
UF Northern Rhodesia
BA Africa
NT Kafue
Lusaka
Nchanga
SA Congo Basin
Lake Kariba
Lake Tanganyika
Zambezi Valley

Zamora
Province on the Protuguese border in NW Spain. Also a city.
BT Spain

Zanga
use Razdan

Zangezur
Mountain range in the Lesser Caucasus extending from E of Lake Sevan S to near the Iranian border. Contains copper and molybdenum ores. Index Soviet republics as applicable. Also search Zangezur Range.
UF Zangezur Range
Zangezurskiy Mountains
BT USSR
SA Armenia
Azerbaidzhan

Zangezur Range
use Zangezur

Zangezurskiy Mountains
use Zangezur

Zanzibar
Island in the Indian Ocean off NE coast of Tanzania. United with Tanganyika in 1964 to form Tanzania.
SA Tanzania

Zao
Volcanic mountain SW of Sendai in N central Honshu.
UF Zao Volcano
BT Honshu
　　Japan

Zao Volcano
use Zao

Zapadna Morava
River which rises in SW Serbia and flows N and then SW joining the Southern Morava River near Stalac to form the Morava River. Called the Moravica River in its upper course.
UF Western Morava River
BT Serbia
　　Yugoslavia

Zaporozhe
City on the left bank of the Dnieper River in Zaporozhe Oblast in S Ukraine.
UF Zaporozhye
BT Ukraine
　　USSR

Zaporozhye
use Zaporozhe

Zaragoza
Term introduced in 1978. Municipality in Mexico, 12 miles NNW of Mexico City. Use Saragossa if referring to Spain.
BT Mexico
SA Saragossa

Zarand Basin
In Zarand area W of Dasht-i-Lut in N Kerman. Also search Zarand.
BT Kerman
　　Iran

Zaria
Town SW of Kano in N central part of country.
BT Nigeria

Zarzis
Town and oasis on the Mediterranean Sea in SE part of country.
BT Tunisia

Zawiercie
City in S part of country. Coal and iron mining nearby.
BT Katowice
　　Poland

Zaysan
use Zaisan

Zaysan Basin
use Zaisan Basin

Zaysan Depression
use Zaisan Basin

Zealand
Largest and easternmost of the major islands of Denmark. Island on which Copenhagen is located. Also search Sjaeland.
UF Seeland
　　Sjaeland
BT Denmark

Zechstein
Composed of evaporites (celestite and rock salt). In Denmark, East Germany, and West Germany. Also search Zechstein sedimentary rocks, and Zechstein evaporites. Includes use on level 3 under age terms(1). See list E.
BA Upper Permian
　　Permian
BT Paleozoic
SA Denmark
　　East Germany
　　West Germany

Zeehan
Mining town in W.
BT Tasmania
　　Australia

Zeeland
Province consisting of several islands on the North Sea coast and part of the mainland S of the Western Scheldt Estuary in SW.
BT Netherlands

Zelten
Village W of Tripoli in extreme NW part of country.
BT Libya

zeolite
BA zeolite group
　　framework silicates
　　silicates
BT minerals
SA apophyllite

zeolite facies
BT facies
SA metamorphic rocks

zeolite group
Includes use in combination with framework silicates (i.e. framework silicates,zeolite group) on level 2 under minerals(1). See list L.
BA framework silicates
　　silicates
BT minerals
NA chabazite
　　clinoptilolite
　　edingtonite
　　epistilbite
　　erionite
　　faujasite
　　ferrierite
　　garronite
　　gmelinite
　　harmotome
　　heulandite
　　laumontite
　　leonhardite
　　mesolite
　　mordenite
　　natrolite
　　offretite
　　phillipsite
　　scolecite
　　stellerite
　　stilbite
　　wairakite
　　yugawaralite
　　zeolite
　　zeolites
SA analcime
　　thomsonite

zeolites
BA zeolite group
　　framework silicates
　　silicates
BT minerals

zeolitization
BA metasomatism
SA hydrothermal alteration

Zeravshan Range
Branch of the Tien Shan extending from Alai Range W for about 200 miles in W Tadzhikistan. Sometimes considered part of the Pamir-Alai system. Also search Zeravshan.
UF Zeravshanskiy Range
BT Tadzhikistan
　　USSR
SA Tien Shan
　　Zeravshan-Hissar

Zeravshan River
Rises at W end of Alai Range and flows W disappearing in the desert near Bukhara. Index Soviet republics as applicable. Also search Zeravshan.
BT USSR
SA Tadzhikistan
　　Uzbekistan

Zeravshan-Hissar
Parallel ranges of the Tien Shan in W Tadzhikistan. The Zeravshan Range lies to the N of the Hissar Range.
BT Tadzhikistan
SA Hissar Range
　　Zeravshan Range

Zeravshanskiy Range
use Zeravshan Range

Zerenda
Village S of Kokchetav in Kokchetav Oblast in N Kazakhstan.
BT Kazakhstan
　　USSR

Zermatt
Resort village in Pennine Mountains in SW part of country.
BT Valais
　　Switzerland

Zetland
use Shetland Islands

Zeya
River in Transbaikalia which rises in the Stanovoi Mountains in E Chita Oblast and flows S and SE into the Amur River.
BT Russian Republic
　　USSR

Zeya-Bureya Basin
Fertile plain between the lower courses of the Zeya and Bureya rivers in SE Amur Oblast in Transbaikalia.
UF Zeya-Bureya Depression
　　Zeya-Bureya Plain
BT Russian Republic

Zeya-Bureya Depression
use Zeya-Bureya Basin

Zeya-Bureya Plain
use Zeya-Bureya Basin

Zhdanov
City in Donetsk Oblast on N shore of Sea of Azov.
UF Mariupol
BT Ukraine
　　USSR

Zhetybay
Village in Uralsk Oblast in extreme W.
BT Kazakhstan
　　USSR

Zhitkovichi
Town WNW of Mozyr in S.
BT Byelorussia
　　USSR

Zhitomir
City W of Kiev on the Teterev River. Also an oblast.
BT Ukraine
　　USSR

Ziaetdin
Station on the Trans-Caspian RR SE of Navoi in S central.
UF Ziatdin
BT Uzbekistan
　　USSR

Ziatdin
use Ziaetdin

Zielona Gora
Province on the East German border in W Poland. Also a city.
BT Poland
NT Nowa Sol
　　Polkowice

Zillertal Alps
Range of Eastern Alps W of the Hohe Tauern in the Tyrol and in Italy's Trentino-Alto Adige. Index countries as applicable. Also search Zillertal.
UF Zillertaler Alps
BT Alps
　　Europe
SA Austria
　　Eastern Alps
　　Italy

Zillertaler Alps
use Zillertal Alps

zinc
Includes use on level 1 and 2 as a commodity term (list C) and as a chemical element (list D); on level 2 under mineral deposits, genesis(1).
UF Zn
SA elements
　　heavy metals
　　lead-zinc deposits
　　mississippi valley-type

zinc blende
use sphalerite

zinc spar
use smithsonite

zinc spinel
use gahnite

zincite
UF red zinc ore
BA oxides
BT minerals

zinckenite
Also search zinkenite.
UF zinkenite
BA sulfantimonites
　　sulfosalts
BT minerals

zinkenite
use zinckenite

zinnwaldite
UF zinwaldite
BZ halides
　　sheet silicates
BT minerals
SA fluorides
　　mica group

zinwaldite
use zinnwaldite

Zirabulak
Station on the Trans-Caspian RR W of Kattakurgan in Samarkand Oblast in S central.
BT Uzbekistan
　　USSR

zircon
Includes use on level 1 and 2 as a commodity term (list C).
BA orthosilicates
　　silicates
BT minerals
SA cyrtolite
　　gems
　　heavy minerals

zirconium
Includes use on level 1 and 2 as a chemical element (list D). Also search Zr.
UF Zr
SA elements
　　Zr-95

zirconolite
BA oxides
BT minerals

zirkelite
Includes use as an igneous rock and

as a mineral.
BZ oxides
pyroclastics and glasses

Zlatibor Massif
use Zlatibor Mountains

Zlatibor Mountains
Range of the S Dinaric Alps in W Serbia.
UF Zlatibor Massif
BT Serbia
Yugoslavia
SA Dinaric Alps

Zloty Stok
Town E of Klodzko in Lower Silesia in SW part of country.
UF Reichenstein
Rowne
BT Wroclaw
Poland

Zmeinogorsk
Town SE of Rubtsovsk in S Altai Kray near the E Kazakhstan border.
BT Russian Republic
USSR

Zn
use zinc

Zoantharia
Subclass. Includes use on level 2 under Coelenterata(1). See list F.
BA Anthozoa
Coelenterata
BT Invertebrata
SA Tetracorallia

Zod
Village in E.
BT Armenia
USSR

zodiacal dust
use cosmic dust

zoisite
BA epidote group
orthosilicates
silicates
BT minerals
SA clinozoisite
tanzanite

Zonal soils
See list M.

BA soils
SA Brown soils
Chernozems
Chestnut soils
Desert soils
Podzols
Red soils
Sierozems
Tundra soils

zonation
A valid term through 1976. Use zoning.

zone, auroral
use auroral zone

zone, Benioff
use Benioff zone

zone of mobility
use asthenosphere

zone, oxidation
use oxidation zone

zones
A valid term through 1977. Use zoning.

zones, fault
use fault zones

zones, fracture
use fracture zones

zones, low-velocity
use low-velocity zones

zones, rift
use rift zones

zones, shear
use shear zones

zones, subduction
use subduction zones

zones, transition
use transition zones

Zonguldak
Province in NW Anatolia. Also a city on the Black Sea.
BT Turkey
Middle East
NT Amasra

zoning
Includes use on level 2 under metamorphism(1); on level 3 under metasomatism(1), under crystal growth(1) and under mineral name. Before 1975, also search mineral zoning. Also search zonation; zones.
SA crystal growth
intrusions
metamorphism
metasomatism

zoogeography
use biogeography

zooplankton
BT plankton

zooxanthellae
SA algae
Protista

Zr
use zirconium

Zr-95
Includes use on level 3 under isotopes(1).
SA isotopes
zirconium

Zubair Formation
Lower Aptian-Hauterivian. In N, central, and S Iraq. In S, comprises over 1200 feet of sandstones, siltstones, and shales.
BT Lower Cretaceous
Cretaceous
SA Aptian
Hauterivian
Iraq
Middle East

Zulia
State encircling almost all of Lake Maracaibo in NW.
BT Venezuela
NT Lake Maracaibo
SA Maracaibo Basin

Zuloaga Limestone
Coahuila and Zacatecas, Mexico. Also search Zuloaga Formation.
BT Upper Jurassic
Jurassic
SA Coahuila
Mexico
Zacatecas

Zululand
Region comprising native reserves of the Zulus in NE.
BT Natal
South Africa

Zungaria
use Dzhungaria

Zuni Mountains
Range in McKinley and Valencia counties in W.
BT New Mexico
United States

zunyite
BZ halides
orthosilicates
BT minerals

Zurich
Canton in N part of country. Also a city at NW end of Lake of Zurich.
BT Switzerland

zussmanite
BA sheet silicates
silicates
BT minerals
SA mica group

Zwickau
City on the Mulde River S of Leipzig in S part of country.
BT Karl-Marx-Stadt
East Germany

Zwischengebirge
use median masses

Zyryanka
Town on the Kolyma River at the mouth of the Zyryanka River in NE Yakutia. A coal mining center.
BT Russian Republic
USSR

Zyryanovsk
City SE of Ust Kamengorsk in Vostochno Kazakhstan Oblast in extreme E.
BT Kazakhstan
USSR

INTRODUCTION TO THE GUIDE TO INDEXING

This Guide is used by the GeoRef editors/indexers for selecting index terms and arranging them into the three-level hierarchical index entries found in the printed products made from the GeoRef database.

Each complete index entry (set) has three parts; first, second, and third-level terms. For example, the five entries below appeared in the 1978 Bibliography and Index of Geology:

(1) 1st level, 2nd level
    3rd level

    Illinois—engineering geology
      petroleum engineering: Low tension waterflood pilot at the Salem Unit, Marion County, Illinois; Part 1, Field implementation and results 10819

(2) 1st level, 2nd level
    3rd level

    Illinois—environmental geology
      land use: Geology for planning in De Witt County, Illinois 10414

(3) 1st level, 2nd level
    3rd level

    Illinois—paleobotany
      pteridophytes: Preliminary report on permineralized Senftenbergia from the Chester Series of Illinois 08870

(4) 1st level, 2nd level
    3rd level

    Illinois—paleontology
      ichnofossils: Some trace fossils from the Clear Creek Limestone (Lower Devonian) of southern Illinois 08889
      problematic fossils: Echiura from the Pennsylvanian Essex fauna of northern Illinois 08962

(5) 1st level, 2nd level
    3rd level

    Illinois—sedimentary petrology
      sediments: Sediments and sedimentary structures of a beach-ridge complex, southwestern shore of Lake Michigan 08628

All of the above documents were indexed with the term Illinois. Then in addition to Illinois document (5), for example, also was indexed with the terms sedimentary petrology (2nd-level term) and sediments (3rd-level term).

This Guide consists of a Main List which is found embedded in the Thesaurus but could be ordered separately, and several special lists of terms.

Main List

The Main List is made up of sets of three-level terms. Each set indicates, for each 1st-level term, specific terms and/or classes of terms which can be used as 2nd- and 3rd-level terms for that 1st-level term.

In this manner, the breakdown of topics under each 1st-level term is regularized.

In addition, each set contains some explanatory notes on the scope of the set and a matrix of other sets related to it in a geological or indexing context. Furthermore, most sets contain lists of common terms which may be used on the 3rd level.

The underlined words and phrases in these sets are terms used in GeoRef indexing. Within an entry for a 1st-level term, the subsequent paragraphs indicate the 2nd-level options; the indentation indicates the 3rd-level options. For example:

21. crust

    topic [age, anomalies, composition, concepts, evolution, genesis, interpretation, observations, processes, properties, structure, theoretical studies, thickness]

    free (no area term)

The above entry indicates that 2nd-level terms under crust will be a topic such as age, composition, etc. For crust, the 3rd-level terms under a topic term can be freely chosen, but should not include any area terms.

Sample entries under crust, according to this breakdown, taken from the 1978 Bibliography and Index of Geology are:

> **crust—anomalies**
> *gravity anomalies:* The effect of the sedimentary layer of the Japan Sea on gravity anomalies and the isostatic state of the Earth's crust 09839
>
> **crust—evolution**
> *mineral deposits, genesis:* Conditions of formation of metamorphic deposits during the evolution of mobile zones of the Earth's crust 11236
> *sial:* Three arguments for continual evolution of sial throughout geologic time 09845
> *transformations:* Tectonosphere, exogene processes, and living matter 09817

The special lists (list A-O)

The special lists are itemized on the contents page. They consist of (1) various categories of terms which are used as 2nd- or 3rd-level terms in index sets with the 1st-level terms in the Main List, as well as (2) classes of 1st-level terms not included in the Main List, e.g. commodity terms (List C), and fossils (List F).

List A

First order terms in GeoRef

This is a listing of all first-level terms, misspellings included, that appear on the SDC file for the years 1961 through 1977. The following codes represent:

G = Geophysical Abstracts Index
N = Bibliography of North American Geology
no letter designation = Geological Society of America Bibliography and Index of Geology

```
Absolute age                                              72 73 74 75 76 77
Absolute age, dates              67 68 69 70 71
Absolute age, dates              61 62 63 64 65 66 67 68 69 70
Absolute age, dates              66 67 68 69 70 71
Absolute age, methods            67 68 69 70 71
Absolute age, methods            61 62 63 64 65 66 67 68 69 70
Absolute age, methods            66 67 68 69 70 71
Absorption spectrophotometry              64
Acoustical exploration                    64
Acoustical logging               61
Actinium                                  69 70    72
Actinium                                     65
Aden                                      68    70
Adriatic Sea                                              72 73 74 75 76 77
Adriatic Sea                              68    70
Adriatic sea                     67 68 69 70 71
Aegean Sea                                                               77
Aeronomy                                                        75 76 77
Afars and Issas                              71
Afars and Issas Territory                                 73 74 75 76 77
Afghanistan                      67 68 69 70 71 72 73 74 75 76 77
Afghanistan                      66 67 68 69 70 71
Africa                           67 68 69 70 71 72 73 74 75 76 77
Africa                           66 67 68 69 70 71
Alabama                                69 70 71 72 73 74 75 76 77
Alabama                          61 62 63 64 65 66 67 68 69 70
Alabama                          66    68 69 70 71
Alaska                                 69 70 71 72 73 74 75 76 77
Alaska                           61 62 63 64 65 66 67 68 69 70
Alaska                           66 67 68 69 70 71
Albania                          67 68 69 70 71 72    74 75 76
Alberta                                69 70 71 72 73 74 75 76 77
Alberta                          61 62 63 64 65 66 67 68 69 70
Alberta                          66 67 68 69 70 71
Algae                            67 68 69 70 71 72 73 74 75 76 77
Algae                            61 62 63 64 65 66 67 68 69 70
Algeria                          67 68 69 70 71 72 73 74 75 76 77
Algeria                          66 67 68 69 70 71
Alkali metals                          68
Alluvial fans                    61 62 63 64 65 66
Alluvial plains                  61
Alpha activation analysis        61
```

	Alps	67	68	69	70	71	72	73	74	75	76	77
G	Alps	66	67	68	69	70	71					
	Aluminum	67	68	69	70	71	72	73	74	75	76	77
N	Aluminum	61	62	63	64	65	66	67	68	69	70	
N	American Samoa			63		65		67	68			
	Americium				69							77
	Amphibia	67	68	69	70	71	72	73	74	75	76	77
N	Amphibia	61	62	63	64	65	66	67	68	69	70	
N	Amphineura				64							
	Andes		68	69	70	71	72	73	74	75	76	77
G	Andes		68		70							
N	Andesite	61										
	Andorra		68			71		73				
	Angiosperms	67	68	69	70	71	72	73	74	75	76	77
N	Angiosperms	61	62	63	64	65	66	67	68	69	70	
	Angola	67	68	69	70	71	72	73	74	75	76	77
G	Angola	66	67		69	70						
N	Anhydrite	61										
N	Anion exchange-spectrochemical analysis	61										
	Annelida	67	68	69	70	71	72	73	74	75	76	77
N	Annelida		62	63			66	67	68	69	70	
	Antarctic Ocean	67	68	69	70	71	72	73	74	75	76	77
G	Antarctic Ocean	66		68								
G	Antarctic region		68	69	70							
	Antarctica	67	68	69	70	71	72	73	74	75	76	77
G	Antarctica	66	67	68	69	70	71					
	Anthozoa	67	68	69	70	71						
N	Anthozoa	61	62	63	64	65	66	67	68	69	70	
N	Anticlines				64							
	Antimony	67	68	69	70	71	72	73	74	75	76	77
N	Antimony	61	62	63	64	65	66	67	68	69	70	
	Apennines	67	68		70	71	72	73	74	75	76	77
N	Appalachian Basin	61										
	Appalachians			69	70	71	72	73	74	75	76	77
N	Appalachians	61	62	63	64	65	66	67	68	69	70	
G	Appalachians		67	68		70	71					
	Arabia	67	68	69								
G	Arabia		68									
	Arabian Peninsula				70	71	72	73	74	75	76	77
	Arabian Sea											77
	Arabian peninsula	67	68	69								
	Arachnida	67	68	69	70	71						
N	Arachnida			63			66			69		
	Archaeocyatha	67	68	69	70	71	72	73	74	75	76	77
N	Archaeocyatha	61	62	63				67		69	70	
N	Arctic		62	63		65	66	67	68	69	70	
G	Arctic	66	67	68		70	71					
N	Arctic America	61			64							
	Arctic Ocean	67	68	69	70	71	72	73	74	75	76	77
N	Arctic Ocean	61	62	63	64	65	66	67	68	69	70	
G	Arctic Ocean	66	67	68	69	70	71					
	Arctic region	67	68	69	70	71	72	73	74	75	76	77
	Argentina	67	68	69	70	71	72	73	74	75	76	77
G	Argentina	66	67	68	69	70	71					
	Argon	67	68	69	70	71	72	73	74	75	76	77
N	Argon	61	62	63	64	65	66	67	68	69	70	
N	Arid regions	61			64							

Arizona			69	70	71	72	73	74	75	76	77
Arizona	61	62	63	64	65	66	67	68	69	70	
Arizona	66	67	68	69	70	71					
Arkansas			69	70	71	72	73	74	75	76	77
Arkansas	61	62	63	64	65	66	67	68	69	70	
Arkansas	66		69	70	71						
Arsenic	67	68	69	70	71	72	73	74	75	76	77
Arsenic		63	64	65	66		68	69	70		
Artesian waters and wells	67	68	69	70	71						
Artesian waters and wells	61	62	63	64	65	66	67	68	69	70	
Arthropoda	67	68	69	70	71	72	73	74	75	76	77
Arthropoda	61	62	63			66	67	68	69		
Artifacts	67	68	69	70	71						
Artifacts	61	62	63	64	65	66	67	68	69	70	
Asbestos	67	68	69	70	71	72	73	74	75	76	77
Asbestos		62	63	64	65	66	67	68	69	70	
Ascension Island						72					
Ascension Island			69								
Ascension island	67			71							
Asia	67	68	69	70	71	72	73	74	75	76	77
Asia	66	67	68	69	70	71					
Asia Minor	67										
Asphalt									70		
Associations	67	68	69	70	71	72	73	74	75	76	77
Associations	61	62	63	64	65	66	67	68	69	70	
Astatine				65							
Asteroidea	67		69	70	71						
Asteroidea							68				
Asteroids			70	71	72	73	74	75	76	77	
Asteroids	66		68	69	70	71					
Asterozoa	68	69	70	71							
Asterozoa	62	63		65	66	67	68	69	70		
Astrophysics and solar physics									76		
Atlantic Coastal Plain		70	71	72	73	74	75	76	77		
Atlantic Coastal Plain	61	62	63	64	65	66	67	68	69	70	
Atlantic Ocean	67	68	69	70	71	72	73	74	75	76	77
Atlantic Ocean	61	62	63	64	65	66	67	68	69	70	
Atlantic Ocean	66	67	68	69	70	71					
Atlantic region									76	77	
Atmosphere	67	68	69	70	71	72	73	74	75	76	77
Atmosphere	61		63		65	66	67	68	69	70	
Atmosphere	66	67	68	69		71					
Atomic energy					66						
Australasia	67	68	69	70	71	72	73	74	75	76	77
Australia	67	68	69	70	71	72	73	74	75	76	77
Australia	66	67	68	69	70	71					
Austria	67	68	69	70	71	72	73	74	75	76	77
Austria	66	67	68	69	70	71					
Automatic data processing	67	68	69	70	71	72	73	74	75	76	77
Automatic data processing		62	63		65	66	67	68	69	70	
Automatic data processing	66	67	68	69	70	71					
Autoradiography			70								
Autoradiography		62		64		66					
Aves	67	68	69	70	71	72	73	74	75	76	77
Aves	61	62	63	64	65	66	67	68	69	70	
Azores			69	70	71	72	73	74	75	76	77
Azores	66	67	68	69	70	71					

	Term										
	Bacteria	67	68	69	70	71	72	73	74	75	76	77
N	Bacteria		62	63	64	65	66	67	68	69	70	
	Bahamas			69	70	71	72	73	74	75	76	77
N	Bahamas	61	62	63	64	65	66	67	68	69	70	
G	Bahamas	66	67	68		70						
	Bahrain											77
	Balearic Islands								74	75	76	77
	Balkan Peninsula						72	73	74	75	76	77
	Balkan peninsula	67			70	71						
	Baltic Sea						72	73	74	75	76	77
G	Baltic Sea		68	69	70	71						
	Baltic region	67	68	69	70	71	72	73	74	75	76	77
G	Baltic region		68									
	Baltic sea	67	68	69	70	71						
G	Banda Sea				70							
	Bangladesh							73	74	75	76	77
	Barbados								74	75	76	77
N	Barbados	61			64		66	67	68	69	70	
G	Barbados			68		70						
G	Barents Sea			69		71						
	Barite	67	68	69	70	71	72	73	74	75	76	77
N	Barite	61	62	63	64	65	66	67	68	69	70	
	Barium	67	68	69	70	71	72	73	74	75	76	77
N	Barium	61	62	63	64	65	66	67	68	69	70	
N	Bars	61			64							
N	Basalt	61						67				
N	Base metals				64							
N	Baselevel				64							
N	Basin and Range									69		
G	Basin and Range Province				70							
N	Basin, structural										70	
	Basins, structural	67	68	69	70	71						
N	Basins, structural	61	62	63	64	65	66	67	68	69	70	
	Basutoland	67										
	Batholiths	67	68	69	70	71						
N	Batholiths	61	62	63	64	65	66	67	68	69	70	
	Bauxite	67	68	69	70	71	72	73	74	75	76	77
N	Bauxite	61	62	63	64	65	66	67	68	69	70	
G	Bay of Biscay				70							
N	Beaches				64							
G	Bechuanaland						66					
	Belgium	67	68	69	70	71	72	73	74	75	76	77
G	Belgium			66	67	68	69	70	71			
	Belize									75	76	77
	Benin										76	77
	Bentonite	67	68	69	70	71	72	73	74	75	76	77
N	Bentonite	61	62	63	64	65	66	67	68	69	70	
	Bering Sea	67					72	73	74	75	76	77
N	Bering Sea								68	69	70	
G	Bering Sea		68	69	70	71						
	Bering sea			69	70	71						
	Bermuda			69	70	71	72	73	74	75	76	77
N	Bermuda	61	62	63	64	65	66	67	68	69	70	
G	Bermuda		68	69	70	71						
G	Bermuda Islands	66										
N	Beryl				64							
	Beryllium	67	68	69	70	71	72	73	74	75	76	77

Beryllium		61	62	63	64	65	66	67	68	69	70
Beryllium										69	
Bhutan										75	76
Bibliography		67	68	69	70	71	72	73	74	75	76	77
Bibliography		61	62	63	64	65	66	67	68	69	70
Bibliography										69	
Biogeochemical prospecting		67	68	69	70	71					
Biogeochemical prospecting			62	63		65	66	67	68	69	70
Biogeochemical surveys											70
Biogeochemistry			62		64					69	
Biogeography				69	70	71	72	73	74	75	76	77
Biogeography		61			64						
Biography		67	68	69	70	71	72	73	74	75	76	77
Biography		61	62	63	64	65	66	67	68	69	70
Bismark Islands										69	
Bismuth		67	68	69	70	71	72	73	74	75	76	77
Bismuth		61	62	63	64	65	66	67	68	69	70
Bitumens		67	68	69	70	71	72	73	74	75	76	77
Bitumens		61	62	63	64	65	66	67	68		70
Bituminous sands		61									
Black Sea		67	68	69	70	71	72	73	74	75	76	77
Black Sea		66	67	68	69	70	71				
Blastoidea		67	68	69	70	71					
Blastoidea		61	62	63	64	65	66	67		69	70
Bogs		61		63							
Bohemia		67									
Bolivia		67	68	69	70	71	72	73	74	75	76	77
Bolivia		66	67	68	69	70	71				
Book reviews								73			76	77
Borates		61				65				69	
Borneo		67	68	69	70	71	72	73			
Borneo					69						
Boron		67	68	69	70	71	72	73	74	75	76	77
Boron		61	62	63	64	65	66	67	68	69	70
Boron					69						
Botswana		67	68	69	70	71	72	73	74	75	76	77
Botswana			68	69							
Boudinage		67									
Boudinage			63		65	66	67	68	69		
Boulders		61	62	63	64	65	66			69	70
Brachiopoda		67	68	69	70	71	72	73	74	75	76	77
Brachiopoda		61	62	63	64	65	66	67	68	69	70
Branchiopoda		67	68	69	70	71					
Branchiopoda		61	62	63			66	67	68	69	70
Brazil		67	68	69	70	71	72	73	74	75	76	77
Brazil		66	67	68	69	70	71				
Breccia		67	68	69	70	71					
Breccia		61	62	63	64	65	66	67	68	69	70
Brines		67	68	69	70	71	72	73	74	75	76	77
Brines		61	62	63	64	65	66	67	68	69	70
British Columbia				69	70	71	72	73	74	75	76	77
British Columbia		61	62	63	64	65	66	67	68	69	70
British Columbia		66	67	68	69	70	71				
British Guiana		66	67								
British Honduras				69	70	71	72	73	74		
British Honduras		61	62	63	64	65	66	67	68	69	70
British Isles		66		68	69	70	71				

	Bromine		68	69	70	71	72	73	74	75	76	77
N	Bromine			63		65	66	67	68	69	70	
N	Brucite								68			
	Bryophytes	67	68	69	70	71	72	73	74	75	76	77
N	Bryophytes							66	67	68	69	
	Bryozoa	67	68	69	70	71	72	73	74	75	76	77
N	Bryozoa	61	62	63	64	65	66	67	68	69	70	
	Bulgaria	67	68	69	70	71	72	73	74	75	76	77
G	Bulgaria	66	67	68	69	70	71					
	Burma	67	68	69	70	71	72	73	74	75	76	77
G	Burma		67		69	70	71					
	Burundi			69	70	71		73		75		77
	Cadmium	67	68	69	70	71	72	73	74	75	76	77
N	Cadmium	61	62	63	64	65		67	68	69	70	
	Calcite		68	69	70	71	72	73	74	75	76	77
N	Calcite			63		65	66		68			
	Calcium	67	68	69	70	71	72	73	74	75	76	77
N	Calcium	61	62	63	64	65	66	67	68	69	70	
N	Calderas	61										
N	Caliche	61			64							
	California			69	70	71	72	73	74	75	76	77
N	California	61	62	63	64	65	66	67	68	69	70	
G	California	66	67	68	69	70	71					
	Californium					71		73				77
N	Californium									69		
	Cambodia	67	68	69	70	71	72	73	74	75	76	77
G	Cambodia			69	70	71						
	Cambrian	67	68	69	70	71	72	73	74	75	76	77
N	Cambrian	61	62	63	64	65	66	67	68	69	70	
	Cameroon	67	68	69	70	71	72	73	74	75	76	77
G	Cameroun	66	67	68	69	70						
	Canada			69	70	71	72	73	74	75	76	77
N	Canada	61	62	63	64	65	66	67	68	69	70	
G	Canada	66	67	68	69	70	71					
N	Canal Zone				64							
	Canary Islands								74	75	76	77
G	Canary Islands		67	68	69	70	71					
	Cape Verde Islands						72	73	74	75	76	77
G	Cape Verde Islands		67	68	69	70	71					
	Cape Verde islands			69	70	71						
	Carbon	67	68	69	70	71	72	73	74	75	76	77
N	Carbon	61	62	63	64	65	66	67	68	69	70	
G	Carbon				68							
N	Carbonate rocks	61										
N	Carbonates	61										
	Carboniferous	67	68	69	70	71	72	73	74	75	76	77
N	Carboniferous	61	62	63	64	65	66	67	68	69	70	
	Caribbean Sea	67					72	73	74	75	76	77
N	Caribbean Sea	61		63	64	65	66	67	68	69	70	
G	Caribbean Sea	66	67	68	69	70	71					
	Caribbean sea		68	69	70	71		74				
	Caribbean region	67	68	69	70	71	72	73	74	75	76	77
N	Caribbean region	61	62	63	64	65	66	67	68	69	70	
G	Caribbean region	66	67	68	69	70	71					
	Caroline islands				68							
	Carpathians	67	68	69	70	71	72	73	74	75	76	77
	Cartography	67	68	69	70	71						

Cartography		61	62	63		65	66	67	68	69	70	
Caspian Sea									72	73	74	75	76	77
Caspian sea					69	70	71					
Catalogs		67		69	70	71	72	73	74	75	76	77
Catalogs		61									70	
Caves		67	68	69	70	71						
Caves		61	62	63	64	65	66	67	68	69	70	
Celebes Sea								72	73			
Celebes sea		67										
Celtic Sea												77
Cenozoic		67	68	69	70	71	72	73	74	75	76	77
Cenozoic		61	62	63	64	65	66	67	68	69	70	
Central African Republic			68	69	70	71	72	73	74	75		77
Central America				69	70	71	72	73	74	75	76	77
Central America		61	62	63	64	65	66	67	68	69	70	
Central America			67	68	69	70	71					
Cephalopoda		67	68	69	70	71						
Cephalopoda		61	62	63	64	65	66	67	68	69	70	
Ceramic materials		67	68	69	70	71	72	73	74	75	76	77
Ceramic materials		61	62	63	64	65	66	67	68	69	70	
Cerium		67	68	69		71	72	73	74	75	76	77
Cerium			62							69		
Cesium		67	68	69	70	71	72	73	74	75	76	77
Cesium		61	62	63	64	65	66	67	68	69	70	
Ceylon		67	68	69	70	71	72	73				
Ceylon		66			69	70						
Chad		67	68	69	70	71	72	73	74	75	76	77
Chad					70							
Chad Republic						71						
Changes of level		67	68	69	70	71	72	73	74	75	76	77
Changes of level		61	62	63	64	65	66	67	68	69	70	
Changes of level		66										
Chelicerata		67			70							
Chemical analyses		61			64							
Chemical analysis		67	68	69	70	71	72	73	74	75	76	77
Chemical analysis			62	63		65	66	67	68	69	70	
Chert					64						70	
Chesapeake Bay							72					
Chesapeake Bay					64		66			69	70	
Chesapeake bay				69	70	71						
Chile		67	68	69	70	71	72	73	74	75	76	77
Chile		66	67	68	69	70	71					
China		67	68	69	70	71	72	73	74	75	76	77
China		66	67	68	69	70						
Chlorine		67	68	69	70	71	72	73	74	75	76	77
Chlorine		61		63	64	65	66	67	68	69	70	
Chordata				69	70	71					76	77
Chordata										69		
Christmas Island		67										
Christmas island						71						
Chromite		67	68	69	70	71	72	73	74	75	76	77
Chromite		61	62	63	64	65	66	67	68	69	70	
Chromium		67	68	69	70	71	72	73	74	75	76	77
Chromium		61	62	63	64	65	66	67	68	69		
Cirripedia		67	68	69	70	71						
Cirripedia		61	62	63	64	65	66	67	68	69		
Clay		61										

	Clay mineralogy	67	68	69	70	71	72	73	74	75	76	77
N	Clay mineralogy	61	62	63	64	65	66	67	68	69	70	
	Clays	67	68	69	70	71	72	73	74	75	76	77
N	Clays		62	63	64	65	66	67	68	69	70	
	Coal	67	68	69	70	71	72	73	74	75	76	77
N	Coal	61	62	63	64	65	66	67	68	69	70	
N	Coal balls	61										
N	Coast Rica		62									
	Cobalt	67	68	69	70	71	72	73	74	75	76	77
N	Cobalt	61	62	63	64	65	66	67	68	69	70	
	Coelenterata	67	68	69	70	71	72	73	74	75	76	77
N	Coelenterata	61	62	63		65	66		68	69		
N	Collapse structures	61										
	Collections	67	68	69	70	71						
N	Collections		62	63		65	66	67	68	69	70	
	Colombia	67	68	69	70	71	72	73	74	75	76	77
G	Colombia		67	68	69	70	71					
	Colorado			69	70	71	72	73	74	75	76	77
N	Colorado	61	62	63	64	65	66	67	68	69	70	
G	Colorado	66	67	68	69	70	71					
N	Colorado Plateau	61	62	63	64	65	66	67	68	69	70	
G	Colorado Plateau	66			70	71						
N	Colorimetric analysis	61										
G	Columbia	66										
	Columbia river			69								
	Columbium						72					
N	Columbium		62									
G	Comets	67	68		70	71						
	Comoro Islands											77
N	Compressibility				64							
	Concretions	67	68	69	70	71						
N	Concretions	61	62	63	64	65	66	67	68	69	70	
N	Conglomerate	61										
	Congo	67	68	69	70	71	72	73	74	75	76	77
G	Congo		67	68	69		71					
G	Congo Republic				70							
	Connate water			69	70	71						
N	Connate water			63		65	66	67	68	69	70	
	Connecticut			69	70	71	72	73	74	75	76	77
N	Connecticut	61	62	63	64	65	66	67	68	69	70	
G	Connecticut	66	67	68	69	70	71					
	Conodonts	67	68	69	70	71	72	73	74	75	76	77
N	Conodonts	61	62	63	64	65	66	67	68	69	70	
	Construction materials	67	68	69	70	71	72	73	74	75	76	77
N	Construction materials	61	62	63	64	65	66	67	68	69	70	
	Continental drift	67	68	69	70	71	72	73	74	75	76	77
N	Continental drift	61	62	63	64	65	66	67	68	69	70	
G	Continental drift	66	67	68	69	70	71					
N	Continental margin							67	68	69	70	
G	Continental margin		67	68	69							
	Continental shelf	67	68	69	70	71	72	73	74	75	76	77
N	Continental shelf	61	62	63	64	65	66					
	Continental slope	67	68	69	70	71	72	73	74	75	76	77
N	Continental slope	61	62	63	64	65	66					
G	Continental slope	66										
	Continents	67	68	69	70	71						
N	Continents	61	62	63	64	65	66	67	68	69	70	

Term											
Continents		66	67	68		70					
Cook Islands				68			71				
Copper		67	68	69	70	71	72	73	74	75	76	77
Copper	61	62	63	64	65	66	67	68	69	70	
Coprolite								68	69		
Coprolites		67	68	69	70	71	72	73	74	75	76	77
Coprolites										70	
Coral Sea											77
Coral Sea						71					
Core		67	68	69	70	71	72	73	74	75	76	77
Core	61	62	63	64	65	66	67	68	69	70	
Core		66	67	68	69	70	71				
Correlation		67	68	69	70	71					
Correlation		62	63		65	66	67	68	69	70	
Corsica				69	70	71	72	73	74	75	76	77
Corsica		66		68	69						
Corundum			68	69	70	71	72		74	75	76	77
Corundum					64	65				69	
Cosmic dust				69	70	71					
Cosmic dust	61				64						
Cosmic-ray methods			68		70						
Cosmogony		66	67	68	69	70					
Costa Rica				69	70	71	72	73	74	75	76	77
Costa Rica	61	62	63	64	65	66	67	68	69	70	
Costa Rica		66	67	68	69	70	71				
Cratering		67	68	69	70	71					
Cratering	61	62	63	64	65	66	67	68	69	70	
Cratering		66	67	68		70	71				
Craters	61			64							
Cretaceous		67	68	69	70	71	72	73	74	75	76	77
Cretaceous	61	62	63	64	65	66	67	68	69	70	
Crete				69							
Crinoidea		67	68	69	70	71					
Crinoidea	61	62	63	64	65	66	67	68	69	70	
Crust		67	68	69	70	71	72	73	74	75	76	77
Crust	61	62	63	64	65	66	67	68	69	70	
Crust		66	67	68	69	70	71				
Crustacea		67	68	69	70	71					
Crustacea	61	62		64	65	66	67	68	69		
Cryogeology				69							
Cryptoexplosion structures		67	68	69	70	71					
Cryptoexplosion structures	61	62	63	64	65	66	67	68	69	70	
Cryptoexplosion structures		66	67	68	69	70	71				
Cryptogams				69							
Crystal chemistry		67	68	69	70	71	72	73	74	75	76	77
Crystal chemistry	61	62	63	64	65	66	67	68	69	70	
Crystal growth						71	72	73	74	75	76	77
Crystal growth	61	62		64							
Crystal structure		67	68	69	70	71	72	73	74	75	76	77
Crystal structure	61	62	63	64	65	66	67	68	69	70	
Crystallography		67	68	69	70	71	72	73	74	75	76	77
Crystallography	61	62	63	64	65	66	67	68	69	70	
Cuba				69	70	71	72	73	74	75	76	77
Cuba	61	62	63	64	65	66	67	68	69	70	
Cuba				69	70	71					
Curium				69							77
Curium		62									

	Term										
	Cyprus	67	68	69	70	71	72	73	74	75	76	77
G	Cyprus		68	69		71						
	Cystoidea	67	68	69	70	71						
N	Cystoidea	61	62	63		65	66	67	68	69	70	
	Czechoslovakia	67	68	69	70	71	72	73	74	75	76	77
G	Czechoslovakia	66	67	68	69	70	71					
N	Dacite	61										
	Dahomey	67		69		71	72		74		76	
G	Dahomey					71						
G	Deception Island				70							
	Deformation	67	68	69	70	71	72	73	74	75	76	77
N	Deformation	61	62	63	64	65	66	67	68	69	70	
G	Deformation	66	67	68	69	70	71					
	Delaware			69	70	71	72	73	74	75	76	77
N	Delaware	61	62	63	64	65	66	67	68	69	70	
G	Delaware				68							
	Deltas	67	68	69	70	71						
N	Deltas	61	62	63	64	65	66	67	68	69	70	
	Denmark	67	68	69	70	71	72	73	74	75	76	77
G	Denmark	66	67	68	69	70	71					
G	Density	66										
N	Desert pavement	61										
N	Deserts	61	62	63	64	65	66					
	Deuterium			69	70	71	72	73	74	75	76	77
	Devonian	67	68	69	70	71	72	73	74	75	76	77
N	Devonian	61	62	63	64	65	66	67	68	69	70	
N	Diabase	61										
	Diagenesis	67	68	69	70	71						
N	Diagenesis	61	62	63	64	65	66	67	68	69	70	
N	Diamond	61			64	65	66					
	Diamonds	67	68	69	70	71	72	73	74	75	76	77
N	Diamonds		62	63			66	67	68	69	70	
	Diapirs	67	68	69	70	71						
N	Diapirs	61	62	63	64	65	66	67	68	69	70	
	Diatomite	67	68	69	70	71	72	73	74	75	76	77
N	Diatomite		62	63	64	65	66	67		69		
	Diatoms	67	68	69	70	71						
N	Diatoms	61	62	63	64	65	66	67	68	69	70	
	Differential thermal analysis	67	68	69	70	71	72	73	74	75	76	77
N	Differential thermal analysis	61	62	63	64	65	66	67	68	69	70	
	Dikes	67	68	69	70	71						
N	Dikes	61	62	63	64	65	66	67	68	69	70	
	District of Columbia			69	70	71	72	73	74	75	76	77
N	District of Columbia		62	63	64	65	66	67	68	69	70	
G	District of Columbia	66				70						
	Dolomite	67	68	69	70	71	72	73	74	75	76	
N	Dolomite	61	62	63	64	65	66	67	68	69	70	
N	Dolomitization	61			64							
	Dolostone											77
N	Domes	61			64							
	Dominican Republic						72	73	74	75	76	77
N	Dominican Republic	61	62	63	64	65	66	67	68	69	70	
N	Drainage changes	61	62	63	64	65	66					
N	Drainage patterns	61	62	63	64	65	66					
N	Dunes	61	62	63	64	65	66					
N	Dysporium			63								
	Dysprosium							73	74			

Term	Years
Earth	67 68 69 70 71 72 73 74 75 76 77
Earth	61 62 63 64 65 66 67 68 69 70
Earth	66 67 68 69 70 71
Earth current exploration	64
Earth currents	61 63 66 68
Earth currents	66 67 68 69 70 71
Earth tides	67 68 69 70 71
Earth tides	61 63 66 67 68 69 70
Earth tides	66 67 68 69 70 71
Earth-current exploration	62 63 65
Earth-current exploration	66
Earth-current methods	67 68 69 70 71
Earth-current methods	66 67 68 69 70
Earth-current methods	67 68 69 70 71
Earth-current surveys	67 68 69 70 71
Earth-current surveys	61 62 63 64 65 66 67 69 70
Earth-current surveys	66 67 68 69 70 71
Earthquakes	67 68 69 70 71 72 73 74 75 76 77
Earthquakes	61 62 63 64 65 66 67 68 69 70
Earthquakes	66 67 68 69 70 71
East China Sea	71
Easter Island	72
Easter island	67 71
Eastern Hemisphere	72 73 74 75 76 77
Eastern hemisphere	70 71
Echinodermata	67 68 69 70 71 72 73 74 75 76 77
Echinodermata	61 62 63 64 65 66 67 68 69 70
Echinoidea	67 68 69 70 71
Echinoidea	61 62 63 65 66 67 68 69 70
Ecology	67 68 69 70 71 72 73 74 75 76 77
Ecology	61 62 63 64 65 66 67 68 69 70
Economic geology	67 68 69 70 71 72 73 74 75 76 77
Economic geology	61 62 63 64 65 66 67 68 69 70
Ecuador	67 69 70 71 72 73 74 75 76 77
Ecuador	67 70
Education	67 68 69 70 71 72 73 74 75 76 77
Education	61 62 63 64 65 66 67 68 69 70
Education	68
Egypt	67 68 69 70 71 72 73 74 75 76 77
Egypt	68 69 70 71
El Salvador	69 70 71 72 73 75 76 77
El Salvador	61 62 63 64 65 66 67 68 69 70
El Salvador	66 68 70 71
Elastic properties	67 68 69 70 71
Elastic properties	61 62 64 65 66 67 68 69 70
Elastic properties	66 67 68 69 70 71
Elastic waves	61 64
Elasticity	61 64
Elba	70
Electric exploration	65
Electrical exploration	61 62 63 64 65 66
Electrical exploration	66
Electrical logging	61 64
Electrical methods	67 68 69 70 71
Electrical methods	66 67 68 69 70
Electrical methods	67 68 69 70 71
Electrical properties	67 68 69 70 71

N	Electrical properties	61 62 63 64 65 66 67 68 69 70
G	Electrical properties	66 67 68 69 70 71
	Electrical surveys	67 68 69 70 71
N	Electrical surveys	61 62 63 64 65 66 67 68 69 70
G	Electrical surveys	66 67 68 69 70 71
	Electron diffraction analysis	71
N	Electron diffraction analysis	61 64 66 67 68
N	Electron-diffraction analysis	65
	Electron microscopy	67 68 69 70 71 72 73 74 75 76 77
N	Electron microscopy	61 62 63 64 65 66 67 68 69 70
N	Electron paramagnetic resonance	68 69 70
N	Electron probe analysis	61 64
N	Electron-probe analysis	66
N	Electron spin resonance	68 69
	Elements	67 68 69 70 71 72 73
N	Elements	61 62 63 64 65 66 67 68 69 70
G	Elements	70
N	Emission specroscopy	61 64
	Energy sources	67 70 71 72 73 74 75 76 77
N	Energy sources	63 66 68
	Engineering geology	67 68 69 70 71 72 73 74 75 76 77
N	Engineering geology	61 62 63 64 65 66 67 68 69 70
	England	67 68 69 70 71 72 73 74 75 76 77
G	England	66 67 68 69 70 71
	English Channel	67 68 69 70 71 72 74 75 76 77
G	English Channel	71
	Environmental geology	70 71 72 73 74 75 76 77
	Eocene	71 72 73 74 75 76 77
N	Eocene	64
	Epeirogenesis	67 68 69 70 71
N	Epeirogenesis	62 63 65 66 67 68 70
	Epeirogeny	72 73 74 75 76 77
G	Eritrea	68
	Erosion	67 68 69 70 71
N	Erosion	61 62 63 64 65 66 67 68 69 70
N	Erosion surfaces	61 62 63 64 65 66
	Estuaries	67 69 70 71
N	Estuaries	61 62 63 65 66 67 68 69 70
	Ethiopia	67 68 69 70 71 72 73 74 75 76 77
G	Ethiopia	66 67 68 70 71
G	Ethiopis	69
	Eurasia	67 68 69 70 71 72 73 74 75 76 77
	Europe	67 68 69 70 71 72 73 74 75 76 77
G	Europe	66 67 68 69 70 71
	Europium	67 69 70 71 72 73 74 75 77
N	Europium	63 66 68 69 70
	Eurypterida	67 68 69 70 71
N	Eurypterida	62 63 64 65 66 69
	Evaporites	67 68 69 70 71 72 73 74 75 76 77
N	Evaporites	61 62 63 64 65 66 67 68 69 70
	Evolution	67 68 69 70 71
N	Evolution	61 62 63 64 65 66 67 68 69 70
	Explosion phenomena	68 69 70 71
N	Explosion phenomena	62 63 65 66 67 68 69 70
G	Explosion phenomena	66 69 70
	Extraterrestrial geology	68 69 70 71
	Faeroe Islands	72 73 74 75 76 77

Faeroe Islands						66				71	
Faeroe islands						67	68	69	70	71	
Faeroes									70		
Falkland Islands						67			72	73	74		76	77
Falkland islands									70	71	
Far East						67	68	69	70	71	72	73	74	75	76	77
Faroe Islands							67	68			
Faults						67	68	69	70	71	72	73	74	75	76	77
Faults	61	62	63	64	65	66	67	68	69	70	
Faults						66	67		69		
Fauna		62									
Feldspar						67		69	70	71	72	73	74	75	76	77
Feldspar		62	63	64	65	66	67	68	69	70	
Fennoscandia						67	68	69	70	71	
Fernando Poo								69			
Fiji						67	68	69	70	71	72	73	74	75	76	77
Fiji						67	68			71	
Finland						67	68	69	70	71	72	73	74	75	76	77
Finland						66	67	68	69	70	71	
Fjords			63								
Flame photometric analysis				64							
Flame photometry	61										
Flora								69			
Flora		62	63								
Florida								69	70	71	72	73	74	75	76	77
Florida	61	62	63	64	65	66	67	68	69	70	
Florida						66	67	68	69	70	71	
Florida keys								69			
Fluid inclusions						67	68	69	70	71	
Fluid inclusions		62	63		65	66	67	68	69	70	
Fluorescence	61			64							
Fluorine						67	68	69	70	71	72	73	74	75	76	77
Fluorine	61	62	63	64	65	66	67	68	69	70	
Fluorite	61			64							
Fluorometric analysis	61										
Fluorspar						67	68	69	70	71	72	73	74	75	76	77
Fluorspar		62	63		65	66	67	68	69	70	
Folds						67	68	69	70	71	72	73	74	75	76	77
Folds	61	62	63	64	65	66	67	68	69	70	
Folds						66	67				
Foliation						67	68	69	70	71	72	73	74	75	76	77
Foliation	61	62	63	64	65	66	67	68	69	70	
Foliations								69			
Foraminifera						67	68	69	70	71	72	73	74	75	76	77
Foraminifera	61	62	63	64	65	66	67	68	69	70	
Formation pressure	61										
Formosa						67	68	69	70	71	72	73	74	75	76	
Formosa							68	69	70		
Fossils, problematic						67	68	69	70	71	72	73	74	75	76	77
Fossils, problematic						66	67	68	69	70	
Fossils, problematical		62	63		65						
Fractures						67	68	69	70	71	72	73	74	75	76	77
Fractures	61	62	63	64	65	66	67	68	69	70	
Fractures						66	67				
France						67	68	69	70	71	72	73	74	75	76	77
France						66	67	68	69	70	71	
Francium					65						

G	Franz Josef Land				69		71				
	French Guiana	67	68	69	70	71	72		74	75	76	77
	Fulgurites		68	69								
N	Fulgurites	61					66		68	69		
N	Fullers' earth				64							
	Fumaroles	67	68	69	70	71						
N	Fumaroles	61	62	63	64		66	67		69	70	
G	Fumaroles					71						
	Fungi	67	68	69	70	71	72	73	74	75	76	77
N	Fungi		62	63		65	66	67	68	69	70	
N	Fusulinidae						66					
N	Gabbro	61										
	Gabon	67	68	69	70	71	72	73	74	75	76	77
G	Gabon	66			69							
	Gadolinium			69	70	71	72	73	74	75		77
N	Gadolinium			63	64			67			70	
	Galapagos Islands						72	73	74	75	76	77
G	Galapagos Islands					71						
	Galapagos islands	67	68	69	70	71						
	Gallium		68	69	70	71	72	73	74	75	76	77
N	Gallium		62			65	66	67	68	69		
	Gambia				70	71		73				
	Garnet					71						
N	Garnet				64							
N	Gas chromatographic analyses				64							
	Gas, natural	67	68	69	70	71	72	73	74	75	76	77
N	Gas, natural	61	62	63	64	65	66	67	68	69	70	
N	Gastroliths										70	
	Gastropoda	67	68	69	70	71						
N	Gastropoda	61	62	63	64	65	66	67	68	69	70	
G	Gegenschein			69								
	Gems	67	68	69	70	71	72	73	74	75	76	77
N	Gems	61	62	63	64	65	66	67	68	69	70	
	General	67	68	69	70	71						
N	General	61	62	63	64	65	66	67	68	69	70	
G	General	66	67	68	69	70	71					
N	Geobotanical prospecting	61										
N	Geochemical exploration				64							
	Geochemical methods					70						
	Geochemical prospecting	67	68	69	70	71						
N	Geochemical prospecting	61	62	63	64	65	66	67	68	69	70	
	Geochemical surveys	67	68	69	70	71						
N	Geochemical surveys		62	63		65	66	67	68	69	70	
	Geochemistry	67	68	69	70	71	72	73	74	75	76	77
N	Geochemistry	61	62	63	64	65	66	67	68	69	70	
	Geochronology	67	68	69	70	71	72	73	74	75	76	77
N	Geochronology	61	62	63	64	65	66	67	68	69	70	
G	Geochronology	66	67	68		70						
N	Geodes	61			64						70	
	Geodesy	67	68	69	70	71	72	73	74	75	76	77
N	Geodesy					65	66	67	68	69	70	
G	Geodesy	66	67	68	69	70	71					
	Geologic barometry		68	69	70	71						
N	Geologic barometry							67		69	70	
G	Geologic barometry					71						
	Geologic cryometry		68									
N	Geologic exploration	61										

Geologic mapping	61		64								
Geologic thermometry	67	68	69	70	71						
Geologic thermometry	61	62	63	64	65	66	67	68	69	70	
Geologic thermometry	66	67	68	69	70	71					
Geological barometry			68								
Geological exploration	67	68	69	70	71						
Geological exploration		62	63	64	65	66	67	68	69	70	
Geology							72	73	74	75	76	77
Geology as a profession	61		64								
Geomorphology	67	68	69	70	71	72	73	74	75	76	77
Geomorphology	61	62	63	64	65	66	67	68	69	70	
Geophysical exploration	61	62	63	64	65						
Geophysical exploration	66										
Geophysical logging			64								
Geophysical methods	67	68	69	70	71	72	73	74	75	76	77
Geophysical methods						66	67	68	69	70	
Geophysical methods	67	68	69	70	71						
Geophysical observations		69									
Geophysical research		69									
Geophysical surveys	67	68	69	70	71	72	73	74	75	76	77
Geophysical surveys	61	62	63	64	65	66	67	68	69	70	
Geophysical surveys	66	67	68	69	70	71					
Geophysics	67	68	69	70	71	72	73	74	75	76	77
Geophysics	61	62	63	64	65	66	67	68	69	70	
Geophysics	66		70								
Georgia		69	70	71	72	73	74	75	76	77	
Georgia	61	62	63	64	65	66	67	68	69	70	
Georgia	66		68	69	70	71					
Geosynclines	67	68	69	70	71	72	73	74	75	76	77
Geosynclines	61	62	63	64	65	66	67	68	69	70	
Geosynclines	66	67	68								
Geotectonics	61										
Geothermal energy	67	68	69	70	71	72	73	74	75	76	77
Geothermal energy	61	62	63	64	65	66	67	68	69	70	
Geothermal energy	66	67	68	69	70	71					
Geothermal gradient	61	62	63	64	65	66					
Geothermal gradient	66										
Geothermal surveys				65				69			
Geothermometry	61										
Germanium	67	68	69	70	71	72	73	74	75	76	77
Germanium		62	63	64		66	67	68	69	70	
Germany	67	68	69	70	71	72	73	74	75	76	77
Germany	66	67	68	69	70	71					
Geysers	67		69	70	71						
Geysers	61		64	65	66	67	68	69	70		
Geysers		68	69	70	71						
Ghana	67	68	69	70	71	72	73	74	75	76	77
Ghana	66	67	68		70	71					
Glacial deposits	61										
Glacial features	61	62	63	64	65						
Glacial geology						72	73	74	75	76	77
Glacial geology	61	62	63	64	65	66					
Glacial lakes	61	62	63	64	65	66					
Glaciation	67	68	69	70	71						
Glaciation	61	62	63	64	65	66	67	68	69	70	
Glaciation	66	67	68	69	70	71					
Glaciers	67	68	69	70	71						

N	Glaciers		61	62	63	64	65	66	67	68	69	70
G	Glaciers		66	67	68	69	70	71				
N	Glaciology							66				
	Glauconite	67							73	74		76	77
N	Glauconite		62	63					69			
	Glossaries	67	68	69	70	71	72	73	74	75	76	77
N	Glossaries		62	63		65	66	67	68	69	70	
N	Glossary	61			64							
	Glossopteris flora			69	70	71						
	Goa	67		69		71	72					
	Gold	67	68	69	70	71	72	73	74	75	76	77
N	Gold	61	62	63	64	65	66	67	68	69	70	
N	Grabens				64					70		
	Granite	67		69	70	71	72	73		75		77
N	Granite	61					67		69	70		
N	Granodiorite	61										
	Graphite	67	68	69	70	71	72	73	74	75	76	77
N	Graphite		62	63	64	65	66	67	68			
N	Graptolites	61			64							
	Graptolithina	67	68	69	70	71	72	73	74	75	76	77
N	Graptolithina		62	63	64	65	66	67	68	69	70	
	Gravel	67	68	69	70	71	72	73	74	75	76	77
N	Gravel		62	63	64	65	66	67	68	69	70	
N	Gravity anomalies	61			64							
N	Gravity exploration	61	62	63	64	65						
G	Gravity exploration	66										
	Gravity field, Earth	67	68	69	70	71						
N	Gravity field, Earth		62	63	64	65	66	67	68	69	70	
G	Gravity field, Earth	66	67	68	69	70	71					
	Gravity field, Moon				70	71						
G	Gravity field, Moon		67	68		70						
	Gravity methods	67	68	69	70	71						
N	Gravity methods						66	67	68	69	70	
G	Gravity methods		67	68	69	70	71					
	Gravity surveys	67	68	69	70	71						
N	Gravity surveys	61	62	63	64	65	66	67	68	69	70	
G	Gravity surveys	66	67	68	69	70	71					
N	Gravity tectonics	61			64							
	Great Britain	67	68	69	70	71	72	73	74	75	76	77
G	Great Britain			68		70						
	Great Lakes			69	70	71	72	73	74	75	76	77
	Great Lakes region			69	70	71	72	73	74	75	76	77
N	Great Lakes region	61	62	63	64	65	66	67	68	69	70	
G	Great Lakes region	66	67	68	69		71					
N	Great Plains							67				
	Greece	67	68	69	70	71	72	73	74	75	76	77
G	Greece	66	67	68	69	70	71					
	Greenland			69	70	71	72	73	74	75	76	77
N	Greenland	61	62	63	64	65	66	67	68	69	70	
G	Greenland	66	67	68	69	70	71					
G	Greenland Sea					71						
	Ground water	67	68	69	70	71	72	73	74	75	76	77
N	Ground water	61	62	63	64	65	66	67	68	69	70	
	Guadalupe	67										
	Guadeloupe			69	70		72	73		75	76	77
N	Guam		62	63	64	65	66	67	68			
G	Guam			69								

Guatemala				69	70	71	72	73	74	75	76	77
Guatemala	61	62	63	64	65	66	67	68	69	70		
Guatemala	66	67		69	70	71						
Guinea			68	69	70	71	72	73	74	75		77
Guinea				69								
Guinea-Bissau										75	76	77
Gulf Coastal Plain				69	70		72	73	74	75	76	77
Gulf Coastal Plain	61	62	63	64	65	66	67	68	69	70		
Gulf Coastal Plain		67	68		70	71						
Gulf Coastal plain					71							
Gulf of Aden	67		69	70	71	72	73	74	75	76	77	
Gulf of Aden		67	68	69	70	71						
Gulf of Alaska			68									
Gulf of Bothnia		67										
Gulf of California				69	70	71	72	73	74	75	76	77
Gulf of California			63	64	65	66	67	68	69	70		
Gulf of California					71							
Gulf of Maine								68				
Gulf of Mexico				69	70	71	72	73	74	75	76	77
Gulf of Mexico	61	62	63	64	65	66	67	68	69	70		
Gulf of Mexico	66	67	68	69	70	71						
Gulf of Saint Lawrence						66		68				
Gulf of Saint Lawrence		67										
Gulf of St. Lawrence							67		69	70		
Gulf of St. Lawrence					71							
Guyana	67	68	69	70	71	72	73	74	75	76	77	
Guyana		68	69	70	71							
Guyots			64									
Gymnosperms	67	68	69	70	71	72	73	74	75	76	77	
Gymnosperms		62	63	64	65	66	67	68	69	70		
Gypsum	67	68	69	70	71	72	73	74	75	76	77	
Gypsum	61	62	63	64	65	66	67	68	69	70		
Hafnium	67	68	69	70	71	72	73	74	75	76	77	
Hafnium	61	62	63	64			68	69	70			
Haiti			69	70	71	72	73	74	75	76	77	
Haiti	61		63	64		66	67	68	69	70		
Halogens		68		71								
Halogens			66	67								
Hawaii			69	70	71	72	73	74	75	76	77	
Hawaii	61	62	63	64	65	66	67	68	69	70		
Hawaii	66	67	68	69	70	71						
Heat flow	67	68	69	70	71	72	73	74	75	76	77	
Heat flow	61				66	67	68	69	70			
Heat flow		67	68	69	70	71						
Heat transfer	66											
Heavy metals	61											
Heavy minerals	67	68	69	70	71	72	73	74	75	76	77	
Heavy minerals	61	62	63	64	65	66	67	68	69	70		
Helium		68	69	70	71	72	73	74	75	76	77	
Helium	61		63	64	65	66	67	68	69	70		
Hemichordata										76	77	
High-pressure research			64									
Himalaya	67	68										
Himalayas		68	69	70	71	72	73	74	75	76	77	
Historical geology			64									
History	67	68	69	70	71							
History	61	62	63	64	65	66	67	68	69	70		

	Holocene										71	72	73	74	75	76	77	
	Holothuroidea						67	68	69	70	71							
N	Holothuroidea		62		64	65	66	67			70							
	Honduras										71	72	73	74	75	76	77	
N	Honduras			63		65		67	68	69	70							
	Hong Kong							68		70	71	72	73	74	75	76	77	
	Hungary						67	68	69	70	71	72	73	74	75	76	77	
G	Hungary					66	67	68	69	70	71							
N	Hydrocarbons	61			64													
	Hydrogen						67	68	69	70	71	72	73	74	75	76	77	
N	Hydrogen	61	62		64	65	66	67	68	69	70							
	Hydrogeology						67	68	69	70	71	72	73	74	75	76	77	
N	Hydrogeology	61	62	63	64	65	66	67	68	69	70							
	Hydrology														75	76	77	
N	Hydrosphere					66		68	69									
	Hydrothermal alteration						67	68	69	70	71							
N	Hydrothermal alteration	61	62	63	64	65	66	67	68	69	70							
N	Hydrothermal solutions	61																
	Hydrozoa						67		69	70	71							
N	Hydrozoa		62	63			66			69								
G	Icarus								69									
N	Ice	61			64	65												
N	Ice ages (ancient)				64													
	Ice ages, ancient						67	68	69	70	71							
N	Ice ages, ancient		62			65	66			69	70							
N	Ice islands	61																
	Ice, non-glacial						67	68	69	70	71							
N	Ice, non-glacial		62			65					70							
N	Ice, nonglacial			63			66	67	68	69								
G	Ice, nonglacial									70	71							
	Iceland						67	68	69	70	71	72	73	74	75	76	77	
G	Iceland					66	67	68	69	70	71							
	Ichnofossils											72	73	74	75	76	77	
	Idaho								69	70	71	72	73	74	75	76	77	
N	Idaho	61	62	63	64	65	66	67	68	69	70							
G	Idaho					66	67	68	69	70	71							
N	Igneous petrology	61			64													
	Igneous rocks						67	68	69	70	71	72	73	74	75	76	77	
N	Igneous rocks	61	62	63	64	65	66	67	68	69	70							
	Illinois								69	70	71	72	73	74	75	76	77	
N	Illinois	61	62	63	64	65	66	67	68	69	70							
G	Illinois					66	67	68	69	70	71							
N	Illinois Basin	61																
N	Ilmenite				64													
N	Impact phenomena						66	67	68	69	70							
G	Impact phenomena						67	68	69	70	71							
N	Impactite		62															
N	Impactites	61																
G	Impactites					66	67											
	Inclusions						67	68	69	70	71	72	73	74	75	76	77	
N	Inclusions	61		63	64	65	66	67	68	69	70							
	India						67	68	69	70	71	72	73	74	75	76	77	
G	India					66	67	68	69	70	71							
	Indian Ocean						67	68	69	70	71	72	73	74	75	76	77	
G	Indian Ocean					66	67	68	69	70	71							
	Indiana								69	70	71	72	73	74	75	76	77	
N	Indiana	61	62	63	64	65	66	67	68	69	70							

Indiana		66	67		69	70					
Indium			67		69	70	71	72	73	74	75	76	77
Indium	61	62						67	68	69		
Indochina			67	68	69	70	71	72	73	74		76	77
Indonesia			67	68	69	70	71	72	73	74	75	76	77
Indonesia		66	67	68	69	70	71					
Industrial materials				68								
Industrial minerals			67	68	69	70	71	72	73	74	75	76	77
Industrial minerals	61	62	63	64	65	66	67	68	69	70		
Infrared exploration					63	64	65					
Infrared exploration		66										
Infrared methods			67	68	69	70	71					
Infrared methods								66	67	68	69	70
Infrared methods			67	68	69	70	71					
Infrared spectroscopy	61	62		64								
Infrared surveys					69	70	71					
Infrared surveys				63		65	66	67	68	69	70	
Infrared surveys		66		68	69	70	71					
Insecta			67	68	69	70	71	72	73	74	75	76	77
Insecta	61	62	63	64	65	66	67	68	69	70		
Interplanetary space											75	76	77
Intrusions			67	68	69	70	71	72	73	74	75	76	77
Intrusions	61	62	63	64	65	66	67	68	69	70		
Invertebrata			67	68	69	70	71	72	73	74	75	76	77
Invertebrata	61	62	63	64	65	66	67	68	69	70		
Iodine			67	68	69	70	71	72	73	74	75	76	77
Iodine	61	62	63	64			67	68	69	70		
Ionian Sea								72	73	74	75	76	77
Ionian sea			67			70	71					
Ionosphere													77
Iowa					69	70	71	72	73	74	75	76	77
Iowa	61	62	63	64	65	66	67	68	69	70		
Iowa		66		68		70						
Iran			67	68	69	70	71	72	73	74	75	76	77
Iran			67	68	69	70	71					
Iraq			67	68	69	70	71	72	73	74	75	76	77
Iraq				68	69	70	71					
Ireland			67	68	69	70	71	72	73	74	75	76	77
Ireland		66	67	68	69	70	71					
Iridium			67	68	69	70	71	72	73	74	75	76	77
Iridium		62	63		65		67	68	69	70		
Irish Sea			67	68		70	71	72	73	74	75	76	77
Irish Sea		66	67	68	69	70	71					
Iron			67	68	69	70	71	72	73	74	75	76	77
Iron	61	62	63	64	65	66	67	68	69	70		
Island arcs	61					66	67	68	69	70		
Isostasy			67	68	69	70	71	72	73	74	75	76	77
Isostasy	61	62	63	64	65	66	67	68	69	70		
Isostasy		66	67	68	69	70	71					
Isotopes			67	68	69	70	71	72	73	74	75	76	77
Isotopes	61	62	63	64	65	66	67	68	69	70		
Isotopes		66	67	68	69	70	71					
Israel			67	68	69	70	71	72	73	74	75	76	77
Israel			67	68	69	70	71					
Italy			67	68	69	70	71	72	73	74	75	76	77
Italy		66	67	68	69	70	71					
Ivory Coast			67	68	69	70	71	72	73	74	75	76	77

G	Ivory Coast					69	70					
	Jamaica					69	70	71	72	73	74	75	76	77
N	Jamaica	61	62	63	64	65	66	67	68	69	70
G	Jamaica	66	67		69	70	71
	Jan Mayen	67			70	71	72	73	74	75	76	77
G	Jan Mayen	66				71
	Japan	67	68	69	70	71	72	73	74	75	76	77
G	Japan	66	67	68	69	70	71
	Japan Sea	67	68	69	70	71	72	73	74	75	76	77
G	Japan Sea	66	67		69	70	71
N	Johnston Island					65
N	Joints	61			64				68
G	Joints					69
	Jordan	67	68	69	70	71	72	73	74	75	76	77
G	Jordan		67		69		71
	Jupiter			69	70	71	72	73	74	75	76	77
G	Jupiter	66	67	68	69	70	71
	Jurassic	67	68	69	70	71	72	73	74	75	76	77
N	Jurassic	61	62	63	64	65	66	67	68	69	70
	Kansas			69	70	71	72	73	74	75	76	77
N	Kansas	61	62	63	64	65	66	67	68	69	70
G	Kansas	66	67	68	69	70	71
	Kaolin	67	68	69	70	71	72	73	74	75	76	77
N	Kaolin	61		63		65	66	67	68	69	70
N	Karst	61	62	63	64	65	66
	Kashmir			69	70	71
	Katanga			69		71	72	73
	Kentucky			69	70	71	72	73	74	75	76	77
N	Kentucky	61	62	63	64	65	66	67	68	69	70
G	Kentucky	66	67		69	70
	Kenya	67	68	69	70	71	72	73	74	75	76	77
G	Kenya	66	67	68		70	71
G	Kerguelen Islands			68	69
G	Kermadec Islands			68
	Korea	67	68	69	70	71	72	73	74	75	76	77
G	Korea		67	68	69	70	71
	Krypton		68	69	70	71	72	73	74	75	76	77
N	Krypton		62			65		67	68	69	70
G	Krypton			68
	Kuril islands	67	68	69
	Kuwait		68	69	70	71	72		74	75	76	77
	Labrador			69	70	71	72	73	74	75	76	77
N	Labrador	61	62	63	64	65	66	67	68	69	70
G	Labrador			69	70	71
N	Laccoliths	61			64
G	Lake Erie				70
G	Lake Ontario			69
G	Lake Superior			69
	Lake Superior region		69	70
N	Lake Superior region	61	62	63	64	65	66	67	68	69	70
G	Lake Superior region	66	67	68	69	70
	Lakes	67	68	69	70	71
N	Lakes	61	62	63	64	65	66	67	68	69	70
	Lakes, extinct	67	68	69	70	71
N	Lakes, extinct	61	62	63	64	65	66	67	68	69	70
N	Lamprophyre	61
N	Landforms		62	63		65

Landslides		61	62	63	64	65	66						
Lanthanides					68	69		71	72				
Lanthanum					68	69	70	71	72	73	74			77
Lanthanum				63		65		67	68	69			
Laos					68	69	70	71	72				76	77
Lapland						69							
Laser methods										68	69	70	
Laser methods					69		71						
Laser surveys											70		
Laterite		61											
Laterites		67	68	69	70	71							
Laterites		62		64	65	66	67	68	69	70			
Lava		67	68	69	70	71	72	73	74	75	76	77	
Lava		61	62	63	64	65	66	67	68	69	70		
Lava flows				64									
Lead		67	68	69	70	71	72	73	74	75	76	77	
Lead		61	62	63	64	65	66	67	68	69	70		
Lead		66											
Lebanon		67	68	69	70	71	72	73	74	75	76	77	
Lebanon		67	68	69									
Lesotho				69	70	71	72	73	74	75	76	77	
Lesser Antilles												77	
Leveling				68									
Leveling networks		66											
Lexicon		61											
Lexicons		67	68	69	70	71	72	73	74	75	76	77	
Lexicons		62	63		65	66	67	68		70			
Liberia		67		69	70	71	72	73	74	75	76	77	
Liberia				70	71								
Libya		67	68	69	70	71	72	73	74	75	76	77	
Libya		67		70									
Lichens				70						76			
Life				68									
Lignite		67	68	69	70	71	72	73	74	75	76	77	
Lignite		61	62	63	64	65	66		68	69	70		
Limestone		67	68	69	70	71	72	73	74	75	76	77	
Limestone		61	62	63	64	65	66	67	68	69	70		
Lineaments			63						69	70			
Lineation		67	68	69	70	71	72	73	74	75	76	77	
Lineation		62	63	64	65	66	67	68	69	70			
Lithium		67	68	69	70	71	72	73	74	75	76	77	
Lithium		61	62	63	64	65	66	67	68	69	70		
Lithium				69									
Lithofacies											70		
Loess		61	62	63		65	66						
Louisiana				69	70	71	72	73	74	75	76	77	
Louisiana		61	62	63	64	65	66	67	68	69	70		
Louisiana		66		68	69	70	71						
Low-temperature analysis				64									
Luminescence		67	68	69	70	71							
Luminescence		61	62	63	64	65	66	67	68		70		
Luminescence		66		69									
Lutetium				69	70		72				76		
Luxembourg		67	68	69	70	71	72	73	74	75	76	77	
Luxembourg				68	69								
Lycopoda				69	70	71							
Macao				69									

	Entry	Years
	Macedonia	69
G	Macquarie Island	68 70
	Madagascar	67 68 69 70 71 72 73 74
G	Madagascar	66 67 68 69 70 71
	Madeira	77
G	Madeira	71
G	Magma	71
	Magmas	67 68 69 70 71 72 73 74 75 76 77
N	Magmas	62 63 65 66 67 68 69 70
G	Magmas	66 67 68 70
N	Magmas and magmatic differentiation	61 64
	Magnesite	67 68 69 70 71 72 73 74 75 76 77
N	Magnesite	62 63 64 66 67 68 70
	Magnesium	67 68 69 70 71 72 73 74 75 76 77
N	Magnesium	61 62 63 64 65 66 67 68 69 70
N	Magnetic anomalies	61 64
N	Magnetic exploration	61 62 63 64 65 66
G	Magnetic exploration	66
N	Magnetic field of the Earth	61 64
	Magnetic field, Earth	69 70 71
N	Magnetic field, Earth	62 63 65 66 67 68 69 70
G	Magnetic field, Earth	66 67 68 69 70 71
G	Magnetic field, Jupiter	67 68
	Magnetic field, Mars	69
G	Magnetic field, Mars	66 68
	Magnetic field, Mercury	69
G	Magnetic field, Mercury	68
	Magnetic field, Moon	69 70 71
G	Magnetic field, Moon	66 67 68 69 70
G	Magnetic field, Sun	66
	Magnetic field, Venus	69
G	Magnetic field, Venus	66 68
	Magnetic field, asteroids	69
	Magnetic field, earth	67 68
G	Magnetic field, interplanetary	67 68 69 70 71
G	Magnetic field, planets	68 69
	Magnetic methods	67 68 69 70 71
N	Magnetic methods	66 67 68 69 70
G	Magnetic methods	67 68 69 70 71
	Magnetic properties	67 68 69 70 71
N	Magnetic properties	61 62 63 64 65 66 67 68 69 70
G	Magnetic properties	66 67 68 69 70 71
	Magnetic surveys	67 68 69 70 71
N	Magnetic surveys	61 62 63 64 65 66 67 68 69 70
G	Magnetic surveys	66 67 68 69 70 71
	Magnetite	71
N	Magnetite	63 67 68 70
	Magnetosphere	75 76 77
N	Magnetotelluric exploration	62 63 65
G	Magnetotelluric exploration	66
	Magnetotelluric methods	67 68 69 70 71
N	Magnetotelluric methods	66 67 68 69 70
G	Magnetotelluric methods	67 68 69 70 71
	Magnetotelluric surveys	67 68 69 70 71
N	Magnetotelluric surveys	63 65 66 67 68 69 70
G	Magnetotelluric surveys	66 67 68 69 70 71
	Maine	69 70 71 72 73 74 75 76 77

Maine		61 62 63 64 65 66 67 68 69 70
Maine		67 69 70 71
Major-element analyses		67 68 69 70 71
Major-element analyses		62 63 64 65 66 67 68 69 70
Malacostraca		67 68 69 70 71
Malacostraca		62 63 64 65 66 67 68 69 70
Malagasy Republic		74 75 76 77
Malawi		67 68 69 70 71 72 74 75 76 77
Malawi		67 68 69 70
Malay Archipelago		71 72 73 74 75 76
Malaya		67 68 69 70 71 72
Malaya		68 70
Malaysia		67 68 69 70 71 72 73 74 75 76 77
Malaysia		66 67 69 71
Maldive Islands		67 75 76
Mali		67 68 69 70 71 72 73 74 75 76 77
Mali		66 67 69 70 71
Malta		73 74 75 76 77
Mammalia		67 68 69 70 71 72 73 74 75 76 77
Mammalia		61 62 63 64 65 66 67 68 69 70
Man, fossil		67 68 69 70 71 72 73 74 75 76 77
Man, fossil		61 62 63 64 65 66 67 68 69 70
Manganese		67 68 69 70 71 72 73 74 75 76 77
Manganese		61 62 63 64 65 66 67 68 69 70
Manitoba		69 70 71 72 73 74 75 76 77
Manitoba		61 62 63 64 65 66 67 68 69 70
Manitoba		66 67 68 69 70 71
Mantle		67 68 69 70 71 72 73 74 75 76 77
Mantle		61 62 63 64 65 66 67 68 69 70
Mantle		66 67 68 69 70 71
Maps		72 73 74 75 76 77
Marble		67 69 70 71 72 73 74 75 76 77
Marble		61 62 63 64 65 66 67 68 70
Mariana Islands		73 74 75 76 77
Mariana islands		67 71
Marine geology		67 68 69 70 71 72 73 74 75 76 77
Marine geology		66 67 68 69 70
Marine geology		67 68 69 70 71
Mars		67 68 69 70 71 72 73 74 75 76 77
Mars		66 67 68 69 70 71
Marshall Islands		72 73 74 75 76 77
Marshall islands		67 70 71
Martinique		69
Maryland		69 70 71 72 73 74 75 76 77
Maryland		61 62 63 64 65 66 67 68 69 70
Maryland		66 68 69 70 71
Mass spectroscopy		61 64
Mass wastage		62 63 65 66
Massachusetts		69 70 71 72 73 74 75 76 77
Massachusetts		61 62 63 64 65 66 67 68 69 70
Massachusetts		66 68 69 70 71
Mathematical geology		71 72 73 74 75 76 77
Mauritania		67 68 69 70 71 72 73 74 75 76 77
Mauritania		67 68 69 70 71
Mauritius		76 77
Mauritius		69
Medical geology		71

	Mediterranean Sea	67	68				72	73	74	75	76	77
G	Mediterranean Sea	66	67	68	69	70	71					
	Mediterranean sea				69	70	71					
	Mediterranean region	67	68	69	70	71	72	73	74	75	76	77
G	Mediterranean region				69							
	Melanesia								74	75		77
	Mercury	67	68	69	70	71	72	73	74	75	76	77
N	Mercury	61	62	63	64	65	66	67	68	69	70	
G	Mercury	66	67	68	69	70	71					
	Mercury Planet									75	76	77
	Merostomata	67		69	70							
N	Merostomata			63	64					69	70	
	Mesozoic	67	68	69	70	71	72	73	74	75	76	77
N	Mesozoic	61	62	63	64	65	66	67	68	69	70	
	Metal		68									
	Metals	67	68	69	70	71	72	73	74	75	76	77
N	Metals		62	63	64	65	66	67	68	69	70	
N	Metamorphic petrology	61										
	Metamorphic rocks	67	68	69	70	71	72	73	74	75	76	77
N	Metamorphic rocks	61	62	63	64	65	66	67	68	69	70	
	Metamorphism	67	68	69	70	71	72	73	74	75	76	77
N	Metamorphism	61	62	63	64	65	66	67	68	69	70	
	Metasomatism	67	68	69	70	71	72	73	74	75	76	77
N	Metasomatism	61	62	63	64	65	66	67	68	69	70	
	Metazoa				69							
	Meteor craters	67	68	69	70	71	72	73	74	75	76	77
N	Meteor craters	61	62	63	64	65	66	67	68	69	70	
G	Meteor craters	66	67	68	69	70	71					
	Meteorites	67	68	69	70	71	72	73	74	75	76	77
N	Meteorites	61	62	63	64	65	66	67	68	69	70	
G	Meteorites	66	67	68	69	70	71					
	Meteorology									75	76	77
G	Meteors	66	67	68		70						
	Mexico			69	70	71	72	73	74	75	76	77
N	Mexico	61	62	63	64	65	66	67	68	69	70	
G	Mexico	66	67	68	69	70	71					
	Mica	67	68	69	70	71	72	73	74	75	76	77
N	Mica	61	62		64	65	66	67	68	69		
	Michigan			69	70	71	72	73	74	75	76	77
N	Michigan	61	62	63	64	65	66	67	68	69	70	
G	Michigan	66	67	68	69	70	71					
	Micronesia	67	68		70	71	72			75	76	77
	Micropaleontology	67	68	69	70	71	72	73	74	75	76	77
N	Micropaleontology	61	62	63	64	65	66	67	68	69	70	
	Microscope methods	67	68	69	70	71						
N	Microscope methods						66	67	68	69	70	
N	Microscope techniques		62	63		65						
	Microscopic methods	67										
	Microseisms	67	68	69	70	71						
N	Microseisms		62	63		65	66	67	68	69	70	
G	Microseisms	66	67	68	69	70	71					
N	Microwave exploration					65						
N	Microwave methods						66	67	68	69	70	
G	Microwave methods		68	69	70	71						
N	Microwave surveys										70	
G	Microwave surveys				70							
	Middle East	67	68	69	70	71	72	73	74	75	76	77

Term	Years
Middle East	69 70 71
Midway Islands	70
Migmatites	61
Military geology	69 70
Military geology	61 62 63 64 65 66 68 69 70
Mineragraphy	67 68 69 70 71
Mineragraphy	61 62 63 64 65 66 67 68 69 70
Mineral collecting	67 68 69 70 71
Mineral collecting	61 62 63 64 65 66 67 68 69 70
Mineral collections	64
Mineral data	67 68 69 70 71
Mineral data	66 67 68 69 70
Mineral deposits, genesis	67 68 69 70 71 72 73 74 75 76 77
Mineral deposits, genesis	61 62 63 64 65 66 67 68 69 70
Mineral descriptions	61 62 63 64 65
Mineral economics	67 68 69 70 71
Mineral economics	62 63 64 65 66 67 68 69 70
Mineral exploration	67 68 69 70 71 72 73 74 75 76 77
Mineral exploration	61 62 63 64 65 66 67 68 69 70
Mineral exploration	67
Mineral resources	67 68 69 70 71 72 73 74 75 76 77
Mineral resources	61 62 63 64 65 66 67 68 69 70
Mineral zoning	67 68 69 70 71
Mineral zoning	62 63 64 65 66 67 68 69 70
Mineralogy	67 68 69 70 71 72 73 74 75 76 77
Mineralogy	61 62 63 64 65 66 67 68 69 70
Minerals	71 72 73 74 75 76 77
Mining geology	67 68 69 70 71 72 73 74 75 76 77
Mining geology	61 62 63 64 65 66 67 68 69 70
Minnesota	69 70 71 72 73 74 75 76 77
Minnesota	61 62 63 64 65 66 67 68 69 70
Minnesota	66 67 68 69 70 71
Minor-element analyses	62
Miocene	68 71 72 73 74 75 76 77
Miocene	64
Mississippi	69 70 71 72 73 74 75 76 77
Mississippi	61 62 63 64 65 66 67 68 69 70
Mississippi	66 67 68 69 70
Mississippi Delta	71
Mississippi River	64
Mississippi Valley	61 62 63 64 65 66 67 68 69 70
Mississippi Valley	67 70
Mississippi embayment	61 67
Mississippian	69 70 71 72 73 74 75 76 77
Mississippian	61 62 63 64 65 66 67 68 69 70
Missouri	69 70 71 72 73 74 75 76 77
Missouri	61 62 63 64 65 66 67 68 69 70
Missouri	66 67 68 69 70 71
Models	64
Mohorovicic discontinuity	67 68 69 70 71 72 73 74 75 76 77
Mohorovicic discontinuity	61 62 63 64 65 66 67 68 69
Mohorovicic discontinuity	66 67 68 69 70
Mollusca	67 68 69 70 71 72 73 74 75 76 77
Mollusca	61 62 63 64 65 66 67 68 69 70
Molybdenum	67 68 69 70 71 72 73 74 75 76 77
Molybdenum	61 62 63 64 65 66 67 68 69 70
Monaco	67 73 76

	Topic	61	62	63	64	65	66	67	68	69	70	71	72	73	74	75	76	77
	Monazite								68	69	70		72	73	74	75	76	77
N	Monazite		62	63									66	67	68	69	70	
	Mongolia							67	68	69	70	71	72	73	74	75	76	77
G	Mongolia						66	67	68	69	70	71						
	Montana									69	70	71	72	73	74	75	76	77
N	Montana	61	62	63	64	65	66	67	68	69	70							
G	Montana						66	67	68	69	70	71						
	Moon							67	68	69	70	71	72	73	74	75	76	77
N	Moon										70							
G	Moon						66	67	68	69	70	71						
	Morocco							67	68	69	70	71	72	73	74	75	76	77
G	Morocco						66	67	68		70	71						
N	Mossbauer analysis				64													
G	Mossbauer effect						66											
	Mozambique							67	68	69	70	71	72	73	74	75	76	77
G	Mozambique						66		68									
	Mud volcanoes							67	68	69	70	71	72	73	74	75	76	77
N	Mud volcanoes	61		63		65		67	68	69	70							
N	Mudflows	61																
	Museums							67	68	69	70	71	72	73	74	75	76	77
N	Museums	61	62	63	64	65	66	67	68	69	70							
	Myriapoda								68			71						
N	Myriapoda			63						69								
N	Natural bridges	61			64													
	Near East							67	68									
	Nebraska									69	70	71	72	73	74	75	76	77
N	Nebraska	61	62	63	64	65	66	67	68	69	70							
G	Nebraska								68									
	Neodymium											71		73	74		76	77
	Neon								68	69	70	71	72	73	74	75	76	77
N	Neon	61			64	65	66	67	68	69	70							
G	Neon						66											
	Nepal							67	68	69	70	71	72	73	74	75	76	77
N	Nepheline syenite	61																
	Neptune									69	70			73	74	75	76	77
G	Neptune								68	69	70	71						
	Neptunium												72	73				77
	Netherlands							67	68	69	70	71	72	73	74	75	76	77
G	Netherlands						66	67	68			71						
	Netherlands Antilles									69	70		72	73	74	75	76	77
N	Netherlands Antilles									69	70							
N	Neutron activation analysis	61			64													
N	Neutron diffraction analysis		62		64	65	66	67	68	69								
	Nevada									69	70	71	72	73	74	75	76	77
N	Nevada	61	62	63	64	65	66	67	68	69	70							
G	Nevada						66	67	68	69	70	71						
G	New Britain										70							
G	New Britain Island											71						
	New Brunswick									69	70	71	72	73	74	75	76	77
N	New Brunswick	61	62	63	64	65	66	67	68	69	70							
G	New Brunswick						66	67	68	69	70	71						
	New Caledonia							67	68	69	70	71	72	73	74	75	76	77
G	New Caledonia								68			71						
	New England									69	70	71	72	73	74	75	76	77
N	New England	61	62	63	64	65	66	67	68	69	70							
G	New England							67	68	69		71						
	New Guinea							67	68	69	70	71	72	73	74	75	76	77

New Guinea			67	68	69	70	71			
New Hampshire				69	70	71	72	73	74	75 76 77
New Hampshire	61	62	63	64	65	66	67	68	69	70
New Hampshire	66			69	70	71				
New Hebrides	67	68	69	70	71	72	73	74	75 76 77	
New Hebrides	66	67	68	69	70	71				
New Jersey				69	70	71	72	73	74	75 76 77
New Jersey	61	62	63	64	65	66	67	68	69	70
New Jersey	66		68	69	70	71				
New Mexico				69	70	71	72	73	74	75 76 77
New Mexico	61	62	63	64	65	66	67	68	69	70
New Mexico	66	67	68	69	70	71				
New South Wales	67	68	69	70	71	72	73	74	75 76 77	
New York				69	70	71	72	73	74	75 76 77
New York	61	62	63	64	65	66	67	68	69	70
New York	66	67	68	69	70	71				
New Zealand	67	68	69	70	71	72	73	74	75 76 77	
New Zealand	66	67	68	69	70	71				
Newfoundland				69	70	71	72	73	74	75 76 77
Newfoundland	61	62	63	64	65	66	67	68	69	70
Newfoundland			67	68	69	70	71			
Nicaragua					70	71	72	73	74	75 76 77
Nicaragua	61	62	63	64	65	66	67	68	69	70
Nicaragua	66	67			70	71				
Nickel	67	68	69	70	71	72	73	74	75 76 77	
Nickel	61	62	63	64	65	66	67	68	69	70
Niger	67	68	69	70	71	72	73	74	75 76 77	
Niger Republic			67							
Nigeria	67	68	69	70	71	72	73	74	75 76 77	
Nigeria	66	67	68	69	70					
Niobium	67	68	69	70	71	72	73	74	75 76 77	
Niobium	61	62	63	64	65	66		68	69	70
Nitrates			68	69	70	71	72			76 77
Nitrates					65			68	69	
Nitrogen	67	68	69	70	71	72	73	74	75 76 77	
Nitrogen		62		64	65	66	67	68	69	70
Noble gases						71	72	73	74	75 76 77
Noble gases				64						70
Nodules	67	68	69	70	71	72	73	74	75 76 77	
Nodules	61	62	63	64	65	66	67	68	69	70
Nonmetals								74	75	76 77
North Africa	67		69	70	71					
North America				69	70	71	72	73	74	75 76 77
North America	61	62	63	64	65	66	67	68	69	70
North America	66	67	68	69	70	71				
North Carolina				69	70	71	72	73	74	75 76 77
North Carolina	61	62	63	64	65	66	67	68	69	70
North Carolina	66	67	68	69	70	71				
North Dakota				69	70	71	72	73	74	75 76 77
North Dakota	61	62	63	64	65	66	67	68	69	70
North Dakota			67		70	71				
North Korea					69					
North Sea	67	68	69	70	71	72	73	74	75 76 77	
North Sea	66	67	68	69	70	71				
North Vietnam				68						
Northern Hemisphere							72	73	74	75 76 77
Northern Ireland	67			70			72	73	74	75 76 77

G	Northern Ireland			67		70					
	Northern Territory		67	68	69	70	71	72	73	74	75	76	77
	Northern hemisphere			68	69	70	71					
	Northwest Territories				69	70	71	72	73	74	75	76	77
N	Northwest Territories	61	62	63	64	65	66	67	68	69	70
G	Northwest Territories	66	67	68	69	70	71				
	Norway	67	68	69	70	71	72	73	74	75	76	77
G	Norway	66	67	68	69	70	71				
G	Norwegian Sea	66		68			71				
	Nova Scotia			69	70	71	72	73	74	75	76	77
N	Nova Scotia	61	62	63	64	65	66	67	68	69	70
G	Nova Scotia	66	67	68	69	70	71				
	Nuclear explosions	67		69	70	71					
N	Nuclear explosions	61	62	63	64	65	66	67	68	69	70
G	Nuclear explosions	66	67	68	69	70	71				
N	Nuclear magnetic resonance								68	69	70
N	Nuclear science	61									
G	Nucleosynthesis	66	67			70					
G	Nutation	66									
	Ocean basins						72	73	74	75	76	77
N	Ocean basins					65	66		68			
G	Ocean basins	66	67								
	Ocean floors						72	73	74	75	76	77
	Ocean waves						72	73	74	75	76	77
	Oceania	67		69	70		73	74	75	76	77
	Oceanography	67	68	69	70	71	72	73	74	75	76	77
N	Oceanography	61	62	63	64	65	66	67	68	69	70
G	Oceanography				68						
N	Oceans	61									
G	Oceans	66	67								
	Oceans, circulation						72	73	74	75	76	77
	Ohio			69	70	71	72	73	74	75	76	77
N	Ohio	61	62	63	64	65	66	67	68	69	70
G	Ohio	66	67	68	69	70	71				
	Oil and gas fields	67	68	69	70	71	72	73	74	75	76	77
N	Oil and gas fields	61	62	63	64	65	66	67	68	69	70
	Oil sand				70	71					
	Oil sands			69			72	73	74	75	76	77
N	Oil sands	61	62	63	64	65	66	67	68	69	70
	Oil shale	67	68	69	70	71	72	73	74	75	76	77
N	Oil shale	61	62	63	64	65	66	67	68	69	70
	Okhotsk Sea						72	73	74	75	76	77
G	Okhotsk Sea				70						
	Okhotsk sea	68		70	71						
	Oklahoma			69	70	71	72	73	74	75	76	77
N	Oklahoma	61	62	63	64	65	66	67	68	69	70
G	Oklahoma	66	67	68	69	70	71				
	Oligocene		68		71	72	73	74	75	76	77
	Oman		68		71	72	73	74	75	76	77
	Ontario			69	70	71	72	73	74	75	76	77
N	Ontario	61	62	63	64	65	66	67	68	69	70
G	Ontario	66	67	68	69	70	71				
N	Oolites	61			64						
N	Optical data processing									69	
	Optical mineralogy	67	68	69	70	71					
N	Optical mineralogy	61	62	63	64	65	66	67	68	69	70
	Optical properties				68						

Ordovician	67	68	69	70	71	72	73	74	75	76	77
Ordovician	61	62	63	64	65	66	67	68	69	70	
Oregon			69	70	71	72	73	74	75	76	77
Oregon	61	62	63	64	65	66	67	68	69	70	
Oregon	66	67	68	69	70	71					
Organic materials	67	68	69	70	71	72	73	74	75	76	77
Organic materials		62	63	64	65	66	67	68	69	70	
Orogeny	67	68	69	70	71	72	73	74	75	76	77
Orogeny	61	62	63	64	65	66	67	68	69	70	
Orogeny	66	67	68		70						
Osmium		68	69	70	71	72	73	74	75	76	77
Osmium			63	64			67	68		70	
Ostracoda	67	68	69	70	71	72	73	74	75	76	77
Ostracoda	61	62	63	64	65	66	67	68	69	70	
Oxygen	67	68	69	70	71	72	73	74	75	76	77
Oxygen	61	62	63	64	65	66	67	68	69	70	
Ozone							73				
Pacific Islands			63			66	67	68	69		
Pacific Islands	66	67	68	69	70	71					
Pacific Ocean	67	68	69	70	71	72	73	74	75	76	77
Pacific Ocean	61	62	63	64	65	66	67	68	69	70	
Pacific Ocean	66	67	68	69	70	71					
Pacific region			69	70	71	72	73	74	75	76	77
Pakistan	67	68	69	70	71	72	73	74	75	76	77
Pakistan	66	67	68	69	70	71					
Paladium					65						
Paleobiogeography	61										
Paleobotany	67	68	69	70	71	72	73	74	75	76	77
Paleobotany	61	62	63	64	65	66	67	68	69	70	
Paleocene					71	72	73	74	75	76	77
Paleoclimatology	67	68	69	70	71	72	73	74	75	76	77
Paleoclimatology	61	62	63	64	65	66	67	68	69	70	
Paleoclimatology	66	67	68	69	70	71					
Paleoecology	67	68	69	70	71	72	73	74	75	76	77
Paleoecology	61	62	63	64	65	66	67	68	69	70	
Paleogeography	67	68	69	70	71	72	73	74	75	76	77
Paleogeography	61	62	63	64	65	66	67	68	69	70	
Paleomagnetism	67	68	69	70	71	72	73	74	75	76	77
Paleomagnetism	61	62	63	64	65	66	67	68	69	70	
Paleomagnetism	66	67	68	69	70	71					
Paleontology	67	68	69	70	71	72	73	74	75	76	77
Paleontology	61	62	63	64	65	66	67	68	69	70	
Paleosalinity		62	63	64							
Paleosalinity	66										
Paleotemperature		67				71					
Paleotemperatures	61			64							
Paleozoic	67	68	69	70	71	72	73	74	75	76	77
Paleozoic	61	62	63	64	65	66	67	68	69	70	
Palladium	67		69	70	71	72	73	74	75	76	77
Palladium		62		64		66	67	68	69	70	
Palynology	67	68	69	70	71	72	73	74	75	76	77
Palynology	61	62	63	64	65	66	67	68	69	70	
Palynomorphs	67	68	69	70	71	72	73	74	75	76	77
Palynomorphs		62	63		65	66	67	68	69	70	
Panama			69	70	71	72	73	74	75	76	77
Panama	61	62	63	64	65	66	67	68	69	70	
Panama			69	70	71						

N	Pantellerite	61										
N	Paper chromatography	61										
	Papua	67	68	69	70	71	72	73	74	75		
G	Papua		67	68		70	71					
	Papua New Guinea									75	76	77
	Paragenesis	67	68	69	70	71	72	73	74	75	76	77
N	Paragenesis	61	62	63	64	65	66	67	68	69	70	
	Paraguay	67		69	70	71	72	73	74	75	76	77
	Patterned ground	67	68	69	70	71						
N	Patterned ground	61	62	63	64	65	66	67	68	69	70	
	Peat	67	68	69	70	71	72	73	74	75	76	77
N	Peat	61	62	63	64	65	66		68	69	70	
	Pebbles	67	68	69	70	71						
N	Pebbles	61	62	63	64	65	66	67	68	69	70	
N	Pediments				64							
	Pegmatite	67	68	69	70	71	72	73	74	75	76	77
N	Pegmatite						66	67	68	69		
N	Pegmatites	61	62	63	64	65					70	
	Pelecypoda	67	68	69	70	71						
N	Pelecypoda	61	62	63	64	65	66	67	68	69	70	
N	Peneplanes				64							
	Pennsylvania			69	70	71	72	73	74	75	76	77
N	Pennsylvania	61	62	63	64	65	66	67	68	69	70	
G	Pennsylvania	66	67	68	69	70	71					
	Pennsylvanian			69	70	71	72	73	74	75	76	77
N	Pennsylvanian	61	62	63	64	65	66	67	68	69	70	
N	Peridotite	61										
N	Periglacial features		62	63		65						
N	Periglacial phenomena	61			64							
N	Perlite	61										
	Permafrost	67	68	69	70	71						
N	Permafrost	61	62	63	64	65	66	67	68	69	70	
G	Permafrost	66										
	Permeability	67	68	69	70	71						
N	Permeability	61	62	63	64	65	66	67	68	69	70	
G	Permeability	66										
	Permian	67	68	69	70	71	72	73	74	75	76	77
N	Permian	61	62	63	64	65	66	67	68	69	70	
G	Permian					71						
	Persian Gulf	67	68	69	70	71	72	73	74	75	76	77
G	Persian Gulf		68									
	Peru	67	68	69	70	71	72	73	74	75	76	77
G	Peru	66	67	68	69	70	71					
	Petrofabrics	67	68	69	70	71						
N	Petrofabrics	61	62	63	64	65	66	67	68	69	70	
G	Petrofabrics		67									
N	Petrogenesis	61										
	Petrography	67										
N	Petrography	61								70		
	Petroleum	67	68	69	70	71	72	73	74	75	76	77
N	Petroleum	61	62	63	64	65	66	67	68	69	70	
N	Petroleum engineering	61										
	Petrology	67	68	69	70	71	72	73	74	75	76	77
N	Petrology	61	62	63	64	65	66	67	68	69	70	
	Phanerozoic						72	73	74	75	76	77
N	Phanerozoic							68	69	70		
	Phase equilibria	67	68	69	70	71	72	73	74	75	76	77

Phase equilibria	61 62 63 64 65 66 67 68 69 70
Phase equilibria	66 67 68 70
Philippine Islands	72 73 74 75 76 77
Philippine Sea	71
Philippine islands	67 68 69 70 71
Philippines	66 67 68 69 70 71
Phosphate	67 68 69 70 71 72 73 74 75 76 77
Phosphate	61 62 63 64 65 66 67 68 69 70
Phosphorescence	61
Phosphorous	64
Phosphorus	67 68 69 70 71 72 73 74 75 76 77
Phosphorus	65 67 68 69 70
Photogeology	67 68 69 70 71
Photogeology	61 62 63 64 65 66 67 68 69 70
Photogeology	70
Photomicroscopy	64
Physical geology	61
Physical properties	64
Physical properties	66 67 70
Phytoliths	69
Pisces	67 68 69 70 71 72 73 74 75 76 77
Pisces	61 62 63 64 65 66 67 68 69 70
Placers	68 69 70 71 72 73 74 75 76 77
Placers	61 62 63 64 65 66 67 68 69 70
Planetology	71 72 73 74 75 76 77
Planets	67 68 69 70 71
Planets	66 67 68 69 70 71
Plantae	72 73 74 75 76 77
Plasticity	61
Plate tectonics	76 77
Platinum	67 68 69 70 71 72 73 74 75 76 77
Platinum	61 62 63 64 65 66 67 68 69 70
Playas	61
Pleistocene	71 72 73 74 75 76 77
Pleistocene	64
Pliocene	71 72 73 74 75 76 77
Pluto	69 74 76 77
Pluto	68 69 70 71
Plutonium	68 69 70 71 72 73 74 75 76 77
Plutonium	67 68 69 70
Plutonium	68
Plutons	64
Poland	67 68 69 70 71 72 73 74 75 76 77
Poland	66 67 68 69 70 71
Polar wandering	66
Polarographic analysis	61
Polonium	71 72 73 74 75 76 77
Polonium	63 65
Polymetallic ores	68 69 70 71 72 73 74 75 76 77
Polymetallic ores	61 62 63 64 65 66 67 68 69 70
Polynesia	74 75 76 77
Popular and elementary geology	67 68 69 70 71 72
Popular and elementary geology	61 62 63 64 65 66 67 68 69 70
Porifera	67 68 69 70 71 72 73 74 75 76 77
Porifera	61 62 63 64 65 66 67 68 69 70
Porosity	67 68 69 70 71
Porosity	61 62 63 64 65 66 67 68 69 70

G	Porosity	66	67								
	Portugal	67	68	69	70	71	72	73	74	75	76	77
G	Portugal	66	67	68	69	70	71				
	Portuguese Guinea	67			70		72	73			
	Portuguese Timor										76	
	Potash	67	68	69	70	71	72	73	74	75	76	77
N	Potash		62	63	64	65	66	67	68	69	70	
	Potassium	67	68	69	70	71	72	73	74	75	76	77
N	Potassium	61	62	63	64	65	66	67	68	69	70	
	Precambrian	67	68	69	70	71	72	73	74	75	76	77
N	Precambrian	61	62	63	64	65	66	67	68	69	70	
N	Primates	61			64							
	Prince Edward Island				70	71	72	73	74		76	77
N	Prince Edward Island	61	62	63	64	65	66	67	68	69	70	
G	Prince Edward Island	66		68								
N	Problematical fossils				64							
	Promethium						72					
	Protactinium			69	70	71	72	73	74			
N	Protactinium					65	66	67		69		
	Proterozoic		68									
	Protista	67	68	69	70	71	72	73	74	75	76	77
N	Protista	61	62	63	64	65	66	67	68	69	70	
	Protozoa	67	68	69	70	71						
N	Protozoa	61	62		64	65	66	67	68		70	
N	Pseudomorphs	61			64							
	Pteridophytes	67	68	69	70	71	72	73	74	75	76	77
N	Pteridophytes		62	63		65	66	67	68	69	70	
	Pterobranchia				70		72				76	
N	Pterobranchia			63							70	
	Pteropoda		68	69	70	71						
N	Pteropoda								68		70	
	Puerto Rico			69	70	71	72	73	74	75	76	77
N	Puerto Rico	61	62	63	64	65	66	67	68	69	70	
G	Puerto Rico	66	67			70	71					
	Pumice				70	71		73	74		76	77
N	Pumice		62	63		65	66			69		
	Pyrenees	67	68	69	70	71	72	73	74	75	76	77
	Pyrite	67	68	69	70	71	72	73	74	75	76	77
N	Pyrite			63	64	65	66	67		69		
	Qatar								74			77
G	Quark		68									
	Quartz crystal	67	68	69	70	71	72	73	74	75	76	77
N	Quartz crystal			63			66					
N	Quartzite	61										
	Quaternary	67	68	69	70	71	72	73	74	75	76	77
N	Quaternary	61	62	63	64	65	66	67	68	69	70	
	Quebec			69	70	71	72	73	74	75	76	77
N	Quebec	61	62	63	64	65	66	67	68	69	70	
G	Quebec	66	67	68	69	70	71					
	Queensland	67	68	69	70	71	72	73	74	75	76	77
N	Radar exploration					65						
N	Radar methods						66	67	68	69	70	
G	Radar methods		67	68	69	70	71					
N	Radar surveys					65	66	67	68	69	70	
G	Radar surveys		67	68	69	70	71					
N	Radioactive-waste disposal	61			64							
	Radioactivity	67	68	69	70	71						

Term	Pages
Radioactivity	61 62 63 64 65 66 67 68 69 70
Radioactivity	66 67 68 69 70 71
Radioactivity exploration	61 62 63 64 65
Radioactivity exploration	66
Radioactivity logging	61 64
Radioactivity methods	67 68 69 70 71
Radioactivity methods	66 67 68 69 70
Radioactivity methods	67 68 69 70 71
Radioactivity surveys	67 68 69 70 71
Radioactivity surveys	61 62 63 64 65 66 67 68 69 70
Radioactivity surveys	66 67 68 69 70 71
Radiochemical analysis	64 67
Radiolaria	67 68 69 70 71 72 73 74 75 76 77
Radiolaria	61 62 63 64 65 66 67 68 69 70
Radiowave methods	68 69
Radiowave surveys	69
Radiowave surveys	70
Radium	67 68 69 70 71 72 73 74 75 76 77
Radium	62 63 64 65 66 67 68 69 70
Radon	67 68 69 70 71 72 73 74 75 76 77
Radon	62 64 65 66 67 68 69 70
Rare earths	67 68 69 70 71 72 73 74 75 76 77
Rare earths	61 62 63 64 65 66 67 68 69 70
Rare gases	71
Rare gases	61 69
Red Sea	67 68 69 70 71 72 73 74 75 76 77
Red Sea	66 67 68 69 70 71
Red Sea region	76 77
Reefs	67 68 69 70 71 72 73 74 75 76 77
Reefs	61 62 63 64 65 66 67 68 69 70
Reflectivity	66
Refractometry	61
Refractory materials	66
Remote-sensing methods	66 67 68 69 70
Remote-sensing methods	67 68 69 70 71
Remote-sensing surveys	68 69 70
Remote-sensing surveys	69 70 71
Reptilia	67 68 69 70 71 72 73 74 75 76 77
Reptilia	61 62 63 64 65 66 67 68 69 70
Reunion	67 68 69 70 71 72 73 74 75 77
Reunion Island	66 67 68 71
Rhenium	68 69 70 71 72 73 74 75 76 77
Rhenium	61 62 64 65 67 69 70
Rhode Island	69 70 71 72 73 74 75 76 77
Rhode Island	61 62 63 64 65 66 67 68 69 70
Rhode Island	66 67 68 69
Rhodesia	68 69 70 71 72 73 74 75 76 77
Rhodesia	68 69 70 71
Rhodium	69 70 71 72 73 74 75 76 77
Rhodium	62 68 69 70
Ripple marks	64
Rivers	67 68 69 70 71
Rivers	61 62 63 64 65 66 67 68 69 70
Rock creep	61
Rock glaciers	65
Rock mechanics	64
Rock mechanics	70 71

		61	62	63	64	65	66	67	68	69	70	71	72	73	74	75	76	77
	Rocky Mountains										70	71	72	73	74	75	76	77
N	Rocky Mountains	61	62	63	64	65	66	67	68	69	70							
G	Rocky Mountains						66	67		69	70	71						
	Rocky mountains									69								
	Romania							67	68	69	70	71	72	73	74	75	76	77
	Rubidium							67	68	69	70	71	72	73	74	75	76	77
N	Rubidium	61	62	63	64	65	66	67	68	69	70							
G	Rumania						66	67	68	69	70	71						
	Ruthenium									69	70	71	72	73	74	75	76	
N	Ruthenium	61		63				67	68									
	Rwanda							67										
G	Rwanda										70	71	72	73	74	75		77
	Ryukyu islands							67	68		70							
	Sahara							67	68	69	70	71	72	73	74	75	76	77
G	Saint Helena								68	69								
N	Saint-Pierre and Miquelon		62						68		70							
	Salt							67	68	69	70	71	72	73	74	75	76	77
N	Salt	61	62	63	64	65	66	67	68	69	70							
	Salt tectonics							67	68	69	70	71	72	73	74	75	76	77
N	Salt tectonics	61	62	63	64	65	66	67	68	69	70							
G	Salt tectonics						66	67		69								
N	Salt water intrusion	61			64													
	Samarium											71	72	73	74	75	76	77
N	Samarium			63							70							
	Samoa								68	69	70	71	72	73	74	75	76	77
	Samoa islands							67										
N	Sampling				64													
	Sand							67	68	69	70	71	72	73	74	75	76	77
N	Sand	61	62	63	64	65	66	67	68	69	70							
	Sand volcanoes							67										
	Sandstone								68	69	70	71	72	73	74	75	76	77
N	Sandstone	61	62	63	64	65	66	67	68	69	70							
	Sarawak							67	68	69	70	71		73		75		
	Sardinia														74	75	76	77
G	Sardinia							67			70	71						
	Saskatchewan									69	70	71	72	73	74	75	76	77
N	Saskatchewan	61	62	63	64	65	66	67	68	69	70							
G	Saskatchewan						66	67	68	69	70	71						
	Saturn									69	70		72	73	74	75	76	77
G	Saturn							67	68	69	70	71						
	Saudi Arabia							67	68	69	70	71	72	73	74	75	76	77
G	Saudi Arabia							67	68	69	70	71						
	Scandinavia							67	68	69	70	71	72	73	74	75	76	77
G	Scandinavia								68	69	70	71						
	Scandium								68	69	70	71	72	73	74	75	76	77
N	Scandium	61	62	63	64	65	66	67	68	69								
	Scaphopoda								68	69	70	71						
N	Scaphopoda		62			65		67	68	69	70							
N	Scolecodonts	61			64													
G	Scotia Arc									69								
G	Scotia Sea								68									
	Scotland							67	68	69	70	71	72	73	74	75	76	77
G	Scotland						66	67	68	69	70	71						
N	Sea mounts								68	69								
	Sea of Japan							67	68									
G	Sea of Japan											71						
	Sea water							67	68	69	70	71	72	73	74	75	76	77

Term										
Sea water		63		65	66	67	68	69	70	
Sea-floor spreading	68	69	70	71	72	73	74	75	76	77
Seamounts		69	70	71						
Seamounts		63		65		67			70	
Seamounts	66	67	68	69	70	71				
Sedimentary petrology	68	69	70	71	72	73	74	75	76	77
Sedimentary petrology	61								70	
Sedimentary rocks	67	68	69	70	71	72	73	74	75	76	77
Sedimentary rocks	61	62	63	64	65	66	67	68	69	70
Sedimentary structures	67	68	69	70	71	72	73	74	75	76	77
Sedimentary structures	61	62	63	64	65	66	67	68	69	70
Sedimentation	67	68	69	70	71	72	73	74	75	76	77
Sedimentation	61	62	63	64	65	66	67	68	69	70
Sediments	67	68	69	70	71	72	73	74	75	76	77
Sediments	61	62	63	64	65	66	67	68	69	70
Seiche				64						
Seismic exploration	61	62	63	64	65					
Seismic exploration	66									
Seismic methods	67	68	69	70	71					
Seismic methods						66	67	68	69	70
Seismic methods		67	68	69	70	71				
Seismic surveys	67	68	69	70	71					
Seismic surveys	61	62	63	64	65	66	67	68	69	70
Seismic surveys	66	67	68	69	70	71				
Seismic waves	61			64						
Seismology	67	68	69	70	71	72	73	74	75	76	77
Seismology	61	62	63	64	65	66	67	68	69	70
Seismology	66	67	68	69	70	71				
Selenium	67	68	69	70	71	72	73	74	75	76	77
Selenium	61	62	63	64	65	66	67	68	69	70
Senegal	67	68	69	70	71	72	73	74	75	76	77
Senegal			69	70						
Serpentine	61									
Shale	61	62		64		66	67	68	69	70
Shatter cones	61									
Shetland Islands	67					72			76	77
Shetland islands		69	70	71						
Shore features	61	62	63	64	65	66				
Shorelines	67	68	69	70	71					
Shorelines	61	62	63	64	65	66	67	68	69	70
Sierra Leone	67	68	69	70	71	72	73	74	75	76	77
Sikkim		68		70	71			74	75		77
Silica				64		66				
Silicate rocks	61									
Siliceous earth				64						
Silicification				64						
Silicon	67	68	69	70	71	72	73	74	75	76	77
Silicon	61	62	63	64	65	66	67	68	69	70
Sills	61			64		66				
Silurian	67	68	69	70	71	72	73	74	75	76	77
Silurian	61	62	63	64	65	66	67	68	69	70
Silver	67	68	69	70	71	72	73	74	75	76	77
Silver	61	62	63	64	65	66	67	68	69	70
Singapore		68	69	70				74	75	76	77
Slate			69	70	71		73			76
Slate	61			64	65		67		69	
Slopes		62								

	Snow		67	68	69	70	71				
N	Snow	61						67	68	69	70	
	Society Islands											76
	Society islands		67	68						75		
	Sodium		67	68	69	70	71	72	73	74	75	76	77
N	Sodium	61	62	63	64	65	66	67	68	69	70	
	Sodium carbonate						71	72	73	74	75	76	77
N	Sodium carbonate			63		65	66	67		69		
	Sodium sulfate				69	70		72	73	74	75	76	77
N	Sodium sulfate		62	63			66		68		70	
	Soils		67	68	69	70	71	72	73	74	75	76	77
N	Soils	61	62	63	64	65	66	67	68	69	70	
	Solar system					70	71						
G	Solar system		66	67	68	69	70	71					
	Solomon Islands							72	73	74	75	76	77
G	Solomon Islands			67		69		71				
	Solomon islands		67	68	69	70	71					
N	Solubility	61			64							
	Somali Republic						71	72	73	74		76	77
	Somalia		67	68	69							
G	Somalia						71					
	South Africa		67	68	69	70	71	72	73	74	75	76	77
G	South Africa	66	67	68	69	70	71					
	South America		67	68	69	70	71	72	73	74	75	76	77
G	South America	66	67	68	69	70	71					
	South Arabia			68									
	South Australia		67	68	69	70	71	72	73	74	75	76	77
	South Carolina				69	70	71	72	73	74	75	76	77
N	South Carolina	61	62	63	64	65	66	67	68	69	70	
G	South Carolina	66	67	68		70	71					
	South Dakota				69	70	71	72	73	74	75	76	77
N	South Dakota	61	62	63	64	65	66	67	68	69	70	
G	South Dakota	66		68	69		71					
G	South Korea						71					
G	South Sandwich Islands		67									
G	South Shetland Islands					70						
G	South West Africa						71					
	South-West Africa		67	68	69	70	71	72	73	74	75	76	77
G	South-West Africa			68	69	70						
	Southern Hemisphere							72	73	74	75	76	77
	Southern Rhodesia	67					72					
G	Southern Rhodesia	66										
	Southern Yemen											77
	Southern hemisphere			68	69	70	71					
	Spain		67	68	69	70	71	72	73	74	75	76	77
G	Spain	66	67	68	69	70	71					
	Spanish Guinea		67			70			73			
	Spanish Sahara		67	68			71		73	74	75	76
G	Spanish Sahara					70						
	Spanish West Africa		67									
	Specific gravity				69	70	71					
N	Specific gravity	61	62	63	64	65	66	67	68	69		
G	Specific gravity	66	67	68	69	70	71					
N	Spectrochemical analysis	61										
N	Spectrophotometric analysis	61			64							
	Spectroscopy		67	68	69	70	71	72	73	74	75	76	77
N	Spectroscopy		62	63		65	66	67	68	69	70	

Spectroscopy	66
Spermatophyta	64
Spitsbergen	67 68 69 70 71 72 73 74 75 76 77
Spitsbergen	67 68 69 70 71
Spitzbergen	66
Springs	67 68 69 70 71 72 73 74 75 76 77
Springs	61 62 63 64 65 66 67 68 69 70
Sri Lanka	74 75 76 77
Standard materials	76 77
Standard rocks	69 70 71 72 73 74 75
Statistical measures	62 63 65 67
Statistical methods	67 68 69 70 71
Statistical methods	62 63 65 66 67 68 69 70
Statistical methods	67 69
Statistics	61 64
Statistics	66
Stocks	67 68 69 70 71
Stocks	61 62 63 64 65 66 67 68 69 70
Stratigraphy	67 68 69 70 71 72 73 74 75 76 77
Stratigraphy	61 62 63 64 65 66 67 68 69 70
Stream capture	61 62 63 64 65
Streams	64
Strength	64
Strength	66 67
Stress	61 64
Stromatolites	67 68 69 70 71
Stromatolites	63 65 66 67 68 69 70
Stromatoporoidea	67 68 69 70 71
Stromatoporoidea	61 62 63 64 65 66 67 68 69 70
Strontium	67 68 69 70 71 72 73 74 75 76 77
Strontium	61 62 63 64 65 66 67 68 69 70
Structural analysis	72 73 74 75 76 77
Structural geology	67 68 69 70 71 72 73 74 75 76 77
Structural geology	61 62 63 64 65 66 67 68 69 70
Stylolites	64
Submarine geology	61 62 63 64 65 66
Submarine geology	66
Subsidence	67 68 69 70 71
Subsidence	61 62 63 64 65 66 67 70
Sudan	67 68 69 70 71 72 73 74 75 76 77
Sudan	66 67 68 70 71
Sulfides	76
Sulfides	61 64
Sulfur	67 68 69 70 71 72 73 74 75 76 77
Sulfur	61 62 63 64 65 66 67 68 69 70
Sun	69 70 71 75 76 77
Sun	68 69 70 71
Surinam	67 68 69 70 71 72 73 74 75 76 77
Surinam	66 68 69 70 71
Surveys	67 68 69 70 71 72 73 74 75 76 77
Surveys	61 62 63 64 65 66 67 68 69 70
Swaziland	67 68 69 70 71 72 73 74 75 76 77
Swaziland	67 70 71
Sweden	67 68 69 70 71 72 73 74 75 76 77
Sweden	66 67 68 69 70 71
Switzerland	67 68 69 70 71 72 73 74 75 76 77
Switzerland	66 67 68 69 70 71

	Entry	Years
	Symposia	68 69 70 71 72 73 74 75 76 77
N	Symposia	68 69 70
	Syria	67 68 69 70 71 72 73 74 75 76 77
G	Syria	67 68 69 70
	Tahiti	67
	Taiwan	71 72 73 74 75 76 77
G	Taiwan	77
	Talc	66 67 68 71
N	Talc	67 68 69 70 71 72 73 74 75 76 77
G	Tanganyika	61 62 63 64 65 66 67 68 69 70
	Tantalum	66 67
N	Tantalum	67 68 69 70 71 72 73 74 75 76 77
	Tanzania	61 62 63 64 65 66 68 69 70
G	Tanzania	67 68 69 70 71 72 73 74 75 76 77
	Tasman Sea	67 68 69 70 71
G	Tasman Sea	77
	Tasmania	66 69 70 71
G	Tasmania	67 68 69 70 71 72 73 74 75 76 77
	Technetium	66 67 68 69 70 71
G	Tectites	76
	Tectonics	70
N	Tectonics	67 68 69 70 71 72 73 74 75 76 77
G	Tectonics	61 62 63 64 65 66 67 68 69 70
	Tectonophysics	66 67 68 69 70 71
	Tektites	71 72 73 74 75 76 77
N	Tektites	67 68 69 70 71 72 73 74 75 76 77
G	Tektites	61 62 63 64 65 66 67 68 69 70
N	Television logging	66 67 68 69 70 71
	Tellurium	64
N	Tellurium	67 68 69 70 71 72 73 74 75 76 77
G	Temperature methods	61 62 63 64 65 66 67 68 69
G	Temperature surveys	68
G	Tenerife	68 69
	Tennessee	66
N	Tennessee	69 70 71 72 73 74 75 76 77
G	Tennessee	61 62 63 64 65 66 67 68 69 70
	Terbium	67 68
N	Terraces	70 71 72
	Tertiary	61 62 63 64 65
N	Tertiary	67 68 69 70 71 72 73 74 75 76 77
	Texas	61 62 63 64 65 66 67 68 69 70
N	Texas	69 70 71 72 73 74 75 76 77
G	Texas	61 62 63 64 65 66 67 68 69 70
	Thailand	66 67 68 69 70 71
G	Thailand	67 68 69 70 71 72 73 74 75 76 77
	Thallium	69 70 71
N	Thallium	68 69 70 71 72 73 74 75 76 77
	Thallophytes	61 62 64 67 69 70
N	Thermal conductivity	71 72 73 74 75 76 77
G	Thermal conductivity	62 63 64 65 66
N	Thermal methods	66
G	Thermal methods	68
N	Thermal properties	69 71
G	Thermal properties	61 64 68
	Thermal springs	66 69 70 71
N	Thermal springs	67
G	Thermal springs	61 62 63 64 65 66 67 68 69 70
		66 67 68 69 70 71

Term											
Thermal surveys							68				
Thermal surveys						71					
Thermal waters		68	69	70	71	72	73	74	75	76	77
Thermal waters	61										
Thermodynamic properties	67	68	69	70	71		73				
Thermodynamic properties		62	63	64	65	66	67	68	69	70	
Thermodynamic properties	66	67	68	69	70						
Thermogravimetric analysis							73	74	75	76	77
Thermogravimetric analysis						66	67		69		
Thermoluminescence	61			64			67	68	69	70	
Thermoluminescence	66	67	68	69	70	71					
Thorium	67	68	69	70	71	72	73	74	75	76	77
Thorium	61	62	63	64	65	66	67	68	69	70	
Thoron			69								
Throium				70							
Thulium					71	72					
Thulium										70	
Till	61										
Tillites	61										
Timor	67	68	69	70		72	73				
Timor Sea		68	69								
Tin	67	68	69	70	71	72	73	74	75	76	77
Tin	61	62	63	64	65	66	67	68	69	70	
Titanium	67	68	69	70	71	72	73	74	75	76	77
Titanium	61	62	63	64	65	66	67	68	69	70	
Togo		68		70		72		74		76	77
Tonga		68		70	71	72	73	74	75	76	77
Tonga Islands					71						
Tonga islands	67		69								
Trace elements	61			64	65	66	67	68	69	70	
Trace-element analyses	67	68	69	70	71						
Trace-element analyses		62	63		65	66	67	68	69	70	
Tracks and trails	67	68	69	70	71						
Tracks and trails	61	62	63	64	65	66	67	68	69	70	
Traps		62									
Triassic	67	68	69	70	71	72	73	74	75	76	77
Triassic	61	62	63	64	65	66	67	68	69	70	
Trieste	67			71							
Trilobita	67	68	69	70	71	72	73	74	75	76	77
Trilobita	61	62	63	64	65	66	67	68	69	70	
Trinidad			69	70	71	72	73	74	75	76	
Trinidad	61	62	63	64	65	66	67	68	69	70	
Trinidad		68									
Trinidad and Tobago											77
Tristan da Cunha		68									
Tritium	67	68	69		71	72	73	74	75	76	77
Tritium	61		63					68		70	
Trucial Coast		68	69	70	71	72	73				
Trucial Oman		68	69	70	71	72					
Tsunami	66	67		69							
Tsunamis	67	68	69	70	71						
Tsunamis		62	63	64	65	66	67	68	69	70	
Tsunamis		68		70	71						
Tuff	61										
Tungsten	67	68	69	70	71	72	73	74	75	76	77
Tungsten	61	62	63	64	65	66	67	68	69	70	
Tunisia	67	68	69	70	71	72	73	74	75	76	77

G	Tunisia	66		68		70	71					
N	Turbidity currents	61			64							
	Turkey		67	68	69	70	71	72	73	74	75	76	77
G	Turkey	66	67	68	69	70	71					
	Tyrrhenian Sea							72	73	74	75	76	77
	Tyrrhenian sea		67			70	71					
G	U.S.S.R.	66	67	68	69	70	71					
	USSR		67	68	69	70	71	72	73	74	75	76	77
	Uganda		67	68	69	70	71	72	73	74	75	76	77
G	Uganda	66		68		70	71					
N	Ultramafic rocks	61										
N	Ultraviolet spectroscopy				64							
	Unconformities		67	68	69	70	71					
N	Unconformities	61	62	63	64	65	66	67	68	69	70	
N	Uniformitarianism	61										
	United Arab Emirates									74	75	76	77
	United Kingdom									74	75	76	77
G	United Kingdom		67	68	69	70	71					
	United States				69	70	71	72	73	74	75	76	77
N	United States	61	62	63	64	65	66	67	68	69	70	
G	United States	66	67	68	69	70	71					
	Uplifts		67	68	69	70	71					
N	Uplifts	61	62	63	64	65	66	67	68	69	70	
	Upper Volta			68	69	70	71	72	73		75	76	77
G	Upper Volta					70						
	Uranium		67	68	69	70	71	72	73	74	75	76	77
N	Uranium	61	62	63	64	65	66	67	68	69	70	
	Uranus					70			73	74	75	76	77
G	Uranus	66	67	68		70	71					
	Uruguay		67	68	69	70	71	72	73	74	75	76	77
G	Uruguay		67			70						
	Utah				69	70	71	72	73	74	75	76	77
N	Utah	61	62	63	64	65	66	67	68	69	70	
G	Utah	66	67	68	69	70	71					
N	Valleys	61	62	63	64	65						
	Vanadium		67	68	69	70	71	72	73	74	75	76	77
N	Vanadium	61	62	63	64	65	66	67	68	69	70	
N	Varves	61			64							
	Veins		67	68	69	70	71					
N	Veins		62	63	64	65	66	67	68	69	70	
	Venezuela		67	68	69	70	71	72	73	74	75	76	77
G	Venezuela	66	67	68	69	70	71					
	Venus			68	69	70	71	72	73	74	75	76	77
G	Venus	66	67	68	69	70	71					
	Vermes				69					74			
	Vermiculite		67	68	69	70	71	72	73	74	75	76	77
N	Vermiculite	61	62	63				67		69			
	Vermont				69	70	71	72	73	74	75	76	77
N	Vermont	61	62	63	64	65	66	67	68	69	70	
G	Vermont	66		68	69		71					
	Vertebrata		67	68	69	70	71	72	73	74	75	76	77
N	Vertebrata	61	62	63	64	65	66	67	68	69	70	
	Victoria		67	68	69	70	71	72	73	74	75	76	77
G	Viet Nam		67			70						
G	VietNam						71					
	Vietnam		67	68	69	70	71	72	73	74	75	76	77
G	Vietnam			68								

Topic											
Virgin Islands		61	62	63	64	65	66	67		69	
Virgin Islands								67			
Virginia							69	70	71	72	73	74	75	76	77
Virginia		61	62	63	64	65	66	67	68	69	70
Virginia		66	67	68	69	70	71				
Volcanism		67	68	69	70	71					
Volcanism		61	62	63	64	65	66	67	68	69	70
Volcanism		66	67	68	69	70	71				
Volcanoes		67	68	69	70	71					
Volcanoes		61	62	63	64	65	66	67	68	69	70
Volcanoes		66	67	68	69	70	71				
Volcanology							72	73	74	75	76	77
Wake Island										70	
Wales		67	68	69	70	71	72	73	74	75	76	77
Wales		66	67	68	69	70	71				
Washington					69	70	71	72	73	74	75	76	77
Washington		61	62	63	64	65	66	67	68	69	70
Washington		66	67	68	69	70	71				
Water		67									
Water		61									
Water falls					64						
Water resources						71	72	73	74	75	76	77
Weathering		67	68	69	70	71	72	73	74	75	76	77
Weathering		61	62	63	64	65	66	67	68	69	70
Well logging			62	63		65	66	67	68	69	70
Well logging		66	67	68	69	70	71				
Well-logging		67	68	69	70	71	72	73	74	75	76	77
Wells and drill holes		67	68	69	70	71					
Wells and drill holes		61	62	63	64	65	66	67	68	69	70
West Africa					69	70	71	72				
West Africa				68		70	71				
West Indies					69	70	71	72	73	74	75	76	77
West Indies		61	62	63	64	65	66	67	68	69	70
West Indies		66	67	68	69	70	71				
West Virginia					69	70	71	72	73	74	75	76	77
West Virginia		61	62	63	64	65	66	67	68	69	70
West Virginia		66	67	68		70	71				
Western Australia		67	68	69	70	71	72	73	74	75	76	77
Western Hemisphere				70		72	73	74	75	76	77
Western hemisphere			69		71						
Williston Basin		61									
Wind work		61	62	63	64	65	66				
Wisconsin					69	70	71	72	73	74	75	76	77
Wisconsin		61	62	63	64	65	66	67	68	69	70
Wisconsin		66	67	68	69	70	71				
World		67									
World								69			
Worms		67	68	69	70	71			75	76	77
Worms		61	62	63	64		66	67	68	69	70
Wyoming					69	70	71	72	73	74	75	76	77
Wyoming		61	62	63	64	65	66	67	68	69	70
Wyoming		66	67	68	69	70	71				
X-ray diffraction			62								
X-ray diffraction analyses		61									
X-ray diffraction analysis		67	68	69	70	71	72	73	74	75	76	77
X-ray diffraction analysis		61	62	63	64	65	66	67	68	69	70
X-ray fluorescence analysis		61			64						

N	X-ray radiography			63		65	66	67		69	70
N	X-ray spectrographic analysis	61			64						
N	Xenoliths	61			64						
	Xenon			68	69	70	71	72	73	74	75	76	77
N	Xenon	61	62	63	64	65	66	67	68	69	70
G	Xenon			68							
G	Yellow Sea						71				
	Yemen		67	68			71	72		74			77
	Ytterbium					70						77
N	Ytterbium									69	
	Yttrium			68	69	70	71	72	73	74	75	76	77
N	Yttrium		62	63	64	65				69	
	Yugoslavia		67	68	69	70	71	72	73	74	75	76	77
G	Yugoslavia	66	67	68	69	70	71					
N	Yukon	61	62	63	64	65	66	67	68	69	70
G	Yukon		67	68	69	70	71					
	Yukon Territory				69	70	71	72	73	74	75	76	77
	Zaire							72	73	74	75	76	77
	Zambia		67	68	69	70	71	72	73	74	75	76	77
G	Zambia			68	69	70	71					
	Zinc		67	68	69	70	71	72	73	74	75	76	77
N	Zinc	61	62	63	64	65	66	67	68	69	70
	Zircon		67		69	70	71	72	73	74		76	77
N	Zircon				64	65	66			69	
	Zirconium		67	68	69	70	71	72	73	74	75	76	77
N	Zirconium		62	63	64		66	67		69	70

List B

Index terms under area terms

The second and third order terms for areas (List O):

<u>areal geology</u> (used for entries that might properly be placed under three or more of the 2nd level headings such as <u>geomorphology</u>, <u>stratigraphy</u>, <u>structural geology</u>)

    topic   [<u>bibliography</u>, <u>explanatory text</u> (for a map), <u>guidebook</u>, <u>maps</u>] or locality

e.g.   (1) <u>France</u>
         (2) <u>areal geology</u>
         (3) <u>Narbonne</u>

<u>economic geology</u>

         commodity (List C)

e.g.   (1) <u>Germany</u>
         (2) <u>economic geology</u>
         (3) <u>coal</u>

<u>engineering geology</u>

    topic   (see Main List, 2nd level terms or <u>dams</u>, <u>earthquakes</u>, <u>foundations</u>, <u>geologic hazards</u>, <u>highways</u>, <u>land subsidence</u>, <u>marine installations</u>, <u>nuclear facilities</u>, <u>reservoirs</u>, <u>rock mechanics</u>, <u>shorelines</u>, <u>slope stability</u>, <u>soil mechanics</u>, <u>tunnels</u>, <u>underground installations</u>, <u>waste disposal</u>, <u>waterways</u>)

e.g.   (1) <u>California</u>
         (2) <u>engineering geology</u>
         (3) <u>earthquakes</u>

<u>environmental geology</u>

    topic   (see Main List, 2nd level terms, or <u>conservation</u>, <u>geologic hazards</u>, <u>impact statements</u>, <u>land use</u>, <u>pollution</u>, <u>reclamation</u>, <u>waste disposal</u>)

e.g.   (1) <u>Idaho</u>
       (2) <u>environmental geology</u>
       (3) <u>land use</u>

<u>geochemistry</u>

           (<u>isotopes</u>) or element(s) or other terms from List D or material(s) or topic (Main List)

e.g.   (1) <u>USSR</u>
       (2) <u>geochemistry</u>
       (3) <u>isotopes</u>

<u>geochronology</u>

    topic   (<u>absolute age</u> or topic from Main List, 2nd level terms) or age (List E)

e.g.   (1) <u>United States</u>
       (2) <u>geochronology</u>
       (3) <u>absolute age</u>

<u>geomorphology</u>

    topic   (see Main List, 2nd level terms, or <u>glacial geology</u>)

e.g.   (1) <u>Sahara</u>
       (2) <u>geomorphology</u>
       (3) <u>eolian features</u>

<u>geophysical surveys</u>

    topic   (see Main List, 2nd level terms under both <u>geophysical surveys</u> and <u>geophysical methods</u>, or <u>geodesy</u>, <u>heat flow</u>, <u>remote sensing</u>, <u>well-logging</u>)

e.g. (1) Oklahoma
(2) geophysical surveys
(3) infrared surveys

general

    topic   (e.g. annual report, bibliography, current research, education, geology, maps, philosophy, research)

e.g. (1) United States
(2) general
(3) current research

hydrogeology

    topic   (ground water, hydrology, springs, thermal waters)

e.g. (1) Louisiana
(2) hydrogeology
(3) ground water

mineralogy

    name of mineral group (List L), in the case of miscellaneous minerals, that term suffices here)

e.g. (1) Italy
(2) mineralogy
(3) framework silicates, feldspar group

oceanography

    topic   (see Main List, 2nd level terms; or continental shelf, continental slope, estuaries, geophysical methods, geophysical surveys, marine geology, nodules, ocean basins, ocean circulation, ocean floors, ocean waves, reefs, sea water, sedimentation, sediments) or locality

e.g. (1) <u>Pacific Ocean</u>
(2) <u>oceanography</u>
(3) <u>marine geology</u>

<u>paleobotany</u>

    fossil group (List F, 1st level terms)

e.g. (1) <u>Arizona</u>
(2) <u>paleobotany</u>
(3) <u>gymnosperms</u>

    Repeat set if more than one equally emphasized fossil group.

<u>paleontology</u> (also used for studies on both <u>fauna</u> and <u>flora</u>)

    fossil group (List F, 1st level terms)

e.g. (1) <u>California</u>
(2) <u>paleontology</u>
(3) <u>foraminifera</u>

    Repeat set if more than one equally emphasized fossil group.

<u>petrology</u> (also used for studies on both crystalline and sedimentary rock <u>complexes</u>)

    topic [<u>breccia</u>, <u>igneous rocks</u>, <u>inclusions</u>, <u>intrusions</u>, <u>magmas</u>, <u>metamorphic rocks</u>, <u>metamorphism</u>, <u>metasomatism</u>, <u>meteorites</u>, <u>phase equilibria</u>, <u>tektites</u>, <u>volcanism</u>, <u>volcanology</u>]

e.g. (1) <u>Colorado</u>
(2) <u>petrology</u>
(3) <u>metamorphism</u>

<u>sedimentary petrology</u>

    topic [<u>breccia</u>, <u>clay mineralogy</u>, <u>diagenesis</u>,

heavy minerals, sedimentary rocks, sedimentary structures, sedimentation, sediments]

e.g. (1) New South Wales
(2) sedimentary petrology
(3) sedimentary rocks

seismology

    topic   (see Main List, 2nd level terms)

e.g. (1) Washington
(2) seismology
(3) earthquakes

soils

    topic   (see Main List, 2nd level terms)

e.g. (1) Alberta
(2) soils
(3) soil group

stratigraphy

    age (List E) biogeography, changes of level, continental drift, paleogeography

e.g. (1) South Carolina
(2) stratigraphy
(3) Cretaceous

    Repeat set if more than one equally emphasized age.

structural geology

    topic   [deformation, epeirogeny, faults, folds, foliation, fractures, geosynclines, lineation, neotectonics, salt tectonics, structural analysis, tectonics]

e.g.   (1) <u>India</u>
            (2) <u>structural geology</u>
            (3) <u>tectonics</u>

<u>tectonophysics</u>

     topic   (see Main List, 2nd level terms; or <u>continental drift</u>, <u>core</u>, <u>crust</u>, <u>heat flow</u>, <u>mantle</u>, <u>plate tectonics</u>, <u>sea-floor spreading</u>]

e.g.   (1) <u>North America</u>
            (2) <u>tectonophysics</u>
            (3) <u>plate tectonics</u>

<u>volcanology</u> (for <u>Quaternary</u> volcanic activity)

         locality

e.g.   (1) <u>Italy</u>
            (2) <u>volcanology</u>
            (3) <u>Mount Etna</u>

List C

Commodity terms

Earth materials of actual or potential value as industrial raw materials are here referred to as "commodities". The following names of commodities, if underlined, are 1st-level terms.

abrasives (general), use under industrial minerals
aluminum
andalusite, use under ceramic materials
anhydrite, use under gypsum
antimony
arsenic
asbestos
asphalt, use under bitumens
barite
base metals
bauxite, (use here only when of economic value)
bentonite, (use here only when of economic value)
beryl, use under beryllium
beryllium
bismuth
bitumens, see also organic materials in Main List
borates, use under boron
boron
brines
bromine, see also brines, salt
brown coal, use lignite
building stone, use construction materials, granite,
    sandstone, limestone, marble
calcite (optical)
cement materials, use under limestone or construction
    materials
ceramic materials
chalk, use under limestone
chromite
clays, see also clay mineralogy in Main List
coal (use only when of economic value)
cobalt
columbium, use niobium
construction materials
copper
corundum
cryolite, use under fluorspar
diamonds
diatomite
dimension stone, use construction materials, granite,
    sandstone, limestone, marble

dolostone
energy sources (a grouper for petroleum, natural gas, coal, uranium, etc.)
evaporites
feldspar
fluorite (as commodity), use under fluorspar
fuller's earth, use under clays
garnet, use under industrial minerals
gas, natural use natural gas
gems
geothermal energy
glauconite
gold
granite
graphite
gravel (includes sand when sand is used as construction material), use also under sediments in Main List
gypsum
heavy minerals, see also sedimentary rocks in Main List; see also placers in Main List (use only when of economic value)
helium
hematite, use under iron
industrial minerals
iodine, use under brines, salt
iron
kaolin
kyanite, use under ceramic materials
lead
lead-zinc deposits
lightweight aggregate, use construction materials
lignite
limestone, see also under sedimentary rocks in Main List
limonite, use under iron
lithium
magnesite
magnetite, use under iron
manganese
marble, see also under metamorphic rocks in Main List
mercury
metals (use when of economic value)
mica
mineral resources (for very general treatments)
molybdenum
monazite, see also heavy minerals, thorium, rare earths
natural gas
nickel
niobium
nitrates
nonmetals (use mainly when of economic value; do not

  include underlined energy sources and water resources)
oil, use petroleum
oil sands
oil shale
peat
pegmatite
perlite, use under construction materials
petroleum, see also oil and gas fields in Main
  List
phosphate
platinum
polymetallic ores
potash
pumice, see also volcanic under igneous rocks in
  Main List
pyrite, see also sulfur
quartz crystal
rare earths
refractory materials, use under ceramic materials
salt
sand [for glass, ceramic, chemical use, etc.]
sandstone, use also under sedimentary rocks in
  Main List
shale (for shale as brick clay, use clays; for bloating
  shale, use under construction materials)
shale oil, use oil shale
silica [as resource], use sand
sillimanite, use under ceramic materials
silver
slate, see also under metamorphic rocks in Main
  List
soapstone, use under talc
sodium carbonate
sodium sulfate
sulfur, see also pyrite
talc
tantalum
tar sands, use oil sands, bitumens
thorium, see also monazite
tin
titanium, see also heavy minerals
trap rock, use under construction materials
trona, use under sodium carbonate
tungsten
uranium
vanadium
vermiculite
water resources
zinc
zircon, see also heavy minerals

These commodity terms may get an automatic area cross-reference or may be used as follows:

commodity term [as in list above]

>topic
>>subtopic e.g. ore deposits (for metals) or deposits (for nonmetals) or abundance, affinities, economics, exploration, genesis, geochemistry, inventory, mineral economics, possibilities, production, properties, reserves, resources, structure

Some of the commodity terms, like "heavy minerals," "industrial minerals," "polymetallic ores," "evaporites," "mineral resources," etc. are general, and include a number of the more specific commodities in this list. In general, articles indexed under the individual commodities would not be also indexed under the broader category term, and vice versa; exception to this rule would include situations where one or two commodities out of a large number treated have received particular attention (e.g., an article containing a particularly elaborate discussion of monazite in both meanings of the terms).

For complete coverage of a commodity appropriate general terms as well as the specific should be consulted.

These commodity names are used as 1st level terms where the principal basis for the article is the economic potential of the material, as in descriptions of deposits. Articles on many of these materials, however, such as clays, granite, limestone, various minerals, etc., may be purely petrologic or mineralogic; these articles are also indexed under their names as 3rd-level terms under igneous rocks, metamorphic rocks, sedimentary rocks, and minerals. Articles containing detailed treatment of the petrology, mineralogy, geochemistry, etc. of mineral deposits are also indexed both ways.

Coal, lignite, and peat are both 1st-level terms in a commodity sense and 3rd-level terms in a sedimentary sense (see also organic materials in Main List).

For chemical elements like aluminum, antimony, arsenic, etc., the terms appropriate to their treatment as commodities are interspersed with the 2nd-level terms appropriate to their treatment as chemical elements; the terms from both sets are combined and listed in alphabetic order in the indices. 2nd-level terms under lead, for example, might be, in succession: abundance, analysis, exploration, geochemistry, etc.

List D

Element sets

2nd- and 3rd-level terms related to elements
(Elements and symbols are listed in the main
body of the Thesaurus)

Other related terms that may be used on the 1st level include:

deuterium

metals

noble gases (use for rare gases or inert gases)

 these are: argon, helium, krypton, neon, radon, xenon

They are all products of radioactive decay.

rare earths (use for lanthanide series or inner
transition elements or lanthanoans)

 these are: cerium, dysprosium, erbium, europium, gadolinium, holmium, lanthanum, lutetium, neodymium, praesodymium, prometheum, samarium, scandium, terbium, thulium, ytterbium, yttrium

tritium

Terms that can be used on 3rd level: major elements, minor elements, trace elements

When one of the element terms or other terms such as metals, noble gases, rare earths is used on the 1st level as a chemical element, it is then followed by:

abundance
 material

<u>geochemistry</u>
    <u>material</u>

<u>analysis</u>
    <u>activation analysis</u>

    <u>alpha-ray spectroscopy</u>

    <u>autoradiography</u>

    <u>chemical analysis</u>

    <u>chromatography</u>

    <u>colorimetry</u>

    <u>electrolytic analysis</u>

    <u>electron diffraction analysis</u>

    <u>emission spectroscopy</u>

    <u>flame photometry</u>

    <u>gamma-ray spectroscopy</u>

    <u>mass spectroscopy</u>

    <u>microwave spectroscopy</u>

    <u>Mossbauer spectroscopy</u>

    <u>neutron diffraction analysis</u>

    <u>optical spectroscopy</u>

    <u>polarography</u>

    <u>radio-frequency spectroscopy</u>

    <u>Raman spectroscopy</u>

    <u>spectroscopy</u>, (and <u>atomic absorption</u>)

    <u>spectroscopy</u>, [and <u>electron paramagnetic resonance</u> (not "electron spin resonance")]

    <u>spectroscopy</u>, (and <u>electron probe</u>)

    <u>spectroscopy</u>, (and <u>ion probe</u>)

    <u>spectroscopy</u>, (and <u>laser probe</u>)

>   <u>spectroscopy</u>, (and <u>nuclear</u> <u>magnetic</u> <u>resonance</u>)
>
>   <u>ultraviolet</u> <u>spectroscopy</u>
>
>   <u>vacuum</u> <u>fusion</u> <u>analysis</u>
>
>   <u>volumetric</u> <u>analysis</u>
>
>   <u>X-ray</u> <u>diffraction</u> <u>analysis</u>
>
>   <u>X-ray</u> <u>fluorescence</u>
>
>   <u>X-ray</u> <u>spectroscopy</u>
>
> <u>isotopes</u>
>> name of isotope or isotope ratio, in chemical symbols (e.g. <u>C-14</u>, <u>D/H</u>, <u>K/Ar</u>, <u>O-18/O-16</u>, <u>S-34/S-32</u>, <u>Sr-87/Sr-86</u>)

Where an element is treated as a mineral resource, the 2nd level heading appropriate for commodities (see List C) should be used. In the index these are interspersed with the 2nd level headings listed here, without distinction.

List E

Geologic ages

First level terms denoting geologic time (epochs, periods and eras) are called "age terms". Only words in the following list are used as age terms on the first and second levels. Stratigraphic terms used with these age terms appear in the main body of the Thesaurus.

Archean	Miocene	Phanerozoic
Cambrian	Mississippian	Pleistocene
Carboniferous	Neogene	Pliocene
Cenozoic	Oligocene	Precambrian
Cretaceous	Ordovician	Proterozoic
Devonian	Paleocene	Quaternary
Eocene	Paleogene	Silurian
Holocene	Paleozoic	Tertiary
Jurassic	Pennsylvanian	Triassic
Mesozoic	Permian	

The 2nd- and 3rd-level index terms following 1st-level age terms are as follows:

topic (when an area term is not relevant; the topic is selected from the 2nd-level terms used under area terms (see List B except for areal geology, engineering geology, environmental geology, and geophysical surveys)

subtopic (e.g. name of fossil group (List F); (no area term)

List F

Fossils

Numbers denote categories in the B.I.G. as follows:

08 paleontology, general
09 paleobotany
10 paleontology, invertebrate
11 paleontology, vertebrate

1st = 1st-level term
2nd = 2nd-level term
3rd = 3rd-level term

M__ indicates micropaleontology category

Acantharina see Radiolaria
acritarchs see palynomorphs
Actiniaria see Anthozoa under Coelenterata
Actinopterygii see Osteichthyes under Pisces
Agnatha see Pisces
Agnostida see Trilobita
Alcyonacea see Octocorallia under Coelenterata

09,M01 algae (1st)

  Charophyta (2nd) (incl. chara)
  Chlorophyta (2nd) (incl. green algae, desmids)
  Chrysophyta (2nd) (incl. golden-brown algae)
M03 Coccolithophoraceae (2nd)
  Cyanophyta (2nd) (incl. blue algae, blue-green algae, Schizophyta)
  Dasycladaceae (2nd)
M05 diatoms (2nd)
M03 nannofossils (2nd) (incl. discoasters, nannoconids)
  Phaeophyta (2nd) (incl. brown algae, seaweed)
  Rhodophyta (2nd) (incl. red algae, Corallinaceae)
  stromatolites (2nd) (also see sedimentary structures, under biogenic structures)
  Tasmanites (2nd)

 Note: calcareous algae is used as such on 3rd level under the appropriate group)

Allogromiina see foraminifera

409

Alveolinellidae see Miliolina under foraminifera
Amblypoda see Mammalia
Ammodiscacea see Textulariina under foraminifera

11   Amphibia (1st)

         Labyrinthodontia (2nd)
         Lepospondyli (2nd)
         Lissamphibia (2nd)

Anapsida see Reptilia

09   angiosperms (1st)

         Dicotyledoneae (2nd)
         Monocotyledoneae (2nd)

10   Annelida (1st) (see also worms)

         Myzostomia (2nd)
         Oligochaetia (2nd)
         Polychaetia (2nd)

Anorplura see Insecta
Anthozoa see Coelenterata
Antipatharia see Ceriantipatharia under Coelenterata
Arachnida see Chelicerata under Arthropoda

10   Archaeocyatha (1st)

Archaeornithes see Aves
Archeocopida see Ostracoda
Archipolypoda see Myriapoda under Arthropoda
Archodonata see Insecta
Archosauria see Reptilia

10   Arthropoda (1st) (exclude Insecta)

         Chelicerata (2nd) (subphylum; use only for taxa not included in Merostomata and Arachnida)

            Arachnida (2nd)
            Merostomata (2nd)

         Crustacea (2nd) (superclass; exclude Ostracoda; use for groups of taxa; also used for Recent classes (Branchiura, Cephalocarida, Mystacocarida)

        Branchiopoda (2nd)
        Cirripedia (2nd)
        Copepoda (2nd)
        Euthycarcinoidea (2nd)
        Malacostraca (2nd)

        Myriapoda (2nd) (superclass; use classes on 3rd level: Archipolypoda, Chilopoda, Diplopoda, Pauropoda, Symphyla)
        Pycnogonida (2nd) subphylum
        Trilobitomorpha (2nd) subphylum; exclude Trilobita but include Trilobitoidea on 3rd level)

Archosauria see Reptilia
Articulata see Brachiopoda
Articulatae see Sphenopsida under pteridophytes
Artiodactyla see Mammalia
Asteroidea see Stelleroidea under Asterozoa under Echinodermata
Asterozoa see Echinodermata
Astrapotheria see Mammalia

11   Aves (1st)

        Archaeornithes (2nd)
        Neornithes (2nd)

09   bacteria (1st) (schizomycetes)

        Bennettitales see gymnosperms
        birds see Aves
        Bivalvia see Mollusca
        Blastoidea see Crinozoa under Echinodermata
        Blattodea see Protorthoptera under Insecta
        blue algae see Cyanophyta under algae
        blue-green algae see Cyanophyta under algae

10   Brachiopoda (1st)

        Articulata (2nd)

            Orthida (2nd)
            Pentamerida (2nd)
            Rhynchonellida (2nd)
            Spiriferida (2nd)
            Strophomenida (2nd)
            Terebratulida (2nd)

Inarticulata (2nd)

Branchiopoda see Crustacea under Arthropoda
Branchiura see Crustacea under Arthropoda
brown algae see Phaeophyta under algae

09 bryophytes (1st) (mosses)

   Hepaticae (2nd)
   Musci (2nd)

10 Bryozoa (1st)

   Cheilostomata (2nd)
   Cryptostomata (2nd)
   Ctenostomata (2nd)
   Cyclostomata (2nd)
   Trepostomata (2nd)

Buliminacea see Rotaliina under foraminifera
Calcispongea see Porifera
Caloneurodea see Protorthoptera under Insecta
Camaroidea see Graptolithina
Camptostromatoidea see Echinozoa under Echinodermata
Carnivora see Mammalia
Carterinacea see Rotaliina under foraminifera
Cassidulinacea see Rotaliina under foraminifera
Caytoniales see gymnosperms
Cephalocarida see Crustacea under Arthropoda
Cephalodiscida see Pterobranchia
Cephalopoda see Mollusca
Ceriantharia see Ceriantipatharia under Coelenterata
Ceriantipatharia see Coelenterata
Cetacea see Mammalia
Chaetognatha see worms
chara see Charophyta under algae
Charophyta see algae
Cheilostomata see Bryozoa
Cheleutoptera see Plecoptera under Insecta
Chelicerata see Arthropoda
Chilopoda see Myriapoda under Arthropoda
Chiroptera see Mammalia
Chitinozoa see palynomorphs
Chlorophyta see algae
Chondrichthyes see Pisces

11 Chordata (1st)

Chrysophyta see algae
Cirripedia see Crustacea under Arthropoda

Coccolithophoraceae see algae

10   Coelenterata (1st)

         Anthozoa (2nd)

            Actiniaria (2nd)
            Corallimorpharia (2nd)
            Heterocorallia (2nd)
            Hexactiniaria (2nd)
            Rugosa (2nd)
            Scleractinia (2nd)
            Tabulata (2nd)
            Zoantharia (2nd)
            Zoanthiniaria (2nd)

         Ceriantipatharia (2nd)

            Antipatharia (2nd)
            Ceriantharia (2nd)

         Hydrozoa (2nd)

         Octocorallia (2nd)

            Alcyonacea (2nd)
            Coenothecacalia (2nd)
            Gorgonacea (2nd)
            Pennatulacea (2nd)
            Stolonifera (2nd)
            Telestacea (2nd)
            Trachypsammiacea (2nd)

         Scyphozoa (2nd)
         Stromatoporoidea (2nd) (could also be used
            under problematic fossils)

Coenothecacalia see Octocorallia under Coelenterata
Coleoptera see Insecta
Collembola see Insecta
Condylarthra see Mammalia
Coniferales see gymnosperms

08,M04   conodonts (1st)

        Copepoda see Crustacea under Arthropoda

10,11   coprolites (1st)

Corallimorpharia see Anthozoa under Coelenterata
Corallinaceae see Rhodophyta under algae
corals see Coelenterata
Cordaitales see gymnosperms
Corynexochida see Trilobita
Creodonta see Mammalia
Crinoidea see Crinozoa under Echinodermata
Crinozoa see Echinodermata
Crustacea see Arthropoda
Cryptostomata see Bryozoa
Ctenostomata see Bryozoa
Cyanophyta see algae
Cycadales see gymnosperms
Cyclocystoidea see Echinozoa under Echinodermata
Cyclostomata see Bryozoa
Cystoidea see Crinozoa under Echinodermata
Dasycladaceae see algae
Demospongea see Porifera
Dendroidea see Graptolithina
Dermaptera see Insecta
desmids see Chlorophyta under algae
Desmostylia see Mammalia
diatoms see algae
Dicotyledoneae see angiosperms
Didymograptina see Graptoloidea under Graptolithina
Dinoflagellata see palynomorphs
dinosaurs see Archosauria under Reptilia
Diplograptina see Graptoloidea under Graptolithina
Diplopoda see Myriapoda under Arthropoda
Diplura see Insecta
Diptera see Insecta
discoasters see nannofossils
     under algae
Discorbacea see Rotaliina under foraminifera
Docodonta see Mammalia
ebridians see Protista

10   Echinodermata (1st)

               Asterozoa (2nd)

               Stelleroidea (2nd)

                   Asteroidea (3rd)
                   Ophiuroidea (3rd)
                   Somasteroidea (3rd)

               Crinozoa (2nd)

                   Blastoidea (2nd)

     <u>Crinoidea</u> (2nd)
     <u>Cystoidea</u> (2nd)
     <u>Edrioblastoidea</u> (2nd)
     <u>Eocrinoidea</u> (2nd)
     <u>Lepidocystoidea</u> (2nd)
     <u>Parablastoidea</u> (2nd)
     <u>Paracrinoidea</u> (2nd)

   <u>Echinozoa</u> (2nd)

     <u>Camptostromatoidea</u> (2nd)
     <u>Cyclocystoidea</u> (2nd)
     <u>Echinoidea</u> (2nd)
     <u>Edrioasteroidea</u> (2nd)
     <u>Helicoplacoidea</u> (2nd)
     <u>Holothuroidea</u> (2nd)
     <u>Ophiocistioidea</u> (2nd)

   <u>Homalozoa</u> (2nd)

     <u>Homoiostelea</u> (2nd)
     <u>Homostelea</u> (2nd)
     <u>Machaeridia</u> (2nd)
     <u>Stylophora</u> (2nd)

<u>Echinoidea</u> see <u>Echinozoa</u> under <u>Echinodermata</u>
<u>Echinozoa</u> see <u>Echinodermata</u>
<u>Ectotropha</u> see <u>Insecta</u>
<u>Edentata</u> see <u>Mammalia</u>
<u>Edrioasteroidea</u> see <u>Echinozoa</u> under <u>Echinodermata</u>
<u>Edrioblastoidea</u> see <u>Crinozoa</u> under <u>Echinodermata</u>
<u>Embidina</u> see <u>Plecoptera</u> under <u>Insecta</u>
<u>Embrithopoda</u> see <u>Mammalia</u>
<u>Eocrinoidea</u> see <u>Crinozoa</u> under <u>Echinodermata</u>
<u>Eopaleodictyoptera</u> see <u>Insecta</u>
<u>Euryapsida</u> see <u>Reptilia</u>
<u>Eurypterida</u> see <u>Merostomata</u> under <u>Arthropoda</u>
<u>Filicales</u> see <u>Filicopsida</u> under <u>pteridophytes</u>
<u>Filicopsida</u> see <u>pteridophytes</u>

10,M06 <u>foraminifera</u> (1st)

   <u>Allogromiina</u> (2nd)

   <u>Fusulinina</u> (2nd)

     <u>Fusulinidae</u> (2nd)

   <u>Miliolina</u> (2nd)

     <u>Alveolinellidae</u> (2nd)
     <u>Miliolacea</u> (2nd)

        Rotaliina (2nd)

            Buliminacea (2nd)
            Carterinacea (2nd)
            Cassidulinacea (2nd)
            Discorbacea (2nd)
            Globigerinacea (2nd)
            Nodosariacea (2nd)
            Orbitoidacea (2nd)
                Orbitoididae (2nd)
            Robertinacea (2nd)
            Rotaliacea (2nd)
                Nummulitidae (2nd)
            Spirillinacea (2nd)

        Textulariina (2nd)

            Ammodiscacea (2nd)
            Lituolacea (2nd)

    fossil wood (3rd) under Plantae or gymnosperms
        or angiosperms

11   fossil man (1st)

09   fungi (1st)

    Fusulinidae see Fusulinina under foraminifera
    Fusulinina see foraminifera
    Gastropoda see Mollusca
    Ginkgoales see gymnosperms
    Globigerinacea see Rotaliina under foraminifera
    Glosselytrodea see Protorthoptera under Insecta
    Glossograptina see Graptoloidea under Graptolithina
    Glossopteris see gymnosperms
    Gnetales see gymnosperms
    golden-brown algae see Chrysophyta under algae
    Gorgonacea see Octocorallia under Coelenterata

10   Graptolithina (1st)

        Camaroidea (2nd)
        Dendroidea (2nd)

        Graptoloidea (2nd)

            Didymograptina (2nd)
            Diplograptina (2nd)
            Glossograptina (2nd)

                    Monograptina (2nd)

            Stolonoidea (2nd)
            Tuboidea (2nd)

    Graptoloidea see Graptolithina
    green algae see Chlorophyta under algae

09   gymnosperms (1st)

            Bennettitales (2nd)
            Caytoniales (2nd)
            Coniferales (2nd)
            Cordaitales (2nd)
            Cycadales (2nd)
            Ginkgoales (2nd)
            Glossopteris (2nd) (for more than the genus)
            Gnetales (2nd)
            Nilssoniales (2nd)
            Pentoxyleae (2nd)
            Pteridospermae (2nd)

    Helicoplacoidea see Echinozoa under Echinodermata

10   Hemichordata (1st)

    Hepaticae see bryophytes
    Heterocorallia see Anthozoa under Coelenterata
    Heteroptera see Insecta
    Hexactiniaria see Anthozoa under Coelenterata
    Holostei see Osteichthyes under Pisces
    Holothuroidea see Echinozoa under Echinodermata
    Homalozoa see Echinodermata
    Homoiostelea see Homalozoa under Echinodermata
    Homoptera see Insecta
    Homostelea see Homalozoa under Echinodermata
    Hyalospongea see Porifera
    Hydropterides see pteridophytes
    Hydrozoa see Coelenterata
    Hymenoptera see Insecta
    hyolithes see problematic fossils
    hystrichosphaerids see acritarchs or Dinoflagellata
            under palynomorphs

08,10,11 ichnofossils (1st)

    Ichthyopterygia see Reptilia
    Inarticulata see Brachiopoda

10   Insecta (1st)

    Anorplura (2nd)
    Archodonata (2nd)
    Coleoptera (2nd)
    Collembola (2nd)
    Dermaptera (2nd)
    Diplura (2nd)
    Diptera (2nd)
    Ectotropha (2nd)
    Eupaleodictyoptera (2nd)
    Eubleptidodea (2nd)
    Heteroptera (2nd)
    Homoptera (2nd)
    Hymenoptera (2nd)
    Lepidoptera (2nd)
    Mallophaga (2nd)

   Mantodea (2nd)

     Isoptera (2nd)
     Protoperlaria (2nd)

   Megaloptera (2nd)

   Odonata (2nd)

     Megasecoptera (2nd)

   Paracoleoptera (2nd)
   Permoneurodea (2nd)
   Plannipennia (2nd)

   Plecoptera (2nd)

     Cheleutoptera (2nd)
     Embidina (2nd)
     Notoptera (2nd)
     Orthoptera (2nd)
     Protelytroptera (2nd)
     Saltatoria (2nd)

   Plectoptera (2nd)

     Protodonata (2nd)

   Protocoleoptera (2nd)
   Protoephemerodea (2nd)
   Protohemiptera (2nd)
   Protomecoptera (2nd)

   Protorthoptera (2nd)

     Blattodea (2nd)
     Caloneurodea (2nd)
     Glosselytrodea (2nd)

     Psocoptera (2nd)
     Raphidiodea (2nd)

     Siphonaptera (2nd)
     Strepsiptera (2nd)
     Syntonopterodea (2nd)
     Sypharopterodea (2nd)
     Thysanoptera (2nd)
     Trichoptera (2nd)

  Insectivora see Mammalia

10 Invertebrata (1st)

  Isoptera see Mantodea under Insecta
  Labyrinthodontia see Amphibia
  Lagomorpha see Mammalia
  Leperditicopida see Ostracoda
  Lepidocystoidea see Crinozoa under Echinodermata
  Lepidoptera see Insecta
  Lepidosauria see Reptilia
  Lepospondyli see Amphibia

09 lichens (1st)

  Lichida see Trilobita
  Lissamphibia see Amphibia
  Litopterna see Mammalia
  Lituolacea see Textulariina under foraminifera
  Lycopsida see pteridophytes
  Machaeridia see Homalozoa under Echinodermata
  Malacostraca see Crustacea under Arthropoda
  Mallophaga see Insecta

11 Mammalia (1st)

    Amblypoda (2nd)
    Artiodactyla (2nd)
    Astrapotheria (2nd)
    Carnivora (2nd)
    Cetacea (2nd)
    Chiroptera (2nd)
    Condylarthra (2nd)
    Creodonta (2nd)
    Desmostylia (2nd)
    Docodonta (2nd)

    Edentata (2nd)
    Embrithopoda (2nd)
    Insectivora (2nd)
    Lagomorpha (2nd)
    Litopterna (2nd)
    Marsupialia (2nd)
    Monotremata (2nd)
    Multituberculata (2nd)
    Notoungulata (2nd)
    Pantotheria (2nd)
    Perissodactyla (2nd)
    Pholidata (2nd)
    Primates (2nd)
    Proboscidea (2nd)
    Ptilodontoidea (2nd)
    Rodentia (2nd)
    Sirenia (2nd)
    Symmetrodonta (2nd)
    Taeniodonta (2nd)
    Taeniolaboidea (2nd)
    Tillodontia (2nd)
    Triconodonta (2nd)
    Tubulidentata (2nd)
    Ungulata (2nd)

Mantodea see Insecta
Marsupialia see Mammalia
Megaloptera see Insecta
Megasecoptera see Odonata under Insecta
megaspores see palynomorphs
Merostomata see Chelicerata under Anthropoda
Miliolacea see Miliolina under foraminifera
   Miliolina see foraminifera
   miospores see palynomorphs

10   Mollusca (1st)

    Aplacophora (2nd)
    Bivalvia (2nd) (incl. rudists under
     Hippuritoida)
    Cephalopoda (2nd)
    Gastropoda (2nd)
    Monoplacophora (2nd)
    Polyplacophora (2nd)
    Rostroconchia (2nd)
    Scaphopoda (2nd)

Monocotyledoneae see angiosperms
Monograptina see Graptoloidea under Graptolithina
Monoplacophora see Mollusca
Monotremata see Mammalia

mosses see bryophytes
Multituberculata see Mammalia
Musci see bryophytes
Myodocopida see Ostracoda
Myriapoda see Arthropoda
Mystacocarida see Crustacea under Arthropoda
Myzostomia see Annelida
nannoconids see nannofossils under algae
nannofossils see algae
Nassellina see Radiolaria
Nematoida see worms
Nematomorpha see worms
Nemerta see worms
Neornithes see Aves
Nilssoniales see gymnosperms
Nodosariacea see Rotaliina under foraminifera
Noeggerathiales see pteridophytes
Notoptera see Plecoptera under Insecta
Notoungulata see Mammalia
Nummulitidae see Rotaliacea under Rotaliina under
    foraminifera
Octocorallia see Coelenterata
Odonata see Insecta
Odontopleurida see Trilobita
Oligochaetia see Annelida
Ophiocistioidea see Echinozoa under Echinodermata
Ophiuroidea see Stelleroidea under Asterozoa
    under Echinodermata
Orbitoidacea see Rotaliina under foraminifera
Orbitodidae see Orbitoidaceae under Rotaliina
    under foraminifera
Orthida see Articulata under Brachiopoda
Orthoptera see Plecoptera under Insecta
Osteichthyes see Pisces

10,M08   Ostracoda (1st)

     Archeocopida (2nd)
     Leperditicopida (2nd)
     Myodocopida (2nd)
     Paleocopida (2nd)
     Podocopida (2nd)

ostracoderms see Agnatha under Pisces
Paleocopida see Ostracoda

09,M09   palynomorphs (1st)

     acritarchs (hystrichosphaerids mostly) (2nd)
     Chitinozoa (2nd)
     Dinoflagellata (peridineans and hystrichosphaerids)

                    (2nd)
                megaspores (2nd)
                miospores (2nd)
                problematic palynomorphs (2nd)

        Pantotheria see Mammalia
        Parablastoidea see Crinozoa under Echinodermata
        Paracoleoptera see Insecta
        Paracrinoidea see Crinozoa under Echinodermata
        Pauropoda see Myriapoda under Arthropoda
        Pelecypoda use Bivalvia under Mollusca
        Pennatulacea see Octocorallia under Coelenterata
        Pentamerida see Articulata under Brachiopoda
        Pentoxyleae see gymnosperms
        peridineans see Dinoflagellata under palynomorphs
        Perissodactyla see Mammalia
        Permoneurodea see Insecta
        Phacopida see Trilobita
        Phaeodarina see Radiolaria
        Phaeophyta see algae
        Pholidata see Mammalia

11   Pisces (1st)

            Agnatha (2nd) (ostracoderms)
            Chondrichthyes (2nd) (selachians, etc.)
            Osteichthyes (2nd) (incl. Actinopterygii, Holostei,
                Teleostei)
            Placodermi (2nd)

        Placodermi see Pisces
        plankton (use under various fossil groups on 3rd)
        Plannipennia see Insecta

09   Plantae (1st)
            (use also vegetation on 3rd level)

        Plecoptera see Insecta
        Plectoptera see Insecta
        Podocopida see Ostracoda
        Polychaetia see Annelida

10   Porifera (1st)

            Calcispongea (2nd)
            Demospongea (2nd)
            Hyalospongea (2nd)

        Primates see Mammalia

422

08,M07   problematic fossils (1st) including hyolithes (3rd)
          and tentaculites (3rd) and
          Stromatoporoidea (3rd)
          problematic palynomorphs see palynomorphs
          Proboscidea see Mammalia
          Protelytroptera see Plecoptera under Insecta

08,09,10,
M10   Protista (1st)

              ebridians (2nd)
              Silicoflagellata (2nd)
              Thecamoeba (2nd)
              Tintinnidae (2nd) (Calpionellidae)

          Protocoleoptera see Insecta
          Protodonata see Plectoptera under Insecta
          Protoephemerodea see Insecta
          Protohemiptera see Insecta
          Protolytroptera see Plecoptera under Insecta
          Protomecoptera see Insecta
          Protoperlaria see Mantodea under Insecta
          Protorthoptera see Insecta
          Protozoa see Protista or other microfossil groups
          Psilophytales see Psilopsida under pteridophytes
          Psilopsida see pteridophytes
          Psocoptera see Insecta
          Pteridophyllen see pteridophytes

09   pteridophytes (1st)

              Filicopsida (2nd) (Filicales)
              Hydropterides (2nd)
              Lycopsida (2nd) (Lycopodiales)
              Noeggerathiales (2nd)
              Psilopsida (2nd) (Psilophytales)
              Pteridophyllen (2nd)
              Sphenopsida (2nd) (Articulatae)

          Pteridospermae see gymnosperms

10   Pterobranchia (1st)

              Cephalodiscida (2nd)
              Rhabdopleurida (2nd)

          Pteropoda used on 3rd level under Gastropoda, Mollusca
          Ptychopariida see Trilobita
          Pycnogonida see Arthropoda

10,M11   Radiolaria (1st)

                Acantharina (2nd)
                Nassellina (2nd)
                Phaeodarina (2nd)
                Spumellina (2nd)

            Raphidiodea see Insecta
            Redlichiida see Trilobita

11   Reptilia (1st)

                Anapsida (2nd)
                Archosauria (2nd) (incl. dinosaurs)
                Euryapsida (2nd)
                Ichthyopterygia (2nd)
                Lepidosauria (2nd)
                Synapsida (2nd)

            Rhabdopleurida see Pterobranchia
            Rhodophyta see algae
            Rhynchonellida see Articulata under Brachiopoda
            Robertinacea see Rotaliina under foraminifera
            Rodentia see Mammalia
            Rostroconchia see Mollusca
            Rotaliacea see Rotaliina under foraminifera
            Rotaliina see foraminifera
            rudists see Bivalvia under Mollusca
            Rugosa see Anthozoa under Coelenterata
            Saltatoria see Plecoptera under Insecta
            Scaphopoda see Mollusca
            schizomycetes see bacteria
            Schizophyta see Cyanophyta under algae
            Scleractinia see Anthozoa under Coelenterata
            Scyphozoa see Coelenterata
            selachians see Chondrichthyes under Pisces
            Silicoflagellata see Protista
            Siphonaptera see Insecta
            Sipunculoida see worms
            Sirenia see Mammalia
            Somasteroidea see Stelleroidea under Asterozoa
                    under Echinodermata
            Sphenopsida see pteridophytes
            Spiriferida see Articulata under Brachiopoda
            Spirillinacea see Rotaliina under foraminifera
            Spongiae use Porifera
            Spumellina see Radiolaria
            Stelleroidea see Asterozoa under Echinodermata
            Stolonifera see Octocorallia under Coelenterata
            Stolonoidea see Graptolithina
            Strepsiptera see Insecta
            stromatolites see algae

Stromatoporoidea see Coelenterata
        or problematic fossils
Strophomenida see Articulata under Brachiopoda
Stylophora see Homalozoa under Echinodermata
Symmetrodonta see Mammalia
Symphyla see Myriapoda under Arthropoda
Synapsida see Reptilia
Syntonopterodea see Insecta
Sypharopterodea see Insecta
Tabulata see Anthozoa under Coelenterata
Taeniodonta see Mammalia
Taeniolaboidea see Mammalia
Tasmanites see algae
Teleostei see Osteichthyes under Pisces
Telestacea see Octocorallia under Coelenterata
tentaculites see problematic fossils
Terebratulida see Articulata under Brachiopoda
Textulariina see foraminifera

09   thallophytes (1st) (group)

Thecamoeba see Protista
Thysanoptera see Insecta
Tillodontia see Mammalia
Tintinnidae see Protista
Trachypsammiacea see Octocorallia under Coelenterata
tracks and trails see ichnofossils
Trepostomata see Bryozoa
Trichoptera see Insecta
Triconodonta see Mammalia

10   Trilobita (1st)

        Agnostida (2nd)
        Corynexochida (2nd)
        Lichida (2nd)
        Odontopleurida (2nd)
        Phacopida (2nd)
        Ptychopariida (2nd)
        Redlichiida (2nd)

Trilobitoidea see Trilobitomorpha under Arthropoda
Trilobitomorpha see Arthropoda
Tuboidea see Graptolithina
Tubulsidentata see Mammalia
Ungulata see Mammalia

11   Vertebrata (1st)

10   worms (1st) (see also Annelida)

                    Chaetognatha (2nd)
                    Nematoida (2nd)
                    Nematomorpha (2nd)
                    Nemerta (2nd)
                    scolecodonts (2nd)
                    Sipunculoida (2nd)

          Zoantharia see Anthozoa under Coelenterata
          Zoanthiniaria see Anthozoa under Coelenterata

The fossil set is made up of one of the 1st level terms above and:

on 2nd level

taxonomic subdivision (allowable 2nd level term from List above; if fossil not yet classified, use miscellanea); for several 2nd level terms use faunal studies or floral studies

or: topic (used when topic is emphasized): e.g. biocenoses, biochemistry, biogeography, biostratigraphy, distribution, ecology, evolution (including phylogeny), fossilization,(including taphonomy), habitat, morphology, paleoecology, occurrence, ontogeny, taxonomy

          on 3rd level

either: age or area or topic (also use in indexing
          the taxon, i.e. name of paleontologic
          subdivision(s) down to genus level or
          species level if new or of importance
          to paper); also use new taxa (whenever
          a new taxonomic group is discussed)

or: subtopic e.g. benthonic taxa, bibliography,
          biometry, catalogs, faunal list, floral
          list, paleoclimatology, phylogeny, planktonic
          taxa, statistical analysis, taphonomy,
          thanatocenoses, zoning (type of material,
          e.g. bones, fossil wood, plankton, shells,
          skeletons, skulls, teeth, tests)

List G

Terms appropriate to a large number of topics

- abundance
- age
- annual report
- anomalies
- applications
- bibliography
- catalogs
- causes
- changes
- classification
- composition
- concepts
- detection
- distribution
- education
- effects
- environment
- evolution
- experimental studies
- extent
- general
- genesis
- history
- identification
- instruments
- interpretation
- manuals
- mechanism
- meetings
- methods
- nomenclature
- objectives
- observations
- occurrence
- origin use genesis
- patterns
- philosophy
- possibilities
- practice
- preparation
- principles
- processes
- properties
- rates
- regional
- review
- structure
- surveys
- symposia
- techniques
- temperature
- textbooks
- textures
- theoretical studies
- varieties

These terms, being general in meaning, should be used sparingly and with some care on the 2nd- or 3rd-level as the case may be.

List H

Igneous rocks

general terms

  <u>acidic composition</u>
  <u>alkalic composition</u>
  <u>calc-alkalic composition</u>
  <u>calcic composition</u>
  <u>hypabyssal rocks</u>
  <u>mafic composition</u>
  <u>plutonic rocks</u>
  <u>volcanic rocks</u>

groups

  1 = <u>basalt family</u>
  2 = <u>alkali basalt family</u>
  3 = <u>lamprophyre and carbonatite family</u>
  4 = <u>trachyte-phonolite family</u>
  5 = <u>andesite-rhyolite family</u>
  6 = <u>ultramafic family</u>
  7 = <u>gabbro family</u>
  8 = <u>alkali gabbro family</u>
  9 = <u>diorite family</u>
 10 = <u>granite-granodiorite family</u>
 11 = <u>syenite family</u>
 12 = <u>pyroclastics and glasses</u>

specific rock; and group

  <u>absarokite</u>: 1
  <u>adamellite</u>: 10
  <u>agpaite</u>: 11
  <u>alaskite</u>: 10
  <u>albitite</u>: 4
  <u>albitophyre</u>: 11
  <u>alkali basalt</u>: 2
  <u>alkali olivine basalt</u>: 1
  <u>alkali syenite</u>: 11
  <u>alkalic granite</u>: 10
  <u>andesite porphyry</u>: 5
  <u>andesitic tuff</u>: 12
  <u>andesite</u>: 5
  <u>ankaramite</u>: 1,6
  <u>ankaratrite</u>: 2
  <u>anorthositic gabbro</u>: 7
  <u>anorthosite</u>: 7

aplite: 10
apogranite: 10
appinite: 11,9
ash-flow tuff: 12
ariegite: 6
augitite: 6
basalt: 1
basaltite: use basalt
basanite: 2
beresite
biotite granite: 10
bostonite: 11
bronzitite: 6
camptonite: 3
carbonatite: 3
catapleiite syenite: 11 use
    syenite and catapleiite
charnockite: 10 (plutonic rocks
    or under metamorphic rocks)
chromitite: 6
clinopyroxenite: 6
columnar basalt: 1
comendite: 5
crinanite: 2
dacite: 5
dacite porphyry: 5
dellenite: 5
diabase: 1
diabase porphyry: 1
diallagite: 6
diorite: 9
diorite porphyry: 9 (volcanic rocks)
dolerite: 1
dunite: 6
echodolite: 4 use phonolite
eleolite syenite: 11 use
    nepheline syenite
enderbite: 10
essexite: 8
eucrite: 7
eulysite: 6
eupholite: 7 use gabbro
farsundite: 10
felsite: 10
ferrodiorite: 9
fluolite: 12 use pitchstone
foyaite: 11,8
gabbro: 7
gabbroic anorthosite: 7
garnet lherzolite: 6
garnet peridotite: 6
garnet pyroxenite: 6
glimmergabbro: 7 use gabbro

granite: 10
granite porphyry: 10
granodiorite: 10
granodiorite porphyry: 10
granophyre: 10 (volcanic rocks)
granosyenite: 10,11
graphic granite: 10
green tuff: 12
griquaite: 6
haplophyre: 10 use granite
hardebank: 6 use kimberlite
hawaiite: 2
hornblende andesite: 5 use
    andesite and hornblende
hornblendite: 6
hyalopsite: 12 use obsidian
hypersthenfels: 7 use norite
ignimbrite: 12
ijolite: 8
kakortokite: 11
kaligranite: 10 use orthogranite
kammgranite: 10 use granite
kankan-ishi: 5 use andesite
keratophyre: 4
kersantite: 3
kimberlite: 6
komatiite: 6
labradoritite: 7
lamproite: 4
lamprophyre: 3
latite: 5
leucitite: 2,6
leucogranite: 10
lherzolite: 6
limburgite: 6
liparite: 5
liparite porphyry: 5
lujavrite: 11
mangerite: 9
meimechite: 6
melaphyre: 1
melilitite: 2,6
melteigite: 8
miaskite: 11
microdiorite: 9
microgabbro: 7
microgranite: 10
micropegmatite: 10
microsyenite: 11
minette: 3
monchiquite: 3
monzodiorite: plutonic rocks
monzonite: 11

mugearite: 1
muscovite granite: 10
naujaite: 11
nepheline basalt: 2
nepheline syenite: 11
nephelinite: 2,6
norite: 7
obsidian: 12
oceanite: 1
olivine basalt: 1
olivine diabase: 1
olivine dolerite: 1
olivine gabbro: 7
olivine nephelinite: 2
olivine tholeiite: 1
olivinite: 6
ophiolite: 6
palagonite: 12
paleophyre: 5 use andesite
pallasite: 6
pantellerite: 5
pegmatite: 10
peridotite: 6
perlite: 12
phonolite: 4
picrite: 1,6
picrite porphyry: 1
pitchstone: 12
pladorit: 10 use granite
plagiogranite: plutonic rocks porphyry
pulaskite: 11
pumice: 12
pyroxene andesite: 5
pyroxenite: 6
quartz andesite: 5 use andesite
    and quartz
quartz diabase: 1
quartz diorite: 9
quartz dolerite: 1
quartzfels: 10 use silexite
quartz keratophyre: 4
quartz latite: 5
quartz monzonite: 10
quartz porphyry: 5
quartz syenite: 11
ramosite: 12 use scoria

rapakivi: 10
rhyodacite: 5
rhyolite: 5
rhyolite porphyry: 5
rhyolite tuff: 12
rodingite: 7
sanakite: 5 use andesite
scorilite: 12 use volcanic glass
shonkinite: 11,8
shoshonite: 1
silexite: 10
soda minette: 3 use minette
sovite: 3
spessartite: 3
spilite: 2
spinel lherzolite: 6
syenite: 11
syenite porphyry: 11
syenodiorite: 9
tephrite: 2
teschenite: 8
tholeiite: 1
tholeiitic basalt: 1
tholeiitic dolerite: 1
tinguaite: 4
tonalite: 9
trachyandesite: 5
trachybasalt: 2
trachydolerite: 2
trachyte: 4
trap rocks: 1
troctolite: 7
trondhjemite: 9
troutstone: 7 use troctolite
tuff: 12
tuffite: 12
uralitite: 1 use diabase
uralite diabase: 1 use diabase
    and uralite
urtite: 8
volcanic ash: 12
volcanic glass: 12
websterite: 6
wehrlite: 6
welded tuff: 12
zirkelite: 12

List I

Sedimentary rocks

groups

1 = <u>carbonate rocks</u>

2 = <u>chemically precipitated rocks</u>

3 = <u>clastic rocks</u>

4 = <u>organic residues</u>

specific rock; and group

<u>algal limestone</u>: 1
<u>anhydrite</u>: 2
<u>anthracite</u>: 4
<u>arenite</u>: 3
<u>argillite</u>: 3
<u>arkose</u>: 3
<u>bauxite</u>: 3
<u>beachrock</u>: 1
<u>bentonite</u>: 3
<u>biocalcarenite</u>: 1
<u>biohermite</u> use <u>klintite</u>
<u>biomicrite</u>: 1
<u>biosparite</u>: 1
<u>bituminous coal</u>: 4
<u>bituminous shale</u>: 3
<u>black shale</u>: 3
<u>boundstone</u>: 1
<u>breccia</u>: 3
<u>brown coal</u>: 4
<u>calcarenite</u>: 1
<u>calcilutite</u>: 1
<u>carbargilite</u>: 4 use <u>coal</u>
<u>caustobiolith</u>: 4
<u>chalk</u>: 1
<u>chert</u>: 2
<u>cinerite</u>: 3
<u>claystone</u>: 3
<u>coal</u>: 4 [<u>macerals</u>: <u>alginite</u>, <u>collinite</u>,
    <u>cutinite</u>, <u>exinite</u>, <u>fusinite</u>, <u>inertinite</u>,
    <u>leptinite</u>, <u>micrinite</u>, <u>resinite</u>, <u>semifusinite</u>,
    <u>sporinite</u>, <u>telinite</u>, <u>vitrinite</u>]
<u>conglomerate</u>: 3

conglomerite use conglomerate
contourite: 3
diatomaceous earth: 3
diatomite: 3
diamictite: 3
dolostone: 1
dropstone use argillite
durain: 4
eolianite: 3
evaporites: 2
fanglomerate: 3
flint: 2
flysch: 3
glauconite use glauconitic
    composition
grainstone: 1
grapestone use as modifier
graywacke: 3
gypsum: 2
halite: 2
hornstein use chert
hornstone use chert
ignimbrite: 3
iron formations: 2 [used with
    iron-rich composition on
    3rd level]
ironstone: 2
itabirite: 2
itacolumite see metamorphic rocks
jasperoid: 2
jaspilite: 2
lignite: 4
limestone: 1
lydite use chert
macerals see under coal
magnesian limestone: 1
marl: 1,3
micrite: 1
microbreccia: 3
mixtite: 3
molasse: 3
mudstone: 3
novaculite: 3
oolitic limestone: 1
orthoquartzite: 3
packstone: 1
paraconglomerate use mudstone,
    conglomeratic texture
pelite use shale
pelodite use varvite

phosphate rocks: 2
phosphorite: 2
porcellanite: 3
protoquartzite: 3 (use sublitharenite)
pyroclastics: 3
radiolarite: 3
red beds: 3
sandstone: 3
saprolite: 3
sapropelite: 4
shale: 3
siliceous sinter: 2
siltstone: 3
sparagmite: 3
spongolite: 3
subarkose: 3
subgraywacke: 3
taconite: 2
tillite: 3
tilloid: 3
tonstein: 3
torbanite: 4
travertine: 2,1
turbidite: 3
tufa: 2,1
tuff: 3
vitrain: 4
wackestone: 1

terms to be used with specific name:

    arenaceous texture
    argillaceous texture
    arkosic composition
    calcareous composition
    conglomeratic texture
    dolomitic composition
    ferruginous composition
    fossiliferous materials
    glauconitic composition
    iron-rich composition
    lithic texture
    oolitic texture
    pebbly texture
    pelletal texture
    pisolitic texture
    siliceous composition

List J

Metamorphic rocks

general terms

metasedimentary rocks
metavolcanic rocks
metaplutonic rocks
metaigneous rocks

groups

1 = hornfels
2 = slates
3 = phyllites
4 = schists
5 = amphibolites
6 = gneisses
7 = granulites
8 = marbles
9 = mylonites
10 = cataclasites
11 = phyllonites
12 = migmatites

specific rock; and group

agmatite: 12
amphibolite: 5
anatexite
augen gneiss: 6
banded gneiss: 6
bardiglio: 8 use marble
basic hornfels: 1 use hornfels and mafic composition
basic schist: 4 use schist and mafic composition
biotite gneiss: 6
biotite schist: 4
blastomylonite: 9
blavierite: 9 use mylonite
blueschist: 4
bookstone: 4 use schist
brucite marble: 8
calcareous schist: 4 use schist and calcareous composition
calciphyre: 8
calc-schist: 4
calc-silicate hornfels: 1 use hornfels and calc-silicate composition
calc-silicate marble: 8 use marble and calc-silicate composition
chiastolite slate: 2 use slate and chiastolite

433

chlorite schist: 4
clay slate: 2 use slate
crystalline limestone: 8 use marble
crystalline schist: 4 use schist
delvin: 2 use slate
desmosite: 2 use adinole
diabase amphibolite: 5 use amphibolite
dictyonite: 12 use migmatite
dolerine: 4 use schist and talc
dolomite marble: 8 use marble and dolomite
eclogite: 7
ectinite: 12
epidiorite: 4
Erianfels: 7 use granulite
fascicular schist: 4 use schist and fascicular texture
fenite
ferruginous quartzite
ferruginous schist: 4 use schist and ferruginous composition
flaser gabbro: 10 use gabbro and flaser structure
flaser gneiss: 10 use gneiss and flaser structure
flaser granite: 10 use granite and flaser structure
Fleckschiefer: 2 use slate
fruited schist: 4 use schist
gabbro schist: 4 use schist and gabbro
Garbenschiefer: 2 use slate
glandular schist: 4 use schist
glaucophane schist: 4
gneiss: 6
gneissic quartzite: 6 use quartzite and gneissic texture
gondite
granite gneiss: 6
granulite: 7
greenschist: 4
greenstone: 4
greisen
griffel schiefer: 3 use phyllite
halleflintgneiss: 7 use leptite
hornblende gneiss: 6 use gneiss and hornblende
hornblende schist: 4 use schist and hornblende
hydromica schist: 4 use schist and hydromica
kinzigite: 7
leptite: 7
leptynite: 7
listwanite
marble: 8
meta-andesite: metaigneous rocks
meta-arkose: metasedimentary rocks
metabasalt: metaigneous rocks
metabasite: metaigneous rocks
metabentonite: metasedimentary rocks
metaconglomerate: metasedimentary rocks
metadiabase: metaigneous rocks
metadiorite: metaigneous rocks

metadolerite: metaigneous rocks
metagabbro: metaigneous rocks
metagranite: metaigneous rocks
metagraywacke: metasedimentary rocks
metalimestone: metasedimentary rocks
metapelite: metasedimentary rocks
metaperidotite: metaigneous rocks
metapyroxenite: metaigneous rocks
metarhyolite: metaigneous rocks
metasandstone: metasedimentary rocks
metasiltstone: metasedimentary rocks
metatuff: metaigneous rocks
mica schist: 4
migmatite: 12
mylonite: 9
ophicalcite: 8
ophite: 8
orthoamphibolite: 5
para-amphibolite: 5
paragneiss: metasedimentary rocks
pelitic gneiss: 6 use gneiss and pelitic texture
pelitic hornfels: 1 use hornfels and pelitic texture
pelitic schist: 4
phyllite: 3
phyllonite: 11
prasinite: 4 use greenschist
predazzite: 8 use pencatite
proteolite: 1 use hornfels
protoclastic gneiss: 6 use gneiss and protoclastic texture
psammitic gneiss: 6 use gneiss and arenaceous texture
psammitic schist: 4 use schist and arenaceous texture
pseudotachylite: 9
pyroxene granulite: 7
quartzite
quartzite schist: 4 use quartzite and schist
ribbon slate use slate
schist: 4
serpentine marble: 8 use verd antique
serpentinite
sillimanite gneiss: 6
slate: 2
slaty quartzite: 2 use quartzite and slaty texture
spotted slate: 2 use slate
staurotile: 4 use schist
stengel gneiss: 6 use gneiss
tonalite gneiss: 6
Winooski marble: 8 use marble

Metasomatic rocks

fenite
greisen
propylite
serpentinite
skarn
tactite
talc rock

List K

Sedimentary structures

Structure type

1 = bedding plane irregularities
2 = biogenic structures
3 = cylindrical structures
4 = planar bedding structures
    (i.e. bedding structures)
5 = primary structures
6 = secondary structures
7 = soft sediment deformation
8 = turbidity current structures

algal banks: 2
algal biscuits: 2
algal mats: 2
algal mounds: 2
antidunes: 1
armored mud balls: 6
ball-and-pillow: 7
banks: 2
bars: 4
bedding: 4
bioturbation: 2
boudinage: 7
burrows: 2
channels: 4
chevron marks: 1
clastic dikes: 7
concretions: 6
cone-in-cone: 6
convoluted beds: 7,8
convoluted deformation: use
  convoluted beds
coprolites: 2
cross-bedding: 4
cross-laminations: 4
cross-stratification: 4
current lineations: use
  parting lineation
current markings: 1
current partings: use
  parting lineation
cut and fill: 4
cyclothems: 4
dunes: 1
flame structures: 7
flaser bedding: 4
flow structures: 7,8
flute casts: 1
frost features: 1

geodes: 6
girvanella: 2
graded bedding: 4,8
grooves: 1
groove casts: 1
ice wedges: 1
ichnofossils: 2
imbrication: 4
laminations: 4
lebensspuren: 2
load casts: 1,7,8
massive bedding: 4
megacyclothems: 4
megaripples: 1
microstylolites: 6
mounds: 1
mudcracks: 1
mud lumps: 1
nodules: 6
olistoliths: 7,8
olistostromes: 7,8
oncolites: 2
parting lineation: 1
pull apart structures: use
  boudinage
rhythmic bedding: 4
ripple drift-cross laminations: 4
ripple-cross-laminations: use
  ripple drift-cross laminations
ripple marks: 1
  (asymmetrical, interference, etc.)
sand bodies: 4
sandstone dikes: 7
sand waves: 1
scour casts: 1
scour marks: 1
septaria: 6
shrinkage cracks: 1
slump structures: 7
sole marks: 1,8
stratification: 4
striations: 1
stromatolites: 2
stylolites: 6
tool marks: 1
tracks: 2
trails: 2
varves: 4

List L

Mineral groups

The following group names are used on the second level of the minerals set. The first term on the third level will be the mineral species. If no specific mineral is given, start the third level with a topic. In cases where the particular mineral species may be referred to more than one group (e.g. kainite: halides, sulfates), multiple entries under the minerals sets should be made.

antimonates and antimonites

antimonides (use under sulfides)

arsenates

arsenides (use under sulfides)

arsenites

artifical minerals - place these in their appropriate group and use the term synthetic materials on the third level

bismuthides (use under sulfides)

borates

carbides (use under native elements and alloys)

carbonates

chlorides (use under halides)

chromates

fluoborates (use under halides)

fluosilicates (use under halides)

halides (includes chlorides, fluoborates, fluosilicates, and fluorides)

iodates

meteorite minerals - place in appropriate group

miscellaneous minerals (for several groups or for minerals of unknown affinity)

molybdates

native elements and alloys (includes carbides, nitrides, phosphides, and silicides)

niobates (use under oxides)

nitrates

nitrides (use under native elements and alloys)

organic compounds

oxides - includes niobates and tantalates

oxysulfides (use under sulfides)

phosphates

phosphides (use under native elements and alloys)

selenates and selenites

selenides (use under sulfides)

silicates - use this term only for very broad treatments of the entire class of minerals; otherwise, use a narrower term below:

1. orthosilicates (neso-subsilicates, sorosilicates and nesosilicates)

2. orthosilicates, epidote group

3. orthosilicates, garnet group

4. orthosilicates, humite group

5. orthosilicates, melilite group

6. orthosilicates, olivine group

7. ring silicates (cyclosilicates)

8. chain silicates (inosilicates)

9. chain silicates, amphibole group

10. chain silicates, pyroxene group

11. sheet silicates (phyllosilicates)

12. <u>sheet silicates</u>, <u>chlorite group</u>
13. <u>sheet silicates</u>, <u>clay minerals</u>
14. <u>sheet silicates</u>, <u>mica group</u>
15. <u>sheet silicates</u>, <u>serpentine group</u>
16. <u>framework silicates</u> (tectosilicates)
17. <u>framework silicates</u>, <u>feldspar group</u>
18. <u>framework silicates</u>, <u>nepheline group</u>
19. <u>framework silicates</u>, <u>scapolite group</u>
20. <u>framework silicates</u>, <u>silica minerals</u>
21. <u>framework silicates</u>, <u>sodalite group</u>
22. <u>framework silicates</u>, <u>zeolite group</u>

<u>silicides</u> (use under <u>native elements and alloys</u>)

<u>sulfates</u>

<u>sulfosalts</u> - includes <u>sulfantimonates</u>, <u>sulfantimonites</u>, <u>sulfarsenates</u>, <u>sulfarsenites</u>, <u>sulfobismuthites</u>, <u>sulfogermanates</u>, <u>sulfostannates</u>, and <u>sulfovanadates</u>

<u>sulfides</u> - includes <u>antimonides</u>, <u>arsenides</u>, <u>bismuthides</u>, <u>oxysulfides</u>, <u>selenides</u>, and <u>tellurides</u>

<u>tantalates</u> (use under <u>oxides</u>)

<u>tellurates</u> <u>and</u> <u>tellurites</u>

<u>tellurides</u> (use under <u>sulfides</u>)

<u>tungstates</u>

<u>vanadates</u>

Sources: 1. For <u>silicates</u> use the classification of Strunz, Hugo, <u>Mineralogische Tabellen</u>, Leipzig, 1970.

        2. For other minerals use Fleischer, Michael, <u>Glossary of Minerals</u>.

List M

Soils

The classification of soils has undergone several changes in American practice. What follows are two of the classifications adopting different criteria. The 1965 classification has been retained fairly intact since.

In addition there follows a comparative table giving equivalent French and German terms for some of the basic soil terms in English.

AMERICAN SOIL CLASSIFICATION SYSTEM - 1964

Zonal soils
- Tundra soils
- Subarctic brown forest soils
- Desert soils
- Red desert soils
- Sierozems
- Brown soils
- Reddish brown soils
- Chestnut soils
- Reddish chestnut soils
- Chernozems
- Brunizems
- Reddish prairie soils
- Noncalcic brown soils
- Podzols
- Brown podzolic soils
- Gray wooded soils
- Sols bruns acides
- Gray-brown podzolic soils
- Red-yellow podzolic soils
- Reddish-brown lateritic soils
- Yellowish-brown lateritic soils
- Latosols

Intrazonal soils
- Solonchak soils (saline soils)
- Solonetz soils
- Soloth soils
- Humic-gley soils
- Alpine meadow soils
- Bog soils
- Low-humic gley soils
- Planosols
- Ground-water podzols
- Ground-water laterite soils

	Brown forest soils
	Rendzinas
	Grumusols
	Calcisols
Azonal soils	Lithosols
	Regosols
	Alluvial soils

AMERICAN SOIL CLASSIFICATION SYSTEM - 1965 et seq.

orders	suborders
Alfisols	Udalfs
	Boralfs
	Ustalfs
	Xeralfs
Aridisols	Argids
	Orthids
Entisols	Aquents
	Fluvents
	Orthents
	Psamments
Histosols	Fibrists
	Folists
	Hemists
	Saprists
Inceptisols	Andepts
	Aquepts
	Ochrepts
	Tropepts
	Umbrepts
Mollisols	Albolls
	Aquolls
	Borolls
	Redolls
	Udolls
	Ustolls
	Xerolls

Oxisols	Aquox
	Humox
	Orthox
	Torrox
	Ustox
Spodosols	Aquods
	Humods
	Orthods
Ultisols	Aquults
	Humults
	Udults
	Ustults
	Xerults
Vertisols	Torrerts
	Uderts
	Usterts
	Xererts

In addition, terms such as loam, laterites, paleosols, placosols, and volcanic ash may be used on the third level.

AMERICAN	GERMAN	FRENCH
Acrisols	Acrisol	Sol-mediterraneen
Albolls		
Alfisols		
Alluvial soils	Auen-Boden	Sol-d'alluvions
Alpine meadow soils	Alpiner Wiesenboden	Sol-hydromorphe
Andepts		
Andosols	Andosol	Sol-peu-evolue
		roche-volcanique
Aqualfs		
Aquents		
Aquepts		
Aquods		
Aquolls		
Aquox		
Aquults		
Arctic tundra soils	Arktische Tundra Boden	Sol-de-toundra
Arenosols	Arenosol	Sol-brut sable
Arents		
Argids		
Aridisols		

Azonal soils	Roh-Boden	Sol-brut
Black earth use Chernozems	Schwarzerde	Chernozem
Bog soils	Moorboden	Tourbe
Boralfs		
Boreal frozen taiga soils		Sol-gele
Boreal taiga and forest soils		Sol
Borolls		Mollisol
Brown desert steppe soils	Burozem	Sierozem
Brown forest soils	Brauner Wald Boden	Sol-brun
Brown loam	Braunlehm	Couche-rouge
Brown podzolic soils	Podsoliger Braunerde	Podzol
Brown soils	Braunerde	Sol-brun
Brunizems	Brunizem	Brunizem
Calcisols		
Cambisols	Cambisol	Sol-brun
Caar	Uebergangsmoor	Tourbe
Chernozems	Chernozem	Chernozem
Chestnut soils	Kastannozem	Sol-chatain
Cryosols	Cryosol	Sol-gele
Desert soils	Wuesten-Boden	Sol-subdesetique
Desert raw soils	Wuesten-roh-Boden	Sol-de-desert
Dy	Dy	Vase
matiere-organique		
Entisols		
Fen		
Ferralites	Eisen-Silikat-Boden	Sol-fersialitique
Ferralsols	Ferralsol	Sol-ferralitique
Ferrods		
Ferruginous soils	Eisenhaltiger-Boden	Sol-ferrugineux
Fibrists		
Fluvents		
Fluviosols		Sol-d'alluvions
Folists		
Frozen ground	Frost-Boden	Sol-gele
Gleys	Gley	Gley
Glensols	Glensol	Glensol
Gray-brown podzolic soils	Fahlerde	Sol-brun lessivage
Gray podzolic soils	Podsolierter Grauer Boden	Podzol
Gray warp soils	Paternia	Sol-peu-evolue or sol-d'alluvions
Gray wooded soils		
Ground-water podzols	Gley-Podsol	Podzol
Ground-water laterite soils	Grundwasser-Laterite	Laterite
Grumosols	Grumosol	Vertisol
Half bog soils	Anmoor	Tourbe
Halomorphic soils	Salz-Boden	Sol-halomorphe

Halosols	Halosol	Sol-halomorphe
Hemists		
High moor	Hochmoor	Tourbe
Histosols		
Humic gley soils	Humus Gley Boden	Sol-hydromorphe
Humic soils	Humus-reiche-Boden	Sol-riche-en-humus
Humods		
Humox		
Hydromorphic soils	Hydromorpher-Boden	Sol-riche-en-humus
Inceptisols		
Intrazonal soils	Intrazonaler Boden	Sol
Kastanozems		
Krasnozems	Krasnozem	Krasnozem
laterites	Laterit-Boden	Sol-ferralitique
Latosols	Latosol	Sol-ferralitique
Lithosols	Gesteins-roh-Boden	Sol-squelettique
Low-humic gley soils		
Luvisols	Luvisols	Sol lessivage
Mediterranean soils	Mediterraner Boden	Sol-mediterraneen
Mollisols		
Mor	Auflagehumus	Humus
Mull	Mull	Humus
Mull soils	Mull-Boden	Sol-a-mull
Muck	Niedermoor	Tourbe
Nitosols	Nitosol	
Noncalcic brown soils		
Ochrepts		
Orthents		
Orthids		
Orthods		
Orthox		
Oxisols		
Parachernozems	Smonitza	Chernozem
sol-d'alluvions		
Paramosols	Paramosol	Sol-chatain
Pararendzina	Borowina	Rendzine
sol-d'alluvions		
Parasierozems	Parasierozem	Sierozem
Peat	Torf	Tourbe
Pelosols	Pelosol	Sol-hydromorphe
Pergelisols	Pergelisol	Sol-gele
Plaggenesch	Plaggenesch	Sol-riche-en-humus action-homme
Plaggepts		
Planosols	Planosol	Couche-rouge
Podzols	Podsol	Podzol
Podzoluvisols	Podzoluvisol	Sol lessivage
Poorly developed soils	Unreife-boden	Sol-peu-developpe
Prairie soils	Brunizem	Brunizem
Protopedons	Protopedon	Sol-brut or
sol-d'alluvions		
Psamments		
Pseudogleys	Pseudogley	Gley

Rambla	Rambla	Sol-brut or
sol-d'alluvions		
Ranker	Ranker	Ranker
Red desert soils	Roter Wuestenboden	Sol-subdesertique
Reddish brown soils	Roetlich-brauner Halbwueste Boden	Sol-subdesertique
Reddish-brown lateritic soils		
Reddish chestnut soils	Roetlich-Kastanienfarbiger Boden	Sol-chatain
Reddish prairie soils	Roetlicher Prairie Boden	Brunizem
Red-yellow podzolic soils	Gelbig roterpodsoliger Boden	Sol-ferralitique
Reg	Reg	Sol-de-desert
Regosols	Regosol	Sol-peu-evolue
Regur	Regur	Vertisol
Rendolls		
Rendzinas	Rendzina	Rendzine
Rhegosols	Rhegosol	Sol-peu-developpe
Rigosols	Rigosol	Sol-brut
Rutmark	Raamark	Sol-brut
zone-froide		
Saprists		
Sapropels	Sapropel	Sierozem
Sierozems	Sierozem	Sierozem
Solonchak soils	Solonchak	Sol-halomorphe
Solonetz soils	Solonetz	Sol-halomorphe
Soloth soils	Solod	Sol-halomorphe
Spodosols		
Stagnogleys	Stagnogley	Gley
Subarctic brown forest soils		
Subboreal desert soils		
Subboreal humid soils		
Subboreal steppe soils		
Subtropical desert soils		
Subtropical dry soils		
Subtropical humid soils		
Syrogleys	Syrogley	Gley
Syrozems	Syrozem	Sol-brut
zone-temperee		
Takyr	Takyr	Sol-de-desert
Terrae calcis	Terrae calcis	Couche-rouge
Terra rossa	Terra rossa	Couche-rouge
Terra fusca	Terra fusca	Couche-rouge
Tir	Tir	Vertisols
Torrox		
Torrerts		
Tropepts		
Tropical desert soils		
Tropical dry soils		
Tropical humid soils		
Tundra soils	Tundra-Boden	Sol-de-toundra

Udalfs
Uderts
Udolls
Udults
Ultisols
Umbrepts
Ustalfs
Ustserts
Ustolls
Ustsults
Vega   Vega   Sol-brun or
sol-d'alluvions
Vertisols
Wet meadow soils   Wiesenboden   Sol-hydromorphe
Xeralfs
Xerests
Xerolls
Xerosols   Xerosol   Sierozem
Xerults
Yeltozems   Yeltozem   Sol-mediterraneen
Yermosols   Ermosol   Sol-de-desert
Zonal soils   Zonaler Boden   Sol

List N

Sediments

groups

1 = carbonate sediments

2 = chemically precipitated sediments

3 = clastic sediments

4 = organic sediments

specific sediment; and group

alluvium: 3
anhydrite: 2
bentonite: 3
boulders: 3
calcrete: 2
caliche: 2
clay: 3
cobbles: 3
colluvium: 3
coquina: 1,3
diamicton: 3
drift: 3
duricrust: 2
dust: 3
eluvium: 3
encrinite use coquina
evaporites: 2
ferricrete: 2
flint clay: 3
flysch: 3
geyserite use siliceous sinter
gravel: 1,3
guano: 4
gypsum: 2
gyttja: 4
halite: 2
iron-rich composition: 2,3
loess: 3
lutite use clay
mud: 1,3
ooze: 3
outwash: 3
peat: 4

pebbles: 3
quartz sand: 3
residuum: 3
sand: 1,3
shingle: 3
siliceous composition: 2,3
siliceous sinter: 2,3
silcrete: 2,3
silt: 3
till: 3
travertine: 1,2
tufa: 1,2
volcanic ash: 3
weathering crust: 2

terms to be used with specific name:

arenaceous texture
argillaceous texture
arkosic composition
calcareous composition
dolomitic composition
ferruginous composition
fossiliferous materials
glauconitic composition
intraclastic texture
lithic texture
oolitic texture
pebbly texture
pelletal texture
pisolitic texture

LIST O

Area terms
(including main area terms)

The following list includes 1st-level terms on the left and right columns.

area term	main area term
Adriatic Sea	Mediterranean Sea
Aegean Sea	Mediterranean Sea
Afghanistan	Asia
Africa	
Alabama	United States
Alaska	United States
Albania	Europe
Alberta	Canada
Algeria	Africa
Alps	Europe
Andes	South America
Andorra	Europe
Angola	Africa
Antarctic Ocean	
Antarctica	
Apennines	Europe
Appalachians	United States or Canada or North America
Arabian Peninsula	Asia
Arabian Sea	Indian Ocean
Arctic Ocean	
Arctic region	
Argentina	South America
Arizona	United States
Arkansas	United States
Asia	
Atlantic Coastal Plain	United States or Canada or North America
Atlantic Ocean	
Atlantic region	
Australasia	
Australia	
Austria	Europe
Azores	Atlantic Ocean
Bahamas	West Indies
Bahrain	Arabian Peninsula
Balearic Islands	Mediterranean Sea
Balkan Peninsula	Europe
Baltic region	
Baltic Sea	Atlantic Ocean
Bangladesh	Asia
Barbados	West Indies
Basin and Range Province	United States
Belgium	Europe
Belize	Central America

448

Benin	Africa
Bering Sea	Pacific Ocean
Bermuda	Atlantic Ocean
Bhutan	Asia
Black Sea	Eurasia
Bolivia	South America
Botswana	Africa
Brazil	South America
British Columbia	Canada
British Honduras see Belize	
Bulgaria	Europe
Burma	Asia
Burundi	Africa
California	United States
Cambodia	Asia
Cameroon	Africa
Canada	
Canadian Shield	Canada
Canary Islands	Atlantic Ocean
Cape Verde Islands	Africa
Caribbean region	
Caribbean Sea	
Carpathians	Europe
Caspian Sea	Eurasia
Celebes Sea	Pacific Ocean
Central African Republic	Africa
Central America	
Ceylon use Sri Lanka	
Chad	Africa
Chile	South America
China	Asia
Colombia	South America
Colorado	United States
Colorado Plateau	United States
Columbia Plateau	United States
Comoro Islands	Indian Ocean
Congo	Africa
Connecticut	United States
Coral Sea	Pacific Ocean
Corsica	Europe or Mediterranean Sea or Mediterranean region
Costa Rica	Central America
Cuba	West Indies
Cyprus	Middle East
Czechoslovakia	Europe
Dahomey use Benin	
Delaware	United States
Denmark	Europe
District of Columbia	United States
Djibouti	Africa
Dominican Republic	West Indies
East China Sea	Pacific Ocean
East Germany	Europe
Eastern U.S.	United States
Ecuador	South America
Egypt	Africa

El Salvador	Central America
England	Europe
English Channel	Atlantic Ocean
Equatorial Guinea	Africa
Ethiopia	Africa
Eurasia	
Europe	
Faeroe Islands	Atlantic Ocean
Falkland Islands	Atlantic Ocean
Far East	Asia
Fernando Po see Equatorial Guinea	
Fiji	Melanesia
Finland	Europe
Florida	United States
Formosa see Taiwan	
France	Europe
French Guiana	South America
Gabon	Africa
Galapagos Islands	Pacific Ocean
Gambia	Africa
Georgia	United States
Germany	Europe
Ghana	Africa
Great Basin	United States
Great Britain	Europe
Great Lakes	United States or Canada or North America
Great Lakes region	United States or Canada or North America
Great Plains	United States or Canada or North America
Greater Antilles	West Indies
Greece	Europe
Greenland	Arctic region
Guadeloupe	West Indies
Guatemala	Central America
Guinea	Africa
Guinea-Bissau	Africa
Gulf Coastal Plain	North America or United States or Mexico
Gulf of Aden	Arabian Sea
Gulf of California	Pacific Ocean
Gulf of Mexico	
Guyana	South America
Haiti	West Indies
Hawaii	United States
Himalayas	Asia
Honduras	Central America
Hong Kong	Asia
Hungary	Europe
Iceland	Atlantic Ocean
Idaho	United States
Illinois	United States
India	Asia
Indian Ocean	
Indiana	United States

Indochina	Asia
Indonesia	Asia
Ionian Sea	Mediterranean Sea
Iowa	United States
Iran	Asia
Iraq	Middle East
Ireland	Europe
Irish Sea	Atlantic Ocean
Israel	Middle East
Italy	Europe
Ivory Coast	Africa
Jamaica	West Indies
Jan Mayen	Arctic Ocean or Atlantic Ocean
Japan	Asia
Japan Sea	Pacific Ocean
Jordan	Middle East
Jura Mountains see France or Switzerland	
Kansas	United States
Katanga see Zaire (Shaba)	
Kentucky	United States
Kenya	Africa
Korea	Asia
Kuwait	Arabian Peninsula
Labrador	Canada
Laos	Asia
Lebanon	Middle East
Lesotho	Africa
Lesser Antilles	West Indies
Liberia	Africa
Libya	Africa
Louisiana	United States
Luxembourg	Europe
Madagascar use Malagasy Republic	
Madeira	Atlantic Ocean
Maine	United States
Malagasy Republic	Africa
Malawi	Africa
Malay Archipelago	Asia
Malaysia	Asia
Maldive Islands	Indian Ocean
Mali	Africa
Malta	Mediterranean Sea
Manitoba	Canada
Mariana Islands	Micronesia
Maritime Provinces	Canada
Marshall Islands	Micronesia
Maryland	United States
Massachusetts	United States
Mauritania	Africa
Mauritius	Indian Ocean
Mediterranean region	
Mediterranean Sea	
Melanesia	Pacific Ocean
Mexico	

Michigan	United States
Micronesia	Pacific Ocean
Middle East	
Midwest	United States
Minnesota	United States
Mississippi	United States
Mississippi Valley	United States
Missouri	United States
Monaco	Europe
Mongolia	Asia
Montana	United States
Morocco	Africa
Mozambique	Africa
Nebraska	United States
Nepal	Asia
Netherlands	Europe
Nertherlands Antilles	Caribbean region
Nevada	United States
New Brunswick	Canada
New Caledonia	Pacific Ocean
New England	United States
New Guinea	Malay Archipelago
New Hampshire	United States
New Hebrides	Melanesia
New Jersey	United States
New Mexico	United States
New South Wales	Australia
New York	United States
New Zealand	Australasia
Newfoundland	Canada
Nicaragua	Central America
Niger	Africa
Nigeria	Africa
North America	
North Carolina	United States
North Dakota	United States
North Sea	Atlantic Ocean
Northern Hemisphere	
Northern Ireland	Europe
Northern Territory	Australia
Northwest Territories	Canada
Norway	Europe
Nova Scotia	Canada
Oceania	Pacific Ocean
Ohio	United States
Okhotsk Sea	Pacific Ocean
Oklahoma	United States
Oman	Arabian Peninsula
Ontario	Canada
Oregon	United States
Pacific Coast	United States
Pacific region	
Pakistan	Asia
Panama	Central America
Papua New Guinea	Australasia
Paraguay	South America
Pennsylvania	United States

Persian Gulf	Indian Ocean
Peru	South America
Philippine Islands	Asia
Philippine Sea	Pacific Ocean
Poland	Europe
Polynesia	Pacific Ocean
Portugal	Europe
Portuguese Guinea see Guinea-Bissau	
Portuguese Timor see Indonesia	
Prince Edward Island	Canada
Puerto Rico	West Indies
Pyrenees	Europe
Qatar	Arabian Peninsula
Quebec	Canada
Queensland	Australia
Red Sea	
Red Sea region	
Reunion	Indian Ocean
Rhode Island	United States
Rhodesia	Africa
Rio Muni see Equatorial Guinea	
Rocky Mountains	United States or North America or Canada
Romania	Europe
Rwanda	Africa
Sahara	Africa
Samoa	Polynesia
Sarawak see Malaysia	
Sardinia	Europe or Mediterranean region or Mediterranean Sea
Saskatchewan	Canada
Saudi Arabia	Arabian Peninsula
Scandinavia	Europe
Scotland	Europe
Senegal	Africa
Seychelles	Indian Ocean
Shetland Islands	Atlantic Ocean
Sierra Leone	Africa
Sikkim see India	
Singapore	Asia
Society Islands	Polynesia
Solomon Islands	Melanesia
Somali Republic	Africa
South Africa	Africa
South America	
South Australia	Australia
South Carolina	United States
South China Sea	Pacific Ocean
South Dakota	United States
Southern Hemisphere	
Southern Yemen (Aden)	Arabian Peninsula
South-West Africa	Africa
Southwestern U.S.	United States

Spain	Europe
Spanish Guinea use Equatorial Guinea	
Spanish Sahara use Sahara and west	
Spitsbergen	Arctic region
Sri Lanka	Asia
Sudan	Africa
Surinam	South America
Swaziland (Kaapvaal Craton)	Africa
Sweden	Europe
Switzerland	Europe
Syria	Middle East
Tahiti	Polynesia
Taiwan	Asia
Tanzania	Africa
Tasman Sea	Pacific Ocean
Tasmania	Australia
Tennessee	United States
Texas	United States
Thailand	Asia
Togo	Africa
Tonga	Polynesia
Trinidad and Tobago	West Indies
Trucial Coast use United Arab Emirates	
Tunisia	Africa
Turkey	Middle East
Tyrrhenian Sea	Mediterranean Sea
Uganda	Africa
United Arab Emirates	Arabian Peninsula
United Kingdom	Europe
United States	
Upper Volta	Africa
Uruguay	South America
USSR	
Utah	United States
Venezuela	South America
Vermont	United States
Victoria	Australia
Vietnam	Asia
Virginia	United States
Wales	Europe
Washington	United States
West Germany	Europe
West Indies	
West Virginia	United States
Western Australia	Australia
Western Hemisphere	
Western Interior	United States or Canada or North America
Western U.S.	United States
Wisconsin	United States
Wyoming	United States
Yellow Sea	Pacific Ocean
Yemen	Arabian Peninsula
Yugoslavia	Europe

Canada
Africa
Africa

Canada
Africa
Africa